Epigenetics for Climate-Smart and Sustainable Agriculture

Epigenetics for Climate-Smart and Sustainable Agriculture

Edited by

Jen-Tsung Chen
Department of Life Sciences, National University of Kaohsiung, Taiwan

CABI

CABI Climate Change Series

Climate change is a major environmental challenge to the world today, with significant threats to ecosystems, food security, water resources and economic stability overall. In order to understand and research ways to alleviate the effects of climate change, scientists need access to information that not only provides an overview of and background to the field, but also keeps them up to date with the latest research findings.

This series addresses many topics relating to climate change, including strategies to develop sustainable systems that minimize impact on climate and/or mitigate the effects of human activity on climate change. Coverage will encompass all areas of environmental and agricultural sciences. Aimed at researchers, upper-level students and policy makers, titles in the series provide international coverage of topics related to climate change, including both a synthesis of facts and discussions of future research perspectives and possible solutions.

Titles Available

1. Climate Change and Crop Production
Edited by Matthew P. Reynolds

2. Crop Stress Management and Climate Change
Edited by José L. Araus and Gustavo A. Slafer

3. Temperature Adaptation in a Changing Climate: Nature at Risk
Edited by Kenneth Storey and Karen Tanino

4. Plant Genetic Resources and Climate Change
Edited by Michael Jackson, Brian Ford-Lloyd and Martin Parry

5. Climate Change Impact and Adaptation in Agricultural Systems
Edited by Jürg Fuhrer and Peter J. Gregory

6. Livestock Production and Climate Change
Edited by Pradeep K. Malik, Raghavendra Bhatta, Junichi Takahashi, Richard A. Kohn and Cadaba S. Prasad

7. Climate Change and Insect Pests
Edited by Christer Björkman and Pekka Niemelä

8. Climate Change and Agricultural Water Management in Developing Countries
Edited by Chu Thai Hoanh, Vladimir Smakhtin and Robyn Johnston

9. Climate Change Challenges and Adaptations at Farm-level: Case Studies from Asia and Africa
Edited by Naveen Prakash Singh, Cynthia. Bantilan, Kattarkandi Byjesh and Swamikannu Nedumaran

10. Climate Change Impacts on Urban Pests
Edited by Partho Dhang

11. Climate Change Impacts and Sustainability: Ecosystems of Tanzania
Edited by Pius Z. Yanda, Claude G. Mung'ong'o and Edmund B. Mabhuye

12. Climate, Ticks and Disease
Edited by Pat Nuttall

13. Climate Change, Adaptation and Gender: Policy, Practice and Methodological Underpinnings
Mamta Mehar and Narayan Prasad

14. Climate Change and Sustainable Agro-ecology in Global Drylands
Edited by Adel El-Beltagy, Rattan Lal and Kauser Malik

15. Climate Change and Food Security
Edited by Sukanta Mondal, Pankaj Singh and Ram Lakhan Singh

16. Epigenetics for Climate-Smart and Sustainable Agriculture
Edited by Jen-Tsung Chen

CABI is a trading name of CAB International

CABI
Nosworthy Way
Wallingford
Oxfordshire OX10 8DE
UK

Tel: +44 (0)1491 832111
E-mail: info@cabi.org
Website: www.cabi.org

CABI
200 Portland Street
Boston
MA 02114
USA

T: +1 (617)682-9015
E-mail: cabi-nao@cabi.org

A catalogue record for this book is available from the British Library, London, UK.

ISBN-13: 9781800626089 (hardback)
9781800626096 (ePDF)
9781800626102 (ePub)

DOI: 10.1079/9781800626102.0000

Commissioning Editor: Ward Cooper
Editorial Assistant: Emma McCann
Production Editor: James Bishop

Typeset by Exeter Premedia Services Pvt Ltd, Chennai, India
Printed and bound in the UK by CPI Group (UK) Ltd, Croydon, CR0 4YY

Contents

Contributors

Naveen Aavula: Division of Genetics, ICAR- Indian Agricultural Research Institute, New Delhi, India. Email: aavulanaveenyadav1997@gmail.com

Thiruvengadam Abarna: Department of Biochemistry, Shrimati Indira Gandhi College, Tiruchirappalli 620002, Tamil Nadu, India. Email: abarnarajkumar@gmail.com

Naglaa A. Abdallah: Department of Genetics Faculty of Agriculture, Cairo University, Cairo, Egypt and National Biotechnology Network of Expertise, ASRT, Egypt. Email: naglaa.abdallah@agr.cu.edu.eg

Reham I. Abdelhamid: Sugar Crops Research Institute, Agricultural Research Center (ARC), Giza12619, Egypt. Email: rehamezz78@gmail.com

Sachin Lalaso Abhang: Department of Agricultural Botany, Dr. Panjabrao Deshmukh Krishi Vidyapeeth, Akola-444104, Maharashtra, India. Email: sachinabhang299@gmail.com

Dipen Nama Adhikari: Department of Agricultural Biotechnology, Assam Agricultural University, Jorhat, Assam, India. Email: dnadhikari77@gmail.com

Khulood Fahad Alabbosh: Department of Biology, College of Science, University of Hail, Hail 2440, Saudi Arabia. Email: k.alabosh@uoh.edu.sa

Malin Paula Alf: Department of Plant Genetics, Institute of Biological Sciences, University of Rostock, Albert-Einstein-Str. 3, Rostock, Germany. Email: malin.alf@uni-rostock.de

Ambika: Department of Genetics and Plant Breeding, UAS, GKVK, Bangalore, Karnataka, India-560 0654. Email: ambikanagaraj1076@gmail.com

Arumugam Vijaya Anand: Department of Human Genetics and Molecular Biology, Bharathiar University, Coimbatore 641046, Tamil Nadu, India. Email: avahgmb@buc.edu.in

Kumari Anjani: Department of Agricultural Biotechnology and Molecular Biology, Dr Rajendra Prasad Central Agricultural University, Pusa-848125, Bihar, India. Email: anjani@rpcau.ac.in

Burcu Arıkan: Istanbul University, Faculty of Science, Department of Molecular Biology and Genetics, Vezneciler, 34134 Istanbul, Türkiye. Email: burcuarikan@istanbul.edu.tr

Amine Assouguem: Faculty of Sciences and Techniques, Sidi Mohammed Ben Abdellah University, Fez, Morocco. Email: assougam@gmail.com

Kousik Atta: Faculty of Agricultural Sciences, Institute of Applied Sciences and Humanities, GLA University, Mathura, Uttar Pradesh, India. Email: kousikatta1995@gmail.com

Sahaya Shibu Babu: Department of Biotechnology, Safi Institute of Advanced Study (S9AS), Via Ramanattukara, Malappuram Dist., Kerala-673663, India. Email: shibubt@gmail.com

Chithra Devi Balasundaram Saraswathy: Department of Botany, PSGR Krishnammal College for Women, Coimbatore-641004, Tamil Nadu. Email: bchitradevi@psgrkcw.ac.in

Sushil Barik: Faculty of Agriculture and Allied Sciences, C.V. Raman Global University, Bhubaneswar-752054, India. Email: susilbarik58@gmail.com

Nagendra Sarma Barua: Department of Plant Breeding and Genetics, Assam Agricultural University, Jorhat, Assam, India. Email: nagendra.sarmabarua@aau.ac.in

P.S. Basavaraj, School of Water Stress Management, ICAR-National Institute of Abiotic Stress Management, Baramati, Maharashtra, India. Email: Basavaraj.ps@icar.gov.in

Sana Basharat: College of Tropical Agriculture and Forestry, Hainan University, Haikou, Hainan, China; Key Laboratory of Tropical Horticultural Crop Quality Regulation, College of Horticulture, Hainan University, Haikou, Hainan, China; Hainan Yazhou Bay Seed Laboratory, Sanya Nanfan Research Institute of Hainan University, Sanya 572025, China; and Fang Zhiyuan Academician Team Innovation Center of Hainan Province, China. Email: sanabasharat143@gmail.com

S. Basker: Department of Botany, Government Arts College, Salem, Tamil Nadu, India. Email: biobasker@gmail.com

Nourhan M. Battor: Nanotechnology & Advanced Nano-Materials Laboratory (NANML), Mycology and Disease Survey Research Department, Plant Pathology Research Institute, Agricultural Research Center, Giza 12619, Egypt. Email: Nourhanbattor@gmail.com

Michael Baum: Department of Biotechnology, International Centre for Agricultural Research in the Dry Areas (ICARDA), Giza, Egypt. Email: m.baum@cgiar.org

Sirajudeen Benazir Begum: PG and Research Department of Zoology, V. O. Chidambaram College, Thoothukudi 628008, Tamil Nadu, India. Email: yasbenz783@gmail.com

Tajudheen Nargis Begum: Department of Biotechnology, Jamal Mohamed College, (Autonomous), Tiruchirapalli 620020, Tamil Nadu, India. Email: nargisalmaas@gmail.com

Zineb Belabess: Plant Protection Laboratory, Regional Center of Agricultural Research of Meknes, National Institute of Agricultural Research, Km 13, Route Haj Kaddour, BP 578, Meknes 50001, Morocco. Email: zineb.belabess@inra.ma

Khadija Benamar: Phytopathology Unit, Department of Plant Protection, École Nationale d'Agriculture de Meknès, Km10, Rte. HajKaddour, BP S/40, Meknes 50001, Morocco; and Department of Biology, Laboratory of Microbial Biotechnology and Bioactive Molecules, Sciences and Technologies Faculty, Sidi Mohamed Ben Abdellah University, P.O Box 2202, Fez, Morocco. Email: khadija.benamar2@usmba.ac.ma

Kawtar Fikri Benbrahim: Department of Biology, Laboratory of Microbial Biotechnology and Bioactive Molecules, Sciences and Technologies Faculty, Sidi Mohamed Ben Abdellah University, PO Box 2202, Fez, Morocco. Email: Kawtar.benbrahim@usmba.ac.ma

H.A. Bhargavi, Division of Vegetable Crops, ICAR- Indian Institute of Horticultural Research, Bengaluru-560089, Karnataka, India. Email: bhargaviha6@gmail.com

Amarjeet Singh Bhogal: Department of Plant Breeding and Genetics, Assam Agricultural University, Jorhat, Assam, India. Email: singhhamarjeet1820@gmail.com

Karma Landup Bhutia, Department of Agricultural Biotechnology and Molecular Biology, Dr Rajendra Prasad Central Agricultural University, Pusa-848125, Bihar, India. Email: klbhutia@rpcau.ac.in

Özgür Çakır: Istanbul University, Faculty of Science, Department of Molecular Biology and Genetics, Vezneciler, 34134 Istanbul, Türkiye. Email: ozgurckr@istanbul.edu.tr

Bandela Chandana: Department of Genetics and Plant Breeding, Kerala Agricultural University, Thiruvananthapuram, Kerala, India. Email: chandanab246@gmail.com

C. A. Cedillo-Jiménez: Center of Applied Research in Biosystems (CARB-CIAB). School of Engineering-Campus Amazcala, Autonomous University of Querétaro, Carr. Amazcala-Chichimequillas, Km 1.0. C.P. 76265, Amazcala, El Marques, Querétaro, México. E-mail: christopher.cedillo@uaq.mx

Chitrita Chatterjee: Shibpur Dinobundhoo Institution College, Howrah-711103, India. Email: chitrita87@gmail.com

Charupriya Chauhan: Department of Genetics and Plant Breeding, College of Agriculture, Acharya Narendra Deva University of Agriculture & Technology, Kumarganj-224229, Ayodhya, Uttar Pradesh, India. Email: charupriyachauhan0@gmail.com

Abhra Abir Chowdhury: PG Department of Botany, Darjeeling Government College, Darjeeling, West Bengal, India, PIN – 734101, India Email: abhrabot23@klyuniv.ac.in

Dibakar Das: Department of Genetics and Plant Breeding, School of Agriculture, GIET University, Gunupur-765022, Rayagada, Odisha, India. Email: dasdibakartlm@gmail.com

Supti Das: Department of Biochemistry, University of Calcutta, 35, Ballygunge Circular Road, Kolkata- 700019, West Bengal, India. Email: suptidas.raiganj@gmail.com

Ilham Dehbi: Phytopathology Unit, Department of Plant Protection, École Nationale d'Agriculture de Meknès, Km10, Rte. HajKaddour, BP S/40, Meknes 50001, Morocco. Email: il.dehbi@edu.umi.ac.ma

Vinay Kumar Dhiman: Department of Basic Sciences, College of Forestry, Dr Y S Parmar University of Horticulture and Forestry, 173230, H.P., India. Email: vinaykrdhiman4@gmail.com

Vivek Kumar Dhiman: Department of Biotechnology, College of Horticulture, Dr Y S Parmar University of Horticulture and Forestry, 173230, H.P., India. Email: vivekkrdhiman@gmail.com

Pratim Dutta: Department of Agricultural Biotechnology, Assam Agricultural University, Jorhat, Assam, India. Email: pratimdutta97@gmail.com

Shreyasi Dutta: Research Center for Natural and Applied Sciences, Raja Narendralal Khan Women's College (Autonomous), Gope Palace, Midnapore- 721102, India Email: shreyasidutta@rnlkwc.ac.in

Tushar K. Dutta: Division of Nematology, ICAR-Indian Agricultural Research Institute, New Delhi 110012. Email: nemaiari@gmail.com

Tuward J. Dweh: Faculty of Agriculture and Allied Sciences, C.V. Raman Global University, Bhubaneswar – 752054, Odisha India. Email: tjosephdweh23@gmail.com

Mayar A. Elnagar: Faculty of Science, Department of Biotechnology, Cairo University. Email: Mayarelnagar289@gmail.com

Sozan E. El-Abeid: Nanotechnology & Advanced Nano-Materials Laboratory (NANML), Mycology and Disease Survey Research Department, Plant Pathology Research Institute, Agricultural Research Center, Giza 12619, Egypt. Email: sozanelabeid@yahoo.com

Mariam A. Abd El-Samea: Faculty of Science, Department of Biotechnology, Cairo University. Email: mariamabdelghaffar262@gmail.com

Rachid Ez-Zouggari: Phytopathology Unit, Department of Plant Protection, École Nationale d'Agriculture de Meknès, Km10, Rte. HajKaddour, BP S/40, Meknes 50001, Morocco; and Laboratory of Biotechnology, Conservation and Valorisation of Natural Resources (LBCVNR), Faculty of Sciences Dhar El Mehraz, Sidi Mohamed Ben Abdallah University, Fez 30000, Morocco. Email: ezzouggarirachid@gmail.com

Abdelaaziz Farhaoui: Phytopathology Unit, Department of Plant Protection, Ecole National of Agriculture Meknes, Km10, Rte Haj Kaddour, BP S/40, Meknes 50001, Morocco. Email: farhaoui.aziz22@gmail.com

Nourhan Fouad: Department of Biotechnology, International Centre for Agricultural Research in the Dry Areas (ICARDA), Giza, Egypt. Email: nourhan.fouad1@gmail.com

Mukul Kumar Gandhi: Department of Agricultural Biotechnology and Molecular Biology, Dr. Rajendra Prasad Central Agricultural University, Pusa-848125, Bihar, India. Email: gandhi.mukul007@gmail.com

Ashish Gautam: Department of Life Science, School of Life Sciences, Central University of Karnataka, Kalaburagi, Karnataka, India. Email: gautam.ashish801@gmail.com

Blassan George: Faculty of Health Sciences, Research Centre Staff Member, University of Johannesburg, South Africa. Email: blassang@uj.ac.za

Shiv Kumar Giri: Department of Biotechnology, Maharaja Agrasen University, Solan, Himachal Pradesh, India. Email: goswamishiv@gmail.com

Rahul Gogoi: Department of Agricultural Biotechnology, Assam Agricultural University, Jorhat, Assam, India. Email: rahulgogoi.biotech@gmail.com

Pemula Gowtham: Department of Human Genetics and Molecular Biology, Bharathiar University, Coimbatore 641046, Tamil Nadu, India. Email: g.pemula96@gmail.com

R.G. Guevara-González: Center of Applied Research in Biosystems (CARB-CIAB). School of Engineering-Campus Amazcala. Autonomous University of Querétaro. Carr. Amazcala-Chichimequillas, Km 1.0. C.P. 76265. Amazcala, El Marques, Querétaro, México. Email: ramon.guevara@uaq.mx

Heena Gulia: Centre for Medical Biotechnology, Maharshi Dayanand University, Rohtak, Haryana, India. Email: guliaheena18@gmail.com

Aladdin Hamwieh: Department of Biotechnology, International Centre for Agricultural Research in the Dry Areas (ICARDA), Giza, Egypt. Email: a.hamwieh@cgiar.org

Seyed Mafia Haniff: Department of Zoology, Jamal Mohamed College, (Autonomous), Tiruchirapalli 620020, Tamil Nadu, India. Email: biomafi@rediffmail.com

Renate Horn: Department of Plant Genetics, Institute of Biological Sciences, University of Rostock, Albert-Einstein-Str. 3, Rostock, Germany. Email: renate.horn@uni-rostock.de

Senthil Kumar Ilakkiya: Department of Biochemistry, Shrimati Indira Gandhi College, Tiruchirappalli 620002, Tamil Nadu, India. Email: ilakkiini@gmail.com

Babar Iqbal: School of Environment and Safety Engineering, Jiangsu University, Zhenjiang 212013, People's Republic of China. Email: babar@ujs.edu.cn

Sarah Jaison: Department of Botany, PSGR Krishnammal College for Women, Coimbatore-641004, Tamil Nadu, India. Email: sarah@psgrkcw.ac.in

Arshad Jalal: The BioActives Lab, Center for Desert Agriculture, Division of Biological and Environmental Sciences, King Abdullah University of Science and Technology, Thuwal 23955-6900, Kingdom of Saudi Arabia. Email: arshad.jalal@kaust.edu.sa

Yunfei Jiang: Department of Plant, Food and Environmental Sciences, Faculty of Agriculture, Dalhousie University, Truro, Nova Scotia, Canada B2N 5E3. E-mail: yunfei.jiang@dal.ca

Anil Namdev Kale: Department of Agricultural Biotechnology and Molecular Biology, Dr Rajendra Prasad Central Agricultural University, Pusa-848125, Bihar, India. Email: anilkale404@gmail.com

Thirunethiran Karpagam: Department of Biochemistry, Shrimati Indira Gandhi College, Tiruchirappalli 620002, Tamil Nadu, India. Email: karpgam@sigc.edu

Salma Kayastha: Faculty of Agriculture and Allied Sciences, C.V. Raman Global University, Bhubaneswar – 752054, Odisha India. Email: salmakayastha3@gmail.com

K. Mohammed Rafiq Khan: Associate Professor, Department of Biotechnology, Indian Academy Degree College – Autonomous, Hennur Cross, Hennur Main Road Kalyan Nagar, Bangalore – 560 043 Karnataka, India. Email: rafiqkhanmd.k@gmail.com; dr.rafiq.biotech@iadc.ac.in

Ankica Kondić-Špika: Institute of Field and Vegetable Crops, National Institute of the Republic of Serbia, Maksima Gorkog 30, Novi Sad, Serbia. Email: ankica.spika@ifvcns.ns.ac.rs

Mohamed Kouighat: National Institute of Agricultural Research (INRA), BO 415, Rabat, Morocco. Email: kouighat84@gmail.com

Anil Kumar: Centre for Medical Biotechnology, Maharshi Dayanand University, Rohtak, Haryana, India. Email: anil.cmbt@mdurohtak.ac.in

Gaurav Kumar: Department of Basic Sciences, College of Horticulture and Forestry, Neri, Dr Y S Parmar University of Horticulture and Forestry, 173230, H.P., India. Email: gauravrana4636@gmail.com

Praveen Kumar: Department of Genetics and Plant Breeding, Agriculture University, Jodhpur, India. Email: bkpraveenars@gmail.com

Rajendra Kumar: Division of Genetics, ICAR-Indian Agricultural Research Institute, New Delhi, India-110012. Email: rajendrak64@yahoo.co.in

Rakesh Kumar: Department of Life Science, School of Life Sciences, Central University of Karnataka, Kalaburagi, Karnataka, India. Email: rakeshkumar@cuk.ac.in

Sarita Kumari: Department of Agricultural Biotechnology and Molecular Biology, Dr. Rajendra Prasad Central Agricultural University, Pusa-848125, Bihar, India. Email: sarita@rpcau.ac.in

Rachid Lahlali: Phytopathology Unit, Department of Plant Protection, École Nationale d'Agriculture de Meknès, Km10, Rte. HajKaddour, BP S/40, Meknes 50001, Morocco. Email: rlahlali@enameknes.ac.ma

P.M. Lakshmi, Environmental Scientist, Directorate of Environment and Climate Change, Government of Kerala, India. Email: drlpm83@gmail.com

Guanlin Li: School of Environment and Safety Engineering, Jiangsu University, Zhenjiang 212013, People's Republic of China. Email: liguanlin@ujs.edu.cn

Pingwu Liu: College of Tropical Agriculture and Forestry, Hainan University, Haikou, Hainan, China; Key Laboratory of Tropical Horticultural Crop Quality Regulation, College of horticulture, Hainan University, Haikou, Hainan, China; Hainan Yazhou Bay Seed Laboratory, Sanya Nanfan Research Institute of Hainan University, Sanya 572025, China; and Fang Zhiyuan Academician Team Innovation Center of Hainan Province, China. Email: hnulpw@hainanu.edu.cn

Irene Luzzi: Department of Agronomy, Food, Natural Resources, Animals and Environment, University of Padova, Viale dell'Università 16, Legnaro (PD), Italy. Email: irene.luzzi@phd.unipd.it

Manaswini Mahapatra: Faculty of Agriculture and Allied Sciences, C.V. Raman Global University, Bhubaneswar-752054, India. Email: manaswini.agri@gmail.com

Aqsa Majgaonkar: Department of Botany, St. Xavier's College (Autonomous), 5, Mahapalika Marg, CST, Fort, Mumbai-400001, Maharashtra, India. Email: aqsa.majgaonkar@xaviers.edu.in

Ramasamy Manikandan: Department of Biochemistry, Shrimati Indira Gandhi College, Tiruchirappalli 620002, Tamil Nadu, India. Email: mani_r_trichy@yahoo.co.in

Arun Kumar Maurya: Department of Botany, Multanimal Modi College, Modinagar, Ghaziabad, Uttar Pradesh 201204, India. Email: akmauryahrc@gmail.com

Dragana Miladinović: Institute of Field and Vegetable Crops, National Institute of the Republic of Serbia, Maksima Gorkog 30, Novi Sad, Serbia. Email: draganavas@yahoo.com dragana.miladinovic@ifvcns.ns.ac.rs

Varucha Misra: ICAR-Indian Institute of Sugarcane Research, Lucknow-226002 (UP), India. Email: misra.varucha@gmail.com

Krishnasreya Mohandas: Department of Botany, PSGR Krishnammal College for Women, Coimbatore-641004, Tamil Nadu, India. Email: sreyamohandas97@gmail.com

Sarmad Moin: Associate Professor, School of Applied Sciences, Suresh Gyan Vihar University, Gyan Vihar Marg, Jagatpura, Jaipur, Rajasthan 302017, India. Email: moinsarmad@gmail.com; sarmad.moin@mygyanvihar.com

Rashmi Mukherjee: Research Center for Natural and Applied Sciences, Raja Narendralal Khan Women's College (Autonomous), Gope Palace, Midnapore-721102, India. Email: rashmimukherjee@rnlkwc.ac.in

Paramita Nandy-Datta, Sister Nibedita Govt. General Degree College for Girls, Kolkata, India Email: paramita_nandy@yahoo.com

Abigna Narla: Department of Genetics and Plant Breeding, Kerala Agricultural University, Thiruvananthapuram, Kerala, India. Email: abigna.narla222@gmail.com

Jagyaswani Nayak: Faculty of Agriculture and Allied Sciences, C.V. Raman Global University, Bhubaneswar – 752054, Odisha India. Email: jagyaswaninayak31@gmail.com

Subharshree Pattnaik: Faculty of Agriculture and Allied Sciences, C.V. Raman Global University, Bhubaneswar – 752054, Odisha India. Email: pattnaiksubhashree77@gmail.com

Madhuri Pradhan: Faculty of Agriculture and Allied Sciences, C.V. Raman Global University, Bhubaneswar-752054, India. Email: madhuripradhan44@gmail.com

Bishun Deo Prasad: Department of Agricultural Biotechnology and Molecular Biology, Dr Rajendra Prasad Central Agricultural University, Pusa-848125, Bihar, India. Email: bdprasad@rpcau.ac.in

Raja Helan Pricilla: Department of Biochemistry, Shrimati Indira Gandhi College, Tiruchirappalli 620002, Tamil Nadu, India. Email: helanpricilla@gmail.com

Anbazhagan Priya: PG and Research Department of Zoology, V. O. Chidambaram College, Thoothukudi 628008, Tamil Nadu, India. Email: priyaanbu88@gmail.com

Elif Pulat: Istanbul University, Institute of Science, Program of Molecular Biology and Genetics, Istanbul, Türkiye. Email: elifpulat001@gmail.com

Aleksandra Radanović: Institute of Field and Vegetable Crops, National Institute of the Republic of Serbia, Maksima Gorkog 30, Novi Sad, Serbia. Email: aleksandra.dimitrijevic@ifvcns.ns.ac.rs

Mohammed Radi: Phytopathology Unit, Department of Plant Protection, École Nationale d'Agriculture de Meknès, Km10, Rte. HajKaddour, BP S/40, Meknes 50001, Morocco; and Environment and Valorization of Microbial and Plant Resources Unit, Faculty of Sciences, Moulay Ismail University, Meknes, Morocco. Email: simodefes13@gmail.com

Khaled Radwan: National Biotechnology Network of Expertise, ASRT, Egypt and Department of Biotechnology, Agricultural Genetic Engineering Research Institute (AGERI), ARC, Giza, Egypt. Email: khradwan@gmail.com

Manoharan Rajesh: Department of Biochemistry, Bharathiar University, Coimbatore 641046, Tamil Nadu, India. Email: rajeshmanoh@gmail.com

Ramkumar Govindarajan: Project Coordinator, BIRAC E-YUVA Centre, PSGR Krishnammal College for Women, Coimbatore-641004, Tamil Nadu, India. Email: ramkumarg@psgrkcw.ac.in

Nalam Sindhu Rani: Department of Genetics and Plant Breeding, Kerala Agricultural University, Thiruvananthapuram, Kerala, India. Email: nsindhurani30@gmail.com

Elakkiya Maruthamuthu Rathinam: Department of Botany, PSGR Krishnammal College for Women, Coimbatore-641004, Tamil Nadu, India. Email: elakkiyarathinam11@gmail.com

Noru Rajasekhar Reddy: Department of Genetics and Plant Breeding, Kerala Agricultural University, Thiruvananthapuram, Kerala, India. Email: rajareddynoru@gmail.com

Lellapalli Rithesh: Department of Plant Pathology, Kerala Agricultural University, Thiruvananthapuram, Kerala, India. Email: rithesh132@gmail.com

Asirvatham Alwin Robert: Department of Endocrinology and Diabetes, Prince Sultan Military Medical City, Ministry of Defense, Riyadh 11159, Kingdom of Saudi Arabia. Email: aalwinrobert@gmail.com

Yasmeen A. Roby: Faculty of Science, Department of Biotechnology, Cairo University. Email: yasmenabdelghany@gstd.sci.cu.edu.eg

Bikram Saha: Post Graduate Dept. of Botany, Darjeeling Govt. College, Darjeeling, West Bengal-734101, India. Email: bikram24@gmail.com

Upala Saha: Sister Nibedita Govt. Department of Botany, General Degree College for Girls, Hastings House, Alipore, Kolkata-700027, India. Email: upala.saha@gmail.com

Jyoti Prakash Sahoo: Faculty of Agriculture and Allied Sciences, C.V. Raman Global University, Bhubaneswar-752054, India. Email: jyotiprakashsahoo2010@gmail.com

Palanisamy Sampathkumar: Department of Chemistry and Biosciences, SASTRA Deemed to be University, Srinivasa Ramanujan Centre, Kumbakonam 612 001, Tamil Nadu, India. Email: pskumar16@yahoo.com

Ajay Samuel Siva Daniel: GRG BIRAC E-YUVA Centre, PSGR Krishnammal College for Women, Coimbatore-641004, Tamil Nadu, India. Email: sjaysamuel@psgrkcw.ac.in

Debojit Sarma: Department of Plant Breeding and Genetics, Assam Agricultural University, Jorhat, Assam, India. Email: debojit.sarma@aau.ac.in

Sitanshu Sekhar Satapathy: Faculty of Agriculture and Allied Sciences, C.V. Raman Global University, Bhubaneswar-752054, India. Email: satapathysitanshu2@gmail.com

Iffat Shaheen: Department of Agriculture, Bahauddin Zakariya University, Multan, Pakistan. Email: iffat.shaheen.malik@gmail.com

Angappan Shanmugapriya: Department of Biochemistry, Shrimati Indira Gandhi College, Tiruchirappalli 620002, Tamil Nadu, India. Email: shanmugapriya@sigc.edu

Gargi Sharma: Department of Plant Breeding and Genetics, Assam Agricultural University, Jorhat, Assam, India. Email: gargi.sharma@aau.ac.in

Kishore Kumar Sharma: Department of Plant Breeding and Genetics, Assam Agricultural University, Jorhat, Assam, India. Email: kishore.sharma@aau.ac.in

Siddhartha Shankar Sharma: Faculty of Agriculture and Allied Sciences, C.V. Raman Global University, Bhubaneswar-752054, India. Email: siddharthashankarsharma77@gmail.com

Vinay Sharma: Department of Genetics and Plant Breeding, Ch. Charan Singh University, Meerut 250004, India. Email: vinaysharma965@gmail.com

Vinay Kumar Sharma: Department of Agricultural Biotechnology and Molecular Biology, Dr. Rajendra Prasad Central Agricultural University, Pusa-848125, Bihar, India. Email: vksharma@rpcau.ac.in

Sanjay Kumar Sheshma: Department of Agricultural Biotechnology and Molecular Biology, Dr. Rajendra Prasad Central Agricultural University, Pusa-848125, Bihar, India. Email: sanjay-sheshma105@gmail.com

Aditya Pratap Singh: Department of Genetics and Plant Breeding, Bidhan Chandra Krishi Viswavidyalaya, Mohanpur-765022, Nadia, West Bengal. Email: adityapratapbckv@gmail.com

Devendra Singh: Faculty of Biotechnology, Institute of Biosciences and Technology, Shri Ramswaroop Memorial University, Barabanki, Uttar Pradesh 225003. Email: devendrasingh.ibst@srmu.ac.in

Gulab Singh: Department of Bio-Nanotechnology, Chaudhary Charan Singh Haryana Agricultural University, Hisar-125001, India. Email: gulab24@gmail.com

Dwaipayan Sinha: Department of Botany, Government General Degree College, Mohanpur, Paschim Medinipur, West Bengal-721436. India. Email: dwaipayansinha@hotmail.com

Ahmed G. Soliman: Faculty of Agriculture, Department of Biotechnology, Ain Shams University. Email: Ahmedgamal_soliman@agr.asu.edu.eg

Sudarshna: Department of Bioscience & Bioengineering, Indian Institute of Technology Jodhpur, Rajasthan 342037, India. Email: sudarshna@iitj.ac.in

Tharani Sureshkumar: Project Assistant, GRG BIRAC EYUVA Centre, PSGR Krishnammal College for Women, Coimbatore-641004, Tamil Nadu, India. Email: tharani@psgrkcw.ac.in

Gunna Sureshbabu Suruthi: Department of Human Genetics and Molecular Biology, Bharathiar University, Coimbatore 641046, Tamil Nadu, India. Email: sruthigunna11@gmail.com

Mohammaed Taoussi: Phytopathology Unit, Department of Plant Protection, Ecole National of Agriculture Meknes, Km10, Rte Haj Kaddour, BP S/40, Meknes 50001, Morocco and Laboratory of Environment and Valorization of Microbial and Plant Resources, Faculty of Sciences, Moulay Ismail University, B.P. 11201, Zitoune, Meknès 50000, Morocco. Email: medtaoussi00@gmail.com

Anamika Thakur: Department of Biotechnology, Dr Y.S. Parmar University of Horticulture and Forestry, Nauni, Solan-173230, Himachal Pradesh, India. Email: inuthakur95@gmail.com

Rakesh Thakur: Department of Fruit Science, College of Horticulture, Dr Y.S. Parmar University of Horticulture and Forestry, 173230, H.P. Email: rakeshthakur_rt@ymail.com

Murugesan Thentral: Department of Biochemistry, Shrimati Indira Gandhi College, Tiruchirappalli 620002, Tamil Nadu, India. Email: thentralmurugesan@gmail.com

Neslihan Turgut Kara: Istanbul University, Faculty of Science, Department of Molecular Biology and Genetics, Vezneciler, 34134 Istanbul, Türkiye. Email: neslihantk@istanbul.edu.tr

Serena Varotto: Department of Agronomy, Food, Natural Resources, Animals and Environment, University of Padova, Viale dell'Università 16, Legnaro (PD), Italy. Email: serena.varotto@unipd.it

Neha Verma: Department of Botany, Baba Mastnath University, Rohtak, Haryana, India. Email: nehabt977@gmail.com

A.L. Villagómez-Aranda: Center of Applied Research in Biosystems (CARB-CIAB). School of Engineering-Campus Amazcala. Autonomous University of Querétaro. Carr. Amazcala-Chichimequillas, Km 1.0. C.P. 76265. Amazcala, El Marques, Querétaro, México. Email: annvillaranda@gmail.com

Muhammad Waseem: College of Tropical Agriculture and Forestry, Hainan University, Haikou, Hainan, China; Key Laboratory of Tropical Horticultural Crop Quality Regulation, College of Horticulture, Hainan University, Haikou, Hainan, China; Yazhou Bay Seed Laboratory, Sanya Nanfan Research Institute of Hainan University, Sanya 572025, China; and Fang Zhiyuan Academician Team Innovation Center of Hainan Province, China. Email: m.waseem.botanist@gmail.com

P. Servin Wesley: Assistant Professor, Department of Biotechnology, SAFI Institute of Advanced Study, Malappuram-673633, Kerala, India. Email: servinwesley@siasindia.org

Sangilimuthu Alagar Yadav: Department of Biotechnology, Safi Institute of Advanced Study (SIAS), Via Ramanattukara, Malappuram Dist., Kerala-673633, India. Email: sangilimuthu.a@kahedu.edu.in

Nesma A. Zakari: Faculty of Science, Department of Biotechnology, Cairo University. Email: nesmaelhawary289@gmail.com

Preface

In the scenario of global climate change, the threat of environmental pollution, and the increasing human population, the issue of nutrition and food security with plant sources is getting more and more critical. Also, biotic stressors including pests, bacteria, viruses, and fungi continue to damage the growth of crops and therefore cause serious loss of food products across the world. Importantly, the threats of biotic (pathogenic) and abiotic (environmental) factors can be even worse when faced with global climate change, particularly rising temperatures. In 2020, it was reported that nearly 690 million people, equal to 8.9% of the global population, are facing the problem of hunger, and it has risen dramatically by nearly 60 million in 5 years. The challenge of global food security is getting severe, as the world may need to produce about 70% more food by 2050 to feed an estimated 9 billion people. In the past few years, the approach of climate-smart agriculture guided actions to transform the agri-food system toward being greener and more climate-resilient, and it supports SDGs (Sustainable Development Goals) by the United Nations and the Paris Agreement to tackle three main objectives, including: (i) sustainably increasing agricultural productivity and incomes; (ii) adapting and building resilience to climate change; and (iii) reducing and/or removing greenhouse gas emissions, where possible.

In agricultural food production, a range of stressors continue to cause huge losses. Fortunately, modern crop breeding approaches have upgraded rapidly, and one of the promising technologies is based on epigenetics, which leads to crop improvement without altering DNA sequences to obtain resilient and stress-tolerant crops. Therefore, the epigenetics system is undoubtedly a powerful tool to support the goal of climate-smart and sustainable agriculture.

The book provides current achievements and future perspectives on epigenetics and epigenomics on climate-smart and sustainable agriculture and it presents promising ways to integrate the resources and technologies with plant functional genomics as well as precision molecular breeding in crops and includes current findings of crop improvement and breeding strategies involving epigenetics strategies to produce future crops with stress-tolerant and climate-smart capacities, which have better performances on growth, and development to achieve enhanced production and better quality of foods. It covers all kinds of critical stressors of temperature, salt, drought, pollutants, pests, fungi, bacteria, viruses, etc. Also, it comprehensively presents the part that crop improvement plays, using epigenetics technologies, in increasing quality and yield and the efficiency of nutrient utilization, and raising ethical and societal issues, and future challenges.

Each year, agricultural production has a huge loss caused by plant diseases that damage global food and nutrition security. The situation can be worse when faced with climate change and increasing human population. For decades, plant molecular biotechnologies have been expected to lead strategies and breeding programs for mitigating the negative impact of plant pests and pathogens. Among them, epigenetics approaches hold promise to release the potential of immunity for improving plant protection by fine-tuning gene expression patterns that are able to control without any modifications to DNA sequences. Using epigenetic strategies, scientists have the opportunity to promote plant disease resilience by controlling epigenetic machinery, manipulating non-coding RNAs, modifying chromatin structures, and editing epigenomes.

This book summarizes ways to conduct epigenetics research, construct epigenomes, and integrate with functional genomics, including approaches of chromatin immunoprecipitation following sequencing (ChIP-Seq), whole-genome bisulfite sequencing (WGBS), DNA methylation analysis, histone modification profiling, and small RNA sequencing, and additionally it introduces publicly available databases for accessing epigenomics data. Current achievements in stress-responsive cell signaling and the underlying key machinery involving epigenetic regulation have been included, and this knowledge might advance crop improvement to combat diverse biotic stressors.

The book illustrates the role of non-coding RNAs in epigenetic regulation and their applications in crop improvement and precision breeding, i.e. by CRISPR/Cas9-mediated epigenome editing. For post-transcriptional gene regulation, plant microRNAs (miRNAs) possess the capacity to regulate plant responses, chiefly via impacts on transcription factors, to a wide range of stressors, including biotic factors, to combat pests and pathogens. Another type of powerful key player belongs to short-interfering RNAs (siRNAs), which engage in plant immunity against biotic stressors by regulating the interaction between plants and pests or pathogens.

The book includes frontier knowledge of plant epigenetics and epigenomics and is highly recommended to students and young researchers as an ideal reference for overviewing one of the burning topics in plant science. Hopefully, it can inspire their future directions for organizing precision crop breeding programs and lead to supporting sustainable agriculture and healthy ecosystems.

This is an ideal complete guide for students, researchers, experts, and professors to efficiently overview such a critical topic covering all aspects of epigenetics, epigenomics, and emerging RNA technologies, from fundamental and advanced tools and current applications in plant sciences, plant functional genomics, plant stress physiology, plant biotechnology, plant pathology, and plant breeding.

The book's editor, Dr Jen-Tsung Chen, appreciates all chapter contributors and the publisher's staff at CABI. Thank you for everything!

1 CRISPR Technologies for Precise Epigenome Editing and Current Applications

Naglaa A. Abdallah[1,2*], Aladdin Hamwieh[3], Khaled Radwan[2,4], Nourhan Fouad[3] and Michael Baum[3]

[1]Department of Genetics Faculty of Agriculture, Cairo University, Cairo, Egypt; [2]National Biotechnology Network of Expertise, ASRT, Cairo, Egypt; [3]Department of Biotechnology, International Centre for Agricultural Research in the Dry Areas (ICARDA), Giza, Egypt; [4]Department of Biotechnology, Agricultural Genetic Engineering Research Institute (AGERI), ARC, Giza, Egypt

Abstract

Gene regulation mediated by epigenetic mechanisms includes DNA methylation, the regulation of non-coding RNAs, the involvement of small RNAs and the histone modification of genes. Epigenetic regulation leads to heritable changes without altering the genome sequence, making it a promising technology for the development of non-transgenic/non-mutagenic modifications. CRISPR technologies are widely used in plant breeding, in understanding and controlling biological functions, and in the development of various epigenome engineering technologies. In recent decades, epigenetics research has evolved, leading to the introduction of epigenome editing technologies for regulating responses to various environmental stresses, regulating growth and development, and treating various diseases. Various efforts have been made to successfully apply epigenome editing *in vivo*, e.g., to improve target specificity, enzymatic activity, and drug delivery for the development of reliable therapeutics. In this chapter, we summarize the recent progress of CRISPR/Cas9 in epigenome editing and introduce the current limitations and future challenges in the application of epigenome editing.

Keywords: CRISPR/Cas, Epigenome editing, DNA methylation/demethylation, Histone acetylation/methylation, Chromatin dynamics, RNA engineering

1.1 CRISPR Technologies

1.1.1 CRISPR/Cas systems: an overview

CRISPR (short for clustered, regularly interspaced short palindromic repeats) and CRISPR-associated (Cas) systems are natural examples of bacterial/archaeal adaptive immunity (Lam and Ye, 2019). They are acquired genetic immunity systems against mobile genetic elements (MGEs), mainly phages and plasmids (He *et al.*, 2018). A CRISPR/Cas system works by storing genetic information about past invasions by adding invader sequences to CRISPR loci (Lam and Ye, 2018). The CRISPR array is made up of short direct repeats that are separated by spacers that

*Corresponding author: naglaa.abdallah@agr.cu.edu.eg

 Epigenetics for Climate-Smart and Sustainable Agriculture (ed. Jen-Tsung Chen)
DOI: 10.1079/9781800626102.0001

come from foreign genetic elements. It also has different Cas genes on either side (Kamruzzaman *et al.*, 2022).

CRISPR/Cas systems provide a breakthrough platform for cheap, programmable and effective sequence-specific DNA targeting (Vora *et al.*, 2016). They have become reliable and powerful genome editing tools (Tang *et al.*, 2019). More importantly, for the optimization of genome editing and other applications, such systems are essential (Mayorga-Ramos *et al.*, 2023). CRISPR/Cas systems provide fitness benefits for bacteria and archaea in the canonical role of defense against phage (Shehreen *et al.*, 2019). Different types of CRISPR/Cas systems exist, with class 1 systems (types I, III, and IV) considered to be older while class 2 systems (types II, V and VI) are commonly utilized for modifications (Sun *et al.*, 2020). A typical CRISPR/Cas system includes an operon, a cluster of Cas genes, a leader sequence and a CRISPR DNA array (Zheng *et al.*, 2020). These systems are made up of CRISPR arrays containing spacer sequences surrounded by repeat sequences along with a number of Cas proteins (Mortensen *et al.*, 2021).

1.1.2 The rise of CRISPR-based technologies

CRISPR technology has revolutionized the processes of genetic engineering and editing. CRISPR and CRISPR-associated (Cas) systems were first found to be a prokaryotic adaptive immune system. They have quickly become powerful tools for editing DNA (Barrangou and Marraffini, 2014). It is also stated that these easy-to-use systems offer fantastic potential and flexibility for genome editing, transcriptional control, and various DNA targeting applications.

In genetic and metabolic engineering, disease modeling, or therapeutic applications, the ability of the CRISPR/Cas system has been exploited to target specific DNA and RNA sequences accurately with high efficiency (Aman *et al.*, 2020). Multiple CRISPR/Cas systems have been developed, such as type II Cas9 and type V Cas12a. This has given rise to huge changes in technology and the same can be used now in many different assessments (Liu *et al.*, 2020).

The increasingly widespread CRISPR/Cas9 technology has become an easily accessible, programmable and precise gene-editing tool with numerous applications, particularly in biomedical research (You *et al.*, 2019). This technology has enabled researchers to generate gene knockouts, knock-ins, and point mutations in eukaryotic genomes with high efficiency (Wu *et al.*, 2019). Furthermore, the development of CRISPR/Cas systems goes beyond genome editing and includes applications in synthetic biology, disease treatment, agriculture, and cancer research (Tian *et al.*, 2019; Riaz *et al.*, 2022; Bandyopadhyay *et al.*, 2023; Ning *et al.*, 2023). CRISPR technologies have enhanced the researcher's ability to specifically control gene expression, facilitated the development of disease-resistant plant varieties, and allowed for a higher ability to detect pathogens (Lin *et al.*, 2017; Mutezo *et al.*, 2021; Mimoune *et al.*, 2022).

The development of CRISPR techniques from a prokaryotic defense mechanism to a manipulable genome editing tool has facilitated a broad range of advances across scientific and medical fields. With advancements in CRISPR methods, there is still a lot of discovery and progress to be made with manipulation.

1.1.3 Mechanisms of CRISPR/Cas systems

Acquired immunity against invasive genetic elements in bacteria and archaea is provided by CRISPR-associated (Cas) and CRISPR to confer acquired immunity to invasive genetic elements (Barrangou and Marraffini, 2014). The CRISPR/Cas systems work by means of a sequence of steps that result in the accurate targeting and degradation of foreign nucleic acids.

The adaptation process, which involves incorporating tiny DNA fragments from invasive elements into CRISPR arrays, is one of the main characteristics of CRISPR/Cas systems (Gasiunas *et al.*, 2014). The acquisition of these spacers is transcribed and further processed into short CRISPR RNAs (crRNAs) that guide the system to target and cleave matching sequences in future invasions (Marraffini and Sontheimer,

2008). It is also argued that the near-perfect identity between CRISPR spacers and target sequences is essential for effective immunity from bacteriophage infections.

CRISPR/Cas systems work by putting together ribonucleoprotein effector complexes made up of Cas proteins and crRNAs. These complexes bind to the target DNA and form an R-loop structure, which then interferes with the intruder's genome (Tuminauskaite *et al.*, 2020). Different types of CRISPR/Cas systems exhibit variations in their mechanisms, such as RNA interference, DNA interference, and cyclic oligoadenylate signaling, which are observed in type III systems (Zink *et al.*, 2020).

In addition, CRISPR/Cas systems have precisely modified genomes in various organisms, with class 2 systems such as Cas9 and Cas12a playing the central role in the redirection, or rather repair, of DNA (Cubbon *et al.*, 2018). These systems rely on host DNA repair accidents to clear the debris of DNA double-strand breaks. The researchers concluded by highlighting the tight relationship that CRISPR/Cas systems appear to have with cellular machinery.

CRISPR/Cas systems have adaptation, processing, and interference modes. In this way, they can mount something like a targeted immune response against an invading foreign genetic element. The more we understand these mechanisms, the better CRISPR technologies can edit genomes to cure disease and advance therapeutic development.

1.2 Epigenome Editing: Principles

1.2.1 Epigenome: definition and importance

We can simply define the epigenome as the process of DNA methylation, histone modifications, and non-coding RNA in cells of an organism. In places where the DNA sequence has not yet changed, it governs the essential gene regulatory system and therefore carries out another normal function without any mutation needed. In this way, various traits will be expressed in future generations (Skaar *et al.*, 2012). It serves as a mediator between the genome and the phenotype, determining exactly what form the genetic data will take. It therefore has a significant impact on the expression of the genotype – or the way our genes tell us what they naturally carry (Villota-Salazar *et al.*, 2016).

Understanding the epigenome is essential since it has such a major impact on processes such as development and differentiation; especially nowadays, shaping it also means contributing significantly to susceptibility to diseases. Epigenome marks are not only inherited but can also be transmitted in both somatic cells through cell division (Konishi and Issa, 2007) and germline too; as a result, there is cellular stability. The epigenome has an important role in cellular identity and response to the environment (Aguilar and Craighead, 2013).

Moreover, the role played by the epigenome in the making of individual traits is demonstrated by response to environmental change. Beyond this, researchers generally agree that epigenetic change is a mechanism by which outside influences such as lifestyle and diet play on gene expression at the cellular level (Sun, 2014). The adaptable nature of the epigenome makes those reversible changes susceptible to influence by our way of life, what we eat, and therapies exercised on us (Maeder and Gersbach, 2016).

Regarding disease, environmental changes have been associated with diseases like cancer, neurodegenerative diseases, and metabolic syndromes. As a result, epigenetic abnormalities in gene expression are capable of serving as a direct mechanism that leads to the development of disease (Jenuwein, 2006). By studying the epigenome, we can not only see how diseases work but also find out how doctors can detect a disease in its earliest stages. So far, nothing has been better at identifying what is going on than the epigenome. The discovery of specific patterns of histone acetylation marks as biomarkers will soon be on almost every screen imaginable and could also open new possibilities for therapeutic intervention (Hariprakash and Ferrari, 2019).

1.2.2 Conventional techniques in epigenome editing

The editing of the epigenome has been known to be the targeted modulation of chromatin structure and gene expression traditionally. Zinc-finger proteins (ZFPs) and transcription activator-like effector proteins (TALEs) have offered key analyses in targeting the genome at single-base pair resolution (Urbano *et al.*, 2019). These technologies have enabled researchers to modulate gene expression by altering the epigenetic landscape at the desired locations in the genome.

In addition, researchers have employed nuclease-free genome editing systems like catalytic dead Cas9 (dCas9) along with effector(s) (Laufer and Singh, 2015). Using dCas9 to guide chromatin modification enzymes, the researchers created a tool that could regulate gene expression without altering the DNA sequences. It is the optimal instrument for research on gene regulation and cell function.

Nevertheless, epigenome editing technologies have now advanced to incorporate CRISPR/Cas systems, particularly CRISPR/Cas9, allowing the epigenome to be manipulated with precision (Gemberling *et al.*, 2021). Utilizing CRISPR-based systems, a gene or non-coding region can be controlled with great exactness; a potent tool for studying basic biological actions through modifications in epigenetics.

In addition, site-specific manipulation of DNA methylation by CRISPR/CRISPR-associated protein 9 (CRISPR/Cas9) has also demonstrated promise for getting epigenetic-based molecular breeding in crop plants under way (Gahlaut *et al.*, 2020). This method enables the customization of targeted DNA methylation marks that can modulate gene expression and other traits in crops.

In the tuning of the epigenome, ZFPs, TALEs and CRISPR/Cas systems are commonly used for site-specific targeted modifications on the epigenetic landscape. These technologies will enable a new world that studies gene regulation and how this impacts diseases and even programs an organism's epigenome, as is being done in agriculture.

1.2.3 Using CRISPR to edit the epigenome

The CRISPR-based epigenome editing toolkit makes a significant contribution to not only genetics but also epigenetics. This generally results in the ability both to control gene expression accurately and to bring about chromatin changes that go far beyond single-gene family modifications. As a result, these systems changed the working practice of everyday biologists. The tools in turn enabled researchers to investigate just what role various kinds of epigenetic changes played in a number of different biological processes and conditions. CRISPR-mediated epigenome editing can be employed for both *in vivo* and *in vitro* studies of gene regulation and disease. A perfect solution is represented in the form of a CRISPR/dCas9 fusion protein that offers a powerful and precise alternative for epigenome editing and transcriptional modulation. It enables the directed manipulation of gene expression and chromatin modifications.

Dead Cas9 (dCas9) has been remodeled for epigenome editing and has the ability to bind DNA to target specific loci without cleavage (Qi *et al.*, 2013). Various epigenome control toolkits have been developed to control transcriptional activation and repression of genes through histone modifications (Chen *et al.*, 2022), DNA methylation, and demethylation (Yang *et al.*, 2018). Nevertheless, Gilbert *et al.* (2013) demonstrated the functionality of dCas9 fusion proteins in mimicking various native protein interactions, such as the KRAB domain of Kox1, VP16 and p65AD.

In addition, dCas9 has been used as CRISPR interference (CRISPRi) to downregulate gene expression through altered epigenetic signaling pathways, including histone acetyltransferases or histone demethylases (Elliott *et al.*, 2021), while CRISPR activation (CRISPRa) has been used to upregulate gene expression. Moreover, Li *et al.* (2018) introduced dCas9 fusion protein-mediated CRISPRi systems such as dCas9-KRAB and dCas9-LSD1 for epigenetic modifications.

In various biomedical applications, the potential of CRISPR/Cas9 has been employed for precise genome editing and epigenome editing (Ferré *et al.*, 2020). In a study by Sugita *et al.* (2021), which was performed *in vivo*, the results confirmed the successful use of CRISPR/Cas systems for epigenome editing, also showing promising effects on treatments of mouse models with diabetes, muscular dystrophy, and acute kidney disease.

Several references have emphasized the importance and versatility of CRISPR-based epigenome editing tools in elucidating gene regulation, disease mechanisms, and potential therapeutic interventions (Lau and Suh, 2018). Vetchinova *et al.* (2021) discussed *in vivo* epigenome editing and transcriptional modulation using CRISPR technology and emphasized the utility of nuclease- dCas9 for binding to target DNA without inducing cleavage. To add to the complexity, Rajaram *et al.* (2023) also revealed an even more precise epigenome editing at allele-specific DNA methylation with CRISPR-based methods. The CRISPR/dCas9 system is utilized with enzymes like methylases and demethylases which mobilize key DNA-modifying domains (Mushtaq *et al.*, 2019).

1.3 Types of Epigenetic Modifications

1.3.1 DNA methyltransferases and demethylases

Gene methyltransferases and demethylases play a crucial role in targeted epigenome editing by enabling precise modifications of DNA methylation models. For epigenetic repression, DNA methylation is used as a modification system by adding a methyl group to the fifth carbon position of cytosine. The enzyme DNA methyltransferases catalyze DNA methylation; at the same time the enzymes TET1–3 remove the methylation modification by attaching a hydroxyl group to the methyl group and thus converting it back to cytosine (Moore *et al.*, 2013).

The fusion of dCas9 with methyltransferases and demethylases is a powerful method for precise epigenome editing since it allows researchers to modulate DNA methylation models and investigate the role of epigenetic modifications in gene regulation and disease mechanisms (Fig. 1.1). Many epigenome editing technologies were developed for DNA methylation by fusion of dCas9 with methyltransferases, such as DNA methyltransferase 3A (Dnmt3a) and DNA methyltransferase 3L (Dnmt3L), or demethylases such as TET1, to enable targeted DNA methylation or demethylation (Katti *et al.*, 2022). To create a targeted DNA methylation control system, these domains are fused to dCas9 and used with synthetic sgRNA molecules specific to target sites of the genome (Stepper *et al.*, 2017; Josipović *et al.*, 2019; Nguyen and Lister, 2021). To avoid off-target effects and achieve more specific and effective methylation of target sites, a dual dCas9-SunTag DNMT3A system was developed, in which the SunTag peptide and the DNMT3A single-chain antibody are fused (Pflueger *et al.*, 2018).

1.3.2 Histone modifications

Histones define chromatin architecture and can up- or downregulate gene transcription; and they can be modified in various ways, including acetylation, methylation, phosphorylation, glycosylation, and ubiquitination. Post-translational histone modifications can regulate gene expression by influencing the binding of histones to DNA. These alterations lead to either activation or repression of target

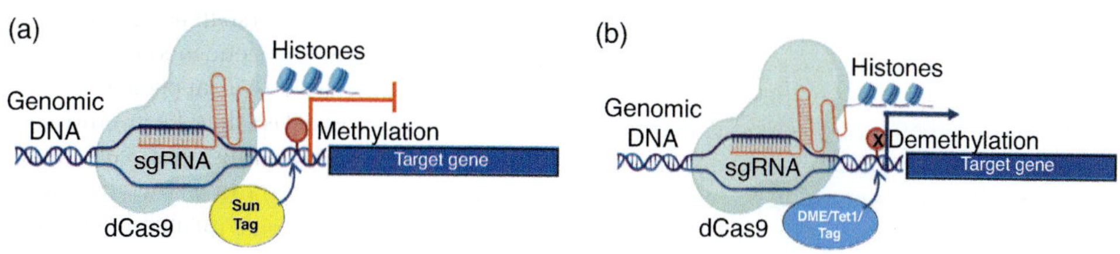

Fig. 1.1. CRISPR/dCas epigenetic regulation via DNA methylation/demethylation (Biorender). (a) The DNA methyl transferases Sun Tag repress gene expression by adding methyl groups to the promoter. (b) Demethyltransferases (such as Tet1 and Tag) remove the methyl group and activate gene expression. Author's own image.

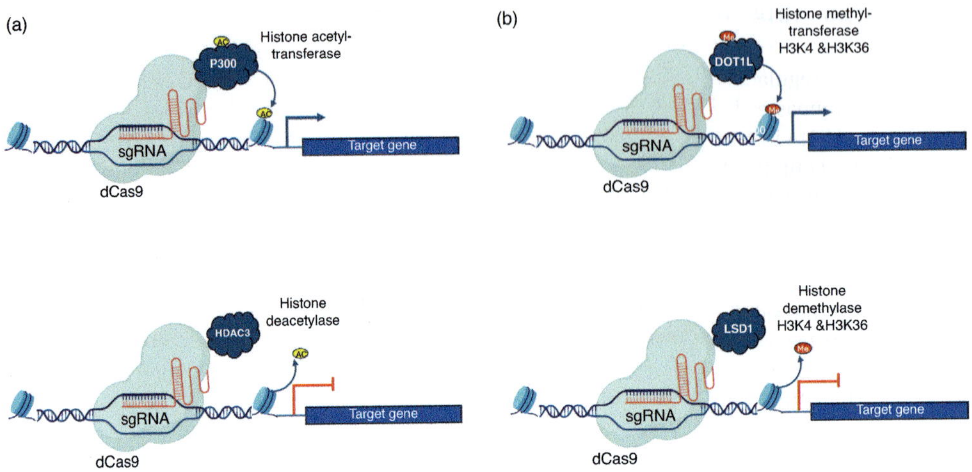

Fig. 1.2. CRISPR/dCas epigenome editing via histone acetylation/methylation (Biorender). (a) Histone acetyltransferases such as P300 at the promoter region activate gene expression, while histone deacetylation causes gene repression. (b) Histone methylation at positions H3K4 and H3K36 using DOT1L activates gene expression while removing methyl groups from these positions represses gene expression. Author's own image

genes, by changing the compacted or relaxed state of chromatin or by recruiting other downstream effectors. Histone acetylation is the transactivation of a particular locus and promotes gene expression by reducing the affinity of binding histones to DNA. However, the crosstalk between histone acetyltransferases and histone deacetylases determines the status of histone acetylation (Hu and Du, 2022). Lysine methylation modifications at H3 histones showed that dimethylation and trimethylation modifications at H3K4 and H3K36 are associated with activation of gene expression; on the other hand, methylation modifications at H3K9 and H3K27 are usually associated with gene repression (Jambhekar *et al.*, 2019).

In general, histone acetylation regulatory sites have higher stability and cell activity than methylations. Epigenome editing by CRISPR/dCas9 histone modifiers is a powerful approach that can precisely direct modifications of histone marks to control gene expression and chromatin structure (Fig. 1.2). Several studies have demonstrated that CRISPR/Cas systems can reprogram chromatin state and transcription by coupling dCas to enzymes that mediate histone acetylation and methylation (Cano-Rodriguez *et al.*, 2016; Fukushima *et al.*, 2019; Chiarella *et al.*, 2020; Li *et al.*, 2020; Zhao *et al.*, 2020; Gemberling *et al.*, 2021; Zhao *et al.*, 2021). In addition, the fusion of dCas9 with histone methyltransferases such as DOT1L or histone demethylases such as LSD1 allows the targeted addition or removal of histone modifications, as reported by Godfrey and Rodriguez-Meira (2022).

The catalytic domains of acetyltransferases such as p300 are frequently used with CRISPR-based editing systems to modulate proximal and distal enhancers for gene regulation. The fused dCas9-p300 proteins acetylate the histone3-lysine 27 residue and lead to the transactivation of endogenous genes (Hilton *et al.*, 2015). To repress the transcription of endogenous promoters, CRISPR/Cas9 histone deacetylase proteins were developed by fusing the dCas9 and HDAC3 proteins. The fusion system represses transcription at endogenous promoters (Kwon *et al.*, 2017). To optimize dCas9-based epigenome editing tools, several effector systems showed better persistent and strong modulation (Li *et al.*, 2020). Thus, the CRISPR/dCas9 histone modifier approach offers promising opportunities for the manipulation of histone marks and chromatin structure, providing insights into gene regulation, disease mechanisms and potential therapeutic interventions through targeted epigenome editing.

1.3.3 RNA engineering

The class of regulatory non-coding RNAs (ncRNAs) encompasses long non-coding RNAs (lncRNAs) and RNA interference (RNAi), among several classes. Long ncRNAs have been shown to be involved in a large number of cell processes, such as transcriptional regulation, post-transcriptional regulation, genomic imprinting, and chromatin remodeling (Engreitz *et al.*, 2016). On the other side is RNAi, which is much more of a post-transcriptional gene silencing methodology and cannot really edit genes in the same way. The CRISPR-based lncRNA editing comprises targeted deletions or mutations and gene activation or repression by dCas9 fused to transcriptional effectors domains KRAB and VP64 (Morelli *et al.*, 2021; Arnan *et al.*, 2022).

In addition, the RNA editing enzyme Cas13 could be used for direct targeting of the lncRNA molecules to degrade the lncRNAs without altering the genome sequences (Gupta *et al.*, 2022). In addition, an inactive Cas13 protein (dCas13) was developed for base editing of the RNA targeting platform without destroying the RNA. Fusion of RNA base-editing proteins as adenosine deaminases is utilized to fuse adenosines to inosines (Tan *et al.*, 2017), resulting in transcript modifications. Recently, Cas7-11 was developed for programming the RNA cleavage system and showed higher specificity and cell toxicity than Cas13 (Ozcan *et al.*, 2021).

In addition, mRNA modification could be used for epigenetic editing. Internal mRNA modification could be achieved with N6-methyladenosine (m6A) to regulate mRNA stability and induce gene repression. The RNA binding system CRISPR/dCas13a was developed by fusion with m6A to bind and edit messenger RNA (mRNA) (Fang *et al.*, 2023). The researchers added that adenosine methylases targeted to specific mRNA by dCAS13a caused mRNA degradation and gene repression in *Arabidopsis*, while for demethylation editing, the m6A demethylase ALKBH5cd was used to activate gene expression in plants. In general, the use of dCas13a showed lower off-target activity compared with dCas9 (Wilson *et al.*, 2020).

1.3.4 Modification of regulatory regions

A characteristic of the genome is that, while containing genes, there are also *cis*-regulatory regions, which are non-coding regions responsible for gene regulation. A decrease in histone density causes it to open chromatin, making it accessible for transcription. Altering the chromatin folding including modulation of the non-coding regions, *cis*-regulatory regions or regions distant from the target gene could be possible through changes in histone conformation as well as through using DNA-binding regulatory proteins, inserting these epigenetic tools and their interactors to control gene transcription and maybe gene silencing at certain genomic location (Splinter and Laat, 2011; Splinter *et al.*, 2013; Zhou *et al.*, 2014).

CRISPR-based regulatory studies have been used to identify target genes in regulatory regions and to alter their expression. The CRISPR/Cas system was developed for epigenetic modifications at the promoters and the proximal and distal genomic enhancer regions (Fig. 1.3). The dCas9-KRAB repressor and the 9Cas9-p300 activator are commonly used for epigenetic modifications and to study loss- and gain-of-function effects (Klann *et al.*, 2017).

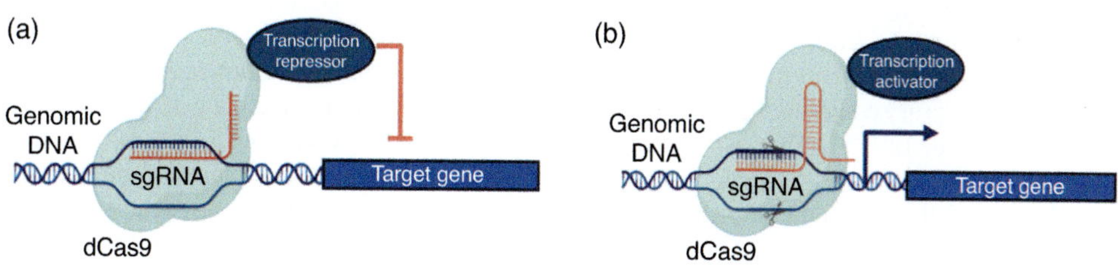

Fig. 1.3. CRISPR/dCas epigenome editing via modification of gene regulation (Biorender). (a) Fusing dCas9 with a transcription repressor will silence gene expression, while (b) fusion dCas9 with a transcription activator will enhance gene expression. Author's own image.

In the epigenetic repressor module, CRISPR interference (CRISPRi) has successfully mediated downregulation and is used to precisely target specific genomic platforms without altering DNA sequences. In this case, the dCas enzyme is fused to the transcriptional repressor domain (TRD) to disrupt gene expression (Xu *et al.*, 2023). A 3_SRDX fused by dCas9 was applied to silence the expression of *TaPDS* in wheat, and it was reported that such silencing could be inherited (Zhou *et al.*, 2022). Also, the applications of TRDs like D144, DLS and MIX have proved active in the repression of transcription in both dicot and monocot plants (Xu *et al.*, 2023).

In the CRISPR/dCas9 activation system (CRISPRa), dCas9 is linked to the target gene with transcriptional activators (TADs) and other transcriptional activators (Fang *et al.*, 2023; Shi *et al.*, 2024), using a complementary guide RNA. The *Streptococcus pyogenes* Cas9 (pcoCas9), fused with transcriptional activation domain VP64, has been widely used for multiple-target gene promoter activation targeting (Liu *et al.*, 2019). Nevertheless, the recruitment of VP128 with multiple TALE-TAD motifs was more efficient at activating multiple genes than VP64 (Wilson *et al.*, 2020).

1.4 Application of Epigenome for Basic Research

1.4.1 Gene regulation and chromatin dynamics

CRISPR essentially lets researchers tweak nature in such a way that they could well do things that had never been done before by being able to successfully edit the epigenetic landscape in a very limited manipulation. Even the capability of repressor elements and CRISPR/Cas9 to change long-range regulatory components was reported by Thakore *et al.* (2015). That is, making it possible to modulate gene expression provides gene flow, even the disabilities brought about by genetic mutation, and explores cause–effect links between repression systems on specific genes and their normal automatic regulation. CRISPR/Cas9 epigenome editing has been used in uncommon imprinting diseases, highlighting its importance for both analysis and potential curative applications in these diseases (Syding *et al.*, 2020).

The assessment highlighted the main CRISPR-based epigenome editors and their position in the consideration of genetic regulation and disease systems. In addition, CRISPR epigenome editing has been used to transactivate functionally equivalent genes and direct gene manifestation in diseases developing from heterozygous dominant haploinsufficiency (Bhokisham *et al.*, 2023). This approach, in turn, offers a powerful means to study genetic regulation and related chromatin dynamics during disease. In addition, Zhao *et al.* (2020) emphasized the revolutionary importance of epigenome editing for controlling gene expression and chromatin modifications more precisely. As a tool, the platform is useful for investigation of how gene regulation functions, and as a source of new epigenome therapies. In summary, CRISPR epigenome editing is commonly used in gene regulation and chromatin dynamics, which may provide the key to understanding how genes are expressed, how diseases develop, and how cells function.

1.4.2 Functional genomics and epigenomics

The development of CRISPR-mediated epigenetic editing provides a new foundation for functional genomics and epigenomics. Rather than just changing fundamental parts of DNA, the system can access a site and generate genes directly at that level, or it could turn off an existing gene. The CRISPR/Cas9 system allows researchers to precisely control the cellular landscape at the epigenetic level. Thus, it is now possible to invent mutations within specific genes, each of them reflecting a different species or race, and all of them free from direct linkage to global populations in the air (Policarpi *et al.*, 2021). This technology is important for the study of functional genomics for regulatory elements. It provides researchers with insight into how epigenetic modifications affect gene expression and impact cellular functions (Dong and Kantor, 2021).

The use of CRISPR-based epigenome editing to study gene regulation, chromatin dynamics, and epigenetic modifications in different biological contexts has already been described. The enzyme activity of the ortho-isoenzyme was demonstrated by the fusion of catalytically inactivated Cas9 molecules with effectors. With this method, quantitative manipulations of individual genes could be performed (Syding *et al.*, 2020). This strategy enables the direct modification of chromatin marks that control gene expression and cell function (Pulecio *et al.*, 2017). These researchers also mentioned that CRISPR-based epigenome editing tools, especially those using the dCas9 approach, offer an exciting way to localize and modify specific chromatin marks associated with cell function or pluripotency based on their previous data. Genomic and epigenomic aberrations have been used to study the pathogenic mechanisms of natural killer/T-cell lymphoma (NKTCL), which is associated with Epstein–Barr virus (EBV) and causes non-Hodgkin's lymphoma. Dysregulation of the expression of EBV-associated epigenetic drivers has been identified in epigenomic studies and in key genes (Tan *et al.*, 2024).

1.4.3 Developmental biology and differentiation research

The ability to perform CRISPR editing of the epigenome is a game changer for developmental biologists and others who are attempting to control cell differentiation. It is used for making a modification that only occurs in specific places of the epigenetic landscape. Although the development of this technology is still in its infancy, the CRISPR/Cas9 system has allowed researchers to explore how specific epigenetic marks alter gene expression at different stages of development and differentiation (Klann *et al.*, 2017). Using CRISPR-based epigenome editing tools such as dCas9 fused with effector enzymes, scientists can make changes in gene expression patterns and investigate what effects these in turn have on cellular differentiation (Fu *et al.*, 2019).

Research has shown that CRISPR-mediated epigenome editing works well during developmental processes to alter gene expression and chromatin behavior. Thus, researchers can use this technique to elucidate the mechanisms behind cell fate determination and differentiation (Dong and Kantor, 2021). This technique is an important way to study how histones and chromatin mediate gene activity during development and differentiation (Policarpi *et al.*, 2021).

In addition, CRISPR-based epigenome editing provides a method to test how specific epigenetic changes affect patterns of gene expression during differentiation. By targeting critical regulatory regions and altering the chromatin state, researchers can trigger and study such epigenetic processes in cell development (Syding *et al.*, 2020). The implementation of this strategy also supports a powerful experimental proof-of-concept for the application of small molecules as epigenetic modulators in biological differentiation research (Clement *et al.*, 2020). Taken together, CRISPR epigenome editing adds another essential to our molecular tool belt and provides a solid foundation for the investigation of developmental biology and differentiation processes via targeted manipulation of the epigenetic network. Finally, it could be used to detect how certain off-genome changes moderated gene expression in development or cellular differentiation which helps understanding of epigenetic regulation in controlling gene expression.

1.5 Therapeutic Applications

1.5.1 Cancer therapy: targeting the epigenome

Another potential approach to cancer treatment is epigenome editing technologies based on CRISPR. It relates to the epigenetic modifications that occur in association with tumorigenesis. By controlling DNA methylation, histone abridgments, or chromatin topology in this way, epigenome manipulation allows scientists to alter gene expression in gene-specific manners by altering the expression of individual genes. This represents a new approach to altering cancer cell gene expression patterns (Urbano *et al.*, 2019).

An important merit of epigenome editing as a cancer therapy is that it has the ability to reprogram the expression of essential genes regardless of whether these target genes harbor mutations (Tanaka and Nakamura, 2023). This offers the possibility to act on gene expression in cancer cells at a highly detailed level. This is a very attractive personalized approach to the treatment of various cancers.

Becoming a versatile tool, epigenome editing technologies are expected for the treatment of genetic and epigenetic diseases through precisely targeting pathogenic epigenetic changes that lead to corrective regulation of expression profiles of disease-related genes. It is an interesting route to making the treatments for many diseases more focused and personalized.

1.5.2 Treatment of genetic and epigenetic disorders

Epigenome editing has been suggested as a potential strategy for treating genetic and epigenetic disorders, especially with the tool sets of CRISPR-based technologies specifically tailored for such commonly observed epigenetic modifications implicated in various types of human diseases. A new approach to modulating gene expression patterns in diseased cells is currently offered by epigenome editing tools that utilize Cas9, which can delete DNA methyltransferases and control the activity of histone acetyltransferases (Majchrzak-Celińska *et al.*, 2021). The power of epigenome editing to correct misregulated gene expression programs that underpin genetic and epigenetic disease is a major advantage in using this strategy to treat these disorders. Researchers could target the disruption of specific regulatory elements and changes in epigenetic markers that lead to the changes in gene expression caused by these disorders, and slow or reverse this process (Vetchinova *et al.*, 2021).

The development of epigenome editing technologies allows the specific targeting of particular epigenetic effector domains to distinct, regulatory genomic loci to modulate specific chromatin marks in specific regions and chromatin contexts. The strategy is targeted and can be personalized for therapeutic intervention of diverse genetic and epigenetic disorders. Moreover, in diseases such as cancer and neurodegenerative or autoimmune disorders, epigenome editing offers a potential to reverse aberrant gene expression patterns that correlate with these pathological conditions. It is these changes in the epigenetic marks that lead to the malfunctioning of the genetic program of a cell, but also hopefully can be reverted when targetable by modulators of these expression patterns.

1.5.3 Epigenome editing in neurological diseases: possible targets and applications

Pathological epigenetic changes occur as a result of neurological disorders and are successfully reversed using epigenome editing. This results in the restoration of the cellular gene transcription state and activities (Christofi and Zaravinos, 2019). In other words, the disease-associated pathological gene expression of SYNGAP1 can be turned off by epigenome editing. By targeting specific regulatory elements and modifying epigenetic marks, such errors could be potentially corrected. Furthermore, epigenome editing is a boon in the treatment of neurological diseases as it can modulate expression at very important targets regardless of the specific mutational status (Krichevsky and Uhlmann, 2019). Interestingly, this mode of treatment provides precise regulation over gene expression in neurons and offers a treatment route for many diseases of the nervous system. Thus, epigenome editing technologies offer great prospects for the treatment of neurological diseases.

1.5.4 Epigenetic reprogramming and regenerative medicine

These targeted epigenetic changes are a major subset of those controlled in regenerative medicine and cellular reprogramming with epigenome editing. Epigenome editing tools will permit the fine-tuning of DNA methylation patterns, histone modifications, and chromatin structure, thus providing a new strategy for the modulation of gene expression in different cell types to be used as a target for regenerative medicine (Yang *et al.*, 2016).

Epigenome technologies make it possible to alter the epigenetic landscape to switch cell identity, i.e. to activate or repress genes that are necessary for differentiation and regeneration of tissue damaged by disease or injury. This recapitulates what nature achieves naturally during development, but never in adulthood, as is the case on Earth today (Cervantes-Gracia *et al.*, 2021). This technique makes it possible to alter the functions of cells and promote regeneration mechanisms in damaged or diseased tissue. One of the main draws to this kind of research is that epigenome editing can be used to influence gene expression patterns in such a way as to coerce repair and regeneration mechanisms inside the body.

It is important that targeted regulatory elements and changes in epigenetic marks (Zhao *et al.*, 2020), can help to make stem cells more robust. The technology of epigenome editing can be used to enforce specified epigenetic changes that drive cellular reprogramming and differentiation, thus resulting in the formation of specialized cell types that are suitable for regenerative medicine applications (Tasca *et al.*, 2020). Hence the technique can also serve as one by which scientists treat a variety of diseases with a new source for highly effective repair options. To sum up, epigenome editing stands to become a significant force in the fields of epigenetic reprogramming and regenerative medicine. It enables researchers to specifically turn on or off single genes or sets of them in an intact cell or tissue, allowing precise control over gene expression patterns and hence cellular activities that drive the process of tissue regeneration and repair.

1.6 Application to Modify Complex Traits

1.6.1 Agricultural biotechnology applications

Several studies successfully demonstrated that CRISPR/Cas systems can currently improve traits of interest in agricultural products such as raising animal productivity and welfare, increasing crop output, and improving seed quality (Maximiano *et al.*, 2021). CRISPR/Cas technology has brought about positive impacts on socioeconomics, human health, and environmental sustainability in agriculture.

Technologies such as CRISPR make it possible to modify a specific genomic site to improve traits such as disease resistance, herbicide tolerance, nutrient quality, etc. in crops. As a result, yields can either be greatly improved or production cost reduced – or both at the same time (better plant quality) – with CRISPR-based genome editing. It is also expected to use CRISPR on a number of other plants in the future, including rice and wheat, which could become the substrate crops for CRISPR ready-to-eat tomatoes, corn, and peas. It has revolutionized crop development for useful traits and altered the conventional breeding method.

A biotechnological tool that this technology offers for crop improvement overcomes any weaknesses that might limit other methods to achieve such results and their findings will soon be put into practice in the form of transformed genetically improved plants. Taken together, advances at the epigenome level could revolutionize agriculture, making it more effective in providing a greater quantity of food with resources that are urgently needed to overcome today's crop challenges.

1.6.2 Epigenetic enhancement in crop plants

Epigenome editing has important applications in agricultural biotechnology politics to improve various traits of crops and animals. The epigenome changes that give rise to desirable traits and epigenome editing tools can re-purpose. It offers an entirely new way of improving agricultural products.

The CRISPR-based epigenetic amplification of crops is also a promising approach to increase the yield. Bioengineered trait generation in crop plants employing tools like epigenome editing to specifically target the epialleles (alleles whose phenotypic effect differs depending on parental origin), linked with complex traits, will permit programmable manipulation of gene expression patterns in crops and thereby genetic agronomic traits. There have been a number of studies demonstrating how CRISPR/Cas systems can be used to safely control gene expression by enhancing the specificity for targets while carving down off-targets in crop improvement. Recently, some recombinant-engineered Cas9 cassettes like Nmcas9, Stcas9, or Sacas9 have been used to enhance target specificity and to reduce off-target cleavage when applied in crops (Jaganathan *et al.*, 2018). This technique of gene editing is sufficient to realize the demands of crop research and agriculture, to make crops grow more efficiently. The landscape of epigenetics in crops is the only tool that enables fine genetic manipulations of crop traits with proven beneficial effects on crop yield (Ahmar *et al.*, 2020).

However, given the fact that it is indeed credible to increase plant stress resistance and has substantially increased overall agriculture yield, CRISPR/Cas systems were developed. Plant researchers have applied soil cultivation experiments to improve crop varieties by adjusting gene expression patterns at the epigenetic level (Li *et al.*, 2021). In cooperation, these results suggest that the use of CRISPR-based technologies for epigenetic crop enhancement is a potent instrument to enhance plant characteristics, improve stress resistance and aggrandize field production. This is a powerful technology that could modernize crop development efforts and help us meet the food security challenges as well.

1.6.3 Animal breeding and trait improvement

Recent advances in animal breeding using epigenetics and CRISPR-based genetic technology to improve traits have enabled a new and powerful instrument for livestock production. By introducing specific epigenetic changes into the gene sequences that control a variety of features, those epigenetic editing tools in animals will affect pattern identification within genes and therefore how they express their qualities. CRISPR/Cas technology has also been successfully applied to animal breeding to improve key agronomic traits in animals and enhance their performance.

Such techniques have allowed the editing of the epigenetic constitution of animals, thereby generating animals that are not only disease-resistant but also fast-growing and better in meat quality (Khwatenge and Nahashon, 2021). CRISPR-associated proteins like CRISPR/Cas9 being used have been a milestone in enhancing various agronomic traits in offspring to non-transgenic mutants for different species. They allow for genome manipulation to control transcription, site-specific targeting and transfection of recombinant DNA, or drug delivery within the animal pulp.

1.7 Ethical, Legal, and Social Implications

1.7.1 Ethical perspectives on epigenome editing

Introducing changes to the epigenome, including those induced by CRISPR-derived technologies, will induce several ethical considerations which call for in-depth discussion. The major ethological concerns on epigenome editing are as follows.

1. **Informed consent:** Before an epigenome editing study can be carried out, it must fairly disclose the potential risks, benefits, and consequences to those who will participate in the study. This approach ensures respect for the autonomy and rights of research participants (Dyke *et al.*, 2019).
2. **Ethical issues on data privacy and data sharing:** While we conducted this analysis, some of the most important ethical considerations were the areas in which epigenetic data privacy and data sharing are concerned (Bunnik *et al.*, 2020).
3. **Equity and access:** The full value of the epigenome editing technology will only be realized if its benefits are available to all elements of society, regardless of geography or socioeconomic status. In the context of epigenetics, for example, efforts are required to improve equity in research to ensure equal access for all people to this technology (Uchiyama *et al.*, 2018).
4. **Potential danger:** While CRISPR-based epigenomic editing technologies allow genome-wide editing abilities, a limited number of libraries can detect a given sequence on the genome and thus possible

damages off-target are still prevalent. Given such off-target potential, there has been a call for caution in developing safe and efficacious interventions, particularly for epigenetic editing.

5. **Inheritance:** Epigenome editing technologies permit inheritance of the induced changes to progeny. Deeper ethical concerns about the impact *per se* of heritable epigenetic changes and the implications this might have for future generations will have to be balanced with great caution (Zeps *et al.*, 2021).
6. **Regulatory management:** Epigenome editing studies can only be carried out under strict regulations and ethical standards. The restriction of responsible science and ethical obligations is required and relevant guidelines, as well as frameworks, must be observed by the conducting studies (Zeps *et al.*, 2021).
7. **Societal and cultural issues:** The use of epigenome editing technologies will also raise a number of societal and cultural concerns and ethical dilemmas about tampering with genomic material. Consideration of these socio-ethical implications by diverse constituencies, including those with organizational responsibilities, designers, ethicists, and policymakers together, is vitally important (Uchiyama *et al.*, 2018).

Whilst the potential to manipulate cellular machinery in this way could offer new possibilities for the treatment of a range of severe and life-limiting conditions, researchers undoubtedly will also need to be alert to these sorts of ethical concerns if they are to know that epigenome editing research is being undertaken responsibly and ethically, with proper regard for the interests and rights of individuals and communities affected by such work.

1.7.2 Policies/regulatory framework

To utilize this technology in a responsible manner, an efficient regulatory framework and policies of epigenome editing are vital tools without which this technology could not be beneficial. Before epigenome editing can be considered for clinical trials, there are a number of important questions that will need to be addressed in the design of the regulatory framework.

1. **Ethical oversight:** Regulatory agencies and ethics committees play a critical role in overseeing epigenome editing research to ensure that ethical standards are met. Limiting ethical oversight may thus enable more equitable access to epigenome editing for different social groups.
2. **Consent provision:** Research workers must get written permission from individuals registered in epigenome editing studies. The ideal procedure is more of a recognition, as the actual process would be contingent upon national laws and likely vary between countries and cultures due to varying legal needs.
3. **Data protection:** For epigenome editing research to be conducted, it must still conform to data protection and security regulations. Ethics oversight boards and IT experts should be consulted in the protection of data from epigenome editing research.
4. **Safety and efficacy:** Regulatory systems must ensure the safety and effectiveness of epigenome editing interventions. Research scientists should perform thorough risk assessments and if necessary, aim to minimize the generation of off-target effects on participants' safety in the studies (Zhao *et al.*, 2020).
5. **Public involvement:** In setting the regulatory framework for epigenome editing, public and stakeholder involvement is of the utmost importance. Public involvement refers to transparency, accountability and inclusivity as soon as decisions about research on epigenome editing are made (Perera *et al.*, 2020).
6. **International guidelines:** The basis of a regulatory framework for epigenome editing is conformity to international guidelines, which encompasses ensuring that the use of these technologies complies with existing regulations. For research to be conducted in a manner that is ethical, researchers and institutions will have to follow existing protocols and standards.

7. **Monitoring and governance:** The regulatory framework should include provisions for monitoring and governing epigenome editing research facilities. This includes the need to create regulatory entities, guidelines and laws for the ethical use of epigenome editing technology (Okada *et al*., 2017).

In short, a good governance framework is crucial to guide the ethically responsible use of epigenome-editing technologies. It helps research to be conducted in compliance with ethical guidelines and protects the rights and welfare of individuals included in studies.

1.7.3 Public perception and acceptance

Public recognition and reception is the product of various factors, including scientific understanding, ethical concerns, and societal attitudes toward genetic manipulation. Here are some key points on public perception and acceptance of the technology.

1. **Scientific understanding:** One confounding factor in the public perception of epigenome editing technology is scientific understanding. This is still not very well understood by the public, and outreach educational events are a critical form of assistance to bring a sense of what is possible and what is not when it comes to editing the epigenome.
2. **Ethical considerations:** Concerns related to privacy, consent, and fairness that arise from epigenome editing technology all go into shaping public opinion. By addressing these ethical issues and then making one's research data practices open to scrutiny, trust from the public can be won.
3. **The perception of risks and benefits for how the public views epigenome editing techniques:** If potential benefits like disease treatment and crop improvement are candidly communicated, and if a record of this effect can be made, then one will generally get public support.
4. **Regulatory oversight:** Epigenome editing regulated under a particular regulatory framework, or another, affects public trust. Strict regulatory controls and adherence to a code of ethics are needed for a controlled safe practice of epigenome editing.
5. **Social/cultural factors:** Attitudes on genetic manipulation, bioethics and the social contribution of science all play a part in affecting public opinion towards epigenome editing.
6. **Participation and dialogue:** Dialogues and engagement with the people, public consultations and participatory decision-making processes can all help allay fears and increase the public's acceptance of epigenome editing technologies.
7. **The media:** Good educational reporting, as well as fair and accurate reporting, can stir the pot on this one.

In brief, it is a blend of scientific, ethical, regulatory, and socio-cultural criteria that have been associated with the recognition and acceptance of this nascent technology. However, the same cannot be said about open direct communication with public discourse to create awareness and get endorsement for editing epigenome.

1.8 Challenges and Limitations

1.8.1 Off-target effects and specificity

This is especially true of new technologies like CRISPR-based tools. Off-target effects and specificity are crucial issues in the field of epigenome editing, however minor they may appear to be at present. Concerning the optimization of CRISPR/Cas9 in terms of on-target specificity, Jamal *et al*. (2018) discussed both improved gRNA design and advanced detection methods for screening off-target sites. They mentioned that these advances have greatly improved the specificity of CRISPR technology as well.

Nomura (2018) emphasized the problem of altering off-target areas in genome editing technologies. It was also stressed that care should be taken to avoid any unexpected side effects resulting from unintentional changes at similar sequences. In epigenome editing approaches the need for precise targeting and minimization of off-target effects is stressed by this. However, CRISPR/Cas-based, targeted epigenome editing tools will have a higher specificity than small molecules, as Psatha *et al.* (2022) pointed out. Against this background, careful analysis of stability and specificity is necessary in order to produce cell-based therapeutics with high therapeutic efficacy and long-lasting effects.

Lau and Suh (2017) elucidated the problem of off-target effects in epigenome editing, along with being able to decode epigenetic signatures inside reprogrammed cells to avert unintended after-effects. They identified the importance of understanding and mitigating off-target effects in epigenome editing applications.

Stover *et al.* (2023) highlighted the necessity of maximizing specificity and minimizing off-target effects in developing a new method for epigenome editing that is essentially required from the standpoint of safety as well as effectiveness. In addition, a comparison of their activity and specificity with those of existing multi-effector targeting systems demonstrated the efficacy and specificity of this novel epigenome editing system.

In other words, the eradication of any off-target effects and increase in specificity are critical factors for creating an epigenome editing solution that is ethical. The development of gene targeting technology, selection methods, and detection tools have greatly advanced the precision of CRISPR-based genome editing and have also given rise to new epigenome editing technologies, which work even more precisely.

1.8.2 Delivery challenges and efficiency

Efficient transfer and high specificity are the main problems in epigenome editing, and especially so when using technologies such as CRISPR-based tools. Syding *et al.* (2020) also stressed that efficient transfer, high specificity, and non-immunogenicity were all essential in epigenome editing. They also stated that the advances in nuclease development, gRNA selection, and detection methods have increased the specificity of CRISPR technology.

Also, Gemberling *et al.* (2021) highlighted the problems of *in vivo* delivery in epigenome editing and stressed the value of targeted gene activation and repression *in vivo*. The majority of the studies that have been reported so far used *ex vivo* cell culture systems, but there are examples of effective *in vivo* applications of epigenome editing as well. Biomolecular engineering strategies are emphasized to maintain the functions of such therapy and overcome limitations of targeting efficiency and cell internalization.

1.8.3 Long-term stability and safety concerns

If these epigenetic changes can be regulated and aimed to behave in the desired manner through a CRISPR-based editing tool, then it is possible for new and highly effective therapies to develop. Nevertheless, seasoned safety is the stake for long-term stability. Bashor *et al.* (2022) emphasized the importance of stability and specificity for epigenome editing in future clinical practices. Even though the specificity of targeted CRISPR/Cas-based epigenome editing tools may be greater than that of disrupting all human epigenomic sites with small molecules, careful analyses of both stability and specificity are critical for tailoring therapeutic efficacy, durability and pharmacokinetic/pharmacodynamic properties.

Besides, Carrillo-Reixach *et al.* (2020) reported a universal epigenetic footprint obscure in human hepatoblastoma (HB). Dysregulation of RNA editing and over-expressed non-coding genes at this oncogenic 14q32 DLK1-DIO3 site are included. Understanding how the long-term effects of changes in epigenetics caused by cancer and other diseases may bear increasingly on their susceptibility to them as well as shape their medical futures.

Aubourg *et al.* (2022) studied the genetics of osteoarthritis and the proteins that are involved. They found that variations related to DNA methylation play a unique part in the development of osteoarthritis. The long-term stability of these epigenetic changes is crucial for the development of targeted therapies and interventions. Also, Li *et al.* (2019) studied the new role that HOX transcription factors may play in predisposition to cancer and in its progression. They found that some current chemotherapy drugs are no longer effective because of long-term repercussions such as deregulating these genes, but that early intervention development requires treating patients based on their unique genetic make-up. In addition, Dong and Kantor (2021) focused on lentiviral vectors for the delivery of gene editing systems of the CRISPR/Cas type. They concluded that long-term stability and safety concerns related to gene-editing tool delivery are key for developing effective and safe therapeutic interventions.

As such, addressing long-term safety and stability for epigenome editing is imperative if these technologies are to effectively translate into the clinic. The evaluation of stability, specificity, and safety profiles of epigenome editing interventions is key to determining their efficacy and safety in diverse disease contexts.

1.8.4 Technical and biological barriers

Epigenome editing has immense potential for the precise regulation of genes and could function as a powerful tool in numerous medical applications, but some technical biological obstacles need to be addressed before some successful implementations. CRISPR/Cas9's versatility was also described by Adli (2018), where the authors showed that CRISPR/Cas9 can be utilized as a tool not only for gene editing but also for gene regulation; epigenome editing; chromatin engineering, and imaging applications. However, technical challenges such as high specificity and efficiency in epigenome editing are still major bottlenecks to wider use. In addition, Policarpi *et al.* (2021) went over unresolved issues in epigenetics that can be addressed by programmable epigenome editing. CRISPR-based technology for epigenetic change must achieve both precision and stability but one technical difficulty that arises from chromatin states having themselves been shaped by context is that researchers simply cannot get a lasting effect.

On the contrary, Xia and Xie (2020) further explored changes in the early mammalian epigenome and pointed out the dynamic nature of such changes. However, they noted the importance of understanding and controlling the intricate processes of epigenetic reprogramming that represent biological barriers to stable and hereditable epigenome editing. Meantime, Kuscu *et al.* (2020) also pointed out that in cases of transplantation, it is the desire to use inactive Cas9, or dead Cas9 (dCas9), for carrying out certain tasks of epigenome editing. High specificity of action and efficient delivery of epigenetic editing tools to target cell populations are technical difficulties that should be removed to proceed successfully with this work.

Nguyen and Karn (2024) discussed the drawbacks of efficient *in vivo* CRISPR delivery and in particular ideas for controlling HIV latency. They noted that epigenome editing for medical treatment, improving the specific target, avoiding off-target effects, and promoting possible delivery routes are the key technical barriers. Also, Tsagaratou (2021) explored TET proteins in T-cell lineage specification and their roles in transformation to malignancy. The complex interaction between epigenetic mechanisms and gene expression is a biological challenge for achieving stable and accurate epigenome editing. Mladenov *et al.* (2021) took the opportunity to review transgenerational plant stress memories and the need to identify transferred epigenetic tags. The technical challenges involved in decoding and manipulating the epigenetic alphabet that hides transgenerational stress memories are all present barriers to achieving steady and inheritable epigenome editing applications for agriculture.

1.9 Future Perspectives

With the help of epigenome editing, we could further clarify the mysteries behind how genes are regulated, what causes disease or death, and even benefit agricultural production. Allis and Jenuwein (2016) expected that genome technology will continue to be harnessed in future control studies of genetic and epigenetic regulation. In their opinion, quantitative CRISPR technology will greatly increase the precision and specificity of future epigenome modifications. Kumar and Mohapatra (2021) also remarked that if we could identify epigenetic pathways in plants and animals, and verify that these have biological functions, then developing strategies for editing the epigenome in future agriculture would become much easier. They also reasoned that this could help agriculture evolve towards climate-resilient and sustainable practices in the face of global climate change.

The availability of advanced epigenome editing tools such as zinc fingers, transcriptional activator-like effectors, and endonuclease-deficient Cas9 proteins was emphasized by Gogolev *et al.* (2021) to affect epigenetic changes at specific sites. Those authors proposed that future applications of these technologies might benefit both food safety and public health through precise editing of the epigenome. Kakoulidou *et al.* (2021) also looked at an important problem facing humanity, in the form of epigenetic modifications for crop improvement: What kinds of markers can provide useful information? How do we evaluate plant performance during growth? They suggested that future advances in the use of epigenome diversity for crop production might change such strategies fundamentally and that improved final crop yields beckon.

Bannister *et al.* (2020) pointed out that epigenetics is becoming increasingly important in our understanding of how the immune system responds to immunizations or diseases. They had the bright idea that advances in methods to manipulate epigenomic features might allow people to establish causality and function, obtaining a deeper understanding of immune responses as well as possible therapeutic interventions. In addition, Ferré *et al.* (2020) proposed that CRISPR/Cas9-empowered technologies hold promise to accomplish precise genome editing and epigenome editing for bioreactors and that these tools offer a bright future with potential applications in all manner of areas from biomedicine to biology and biotechnology.

Also, Han *et al.* (2020) had a hypothesis that, one day, histone modifications and other epigenetic changes could be used to improve diagnosis or monitor allergic inflammation. They proposed that in the future, if we could better understand epigenetic mechanisms, this would offer new possibilities for the prevention of allergic diseases. The versatility of CRISPR/Cas9 was noted by Zhou *et al.* (2014) who reported that this editing tool is flexible as it is used in different types of other gene-activated suppressors, epigenome editors, and even insertions or deletions. The authors expected that advances in both CRISPR technology and analysis strategies will continue to fuel exciting new developments in gene regulation and genome editing.

The coming era of epigenetic editing heralds enormous promise for advancing research as well as pharmaceuticals, agriculture, and biotechnology. The expectation is that innovation or change of a transformative nature will take place within an array of areas, as technology and research development will be complemented by advances in deployment strategies.

1.10 Conclusion

Using CRISPR technology for editing the epigenome is a powerful way to achieve precise gene regulation, elegant medical treatment, and revolutionary crops. This chapter has introduced the various applications of CRISPR-inspired epigenome editing, suggesting that it can turn a corner in more than one discipline. As a result of this system with CRISPR/dCas9 fusions, people can change DNA methylation patterns and histone modifications where they want. This has given gene regulation and disease mechanism researchers the advantage of recruiting chromatin-modifying enzymes to specific genomic loci and unveiled the role of epigenetic changes in backgrounds as diverse as biological function. However, with all these great progressions, there remain many hurdles to clear before epigenome editing can be brought to clinical use. High specificity and efficacy, stable gene expression and effective methods for gene delivery are the paramount obstacles that require to be overcome. An additional factor may be whether it raises ethical concerns when used in people or different life forms on the face of our planet.

Future work on CRISPR tools and techniques will allow for increased options when editing the epigenome. By developing more sophisticated techniques together with sophisticated tools, we will be able to become better epigenome manipulators – maybe in a way that could even revolutionize medicine, agriculture, or biotechnology. We learn that the prospect of CRISPR-based epigenome editing is huge; it allows us to study gene regulation in new ways and offers novel therapeutic solutions to old problems for which none is currently available. Those engaged in the long-term development of this field, which we assume will be of interest to employers and investors, will continue striving for the realization of the technology's whole potential despite obstacles faced in its path.

References

Adli, M. (2018) The CRISPR tool kit for genome editing and beyond. *Nature Communications* 9(1).

Aguilar, C.A. and Craighead, H.G. (2013) Micro- and nanoscale devices for the investigation of epigenetics and chromatin dynamics. *Nature Nanotechnology* 8(10), 709–718.

Ahmar, S., Gill, R.A., Jung, K.-H., Faheem, A., Qasim, M.U. *et al.* (2020) Conventional and molecular techniques from simple breeding to speed breeding in crop plants: Recent advances and future outlook. *International Journal of Molecular Sciences* 21(7), 2590.

Allis, C.D. and Jenuwein, T. (2016) The molecular hallmarks of epigenetic control. *Nature Reviews. Genetics* 17(8), 487–500.

Aman, R., Mahas, A. and Mahfouz, M. (2020) Nucleic acid detection using CRISPR/Cas biosensing technologies. *ACS Synthetic Biology* 9(6), 1226–1233.

Arnan, C., Ullrich, S., Pulido-Quetglas, C., Nurtdinov, R., Esteban, A. *et al.* (2022) Paired guide RNA CRISPR-Cas9 screening for protein-coding genes and lncRNAs involved in transdifferentiation of human B-cells to macrophages. *BMC Genomics* 23(1), 402.

Aubourg, G., Rice, S.J., Bruce-Wootton, P. and Loughlin, J. (2022) Genetics of osteoarthritis. *Osteoarthritis and Cartilage* 30(5), 636–649.

Bandyopadhyay, A., Köpke, M. and Moon, T.S. (2023) CRISPR-aided bioengineering for value-added product development. *Frontiers in Bioengineering and Biotechnology* 11, 1340377.

Bannister, S., Messina, N.L., Novakovic, B. and Curtis, N. (2020) The emerging role of epigenetics in the immune response to vaccination and infection: A systematic review. *Epigenetics* 15(6–7), 555–593.

Barrangou, R. and Marraffini, L.A. (2014) CRISPR-Cas systems: Prokaryotes upgrade to adaptive immunity. *Molecular Cell* 54(2), 234–244.

Bashor, C.J., Hilton, I.B., Bandukwala, H., Smith, D.M. and Veiseh, O. (2022) Engineering the next generation of cell-based therapeutics. *Nature Reviews Drug Discovery* 21(9), 655–675.

Bhokisham, N., Laudermilch, E., Traeger, L.L., Bonilla, T.D., Ruiz-Estevez, M. *et al.* (2023) CRISPR-Cas system: The current and emerging translational landscape. *Cells* 12(8), 1103.

Bunnik, E.M., Timmers, M. and Bolt, I.L. (2020) Ethical issues in research and development of epigenome-wide technologies. *Epigenetics Insights* 13, 2516865720913253.

Cano-Rodriguez, D., Gjaltema, R.A.F., Jilderda, L.J., Jellema, P., Dokter-Fokkens, J. *et al.* (2016) Writing of h3k4me3 overcomes epigenetic silencing in a sustained but context-dependent manner. *Nature Communications* 7(1), 12284.

Carrillo-Reixach, J., Torrens, L., Simon-Coma, M., Royo, L., Domingo-Sàbat, M. *et al.* (2020) Epigenetic footprint enables molecular risk stratification of hepatoblastoma with clinical implications. *Journal of Hepatology* 73(2), 328–341.

Cervantes-Gracia, K., Gramalla-Schmitz, A., Weischedel, J. and Chahwan, R. (2021) APOBECs orchestrate genomic and epigenomic editing across health and disease. *Trends in Genetics* 37(11), 1028–1043.

Chen, E., Lin-Shiao, E., Trinidad, M., Saffari Doost, M., Colognori, D. *et al.* (2022) Decorating chromatin for enhanced genome editing using CRISPR-Cas9. *Proceedings of the National Academy of Sciences of the United States of America* 119(49), e2204259119.

Chiarella, A.M., Butler, K.V., Gryder, B.E., Lu, D., Wang, T.A. *et al.* (2020) Dose-dependent activation of gene expression is achieved using CRISPR and small molecules that recruit endogenous chromatin machinery. *Nature Biotechnology* 38(1), 50–55.

Christofi, T. and Zaravinos, A. (2019) RNA editing in the forefront of epitranscriptomics and human health. *Journal of Translational Medicine* 17(1), 319.

Clement, K., Hsu, J.Y., Canver, M.C., Joung, J.K. and Pinello, L. (2020) Technologies and computational analysis strategies for CRISPR applications. *Molecular Cell* 79(1), 11–29.

Cubbon, A., Ivancic-Bace, I. and Bolt, E.L. (2018) CRISPR-Cas immunity, DNA repair and genome stability. *Bioscience Reports* 38(5), BSR20180457.

Dong, W. and Kantor, B. (2021) Lentiviral vectors for delivery of gene-editing systems based on CRISPR/Cas: Current state and perspectives. *Viruses* 13(7), 1288.

Dyke, S.O.M., Saulnier, K.M., Dupras, C., Webster, A.P., Maschke, K. *et al.* (2019) Points-to-consider on the return of results in epigenetic research. *Genome Medicine* 11(1), 9.

Elliott, E.K., Haupt, L.M. and Griffiths, L.R. (2021) Mini review: Genome and transcriptome editing using CRISPR-Cas systems for haematological malignancy gene therapy. *Transgenic Research* 30(2), 129–141.

Engreitz, J.M., Haines, J.E., Perez, E.M., Munson, G., Chen, J. *et al.* (2016) Local regulation of gene expression by lncRNA promoters, transcription and splicing. *Nature* 539(7629), 452–455.

Fang, R., Chen, X., Shen, J. and Wang, B. (2023) Targeted mRNA demethylation in *Arabidopsis* using plant m6a editor. *Plant Methods* 19(1), 81.

Ferré, L.B., Kjelland, M.E., Strøbech, L.B., Hyttel, P., Mermillod, P. *et al.* (2020) Review: Recent advances in bovine *in vitro* embryo production: Reproductive biotechnology history and methods. *Animal* 14(5), 991–1004.

Fu, B.X.H., Smith, J.D., Fuchs, R.T., Mabuchi, M., Curcuru, J. *et al.* (2019) Target-dependent nickase activities of the CRISPR-Cas nucleases cpf1 and cas9. *Nature Microbiology* 4(5), 888–897.

Fukushima, H.S., Takeda, H. and Nakamura, R. (2019) Targeted *in vivo* epigenome editing of H3K27me3. *Epigenetics & Chromatin* 12(1), 12.

Gahlaut, V., Zinta, G., Jaiswal, V. and Kumar, S. (2020) Quantitative epigenetics: A new avenue for crop improvement. *Epigenomes* 4(4), 25.

Gasiunas, G., Sinkunas, T. and Siksnys, V. (2014) Molecular mechanisms of CRISPR-mediated microbial immunity. *Cellular and Molecular Life Sciences* 71, 449–465.

Gemberling, M.P., Siklenka, K., Rodriguez, E., Tonn-Eisinger, K.R., Barrera, A. *et al.* (2021) Transgenic mice for *in vivo* epigenome editing with CRISPR-based systems. *Nature Methods* 18(8), 965–974.

Gilbert, L.A., Larson, M.H., Morsut, L., Liu, Z., Brar, G.A. *et al.* (2013) CRISPR-mediated modular RNA-guided regulation of transcription in eukaryotes. *Cell* 154(2), 442–451.

Godfrey, L.C. and Rodriguez-Meira, A. (2022) Viewing AML through a new lens: Technological advances in the study of epigenetic regulation. *Cancers* 14(23), 5989.

Gogolev, Y.V., Ahmar, S., Akpinar, B.A., Budak, H., Kiryushkin, A.S. *et al.* (2021) OMICs, epigenetics, and genome editing techniques for food and nutritional security. *Plants* 10(7), 1423.

Gupta, R., Ghosh, A., Chakravarti, R., Singh, R., Ravichandiran, V. *et al.* (2022) Cas13d: A new molecular scissor for transcriptome engineering. *Frontiers in Cell and Developmental Biology* 10, 866800.

Han, X., Krempski, J.W. and Nadeau, K. (2020) Advances and novel developments in mechanisms of allergic inflammation. *Allergy* 75(12), 3100–3111.

Hariprakash, J.M. and Ferrari, F. (2019) Computational biology solutions to identify enhancers-target gene pairs. *Computational and Structural Biotechnology Journal* 17, 821–831.

He, Y., Wang, M., Liu, M., Huang, L., Liu, C. *et al.* (2018) Cas1 and cas2 from the type II-C CRISPR-Cas system of *Riemerella anatipestifer* are required for spacer acquisition. *Frontiers in Cellular and Infection Microbiology* 8, 195.

Hilton, I.B., D'Ippolito, A.M., Vockley, C.M., Thakore, P.I., Crawford, G.E. *et al.* (2015) Epigenome editing by a CRISPR-Cas9-based acetyltransferase activates genes from promoters and enhancers. *Nature Biotechnology* 33(5), 510–517.

Hu, H. and Du, J. (2022) Structure and mechanism of histone methylation dynamics in *Arabidopsis*. *Current Opinion in Plant Biology* 67, 102211.

Jaganathan, D., Ramasamy, K., Sellamuthu, G., Jayabalan, S. and Venkataraman, G. (2018) CRISPR for crop improvement: An update review. *Frontiers in Plant Science* 9, 985.

Jamal, M., Ullah, A., Ahsan, M., Tyagi, R., Habib, Z.R.K. *et al.* (2018) Improving CRISPR-Cas9 on-target specificity. *Current Issues in Molecular Biology* 26(1), 65–80.

Jambhekar, A., Dhall, A. and Shi, Y. (2019) Roles and regulation of histone methylation in animal development. *Nature Reviews. Molecular Cell Biology* 20(10), 625–641.

Jenuwein, T. (2006) The epigenetic magic of histone lysine methylation: Delivered on 6 July 2005 at the 30th FEBS congress in Budapest, Hungary. *The FEBS Journal* 273(14), 3121–3135.

Josipović, G., Tadić, V., Klasić, M., Zanki, V., Bečeheli, I. *et al.* (2019) Antagonistic and synergistic epigenetic modulation using orthologous CRISPR/dCas9-based modular system. *Nucleic Acids Research* 47(18), 9637–9657.

Kakoulidou, I., Avramidou, E.V., Baránek, M., Brunel-Muguet, S., Farrona, S. *et al.* (2021) Epigenetics for crop improvement in times of global change. *Biology* 10(8), 766.

Kamruzzaman, M., Yan, A. and Castro-Escarpulli, G. (2022) CRISPR-Cas systems in bacteria and archaea. *Frontiers in Microbiology* 13, 887778.

Katti, A., Diaz, B.J., Caragine, C.M., Sanjana, N.E. and Dow, L.E. (2022) CRISPR in cancer biology and therapy. *Nature Reviews. Cancer* 22(5), 259–279.

Khwatenge, C.N. and Nahashon, S.N. (2021) Recent advances in the application of CRISPR/Cas9 gene editing system in poultry species. *Frontiers in Genetics* 12, 627714.

Klann, T.S., Black, J.B., Chellappan, M., Safi, A., Song, L. *et al.* (2017) CRISPR-Cas9 epigenome editing enables high-throughput screening for functional regulatory elements in the human genome. *Nature Biotechnology* 35(6), 561–568.

Konishi, K. and Issa, J.P.J. (2007) Targeting aberrant chromatin structure in colorectal carcinomas. *Cancer Journal* 13(1), 49–55.

Krichevsky, A.M. and Uhlmann, E.J. (2019) Oligonucleotide therapeutics as a new class of drugs for malignant brain tumors: Targeting mRNAs, regulatory RNAs, mutations, combinations, and beyond. *Neurotherapeutics* 16(2), 319–347.

Kumar, S. and Mohapatra, T. (2021) Dynamics of DNA methylation and its functions in plant growth and development. *Frontiers in Plant Science* 12, 596236.

Kuscu, C., Kuscu, C., Bajwa, A., Eason, J.D., Maluf, D. *et al.* (2020) Applications of CRISPR technologies in transplantation. *American Journal of Transplantation* 20(12), 3285–3293.

Kwon, D.Y., Zhao, Y.T., Lamonica, J.M. and Zhou, Z. (2017) Locus-specific histone deacetylation using a synthetic CRISPR-Cas9-based HDAC. *Nature Communications* 8(1), 15315.

Lam, T.J. and Ye, Y. (2018) CRISPRs for strain tracking and its application to microbiota transplantation data analysis. *Bioinformatics* 344358.

Lam, T.J. and Ye, Y. (2019) CRISPRs for strain tracking and their application to microbiota transplantation data analysis. *The CRISPR Journal* 2(1), 41–50.

Lau, C.H. and Suh, Y. (2017) Genome and epigenome editing in mechanistic studies of human aging and aging-related disease. *Gerontology* 63(2), 103–117.

Lau, C.H. and Suh, Y. (2018) *In vivo* epigenome editing and transcriptional modulation using CRISPR technology. *Transgenic Research* 27, 489–509.

Laufer, B.I. and Singh, S.M. (2015) Strategies for precision modulation of gene expression by epigenome editing: An overview. *Epigenetics & Chromatin* 8(1), 34.

Li, B., Huang, Q. and Wei, G.H. (2019) The role of HOX transcription factors in cancer predisposition and progression. *Cancers* 11(4), 528.

Li, C., Brant, E., Budak, H. and Zhang, B. (2021) CRISPR/Cas: A Nobel prize award-winning precise genome editing technology for gene therapy and crop improvement. *Journal of Zhejiang University. SCIENCE. B* 22(4), 253–284.

Li, K., Liu, Y., Cao, H., Zhang, Y., Gu, Z. *et al.* (2020) Interrogation of enhancer function by enhancer-targeting CRISPR epigenetic editing. *Nature Communications* 11(1), 485.

Li, L., Hu, S. and Chen, X. (2018) Non-viral delivery systems for CRISPR/Cas9-based genome editing: Challenges and opportunities. *Biomaterials* 171, 207–218.

Lin, J., Zhou, Y., Liu, J., Chen, J., Chen, W. *et al.* (2017) Progress and application of CRISPR/Cas technology in biological and biomedical investigation. *Journal of Cellular Biochemistry* 118(10), 3061–3071.

Liu, X.M., Zhou, J., Mao, Y., Ji, Q. and Qian, S.B. (2019) Programmable RNA N6-methyladenosine editing by CRISPR-Cas9 conjugates. *Nature Chemical Biology* 15(9), 865–871.

Liu, Z., Dong, H., Cui, Y., Cong, L. and Zhang, D. (2020) Application of different types of CRISPR/Cas-based systems in bacteria. *Microbial Cell Factories* 19(1), 172.

Maeder, M.L. and Gersbach, C.A. (2016) Genome-editing technologies for gene and cell therapy. *Molecular Therapy* 24(3), 430–446.

Majchrzak-Celińska, A., Warych, A. and Szoszkiewicz, M. (2021) Novel approaches to epigenetic therapies: From drug combinations to epigenetic editing. *Genes* 12(2), 208.

Marraffini, L.A. and Sontheimer, E.J. (2008) CRISPR interference limits horizontal gene transfer in staphylococci by targeting DNA. *Science* 322(5909), 1843–1845.

Maximiano, M.R., Távora, F.T.P.K., Prado, G.S., Dias, S.C., Mehta, A. *et al.* (2021) CRISPR genome editing technology: A powerful tool applied to developing agribusiness. *Journal of Agricultural and Food Chemistry* 69(23), 6379–6395.

Mayorga-Ramos, A., Zúñiga-Miranda, J., Carrera-Pacheco, S.E., Barba-Ostria, C. and Guamán, L.P. (2023) CRISPR-Cas-based antimicrobials: Design, challenges, and bacterial mechanisms of resistance. *ACS Infectious Diseases* 9(7), 1283–1302.

Mimoune, N., Bahouh, M.W., Boukhechem, S., Khelef, D., Kaidi, R. *et al.* (2022) CRISPR/Cas9 and cancer. *Veterinarska Stanica* 53(2), 127–139.

Mladenov, V., Fotopoulos, V., Kaiserli, E., Karalija, E., Maury, S. *et al.* (2021) Deciphering the epigenetic alphabet involved in transgenerational stress memory in crops. *International Journal of Molecular Sciences* 22(13), 7118.

Moore, L.D., Le, T. and Fan, G. (2013) DNA methylation and its basic function. *Neuropsychopharmacology* 38(1), 23–38.

Morelli, E., Gulla', A., Amodio, N., Taiana, E., Neri, A. *et al.* (2021) CRISPR interference (CRISPRi) and CRISPR activation (CRISPRa) to explore the oncogenic lncRNA network. In: Navarro, A. (ed.) *Long Non-Coding RNAs in Cancer*. Humana New York (Springer Nature), New York, pp. 189–204.

Mortensen, K., Lam, T.J. and Ye, Y. (2021) Comparison of CRISPR-Cas immune systems in healthcare-related pathogens. *Frontiers in Microbiology* 12, 758782.

Mushtaq, M., Sakina, A., Wani, S.H., Shikari, A.B., Tripathi, P. *et al.* (2019) Harnessing genome editing techniques to engineer disease resistance in plants. *Frontiers in Plant Science* 10, 550.

Mutezo, W., Sedibe, M.M., Mofokeng, A., Shargie, N. and Soko, T. (2021) The application of CRISPR/Cas9 technology in the management of genetic and nongenetic plant traits. *International Journal of Agronomy* 2021(1), 9993784.

Nguyen, K. and Karn, J. (2024) The sounds of silencing: Dynamic epigenetic control of HIV latency. *Current Opinion in HIV and AIDS* 19(3), 102–109.

Nguyen, T.V. and Lister, R. (2021) Genomic targeting of TET activity for targeted demethylation using CRISPR/Cas9. In: Bogdanovic, O. and Vermeulen, M. (eds) *TET Proteins and DNA Demethylation: Methods and Protocols*. Humana New York (Springer Nature), New York, pp. 181–194.

Ning, L., Xi, J., Zi, Y., Chen, M., Zou, Q. *et al.* (2023) Prospects and challenges of CRISPR/Cas9 gene-editing technology in cancer research. *Clinical Genetics* 104(6), 613–624.

Nomura, W. (2018) Development of toolboxes for precision genome/epigenome editing and imaging of epigenetics. *Chemical Record* 18(12), 1717–1726.

Okada, M., Kanamori, M., Someya, K., Nakatsukasa, H. and Yoshimura, A. (2017) Stabilization of foxp3 expression by CRISPR-dCas9-based epigenome editing in mouse primary T cells. *Epigenetics & Chromatin* 10, 24.

Özcan, A., Krajeski, R., Ioannidi, E., Lee, B., Gardner, A. *et al.* (2021) Programmable RNA targeting with the single-protein CRISPR effector Cas7-11. *Nature* 597(7878), 720–725.

Perera, B.P.U., Faulk, C., Svoboda, L.K., Goodrich, J.M. and Dolinoy, D.C. (2020) The role of environmental exposures and the epigenome in health and disease. *Environmental and Molecular Mutagenesis* 61(1), 176–192.

Pflueger, C., Tan, D., Swain, T., Nguyen, T., Pflueger, J. *et al.* (2018) A modular dCas9-suntag DNMT3A epigenome editing system overcomes pervasive off-target activity of direct fusion dCas9-DNMT3A constructs. *Genome Research* 28(8), 1193–1206.

Policarpi, C., Dabin, J. and Hackett, J.A. (2021) Epigenetic editing: Dissecting chromatin function in context. *BioEssays* 43(5), 2000316.

Psatha, N., Paschoudi, K., Papadopoulou, A. and Yannaki, E. (2022) *In vivo* hematopoietic stem cell genome editing: Perspectives and limitations. *Genes* 13(12), 2222.

Pulecio, J., Verma, N., Mejía-Ramírez, E., Huangfu, D. and Raya, A. (2017) CRISPR/Cas9-based engineering of the epigenome. *Cell Stem Cell* 21(4), 431–447.

Qi, L.S., Larson, M.H., Gilbert, L.A., Doudna, J.A., Weissman, J.S. *et al.* (2013) Repurposing CRISPR as an RNA-guided platform for sequence-specific control of gene expression. *Cell* 152(5), 1173–1183.

Rajaram, N., Kourouklі, A.G., Bens, S., Bashtrykov, P. and Jeltsch, A. (2023) Development of super-specific epigenome editing by targeted allele-specific DNA methylation. *Epigenetics & Chromatin* 16(1), 41.

Riaz, A., Kanwal, F., Ahmad, I., Ahmad, S., Farooq, A. *et al.* (2022) New hope for genome editing in cultivated grasses: CRISPR variants and application. *Frontiers in Genetics* 13, 866121.

Shehreen, S., Chyou, T.-Y., Fineran, P.C. and Brown, C.M. (2019) Genome-wide correlation analysis suggests different roles of CRISPR-Cas systems in the acquisition of antibiotic resistance genes in diverse species. *Philosophical Transactions of the Royal Society of London. Series B, Biological Sciences* 374(1772), 20180384.

Shi, C., Zou, W., Liu, X., Zhang, H., Li, X. *et al.* (2024) Programmable RNA n^6-methyladenosine editing with CRISPR/dCas13a in plants. *Plant Biotechnology Journal* 22(7), 1867–1880.

Skaar, D.A., Li, Y., Bernal, A.J., Hoyo, C., Murphy, S.K. *et al.* (2012) The human imprintome: Regulatory mechanisms, methods of ascertainment, and roles in disease susceptibility. *ILAR Journal* 53(3–4), 341–358.

Splinter, E. and Laat, W. (2011) The complex transcription regulatory landscape of our genome: Control in three dimensions. *The EMBO Journal* 30(21), 4345–4355.

Splinter, K.D., Turner, I.L. and Davidson, M.A. (2013) How much data is enough? The importance of morphological sampling interval and duration for calibration of empirical shoreline models. *Coastal Engineering* 77, 14–27.

Stepper, P., Kungulovski, G., Jurkowska, R.Z., Chandra, T., Krueger, F. *et al.* (2017) Efficient targeted DNA methylation with chimeric dCas9-Dnmt3a-Dnmt3L methyltransferase. *Nucleic Acids Research* 45(4), 1703–1713.

Stover, J.D., Trone, M.A., Lawrence, B. and Bowles, R.D. (2023) Multiplex epigenome editing of ion channel expression in nociceptive neurons abolished degenerative IVD-conditioned media-induced mechanical sensitivity. *JOR Spine* 6(2), e1253.

Sugita, E., Hayashi, K., Hishikawa, A. and Itoh, H. (2021) Epigenetic alterations in podocytes in diabetic nephropathy. *Frontiers in Pharmacology* 12, 759299.

Sun, Y.V. (2014) The influences of genetic and environmental factors on methylome-wide association studies for human diseases. *Current Genetic Medicine Reports* 2(4), 261–270.

Sun, D., Mao, X., Fei, M., Chen, Z., Zhu, T. *et al.* (2020) Histone-like nucleoid-structuring protein (H-NS) paralogue stpa activates the type i-e CRISPR-Cas system against natural transformation in *Escherichia coli*. *Applied and Environmental Microbiology* 86(14), e00731–20.

Syding, L.A., Nickl, P., Kasparek, P. and Sedlacek, R. (2020) CRISPR/Cas9 epigenome editing potential for rare imprinting diseases: A review. *Cells* 9(4), 993.

Tan, M.H., Li, Q., Shanmugam, R., Piskol, R., Kohler, J. *et al.* (2017) Dynamic landscape and regulation of RNA editing in mammals. *Nature* 550(7675), 249–254.

Tan, Y., Küçük, C. and Cui, L. (2024) Genomics and epigenomics of natural killer/T cell lymphoma: Diagnostic, prognostic, and therapeutic implications. In: Rezaei, N. (ed.) *Interdisciplinary Cancer Research*. Springer, Cham.

Tanaka, M. and Nakamura, T. (2023) Targeting epigenetic aberrations of sarcoma in CRISPR era. *Genes, Chromosomes & Cancer* 62(9), 510–525.

Tang, Z., Chen, S., Chen, A., He, B., Zhou, Y. *et al.* (2019) CasPDB: An integrated and annotated database for Cas proteins from bacteria and archaea. *Database* baz093.

Tasca, F., Wang, Q. and Gonçalves, M.A.F.V. (2020) Adenoviral vectors meet gene editing: A rising partnership for the genomic engineering of human stem cells and their progeny. *Cells* 9(4), 953.

Thakore, P.I., D'Ippolito, A.M., Song, L., Safi, A., Shivakumar, N.K. *et al.* (2015) Highly specific epigenome editing by CRISPR-Cas9 repressors for silencing of distal regulatory elements. *Nature Methods* 12(12), 1143–1149.

Tian, X., Gu, T., Patel, S., Bode, A.M., Lee, M.H. *et al.* (2019) CRISPR/Cas9 – an evolving biological tool kit for cancer biology and oncology. *NPJ Precision Oncology* 3(1), 8.

Tsagaratou, A. (2021) Deciphering the multifaceted roles of TET proteins in T-cell lineage specification and malignant transformation. *Immunological Reviews* 300(1), 22–36.

Tuminauskaite, D., Norkunaite, D., Fiodorovaite, M., Tumas, S., Songailiene, I. *et al.* (2020) DNA interference is controlled by R-loop length in a type I-F1 CRISPR-Cas system. *BMC Biology* 18(1), 65.

Uchiyama, M., Nagai, A. and Muto, K. (2018) Survey on the perception of germline genome editing among the general public in Japan. *Journal of Human Genetics* 63(6), 745–748.

Urbano, A., Smith, J., Weeks, R.J. and Chatterjee, A. (2019) Gene-specific targeting of DNA methylation in the mammalian genome. *Cancers* 11(10), 1515.

Vetchinova, A.S., Fedotova, E.Y. and Illarioshkin, S.N. (2021) Editing the epigenome in neurodegenerative diseases. *Neurochemical Journal* 15(4), 359–366.

Villota-Salazar, N.A., Mendoza-Mendoza, A. and González-Prieto, J.M. (2016) Epigenetics: From the past to the present. *Frontiers in Life Science* 9(4), 347–370.

Vora, S., Tuttle, M., Cheng, J. and Church, G. (2016) Next stop for the CRISPR revolution: RNA-guided epigenetic regulators. *The FEBS Journal* 283(17), 3181–3193.

Wilson, C., Chen, P.J., Miao, Z. and Liu, D.R. (2020) Programmable m6a modification of cellular RNAs with a Cas13-directed methyltransferase. *Nature Biotechnology* 38(12), 1431–1440.

Wu, W., Yang, Y. and Lei, H. (2019) Progress in the application of CRISPR: From gene to base editing. *Medicinal Research Reviews* 39(2), 665–683.

Xia, W. and Xie, W. (2020) Rebooting the epigenomes during mammalian early embryogenesis. *Stem Cell Reports* 15(6), 1158–1175.

Xu, L., Sun, B., Liu, S., Gao, X., Zhou, H. *et al.* (2023) The evaluation of active transcriptional repressor domain for crispri in plants. *Gene* 851, 146967. DOI: 10.1016/j.gene.2022.146967.

Yang, D., Scavuzzo, M.A., Chmielowiec, J., Sharp, R., Bajic, A. *et al.* (2016) Enrichment of G2/M cell cycle phase in human pluripotent stem cells enhances HDR-mediated gene repair with customizable endonucleases. *Scientific Reports* 6(1), 21264.

Yang, J., Meng, X., Pan, J., Jiang, N., Zhou, C. *et al.* (2018) CRISPR/Cas9-mediated noncoding RNA editing in human cancers. *RNA Biology* 15(1), 35–43.

You, L., Tong, R., Li, M., Liu, Y., Xue, J. *et al.* (2019) Advancements and obstacles of CRISPR-Cas9 technology in translational research. *Molecular Therapy. Methods & Clinical Development* 13, 359–370.

Zeps, N., Lysaght, T., Chadwick, R., Erler, A., Foo, R. *et al.* (2021) Ethics and regulatory considerations for the clinical translation of somatic cell human epigenetic editing. *Stem Cell Reports* 16(7), 1652–1655.

Zhao, W., Wang, Y. and Liang, F.S. (2020) Chemical and light inducible epigenome editing. *International Journal of Molecular Sciences* 21(3), 998.

Zhao, W., Xu, Y., Wang, Y., Gao, D., King, J. *et al.* (2021) Investigating crosstalk between H3K27 acetylation and H3K4 trimethylation in CRISPR/dCas-based epigenome editing and gene activation. *Scientific Reports* 11(1), 15912.

Zheng, Z., Zhang, Y., Liu, Z., Dong, Z., Xie, C. *et al.* (2020) The CRISPR-Cas systems were selectively inactivated during evolution of bacillus cereus group for adaptation to diverse environments. *The ISME Journal* 14(6), 1479–1493.

Zhou, H.Y., Katsman, Y., Dhaliwal, N.K., Davidson, S., Macpherson, N.N. *et al.* (2014) A sox2 distal enhancer cluster regulates embryonic stem cell differentiation potential. *Genes & Development* 28(24), 2699–2711.

Zhou, H., Xu, L., Li, F. and Li, Y. (2022) Transcriptional regulation by CRISPR/dCas9 in common wheat. *Gene* 807, 145919. DOI: 10.1016/j.gene.2021.145919.

Zink, I.A., Wimmer, E. and Schleper, C. (2020) Heavily armed ancestors: CRISPR immunity and applications in archaea with a comparative analysis of CRISPR types in sulfolobales. *Biomolecules* 10(11), 1523.

2 Epigenetics and Epigenomics for Climate-Smart and Sustainable Agriculture: Unlocking Potential for Crop Improvement

Gunna Sureshbabu Suruthi[1], Arumugam Vijaya Anand[1], Manoharan Rajesh[2], Tajudheen Nargis Begum[3], Sirajudeen Benazir Begum[4], Anbazhagan Priya[4], Palanisamy Sampathkumar[5] and Asirvatham Alwin Robert[6]*

[1]Department of Human Genetics and Molecular Biology, Bharathiar University, Coimbatore, Tamil Nadu, India; [2]Department of Biochemistry, Bharathiar University, Coimbatore, Tamil Nadu, India; [3]Department of Biotechnology, Jamal Mohamed College, (Autonomous), Tiruchirapalli, Tamil Nadu, India; [4]PG and Research Department of Zoology, V.O. Chidambaram College, Thoothukudi, Tamil Nadu, India; [5]Department of Chemistry and Biosciences, SASTRA Deemed to be University, Srinivasa Ramanujan Centre, Kumbakonam, Tamil Nadu, India; [6]Department of Endocrinology and Diabetes, Prince Sultan Military Medical City, Ministry of Defense, Riyadh, Kingdom of Saudi Arabia

Abstract

The rising challenges posed by climate modification necessitate inventive techniques in agriculture that enhance crop resilience and sustainability. Epigenetics and epigenomics offer promising avenues for crop improvement by regulating gene expression and plant responses to environmental stressors without altering the DNA sequence. This chapter explores the role of epigenetic processes, like methylation of DNA, modifications of histone, and non-coding RNAs, in plant development and stress responses. It delves into how climate change impacts crop epigenomes and how epigenomic tools can be leveraged to enhance crop resilience. The potential of tolerable agricultural methods, like crop rotation and intercropping, to influence plant epigenomes and improve productivity is also discussed. Furthermore, the chapter highlights the integration of epigenetic tools with traditional breeding methods to develop climate-smart crops. Epigenetic breeding strategies, which utilize epigenetic variations for crop improvement, present a novel approach to selecting and stabilizing desirable traits. The chapter explores the challenges and prospects of applying epigenetic knowledge to achieve sustainable agricultural practices and crop enhancement in the face of climate differences.

Keywords: Epigenetics, Epigenomics, Sustainable agriculture, Climate-smart agriculture, Crop improvement

*Corresponding author: aalwinrobert@gmail.com

DOI: 10.1079/9781800626102.0002

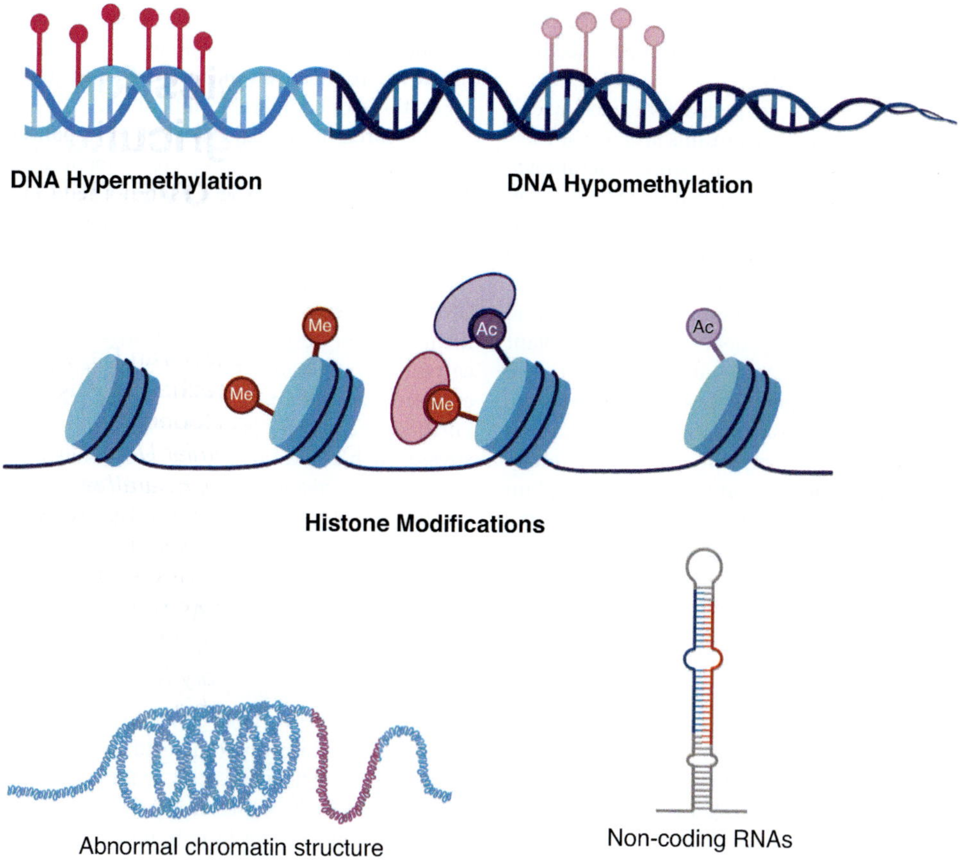

Fig. 2.1. Epigenetic modifications. Author's own image.

2.1 Introduction to Epigenetics in Plants

Epigenetics in plants involves a complex network of gene regulation mechanisms beyond the traditional genetic code, influencing the expression of genes and phenotype without modifying the sequence of DNA (Chen *et al.*, 2010). The significance of this topic in several biological processes, such as development, stress reactions, and adaptability, has drawn attention. Because of the variety of epigenetic processes governing plant development and stress responses, plants are helpful models for researching epigenetic phenomena (Munshi *et al.*, 2015). In plants, methylation of DNA, modifications in histone, and non-coding RNAs are the main epigenetic processes that regulate the expression of genes and functions of cells (Fig. 2.1) (Munshi *et al.*, 2015).

Methylation of DNA modifies the patterns of the expression genes by introducing a methyl group to DNA, usually at cytosine residues. Histone alterations affect transcription and gene accessibility by changing the structure of the chromatin. Non-coding RNAs such as microRNAs control the expression of post-transcriptional genes by directing particular mRNAs toward translational suppression or destruction (Munshi *et al.*, 2015). Plants can inherit epigenetic changes from one generation to the subsequent generation, which helps them adjust to their ever-changing surroundings. According to Abobatta (2018), epigenetic memory improves reactions to upcoming stressors and raises adaptive capacity. Based on studies, morphological features and differences in gene expression in plants and crops can be stably inherited due to epigenetic modifications (Disi *et al.*, 2012).

Plant genomes are perfect for exploring epigenetic regulatory mechanisms because of their plasticity and resilience to stress (Avramova, 2011). As well as affecting gene expression, epigenetic changes are essential for plants' stress tolerance. Multiple epigenomes can be established within a single plant in response to developmental and environmental alerts, as environmental factors can cause epigenetic alterations in plant genomes (Chinnusamy and Zhu, 2009). The significance of epigenetic systems in preserving plant health and resilience has been emphasized by the relationship between epigenetic diversity and increased production and stability in plant populations (Latzel *et al.*, 2013). The functional diversity found in plant populations results from the relations between genetic and epigenetic variables, with epigenetic changes affecting traits associated with stress responses and plant growth (Medrano *et al.*, 2014).

It has been suggested that plant adaptation and speciation are fueled by epigenetic inheritance. Plants can adapt to their surroundings and flourish in a diverse environment because of the ability of environmental stimuli to cause epigenetic modifications, like DNA methylation and histone modifications, that last for generations (Minow and Colasanti, 2020). Plants can quickly adapt to shifting ecological stresses because of this environmentally induced epigenetic transgenerational inheritance, which is important for evolutionary processes (Skinner and Nilsson, 2021).

2.2 Epigenetic Regulation of Plant Stress Responses

Epigenetic regulation is a critical factor in plant stress responses, influencing how plants adapt to different environmental challenges like drought, salinity, temperature extremes, pests, and diseases. Enhancing plant stress toleration requires an understanding of the processes behind epigenetic regulation in response to abiotic stresses (Kim, 2021). According to Gümüş *et al.* (2023), epigenetic changes can control gene expression under stress. According to Chu *et al.* (2020), these alterations serve as a type of stress memory, enabling plants to react to recurrent stresses more effectively and prepare subsequent generations for comparable difficulties.

Research has revealed that plant responses to abiotic stressors such as severe temperatures, droughts, salt, and deficiencies in nutrients are strongly influenced by epigenetic modifications, such as DNA methylation patterns (Fig. 2.2) (Miryeganeh, 2021). For instance, differences in DNA methylation can affect how genes express themselves in response to salt stress, which can alter how various rice genotypes adapt to environmental changes (Akhter *et al.*, 2021). Furthermore, during plant responses to abiotic stresses such as low phosphorus conditions, epigenetic mechanisms like chromatin remodeling and histone alterations may control gene expression (Mierziak and Wojtasik, 2024).

Epigenetic control also plays a necessary function in plant defense against biotic stressors such as pests and diseases. Investigations have highlighted the involvement of epigenetic processes in plant defense mechanisms, like the regulation of stress-related genes and the activation of defense pathways (Lucibelli *et al.*, 2022). Epigenetic modifications can influence gene expression, which is involved in plant defense reactions, contributing to enhanced resistance against pathogens (Shahrajabian *et al.*, 2019). Furthermore, chromatin remodelers and non-coding RNAs have been recognized as important parts of plant defense against biotic stressors.

The interplay between epigenetic processes and stress reactions in plants is complex and multifaceted. Epigenetic modifications not only influence gene expression in response to stress but also contribute to the acquisition of stress tolerance and the priming of plants for future challenges. Comprehending the epigenetic processes underlying stress responses in plants is crucial for conceiving methods to improve crop stability and agricultural production in the face of modifying environmental circumstances.

2.3 Impact of Climate Change on Crop Epigenomes

Climate difference affects crop production by affecting plant epigenomes through various

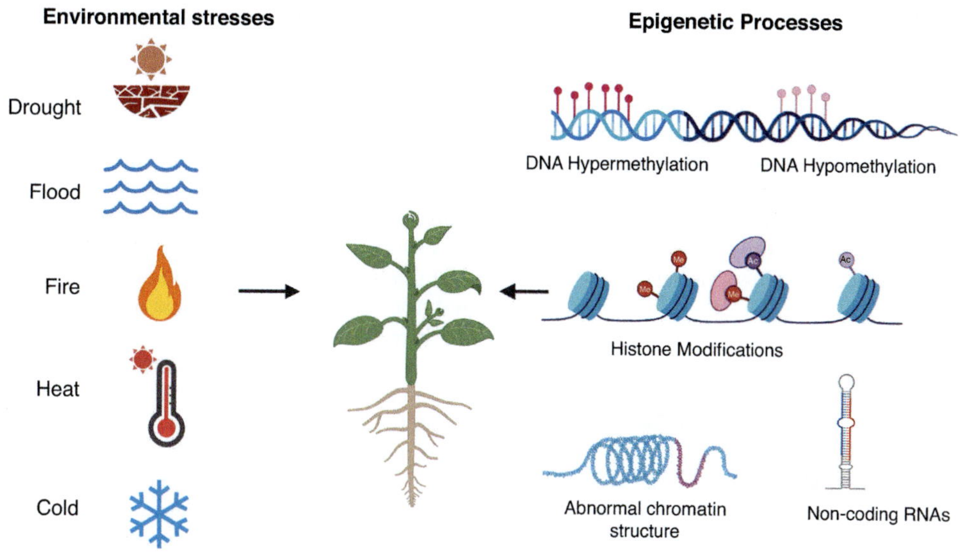

Fig. 2.2. Epigenetic mechanisms against abiotic stresses in plants. Author's own image.

environmental stressors like temperature fluctuations, drought, salinity, and other abiotic factors. The epigenetic processes of plants to these stressors are crucial for their adaptation and survival in changing environmental conditions (Chinnusamy *et al.*, 2013). Research has shown that abiotic stressors can lead to changes in DNA methylation imprints, histone modifications, and non-coding RNA expression in plants, thereby altering gene expression profiles and phenotypic traits (Yung *et al.*, 2023). For example, salinity and low nitrogen availability can restrict crop growth, prompting plants to undergo epigenetic modifications to cope with these stressors (Agarwal and Kumar, 2021).

Studies on grapevines have illustrated that irrigation and grafting practices can impact the epigenomic patterns of plants, emphasizing the consequence of environmental factors on the plant epigenome (Williams *et al.*, 2020). Similarly, the preference for enhanced energy use efficiency and drought toleration in canola has resulted in different transcriptome and epigenome modifications, demonstrating the plant's capacity to adjust to adverse growth requirements (Verkest *et al.*, 2015). The dynamic chromatin regulatory programs during the embryogenesis of hexaploid wheat offer insights into the epigenomic regulation of crop development, highlighting the significance of epigenetic mechanisms in shaping plant phenotypes (Zhao *et al.*, 2023). Additionally, the importance of plant growth-promoting rhizobacteria in enhancing salt stress tolerance in crops emphasizes the intricate association between plant epigenomes and microbes (Giannelli *et al.*, 2023).

Epigenetic changes induced by environmental stressors can have transgenerational effects on plant traits, as observed in winter oilseed rape revealed to drought stress conditions (Hatzig *et al.*, 2018). Late planting of short-season cotton in saline areas and the impacts of soil saltiness on cotton yield and leaf senescence further emphasize the complex relationship between environmental stressors, plant growth, and epigenetic modifications (Zhang *et al.*, 2012). Examinations on soybeans have demonstrated that salinity can affect nitrogen assimilation and seed yield, suggesting that epigenetic changes mediate plant responses to stress due to salinity (Rabie and Kumazawa, 1988). Additionally, the interaction of sodium chloride and temperature on alfalfa germination highlights the intricate interplay between environmental factors and plant physiological responses (Stone *et al.*, 1979). The effects of boron and saltiness on grafted and non-grafted melon plants suggest that rootstock interactions can influence the plant's response to environmental stressors, potentially leading to differential epigenetic modifications (Edelstein *et al.*, 2005). Furthermore, stress memory in plants subjected to repeated drought episodes can impact physiological processes and stress responses, underscoring the long-term effects of environmental stress on plant epigenomes (Fleta-Soriano and Munné-Bosch, 2016).

2.4 Epigenomics for Enhancing Crop Resilience

Epigenomics, the investigation of modifications in gene expression or cellular phenotype that do not concern the alterations to the underlying DNA sequence, has emerged as a promising field for enhancing crop resilience in the face of climate modification and environmental stresses. Several studies have highlighted the consequence of epigenetic mechanisms in enhancing stress toleration in plants, which can eventually direct the development of climate-smart crops (Kumar and Rani, 2023). By understanding how epigenetic modifications influence the expression of genes in response to stress, investigators can determine epigenetic markers associated with stress resilience in crops (Tan and Rodríguez López, 2023).

Breeding strategies that incorporate epigenetic information have the potential to revolutionize crop improvement programs. Epigenetic variations can be harnessed to develop climate-resilient crops with improved adaptability to altering environmental circumstances (Tharun *et al.*, 2024). Integrating epigenomic data into genetic engineering and genome editing methods like clustered regularly interspaced short palindromic repeats/CRISPR-associated protein 9 (CRISPR/Cas9) can further enhance the precision and efficiency of creating stress-resistant crop varieties (Mushtaq *et al.*, 2019). By targeting the plant epigenome, researchers can boost genetic resilience in crops, leading to improved yields and sustainability (Mushtaq *et al.*, 2019).

Epigenetic analysis not only aids in understanding stress responses in plants but also provides valuable insights into processes of stress priming, which can improve a crop's capacity to tolerate environmental strain without compromising yield (Kakoulidou *et al.*, 2021). The application of epigenetics in crop improvement extends to the generation of more resilient grapevines, highlighting the versatility of epigenomic tools in enhancing different crop species (Tan and Rodríguez López, 2023). Furthermore, the exploitation of epigenetic variations in crop wild relatives can contribute to both resilience and crop security, offering an additional strategy for breeding climate-resilient crops (Varotto *et al.*, 2022). Incorporating epigenetic markers into breeding programs can facilitate the development of elite crops with enhanced agronomic characteristics and stress stability (Tharun *et al.*, 2024). By identifying and manipulating epigenomic variability, breeders can create novel crop varieties that exhibit broad-spectrum resistance to abiotic and biotic stresses (Tirnaz and Batley, 2019). These epigenetically enhanced crops hold the prospects to address global food safety challenges by increasing yield potential and tolerance to stressful environments (Raza *et al.*, 2023).

Omics-driven strategies, like metabolomics and transcriptomics, have been instrumental in elucidating stress responses and tolerance mechanisms in crops, paving the route for the progression of temperature-resilient and drought-smart plants (Raza *et al.*, 2024). By leveraging these technologies, researchers can accelerate the breeding of climate-resilient crops with superior nutritive value and enhanced abiotic stress tolerance (Zenda *et al.*, 2021). Additionally, integrating genomics and molecular breeding approaches can provide valuable insights for selecting molecular tags that regulate complex traits, expediting the progression of high-yielding stress-tolerant crop varieties (Kujur *et al.*, 2013).

2.5 Sustainable Agricultural Practices and Epigenetics

Sustainable agricultural practices are essential for ensuring food security and promoting environmental sustainability. One significant aspect of sustainable agriculture is the impact of these practices on crop epigenomes. Organic farming and reduced chemical usage have been shown to have implications for the epigenetic modification of plants, potentially affecting the expression of gene patterns and crop traits (Pikaard and Scheid, 2014). This emphasizes the significance of considering the epigenetic consequences of agricultural practices when aiming for sustainable production systems.

Crop rotation, intercropping, and agroforestry are sustainable practices that have been recognized for their epigenetic benefits. These practices not only contribute to soil health and biodiversity but also influence the epigenomes of crops. Crop rotations, where different crops are planted in a sequence over time, have been shown to have positive effects on crop epigenomes by potentially altering gene expression patterns and enhancing stress tolerance (Marques *et al.*, 2020). Intercropping, the simultaneous cultivation of two or more plants in proximity, has also been associated with beneficial epigenetic effects, potentially leading to improved crop resilience and productivity. Additionally, agroforestry systems, which integrate trees and shrubs into agricultural landscapes, can influence plant epigenomes through interactions with soil microbiomes and root exudates, contributing to enhanced crop performance (Wang *et al.*, 2021).

The position of soil microbiomes in shaping plant epigenomes is a critical aspect of sustainable agriculture. Soil microbiomes, comprising diverse communities of microorganisms, play a significant role in nutrient cycling, suppressing disease, and plant development. These microbial communities interact with plant roots and influence gene expression in crops, thereby impacting their epigenomes (Wang *et al.*, 2021). The composition and diversity of soil microbiomes can be impacted by agricultural approaches like crop rotation and intercropping, which in turn can modulate the epigenetic regulation of plants (Wahbi *et al.*, 2016). Comprehending the intricate relationship between soil microbiomes and plant epigenomes is essential for optimizing sustainable agricultural practices and enhancing crop productivity in a changing environment.

In the context of endurable agricultural practices, the adoption of conservation agriculture (CA) principles has gained prominence. CA emphasizes minimal soil disruption, soil cover, and crop rotation as key components for sustainable crop production. These traditions not only contribute to soil health and water conservation but also have implications for crop epigenetics. By maintaining soil structure and microbial diversity, CA practices can influence the epigenetic regulation of crops, potentially leading to improved stress tolerance and yield stability (Ikazaki *et al.*, 2018). The integration of CA principles with a focus on crop epigenetics can enhance the resilience of agricultural systems to environmental challenges.

Furthermore, the economic implications of assuming tolerable agricultural practices on crop yield and household earnings are essential considerations for smallholder farmers. Studies have shown that the adoption of endurable agricultural practices can have a positive impact on farm products and household revenues among smallholder farmers (Mugula *et al.*, 2023). By implementing practices like crop rotation, intercropping, and reduced chemical usage, farmers can not only improve crop productivity but also enhance their economic well-being. These findings underscore the importance of promoting sustainable agricultural practices that not only benefit the environment and crop production but also contribute to the socio-economic development of farming communities.

2.6 Epigenetic Tools and Techniques for Sustainable Agriculture

Epigenetic tools and techniques are instrumental in advancing sustainable agriculture, especially in the realm of creating climate-smart crops. The use of epigenetic editing tools, like CRISPR/dCas9 and RNA interference, has transformed crop improvement by allowing precise changes in the epigenetic status. These techniques empower researchers to explore the connection between epigenetic modifications and gene expression, offering a potent means to adjust gene activity and control specific traits in crops (Fukushima *et al.*, 2019). The employment of epigenetic editing tools in creating climate-smart crop varieties shows great potential in bolstering agricultural sustainability in the face of climate modification and environmental challenges. By utilizing tools like CRISPR/dCas9, researchers can pinpoint specific epigenetic marks linked to stress tolerance and resilience, thereby engineering crops that are more resilient to adverse environmental circumstances. This targeted epigenome editing strategy allows for precise modulation of gene expression patterns, providing a pathway to develop crop varieties with enhanced adaptability and productivity under changing climatic conditions (Waryah *et al.*, 2018).

Epigenomics has emerged as a crucial factor in enhancing breeding approaches for developing climate-resilient crops. By harnessing epigenetic mechanisms, researchers can enhance traditional breeding strategies to create crop varieties with improved resilience to environmental stresses. Epigenetic tools not only facilitate the modification of specific epigenetic marks associated with stress response but also offer a way to fine-tune gene expression for desired agronomic traits, paving the way for the progression of climate-smart crops (Chandana *et al.*, 2022). In the pursuit of sustainable agriculture, integrating epigenetic editing tools with conventional breeding methods has the potential to revolutionize crop improvement programs. These tools enable precise epigenome editing, allowing researchers to adjust gene expression and regulatory elements to enhance desired traits in crops. By combining the power of epigenetic editing with traditional breeding practices, scientists can expedite the action of climate-resilient crops with improved yield prospects and stress tolerance (Thakore *et al.*, 2015).

The progression of climate-smart crops is crucial to ensure food safety and sustainable crop production under changing environmental conditions. Epigenetic tools provide a means to deepen physiological phenotyping and enhance crop productivity by creating new crops with improved yield and environmental resilience (Table 2.1). By leveraging epigenetic editing tools, researchers can target specific epigenetic modifications that influence crop performance under stress conditions, directing the production of resilient crop varieties capable of thriving in challenging environments (Hirayama and Mochida, 2022).

2.7 Epigenetic Breeding

Epigenetic breeding strategies provide a novel approach to selecting and breeding crops with desirable epigenetic traits, potentially revolutionizing traditional plant breeding methods. By utilizing epigenetic variation, researchers can manipulate the expression of gene patterns without changing the underlying DNA sequence, offering a powerful tool for enhancing crop characteristics (Mladenov *et al.*, 2021). The concept of epigenetic breeding, also known as epi-breeding, involves the examination of epigenetic marks across generations to improve traits in crops (Gupta and Salgotra, 2022). This procedure permits the targeted modification of epigenetic states to select specific traits, providing a more precise and efficient method of crop improvement.

The integration of epigenetics in crop breeding has shown promise in enhancing crop resilience to environmental stressors, such as abiotic stress and climate change. Epigenomics has been recognized as a valuable tool for developing climate-resilient crops by modulating plant responses to environmental stimuli through epigenetic modifications (Tan and Rodríguez López, 2023). By comprehending the epigenetic processes concerned with stress memory and adaptation, researchers can develop crops with increased resilience and adaptability to varying environmental situations (Yang *et al.*, 2015). Epigenetic breeding has been demonstrated to be feasible in crop plants, providing evidence of its potential to improve crop traits and enhance productivity (Fang *et al.*, 2023).

One of the essential benefits of epigenetic breeding is the capability to induce and exploit epigenetic variations to drive crop enhancement. Locus-specific manipulation of DNA methylation using devices like CRISPR/Cas9 can promote molecular breeding of important crops by modulating gene expression patterns (Gahlaut *et al.*, 2020). The exploitation of epigenetic variations, particularly in crop wild relatives, offers a valuable source of variability that can be presented in crops through epigenetic breeding, contributing to agrobiodiversity preservation and crop improvement (Varotto *et al.*, 2022). Epigenetic breeding, focusing on genetic breeding for epigenetic modifications, presents a unique way to enhance crop resistance to diseases and environmental stressors (Zhi and Chang, 2021).

Moreover, epigenetic breeding strategies can be instrumental in selecting and stabilizing

Table 2.1. Epigenetic tools and techniques for enhancing crop resilience in sustainable agriculture. Author's own table.

Epigenetic tool/technique	Description	Application in sustainable agriculture	Reference
CRISPR/dCas9	Epigenome editing tool for precise modifications	Enhances stress tolerance and adaptability in crops	Fukushima *et al.* (2019)
RNA interference	Technique to suppress gene expression	Modulates specific traits in crops for improved performance	Fukushima *et al.* (2019)
Epigenetic markers	Identifiable epigenetic modifications	Facilitates selection of desirable traits in breeding programs	Tharun *et al.* (2024)
Epigenome-wide association studies	Analysis of epigenetic variations across genomes	Identifies epigenetic markers associated with stress resilience	Tan and Rodríguez López (2023)
Integrative omics approaches	Combines genomics, proteomics, transcriptomics, and metabolomics with epigenomics	Elucidates stress responses for developing temperature-resilient crops	Raza *et al.* (2024)
Epigenetic breeding	Selection based on epigenetic variations	Develops crop varieties with improved agronomic performance	Tharun *et al.* (2024)
Locus-specific DNA methylation manipulation	Targeted modification of DNA methylation patterns	Facilitates molecular breeding of important crop plants	Gahlaut *et al.* (2020)
Transgenerational memory induction	Techniques to induce heritable epigenetic changes	Enhances long-term stress tolerance in crops	Rajnović *et al.* (2020)
Integration with conventional breeding	Combines epigenetic tools with traditional methods	Accelerates development of climate-resilient crop varieties	Thakore *et al.* (2015)
Artificial epigenome editing	Targeted modification of epigenetic states	Generates novel crop varieties with improved traits	Verkest *et al.* (2015)

desirable traits in crops. The heritability of epigenetic variations over multiple generations has been observed in soybean breeding, highlighting the potential for long-term trait improvement through epigenetic mechanisms (Ketumile *et al.*, 2022). Induced DNA methylation has been proposed as a tool to introduce desirable characteristics in crops within a single generation, with lasting effects in subsequent generations (Rajnović *et al.*, 2020). By selecting favorable epigenetic states, creating new epialleles, and regulating gene expression, plant breeders can leverage epigenetic information to develop crop improvement strategies (Ostrowska-Mazurek *et al.*, 2020).

The integration of epigenetics into plant breeding programs offers a new dimension to traditional breeding approaches, providing additional resources for enhancing crop performance and sustainability. Epigenetic modifications can be targeted to improve traits associated with energy usage efficiency, drought tolerance, and stress resistance in crops (Kakoulidou *et al.*, 2021). The identification of epialleles and functional epigenetic regulatory strategies can lead to diverse strategies for epigenetic breeding of crops, including the benefit of mutant lines and recurrent epi-selection (Jeynes-Cupper and Catoni, 2023). Understanding the dynamics of DNA methylation and exploring artificial epigenome editing techniques can further enhance crop breeding efforts by generating novel varieties with improved agronomic traits (Verkest *et al.*, 2015).

2.8 Epigenetic Memory in Plants

Epigenetic memory in plants refers to the capability of plants to keep details about past environmental conditions through epigenetic marks, allowing them to adapt and respond more effectively to future stresses. This phenomenon has been extensively studied in the context of both biotic and abiotic stresses. Research by Ashapkin *et al.* (2020) highlighted how epigenetic mechanisms can influence plant adaptation to rapidly varying environments, particularly in response to confrontational abiotic and biotic strains. Similarly, Müller-Xing *et al.* (2014) discussed how plants exhibit a form of 'plant memory' where exposure to various stresses is memorized in the cells through epigenetic mechanisms.

Miryeganeh (2021) reported that epigenetic marks deliver a mechanical cause for stress memory in crops, allowing them to react more efficiently to frequent stresses and prepare future generations for conceivable challenges. Additionally, studies by Latzel *et al.* (2016) and Sun *et al.* (2021) emphasized how environmentally caused epigenetic changes enable plants to recognize past interactions and develop heritable stress memory for improved responses against various stresses.

Transcriptional memory, as explored by Wei *et al.* (2021), refers to plants' ability to remember previously experienced stresses and acquire enhanced tolerance to similar or different stressors. This memory mechanism plays a vital role in priming plants for future challenges. Understanding these mechanisms is crucial for leveraging epigenetic memory in crop improvement efforts, as emphasized by Duarte-Aké *et al.* (2023). The function of stress memory in enhancing drought toleration in plants is a key focus of research, as highlighted by Liu *et al.* (2022a). By examining stress memory reactions and seed priming, researchers aim to improve plant resilience to drought through physiological, biochemical, molecular, and epigenetic perspectives. Leveraging epigenetic variation in crop wild relatives for agrobiodiversity preservation and crop enhancement (Varotto *et al.*, 2022) showcases the potential of epigenetics in shaping agronomically significant traits for enhanced crop productivity. Kinoshita and Seki (2014) provided proof reinforcing the essential role of epigenetic mechanisms in plant stress memory and adaptation, further underlining the importance of epigenetics in plant reactions to environmental challenges.

The intricate relationship between stress rescue, RNA turnover, and epigenetics in plant memory was discussed by Crisp *et al.* (2016), shedding light on how epigenetic memory complements genetic selection in facilitating plant adaptation to changing environments. Charng *et al.* (2023) emphasized the importance of maintaining abiotic stress memory in plants to facilitate acclimation to future environmental changes. Research by Lämke and Bäurle (2017) delved into the epigenetic and chromatin-based processes involved in environmental stress adaptation and memory in plants, providing insights into how these modifications enable plants to react more efficiently to regular stresses. The multifaceted function of RNA-based regulation in plant stress memory, as discussed by Xu *et al.* (2024), further elucidates the intricate molecular mechanisms underlying epigenetic memory in plants. Abdulraheem *et al.* (2024) highlighted the ongoing discoveries in plant epigenetic regulation in reaction to stress and the importance of future applications in crop enhancement.

Epigenetic regulation, particularly DNA methylation, is crucial for plants to react to environmental stresses, as discussed by Lanciano and Mirouze (2017). The concept of plant environmental memory, as explored by Auge *et al.* (2023), underscores the adaptive responses of plants to different environments through memory mechanisms. Avramova (2015) delved into the transcriptional 'memory' of stress in plants, linking short- and long-term memory responses to plant acclimation and adaptation prospects. Moreover, recent advancements in understanding plant stress memory, as highlighted by Liu *et al.* (2022b), offer promising opportunities for rewiring stress memory to enhance crop resilience.

2.9 Transgenerational Epigenetic Inheritance in Plants

Transgenerational epigenetic inheritance in plants is a captivating area of research that explores how epigenetic traits are transmitted across generations, influencing plant health, adaptation, and responses to environmental stimuli. The information of epigenetic changes, via the germline, serves as a molecular mechanism for the potential impacts of various environmental exposures on plant health. This phenomenon, known as transgenerational epigenetic inheritance, concerns the transmission of epigenetic information via the germline (Horsthemke, 2018). The part of epigenetics in mediating environmental influences on phenotype, including transgenerational epigenetic inheritance, has been considerably investigated and documented (Morgan and Whitelaw, 2009).

Epigenetic transgenerational inheritance, often described as the germline transmission of epigenetic transmission between generations in the lack of direct exposures or genetic manipulations, has been a subject of significant interest in the field of plant biology (Yan, 2014). Studies have shown that environmentally induced epigenetic changes in plants, such as methylation of DNA patterns caused by heavy metal stress, can be readily inheritable across organismal generations, reinforcing the concept of transgenerational epigenetic inheritance in plants (Ou *et al.*, 2012). Similarly, the investigation has highlighted the involvement of RNA-mediated strategies and modifications in chromatin states in transgenerational epigenetic inheritance in plants (Neeb and Nowacki, 2018).

The effect of epigenetic transgenerational inheritance on plant healthiness and disease susceptibility is profound. Investigations have demonstrated the inheritance of altered methylation of DNA, non-coding RNA, and histone retention in plants, indicating the involvement of multiple epigenetic processes in transgenerational inheritance. Additionally, observations suggest an integration of methylation of DNA, non-coding RNA, and histone retention in part involving RNA-directed methylation of DNA and DNA methylation-led histone retention in epigenetic transgenerational inheritance. The environmentally generated epigenetic transgenerational inheritance of epimutations has been shown to promote genetic mutations, further emphasizing the intricate interplay between environmental exposures, epigenetic modifications, and genetic outcomes in crops. The potential influence of epigenetic transgenerational inheritance on disease etiology, variation in phenotype, and evolution in plants underscores the significance of understanding and unraveling the mechanisms underlying this phenomenon. Moreover, the developmental origins of epigenetic transgenerational inheritance have been investigated, shedding light on the intricate processes that contribute to the dissemination of epigenetic details across generations in plants and other organisms (Hanson and Skinner, 2016).

2.10 Challenges and Limitations

Epigenetics and epigenomics offer promising avenues for enhancing climate-smart and sustainable agriculture by leveraging the dynamic nature of DNA methylation and epigenetic regulation in plants. These advancements in epigenomic tools hold the potential to enable the manipulation of plant phenotypic plasticity, facilitating the development of climate-resilient crops for sustainable agricultural practices. Understanding the processes of plant epigenetic modification in response to strain is crucial for enhancing stress tolerance in crops, revolutionizing breeding strategies, and ensuring sustainable food production amidst climate change and environmental challenges (Abdulraheem *et al.*, 2024). Climate-smart agriculture (CSA) has emerged as a critical strategy to enhance agricultural productivity sustainably. By prioritizing food security and considering mitigation benefits, CSA aims to manage the challenges posed by climate modification and environmental stresses in agriculture (Managa and Nkobole-Mhlongo, 2016). However, the implementation of CSA has been hindered by conceptual ambiguities and institutional challenges, highlighting the need for clearer frameworks and mechanisms for its effective application (Akamani, 2021).

Epigenetic processes play a substantial part in enhancing crop productivity, quality, and resilience to environmental stressors, thereby contributing to sustainable agriculture (Duarte-Aké *et al.*, 2023). Methylation of DNA and modifications of histone are key epigenetic alterations that enable plants to acclimatize to challenges like drought, emphasizing the importance of epigenetic and hormonal modulation in promoting resilience in agricultural systems (Kaya *et al.*, 2023). Integrating epigenetic insights into crop breeding and genetic engineering strategies can offer novel solutions to address the consequences of climate modification on agriculture (Duarte-Aké *et al.*, 2023). Agriculture faces multifaceted challenges due to environmental strains, including abiotic and biotic factors, which limit plant growth, reduce yields, and contribute to global food crises (Sghaier-Hammami and Hammami, 2023). The complexities of climate change further exacerbate these challenges, necessitating innovative approaches such as nanotechnology and ecosystem-based strategies to manage plant stress effectively. By exploring nanotechnological interventions and ecosystem-based solutions, agriculture can enhance its resilience to environmental stressors and improve food security (Sghaier-Hammami and Hammami, 2023).

The concept of CSA aligns with the broader goal of increasing food security, promoting adaptation, and mitigating climate change impacts sustainably (Neufeldt *et al.*, 2013). To achieve these objectives, it is essential to go beyond traditional agricultural practices and embrace innovative approaches that integrate environmental sustainability, social equity, and economic viability. Organic farming practices, agroecology, and sustainable agricultural systems offer pathways to manage the interconnected challenges of environmental degradation, food safety, and socio-economic well-being (Hazmi *et al.*, 2023). Sustainable agriculture necessitates a holistic approach that balances food production with environmental protection and socio-economic well-being (Choudhary *et al.*, 2023). By integrating agricultural and environmental sustainability across generations, it is possible to navigate the complexities of climate change, mitigate greenhouse gas emissions, and improve carbon sequestration in agricultural systems. However, achieving agricultural sustainability requires continuous innovation, knowledge exchange, and policy support to address the evolving challenges faced by the agricultural sector (Musvoto *et al.*, 2015). The embracement of sustainable agricultural techniques is crucial for mitigating the impacts of insect pests, enhancing environmental stewardship, and promoting economically viable farming systems. Integrated pest management (IPM) principles offer effective strategies to manage insect pests while minimizing environmental risks and supporting sustainable agriculture (Tiwari, 2024). By promoting the dissemination of climate-smart agricultural learning through agriculturalist field schools and community engagement, smallholder farmers can enhance their resilience to climate change and adopt sustainable practices (Gichuki *et al.*, 2023).

2.11 Future Prospects

Prospects of epigenetics and epigenomics in the context of climate-smart and sustainable agriculture hold significant promise for revolutionizing agricultural practices to meet the challenges posed by climate modification and ensure food safety. By comprehending the molecular processes underlying epigenetic pathways in plants and their responses to environmental stressors, investigators can pave the way for inventive approaches such as epigenome editing to develop climate-resilient crops. This approach aligns with the urgent need to enhance agricultural productivity while mitigating the impacts of climate change on global food systems (Kumar and Mohapatra, 2021).

The integration of digitalization and climate-smart agricultural techniques presents a compelling opportunity to enhance the efficiency and sustainability of the agricultural sector, thereby contributing to food and environmental safety by lowering greenhouse gas emissions (Vărzaru, 2024). By leveraging epigenetic insights and adopting climate-smart agricultural technologies, it is possible to improve agricultural productivity, increase resilience to climate extremes, and lower greenhouse gas emissions, thus promoting sustainable development and food security. These advancements underscore the potential of epigenetics and epigenomics to drive transformative changes in agricultural systems towards greater sustainability and climate resilience (Zeleke *et al.*, 2024).

CSA emerges as a key process for sustainably improving agricultural productivity, improving resilience to climate modification, and lowering greenhouse gas emissions (Kumar, 2022). The adoption of CSA practices, including sustainable farming techniques and efficient plant nutrient management, offers a pathway to manage the challenges posed by climate modification and confirm the long-term sustainability of agricultural systems (Kottegoda *et al.*, 2023). By communicating climate-smart agricultural strategies and disseminating information among farmers, it is possible to enhance food security, increase yields, and alleviate poverty in vulnerable agricultural communities (Nagar *et al.*, 2023).

The future of epigenetics and epigenomics in agriculture lies in their potential to contribute to sustainable intensification and climate-smart agricultural practices. By focusing on soil carbon sequestration, climate-resilient crop selection, and technology integration, climate-smart landscapes can offer integrated solutions that benefit food security, biodiversity conservation, and poverty alleviation (Bai *et al.*, 2019). The function of biosensors in climate-smart organic agriculture further underscores the importance of monitoring and optimizing agricultural practices to enhance sustainability and environmental stewardship (Ukhurebor, 2020). Measuring the implementation of farms concerning CSA objectives through a set of indicators can provide a useful understanding of the effectiveness of sustainable agricultural practices (Selbonne *et al.*, 2023). By scaling up climate-smart agricultural techniques and practices in sub-Saharan Africa, it is possible to achieve sustainable development goals, enhance food security, and mitigate the influences of climate modification on agricultural systems (Phiri *et al.*, 2022). Evaluating the role of CSA in promoting sustainable livelihoods underscores the importance of continued support from governmental and non-governmental sectors to drive the adoption of climate-resilient practices in rural communities (Muzorewa and Chitakira, 2022).

2.12 Conclusion

Epigenetics and epigenomics offer exciting new avenues for advancing climate-smart and sustainable agriculture. By harnessing our growing understanding of epigenetic mechanisms in plants, investigators and breeders can create crops with enhanced resilience to environmental strains and improved productivity under changing climate conditions. Key areas of promise include utilizing epigenetic breeding strategies to select for beneficial traits without altering DNA sequences, leveraging epigenetic memory and transgenerational inheritance to create crops with 'stress memory' and enhanced adaptability, and employing epigenome editing tools like CRISPR/dCas9 to precisely modulate gene expression for desired traits. Additionally, integrating epigenomic data into crop improvement programs can accelerate the development of climate-resilient varieties, while exploring how sustainable agricultural practices influence crop epigenomes can help optimize farming systems.

However, challenges remain in fully understanding complex epigenetic mechanisms and translating epigenomic insights into practical applications. Continued research is needed to elucidate epigenetic regulation of stress responses and develop robust epigenetic breeding technologies. As climate change intensifies pressures on global agriculture, epigenetics and epigenomics will likely play an increasingly vital role in crop adaptation and food security. By combining epigenetic approaches with other innovative technologies and sustainable practices, the agricultural sector can work towards greater resilience, productivity and environmental stewardship in the face of mounting challenges. Realizing this potential will require sustained investment in research, interdisciplinary collaboration, and effective knowledge transfer to farming communities worldwide. As we persist in unravelling the complexities of plant epigenomes, the future of agriculture looks poised to benefit from these molecular insights, offering hope for more sustainable and climate-resilient food production systems.

References

Abdulraheem, M.I., Xiong, Y., Moshood, A.Y., Cadenas-Pliego, G., Zhang, H. *et al.* (2024) Mechanisms of plant epigenetic regulation in response to plant stress: Recent discoveries and implications. *Plants* 13(2), 163.

Abobatta, W.F. (2018) Epigenetics in the agricultural sector. *Annals of Reviews & Research* 2(5), 1–4.

Agarwal, S. and Kumar, A. (2021) Ameliorative effects of nitrogen application on vegetative growth and productivity in genotypes of ragi (*Eleusine coracana* Gaertn.) under saline conditions. *Indian Journal of Agricultural Research* 22 September 2021, 5 pp.

Akamani, K. (2021) An ecosystem-based approach to climate-smart agriculture with some considerations for social equity. *Agronomy* 11(8), 1564.

Akhter, Z., Bi, Z., Ali, K., Sun, C., Fiaz, S. *et al.* (2021) In response to abiotic stress, DNA methylation confers epigenetic changes in plants. *Plants* 10(6), 1096.

Ashapkin, V.V., Kutueva, L.I., Aleksandrushkina, N.I. and Vanyushin, B.F. (2020) Epigenetic mechanisms of plant adaptation to biotic and abiotic stresses. *International Journal of Molecular Sciences* 21(20), 7457.

Auge, G., Hankofer, V., Groth, M., Antoniou-Kourounioti, R., Ratikainen, I. *et al.* (2023) Plant environmental memory: Implications, mechanisms and opportunities for plant scientists and beyond. *AoB PLANTS* 15(4), lad032.

Avramova, Z. (2011) Epigenetic regulatory mechanisms in plants. In: Tollefsbol, T. (ed.) *Handbook of Epigenetics*. Academic Press, San Diego, California, pp. 251–278.

Avramova, Z. (2015) Transcriptional 'memory' of a stress: transient chromatin and memory (epigenetic) marks at stress-response genes. *The Plant Journal* 83(1), 149–159.

Bai, X., Huang, Y., Ren, W., Coyne, M., Jacinthe, P.-A. *et al.* (2019) Responses of soil carbon sequestration to climate-smart agriculture practices: A meta-analysis. *Global Change Biology* 25(8), 2591–2606.

Chandana, B.S., Mahto, R.K., Singh, R.K., Ford, R., Vaghefi, N. *et al.* (2022) Epigenomics as potential tools for enhancing magnitude of breeding approaches for developing climate resilient chickpea. *Frontiers in Genetics* 13, 900253.

Charng, Y.Y., Mitra, S. and Yu, S.J. (2023) Maintenance of abiotic stress memory in plants: Lessons learned from heat acclimation. *The Plant Cell* 35(1), 187–200.

Chen, M., Lv, S. and Meng, Y. (2010) Epigenetic performers in plants. *Development, Growth & Differentiation* 52(6), 555–566.

Chinnusamy, V. and Zhu, J.K. (2009) Epigenetic regulation of stress responses in plants. *Current Opinion in Plant Biology* 12(2), 133–139.

Chinnusamy, V., Dalal, M. and Zhu, J.K. (2013) Epigenetic regulation of abiotic stress responses in plants. In: Jenks, M.A. and Hasegawa, P.M. (eds) *Plant Abiotic Stress*, 2nd edn. John Wiley & Sons, Hoboken, New Jersey, pp. 203–229.

Choudhary, S., Boruah, A., Ram, N., Gulaiya, S., Choudhary, C.S. *et al.* (2023) Millet's role in sustainable agriculture: A comprehensive review. *International Journal of Plant & Soil Science* 35(22), 556–568.

Chu, S., Zhang, X., Yu, K., Lv, L., Sun, C. *et al.* (2020) Genome-wide analysis reveals dynamic epigenomic differences in soybean response to low-phosphorus stress. *International Journal of Molecular Sciences* 21(18), 6817.

Crisp, P.A., Ganguly, D., Eichten, S.R., Borevitz, J.O. and Pogson, B.J. (2016) Reconsidering plant memory: Intersections between stress recovery, RNA turnover, and epigenetics. *Science Advances* 2(2), e1501340.

Disi, J.O., Wei, D., Mei, J. and Qian, W. (2012) Understanding epigenetic effects in crop species. *African Journal of Plant Science* 6(14), 355–363.

Duarte-Aké, F., Us-Camas, R. and De-la-Peña, C. (2023) Epigenetic regulation in heterosis and environmental stress: The challenge of producing hybrid epigenomes to face climate change. *Epigenomes* 7(3), 14.

Edelstein, M., Ben-Hur, M., Cohen, R., Burger, Y. and Ravina, I. (2005) Boron and salinity effects on grafted and non-grafted melon plants. *Plant and Soil* 269, 273–284.

Fang, W., Fasano, C. and Perrella, G. (2023) Unlocking the secret to higher crop yield: The potential for histone modifications. *Plants* 12(8), 1712.

Fleta-Soriano, E. and Munné-Bosch, S. (2016) Stress memory and the inevitable effects of drought: A physiological perspective. *Frontiers in Plant Science* 7, 143.

Fukushima, H.S., Takeda, H. and Nakamura, R. (2019) Targeted *in vivo* epigenome editing of H3K27me3. *Epigenetics & Chromatin* 12, 1–12.

Gahlaut, V., Zinta, G., Jaiswal, V. and Kumar, S. (2020) Quantitative epigenetics: A new avenue for crop improvement. *Epigenomes* 4(4), 25.

Giannelli, G., Potestio, S. and Visioli, G. (2023) The contribution of PGPR in salt stress tolerance in crops: Unravelling the molecular mechanisms of cross-talk between plant and bacteria. *Plants* 12(11), 2197.

Gichuki, C., Osewe, M. and Ndiritu, S.W. (2023) Dissemination of climate smart agricultural knowledge through farmer field schools (FFS): Analyzing the application CAS knowledge by smallholder farmers. *International Journal of Development Issues* 22(3), 399–417.

Gümüş, T., Aydın, T., Gündüz, B., Meriç, S., Ayan, A. *et al.* (2023) Stress memory and priming applications in plants: Potential for breeders. In: El-Esawi, M.A. (ed.) *Recent Trends in Plant Breeding and Genetic Improvement*. IntechOpen, London, pp. 71–94.

Gupta, C. and Salgotra, R.K. (2022) Epigenetics and its role in effecting agronomical traits. *Frontiers in Plant Science* 13, 925688.

Hanson, M.A. and Skinner, M.K. (2016) Developmental origins of epigenetic transgenerational inheritance. *Environmental Epigenetics* 2(1), dvw002.

Hatzig, S.V., Nuppenau, J.N., Snowdon, R.J. and Schießl, S.V. (2018) Drought stress has transgenerational effects on seeds and seedlings in winter oilseed rape (*Brassica napus* L.). *BMC Plant Biology* 18, 1–13.

Hazmi, M., Suryaningrum, D.A., Umarie, I. and Swasono, M.A. (2023) Emerging trends and future directions in organic agriculture and environmentally-friendly farming practices: A bibliometric analysis. *West Science Interdisciplinary Studies* 1(7), 438–447.

Hirayama, T. and Mochida, K. (2022) Plant hormonomics: A key tool for deep physiological phenotyping to improve crop productivity. *Plant and Cell Physiology* 63(12), 1826–1839.

Horsthemke, B. (2018) A critical view on transgenerational epigenetic inheritance in humans. *Nature Communications* 9(1), 2973.

Ikazaki, K., Nagumo, F., Simporé, S. and Barro, A. (2018) Are all three components of conservation agriculture necessary for soil conservation in the Sudan Savanna? *Soil Science and Plant Nutrition* 64(2), 230–237.

Jeynes-Cupper, K. and Catoni, M. (2023) Long distance signalling and epigenetic changes in crop grafting. *Frontiers in Plant Science* 14, 1121704.

Kakoulidou, I., Avramidou, E.V., Baránek, M., Brunel-Muguet, S., Farrona, S. *et al.* (2021) Epigenetics for crop improvement in times of global change. *Biology* 10(8), 766.

Kaya, C., Uğurlar, F. and Adamakis, I.D.S. (2023) Epigenetic and hormonal modulation in plant-plant growth-promoting microorganism symbiosis for drought-resilient agriculture. *International Journal of Molecular Sciences* 24(22), 16064.

Ketumile, D., Yang, X., Sanchez, R., Kundariya, H., Rajewski, J. *et al.* (2022) Implementation of epigenetic variation in sorghum selection and implications for crop resilience breeding. *Frontiers in Plant Science* 12, 798243.

Kim, J.H. (2021) Multifaceted chromatin structure and transcription changes in plant stress response. *International Journal of Molecular Sciences* 22(4), 2013.

Kinoshita, T. and Seki, M. (2014) Epigenetic memory for stress response and adaptation in plants. *Plant and Cell Physiology* 55(11), 1859–1863.

Kottegoda, N., Nimanka, S., Fernando, N., Silva, M. and Munaweera, I. (2023) Climate smart agriculture: The role of fertilizer innovations and efficient plant nutrient management. *Vidyodaya Journal of Science* 1(s1).

Kujur, A., Saxena, M.S., Bajaj, D. and Laxmi Parida, S.K. (2013) Integrated genomics and molecular breeding approaches for dissecting the complex quantitative traits in crop plants. *Journal of Biosciences* 38(5), 971–987.

Kumar, V. (2022) Climate-smart agriculture: Need of the hour. *Journal of Agriculture Research and Technology* 47, 32–36.

Kumar, M. and Rani, K. (2023) Epigenomics in stress tolerance of plants under the climate change. *Molecular Biology Reports* 50(7), 6201–6216.

Kumar, S. and Mohapatra, T. (2021) Dynamics of DNA methylation and its functions in plant growth and development. *Frontiers in Plant Science* 12, 596236.

Lämke, J. and Bäurle, I. (2017) Epigenetic and chromatin-based mechanisms in environmental stress adaptation and stress memory in plants. *Genome Biology* 18, 1–11.

Lanciano, S. and Mirouze, M. (2017) DNA methylation in rice and relevance for breeding. *Epigenomes* 1(2), 10.

Latzel, V., Rendina González, A.P. and Rosenthal, J. (2016) Epigenetic memory as a basis for intelligent behavior in clonal plants. *Frontiers in Plant Science* 7, 1354.

Latzel, V.I.T., Allan, E., Bortolini Silveira, A., Colot, V., Fischer, M. *et al.* (2013) Epigenetic diversity increases the productivity and stability of plant populations. *Nature Communications* 4(1), 2875.

Liu, H., Able, A.J. and Able, J.A. (2022a) Priming crops for the future: Rewiring stress memory. *Trends in Plant Science* 27(7), 699–716.

Liu, X., Quan, W. and Bartels, D. (2022b) Stress memory responses and seed priming correlate with drought tolerance in plants: An overview. *Planta* 255(2), 45.

Lucibelli, F., Valoroso, M.C. and Aceto, S. (2022) Plant DNA methylation: An epigenetic mark in development, environmental interactions, and evolution. *International Journal of Molecular Sciences* 23(15), 8299.

Managa, L.R. and Nkobole-Mhlongo, N. (2016) Towards climate-smart agricultural approach: Prospect for smallholder farmers in semi-arid regions. *Journal of Agriculture and Environmental Sciences* 5(2), 36–46.

Marques, E., Kur, A., Bueno, E. and Wettberg, E. (2020) Defining and improving the rotational and intercropping value of a crop using a plant–soil feedbacks approach. *Crop Science* 60(5), 2195–2203.

Medrano, M., Herrera, C.M. and Bazaga, P. (2014) Epigenetic variation predicts regional and local intraspecific functional diversity in a perennial herb. *Molecular Ecology* 23(20), 4926–4938.

Mierziak, J. and Wojtasik, W. (2024) Epigenetic weapons of plants against fungal pathogens. *BMC Plant Biology* 24(1), 175.

Minow, M.A. and Colasanti, J. (2020) Does variable epigenetic inheritance fuel plant evolution. *Genome* 63(5), 253–262.

Miryeganeh, M. (2021) Plants' epigenetic mechanisms and abiotic stress. *Genes* 12(8), 1106.

Mladenov, V., Fotopoulos, V., Kaiserli, E., Karalija, E., Maury, S. *et al.* (2021) Deciphering the epigenetic alphabet involved in transgenerational stress memory in crops. *International Journal of Molecular Sciences* 22(13), 7118.

Morgan, D.K. and Whitelaw, E. (2009) The role of epigenetics in mediating environmental effects on phenotype. In: Kalhan, S.C., Prentice, A.M. and Yainik, C.S. (eds) *Emerging Societies – Coexistence of Childhood Malnutrition and Obesity, Nestlé Nutrition Institute Workshop Series*, Vol. 63. Karger Publishers, Basel, Switzerland, pp. 109–119.

Mugula, J., Ahmad, A.K., Msinde, J. and Kadigi, M. (2023) Economic impact of adopting bundled SAPs on crop yields and household incomes among smallholder maize farmers in Morogoro region, Tanzania. *International Journal of Research in Business and Social Science* 12(8), 261–273.

Müller-Xing, R., Xing, Q. and Goodrich, J. (2014) Footprints of the sun: Memory of UV and light stress in plants. *Frontiers in Plant Science* 5, 474.

Munshi, A., Ahuja, Y.R. and Bahadur, B. (2015) Epigenetic mechanisms in plants: An overview. *Plant Biology and Biotechnology* II, 265–278.

Mushtaq, M., Sakina, A., Wani, S.H., Shikari, A.B., Tripathi, P. *et al.* (2019) Harnessing genome editing techniques to engineer disease resistance in plants. *Frontiers in Plant Science* 10, 550.

Musvoto, C., Nortje, K., Wet, B., Mahumani, B.K. and Nahman, A. (2015) Imperatives for an agricultural green economy in South Africa. *South African Journal of Science* 111(1–2), 1–8.

Muzorewa, W. and Chitakira, M. (2022) Evaluating the role of climate smart agriculture towards sustainable livelihoods in Mutare district, Zimbabwe. *Significances of Bioengineering & Biosciences* 4(5), 417–425.

Nagar, A.K., Namdeo, S., Verma, B., Shrivastava, S. and Sarkar, R. (2023) Role of climate-smart agriculture towards sustainable food production. *International Journal of Environment and Climate Change* 13(12), 465–472.

Neeb, Z.T. and Nowacki, M. (2018) RNA-mediated transgenerational inheritance in ciliates and plants. *Chromosoma* 127(1), 19–27.

Neufeldt, H., Jahn, M., Campbell, B.M., Beddington, J.R., DeClerck, F. *et al.* (2013) Beyond climate-smart agriculture: Toward safe operating spaces for global food systems. *Agriculture & Food Security* 2, 1–6.

Ostrowska-Mazurek, A., Kasprzak, P., Kubala, S., Zaborowska, M. and Sobieszczuk-Nowicka, E. (2020) Epigenetic landmarks of leaf senescence and crop improvement. *International Journal of Molecular Sciences* 21(14), 5125.

Ou, X., Zhang, Y., Xu, C., Lin, X., Zang, Q. *et al.* (2012) Transgenerational inheritance of modified DNA methylation patterns and enhanced tolerance induced by heavy metal stress in rice (*Oryza sativa* L.). *PLoS ONE* 7(9), e41143.

Phiri, A.T., Charimbu, M., Edewor, S.E. and Gaveta, E. (2022) Sustainable scaling of climate-smart agricultural technologies and practices in Sub-Saharan Africa: The case of Kenya, Malawi, and Nigeria. *Sustainability* 14(22), 14709.

Pikaard, C.S. and Scheid, O.M. (2014) Epigenetic regulation in plants. *Cold Spring Harbor Perspectives in Biology* 6(12), a019315.

Rabie, R.K. and Kumazawa, K. (1988) Effect of salinization and desalinization on the uptake, distribution, and assimilation of fertilizer nitrogen by nodulated soybeans. *Soil Science and Plant Nutrition* 34(4), 493–498.

Rajnović, T., Vokurka, A. and Bolarić, S. (2020) Epigenetics in plant breeding. *Journal of Central European Agriculture* 21(1), 56–61.

Raza, A., Mubarik, M.S., Sharif, R., Habib, M., Jabeen, W. *et al.* (2023) Developing drought-smart, ready-to-grow future crops. *The Plant Genome* 16(1), e20279.

Raza, A., Bashir, S., Khare, T., Karikari, B., Copeland, R.G.R. *et al.* (2024) Temperature-smart plants: A new horizon with omics-driven plant breeding. *Physiologia Plantarum* 176(1), e14188.

Selbonne, S., Guindé, L., Causeret, F., Chopin, P., Sierra, J. *et al.* (2023) How to measure the performance of farms with regard to climate-smart agriculture goals? A set of indicators and its application in Guadeloupe. *Agriculture* 13(2), 297.

Sghaier-Hammami, B. and Hammami, S.B. (2023) New insights in nanotechnology for plant stress management. *Frontiers in Plant Science* 14, 1319936.

Shahrajabian, M.H., Sun, W. and Cheng, Q. (2019) DNA methylation as the most important content of epigenetics in traditional Chinese herbal medicine. *Journal of Medicinal Plants Research* 13(16), 357–369.

Skinner, M.K. and Nilsson, E.E. (2021) Role of environmentally induced epigenetic transgenerational inheritance in evolutionary biology: Unified evolution theory. *Environmental Epigenetics* 7(1), dvab012.

Stone, J.E., Marx, D.B. and Dobrenz, A.K. (1979) Interaction of sodium chloride and temperature on germination of two Alfalfa cultivars 1. *Agronomy Journal* 71(3), 425–427.

Sun, C., Ali, K., Yan, K., Fiaz, S., Dormatey, R. *et al.* (2021) Exploration of epigenetics for improvement of drought and other stress resistance in crops: A review. *Plants* 10(6), 1226.

Tan, J.W. and Rodríguez López, C.M. (2023) Epigenomics: A new tool for the generation of climate resilient grapevines. *Frontiers in Horticulture* 2, 1116866.

Thakore, P.I., D'Ippolito, A.M., Song, L., Safi, A., Shivakumar, N.K. *et al.* (2015) Highly specific epigenome editing by CRISPR-cas9 repressors for silencing of distal regulatory elements. *Nature Methods* 12(12), 1143–1149.

Tharun, P.S., Talekar, N. and Akkati, V.R. (2024) Applications of molecular marker implementation for enhanced oilseed breeding through marker-assisted selection (Mas). *Journal of Experimental Agriculture International* 46(5), 865–876.

Tirnaz, S. and Batley, J. (2019) Epigenetics: Potentials and challenges in crop breeding. *Molecular Plant* 12(10), 1309–1311.

Tiwari, A.K. (2024) Insect pests in agriculture identifying and overcoming challenges through IPM. *Archives of Current Research International* 24(3), 124–130.

Ukhurebor, K.E. (2020) The role of biosensor in climate smart organic agriculture toward agricultural and environmental sustainability. In: Meena, R.S. (ed.) *Agrometeorology*. IntechOpen, London.

Varotto, S., Krugman, T., Aiese Cigliano, R., Kashkush, K., Kondić-Špika, A. *et al.* (2022) Exploitation of epigenetic variation of crop wild relatives for crop improvement and agrobiodiversity preservation. *Theoretical and Applied Genetics* 135(11), 3987–4003.

Vărzaru, A.A. (2024) Assessing agricultural impact on greenhouse gases in the European union: A climate-smart agriculture perspective. *Agronomy* 14(4), 821.

Verkest, A., Byzova, M., Martens, C., Willems, P., Verwulgen, T. *et al.* (2015) Selection for improved energy use efficiency and drought tolerance in canola results in distinct transcriptome and epigenome changes. *Plant Physiology* 168(4), 1338–1350.

Wahbi, S., Prin, Y., Thioulouse, J., Sanguin, H., Baudoin, E. *et al.* (2016) Impact of wheat/faba bean mixed cropping or rotation systems on soil microbial functionalities. *Frontiers in Plant Science* 7, 1364.

Wang, G., Bei, S., Li, J., Bao, X., Zhang, J. *et al.* (2021) Soil microbial legacy drives crop diversity advantage: Linking ecological plant–soil feedback with agricultural intercropping. *Journal of Applied Ecology* 58(3), 496–506.

Waryah, C.B., Moses, C., Arooj, M. and Blancafort, P. (2018) Zinc fingers, TALEs, and CRISPR systems: A comparison of tools for epigenome editing. In: Jeltsch, A. and Rots, M.G. (eds) *Epigenome Editing: Methods and Protocols*. Methods in Molecular Biology Series, Humana New York, pp. 19–63.

Wei, X.I.E., Qianqian, T.A.N.G., Fei, Y.A.N. and Zeng, T.A.O. (2021) Transcriptional memory and response to adverse temperatures in plants. *Journal of Zhejiang University. Science. B* 22(10), 791–804.

Williams, B.R., Edwards, C.E., Kwasniewski, M.T. and Miller, A.J. (2020) Epigenomic patterns reflect irrigation and grafting in the grapevine clone 'Chambourcin'. *BioRxiy*.

Xu, W.B., Cao, F., Liu, P., Yan, K. and Guo, Q.H. (2024) The multifaceted role of RNA-based regulation in plant stress memory. *Frontiers in Plant Science* 15, 1387575.

Yan, W. (2014) Potential roles of noncoding RNAs in environmental epigenetic transgenerational inheritance. *Molecular and Cellular Endocrinology* 398(1–2), 24–30.

Yang, X., Kundariya, H., Xu, Y.Z., Sandhu, A., Yu, J. *et al.* (2015) MutS HOMOLOG1-derived epigenetic breeding potential in tomato. *Plant Physiology* 168(1), 222–232.

Yung, W.-S., Ting-Fung, C., Kong, F. and Lam, H.M. (2023) The plant genome special section: Epigenome and epitranscriptome in plant-environment interactions. *The Plant Genome* 16(4).

Zeleke, G., Teshome, M. and Ayele, L. (2024) Determinants of smallholder farmers' decisions to use multiple climate-smart agricultural technologies in North Wollo Zone, Northern Ethiopia. *Sustainability* 16(11), 4560.

Zenda, T., Liu, S., Dong, A., Li, J., Wang, Y. *et al.* (2021) Omics-facilitated crop improvement for climate resilience and superior nutritive value. *Frontiers in Plant Science* 12, 774994.

Zhang, H.J., Dong, H.Z., Li, W.J. and Zhang, D.M. (2012) Effects of soil salinity and plant density on yield and leaf senescence of field-grown cotton. *Journal of Agronomy and Crop Science* 198(1), 27–37.

Zhao, L., Yang, Y., Chen, J., Lin, X., Zhang, H. *et al.* (2023) Dynamic chromatin regulatory programs during embryogenesis of hexaploid wheat. *Genome Biology* 24(1), 7.

Zhi, P. and Chang, C. (2021) Exploiting epigenetic variations for crop disease resistance improvement. *Frontiers in Plant Science* 12, 692328.

3 The Role of Plant Epigenetic Memory Mechanisms in Abiotic Stress Tolerance

Neslihan Turgut Kara[1]*, Burcu Arıkan[1], Elif Pulat[2,3] and Özgür Çakır[1]
[1]Istanbul University, Faculty of Science, Department of Molecular Biology and Genetics, Vezneciler, Istanbul, Türkiye; [2]Istanbul University, Institute of Science, Program of Molecular Biology and Genetics, Istanbul, Türkiye; [3]Acıbadem University, Faculty of Engineering and Natural Sciences, Department of Molecular Biology and Genetics, Istanbul, Türkiye

Abstract
Increasing knowledge of how plants perceive environmental stimuli, including the mechanisms of cellular signaling and gene expression cascades, provides evidence that plants can remember past environmental events. The use of these memories to enhance responses in situations where developmental stages or stress conditions must be repeated and their ability to be inherited through cell divisions reveals a potential mechanism for stress memory in plants. Recent evidence indicates that stress-associated epigenetic mechanisms are essential for plant stress memories and various adaptation mechanisms. This chapter discusses the importance of turning on and off the transcriptional profiles of stress-related genes, particularly in understanding various types of plant abiotic stress memories regulated by epigenetic changes. In addition, the chapter draws attention to the lack of information on how stress memory is transferred and points out that the regulatory mechanisms in plants can only be understood by deciphering the memory mechanisms.

Keywords: Epigenetic, Memory, Abiotic stress, Epigenetic modification, Gene expression

3.1 Introduction

Agriculture is essential for feeding the rapidly growing global population, relying heavily on major crops like cotton, maize, sugarcane, rice, barley, wheat, and soybean for fiber, fuel, and food. To meet the needs of a growing global population, agricultural output must double by 2050. However, the effects of climate change disrupt the plant's physiological and biochemical processes, leading to reduced growth and productivity, making this task even more difficult (Janni *et al.*, 2024). Recent advancements, including affordable genetic profiling systems and genome editing techniques, offer promising avenues for improving crop resilience and productivity (Kumari *et al.*, 2022). Feeding the growing world population would require utilizing the latest discoveries in plant science, especially plant epigenetics. Epigenetics is increasingly recognized as a significant contributor to vital traits in various plant species. This field encompasses processes such as DNA methylation, histone modification, and ncRNA modification, which regulate gene expression without modifying the DNA sequence. Understanding plant epigenetics

*Corresponding author: neslihantk@istanbul.edu.tr

DOI: 10.1079/9781800626102.0003

is crucial for comprehending how plants adapt to extreme stress conditions by adjusting their phenotypes (Saeed *et al.*, 2022).

Conrad Hal Waddington published his famous epigenetic landscape drawing in 1957, depicting the cell's path through development from the undifferentiated state, providing us with a better understanding of how differentiated cell fate actually occurs (Waddington, 1957). Epigenetics has a broad meaning defined by Waddington and a narrow meaning related to cellular memory and cellular inheritance. It generally refers to the biology branch investigating the causal interactions between genes and the products that create the phenotype. In its narrower but modern sense, it involves the study of developmental processes in both prokaryotes and eukaryotes that lead to or can sustain changes that persist even in the absence of the original triggering input underlying an organisms' relationship with memory mechanisms (Waddington, 1975; Jablonka, 2017).

In plants, DNA methylation typically involves the addition of methyl groups to cytosine bases [CG (or CpG), CHG, or CHH (where H corresponds to A, T, or C)] and can lead to gene silencing or activation, depending on the location and extent of methylation. Histone modifications such as acetylation, methylation, and phosphorylation affect gene expression patterns by changing the structure and accessibility of chromatin. Additionally, small RNAs, including microRNAs (miRNAs) and small interfering RNAs (siRNAs), can guide gene silencing through sequence-specific interactions with target mRNAs, leading to translational repression or mRNA degradation. These epigenetic modifications can persist across cell divisions and sometimes even across generations, contributing to establishing and maintaining cellular memory. Moreover, environmental stimuli such as abiotic stresses can also induce specific epigenetic changes that confer adaptive responses and memory in plants, enabling plants to better cope with repetitive stressors (Pikaard and Mittelsten Scheid, 2014).

The scope of plant epigenetics has been expanded to include transgenerational inheritance. Epigenetic modifications acquired in response to environmental stimuli can be transmitted to subsequent generations, thus affecting offspring phenotypes. This phenomenon underlines the tremendous flexibility of plants to respond dynamically to their environment and adapt to changing environmental conditions, and reveals the difference between this group of living organisms. It also provides clues to the evolutionary forces that enable plant species to diversify and acquire adaptive traits. Elucidating the epigenetic mechanisms underlying plant development and environmental responses may enable us to uncover new strategies to increase crop productivity, resilience, and sustainability in the face of agricultural challenges.

In summary, studies on plant epigenetic memory may provide promising clues toward unraveling the molecular mechanisms that support plant development, adaptation, and evolution. Plant epigenetics not only advances our understanding of plant biology by bridging the gap between genotype and phenotype but also enables us to identify genes and gene products that can serve as targets for biotechnological advances by understanding the molecular scenarios of stress memory consistently provided by plants.

This chapter discusses the importance of regulating transcriptional profiles of stress-related genes, especially through epigenetic changes, in understanding plant abiotic stress memories for better tolerance mechanisms. At the same time, it tries to explain the importance of understanding the regulatory mechanisms of plant stress responses and deciphering the memory mechanisms by emphasizing that we need to have knowledge about stress memory transfer.

3.2 Transfer of Stress Memory

After exposure to stress, plants can give repetitive responses called memory. In particular, the reactions of plants to environmental stress are regulated through changes in transcription, and in this way, the molecular markers of their initial encounter with stress can be preserved through transcriptional regulation. Later, when they encounter similar stresses, they can combat stress by initiating stress response and adaptation mechanisms instantly and more effectively. This plant response, stress priming, can activate short-term or long-term stress memory (Liu *et al.*, 2022). These types of responses are reversible, as they do not require DNA modification and typically affect the plant phenotype, and the effect of this memory persists for days or weeks and can be transmitted to subsequent generations (Zuo *et al.*, 2023). D'Urso and Brickner (2014) classified epigenetic memory first as cellular memories, which are transcriptional states that can be mitotically inherited in response to cues during development. The second is that transcriptional memories are mitotically inherited changes in the organism's response to environmental changes due to previous experiences. The last is transgenerational memory, which is meiotically inherited changes reflected in the phenotype along with gene expression according to the experiences of previous generations. Lämke and Bäurle (2017) classified stress memory in plants as somatic, intergenerational, and transgenerational stress memory. Somatic stress memory (where memory traces are inherited mitotically) is also known as mitotic stress memory and is limited to the duration of a single generation. Early-stage stress during plant growth and development can trigger this short-term stress memory. Stress experienced by parent plants during the reproductive stage and affecting the resulting seeds leads to intergenerational stress memory mediated by molecular signals transmitted by the parent plant to the seed or embryo. Stress memory in a generation that is not directly exposed to stress due to ancestral exposure during the reproductive phase is called transgenerational memory (Lämke and Bäurle, 2017). In response to stress, plants can inherit epigenetic markers across multiple generations through transgenerational epigenetic inheritance and stress memory (Hauser *et al.*, 2011).

3.3 Mechanisms of Epigenetic Stress Memory

There has been a growing number of reports detailing plant stress memory, which refers to the lasting influence of stress. This effect entails the significant impact of past exposure to abiotic or biotic stress on how plants respond to subsequent stress in either the current or future generations (Liu *et al.*, 2022). Epigenetic mechanisms are vital for recording and retaining memories of environmental shifts and for precisely adjusting gene expression (Nishio *et al.*, 2024). Epigenetic modifications resulting from the actions of DNA or histone methylation, other histone modifications, microRNAs, and long non-coding RNAs are quite important as they regulate genome-wide transcription activities for the formation of plant stress responses and tolerance. At the chromosome level, structure, location, and histone variant alterations all have a role in how plants respond to stress, and these modifications can be inherited through both mitotic and meiotic divisions (Kinoshita and Seki, 2014).

Re-encounters with stress factors or priming practices help plants adapt to recurring stress events and increase their resilience by interacting with these epigenetic mechanisms. In other words, to protect themselves against repeated stress events, plants have developed adaptive mechanisms that involve a process called priming, in which they remember previous stress exposures. Priming is a phenomenon where plants are able to respond more successfully to repeated stress events when they have experienced stress before. After priming, plants spend a stress-free period between two stresses and then respond more prepared to the second stress stimulus (Turgut Kara *et al.*, 2020; Liu *et al.*, 2022). After the priming event, this behavior, also called 'ready response' or 'memory', occurs with the methylation states of genes and histones, other types of modifications of histones, the actions of ncRNAs, and other components that function in chromatin remodeling.

3.3.1 Stress memory and DNA methylation

Certain cytosine bases in CG, CHG, and CHH sites undergo methylation in higher plants, with 5-methylcytosine (5mC) being the most prevalent form. In addition to cytosine, there is adenine methylation, known as 6mA in DNA and m6A in RNA, an epigenetic modification crucial for various cellular processes. This reversible modification involves the covalent binding of a methyl group, typically at the N6 position of the purine ring, influencing nucleic acid stability and molecular interactions. DNA 6mA plays significant roles in genome stability, gene expression, DNA replication, repair mechanisms, and stress responses (Jiménez-Ramírez *et al.*, 2022). Studies have mapped DNA methylation across the *Arabidopsis* genome, revealing its presence in large portions of genes and transposable elements (TEs) (Zilberman *et al.*, 2007). In plants, *de novo* DNA methylation is facilitated by the RNA-directed DNA methylation (RdDM) pathway, where small interfering RNAs (siRNAs) produced by RNA polymerase IV (Pol IV) initiate DNA methylation at homologous Pol V-transcribed loci, leading to transcriptional silencing at repetitive DNA sequences, including various types of TEs. Maintenance of methylation and active demethylation in promoter and gene body regions involves various methyltransferases (methyltransferase MET1 (homolog of mammalian DNMT1) maintains DNA methylation in the CG context, while two conserved plant-specific methyltransferases, Chromomethylase 3 (CMT3) and CMT2, help maintain CHG and CHH methylation, respectively) (Bewick *et al.*, 2017) and demethylase [Repressor of silencing 1 (ROS1), Demeter (DME), Demeter-like 2 (DML2), and Demeter-like 3 (DML3)] (Choi *et al.*, 2002). DNA methylation plays a crucial role in maintaining genome stability and regulating gene transcription, with methylation in gene promoter regions generally suppressing transcription, while methylation in the gene body may enhance transcription (Kumar and Mohapatra, 2021).

In the rice genome, Zheng *et al.* (2013) showed that differentially methylated regions (DMRs) occur mainly in gene promoter and exon regions, which has implications for gene expression regulation. Also, DMRs have been observed to affect the expression of other genes near the genes. Rice plants with drought susceptibility primarily displayed increased methylation levels, while those with drought tolerance exhibited decreased methylation levels. They also noted the presence of numerous DMRs across all 12 rice chromosomes, predominantly located in gene promoter and exon regions. Additionally, the tolerant genotypes not only showed reduced methylation levels but also demonstrated a greater capacity for adjusting DNA methylation levels. Active DNA demethylation, mediated by enzymes like ROS1, plays a role in regulating gene expression under stress conditions, with a subset of ABA-inducible genes being regulated by ROS1-dependent DNA demethylation (Zheng *et al.*, 2013; Kim *et al.*, 2019). According to Wang *et al.* (2015), a sizable percentage of salt-induced DNA methylation or demethylation changes in rice continue even after the stress is eliminated, suggesting that these modifications can store memory of environmental salt stress and transmit the stress-induced epigenetic states to daughter cells during mitotic cell divisions in the current generation. Additionally, it is possible that specific variations in the 5mC or 6mA genomic locations influence how plants respond somatically to abiotic stressors (Liu and He, 2020).

3.3.2 Stress memory and histone modifications

In eukaryotic cells, DNA is packed into the nucleus by wrapping around histone octamers, forming nucleosomes. Modifications like acetylation, methylation, phosphorylation, and ubiquitination on histone tails can alter gene activity. These modifications have diverse effects depending on the type and residue modified. Typically, deacetylation of certain lysine residues tightens DNA-histone interaction, suppressing gene transcription. The response of plants to abiotic stress is influenced by the interaction between metabolic activity and the dynamics of histone methylation. Relatedly, stress memory formation is also closely linked to histone methylation (Shi *et al.*, 2024). Studies demonstrate how histone modification complexes regulate gene expression, particularly in response to stress (Allis and Jenuwein, 2016; Kumar and Mohapatra, 2024). For instance, the abscisic acid (ABA) receptor genes are repressed by a complex involving a histone deacetylase (HDA) and the histone-binding WD40-repeat protein MSI1, which impacts *Arabidopsis's* susceptibility to this phytohormone. A histone deacetylation complex comprising HDA19 was found to suppress the expression of ABA receptor genes (Mehdi *et al.*, 2016). When ABA4 interacts with HDA9, a REDUCED Potassium Dependency Protein 3 (RPD3)/HDA1-type class 1 histone deacetylase, and Powerdress (PWR), a homolog of human NcoR1, a chromatin remodeling complex is formed. PWR and HDA9 interact with ABI4 to control plant drought signaling through epigenetically regulated expression (Khan *et al.*, 2020).

Studies on stress-responsive gene regulation reveal changes in histone modifications like H3K4me3, H3K36me3, H3K9me2, and H3K27me3 influencing gene expression. Generally, H3K4me3 and H3K36me3 are associated with transcriptional activation, while H3K9me2 and H3K27me3 are correlated with transcriptional silencing (Xiao *et al.*, 2016; Imran *et al.*, 2022). Specifically, H3K4me3 plays a significant role in transcriptional memory for various abiotic stress-responsive genes. For instance, upon the onset of dehydration, the H3K4me3 level at the dehydration memory genes increases and persists for a while. Moreover, it was shown that following repeated treatments with drought stress, these memory-trained genes had higher levels of H3K4me3 than untrained genes. This finding suggests that H3K4me3 plays a role in maintaining drought stress memory (Ding *et al.*, 2012).

Similar to this, the expression of Δ1-pyrroline-5-carboxylate synthetase 1 (P5CS1), which encodes an enzyme involved in proline biosynthesis, is raised by primary salt stress, decreases to a basal level during recovery and increases in response to recurrent salt stress. Subsequent research revealed that the preservation of an elevated H3K4me3 level at the P5CS1 locus during the recovery phase following salt stress is connected to the salt-induced transcriptional memory of P5CS1 (Feng *et al.*, 2016). Also, plants under heat stress have been shown to accumulate H3K4me2/3. After the heat wave passed, the heat memory genes showed H3K4 hypermethylation that persisted for many days (Lämke *et al.*, 2016; Liu *et al.*, 2018).

Exposure to cold stress reduced the abundance of H3K27me3 in genes responsive to such stress, such as COLD-REGULATED 15A (COR15A) and *galactinol synthase 3* (GOLS3). This reduction persisted for several days even after plants returned to their regular conditions, indicating that H3K27me3 functions as a memory marker for cold stress (Kwon *et al.*, 2009). Hence, histone methylation, specifically H3K4me3 and H3K27me3, is therefore essential for providing plants with transcriptional memory of abiotic stressors (Shi *et al.*, 2024).

3.3.3 Stress memory and non-coding RNAs

Non-coding RNAs (ncRNAs) are transcribed from various genome regions, constituting over 90% of all RNA transcripts (Chekanova *et al.*, 2007). Despite not encoding proteins, ncRNAs play crucial regulatory roles in diverse biological processes. Specifically, they have been demonstrated to be essential in controlling the expression of genes at the epigenetic, transcriptional, posttranscriptional, and post-translational levels (Chen *et al.*, 2021). In plants, thousands of ncRNAs have been identified, including microRNAs (miRNAs), TE-derived small interfering RNAs (siRNAs), and long non-coding RNAs (lncRNAs), which exert their effects through different mechanisms (Ma *et al.*, 2022). TE-derived siRNAs, on the other hand, participate in the transcriptional silencing of TEs through the RdDM pathway. RNA-directed DNA methylation (RdDM) is a biological mechanism where non-coding RNA molecules guide the addition of DNA methylation to particular DNA sequences. While the RdDM pathway is exclusive to plants, alternative mechanisms of RNA-directed chromatin modification have been identified in fungi and animals (Erdmann and Picard, 2020). Though they cannot translate into proteins, ncRNAs have many important roles in many cellular and biological processes, including plant growth, development, senescence, and responses to biotic and abiotic stresses. These functions are frequently linked to transcription, turnover, editing, chromatin remodeling, epigenetic memory, RNA splicing, transcription, and post-transcriptional control of gene expression (Yoon *et al.*, 2013; Waititu *et al.*, 2020; Sebastian-DelaCruz *et al.*, 2021; Yadav *et al.*, 2024).

miRNAs and lncRNAs regulate genes involved in responses to stressors like drought by targeting pathways related to ABA-mediated responses, osmoprotection, and antioxidant defense. Stress-responsive elements in TE promoters lead to the production of ncRNAs under stress conditions. Transcriptomic studies have revealed significant changes in miRNA expression in response to drought stress in plants like *Arabidopsis* and rice. For instance, in rice, miR820 derived from TEs targets *de novo* DNA methyltransferase transcripts involved in epigenetic silencing under stress (Nosaka *et al.*, 2012).

3.3.4 Stress memory and chromatin remodeling complexes

In plants, chromatin remodeling complexes play a crucial role in regulating gene expression epigenetically. These complexes alter the structure of chromatin, which affects the accessibility of DNA to transcriptional machinery and regulatory proteins (Pandey *et al.*, 2016; Abdulraheem *et al.*, 2024).

Mitotically inherited epigenetic or dynamic chromatin state changes affect the plant's response to abiotic stress. Chromatin remodeling complexes, such as the SWI/SNF complex, are the universal epigenetic inheritance currency, persistently keeping chromatin open for the function of enhancers and promoters (Ahmad *et al.*, 2024). The SNF2 family protein homolog of Brahma, the ATPase of the Drosophila SWITCHING DEFECTIVE2/SUCROSE NON-FERMENTING (SWI/SNF) complex involved in chromatin remodeling during transcription, is *Arabidopsis thaliana* BRAHMA (BRM, also known as AtBRM) (Hurtado *et al.*, 2006). It utilizes ATP to alter the interaction between nucleosomes and DNA, thus influencing their positioning and occupancy. BRM contains various functional domains, including those involved in protein–protein interactions, ATP binding for transcriptional activation, and DNA binding. Through whole-genome mapping and transcriptome analyses in *Arabidopsis*, it has been revealed that the BRM complex is present at numerous genomic sites, most of which are located within or close to genes, impacting gene transcription positively or negatively (Archacki *et al.*, 2017).

The mechanism by which BRM-containing SWI/SNF complexes access and occupy their target sites has been partially elucidated. Studies suggest that physical interactions with specific proteins, such as the plant-specific H3K27 demethylase REF6 or various transcription factors, facilitate their recruitment to genomic loci (Li *et al.*, 2016). Furthermore, BRM's role in the ABA signaling pathway has been highlighted, where it modulates ABA responses and drought tolerance by regulating nucleosome stability and gene transcription. Post-translational modifications of BRM, such as phosphorylation by SnRK2.6/OST1 and dephosphorylation by PP2CA, also play a crucial role in regulating its function. These modifications serve as a 'switch' controlling BRM-mediated chromatin remodeling and the transcription of ABA-responsive genes. Additionally, studies have shown that under stress conditions like salt stress, there is enhanced expression of PP2C genes in plants lacking BRM, further implicating BRM in stress response pathways (Peirats-Llobet *et al.*, 2016).

Chromatin looping is another regulatory mechanism influencing gene expression. Coordinating transcriptional activities is aided by the determination of chromatin domains by long-range interactions both within and across chromosomes. Conversely, short-range chromatin interactions have been identified as dynamic processes controlling the expression of particular loci (Gagliardi and Manavella, 2020). Chromatin loops are influenced by factors like environmental signals and chromatin remodeling complexes. Then, interactions at the local scale that affect transcriptional memory and gene expression are arranged in the genome into chromatin loops (Rodriguez-Granados *et al.*, 2016). According to studies, alterations in nuclear structure and stress circumstances may be functionally related. Given the link between chromatin changes and somatic memory, there may be a molecular foundation for stress memory (Bhadouriya *et al.*, 2021).

3.4 Epigenetic Memory-Related Genes After Abiotic Stress

Abiotic stresses, such as drought, salinity, and extreme temperatures, pose significant challenges to crop productivity worldwide. Limitation in plant mobility has enhanced their adaptability to a broader spectrum of stress conditions. Recent advancements have shed

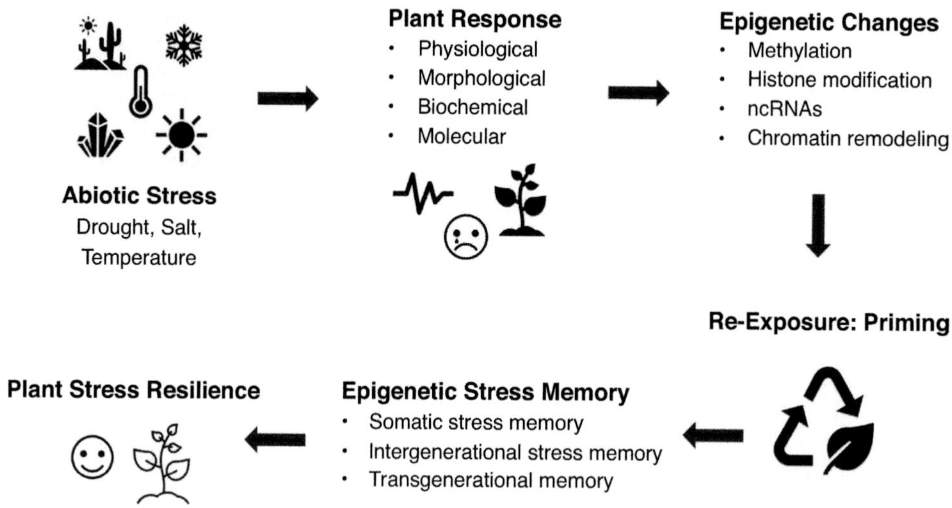

Fig. 3.1. Abiotic stress is remembered through epigenetic modifications, and the plant gains resilience through priming. Author's own image.

light on the underlying molecular mechanisms enabling plants to withstand stress. Both genetic and epigenetic evidence indicate that plants have the ability to 'remember' past stress events and respond more effectively to subsequent exposures. This phenomenon, known as abiotic stress memory, offers promising avenues for improving crop resilience and yield stability. Abiotic stress memory involves the plant's ability to retain information about past stress events and modulate its physiological and molecular responses accordingly.

Environmental factors and developmental stages influence plants' establishment and persistence of abiotic stress memory. Priming, a phenomenon where prior exposure to mild stress enhances subsequent stress tolerance, is an important aspect of stress memory. Environmental cues, such as light, temperature, and nutrient availability, can modulate the strength and duration of stress memory. Furthermore, stress experienced during critical developmental stages, such as germination and flowering, can have long-lasting effects on plant resilience (Fig. 3.1). In recent years, many studies have been conducted to identify genes that play a role in the response mechanisms that result in resilience and memory formation against drought, salinity, and heat stress (Table 3.1).

3.4.1 Drought-responsive genes

In order to overcome stress conditions such as drought, plants develop strategies to withstand adverse conditions by activating their tolerance genes and regulating their biological adaptations (Kim and Lee, 2023). Central to this response is the regulation of endogenous ABA, an important plant hormone. ABA manages stomatal closure, reduces transpiration to preserve water, and helps the plant increase its resistance to stress by synthesizing osmoprotectants. Complex remodeling of chromatin architecture under epigenetic regulation plays a role in initiating such responses. During this epigenetic reorganization, repressive chromatin is converted to an active state to facilitate gene transcription (Seki *et al.*, 2002; Hoth *et al.*, 2002; Sah *et al.*, 2016).

When plants were subjected to double stress as opposed to a first-time drought challenge, ABA revealed variations. Following the first drought, ABA levels soared 40-fold before eventually returning to their pre-drought levels. However, ABA levels rose greater in double-stressed plants during the second stress period than in plants that were being challenged by drought for the first time. Through reprogramming gene expression, ABA provides protection

Table 3.1. Genes determined to be responsible for abiotic stress memory establishment and their related pathways. Author's own table.

Plant	Stress	Epigenetic Memory Mechanism	Gene	Function /Pathway	Reference
Arabidopsis thaliana *Solanum tuberosum*	Drought Salinity	Transcriptional memory	LTP2, LTP3, LTP4 (Lipid transfer protein)	Calmodulin binding lipid transport protein, localized in cell wall	Liu *et al.*, 2014; Chen *et al.*, 2019
Arabidopsis thaliana *Oryza sativa*	Salinity Drought	Transcriptional memory, lncRNA regulation	P5CS(Δ1-PYRROLINE-5-CARBOXYLATE SYNTHETASE 1)	Proline accumulation, stress response.	Feng *et al.*, 2016; Li *et al.*, 2019
Oryza sativa *Arabidopsis thaliana*	Drought	Dosage memory genes, ABA-regulation, lncRNA, or DNA methylation	SnRK2 (SAPK2, SAPK7, SAPK6, SAPK9, and SAPK10) CYP76C2, CYP707A3	Serine/threonine protein kinase Allene oxide synthase	Li *et al.*, 2019
Arabidopsis thaliana	Drought Cold	Chromatin/ Transcriptional memory	RESPONSIVE TO DESICCATION 29A (RD29A), LTI	Stress response protein	Liu *et al.*, 2023
Arabidopsis thaliana	Drought Salinity Cold	Chromatin/ Transcriptional memory	RESPONSIVE TO DESICCATION 29B (RD29B), LTI	Stress response protein, ABA-responsive, signaling	Liu *et al.*, 2023
Arabidopsis thaliana *Nicotiana tabacum*	Salinity	DNA methylation, Transcriptional memory	ROS1 (repressor of silencing), (DEMETER-LIKE 1)	Repressor of transcriptional silencing. Activating stress response genes. Reduces DNA methylations	Bharti *et al.*, 2015; Zuo *et al.*, 2023
Arabidopsis thaliana	Heat	Transcriptional memory	HSFA2 (heat shock transcription factor A2) HSFA1, HSFA3	DNA-binding transcription factor activity	Liu *et al.*, 2019; Kappel *et al.*, 2023
Arabidopsis thaliana	Heat	Histone demethylation	REF6 (RELATIVE OF EARLY FLOWERING 6)	H3K27 histone demethylase	Liu *et al.*, 2019
Arabidopsis thaliana	Heat	Chromatin remodeling	FGT1, 2 and 3 (FORGETTER)	Nucleosome remodeling	Brzezinka *et al.*, 2016; Castellanos *et al.*, 2020

Arabidopsis thaliana	Heat	Transcriptional memory	HSP17.6A and FBA6 (Fructose 1,6-bisphosphate aldolase)	Small heat shock protein and carbohydrate metabolism	Olas *et al*., 2021; Yamaguchi *et al*., 2021
Arabidopsis thaliana	Heat	Transcriptional memory, post-transcriptional regulation	HSP101-HSA32 HSP21 HSP22 HSP18.2	Heat shock proteins/protein folding and heat stress associated protein, post-transcriptional regulation	Yamaguchi *et al*., 2021
Arabidopsis thaliana	Heat	Histone demethylation	JMJ (JUMONJI)	H3K27 histone demethylase	Yamaguchi *et al*., 2021
Arabidopsis thaliana	Heat	Histone methylation	SDG25, ATX1 (trithorax family proteins)	H3K4 methyltransferase controls flowering	Song *et al*., 2021
Arabidopsis thaliana	Heat	Transcriptional memory	CLV1 (CLAVATA1) CLV3	Receptor serine/threonine kinase activity, meristem size control	Olas *et al*., 2021
Arabidopsis thaliana	Cold	DNA methylation, Transcriptional regulation	PICKLE (PKL)	Putative CHD3-type chromatin remodeling factor, RdDM, and growth regulation	Yang *et al*., 2019

during periods of drought (Fleta-Soriano *et al.*, 2015).

Abiotic stress (mainly drought or high-salinity) conditions trigger changes in gene expression, with more than half of these genes responding to ABA application (Todaka *et al.*, 2019; Nguyen *et al.*, 2022). ABA is produced in vascular tissues and transported to its targets, including guard cells, through various transporters. In *Arabidopsis*, three transporter families have been identified: ATP binding cassette (ABC), detoxification efflux carriers (DTX)/multidrug and toxic compound extrusion (MATE) proteins, and nitrate transporter 1/peptide transporter family (NPF) (Shimizu *et al.*, 2022). ABA promotes the expression of genes that produce osmoprotectants like trehalose and late embryogenesis-abundant proteins and regulates genes involved in ABA metabolism, transport, and signaling (Cheong, 2022). The promoters of ABA-responsive genes contain a conserved *cis*-acting element called ABRE (ABA-responsive element) (Uno *et al.*, 2000). However, genes induced by drought do not respond to ABA, suggesting ABA-independent pathways (Shinozaki and Yamaguchi-Shinozaki, 2007). These genes' promoters contain the dehydration-responsive element/C-repeat (DRE/CRT) element, which mediates ABA-independent gene expression. Transcription factors like C-repeat/dehydration-responsive element binding factors (CBF/DREB1) and DREB2 bind to DRE/CRT elements and regulate gene expression under drought stress, which are also implicated in tolerance to other abiotic stresses. Both dicotyledonous and monocotyledonous plants possess similar transcription factors for drought stress tolerance. The *Arabidopsis* RD29A gene has both ABRE and DRE/CRT elements in its promoter region (Wang *et al.*, 2022).

Drought conditions are known to have detrimental effects on plant physiology and metabolism. Stress memory, which is highly correlated with antioxidant and osmotic regulations, can result from differential metabolite or ROS accumulation or antioxidant enzyme activity in addition to the transcriptional regulation of drought-responsive genes (Liu *et al.*, 2022).

The drought priming effect on transcriptional responses of stress memory has been previously shown in *Arabidopsis*. Induced differential expression of stress-response genes was found to be associated with such memory. These genes are transcribed at normal levels when conditions return to normal while staying bound to Ser5P Polymerase II. Increased levels of these stalled RNA polymerases and H3K4me3 marks were found to be responsible for making their associated genes remember the drought conditions. Trained plants start to lose less water during drought after multiple exposures, indicating the physiological correspondence of the memory (Ding *et al.*, 2012).

An example of lncRNAs taking part in drought memory is observed in rice plants. Target analysis of three individual lncRNAs revealed transcripts of a serine-threonine protein kinase, PAL (phenylalanine ammonia-lyase) gene, and ferredoxin, which were previously found to be drought memory-related. The same study analyzed the regulation of drought memory by DNA methylations on associated genes. Out of 5373 candidate memory transcripts, expressions of 3064 were significantly related to DNA methylations, most of them being CHH methylations. Rice drought memory was thereby hypothesized to be possibly related to DNA methylations of drought-response genes as well (Li *et al.*, 2019).

In *Glycine max* L., drought-induced genes associated with gene ontology terms, such as response to cold, heat acclimation, and hyperosmotic salinity response, were determined. Drought-induced memory transcripts were determined by comparison of post- and pre-treatment with secondary stress and validated by qPCR. Among these, transcription factors (WRKY, MYB, NAC), osmotin, ABA-responsive stress proteins (SMP (stress management protein), LEA (late-embryogenesis abundant)), and dehydration responsive element AP2 were validated to have higher expressions after recurring drought stress, and associated with drought memory (Kim *et al.*, 2020).

Many genes taking part in epigenetic mechanisms are identified to be involved in creating memory. ATX-1 (TRITHORAX-LIKE 1), as a histone methyltransferase, creates previously mentioned H3K4me3 marks, whereas, in *atx1* mutants, memory response faltered, pointing out its role in these circumstances. ABA-responsive proteins such as AREB1, AREB2, and ABF3 take part in the induction of memory-associated genes (Ding *et al.*, 2012). Memory genes such as pre-mRNA-splicing factor 38A, vacuolar amino acid transporter 1-like, and UDP-sugar pyrophosphorylase are regulated by promoter methylations, indicating the aforementioned DNA methylation participation to be valid.

Adding to the epigenetic regulation mechanism responsible for memory, small RNAs are determined to contribute as much as DNA methylations and histone-modifying enzymes. Previously studied *ago1* and *dcl1* mutants were associated with heat tolerance (Stief *et al.*, 2014). The aforementioned acquisition of heat memory through miR156-regulated SPL2 repression, with higher expression of miR156 strengthening the memory. The same situation was observed in *Medicago sativa*'s drought tolerance and then reported in other plants (Arshad *et al.*, 2017). Whether this tolerance creates a drought memory in a similar manner to the heat memory of *Arabidopsis* is still being debated (Jacques *et al.*, 2021).

He *et al.* (2024) studied cotton seedling responses to repetitive dehydration stress, examining physiological changes, gene expression patterns, and histone modification levels. Their findings revealed the development of short-term dehydration stress memory in cotton seedlings. Chromatin immunoprecipitation-qPCR results suggested that increased H3K4me3 levels during recovery may contribute to transcriptional behavior in dehydration memory genes. Additionally, their virus-induced gene silencing assay implicated the histone demethylase gene GhJMJ17-A/D in regulating the transcriptional behavior of memory genes in cotton.

3.4.2 Salt-responsive genes

Salinity stress is a significant factor influencing soil quality and plant productivity (Acosta-Motos *et al.*, 2017; Zuo *et al.*, 2023). Plants have developed various mechanisms to cope with and mitigate the effects of salt stress, including modulating ion homeostasis, ion compartmentalization and export, and the biosynthesis of osmoprotectants (Balasubramaniam *et al.*, 2023). Furthermore, plants exhibit transcriptional memory in response to salt stress (do Amaral *et al.*, 2020). Long-term somatic memory induced by salt stress has been observed in *Arabidopsis* seedlings, with pre-treated plants absorbing less salt and exhibiting higher survival rates upon second exposure to stress compared with controls. Not only has salt conditioning been shown to increase plant survival under salt-stress conditions significantly, but it has also increased the plant's resilience to drought (Rossatto *et al.*, 2023). Evidence suggests that the transcriptional memory of plants under salt stress is linked to epigenetic changes, including DNA methylation and histone protein modification, regulating gene expression levels (Singroha *et al.*, 2022). In Nonabokra rice varieties, low H3K27me3 and high H3K4me3 were seen in control plants, accompanied by loss of DNA methylation at the OsBZ8 gene. Elevated DNA methylation and enriched H3K27me3 are observed in control IR64 plants. This indicates a significant difference in chromatin modifications among rice varieties that regulate the differential expression of the OsBZ8 gene during salt stress (Paul *et al.*, 2017).

Yang *et al.* (2022) showed that transcriptionally activating the production of accelerated cell death 6 (ACD6), aconitate hydratase 3 (ACO3), and glutathione S-transferase family 14 (GSTF14) by demethylation of repetitions in their promoters caused by abiotic stimuli such as cold, salt, and water scarcity. In tobacco, overexpression of *Arabidopsis* ROS1 reduced DNA methylation in the promoter region of genes encoding antioxidant pathway enzymes and flavonoid production, increasing the plant's resistance to salt stress (Bharti *et al.*, 2015). Transcriptional regulation of salt stress-responsive genes involves histone methylation, with H3K4me3 playing a pivotal role. This modification is crucial for genes like RD29A, RD29B, COR15A, and COR15B, which are involved in abiotic stress responses (Shen *et al.*, 2022). Additionally, histone methylation/demethylation enzymes such as JMJ15 in *Arabidopsis* and SDG721 in rice modulate salt stress response through H3K4me3 and H3K27me3 modifications (Shen *et al.*, 2014; Liu *et al.*, 2021).

In addition to H3K4me3 and H3K27me3 modifications, changes in H3K4me2 and H3K9ac levels under salt stress indicate a dynamic process, suggesting the temporal nature of transcriptional memory in plants. Following salinity application, alterations in the histone modifications indicate the complicated regulation of chloride ion homeostasis during salt stress priming (Yung *et al.*, 2022). Salinity treatment raised MYB84 expression by reducing DNA methylation at the MYB84 promoter. The ability of soybean plants to withstand salt was improved when the MYB84 protein was found to bind to a *cis*-element in the promoter of AKT1, a homolog of *Arabidopsis* K + transporter 1 (AKT1) implicated in the salt stress response (Zhang *et al.*, 2020).

3.4.3 Temperature-responsive genes

Among the pivotal environmental factors influencing plant growth and development, temperature emerges. While fluctuations in temperature are not confined to seasonal changes or the day–night cycle, heat stress emerges as a critical concern. Heat stress occurs when temperatures exceed a plant's optimal range for growth, hindering its ability to achieve genetic potential (Saeed *et al.*, 2022). This detrimental condition substantially threatens plant growth and development, sometimes culminating in severe damage and, in extreme cases, plant mortality. Numerous studies have elucidated regulatory networks that activate stress-responsive genes, fostering thermotolerance and equipping plants with the capacity to adapt to heat stress. Heat stress, being one of the most studied abiotic stress factors, disrupts the membrane thermostability of photosynthetic processes while affecting the plant's respiration, transpiration, and osmotic regulation. The molecular genetic response associated with these changes usually entails the expression regulation of heat shock factors (HSFs) and heat shock proteins (HSPs) (Ohama *et al.*, 2017). Certain small non-coding RNAs (ncRNAs) also help in regulating the expression of such factors, including miRNAs, siRNAs, lncRNAs, and circRNAs, with many examples detected in plants (Wang *et al.*, 2012; Guan *et al.*, 2013; Stief *et al.*, 2014; Zhao *et al.*, 2016). All role takers of the heat stress response (HSR) should be tightly regulated to be associated with adverse conditions. This regulation entails a complex network of epigenetic regulation and stress memory.

Epigenetic changes during the response to high-temperature conditions generally comprise DNA methylation, histone modifications, or chromatin remodeling of heat stress-inducible genes, aside from the ncRNA regulation of gene expression. What makes up the heat stress memory, though, is an intricate network of transcription factors and transposable elements regulating the specific histone modifications of HSR elements (mostly for memory, H3K4me2/3 or H3K27me3 demethylation) (Roudier *et al.*, 2011), and also. the participation of a particular retrotransposon in heat stress memory named 'ONSEN', intertwined with the RdDM pathway. Usually, epigenetically silenced ONSEN can become transcriptionally active during heat stress conditions. This phenomenon was found to be correlated with the siRNAs partaking in RdDM silencing of ONSEN, which is impaired during said adverse conditions. ONSEN is also able to form extrachromosomal circular DNA (eccDNA) molecules in the cell when transcribed under stress conditions, just as other LTR-RT elements. These eccDNA molecules and their transcripts are degraded as the stress subsides. Mutants with inactive siRNA regulation accumulated ONSEN insertions in their genomes, with a pattern indicating that this transposition event happened during flowering. Stress memory remained during development in mutants, allowing ONSEN to relocate its copies during differentiation. Insertion events of ONSEN were shown to provide heat responsiveness to nearby genes, making its potential role highly significant for heat stress memory (Ito *et al.*, 2011).

Histone variant exchange is another factor contributing to epigenetic memory. Specifically, in high-temperature conditions, the H2A.Z histone variant has been shown to play an active role in *Arabidopsis* plants (Kumar and Wigge, 2010). H2A.Z nucleosomes act by regulating the thermosetting during flowering, exemplifying the importance of histone variant action for gene activation in a timely fashion. FT (FLOWERING LOCUS T) promoter is able to be bound by PIF4 (PHYTOCHROME INTERACTING FACTOR 4) in the presence of H2A.Z, resulting in flowering (Kumar *et al.*, 2012). PIF4 also regulates other gene expressions related to morphological changes under heat conditions such as YUC8, TAA1, or SAUR19-24 (Liu *et al.*, 2015). When conditions become adverse, epigenetic regulation changes accordingly.

During harsher temperatures causing stress, ASF1A/B removes H3K56ac from HSR genes, representing only one of the histone modifications occurring under stress. One of the prominent examples of histone modification-regulated transcription factors is HSFA2, which is responsible for heat stress memory. H3K4me2/3 modification was found to be associated with said memory in a study where priming with heat shock resulted in higher methylation levels (Lämke *et al.*, 2016). H3K4 methylation was also found to be dependent on HSFA2 in creating this transcriptional memory of heat stress, methylations even remaining after the detachment of HSFA2. Adding to the H3K4 mark, the long-term epigenetic memory of high temperatures was also found to be maintained by the positive feedback loop of HSFA2 and an H3K27me3 demethylase RELATIVE OF EARLY FLOWERING 6 (REF6), making H3K27me3 demethylation another important heat stress memory mark in plants (Liu *et al.*, 2019). FGT1 (FORGETTER 1) action remodels nucleosomes, contributing to heat stress memory via HSFA2 activity (Brzezinka *et al.*, 2016).

Another path for HSFA2 is the induction of carbohydrate metabolism genes HSP17.6A and FRUCTOSE BISPHOSPHATE ALDOLASE6 (FBA6) in shoot apical meristems, which occur by promoter binding. When encountered with heat twice, HSP17.6A and FBA6 show higher expression, indicating the existence of a transcriptional memory (Olas *et al.*, 2021). Additionally, FGT2 and FGT3 had similar effects on heat stress memory with similar modes of action (Friedrich *et al.*, 2021).

HSP101-HSA32 regulation represents another example of genes responsible for heat stress memory along with many other examples such as HSP21, HSFA1, miR156-regulated SQUAMOSA-PROMOTER BINDING-LIKE (SPL) TFs, HSP22, ACORBATE PEROXIDASE2 (APX2), and HSP18.2 (Nishio *et al.*, 2024). Also, though indirectly, genes encoding chromatin remodeling enzymes, such as JUMONJI (JMJ) histone demethylases, create chromatin changes affecting related gene expression, therefore having influential gene expressions themselves for the heat stress memory. JUMONJI is responsible for deleting H3K27me3 modifications and regulating genes such as HSP22 and HSP17.6C, resulting in their activation. Knockout mutants of JMJ cannot sustain a memory of the heat condition, pointing to the significance of H3K27me3 demethylation of the mentioned loci during heat stress memory (Yamaguchi *et al.*, 2021). Many mechanisms such as these are still being identified, making it possible to expand our knowledge of how high-temperature responses in plants are regulated.

Opposingly, low-temperature memory is best represented by FLC/FWA systems, with FLC using the H3K27me3 mark, though some other memory mechanisms were also determined. Using the RdDM pathway, the ALN (ALLANTOINASE) gene negatively regulates seed dormancy, promoting germination during cold temperatures. Demethylation of 5mC is another regulator activating the expression of target genes such as CKA2 and CKB4 regulating the circadian rhythm, thus flowering time of the plant according to temperature (Gallusci *et al.*, 2023).

Aside from such regulations and memory in response to changing environments, cold stress also has memory mechanisms associated with it. ICE1 (INDUCER OF CBF EXPRESSION 1) activated CBFs (C-repeat binding factor/dehydration-responsive-element-binding protein) trigger the expression of COR (cold-regulated) genes, as demonstrated in *Arabidopsis* plants. Cold tolerance generated by this mechanism is also achieved in rice plants, which is highly important for this cold-susceptible plant. Bioinformatic analysis of rice ICE1-CBF-COR elements in comparison with *Arabidopsis thaliana* represented close relations in structure and expression profiles (Deng *et al.*, 2017).

Cold memory in *Arabidopsis* has been demonstrated in a study where a freezing-tolerant ecotype N14 was used. Cold-primed plants were more tolerant to the stress when exposed later, indicating a memory of the cold, extending to 7 days after the first exposure. Interestingly, the tolerance resulted from a novel transcriptomic profile with varying metabolite productions and not due to higher expression of cold response genes, as usually observed in priming events, pointing to the possible role of metabolome profiles in creating cold tolerance memory (Zuther *et al.*, 2019).

From previously mentioned COR systems, COR15A and CHS were possibly associated with cold memory generation in an *Arabidopsis* study, where the cold memory was found to be regulated by plant age. COR expressions were reduced after reoccurring exposures to stress, which encourages them to be classified as related to stress memory due to previous examples of such stress-associated genes being up- or downregulated after repetitive exposure to stress. Stress caused H3K4me3 histone modifications, resulting in higher transcript levels at initial stress, whereas recurring stress lowered COR15A and CHS levels, leading them to act like memory genes. COR15A, being an H3K27me3 target, is suspected to be downregulated by such an inhibitory modification during repeating stress, though it is still only hypothesized (Leuendorf *et al.*, 2020).

Both adverse temperature conditions trigger epigenetic regulation mechanisms, resulting in a memory of the stress experienced by the plant, which may be inherited by the next generation. These sensing of environmental changes and the following epigenetic changes can also provide certain mechanisms for plants to thermoregulate their growth and developmental processes.

3.5 Conclusion and Future Prospects

The abiotic stress memory in plants offers an intriguing perspective on adaptability and resilience to environmental challenges. By deciphering the molecular mechanisms of stress memory and utilizing this understanding in plant improvement efforts, researchers can make significant contributions to global food security and sustainability. Continuous interdisciplinary research efforts are crucial to fully exploit the potential of abiotic stress memory in supporting plant adaptation and productivity amidst changing climates and resource constraints.

By harnessing natural diversity and utilizing genetic and epigenetic engineering techniques, researchers can enhance stress memory in plants, thereby improving their adaptation to harsh environments. Breeding programs can integrate stress memory traits to develop resilient plant varieties characterized by high-yield stability and sustainability. Additionally, combining precision farming methods with advances in omics technologies enables targeted manipulation of stress memory mechanisms for customized plant management strategies. Increasing studies aimed at understanding epigenetic memory suggest that changes in the methylome and other epigenetic system components may represent hereditary epigenetic adaptation over time, along with genetic adaptation. Genetic methods for site-specific epigenetic alterations, including CRISPR/Cas systems and RNAi-based targeting, can be used to modify the epigenetic components of plants as they are more precise and reversible to achieve the epi-breeding goal of increasing agriculturally important traits such as yield, quality, and disease resistance. It is used to increase stress tolerance (Gaude *et al.*, 2024).

Recent advancements in plant priming include the utilization of natural metabolites, synthetic chemical compounds, and nanotechnology and its tools. Chemical priming has emerged as an effective technology to enhance tolerance to stressors. For instance, nanoparticles have significantly protected plants against adverse environmental conditions. These roles are often attributed to their nano-scale dimensions, which confer various physicochemical properties such as high solubility, reactivity, and biochemical activity. However, these properties also make them ideal candidates for serving as smart nanocarriers, ensuring targeted delivery and optimal priming activity of chemical priming agents while mitigating their potential environmental impacts (Gohari *et al.*, 2024)

Sequencing technologies currently used provide the ability to sequence long DNA fragments perfectly. Processing and interpreting complex epigenetic modification data with the acquired sequence information will be possible with the power of machine learning, an application of artificial intelligence. Machine learning can enhance the annotation of transcriptomics, epigenomics, and proteomics and accurately classify differentially expressed genes, proteins, pseudogenes, and transposon transcripts. It can also accurately predict tissue-, condition-, and stress-type-specific DNA methylation and histone modification profiles and stress memory-associated gene expression patterns in plants (Libbrecht and Noble, 2015; Sartor *et al.*, 2019; N'diaye *et al.*, 2020; Wang *et al.*, 2021).

Although the primary focus is on plant epigenetic memory's role in adapting to abiotic stress, it is crucial to underscore that, in natural environments, the interplay between epigenetic and genetic inheritance is inexorable over extended durations. Consequently, no 'purely genetic' or 'purely epigenetic' alteration exists. Nevertheless, owing to the context-sensitive nature of epigenetic variations and their higher occurrence compared with genetic variations, they frequently instigate, impede, and expedite evolutionary shifts (Jablonka, 2017).

It is now known that epigenetic mechanisms play a critical role in neural learning and, therefore, in the intergenerational transmission of learned behaviors in mammalian species. Another crucial aspect reminding us of the necessity to understand epigenetic memory is the requirement to feed the human population, which is projected to reach 9.3 billion by 2050 despite increasingly adverse environmental conditions. Hence, integrating epigenetic data into breeding programs holds promise for developing stress-tolerant plants with enhanced epigenetic memory (Bronfman *et al.*, 2016). In conclusion, accessing key genes involved in plant epigenetic memory formation will be more guided with new technologies, enabling correct and rapid steps toward producing stress-resistant crops.

References

Abdulraheem, M.I., Xiong, Y., Moshood, A.Y., Cadenas-Pliego, G., Zhang, H. *et al.* (2024) Mechanisms of plant epigenetic regulation in response to plant stress: Recent discoveries and implications. *Plants* 13(2), 163.

Acosta-Motos, J., Ortuño, M., Bernal-Vicente, A., Diaz-Vivancos, P., Sanchez-Blanco, M. *et al.* (2017) Plant responses to salt stress: Adaptive mechanisms. *Agronomy* 7, 18.

Ahmad, K., Brahma, S. and Henikoff, S. (2024) Epigenetic pioneering by SWI/SNF family remodelers. *Molecular Cell* 84, 194–201.

Allis, C.D. and Jenuwein, T. (2016) The molecular hallmarks of epigenetic control. *Nature Reviews Genetics* 2016 17, 487–500.

Archacki, R., Yatusevich, R., Buszewicz, D., Krzyczmonik, K., Patryn, J. *et al.* (2017) *Arabidopsis* SWI/SNF chromatin remodeling complex binds both promoters and terminators to regulate gene expression. *Nucleic Acids Research* (6), 3116–3129.

Arshad, M., Feyissa, B.A., Amyot, L., Aung, B. and Hannoufa, A. (2017) MicroRNA156 improves drought stress tolerance in alfalfa (Medicago sativa) by silencing SPL13. *Plant Science* 258, 122–136.

Balasubramaniam, T., Shen, G., Esmaeili, N. and Zhang, H. (2023) Plants' response mechanisms to salinity stress. *Plants (Basel)* 12(12), 2253.

Bewick, A.J., Niederhuth, C.E., Ji, L., Rohr, N.A., Griffin, P.T. *et al.* (2017) The evolution of CHROMOMETHYLASES and gene body DNA methylation in plants. *Genome Biology* 18, 65.

Bhadouriya, S.L., Mehrotra, S., Basantani, M.K., Loake, G.J. and Mehrotra, R. (2021) Role of chromatin architecture in plant stress responses: An update. *Frontiers in Plant Science* 12 11, 603380.

Bharti, P., Mahajan, M., Vishwakarma, A.K., Bhardwaj, J. and Yadav, S.K. (2015) AtROS1 overexpression provides evidence for epigenetic regulation of genes encoding enzymes of flavonoid biosynthesis and antioxidant pathways during salt stress in transgenic tobacco. *Journal of Experimental Botany* 66, 5959–5969.

Bronfman, Z., Ginsburg, S. and Jablonka, E. (2016) The epigenetics of neural learning. In: Murphy, R. and Honey, R. (eds) *The Wiley Blackwell Handbook on the Cognitive Neuroscience of Learning*. Wiley, New York, pp. 136–176.

Brzezinka, K., Altmann, S., Czesnick, H., Nicolas, P., Gorka, M. *et al.* (2016) *Arabidopsis* FORGETTER1 mediates stress-induced chromatin memory through nucleosome remodeling. *ELife* 5, e17061.

Castellanos, U.R., Friedrich, T., Petrovic, N., Altmann, S., Brzezinka, K. *et al.* (2020) FORGETTER2 protein phosphatase and phospholipase D modulate heat stress memory in *Arabidopsis*. *The Plant Journal* 104, 7–17.

Chekanova, J.A., Gregory, B.D., Reverdatto, S.V., Chen, H., Kumar, R. *et al.* (2007) Genome-wide high-resolution mapping of exosome substrates reveals hidden features in the *Arabidopsis* transcriptome. *Cell* 131, 1340–1353.

Chen, Q., Liu, K., Yu, R., Zhou, B., Huang, P. *et al.* (2021) From "Dark Matter" to "Star": Insight into the regulation mechanisms of plant functional long non-coding RNAs. *Frontiers in Plant Science* 12, 650926.

Chen, Y., Li, C., Yi, J., Yang, Y., Lei, C. *et al.* (2019) Transcriptome response to drought, rehydration and re-dehydration in potato. *International Journal of Molecular Sciences* 21, 159.

Cheong, J.J. (2022) Modulation of abscisic acid signaling for stomatal operation under salt stress conditions. *Advances in Botanical Research* 103, 89–121.

Choi, Y., Gehring, M., Johnson, L., Hannon, M., Harada, J.J. *et al.* (2002) DEMETER, a DNA glycosylase domain protein, is required for endosperm gene imprinting and seed viability in *Arabidopsis*. *Cell* 110, 33–42.

Deng, C., Ye, H., Fan, M., Pu, T. and Yan, J. (2017) The rice transcription factors osice confer enhanced cold tolerance in transgenic *Arabidopsis*. *Plant Signal Behavior* 12(5), e1316442.

Ding, Y., Fromm, M. and Avramova, Z. (2012) Multiple exposures to drought 'train' transcriptional responses in *Arabidopsis*. *Nature Communications* 3, 1–9.

Do Amaral, M.N., Arge, L.W.P., Auler, P.A., Rossatto, T., Milech, C. *et al.* (2020) Long-term transcriptional memory in rice plants submitted to salt shock. *Planta* (6), 111.

D'Urso, A. and Brickner, J.H. (2014) Mechanisms of epigenetic memory. *Trends in Genetics* 30(6), 230–236.

Erdmann, R.M. and Picard, C.L. (2020) RNA-directed DNA methylation. *PLoS Genetics* 16(10), e1009034.

Feng, X.J., Li, J.R., Qi, S.L., Lin, Q.F., Jin, J.B. *et al.* (2016) Light affects salt stress-induced transcriptional memory of P5CS1 in *Arabidopsis*. *Proceedings of the National Academy of Sciences of the United States of America* 113(51), E8335–E8343.

Fleta-Soriano, E., Pintó-Marijuan, M. and Munné-Bosch, S. (2015) Evidence of drought stress memory in the facultative CAM, *Aptenia cordifolia*: Possible role of phytohormones. *PloS ONE* 10(8), e0135391.

Friedrich, T., Oberkofler, V., Trindade, I., Altmann, S., Brzezinka, K. *et al.* (2021) Heteromeric HSFA2/HSFA3 complexes drive transcriptional memory after heat stress in *Arabidopsis*. *Nature Communications* 12, 3426.

Gagliardi, D. and Manavella, P.A. (2020) Short-range regulatory chromatin loops in plants. *The New Phytologist* 228, 466–471.

Gallusci, P., Agius, D.R., Moschou, P.N., Dobránszki, J., Kaiserli, E. *et al.* (2023) Deep inside the epigenetic memories of stressed plants. *Trends in Plant Science* 28, 142–153.

Gaude, A.A., Siqueira, R.H., Botelho, S.B. and Jalmi, S.K. (2024) Epigenetic arsenal for stress mitigation in plants. *Biochimica et Biophysica Acta (BBA) – General Subjects* 130620.

Gohari, G., Jiang, M., Manganaris, G.A., Zhou, J. and Fotopoulos, V. (2024) Next generation chemical priming: With a little help from our nanocarrier friends. *Trends in Plant Science* 29(2), 150–166.

Guan, Q., Lu, X., Zeng, H., Zhang, Y. and Zhu, J. (2013) Heat stress induction of miR398 triggers a regulatory loop that is critical for thermotolerance in *Arabidopsis*. *Plant Journal* 74, 840–851.

Hauser, M.T., Aufsatz, W., Jonak, C. and Luschnig, C. (2011) Transgenerational epigenetic inheritance in plants. *Biochimica et Biophysica Acta* 1809(8), 459–468.

He, S., Zhang, P., Wang, Y., Zheng, R., Li, Y. *et al.* (2024) Transcriptional patterns and histone modification signatures reveal dehydration memory behaviour in seedlings of *Gossypium hirsutum*. *Environmental and Experimental Botany* 220, 105663.

Hoth, S., Morgante, M., Sanchez, J.P., Hanafey, M.K., Tingey, S.V. *et al.* (2002) Genome-wide gene expression profiling in *Arabidopsis thaliana* reveals new targets of abscisic acid and largely impaired gene regulation in the abi1-1 mutant. *Journal of Cell Science* 115, 4891–4900.

Hurtado, L., Farrona, S. and Reyes, J.C. (2006) The putative SWI/SNF complex subunit BRAHMA activates flower homeotic genes in *Arabidopsis thaliana*. *Plant Molecular Biology* 62(1–2), 291–304.

Imran, M., Shafiq, S., Ilahi, S., Ghahramani, A., Bao, G. *et al.* (2022) Post-transcriptional regulation of 2-acetyl-1-pyrroline (2-AP) biosynthesis pathway, silicon, and heavy metal transporters in response to Zn in fragrant rice. *Frontiers in Plant Science* 17 13, 948884.

Ito, H., Gaubert, H., Bucher, E., Mirouze, M., Vaillant, I. *et al.* (2011) An siRNA pathway prevents transgenerational retrotransposition in plants subjected to stress. *Nature* 472(7341), 115–119.

Jablonka, E. (2017) Evolutionary consequences of epigenetic inheritance. *Interface Focus* 7, 20160135.

Jacques, C., Salon, C., Barnard, R.L., Vernoud, V. and Prudent, M. (2021) Drought stress memory at the plant cycle level: A review. *Plants* 10, 1873.

Janni, M., Maestri, E., Gullì, M., Marmiroli, M. and Marmiroli, N. (2024) Plant responses to climate change, how global warming may impact on food security: A critical review. *Frontiers in Plant Science* 5 14, 1297569.

Jiménez-Ramírez, I.A., Pijeira-Fernández, G., Moreno-Cálix, D.M. and De-la-Peña, C. (2022) Same modification, different location: The mythical role of n6-adenine methylation in plant genomes. *Planta* 256(1), 9.

Kappel, C., Friedrich, T., Oberkofler, V., Jiang, L., Crawford, T. *et al.* (2023) Genomic and epigenomic determinants of heat stress-induced transcriptional memory in *Arabidopsis*. *Genome Biology* 24, 1–23.

Khan, I.U., Ali, A., Khan, H.A., Baek, D., Park, J. *et al.* (2020) PWR/HDA9/ABI4 complex epigenetically regulates ABA dependent drought stress tolerance in *Arabidopsis*. *Frontiers in Plant Science* 11, 623.

Kim, K.H. and Lee, B.M. (2023) Effects of climate change and drought tolerance on maize growth. *Plants (Basel)* 12 12(20), 3548.

Kim, Y.K., Chae, S., Oh, N.I., Nguyen, N.H. and Cheong, J.J. (2020) Recurrent drought conditions enhance the induction of drought stress memory genes in *Glycine max* L. *Frontiers in Genetics* 11, 576086.

Kim, J.S., Lim, J.Y., Shin, H., Kim, B.G., Yoo, S.D. *et al.* (2019) ROS1-dependent DNA demethylation is required for ABA-inducible NIC3 expression. *Plant Physiology* 179(4), 1810–1821.

Kinoshita, T. and Seki, M. (2014) Epigenetic memory for stress response and adaptation in plants. *Plant Cell Physiology* 55(11), 1859–1863.

Kumar, S.V. and Wigge, P.A. (2010) H2A.Z-containing nucleosomes mediate the thermosensory response in *Arabidopsis*. *Cell* 140, 136–147.

Kumar, S.V., Lucyshyn, D., Jaeger, K.E., Alós, E., Alvey, E. *et al.* (2012) PHYTOCHROME INTERACTING FACTOR4 controls the thermosensory activation of flowering. *Nature* 484, 242.

Kumari, A., Lakshmi, G.A., Krishna, G.K., Patni, B., Prakash, S. *et al.* (2022) Climate change and its impact on crops: A comprehensive investigation for sustainable agriculture. *Agronomy* 12, 3008.

Kumar, S. and Mohapatra, T. (2021) Dynamics of DNA methylation and its functions in plant growth and development. *Frontiers in Plant Science* 12, 596236.

Kumar, S. and Mohapatra, T. (2024) Epigenetic modifications in genome help remembering the stress tolerance strategy adopted by the plant. *Frontiers in Bioscience* 29(3), 126.

Kwon, C.S., Lee, D., Choi, G. and Chung, W.I. (2009) Histone occupancy-dependent and -independent removal of H3K27 trimethylation at cold-responsive genes in *Arabidopsis*. *The Plant Journal* 60(1), 112–121.

Lämke, J. and Bäurle, I. (2017) Epigenetic and chromatin-based mechanisms in environmental stress adaptation and stress memory in plants. *Genome Biology* 18(1), 124.

Lämke, J., Brzezinka, K., Altmann, S. and Bäurle, I. (2016) A hit-and-run heat shock factor governs sustained histone methylation and transcriptional stress memory. *The EMBO Journal* 35(2), 162–175.

Leuendorf, J.E., Frank, M. and Schmülling, T. (2020) Acclimation, priming and memory in the response of *Arabidopsis thaliana* seedlings to cold stress. *Scientific Reports* 10.

Li, C., Gu, L., Gao, L., Chen, C., Wei, C.-Q. *et al.* (2016) Concerted genomic targeting of H3K27 demethylase REF6 and chromatin-remodeling atpase BRM in *Arabidopsis*. *Nature Genetics* 48(6), 687–693.

Li, P., Yang, H., Wang, L., Liu, H., Huo, H. *et al.* (2019) Physiological and transcriptome analyses reveal short-term responses and formation of memory under drought stress in rice. *Frontiers in Genetics* 10, 433860.

Libbrecht, M.W. and Noble, W.S. (2015) Machine learning applications in genetics and genomics. *Nature Reviews Genetics* 16, 321–332.

Liu, J. and He, Z. (2020) Small DNA methylation, big player in plant abiotic stress responses and memory. *Frontiers in Plant Science* 11, 595603.

Liu, N., Fromm, M. and Avramova, Z. (2014) H3K27me3 and H3K4me3 chromatin environment at super-induced dehydration stress memory genes of *Arabidopsis thaliana*. *Molecular Plant* 7, 502–513.

Liu, H., Able, A.J. and Able, J.A. (2022) Priming crops for the future: Rewiring stress memory. *Trends in Plant Science* 27(7), 699–716.

Liu, H.C., Lämke, J., Lin, S.Y., Hung, M.J., Liu, K.M. *et al.* (2018) Distinct heat shock factors and chromatin modifications mediate the organ-autonomous transcriptional memory of heat stress. *The Plant Journal* 95(3), 401–413.

Liu, H.H., Tian, X., Li, Y.J., Wu, C.A. and Zheng, C.C. (2019) Microarray-based analysis of stress-regulated microRNs in *Arabidopsis thaliana*. *RNA* 14, 836–843.

Liu, J., Feng, L., Li, J. and He, Z. (2015) Genetic and epigenetic control of plant heat responses. *Frontiers in Plant Science* 6.

Liu, W., Thapa, P. and Park, S.W. (2023) RD29A and RD29B rearrange genetic and epigenetic markers in priming systemic defense responses against drought and salinity. *Plant Science* 337, 111895.

Liu, Y., Chen, X., Xue, S., Quan, T., Cui, D. *et al.* (2021) SET DOMAIN GROUP 721 protein functions in saline-alkaline stress tolerance in the model rice variety Kitaake. *Plant Biotechnology Journal* 19(12), 2576–2588.

Ma, X., Zhao, F. and Zhou, B. (2022) The characters of non-coding RNAs and their biological roles in plant development and abiotic stress response. *International Journal of Molecular Sciences* 23, 4124.

Mehdi, S., Derkacheva, M., Ramström, M., Kralemann, L., Bergquist, J. *et al.* (2016) The WD40 domain protein MSI1 functions in a histone deacetylase complex to fine-tune abscisic acid signaling. *The Plant Cell* 28, 42–54.

N'Diaye, A., Byrns, B., Cory, A.T., Nilsen, K.T., Walkowiak, S. *et al.* (2020) Machine learning analyses of methylation profiles uncovers tissue-specific gene expression patterns in wheat. *The Plant Genome* 13, e20027.

Nguyen, N.H., Vu, N.T. and Cheong, J.J. (2022) Transcriptional stress memory and transgenerational inheritance of drought tolerance in plants. *International Journal of Molecular Sciences* 23, 12918.

Nishio, H., Kawakatsu, T. and Yamaguchi, N. (2024) Beyond heat waves: Unlocking epigenetic heat stress memory in *Arabidopsis*. *Plant Physiology* 1934(4), 1934–1951.

Nosaka, M., Itoh, J., Nagato, Y., Ono, A., Ishiwata, A. *et al.* (2012) Role of transposon-derived small RNAs in the interplay between genomes and parasitic DNA in rice. *PLoS Genetics* 8(9), e1002953.

Ohama, N., Sato, H., Shinozaki, K. and Yamaguchi-Shinozaki, K. (2017) Transcriptional regulatory network of plant heat stress response. *Trends in Plant Science* 22, 53–65.

Olas, J.J., Apelt, F., Annunziata, M.G., John, S., Richard, S.I. *et al.* (2021) Primary carbohydrate metabolism genes participate in heat-stress memory at the shoot apical meristem of *Arabidopsis thaliana*. *Molecular Plant* 14, 1508–1524.

Pandey, G., Sharma, N., Sahu, P.P. and Prasad, M. (2016) Chromatin-based epigenetic regulation of plant abiotic stress response. *Current Genomics* 17, 490–498.

Paul, A., Dasgupta, P., Roy, D. and Chaudhuri, S. (2017) Comparative analysis of histone modifications and DNA methylation at OsBZ8 locus under salinity stress in IR64 and Nonabokra rice varieties. *Plant Molecular Biology* 95(1–2), 63–88.

Peirats-Llobet, M., Han, S.K., Gonzalez-Guzman, M., Jeong, C.W., Rodriguez, L. *et al.* (2016) A direct link between abscisic acid sensing and the chromatin-remodeling ATPase BRAHMA via core ABA signaling pathway components. *Molecular Plant* (1), 136–147.

Pikaard, C.S. and Mittelsten Scheid, O. (2014) Epigenetic regulation in plants. *Cold Spring Harbor Perspectives in Biology* 6(12), a019315.

Rodriguez-Granados, N.Y., Ramirez-Prado, J.S., Veluchamy, A., Latrasse, D., Raynaud, C. *et al.* (2016) Put your 3D glasses on: Plant chromatin is on show. *Journal of Experimental Botany* 67(11), 3205–3221.

Rossatto, T., Souza, G.M., Amaral, M.N., Auler, P.A., Pérez-Alonso, M.M. *et al.* (2023) Cross-stress memory: Salt priming at vegetative growth stages improves tolerance to drought stress during grain-filling in rice plants. *Environmental and Experimental Botany* 206, 105187.

Roudier, F., Ahmed, I., Bérard, C., Sarazin, A., Mary-Huard, T. *et al.* (2011) Integrative epigenomic mapping defines four main chromatin states in *Arabidopsis*. *The EMBO Journal* 30(10), 1928–1938.

Saeed, F., Chaudhry, U.K., Bakhsh, A., Raza, A., Saeed, Y. *et al.* (2022) Moving beyond DNA sequence to improve plant stress responses. *Frontiers in Genetics* 13, 874648.

Sah, S.K., Reddy, K.R. and Li, J. (2016) Abscisic acid and abiotic stress tolerance in crop plants. *Frontiers in Plant Science* 7, 571.

Sartor, R.C., Noshay, J., Springer, N.M. and Briggs, S.P. (2019) Identification of the expressome by machine learning on omics data. *Proceedings of the National Academy of Sciences of the United States of America* 116(36), 18119–18125.

Sebastian-delaCruz, M., Gonzalez-Moro, I., Olazagoitia-Garmendia, A., Castellanos-Rubio, A. and Santin, I. (2021) Non-coding RNA. The role of lncrnas in gene expression regulation through MRNA stabilization. *Non-Coding RNA* 7(1), 3.

Seki, M., Ishida, J., Narusaka, M., Fujita, M., Nanjo, T. *et al.* (2002) Monitoring the expression pattern of around 7,000 *Arabidopsis* genes under ABA treatments using a full-length cDNA microarray. *Functional & Integrative Genomics* 2(6), 282–291.

Shen, Y., Conde, E.S.N., Audonnet, L., Servet, C., Wei, W. *et al.* (2014) Over-expression of histone H3K4 demethylase gene JMJ15 enhances salt tolerance in *Arabidopsis*. *Frontiers in Plant Science* 5, 290.

Shen, Y., Chi, Y., Lu, S., Lu, H. and Shi, L. (2022) Involvement of JMJ15 in the dynamic change of genome-wide h3k4me3 in response to salt stress. *Frontiers in Plant Science* 13, 1009723.

Shi, L., Cui, X. and Shen, Y. (2024) The roles of histone methylation in the regulation of abiotic stress responses in plants. *Plant Stress* 11, 100303.

Shimizu, T., Kanno, Y., Watanabe, S. and Seo, M. (2022) *Arabidopsis* NPF5.1 regulates ABA homeostasis and seed germination by mediating ABA uptake into the seed coat. *Plant Signaling & Behavior* 17(1), 2095488.

Shinozaki, K. and Yamaguchi-Shinozaki, K. (2007) Gene networks involved in drought stress response and tolerance. *Journal of Experimental Botany* 58(2), 221–227.

Singroha, G., Kumar, S., Gupta, O.P., Singh, G.P. and Sharma, P. (2022) Uncovering the epigenetic marks involved in mediating salt stress tolerance in plants. *Frontiers in Genetics* 13, 811732.

Song, Z.T., Zhang, L.L., Han, J.J., Zhou, M. and Liu, J.X. (2021) Histone H3K4 methyltransferases SDG25 and ATX1 maintain heat-stress gene expression during recovery in *Arabidopsis*. *The Plant Journal* 105(5), 1326–1338.

Stief, A., Altmann, S., Hoffmann, K., Pant, B.D., Scheible, W.R. *et al.* (2014) *Arabidopsis* miR156 regulates tolerance to recurring environmental stress through SPL transcription factors. *The Plant Cell* 26(4), 1792–1807.

Todaka, D., Takahashi, F., Yamaguchi-Shinozaki, K. and Shinozaki, K. (2019) ABA-responsive gene expression in response to drought stress: Cellular regulation and long-distance signaling. *Advances in Botanical Research* 92, 83–113.

Turgut Kara, N., Arikan, B. and Celik, H. (2020) Epigenetic memory and priming in plants. *Genetica* 148(2), 47–54.

Uno, Y., Furihata, T., Abe, H., Yoshida, R., Shinozaki, K. *et al.* (2000) *Arabidopsis* basic leucine zipper transcription factors involved in an abscisic acid-dependent signal transduction pathway under drought and high-salinity conditions. *Proceedings of the National Academy of Sciences* 97(21), 11632–11637.

Waddington, C.H. (1957) *Strategy of Genes*. George Allen and Unwin, London.

Waddington, C.H. (1975) *The Evolution of an Evolutionist*. Edinburgh University Press, Edinburgh.

Waititu, J.K., Zhang, C., Liu, J. and Wang, H. (2020) Plant non-coding RNAs: Origin, biogenesis, mode of action and their roles in abiotic stress. *International Journal of Molecular Sciences* 21, 8401.

Wang, H., Lu, S., Guan, X., Jiang, Y., Wang, B. *et al.* (2022) Dehydration-responsive element binding protein 1C, 1E, and 1G promote stress tolerance to chilling, heat, drought, and salt in rice. *Frontiers in Plant Science* 13, 851731.

Wang, W., Huang, F., Qin, Q., Zhao, X., Li, Z. *et al.* (2015) Comparative analysis of DNA methylation changes in two rice genotypes under salt stress and subsequent recovery. *Biochemical and Biophysical Research Communications* 465(4), 790–796.

Wang, Y., Sun, F., Cao, H., Peng, H., Ni, Z. *et al.* (2012) TamiR159 directed wheat tagamyb cleavage and its involvement in anther development and heat response. *PLoS ONE* 7(11), 30600392.

Wang, Y., Zhang, P., Guo, W., Liu, H., Li, X. *et al.* (2021) A deep learning approach to automate whole-genome prediction of diverse epigenomic modifications in plants. *New Phytologist* 232(2), 880–897.

Xiao, J., Lee, U.S. and Wagner, D. (2016) Tug of war: Adding and removing histone lysine methylation in *Arabidopsis*. *Current Opinion in Plant Biology* 34, 41–53.

Yadav, A., Mathan, J., Dubey, A.K. and Singh, A. (2024) The emerging role of non-coding rnas (ncRNAs) in plant growth, development, and stress response signaling. *Non-Coding RNA* 10(1), 13.

Yamaguchi, N., Matsubara, S., Yoshimizu, K., Seki, M., Hamada, K. *et al.* (2021) H3K27me3 demethylases alter HSP22 and HSP17.6C expression in response to recurring heat in *Arabidopsis*. *Nature Communications* 12, 3480.

Yang, L., Lang, C., Wu, Y., Meng, D., Yang, T. *et al.* (2022) ROS1-mediated decrease in DNA methylation and increase in expression of defense genes and stress response genes in *Arabidopsis* thaliana due to abiotic stresses. *BMC Plant Biology* (1), 104.

Yang, R., Hong, Y., Ren, Z., Tang, K., Zhang, H. *et al.* (2019) A role for PICKLE in the regulation of cold and salt stress tolerance in *Arabidopsis*. *Frontiers in Plant Science* 10, 458456.

Yoon, J.H., Abdelmohsen, K. and Gorospe, M. (2013) Posttranscriptional gene regulation by long noncoding RNA. *Journal of Molecular Biology* 425(19), 3723–3730.

Yung, W.S., Wang, Q., Huang, M., Wong, F.L., Liu, A. *et al.* (2022) Priming-induced alterations in histone modifications modulate transcriptional responses in soybean under salt stress. *The Plant Journal* 109(6), 1575–1590.

Zhang, W., Wang, N., Yang, J., Guo, H., Liu, Z. *et al.* (2020) The salt-induced transcription factor GmMYb84 confers salinity tolerance in soybean. *Plant Science* 291, 110326.

Zhao, J., He, Q., Chen, G., Wang, L. and Jin, B. (2016) Regulation of non-coding RNAs in heat stress responses of plants. *Frontiers Plant Science* 7, 1213.

Zheng, X., Chen, L., Li, M., Lou, Q., Xia, H. *et al.* (2013) Transgenerational variations in DNA methylation induced by drought stress in two rice varieties with distinguished difference to drought resistance. *PLoS ONE* 8(11), e80253.

Zilberman, D., Gehring, M., Tran, R.K., Ballinger, T. and Henikoff, S. (2007) Genome-wide analysis of *Arabidopsis thaliana* DNA methylation uncovers an interdependence between methylation and transcription. *Nature Genetics* 39, 61–69.

Zuo, D.D., Ahammed, G.J. and Guo, D.L. (2023) Plant transcriptional memory and associated mechanism of abiotic stress tolerance. *Plant Physiology and Biochemistry* 201, 107917.

Zuther, E., Schaarschmidt, S., Fischer, A., Erban, A., Pagter, M. *et al.* (2019) Molecular signatures associated with increased freezing tolerance due to low temperature memory in *Arabidopsis*. *Plant Cell Environment* 42, 854–873.

4 Epigenetics for Developing Abiotic Stress-Tolerant Plants: Updated Methods and Current Achievements

Dwaipayan Sinha[1]*, Aqsa Majgaonkar[2], Arun Kumar Maurya[3], Supti Das[4], Rashmi Mukherjee[5], Upala Saha[6], Bikram Saha[7], Shreyasi Dutta[5], Abhra Abir Chowdhury[7], Chitrita Chatterjee[8] and Paramita Nandy-Datta[6]*

[1]Department of Botany, Government General Degree College, Mohanpur, Paschim Medinipur, West Bengal, India; [2]Department of Botany, St Xavier's College (Autonomous), Mumbai, Maharashtra, India; [3]Department of Botany, Multanimal Modi College, Modinagar, Ghaziabad, Uttar Pradesh, India; [4]Department of Biochemistry, University of Calcutta, Kolkata, West Bengal, India; [5]Research Center for Natural and Applied Sciences, Raja Narendralal Khan Women's College (Autonomous), Gope Palace, Midnapore, India; [6]Sister Nibedita Govt. General Degree College for Girls, Hastings House, Alipore, Kolkata, India; [7]Post Graduate Dept. of Botany, Darjeeling Govt. College, Darjeeling, West Bengal, India; [8]Shibpur Dinobundhoo Institution College, Howrah, India

Abstract

Epigenetics can be defined as the modulation of certain chemical programming that influences the DNA transcription mechanism, leading to altered gene expression without interfering with the reading frame. DNA at its core undergoes hyper- or hypo-methylation at its fifth carbon atom and encounters histone modification and non-coding RNA-regulated silencing to acquire epigenetic adaptations. Despite being inherited, the epigenomes differ from the polymorphic DNA; having been induced by stress memory, epigenetic modifications do not involve the sequence of structural genes, unlike DNA polymorphisms that are rather random in origin and demand base sequence alteration. Thus epigenetic inheritance explains the Lamarckian theory of acquired phenotype in the offspring due to certain experiences that came across from the parental generation. In plant kingdoms, naturally occurring epigenetic modifications bring about tolerance against adverse environmental conditions like extreme heat, cold, salinity, and heavy metal pollution in the substrate. Conventional breeding strategies are amalgamated with induced epigenetic alterations in view of acquiring desirable traits, especially abiotic stress tolerance in crops. In recent decades CRISPR/Cas technology has emerged as an optimized tool for plant epi-breeding that aims to improve growth and yield as well as to develop stress-tolerant varieties of crops and medicinal species. The present article covers in detail the molecular mechanisms of epigenetic manipulations in plant genomes and present-day epi-breeding approaches taking on CRISPR/Cas technology.

Keywords: Epigenetics, CRISPR, Stress, Methylation, Acetylation, Toxicity

---

*Corresponding authors: paramita_nandy@yahoo.com and dwaipayansinha@hotmail.com

 Epigenetics for Climate-Smart and Sustainable Agriculture (ed. Jen-Tsung Chen)
DOI: 10.1079/9781800626102.0004

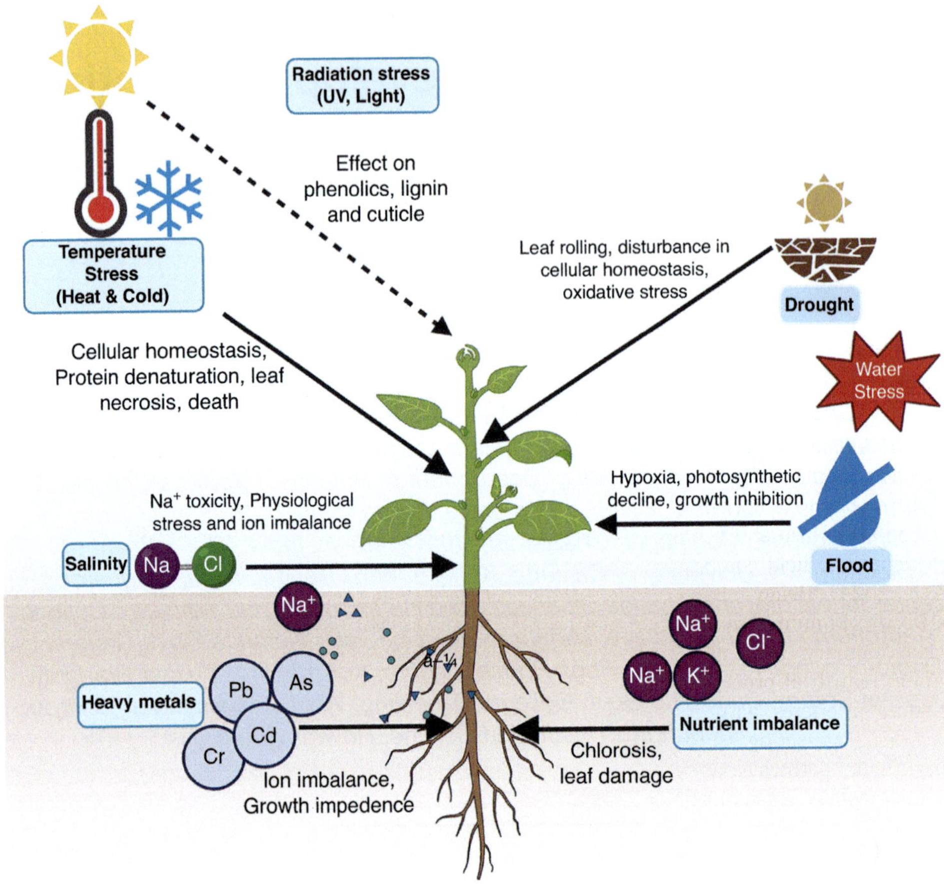

Fig. 4.1. Pictorial overview of the different types of abiotic stressors and their various effects on plants. Author's own image.

4.1 Introduction

4.1.1 Overview of abiotic stress factors in plants

Extreme temperatures, drought, heavy metal toxicity, salinity, UV irradiation, and nutrient imbalance are abiotic stressors that impede vegetative growth, maturation, and crop yield, posing threats to overall food security and sustainability (Mittal *et al.*, 2023; Sarma *et al.*, 2023), as depicted in Fig. 4.1. Grasping the fundamental processes behind plant responses to these stresses is essential for developing resilient crops capable of thriving in harsh conditions (Shah *et al.*, 2021; Verslues *et al.*, 2023).

Drought significantly impacts global plant production, which is compounded by changes in evaporation, transpiration, and irregular rainfall patterns caused by climate change (Seleiman *et al.*, 2021). This stress causes physiological and molecular reactions in plants, including decreased reproduction, altered hormone communication, reduced photosynthesis, stomatal closure, and growth inhibition, thus severely affecting agricultural productivity (Mohapatra *et al.*, 2023). Drought-induced xylem embolism affects water transport, potentially causing plant death via hydraulic collapse (Cotrina Cabello *et al.*, 2023). Plant growth is hindered by salt stress due to the build-up of large concentrations of salts such as Na^+, Cl^-, K^+, and SO_4^{2-} in the root zone, resulting in ionic and osmotic imbalances that limit water intake (Balasubramaniam *et al.*, 2023). This stress causes oxidative damage and slows growth, forcing plants to engage defensive mechanisms such as ion homeostasis, osmolyte and phytohormone production, and systems for antioxidant defense (Singh *et al.*, 2022). Extreme temperatures, including cold and heat stress, are major environmental stresses on crop plants (Hasanuzzaman *et al.*, 2013). Beyond adaptive limitations, thermal stress significantly impacts plants' ability to survive, develop, metabolize and produce, resulting in oxidative stress (Devireddy *et al.*, 2021). The level of damage is determined by the period of stress, plant genotype, and growth phase. Plants respond to temperature-induced damage by generating heat shock proteins and changing membrane lipid composition (Pandey and Singh, 2024). Heavy-metal contamination of soil (lead, cadmium, arsenic, chromium, copper, nickel, mercury, zinc, aluminum, etc.) from mining, industrial processes, and agricultural practices interferes with vital plant cellular functions, impacting metabolism, growth, biosynthetic pathways and altering secondary metabolites concentrations (Asiminicesei *et al.*, 2024; Taghavi *et al.*, 2024). In response, plants employ defense mechanisms such as metal chelation, compartmentalization, and efflux transport. These defense mechanisms are backed by intricate signaling networks that comprise distinct proteins, molecular messengers, antioxidants (enzymatic and non-enzymatic), root exudates, and reinforced cell walls (Ningombam *et al.*, 2024). Nutrient imbalances can impede plant growth, cause damage, and limit yield (Farhan *et al.*, 2024). Inadequate nutrient availability disturbs metabolic pathways, which affects plant physiology. To cope with nutritional stress, plants adapt root development, remobilize minerals inside cells, secrete organic compounds, and modify nutrient uptake and assimilation mechanisms (Singh *et al.*, 2024).

Plant responses to various abiotic stressors entail extensive adaptation mechanisms controlled by complex signaling networks and gene regulatory pathways, which are critical for acclimatization and survival. Epigenetic studies present a promising way to understand the molecular basis of these reactions, with the perspective of improving crop resistance, thus guaranteeing global food security.

4.1.2 Epigenetics and abiotic stress response

Epigenetics, the exploration of inheritable modifications in gene countenance without alterations in the DNA sequence, has become a powerful tool in unraveling plant stress response mechanisms (Ball and Sadhukhan, 2023). Recent advances in epigenetics research studies and reversible control over inheritance and gene expression suggest that plants adopt a short-term approach for survival under abiotic stress factors. It is well established that species with reversible gene expression have an evolutionary advantage since they can inhibit unwarranted rearrangements and population diversity (Boyko and Kovalchuk, 2008). Several unknowns exist regarding the nature of the biochemical signals that cause alterations in the plant epigenome during environmental stress situations. However, there seems to be a strong link between epigenetic changes and phytohormone signaling (Ohama *et al.*, 2017). Dynamic epigenetic alterations are exhibited by plants in response to hostile environmental conditions, which regulate gene expression and facilitate adaptation and survival (Gallusci *et al.*, 2023). Epigenetic processes like histone changes, DNA methylation, and small RNA-facilitated gene silencing are instrumental in coordinating plant reactions to stress (Fig. 4.2) (Akhter *et al.*, 2021). Elevated capacity of floras to tolerate stress is connected to variations in DNA methylation at stress-responsive gene loci and affects stomatal development, pollen tube formation, fruit ripening, embryogenesis, and nodule formation (Garg *et al.*, 2015). Acetylation and methylation of histones are vital in controlling the transcriptional activity of genes linked to stress and the accessibility of chromatin, which in turn modulates how plants respond to their abiotic stress factors (Bhadouriya *et al.*, 2021). Post-transcriptional gene regulation is a function of short RNAs, viz. microRNAs (miRNAs) and small interfering RNAs (siRNAs) that regulate stress-responsive gene expression networks in plants (Guerra

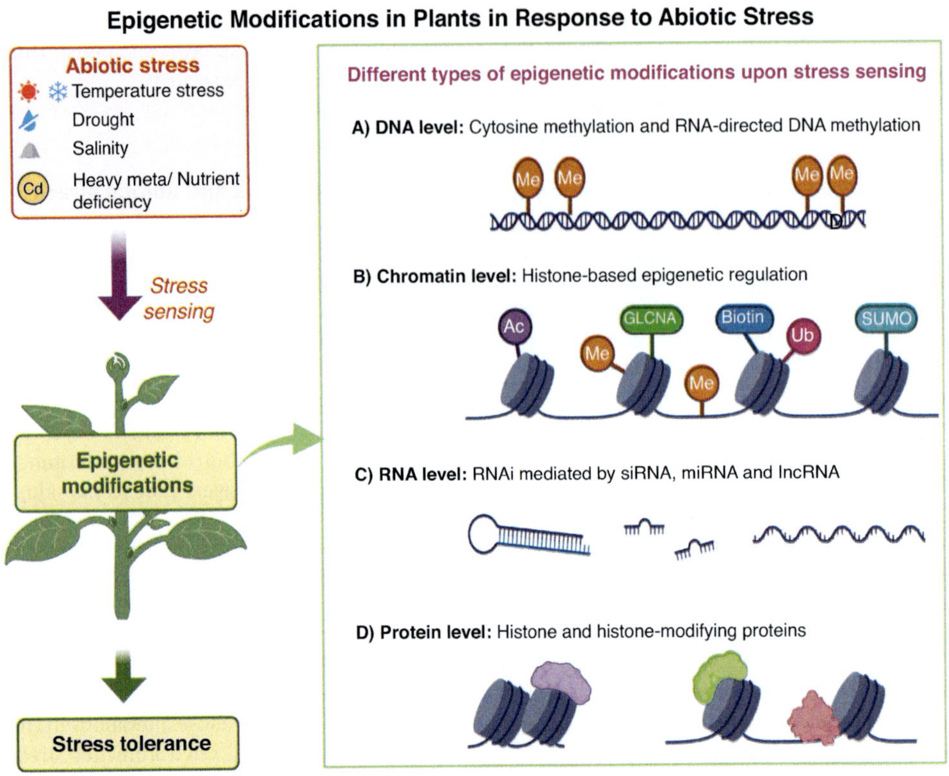

Fig. 4.2. Illustration depicting the epigenetic modifications that occur in response to abiotic stress in plants. Author's own image.

et al., 2015). Epigenetic alterations also influence bacterial symbiosis and plant immunological responses to pathogens (Al Akeel, 2013). The significance of epigenetic control for additional agriculturally significant traits like vernalization, hybrid vigor, and self-incompatibility has been well demonstrated (Kumar, 2018). These findings highlight the crucial role of epigenetics in influencing plant–environment interactions and maintaining agricultural sustainability. This chapter aims to elaborate on the various epigenetic modifications in plants in response to abiotic stress.

4.2 Understanding Epigenetic Mechanisms

Stress elicits plant epigenetic responses that significantly alter gene expression, ensuring the reprogramming of the plant's transcriptome. Epigenetic modifications, such as DNA methylation (DM), histone modification (HM), and non-coding short RNAs, cause changes in the structure of chromatin. Altercations to the DNA and/or associated protein structure impact gene activity (Jogam *et al.*, 2022) which can be constant and transmissible to future generations, enabling plants to transmit stress experiences across subsequent generations (Sudan *et al.*, 2018).

4.2.1 DNA methylation: impact on stress tolerance

DNA methylation is the process of adding a methyl moiety to the C5 location on the cytosine, which leads to the formation of 5mC (Lister *et al.*, 2008). DNA methylation is not limited to the promoter regions of genes but also occurs in

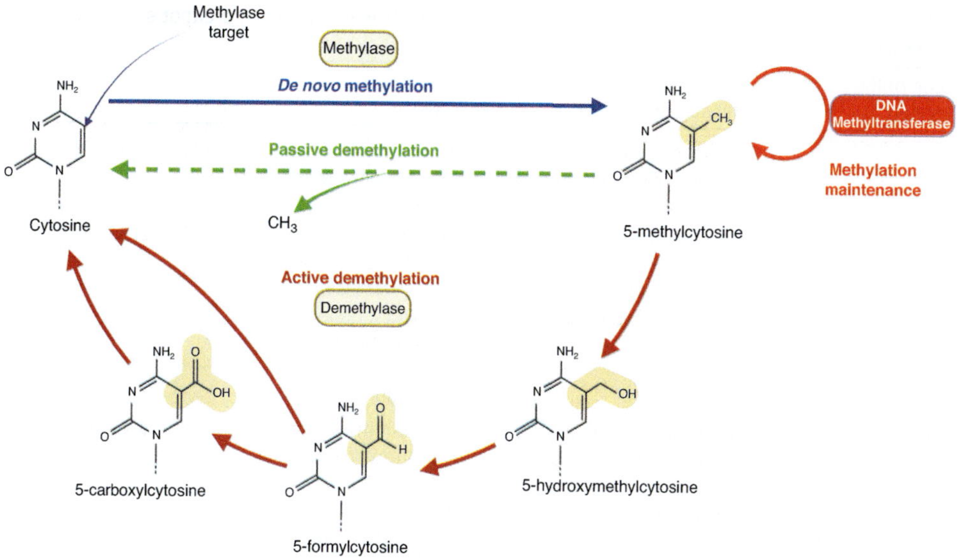

Fig. 4.3. Pictorial representation of DNA methylation process during epigenetic modifications. Author's own image.

the coding sections. Plants exhibit three distinct mechanisms of DNA methylation (Fig. 4.3):

1. maintenance of DNA methylation (methylation of hemimethylated symmetrical sequences)
2. initiation of DNA methylation (methylation at previously unmethylated sites); and
3. DNA demethylation (where methylation status can be upturned).

During DNA replication and cell division in plants, the methylation mark is mostly conserved by DNA methyltransferases, which include maintenance methyltransferases and *de novo* methyltransferases. (van Dijk *et al.*, 2010). Modern investigations indicate that DNA methylation serves as a mechanism through which plants mitigate abiotic stresses that jeopardize plant growth, reproduction, and yield along with species diversity (Ashapkin *et al.*, 2020; Akhter *et al.*, 2021).

DNA methylation and drought stress

Drought is the most destructive non-living factor that negatively impacts plant productivity (Lobell and Gourdji, 2012; Chaudhry and Sidhu, 2022). Among the various intricate mechanisms which plants have evolved to respond to drought stress, DNA methylation prominently regulates gene expression (Table 4.1). Under drought stress, plants such as mulberry (*Morus alba*) show an 8.64% increase, and *Populus trichocarpa* exhibits a 2.29% rise, in overall methylation levels compared with control plants (Liang *et al.*, 2014; Li *et al.*, 2020). Drought-resistant plants exhibit a more consistent methylome under water-deficit conditions, with genes associated primarily with stress reaction, programmed cell death, and other processes identified as differentially methylated regions (DMRs) in rice (*Oryza sativa*) (Wang *et al.*, 2016), mulberry (*Morus alba*) (Ackah *et al.*, 2022), mungbean (*Vigna radiata*) (Zhao *et al.*, 2022), and maize (*Zea mays*) (Wang *et al.*, 2021a). In a separate study, rice varieties IR 20 and CO 43, alongside drought-resistant genotypes PL and PMK3, as well as their F1 hybrids, were employed to examine variations in cytosine methylation in reaction to drought strain (Gayacharan and Joel, 2013). Elevated levels of 5mC in the leaves of drought-resistant rice plants explain the significance of DNA methylation in the drought-resistant rice cultivar; during recovery, the elevated 5-mC level,

Table 4.1. Selected investigations on DNA methylation in plants with respect to drought stress. Author's own table.

Name of the plant exposed	Epigenetic response investigated	Outcomes	References
Rice	Variance in the manifestation of cytosine methylation	Methylation percentage correlated positively with spikelet sterility	Gayacharan and Joel, 2013
Tomato	Investigation of epigenetic markers in the tomato root is essential for detecting drought stress	i. Demethylation in the controlling section of ABA stress and ripening 2 genes ii. Loss of H3K9me2 mark after stress imposition	González *et al.*, 2013
Arabidopsis	H3 lysine 4 mono-, di-, and tri-methylation pattern in chromatin of *Arabidopsis* leaves subjected to water stress	i. The H3K4me1 and H3K4me2 marks exhibit a greater extent of distribution compared to the H3K4me3 marks (62%). However, the levels of H3K4me2 and H3K4me1 only saw slight changes after dehydration stress. ii. The abundance of H3K4me3 exhibited significant changes in response to alterations in transcript levels of target genes.	van Dijk *et al.*, 2010

even after 10 days of stress withdrawal, indicates the essence of epigenetic modification in 'stress memory' (Kumar *et al.*, 2023). A crucial quantitative trait locus on the rice chromosome, qDTY1, played a vital role in suitable grain yield under water scarcity stress (Vikram *et al.*, 2011; Dhawan *et al.*, 2021). Demethylation-related epigenetic changes persisted even up to the fifth generation in clonally propagated *Trifolium repens* and *Fragaria vesca* (Xu *et al.*, 2016; Rendina González *et al.*, 2018). Transcriptome and epigenome analysis revealed that *Brassica napus* can resist drought and efficiently utilize nitrogen for up to three subsequent generations (Verkest *et al.*, 2015). Furthermore, it is assumed that hypermethylation may indicate drought susceptibility, while hypomethylation may indicate drought tolerance (Gayacharan and Joel, 2013). Methylation polymorphism can be effectively utilized in programs aimed at screening for drought tolerance.

The stress caused by drought leads to the modification of H3K4me3 by the histone methyltransferase, ATX1. This modification then triggers the activation of NCED3, leading to the increased manifestation of genes linked to drought and ABA (Kim *et al.*, 2015). Numerous drought-responsive genes had enriched active histone marks, namely H3K4me3 and H3K9ac (Ost *et al.*, 2023). During recovery, the removal of H3K9ac was rapid, whereas H3K4me3 declined slowly, indicating its involvement in epigenetic memory (Lim *et al.*, 2015). In *Arabidopsis thaliana*, a gene named ATDM1 has been reported, which aids in overcoming drought stress by methylation of relevant genes (Lee *et al.*, 2023). Generally, plants remember the first exposure to drought stress as the 'first memory' and combat it more efficiently later within the same environmental exposure (Chinnusamy and Zhu, 2009; Kinoshita and Seki, 2014; Marcos *et al.*, 2018; Tombesi *et al.*, 2018). The epigenetic alteration can also be passed on to the next generation through meiotic and mitotic mediation (Meins and Thomas, 2003; Iwasaki and Paszkowski, 2014). HATs and HDACs work together to produce dynamic histone acetylation levels (Boycheva *et al.*, 2014). Expression of some HAT genes like OsHAM701, OsHAG702, and OsHAG703 in rice; BraHAC1, BraHAC3, BraHAC2, and BraHAC7 in Chinese cabbage significantly enhanced histone acetylation after drought exposure (Hou *et al.*, 2021) (Eom and Hyun, 2018) (Li *et al.*, 2021).

Cytosine-5-methyltransferases and demethylases are crucial enzymes that actively regulate the DNA methylation state of plant genomes during water scarcity stress (Sun *et al.*, 2022). Studies indicate that plants, including apples (Xu *et al.*, 2018), tomatoes (Kumar *et al.*, 2016), and eggplant (Moglia *et al.*, 2019), exhibited an increase in the regulation of 5mC methyltransferase and demethylase as a reaction to drought stress. The plant DNA methylation varies in degree, quantity, and polymorphism when subjected to water-deficient circumstances. These variations were shown to be specific to different tissues and genotypes. The drought-tolerant wheat genotype AK58 displayed higher levels of DNA methylation and demethylation, along with a rise in methylation polymorphisms, in comparison with the typical wheat genotype MinMai 13 (Duan *et al.*, 2020). Moreover, the methylation polymorphisms in the root exhibited a higher level than those in the foliage under drought-like conditions. This difference was particularly pronounced in AK58, suggesting that the observed changes in DNA methylation might be instrumental in AK58's faster response to water deprivation (Duan *et al.*, 2020).

DNA methylation and heat stress

Heat-sensitive genotypes had more methylation after heat treatment than heat-resistant ones. Heat-resistant genotypes demethylated DNA more than heat-sensitive genotypes (Gao *et al.*, 2014). Based on evidence, it has been demonstrated that in the majority of situations, plants subjected to high temperatures have a decreased overall level of methylation compared with control plants. This phenomenon is exhibited by the species *Populus simonii* and *Brassica napus* (Ci *et al.*, 2015; Li *et al.*, 2016). The presence of DNA methylases and demethylase genes in the *Brassica napus* genome is crucial to sustaining the balance of DNA methylation amid heat stress in plant genomes (Fan *et al.*, 2020). A total of 325 genes exhibiting differential methylation and responsive to heat stress at a temperature of 42°C have been identified in maize seedlings (Qian *et al.*, 2019) Remarkably when exposed to heat stress, it was noted that 9-dimethylguanosine (DMG) molecules linked with spliceosomes displayed a reduced amount of methylation (Qian *et al.*, 2019). Moreover, when exposed to heat stress ranging from 37°C to 45°C, the heat-sensitive variant of *Brassica napus* exhibited a greater level of DNA methylation compared with its heat-tolerant variant (Gao *et al.*, 2014). RNA-directed DNA methylation (RdDM) in plants encompasses several modules, including siRNAs and the DNA methyltransferase DRM2 (Law and Jacobsen, 2010). DNA methylation happens at particular locations: asymmetric CpNpN and symmetric CpG/CpNpG patterns. Chromomethylase (CMT) controls the methylation of asymmetric cytosine (CpNpG) in plants (Bartee *et al.*, 2001). Through epigenetic regulation, heat exposure can cause repetitive elements to become active in *Arabidopsis* seedlings (Pecinka *et al.*, 2010). Exposure of *Arabidopsis* plants to elevated temperatures activates the crucial genes involved in DNA methylation, including domains rearranged methyltransferase 2 (DRM2), nuclear RNA polymerase D1 (NRPD1), and nuclear RNA polymerase E 1 (NRPE1) (Naydenov *et al.*, 2015; Borges *et al.*, 2021). The transcriptome analysis in heat-treated cotton anthers revealed the presence of many genes associated with histone modification and DNA methylation. During heat stress, the cotton line that can tolerate high temperatures generates anthers and normal pollen. On the other hand, the cotton line sensitive to heat has a problem with anther dehiscence and cannot produce viable pollen. Heat stress caused a reduction in the levels of DNA cytosine-5-methyltransferase1 (DNMT1) and S-adenosyl-l-methionine-dependent methyltransferase (SAM MTase) through the various phases of tetrad and tapetum disintegration in thermal-sensitive cotton anthers. However, in heat-resilient cotton anthers, the expression of these genes remained similar, except for an increase in the manifestation of DRM3 during the tetrad phase (Min *et al.*, 2014). In heat-sensitive cotton anthers, heat stress inhibits the manifestation of three proteins necessary for proper DNA methylation, viz. nuclear RNA polymerase D1B (NRPD1B), *S*-adenosyl-L-homocysteine hydrolase 1 (SAHH1), and needed for RDR2 independent DNA methylation (NERD) (Min *et al.*, 2014). Heat stress induces a reduction in DNA methylation levels in heat-susceptible cotton anthers, while heat-tolerant cotton anthers have elevated DNA methylation levels. Furthermore, it is plausible that the heat-sensitive cotton undergoes hypomethylation, resulting in pollen sterility and anomalies in anther dehiscence (Ma *et al.*, 2018). Treatment of soybean root hairs and roots without root hairs to heat stress led to a decrease in methylation of CHH (Hossain *et al.*, 2017). Additionally, CG and CHG become hypomethylated under heat exposure in *Brassica napus* microspore cultures (Li *et al.*, 2016). Table 4.2 displays a compilation of specific research studies investigating DNA methylation in plants as a reaction to heat stress.

DNA methylation and cold stress

Alterations in DNA methylation are pivotal for a plant's capacity to control gene expression when confronted with low temperatures or

Table 4.2. Selected investigations on DNA methylation in plants with respect to heat stress. Author's own table.

Name of the plant exposed	Epigenetic response investigated	Outcome of the experiment	References
Brassica spp.	Investigation of methylation levels and changes in cytosine methylation trends in seedlings of heat-resilient and heat-susceptible rapeseed types.	i. The heat-susceptible genotype exhibited a more significant augmentation in methylation levels under heat exposure in comparison to the heat-resilient genotype. ii. The genotype that is resistant to heat saw a greater number of DNA demethylation events, while the genotype that is susceptible to heat showed a higher occurrence of DNA methylation.	Gao *et al.*, 2014
Arabidopsis	The impact of heat stress on the activation of genes responsible for encoding crucial components involved in DNA methylation.	i. Findings indicate a synchronized increase in the expression of DRM2, NRPD1, and NRPE1 in response to elevated temperature. ii. The initiation of DRM2 manifestation in heat stress may necessitate the involvement of PolIV and/or PolV.	Naydenov *et al.*, 2015

freezing stress (Fan *et al.*, 2013; Thiebaut *et al.*, 2019). For instance, exposure to cold stress led to significant reductions in DNA methylation intensities in sugar beet, particularly under high-chill conditions (Gutschker *et al.*, 2022). In contrast, apples showed no notable decrease in methylation levels under low-chill conditions (Kumar *et al.*, 2016). Cold acclimatization involves both methylation and demethylation processes. For example, in *Brassica* species, 1562 genes exhibited differential methylation patterns during cold acclimation, resulting in demethylation of their promoters and increased transcriptional activity (Liu *et al.*, 2017). Furthermore, DNA demethylation has been observed in cold adaptation processes in species such as rubber plants (Tang *et al.*, 2018), *Brassica rapa* (Liu *et al.*, 2017), tomato (Zhang *et al.*, 2016), and sweet orange (*Citrus sinensis*), in which it promotes the activation of cold-linked and cold-reactive genes by demethylating their promoters (Sicilia *et al.*, 2020). Table 4.3 displays a collection of

Table 4.3. Selected investigations on DNA methylation and transcriptional activation in plants with respect to cold stress. Author's own table.

Name of the plant exposed	Epigenetic response investigated	Outcome of the experiment	References
Maize	An examination of DNA methylation in maize nucleosomes and the reversal of methylation under environmental stress	The exposure to cold stress caused significant demethylation in the core, but the linker remained comparatively unaffected.	Steward *et al.*, 2002
Rice	Study on the epigenetic modifications that occur when the OsDREB1b genes are expressed in rice under cold stress	i. There is a significant change in the arrangement of nucleosomes in the region upstream of the cis-elements and TATA box at the core promoter of OsDREB1b genes. This change is accompanied by an increase in sensitivity to DNaseI or MNase digestion. ii. Upon returning stressed plants to normal temperature, it was shown that the elevated transcript level of OsDREB1b produced by cold and the acetylation of histone H3 in the upstream region would diminish.	Roy *et al.*, 2014
Beta vulgaris	The impact of freezing on two *Beta vulgaris* accessions' gene expression and genome-wide DNA methylation	Noteworthy decrease in the levels of DNA methylation, precisely in the CHH scenario (and to a lesser degree in CHG). This decrease was accompanied by a decrease in the expression of the chromomethyltransferase CMT2 and an increased manifestation of many genes involved in dynamic DNA demethylation.	Gutschker *et al.*, 2022

research that examines the relationship between DNA methylation and the activation of gene transcription in plants when exposed to cold stress.

DNA methylation and salinity stress

The high salt content in soil is a major agricultural hindrance in the path of future food security (Rahman *et al.*, 2021) because high salinity disrupts osmotic pressure and ion balance, as well as protein synthesis, photosynthesis, energy metabolism, and lipid metabolism in plants (Muchate *et al.*, 2016). Salinity stress induces alterations in DNA methylation patterns in the genomes of different plant species (Table 4.4). On the one hand, exposure to high levels of salt caused an augmentation in the methylome composition of *Medicago* species (Al-Lawati *et al.*, 2016), whereas, on the other hand, DNA demethylation plays an integral part in salinity stress as well. Under salinity stress, gene expression is regulated depending on the position of methylated cytosine (Moore *et al.*, 2013). Methylation in the promoter area ceases expression of the downstream structural gene, whereas that in the exon promotes gene manifestation (Baek *et al.*, 2010). In salinity-tolerant cultivars of olive, DNA methylation enhances following exposure to Na^+ (Bazakos *et al.*, 2015). Prior exposure to low salt treatment can trigger a lasting memory that enables a quick reaction to subsequent stress caused by low levels of H3K27me3 in HKT, a protein that controls the distribution of sodium ions (Kim *et al.*, 2015). Methylation in the sRNA segment located upstream from the salt-resistant AtHKT1 gene triggers its transcription, thus significantly affecting salinity tolerance (Baek *et al.*, 2010). When *Arabidopsis thaliana* was exposed to salt stress, epigenetic regulation influenced the flavonoid and antioxidant pathways (Bharti *et al.*, 2015). In *Nicotiana tabacum* genes that are related to flavonoid biosynthesis, like chalcone, flavone-3-hydroxylase, flavonol synthase, and those involved in antioxidant scavenging, comprising glutathione peroxidase, *S*-transferase get demethylated at both coding and promoter regions by overexpression of AtROS1 (Repressor of silencing1) at the time of salinity exposure (Bharti *et al.*, 2015).

DNA methylation and heavy metal stress

Plants experience a notable interaction between their physiological reactions, gene manifestation levels, and DNA methylation trends when subjected to intense heavy metal stress (Ding *et al.*, 2014). Plant reaction to heavy metals (HMs) may be regulated by DNA methylation through two main pathways. Firstly, hypermethylation acts as a defensive strategy that can safeguard DNA from potential injury by HMs, thereby preventing genomic instability and enhancing plant survival in challenging environments. Secondly, DNA methylation regulates gene expression not only in the promoter regions but also in the coding locations of genes (Sun *et al.*, 2022).

For instance, rice plants stressed with cadmium (Cd) (Feng *et al.*, 2016a), radish (Tang *et al.*, 2021), and soybean (Sun *et al.*, 2021) exhibited increased methylation levels following Cd exposure compared with control plants. Table 4.5 represents a few investigations on DNA methylation and transcriptional activation in plants concerning heavy metal stress.

4.2.2 Histone modifications: regulation of stress-responsive genes

Basic histone proteins undergo various post-translational modifications (PTMs) that impact chromatin structure and function, influencing processes such as transcription, DNA repair, replication, and recombination (Bannister and Kouzarides, 2011). Key histone PTMs include methylation, acetylation, phosphorylation, ubiquitination, and sumoylation, catalyzed by specific histone-modifying enzymes at distinct genomic sections (Kouzarides, 2007).

Acetylation of lysine and arginine residues is generally believed to reduce DNA–histone interactions, thereby relaxing chromatin assembly and enhancing DNA accessibility for transcription (Allis and Jenuwein, 2016). In contrast, histone methylation influences histone–DNA associations without directly affecting nucleosome stability due to its neutral nature, yet it does impact local hydrophobicity (Xiao *et al.*, 2016).

Table 4.4. Selected investigations on DNA methylation and transcriptional activation in plants with respect to salinity stress. Author's own table.

Name of the plant exposed	Epigenetic response investigated	Outcome of the experiment	References
Chloris virgata	Investigation of root DNA methylation pattern	i. Cytosine methylation in leaves and roots. ii. Variations in DNA methylation were brought on by both salt and alkaline stress, mainly in the roots. iii. Root DNA methylation may help acquire and inherit salt and alkaline stress resistance.	Cao *et al.*, 2012
Mesembryanthemum crystallinum	Investigation on satellite DNA methylation of the CpNpG type in plants grown in both normal and salt-stressed environments	Switch from the C3 photosynthetic cycle to CAM mode.	Dyachenko *et al.*, 2006
Rice	Change in methylation patterns in 'Pokkali' cultivar	Pokkali, a rice variety that can tolerate high levels of salt, showed a significant and quick reduction in DNA methylation when exposed to salt stress. On the other hand, the salt-sensitive variety 'IR29' had a limited capacity to adjust DNA methylation levels in response to salt stress.	Ferreira *et al.*, 2015
Triticum aestivum	Methylation status of CCGG sequence of nuclear genome of root DNA in cultivars of wheat	Plants displayed a clear alteration in DNA methylation when exposed to salt stress, and the extent of this change was directly determined by the dosage.	Lan *et al.*, 2009
Arabidopsis	Methylation patterns in the offspring of *Arabidopsis thaliana*	i. The primary cause of hypermethylation in the offspring of stressed plants was alterations in the 5′ and 3′ regions, as well as in exons, rather than introns. ii. Connection between hypermethylation and reduced gene manifestation and an elevation of H3K9me2 and reduction of H3K9ac histones.	Bilichak *et al.*, 2012
Maize	Methylation-sensitive amplified polymorphism was employed to screen salt and water stress-induced methylation patterns in maize	i. The expression of root zmPP2C was dramatically downregulated by salt-induced intron methylation, whereas the expression of leaf zmGST was moderately increased by salt-induced demethylation.	Tan, 2010

Continued

Table 4.4. Continued

Name of the plant exposed	Epigenetic response investigated	Outcome of the experiment	References
Rapeseed	Using methylation sensitive amplified polymorphism, salinity-tolerant and sensitive rapeseed cultivars were assessed for cytosine methylation following salt stress.	i. Salinity influenced DNA methylation. Methylation was reduced in the salt-resilient cultivar and increased in the salt-sensitive cultivar.	Marconi *et al.*, 2013

Phosphorylation of amino acid residues like threonine, serine, and tyrosine introduces additional negative charges to the polypeptide chain, thereby weakening DNA–histone interactions. Ubiquitination of lysine residues involves attaching small amino acid chains to histone tails, which can decrease nucleosome stability. However, the precise mechanisms of transcriptional regulation by ubiquitination remain unclear (Zhou *et al.*, 2017, Zhou *et al.*, 2018).

Sumoylation, involving the attachment of the SUMO moiety (a small ubiquitin-related modifier), is another crucial post-translational modification in response to various plant stress (Miller *et al.*, 2013; Elrouby, 2017). It is crucial in altering mRNA profiles and functions as a modifier in RNA processing and metabolism. Sumoylation interacts with RNA binding proteins and elements involved in 3' pre-mRNA processing, RNA editing, transcription termination, and mRNA export (Richard *et al.*, 2013) (Lamoliatte *et al.*, 2014).

Histone modification in drought stress response

Robust histone deacetylation and nucleosome replacement are necessary (Greco *et al.*, 2012) to fully suppress the expression of stress-induced genes and restore chromatin status under normal (non-drought) conditions. Therefore, histone acetylation is associated with drought stress and ABA responses in plants, where the acetylation levels increase on drought-responsive genes such as RD20, RD29A, and RD29B (Kim *et al.*, 2008). Additionally, there is a rapid enhancement of H3K9ac in these gene regions.

Histone modification in salt stress response

Key mechanisms involved in salt stress responses include acetylation, methylation, and phosphorylation. Histone acetylation typically correlates with increased gene activity and is controlled by the contrasting actions of histone acetyltransferases (HATs) and histone deacetylases (HDACs). Both tobacco BY-2 and *Arabidopsis* T87 culture cells experience a quick and temporary increase in the phosphorylation of histone H3 at Ser-10, which is then followed by the phosphoacetylation of H3 and the acetylation of histone H4 (Kim *et al.*, 2015).

Histone acetylation-based epigenetic mechanisms control the manifestation of genes, including ABI1 and ABI2, implicated in the ABA signaling pathway, during salt stress (Lim *et al.*, 2015). To maintain plant physiology through epigenetic control, it is necessary to achieve a delicate equilibrium between positive and negative regulators (Kaldis *et al.*, 2010). The transcriptional adaptor 2b protein (ADA2b) influences acetyltransferase activity in *Arabidopsis*; mutant plants with altered ADA2b show reduced salinity tolerance (Kaldis *et al.*, 2010).

Mutations in HD2C, HDA6, and HDA19 in *Arabidopsis* disrupt acetylation levels and lead to increased sensitivity to salinity in mutant plants (Luo *et al.*, 2012). Epigenetic modifications contribute to the long-term somatic memory of salinity sensitivity (Sani *et al.*, 2013).

Histone modification in heat stress response

Histone changes, such as acetylation and Sumoylation, have a role in the cellular response

Table 4.5. Selected investigations on DNA methylation and transcriptional activation in plants with respect to heavy metal stress. Author's own table.

Name of the plant exposed	Epigenetic response investigated	Outcome of the experiment	References
Rice	An analysis of the DNA methylation pattern in the leaves of rice plants exposed to HMs, as well as in their three subsequent generations.	i. DNA methylation patterns changed in stressed plants, and these changes were limited to CHG hypomethylation alone. ii. The CHG-demethylated states could be inherited through both the maternal and paternal germline, but they commonly occur alongside further hypomethylation. iii. The stress triggered changes in the manifestation of genes responsible for DNA methyltransferases, DNA glycosylase, and SWI/SNF chromatin remodeling factor (DDM1).	Ou *et al.*, 2012
Populus alba L. (white poplar)	Investigation of epigenetic modifications that arise during the development of tolerance to HMs in mycorrhizal and non-mycorrhizal poplar.	The expression of genes in treated samples differed from that of controls in response to HMs and/or arbuscular mycorrhizal fungi (AMF) inoculation. More precisely, the levels of gene transcripts related to RNA processing, cell wall, and amino acid metabolism were increased when arbuscular mycorrhizal fungi (AMF) were present, regardless of the presence of HMs.	Cicatelli *et al.*, 2013

Continued

Table 4.5. Continued

Name of the plant exposed	Epigenetic response investigated	Outcome of the experiment	References
Posidonia oceanic	Investigation of the effect of cadmium treatment on DNA methylation patterns and its influence on chromatin reconfiguration.	i. Cadmium administration caused an increase in DNA hypermethylation, as well as an up-regulation of chromomethylase. ii. High cadmium stress resulted in a gradual formation of heterochromatin in the nucleus during the interphase. iii. Apoptotic figures were also detected following prolonged exposure to the substance.	Greco *et al.*, 2012
Rapeseed	An investigation of the impact of selenium on the growth of rape seedlings under conditions of cadmium-induced stress.	Selenium inhibited alterations in the DNA methylation pattern induced in rape seedlings by high doses of cadmium.	Filek *et al.*, 2008
Noccaea caerulescens	A study into the association between epigenetic alterations and the response to elevated levels of nickel in plant foliage.	The expression of MET1, DRM2, and HDA8 genes, which are associated with the alteration of epigenetic DNA and histones, was increased in the existence of elevated levels of nickel in the leaves.	Gullì *et al.*, 2018

to heat stress. Recent research has shown that the presence of certain histone variations, specifically H2A and H3, not only controls gene expression but also has important functions in DNA repair and the creation of chromosome centromeres in eukaryotes (Mizuguchi *et al.*, 2004; Lu *et al.*, 2009; Choi *et al.*, 2013).

Histone modification in cold stress response

During periods of low-temperature stress, there is a noticeable rise in the levels of histone acetylation at specific genes that respond to cold, such as COR15A and AtGolS3. This indicates that the changes in gene expression caused by cold stress are likely influenced by epigenetic modifications, which in turn enhance the ability to tolerate the stress (Kim *et al.*, 2015). In maize, histone acetylation at cold-responsive genes like ZmDREB1 and ZmCOR413 is also heightened (Hu *et al.*, 2011). Additionally, cold stress enhances histone acetylation of OsDREB1b in rice (Roy *et al.*, 2014). Conversely, DNA methylation levels have been observed to decrease under cold stress.

4.2.3 Role of small RNAs in stress adaptation

Small RNAs can regulate gene expression at both the transcriptional and post-transcriptional levels. Currently, small RNAs are classified into miRNAs and three classes of endogenous siRNAs: *trans*-acting siRNAs (ta-siRNAs), heterochromatic siRNAs (hc-siRNAs), and natural antisense siRNAs (nat-siRNAs) (Phillips *et al.*, 2007).

Small RNAs, specifically miRNAs and siRNAs, have a crucial function in precisely adjusting the manifestation of stress- or development-related genes employing post-transcriptional gene silencing (Mallory and Vaucheret, 2006). Epigenetic regulatory systems have certain benefits over other mechanisms. They can quickly adapt to various environmental and developmental signals and function as mobile signaling molecules to regulate the expression levels of certain genes (Ravichandran *et al.*, 2019). Research indicates that plant miRNA-mediated pathways contribute to stress memory, aiding plants in adapting to recurring stress (Stief *et al.*, 2014), and facilitating the plant's acclimatization under stress conditions.

Role of miRNAs in stress responses

Environmental concentrations of phosphate control the expression of miR399 (Fujii *et al.*, 2005; Chiou *et al.*, 2006), with low phosphate levels strongly inducing miR399 expression. miR393 targets AUX/IAA proteins for proteolysis indirectly through SCF E3 ubiquitin ligases in an auxin-dependent manner, crucial for auxin-induced growth procedures (Vierstra, 2003), and also associated with abiotic stress responses. ABA plays a vital role in osmotic stress responses and seed maturation. In germinating *Arabidopsis* seeds, ABA induces the build-up of miR159, which targets the seed-specific transcription factor ABI3. Subsequently, miR159 mediates the cleavage of MYB101 and MYB33 transcripts, which are positive regulators of ABA responses (Reyes and Chua, 2007).

miR398 targets two closely related Cu/Zn-superoxide dismutase genes, CSD1 and CSD2, involved in scavenging superoxide radicals (Jones-Rhoades and Bartel, 2004). It cleaves the transcripts of CSD1 and CSD2 as well. The process of oxidative stress leads to the transcriptional downregulation of miR398, which is essential for the accumulation of CSD1 and CSD2 mRNAs at the post-transcriptional level. This accumulation plays a vital role in countering oxidative stress.

Twenty-two novel miRNAs have been identified in salt-stressed *Raphanus sativus* (Sun *et al.*, 2015). In maize, downregulated miRNAs such as mir17, mir205, and mir250 target genes involved in enhancing survival strategies, including casein kinase II, glutathione peroxidase, pyrroline-5-carboxylate synthase, and translation initiation factors (IF1) (Fu *et al.*, 2017). The type of stress and plant species influence the expression of miRNAs in *Saccharum* species. Selected studies in which sugarcane modulates miRNA expression in response to salt stress are summarized in Table 4.6.

AFB2,3 (auxin signaling F box protein 2,3-producing gene) and transport inhibitor response protein 1 (TIR1) are examples of auxin-responsive genes that are potential targets of miR319; both genes, when decreased in expression, may aid plants in withstanding salt stress (Roy and Soni, 2022). In *Arabidopsis*, heightened expression of the miR393-resistant TIR1 gene bolstered tolerance to salinity (Roy and Soni, 2022). Eleven miRNAs responsive to salt stress were identified in Suaeda maritime (Gharat and Shaw, 2015). Members of the miR172 family are instrumental in stress tolerance in *Arabidopsis* and *Glycine max* (Bai *et al.*, 2018).

After subjecting the organisms to heat stress, the expression of miR156, which specifically acts on the SPL gene, is observed to increase (Li *et al.*, 2016). *Arabidopsis* ago1 mutants exhibit anomalies in heat acclimation, indicating that miR156-mediated post-transcriptional regulation of SPL genes contributes to heat stress memory. Heat stress activates the ONSEN retrotransposon (Nozawa *et al.*, 2022), while RdDM suppresses ONSEN's transpositional and transcriptional activation (Hayashi *et al.*, 2020). Genes near RdDM

Table 4.6. Modulation of miRNA in sugarcane in response to salt stress. Author's own table.

No.	miRNA modulated	Type of change	Reference
1.	miR156	Upregulation	Yang *et al.*, 2017
2.	miR159	Upregulation	Patade and Suprasanna, 2010
3.	miR166, 168,169, 396, 397, 398	Upregulation	Carnavale Bottino *et al.*, 2013

targets exhibit altered transcription due to heat stress-induced DNA methylation; normal transcription is restored by inducing DNA methylation (Kim *et al.*, 2015). In rapeseed seedlings, heat stress significantly reduced the expressivity of miRNAs bramiR1885b.3 and bra-miR5716 (Yu *et al.*, 2012). Heat stress also increased the expressivity of miR160a, 166a, 167 hour, and 5175a in barley, while decreasing the expression of their target genes, including Auxin response transcription factors (ARFs) (Kruszka *et al.*, 2014). Similarly, heat stress in *Arabidopsis* rapidly upregulated miR398 expression while downregulating target genes such as CSD (encoding copper/zinc SOD) and CCS (encoding a chaperone for CSD) (Guan *et al.*, 2013). These findings emphasize the important function of miRNAs in plant reactions to heat stress, and the function of siRNAs and other regulatory RNA molecules in the process of stress adaption.

The nat-siRNA, responsible for regulating proline metabolism, is a well-studied small RNA molecule that is distinct from miRNA and has a role in adapting to the environment (Borsani *et al.*, 2005). Salt stress induces an elevation in the concentration of nat-siRNA, resulting in the build-up of proline, which is essential for salt tolerance (Borsani *et al.*, 2005). Exposure to abiotic stress builds stress resistance in plants by non-coding up- and downregulated RNAs (siRNA and miRNA), which upregulate effective regulators like plant hormones, transcription factors, reactive oxygen species (ROS) and downregulate the antipathetic regulators (Banerjee *et al.*, 2017; Zhang *et al.*, 2019).

4.3 Techniques for Studying Epigenetic Changes

4.3.1 Bisulfite sequencing: profiling DNA methylation patterns

One of the most extensively researched epigenetic indicators is DNA methylation, which involves adding methyl groups to cytosine bases in CpG dinucleotides (Bellacosa and Drohat, 2015; Kim and Costello, 2017). Bisulfite sequencing has emerged as a pivotal technique to decipher methylation patterns across the genomic DNA, offering insights into gene regulation mechanisms in health and disease (Fig. 4.4) (Wreczycka *et al.*, 2017; Barros-Silva *et al.*, 2018; Gouil and Keniry, 2019).

Bisulfite sequencing is based on the chemical transformation of unmethylated cytosines to uracils (which are subsequently read as thymines during sequencing), while methylated cytosines remain unchanged (Reinders and Paszkowski, 2010; Berney and McGouran, 2018; Gouil and Keniry, 2019). This differential treatment allows for the discrimination between methylated and unmethylated cytosines at single-nucleotide resolution (Singer, 2019). The method commences with the fragmentation of genomic DNA, which is subsequently subjected to bisulfite treatment. During this treatment, sodium bisulfite transforms cytosines that are not methylated into uracil, while cytosines that are methylated stay as cytosines (Clark *et al.*, 2006; Hayatsu *et al.*, 2008; Patterson *et al.*, 2011). Following bisulfite treatment, PCR amplification of the transformed DNA fragments is performed, followed by high-throughput sequencing to ascertain the methylation state at each CpG site (Sun *et al.*, 2014; Pajares *et al.*, 2021).

Bisulfite sequencing has revolutionized our understanding of DNA methylation dynamics in various biological contexts (Ambrosi *et al.*, 2017) (Harrison and Parle-McDermott, 2011). It is particularly valuable in studying tissue-specific methylation patterns, developmental processes, and disease states (Liang *et al.*, 2011; Mendizabal and Yi, 2016; Yong *et al.*, 2016). Understanding these patterns is essential because DNA methylation can help in gene regulation in response to environmental stressors such as drought, salt, and extreme temperatures (Garg *et al.*, 2015; Viggiano and de Pinto, 2017; Akhter *et al.*, 2021). Li *et al.* (2020) studied the methylation patterns in *Morus alba* under drought stress. By employing whole-genome bisulfite sequencing, a thorough map of methylated cytosines was generated and the effects of these alterations on gene expression were examined using transcriptomics. The results revealed that the percentage of methylated

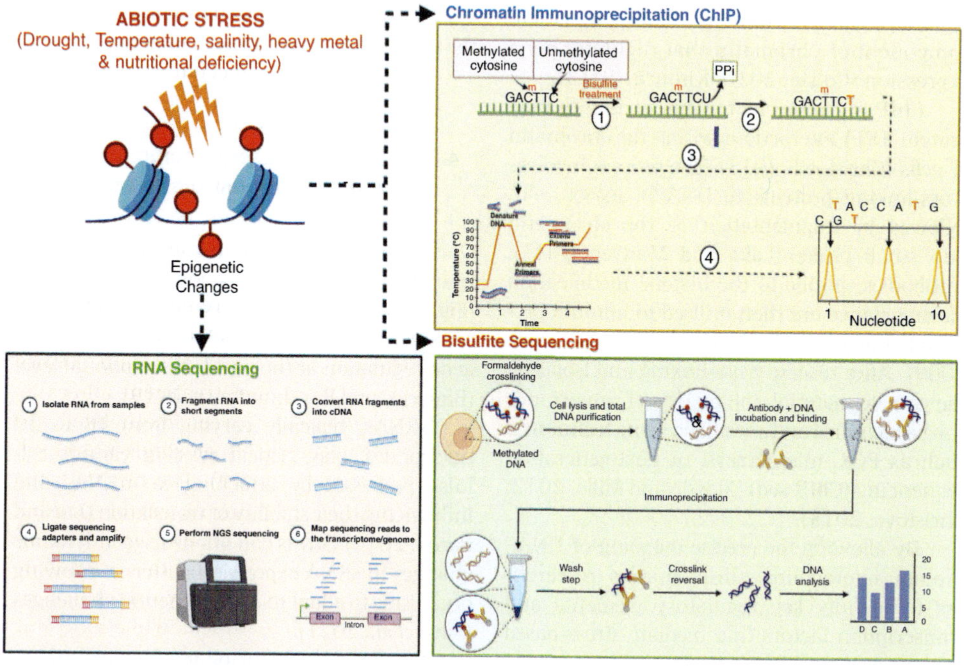

Fig. 4.4. Pictorial representation of selected techniques involved in the study of epigenetic changes. Author's own image.

cytosines varied based on sequence context and treatment conditions. Under drought stress, methylation levels increased by 8.64% compared with control, primarily due to a 6.24% rise in mCG methylation. These methylated genes were enhanced in gene ontology (GO) subcategories such as catalytic activity, cellular processes, metabolic processes, response to stimuli, and regulation of biological processes (Li *et al.*, 2020). By identifying specific methylation changes associated with stress-responsive genes, researchers can elucidate the epigenetic mechanisms plants use to adapt to adverse conditions, potentially leading to the improvement of crops with improved stress tolerance and resilience (Kumar, 2018; Sun *et al.*, 2021).

One key advantage of bisulfite sequencing is its high resolution, enabling the detection of methylation patterns at single-base resolution. This precision is crucial for identifying subtle changes in DNA methylation that may be missed by less sensitive techniques (Clark *et al.*, 2017; Gouil and Keniry, 2019). Moreover, bisulfite sequencing can analyze large genomic regions or even whole genomes, providing a comprehensive view of epigenetic landscapes (Mendizabal and Yi, 2016; Wreczycka *et al.*, 2017).

However, bisulfite sequencing also presents challenges. It requires a substantial amount of DNA, which can be limiting when studying scarce samples or single cells (Hodges *et al.*, 2009). Additionally, bisulfite treatment introduces biases, such as incomplete conversion of unmethylated cytosines or DNA degradation. These factors necessitate careful experimental design and data analysis to ensure accurate interpretation of results (Patterson *et al.*, 2011; Olova *et al.*, 2018).

4.3.2 Chromatin immunoprecipitation (ChIP) and analyzing histone modifications

Techniques like chromatin immunoprecipitation (ChIP) are central to epigenetics (Fig. 4.4), which allows researchers to investigate modifications to histone proteins – a critical component of chromatin that regulates gene expression (Collas, 2010; Kimura, 2013).

ChIP is a powerful technique used to map protein–DNA interactions within the chromatin of cells (Christova, 2013). The process involves cross-linking proteins to DNA in intact cells, followed by fragmentation of the chromatin into small pieces (Lake and Minteer, 2017). Antibodies specific to the histone modification of importance are then utilized to immunoprecipitate the protein-DNA complexes (Milne *et al.*, 2009). After reverse cross-linking and isolating the genetic material, the regions of importance can be identified and quantified using techniques such as PCR, microarrays, or next-generation sequencing (ChIP-seq) (Massie and Mills, 2012; Christova, 2013).

By allowing the precise mapping of DNA–protein interactions, ChIP enables researchers to identify key regulatory elements and transcription factors that mediate stress-based responses at the chromatin level (Kumar *et al.*, 2021). Understanding the intricate signaling pathways and gene networks that drive plant response to environmental challenges like drought, salt, and severe temperatures is of utmost importance (Thanmalagan *et al.*, 2022; Adel and Carels, 2023). A study utilized ChIP-seq, a technique combining ChIP with high-throughput DNA sequencing, to analyze stressed tomato plant leaves (Ricardi *et al.*, 2014). The study employed a high-quality anti-ASR1 antibody. This method facilitated the discovery of a novel group of target genes, several of which have significant involvement in the reaction to drought-induced stress. The ASR1-enriched genomic regions contain genes that encode enzymes responsible for cell wall formation and remodeling, as well as channels that are essential for the transfer of water and solutes, such as aquaporins. In addition, they discovered a prominent agreement in the ASR1-binding DNA motif (Ricardi *et al.*, 2014). Through ChIP, scientists can elucidate how histone modifications, DNA methylation, and other chromatin remodeling events alter gene expression profiles in stressed plants, providing insights that are crucial for developing crops with enhanced stress tolerance (Adel and Carels, 2023; Sharma *et al.*, 2023; Abdulraheem *et al.*, 2024). This knowledge is essential for improving agricultural productivity and sustainability in the face of changing climate conditions.

4.3.3 Small RNA sequencing: profiling stress-responsive sRNAs

One of the pivotal techniques in epigenetic research is small RNA sequencing (sRNA-seq) (Fig. 4.4), which enables the profiling of stress-responsive small RNAs (sRNAs) implicated in gene regulation at the post-transcriptional level (Banerjee and Roychoudhury, 2019).

sRNAs, typically varying from 18 to 30 nucleotides, play critical roles in various cellular processes by attaching to mRNAs and influencing their stability or translation (Liu and Paroo, 2010). Stress conditions trigger dynamic changes in sRNA expression patterns, allowing organisms to adapt to environmental challenges (Silva *et al.*, 2021).

Small RNA sequencing involves several key steps to capture and analyze these short RNA molecules from biological samples. Initially, total RNA is extracted and optimized for sRNAs using size selection methods (Buschmann *et al.*, 2016). Subsequently, adaptors are ligated to the ends of the sRNAs, succeeded by reverse transcription and PCR amplification to generate a cDNA library (Hafner *et al.*, 2012). High-throughput sequencing systems, such as Illumina or Ion Torrent, subsequently analyze millions of tiny RNA molecules in the library by sequencing them (Persson *et al.*, 2017). Differential expression analysis identifies stress-responsive sRNAs by comparing samples under stress vs control conditions (Azahar *et al.*, 2020). One sRNA transcriptome study was aimed toward the identification and characterization of miRNAs responding to submergence to elucidate the response process of corn seedlings under submerged conditions (Azahar *et al.*, 2020). They constructed two sRNA libraries from control and submerged corn roots for this investigation.

Small RNA sequencing has been instrumental in uncovering stress-responsive sRNA profiles across diverse organisms, from plants to animals. A recent study has discovered a new ta-siRNA-generating locus (TAS) in wheat. This locus is derived from a section of 28S ribosomal RNA and exhibits varying levels of expression when the plant is infested with leaf rust (Dutta *et al.*, 2017a). Scientists generated four short RNA libraries from wheat near-isogenic lines, which were subjected to treatment with the leaf rust pathogen, as well as untreated controls, to make a comparison. This analysis revealed a TAS capable of generating four ta-siRNAs through phasing events, with miRNA TamiR16 acting as the phase initiator. The ta-siRNAs target several genes, including α-gliadin, leucine-rich repeat, trans-membrane proteins, glutathione-S-transferase, and fatty acid desaturase, which are crucial for stress response and normal plant development.

The study examined the expression of the TAS gene, its ta-siRNAs, and their target genes at five distinct time points after introducing a pathogen into both susceptible and resistant wheat isolines. These results were then compared with control samples that were not infected with the pathogen. The comparison analysis showed clear and opposite expression patterns between the vulnerable and resistant isolines. The expression patterns of the target genes suggested a tendency towards effector-triggered vulnerability, which promotes disease. This study offers significant insights into the regulatory mechanisms of wheat genes that are influenced by ta-siRNAs in response to leaf rust (Dutta *et al.*, 2017b; Goswami *et al.*, 2019).

Small RNA sequencing represents a powerful tool for studying stress-responsive epigenetic changes. Its ability to profile sRNA expression profiles underpins discoveries in environmental adaptation, epigenetic inheritance, and disease mechanisms (Goswami *et al.*, 2019).

4.4 Stress Memory

Research findings widely accept the idea of stress memory in plants, which helps them better survive under similar stressors after mild exposure (Sun *et al.*, 2022). The first stress exposure leaves an imprint on the plant that influences the plant to alter its epigenetic machinery to adapt to and withstand future stress in various ways, affecting its responses in the same or following generations (Kambona *et al.*, 2023). Hence, stress-based memory is the ability of a plant to improve its resilience to later stresses by storing prior stress information so that it can respond differently when tested under the same stress in the future (Kambona *et al.*, 2023). Plants with short-term memory can protect themselves from a second stress but cannot pass this memory to their offspring (Sun *et al.*, 2022). On the other hand, plants that achieve long-term stress memory, including intergenerational memory, can transmit this quality to the first stress-free generation of offspring, and transgenerational memory can be stably inherited by at least two stress-free subsequent generations (Sun *et al.*, 2022). DNA methylation is heritable and significantly affects the long-term stress memory of various plants, including rice, *Arabidopsis* (Wibowo *et al.*, 2016), oilseed rape (Hatzig *et al.*, 2018), soybean (Chen *et al.*, 2019), and maize under different abiotic stresses (Tan, 2010). In upland paddy fields, the persistence of drought-induced DNA methylation across multiple offspring generations demonstrates a transgenerational stress memory (Zheng *et al.*, 2017). If the parental generation of *Polygonum persicaria* is subjected to drought stress, their descendants respond by growing longer roots and achieving higher biomass than the progeny of parents who were adequately watered, all within the same genetic lineage (Herman and Sultan, 2016). Numerous studies have shown that histone alterations play a part in preserving the transcriptional memory of stress-responsive genes. For example, the repressive marker H3K27me3 has been linked to salt stress memory, whereas the active transcription marker H3K4me3 has been shown after chronic dehydration. In one case, the *Arabidopsis* P5CS1 gene demonstrated salt-induced preservation of genetic information linked to the synthesis of proline biosynthetic enzymes (Feng *et al.*, 2016b). Increased proline build-up results from this retention, which acts as an adaptive tactic for salt tolerance (Feng *et al.*, 2016b). After the initial salt stress, P5CS1 expression decreased to a negligible level during the recovery period and elevated to a greater level after another salt stress (Feng *et al.*, 2016b). Because of its function in maintaining H3K4me3 levels at P5CS1, the light-responsive transcription factor HY5 is necessary for long-lasting transcriptional action on P5CS1 (Feng *et al.*, 2016b). Accessibility of sunlight during the recovery phase is essential for this process (Feng *et al.*, 2016b). Other research findings on stomatal clones of grapevines revealed that 40% of epigenetic changes disappeared within a year after the stress conditions were removed (Widiez *et al.*, 2014; Pandey *et al.*, 2016). In the moss *Physcomitrella patens*, the three activating histone marks (H3K4me3, H3K27Ac, and H3K9Ac) show conspicuous alterations when the species progresses from juvenile to adulthood and experiences drought stress. This finding suggests that genes expressed during drought stress in adult gametophores respond to desiccation (Widiez *et al.*, 2014). Transmitting stress-induced epigenetic modifications can be considered a tactic for growing crop types with increased stress tolerance (Pandey *et al.*, 2016).

4.5 Epigenetic Breeding Approaches for Crop Improvement – Role of CRISPR

The present era of genome editing is defined by the binding of nuclease to specific DNA sequences and cutting and repairing specific sequences through non-homologous end joining (Gaj *et al.*, 2013). The use of meganucleases such as ZFNs and TALENs for genome editing is the first-generation genome editing technology. ZFNs, composed of the zinc fingers recognition domain and the FokI restriction enzyme domain, induce genome editing by deletions, insertions, inversions, translocations, and point mutations (Carroll, 2011). TALENs consist of a nuclease that cuts DNA in cells and a TALE domain that is designed to mimic the natural transcription activator-like effector proteins (Joung and Sander, 2013). TALENs are more user-friendly than ZFNs because they have higher binding specificity. TALENs have demonstrated efficacy in plant species such as *Arabidopsis* spp., tobacco, rice, and *Brachypodium* spp. (Cermak *et al.*, 2011; Li *et al.*, 2012). Clustered regularly interspaced short palindromic repeats (CRISPR)/CRISPR-associated protein 9 (Cas9) is the second-generation genome editing technique that is easy to design, less time-consuming, and cost-effective. Recent genome editing technology has allowed breeders to introduce desired characters in crops (Jaganathan *et al.*, 2018); CRISPR/Cas employs Cas9 and guided RNAs (gRNAs) to selectively target numerous locations in the genome. Genome-edited crops surpass transgenic plants by transmitting modified DNA for targeted features (Malzahn *et al.*, 2017; Afzal *et al.*, 2023).

Resistance of plants to stress is regulated by several coding and non-coding DNA stretches involving epigenetic markers. RNA-targeting Cas proteins perform significant functions such as pathogen recognition, RNA degradation, RNA base editing, RNA/DNA detection, and live RNA imaging (Shelake *et al.*, 2019a; Pramanik *et al.*, 2021). The ability of the CRISPR technique to trigger chromosomal rearrangements like translocation, inversion, duplication, or deletion leading to the modification of linkage groups and reformation or disruption of reading frames that control essential agronomical traits is one of the technique's major applications in epigenetic breeding initiatives (Shelake *et al.*, 2019b; Huang and Puchta, 2021).

In the model plant *Arabidopsis*, drought tolerance was achieved by applying CRISPR/Cas9-based techniques such as a mutation in the *Arabidopsis* OST2 gene, knockout of *Arabidopsis* gene miR169a, activation of ABA-responsive element binding gene (AREB1) and suppression of trehalose (TRE1) gene (Osakabe *et al.*, 2016; Zhao *et al.*, 2016; Park *et al.*, 2017; Roca Paixão *et al.*, 2019; Nuñez-Muñoz *et al.*, 2021). The drought tolerance in wheat was enhanced by utilizing CRISPR/Cas9 technology to modify the TaDREB2 and TaERF3 genes. This resulted in the alteration of the SAL1 gene, which is responsible for negatively regulating drought tolerance. Additionally, the SINPR1 gene was deactivated through deletion (Kim *et al.*, 2018; Li *et al.*, 2019a; Abdallah *et al.*, 2022). The salt tolerance of *Oryza sativa* was enhanced by removing the OsbHLH024 gene and activating the expression of ion transporter genes including OsHKT1;3, OsHAK7, and OsSOS1, and altering the OsRAV2 gene using CRISPR/Cas9 technology (Liu *et al.*, 2020; Alam *et al.*, 2022). The mutation in the heat-tolerant BRZ1 gene has shown its function in the generation of reactive oxygen species (ROS) in tomato plants. The same approach was used to create tomato mutants called CRISPR/Cas-mediated heat-stress-sensitive albino 1 (HSA1), which exhibited increased susceptibility to temperature stress (Qiu *et al.*, 2018; Yin *et al.*, 2018). The CRISPR technique was used to modify the thermosensitive male sterile gene in maize, to enhance the production of plants that are vulnerable to temperature-induced male sterility (Li *et al.*, 2017). Another study found that the disruption of the NCED4 gene in lettuce, which is responsible for producing the important regulatory enzyme ABA, resulted in increased seed germination at higher temperatures. Consequently, the LsNCED4 mutants could have practical value in manufacturing settings that experience elevated temperatures (Bertier *et al.*, 2018). Table 4.7 summarizes the successful application of CRISPR/Cas9-based genome editing in plants to improve their tolerance to abiotic stress.

4.6 CRISPR-Based Epigenome Editing Strategies

The concept of genome editing, as introduced by Capecchi in 1980, involves the targeted modification of genetic material at specified locations within the genome (Fig. 4.5) (Capecchi, 1980; Chen *et al.*, 2019; Sedeek *et al.*, 2019). The natural example of genome alteration that existed on the Earth for thousands of years is the present-day modern corn, which is quite different from its wild antecedent (Holmes *et al.*, 2016; Sedeek *et al.*, 2019). Based on onsite-specific endonucleases, plant genome editing techniques have been classified into four major types: (i) meganucleases; (ii) zinc-finger nucleases (ZFNs); (iii) transcription activator-like effector nucleases (TALENs); and (iv) clustered regularly interspaced short palindromic repeats (CRISPR)/CRISPR-associated protein 9 (Cas9) (Adli, 2018; Gao, 2021). The CRISPR system is characterized by its simplicity, versatility, and affordability. Consequently, it has been extensively employed to enhance various characteristics across virtually all biological domains (Bortesi and Fischer, 2015) making it the most usable method by researchers (Scheben *et al.*, 2016; Minkenberg *et al.*, 2017).

The CRISPR/Cas9 system, initially used in *Streptococcus pyogenes* against invading viruses, became popularized as the most extensively accepted and applied system as a genome editing method against abiotic stresses (Jinek *et al.*, 2012; Bortesi and Fischer, 2015). The CRISPR system comprises two components: a single chimeric RNA called guide RNA (gRNA) and the enzyme Cas9. These short guide RNAs identify the foreign genetic materials and recruit the enzymatic Cas proteins, to cut and remove the foreign invading genetic material. The nuclease activity of Cas9 is to cut genomic DNA. This is done by binding through Watson and Crick base pairing and supported by CRISPR RNA (crRNA) and *trans*-acting CRISPR RNA-composed gRNA. Scientists are currently developing a modified version of gRNA that can attach to a specific sequence of 20 nucleotides and induce precise Cas9-mediated cutting of both DNA strands at a predetermined site in the target. This demonstrates the effectiveness of this technology in inhibiting or activating gene activity. The combination of RNA and enzymes allows for the precise modification of targeted genomes, hence facilitating desired alterations in gene engineering (Jinek *et al.*, 2012). However, for the action of Cas endonuclease, protospacer adjacent motif (PAM), a short DNA sequence, is required downstream to the given target site that must be compatible with the associated Cas protein (Deveau *et al.*, 2008; Garneau *et al.*, 2010; Sapranauskas *et al.*, 2011; Gasiunas *et al.*, 2012; Jinek *et al.*, 2012).

CRISPR/Cas9 technology involves five steps, starting with target gene selection followed by the designing of sgRNA specific to the target gene, Cas9 and sgRNA assemblage, targeted plant transformation, and finally regeneration and screening of the plant lines (Erdoğan *et al.*, 2023).

4.7 Conclusion

To conclude, epigenetics provides prospective pathways for improving abiotic stress tolerance in plants, as indicated by the recent advances and approaches mentioned. The research reviewed demonstrates the dynamic interplay between epigenetic regulation and plant responses to environmental challenges, emphasizing both the complexity and the possibility of targeted modification.

One of the important findings from recent research is the function of DNA methylation and histone changes in modulating stress responses.

Table 4.7. Utilization of the CRISPR-based genome editing technique in plants to enhance abiotic stress tolerance. Author's own table.

No.	Name of the plant	Name of the target gene or site of action	Methodology of transformation	Target trait/ Improved trait	Reference
1	Wheat	TaDREB2 and TaERF3	CRISPR/Cas9-based genome editing.	TaDREB2 and TaERF3 expression increased and caused longer incubation of seedlings under dehydrated conditions.	Kim *et al.*, 2018
2	*Oryza sativa* subsp. *indica*	OsDST gene	Deletion mutation in DST gene by CRISPR/Cas9.	Mutant produced leaves with broader width and reduced stomatal density thus enhancing leaf water retention capacity under drought stress.	Santosh Kumar *et al.*, 2020
3	*Arabidopsis thaliana*	OST2	CRISPR/Cas9 generates new alleles for OST2, a proton pump in *Arabidopsis*.	OST2 increased stomatal response.	Osakabe *et al.*, 2016
4	Papaya	CpDreb2-type gene	Disruption of gene	Elevated drought tolerance	Arroyo-Herrera *et al.*, 2016
5	*Solanum lycopersicum*	SlNPR1	Generated slnpr1 mutants by CRISPR/Cas9.	SlNPR1 mutants are weaker than of wild plants in respect to drought tolerance.	Li *et al.*, 2019b
6	*Arabidopsis thaliana*	AtWRKY, AtWRKY4	Loss of function of AtWRKY, AtWRKY4 genes by CRISPR/Cas9.	AtWRKY3 and AtWRKY4 genes upregulated under salt and Me-JA stresses.	Li *et al.*, 2021
7	*Glycine max*	GmAITR	CRISPR/Cas9 induced mutation of GmAITR genes.	Greater salinity tolerance in soybean.	Wang *et al.*, 2021b
8	*O. sativa*	OsRAV2	OsRAV2 gene mutation by CRISPR/Cas9.	GT-1 element directly controls the response of salt of OsRAV2.	Duan *et al.*, 2016

Continued

Table 4.7. Continued

No.	Name of the plant	Name of the target gene or site of action	Methodology of transformation	Target trait/ Improved trait	Reference
9	*Solanum tuberosum*	Coilin	CRISPR/Cas9-based genome editing.	Can tolerate osmotic and salt stress; also develop resistance against potato virus Y	Makhotenko *et al.*, 2019
10	*S. lycopersicum*	SlHyPRP1	Elimination of SlHyPRP1 domain(s).	Can tolerate high salinity during seed germination and vegetative growth stages	Tran *et al.*, 2021
11	*Lactuca sativa*	NCED4	CRISPR/ Cas9-induced modifications of NCED4.	Germination was possible at 37°C.	Bertier *et al.*, 2018
12	*O. sativa*	OsPDS	Induction of heat shock protein by CRISPR/Cas9.	Heat-resistant seedlings.	Nandy *et al.*, 2019
13	*S. lycopersicum*	SlAGL6	Knockout of SlAGL6 gene by CRISPR/Cas9	Parthenocarpic fruit production under heat-stress.	Klap *et al.*, 2017
14	*O. sativa*	OsAnn3	CRISPR/Cas9-based knockout of OsAnn3	Cold tolerance in rice	Shen *et al.*, 2017
15	*O. sativa*	GS3, OsMYB30, OsPIN5b	CRISPR/Cas9 based editing of OsPIN5b, GS3, OsMYB30	Increased yield of rice and cold tolerance.	Zeng *et al.*, 2020
16	*S. lycopersicum*	SlCBF1	slcbf1 mutants applying CRISPR/ Cas9	Tomato plants were able to tolerate cold stress.	Li *et al.*, 2018

These epigenetic marks not only regulate gene expression but also influence the heritability of stress memory over generations. Understanding the processes by which stress-induced epigenetic alterations are inherited allows researchers to create techniques for inducing and stabilizing positive epigenetic modifications in crops.

The introduction of high-throughput sequencing technologies has transformed epigenetic research by allowing genome-wide characterization of DNA methylation patterns and histone changes under stress situations. Integrative omics techniques have revealed unparalleled insights into the biochemical pathways and regulatory networks that are influenced by epigenetic mechanisms during stress adaptation. Such comprehensive datasets are useful for identifying major epigenetic regulators and prospective genetic engineering targets.

Furthermore, the concept of epigenetic priming represents a promising technique for improving crop stress tolerance. Researchers can create adaptive epigenetic modifications in plants by exposing them to mild stress conditions beforehand, preparing them for subsequent severe stress events. This priming effect improves both immediate stress responses and

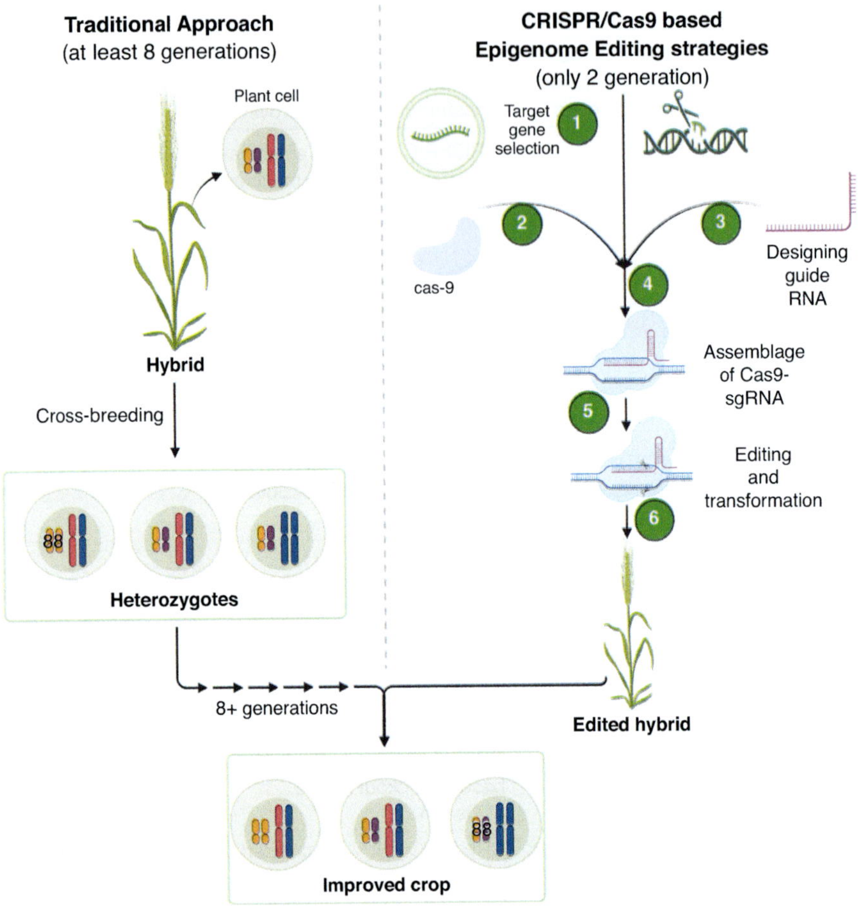

Fig. 4.5. Traditional approach *vis-à-vis* CRISPR-based genome editing strategies for plant improvement. Step 1. Selection of desired gene; 2. Recruitment of Cas9; 3. Selection/designing of guide RNA (gRNA); 4. Assembly of Cas9 and gRNA; 5. Editing; 6. Transformation of plant species.

transgenerational stress resilience, providing a long-term strategy for crop development.

In addition to DNA methylation and histone modifications, non-coding RNAs (ncRNAs) have emerged as important participants in the epigenetic regulation of stress responses. Small RNAs, including miRNAs and siRNAs, have a role in post-transcriptional gene silencing and are thought to regulate stress-responsive gene expression. Understanding the involvement of ncRNAs in stress adaptation may open up new options for developing RNA-based biotechnological therapies to alleviate abiotic stress effects on crops.

Translating basic epigenetic research into practical applications remains difficult. Effective delivery strategies for epigenome editing tools, as well as the durable inheritance of induced epigenetic alterations in crop progeny, are crucial issues that require additional investigation. Furthermore, ethical concerns about the long-term impact of epigenetic changes on plant fitness and ecosystem interactions require careful examination.

Significant progress has been achieved in understanding the function of epigenetics in plant stress responses, applying this information to the development of abiotic stress-tolerant crops. Future research should focus on improving epigenome editing technologies, investigating the potential of ncRNAs as biotechnological tools, and addressing regulatory and ethical concerns. By combining epigenetic approaches with standard breeding methods, we can speed up the production of crop varieties that can maintain agricultural productivity in the face of changing environmental conditions. Epigenetics has enormous promise as a revolutionary instrument for ensuring agricultural sustainability in the 21st century and beyond.

References

Abdallah, N.A., Elsharawy, H., Abulela, H.A., Thilmony, R., Abdelhadi, A.A. *et al.* (2022) Multiplex CRISPR/Cas9-mediated genome editing to address drought tolerance in wheat. *GM Crops & Food* 1–17. DOI: 10.1080/21645698.2022.2120313.

Abdulraheem, M.I., Xiong, Y., Moshood, A.Y., Cadenas-Pliego, G., Zhang, H. *et al.* (2024) Mechanisms of plant epigenetic regulation in response to plant stress: Recent discoveries and implications. *Plants (Basel)* 13(2), 163. DOI: 10.3390/plants13020163.

Ackah, M., Guo, L., Li, S., Jin, X., Asakiya, C. *et al.* (2022) DNA methylation changes and its associated genes in mulberry (*Morus alba* L.) yu-711 response to drought stress using methylrad sequencing. *Plants* 11, 190. DOI: 10.3390/plants11020190.

Adel, S. and Carels, N. (2023) Plant tolerance to drought stress with emphasis on wheat. *Plants (Basel)* 12, 2170. DOI: 10.3390/plants12112170.

Adli, M. (2018) The CRISPR tool kit for genome editing and beyond. *Nature Communications* 9, 1911. DOI: 10.1038/s41467-018-04252-2.

Afzal, M., Hindawi, S.E.S., Alghamdi, S.S., Migdadi, H.H., Khan, M.A. *et al.* (2023) Potential breeding strategies for improving salt tolerance in crop plants. *Journal of Plant Growth Regulation* 42, 3365–3387. DOI: 10.1007/s00344-022-10797-w.

Akhter, Z., Bi, Z., Ali, K., Sun, C., Fiaz, S. *et al.* (2021) In response to abiotic stress, DNA methylation confers epigenetic changes in plants. *Plants (Basel)* 10, 1096. DOI: 10.3390/plants10061096.

Al Akeel, R. (2013) Role of epigenetic reprogramming of host genes in bacterial pathogenesis. *Saudi Journal of Biological Sciences* 20, 305–309. DOI: 10.1016/j.sjbs.2013.05.003.

Alam, M.S., Kong, J., Tao, R., Ahmed, T., Alamin, M. *et al.* (2022) CRISPR/Cas9 mediated knockout of the *OsBHLHH024* transcription factor improves salt stress resistance in rice (*Oryza sativa* L). *Plants (Basel)* 11, 1184. DOI: 10.3390/plants11091184.

Al-Lawati, A., Al-Bahry, S., Victor, R., Al-Lawati, A.H. and Yaish, M.W. (2016) Salt stress alters DNA methylation levels in alfalfa (*Medicago* spp.). *Genetics and Molecular Research* 15. DOI: 10.4238/gmr.15018299.

Allis, C.D. and Jenuwein, T. (2016) The molecular hallmarks of epigenetic control. *Nature Reviews. Genetics* 17, 487–500. DOI: 10.1038/nrg.2016.59.

Ambrosi, C., Manzo, M. and Baubec, T. (2017) Dynamics and context-dependent roles of DNA methylation. *Journal of Molecular Biology* 429, 1459–1475. DOI: 10.1016/j.jmb.2017.02.008.

Arroyo-Herrera, A., Figueroa-Yáñez, L., Castaño, E., Santamaría, J., Pereira-Santana, A. *et al.* (2016) A novel Dreb2-type gene from *Carica papaya* confers tolerance under abiotic stress. *Plant Cell, Tissue and Organ Culture* 125, 119–133. DOI: 10.1007/s11240-015-0934-9.

Ashapkin, V.V., Kutueva, L.I., Aleksandrushkina, N.I. and Vanyushin, B.F. (2020) Epigenetic mechanisms of plant adaptation to biotic and abiotic stresses. *International Journal of Molecular Sciences* 21, 7457. DOI: 10.3390/ijms21207457.

Asiminicesei, D.M., Fertu, D.I. and Gavrilescu, M. (2024) Impact of heavy metal pollution in the environment on the metabolic profile of medicinal plants and their therapeutic potential. *Plants* 13, 913. DOI: 10.3390/plants13060913.

Azahar, I., Ghosh, S., Adhikari, A., Adhikari, S., Roy, D. *et al.* (2020) Comparative analysis of maize root sRNA transcriptome unveils the regulatory roles of miRNAs in submergence stress response mechanism. *Environmental and Experimental Botany* 171, 103924. DOI: 10.1016/j.envexpbot.2019.103924.

Baek, D., Jiang, J., Chung, J.-S., Wang, B., Chen, J. *et al.* (2010) Regulated *AtHKT1* gene expression by a distal enhancer element and DNA methylation in the promoter plays an important role in salt tolerance. *Plant and Cell Physiology* 52, 149–161. DOI: 10.1093/pcp/pcq182.

Bai, Q., Wang, X., Chen, X., Shi, G., Liu, Z. *et al.* (2018) Wheat miRNA TaemiR408 acts as an essential mediator in plant tolerance to Pi deprivation and salt stress via modulating stress-associated physiological processes. *Frontiers in Plant Science* 9, 499. DOI: 10.3389/fpls.2018.00499.

Balasubramaniam, T., Shen, G., Esmaeili, N. and Zhang, H. (2023) Plants' response mechanisms to salinity stress. *Plants* 12, 2253. DOI: 10.3390/plants12122253.

Ball, K. and Sadhukhan, S. (2023) Epigenetics – the molecular tool in understanding abiotic stress response in plants. In: Roychoudhury, A. (ed.) *Biology and Biotechnology of Environmental Stress Tolerance in Plants*. Apple Academic Press, New York, pp. 349–400. DOI: 10.1201/9781003346401-15.

Banerjee, A. and Roychoudhury, A. (2019) The regulatory signaling of gibberellin metabolism and its crosstalk with phytohormones in response to plant abiotic stresses. *Plant Signaling Molecules* 333–339. DOI: 10.1016/B978-0-12-816451-8.00020-4.

Banerjee, S., Sirohi, A., Ansari, A.A. and Gill, S.S. (2017) Role of small RNAs in abiotic stress responses in plants. *Plant Gene* 11, 180–189. DOI: 10.1016/j.plgene.2017.04.005.

Bannister, A.J. and Kouzarides, T. (2011) Regulation of chromatin by histone modifications. *Cell Research* 21, 381–395. DOI: 10.1038/cr.2011.22.

Barros-Silva, D., Marques, C.J., Henrique, R. and Jerónimo, C. (2018) Profiling DNA methylation based on next-generation sequencing approaches: New insights and clinical applications. *Genes* 9, 429. DOI: 10.3390/genes9090429.

Bartee, L., Malagnac, F. and Bender, J. (2001) *Arabidopsis* cmt3 chromomethylase mutations block non-CG methylation and silencing of an endogenous gene. *Genes & Development* 15, 1753–1758. DOI: 10.1101/gad.905701.

Bazakos, C., Manioudaki, M.E., Sarropoulou, E., Spano, T. and Kalaitzis, P. (2015) 454 pyrosequencing of olive (*Olea europaea* L.) transcriptome in response to salinity. *PloS ONE* 10, e0143000. DOI: 10.1371/journal.pone.0143000.

Bellacosa, A. and Drohat, A.C. (2015) Role of base excision repair in maintaining the genetic and epigenetic integrity of CpG sites. *DNA Repair* 32, 33–42. DOI: 10.1016/j.dnarep.2015.04.011.

Berney, M. and McGouran, J.F. (2018) Methods for detection of cytosine and thymine modifications in DNA. *Nature Reviews Chemistry* 2, 332–348. DOI: 10.1038/s41570-018-0044-4.

Bertier, L.D., Ron, M., Huo, H., Bradford, K.J., Britt, A.B. *et al.* (2018) High-resolution analysis of the efficiency, heritability, and editing outcomes of CRISPR/Cas9-induced modifications of *NCED4* in lettuce (*Lactuca sativa*). *Genetics* 8, 1513–1521. DOI: 10.1534/g3.117.300396.

Bhadouriya, S.L., Mehrotra, S., Basantani, M.K., Loake, G.J. and Mehrotra, R. (2021) Role of chromatin architecture in plant stress responses: An update. *Frontiers in Plant Science* 11. DOI: 10.3389/fpls.2020.603380.

Bharti, P., Mahajan, M., Vishwakarma, A.K., Bhardwaj, J. and Yadav, S.K. (2015) AtROS1 overexpression provides evidence for epigenetic regulation of genes encoding enzymes of flavonoid biosynthesis and antioxidant pathways during salt stress in transgenic tobacco. *Journal of Experimental Botany* 66, 5959–5969. DOI: 10.1093/jxb/erv304.

Bilichak, A., Ilnystkyy, Y., Hollunder, J. and Kovalchuk, I. (2012) The progeny of *Arabidopsis thaliana* plants exposed to salt exhibit changes in DNA methylation, histone modifications and gene expression. *PloS ONE* 7, e30515. DOI: 10.1371/journal.pone.0030515.

Borges, F., Donoghue, M.T.A., LeBlanc, C., Wear, E.E., Tanurdžić, M. *et al.* (2021) Loss of small-RNA-directed DNA methylation in the plant cell cycle promotes germline reprogramming and somaclonal variation. *Current Biology* 31, 591–600. DOI: 10.1016/j.cub.2020.10.098.

Borsani, O., Zhu, J., Verslues, P.E., Sunkar, R. and Zhu, J.-K. (2005) Endogenous siRNAs derived from a pair of natural *cis*-antisense transcripts regulate salt tolerance in *Arabidopsis*. *Cell* 123, 1279–1291. DOI: 10.1016/j.cell.2005.11.035.

Bortesi, L. and Fischer, R. (2015) The CRISPR/Cas9 system for plant genome editing and beyond. *Biotechnology Advances* 33, 41–52. DOI: 10.1016/j.biotechadv.2014.12.006.

Boycheva, I., Vassileva, V. and Iantcheva, A. (2014) Histone acetyltransferases in plant development and plasticity. *Current Genomics* 15, 28–37. DOI: 10.2174/138920291501140306112742.

Boyko, A. and Kovalchuk, I. (2008) Epigenetic control of plant stress response. *Environmental and Molecular Mutagenesis* 49, 61–72. DOI: 10.1002/em.20347.

Buschmann, D., Haberberger, A., Kirchner, B., Spornraft, M., Riedmaier, I. *et al.* (2016) Toward reliable biomarker signatures in the age of liquid biopsies – how to standardize the small RNA-seq workflow. *Nucleic Acids Research* 44, 5995–6018. DOI: 10.1093/nar/gkw545.

Cao, D., Gao, X., Liu, J., Wang, X., Geng, S. *et al.* (2012) Root-specific DNA methylation in *Chloris virgata*, a natural alkaline-resistant halophyte, in response to salt and alkaline stresses. *Plant Molecular Biology Reporter* 30, 1102–1109. DOI: 10.1007/s11105-012-0420-z.

Capecchi, M.R. (1980) High efficiency transformation by direct microinjection of DNA into cultured mammalian cells. *Cell* 22, 479–488. DOI: 10.1016/0092-8674(80)90358-x.

Carnavale Bottino, M., Rosario, S., Grativol, C., Thiebaut, F., Rojas, C.A. *et al.* (2013) High-throughput sequencing of small RNA transcriptome reveals salt stress regulated microRNAs in sugarcane. *PLoS ONE* 8(3), e59423. DOI: 10.1371/journal.pone.0059423.

Carroll, D. (2011) Genome engineering with zinc-finger nucleases. *Genetics* 188, 773–782. DOI: 10.1534/genetics.111.131433.

Cermak, T., Doyle, E.L., Christian, M., Wang, L., Zhang, Y. *et al.* (2011) Efficient design and assembly of custom TALEN and other TAL effector-based constructs for DNA targeting. *Nucleic Acids Research* 39, e82–e82. DOI: 10.1093/nar/gkr218.

Chaudhry, S. and Sidhu, G.P.S. (2022) Climate change regulated abiotic stress mechanisms in plants: A comprehensive review. *Plant Cell Reports* 41, 1–31. DOI: 10.1007/s00299-021-02759-5.

Chen, K., Wang, Y., Zhang, R., Zhang, H. and Gao, C. (2019) CRISPR/Cas genome editing and precision plant breeding in agriculture. *Annual Review of Plant Biology* 70, 667–697. DOI: 10.1146/annurev-arplant-050718-100049.

Chinnusamy, V. and Zhu, J.-K. (2009) Epigenetic regulation of stress responses in plants. *Current Opinion in Plant Biology* 12, 133–139. DOI: 10.1016/j.pbi.2008.12.006.

Chiou, T.-J., Aung, K., Lin, S.-I., Wu, C.-C., Chiang, S.-F. *et al.* (2006) Regulation of phosphate homeostasis by microRNA in *Arabidopsis*. *The Plant Cell* 18, 412–421. DOI: 10.1105/tpc.105.038943.

Choi, K., Zhao, X., Kelly, K.A., Venn, O., Higgins, J.D. *et al.* (2013) *Arabidopsis meiotic* crossover hot spots overlap with H2A.Z nucleosomes at gene promoters. *Nature Genetics* 45, 1327–1336. DOI: 10.1038/ng.2766.

Christova, R. (2013) Detecting DNA–protein interactions in living cells – ChIP approach. *Protein-Nucleic Acids Interactions* 101–133. DOI: 10.1016/B978-0-12-411637-5.00004-4.

Ci, D., Song, Y., Tian, M. and Zhang, D. (2015) Methylation of miRNA genes in the response to temperature stress in *Populus simonii*. *Frontiers in Plant Science* 6, 921. DOI: 10.3389/fpls.2015.00921.

Cicatelli, A., Todeschini, V., Lingua, G., Biondi, S., Torrigiani, P. *et al.* (2013) Epigenetic control of heavy metal stress response in mycorrhizal versus non-mycorrhizal poplar plants. *Environmental Science and Pollution Research* 21, 1723–1737. DOI: 10.1007/s11356-013-2072-4.

Clark, S.J., Statham, A., Stirzaker, C., Molloy, P.L. and Frommer, M. (2006) DNA methylation: Bisulphite modification and analysis. *Nature Protocols* 1, 2353–2364. DOI: 10.1038/nprot.2006.324.

Clark, S.J., Smallwood, S.A., Lee, H.J., Krueger, F., Reik, W. *et al.* (2017) Genome-wide base-resolution mapping of DNA methylation in single cells using single-cell bisulfite sequencing (scbs-seq). *Nature Protocols* 12, 534–547. DOI: 10.1038/nprot.2016.187.

Collas, P. (2010) The current state of chromatin immunoprecipitation. *Molecular Biotechnology* 45, 87–100. DOI: 10.1007/s12033-009-9239-8.

Cotrina Cabello, G.G., Ruiz Rodriguez, A., Husnain Gondal, A., Areche, F.O., Flores, D.D.C. *et al.* (2023) Plant adaptability to climate change and drought stress for crop growth and production. *CABI Reviews*. DOI: 10.1079/cabireviews.2023.0004.

Deveau, H., Barrangou, R., Garneau, J.E., Labonté, J., Fremaux, C. *et al.* (2008) phage response to CRISPR-encoded resistance in *Streptococcus thermophilus*. *Journal of Bacteriology* 190, 1390–1400. DOI: 10.1128/JB.01412-07.

Devireddy, A.R., Tschaplinski, T.J., Tuskan, G.A., Muchero, W. and Chen, J.-G. (2021) Role of reactive oxygen species and hormones in plant responses to temperature changes. *International Journal of Molecular Sciences* 22, 8843. DOI: 10.3390/ijms22168843.

Dhawan, G., Kumar, A., Dwivedi, P., Gopala Krishnan, S., Pal, M. *et al.* (2021) Introgression of qdty1.1 governing reproductive stage drought tolerance into an elite basmati rice variety “Pusa basmati 1” through marker assisted backcross breeding. *Agronomy* 11, 202. DOI: 10.3390/agronomy11020202.

Ding, H., Gao, J., Qin, C., Ma, H., Huang, H. *et al.* (2014) The dynamics of DNA methylation in maize roots under Pb stress. *International Journal of Molecular Sciences* 15, 23537–23554. DOI: 10.3390/ijms151223537.

Duan, H., Li, J., Zhu, Y., Jia, W., Wang, H. *et al.* (2020) Responsive changes of DNA methylation in wheat (*Triticum aestivum*) under water deficit. *Scientific Reports* 10. DOI: 10.1038/s41598-020-64660-7.

Duan, Y.B., Li, J., Qin, R.Y., Xu, R.F., Li, H. *et al.* (2016) Identification of a regulatory element responsible for salt induction of rice *OsRAV2* through *ex situ* and *in situ* promoter analysis. *Plant Molecular Biology* 90(1–2), 49–62. DOI: 10.1007/s11103-015-0393-z.

Dutta, S., Kumar, D., Jha, S., Prabhu, K.V., Kumar, M. *et al.* (2017a) Identification and molecular characterization of a trans-acting small interfering RNA producing locus regulating leaf rust responsive gene expression in wheat (*Triticum aestivum* L.). *Planta* 246, 939–957. DOI: 10.1007/s00425-017-2744-2.

Dutta, S., Kumar, D., Jha, S., Prabhu, K.V., Kumar, M. *et al.* (2017b) Identification and molecular characterization of a trans-acting small interfering RNA producing locus regulating leaf rust responsive gene expression in wheat (*Triticum aestivum* L.). *Planta* 246, 939–957. DOI: 10.1007/s00425-017-2744-2.

Dyachenko, O.V., Zakharchenko, N.S., Shevchuk, T.V., Bohnert, H.J., Cushman, J.C. *et al.* (2006) Effect of hypermethylation of CCWGG sequences in DNA of *Mesembryanthemum crystallinum* plants on their adaptation to salt stress. *Biochemistry (Moscow)* 71, 461–465. DOI: 10.1134/S000629790604016X.

Elrouby, N. (2017) Regulation of plant cellular and organismal development by SUMO 227–247. DOI: 10.1007/978-3-319-50044-7_14.

Eom, S.H. and Hyun, T.K. (2018) Histone acetyltransferases (hats) in Chinese cabbage: Insights from histone H3 acetylation and expression profiling of hats in response to abiotic stresses. *Journal of the American Society for Horticultural Science* 143, 296–303. DOI: 10.21273/JASHS04436-18.

Erdoğan, İ., Cevher-Keskin, B., Bilir, Ö., Hong, Y. and Tör, M. (2023) Recent developments in CRISPR/Cas9 genome-editing technology related to plant disease resistance and abiotic tress tolerance. *Biology* 12, 1037. DOI: 10.3390/biology12071037.

Fan, H.H., Wei, J., Li, T.C., Li, Z.P., Guo, N. *et al.* (2013) DNA methylation alterations of upland cotton (*Gossypium hirsutum*) in response to cold stress. *Acta Physiologiae Plantarum* 35, 2445–2453. DOI: 10.1007/s11738-013-1278-x.

Fan, S., Liu, H., Liu, J., Hua, W., Xu, S. *et al.* (2020) Systematic analysis of the DNA methylase and demethylase gene families in rapeseed (*Brassica napus* L.) and their expression variations after salt and heat stresses. *International Journal of Molecular Sciences* 21, 953. DOI: 10.3390/ijms21030953.

Farhan, M., Sathish, M., Kiran, R., Mushtaq, A., Baazeem, A. *et al.* (2024) Plant nitrogen metabolism: Balancing resilience to nutritional stress and abiotic challenges. *Phyton (B Aires)* 93, 581–609. DOI: 10.32604/phyton.2024.046857.

Feng, S.J., Liu, X.S., Tao, H., Tan, S.K., Chu, S.S. *et al.* (2016a) Variation of DNA methylation patterns associated with gene expression in rice (*Oryza sativa*) exposed to cadmium. *Plant, Cell and Environment* 39, 2629–2649. DOI: 10.1111/pce.12793.

Feng, X.J., Li, J.R., Qi, S.L., Lin, Q.F., Jin, J.B. *et al.* (2016b) Light affects salt stress-induced transcriptional memory of P5CS1 in *Arabidopsis*. *Proceedings of the National Academy of Sciences of the United States of America* 113, E8335–8343. DOI: 10.1073/pnas.1610670114.

Ferreira, L.J., Azevedo, V., Maroco, J., Oliveira, M.M. and Santos, A.P. (2015) Salt tolerant and sensitive rice varieties display differential methylome flexibility under salt stress. *PloS ONE* 10, e0124060. DOI: 10.1371/journal.pone.0124060.

Filek, M., Keskinen, R., Hartikainen, H., Szarejko, I., Janiak, A. *et al.* (2008) The protective role of selenium in rape seedlings subjected to cadmium stress. *Journal of Plant Physiology* 165, 833–844. DOI: 10.1016/j.jplph.2007.06.006.

Fu, R., Zhang, M., Zhao, Y., He, X., Ding, C. *et al.* (2017) Identification of salt tolerance-related microRNAs and their targets in maize (*Zea mays* l.) using high-throughput sequencing and degradome analysis. *Frontiers in Plant Science* 8, 864. DOI: 10.3389/fpls.2017.00864.

Fujii, H., Chiou, T.-J., Lin, S.-I., Aung, K. and Zhu, J.-K. (2005) A miRNA involved in phosphate-starvation response in *Arabidopsis*. *Current Biology* 15, 2038–2043. DOI: 10.1016/j.cub.2005.10.016.

Gaj, T., Gersbach, C.A. and Barbas, C.F. (2013) ZFN, TALEN, and CRISPR/Cas-based methods for genome engineering. *Trends in Biotechnology* 31, 397–405. DOI: 10.1016/j.tibtech.2013.04.004.

Gallusci, P., Agius, D.R., Moschou, P.N., Dobránszki, J., Kaiserli, E. *et al.* (2023) Deep inside the epigenetic memories of stressed plants. *Trends in Plant Science* 28, 142–153. DOI: 10.1016/j.tplants.2022.09.004.

Gao, C. (2021) Genome engineering for crop improvement and future agriculture. *Cell* 184, 1621–1635. DOI: 10.1016/j.cell.2021.01.005.

Gao, G., Li, J., Li, H., Li, F., Xu, K. *et al.* (2014) Comparison of the heat stress induced variations in DNA methylation between heat-tolerant and heat-sensitive rapeseed seedlings. *Breeding Science* 64, 125–133. DOI: 10.1270/jsbbs.64.125.

Garg, R., Narayana Chevala, V., Shankar, R. and Jain, M. (2015) Divergent DNA methylation patterns associated with gene expression in rice cultivars with contrasting drought and salinity stress response. *Scientific Reports* 5, 14922. DOI: 10.1038/srep14922.

Garneau, J.E., Dupuis, M.-È., Villion, M., Romero, D.A., Barrangou, R. *et al.* (2010) The CRISPR/Cas bacterial immune system cleaves bacteriophage and plasmid DNA. *Nature* 468, 67–71. DOI: 10.1038/nature09523.

Gasiunas, G., Barrangou, R., Horvath, P. and Siksnys, V. (2012) Cas9-crrna ribonucleoprotein complex mediates specific DNA cleavage for adaptive immunity in bacteria. *Proceedings of the National Academy of Sciences of the United States of America* 109. DOI: 10.1073/pnas.1208507109.

Gayacharan and Joel, A.J. (2013) Epigenetic responses to drought stress in rice (*Oryza sativa* L.). *Physiology and Molecular Biology of Plants* 19, 379–387. DOI: 10.1007/s12298-013-0176-4.

Gharat, S.A. and Shaw, B.P. (2015) Novel and conserved miRNAs in the halophyte *Suaeda maritima* identified by deep sequencing and computational predictions using the ESTs of two mangrove plants. *BMC Plant Biology* 15, 301. DOI: 10.1186/s12870-015-0682-3.

González, R.M., Ricardi, M.M. and Iusem, N.D. (2013) Epigenetic marks in an adaptive water stress-responsive gene in tomato roots under normal and drought conditions. *Epigenetics* 8, 864–872. DOI: 10.4161/epi.25524.

Goswami, K., Tripathi, A., Gautam, B. and Sanan-Mishra, N. (2019) Impact of next-generation sequencing in elucidating the role of microRNA related to multiple abiotic stresses. *Molecular Plant Abiotic Stress* 389–426. DOI: 10.1002/9781119463665.

Gouil, Q. and Keniry, A. (2019) Latest techniques to study DNA methylation. *Essays in Biochemistry* 63, 639–648. DOI: 10.1042/EBC20190027.

Greco, M., Chiappetta, A., Bruno, L. and Bitonti, M.B. (2012) In *Posidonia oceanica* cadmium induces changes in DNA methylation and chromatin patterning. *Journal of Experimental Botany* 63, 695–709. DOI: 10.1093/jxb/err313.

Guan, Q., Lu, X., Zeng, H., Zhang, Y. and Zhu, J. (2013) Heat stress induction of miR398 triggers a regulatory loop that is critical for thermotolerance in *Arabidopsis*. *The Plant Journal* 74, 840–851. DOI: 10.1111/tpj.12169.

Guerra, D., Crosatti, C., Khoshro, H.H., Mastrangelo, A.M., Mica, E. *et al.* (2015) Post-transcriptional and post-translational regulations of drought and heat response in plants: A spider's web of mechanisms. *Frontiers in Plant Science* 6. DOI: 10.3389/fpls.2015.00057.

Gullì, M., Marchi, L., Fragni, R., Buschini, A. and Visioli, G. (2018) Epigenetic modifications preserve the hyperaccumulator *Noccaea caerulescens* from Ni geno-toxicity. *Environmental and Molecular Mutagenesis* 59, 464–475. DOI: 10.1002/em.22191.

Gutschker, S., Corral, J.M., Schmiedl, A., Ludewig, F., Koch, W. *et al.* (2022) Multi-omics data integration reveals link between epigenetic modifications and gene expression in sugar beet (*Beta vulgaris* subsp. *vulgaris*) in response to cold. *BMC Genomics* 23, 144. DOI: 10.1186/s12864-022-08312-2.

Hafner, M., Renwick, N., Farazi, T.A., Mihailović, A., Pena, J.T.G. *et al.* (2012) Barcoded cDNA library preparation for small RNA profiling by next-generation sequencing. *Methods* 58, 164–170. DOI: 10.1016/j.ymeth.2012.07.030.

Harrison, A. and Parle-McDermott, A. (2011) DNA methylation: A timeline of methods and applications. *Frontiers in Genetics* 2, 74. DOI: 10.3389/fgene.2011.00074.

Hasanuzzaman, M., Nahar, K. and Fujit, M. (2013) Extreme temperature responses, oxidative stress and antioxidant defense in plants. In: Vahdati, K. and Leslie, C. (eds) *Abiotic Stress – Plant Responses and Applications in Agriculture*. InTech. Rijeka, Croatia, pp. 169–205. DOI: 10.5772/54833.

Hatzig, S.V., Nuppenau, J.-N., Snowdon, R.J. and Schießl, S.V. (2018) Drought stress has transgenerational effects on seeds and seedlings in winter oilseed rape (*Brassica napus* L). *BMC Plant Biology* 18, 297. DOI: 10.1186/s12870-018-1531-y.

Hayashi, Y., Takehira, K., Nozawa, K., Suzuki, T., Masuta, Y. *et al.* (2020) Onsen shows different transposition activities in RDDM pathway mutants. *Genes and Genetic Systems* 95, 183–190. DOI: 10.1266/ggs.20-00019.

Hayatsu, H., Shiraishi, M. and Negishi, K. (2008) Bisulfite modification for analysis of DNA methylation. *Current Protocols in Nucleic Acid Chemistry* 33. DOI: 10.1002/0471142700.nc0610s33.

Herman, J.J. and Sultan, S.E. (2016) DNA methylation mediates genetic variation for adaptive transgenerational plasticity. *Proceedings. Biological Sciences* 283, 20160988. DOI: 10.1098/rspb.2016.0988.

Hodges, E., Smith, A.D., Kendall, J., Xuan, Z., Ravi, K. *et al.* (2009) High definition profiling of mammalian DNA methylation by array capture and single molecule bisulfite sequencing. *Genome Research* 19, 1593–1605. DOI: 10.1101/gr.095190.109.

Holmes, M.W., Hammond, T.T., Wogan, G.O.U., Walsh, R.E., LaBarbera, K. *et al.* (2016) Natural history collections as windows on evolutionary processes. *Molecular Ecology* 25, 864–881. DOI: 10.1111/mec.13529.

Hossain, M.S., Kawakatsu, T., Kim, K.D., Zhang, N., Nguyen, C.T. *et al.* (2017) Divergent cytosine DNA methylation patterns in single-cell, soybean root hairs. *New Phytologist* 214(2), 808–819. DOI: 10.1111/nph.14421.

Hou, J., Ren, R., Xiao, H., Chen, Z., Yu, J. *et al.* (2021) Characteristic and evolution of HAT and HDAC genes in gramineae genomes and their expression analysis under diverse stress in *Oryza sativa*. *Planta* 253. DOI: 10.1007/s00425-021-03589-1.

Hu, Y., Zhang, L., Zhao, L., Li, J., He, S. *et al.* (2011) Trichostatin A selectively suppresses the cold-induced transcription of the ZmDREB1 gene in maize. *PloS ONE* 6, e22132. DOI: 10.1371/journal.pone.0022132.

Huang, T.-K. and Puchta, H. (2021) Novel CRISPR/Cas applications in plants: From prime editing to chromosome engineering. *Transgenic Research* 30, 529–549. DOI: 10.1007/s11248-021-00238-x.

Iwasaki, M. and Paszkowski, J. (2014) Epigenetic memory in plants. *The EMBO Journal* 33, 1987–1998. DOI: 10.15252/embj.201488883.

Jaganathan, D., Ramasamy, K., Sellamuthu, G., Jayabalan, S. and Venkataraman, G. (2018) CRISPR for crop improvement: An update review. *Frontiers in Plant Science* 17(9), 985. DOI: 10.3389/fpls.2018.00985.

Jinek, M., Chylinski, K., Fonfara, I., Hauer, M., Doudna, J.A. *et al.* (2012) A programmable dual-RNA-guided DNA endonuclease in adaptive bacterial immunity. *Science (1979)* 337, 816–821. DOI: 10.1126/science.1225829.

Jogam, P., Sandhya, D., Alok, A., Peddaboina, V., Allini, V.R. *et al.* (2022) A review on CRISPR/Cas-based epigenetic regulation in plants. *International Journal of Biological Macromolecules* 219, 1261–1271. DOI: 10.1016/j.ijbiomac.2022.08.182.

Jones-Rhoades, M.W. and Bartel, D.P. (2004) Computational identification of plant microRNAs and their targets, including a stress-induced miRNA. *Molecular Cell* 14, 787–799. DOI: 10.1016/j.molcel.2004.05.027.

Joung, J.K. and Sander, J.D. (2013) TALENs: A widely applicable technology for targeted genome editing. *Nature Reviews Molecular Cell Biology* 14, 49–55. DOI: 10.1038/nrm3486.

Kaldis, A., Tsementzi, D., Tanriverdi, O. and Vlachonasios, K.E. (2010) *Arabidopsis thaliana* transcriptional co-activators ADA2b and SGF29a are implicated in salt stress responses. *Planta* 233, 749–762. DOI: 10.1007/s00425-010-1337-0.

Kambona, C.M., Koua, P.A., Léon, J. and Ballvora, A. (2023) Stress memory and its regulation in plants experiencing recurrent drought conditions. *Theoretical and Applied Genetics* 136, 26. DOI: 10.1007/s00122-023-04313-1.

Kim, M. and Costello, J. (2017) DNA methylation: An epigenetic mark of cellular memory. *Experimental and Molecular Medicine* 49, e322. DOI: 10.1038/emm.2017.10.

Kim, D., Alptekin, B. and Budak, H. (2018) CRISPR/Cas9 genome editing in wheat. *Functional & Integrative Genomics* 18, 31–41. DOI: 10.1007/s10142-017-0572-x.

Kim, J.-M., To, T.K., Ishida, J., Morosawa, T., Kawashima, M. *et al.* (2008) Alterations of lysine modifications on the histone H3 N-tail under drought stress conditions in *Arabidopsis thaliana*. *Plant & Cell Physiology* 49, 1580–1588. DOI: 10.1093/pcp/pcn133.

Kim, J.-M., Sasaki, T., Ueda, M., Sako, K. and Seki, M. (2015) Chromatin changes in response to drought, salinity, heat, and cold stresses in plants. *Frontiers in Plant Science* 6. DOI: 10.3389/fpls.2015.00114.

Kimura, H. (2013) Histone modifications for human epigenome analysis. *Journal of Human Genetics* 58, 439–445. DOI: 10.1038/jhg.2013.66.

Kinoshita, T. and Seki, M. (2014) Epigenetic memory for stress response and adaptation in plants. *Plant and Cell Physiology* 55, 1859–1863. DOI: 10.1093/pcp/pcu125.

Klap, C., Yeshayahou, E., Bolger, A.M., Arazi, T., Gupta, S.K. *et al.* (2017) Tomato facultative parthenocarpy results from slagamous-LIKE 6 loss of function. *Plant Biotechnology Journal* 15, 634–647. DOI: 10.1111/pbi.12662.

Kouzarides, T. (2007) Chromatin modifications and their function. *Cell* 128, 693–705. DOI: 10.1016/j.cell.2007.02.005.

Kruszka, K., Pacak, A., Swida-Barteczka, A., Nuc, P., Alaba, S. *et al.* (2014) Transcriptionally and post-transcriptionally regulated microRNAs in heat stress response in barley. *Journal of Experimental Botany* 65, 6123–6135. DOI: 10.1093/jxb/eru353.

Kumar, G., Rattan, U.K. and Singh, A.K. (2016) Chilling-mediated DNA methylation changes during dormancy and its release reveal the importance of epigenetic regulation during winter dormancy in apple (*Malus x Domestica Borkh*). *PLoS ONE* 11, e0149934. DOI: 10.1371/journal.pone.0149934.

Kumar, S. (2018) Epigenomics of plant responses to environmental stress. *Epigenomes* 2, 6. DOI: 10.3390/epigenomes2010006.

Kumar, S., Kaur, S., Seem, K., Kumar, S. and Mohapatra, T. (2021) Understanding 3D genome organization and its effect on transcriptional gene regulation under environmental stress in plant: A chromatin perspective. *Frontiers in Cell and Developmental Biology* 9, 774719. DOI: 10.3389/fcell.2021.774719.

Kumar, S., Seem, K. and Mohapatra, T. (2023) Biochemical and epigenetic modulations under drought: Remembering the stress tolerance mechanism in rice. *Life (Basel)* 13, 1156. DOI: 10.3390/life13051156.

Lake, E.C. and Minteer, C.R. (2017) A review of the integration of classical biological control with other techniques to manage invasive weeds in natural areas and rangelands. *BioControl* 63(1), 71–86. DOI: 10.1007/s10526-017-9853-5.

Lamoliatte, F., Caron, D., Durette, C., Mahrouche, L., Maroui, M.A. *et al.* (2014) Large-scale analysis of lysine sumoylation by SUMO remnant immunoaffinity profiling. *Nature Communications* 5, 5409. DOI: 10.1038/ncomms6409.

Lan, Z., Yan hao, X. and Jian bo, W. (2009) DNA-methylation changes induced by salt stress in wheat *triticum aestivum*. *African Journal of Biotechnology* 8, 6201–6207. DOI: 10.5897/AJB09.1058.

Law, J.A. and Jacobsen, S.E. (2010) Establishing, maintaining and modifying DNA methylation patterns in plants and animals. *Nature Reviews Genetics* 11, 204–220. DOI: 10.1038/nrg2719.

Lee, S., Choi, J., Park, J., Hong, C.P., Choi, D. *et al.* (2023) DDM1-mediated gene body DNA methylation is associated with inducible activation of defense-related genes in *Arabidopsis*. *Genome Biology* 24, 106. DOI: 10.1186/s13059-023-02952-7.

Li, J., Huang, Q., Sun, M., Zhang, T., Li, H. *et al.* (2016) Global DNA methylation variations after short-term heat shock treatment in cultured microspores of *Brassica napus* cv. *Topas. Scientific Reports* 6, 38401. DOI: 10.1038/srep38401.

Li, J., Zhang, H., Si, X., Tian, Y., Chen, K. *et al.* (2017) Generation of thermosensitive male-sterile maize by targeted knockout of the ZmTMS5 gene. *Journal of Genetics and Genomics* 44, 465–468. DOI: 10.1016/j.jgg.2017.02.002.

Li, L., Du, Y., He, C., Dietrich, C.R., Li, J. *et al.* (2019a) Maize glossy6 is involved in cuticular wax deposition and drought tolerance. *Journal of Experimental Botany* 70, 3089–3099. DOI: 10.1093/jxb/erz131.

Li, R., Liu, C., Zhao, R., Wang, L., Chen, L. *et al.* (2019b) CRISPR/Cas9-mediated SlNPR1 mutagenesis reduces tomato plant drought tolerance. *BMC Plant Biology* 19, 38. DOI: 10.1186/s12870-018-1627-4.

Li, R., Zhang, L., Wang, L., Chen, L., Zhao, R. *et al.* (2018) Reduction of tomato-plant chilling tolerance by CRISPR–Cas9-mediated SlCBF1 mutagenesis. *Journal of Agricultural and Food Chemistry* 66, 9042–9051. DOI: 10.1021/acs.jafc.8b02177.

Li, R., Hu, F., Li, B., Zhang, Y., Chen, M. *et al.* (2020) Whole genome bisulfite sequencing methylome analysis of mulberry (*Morus alba*) reveals epigenome modifications in response to drought stress. *Scientific Reports* 10, 8013. DOI: 10.1038/s41598-020-64975-5.

Li, S., He, X., Gao, Y., Zhou, C., Chiang, V.L. *et al.* (2021) Histone acetylation changes in plant response to drought stress. *Genes (Basel)* 12, 1409. DOI: 10.3390/genes12091409.

Li, T., Liu, B., Spalding, M.H., Weeks, D.P. and Yang, B. (2012) High-efficiency TALEN-based gene editing produces disease-resistant rice. *Nature Biotechnology* 30, 390–392. DOI: 10.1038/nbt.2199.

Liang, D., Zhang, Z., Wu, H., Huang, C., Shuai, P. *et al.* (2014) Single-base-resolution methylomes of *Populus trichocarpa* reveal the association between DNA methylation and drought stress. *BMC Genetics* 15, S9–S9. DOI: 10.1186/1471-2156-15-S1-S9.

Liang, P., Song, F., Ghosh, S., Morien, E., Qin, M. *et al.* (2011) Genome-wide survey reveals dynamic widespread tissue-specific changes in DNA methylation during development. *BMC Genomics* 12, 231. DOI: 10.1186/1471-2164-12-231.

Lim, C.W., Baek, W., Jung, J., Kim, J.-H. and Lee, S.C. (2015) Function of ABA in stomatal defense against biotic and drought stresses. *International Journal of Molecular Sciences* 16, 15251–15270. DOI: 10.3390/ijms160715251.

Lister, R., O'Malley, R.C., Tonti-Filippini, J., Gregory, B.D., Berry, C.C. *et al.* (2008) Highly integrated single-base resolution maps of the epigenome in *Arabidopsis*. *Cell* 133, 523–536. DOI: 10.1016/j.cell.2008.03.029.

Liu, Q. and Paroo, Z. (2010) Biochemical principles of small RNA pathways. *Annual Review of Biochemistry* 79(1), 295–319. DOI: 10.1146/annurev.biochem.052208.151733.

Liu, T., Li, Y., Duan, W., Huang, F. and Hou, X. (2017) Cold acclimation alters DNA methylation patterns and confers tolerance to heat and increases growth rate in *Brassica rapa*. *Journal of Experimental Botany* 68, 1213–1224. DOI: 10.1093/jxb/erw496.

Liu, X., Wu, D., Shan, T., Xu, S., Qin, R. *et al.* (2020) The trihelix transcription factor OsGTγ-2 is involved adaption to salt stress in rice. *Plant Molecular Biology* 103, 545–560. DOI: 10.1007/s11103-020-01010-1.

Lobell, D.B. and Gourdji, S.M. (2012) The influence of climate change on global crop productivity. *Plant Physiology* 160, 1686–1697. DOI: 10.1104/pp.112.208298.

Lu, P.Y.T., Lévesque, N. and Kobor, M.S. (2009) NuA4 and SWR1-C: Two chromatin-modifying complexes with overlapping functions and components. *Biochemistry and Cell Biology* 87, 799–815. DOI: 10.1139/O09-062.

Luo, M., Wang, Y.-Y., Liu, X., Yang, S., Lu, Q. *et al.* (2012) HD2C interacts with HDA6 and is involved in ABA and salt stress response in *Arabidopsis*. *Journal of Experimental Botany* 63, 3297–3306. DOI: 10.1093/jxb/ers059.

Ma, Y., Min, L., Wang, M., Wang, C., Zhao, Y. *et al.* (2018) Disrupted genome methylation in response to high temperature has distinct effects on microspore abortion and anther indehiscence. *The Plant Cell* 30, 1387–1403. DOI: 10.1105/tpc.18.00074.

Makhotenko, A.V., Khromov, A.V., Snigir, E.A., Makarova, S.S., Makarov, V.V. *et al.* (2019) Functional analysis of coilin in virus resistance and stress tolerance of potato *Solanum tuberosum* using CRISPR-Cas9 editing. *Doklady Biochemistry and Biophysics* 484, 88–91. DOI: 10.1134/S1607672919010241.

Mallory, A.C. and Vaucheret, H. (2006) Functions of microRNAs and related small RNAs in plants. *Nature Genetics* 38, S31–S36. DOI: 10.1038/ng1791.

Malzahn, A., Lowder, L. and Qi, Y. (2017) Plant genome editing with TALEN and CRISPR. *Cell and Bioscience* 7, 21. DOI: 10.1186/s13578-017-0148-4.

Marconi, G., Pace, R., Traini, A., Raggi, L., Lutts, S. *et al.* (2013) Use of MSAP markers to analyse the effects of salt stress on DNA methylation in rapeseed (*Brassica napus* var. *oleifera*). *PloS ONE* 8, e75597. DOI: 10.1371/journal.pone.0075597.

Marcos, F.C.C., Silveira, N.M., Marchiori, P.E.R., Machado, E.C., Souza, G.M. *et al.* (2018) Drought tolerance of sugarcane propagules is improved when origin material faces water deficit. *PLoS ONE* 13, e0206716. DOI: 10.1371/journal.pone.0206716.

Massie, C.E. and Mills, I.G. (2012) Mapping protein-DNA interactions using chip-sequencing. *Methods in Molecular Biology* 809, 157–173. DOI: 10.1007/978-1-61779-376-9_11.

Meins, F. and Thomas, M. (2003) Meiotic transmission of epigenetic changes in the cell-division factor requirement of plant cells. *Development* 130, 6201–6208. DOI: 10.1242/dev.00856.

Mendizabal, I. and Yi, S.V. (2016) Whole-genome bisulfite sequencing maps from multiple human tissues reveal novel CpG islands associated with tissue-specific regulation. *Human Molecular Genetics* 25(1), 69–82. DOI: 10.1093/hmg/ddv449.

Miller, M.J., Scalf, M., Rytz, T.C., Hubler, S.L., Smith, L.M. *et al.* (2013) Quantitative proteomics reveals factors regulating RNA biology as dynamic targets of stress-induced sumoylation in *Arabidopsis*. *Molecular and Cellular Proteomics* 12, 449–463. DOI: 10.1074/mcp.M112.025056.

Min, L., Li, Y., Hu, Q., Zhu, L., Gao, W. *et al.* (2014) Sugar and auxin signaling pathways respond to high-temperature stress during anther development as revealed by transcript profiling analysis in cotton. *Plant Physiology* 164, 1293–1308. DOI: 10.1104/pp.113.232314.

Milne, T.A., Zhao, K. and Hess, J.L. (2009) Chromatin immunoprecipitation (ChIP) for analysis of histone modifications and chromatin-associated proteins. *Methods in Molecular Biology* 538, 409–423. DOI: 10.1007/978-1-59745-418-6_21.

Minkenberg, B., Wheatley, M. and Yang, Y. (2017) CRISPR/Cas9-enabled multiplex genome editing and its application. *Progress in Molecular Biology and Translational Science* 111–132. DOI: 10.1016/bs.pmbts.2017.05.003.

Mittal, U., Kumar, V., Kukreja, S., Singh, B., Pandey, N.K. *et al.* (2023) Role of beneficial elements in developing resilience to abiotic and biotic stresses in plants: Present status and future prospects. *Journal of Plant Growth Regulation* 42, 3789–3813. DOI: 10.1007/s00344-022-10840-w.

Mizuguchi, G., Shen, X., Landry, J., Wu, W.-H., Sen, S. *et al.* (2004) ATP-driven exchange of histone H2AZ variant catalyzed by SWR1 chromatin remodeling complex. *Science* 303, 343–348. DOI: 10.1126/science.1090701.

Moglia, A., Gianoglio, S., Acquadro, A., Valentino, D., Milani, A.M. *et al.* (2019) Identification of DNA methyltransferases and demethylases in *Solanum melongena* L., and their transcription dynamics during fruit development and after salt and drought stresses. *PLoS ONE* 14. DOI: 10.1371/journal.pone.0223581.

Mohapatra, M.D., Poosapati, S., Sahoo, R.K. and Swain, D.M. (2023) Helicase: A genetic tool for providing stress tolerance in plants. *Plant Stress* 9, 100171. DOI: 10.1016/j.stress.2023.100171.

Moore, L.D., Le, T. and Fan, G. (2013) DNA methylation and its basic function. *Neuropsychopharmacology* 38, 23–38. DOI: 10.1038/npp.2012.112.

Muchate, N.S., Nikalje, G.C., Rajurkar, N.S., Suprasanna, P. and Nikam, T.D. (2016) Plant salt stress: Adaptive responses, tolerance mechanism and bioengineering for salt tolerance. *The Botanical Review* 82, 371–406. DOI: 10.1007/s12229-016-9173-y.

Nandy, S., Pathak, B., Zhao, S. and Srivastava, V. (2019) Heat-shock-inducible CRISPR/Cas9 system generates heritable mutations in rice. *Plant Direct* 3. DOI: 10.1002/pld3.145.

Naydenov, M., Baev, V., Apostolova, E., Gospodinova, N., Sablok, G. *et al.* (2015) High-temperature effect on genes engaged in DNA methylation and affected by DNA methylation in *Arabidopsis*. *Plant Physiology and Biochemistry* 87, 102–108. DOI: 10.1016/j.plaphy.2014.12.022.

Ningombam, L., Hazarika, B.N., Yumkhaibam, T., Heisnam, P. and Singh, Y.D. (2024) Heavy metal priming plant stress tolerance deciphering through physiological, biochemical, molecular and omics mechanism. *South African Journal of Botany* 168, 16–25. DOI: 10.1016/j.sajb.2024.02.032.

Nozawa, K., Masuda, S., Saze, H., Ikeda, Y., Suzuki, T. *et al.* (2022) Epigenetic regulation of ecotype-specific expression of the heat-activated transposon ONSEN. *Frontiers in Plant Science* 899105. DOI: 10.3389/fpls.2022.899105.

Nuñez-Muñoz, L., Vargas-Hernández, B., Hinojosa-Moya, J., Ruiz-Medrano, R. and Xoconostle-Cázares, B. (2021) Plant drought tolerance provided through genome editing of the trehalase gene. *Plant Signaling and Behavior* 16, 1877005. DOI: 10.1080/15592324.2021.1877005.

Ohama, N., Sato, H., Shinozaki, K. and Yamaguchi-Shinozaki, K. (2017) Transcriptional regulatory network of plant heat stress response. *Trends in Plant Science* 22(1), 53–65. DOI: 10.1016/j.tplants.2016.08.015.

Olova, N., Krueger, F., Andrews, S., Oxley, D., Berrens, R.V. *et al.* (2018) Comparison of whole-genome bisulfite sequencing library preparation strategies identifies sources of biases affecting DNA methylation data. *Genome Biology* 19, 33. DOI: 10.1186/s13059-018-1408-2.

Osakabe, Y., Watanabe, T., Sugano, S.S., Ueta, R., Ishihara, R. *et al.* (2016) Optimization of CRISPR/Cas9 genome editing to modify abiotic stress responses in plants. *Scientific Reports* 6, 26685. DOI: 10.1038/srep26685.

Ost, C., Cao, H.X., Nguyen, T.L., Himmelbach, A., Mascher, M. *et al.* (2023) Drought-stress-related reprogramming of gene expression in barley involves differential histone modifications at ABA-related genes. *International Journal of Molecular Sciences* 24. DOI: 10.3390/ijms241512065.

Ou, X., Zhang, Y., Xu, C., Lin, X., Zang, Q. *et al.* (2012) Transgenerational inheritance of modified DNA methylation patterns and enhanced tolerance induced by heavy metal stress in rice (*Oryza sativa* L). *PloS ONE* 7(9), e41143. DOI: 10.1371/journal.pone.0041143.

Pajares, M.J., Palanca-Ballester, C., Urtasun, R., Alemany-Cosme, E., Lahoz, A. *et al.* (2021) Methods for analysis of specific DNA methylation status. *Methods* 187, 3–12. DOI: 10.1016/j.ymeth.2020.06.021.

Pandey, V. and Singh, S. (2024) Plant adaptation and tolerance to heat stress: Advance approaches and future aspects. *Combinatorial Chemistry and High Throughput Screening* 27, 1701–1715. DOI: 10.2174/0113862073300371240229100613.

Pandey, G., Sharma, N., Sahu, P.P. and Prasad, M. (2016) Chromatin-based epigenetic regulation of plant abiotic stress response. *Current Genomics* 17, 490–498. DOI: 10.2174/1389202917666160520103914.

Park, J.-J., Dempewolf, E., Zhang, W. and Wang, Z.-Y. (2017) RNA-guided transcriptional activation via CRISPR/dCas9 mimics overexpression phenotypes in *Arabidopsis*. *PLoS ONE* 12, e0179410. DOI: 10.1371/journal.pone.0179410.

Patade, V.Y. and Suprasanna, P. (2010) Short-term salt and PEG stresses regulate expression of microRNA, miR159 in sugarcane leaves. *Journal of Crop Science and Biotechnology* 13, 177–182. DOI: 10.1007/s12892-010-0019-6.

Patterson, K., Molloy, L., Qu, W. and Clark, S. (2011) DNA methylation: Bisulphite modification and analysis. *Journal of Visualized Experiments* 3170. DOI: 10.3791/3170.

Pecinka, A., Dinh, H.Q., Baubec, T., Rosa, M., Lettner, N. *et al.* (2010) Epigenetic regulation of repetitive elements is attenuated by prolonged heat stress in *Arabidopsis*. *The Plant Cell* 22, 3118–3129. DOI: 10.1105/tpc.110.078493.

Persson, H., Søkilde, R., Pirona, A.C. and Rovira, C. (2017) Preparation of highly multiplexed small RNA sequencing libraries. *BioTechniques* 63, 57–64. DOI: 10.2144/000114574.

Phillips, J.R., Dalmay, T. and Bartels, D. (2007) The role of small RNAs in abiotic stress. *FEBS Letters* 581, 3592–3597. DOI: 10.1016/j.febslet.2007.04.007.

Pramanik, D., Shelake, R.M., Kim, M.J. and Kim, J.-Y. (2021) CRISPR-mediated engineering across the central dogma in plant biology for basic research and crop improvement. *Molecular Plant* 14, 127–150. DOI: 10.1016/j.molp.2020.11.002.

Qian, Y., Hu, W., Liao, J., Zhang, J. and Ren, Q. (2019) The dynamics of DNA methylation in the maize (*Zea mays* L.) inbred line B73 response to heat stress at the seedling stage. *Biochemical and Biophysical Research Communications* 512, 742–749. DOI: 10.1016/j.bbrc.2019.03.150.

Qiu, Z., Kang, S., He, L., Zhao, J., Zhang, S. *et al.* (2018) The newly identified *heat-stress sensitive albino 1* gene affects chloroplast development in rice. *Plant Science* 267, 168–179. DOI: 10.1016/j.plantsci.2017.11.015.

Rahman, M.M., Mostofa, M.G., Keya, S.S., Siddiqui, M.N., Ansary, M.M.U. *et al.* (2021) Adaptive mechanisms of halophytes and their potential in improving salinity tolerance in plants. *International Journal of Molecular Sciences* 22, 10733. DOI: 10.3390/ijms221910733.

Ravichandran, S., Ragupathy, R., Edwards, T., Domaratzki, M. and Cloutier, S. (2019) MicroRNA-guided regulation of heat stress response in wheat. *BMC Genomics* 20, 488. DOI: 10.1186/s12864-019-5799-6.

Reinders, J. and Paszkowski, J. (2010) Bisulfite methylation profiling of large genomes. *Epigenomics* 2, 209–220. DOI: 10.2217/epi.10.6.

Rendina González, A.P., Preite, V., Verhoeven, K.J.F. and Latzel, V. (2018) Transgenerational effects and epigenetic memory in the clonal plant *Trifolium repens*. *Frontiers in Plant Science* 871, 1677. DOI: 10.3389/fpls.2018.01677.

Reyes, J.L. and Chua, N.-H. (2007) ABA induction of miR159 controls transcript levels of two MYB factors during *Arabidopsis* seed germination. *The Plant Journal* 49, 592–606. DOI: 10.1111/j.1365-313X.2006.02980.x.

Ricardi, M.M., González, R.M., Zhong, S., Domínguez, P.G., Duffy, T. *et al.* (2014) Genome-wide data (ChIP-seq) enabled identification of cell wall-related and aquaporin genes as targets of tomato ASR1, a drought stress-responsive transcription factor. *BMC Plant Biology* 14, 29. DOI: 10.1186/1471-2229-14-29.

Richard, P., Feng, S. and Manley, J.L. (2013) A SUMO-dependent interaction between senataxin and the exosome, disrupted in the neurodegenerative disease AOA2, targets the exosome to sites of transcription-induced DNA damage. *Genes & Development* 27, 2227–2232. DOI: 10.1101/gad.224923.113.

Roy, S. and Soni, P. (2022) Unraveling the epigenetic landscape for salt tolerance in plants. *International Journal of Plant Biology* 13, 443–462. DOI: 10.3390/ijpb13040036.

Roca Paixão, J.F., Gillet, F.-X., Ribeiro, T.P., Bournaud, C., Lourenço-Tessutti, I.T. *et al.* (2019) Improved drought stress tolerance in *Arabidopsis* by CRISPR/dCas9 fusion with a histone acetyltransferase. *Scientific Reports* 9, 8080. DOI: 10.1038/s41598-019-44571-y.

Roy, D., Paul, A., Roy, A., Ghosh, R., Ganguly, P. *et al.* (2014) Differential acetylation of histone H3 at the regulatory region of *OsDREB1b* promoter facilitates chromatin remodelling and transcription activation during cold stress. *PloS ONE* 9, e100343. DOI: 10.1371/journal.pone.0100343.

Sani, E., Herzyk, P., Perrella, G., Colot, V. and Amtmann, A. (2013) Hyperosmotic priming of *Arabidopsis* seedlings establishes a long-term somatic memory accompanied by specific changes of the epigenome. *Genome Biology* 14, R59–R59. DOI: 10.1186/gb-2013-14-6-r59.

Santosh Kumar, V.V., Verma, R.K., Yadav, S.K., Yadav, P., Watts, A. *et al.* (2020) CRISPR-cas9 mediated genome editing of drought and salt tolerance (OsDST) gene in indica mega rice cultivar MTU1010. *Physiology and Molecular Biology of Plants* 26, 1099–1110. DOI: 10.1007/s12298-020-00819-w.

Sapranauskas, R., Gasiunas, G., Fremaux, C., Barrangou, R., Horvath, P. *et al.* (2011) The *Streptococcus thermophilus* CRISPR/Cas system provides immunity in *Escherichia coli*. *Nucleic Acids Research* 39(21), 9275–9282. DOI: 10.1093/nar/gkr606.

Sarma, B., Kashtoh, H., Lama Tamang, T., Bhattacharyya, P.N., Mohanta, Y.K. *et al.* (2023) Abiotic stress in rice: Visiting the physiological response and its tolerance mechanisms. *Plants (Basel)* 12(23), 3948. DOI: 10.3390/plants12233948.

Scheben, A., Yuan, Y. and Edwards, D. (2016) Advances in genomics for adapting crops to climate change. *Current Plant Biology* 6, 2–10. DOI: 10.1016/j.cpb.2016.09.001.

Sedeek, K.E.M., Mahas, A. and Mahfouz, M. (2019) Plant genome engineering for targeted improvement of crop traits. *Frontiers in Plant Science* 10, 114. DOI: 10.3389/fpls.2019.00114.

Seleiman, M.F., Al-Suhaibani, N., Ali, N., Akmal, M., Alotaibi, M. *et al.* (2021) Drought stress impacts on plants and different approaches to alleviate its adverse effects. *Plants* 10(2), 259. DOI: 10.3390/plants10020259.

Shah, A., Nazari, M., Antar, M., Msimbira, L.A., Naamala, J. *et al.* (2021) PGPR in agriculture: A sustainable approach to increasing climate change resilience. *Frontiers in Sustainable Food Systems* 5. DOI: 10.3389/fsufs.2021.667546.

Sharma, M., Sidhu, A.K., Samota, M.K., Gupta, M., Koli, P. *et al.* (2023) Post-translational modifications in histones and their role in abiotic stress tolerance in plants. *Proteomes* 11, 38. DOI: 10.3390/proteomes11040038.

Shelake, R.M., Pramanik, D. and Kim, J.-Y. (2019a) Evolution of plant mutagenesis tools: A shifting paradigm from random to targeted genome editing. *Plant Biotechnology Reports* 13(5), 423–445. DOI: 10.1007/s11816-019-00562-z.

Shelake, R.M., Pramanik, D. and Kim, J.-Y. (2019b) Exploration of plant-microbe interactions for sustainable agriculture in CRISPR era. *Microorganisms* 7(8), 269. DOI: 10.3390/microorganisms7080269.

Shen, C., Que, Z., Xia, Y., Tang, N., Li, D. *et al.* (2017) Knock out of the annexin gene osann3 via CRISPR/Cas9-mediated genome editing decreased cold tolerance in rice. *Journal of Plant Biology* 60(6), 539–547. DOI: 10.1007/s12374-016-0400-1.

Sicilia, A., Scialò, E., Puglisi, I. and Lo Piero, A.R. (2020) Anthocyanin biosynthesis and DNA methylation dynamics in sweet orange fruit [*Citrus sinensis* L. (Osbeck)] under cold stress. *Journal of Agricultural and Food Chemistry* 68(26), 7024–7031. DOI: 10.1021/acs.jafc.0c02360.

Silva, W.T.A.F., Otto, S.P. and Immler, S. (2021) Evolution of plasticity in production and transgenerational inheritance of small RNAs under dynamic environmental conditions. *PLoS Genetics* 17(5), e1009581. DOI: 10.1371/journal.pgen.1009581.

Singer, B.D. (2019) A practical guide to the measurement and analysis of DNA methylation. *American Journal of Respiratory Cell and Molecular Biology* 61(4), 417–428. DOI: 10.1165/rcmb.2019-0150TR.

Singh, B., Singh, S., Singh, B. and Chattu, V.K. (2022) India's neighbourhood vaccine diplomacy during COVID-19 pandemic: Humanitarian and geopolitical perspectives. *Journal of Asian and African Studies* 58, 1021–1037. DOI: 10.1177/00219096221079310.

Singh, K., Gupta, S. and Singh, A.P. (2024) Review: Nutrient-nutrient interactions governing underground plant adaptation strategies in a heterogeneous environment. *Plant Science* 342, 112024. DOI: 10.1016/j.plantsci.2024.112024.

Steward, N., Ito, M., Yamaguchi, Y., Koizumi, N. and Sano, H. (2002) Periodic DNA methylation in maize nucleosomes and demethylation by environmental stress. *The Journal of Biological Chemistry* 277(40), 37741–37746. DOI: 10.1074/jbc.M204050200.

Stief, A., Altmann, S., Hoffmann, K., Pant, B.D., Scheible, W.-R. *et al.* (2014) *Arabidopsis* miR156 regulates tolerance to recurring environmental stress through SPL transcription factors. *The Plant Cell* 26(4), 1792–1807. DOI: 10.1105/tpc.114.123851.

Sudan, J., Raina, M. and Singh, R. (2018) Plant epigenetic mechanisms: Role in abiotic stress and their generational heritability. *3 Biotech* 8, 172. DOI: 10.1007/s13205-018-1202-6.

Sun, C., McAndrew, T., Smith, B.C., Chen, Z., Frimer, M. *et al.* (2014) Characterization of HPV DNA methylation of contiguous CpG sites by bisulfite treatment and massively parallel sequencing-the FRAGMENT approach. *Frontiers in Genetics* 5, 150. DOI: 10.3389/fgene.2014.00150.

Sun, D.-D., Sun, J.-W., Huang, L.-Y., Chen, N. and Wang, Q.-W. (2021) Effects of cadmium stress on DNA methylation in soybean. *Biotechnology & Biotechnological Equipment* 35(1), 1696–1705. DOI: 10.1080/13102818.2021.1980107.

Sun, M., Yang, Z., Liu, L. and Duan, L. (2022) DNA methylation in plant responses and adaption to abiotic stresses. *International Journal of Molecular Sciences* 23(13), 6910. DOI: 10.3390/ijms23136910.

Sun, X., Xu, L., Wang, Y., Yu, R., Zhu, X. *et al.* (2015) Identification of novel and salt-responsive miRNAs to explore miRNA-mediated regulatory network of salt stress response in radish (*Raphanus sativus* L). *BMC Genomics* 16(1), 197. DOI: 10.1186/s12864-015-1416-5.

Taghavi, M., Bakhshi, K., Zarei, A., Hoseinzadeh, E. and Gholizadeh, A. (2024) Soil pollution indices and health risk assessment of metal(loid)s in the agricultural soil of pistachio orchards. *Scientific Reports* 14(1), 8971. DOI: 10.1038/s41598-024-59450-4.

Tan, M. (2010) Analysis of DNA methylation of maize in response to osmotic and salt stress based on methylation-sensitive amplified polymorphism. *Plant Physiology and Biochemistry* 48(1), 21–26. DOI: 10.1016/j.plaphy.2009.10.005.

Tang, M., Xu, L., Wang, Y., Dong, J., Zhang, X. *et al.* (2021) Melatonin-induced DNA demethylation of metal transporters and antioxidant genes alleviates lead stress in radish plants. *Horticulture Research* 8(1), 124. DOI: 10.1038/s41438-021-00561-8.

Tang, X., Wang, Q., Yuan, H. and Huang, X. (2018) Chilling-induced DNA demethylation is associated with the cold tolerance of *Hevea brasiliensis*. *BMC Plant Biology* 18(1), 70. DOI: 10.1186/s12870-018-1276-7.

Thanmalagan, R.R., Jayaprakash, A., Roy, A., Arunachalam, A. and Lakshmi, P.T.V. (2022) A review on applications of plant network biology to understand the drought stress response in economically important cereal crops. *Plant Gene* 29, 100345. DOI: 10.1016/j.plgene.2021.100345.

Thiebaut, F., Hemerly, A.S. and Ferreira, P.C.G. (2019) A role for epigenetic regulation in the adaptation and stress responses of non-model plants. *Frontiers in Plant Science* 10, 246. DOI: 10.3389/fpls.2019.00246.

Tombesi, S., Frioni, T., Poni, S. and Palliotti, A. (2018) Effect of water stress "memory" on plant behavior during subsequent drought stress. *Environmental and Experimental Botany* 150, 106–114. DOI: 10.1016/j.envexpbot.2018.03.009.

Tran, M.T., Doan, D.T.H., Kim, J., Song, Y.J., Sung, Y.W. *et al.* (2021) CRISPR/Cas9-based precise excision of slhyprp1 domain(s) to obtain salt stress-tolerant tomato. *Plant Cell Reports* 40(6), 999–1011. DOI: 10.1007/s00299-020-02622-z.

van Dijk, K., Ding, Y., Malkaram, S., Riethoven, J.-J.M., Liu, R. *et al.* (2010) Dynamic changes in genome-wide histone H3 lysine 4 methylation patterns in response to dehydration stress in *Arabidopsis thaliana*. *BMC Plant Biology* 10, 238. DOI: 10.1186/1471-2229-10-238.

Verkest, A., Byzova, M., Martens, C., Willems, P., Verwulgen, T. *et al.* (2015) Selection for improved energy use efficiency and drought tolerance in canola results in distinct transcriptome and epigenome changes. *Plant Physiology* 168(4), 1338–1350. DOI: 10.1104/pp.15.00155.

Verslues, P.E., Bailey-Serres, J., Brodersen, C., Buckley, T.N., Conti, L. *et al.* (2023) Burning questions for a warming and changing world: 15 unknowns in plant abiotic stress. *The Plant Cell* 35(1), 67–108. DOI: 10.1093/plcell/koac263.

Vierstra, R.D. (2003) The ubiquitin/26S proteasome pathway, the complex last chapter in the life of many plant proteins. *Trends in Plant Science* 8(3), 135–142. DOI: 10.1016/S1360-1385(03)00014-1.

Viggiano, L. and Pinto, M.C. (2017) Dynamic DNA methylation patterns in stress response. *Plant Epigenetics* 281–302. DOI: 10.1007/978-3-319-55520-1_15.

Vikram, P., Swamy, B.P.M., Dixit, S., Ahmed, H.U., Teresa Sta Cruz, M. *et al.* (2011) $QDTY_{1.1}$, a major QTL for rice grain yield under reproductive-stage drought stress with a consistent effect in multiple elite genetic backgrounds. *BMC Genetics* 12, 89. DOI: 10.1186/1471-2156-12-89.

Wang, Q., Xu, J., Pu, X., Lv, H., Liu, Y. *et al.* (2021a) Maize DNA methylation in response to drought stress is involved in target gene expression and alternative splicing. *International Journal of Molecular Sciences* 22(15), 8285. DOI: 10.3390/ijms22158285.

Wang, T., Xun, H., Wang, W., Ding, X., Tian, H. *et al.* (2021b) Mutation of *gmaitr* genes by CRISPR/Cas9 genome editing results in enhanced salinity stress tolerance in soybean. *Frontiers in Plant Science* 12, 779598. DOI: 10.3389/fpls.2021.779598.

Wang, W., Qin, Q., Sun, F., Wang, Y., Xu, D. *et al.* (2016) Genome-wide differences in DNA methylation changes in two contrasting rice genotypes in response to drought conditions. *Frontiers in Plant Science* 7, 1675. DOI: 10.3389/fpls.2016.01675.

Wibowo, A., Becker, C., Marconi, G., Durr, J., Price, J. *et al.* (2016) Hyperosmotic stress memory in *Arabidopsis* is mediated by distinct epigenetically labile sites in the genome and is restricted in the male germline by DNA glycosylase activity. *eLife* 5, e13546. DOI: 10.7554/eLife.13546.

Widiez, T., Symeonidi, A., Luo, C., Lam, E., Lawton, M. *et al.* (2014) The chromatin landscape of the moss *Physcomitrella patens* and its dynamics during development and drought stress. *The Plant Journal* 79(1), 67–81. DOI: 10.1111/tpj.12542.

Wreczycka, K., Gosdschan, A., Yusuf, D., Grüning, B., Assenov, Y. *et al.* (2017) Strategies for analyzing bisulfite sequencing data. *Journal of Biotechnology* 261, 105–115. DOI: 10.1016/j.jbiotec.2017.08.007.

Xiao, J., Lee, U.-S. and Wagner, D. (2016) Tug of war: Adding and removing histone lysine methylation in *Arabidopsis*. *Current Opinion in Plant Biology* 34, 41–53. DOI: 10.1016/j.pbi.2016.08.002.

Xu, J., Tanino, K.K. and Robinson, S.J. (2016) Stable epigenetic variants selected from an induced hypomethylated *Fragaria vesca* population. *Frontiers in Plant Science* 7, 1768. DOI: 10.3389/fpls.2016.01768.

Xu, J., Zhou, S., Gong, X., Song, Y., Nocker, S. *et al.* (2018) Single-base methylome analysis reveals dynamic epigenomic differences associated with water deficit in apple. *Plant Biotechnology Journal* 16(2), 672–687. DOI: 10.1111/pbi.12820.

Yang, Y., Zhang, X., Su, Y., Zou, J., Wang, Z. *et al.* (2017) MiRNA alteration is an important mechanism in sugarcane response to low-temperature environment. *BMC Genomics* 18(1), 833. DOI: 10.1186/s12864-017-4231-3.

Yin, Y., Qin, K., Song, X., Zhang, Q., Zhou, Y. *et al.* (2018) BZR1 transcription factor regulates heat stress tolerance through FERONIA receptor-like kinase-mediated reactive oxygen species signaling in tomato. *Plant & Cell Physiology* 59(11), 2239–2254. DOI: 10.1093/pcp/pcy146.

Yong, W.-S., Hsu, F.-M. and Chen, P.-Y. (2016) Profiling genome-wide DNA methylation. *Epigenetics & Chromatin* 9(1), 26. DOI: 10.1186/s13072-016-0075-3.

Yu, X., Wang, H., Lu, Y., de Ruiter, M., Cariaso, M. *et al.* (2012) Identification of conserved and novel micrornas that are responsive to heat stress in *brassica rapa*. *Journal of Experimental Botany* 63(2), 1025–1038. DOI: 10.1093/jxb/err337.

Zeng, Y., Wen, J., Zhao, W., Wang, Q. and Huang, W. (2020) Rational improvement of rice yield and cold tolerance by editing the three genes ospin5b, GS3, and osmyb30 with the CRISPR–Cas9 system. *Frontiers in Plant Science* 10. DOI: 10.3389/fpls.2019.01663.

Zhang, A., Liu, Y., Wang, F., Li, T., Chen, Z. *et al.* (2019) Enhanced rice salinity tolerance via CRISPR/Cas9-targeted mutagenesis of the *osrr22* gene. *Molecular Breeding* 39, 47. DOI: 10.1007/s11032-019-0954-y.

Zhang, B., Tieman, D.M., Jiao, C., Xu, Y., Chen, K. *et al.* (2016) Chilling-induced tomato flavor loss is associated with altered volatile synthesis and transient changes in DNA methylation. *Proceedings of the National Academy of Sciences* 113(44), 12580–12585. DOI: 10.1073/pnas.1613910113.

Zhao, P., Ma, B., Cai, C. and Xu, J. (2022) Transcriptome and methylome changes in two contrasting mungbean genotypes in response to drought stress. *BMC Genomics* 23(1), 80. DOI: 10.1186/s12864-022-08315-z.

Zhao, Y., Zhang, C., Liu, W., Gao, W., Liu, C. *et al.* (2016) An alternative strategy for targeted gene replacement in plants using a dual-sgRNA/Cas9 design. *Scientific Reports* 6, 23890. DOI: 10.1038/srep23890.

Zheng, X., Chen, L., Xia, H., Wei, H., Lou, Q. *et al.* (2017) Transgenerational epimutations induced by multi-generation drought imposition mediate rice plant's adaptation to drought condition. *Scientific Reports* 7, 39843. DOI: 10.1038/srep39843.

Zhou, J., Liu, D., Wang, P., Ma, X., Lin, W. *et al.* (2018) Regulation of *Arabidopsis* brassinosteroid receptor BRI1 endocytosis and degradation by plant U-box PUB12/PUB13-mediated ubiquitination. *Proceedings of the National Academy of Sciences of the United States of America* 115, E1906–E1915. DOI: 10.1073/pnas.1712251115.

Zhou, Y., Romero-Campero, F.J., Gómez-Zambrano, Á., Turck, F. and Calonje, M. (2017) H2A monoubiquitination in *Arabidopsis thaliana* is generally independent of LHP1 and PRC2 activity. *Genome Biology* 18(1), 69. DOI: 10.1186/s13059-017-1197-z.

5 Plant Single-Cell Epigenomics: Updated Methods and Current Applications

Gunna Sureshbabu Suruthi[1], Arumugam Vijaya Anand[1], Manoharan Rajesh[2], Pemula Gowtham[1], Tajudheen Nargis Begum[3], Seyed Mafia Haniff[4], Palanisamy Sampathkumar[5] and Asirvatham Alwin Robert[6]*

[1]Department of Human Genetics and Molecular Biology, Bharathiar University, Coimbatore, Tamil Nadu, India; [2]Department of Biochemistry, Bharathiar University, Coimbatore, Tamil Nadu, India; [3]Department of Biotechnology, Jamal Mohamed College, (Autonomous), Tiruchirapalli, Tamil Nadu, India; [4]Department of Zoology, Jamal Mohamed College, (Autonomous), Tiruchirapalli, Tamil Nadu, India; [5]Department of Chemistry and Biosciences, SASTRA Deemed to be University, Srinivasa Ramanujan Centre, Kumbakonam, Tamil Nadu, India; [6]Department of Endocrinology and Diabetes, Prince Sultan Military Medical City, Ministry of Defense, Riyadh, Kingdom of Saudi Arabia

Abstract

Plant single-cell epigenomics has emerged as a transformative field, providing high-resolution insights into gene regulation, cellular identity, and epigenetic dynamics in plants. This chapter delves into the updated methods and current applications of single-cell epigenomics in plant biology. It begins with an introduction to single-cell epigenomics, highlighting historical milestones and the evolution of techniques. Key methods discussed include single-cell isolation techniques, epigenomic assays, and advanced computational methods for data research. The chapter explores the significant applications of these methods in understanding plant development, stress responses, and epigenomic variation. It also addresses the challenges and limitations in this field and proposes prospects for advancing plant single-cell epigenomics. By integrating multi-omics approaches and innovative technologies, researchers can unlock unprecedented insights into the complex regulatory networks and molecular mechanisms governing plant biology, paving the way for sustainable agricultural practices and crop improvement.

Keywords: Single-cell epigenomics, Gene regulation, Epigenomics, Crop improvement, Epigenetic dynamics

5.1 Introduction to Single-Cell Epigenomics

Single-cell epigenomics in plants is an effective tool for understanding gene regulation and cell identity at a high resolution (Clark *et al.*, 2016). This field faces unique challenges in plant research, including analyzing plant single-cell epigenomic information and biological challenges specific to plants (Mendieta *et al.*,

*Corresponding author: aalwinrobert@gmail.com

 Epigenetics for Climate-Smart and Sustainable Agriculture (ed. Jen-Tsung Chen)

DOI: 10.1079/9781800626102.0005

2023). Epigenetic modifications in plants can be induced by developmental and environmental signals, leading to the generation of multiple epigenomes from a single genome in response to various cues. Understanding the epigenetic regulation of plant stress responses is crucial for comprehending how plants adapt to changing environmental conditions (Chinnusamy and Zhu, 2009). Single-cell epigenomic techniques offer the potential to revolutionize our knowledge of gene regulation and provide insights into cellular identity, heterogeneity, and regulatory mechanisms (Clark *et al.*, 2016).

In the realm of single-cell epigenomics, advancements in technology have enabled the concurrent analysis of multiple layers of genomic information at the single-cell level (Kelsey *et al.*, 2017). Techniques such as single-cell combinatorial-indexing multi-target chromatin integration labeling observed by sequencing (sci-mtChIL-seq) allow for the investigation of RNA polymerase II attaching to chromatin and other epigenomic characteristics in individual cells. These methods empower researchers to conduct genome-wide transcription and epigenetic analyses at a resolution previously unattainable (Ohkawa *et al.*, 2024). Moreover, the spatially determined profiling of epigenetic actions in complex tissues has been made possible through innovative techniques like epigenomic multiplexed error-robust fluorescence *in situ* hybridization, which retains the spatial context of cells (Lu *et al.*, 2022).

The ability to probe various layers of epigenetic principles in a single cell provides a comprehensive view of molecular circuitry underlying cell identity, state, and gene regulatory networks. Furthermore, the expansion of single-cell multi-omics sequencing technologies like single-cell chromatin overall omic-scale landscape sequencing has enabled the simultaneous analysis of chromatin state, methylation of DNA, copy number variation, and ploidy in individual mammalian cells (Guo *et al.*, 2017). The integration of single-cell epigenomic profiles with transcriptomic data enhances our experience of gene regulatory activities and provides a more holistic perspective of cellular processes (Lee *et al.*, 2018). By profiling chromatin availability at the single-cell status, researchers can gain insights into the regulatory landscape and developmental processes (Ma and Zhang, 2020). Approaches like single-cell cleavage under targets and tagmentation permits the probing of modifications of histone at a single-cell level, enabling the study of epigenetic dynamics in thousands of cells (Bartosovic and Castelo-Branco, 2023). Additionally, the characterization and optimization of multi-omic single-cell epigenomic profiling platforms like single-nucleus assay for transposase-accessible chromatin sequencing and single-nucleus RNA sequencing offer a comprehensive approach to studying open chromatin and expression of genes with single-cell resolution (Sandoval *et al.*, 2023).

Single-cell epigenomics in plants represents a cutting-edge field that offers unprecedented insights into gene regulation, cell identity, and epigenetic dynamics. By overwhelming biological and computational challenges unique to plants, researchers can unravel the complexities of plants at a level of detail never before possible. The incorporation of single-cell epigenomic information with other omics layers provides a comprehensive view of cellular processes and regulatory networks, paving the way for transformative discoveries in plant research.

5.2 Historical Perspective and Milestones

The specialization of plant single-cell epigenomics has seen noteworthy progress marked by key discoveries that have enhanced our comprehension of gene regulation and cell identity at a cellular level. Initially, researchers encountered challenges in investigating plant single-cell epigenomic information due to the distinct biological characteristics of plants (Mendieta *et al.*, 2023). Nevertheless, the introduction of single-cell omics technologies has transformed plant biology by allowing scientists to examine the molecular properties of particular plant cells, including their epigenomes and transcriptomes (Thibivilliers and Libault, 2021).

A pivotal advancement in the field was the introduction of single-cell multi-epigenomics approaches, such as sci-mtChIL-seq, which enabled the simultaneous research of RNA polymerase II binding to chromatin and epigenomic characteristics in plant cells (Ohkawa *et al.*,

2024). This technological breakthrough facilitated the monitoring of chromatin structure changes at a single-cell level, offering insights into the dynamic regulation of gene expression in plants. Additionally, the innovation of single-cell DNA methylome profiling has facilitated the examination of epigenomic heterogeneity in complex plant tissues and through cellular reprogramming processes (Luo *et al.*, 2018).

Over time, the enhancement of single-cell epigenomic library construction strategies have played a crucial role in elucidating the epigenetic variety and causal relationships between the epigenome and transcriptome at a single-cell status (Park *et al.*, 2022). These methodological advancements have paved the way for studying cellular differentiation and disease occurrence with unprecedented resolution. Furthermore, the integration of single-cell epigenomics and transcriptomics from the same kind of cell has allowed researchers to explore the intricate association between chromatin components and transcriptional levels (Pinheiro *et al.*, 2021). The specialization of plant single-cell epigenomics has also benefited from improvements in computational tools and analysis pipelines. Tools like epigenomics single-cell analysis in Python (EpiScanpy) have facilitated integrated single-cell epigenomic analysis, providing a comprehensive range of analysis options for different types of single-cell epigenomic information (Danese *et al.*, 2021). These tools have been essential in managing the considerable volume of data developed by single-cell epigenomic studies and extracting meaningful biological insights from complex datasets.

Moreover, the development of combinatorial indexing strategies for epigenomic profiling of plant single cells has eliminated technical barriers and improved the regular examination of single-cell epigenomes (SCEs). These strategies have brought new insights into the growth and development of the plant, as well as their relations with the environment. Additionally, the incorporation of single-cell epigenomic variability studies has revealed functional heterogeneity in plant cells, shedding light on the regulatory processes governing cellular behavior (Tu *et al.*, 2021). Looking ahead, the future of plant single-cell epigenomics holds great promise, with ongoing advancements in technology and methodologies poised to further unravel the complexity of gene regulation and cell individuality in plants. The continuous refinement of single-cell epigenomic approaches, connected with the integration of multi-omics approaches, is set to provide unparalleled insights into plant biology at a cellular level. By leveraging these state-of-the-art tools and analytical strategies, researchers are positioned to unlock new dimensions of plant epigenetic regulation and cellular dynamics.

5.3 Single-Cell Epigenomics in Plant Developmental Stages

Seed germination is an essential developmental phase in plant growth, characterized by dynamic epigenomic changes that regulate gene expression and cellular processes. Throughout seed development and germination, there is a significant reconfiguration of DNA methylation patterns, with notable gains and losses of methylation observed, indicating a complex epigenetic landscape during this phase (Kawakatsu *et al.*, 2017). Moreover, comprehensive transcriptomic and epigenomic remodeling ensues during seed germination, emphasizing the interconnected regulatory mechanisms that govern this critical developmental transition. The integrated investigation of high-resolution RNA sequencing, small RNA sequencing, and methylcytosine sequencing reveals profound transformations in both the transcriptome and epigenome associated with kernel germination (Narsai *et al.*, 2017).

Additionally, the germination process involves key physiological processes orchestrated by changes in gene expression and protein dynamics, as supported by combined transcriptome and proteome analyses. In response to environmental cues such as decreased salinity, seeds activate mechanisms to mobilize energy reserves, inhibit dormancy-related pathways, and prepare for germination by modulating hormone signaling and antioxidant systems (Zhang *et al.*, 2023c). These findings highlight the intricate interplay between transcriptional and proteomic regulation during kernel germination, underscoring the importance of understanding the molecular processes underlying this developmental stage.

Furthermore, epigenetic regulation plays an important function in modulating growth and developmental pathways during seed germination, as exemplified by the MicroRNA159-regulated gibberellic acid – myeloblastosis (GAMYB) – like genes in *Arabidopsis*. MicroRNAs, such as miR159, target specific transcription factors like GAMYB-associated genes to inhibit the development and facilitate programmed cell death, influencing the developmental fate of germinating seeds (Alonso-Peral *et al.*, 2010). This regulatory network emphasizes the significance of epigenetic mechanisms in fine-tuning gene expression patterns and cellular processes essential for successful kernel germination and earlier seedling facility.

The epigenomic dynamics during seed germination are characterized by extensive DNA methylation reconfiguration, transcriptomic and proteomic remodeling, and microRNA-mediated regulatory pathways that impact gene expression and cellular responses. Understanding the intricate interplay between epigenetic modifications, gene regulation, and protein dynamics during seed germination is essential for unraveling the molecular mechanisms that govern this critical developmental stage in plant growth and development. By elucidating the epigenetic landscape and regulatory networks underlying seed germination, researchers can gain insights into how plants adapt to environmental cues, optimize resource utilization, and ensure successful seedling emergence and establishment.

5.4 Environmental Influences on Plant Epigenomes

The influence of climate modification on plant epigenomes is an essential area of study that sheds light on how environmental strains influence the epigenetic tools of plants. Unpredictable climate changes expose plants to various abiotic strains like excessive temperatures, dehydration, high salinity, and heavy metal toxicity, affecting plant productivity (Fig. 5.1) (Miryeganeh, 2021). These stressors trigger epigenetic responses in plants, directing to modifications in gene expression and chromatin modifications to cope with challenging environmental conditions. Climate change, characterized by rapid temperature fluctuations and increasing atmospheric CO_2 concentrations, affects plant cold adaptation and freezing toleration, highlighting the

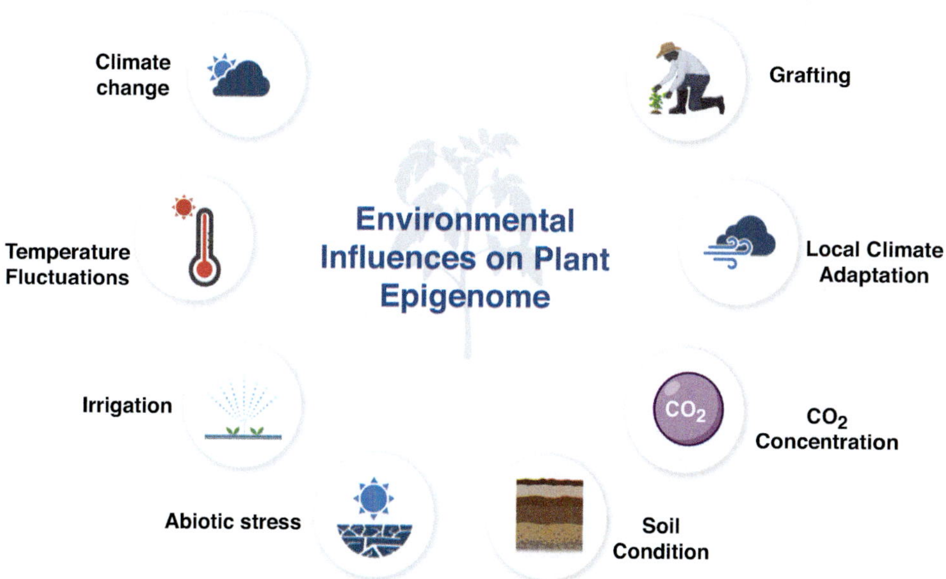

Fig. 5.1. Environmental influences on plant epigenome. Author's own image.

importance of understanding the epigenetic pathways involved in plant responses to changing climatic conditions (Liu *et al.*, 2019).

Epigenomic patterns in plants reflect environmental influences such as irrigation and grafting, indicating that the epigenome serves as a mechanism through which plants respond at a molecular level to their surroundings (Williams *et al.*, 2020). The epigenome's plasticity allows plants to adapt to environmental changes by modulating gene expression through epigenetic modifications. Moreover, temperature, a key ecological factor, limits the geographic distribution of plants, and climate warming alters plant distribution boundaries, emphasizing the profound impact of climate instability on plant ecosystems (Wang *et al.*, 2020). These changes in plant distribution patterns underscore the need to investigate how epigenetic mechanisms mediate plant responses to shifting environmental conditions.

Epigenetic modifications play a crucial part in controlling plant gene expression in reaction to environmental stresses, with studies suggesting that abiotic stressors can induce positive effects on plant growth through epigenetic alterations. The ability of plants to exhibit clonal success across diverse environments and climatic changes underscores the significance of epigenetics as a potential mechanism for plant adaptation and survival (Douhovnikoff and Dodd, 2015). Understanding how epigenetic changes contribute to plant resilience in the face of environmental challenges is essential for creating methods to enhance crop productivity and sustainability in a modifying climate.

The genomic and epigenetic footprint of local transformation to inconsistent climates in kiwifruit highlights the correlation between epigenetic variations and climatic variables, indicating that epigenetic modifications are involved in plant reactions to abiotic strains (Zhang *et al.*, 2023b). DNA methylation dynamics during stress responses in woodland strawberries demonstrate the intricate relationship between environmental pressures and epigenetic modifications, emphasizing the role of epigenetic mechanisms in plant transformation to adverse conditions (López *et al.*, 2022). These studies underscore the importance of epigenomic regulation in mediating plant responses to environmental stressors and deliver an understanding of the molecular mechanisms of basic plant resilience.

Epigenomic studies have the potential to enhance breeding approaches for developing climate-resilient crops by leveraging epigenetic mechanisms to improve stress tolerance and performance stability in plants (Chandana *et al.*, 2022). The moldability of chromatin through environmental perturbations indicates that targeting chromatin regulators and enzymes could facilitate the development of climate-resilient crops through epigenetic engineering (Kumar, 2018). By understanding the epigenetic responses of plants to environmental stress, researchers can identify key regulatory pathways that govern plant adaptation and resilience in changing climatic conditions.

5.5 Methods and Technologies

The specialization of single-cell epigenomics in plants has been revolutionized by emerging techniques for grasping methylation of DNA, chromatin accessibility, modifications of histone, chromosome conformation, and replication dynamics (Schwartzman and Tanay, 2015). Multiplex single-cell outlining of chromatin attainability using combinatorial cellular indexing has accelerated the description of complex plant tissues having myriad cell types and dynamic functions like differentiation (Cusanovich *et al.*, 2015). These technologies offer a comprehensive view of the epigenomic landscape of plant cells, providing insights into gene regulation, cellular identity, and chromatin dynamics at a single-cell decision.

5.5.1 Single-cell isolation techniques

Single-cell isolation techniques have significantly advanced fields including biology, medicine, and materials science by enabling the study of individual cells in detail. Among the diverse methods available, microfluidics has emerged as a powerful tool for high-throughput single-cell segregation and culture (Lin *et al.*, 2016). Lin *et al.* (2016) introduced a microfluidic platform that can achieve increased efficiency of single-cell isolation (~77%) and support great single-cell culture for up to 7 days, allowing researchers to conduct cellular heterogeneity analysis using clonogenic assays (Lin *et al.*, 2016). This technology not only improves the throughput of single-cell studies but also facilitates the investigation of cellular behaviors over extended periods.

Droplet-based microfluidics, as highlighted by Lin *et al.* (2022), offers another promising approach for single-cell isolation, attracting significant attention due to its high throughput, biocompatibility, and ease of operation. By leveraging 3D-printed microfluidics for droplet generation, researchers can encapsulate single cells for subsequent optical imaging, providing a versatile platform for various applications in single-cell investigation (Lin *et al.*, 2022). The simplicity and scalability of droplet-based technologies make them valuable tools for studying particular cells in a high-throughput technique.

In the realm of nuclei isolation, Li *et al.* (2022b) presented a nuclei isolation procedure tailored for various angiosperm species, shedding light on the importance of plant nuclei isolation for single-cell genomics studies. This protocol is a beneficial aid for researchers aiming to delve into the genomics of plant cells at the single-cell status, offering insights into the intricacies of plant cellular composition and function (Li *et al.*, 2022b). By optimizing methods for nuclei isolation, investigators can unlock a more serious knowledge of plant biology and genetic regulation.

Moreover, the isolation of nuclei labelled in distinctive cell types procedure has revolutionized cell type-specific expression of genes and chromatin profiling in *Arabidopsis thaliana*. This innovative approach enables the affinity-based isolation of nuclei from individual cell types within tissues, circumventing challenges associated with traditional mechanical purification techniques (Deal and Henikoff, 2011). By selectively isolating nuclei from specific cell types, researchers can unravel the complexity of gene expression and chromatin dynamics at a cellular level, paving the way for in-depth molecular studies.

Furthermore, the outcome of high-throughput single-cell manipulation methods, as demonstrated by Arakawa *et al.* (2011), has significantly advanced the field of single-cell isolation by enabling the isolation of a large amount of target cells with high efficiency. By leveraging microfluidics and microchannel technologies, researchers can accomplish high-throughput single-cell isolation, opening up new possibilities for investigating cellular heterogeneity and functional diversity at the single-cell level (Arakawa *et al.*, 2011). These advancements in the manipulation of single-cell systems have broad implications for various research areas, including cell biology, medicine, and biotechnology.

In the context of nuclei isolation for RNA sequencing, Thibivilliers *et al.* (2020) outlined a method for isolating root nuclei to facilitate single-cell RNA sequencing studies (Fig. 5.2). This protocol offers a valuable resource for researchers interested in exploring the transcriptomic landscape of plant cells at a single-cell resolution, providing insights into gene expression dynamics and regulatory networks within plant tissues (Thibivilliers *et al.*, 2020). By isolating plant root nuclei, researchers can unravel the molecular processes underpinning plant growth and responses to environmental cues at a cellular level.

5.5.2 Epigenomic assays for single cells

Epigenomic assays for single cells in plants have greatly progressed our knowledge of gene regulation and chromatin dynamics at a cellular level. Single-cell assay for transposase-accessible chromatin sequencing (scATAC-seq) has appeared as a strong tool for profiling available chromatin areas in particular plant cells, enabling the recognition of open chromatin areas and transcription factor binding sites across plant genomes. This technique has been instrumental in delineating gene regulations and control modules in various plant species, shedding light on the intricate mechanisms governing gene expression at a single-cell resolution (Maher *et al.*, 2018). Additionally, the application of scATAC-seq in *Arabidopsis thaliana* roots has revealed thousands of distinctive available sites, allowing for the complete characterization of major cell types within the root system (Dorrity *et al.*, 2021).

Single-cell DNA methylation sequencing methods, including single-cell bisulfite

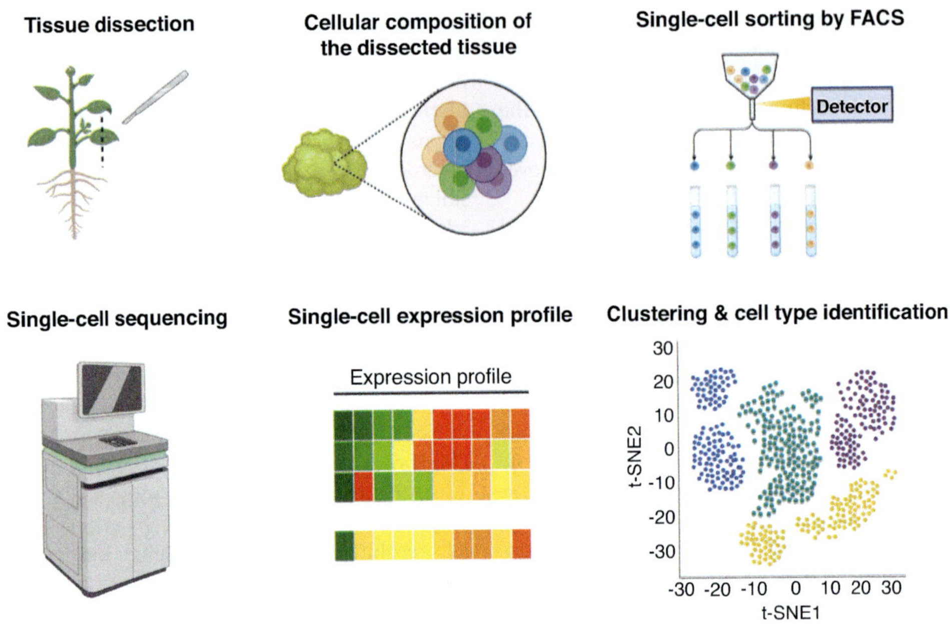

Fig. 5.2. Single-cell analysis. Author's own image.

sequencing (scBS-seq) and single-cell reduced representation bisulfite sequencing (scRRBS), offers insights into the methylation of the DNA landscape of particular plant cells. By profiling DNA methylation patterns at a single-cell level, researchers can unravel the epigenetic modifications that control the expression of genes and cellular identity in plants. These techniques provide a nuanced understanding of the dynamic epigenetic changes occurring within plant cells, offering valuable information for studying developmental processes and environmental responses (Marand *et al.*, 2021). Furthermore, single-cell chromatin immunoprecipitation sequencing (scChIP-seq) facilitates the investigation of chromatin modifications and protein–DNA interactions in individual plant cells. By profiling chromatin immunoprecipitation at a single-cell resolution, researchers can elucidate the regulatory networks governing the expression of genes and chromatin organization in plants. This approach has been pivotal in unraveling the executive landscape of *Arabidopsis thaliana* roots, providing an understanding of the epigenetic mechanisms underlying tissue heterogeneity and cell-specific gene regulation (Dorrity *et al.*, 2021).

Multi-omics approaches, such as transcriptome sequencing, single-cell methylome and single-cell nucleosome, methylation and transcription sequencing, have revolutionized our ability to incorporate various omics data at a single-cell level in plants. These methodologies allow for the comprehensive analysis of multiple molecular layers, including proteomics, transcriptomics, genomics, and epigenomics, within individual plant cells. By combining multi-omics data, researchers can unravel complex regulatory networks, identify key markers for different cell populations, and acquire a holistic knowledge of cellular events in plants (Zhang *et al.*, 2022). This integrative approach holds enormous assurance for increasing our understanding of plant biology and revving the development of sustainable agricultural practices (Kumar *et al.*, 2023). Thus, epigenomic assays for single cells in plants, including scATAC-seq, scBS-seq, scChIP-seq, and multi-omics approaches, have transformed our ability to dissect the molecular intricacies of plant cells at a single-cell resolution. These cutting-edge techniques provide unprecedented insights into gene regulation, chromatin dynamics, and epigenetic modifications within individual plant cells, paving the way for groundbreaking discoveries in plant biology and agriculture.

5.5.3 Computational tools and data analysis

Single-cell epigenomic analysis in plants involves a sophisticated array of computational methods and data investigation approaches to unravel the intricacies of the epigenome at a single-cell level. The exploration of single-cell epigenomic data, such as methylation of DNA and chromatin accessibility, necessitates advanced methodologies to preprocess and ensure data quality (Danese *et al.*, 2021). Tools like EpiScanpy and single-cell epigenomics tools have been conceived to address the challenges specific to single-cell epigenomic data analysis, offering integrated solutions for studying the epigenetic landscape at a cellular resolution (Gao *et al.*, 2023). These tools can enable researchers to delve into the regulatory mechanisms that govern gene expression through epigenetic modifications.

Dimensionality reduction and clustering methods are essential for making sense of the vast amount of single-cell epigenomic information. Techniques like Single-Cell Analysis of Transcriptome and Epigenome for scATAC-seq signal extraction and tools like Multimodal VeloGraph for inferring dynamic gene regulation from multi-omic data aid in reducing the complexity of the data while retaining crucial information (Li *et al.*, 2022a). By employing these methods, researchers can identify patterns, cell subpopulations, and regulatory relationships within the epigenomic data, paving the path for a more in-depth knowledge of plant epigenetics.

Integrative exploration of single-cell epigenomic information with other single-cell datasets, such as transcriptomics and proteomics, is a powerful approach to gaining comprehensive insights into cellular processes. Tools like single-cell aggregation and integration (scAI) and Single-cell Heterogeneity Analysis by Regression and Topic modeling offer solutions for the integrative research of parallel single-cell transcriptomic and epigenomic parameters, enabling researchers to examine cellular heterogeneity and understand transcriptional regulatory processes (Kazwini and Sanguinetti, 2024). These integrative analyses provide a holistic view of cellular function by combining information from different molecular layers, enhancing our understanding of plant cell biology.

Challenges in data interpretation and integration are inherent in single-cell epigenomic studies, especially when dealing with large-scale multi-omic datasets. Computational frameworks like Single-cell Spatial Transcriptomics and Single-cell Transcriptomics Integration, Differentiation, and Epigenomic Analysis with Single-cell have been created to facilitate interactive exploration of single-cell multi-omics information and to standardize and integrate epigenomic datasets across various types of cells (Xiang *et al.*, 2021). These tools address the complexities of data interpretation by providing users with interactive platforms and algorithms to streamline the analysis of diverse epigenomic datasets, ultimately aiding in the elucidation of complex biological processes in plants.

Moreover, the spatially determined epigenomic information of single cells in complex plant tissues, as proposed, offers a unique opportunity to comprehend how the epigenome influences cell fate and function within their native environment (Lu *et al.*, 2022). By profiling epigenetic changes at a single-cell resolution in complex tissues, researchers can unravel the regulatory networks that govern cell differentiation and development in plants. This spatially resolved approach provides a more nuanced understanding of epigenetic principles in the context of tissue architecture and cellular interactions.

The specialization of single-cell epigenomics in plants is rapidly emerging, driven by the outcome of advanced computational devices and data investigation procedures. From preprocessing and quality control to integrative investigation with other single-cell data types, researchers are equipped with a diverse array of tools to explore the epigenetic landscape at a cellular level. Despite the challenges in data interpretation and integration, the continuous innovation in computational methodologies is expanding our understanding of plant epigenetics and paving the path for groundbreaking findings in the field.

5.6 Current Applications in Plant Biology

Single-cell epigenomics has evolved as a valuable device in plant biology, offering a detailed insight into the epigenetic landscape at a cellular level. The utilization of SCE analysis in plants has significant implications across various research domains.

5.6.1 Plant development and differentiation

Single-cell epigenomics has evolved as a valuable device in plant biology, providing insights into plant growth and differentiation. The use of SCE analysis in plant biology has been steadily progressing, focusing on understanding the regulatory mechanisms governing gene expression and cell fate determination. Technologies like chromatin immunoprecipitation sequencing (ChIP-seq) and Assay for Transposase-Accessible Chromatin sequencing (ATAC-seq) have changed the field by permitting investigators to profile the chromatin regulatory terrain at a single-cell resolution, offering crucial information on the epigenetic modifications that steer plant development (Ma and Zhang, 2020).

Single-cell transcriptomics has also been crucial in plant research, enabling the comprehensive analysis of the expression gene patterns at the individual cell. By combining single-cell transcriptomic data with epigenomic information, researchers can untangle the complex regulatory networks controlling plant development and differentiation. These technologies have been pivotal in studying various plant species, elucidating the molecular mechanisms underlying essential biological processes (Seyfferth *et al.*, 2021). The potential of single-cell epigenomics to overcome the limitations of bulk epigenomic analyses and explore new avenues in plant biology is evident. By capturing the epigenetic profiles of individual cells, researchers can reveal the dynamic and heterogeneous nature of cell compositions during plant development. This approach provides more knowledge of epigenetic principles in plants, especially in developing tissues with complex cell populations (Schwartzman and Tanay, 2015).

The use of single-cell multi-omics strategies in horticulture investigation has offered a useful understanding of the molecular processes governing the development of fruits, vegetables, and ornamental crops. By integrating data from single-cell transcriptomics, epigenomics, and other omics layers, researchers can obtain a comprehensive view of cellular processes in plants, leading to advancements in horticultural practices and crop enhancement (Zhang *et al.*, 2023a). Advancements in single-cell omics methods have been transformative for plant biology, with applications spanning epigenomics, DNA methylation, transcriptomics, metabolomics, and proteomics. These technologies have been instrumental in studying plant cell identity, differentiation, and development, providing a clear insight into the molecular functions steering plant growth and responses to environmental stimuli (Mo and Jiao, 2022).

The progress of computational methods for single-cell omics data analysis has further improved our ability to extract meaningful insights from complex datasets. By utilizing computational tools, researchers can integrate multi-omics data, conduct spatiotemporal clustering of the epigenome, and uncover dynamic gene regulatory networks in plants (Efremova and Teichmann, 2020). These computational approaches are essential for deciphering the intricate relationships between different molecular layers in plant cells. The spatially resolved profiling of SCE in plants has offered a unique perspective on how epigenetic modifications influence cell fate and function within complex tissues. By mapping the epigenome at a single-cell resolution, researchers can elucidate the spatial organization of chromatin states and gene expression patterns, providing a valuable understanding of the regulatory mechanisms governing plant growth and differentiation (Farlik *et al.*, 2015).

5.6.2 Stress responses and adaptation

Epigenetic mechanisms can respond to environmental stresses, aiding in their adaptation to varying situations (Chang *et al.*, 2020). By understanding these mechanisms, researchers can unravel the intricate processes that govern how plants cope with stressors and adapt to their surroundings (Cuperus, 2022). Single-cell epigenomics has appeared as a novel method in studying plant stress responses, allowing for a detailed examination of the genetic and epigenetic landscapes at a cellular level. This approach enables the identification of active regulatory spots and the underlying genome principles in different cell types and tissues, shedding light on stress adaptation (Mishra *et al.*, 2023). Single-cell transcriptomics has been instrumental in dissecting the coordinated stress responses in plants, providing insights into how gene expression patterns vary across different cell types under stress conditions. This technology offers a comprehensive view of the molecular changes that occur in plants when faced with environmental challenges, highlighting the interconnectedness of cellular responses in shaping the overall plant stress response (Wang *et al.*, 2024). Moreover, single-cell studies at the genome hierarchy have revealed previously unknown heterogeneity in cellular states, offering a more in-depth knowledge of the regulatory responses concerned in stress and developmental transitions (Rhee *et al.*, 2019).

Epigenetic changes, like methylation of DNA and modifications in histone, play a key role in controlling gene expression in response to stress in plants. These modifications create specific chromatin states that influence the expression of stress-related genes, thereby facilitating plant adaptation to adverse environmental conditions (Chang *et al.*, 2020). The dynamic nature of DNA methylation during stress responses has been shown to impact gene expression, highlighting the intricate interplay between epigenetic modifications and stress adaptation in plants. Understanding these epigenetic dynamics is essential for deciphering the molecular mechanisms that govern stress responses in plants (López *et al.*, 2022).

The application of single-cell omics technologies in plant research has furnished a valuable understanding of the molecular processes due to stress responses and adaptation. By integrating epigenomic data with other omics layers, researchers can gain a comprehensive understanding of how epigenetic regulation influences cellular reactions to stress in plants (Mo and Jiao, 2022). This holistic approach permits the identification of key regulatory elements that drive stress adaptation mechanisms in plants, paving the path for the development of processes to enhance stress tolerance in crops. Epigenetic memory, a phenomenon where epigenetic alterations are inherited and retained across cell divisions, plays a crucial role in plants' ability to cope with recurring stresses. This memory allows plants to respond more efficiently to subsequent stressors, enhancing their resilience and adaptive capacity. By unraveling the epigenetic mechanisms involved in stress memory, researchers can uncover novel targets for improving stress tolerance in crops and enhancing agricultural productivity (Mo and Jiao, 2022).

5.6.3 Evolution and epigenomic variation

The approach of single-cell epigenomics in studying plant evolution and epigenomic variation furnishes a thorough knowledge of the molecular mechanisms driving these functions. By exploring the epigenetic landscapes of individual plant cells, researchers can reveal the dynamic and heterogeneous nature of epigenetic modifications contributing to evolutionary changes and epigenomic diversity (Schwartzman and Tanay, 2015). This approach facilitates the identification of epigenetic variations within and between cellular clusters, shedding light on the intricate regulatory networks shaping plant evolution (He *et al.*, 2011).

Studies have highlighted the effectiveness of single-cell epigenomics in uncovering the epigenetic modification of gene expression during plant development. Genome-wide analyses have elucidated the part of DNA methylation in controlling gene imprinting in plants, offering insights into the epigenetic control of developmental processes. The dynamic alterations in DNA methylation patterns during seed development underscore the complex interplay between the epigenome and plant development, emphasizing the significance of epigenetic mechanisms in influencing plant phenotypes (He *et al.*, 2011).

Additionally, single-cell epigenomic profiling has been pivotal in deciphering the chromatin regulatory landscape of plant cells. Technologies like ChIP-seq and analysis for transposase-accessible chromatin with high-throughput sequencing (ATAC-seq) have enabled investigators to map specific chromatin states and transcription factor binding sites, offering a complete view of the regulatory mechanisms governing gene expression in plants. These methodologies provide a helpful understanding of the spatial and temporal dynamics of chromatin modifications, contributing to our comprehension of epigenomic variation in plant cells (Ma and Zhang, 2020). Epigenomic variations, including DNA methylation patterns and chromatin states, significantly influence plant diversity and adaptation. These epigenetic modifications can impact phenotypic traits and provide for the evolutionary conquest of plant species. Through the study of epigenomic polymorphisms, researchers can deepen their understanding of the factors driving diversity and heterogeneity in plant populations, providing new perspectives on the role of epigenetic variations in plant evolution (Dar *et al.*, 2022).

The incorporation of single-cell transcriptomic and epigenomic profiles has offered a holistic perspective of cellular phenotypes and regulatory networks in plants. The scAI technique has facilitated the simultaneous analysis of transcriptomic and epigenomic data from individual cells, offering an understanding of the coordination between gene expression and epigenetic modifications. This integrative analysis has the potential to unveil novel regulatory mechanisms underlying plant evolution and epigenomic variation (Jin *et al.*, 2020). Furthermore, single-cell epigenomic methods can elucidate the dynamics of epigenetic changes during plant evolution. By deconvoluting single-cell chromatin accessibility information, researchers can unravel the epigenomic characteristics of genetic subclones in plants, providing an understanding of the processes of transcriptional control and cellular heterogeneity. These approaches serve as powerful tools for studying the evolution of epigenomic landscapes in plants and understanding the impact of epigenetic modifications on the expression of genes and the function of cells (Nikolic *et al.*, 2021).

5.7 Challenges and Limitations

Single-cell epigenomics in plants encounters numerous challenges and limitations that must be addressed to propel research in this field. The research of plant single-cell epigenomic information is notably intricate due to the unique biological hurdles specific to plants (Mendieta *et al.*, 2023). While single-cell epigenomics has the potential to unveil chromatin remodeling in various cells within the reproductive lineage, offering a better comprehensive understanding of reprogramming possibilities during the formation of gamete and development of seed (Fang *et al.*, 2021), there are still significant hurdles to overcome. The sensitivity and throughput of single-cell strategies for investigating histone modifications in plants are presently limited (Bartosovic *et al.*, 2021), underscoring the necessity for advancements in this area (Table 5.1).

One of the critical challenges in plant single-cell epigenomics is the historical and foundational understanding of this research area (Mendieta *et al.*, 2023). Additionally, the computational challenges related to investigating plant single-cell epigenomic data are substantial, necessitating the development of robust tools and methodologies to extract meaningful insights (Gao *et al.*, 2023). Despite these challenges, single-cell epigenomics holds immense assurance for increasing our understanding of plant biology. By revealing chromatin structure changes at the single-cell level, this approach can furnish valuable insights into lineage determination and developmental processes (Ohkawa *et al.*, 2024). Moreover, single-cell epigenomics has the prospect of uncovering the variance in the organization of chromatin and regulation of genes within complex cell populations, shedding light on the regulatory mechanisms governing plant development (de Boer and Regev, 2018).

To overcome the challenges and limitations in plant single-cell epigenomics, ongoing research is focused on developing innovative methodologies and tools. Efforts are being made to enhance the quality control steps in SCE data analysis, addressing issues such as amplification noise that can arise in these experiments (Zhao *et al.*, 2017). Furthermore, the progress in single-cell epigenomic library construction approaches is crucial for elucidating the epigenetic diversity and causal relationships between the epigenome and transcriptome at a single-cell status (Park *et al.*, 2022).

Table 5.1. Challenges and limitations in plant single-cell epigenomics. Author's own table.

Challenge/Limitation	Description	Impact	Reference
Biological complexity	Unique biological hurdles specific to plants, such as cell wall barriers	Complicates sample preparation and analysis	Mendieta *et al.* (2023)
Limited sensitivity	Current methods for analyzing histone modifications in plants lack sensitivity	Restricts detection of subtle epigenetic changes	Bartosovic *et al.* (2021)
Low throughput	Existing techniques have limited throughput for plant single-cell epigenomic analysis	Constrains the scale of experiments and statistical power	Bartosovic *et al.* (2021)
Computational challenges	Difficulties in analyzing and interpreting plant single-cell epigenomic data	Hinders extraction of meaningful insights from complex datasets	Gao *et al.* (2023)
Technical noise	Issues such as amplification noise in single-cell experiments	Affects data quality and interpretation	Zhao *et al.* (2017)
Integration of multi-omics data	Challenges in integrating epigenomic data with other omics layers	Limits comprehensive understanding of cellular processes	Mo and Jiao (2022)
Tissue-specific barriers	Difficulty in isolating single cells from complex plant tissues	Restricts analysis of certain plant cell types or developmental stages	Fang *et al.* (2021)
Limited historical context	Lack of foundational understanding in some areas of plant epigenomics	Impacts interpretation and contextualization of new findings	Mendieta *et al.* (2023)

5.8 Future Prospects

Single-cell epigenomics in plants offers significant potential for advancing our understanding of gene regulation, cell identity, and developmental processes. The field is rapidly progressing, with the continuous blossoming of the latest technologies and methodologies to enable more complete analyses of plant epigenomes at the single-cell status. The introduction of single-cell RNA sequencing has already transformed our ability to explore molecular associations among particular cells (Ryu *et al.*, 2019). This technology has opened avenues for investigating lineage determination and chromatin structure changes in plants (Ohkawa *et al.*, 2024). As emphasized, emerging single-cell epigenomic methods can improve our knowledge of gene regulation by providing unparalleled insights into the epigenetic landscape of cells (Clark *et al.*, 2016).

Furthermore, the utilization of single-cell epigenomic data is set to revolutionize our comprehension of the significance of *cis*-regulatory elements in plant genomes (Mendieta *et al.*, 2023). Through the incorporation of single-cell multimodal epigenomic data, researchers can now assess considerable histone transformations and chromatin accessibility within the same plant cell, offering a more holistic view of the epigenome (Gao and Welch, 2024). Additionally, the refinement of single-cell epigenomic library construction techniques, focusing on DNA methylation, chromatin accessibility, and histone modifications, is pivotal for elucidating epigenetic diversity and causal relationships at a single-cell status (Park *et al.*, 2022).

Single-cell approaches have illuminated the epigenetic diversity of plant tissues, enabling the profiling of the modifications of histone and transcription factors in intricate tissues (Bartosovic *et al.*, 2021). The association of single-cell epigenomics and transcriptomics from the same cell empowers researchers to address longstanding queries regarding the correlation between chromatin features and transcriptional positions (Pinheiro *et al.*, 2021). Moreover, the joint detection of chromatin possession and transcriptome at a single-cell level delivers a more detailed interpretation of chromatin dynamics in plant cells (Xiong *et al.*, 2021).

In the realm of plant development, single-cell and single-cell-type approaches are indispensable for generating precise profiles of the metabolome, proteome, transcriptome, and epigenome during processes like nodule development (Libault, 2018). The amalgamation of single-cell RNA sequencing and epigenomic analyses will be pivotal in unraveling early cell lineage specification and epigenetic mechanisms in plant development and diseases. Single-cell DNA methylome character has facilitated the examination of epigenomic heterogeneity in plant tissues and cellular reprogramming processes (Luo *et al.*, 2018).

Looking ahead, the future of single-cell epigenomics in plants lies in its ability to document the past and predict the future of genome function and cell identity (Kelsey *et al.*, 2017). Multi-omics studies capturing the epigenome and transcriptome within single plant cells during development will offer substantial insights into the regulatory mechanisms steering cellular differentiation. The continuous enhancement of SCE sequencing techniques will further augment our capacity to decipher the complexities of plant epigenetic regulation.

5.9 Conclusion

Single-cell epigenomics has appeared as a groundbreaking field in plant biology, offering unprecedented insights into gene regulation, cell identity, and developmental processes at a cellular resolution. This cutting-edge approach has revolutionized our understanding of plant epigenetics by enabling the investigation of chromatin dynamics, DNA methylation patterns, and other epigenetic modifications at the individual cell level. Advanced technologies and methodologies, such as scATAC-seq, scBS-seq, and multi-omics approaches, have significantly enhanced our ability to probe the epigenetic landscape of plant cells. These techniques have been instrumental in unraveling the complexities of plant development, stress responses, and evolutionary processes, providing a more nuanced understanding of how epigenetic modifications influence cellular function and adaptation. The incorporation of single-cell epigenomic information with other omics layers, particularly transcriptomics, has offered a holistic perspective of cellular processes and regulatory networks in plants. Computational tools and data analysis methods have been conceived to address the outstanding challenges posed by single-cell epigenomic data in plants, facilitating more robust and comprehensive analyses.

Despite the significant progress, the field still faces challenges, including technical limitations in sensitivity and throughput, as well as computational hurdles in data interpretation and integration. However, ongoing research and technological advancements are continuously addressing these issues. Looking to the future, single-cell epigenomics in plants holds immense promise for further elucidating the intricate relationships between epigenetic modifications, gene expression, and cellular identity. This field is poised to drive transformative discoveries in plant biology, potentially leading to innovations in crop improvement, stress tolerance, and sustainable agriculture.

References

Alonso-Peral, M.M., Li, J., Li, Y., Allen, R.S., Schnippenkoetter, W. *et al.* (2010) The microrna159-regulated GAMYB-like genes inhibit growth and promote programmed cell death in *Arabidopsis*. *Plant Physiology* 154(2), 757–771. DOI: 10.1104/pp.110.160630.

Arakawa, T., Noguchi, M., Sumitomo, K., Yamaguchi, Y. and Shoji, S. (2011) High-throughput single-cell manipulation system for a large number of target cells. *Biomicrofluidics* 5(1). DOI: 10.1063/1.3567101.

Bartosovic, M. and Castelo-Branco, G. (2023) Multimodal chromatin profiling using nanobody-based single-cell CUT&tag. *Nature Biotechnology* 41(6), 794–805. DOI: 10.1038/s41587-022-01535-4.

Bartosovic, M., Kabbe, M. and Castelo-Branco, G. (2021) Single-cell CUT&tag profiles histone modifications and transcription factors in complex tissues. *Nature Biotechnology* 39(7), 825–835. DOI: 10.1038/s41587-021-00869-9.

Chandana, B.S., Mahto, R.K., Singh, R.K., Ford, R., Vaghefi, N. *et al.* (2022) Epigenomics as potential tools for enhancing magnitude of breeding approaches for developing climate resilient chickpea. *Frontiers in Genetics* 13, 900253. DOI: 10.3389/fgene.2022.900253.

Chang, Y.-N., Zhu, C., Jiang, J., Zhang, H., Zhu, J.-K. *et al.* (2020) Epigenetic regulation in plant abiotic stress responses. *Journal of Integrative Plant Biology* 62(5), 563–580. DOI: 10.1111/jipb.12901.

Chinnusamy, V. and Zhu, J.K. (2009) Epigenetic regulation of stress responses in plants. *Current Opinion in Plant Biology* 12(2), 133–139. DOI: 10.1016/j.pbi.2008.12.006.

Clark, S.J., Lee, H.J., Smallwood, S.A., Kelsey, G. and Reik, W. (2016) Single-cell epigenomics: Powerful new methods for understanding gene regulation and cell identity. *Genome Biology* 17, 1–10. DOI: 10.1186/s13059-016-0944-x.

Cuperus, J.T. (2022) Single-cell genomics in plants: Current state, future directions, and hurdles to overcome. *Plant Physiology* 188(2), 749–755. DOI: 10.1093/plphys/kiab478.

Cusanovich, D.A., Daza, R., Adey, A., Pliner, H.A., Christiansen, L. *et al.* (2015) Multiplex single cell profiling of chromatin accessibility by combinatorial cellular indexing. *Science* 348(6237), 910–914. DOI: 10.1126/science.aab1601.

Danese, A., Richter, M.L., Chaichoompu, K., Fischer, D.S., Theis, F.J. *et al.* (2021) EpiScanpy: Integrated single-cell epigenomic analysis. *Nature Communications* 12(1), 5228. DOI: 10.1038/s41467-021-25131-3.

Dar, T.U.H., Akhter, N. and Dar, S.A. (2022) Editorial: Epigenomic polymorphisms: The drivers of diversity and heterogeneity. *Frontiers in Genetics* 13, 1008178. DOI: 10.3389/fgene.2022.1008178.

Deal, R.B. and Henikoff, S. (2011) The INTACT method for cell type-specific gene expression and chromatin profiling in *Arabidopsis thaliana*. *Nature Protocols* 6(1), 56–68. DOI: 10.1038/nprot.2010.175.

de Boer, C.G. and Regev, A. (2018) BROCKMAN: Deciphering variance in epigenomic regulators by k-mer factorization. *BMC Bioinformatics* 19(1), 13. DOI: 10.1186/s12859-018-2255-6.

Dorrity, M.W., Alexandre, C.M., Hamm, M.O., Vigil, A.-L., Fields, S. *et al.* (2021) The regulatory landscape of *Arabidopsis thaliana* roots at single-cell resolution. *Nature Communications* 12(1), 3334. DOI: 10.1038/s41467-021-23675-y.

Douhovnikoff, V. and Dodd, R.S. (2015) Epigenetics: A potential mechanism for clonal plant success. *Plant Ecology* 216, 227–233. DOI: 10.1007/s11258-014-0430-z.

Efremova, M. and Teichmann, S.A. (2020) Computational methods for single-cell omics across modalities. *Nature Methods* 17(1), 14–17. DOI: 10.1038/s41592-019-0692-4.

Fang, H., Shao, Y. and Wu, G. (2021) Reprogramming of histone H3 lysine methylation during plant sexual reproduction. *Frontiers in Plant Science* 12, 782450. DOI: 10.3389/fpls.2021.782450.

Farlik, M., Sheffield, N.C., Nuzzo, A., Datlinger, P., Schönegger, A. *et al.* (2015) Single-cell DNA methylome sequencing and bioinformatic inference of epigenomic cell-state dynamics. *Cell Reports* 10(8), 1386–1397. DOI: 10.1016/j.celrep.2015.02.001.

Gao, C. and Welch, J.D. (2024) Integrating single-cell multimodal epigenomic data using 1D-convolutional neural networks. *bioRxiv* 02(16), 580655. DOI: 10.1101/2024.02.16.580655.

Gao, Z., Chen, X., Li, Z., Cui, X., Chen, S. *et al.* (2023) ScEpiTools: A database to comprehensively interrogate analytic tools for single-cell epigenomic data. *Bioinformatics* 04. DOI: 10.1101/2023.04.27.538652.

Guo, F., Li, L., Li, J., Wu, X., Hu, B. *et al.* (2017) Single-cell multi-omics sequencing of mouse early embryos and embryonic stem cells. *Cell Research* 27(8), 967–988. DOI: 10.1038/cr.2017.82.

He, G., Elling, A.A. and Deng, X.W. (2011) The epigenome and plant development. *Annual Review of Plant Biology* 62(1), 411–435. DOI: 10.1146/annurev-arplant-042110-103806.

Jin, S., Zhang, L. and Nie, Q. (2020) ScAI: An unsupervised approach for the integrative analysis of parallel single-cell transcriptomic and epigenomic profiles. *Genome Biology* 21(1), 19. DOI: 10.1186/s13059-020-1932-8.

Kawakatsu, T., Nery, J.R., Castanon, R. and Ecker, J.R. (2017) Dynamic DNA methylation reconfiguration during seed development and germination. *Genome Biology* 18(1), 171. DOI: 10.1186/s13059-017-1251-x.

Kazwini, N.E. and Sanguinetti, G. (2024) SHARE-topic: Bayesian interpretable modeling of single-cell multi-omic data. *Genome Biology* 25(1), 55. DOI: 10.1186/s13059-024-03180-3.

Kelsey, G., Stegle, O. and Reik, W. (2017) Single-cell epigenomics: Recording the past and predicting the future. *Science* 358(6359), 69–75. DOI: 10.1126/science.aan6826.

Kumar, S. (2018) Epigenomics of plant responses to environmental stress. *Epigenomes* 2(1), 6. DOI: 10.3390/epigenomes2010006.

Kumar, U., Raj, S., Sreenikethanam, A., Maddheshiya, R., Kumari, S. *et al.* (2023) Multi-omics approaches in plant–microbe interactions hold enormous promise for sustainable agriculture. *Agronomy* 13(7), 1804. DOI: 10.3390/agronomy13071804.

Lee, D.-S., Luo, C., Zhou, J., Chandran, S., Rivkin, A. *et al.* (2018) Single-cell multi-omic profiling of chromatin conformation and DNA methylome. *BioRxiv* 503235. DOI: 10.1101/503235.

Li, C., Virgilio, M., Collins, K.L. and Welch, J.D. (2022a) Single-cell multi-omic velocity infers dynamic and decoupled gene regulation. *Bioinformatics*. DOI: 10.1101/2021.12.13.472472.

Li, C., Wood, J.C., Deans, N.C., Jarrell, A.F., Martin, D. *et al.* (2022b) Nuclei isolation protocol from diverse angiosperm species. *BioRxiv* 11. DOI: 10.1101/2022.11.03.515090.

Libault, M. (2018) Transcriptional reprogramming of legume genomes: Perspective and challenges associated with single-cell and single cell-type approaches during nodule development. *Frontiers in Plant Science* 9, 1600. DOI: 10.3389/fpls.2018.01600.

Lin, C.H., Chang, H.C. and Hsu, C.H. (2016) A microfluidic platform for high-throughput single-cell isolation and culture. *Journal of Visualized Experiments* 112, 54105. DOI: 10.3791/54105.

Lin, Z., Zhang, W., Liu, T., Luo, S., Wu, Y. *et al.* (2022) 3D printed microfluidics-based droplet generation for single cell encapsulation and optical imaging. *Optics in Health Care and Biomedical Optics XII* 2320, 311–317. DOI: 10.1117/12.2644130.

Liu, Y., Dang, P., Liu, L. and He, C. (2019) Cold acclimation by the CBF-COR pathway in a changing climate: lessons from *Arabidopsis thaliana*. *Plant Cell Reports* 38, 511–519. DOI: 10.1007/s00299-019-02376-3.

López, M.E., Roquis, D., Becker, C., Denoyes, B. and Bucher, E. (2022) DNA methylation dynamics during stress response in woodland strawberry (*Fragaria vesca*). *Horticulture Research* 9, uhac174. DOI: 10.1093/hr/uhac174.

Lu, T., Ang, C.E. and Zhuang, X. (2022) Spatially resolved epigenomic profiling of single cells in complex tissues. *Cell* 185(23), 4448–4464. DOI: 10.1016/j.cell.2022.09.035.

Luo, C., Rivkin, A., Zhou, J., Sandoval, J.P., Kurihara, L. *et al.* (2018) Robust single-cell DNA methylome profiling with snmc-seq2. *Nature Communications* 9(1), 3824. DOI: 10.1038/s41467-018-06355-2.

Maher, K.A., Bajic, M., Kajala, K., Reynoso, M., Pauluzzi, G. *et al.* (2018) Profiling of accessible chromatin regions across multiple plant species and cell types reveals common gene regulatory principles and new control modules. *The Plant Cell* 30(1), 15–36. DOI: 10.1105/tpc.17.00581.

Marand, A.P., Zhang, X., Nelson, J., Braga Dos Reis, P.A. and Schmitz, R.J. (2021) Profiling single-cell chromatin accessibility in plants. *STAR Protocols* 2(3), 100737. DOI: 10.1016/j.xpro.2021.100737.

Ma, S. and Zhang, Y. (2020) Profiling chromatin regulatory landscape: Insights into the development of ChIP-seq and ATAC-seq. *Molecular Biomedicine* 1(1), 9. DOI: 10.1186/s43556-020-00009-w.

Mendieta, J.P., Sangra, A., Yan, H., Minow, M.A.A. and Schmitz, R.J. (2023) Exploring plant cis-regulatory elements at single-cell resolution: Overcoming biological and computational challenges to advance plant research. *The Plant Journal* 115(6), 1486–1499. DOI: 10.1111/tpj.16351.

Miryeganeh, M. (2021) Plants' epigenetic mechanisms and abiotic stress. *Genes* 12(8), 1106. DOI: 10.3390/genes12081106.

Mishra, S., Pandey, N., Chawla, S., Sharma, M., Chandra, O. *et al.* (2023) Matching queried single-cell open-chromatin profiles to large pools of single-cell transcriptomes and epigenomes for reference supported analysis. *Genome Research* 33(2), 218–231. DOI: 10.1101/gr.277015.122.

Mo, Y. and Jiao, Y. (2022) Advances and applications of single-cell omics technologies in plant research. *The Plant Journal* 110(6), 1551–1563. DOI: 10.1111/tpj.15772.

Narsai, R., Gouil, Q., Secco, D., Srivastava, A., Karpievitch, Y.V. *et al.* (2017) Extensive transcriptomic and epigenomic remodelling occurs during *Arabidopsis thaliana* germination. *Genome Biology* 18(1), 18. DOI: 10.1186/s13059-017-1302-3.

Nikolic, A., Singhal, D., Ellestad, K., Johnston, M., Shen, Y. *et al.* (2021) Copy-scat: Deconvoluting single-cell chromatin accessibility of genetic subclones in cancer. *Science Advances* 7(42), eabg6045. DOI: 10.1126/sciadv.abg6045.

Ohkawa, Y., Harada, A., Fujii, T., Tomimatsu, K., Kato, M. *et al.* (2024) *Tracking chromatin structure changes by single-cell multi-epigenomics with RNA polymerase II binding profiles*. DOI: 10.21203/rs.3.rs-4503255/v1.

Park, K., Jeon, M.C., Kim, B., Cha, B. and Kim, J.I. (2022) Experimental development of the epigenomic library construction method to elucidate the epigenetic diversity and causal relationship between epigenome and transcriptome at a single-cell level. *Genomics & Informatics* 20(1). DOI: 10.5808/gi.21078.

Pinheiro, I., Torres-Padilla, M.E. and Almouzni, G. (2021) Epigenomics in the single cell era, an important read out for genome function and cell identity. *Epigenomics* 13(13), 981–984. DOI: 10.2217/epi-2021-0153.

Rhee, S.Y., Birnbaum, K.D. and Ehrhardt, D.W. (2019) Towards building a plant cell atlas. *Trends in Plant Science* 24(4), 303–310. DOI: 10.1016/j.tplants.2019.01.006.

Ryu, K.H., Huang, L., Kang, H.M. and Schiefelbein, J. (2019) Single-cell RNA sequencing resolves molecular relationships among individual plant cells. *Plant Physiology* 179(4), 1444–1456. DOI: 10.1104/pp.18.01482.

Sandoval, L., Mohammed Ismail, W., Mazzone, A., Dumbrava, M., Fernandez, J. *et al.* (2023) Characterization and optimization of multiomic single-cell epigenomic profiling. *Genes* 14(6), 1245. DOI: 10.3390/genes14061245.

Schwartzman, O. and Tanay, A. (2015) Single-cell epigenomics: Techniques and emerging applications. *Nature Reviews. Genetics* 16(12), 716–726. DOI: 10.1038/nrg3980.

Seyfferth, C., Renema, J., Wendrich, J.R., Eekhout, T., Seurinck, R. *et al.* (2021) Advances and opportunities in single-cell transcriptomics for plant research. *Annual Review of Plant Biology* 72(1), 847–866. DOI: 10.1146/annurev-arplant-081720-010120.

Thibivilliers, S. and Libault, M. (2021) Enhancing our understanding of plant cell-to-cell interactions using single-cell omics. *Frontiers in Plant Science* 12, 696811. DOI: 10.3389/fpls.2021.696811.

Thibivilliers, S., Anderson, D. and Libault, M. (2020) Isolation of plant root nuclei for single cell RNA sequencing. *Current Protocols in Plant Biology* 5(4), e20120. DOI: 10.1002/cppb.20120.

Tu, X., Marand, A.P., Schmitz, R.J. and Zhong, S. (2021) A combinatorial indexing strategy for epigenomic profiling of plant single cells. *BioRxiv*. DOI: 10.1101/2021.11.07.467612.

Wang, G., Ryu, K.H., Dinneny, A., Carlson, J., Goodstein, D. *et al.* (2024) Diversification of gene expression across extremophytes and stress-sensitive species in the Brassicaceae. *Plant Biology*. DOI: 10.1101/2024.06.21.599952.

Wang, R., Jiang, C., Huang, T., Zhang, Z., Wang, M. *et al.* (2020) A simulation study of the geographical distribution of *Actinidia arguta* in *China*. *Polish Journal of Environmental Studies* 29(2), 1889–1898. DOI: 10.15244/pjoes/109306.

Williams, B.R., Edwards, C.E., Kwasniewski, M.T. and Miller, A.J. (2020) Epigenomic patterns reflect irrigation and grafting in the grapevine clone 'chambourcin'. *Genomics*. DOI: 10.1101/2020.09.09.290072.

Xiang, G., Giardine, B.M., Mahony, S., Zhang, Y. and Hardison, R.C. (2021) S3V2-IDEAS: A package for normalizing, denoising and integrating epigenomic datasets across different cell types. *Bioinformatics* 37(18), 3011–3013. DOI: 10.1093/bioinformatics/btab148.

Xiong, H., Luo, Y., Wang, Q., Yu, X. and He, A. (2021) Single-cell joint detection of chromatin occupancy and transcriptome enables higher-dimensional epigenomic reconstructions. *Nature Methods* 18(6), 652–660. DOI: 10.1038/s41592-021-01129-z.

Zhang, J., Ahmad, M. and Gao, H. (2023a) Application of single-cell multi-omics approaches in horticulture research. *Molecular Horticulture* 3(1), 18. DOI: 10.1186/s43897-023-00067-y.

Zhang, R., Zhang, C., Yu, C., Dong, J. and Hu, J. (2022) Integration of multi-omics technologies for crop improvement: Status and prospects. *Frontiers in Bioinformatics* 2, 1027457. DOI: 10.3389/fbinf.2022.1027457.

Zhang, X., Guo, R., Shen, R., Landis, J.B., Jiang, Q. *et al.* (2023b) The genomic and epigenetic footprint of local adaptation to variable climates in kiwifruit. *Horticulture Research* 10(4), uhad031. DOI: 10.1093/hr/uhad031.

Zhang, Y., Yue, S., Liu, M., Wang, X., Xu, S. *et al.* (2023c) Combined transcriptome and proteome analysis reveal the key physiological processes in seed germination stimulated by decreased salinity in the seagrass *Zostera marina* L. *BMC Plant Biology* 23(1), 605. DOI: 10.1186/s12870-023-04616-x.

Zhao, C., Hu, S.E., Huo, X. and Zhang, Y. (2017) Dr.seq2: A quality control and analysis pipeline for parallel single cell transcriptome and epigenome data. *PloS ONE* 12(7), e0180583. DOI: 10.1371/journal.pone.0180583.

6 Epigenetics Strategies Based on the CRISPR/Cas System for Breeding of Climate-Resilient Crops

Thiruvengadam Abarna[1*], Thirunethiran Karpagam[1], Angappan Shanmugapriya[1], Murugesan Thentral[1], Senthil Kumar Ilakkiya[1], Raja Helan Pricilla[1], Ramasamy Manikandan[1] and Arumugam Vijaya Anand[2]

[1]Department of Biochemistry, Shrimati Indira Gandhi College, Tiruchirappalli, Tamil Nadu, India; [2]Department of Human Genetics and Molecular Biology, Bharathiar University, Coimbatore, Tamil Nadu, India

Abstract

New avenues for functioning genomics and plant enhancement have been made possible by recent advancements in genome editing. A potent technique for genome editing is the CRISPR (clustered regularly interspaced short palindromic repeat)/Cas9 system. There is no other way to overlook the advantages of cutting-edge crop advancement tools like genome editing for climate-smart endurable agriculture, which will be required to provide a world population that is exponentially developing and will require more food with adequate nutrition, enhanced stress tolerance, and improved disease resistance. Contrary to genome editing procedures like zinc-finger nucleases (ZFNs) and transcription activator-like effector nucleases (TALENs), CRISPR/Cas9 offers cost-effectiveness, versatility in the use of different guide RNAs, and ease of cloning. Genome editing has been satisfactorily operated on several crops, and this method could enable the selective transformation of virtually any sequence of crop genome to provide unexplored variation and expedite breeding endeavours. Compared with conventional breeding, the CRISPR/Cas9 method will hasten the breeding of superior crop cultivars and pave the road for the elimination of hunger. This chapter focuses on the role of the CRISPR/Cas system in the breeding of climate-resilient crops.

Keywords: Clustered regularly interspaced short palindromic repeat, Sequence-specific nucleases, Transcription activator-like effector nucleases, Zinc-finger nucleases, Genome editing

6.1 Introduction

Changes in weather like rainfall and temperature affect how well crops grow. When it is too dry or too hot, crops can suffer, and even when it is warmer, they might mature faster. This can mess up their normal growth stages like flowering and fruiting. It also creates notable threats to the production of agriculture and the capability to guarantee adequate global nourishment to the growing population, hence there is a serious need to expand crop productivity to fulfill the nutritional requirements of the growing population. Developing new crop varieties together

*Corresponding author: abarnarajkumar@gmail.com

 Epigenetics for Climate-Smart and Sustainable Agriculture (ed. Jen-Tsung Chen)
DOI: 10.1079/9781800626102.0006

with new governance approaches might help to address these problems. Through the use of newly evolved breeding strategies, and advanced agriculture techniques, crop types could be made with enhanced stability and increased productivity and nutrition, sustaining forthcoming global food needs. In light of current biotechnology knowledge, novel technologies will be utilized to mitigate these environmental hazards and potentially achieve sustainable industrial manufacturing production (Camerlengo *et al.*, 2022).

Many strategies are used by plants to adjust to these changing environmental conditions, such as altered floral rewards, accelerated or paused growth, and altered flowering and fruiting timing. However, all of these adaptive techniques result in a considerable decrease in the plant's yield. Agricultural plants are especially sensitive to the effects of unfavorable weather, since most of them have evolved to produce more while being less tolerant of environmental challenges. Therefore, to allow crops to adapt to changing climatic circumstances, it is important to evolve crops that are high-yielding, disease-resistant, and stress-tolerant. Thus, besides improving to satisfy the needs of population growth, agricultural yields must be improved in the face of environmental constraints by creating and utilizing crop types that are stress-tolerant. Recent advances in molecular techniques like genomics-assisted breeding, high-resolution phenotyping, and advanced genome engineering can be exploited for developing advanced, climate-resilient crops. Resilient crops are those that can thrive in extreme environmental circumstances, including drought, flooding, extreme temperatures, and saltiness. They are critical for ensuring food security in a world encountering climate adaptation and a growing number of inhabitants (Acevedo *et al.*, 2020).

6.1.1 Climate change challenges

Increased droughts and water scarcity

Climate change is leading to an increase in the frequency and severity of droughts worldwide (Cook *et al.*, 2021). Traditional crops may not survive these dry periods, leading to crop failures and food insecurity. Resilient crops, with their ability to tolerate drier conditions, offer a more reliable food source (Rosenzweig *et al.*, 2001).

Extreme weather events

Changing weather patterns also bring a rise in extreme weather events like floods, heatwaves, and storms. These events can devastate crops that are not adapted to such harsh conditions. Resilient crops have a better chance of withstanding these challenges and producing yields (Sapakhova *et al.*, 2023).

Rising temperatures

As global temperatures climb, some areas might become unsuitable for traditionally grown crops. Resilient crops, bred to tolerate higher temperatures, can ensure that food production continues even in these changing environments (Moore and Lobell, 2014).

6.1.2 Growing population demands

Increased food needs

The global population is expected to be almost 10 billion by 2050. This means a significant increase in food production is needed (Godfray *et al.*, 2010). Resilient crops with higher yields and shorter growing seasons can help meet this growing demand (Long *et al.*, 2006).

Land degradation

Intensive agriculture and population growth can lead to land degradation, reducing the amount of arable land available (Lambin *et al.*, 2003). Resilient crops can thrive in less fertile or marginal lands, helping to maximize food production on the land we have (Hobbs *et al.*, 2008).

Food security

By ensuring reliable crop yields even in the beginning of climate constraints, resilient crops contribute to greater food security, especially for vulnerable populations.

Sustainable agriculture

Resilient crops often require less water and other resources, promoting more sustainable agricultural practices (Tilman *et al.*, 2002).

Improved nutrition

Many resilient crops are also highly nutritious, offering essential vitamins and minerals to combat malnutrition (Jones and Harrison, 2013).

6.1.3 Traditional breeding

There are several approaches to developing resilient crops. Traditional breeding relies on selecting plants with desirable characteristics (e.g. disease resistance and drought tolerance) and crossing them to create offspring that inherit those traits. Over generations, breeders can progressively improve these traits in a population.

The benefits of this approach are that there are well-established and safe techniques with a long history of success. Consumers generally have greater acceptance of traditionally bred crops. Traditional breeding maintains genetic diversity within a species.

The limitation is that this method is a slow process, taking many generations to achieve desired results. It relies on existing genetic variation within the species, and identifying and isolating the desired genes can be challenging.

6.1.4 Genetic modification (GM)

Genetic modification (GM) directly manipulates a plant's DNA to introduce typical genes related to desired traits. Techniques like CRISPR/Cas9 offer more precise control over genetic changes. The benefits of this approach are faster and more targeted compared with traditional breeding. It can introduce genes from unrelated species, overcoming the limitations of natural variation. It may introduce entirely new traits not found naturally. The limitation is that this method is a newer technology with potential long-term effects that need ongoing research. Public concerns exist regarding the safety and potential environmental impacts of GM crops. Regulatory hurdles can slow down the development and adoption of GM crops.

6.1.5 Combining approaches

Breeders can utilize markers associated with desirable traits identified through genetic analysis to select plants for breeding programs, accelerating the process. Once a desirable gene is identified through traditional breeding, GM techniques may be used to directly introduce it into other crop varieties.

CRISPR/Cas (clustered regularly interspaced short palindromic repeat/CRISPR-associated system) techniques are broadly categorized into two major categories: Class I and Class II. These two classes are additionally separated into six types, which are then subdivided into 33 subtypes. Each subtype is characterized by its specific CRISPR array structure and unique gene composition.

- **Class I** systems utilize a complex of multiple Cas proteins, comprehended as CASCADE (CRISPR-based complex for antiviral protection), to target and neutralize viral DNA. The subtypes within Class I include type I (Cas3), type III (Cas10), and type IV (Csf1).
- **Class II** systems, on the other hand, rely on a single, large Cas protein with multiple functional regions to achieve their antiviral activity. The subtypes within Class II include type II (Cas9), type V (Cas12a, Cas12b, etc.), and type VI (Cas13a, Cas13b, etc.).

6.1.6 CRISPR locus structure

The typical structure of a CRISPR locus comprises:

- an AT-rich leader sequence;
- repeating short sequences (direct repeats); and
- unique spacer sequences of similar length.

A single CRISPR array generally contains fewer than 50 of these repeat-spacer units. Additionally, small groups of Cas genes are typically found near the CRISPR collections, which are crucial for the system's function. This organization allows the CRISPR/Cas system to identify and cut precise sequences of foreign DNA, providing a form of adaptive immunity for bacteria and archaea. A significant area of study for crop enhancement in light of ongoing climate change is epigenetics. Gene expression changes and transmitted phenotypic diversity have been linked to heritable epigenetic modifications (EGM), which can take place without modifying DNA sequences. In an era contested by world climate instability, epigenetic variety, genetically, chemically, or environmentally impacted, assists in optimizing crop attributes by regulating plant growth and physiological responses to environmental constraints (Agarwal *et al.*, 2020). Beyond variations in DNA sequence, epigenetic alterations could aid in breeding by provisioning beneficial signals and enabling the usage of epigenome variety to forecast plant performance and promote agricultural yield. Many strategies have been designed in response to the challenges of translating model plant epigenetic systems to crops. Using epigenetics for *in vitro* rejuvenation to speed up crop breeding, modeling frameworks for predicting epigenetically regulated adaptive traits, and modifications to certain epigenetic marks that regulate the expression of desired-feature genes are a few of these tactics.

The main challenge facing agriculture in the 21st century is to boost crop yield by accelerating the creation of hardy crop varieties. Thus, epigenetics delivers basic molecular details with possible direct applications in crop enrichment, toleration, and adaptation within the context of climate modification (Sow *et al.*, 2018). In the context of the growing demand for global food safety, one of the main challenges confronting agriculture in the 21st century is to accelerate crop breeding and the growth of novel plants with increased yields, best agronomic features, and more stability to climate alteration. Improvement programs in forestry aim to find trees that can withstand biotic and abiotic stress situations better when temperatures rise. Modern breeding methods require the benefit of *in vitro* plant rejuvenation systems because they enable the clonal propagation of favoured genotypes via somatic embryogenesis, the production of double-haploids with novel genetic variability via microspore embryogenesis, and the performance of gene editing (Kapazoglou *et al.*, 2020).

Selecting genes linked to critical agronomic traits has helped breeders generate the highest cultivars with increased yield by 'crossing the elite with the nobility and waiting for the most useful' in crops. To select for good germplasm, traditional plant breeding techniques are more challenging and require a lot of work. However, by flawlessly incorporating a foreign gene into a particular location of the genome, modern genome editing (GE) techniques allow one to effectively replace an existing allele with a different one (Ahmad, 2023). GE delivers predictable and heritable mutations in selected genomic sites with minor off-target impacts. These mutations do not entail the integration of foreign gene sequences. Large fragment replacements, single-nucleotide substitutions, deletions, and insertions are examples of GE-mediated DNA alterations. Systems including transcription activator-like effector nucleases (TALENs), zinc-finger nucleases (ZFNs), and homing endonucleases (HEs) were utilized as GE devices before the discovery of CRISPR-mediated protein (Cas) (Kumar *et al.*, 2022; Haroon *et al.*, 2022).

The foundation of crop breeding and the evolution of new genetic resources is mutation. Physical and chemical mutagens, RNA interference, ZFNs, TALENs, and CRISPR/Cas are some of the strategies employed in plant mutagenesis. The CRISPR/Cas method's implementation offers a favorable platform for site-specific GE and ushers in a new era in which scientists can quickly and accurately modify key genes to produce novel germplasms with particular agronomic traits (Wang and Doudna, 2023). Based on the originally engineered CRISPR/Cas9, various CRISPR devices have been designed for specific and convincing GE across all layers of the central dogma methods (Pramanik *et al.*, 2021). Gene knockout, deletion of the gene, base editing, insertion of gene or replacement, selected random mutagenesis, epigenetic modification, transcriptional principles, and RNA editing are all made possible by the development of the CRISPR toolkit (Shelake *et al.*, 2022). The third agricultural green revolution will result from the CRISPR/Cas technique's iterative advancements, which have given rise to previously unheard-of possibilities for gene position assessments and the development of desirable germplasms in various plants (Fernie and Yan, 2019). Crop breeding will be substantially aided by the association of the CRISPR/Cas method with conventional breeding techniques, molecular marker-aided preference, double-haploid approach, sterility, and asexual reproduction. It predominantly points out the applications of CRISPR/Cas technology in gene function analysis, new germplasm formation, crop *de novo* propagation, asexual reproduction, double-haploid breeding, conventional hybrid seed production, and decoupling the trade-off effect (Nerkar *et al.*, 2022).

6.2 CRISPR/Cas System: An Overview

The science of genetic engineering has encountered a revolution attributed to the GE approach known as the CRISPR/Cas method (Fig. 6.1). It was first identified as an element of bacterial adaptive immune systems that facilitates precise, targeted alterations to DNA. It was discovered that bacteria use the CRISPR/Cas method as a defense mechanism against viruses, or phages. It is constructed from a collection of CRISPR-mediated (Cas) genes that express Cas proteins with endonuclease action

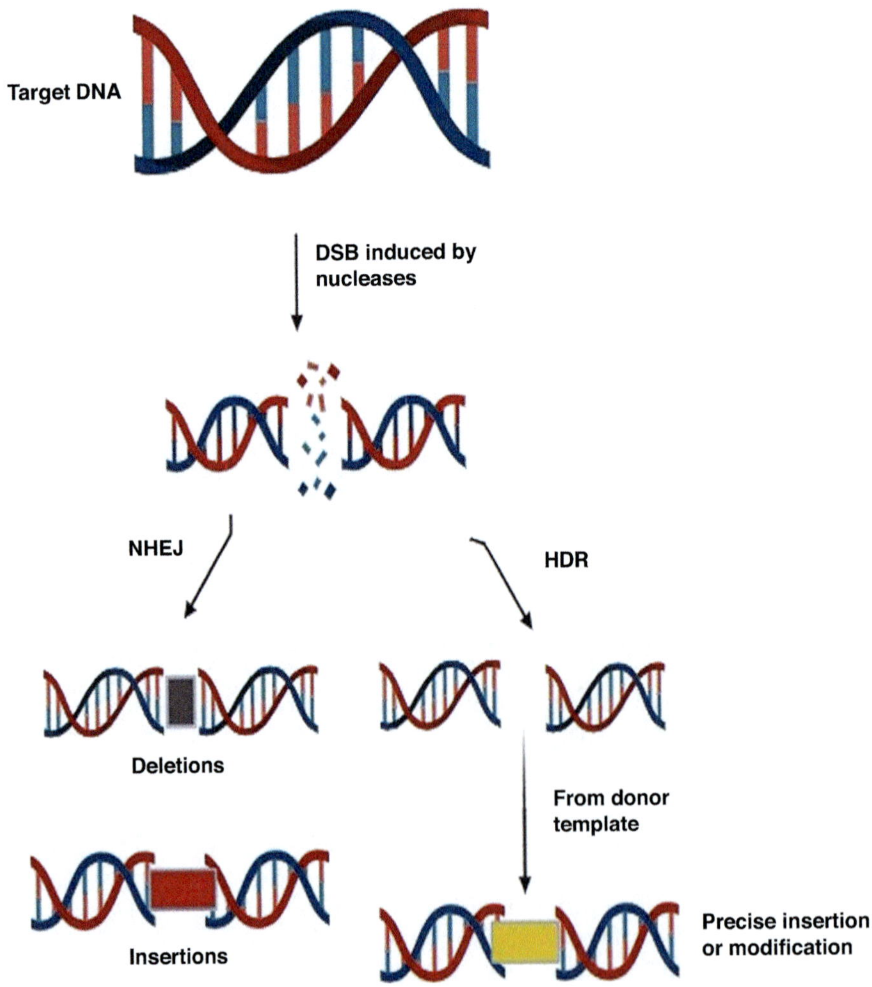

Fig. 6.1. CRISPR/Cas9 genome editing. Author's own image.

and CRISPR repeat-spacer arrays that also translate into CRISPR RNA (crRNA) and *trans*-activating CRISPR RNA (tracrRNA) (Koonin and Makarova, 2009). When a foreign genetic substance infiltrates prokaryotes, Cas proteins can cut the foreign DNA into brief segments, which are subsequently incorporated into the CRISPR array as new spacers (Makarova *et al.*, 2011). When the same invader returns, the host is protected because crRNA promptly identifies and pairs with the foreign DNA, directing the Cas protein to break target sequences of the foreign DNA (Makarova *et al.*, 2011).

One of the most well-characterized and widely utilized categories in many CRISPR-Cas methods is the CRISPR-Cas9 technique, which is derived from *Streptococcus pyogenes* (SpCas9) (Jiang and Doudna, 2017). RNA-guided Cas9 endonuclease and a single-guide RNA (sgRNA) are the two primary parts of the CRISPR-Cas9 system (Ran *et al.*, 2013). Each of the two nuclease domains that the Cas9 protein includes (HNH and RuvC) cleaves one strand of the target DNA (Chen *et al.*, 2014). A condensed form of crRNA and tracrRNA is called a sgRNA (Deltcheva *et al.*, 2011). A Cas9 ribonucleoprotein, which is composed of the Cas9 nuclease and sgRNA, can attach to and split a particular DNA target (Deltcheva *et al.*, 2011). Additionally, the association of the Cas9 protein to the target DNA needs the existence of a protospacer adjacent motif (PAM) sequence (Ran *et al.*, 2013). The two endogenous self-repair mechanisms, the error-prone non-homologous end joining (NHEJ) pathway and the homology directed repair (HDR) pathway, can repair double-strand breaks (DSBs) caused by sgRNA recruiting Cas9 endonuclease to a distinct place in the genome during the GE approach (Ceccaldi *et al.*, 2016).

6.3 Epigenetics Strategies for Developing Climate-Resilient Crops

Epigenetics directs to heritable differences in gene expression that do not affect modifications to the underlying DNA sequence. In the context of crop resilience, EGM can influence plant responses to environmental stresses. Epigenetics offers innovative strategies for enhancing crop strength against environmental pressures including drought, extreme temperatures, saltiness, and pathogen attacks (Jiang and Doudna, 2017). These strategies involve modifying gene expression without changing the DNA sequence, primarily through methylation of DNA, modifications of histone, and non-coding RNAs.

6.3.1 DNA methylation

DNA methylation affects adding a methyl group to the cytosine bases in DNA, commonly at CpG sites, which can alter gene expression. Modulating methylation can alter gene expression to enhance stress responses in two ways: targeted demethylation, and targeted methylation.

- **Targeted demethylation** uses deactivated Cas9 (dCas9) fused with demethylases (e.g. TET proteins) to remove methyl groups from stress-responsive genes, potentially activating them to improve stress tolerance.
- **Targeted methylation** uses dCas9 fused with methyltransferases (e.g. DNMT3A) to add methyl groups and silence genes that negatively affect stress responses.

Modifying methylation patterns of genes involved in stress responses can enhance plant resilience (Sow *et al.*, 2021). For example, demethylating genes that promote drought tolerance can activate these genes under water-deficit conditions. It also induces transgenerational memory that the plants can sometimes pass on stress-induced methylation patterns to their offspring, providing them with a 'memory' of stress and enhancing their resilience.

6.3.2 Histone modification

Histones are proteins near which DNA winds, and their modification (e.g. methylation, acetylation) can influence gene expression by altering chromatin structure. Modifications such as acetylation and methylation of histones can influence gene expression.

- **Histone acetylation:** Fusing dCas9 with histone acetyltransferases (e.g. p300) can enhance the expression of genes concerned with stress reactions.
- **Histone methylation:** Fusing dCas9 with histone methyltransferases (e.g. SET domain-containing proteins) can either trigger or suppress genes, depending on the precise modification.

Acetylation of histones generally promotes gene expression, which can be used to activate stress-responsive genes. Conversely, histone methylation can either suppress or activate genes, depending on the distinctive marks (Juma *et al.*, 2022). Plants can dynamically adjust histone modifications in response to environmental changes, improving their ability to cope with stress.

6.3.3 Non-coding RNAs

Non-coding RNAs, including microRNAs (miRNAs) and long non-coding RNAs (lncRNAs), play important parts in controlling gene expression in the transcriptional and post-transcriptional groups. These small RNAs can target mRNAs for degradation or inhibit their translation, thereby modulating the expression of stress-responsive genes (Seifert *et al.*, 2018). They can interfere with DNA, RNA, or proteins to control gene expression and chromatin state, aiding in the plant's stress response.

6.3.4 Epigenome editing

CRISPR/dCas9-based systems utilize dead Cas9 (dCas9) fused with epigenetic modifiers, like DNA methyltransferases (DNMTs) or histone acetyltransferases (HATs), to target specific genomic loci. dCas9-DNMT can be used to methylate specific genes to suppress their expression, while dCas9-TET can demethylate genes to activate them. This precise targeting allows for the manipulation of stress-responsive genes to enhance resilience (Ntui *et al.*, 2020).

Drought tolerance

Epigenetic activation of genes implicated in abscisic acid signaling pathways can enhance drought tolerance. For instance, activating genes like DREB (dehydration-responsive element-binding proteins) can improve water retention and stress response (Abdallah *et al.*, 2022).

Heat tolerance

Modifying the epigenetic landscape of heat shock proteins (HSPs) can help plants better withstand high temperatures. Histone acetylation near HSP genes can increase their expression during heat stress.

Salt tolerance

Targeting genes involved in ion transport and osmotic balance, such as salt overly sensitive 1, through demethylation or histone modification can enhance salinity tolerance.

6.3.5 Epigenetic priming

Treating seeds or plants with chemicals (e.g. β-aminobutyric acid, salicylic acid) or mild environmental stress (e.g. mild drought) that induce beneficial EGM, like DNA methyltransferase inhibitors or histone deacetylase inhibitors, can 'prime' the plants to better withstand stress. For example, treating soybean plants with azacitidine, a DNA methyltransferase inhibitor, has been shown to enhance drought tolerance by modulating stress-responsive gene expression (Jacobs *et al.*, 2015).

6.3.6 Epigenetic memory and transgenerational inheritance

Plants can sometimes 'remember' environmental stresses through epigenetic marks, which can be given to succeeding generations, enhancing their resilience. Plants exposed to stress can retain epigenetic marks that prepare them for future stress events. This memory can be leveraged in breeding programs to develop more resilient crops (Gevers *et al.*, 2019). For example, *Arabidopsis* plants exposed to mild drought stress can pass on epigenetic marks to their offspring, resulting in improved drought tolerance in the next generation.

6.4 CRISPR-Mediated Genome Editing

6.4.1 In rice

For a significant section of the global population, rice is a crucial crop because it serves as both their primary food source and their primary source of income from agriculture. It seems logical that the primary focus of recent research has been on GE to improve the quality of rice. According to studies using the CRISPR GE approach, it is now feasible to target and tweak rice genes to create superior kinds. In addition to facilitating site-specific integration events, the effective use and effect of CRISPR devices for rice GE will aid in the regulation of gene expression and discovery, rice functional genomics, and the creation of unique, enhanced rice features (Mazumdar *et al.*, 2016). The rice genome has a significant amount of undiscovered genes that could be utilized to boost stability to climate modification. Sequence-specific nucleases (SSNs) have been effectively utilized to target and modify many genes. The CRISPR/Cas9 method has shown the greatest promise in terms of precision, efficiency, and cost-effectiveness for producing crops with selected traits when compared with alternative systems like TALENs and ZFNs (Biswal *et al.*, 2019). The most widely utilized genetic engineering technology (GET) to enhance multiple rice agronomic features, including grain production, abiotic stress toleration, resistance to herbicide and disease, and quality of rice, is the CRISPR/Cas9 technique. In contrast to classical breeding, which takes 6–7 years to select progenies, the CRISPR/Cas9 method attains a sufficient level of homozygosity in less than a year, creating a strong tool for breeding of plants (Tabassum *et al.*, 2021).

To create rice plants with the CRISPR/Cas9 technique (Fig. 6.2), an expression vector comprising the gRNA selected to OsLip1 has to be created. An oligo duplex was generated

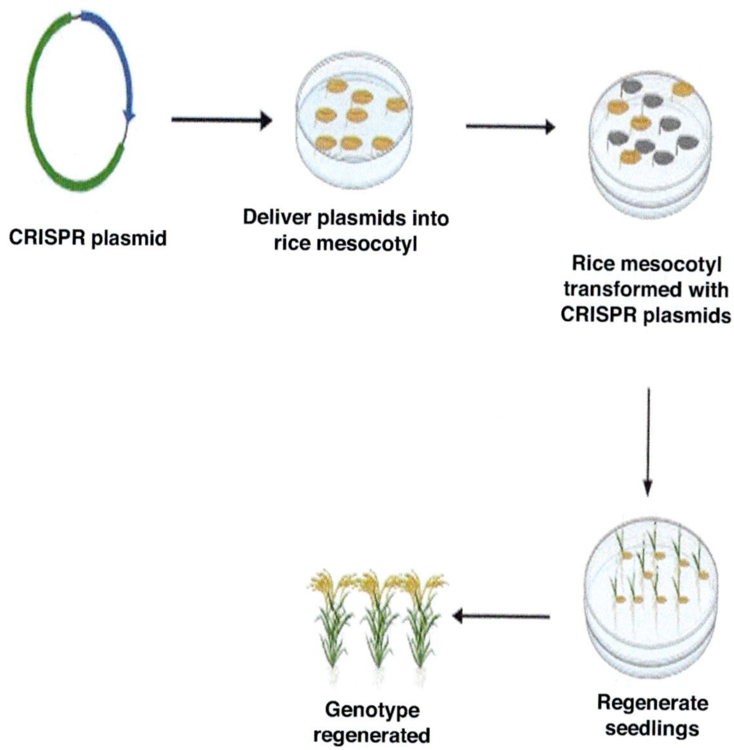

Fig. 6.2. CRISPR-based genome editing in rice.

utilizing primers precise to gRNA, and it was subsequently ligated using the BsaI-digested binary vector pRGEB31. Plasmids were recovered from the isolated colonies following the transformation of the ligated product in DH5α *Escherichia coli*. To ensure the gRNA sequence, the constructs were Sanger sequenced utilizing the M13 reverse primer. The CRISPR/Cas procedure has great promise for exactly and successfully modifying the genetic make-up of plants. The capability to alter specific genes that control a variety of applications in crop development has been greatly enhanced by inserting indels (insertion/deletions) into genomes (Thakur *et al.*, 2022). The two types of GE methods are gene targeting and targeted mutagenesis. SSNs, such as CRISPR/Cas9, are employed in targeted mutagenesis to cause DNA DSBs at the selected sequence or sequences. The gene targeting leverages the homologous recombination path, which permits donor DNA supplied from outside to repair DSBs that are either naturally occurring or specifically brought on by SSNs.

A ubiquitous plant gene-targeting strategy based on positive/negative selection has only been developed in rice. Marker-free gene-targeted rice that only has the right mutation in the targeted locus may be generated by carefully deleting the positive option marker gene utilizing the piggyBac transposon approach. Moreover, the CRISPR/Cas method was initially invented in rice (Endo and Toki, 2020). Furthermore, rice genes can be modified through GE to create non-toxic and healthier varieties. Consuming rice as a staple food puts people at risk for serious health problems due to the heavy chemical cadmium (Cd), which is extremely poisonous. Developing unique methods for Cd-free rice is essential for individual health since it is difficult to generate low-Cd rice cultivars using traditional methods. CRISPR/Cas9 editing has recently been employed to evolve Indica rice with decreased Cd accumulation by deleting the gene *OsNramp5* (Mishra *et al.*, 2018).

6.4.2 In wheat

Common wheat provides substantial amounts of protein and carbohydrates to about 60% of the global human population. Prolonged economic development is driving up demand for premium grain quality in high-yielding cultivars. In hexaploid wheat, grain quality is a multigenic quality that is more complex and controlled by multiple factors concurrently (Fig. 6.3). CRISPR/Cas9-mediated GE can produce more allelic variations more quickly and accurately (Zhang *et al.*, 2021). The benefit of the CRISPR/Cas method has become widespread in current years as a quick, accurate, and reasonably priced way to modify specific DNA sequences. GE can be utilized for gene elimination, single-base substitution, and gene/allele substitute. To achieve the best design for gRNA and Cas expression constructs, and an appropriate modification and rejuvenation system for providing the transgenes, appropriate mutation screening, and genome sequence screening using *in silico* tools is necessary. Three mechanisms exist for cells to repair breaks in their DNA caused by the CRISPR/Cas system: HDR, NHEJ pathway, or both. The plant genome is complex and redundant; thus changing just one of the mutations may require changing the others to affect cellular activity. The CRISPR/Cas system allows for the editing of many genes by introducing gRNA cassettes with one or more promoters into a single vector method. During the expression of multiple gRNA cassettes, arrays of gRNA may form, which would then be cut by endogenous RNase to liberate the gRNAs (Abdallah *et al.*, 2022). *Arabidopsis* and rice are the two plants for which CRISPR GE techniques have been most extensively studied. Wheat is weakening in the benefit of CRISPR-established genetic changes. The hexaploid genome, size, and tissue culture reluctance of wheat are the main obstacles to the effective use of CRISPR. The recently released high-grade reference genome for wheat from the International Wheat Genome Sequencing Consortium (IWGSC) will significantly accelerate the benefit of CRISPR-mediated genome engineering in wheat and help find a solution to the global food safety issue in the forthcoming decades.

Productivity will be increased in abiotic and biotic stress settings by employing the CRISPR method to develop exogenous DNA-free enhanced mutants with desirable traits. To solve multigene complex traits, newly designed multiplex GE toolkits can be employed (Kumar *et al.*, 2019). By utilizing particle bombardment, CRISPR/Cas9 DNA or RNA is introduced into

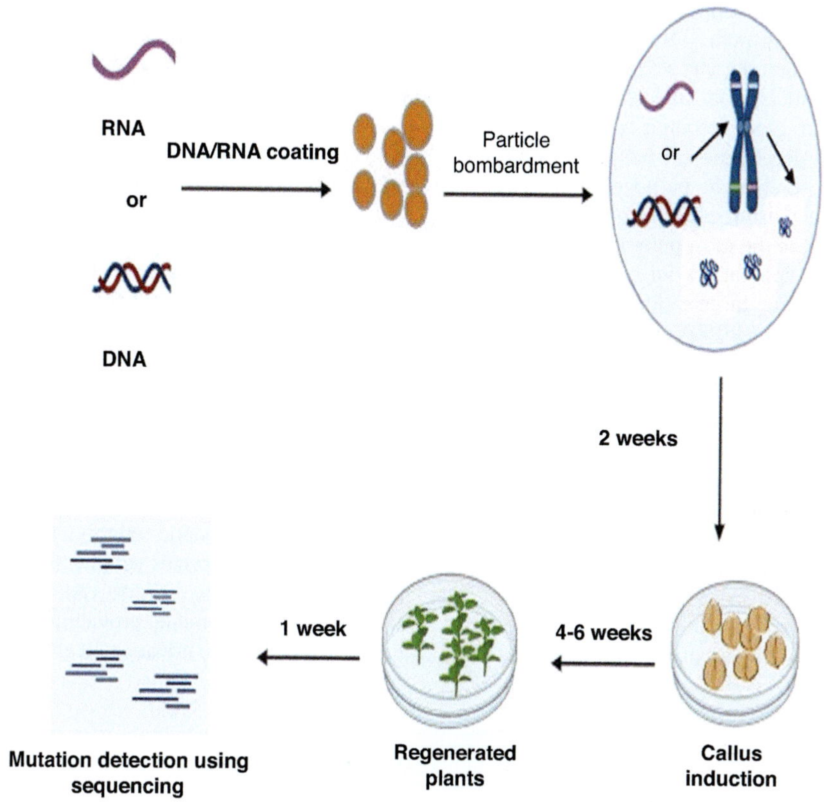

Fig. 6.3. CRISPR-mediated genome editing in wheat.

immature embryos of wheat. Although the damaged embryos produce callus cells that help seedlings grow back, the DNA or RNA of CRISPR/Cas9 breaks down after a brief period of expression and activity. An easy yet convincing technique of GE that enables mutant plants to regenerate either by transiently expressing CRISPR/Cas9 DNA or by transiently expressing Cas9-coding sequence and guide RNA *in vitro* transcripts into wheat callus cells is the way targeted mutants are identified from analysis of the T0 seedlings using DNA sequencing and the PCR-RE test. Furthermore, our tissue culture methods do not include expensive, time-consuming, or labor-intensive pesticide selection; instead, homozygous mutant plants with no noticeable transgenes are recognized in T0 populations. For tetraploid durum wheat, these genome-editing methods are quite successful (Zhang *et al.*, 2016).

6.5 CRISPR-Mediated Disease Resistance

Pathogenic bacteria are responsible for significant declines in crop output worldwide. Using disease-resistant crops can be an endurable process to satisfy the growing global population's food needs. Unfortunately, the labor- and time-intensive nature of traditional plant breeding prevents many crops from being improved effectively. The effect of GE technologies, particularly the CRISPR/Cas9 system, has made it possible to introduce improved crop traits more quickly and efficiently (Thomazella *et al.*, 2016). Plant breeding has been shown to be the most prosperous method for designing new crop forms since agricultural domestication, leading to major improvements in the world's food supply and the growth of society. Increasing plant resilience is essential for adjusting crop yield to match global population growth. Numerous pathogens, including bacteria, viruses, and fungi, can lead to substantial financial losses in agricultural production. New breeding techniques (NBTs), including developmental biology, abiotic stress tolerance, and plant–pathogen resistance, are gaining increasing interest from plant scientists. NBTs include the most potent and sophisticated molecular approaches for precise GM of one or more gene targets. Meganucleases, ZFNs, TALENs and CRISPR/Cas9 are the four types of nucleases used in GE. An effective technique for improving agricultural agronomic characteristics in the future is CRISPR/Cas9 editing (Borrelli *et al.*, 2018).

The CRISPR/Cas9 system has evolved into a powerful new tool against viruses in plants. Reports of the evolution of viral resistance in plants caused by CRISPR/Cas9 can be categorized using two main methods. The first approach aims to modify the genome of viruses; the second strategy targets the host plant components that initiate the viral process to modify the genome of plants. Tobacco and *Arabidopsis* are two model species where the viral genes necessary for replication are targeted, and these species have been the main targets of CRISPR/Cas9-driven virus resistance in plants up to this point (Khatodia *et al.*, 2017). Resistance to bacteria, fungi, and RNA viruses is made possible by focusing on host susceptibility factors (Fig. 6.4). The viral genome aims to develop resistance against RNA- and DNA-based viruses (Mushtaq *et al.*, 2019).

6.6 Advantages and Challenges

CRISPR-based epigenetic editing allows for precise targeting of specific genes without altering the DNA sequence. Unlike GM, epigenetic changes can be reversible, providing flexibility in breeding programs. Advances in gRNA design

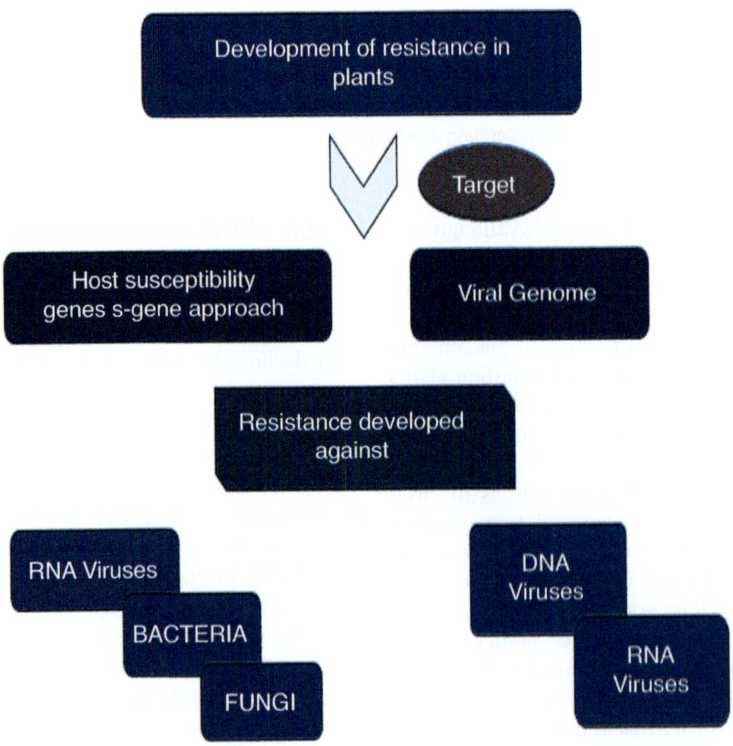

Fig. 6.4. CRISPR-mediated disease resistance.

and dCas9 technology have minimized off-target effects, making the process safer. Active delivery of CRISPR/dCas9 elements into plant cells remains a technical affirmation. The redundancy and complexity of epigenetic networks in plants require a comprehensive understanding to avoid unintended consequences (Ayanoğlu *et al.*, 2020). EGM, although not altering the DNA sequence, still faces regulatory scrutiny and public perception issues similar to those of GM.

6.7 Future Prospects

Throughout the realm of science, CRISPR/Cas9 is an extensively used, extremely potent gene-editing method. The CRISPR/Cas9 method was developed organically by archaea and bacteria as a defense tool in opposition to phage invasion and plasmid transfer. Transcription of the CRISPR region starts when homologous DNA is introduced into the bacteria again. A series of maturation and processing stages results in the production of sgRNA. The homologous spacer region is broken by the DNA strand that Cas9 is directed to shear by the sgRNA. PAMs are short, guanine-enhanced sequences that are needed for the identification of the sgRNA. NGG is preferred as a PAM by SpCas9 since it is found in the genomes of many organisms. This facilitates the application of the CRISPR approach in the areas of medicine, horticulture, and animal research. CRISPR/Cas9 modifies the nucleotide series.

However, not all regions of the genome that are highly chromatinized can be available to CRISPR/Cas9. Favorable applications of the methods include neurological illnesses, sickle cell anemia, cancer, and cardiac issues. Damage to a gene's base sequence can result in inactivation from a single damaging gene, but this inactivation is insufficient to address the several processes involved in each sickness episode. Therefore, scientists explored probable ways to modify the enzyme by determining the physicochemical structure of Cas9, how it cleaves double chains and different effects. They provided Cas9 additional capabilities by introducing effectors and changing its structural domain. These comprised transcriptional regulatory devices like dead Cas9 effectors and single-base substitution devices, including cytosine base editors, adenine base editors, and prime editors. Furthermore, RNA identification and cleavage are possible using Cas13a, which was isolated from *Leptotrichia shahii* (Li *et al.*, 2023).

Although CRISPR/Cas9 has a promising future in illness treatment, its safety may be jeopardized because it depends on DSBs to start the gene editing process. DNA breaks caused by CRISPR/Cas9 have the power to eliminate thousands of base pairs and produce unique genotypes and could be deleterious to mitotically active cells. CRISPR/Cas9 genome engineering in human pluripotent stem cells may result in p53 alterations, which may restrict the application of this method in cell replacement therapy. Furthermore, only 38% of HEK293T cells show CRISPR-mediated HDR and the high incidence of indel mutations limits the effectiveness of gene editing. However, single nucleotide polymorphisms (SNPs) are primarily responsible for most known genetic disorders (Wang *et al.*, 2022). Since CRISPR/Cas9 is a flexible, user-friendly, useful, and efficient technology, its future in cancer biology is bright. It will have an impact on cancer biology in general in the future, provided that researchers have developed efficient techniques and tools for delivering the technology to the intended cell or tissue, as well as guidelines and control measures for minimizing any off-target effects of the technology (Tiruneh *et al.*, 2021). This process offers a novel approach to cancer treatment by enabling the previously unattainable modification of the target cells' DNA.

The technology's potency, flexibility, and adaptability will make it the most effective cancer treatment available in the future. CRISPR/Cas methods have revolutionized synthetic biology and are now an essential tool for a variety of applications. Every day, advances in technology expand the possibilities for precise gene editing in a range of prokaryotic and eukaryotic organisms. Nevertheless, several obstacles remain with the technique that hinders its effective implementation. This technology sparked ethical concerns because it has also been used to modify human embryos. Moreover, the multigene activation approach helps examine the functional area of complex proteins, and the simultaneous activation of multiple genes is required to boost protein compliance. Ultimately, strategically activating several defense genes in response to pathogen infection may be one way to increase plant immunity without influencing its characteristics. Plants may be able to identify the molecules of grazing insects and respond properly and defensively. However, the effectors that insects create hinder the host plant's defensive reaction. This creates wide-spectrum insect-resistant plant materials and stops insect effectors from damaging plant defense mechanisms by employing the CRISPR-activation system (Thomazella *et al.*, 2016).

6.8 Conclusion

Combining the CRISPR/Cas technique with epigenetic strategies offers a promising avenue for developing climate-resilient crops. By precisely modulating gene expression through targeted epigenetic changes, researchers can enhance crop tolerance to environmental stresses, ensuring agricultural productivity and food security in the face of climate modification. Continuous advancements in delivery methods, knowledge of epigenetic mechanisms, and addressing regulatory challenges will be crucial for the successful application of these technologies.

References

Abdallah, N.A., Elsharawy, H., Abulela, H.A., Thilmony, R., Abdelhadi, A.A. *et al.* (2022) Multiplex CRISPR/Cas9-mediated genome editing to address drought tolerance in wheat. *GM Crops & Food* 1–17. DOI: 10.1080/21645698.2022.2120313.

Acevedo, M., Pixley, K., Zinyengere, N., Meng, S., Tufan, H. *et al.* (2020) A scoping review of adoption of climate-resilient crops by small-scale producers in low- and middle-income countries. *Nature Plants* 6(10), 1231–1241. DOI: 10.1038/s41477-020-00783-z.

Agarwal, G., Kudapa, H., Ramalingam, A., Choudhary, D., Sinha, P. *et al.* (2020) Epigenetics and epigenomics: Underlying mechanisms, relevance, and implications in crop improvement. *Functional & Integrative Genomics* 20(6), 739–761. DOI: 10.1007/s10142-020-00756-7.

Ahmad, M. (2023) Plant breeding advancements with "CRISPR-Cas" genome editing technologies will assist future food security. *Frontiers in Plant Science* 14, 1133036. DOI: 10.3389/fpls.2023.1133036.

Ayanoğlu, F.B., Elçin, A.E. and Elçin, Y.M. (2020) Bioethical issues in genome editing by CRISPR-Cas9 technology. *Turkish Journal of Biology* 44(2), 110–120. DOI: 10.3906/biy-1912-52.

Biswal, A.K., Mangrauthia, S.K., Reddy, M.R. and Yugandhar, P. (2019) CRISPR mediated genome engineering to develop climate smart rice: Challenges and opportunities. *Seminars in Cell & Developmental Biology* 96, 100–106. DOI: 10.1016/j.semcdb.2019.04.005.

Borrelli, V.M.G., Brambilla, V., Rogowsky, P., Marocco, A. and Lanubile, A. (2018) The enhancement of plant disease resistance using CRISPR/Cas9 technology. *Frontiers in Plant Science* 9, 1245. DOI: 10.3389/fpls.2018.01245.

Camerlengo, F., Frittelli, A. and Pagliarello, R. (2022) CRISPR towards a sustainable agriculture. *Encyclopedia* 2(1), 538–558. DOI: 10.3390/encyclopedia2010036.

Ceccaldi, R., Rondinelli, B. and D'Andrea, A.D. (2016) Repair pathway choices and consequences at the double-strand break. *Trends in Cell Biology* 26(1), 52–64. DOI: 10.1016/j.tcb.2015.07.009.

Chen, H., Choi, J. and Bailey, S. (2014) Cut site selection by the two nuclease domains of the Cas9 RNA-guided endonuclease. *The Journal of Biological Chemistry* 289(19), 13284–13294. DOI: 10.1074/jbc.M113.539726.

Cook, B.I., Mankin, J.S., Williams, A.P., Marvel, K.D., Smerdon, J.E. *et al.* (2021) Uncertainties, limits, and benefits of climate change mitigation for soil moisture drought in Southwestern North America. *Earth's Future* 9(9), e2021EF002014. DOI: 10.1029/2021EF002014.

Deltcheva, E., Chylinski, K., Sharma, C.M., Gonzales, K., Chao, Y. *et al.* (2011) CRISPR RNA maturation by *trans*-encoded small RNA and host factor RNAse III. *Nature* 471(7340), 602–607. DOI: 10.1038/nature09886.

Endo, M. and Toki, S. (2020) Genome editing in rice. *Rice (New York)* (1), 27. DOI: 10.1186/s12284-020-00384-6.

Fernie, A.R. and Yan, J. (2019) De novo domestication: An alternative route toward new crops for the future. *Molecular Plant* 12(5), 615–631. DOI: 10.1016/j.molp.2019.03.016.

Gevers, C., van Rijswick, H.F.M.W. and Swart, J. (2019) Peasant seeds in France: Fostering a more resilient agriculture. *Sustainability* 11(11), 3014. DOI: 10.3390/su11113014.

Godfray, J., Charles, H., Beddington, R., Crute, R., Lawrence, H. *et al.* (2010) Food security: The challenge of feeding 9 billion people. *Science* 327(5967), 812–818. DOI: 10.1126/science.1185383.

Haroon, M., Wang, X., Afzal, R., Zafar, M.M., Idrees, F. *et al.* (2022) Novel plant breeding techniques shake hands with cereals to increase production. *Plants* 11(8), 1052. DOI: 10.3390/plants11081052.

Hobbs, P.R., Sayre, K. and Gupta, R. (2008) The role of conservation agriculture in sustainable agriculture. *Philosophical Transactions of the Royal Society of London. Series B, Biological Sciences* 363(1491), 543–555. DOI: 10.1098/rstb.2007.2169.

Jacobs, T.B., LaFayette, P.R., Schmitz, R.J. and Parrott, W.A. (2015) Targeted genome modifications in soybean with CRISPR/Cas9. *BMC Biotechnology* 15, 1–10. DOI: 10.1186/s12896-015-0131-2.

Jiang, F. and Doudna, J.A. (2017) CRISPR-Cas9 structures and mechanisms. *Annual Review of Biophysics* 46(1), 505–529. DOI: 10.1146/annurev-biophys-062215-010822.

Jones, I.S.F. and Harrison, D.P. (2013) The enhancement of marine productivity for climate stabilization and food security. In: Richmond, A. and Hu, Q. (eds) *Handbook of Microalgal Culture: Applied Phycology and Biotechnology*, 2nd edn. Wiley-Blackwell, Chichester, UK, pp. 692–703. DOI: 10.1002/9781118567166.

Juma, B.S., Mukami, A., Mweu, C., Ngugi, M.P. and Mbinda, W. (2022) Targeted mutagenesis of the *CYP79D1* gene via CRISPR/Cas9-mediated genome editing results in lower levels of cyanide in cassava. *Frontiers in Plant Science* 13, 1009860. DOI: 10.3389/fpls.2022.1009860.

Kapazoglou, A., Tani, E., Avramidou, E.V., Abraham, E.M., Gerakari, M. *et al.* (2020) Epigenetic changes and transcriptional reprogramming upon woody plant grafting for crop sustainability in a changing environment. *Frontiers in Plant Science* 11, 613004. DOI: 10.3389/fpls.2020.613004.

Khatodia, S., Bhatotia, K. and Tuteja, N. (2017) Development of CRISPR/Cas9 mediated virus resistance in agriculturally important crops. *Bioengineered* 8(3), 274–279. DOI: 10.1080/21655979.2017.1297347.

Koonin, E.V. and Makarova, K.S. (2009) CRISPR-Cas: An adaptive immunity system in prokaryotes. *F1000 Biology Reports* 1, 95. DOI: 10.3410/B1-95.

Kumar, D., Yadav, A., Ahmad, R., Dwivedi, U.N. and Yadav, K. (2022) CRISPR-based genome editing for nutrient enrichment in crops: A promising approach toward global food security. *Frontiers in Genetics* 13, 932859. DOI: 10.3389/fgene.2022.932859.

Kumar, R., Kaur, A., Pandey, A., Mamrutha, H.M. and Singh, G.P. (2019) CRISPR-based genome editing in wheat: A comprehensive review and future prospects. *Molecular Biology Reports* 46(3), 3557–3569. DOI: 10.1007/s11033-019-04761-3.

Lambin, E.F., Geist, H.J. and Lepers, E. (2003) Dynamics of land-use and land-cover change in tropical regions. *Annual Review of Environment and Resources* 28(1), 205–241. DOI: 10.1146/annurev.energy.28.050302.105459.

Li, T., Yang, Y., Qi, H., Cui, W., Zhang, L. *et al.* (2023) CRISPR/Cas9 therapeutics: Progress and prospects. *Signal Transduction and Targeted Therapy* 8(1), 36. DOI: 10.1038/s41392-023-01309-7.

Long, S.P., Zhu, X.G., Naidu, S.L. and Ort, D.R. (2006) Can improvement in photosynthesis increase crop yields? *Plant Cell & Environment* 29(3), 315–330. DOI: 10.1111/j.1365-3040.2005.01493.x.

Makarova, K.S., Haft, D.H., Barrangou, R., Brouns, S.J.J., Charpentier, E. *et al.* (2011) Evolution and classification of the CRISPR-Cas systems. *Nature Reviews. Microbiology* 9(6), 467–477. DOI: 10.1038/nrmicro2577.

Mazumdar, S., Quick, W.P. and Bandyopadhyay, A. (2016) CRISPR-Cas9 mediated genome editing in rice, advancements and future possibilities. *Indian Journal of Plant Physiology* 21(4), 437–445. DOI: 10.1007/s40502-016-0252-1.

Mishra, R., Joshi, R.K. and Zhao, K. (2018) Genome editing in rice: recent advances, challenges, and future implications. *Frontiers in Plant Science* 9, 1361. DOI: 10.3389/fpls.2018.01361.

Moore, F. and Lobell, D. (2014) Adaptation potential of European agriculture in response to climate change. *Nature Climate Change* 4(7), 610–614. DOI: 10.1038/nclimate2228.

Mushtaq, M., Sakina, A., Wani, S.H., Shikari, A.B., Tripathi, P. *et al.* (2019) Harnessing genome editing techniques to engineer disease resistance in plants. *Frontiers in Plant Science* 10, 550. DOI: 10.3389/fpls.2019.00550.

Nerkar, G., Devarumath, S., Purankar, M., Kumar, A., Valarmathi, R. *et al.* (2022) Advances in crop breeding through precision genome editing. *Frontiers in Genetics* 13, 880195. DOI: 10.3389/fgene.2022.880195.

Ntui, V.O., Tripathi, J.N. and Tripathi, L. (2020) Robust CRISPR/Cas9 mediated genome editing tool for banana and plantain (*Musa* spp.). *Current Plant Biology* 21, 100128. DOI: 10.1016/j.cpb.2019.100128.

Pramanik, D., Shelake, R.M., Kim, M.J. and Kim, J.Y. (2021) CRISPR-mediated engineering across the central dogma in plant biology for basic research and crop improvement. *Molecular Plant* 14(1), 127–150. DOI: 10.1016/j.molp.2020.11.002.

Ran, F.A.F.A., Hsu, P.D., Wright, J., Agarwala, V., Scott, D.A. *et al.* (2013) Genome engineering using the CRISPR-Cas9 system. *Nature Protocols* 8(11), 2281–2308. DOI: 10.1038/nprot.2013.143.

Rosenzweig, C., Iglesias, A., Yang, X.B., Epstein, P.R. and Chivian, E. (2001) Climate change and extreme weather events. *Global Change and Human Health* 2(2), 90–104. DOI: 10.1023/A:1015086831467.

Sapakhova, Z., Raissova, N., Daurov, D., Zhapar, K., Daurova, A. *et al.* (2023) Sweet potato as a key crop for food security under the conditions of global climate change: a review. *Plants (Basel)* (13), 2516. DOI: 10.3390/plants12132516.

Seifert, F., Thiemann, A., Schrag, T.A., Rybka, D., Melchinger, A.E. *et al.* (2018) Small RNA-based prediction of hybrid performance in maize. *BMC Genomics* 19(1), 14. DOI: 10.1186/s12864-018-4708-8.

Shelake, R.M., Kadam, U.S., Kumar, R., Pramanik, D., Singh, A.K. *et al.* (2022) Engineering drought and salinity tolerance traits in crops through CRISPR-mediated genome editing: Targets, tools, challenges, and perspectives. *Plant Communications* 3(6), 100417. DOI: 10.1016/j.xplc.2022.100417.

Sow, M., Nguyen, T. and Li, S.J. (2018) Crop boom as a trigger of smallholder livelihood and land use transformations: The case of coffee production in the northern mountain region of Vietnam. *Land* 9(2), 56. DOI: 10.3390/land9020056.

Sow, M.D., Le Gac, A.-L., Fichot, R., Lanciano, S., Delaunay, A. *et al.* (2021) RNAi suppression of DNA methylation affects the drought stress response and genome integrity in transgenic poplar. *The New Phytologist* 232(1), 80–97. DOI: 10.1111/nph.17555.

Tabassum, J., Ahmad, S., Hussain, B., Mawia, A.M., Zeb, A. *et al.* (2021) Applications and potential of genome-editing systems in rice improvement: Current and future perspectives. *Agronomy* 11(7), 1359. DOI: 10.3390/agronomy11071359.

Thakur, T., Sinha, K., Kaur, T., Kapoor, R., Kumar, G. *et al.* (2022) Efficient genetic transformation of rice for CRISPR/Cas9 mediated genome-editing and stable overexpression studies: A case study on rice lipase 1 and galactinol synthase encoding genes. *Agronomy* 12(1), 179. DOI: 10.3390/agronomy12010179.

Thomazella, D., Brail, Q., Dahlbeck, D. and Staskawicz, B. (2016) CRISPR-Cas9 mediated mutagenesis of a *DMR6* ortholog in tomato confers broad-spectrum disease resistance. *bioRxiv* 064824. DOI: 10.1101/064824.

Tilman, D., Cassman, K.G., Matson, P.A., Naylor, R. and Polasky, S. (2002) Agricultural sustainability and intensive production practices. *Nature* 418(6898), 671–677. DOI: 10.1038/nature01014.

Tiruneh, G., Medhin, M., Chekol Abebe, E., Sisay, T., Berhane, N. *et al.* (2021) Current applications and future perspectives of CRISPR-Cas9 for the treatment of lung cancer. *Biologics* 15, 199–204. DOI: 10.2147/BTT.S310312.

Wang, J.Y. and Doudna, J.A. (2023) CRISPR technology: A decade of genome editing is only the beginning. *Science* 379(6629), eadd8643. DOI: 10.1126/science.add8643.

Wang, S.W., Gao, C., Zheng, Y.M., Yi, L., Lu, J.C. *et al.* (2022) Current applications and future perspective of CRISPR/Cas9 gene editing in cancer. *Molecular Cancer* 21(1), 57. DOI: 10.1186/s12943-022-01518-8.

Zhang, S., Zhang, R., Gao, J., Song, G., Li, J. *et al.* (2021) CRISPR/Cas9-mediated genome editing for wheat grain quality improvement. *Plant Biotechnology Journal* 19(9), 1684–1686. DOI: 10.1111/pbi.13647.

Zhang, Y., Liang, Z., Zong, Y., Wang, Y., Liu, J. *et al.* (2016) Efficient and transgene-free genome editing in wheat through transient expression of CRISPR/Cas9 DNA or RNA. *Nature Communications* 7(1), 12617. DOI: 10.1038/ncomms12617.

7 Epigenetics for Combating Multi-Stressors in Plants under Global Climate Change: Updated Methods and Current Achievements

Ashish Gautam[1]*, Anamika Thakur[2], Charupriya Chauhan[3], Vinay Sharma[4]* and Rakesh Kumar[1]

[1]Department of Life Science, School of Life Sciences, Central University of Karnataka, Kalaburagi, Karnataka, India; [2]Department of Biotechnology, Dr. Y.S. Parmar University of Horticulture and Forestry, Nauni Solan, Himachal Pradesh, India; [3]Department of Genetics and Plant Breeding, College of Agriculture, Acharya Narendra Deva University of Agriculture & Technology, Ayodhya, Uttar Pradesh, India; [4]Department of Genetics and Plant Breeding, Ch. Charan Singh University, Meerut, India

Abstract

In the face of escalating global climate change, plants are confronted with an array of environmental stressors, such as drought, heat, salinity, and pathogens, which severely impact agricultural productivity and food security. Epigenetic mechanisms have emerged as pivotal regulators enabling plants to adapt and respond to these multifaceted challenges. This chapter delves into the latest advancements and methodologies in utilizing epigenetic approaches to enhance plant resilience against various stressors. Drawing upon recent research, this comprehensive chapter explores the intricate interplay between epigenetic modifications, gene expression, and stress response pathways. Also, it provides an in-depth analysis of state-of-the-art techniques, including genome-wide DNA methylation profiling, histone modification assays, and small RNA sequencing, shedding light on their applicability in deciphering epigenetic changes triggered by environmental stressors. Additionally, the text discusses the role of non-coding RNAs, chromatin remodelling complexes, and epigenome editing tools in manipulating plant epigenetic regulation for stress tolerance. Furthermore, this gives an overview of synthesizing current achievements in the field, highlighting successful examples of epigenetic interventions that have enhanced stress resilience in diverse plant species. Case studies elucidate how targeted epigenetic modifications have improved drought tolerance, disease resistance, and overall growth under adverse conditions. This chapter is a valuable resource for researchers, agronomists, and policymakers, providing a roadmap for harnessing epigenetic mechanisms in the ongoing battle against climate change-induced plant stressors. This work contributes significantly to advancing sustainable agriculture and global food security in an ever-changing climate by bridging the gap between cutting-edge methodologies and practical applications.

Keywords: Epigenetics, Multi-stressors, Global climate change, Plant resilience, Environmental stress response

*Corresponding authors: gautam.ashish801@gmail.com and vinaysharma965@gmail.com

 Epigenetics for Climate-Smart and Sustainable Agriculture (ed. Jen-Tsung Chen)
DOI: 10.1079/9781800626102.0007

7.1 Introduction

Plants, being sessile in nature, are constantly subjected to varying environmental conditions that can severely impact their growth, development, and overall growth (Zhang and Bartels, 2018). Abiotic stresses, such as drought, salinity, extreme temperatures, and nutrient deficiencies, pose significant challenges to global agricultural productivity (Wang *et al.*, 2016). With the ongoing climate change, the incidence and severity of these stresses are predicted to escalate, making it essential to develop crops with enhanced resilience to such adverse conditions (Liu *et al.*, 2019). While traditional breeding and genetic engineering have led to significant improvements in plant stress tolerance, the complexity of stress responses – which are often governed by multiple genes and influenced by environmental factors – calls for new and innovative approaches to further enhance plant resilience (Springer and Schmitz, 2017). In this regard, epigenetics has emerged as a promising area of study, providing new insights and potential applications in the field of plant stress biology (Song and Zhang, 2018).

Epigenetics involves the study of heritable changes in gene expression that occur without alterations in the DNA sequence. This field encompasses various molecular mechanisms such as DNA methylation, histone modification, and non-coding RNA-mediated regulation of gene expression (Gallego-Bartolomé, 2020). These epigenetic modifications are essential for regulating gene expression in response to environmental stimuli, enabling plants to adapt to challenging conditions (Miryeganeh, 2025). Unlike genetic changes, which are permanent, epigenetic modifications are reversible and dynamic, offering plants a flexible and rapid means to adjust gene expression in response to stress (Varotto *et al.*, 2021). Recent advances in epigenetics have shown that these modifications play a critical role in controlling the expression of stress-responsive genes, thereby influencing the plant's capacity to withstand and recover from abiotic stresses (Lämke and Bäurle, 2017). Plant responses to abiotic stresses are regulated by epigenetic mechanisms, such as DNA methylation, histone modifications, and non-coding RNAs. DNA methylation involves adding a methyl group to the 5th carbon of cytosine residues in CpG dinucleotides, and is one of the most studied epigenetic modifications in plants. This process is crucial for gene silencing, transposon repression, and genomic imprinting. Under abiotic stress conditions, alterations in DNA methylation patterns have been observed across various plant species. For example, drought stress can modify the methylation status of key stress-responsive genes, leading to their activation or repression, which subsequently influences the plant's stress response (Liu *et al.*, 2019). Similarly, salinity stress triggers genome-wide changes in DNA methylation, which can either enhance or suppress the expression of genes involved in ion transport, osmotic balance, and oxidative stress management (Miryeganeh, 2025). Also, the response alters with histone modification. Histones, the protein components of chromatin, undergo various post-translational modifications, such as acetylation, methylation, phosphorylation, and ubiquitination. These modifications affect chromatin structure and, consequently, gene expression. Histone acetylation, often associated with transcriptional activation, is typically enhanced in response to abiotic stress, facilitating the expression of stress-responsive genes. Conversely, histone deacetylation can lead to gene repression under stress conditions (Miryeganeh, 2025). Histone methylation, depending on the specific amino acid residue and the degree of methylation, can either activate or repress gene expression (Zhang *et al.*, 2017). For instance, trimethylation of histone H3 lysine 4 (H3K4me3) is associated with active transcription, while H3K27me3 is linked to gene silencing (Lämke and Bäurle, 2017). Alterations in these histone marks are involved in regulating genes related to stress perception, signal transduction, and adaptive responses (Lämke and Bäurle, 2017). Non-coding RNAs, including microRNAs (miRNAs) and long non-coding RNAs (lncRNAs), play a crucial role in regulating gene expression in plants. These small RNA molecules can guide the RNA-induced silencing complex (RISC) to target specific mRNAs for degradation or translational repression (Sharma *et al.*, 2020). Under abiotic stress conditions, the expression of various miRNAs is altered, leading to the regulation of genes involved in stress responses. For example, miR398 is downregulated under oxidative stress, leading to the accumulation of its target mRNAs, which encode proteins involved in reactive oxygen species (ROS) detoxification. Similarly, lncRNAs have been shown to act as molecular scaffolds, decoys, or guides, influencing chromatin remodeling and gene expression in response to stress (Zhang *et al.*, 2017).

Another intriguing aspect of plant epigenetics is transgenerational epigenetic inheritance, where epigenetic modifications induced by environmental stresses can be passed down to subsequent generations. This inheritance mechanism allows progeny to 'remember' the stress encountered by their parents and respond more effectively when faced with similar conditions. Research has shown that plants subjected to stresses such as drought, salinity, or cold can exhibit altered DNA methylation and histone modification patterns that are retained in their offspring, even in the absence of the initial stress. This epigenetic memory provides a selective advantage in fluctuating environments, contributing to the long-term adaptation and survival of plant populations. The potential applications of epigenetics in plant breeding and biotechnology are immense (Springer and Schmitz, 2017). By identifying and manipulating key epigenetic marks associated with stress tolerance, it may be possible to develop crops with enhanced resilience to abiotic stresses (Gallego-Bartolomé, 2020). For instance, epigenetic editing tools like CRISPR/dCas9-based systems can be utilized to target specific DNA methylation or histone modification sites, enabling precise regulation of stress-responsive genes. Furthermore, understanding the epigenetic basis of stress responses could lead to the development of epigenetic biomarkers for screening and selecting stress-tolerant genotypes in breeding programs (Zhang *et al.*, 2017). However, the field of plant epigenetics is still emerging, and numerous challenges remain to be addressed (Song and Zhang, 2018).

7.2 Plant Epigenetic Regulation in Response to Stress

Abiotic stresses such as drought, salinity, extreme temperatures, and cold are among the most severe challenges, often leading to substantial losses in crop yields. These stressors are becoming increasingly frequent and severe due to climate change, posing a growing threat to global food security (IPCC, 2014). In addition to abiotic factors, plants must also contend with biotic stresses, including attacks from pathogens, pests, and other harmful organisms. The combined effects of these stresses can drastically alter plant physiology and development, forcing plants to deploy complex and adaptive strategies to survive (Deryng *et al.*, 2014; Loreti *et al.*, 2016; Kapazoglou *et al.*, 2018). To cope with these challenges, plants have evolved sophisticated mechanisms to sense and respond to environmental stressors. A key aspect of this adaptive response is the reprogramming of gene expression, which allows plants to modulate their growth and development according to the prevailing conditions. This reprogramming is not just limited to the transcriptional level; it also involves extensive post-transcriptional regulation and epigenetic modifications (Arnholdt-Schmitt, 2004; Etalo *et al.*, 2013; Toljamo *et al.*, 2016). Epigenetic changes in plants are mediated by modifications in the chromatin structure and the regulation of gene expression without altering the underlying DNA sequence, thereby optimizing their physiological and metabolic processes to cope with adverse conditions. These modifications include DNA methylation, histone modifications, and RNA-directed DNA methylation (RdDM), each contributing to the regulation of stress-responsive genes (Gutzat and Mittelsten Scheid, 2012; Annacondia *et al.*, 2018; Baluska *et al.*, 2018). Epigenetic regulation is increasingly recognized as a key component of plant responses to environmental stress (Mirouze and Paszkowski, 2011;

Koc *et al.*, 2020). For example, exposure to heat stress can disrupt the organization of chromocenters and lead to widespread alterations in nucleosome composition (Pecinka *et al.*, 2010; Santos *et al.*, 2011; Tiwari *et al.*, 2023). Similarly, abiotic stresses like salinity and drought are known to trigger histone modifications, such as acetylation and methylation, in genes involved in stress responses (van Dijk *et al.*, 2010; Ding *et al.*, 2012b; Kim *et al.*, 2012). The importance of these epigenetic modifications is further underscored by studies showing that mutations in demethylase genes result in the downregulation of stress-responsive genes, highlighting the crucial role of DNA methylation in regulating stress responses (Ding *et al.*, 2012a; Shen *et al.*, 2014). Salinity stress has also been linked to the demethylation of specific genomic loci, demonstrating the dynamic nature of DNA methylation in response to environmental challenge (Wibowo *et al.*, 2016). Additionally, stress priming has been shown to induce specific histone methylation patterns in several stress-related genes, reinforcing the concept of epigenetic memory in plants (Sani *et al.*, 2013; Lämke *et al.*, 2016). While the idea of plant stress memory has often been treated as a theoretical concept, accumulating evidence supports the view that it is indeed a real biological phenomenon, as reviewed by Galviz *et al.* (2020). Stress-induced epigenetic modifications can have long-lasting effects, contributing to transgenerational inheritance and epigenetic variation. These changes can lead to the formation of epialleles, which may be stable and inherited across generations, thus playing a role in plant evolution and adaptation. The stability of these epigenetic marks and their heritability are influenced by various factors, including the nature of the stress, the developmental stage of the plant, and the specific genetic and epigenetic context. For example, certain epigenetic marks established during stress conditions can be maintained through cell divisions and even transmitted to progeny, thereby providing a form of 'stress memory' that enhances the resilience of future generations. This transgenerational inheritance of epigenetic marks allows plants to 'remember' past stress events and mount a more efficient response when encountering similar stresses in the future. Understanding these processes is essential for leveraging epigenetic regulation in crop improvement. By creating and identifying epigenetic variation and epialleles, researchers can develop crops that are more resilient to environmental stresses. This approach, known as epigenomics-assisted breeding, involves using advanced techniques like CRISPR/Cas9 and dCas to introduce targeted epigenetic modifications. These technologies allow precise control over gene expression, enabling the development of crop varieties with enhanced stress tolerance. Epigenetic regulation also plays a role in the plant's response to nutrient availability and other environmental cues. For example, nutrient deficiency or excess can trigger epigenetic changes that modulate the expression of genes involved in nutrient uptake, transport, and assimilation. These modifications enable the plant to adjust its nutrient acquisition and utilization strategies in response to changes in nutrient availability, thereby maintaining optimal growth and development. Additionally, environmental factors such as light quality and intensity can influence epigenetic regulation, affecting processes such as photomorphogenesis and circadian rhythms. The ability of plants to sense and respond to environmental cues through epigenetic modifications is a key aspect of their adaptability and resilience. Overall, the study of plant epigenetic regulation in response to stress provides valuable insights into the mechanisms that underpin plant adaptation and resilience. The ability to modulate gene expression through epigenetic modifications allows plants to fine-tune their physiological and metabolic processes in response to a wide range of environmental challenges. By understanding these mechanisms, researchers can develop innovative strategies to enhance crop resilience and productivity, ultimately contributing to sustainable agriculture and food security in the face of global climate change. The integration of epigenetic knowledge into crop breeding programs holds great promise for the development of stress-tolerant crop varieties that can thrive in diverse and changing environments, ensuring the sustainability and productivity of agricultural systems worldwide. Ultimately, the ability of plants to adapt to environmental stress through epigenetic regulation underscores the importance of understanding these mechanisms for agricultural improvement. As we continue to face the challenges of climate change, leveraging the natural stress response mechanisms of plants could prove vital in developing strategies to enhance crop resilience and sustain agricultural productivity. By exploring the intricate relationships between stress, epigenetic regulation, and plant development, researchers can uncover new ways to breed crops that are better suited to withstand the multifaceted challenges of the future.

7.3 Abiotic Stress

7.3.1 Cold stress

Cold stress triggers a cascade of molecular events in plants, culminating in enhanced freeze tolerance through a process called cold acclimation. This involves the upregulation of C-repeat binding factor (CBF) proteins, master regulators of cold-responsive genes (Chinnusamy *et al.*, 2007). PICKLE (PKL), a chromatin remodeler, fine-tunes CBF3 activity, impacting the expression of downstream cold-responsive genes like RD29A and COR15A/B (Agius *et al.*, 2013; Kim *et al.*, 2015). PKL's involvement in RdDM and H3K27me3 pathways suggests these epigenetic marks may serve as 'memory' for cold-induced tolerance (Agius *et al.*, 2013). The ubiquitin-proteasome system also plays a crucial role. In *Arabidopsis*, HOS15, a WD40 repeat protein, targets HISTONE DEACETYLASE 2C (HD2C) for degradation, thereby relieving HD2C's repressive effect on cold acclimation genes (Lee *et al.*, 2009). This intricate interplay is evidenced by the contrasting phenotypes of hos15 and hd2c mutants: hos15 mutants are cold-sensitive, while hd2c mutants exhibit enhanced freezing tolerance even without acclimation (Lee *et al.*, 2009). Mechanistically, HOS15 promotes CBF binding to COR gene promoters by removing HD2C, highlighting a dynamic regulatory network governing plant cold acclimation (Lee *et al.*, 2009).

7.3.2 Heat stress

Plants deploy a sophisticated heat shock (HS) response orchestrated by heat shock transcription factors A1 (HsfA1s), the conductors of a multi-gene symphony. These master regulators, when silenced, leave plants vulnerable to heat, underscoring their critical role (Schramm *et al.*, 2008). HsfA1s directly activate a chorus of HS-responsive transcription factors, including DREB2A, HsfA2, HsfA7a, HsfBs, and MBF1C, each with unique expression patterns and functions (Schramm *et al.*, 2008). While HsfA1s are essential, their influence on HS-inducible genes is subtle compared with other HsfAs like HsfA2 and HsfA3, likely due to intricate post-translational modifications and interactions with regulatory partners (Schramm *et al.*, 2008). Epigenetic players also contribute to basal heat tolerance. RdDM and the histone deacetylase HDA6, acting independently, shape this resilience (Kumar *et al.*, 2012). Intriguingly, nearby transposons can modulate heat-dependent gene expression, adding another layer of complexity. HS triggers a surge in H3K9Ac and H3K4me3 on heat shock protein genes, highlighting the dynamic interplay between histone modifications and DNA methylation in both intergenerational and intragenerational stress memory (Kumar *et al.*, 2012). This intricate epigenetic landscape will be explored further in subsequent sections.

7.3.3 Salt stress

Salt stress, a formidable challenge for plants, unleashes a cascade of detrimental effects, from sodium ion toxicity to osmotic imbalance and oxidative damage. DNA methylation, a key epigenetic mark, plays a dynamic role in plant responses to salinity (Bressan *et al.*, 2002). Studies have revealed contrasting patterns of methylation changes in salt-tolerant and susceptible rice varieties. In the salt-tolerant Pokkali rice, a dramatic reduction in global DNA methylation was observed, while the susceptible IR29 exhibited a much smaller decrease (Bressan *et al.*, 2002). Interestingly, these methylation changes were largely stable even after the stress was removed, suggesting a potential role in long-term salt memory. Further investigations have highlighted the complexity of salt-induced methylation changes, with some regions showing persistent alterations while others revert to their original state upon stress relief. The HKT1 gene, a crucial player in sodium transport, provides a compelling example of methylation-mediated regulation. In *Arabidopsis*, heavy methylation of the HKT1 promoter silences its expression, while a reduction in non-CG methylation fine-tunes its activity, highlighting the intricate balance required for salt tolerance. Histone modifications, another layer of epigenetic control, also contribute to salt adaptation. For instance, histone acetylation and deacetylation, mediated by HAT and HDAC complexes, respectively, modulate gene expression in response to salinity. Remarkably, priming plants with mild salt stress can induce long-lasting changes in histone methylation profiles, leading to enhanced salt tolerance upon subsequent exposure. This 'memory' effect is associated with specific alterations in H3K27me3, a repressive histone mark. The R2R3-MYB transcription factor family, known for its role in stress responses, also exhibits methylation-dependent regulation. MYB74, a member of this family, is upregulated by salt stress, leading to the activation of downstream stress-responsive genes. This induction is accompanied by a decrease in DNA methylation and siRNA accumulation in the MYB74 promoter, underscoring the intricate interplay between epigenetic modifications and gene expression in plant salt adaptation.

7.3.4 Water stress

Water deficit stress is a recurring challenge for most plants, prompting them to develop various adaptive mechanisms, including stress memory, to cope with drought conditions (Avramova, 2015; Avramova, 2019). A key player in this process is abscisic acid (ABA), which regulates the expression of numerous drought-responsive genes. In *Arabidopsis thaliana*, repeated dehydration episodes lead to an upregulation of several ABA-induced genes, alongside a persistent partial closure of stomata during the recovery phase (Ding *et al.*, 2011; Ding *et al.*, 2012b; Ding *et al.*, 2013; Virlouvet and Fromm, 2015). The ATX1 methylase is crucial for this response, as it enhances the transcription of multiple stress-related genes by modifying histone markers, such as H3K4me3, particularly at the promoter of the NCED3 gene, which is vital for ABA biosynthesis (Ding *et al.*, 2011). Mutations in ATX1 result in heightened drought sensitivity due to reduced ABA levels and higher stomatal apertures, which accelerate water loss. Furthermore, the epigenetic landscape of drought stress-responsive genes is modulated by changes in histone modifications, such as increases in H3K4me3 and H3K9ac, which correlate with the severity of dehydration stress. During recovery, these modifications are reversed, with H3K9ac levels rapidly declining. Additionally, LHP1, part of the PRC1 complex, contributes to the repression of ABA-responsive genes through increased H3K27me3 marks, negatively regulating drought tolerance (Ramirez-Prado *et al.*, 2019). Beyond histone modifications, DNA methylation patterns also play a role in drought stress responses, with multigenerational studies in rice showing that drought stress can induce stable, heritable changes in DNA methylation (Zheng *et al.*, 2013). In *Populus trichocarpa*, drought-induced DNA methylation changes were observed to broadly affect gene expression, suggesting that epigenetic modifications are a key regulatory mechanism in the plant's response to water deficit stress (Liang *et al.*, 2014).

7.3.5 Multiple stress

Investigating the epigenetic imprints of multiple abiotic stressors reveals a fascinating tapestry of dynamic and stress-specific responses. A study utilizing 60 clones of a resilient poplar genotype (*Populus simonii* 'QL9') subjected to salinity, osmotic, heat, and cold stress illuminated the intricate dance of DNA methylation in response to environmental duress (Zhang *et al.*, 2017). Intriguingly, all four stressors elicited a rapid increase in global DNA methylation levels, with heat shock wielding the most potent initial effect. However, as the stress duration extended, osmotic and cold stress ultimately surpassed heat shock in their methylation-inducing prowess, highlighting the temporal dimension of epigenetic modifications. Delving deeper, genome-wide methylation profiling unveiled a complex landscape of stress-specific differentially methylated regions (DMRs). Heat and osmotic stress displayed a surprising degree of concordance in their methylation signatures, while cold stress carved its own distinct epigenetic path. These DMRs, enriched in genes involved in crucial regulatory functions like transcription, protein modification, and signal transduction, underscore the profound influence of stress-induced methylation changes on cellular processes. Intriguingly, the study also uncovered stress-induced methylation alterations in non-coding RNA genes, including miRNAs and lncRNAs, suggesting an additional layer of epigenetic regulation (Zhang *et al.*, 2017). Furthermore, these stress-induced methylation changes, though initially transient, demonstrated a remarkable degree of persistence, with a subset of DMRs enduring even after 6 months. This epigenetic memory hints at potential mechanisms for acclimation and future stress resilience. Further analysis of genes involved in the methylation machinery itself revealed a complex interplay between stress and gene expression. While osmotic stress primarily boosted the expression of methylation-related genes, heat shock elicited a mixed response, upregulating some while downregulating others. Cold stress, in contrast, stimulated the expression of both methylation and demethylation-related genes, suggesting a dynamic balancing act between these opposing enzymatic forces.

7.3.6 Nutrient deficit stress

Plants, masters of adaptation, have evolved intricate mechanisms to navigate the fluctuating availability of essential nutrients. Nitrogen, a key building block for life, exerts its influence through a complex interplay with epigenetic modifications. High nitrogen levels trigger a negative feedback loop, silencing the expression of NRT2.1, a crucial nitrogen transporter (Zhang *et al.*, 2017). This silencing is orchestrated by HNI9, a protein that directs the deposition of repressive H3K27me3 marks, effectively silencing the gene (Zhang *et al.*, 2017). Iron, another vital nutrient, is regulated by a symphony of epigenetic players. PRMT5, an enzyme responsible for adding methyl groups to histone tails, plays a critical role in iron homeostasis. In its absence, plants accumulate more iron and exhibit enhanced tolerance to iron deficiency (Zhang *et al.*, 2017). GCN5, a histone acetyltransferase, emerges as a key regulator of iron translocation. Its expression surges under iron deficiency, highlighting its role in orchestrating iron movement within the plant (Zhang *et al.*, 2017). Phosphate, a cornerstone of energy transfer and nucleic acid synthesis, also leaves its mark on the epigenome. Phosphate starvation triggers a cascade of changes in DNA methylation, particularly near genes involved in phosphate uptake and utilization (Zhang *et al.*, 2017). Zinc, a trace element crucial for enzyme function, also elicits epigenetic responses. Prolonged zinc deficiency leads to widespread changes in DNA methylation, influencing the expression of zinc transporters and other genes involved in zinc homeostasis (Zhang *et al.*, 2017). These examples highlight the intricate dance between nutrient availability and epigenetic modifications. Plants, through these epigenetic mechanisms, fine-tune their response to nutrient fluctuations, ensuring survival in a constantly changing environment.

7.4 Biotic Stress

7.4.1 Viruses

Plants, lacking the adaptive immunity of animals, have instead cultivated a sophisticated RNA-centric defense system to counter viral invaders. This intricate network deploys a multifaceted arsenal, including epigenetic mechanisms like post-transcriptional gene silencing (PTGS) and transcriptional gene silencing (TGS), to combat viral infection and fine-tune their own gene expression (Voinnet, 2009). PTGS, a rapid and highly specific defense mechanism, targets RNA viruses for destruction. Upon recognizing viral double-stranded RNA, plants unleash DICER-like (DCL) endoribonucleases, particularly DCL2 and DCL4, to dice these foreign molecules into small interfering RNAs (siRNAs) (Voinnet, 2009). These siRNAs, guided by Argonaute proteins, then target viral RNA for degradation, effectively halting viral replication in its tracks (Voinnet, 2009). In contrast, TGS provides a more enduring defense strategy, primarily targeting DNA viruses. This pathway hinges on the RNA-directed DNA methylation (RdDM) pathway, which ultimately silences viral gene expression by introducing DNA methylation marks into viral DNA sequences (Voinnet, 2009). Adding another layer of complexity to this epigenetic arms race, recent discoveries have unveiled the importance of RNA methylation in viral infection dynamics. The alfalfa mosaic virus (AMV), for instance, exploits m6A methylation of its RNA genome to evade the host's defenses (Li *et al.*, 2018). However, *Arabidopsis* has evolved a counter-defense mechanism in the form of ALKBH9B, an m6A demethylase. This enzyme, residing within cytoplasmic siRNA bodies, can efficiently remove m6A marks from viral RNA, thereby impairing AMV's ability to systemically infect the plant (Li *et al.*, 2018). Interestingly, this defense mechanism appears to be target-specific, as ALKBH9B does not impact the infectivity of the cucumber mosaic virus (CMV), highlighting the intricate coevolutionary dance between plants and their viral pathogens (Li *et al.*, 2018).

7.4.2 Microbes

Plants engage in a multifaceted molecular dialogue with the microbial world, employing an intricate network of defense mechanisms to ward off pathogens while fostering beneficial interactions. Central to this ongoing conversation is the dynamic language of epigenetics, with DNA methylation emerging as a key mediator of plant–microbe interactions. The plant immune system, unlike its animal counterpart, relies on innate mechanisms, including the broadly effective PAMP-triggered immunity (PTI) and the highly specific effector-triggered immunity (ETI) (Luna *et al.*, 2012). Intriguingly, both PTI and ETI engage epigenetic modifications to orchestrate a complex transcriptional reprogramming, fine-tuning the plant's defense arsenal (Luna *et al.*, 2012). DNA methylation, a potent epigenetic mark, plays a multifaceted role in this immune response. Studies in the model plant *Arabidopsis* infected with the bacterial pathogen *Pseudomonas syringae* have revealed a surprising twist: mutants deficient in DNA methylation, typically more susceptible to abiotic stress, display enhanced resistance to bacterial infection (Luna *et al.*, 2012). This counterintuitive observation suggests that pathogens may exploit DNA methylation machinery to suppress the host's defenses. Indeed, infection triggers distinct DNA methylation changes, with different methylation contexts (CG, CHG, CHH) exhibiting unique responses to specific pathogens or defense signaling molecules like salicylic acid (SA). These findings highlight the intricate dance between methylation dynamics and the plant's immune surveillance system. Further investigation has unveiled the involvement of specific DNA methylation pathways in shaping immune responses. Flagellin, a bacterial protein recognized by the plant immune receptor FLS2, triggers a transient suppression of the RdDM pathway. This downregulation of RdDM, often associated with gene silencing, allows the activation of key defense genes, enabling a rapid and robust immune response (Luna *et al.*, 2012). Conversely, the DNA demethylase ROS1 appears to play a contrasting role. Mutants lacking ROS1 exhibit heightened susceptibility to bacterial infection, suggesting that active DNA demethylation is crucial for mounting an effective defense (Luna *et al.*, 2012). Intriguingly, DNA methylation not only orchestrates immediate immune responses but also contributes to long-term immunity through a phenomenon known as priming (Luna *et al.*, 2012). This heightened state of defense readiness, often induced by prior pathogen exposure or treatment with defense-priming agents, enables a faster and more robust response to subsequent attacks. Studies have revealed that DNA hypomethylation, often associated with a more permissive chromatin state, can prime plants for enhanced defense, while hypermethylation appears to suppress defense responses. Further unraveling this epigenetic landscape, researchers have identified specific histone methyltransferases (HMTs), enzymes responsible for depositing methylation marks on histone proteins, as crucial players in plant immunity. Mutants lacking certain HMTs, such as SDG8 and SDG25, display compromised immunity to both fungal and bacterial pathogens, highlighting the importance of histone methylation in fine-tuning defense gene expression (Luna *et al.*, 2012).

7.4.3 Pests

The interaction between plants and pests presents a fascinating battleground of coevolution, where each combatant continuously adapts to gain an advantage. Increasingly, researchers are uncovering the intricate role of epigenetics, particularly DNA methylation, in this ongoing arms race (Wang *et al.*, 2017). The soybean cyst nematode (SCN), a microscopic yet formidable pest, provides a compelling case study for understanding how methylation dynamics shape both plant susceptibility and resistance (Wang *et al.*, 2017). SCN infects soybean roots, hijacking the plant's cellular machinery to construct a specialized feeding site known as a syncytium. This parasitic invasion triggers a cascade of molecular changes, including a reprogramming of the plant's epigenome (Wang *et al.*, 2017). Whole-genome bisulfite sequencing (WGBS) analysis has revealed that SCN infection induces both hypermethylation and hypomethylation events across the soybean genome, with hypomethylation emerging as the dominant trend. These methylation changes often target genes involved in defense responses, highlighting the nematode's strategy to disarm the plant's immune system (Wang *et al.*, 2017). Intriguingly, the extent and nature of DNA methylation reprogramming differ significantly between SCN-resistant and SCN-susceptible soybean lines (Wang *et al.*, 2017). Resistant lines exhibit a more measured response, with fewer differentially methylated regions (DMRs) compared with their susceptible counterparts. This observation suggests that resistant varieties may possess a more tightly regulated epigenetic landscape, enabling a more targeted and effective defense against SCN invasion (Wang *et al.*, 2017). Further analysis has revealed a strong correlation between DNA methylation changes and gene expression alterations in SCN-infected roots (Wang *et al.*, 2017). Many genes displaying differential methylation also exhibit altered expression levels, suggesting that SCN manipulates the plant's epigenome to modulate gene activity for its benefit (Wang *et al.*, 2017). This finding underscores the intricate interplay between DNA methylation and transcriptional regulation in the context of plant-pest interactions (Wang *et al.*, 2017). Remarkably, some methylation changes induced by SCN infection appear to be heritable, potentially contributing to transgenerational epigenetic inheritance (Wang *et al.*, 2017). Resistant soybean lines and their progeny exhibit specific DMRs that are absent in susceptible lines, suggesting that these epigenetic marks may be passed down through generations, providing a sort of 'molecular memory' of past infections. This intriguing possibility raises the prospect of developing novel pest control strategies that exploit epigenetic inheritance to enhance crop resilience.

7.4.4 Parasitic plants

Parasitic plants, masters of subterfuge in the botanical world, deploy specialized structures called haustoria to surreptitiously tap into the resources of their unsuspecting host plants. These molecular siphons not only extract vital nutrients and water but also serve as conduits for a clandestine exchange of molecular information, including RNA and small RNAs (sRNAs), which can profoundly influence the epigenetic landscape of the host–parasite interaction (Honaas *et al.*, 2016). The parasitic plant *Cuscuta pentagona*, commonly known as dodder, exemplifies this intricate molecular cross-talk. Transcriptome analysis has revealed a bidirectional flow of messenger RNA (mRNA) between the dodder and its host plants, *Arabidopsis* and tomato. Intriguingly, the volume of this mRNA traffic is heavily influenced by the host, with *Arabidopsis* exhibiting a significantly greater propensity for mRNA exchange compared with tomato (Honaas *et al.*, 2016). This disparity hints at the presence of active defense mechanisms in tomato, potentially limiting the parasite's ability to manipulate its gene expression. While the functional significance of this interspecies mRNA exchange remains an enigma, the persistence of some host mRNAs within the parasite's tissues suggests a potential role in modulating host defenses or hijacking cellular processes for the parasite's benefit. Adding further intrigue to this molecular dialogue, recent studies have uncovered a fascinating role for sRNAs, particularly (miRNAs), in the parasitic arsenal. Dodder species have been shown to release miRNAs into their hosts, where they can silence specific host genes. Remarkably, these mobile miRNAs appear to have evolved to specifically target host mRNAs, suggesting a sophisticated mechanism for manipulating host defenses to the parasite's advantage. For instance, dodder miRNAs targeting the *Arabidopsis* genes SEOR1 and AFB3, both implicated in plant defense, effectively cripple the host's ability to resist parasitic invasion.

7.5 Role of Epialleles in Crop Improvement

Reprogramming is essential to eliminate epigenetic fingerprints acquired during development or caused by the environment, to ensure that the subsequent development of the embryo accurately represents the genetic blueprint specific to each species. When reprogramming fails, epigenetic signatures possess a chance to be passed meiotically. Epigenetic mutations, known as epimutations, can result in the formation of epigenetic alleles, called epialleles. Epialleles are two alleles that occupy the same locus and share the same DNA sequence but reside in distinct epigenetic contexts. Therefore, this epigenetic background serves as an additional level of information that can aid in the understanding of phenotypic variation. These epimutations can have either a neutral or deleterious impact. Additionally, the transgenerational inheritance of epigenetic traits has the potential to be advantageous and may exhibit responsiveness to environmental stressors. This has significant implications for heredity, breeding, and evolution. Researchers have identified several epialleles in various species of plants. These genes govern several phenotypic traits, such as flower structure, flowering duration, maturation of fruits, plant structure, length of root biomass, sex determination, vitamin E accumulation, and more. In rice *Epi-d1* and *Epi-df* exhibit a dwarf stature. DNA hypermethylation in the *DWARF1* promoter region triggers downregulation in the case of *Epi-d1* (Miura *et al.*, 2009), and hypomethylation in the 5′ region of *FERTILIZATION-INDEPENDENT ENDOSPERM1* leads to ectopic expression in *Epi-df* (Zhang *et al.*, 2012). In *Epi-rav6*, hypomethylation of an MITE (Miniature Inverted-Repeat TE) in the *RELATED TO ABSCISIC ACID INSENSITIVE3 (ABI3)/VIVIPAROUS1 (VP1) 6 (RAV6)* promoter results in the gene being over-expressed, which leads to a wider leaf angle and a smaller grain size (Zhang *et al.*, 2015). A *clark kent* (*clk*) epiallele, which is a well-known case of an epiallele, was identified in *Arabidopsis*. Other notable epimutants in *Arabidopsis*, including *SUPERMAN* (*SUP*), *Phosphoribosyl Anthranilate Isomerise* (*PAI2*), *AGAMOUS* (*AG*), *Flowering Wageningen* (*FWA*), *BAL1*, *BONSAI* (*BNS*), and *Folate transporter 1* (*FOLT1*), were found to be related with various traits, such as the time of flowering, plant stature, starch accumulation, etc. In maize, four epialleles – *red1* (*r1*), *booster 1* (*b1*), *purple plant* (*pl1*), and *pericarp colour* (*p1*) – were involved in pigmentation (Brink, 1956; Patterson *et al.*, 1993; Hollick *et al.*, 1995; Cocciolone *et al.*, 2001). In rice, short panicle results from two epialleles: *Squamosa promoter binding protein-like* (*SPL14*) and *Epigenetic short panicle* (*ESP*) (Miura *et al.*, 2010; Luan *et al.*, 2019). In tomato, two notable epigenetic variations were identified; *colorless non-maturation* (*CNR*), associated with the ripening of fruit (Manning *et al.*, 2006), and *Vitamin E3* (*VTE3*), involved in vitamin E content (Quadrana *et al.*, 2014). In *Brassica rapa*, an epigenetic mutation in the *Pollen S-determinant* gene (*SP11*) resulted in self-incompatibility (Shiba *et al.*, 2006), while in *Cucumis melo*, a mutation in the *WASP/N--WASP-interacting protein 1* (*WIP 1*) developed female flowers (Martin *et al.*, 2009). Traits are also affected by epialleles involving transposable elements. Hypomethylation in Karma retro TE resulted in aberrant DEF gene splicing in oil palm, leading to parthenocarpic fruit and reduced yield (Ong-Abdullah *et al.*, 2015). These traits will be helpful in future breeding pipelines. Additionally, new epigenomic methodologies made it feasible to find epiallelic variants, which are likely to be involved in understanding the function they play in plants. Such epigenetic changes identified in various crops, alongside other gene pools, might provide exciting possibilities for improvements in plants.

7.6 Crop Improvement Through Epigenome-Wide Association Studies (EWAS)

Genome-wide association studies (GWAS) remain an effective tool in plant research, facilitating the exploration of the genetic blueprint underlying controls of complex traits and aiding the advancement of crop improvement programs. Epigenome-wide association studies (EWAS) can also serve as an important tool for gaining insight into the underlying causes of phenotypic variation. Upon gaining complete epigenome profiles, the researcher can execute EWAS to discover a connection between a specific phenotypic trait and variations in epigenetic patterns. EWAS consists of the following main phases: (1) sequencing of the epigenome; (2) phenotyping for the trait of interest; and (3) employing bioinformatics and statistical testing to establish a genome-scale association between both variables. Despite the extensive use of EWAS in human populations to identify epigenetic variants associated with diseases, its deployment in plants is limited. In oil palm, the first EWAS study identified epigenetic variations in the MANTLED gene at the Karma splice site (Ong-Abdullah *et al.*, 2015). While the Bad Karma epiallele is associated with fruit deformities linked to reduced yield, the Good Karma epiallele predicts a wild-type fruit set. EWAS was also used to determine the correlation between DNA methylation pattern and climate gradient in *Quercus lobata* (Gugger *et al.*, 2016). With innovations in the field of sequencing, the increasing availability of whole-epigenome resources will facilitate the execution of EWAS. Consequently, this will enhance the advancement of EWAS research in crops and contribute to the current understanding of epigenetic influences. Additionally, a pipeline has been established to investigate epidiversity and EWAS in plants (Can *et al.*, 2021). The plant offers great possibility for investigating epigenetic signatures in wild populations and their influence on phenotypic variations due to their high propensity for epimutations. Epibreeding strategies for crop development can benefit from the defined epigenetic variation. Table 7.1 provides a summary of the epialleles and epigenetic mechanisms identified in various plants. Fig. 7.1 presents a schematic illustration of epigenome-wide association mapping (EWAS).

7.7 Challenges in Epigenetics Research

Research in plant epigenetics faces several significant challenges that complicate our understanding and application of these mechanisms. One major issue is the intricate nature of epigenetic modifications, including DNA methylation, histone modifications, and non-coding RNAs, and their interactions within the plant genome (Gallego-Bartolomé, 2020). Identifying the functional significance of specific epigenetic marks and their precise roles in stress responses demands sophisticated and often expensive technologies, such as high-resolution sequencing and chromatin accessibility assays (Liu *et al.*, 2023). Additionally, the dynamic nature of epigenetic changes makes it difficult to establish consistent patterns across various developmental stages and environmental conditions (Lämke and Bäurle, 2017; Varotto *et al.*, 2021). Integrating

Table 7.1. Summary of the epialleles and epigenetic mechanisms in various plants. Author's own table.

Crop	Gene/loci	Target traits	Epigenetic mechanism	Reference
Rice	*OsSPL14*	Enhances panicle branching and grain yield	Hypomethylation in the 2.6-kb upstream region	Miura *et al.*, 2010
	OsFIE1	Dwarf stature and floral defects	Hypomethylation in the 5′ region of FIE1	Zhang *et al.*, 2012
	D1 (DWARF1)	Dwarf stature	Repressive histone modifications and DNA methylation marks in the D1 promoter region	Miura *et al.*, 2009
	RAV6	Influences the size of seeds and the angle of the leaf	Hypomethylation in the promoter region of gene resulted in an atypical expression	Zhang *et al.*, 2015
	OsAK1	Impacts the ability for photosynthesis.	At the promoter region, hypermethylation resulted in silencing of OsAK1	Wei *et al.*, 2017
	RIZBZ1, RPBF	Thickened aleurone and enhanced nutritional value	*OsROS1*-mediated DNA demethylation	Liu *et al.*, 2018
	ESP	Influences panicle architecture	Hypomethylation in the transcriptional termination region	Luan *et al.*, 2019
Maize	*lpa1-241*	Decreased phytic acid and increased inorganic P in the seed	Paramutation	Pilu *et al.*, 2009
	b1	Affects anthocyanin pigmentation	Paramutation	Patterson *et al.*, 1993
	pl1	Affects anthocyanin pigmentation	Paramutation	Hollick *et al.*, 1995
	p1	Affects pigmentation	Paramutation	Cocciolone *et al.*, 2001
Sorghum bicolor	QTLs	Number of tillers, plant height, flowering duration, and crop yield	Expression of MSH1 in lines with the Tx430 genetic background was suppressed to induce *de novo* epigenetic variation	Ketumile *et al.*, 2022
Glycine max (L.) Merr.	QTLs	Affects growth rate, male sterility, increased branching, influenced leaf and flower morphology	Expression of MSH1 in lines was suppressed to induce *de novo* epigenetic variation	Raju, 2017

Continued

Table 7.1. Continued

Crop	Gene/loci	Target traits	Epigenetic mechanism	Reference
Solanum lycopersicum	*LeSPL-CNR*	Colorless fruits	Increased DNA methylation in the promoter region results in the silencing of expression	Manning *et al.*, 2006
	QTLs	Alterations in the morphology of leaves, variegation, dwarfing, male sterility, floral growth and development, and flowering duration	Use of the MSH1 silencing line induces *de novo* epigenetic variation	Yang *et al.*, 2015
	SLTAB2	Influencing the production of pigments	Paramutation	Gouil *et al.*, 2016
	CNR, VTE3	Fruit development and ripening	SIDML2 facilitates demethylation during fruit ripening.	Gallusci *et al.*, 2016
Fragaria ananassa	Ripening-related genes	Fruit ripening	Hypomethylation during ripening	Cheng *et al.*, 2018
Sweet orange	Ripening-related genes	Fruit development and ripening	Hypermethylation during fruit-ripening	Huang *et al.*, 2019
Cotton	*COL2D*	Photoperiod sensitivity	Hypomethylation on the 5' region of gene	Song *et al.*, 2017
Arabidopsis thaliana	*FWA*	Delay flowering	Hypomethylation of two direct repeats in the 5′ regions of the gene	Soppe *et al.*, 2000
	DWF4	Loss of elongation in petioles and branches	Insertion of CAC1 activated by ddm1 hypomethylation	Miura *et al.*, 2001
	Bns	Stunted growth	Hypermethylation	Saze and Kakutani, 2007

epigenetic data with genomic and phenotypic information to fully understand how these changes impact plant traits presents another challenge (Song and Zhang, 2018). Moreover, applying epigenetic knowledge to practical uses, such as crop improvement and stress management, involves overcoming technical challenges and addressing potential off-target effects in epigenetic editing techniques (Springer and Schmitz, 2017; Zhang *et al.*, 2023). Addressing these issues is essential for advancing our grasp of epigenetics and leveraging its potential in plant science and agriculture.

7.8 Conclusion

Epigenetics is a key area of investigation that is now gaining importance as a result of the evidence that elucidates how it is involved in the tolerance of biotic and abiotic stress in plants. Interestingly, over a period many studies have

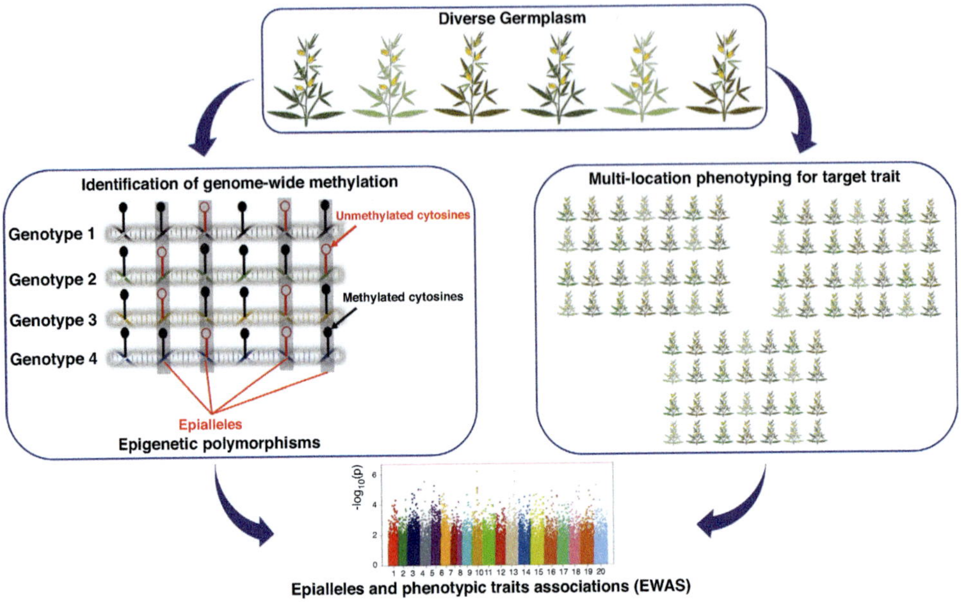

Fig. 7.1. Epigenome-wide association mapping (EWAS) schematic illustration. Author's own image.

been documented that provide an in-depth look at the importance of epigenetic control in the adaptation to biotic and abiotic stress in cereal crops, including wheat, barley, as well as other crops. Thus, there are plenty of opportunities to investigate this arena in cereals with highly complex genomes, such as wheat, and the advancements in genomics technology will undoubtedly facilitate this exploration. In wheat, the role of epigenetic components such as DNA methylation, chromatin modifications, and sRNA in regulating gene expression have been reported. Additionally, the idea of epialleles has taken on value, and there exists some proof of epiallele discovery in crops such as *Arabidopsis*, maize, and rice. An effective strategy would involve developing epi-markers with genetic markers to identify quantitative traits. Several traits have been effectively associated with particular epigenetic signatures, even in instances in which there were not any genetic markers. An effective strategy would involve developing epi-markers to identify quantitative traits. Several traits have been effectively associated with particular epigenetic signatures, even in instances in which there were not any genetic markers. Additional investigation is required to determine the suitability of this strategy for selecting desirable traits in crops and so it may be further deployed. It involves carrying out further investigations on the stability of epialleles by examining segregating populations. After discovering and developing stable epialleles that are associated with the presence of a particular trait, they could potentially be utilized as epi-markers in the breeding pipeline. After being found, such epi-markers may be targeted for epigenomic editing to deliberately induce certain epi-mutations and generate an ideal phenotype. Epigenetics provides new opportunities for employing approaches in crop breeding to select and modify epialleles and traits.

References

Agius, F., Kapoor, A., Ríos, G., Koenig, D., Grossniklaus, U. *et al.* (2013) PICKLE is required for epigenetic regulation of cold-responsive gene expression in *Arabidopsis*. *PLoS Genetics* 9(4), e1003466.

Annacondia, M.L., Magerøy, M.H. and Martinez, G. (2018) Stress response regulation by epigenetic mechanisms: Changing of the guards. *Physiologia Plantarum* 162(2), 239–250.

Arnholdt-Schmitt, B. (2004) Stress-induced cell reprogramming. A role for global genome regulation? *Plant Physiology* 136(1), 2579–2586.

Avramova, Z. (2015) Transcriptional "memory" of a stress: Transient chromatin and memory (epigenetic) marks at stress-response genes. *The Plant Journal* 83(1), 149–159.

Avramova, Z. (2019) Defence-related priming and responses to recurring drought: Two manifestations of plant transcriptional memory mediated by the ABA and JA signalling pathways. *Plant, Cell & Environment* 42(3), 983–997.

Baluska, F., Gagliano, M. and Witzany, G. (eds) (2018) *Memory and Learning in Plants*, Signaling and Communication in Plants. Springer International Publishing, Cham, Switzerland.

Bressan, R.A., Zhu, J.-K., Bohnert, H.J. and Garg, A.K. (2002) Plant responses to water stress. *Plant, Cell & Environment* 25(1), 25–38.

Brink, R.A. (1956) A genetic change associated with the R locus in maize which is directed and potentially reversible. *Genetics* 41(6), 872–889.

Can, S.N., Nunn, A., Galanti, D., Langenberger, D., Becker, C. *et al*. (2021) The EpiDiverse plant epigenome-wide association studies (EWAS) pipeline. *Epigenomes* 5, 12.

Cheng, J., Niu, Q., Zhang, B., Chen, K., Yang, R. *et al*. (2018) Downregulation of rddm during strawberry fruit ripening. *Genome Biology* 19, 1–14.

Chinnusamy, V., Ohta, M., Kanrar, S., Lee, B.H., Hong, X. *et al*. (2007) ICE1: A regulator of cold-induced transcriptome changes and freezing tolerance in *Arabidopsis*. *Genes & Development* 21(10), 1043–1054.

Cocciolone, S.M., Chopra, S., Flint-Garcia, S.A., McMullen, M.D. and Peterson, T. (2001) Tissue-specific patterns of a maize myb transcription factor are epigenetically regulated. *The Plant Journal* 27(5), 467–478.

Deryng, D., Conway, D., Ramankutty, N., Price, J. and Warren, R. (2014) Global crop yield response to extreme heat stress under multiple climate change futures. *Environmental Research Letters* 9(3), 034011.

Ding, B., Bellizzi, M. del R., Ning, Y., Meyers, B.C. and Wang, G.-L. (2012a) HDT701, a histone H4 deacetylase, negatively regulates plant innate immunity by modulating histone H4 acetylation of defense-related genes in rice. *The Plant Cell* 24(9), 3783–3794.

Ding, Y., Avramova, Z. and Fromm, M. (2011) The *Arabidopsis trithorax*-like factor ATX1 functions in dehydration stress responses via ABA-dependent and ABA-independent pathways. *The Plant Journal* 66(5), 735–744.

Ding, Y., Fromm, M. and Avramova, Z. (2012b) Multiple exposures to drought "train" transcriptional responses in *Arabidopsis*. *Nature Communications* 3(1), 740.

Ding, Y., Liu, N., Virlouvet, L., Riethoven, J.J., Fromm, M. *et al*. (2013) Four distinct types of dehydration stress memory genes in *Arabidopsis thaliana*. *BMC Plant Biology* 13, 229.

Etalo, D.W., Stulemeijer, I.J.E., Esse, H.P., Vos, R.C.H., Bouwmeester, H.J. *et al*. (2013) System-wide hypersensitive response-associated transcriptome and metabolome reprogramming in tomato. *Plant Physiology* 162(3), 1599–1617.

Gallego-Bartolomé, J. (2020) DNA methylation in plants: Mechanisms and tools for targeted manipulation. *New Phytologist* 227(1), 38–44.

Gallusci, P., Hodgman, C., Teyssier, E. and Seymour, G.B. (2016) DNA methylation and chromatin regulation during fleshy fruit development and ripening. *Frontiers in Plant Science* 7, 190266.

Galviz, Y.C.F., Ribeiro, R.V. and Souza, G.M.Y. (2020) Yes, plants do have memory. *Theoretical and Experimental Plant Physiology* 32(3), 195–202.

Gouil, Q., Novák, O. and Baulcombe, D.C. (2016) SLTAB2 is the paramutated SULFUREA locus in tomato. *Journal of Experimental Botany* 67(9), 2655–2664.

Gugger, P.F., Fitz-Gibbon, S., PellEgrini, M. and Sork, V.L. (2016) Species-wide patterns of DNA methylation variation in *Quercus lobata* and their association with climate gradients. *Molecular Ecology* 25(8), 1665–1680.

Gutzat, R. and Mittelsten Scheid, O. (2012) Epigenetic responses to stress: Triple defense? *Current Opinion in Plant Biology* 15(5), 568–573.

Hollick, J.B., Patterson, G.I., Coe, E.H., Cone, K.C. and Chandler, V.L. (1995) Allelic interactions heritably alter the activity of a metastable maize pl allele. *Genetics* 141(2), 709–719.

Honaas, L.A., Thieme, C.A., Smith, J.A. and McNeal, J.R. (2016) Dodder (*Cuscuta pentagona*) alters the transcriptome of its host, *Arabidopsis thaliana*, and transfers mRNA into host tissues. *Planta* 244(2), 399–412.

Huang, H., Liu, R., Niu, Q., Tang, K., Zhang, B. *et al.* (2019) Global increase in DNA methylation during orange fruit development and ripening. *Proceedings of the National Academy of Sciences* 116(4), 1430–1436.

IPCC (2014) *IPCC Panel Climate Change: Synthesis Report*. IPCC, Geneva, Switzerland.

Kapazoglou, A., Ganopoulos, I., Tani, E. and Tsaftaris, A. (2018) Epigenetics, epigenomics and crop improvement. *Advances in Botanical Research* 86, 287–324.

Ketumile, D., Yang, X., Sanchez, R., Kundariya, H., Rajewski, J. *et al.* (2022) Implementation of epigenetic variation in sorghum selection and implications for crop resilience breeding. *Frontiers in Plant Science* 12, 798243.

Kim, J.M., To, T.K., Ishida, J., Matsui, A., Kimura, H. *et al.* (2012) Transition of chromatin status during the process of recovery from drought stress in *Arabidopsis thaliana*. *Plant & Cell Physiology* 53(5), 847–856.

Kim, S.Y., Lee, J.H., Kim, J.H., Lee, I. and Kim, J.Y. (2015) PICKLE affects CBF3 activity and cold acclimation in *Arabidopsis*. *Plant Physiology* 167(2), 632–643.

Koc, A., Markovic, D., Ninkovic, V. and Martinez, G. (2020) Molecular mechanisms regulating priming and stress memory. In: Hossain, M.A., Liu, F., Burritt, D., Fujita, M. and Huang, B. (eds) *Priming-Mediated Stress and Cross-Stress Tolerance in Crop Plants*. Academic Press, Cambridge, Massachusetts, pp. 247–265.

Kumar, S., Arif, I.A. and Kumar, P. (2012) Epigenetic modifications in response to heat stress in plants. *Plant Signaling & Behavior* 7(11), 1442–1458.

Lämke, J. and Bäurle, I. (2017) Epigenetic and chromatin-based mechanisms in environmental stress adaptation and stress memory in plants. *Genome Biology* 18(1), 124.

Lämke, J., Brzezinka, K., Altmann, S. and Bäurle, I. (2016) A hit-and-run heat shock factor governs sustained histone methylation and transcriptional stress memory. *The EMBO Journal* 35(2), 162–175.

Lee, B.H., Henderson, J.A., Zhu, J.-K. and Gong, Z. (2009) HOS15, a WD40 repeat protein, is involved in cold acclimation and interacts with ICE1 and HD2C in *Arabidopsis*. *Plant Physiology* 150(2), 1014–1026.

Li, G., Zhang, Y., Yang, R., Wang, H. and Chen, X. (2018) The m6A demethylase ALKBH9B suppresses alfalfa mosaic virus infection in *Arabidopsis*. *Nature Communications* 9(1), 1–11.

Liang, D., Zhang, Z., Wu, H., Huang, C., Shuai, P. *et al.* (2014) Single-base-resolution methylomes of *Populus trichocarpa* reveal the association between DNA methylation and drought stress. *BMC Genetics* 15, S9.

Liu, C., Ding, Y. and Guo, H. (2019) DNA methylation in plant stress responses. *Frontiers in Plant Science* 10, 902.

Liu, J., Wu, X., Yao, X., Yu, R., Larkin, P.J. *et al.* (2018) Mutations in the DNA demethylase OsROS1 result in a thickened aleurone and improved nutritional value in rice grains. *Proceedings of the National Academy of Sciences* 115(44), 11327–11332.

Liu, S., Chen, C. and Wang, X. (2023) Advances in high-resolution sequencing technologies for epigenetic research. *Journal of Experimental Botany* 74(8), 2341–2354.

Loreti, E., Veen, H. and Perata, P. (2016) Plant responses to flooding stress. *Current Opinion in Plant Biology* 33, 64–71.

Luan, X., Liu, S., Ke, S., Dai, H., Xie, X.M. *et al.* (2019) Epigenetic modification of ESP, encoding a putative long noncoding RNA, affects panicle architecture in rice. *Rice* 12, 20.

Luna, E., Pastor, V., Robles, P. and Pieterse, C.M.J. (2012) Plants under attack: Epigenetic regulation of defense responses. *Current Opinion in Plant Biology* 15(5), 483–489.

Manning, K., Tör, M., Poole, M., Hong, Y., Thompson, A.J. *et al.* (2006) A naturally occurring epigenetic mutation in A gene encoding an SBP-box transcription factor inhibits tomato fruit ripening. *Nature Genetics* 38(8), 948–952.

Martin, A., Troadec, C., Boualem, A., Rajab, M., Fernandez, R. *et al.* (2009) A transposon-induced epigenetic change leads to sex determination in melon. *Nature* 461(7267), 1135–1138.

Mirouze, M. and Paszkowski, J. (2011) Epigenetic contribution to stress adaptation in plants. *Current Opinion in Plant Biology* 14(3), 267–274.

Miryeganeh, M. (2025) Epigenetic mechanisms driving adaptation in tropical and subtropical plants: Insights and future directions. *Plant, Cell & Environment* 48, 3487–3499. DOI: 10.1111/pce.15370.

Miura, A., Yonebayashi, S., Watanabe, K., Toyama, T., Shimada, H. *et al.* (2001) Mobilization of transposons by a mutation abolishing full DNA methylation in *Arabidopsis*. *Nature* 411(6834), 212–214.

Miura, K., Agetsuma, M., Kitano, H., Yoshimura, A., Matsuoka, M. *et al.* (2009) A metastable DWARF1 epigenetic mutant affecting plant stature in rice. *Proceedings of the National Academy of Sciences* 106(27), 11218–11223.

Miura, K., Ikeda, M., Matsubara, A., Song, X.J., Ito, M. *et al.* (2010) OsSPL14 promotes panicle branching and higher grain productivity in rice. *Nature Genetics* 42(6), 545–549.

Ong-Abdullah, M., Ordway, J.M., Jiang, N., Ooi, S.E., Kok, S.Y. *et al.* (2015) Loss of karma transposon methylation underlies the mantled somaclonal variant of oil palm. *Nature* 525(7570), 533–537.

Patterson, G.I., Thorpe, C.J. and Chandler, V.L. (1993) Paramutation, an allelic interaction, is associated with a stable and heritable reduction of transcription of the maize B regulatory gene. *Genetics* 135(3), 881–894.

Pecinka, A., Dinh, H.Q., Baubec, T., Rosa, M., Lettner, N. *et al.* (2010) Epigenetic regulation of repetitive elements is attenuated by prolonged heat stress in *Arabidopsis*. *Plant Cell* 22(9), 3118–3129.

Pilu, R., Panzeri, D., Cassani, E., Badone, F.C., Landoni, M. *et al.* (2009) A paramutation phenomenon is involved in the genetics of maize low phytic acid1-241 (lpa1-241) trait. *Heredity* 102(3), 236–245.

Quadrana, L., Almeida, J., Asís, R., Duffy, T., Dominguez, P.G. *et al.* (2014) Natural occurring epialleles determine vitamin E accumulation in tomato fruits. *Nature Communications* 5, 3027.

Raju, S.K.K. (2017) Utilizing muts HOMOLOG 1 (MSH1) derived epigenetic variation in breeding for yield and stress adaptability. Thesis, University of Nebraska – Lincoln ProQuest.

Ramirez-Prado, J.S., Latrasse, D., Rodriguez-Granados, N.Y., Huang, Y., Manza-Mianza, D. *et al.* (2019) The polycomb protein LHP1 regulates *Arabidopsis thaliana* stress responses through the repression of the MYC2-dependent branch of immunity. *The Plant Journal* 100(6), 1118–1131.

Sani, E., Herzyk, P., Perrella, G., Colot, V. and Amtmann, A. (2013) Hyperosmotic priming of *Arabidopsis* seedlings establishes a long-term somatic memory accompanied by specific changes of the epigenome. *Genome Biology* 14(6), R59.

Santos, A.P., Ferreira, L., Maroco, J. and Oliveira, M.M. (2011) Abiotic stress and induced DNA hypomethylation cause interphase chromatin structural changes in rice rDNA loci. *Cytogenetic and Genome Research* 132(4), 297–303.

Saze, H. and Kakutani, T. (2007) Heritable epigenetic mutation of a transposon-flanked *Arabidopsis* gene due to lack of the chromatin-remodeling factor DDM1. *The EMBO Journal* 26(15), 3641–3652.

Schramm, F., Ganguli, A. and Scharf, K.D. (2008) Transcriptional regulation of the heat shock response in *Arabidopsis thaliana*. *Plant, Cell & Environment* 31(6), 582–594.

Shen, Y., Conde e Silva, N., Audonnet, L., Servet, C., wei, W. *et al.* (2014) Over-expression of histone H3K4 demethylase gene JMJ15 enhances salt tolerance in *Arabidopsis*. *Frontiers in Plant Science* 5, 290.

Shiba, H., Kakizaki, T., Iwano, M., Tarutani, Y., Watanabe, M. *et al.* (2006) Dominance relationships between self-incompatibility alleles controlled by DNA methylation. *Nature Genetics* 38(3), 297–299.

Song, Q. and Zhang, T. (2018) Epigenetic regulation of gene expression in response to abiotic stress in plants. *Journal of Integrative Plant Biology* 60(9), 1004–1016. DOI: 10.1111/jipb.12654.

Song, Q., Zhang, T., Stelly, D.M. and Chen, Z.J. (2017) Epigenomic and functional analyses reveal roles of epialleles in the loss of photoperiod sensitivity during domestication of allotetraploid cottons. *Genome Biology* 18, 1–4.

Soppe, W.J., Jacobsen, S.E., Alonso-Blanco, C., Jackson, J.P., Kakutani, T. *et al.* (2000) The late flowering phenotype of fwa mutants is caused by gain-of-function epigenetic alleles of a homeodomain gene. *Molecular Cell* 6(4), 791–802.

Springer, N.M. and Schmitz, R.J. (2017) Exploiting induced and natural epigenetic variation for crop improvement. *Nature Reviews. Genetics* 18(9), 563–575. DOI: 10.1038/nrg.2017.45.

Tiwari, A., Pandey-Rai, S., Rai, K.K., Tiwari, A. and Pandey, N. (2023) Molecular and epigenetic basis of heat stress responses and acclimatization in plants. *Nucleus* 66, 69–79.

Toljamo, A., Blande, D., Kärenlampi, S. and Kokko, H. (2016) Reprogramming of strawberry (*Fragaria vesca*) root transcriptome in response to *Phytophthora cactorum*. *PloS ONE* 11, e0161078.

van Dijk, K., Ding, Y., Malkaram, S., Riethoven, J.J.M., Liu, R. *et al.* (2010) Dynamic changes in genome-wide histone H3 lysine 4 methylation patterns in response to dehydration stress in *Arabidopsis thaliana*. *BMC Plant Biology* 10, 238.

Varotto, S., Tani, E., Abraham, E. *et al.* (2021) Epigenetics: A key player in plants adaptation to environmental stresses. *Plants* 10(3), 452.

Virlouvet, L. and Fromm, M. (2015) Physiological and transcriptional memory in guard cells during repetitive dehydration stress. *The New Phytologist* 205(2), 596–607.

Voinnet, O. (2009) Induction and suppression of RNA silencing: Insights from viral infections. *Nature Reviews Genetics* 10(3), 208–220.

Wang, W., Wu, Y. and Yang, C. (2016) Genome-wide analysis of DNA methylation in *Brachypodium distachyon* reveals differences between desiccation-tolerant and non-tolerant plants. *Plant Biotechnology Journal* 14(1), 55–66.

Wang, Z., Li, B., Song, S., Sun, S. and Huang, J. (2017) Epigenetic regulation in plant-nematode interactions. *Frontiers in Plant Science* 8, 1184.

Wei, X., Song, X., Wei, L., Tang, S., Sun, J. *et al.* (2017) An epiallele of rice AK1 affects photosynthetic capacity. *Journal of Integrative Plant Biology* 59(3), 158–163.

Wibowo, A., Becker, C., Marconi, G., Durr, J., Price, J. *et al.* (2016) Hyperosmotic stress memory in *Arabidopsis* is mediated by distinct epigenetically labile sites in the genome and is restricted in the male germline by DNA glycosylase activity. *eLife* 5, e13546.

Yang, X., Kundariya, H., Xu, Y.Z., Sandhu, A., Yu, J. *et al.* (2015) MutS HOMOLOG1-derived epigenetic breeding potential in tomato. *Plant Physiology* 168(1), 222–232.

Zhang, Q. and Bartels, D. (2018) Molecular responses to dehydration and desiccation in desiccation-tolerant angiosperm plants. *Journal of Experimental Botany* 69(13), 3211–3222.

Zhang, H., Zhou, H., Yang, J., Liu, X., Zhang, H. *et al.* (2017) Genome-wide analysis reveals the dynamic and stress-specific DNA methylation reprogramming during multiple abiotic stress responses in poplar. *Plant Biotechnology Journal* 15(12), 1411–1421.

Zhang, J., Li, C. and Jiang, W. (2023) Advances in CRISPR/Cas-based epigenetic editing tools for plant research and improvement. *Plant Biotechnology Journal* 21(6), 1121–1134.

Zhang, L., Cheng, Z., Qin, R., Qiu, Y., Wang, J.L. *et al.* (2012) Identification and characterization of an epi-allele of FIE1 reveals a regulatory linkage between two epigenetic marks in rice. *The Plant Cell* 24(11), 4407–4421.

Zhang, X., Sun, J., Cao, X. and Song, X. (2015) Epigenetic mutation of RAV6 affects leaf angle and seed size in rice. *Plant Physiology* 169(3), 2118–2128.

Zheng, X., Chen, L., Li, M., Lou, Q., Xia, H. *et al.* (2013) Transgenerational variations in DNA methylation induced by drought stress in two rice varieties with distinguished difference to drought resistance. *PloS ONE* 8(11), e80253.

8 Epigenetics for Combating Heavy Metal and Metalloid Stresses in Plants: Updated Methods and Current Achievements

Babar Iqbal[1]*, Khulood Fahad Alabbosh[2]*, Guanlin Li[1]* and Arshad Jalal[3]*

[1]School of Environment and Safety Engineering, Jiangsu University, Zhenjiang, People's Republic of China; [2]Department of Biology, College of Science, University of Hail, Hail, Saudi Arabia; [3]The BioActives Lab, Center for Desert Agriculture, King Abdullah University of Science and Technology, Thuwal, Kingdom of Saudi Arabia

Abstract

Pollution by heavy metals and metalloids poses a severe danger to plant development and agricultural output. This chapter overviews current strategies and accomplishments in using epigenetic pathways to improve plant tolerance to toxic heavy metals and metalloid stressors. The responses of plants to environmental stresses are mediated by epigenetic alterations such as DNA methylation, histone modifications, and small RNA control. Researchers have been able to disentangle the complexities of epigenetic regulation in response to contamination by heavy metals and metalloids due to technological developments like high-throughput genome sequencing and the study of functional genomics. Contemporary accomplishments include the recognition of particular epigenetic markers linked with stress tolerance, epigenome editing tools, and the use of epigenetic data in stress-resistant crop breeding programming. Thus, this emphasizes the emerging importance of epigenetics in plant stress biology and its ability to minimize the negative impacts of pollution by heavy metals and metalloids on global agriculture.

Keywords: Epigenetics, Heavy metal stress, Metalloid stress, Plant stress tolerance, Bioremediation

8.1 Introduction

Plant epigenetics is critical in reacting to environmental pressures such as pollution by heavy metals. The word 'epigenetics' is used to describe heritable variations in the expression of genes caused by changes in the DNA sequence (Liu *et al.*, 2020). Various environmental variables, particularly heavy metal stress, might impact these alterations. Heavy metal stressors in plants are a major concern because they harm growth, development, and, eventually, crop output (Iqbal *et al.*, 2023a). These metals can build in the soil as a result of industrial, mining, and agricultural processes, presenting a hazard to plant health and, by extension, human health via the food chain (Iqbal *et al.*, 2023b). Plant adaptations to heavy metal stressors are influenced by epigenetic processes such as DNA methylation, histone

*Corresponding authors: babar@ujs.edu.cn, k.alabosh@uoh.edu.sa, liguanlin@ujs.edu.cn and arshad.jalal@kaust.edu.sa

 Epigenetics for Climate-Smart and Sustainable Agriculture (ed. Jen-Tsung Chen)
DOI: 10.1079/9781800626102.0008

changes, and small RNA-mediated gene transcription (Fasani *et al.*, 2023).

Heavy metal stress can cause alterations in plant DNA methylation patterns. DNA methylation is the process of adding a methyl group to DNA, which frequently results in altered gene expression (Thakur *et al.*, 2022). Changes in DNA methylation patterns have been demonstrated in studies to alter the functioning of genes associated with metal ion transport, detoxification, and stress-related pathways in plants (Cong *et al.*, 2019). Heavy metals may additionally affect histone changes, which affect the make-up of chromatin and gene expression. Histone changes like acetylation, methylation, and phosphorylation can influence gene expression in stress responses. Histone acetylation, for example, is related to the active expression of genes and has been connected with crop reactions when exposed to stress caused by heavy metals. Small RNAs, such as microRNAs (miRNAs) and small interfering RNAs (siRNAs), regulate gene expression. Specific miRNAs and siRNAs have been discovered to have been responsible for modulating genes linked with heavy metal detoxifying and physiological responses to stress in crops (Srivastava and Suprasanna, 2021).

Understanding the epigenetic processes underpinning crop responses when faced with heavy metal stressors is critical for enhancing stress tolerance, and these systems might be manipulated to generate crops that are more resistant to contaminants like heavy metals (Cui *et al.*, 2020). Using epigenetic information can contribute to the development of phytoremediation systems, in which plants are utilized to remove, maintain, and eliminate heavy metals from polluted soil. Epigenetic changes can be used to breed or create cultivars that are more resistant to stress from heavy metals, thus assuring the safety and security of food in polluted areas (Gogolev *et al.*, 2021).

This research is ongoing and shows promise for minimizing the negative impacts of heavy metal stressors on plant development and agricultural output while also benefiting from achieving environmental sustainability. Significant advances and promising results have been reported regarding studies into epigenetics and its involvement in alleviating metalloid stressors in plants. Metalloids, which include arsenic and selenium, provide distinct concerns due to their hazardous qualities that can impair human health across the food chain. Understanding how epigenetic pathways impact the response of plants to these metalloids is critical for creating effective countermeasures.

Researchers may now investigate patterns of DNA methylation and histone changes related to metalloid stress using methods such as whole-genome bisulfite sequencing (WGBS) and chromatin immunoprecipitation sequencing (ChIP-seq) (Peng and Zhang, 2009). The use of CRISPR/Cas-based epigenome programming allows for precise alterations to epigenetic markers, making it easier to analyze individual epigenetic modifications and their consequences on crop reactions when exposed to metalloids (Lee *et al.*, 2020). Integrating data from genomes, transcriptomics, proteomics, and metabolomics yields a comprehensive picture of how metalloid stress affects epigenetic control and gene expression in plants. Researchers identified epigenetic changes that control genes associated with metalloid absorption, confinement, and detoxification within plants (Fig. 8.1). Identifying various regulatory processes can help improve the resilience of plants to metalloid stress. Plants can pass on epigenetic modifications caused by metalloid stress to later generations, potentially providing adaptive advantages, according to research into transgenerational epigenetic inheritance (Fasani *et al.*, 2023). The study of epigenetic modifications in response to metalloid stress sheds light on how plants adapt to difficult circumstances, thereby leading to our current comprehension of plant biology and evolution. In the face of rising environmental concerns, additional investigation into epigenetics and metalloid stresses in plants has enormous promise in promoting agricultural sustainability, conserving the environment, and maintaining global food security.

The application of epigenetics to combat heavy metal and metalloid stressors in plants entails a multidisciplinary approach to studying, modifying, and utilizing epigenetic pathways for adaptation to stress and environmental sustainability. The goals are: (1) to identify specific epigenetic markers associated with heavy metal and metalloid stress in plants, such as changes in DNA methylation patterns and histone modifications; (2) to decipher the epigenetic

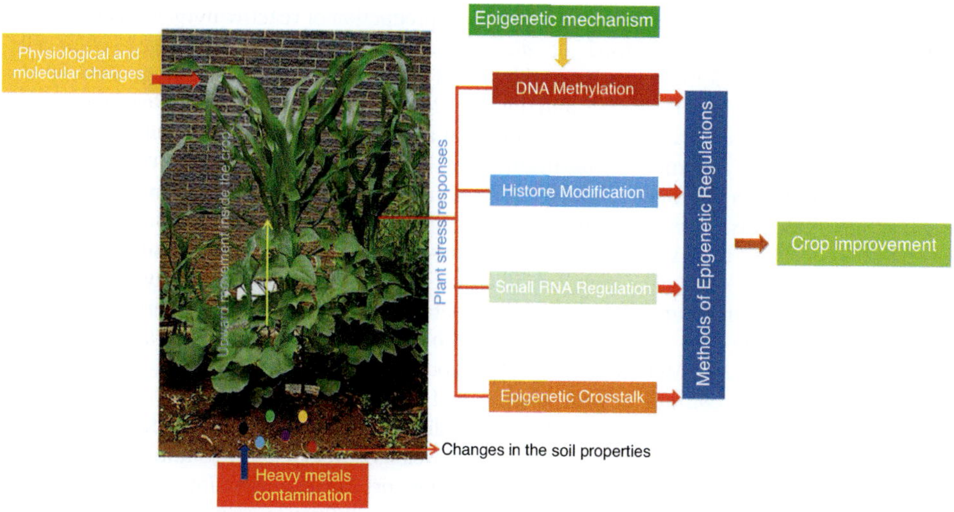

Fig. 8.1. Schematic diagram of heavy metal contamination and pathways to improve crop plants. Author's own image.

modification regulatory mechanisms that regulate genes responsible for metalloid uptake, transport, sequestration, and detoxification in plants; and (3) to investigate the inheritance of stress-induced epigenetic changes across plant generations and assess their role in long-term adaptation.

8.2 Heavy Metal and Metalloid Stress in Plants

8.2.1 Sources and types of heavy metals and metalloids

Heavy metals and metalloids are elements with large atomic weights that, even in tiny concentrations, can be poisonous or harmful to living creatures (Iqbal *et al.*, 2023a). These belong to naturally occurring elements, but they may also enter the environment as a result of human activity, causing significant environmental and health dangers (Iqbal *et al.*, 2023c). There are several causes of toxic metals and metalloid contamination, the majority of which are natural sources that exist within the outermost layer of the Earth and also might be dispersed into the surrounding atmosphere through rock weathering and soil erosion (Edelstein and Ben-Hur, 2018). Similarly, volcanic eruptions can also cause heavy metals and metalloids to be released into the environment and surrounding soil (Edelstein and Ben-Hur, 2018). Anthropogenic sources, on the other hand, include industrial operations that emit heavy metals and metalloids into the air, soil, and water, such as mining, smelting, metal refining, and manufacturing (Briffa *et al.*, 2020). Furthermore, agricultural methods such as the incorporation of chemical fertilizers, insecticides, and sewage sludge leach heavy metals into the soil, damaging crops and water supplies (Lin *et al.*, 2022). Similarly, inappropriate disposal of electronic garbage, batteries, and heavy metal-containing home items can pollute soil and water. Finally, urban run-off, automobile carbon emissions, and industrial discharges all play a role in heavy metal and metalloid pollution in cities (Briffa *et al.*, 2020).

Many heavy metals and metalloids are discharged into the environment, including lead (Pb), which is widely found in batteries, paints, and pipes (Edelstein and Ben-Hur, 2018). Excessive exposure to lead is associated with neurological and developmental disorders. Mercury (Hg) is present in industrial processes as well as some seafood; mercury can impair the brain system and is especially dangerous to fetuses (Balali-Mood *et al.*, 2021). Cadmium (Cd) is found in rechargeable batteries, plastics, and cigarette smoke, and the element is a recognized carcinogen that can cause kidney damage. Arsenic (As) naturally occurs in groundwater, pesticides, and some construction materials; it is very poisonous and has been related to a variety of health problems. Chromium (Cr) is employed in industrial operations and comes in numerous forms, including some that are carcinogenic (Balali-Mood *et al.*, 2021). Metalloids, on the other hand, primarily contain arsenic (As), which is classified as a heavy metal and shows metalloid features. It occurs naturally in the Earth's crust and can pollute groundwater, providing serious health dangers (Fan *et al.*, 2021). Similarly, selenium (Se) is found in soil and rocks and is necessary in tiny levels but harmful in large ones, affecting aquatic ecosystems and animal health (Malisa, 2001). Understanding the origins and kinds of heavy metals and metalloids is critical for putting in place effective measures to decrease their environmental impact, avoid human exposure, and create remediation plans for polluted locations. Sustainable methods and tougher controls are being implemented to reduce the discharge and exposure to these harmful components.

8.2.2 Impact on plant growth and productivity

Heavy metal and metalloid stress have a profound influence on plant development and production, providing significant obstacles to agriculture and the sustainability of ecosystems. The presence of metalloids and heavy metals can obstruct the absorption and delivery of key elements such as iron, magnesium, and calcium, resulting in nutritional shortages in plants (Noor *et al.*, 2022). Similarly, these components can limit cell division and elongation, resulting in slowed root and shoot growth and total plant size reduction. Heavy metals and metalloids have the potential to disrupt chlorophyll formation and photosynthesis, limiting the plant's capacity for converting sunlight into energy (Berni *et al.*, 2019). Heavy metals can also cause the production of reactive oxygen species (ROS), which can cause oxidative damage to plant cells and impair biological functioning (Feki *et al.*, 2021). Heavy metals can alter plant physiological and biochemical characteristics by disturbing ion homeostasis, resulting in an imbalance of critical ions such as potassium, calcium, and magnesium (Berni *et al.*, 2019). Furthermore, these metals may be absorbed and accumulated in plant tissues, causing toxicity and altering metabolic processes. Heavy metal stress may also disrupt enzyme activity and metabolic processes in plants, influencing glucose, lipid, and protein metabolism (Mir *et al.*, 2018). In addition, these metals can impede reproduction as well as development by limiting seed germination or causing aberrant seedling growth, which can influence crop establishment (Iqbal *et al.*, 2024). Heavy metals and metalloids can interfere with floral growth, pollen viability, and fertilization, resulting in decreased seed set and crop output. On the other hand, heavy metals can have long-term impacts on plants, creating chronic toxicity and harming their long-term development, health, and survival (Noor *et al.*, 2022). Similarly, heavy metal and metalloid deposition in soil can remain for long periods, rendering land unsuitable for cultivation and posing dangers to the environment and human health via the food chain (Kumar *et al.*, 2019).

There are several strategies to reduce heavy metal toxicity, the most important of which is the plant's capacity to tolerate and collect heavy metals and metalloids, making them ideal for phytoremediation – the use of plants to remove, stabilize, or detoxify polluted soil. Additionally, via selective breeding or genetic engineering, plant types with tolerance or decreased absorption of certain elements can be developed. Similarly, soil additives such as organic matter or certain minerals in the soil can assist in reducing heavy metal bioavailability and absorption by plants (Fig. 8.2).

8.2.3 Mechanisms of plant response to stress

To respond to and cope with heavy metal and metalloid stress, plants have evolved several mechanisms. These mechanisms encompass a

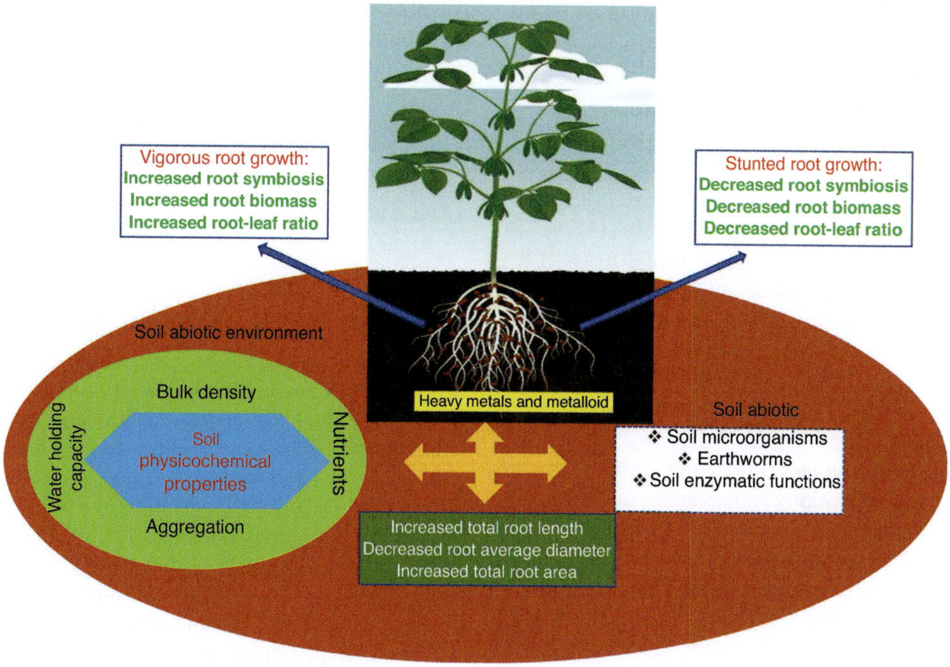

Fig. 8.2. Response of heavy metals and metalloids to the soil and crops. Author's own image.

variety of physiological and biochemical processes, and molecular mechanisms that try to detoxify, sequester, and tolerate substances. The heavy metals and metalloids are sequestered in vacuoles by plants, lowering their harmful effects in the cytoplasm (Arif *et al.*, 2016). However, tonoplast transporters help to transfer these components into vacuoles. Similarly, specific transport proteins in the cell membrane assist heavy metal and metalloid outflow, minimizing intracellular build-up. Furthermore, chelating compounds such as phytochelatins and metallothionein bind and sequester heavy metals, reducing their harmful effects on the cytosol (Faizan *et al.*, 2024).

Heavy metals can also attach to cell wall components such as pectin and cellulose, restricting their entrance into the protoplast and enhancing their toxicity (Kosakivska *et al.*, 2021). Metal stress activates antioxidant mechanisms comprising enzymes (e.g. superoxide dismutase, catalase, peroxidases) and non-enzymatic antioxidants (e.g. glutathione, ascorbate) to neutralize reactive oxygen species (Kamran *et al.*, 2021). Metabolic pathways must be modified to reduce the harmful effects of heavy metals; for example, rerouting energy resources or modifying secondary metabolite synthesis. Furthermore, modification of transporter proteins that are involved in nutrient intake restricts heavy metal entrance while maintaining necessary nutrient uptake (Khalid *et al.*, 2019). Metal transportation, detoxification, and responses to stress pathways are all influenced by genetic and epigenetic control, which takes into account variations in gene expression (Kosakivska *et al.*, 2021). By enhancing nutrient absorption and changing root architecture, symbiotic partnerships with mycorrhizal fungi can improve metal tolerance. Some plant species may hyper-accumulate heavy metals and metalloids, pulling them from the soil into their tissues for removal (Kumar *et al.*, 2019). Heavy metals and metalloids are targeted and localized into certain compartments within cells to minimize interfering with important cellular functions. Integrating these systems reveals how plants adapt and live in metal-contaminated environments, opening up possibilities for agricultural sustainability and environmental protection.

8.3 Epigenetic Mechanisms in Plant Stress Responses

8.3.1 DNA methylation

DNA methylation is an important epigenetic process in plant stress responses, particularly those to heavy metal and metalloid stressors. It entails adding a methyl group to the DNA molecule, especially at cytosine residues, which results in alterations in gene expression without changing the underlying DNA sequence (Moore *et al.*, 2013). Depending on where it occurs in the genome, DNA methylation may either stimulate or suppress gene expression. Methylation of gene promoters frequently results in gene silence by blocking transcription factor binding or attracting proteins that impede transcriptional machinery (Anastasiadi *et al.*, 2018). Methylation inside gene bodies or enhancer regions, on the other hand, may occasionally correspond with active gene expression. Plants dynamically adjust their DNA methylation patterns in response to stress to regulate the expression of stress-responsive genes (Lea *et al.*, 2016).

Changes in DNA methylation assist plants in adapting to environmental stressors such as heavy metal and metalloid stresses by changing the expression of genes involved in stress signaling, detoxification, and defense processes (Gallego-Bartolomé, 2020). Stress-induced changes in DNA methylation patterns can be passed down through generations, leading to plant adaptive responses. Inherited stress-induced DNA methylation modifications allow plants to prepare offspring for comparable stress conditions faced by the parent plant (Liu and He, 2020). Researchers can uncover stress-induced alterations in DNA methylation patterns across the plant genome using high-throughput sequencing and epigenome mapping tools (Chen *et al.*, 2019). Plant-specific DNA methyltransferases and demethylases control the dynamics of DNA methylation, reacting to stress signals and affecting gene expression patterns (Conde *et al.*, 2017). Understanding the significance of DNA methylation in responding to stress offers possibilities for designing plants with increased stress tolerance by modifying certain methylation patterns (Gallego-Bartolomé, 2020). Targeted epigenome editing approaches have the potential to accurately change DNA methylation marks to improve plant resistance to heavy metal and metalloid stressors. DNA methylation is a constantly changing and complicated mechanism in plant reactions to environmental challenges, considerably contributing to their capacity to adapt and endure harsh situations (Chen *et al.*, 2019).

8.3.2 Histone modifications

Histone modifications are critical epigenetic processes that regulate gene expression and stress-related plant responses, including the response to heavy metals and metalloids (Ueda and Seki, 2020). Histone proteins, which wrap DNA to create chromatin, may be chemically modified in a variety of ways, including acetylation, methylation, phosphorylation, and ubiquitination (Fu *et al.*, 2018). Histone alterations affect chromatin shape, impacting DNA accessibility to transcriptional machinery. Histone acetylation, for example, is often associated with accessible chromatin and active gene expression, whereas methylation may possess stimulating or regulatory effects depending on the individual histone residue and degree of methylation (Ueda and Seki, 2020).

In response to environmental pressures, particularly heavy metal and metalloid stresses, plants dynamically change histone marks. Changes in histone modifications caused by stress can influence the expression of stress-responsive genes, such as those that participate in detoxification, stress signaling, and defense systems (Kumar and Mohapatra, 2021). In response to stress, certain histone changes at the promoter region or regulatory domains of stress-responsive genes may either stimulate or inhibit their expression (Fu *et al.*, 2018). Changes in the regulation of histone acetylation or methylation patterns, for example, can influence the functioning of genes that regulate metal transportation, chelation, and antioxidant defense. Stress-induced histone changes can occasionally be passed down through generations, contributing to plant epigenetic memory (Kumar and Mohapatra, 2021). These heritable changes may predispose future generations to better stress responses or modify their vulnerability to environmental stresses. Researchers may map and evaluate histone modification patterns across the plant genome using high-throughput DNA sequencing and chromatin immunoprecipitation methods.

8.3.3 Small RNA regulation

Small RNA-mediated regulation is an important epigenetic mechanism in plant stress responses, including heavy metal and metalloid stress responses. Small RNAs, which are generally 20–24 nucleotides long, influence gene expression by degrading or inhibiting messenger RNAs (mRNAs). Small RNAs play a part in gene silencing, antiviral defense, and transposon control. Small RNAs have an important role in controlling the responses of plants to environmental stressors such as heavy metals and metalloids (Wang *et al.*, 2016). Plants generate stress-responsive short RNAs that target particular genes involved in stress adaption and defense when stressed (Iwasaki *et al.*, 2019). Stress-responsive short RNAs can influence the expression of target genes post-transcriptionally by regulating mRNA disintegration or inhibiting translation, hence changing the amounts of proteins implicated in stress responses (Mousavi *et al.*, 2019). Small RNAs can also play a role in epigenetic control by directing DNA methylation and histone changes at target sites, resulting in transcriptional gene silence (Simon and Meyers, 2011). RNA-directed DNA methylation (RdDM) is a mechanism in which siRNAs drive DNA methylation and gene silence in response to stress. Small RNAs generated by stress can help plants adapt to environmental challenges by modulating variations in gene expression and stress-responsive pathways. Some stress-induced short RNAs may be passed down through generations, possibly impacting plant stress responses in later generations (Wang *et al.*, 2016). Stress-responsive short RNAs in plants subjected to heavy metal and metalloid stressors may be identified and profiled using high-throughput sequencing technology (Ueda and Seki, 2020).

Understanding the roles of DNA methylation, histone changes, and small RNA-mediated control in plant stress responses reveals molecular pathways underpinning stress adaption. This manipulation provides prospects for stress-tolerant crop engineering and agricultural resilience in settings contaminated with heavy metals and metalloids.

8.3.4 Epigenetic crosstalk

The connection and communication between distinct epigenetic systems inside a cell or organism is referred to as epigenetic crosstalk. Epigenetic crosstalk in the context of plant stress responses, especially the responses to heavy metal and metalloid stressors, refers to the synchronized interactions and integration of several epigenetic mechanisms to regulate gene expression and stress adaption (Chang *et al.*, 2020). Under stress, crosstalk occurs when DNA methylation, histone changes, and small RNA pathways collaborate to control gene expression. Epigenetic crosstalk allows multiple epigenetic processes to work together to alter the expression of stress-responsive genes. DNA methylation patterns, for example, may alter histone changes at certain gene loci implicated in stress response pathways. Epigenetic changes can result in feedback loops in which one change impacts the emergence or removal of another epigenetic mark (Sun *et al.*, 2013). This interaction can result in the amplification or reinforcement of stress-related epigenetic alterations, which contribute to strong and long-lasting stress responses.

Crosstalk between distinct epigenetic pathways may contribute to stress-induced epigenetic changes being passed down through generations (Wu *et al.*, 2023). Stress-induced alterations in DNA methylation, histone modifications, or small RNA pathways may be passed down through generations, impacting stress responses in the offspring. Plants may perceive environmental signals, such as heavy metal and metalloid stressors, and develop adaptive responses by integrating numerous epigenetic mechanisms (Mathur *et al.*, 2022). Crosstalk across epigenetic pathways aids plants in fine-tuning their stress responses to maximize survival in changing conditions.

The intricacy of connections between multiple epigenetic pathways makes studying epigenetic crosstalk difficult. Understanding and regulating these interactions, on the other hand, provides potential for designing ways to improve plant resistance to environmental challenges. Epigenetic crosstalk is a complex and dynamic mechanism that orchestrates gene expression regulation and stress responses in plants. Investigating the interaction of diverse epigenetic pathways provides insights into the molecular basis of stress adaptation and opens options for boosting crop resilience in demanding settings contaminated with heavy metals and metalloids.

8.4 Methods for Studying Epigenetic Regulation in Stressed Plants

8.4.1 Epigenome sequencing technologies

Epigenome sequencing methods allow for the full mapping and investigation of epigenetic changes throughout the whole genome of stressed plants. These methods shed light on how stress-induced epigenetic alterations affect gene expression and stress responses. Whole-genome bisulfite sequencing (WGBS) is a strong approach for mapping DNA methylation patterns down to the single nucleotide level. WGBS detects variations in DNA methylation in stressed plants, demonstrating stress-induced changes in methylation patterns across the genome (Feng *et al.*, 2020). Using antibodies that identify particular histone marks, chromatin immunoprecipitation sequencing (ChIP-seq) enables genome-wide mapping of histone modifications. ChIP-seq reveals changes in histone modifications in stressed plants, showing stress-responsive variations in chromatin structure and gene regulation (Kim *et al.*, 2015). RNA sequencing (RNA-seq) aids in the identification of tiny RNA populations and their targets. RNA-seq in stressed plants elucidates stress-responsive short RNAs and their regulatory functions in gene expression under stress conditions (Li *et al.*, 2018). ATAC-seq (assay for transposase-accessible chromatin using sequencing) is a technique for profiling open chromatin areas by sequencing accessible DNA sections using a hyperactive Tn5 transposase. ATAC-seq detects modifications to chromatin accessibility in stressed plants, identifying areas of the genome experiencing transcriptional oversight in response to stress (Wang *et al.*, 2022b). The precision, reliability, and throughput of epigenome sequencing approaches are constantly improving as sequencing technology evolves. These advances allow for more thorough investigations of epigenetic alterations in stressed plants, making it easier to identify stress-responsive epigenetic changes. Integrating epigenomic data with other omics datasets (genomics, transcriptomics, and proteomics) allows for a more complete understanding of how epigenetic modifications impact gene expression and stress responses in stressed plants. In stressed plants, epigenome sequencing tools provide crucial insights into the molecular processes underpinning stress responses. They aid in the identification of stress-induced epigenetic alterations, the discovery of regulatory mechanisms, and the creation of methods to improve plant resistance to a variety of environmental stressors, including heavy metal and metalloid stresses (Lämke and Bäurle, 2017).

8.4.2 Functional genomics approaches

Functional genomics techniques in stressed plants attempt to provide a comprehensive understanding of how genes operate and respond to stress conditions. These techniques investigate gene activity, relationships, and networks to give information on stress adaption processes (Yampolsky *et al.*, 2014). Similarly, transcriptomics is mostly investigated using RNA-seq to assess gene expression changes in stressed plants, while microarrays are used to examine gene expression patterns under stressful circumstances to uncover stress-responsive genes and pathways (Jain, 2015). Additionally, proteomics, which involves mass spectrometry, detects and measures proteins expressed in stressed plants, revealing changes in protein abundance, alterations, and interactions. Similarly, two-dimensional gel electrophoresis (2D-PAGE) involves the separation and visualization of proteins to compare their expression patterns in stressed versus non-stressed plants (Budak *et al.*, 2015). Metabolomics is the use of mass spectrometry and nuclear magnetic resonance (NMR) techniques for the profiling and quantification of metabolites in stressed plants, revealing alterations in metabolic pathways and stress-related compounds. Furthermore, integrative analysis has demonstrated that data from several omics techniques (genomics, transcriptomics, proteomics, metabolomics, and epigenomics) could be merged to generate complete models of plant stress responses (Choudhury *et al.*, 2021). Systems biology techniques, on the other hand, employ computer models and network analysis to comprehend the interconnection of genes, proteins, and metabolites in stressed plants (Cramer *et al.*, 2011).

Finally, functional genomics techniques give a comprehensive knowledge of how plants respond to stressors such as heavy metals and metalloids. These technologies discover essential genes, proteins, and stress responses, allowing for the creation of stress-tolerant crops and strategies for sustainable agriculture in demanding conditions (Budak *et al.*, 2015).

8.4.3 Bioinformatics tools and resources

Bioinformatics techniques and resources are critical for evaluating, interpreting, and drawing insights from vast amounts of data generated by stressed plants. BLAST (Basic Local Alignment Search Tool) is a sequence alignment and analysis tool that identifies sequence similarities between nucleotide or protein sequences, assisting in gene identification and annotation (Lobo, 2008). Bowtie, or BWA, is an alignment tool used to map sequencing reads to a reference genome, which is essential for RNA-seq and ChIP-seq research. Similarly, SOAPdenovo and SPAdes Tools for *de novo* assembly of genomes from short-read sequencing data, which is required for non-model plants without reference genomes (Schaarschmidt *et al.*, 2020; Musich *et al.*, 2021). Augustus, on the other hand, is a tool for genome annotation, gene prediction, and finding functional components in genomes (Stanke and Morgenstern, 2005). Furthermore, the functional annotation and pathway analysis programs include DAVID and Enrichr, and this recognizes disproportionately represented operational terms and pathways in gene sets derived from transcriptomic or proteomic data, and KEGG and Reactome, providing pathway data necessary for comprehending how life functions are affected by stress in plants (Lobo, 2008). Similarly, network analysis and visualization tools include Cytoscape and STRING for displaying and evaluating molecular communication coalitions generated by omics data, demonstrating connections among genes, proteins, and pathways; and VisANT and Gephi for building and examining biological networks and pathways. Bismark and MethylKit are epigenomics tools for processing and evaluating DNA methylation data from WGBS or reduced representation bisulfite sequencing (RRBS). MACS and SICER are tools for finding histone modification peaks and areas in ChIP-seq data from stressed plants. Finally, metabolomics data analysis is performed (Prathibha *et al.*, 2023). MetaboAnalyst and XCMS are used to process and analyze metabolomics data, discovering compounds and pathways that are impacted by stress.

8.5 Epigenetic Engineering and Crop Improvement

8.5.1 Epigenome editing techniques

The change of epigenetic markers inside an organism's DNA to obtain certain features or desired qualities is known as epigenetic engineering. This method of crop enhancement tries to increase qualities like stress tolerance, yield, and nutritional content. In this method, epigenome modification methods are used to accurately change epigenetic markers. CRISPR/Cas-based systems allow for the targeted modification of DNA methylation or histone modifications at specific genomic loci; however, CRISPR/dCas9 fusion proteins by catalytically inactive Cas9 (dCas9) can be fused to epigenetic modifiers such as DNA methyltransferases or histone acetyltransferases/deacetylases, enabling precise control over targeted epigenetic changes (Wang *et al.*, 2022a). Second, the DNMT3A/dCas9 fusion protein connects the catalytic domain of DNA methyltransferase 3A (DNMT3A) to dCas9, allowing for targeted DNA methylation. By adding DNA methylation marks, this method may be utilized to mute certain genes or regulatory regions (Matzke and Mosher, 2014). Finally, the CRISPR/dCas9 histone acetyltransferase (HAT) or histone methyltransferase (HMT) fusion protein connects dCas9 to HAT or HMT enzymes, allowing researchers to precisely change histone acetylation or methylation at specific genomic loci, impacting gene expression and chromatin structure (Zhou *et al.*, 2021). To improve agricultural stress tolerance, epigenetic engineering can be used to modulate the expression of stress-responsive genes by targeted DNA methylation or histone modification. As a result, altering epigenetic marks linked with genes that regulate the process of development and growth can help enhance crop output. It may also be used to change the expression of genes that contribute to nutrient production, resulting in crops with better nutritional profiles (Lieberman-Lazarovich *et al.*, 2022).

As a result, selectivity in epigenome editing is required to minimize unintentional changes at non-targeted genomic sites. In contrast, the use of epigenetic engineering in crops presents ethical considerations, and regulatory frameworks to oversee its usage must be developed. Progress in epigenome editing tools and our understanding of plant epigenetics bode well for more accurate and efficient crop modification efforts. Epigenetic engineering might be critical in generating crops that are resistant to environmental challenges, thereby contributing to global food security (Dalakouras and Vlachostergios, 2021). By carefully altering epigenetic markers, epigenetic engineering, enabled by epigenome editing methods, provides a formidable toolkit for crop improvement. While there are limitations, current research is opening the road for creative agricultural uses to meet the demands of a changing environment and rising population.

8.5.2 Incorporating epigenetic information into breeding programs

Incorporating epigenetic information into breeding programs entails using epigenetic variants and alterations to improve crop breeding tactics. Consider how epigenetic differences among populations might lead to a phenotypic variety. Identifying and utilizing these differences may help in the selection of desired features in breeding programs. However, like GWAS (Genome-Wide Association Studies), Epigenome-Wide Association Studies (EWAS) relate epigenetic alterations to phenotypic variables of relevance (Campagna *et al.*, 2021). This information helps breeders to make breeding decisions by linking certain epigenetic markers with desired features. Furthermore, epigenetic memory and transgenerational effects may be achieved through the use of transgenerational epigenetic inheritance, therefore understanding and harnessing stress-induced epigenetic modifications that persist across generations might assist in the breeding of stress-tolerant crops. Similarly, unique epigenetic marks linked to desired features can be used as selection markers in breeding programs, assisting in the production of improved crop types. Epigenetic factors influencing hybrid vigor may be investigated to improve hybrid breeding tactics, resulting in increased crop production.

8.6 Challenges and Considerations

It is critical to ensure the long-term viability and repeatability of desired epigenetic alterations over generations for successful inclusion into breeding programs. Similarly, high-throughput screening and characterization of epigenetic indicators for breeding applications require reliable and cost-effective methodologies. Introducing epigenetic knowledge into breeding programs has the potential to speed up the production of crops with improved qualities, including stress tolerance, yield, and nutritional content (Wang *et al.*, 2022a). Further study into epigenetic processes and inheritance patterns will develop ways for efficiently employing epigenetic data during crop modification. In a nutshell, adding epigenetic information into breeding programs has the potential to transform crop development by increasing breeders' genetic toolset (Wang *et al.*, 2022a). Using epigenetic changes and understanding their function in plant attributes might help design more robust, productive, and adaptive crop varieties to tackle global food security issues.

8.7 Case Studies and Current Achievements

8.7.1 Examples of successful epigenetic approaches

Although specific examples in agricultural operations are still emerging, viable epigenetic techniques for mitigating heavy metal and metalloid stressors in plants have been extensively investigated in recent years. Researchers have found unique epigenetic variants (epialleles) related to heavy metal tolerance in various plants through epigenetic variation studies. These epigenetic modifications might be passed down through generations and contribute to stress resistance (Boquete *et al.*, 2022). Conversely, regarding epigenome alteration for stress tolerance, the alteration of DNA methylation patterns has been examined through numerous experimental investigations employing CRISPR/Cas-based technology, and it has demonstrated the capacity to modify DNA methylation patterns in plants (Tonosaki *et al.*, 2022). In certain cases, targeted alterations in DNA methylation have been linked to enhanced tolerance to heavy metals. Pre-exposure of plants to sublethal levels of heavy metals has also been shown to cause alterations in DNA methylation and histone modifications. This priming effect may impart increased tolerance to future heavy metal exposures.

Some studies have demonstrated that some stress-induced epigenetic changes can be passed down through generations. This transgenerational transmission of epigenetic modifications might lead to the development of stress-tolerant crops (Ou *et al.*, 2012). Changes in histone acetylation or methylation, on the other hand, have been connected to the regulation of stress-responsive genes implicated in detoxification and metal tolerance pathways (Boquete *et al.*, 2022). In light of these instances, it is critical to note that the use of epigenetic techniques in agriculture, especially for countering heavy metal and metalloid stressors in crops, is still in the experimental and research stages. There is still much to learn about the durability, heredity, and broader application of induced epigenetic alterations in agricultural contexts. Before the actual adoption of epigenetic alterations in crops, ethical and regulatory implications must be carefully considered.

8.7.2 Crops with enhanced heavy metal tolerance

Several crop species have been researched for their ability to withstand heavy metal stressors, and research has yielded encouraging results in several situations. Genetic changes targeting genes involved in metal transport, chelation, and detoxification were used to produce improved tolerance in *Arabidopsis thaliana* variants (Boquete *et al.*, 2022). Some rice varieties are also naturally tolerant or accumulate heavy elements such as cadmium. As a result, selective breeding and genetic engineering have been used to improve metal tolerance while reducing metal deposition in edible sections (Thongnok *et al.*, 2022). Previous studies have shown that certain maize cultivars are resistant to heavy metal stressors. As a result, breeding efforts are focused mostly on discovering and selecting cultivars that accumulate fewer heavy metals in grain while maintaining yield under stressful circumstances (Rizvi *et al.*, 2022; Rahman *et al.*, 2023). Similarly, researchers carried out several trials on barley cultivars that have been discovered to have an intrinsic resistance to heavy metals. Thus, the goal of this research is to better understand the genetic features related to tolerance and to use these qualities in breeding programs (Tanwar *et al.*, 2023). Brassica species, on the other hand, are recognized for their propensity to hyperaccumulate specific heavy metals, and research is being conducted to investigate their potential in phytoremediation, or the use of these plants to recover heavy metals from polluted soils (Bortoloti and Baron, 2022.). Sunflower (*Helianthus annuus*) cultivars that are resistant to heavy metals such as lead can also be utilized in phytoremediation. As a result, research focuses on understanding their tolerance mechanisms and prospective applications in polluted environments (Waseem *et al.*, 2024). These crops have varied degrees of inherent tolerance or the potential for higher resistance through breeding or methods of genetic engineering.

8.8 Challenges and Future Directions

8.8.1 Ethical and environmental considerations

Several ethical and environmental considerations come into play when using epigenetics to address heavy metal and metalloid stresses in plants, including unintended consequences from manipulating epigenome implications for plant characteristics or interactions with the environment, raising concerns about unintended consequences on ecosystems. Developing strong regulatory frameworks and norms is critical to ensuring the safe and ethical use of epigenetic modification in agriculture. As a result, engaging stakeholders such as farmers, consumers, and politicians to address concerns and maintain openness about the use of epigenetics in crop enhancement is essential. Environmental issues, on the other hand, primarily entail the ecological impact of changing the epigenetic landscape of crops, which may influence interactions with other species, potentially affecting biological diversity, ecosystems, and natural habitats. It also has long-term implications since it evaluates the environmental repercussions of epigenetic alterations, such as their durability and possible effects on soil microbiota and ecological processes. Sustainable agriculture entails combining the benefits of increased stress tolerance with larger agricultural goals such as decreasing chemical inputs and encouraging agroecological techniques. While plants can be utilized for phytoremediation, increasing their metal tolerance may accidentally spread toxins or negatively impact local ecosystems if not carefully controlled. It is crucial to address these moral and ecological issues when using epigenetics to counteract heavy metal and metalloid stressors in plants. Implementing thorough risk assessments, encouraging responsible research and innovation, and participating in open communication with stakeholders may all assist in navigating these challenges and ensuring the safe and sustainable application of epigenetic techniques in agriculture.

8.8.2 Unresolved questions and research gaps

Despite significant advances, major research gaps and unanswered concerns remain in the field of employing epigenetics to counteract heavy metal and metalloid stressors in plants. Understanding the durability and heredity of induced epigenetic alterations over generations is still a major scientific challenge. It is critical to determine the transgenerational stability of stress-induced epigenetic changes and their propensity to persist in following generations. It is a huge problem to improve the specificity of epigenome editing techniques to target exact genomic locations while avoiding off-target effects. Unintended alterations in non-targeted locations must be addressed for safe and predictable applications. It is critical to validate the functional effects of induced epigenetic changes. Comprehensive functional investigations are required to establish direct causal links between particular epigenetic changes and stress tolerance features. It is also critical to comprehend the larger ecological ramifications of changed epigenetic characteristics in plants. An essential area of study is determining how modified plants interact with ecosystems, including possible effects on nearby flora and wildlife, along with investigating how epigenetic alterations may be used to supplement or integrate traditional breeding tactics for generating stress-tolerant crops. Further research is needed to optimize the synergy between genetic and epigenetic techniques in crop development. Multi-omics data may be used to decipher complicated epigenetic regulation networks in response to heavy metal stressors. It is still difficult to develop computational tools and models for understanding large-scale epigenomic data. Other possibilities include: (i) developing extensive ethical norms and regulatory frameworks for the use of epigenetics in agriculture; (ii) addressing social issues while assuring the appropriate and ethical application of epigenetic technology in agricultural development; and (iii) converting laboratory-based results into practical applications in the field. It is critical to evaluate the efficacy, safety, and scalability of epigenetic techniques in agricultural contexts. To promote the effective and responsible application of epigenetic techniques in boosting plant tolerance to heavy metal and metalloid stressors, these research gaps must be filled. Bridging these gaps will help to develop sustainable agricultural techniques and stress-tolerant crop types for a changing climate.

8.8.3 Future prospects in epigenetics for plant stress mitigation

By considering epigenetic approaches that provide accuracy in regulating stress-responsive genes, permitting tailored responses to diverse stressors, including heavy metals and metalloids, and recognizing and exploiting epigenetic transmission from generation to generation to confer stress resilience across plant generations, the foreseeable potential of epigenetics in mitigating plant stress is promising and multifaceted. Other possibilities are synergy with genetic approaches, phytoremediation efforts, and using plants to detoxify contaminated soils by improving their contaminant tolerance, making it possible to focus on improvements of particular stress-related traits without first integrating foreign DNA, potentially expediting crop improvement, advances in implementing multi-omics data to unravel complex epigenetic regulatory networks, and the development of robust ethical and regulatory frameworks to guide responsible research. The future of epigenetics in plant stress reduction offers enormous promise, with novel answers to the complex issues faced by environmental pressures. Continued research, technical breakthroughs, ethical concerns, and collaborative efforts are critical to realizing the full benefits of epigenetic techniques in global agriculture and food production.

8.9 Conclusion

In the quest to improve plant resistance to heavy metal and metalloid stressors, epigenetics appears to be a promising field. Epigenetic research approaches that have been updated provide insights into how plants respond and adapt to these difficulties. Current successes demonstrate advances in understanding stress-induced epigenetic alterations, providing views into targeted stress tolerance adjustments. However, unanswered problems about the durability of induced alterations, ethical implications, and practical uses in agricultural contexts remain. Integrating epigenetic insights into breeding and precision editing has considerable potential for developing stress-resistant crops, but careful attention and more study are required to fully realize the potential of epigenetics in sustainable agriculture.

References

Anastasiadi, D., Esteve-Codina, A. and Piferrer, F. (2018) Consistent inverse correlation between DNA methylation of the first intron and gene expression across tissues and species. *Epigenetics & Chromatin* 11(1), 37. DOI: 10.1186/s13072-018-0205-1.

Arif, N., Yadav, V., Singh, S., Singh, S., Ahmad, P. *et al.* (2016) Influence of high and low levels of plant-beneficial heavy metal ions on plant growth and development. *Frontiers in Environmental Science* 4, 69. DOI: 10.3389/fenvs.2016.00069.

Balali-Mood, M., Naseri, K., Tahergorabi, Z., Khazdair, M.R. and Sadeghi, M. (2021) Toxic mechanisms of five heavy metals: Mercury, lead, chromium, cadmium, and arsenic. *Frontiers in Pharmacology* 12, 643972. DOI: 10.3389/fphar.2021.643972.

Berni, R., Luyckx, M., Xu, X., Legay, S., Sergeant, K. *et al.* (2019) Reactive oxygen species and heavy metal stress in plants: Impact on the cell wall and secondary metabolism. *Environmental and Experimental Botany* 161, 98–106. DOI: 10.1016/j.envexpbot.2018.10.017.

Boquete, M.T., Schmid, M.W., Wagemaker, N.C.A.M., Carey, S.B., McDaniel, S.F. *et al.* (2022) Molecular basis of intraspecific differentiation for heavy metal tolerance in the copper moss *Scopelophila cataractae*. *Environmental and Experimental Botany* 201, 104970. DOI: 10.1016/j.envexpbot.2022.104970.

Bortoloti, G.A. and Baron, D. (2022) Phytoremediation of toxic heavy metals by brassica plants: A biochemical and physiological approach. *Environmental Advances* 8, 100204. DOI: 10.1016/j.envadv.2022.100204.

Briffa, J., Sinagra, E. and Blundell, R. (2020) Heavy metal pollution in the environment and their toxicological effects on humans. *Heliyon* 6(9), e04691. DOI: 10.1016/j.heliyon.2020.e04691.

Budak, H., Hussain, B., Khan, Z., Ozturk, N.Z. and Ullah, N. (2015) From genetics to functional genomics: Improvement in drought signaling and tolerance in wheat. *Frontiers in Plant Science* 6, 1012. DOI: 10.3389/fpls.2015.01012.

Campagna, M.P., Xavier, A., Lechner-Scott, J., Maltby, V., Scott, R.J. *et al.* (2021) Epigenome-wide association studies: Current knowledge, strategies and recommendations. *Clinical Epigenetics* 13(1), 214. DOI: 10.1186/s13148-021-01200-8.

Chang, Y.N., Zhu, C., Jiang, J., Zhang, H., Zhu, J.K. *et al.* (2020) Epigenetic regulation in plant abiotic stress responses. *Journal of Integrative Plant Biology* 62(5), 563–580. DOI: 10.1111/jipb.12901.

Chen, R., Li, M., Zhang, H., Duan, L., Sun, X. *et al.* (2019) Continuous salt stress-induced long non-coding rnas and DNA methylation patterns in soybean roots. *BMC Genomics* 20(1), 730. DOI: 10.1186/s12864-019-6101-7.

Choudhury, S., Sharma, P., Moulick, D. and Mazumder, M.K. (2021) Unrevealing metabolomics for abiotic stress adaptation and tolerance in plants. *Journal of Crop Science and Biotechnology* 24(5), 479–493. DOI: 10.1007/s12892-021-00102-8.

Conde, D., Moreno-Cortés, A., Dervinis, C., Ramos-Sánchez, J.M., Kirst, M. *et al.* (2017) Overexpression of DEMETER, a DNA demethylase, promotes early apical bud maturation in poplar. *Plant, Cell & Environment* 40(11), 2806–2819. DOI: 10.1111/pce.13056.

Cong, W., Miao, Y., Xu, L., Zhang, Y., Yuan, C. *et al.* (2019) Transgenerational memory of gene expression changes induced by heavy metal stress in rice (*Oryza sativa* L.). *BMC Plant Biology* 19(1), 282. DOI: 10.1186/s12870-019-1887-7.

Cramer, G.R., Urano, K., Delrot, S., Pezzotti, M. and Shinozaki, K. (2011) Effects of abiotic stress on plants: A systems biology perspective. *BMC Plant Biology* 11, 163. DOI: 10.1186/1471-2229-11-163.

Cui, J., Li, Y., Jin, Q. and Li, F. (2020) Silica nanoparticles inhibit arsenic uptake into rice suspension cells *via* improving pectin synthesis and the mechanical force of the cell wall. *Environmental Science: Nano* 7(1), 162–171. DOI: 10.1039/C9EN01035A.

Dalakouras, A. and Vlachostergios, D. (2021) Epigenetic approaches to crop breeding: current status and perspectives. *Journal of Experimental Botany* 72(15), 5356–5371. DOI: 10.1093/jxb/erab227.

Edelstein, M. and Ben-Hur, M. (2018) Heavy metals and metalloids: Sources, risks and strategies to reduce their accumulation in horticultural crops. *Scientia Horticulturae* 234, 431–444. DOI: 10.1016/j.scienta.2017.12.039.

Faizan, M., Alam, P., Hussain, A., Karabulut, F., Tonny, S.H. *et al.* (2024) Phytochelatins: A key regulator against heavy metal toxicity in plants. *Plant Stress* 100355. DOI: 10.1016/j.stress.2024.100355.

Fan, X., Lu, X., Yu, B., Zuo, L., Fan, P. *et al.* (2021) Risk and sources of heavy metals and metalloids in dust from university campuses: A case study of Xi'an, China. *Environmental Research* 202, 111703. DOI: 10.1016/j.envres.2021.111703.

Fasani, E., Giannelli, G., Varotto, S., Visioli, G., Bellin, D. *et al.* (2023) Epigenetic control of plant response to heavy metals. *Plants* 12(18), 3195. DOI: 10.3390/plants12183195.

Feki, K., Tounsi, S., Mrabet, M., Mhadhbi, H. and Brini, F. (2021) Recent advances in physiological and molecular mechanisms of heavy metal accumulation in plants. *Environmental Science and Pollution Research* 28(46), 64967–64986. DOI: 10.1007/s11356-021-16805-y.

Feng, S., Zhong, Z., Wang, M. and Jacobsen, S.E. (2020) Efficient and accurate determination of genome-wide DNA methylation patterns in *Arabidopsis thaliana* with enzymatic methyl sequencing. *Epigenetics & Chromatin* 13(1), 42. DOI: 10.1186/s13072-020-00361-9.

Fu, Y., Ma, H., Chen, S., Gu, T. and Gong, J. (2018) Control of proline accumulation under drought via a novel pathway comprising the histone methylase CAU1 and the transcription factor ANAC055. *Journal of Experimental Botany* 69(3), 579–588. DOI: 10.1093/jxb/erx419.

Gallego-Bartolomé, J. (2020) DNA methylation in plants: mechanisms and tools for targeted manipulation. *The New Phytologist* 227(1), 38–44. DOI: 10.1111/nph.16529.

Gogolev, Y.V., Ahmar, S., Akpinar, B.A., Budak, H., Kiryushkin, A.S. *et al.* (2021) Omics, epigenetics, and genome editing techniques for food and nutritional security. *Plants* 10(7), 1423. DOI: 10.3390/plants10071423.

Iqbal, B., Javed, Q., Khan, I., Tariq, M., Ahmad, N. *et al.* (2023a) Influence of soil microplastic contamination and cadmium toxicity on the growth, physiology, and root growth traits of *Triticum aestivum* L. *South African Journal of Botany* 160, 369–375. DOI: 10.1016/j.sajb.2023.07.025.

Iqbal, B., Zhao, T.T., Yin, W.Q., Zhao, X., Xie, Q.J. *et al.* (2023b) Impacts of soil microplastics on crops: A review. *Applied Soil Ecology* 18, 104680. DOI: 10.1016/j.apsoil.2022.104680.

Iqbal, B., Li, G., Alabbosh, K.F., Hussain, H., Khan, I. *et al.* (2023c) Advancing environmental sustainability through microbial reprogramming in growth improvement, stress alleviation, and phytoremediation. *Plant Stress* 10, 100283. DOI: 10.1016/j.stress.2023.100283.

Iqbal, B., Zhao, X., Khan, K.Y., Javed, Q., Nazar, M. *et al.* (2024) Microplastics meet invasive plants: Unraveling the ecological hazards to agroecosystems. *The Science of the Total Environment* 906, 167756. DOI: 10.1016/j.scitotenv.2023.167756.

Iwasaki, M., Hyvärinen, L., Piskurewicz, U. and Lopez-Molina, L. (2019) Non-canonical RNA-directed DNA methylation participates in maternal and environmental control of seed dormancy. *eLife* 8, e37434. DOI: 10.7554/eLife.37434.

Jain, M. (2015) Function genomics of abiotic stress tolerance in plants: A CRISPR approach. *Frontiers in Plant Science* 6, 375. DOI: 10.3389/fpls.2015.00375.

Kamran, M., Danish, M., Saleem, M.H., Malik, Z., Parveen, A. *et al.* (2021) Application of abscisic acid and 6-benzylaminopurine modulated morpho-physiological and antioxidative defense responses of tomato (*Solanum lycopersicum* L.) by minimizing cobalt uptake. *Chemosphere* 263, 128169. DOI: 10.1016/j.chemosphere.2020.128169.

Khalid, N., Aqeel, M. and Noman, A. (2019) System biology of metal tolerance in plants: An integrated view of genomics, transcriptomics, metabolomics, and phenomics. In: Sablok, G. (ed.) *Plant Metallomics and Functional Omics*. Springer Nature Switzerland AG, pp. 107–144. DOI: 10.1007/978-3-030-19103-0_6.

Kim, J.-M., Sasaki, T., Ueda, M., Sako, K. and Seki, M. (2015) Chromatin changes in response to drought, salinity, heat, and cold stresses in plants. *Frontiers in Plant Science* 6, 114. DOI: 10.3389/fpls.2015.00114.

Kosakivska, I.V., Babenko, L.M., Romanenko, K.O., Korotka, I.Y. and Potters, G. (2021) Molecular mechanisms of plant adaptive responses to heavy metals stress. *Cell Biology International* 45(2), 258–272. DOI: 10.1002/cbin.11503.

Kumar, S. and Mohapatra, T. (2021) Dynamics of DNA methylation and its functions in plant growth and development. *Frontiers in Plant Science* 12, 596236. DOI: 10.3389/fpls.2021.596236.

Kumar, S., Prasad, S., Yadav, K.K., Shrivastava, M., Gupta, N. *et al.* (2019) Hazardous heavy metals contamination of vegetables and food chain: Role of sustainable remediation approaches-A review. *Environmental Research* 179, 108792. DOI: 10.1016/j.envres.2019.108792.

Lämke, J. and Bäurle, I. (2017) Epigenetic and chromatin-based mechanisms in environmental stress adaptation and stress memory in plants. *Genome Biology* 18(1), 124. DOI: 10.1186/s13059-017-1263-6.

Lea, A.J., Altmann, J., Alberts, S.C. and Tung, J. (2016) Resource base influences genome-wide DNA methylation levels in wild baboons (*Papio cynocephalus*). *Molecular Ecology* 25(8), 1681–1696. DOI: 10.1111/mec.13436.

Lee, J.H., Mazarei, M., Pfotenhauer, A.C., Dorrough, A.B., Poindexter, M.R. *et al.* (2020) Epigenetic footprints of CRISPR/Cas9-mediated genome editing in plants. *Frontiers in Plant Science* 10, 1720. DOI: 10.3389/fpls.2019.01720.

Li, J.R., Liu, C.C., Sun, C.H. and Chen, Y.T. (2018) Plant stress RNA-seq nexus: A stress-specific transcriptome database in plant cells. *BMC Genomics* 19(1), 966. DOI: 10.1186/s12864-018-5367-5.

Lieberman-Lazarovich, M., Kaiserli, E., Bucher, E. and Mladenov, V. (2022) Natural and induced epigenetic variation for crop improvement. *Current Opinion in Plant Biology* 70, 102297. DOI: 10.1016/j.pbi.2022.102297.

Lin, H., Wang, Z., Liu, C. and Dong, Y. (2022) Technologies for removing heavy metal from contaminated soils on farmland: A review. *Chemosphere* 305, 135457. DOI: 10.1016/j.chemosphere.2022.135457.

Liu, J. and He, Z. (2020) Small DNA methylation, big player in plant abiotic stress responses and memory. *Frontiers in Plant Science* 11, 595603. DOI: 10.3389/fpls.2020.595603.

Liu, J., Dou, X., Chen, C., Chen, C., Liu, C. *et al.* (2020) N6-methyladenosine of chromosome-associated regulatory RNA regulates chromatin state and transcription. *Science* 367(6477), 580–586. DOI: 10.1126/science.aay6018.

Lobo, I. (2008) Basic local alignment search tool (BLAST). *Nature Education* 1(1), 215.

Malisa, E.P. (2001) The behaviour of selenium in geological processes. *Environmental Geochemistry and Health* 23(2), 137–158. DOI: 10.1023/A:1010908615486.

Mathur, P., Tripathi, D.K., Baluška, F. and Mukherjee, S. (2022) Auxin-mediated molecular mechanisms of heavy metal and metalloid stress regulation in plants. *Environmental and Experimental Botany* 196, 104796. DOI: 10.1016/j.envexpbot.2022.104796.

Matzke, M. and Mosher, R. (2014) RNA-directed DNA methylation: An epigenetic pathway of increasing complexity. *Nature Reviews. Genetics* 15(6), 394–408. DOI: 10.1038/nrg3683.

Mir, M.A., John, R., Alyemeni, M.N., Alam, P. and Ahmad, P. (2018) Jasmonic acid ameliorates alkaline stress by improving growth performance, ascorbate glutathione cycle and glyoxylase system in maize seedlings. *Scientific Reports* 8(1), 2831. DOI: 10.1038/s41598-018-21097-3.

Moore, L., Le, T. and Fan, G. (2013) DNA methylation and its basic function. *Neuropsychopharmacology* 38(1), 23–38. DOI: 10.1038/npp.2012.112.

Mousavi, S., Regni, L., Bocchini, M., Mariotti, R., Cultrera, N.G.M. *et al.* (2019) Physiological, epigenetic and genetic regulation in some olive cultivars under salt stress. *Scientific Reports* 9(1), 1093. DOI: 10.1038/s41598-018-37496-5.

Musich, R., Cadle-Davidson, L. and Osier, M.V. (2021) Comparison of short-read sequence aligners indicates strengths and weaknesses for biologists to consider. *Frontiers in Plant Science* 12, 657240. DOI: 10.3389/fpls.2021.657240.

Noor, I., Sohail, H., Sun, J., Nawaz, M.A., Li, G. *et al.* (2022) Heavy metal and metalloid toxicity in horticultural plants: Tolerance mechanism and remediation strategies. *Chemosphere* 303(3), 135196. DOI: 10.1016/j.chemosphere.2022.135196.

Ou, X., Zhang, Y., Xu, C., Lin, X., Zang, Q. *et al.* (2012) Transgenerational inheritance of modified DNA methylation patterns and enhanced tolerance induced by heavy metal stress in rice (*Oryza sativa* L). *PloS One* 7(9), e41143. DOI: 10.1371/journal.pone.0041143.

Peng, H. and Zhang, J. (2009) Plant genomic DNA methylation in response to stresses: Potential applications and challenges in plant breeding. *Progress in Natural Science* 19, 1037–1045. DOI: 10.1016/j.pnsc.2008.10.014.

Prathibha, M.D., Harsha, S.G., Geetha, G.A., Lokesha, A.N. and Shivashankara, K.S. (2023) Metabolomics as a selection tool for abiotic stress tolerance in crops. In: Harohalli Masthigowda, M., Gopalareddy, K., Khobra, R., Singh, G. and Pratap Singh, G. (eds) *Translating Physiological Tools to Augment Crop Breeding*. Springer, Singapore. DOI: 10.1007/978-981-19-7498-4_14.

Rahman, S., Ahmad, I. and Nafees, M. (2023) Mitigation of heavy metal stress in maize (*Zea mays* L.) through application of silicon nanoparticles. *Biocatalysis and Agricultural Biotechnology* 50, 102757. DOI: 10.1016/j.bcab.2023.102757.

Rizvi, A., Ahmed, B., Khan, S.M., Rajput, V.D., Umar, S. *et al.* (2022) Maize associated bacterial microbiome linked mitigation of heavy metal stress: A multidimensional detoxification approach. *Environmental and Experimental Botany* 200, 104911. DOI: 10.1016/j.envexpbot.2022.104911.

Schaarschmidt, S., Fischer, A., Zuther, E. and Hincha, D.K. (2020) Evaluation of seven different RNA-seq alignment tools based on experimental data from the model plant *Arabidopsis thaliana*. *International Journal of Molecular Sciences* 21(5), 1720. DOI: 10.3390/ijms21051720.

Simon, S.A. and Meyers, B.C. (2011) Small RNA-mediated epigenetic modifications in plants. *Current Opinion in Plant Biology* 14(2), 148–155. DOI: 10.1016/j.pbi.2010.11.007.

Srivastava, S. and Suprasanna, P. (2021) MicroRNAs: Tiny, powerful players of metal stress responses in plants. *Plant Physiology and Biochemistry* 166, 928–938. DOI: 10.1016/j.plaphy.2021.07.004.

Stanke, M. and Morgenstern, B. (2005) AUGUSTUS: A web server for gene prediction in eukaryotes that allows user-defined constraints. *Nucleic Acids Research* 33, W465–W467. DOI: 10.1093/nar/gki458.

Sun, X., He, Y., Huang, C., Ma, T.T. and Li, J. (2013) The epigenetic feedback loop between DNA methylation and microRNAs in fibrotic disease with an emphasis on DNA methyltransferases. *Cellular Signalling* 25(9), 1870–1876. DOI: 10.1016/j.cellsig.2013.05.013.

Tanwar, U.K., Stolarska, E., Rudy, E., Paluch-Lubawa, E., Grabsztunowicz, M. *et al.* (2023) Metal tolerance gene family in barley: An *in silico* comprehensive analysis. *Journal of Applied Genetics* 64(2), 197–215. DOI: 10.1007/s13353-022-00744-6.

Thakur, M., Praveen, S., Divte, P.R., Mitra, R., Kumar, M. *et al.* (2022) Metal tolerance in plants: Molecular and physicochemical interface determines the “not so heavy effect” of heavy metals. *Chemosphere* 287, 131957. DOI: 10.1016/j.chemosphere.2021.131957.

Thongnok, S., Siripornadulsil, W. and Siripornadulsil, S. (2022) Responses to arsenic stress of rice varieties coinoculated with the heavy metal-resistant and rice growth-promoting bacteria *Pseudomonas stutzeri* and *Cupriavidus taiwanensis*. *Plant Physiology and Biochemistry* 191, 42–54. DOI: 10.1016/j.plaphy.2022.09.014.

Tonosaki, K., Fujimoto, R., Dennis, E.S., Raboy, V. and Osabe, K. (2022) Will epigenetics be a key player in crop breeding? *Frontiers in Plant Science* 13, 958350. DOI: 10.3389/fpls.2022.958350.

Ueda, M. and Seki, M. (2020) Histone modifications form epigenetic regulatory networks to regulate abiotic stress response. *Plant Physiology* 182(1), 15–26. DOI: 10.1104/pp.19.00988.

Wang, Q., Liu, N., Yang, X., Tu, L. and Zhang, X. (2016) Small RNA-mediated responses to low- and high-temperature stresses in cotton. *Scientific Reports* 6(1), 35558. DOI: 10.1038/srep35558.

Wang, S., Yan, J., Hu, B., Wang, R. and Xu, J. (2022a) Advanced epigenomic engineering in crop quality improvement. *Current Opinion in Food Science* 46, 100843. DOI: 10.1016/j.cofs.2022.100843.

Wang, X., Chen, C., He, C., Chen, D. and Yan, W. (2022b) Mapping open chromatin by ATAC-seq in bread wheat. *Frontiers in Plant Science* 13, 1074873. DOI: 10.3389/fpls.2022.1074873.

Waseem, M., Khilji, S.A., Tariq, S., Jamal, A., Alomrani, S.O. *et al*. (2024) Phytoremediation of heavy metals from industrially contaminated soil using sunflower (*Helianthus annus* L.) by inoculation of two indigenous bacteria. *Plant Stress* 11, 100297. DOI: 10.1016/j.stress.2023.100297.

Wu, Z., Qu, J., Zhang, W. and Liu, G.H. (2023) Stress, epigenetics, and aging: Unraveling the intricate crosstalk. *Molecular Cell*. DOI: 10.1016/j.molcel.2023.10.006.

Yampolsky, L.Y., Zeng, E., Lopez, J., Williams, P.J., Dick, K.B. *et al*. (2014) Functional genomics of acclimation and adaptation in response to thermal stress in *Daphnia*. *BMC Genomics* 15(1), 859. DOI: 10.1186/1471-2164-15-859.

Zhou, S., Xing, M., Zhao, Z., Gu, Y., Xiao, Y. *et al*. (2021) DNA methylation modification in heterosis initiation through analyzing rice hybrid contemporary seeds. *The Crop Journal* 9(5), 1179–1190. DOI: 10.1016/j.cj.2020.12.003.

9 Epigenetics for Combating Drought Stress in Plants: Updated Methods and Current Achievements

Vinay Kumar Dhiman[1], Sudarshna[2], Rakesh Kumar[3], Gaurav Kumar[4], Vivek Kumar Dhiman[5] and Devendra Singh[6*]

[1]Department of Basic Sciences, College of Forestry, Dr Y S Parmar University of Horticulture and Forestry, Nauni, Solan, H.P., India; [2]Department of Bioscience & Bioengineering, Indian Institute of Technology, Jodhpur, Rajasthan, India; [3]Department of Fruit Science, College of Horticulture and Forestry, Dr Y S Parmar University of Horticulture and Forestry, Nauni, Solan, H.P., India; [4]Department of Basic Sciences, College of Horticulture and Forestry, Neri, Dr Y S Parmar University of Horticulture and Forestry, Nauni, Solan, H.P., India; [5]Department of Biotechnology, College of Horticulture, Dr Y S Parmar University of Horticulture and Forestry, Nauni, Solan, H.P., India; [6]Faculty of Biotechnology, Institute of Biosciences and Technology, Shri Ramswaroop Memorial University, Barabanki, Uttar Pradesh, India

Abstract

Plants can survive constant vulnerability against stress in the environment with their ability to change genetics over time. The modification in their genes to adapt to different types of stress can be observed in specific genes that are developed through certain stresses. The expression of such genes from genetic molecules (DNA or RNA) to proteins to specific functioning tissues at the desired period is triggered by certain phytohormones. Also, the formation of specific modifications in different plants in relation to the environment is crucial. Epigenetic regulation takes place in the form of modification of DNA with methyl group in cytosine. The modification of histone with methyl, acetyl, or phosphoryl groups interferes with the activation or repression of many genes causing upregulation or downregulation of genes. Also, there is regulation of post-transcriptional genes silencing by non-coding RNA (sRNA or miRNA) that are expressed in gene regulation. In different stress conditions, these epigenetic changes are expressed differently in different plant species along with morpho-physiological characters. The creation of memory in stress is achieved by transferring epigenomic alleles formed due to environmental conditions for survival. These alleles are not genetically passed on to the next generation, but they are necessary for evolution. In drought, plants respond to stress by governing their hormonal changes, introducing physiological or morphological advantages, or through epigenomic changes.

Keywords: Plant, Stress, Epigenomic changes, Methylation, Adaptation

*Corresponding author: devendrasingh.ibst@srmu.ac.in

 Epigenetics for Climate-Smart and Sustainable Agriculture (ed. Jen-Tsung Chen)
DOI: 10.1079/9781800626102.0009

9.1 Introduction

Plants are able to monitor changes in day/night duration and also in seasonal variations to adapt their metabolic rate for growth and development with the environment. Also, plants are sessile and are restricted to the area and environment around them. They have to establish themselves against countless physical attacks by insects and animals, weather fluctuations, scavenging soil for nutrients, competition against other invasive species, and so on (Osnato *et al.*, 2021; Venkat and Muneer, 2022). Therefore, for survival, they have to alter or modify their physical defense and also counteract external forces. Another significant discovery is the notion of 'stress memory', which suggests that plants may save genetic memory records of previous stressful encounters and use them to optimize their stress responses in subsequent instances.

The phenomenon of stress memory in plants can be seen as a mechanism that boosts their ability to withstand challenging conditions. The persistence of plasticity requires the maintenance of responses and the activation of signal pathways for gene regulation and expression (Lagiotis *et al.*, 2023). This effect has been shown in *Arabidopsis thaliana*, where exposure to drought stress results in enduring alterations in DNA methylation patterns at the upregulated regions of stress-associated genes, even after the stress has been alleviated. Plants are consistently subjected to unpredictable variations in the environment, including fluctuations in temperature, intensity of light, and the availability of nutrients and water. Furthermore, plants face additional challenges from a variety of diseases and insects, in addition to abiotic pressures. In order to endure these challenges, plants have evolved intricate sensing systems that are facilitated by signaling cascades and gene transcription networks. These processes enable plants to react to environmental inputs (Pissolato *et al.*, 2024).

Plants are immobile and must endure and thrive in difficult non-living (such as extreme temperatures, light levels, radiation, drought, flooding, saltiness, lack or excess of nutrients, and presence of heavy metals) and living (such as viruses, fungi, bacteria, nematodes, pests that suck or chew, and herbivores) stressful conditions by detecting the causes of stress and making molecular adaptations (Negi *et al.*, 2021; Singh *et al.*, 2022b; Tyagi *et al.*, 2024). The complex regulatory system, driven by epigenetic processes, controls the changes in plants' reactivity to extreme environmental circumstances. Plants often retain a memory of a particular environmental stress that they encountered throughout their lifespan. The phenomenon in plants that allows them to retain information and respond well to future challenges is known as plant memory or epigenetic memory (Fig. 9.1).

These epigenetic mechanisms allow changes that persist in cell division even in the absence of stimulus, thereby providing a memory that reinforces the maintenance of responses. The effects of stress memory-related mechanisms that cause changes in gene expression are connected to physiological, biochemical, metabolomic, and morphological alterations in plants. When epigenetic modifications are accelerated by the environment, the resulting phenotype primarily reflects specific environmental interactions. The changes plants

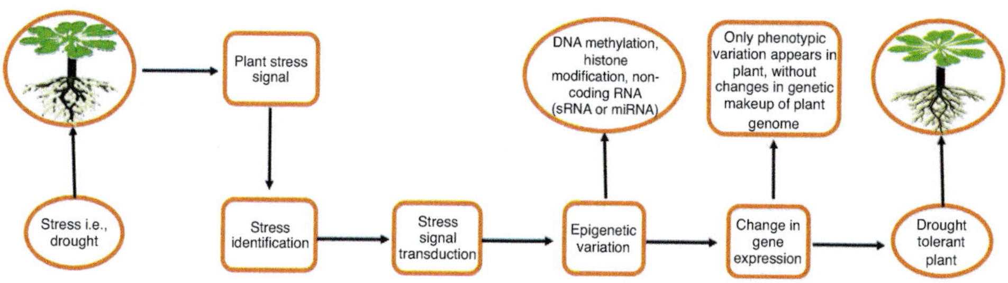

Fig. 9.1. A pictorial representation of the epigenetic processes involved in plants' response to drought stress. Author's own image.

undergo during a particular period can become permanent and heritable for future generations (Godwin and Farrona, 2020).

Stress experiences pose challenges to multicellular plants, resulting in a metabolic adaptation at all levels of the organization, ranging from cellular to the whole plant. Extensive research has been conducted on the impact of drought on the metabolic reaction of plants. However, our understanding of stress memory, as shown by the behavior of plants exposed to repeated drought, remains restricted. The molecular response of plants to future drought exposure may be characterized by an augmented, more effective, and faster reply. Various methods of drought stress imprint might be examined. These pathways rely on the plant's reaction at the metabolomics, proteomics, and transcriptomics levels. The reaction and adaptation of plants to drought stress, as well as other stressors that result in a lack of water in the soil, may be controlled by the influence of molecules and metabolites, which include protective chemicals and hormones (Ahluwalia *et al.*, 2021). The production of these molecules and metabolites may contribute to the process of stress tolerance and acclimation by facilitating a stronger response to stress stimuli upon future exposure (Singh *et al.*, 2022a). Various plant-derived chemicals have been recognized as important in their ability to respond to drought stress. The ability of plants to withstand drought is associated with the build-up of certain substances, including abscisic acid (ABA), ferulic acid, carbohydrates, amino acids, and proteins. Both ABA and jasmonic acid have a role in helping switchgrass plants (*Panicum virgatum*) adapt to secondary dehydration stress by controlling the interaction and communication between ABA and jasmonic acid, as well as the transmission of signals (Hayford *et al.*, 2022). In another study, double-stressed *Aptenia cordifolia* plants exhibited drought stress memory, with gibberellins and abscisic acid potentially influencing plant response and abscisic acid levels potentially modulating plant cellular redox state (Fleta-Soriano *et al.*, 2015).

Drought is a significant environmental stressor that severely affects plant growth and overall productivity. Plants have evolved with complex mechanisms to survive drought stress, including the development of 'drought stress memory'. This memory allows plants to respond more effectively to subsequent drought stress, enhancing their survival and productivity (Wojtyla *et al.*, 2020). The combination of genetic and epigenetic factors makes such a response successful in long-term resistance to such stress (Saha *et al.*, 2022). Epigenetic regulation is based on the modification of DNA through methylation or histone modification after translation of methylation and acetylation, and non-coding RNA-mediated regulation which does not code for proteins but expresses gene regulation (Wang *et al.*, 2022). Throughout the drought stress period and post-stress recovery period from drought, methylated histones are formed in higher amounts and deposited at the drought tolerance genes. Following drought stress, the RNA polymerase experiences a momentary halt and only restarts its activity when a pause-breaking element is introduced during the succeeding drought phase (Sadhukhan *et al.*, 2022). The regulation of short-term memory during drought is controlled by signaling from abscisic acid (ABA) and other phytohormones, which effectively influence the transcriptional response of many genes after a period of drought (Chauhan *et al.*, 2023). However, a prolonged period of recuperation accompanied by watering leads to the loss of the ability to withstand drought conditions, which is helped by specialized enzymes that remove methyl groups from DNA and variables that impact the accessibility of chromatin.

9.2 Epigenetic Inheritance of Phenotype Across Generations

Epigenetic inheritance of phenotype is not contingent upon alterations in DNA sequences. Plants undergo epigenetic modifications to adapt to challenging environmental circumstances and ensure their survival. The phenomenon known as 'plant stress memory' is responsible for the ability of plants to adapt to and withstand stress in challenging environments, as well as pass on this ability to future generations. This mechanism is the underlying cause of the observed modifications in plants. These organisms are immobile and thrive in a specific habitat that is consistently subjected to changes in the environment, posing a direct challenge to their ability to adapt ecologically. Fluctuations in light intensity impact the fluctuations in temperature during the day. Additionally, the altitude and location significantly influence the magnitude of temperature, resulting in either warm or cold environments. The involvement of mitotic and meiotic transmission of epigenetic changes in plants has been suggested. Recent research has shown that plants possess the ability to retain and use previously employed coping methods to effectively and promptly react to stress, hence safeguarding themselves against its recurrence.

The term epigenetics refers to the 'addition of' or 'on top of' genetics and is described as a phenomenon in which phenotypic differences occur in the genome of genetically identical cells or organisms over time (Hyde *et al.*, 2019). Epigenetic inheritance of phenotypes across generations does not depend on normal genetic modifications of DNA sequences. It is a weaker form of genetic inheritance, which is not through the germline but is perpetuated through mitosis or cell division so that the offspring of that cell inherit that expression state. Intra- and inter-individual variations in plants are determined via covalent modification of DNA and chromatin, which are the basis of genome structure, and these changes do not alter the genome sequence (Senaldi and Smith-Raska, 2020). The plant cells are plastic due to epigenome flexibility, which is responsible for triggering internal and external stimuli in developmental stages or with environmental variations, thereby creating memories related to stress (Köhler and Springer, 2017).

9.3 Effect of Drought and Plant Response Through Epigenetics

Stress has always been a greater challenge for the existence of plants, and water-deprived conditions are a greater threat to many species. The adjustment in metabolic rate and organization of cellular systems in plants is remarkable, as mild or shorter periods of water-deprived conditions can enhance stress tolerance to some extent. Only a severe shortage of water in the soil is lethal to plants (Seleiman *et al.*, 2021). In plants, damage caused by drought is prevented by altering their physical structure in response to drought. This includes development of deeper root systems to access more water, reduction of the number of leaves to limit water loss, or growing smaller, thicker leaves to reduce evaporation, succulent stems, tissue elasticity, leaf shedding, reduction in leaves size, and lowering of xylem hydraulic conductance (Filipović, 2020). There may also be reduction in usage of water via closing stomata (pores on the leaf surface) to reduce water loss, altering photosynthetic rates to conserve water, and increasing the synthesis of abscisic acid, a plant hormone that helps to regulate water (Pirasteh-Anosheh *et al.*, 2016). Rhizosphere bacterial and fungal associations also help in reducing nutrient and stress tolerance, either by modification of the root system (nodules and mycorrhizal associations) or by production of growth-promoting metabolites (Ahluwalia *et al.*, 2021; Dhiman *et al.*, 2022; Sudarshna and Sharma, 2024). Plants can accumulate certain solutes like proline, sugars, and amino acids that help to maintain cell turgor pressure, acting as osmoprotectants and protecting cells from dehydration. Stress memory induction is responsible for stress priming, i.e. hardening by stress, training (high-stress and low-stress training), stress acclimation, and conditioning in plants (Khan *et al.*, 2022). Plants are able to adjust the patterns of expression of genes that are responsive to such conditions for the regulation of genes under repeated stress. Memory genes are expressed significantly at higher or lower levels over dehydration in a responsive and efficient manner during drought stress. At the molecular level, the plant responds through genomic (gene level), transcriptomic (mRNA), proteomics (proteins), and metabolomics (metabolites) (Abdulraheem *et al.*, 2024).

Stress occurrences in early plant growth stages cause short-term memory for the stress for a couple of weeks. They cause a temporary effect on morphology and trigger the regulation of specific metabolites; after the end of the short-term stress period, plants return to the growing stage just like earlier, with no long-term memory. Long-term stress plays a crucial role in adapting to the environment and further evolution as the epigenetic changes occur during the lifetime of the plant and even further to its offspring (Jacques *et al.*, 2021). Long-term stress memory enhances drought resistance

Table 9.1. Epigenetic response in different crops in relation to drought and other stress tolerance. Author's own table.

Plant species	Epigenetic mechanism	Impact on drought and other stress tolerance	Example study
Arabidopsis thaliana	DNA methylation	Enhances drought tolerance	Sun *et al*., 2022a
Oryza sativa (rice)	Histone modifications	Regulates stress-responsive genes	Kumar *et al*., 2023
Zea mays (maize)	Small RNA-mediated pathways	Modulates stress response	Zhang *et al*., 2019
Triticum aestivum (wheat)	Chromatin remodeling	Facilitates stress adaptation	Concia *et al*., 2020
Glycine max (soybean)	Non-coding RNA regulation	Alters gene expression	Cadavid *et al*., 2023
Solanum lycopersicum (tomato)	DNA demethylation	Improves water use efficiency	Kumar *et al*., 2016
Hordeum vulgare (barley)	H3K4me3 histone modification	Enhances root development	Baker *et al*., 2015
Sorghum bicolor (sorghum)	Small RNA-mediated pathways	Modulates stomatal closure	Liu *et al*., 2023
Gossypium herbaceum (cotton)	Chromatin remodeling	Increases ROS scavenging	Zafar *et al*., 2023
Pisum sativum (pea)	Histone acetylation	Regulates ABA signaling	Smolikova *et al*., 2024

in plants. A few examples, such as *Arabidopsis thaliana*, barley (*Hordeum vulgare*), cotton (*Gossypium* spp.), sorghum (*Sorghum bicolor*), sunflower (*Helianthus annuus*), and eucalyptus (*Eucalyptus* spp.), showed resistance to drought exposure (Pandian, 2022). Plants retain information about stressful conditions and utilize this memory to better adapt to future stressors. The use of such stress memory is of immense importance in agriculture, as climate variations are major factors that affect agriculture due to extreme weather phenomena (Table 9.1).

9.4 Molecular Mechanism of Stress Memory in Plants

The molecular mechanisms involved in plant stress memory involve epigenetic regulation and transcriptomic and metabolomic alterations. Epigenetic regulation comprises DNA methylation (-CH3), covalent modification of histone proteins, chromatin rearrangement, changes in nucleosome positioning, and non-coding RNA-mediated gene regulation (Dar *et al*., 2022). The changes in genes involve methylation (-CH3) of DNA and histone modifications associated with responsive memory genes and subsequent enhanced transcriptional in drought stress (Sun *et al*., 2022b). Transcriptomic changes are mediated during stress-induced memory involved in various transcription factors. For example, WRKY and AREB are among the transcription factors that have been associated with stress memory (Chauhan *et al*., 2023). Stress memory is also mediated by metabolomic changes, including alterations in the levels of amino acids, sugars, and other metabolites. These changes can help plants adapt and survive unpredictable fluctuations in the environment. Non-coding RNA such as small RNAs, particularly microRNAs (miRNAs), are vital in regulating plant stress memory. These miRNAs control the expression of target genes associated with stress response, thereby increasing the plant's ability to withstand stress. The upregulation of the drought memory-associated miRNA, tae-miR531_L-2, greatly enhanced the drought resistance of genetically modified *Arabidopsis thaliana* (Yadav *et al.*, 2024). Thus, unraveling the molecular mechanisms, including regulatory RNA's role in controlling stress memory in plants, is of paramount importance for sustainable agriculture in the face of increasing climate variability.

9.5 Significance of Phytohormones in Drought Stress Memory

Abscisic acid (ABA) serves a vital role in aiding plants in withstanding drought stress. It acts as a pivotal mediator in signaling response by triggering adaptive reactions to environmental stresses. ABA triggers a multitude of physiological processes, including bud dormancy, regulation of stomatal opening and closure, seed germination, and modulation of gene expression under stress conditions (Vishwakarma *et al.*, 2017; Ali *et al.*, 2022). Moreover, in response to drought stress, the most rapid response regulated by ABA in plants is the closure of stomata. This mechanism reduces water loss by limiting transpiration (Singh *et al.*, 2022c). When exposed to ABA, guard cells undergo a size reduction, leading to the closure of the pore through which air passes. Additionally, ABA plays a role in facilitating the production and transport of other signaling molecules. The intricate signaling network of ABA is essential, involving various sites for ABA synthesis, transporter proteins in the plasma membrane, and numerous components of ABA-dependent signaling pathways (Kumar *et al.*, 2019). ABA-responsive genes are crucial in detecting stress and adjusting hormonal pathways. As water availability reduces, ABA levels rise, inhibiting enzymes responsible for breaking dormancy and promoting germination (Ali *et al.*, 2022).

The plant hormone ABA, produced in response to drought stress, plays a crucial role in increasing drought resistance. ABA regulates several developmental stages, including germination, stomatal closure, and root growth. It also coordinates growth inhibition in seedlings by causing reversible changes in transcriptional expression patterns. The regulatory route of ABA signaling has been previously clarified with a thorough investigation into the epigenetic processes that underlie ABA stress responses. Research has shown that there is a strong connection between changes in histone dynamics and gene expression due to drought stress. Histone alterations play a vital role in the response to ABA throughout the development of seedlings.

For example, when BR-activated BES1 combines with TPL-HDA19, it creates a complex that directly promotes the removal of acetyl groups from the chromatin of ABI3. This results in the suppression of ABI3 and, subsequently, ABI5, which are important regulators of ABA signaling during the early development of seedlings (Zhao *et al.*, 2019). Histone demethylases are involved in the precise regulation of the SnRK2.8 kinase via the ABI3 transcription factor. Wu *et al.* (2019) showed that the inhibition of growth in *A. thaliana* by ABA is regulated by histone demethylases JMJ30 and JMJ32. ABI3 triggers the production of JMJ30 during the post-germination phase in response to ABA. This results in the elimination of the restrictive histone marker H3K27me3 from the promoter region of SnRK2.8 and the subsequent activation of SnRK2.8 expression. SnRK2.8 kinase, in turn, stimulates the production of ABI3 and facilitates growth cessation (Hasan *et al.*, 2022).

9.6 RNA Regulation in Plant Stress Memory

Plants have an impressive ability to remember drought stress, both in the immediate term and over multiple generations, due to epigenetic mechanisms that control gene expression. This allows them to react more efficiently to subsequent drought exposure, which is frequently recurring rather than isolated, marked by occasional intervals of moisture recovery. During the periods of recovery, several genes linked to drought tolerance experience histone methylation, which serves as a marker for their activation when necessary (Kim *et al.*, 2015; Kumar and Singh, 2016). Pause-breaking factors activate RNA polymerase, which was first halted, in response to further drought events. During a drought, there is a phenomenon called

DNA demethylation that takes place near genes that are involved in responding to drought. This process is carefully regulated by genetic factors. The progeny of plants exposed to drought inherit methylation patterns that boost their ability to respond to drought (Verma *et al.*, 2018; Lephatsi *et al.*, 2021).

Extended periods of adequate hydration, on the other hand, lead to the depletion of the ability to withstand dry conditions, which is aided by demethylases and variables that affect the accessibility of chromatin. Small RNAs are essential for controlling drought memory by regulating the expression of genes involved in responding to drought. Current research is focused on understanding the genetic basis of drought memory and the complex interaction between genetic and epigenetic factors in its inheritance (Zuo *et al.*, 2021). Essentially, the recollection of drought stress in plants is a complex process influenced by genetic and epigenetic factors, including the coordination of regulatory RNAs.

MicroRNAs (miRNAs) are crucial for enhancing plant resilience in the face of difficult environmental circumstances, including drought, salt, severe temperatures, and nutrient deprivation. RNA molecules regulate gene expression after transcription, influencing plants' various physiological and biochemical processes. MicroRNAs control crucial physiological functions, including stress responses, ion balance, antioxidant defense, and hormone signaling, by specifically targeting transcripts that encode transcription factors and signaling molecules (Zuo *et al.*, 2021; Islam *et al.*, 2024). In addition, they have an impact on physical characteristics such as the structure of the roots and the regulation of stomata, which are important for improving the ability of plants to withstand environmental challenges. Utilizing genetic engineering and breeding approaches can enhance our understanding of plant systems and boost agricultural output (Raza *et al.*, 2023).

9.7 Role of Histone Modification in Plant Drought Stress Tolerance

The modification of histone structure plays a crucial role in the establishment of drought memory in plants. In particular, alterations such as methylation are strongly linked to genes that have a lasting recollection of drought stress. This leads to an increased activation of transcriptional responses when exposed to drought repeatedly (Banerjee and Roychoudhury, 2017). Following a period of drought, there is a notable increase in the presence of methylated histones in the genes that play a crucial role in drought tolerance. During this period, the activity of RNA polymerase comes to a halt. However, it is later reignited by a stimulus that ends the pause caused by the subsequent drought. This process plays a vital role in the plant's ability to remember and adapt to arid conditions, and the signaling of these molecules is regulated by ABA and other phytohormones (Mayer *et al.*, 2017).

In addition, drought conditions trigger DNA demethylation near genes associated with drought response, a process controlled by genetics. The offspring of plants that have experienced drought conditions display a heightened sensitivity to drought, due to the inheritance of specific methylation patterns. Nevertheless, excessive watering during the recovery phase can result in the depletion of drought memory, which is aided by demethylases and chromatin accessibility factors (Lata *et al.*, 2015; Sadhukhan *et al.*, 2022).

Acetylation and deacetylation reactions of histone tails are carried out by enzyme acetylases and histone deacetylases. Multiple studies have emphasized the crucial function of histone acetyltransferase (HAT) in plants growth and stress adaptation. Histone deacetylases (HDACs) consist of three separate groups, with HD2 being particularly notable as deacetylases unique to plants. HDACs facilitate gene silence by enhancing chromatin compaction and eliminating acetyl groups from histone proteins (Jiang *et al.*, 2020; Kumar *et al.*, 2021).

Phosphorylation is a crucial alteration of histones that is necessary for activating gene transcription, repairing DNA, condensing and segregating chromosomes throughout the cell cycle, and inducing death. Both serine (S) and threonine (T) may be phosphorylated (Rossetto *et al.*, 2012). Ubiquitination is a histone modification that entails the addition of a tiny, conserved protein called ubiquitin, which is composed of 75 amino acids, to histone proteins. This procedure specifically modifies linker histones and nucleosome components. Ubiquitin forms an isopeptide bond with certain lysines, which is a reversible process that also acts as a signal for protein destruction in proteasomes. Moreover, ubiquitination has the ability to impact the intracellular distribution and metabolic functionality of the protein being targeted (Vaughan *et al.*, 2021). Sumoylation, a histone modification process, involves the SUMO (small ubiquitin-related modifier) protein. Similar to ubiquitin, SUMO protein forms covalent bonds with other proteins through specialized enzymes (Sahin *et al.*, 2022). This sumoylation process is pivotal in regulating transcription and managing the stress response in plants.

9.8 Role of DNA Methylation in Plant Drought Stress Tolerance

DNA methylation significantly contributes to plant drought stress tolerance, serving as an extensively studied mechanism for epigenetic control of gene expression. This process is crucial in imparting tolerance against biotic and abiotic stressors, including drought. In plants, the methylation levels of certain genes shift throughout growth and development in response to environmental stimuli (Sharma *et al.*, 2022). DNA methylation aids in plant survival under stress by preventing unfavorable genetic alterations at specific loci. Drought stress can trigger alterations in DNA methylation levels, inducing drought tolerance in plants, particularly through changes in the methylation patterns of key transcription factor genes (Akhter *et al.*, 2021). Thus, epigenetic modifications in the genome represent a vital regulatory mechanism for plant adaptation to drought and potentially other environmental stressors. Additionally, DNA methylation is associated with gene body methylation (gbM), influencing both transcript abundance and stability (Williams *et al.*, 2023). The availability of mutants for genes encoding methylation enzymes has facilitated the elucidation of the biological significance of methylation. DNA methylation is a key mechanism that helps plants adapt to drought stress. It regulates gene expression, prevents unfavorable genetic rearrangements, and contributes to the stability of transcripts, thereby enhancing the plant's drought tolerance.

9.9 Non-Coding RNAs Regulate Plant Drought Stress Tolerance

Non-coding RNAs (ncRNAs), including small RNAs and long non-coding RNAs (lncRNAs), play a crucial role in regulating drought memory in plants. Small RNAs have a crucial function in controlling the ability of plants to remember and respond to drought conditions. They do this by adjusting the levels of certain genes that are involved in the plant's response to drought (Jha *et al.*, 2023). These ribonucleic acids (RNAs) play a role in creating complex systems of control, which have the capacity to affect a large number of genes and, hence, determine the overall reaction of plants. Long ncRNAs are over 200 nt in length and have no protein-coding ability. They are produced by the transcription of RNA polymerase II and exert their functions through an epigenetic mechanism, transcriptional regulation, post-transcriptional modification, and translation (Mattick *et al.*, 2023). Long ncRNAs are involved in both the growth and development of plants, as well as their ability to respond to abiotic stressors (Dykes and Emanueli, 2017). They interact with their targets to establish regulatory networks that govern the activity of numerous genes, hence directing the overall reaction of plants to drought and other non-living environmental pressures. The important genetic controls facilitated by ncRNAs have emerged as transformative factors in modifying plants' responses to drought and other non-living environmental stimuli (Yadav *et al.*, 2024). Non-coding RNAs play a significant role in regulating drought memory in plants, contributing to their adaptive responses to drought stress.

9.10 Plant Epigenetic Regulation: Recent Findings and Studies of Role in Drought Stress

Proline accumulation is believed to have a substantial impact on drought memory in several plant species. Crops have significant challenges throughout their growth cycle when they are subjected to prolonged or frequent drought occurrences. To counteract these situations, plants have developed a variety of tactics, which include making changes to their physiology and development through the regulation of gene expression (Kapoor *et al*., 2020). During periods of drought, a large number of genes that respond to drought are activated, and their levels of RNA return to normal levels when the plants are watered again. Therefore, as a result of repeated periods of dehydration stress, the levels of transcripts display recurring variations. This phenomenon has been referred to as memory, imprinting, priming, training, and adaptation to stress (Kim *et al*., 2020). *Arabidopsis* plants that had been previously subjected to dehydration withered at a much slower pace in comparison with plants that had not been treated to dehydration (Huang *et al*., 2008).

Overall, the ability of plants to remember and respond to drought stress has the potential to provide protection against future droughts in crops. This is achieved by a combination of changes in gene activity that are not caused by changes in the DNA sequence (epigenetic alterations), responses to plant hormones, and the control of mechanisms that guard against damage caused by reactive oxygen species (antioxidative protection) (Liu *et al*., 2022). Drought priming allows plants to regulate gene responses in order to tolerate recurrent drought, hence enhancing their capacity to resist natural, unpredictable drought cycles when grown in the field. Gaining a more profound comprehension of the regulatory systems and identifying the most effective levels of stress will assist in choosing drought-resistant cultivars (Marthandan *et al*., 2020).

Scientists have investigated the capacity of stress memory to improve agricultural output during periods of drought. Zaki and Radwan (2022) identified five potato cultivars sourced from the Middle East that exhibited high tolerance to drought, while 11 others showed moderate tolerance. This tolerance was associated with the upregulation of stress-related genes such as DRO, ERECTA, ERF, DREB, and StMYB. Sun *et al*. (2022a) used the functional potato gene StDRO1 in Unica, a drought-tolerant cultivar of *Arabidopsis*. They observed an enhanced biochemical response, including increased levels of peroxidase, catalase, superoxide dismutase, and proline content, which provided its potential role in potatoes under drought stress.

Nevertheless, the precise molecular pathways responsible for stress memory still need to be better understood. Epigenetic changes in the chromatin structure of memory genes, such as DNA methylation at the promoter region, modifications of histones, or formation of short RNA molecules, are believed to be the main factors, according to the current view. Epigenetic alterations are thought to be transferred mitotically to newly generated cells during cell division Furthermore, it is possible to transfer the epigenetic memory state to future generations of plants via meiosis, which is known as transgenerational epigenetic inheritance (Rissman and Adli, 2014). The possibility of inheriting learned features across generations has improved agricultural yield. Virdi *et al*. (2015) studied heritable changes in *Arabidopsis* due to MSH1 mutation, and these mutations were observed to impact plant development and stress responses. Tabassum *et al*. (2018) found that priming wheat seeds enhanced crop performance during drought stress by improving tissue water status, accumulating osmolytes, and reducing lipid peroxidation. Drought priming allows plants to regulate gene responses, enhancing their capacity to resist many droughts and improving their resilience to unpredictable drought cycles in field cultivation. Raju *et al*. (2018) created a breeding method for soybeans that uses epigenetics to improve yield and stability. This is done by controlling the pathways related to development, defense, phytohormones, and response to abiotic stress (Rathva *et al*., 2024).

9.11 Chromatin Alterations Caused by Histone Modifications and DNA Methylation

Chromatin alterations caused by histone modifications and DNA methylation are essential for enhancing drought tolerance in plants. Chromatin regulators play a crucial role in regulating stress-responsive gene networks under abiotic stress conditions, with specific histone modification sites and histone modifiers identified through genetic and biochemical approaches. DNA methylation plays a crucial role in controlling gene expression and maintaining the integrity of the plant genome

(Breiling and Lyko, 2015). The addition of a methyl group to the cytosine residue of DNA results in the formation of 5-methylcytosine (5mC). The 5mC molecule has a role in several biological processes, including maintaining the integrity of the genome, regulating gene expression, controlling development, and responding to environmental stress (Kumar and Mohapatra, 2021). It functions as a suppressive indicator that inhibits the expression of genes, and its levels are controlled by both methylation and demethylation processes. DNA methylation can improve crop productivity, increase resistance to diseases, and boost tolerance to environmental challenges. Epigenetic markers may be used in crop development programs using CRISPR/dCas9-based targeted DNA (de)methylation editing, which has previously been exploited to produce an inheritable late-flowering characteristic. This presents a novel approach to genetically modifying plants by using the epi-allele/epimark during the production of climate-smart agricultural plants.

9.12 Conclusion

Drought stress memory, when combined with epigenetic alterations, phytohormonal responses, and modulation of antioxidative protection, may provide crops with resilience against repeated droughts. Additionally, DNA methylation has a role in enhancing plants' ability to withstand drought conditions. Although there have been many studies investigating the effects of drought stress on DNA methylation levels in annual herbaceous plants, there needs to be more research on this topic for perennial woody plants. The levels of methylation of cytosines throughout the whole genome rise dramatically in response to drought stress, and this increase is positively associated with gene expression. When methylation in DNA significantly increases, it results in the silencing of genes. Genes responsible for some transcription factors show a decrease in DNA methylation and transcriptome expression after drought treatment, whereas genes encoding other transcription factors exhibit an increase in DNA methylation and gene expression. Hence, alterations in DNA methylation control the activation of drought-response genes on a comprehensive level across the genome. Histone acts as spools that cover DNA and wind it; the modification of histone with the addition of methyl, acetyl, or phosphoryl groups leads to the downregulation or upregulation of genes. Essentially, the modifications of histones, specifically methylation and demethylation, play a crucial role in the establishment and regulation of drought memory in plants. They facilitate plant adaptation to changes in the environment, improving their chances of survival during periods of drought. Non-coding RNA leads to the silencing of genes during post-transcriptional events and merely regulates the expression of genes.

References

Abdulraheem, M.I., Xiong, Y., Moshood, A.Y., Cadenas-Pliego, G., Zhang, H. *et al.* (2024) Mechanisms of plant epigenetic regulation in response to plant stress: Recent discoveries and implications. *Plants* 13(2), 163.

Ahluwalia, O., Singh, P.C. and Bhatia, R. (2021) A review on drought stress in plants: Implications, mitigation and the role of plant growth promoting rhizobacteria. *Resources, Environment and Sustainability* 5, 100032.

Akhter, Z., Bi, Z., Ali, K., Sun, C., Fiaz, S. *et al.* (2021) In response to abiotic stress, DNA methylation confers epigenetic changes in plants. *Plants* 10(6), 1096.

Ali, F., Qanmber, G., Li, F. and Wang, Z. (2022) Updated role of ABA in seed maturation, dormancy, and germination. *Journal of Advanced Research* 35, 199–214.

Baker, K., Dhillon, T., Colas, I., Cook, N., Milne, I. *et al.* (2015) Chromatin state analysis of the barley epigenome reveals a higher-order structure defined by H3K27me1 and H3K27me3 abundance. *The Plant Journal* 84(1), 111–124.

Banerjee, A. and Roychoudhury, A. (2017) Epigenetic regulation during salinity and drought stress in plants: Histone modifications and DNA methylation. *Plant Gene* 11, 199–204.

Breiling, A. and Lyko, F. (2015) Epigenetic regulatory functions of DNA modifications: 5-methylcytosine and beyond. *Epigenetics & Chromatin* 8(1), 9.

Cadavid, I.C., Balbinott, N. and Margis, R. (2023) Beyond transcription factors: More regulatory layers affecting soybean gene expression under abiotic stress. *Genetics and Molecular Biology* 46, e20220166.

Chauhan, D., Singh, D., Pandey, H., Khan, Z., Srivastava, R. *et al.* (2023) Impact of transcription factors in plant abiotic stress: A recent advancement for crop improvement. In: *Plant Transcription Factors*. Academic Press, pp. 271–286.

Concia, L., Veluchamy, A., Ramirez-Prado, J.S., Martin-Ramirez, A., Huang, Y. *et al.* (2020) Wheat chromatin architecture is organized in genome territories and transcription factories. *Genome Biology* 21, 1–20.

Dar, F.A., Mushtaq, N.U., Saleem, S., Rehman, R.U., Dar, T.U.H. *et al.* (2022) Role of epigenetics in modulating phenotypic plasticity against abiotic stresses in plants. *International Journal of Genomics*. DOI: 10.1155/2022/1092894.

Dhiman, V.K., Rana, N., Dhiman, V.K., Pandey, H., Verma, P. *et al.* (2022) Effect of rhizobial isolates and nitrogen fertilizers on nursery performance, nodulation behavior and nitrogenase activity of *Dalbergia sissoo* Roxb. seedlings. *Plant Stress* 4, 100080.

Dykes, I.M. and Emanueli, C. (2017) Transcriptional and post-transcriptional gene regulation by long non-coding RNA. *Genomics, Proteomics & Bioinformatics* 15(3), 177–186.

Filipović, A. (2020) Water plant and soil relation under stress situations. *Soil Moisture Importance* 73.

Fleta-Soriano, E., Pintó-Marijuan, M. and Munné-Bosch, S. (2015) Evidence of drought stress memory in the facultative CAM, *Aptenia cordifolia*: Possible role of phytohormones. *PloS ONE* 10(8), e0135391.

Godwin, J. and Farrona, S. (2020) Plant epigenetic stress memory induced by drought: A physiological and molecular perspective. *Plant Epigenetics and Epigenomics: Methods and Protocols* 243–259.

Hasan, M.M., Liu, X.-D., Waseem, M., Guang-Qian, Y., Alabdallah, N.M. *et al.* (2022) ABA activated SnRK2 kinases: An emerging role in plant growth and physiology. *Plant Signaling & Behavior* 17(1), 2071024.

Hayford, R.K., Serba, D.D., Xie, S., Ayyappan, V., Thimmapuram, J. *et al.* (2022) Global analysis of switchgrass (*Panicum virgatum* L.) transcriptomes in response to interactive effects of drought and heat stresses. *BMC Plant Biology* 22(1), 107.

Huang, D., Wu, W., Abrams, S.R. and Cutler, A.J. (2008) The relationship of drought-related gene expression in *Arabidopsis thaliana* to hormonal and environmental factors. *Journal of Experimental Botany* 59(11), 2991–3007.

Hyde, L.K., Friso, S. and Choi, S.W. (2019) Introduction to epigenetics. In: Caterina, R., Martinez, J.A. and Kohlmeir, M. (eds) *Principles of Nutrigenetics and Nutrigenomics: Fundamentals of Individualized Nutrition*. Academic Press, pp. 129–139.

Islam, W., Adnan, M., Alomran, M.M., Qasim, M., Waheed, A. *et al.* (2024) Plant responses to temperature stress modulated by microRNAs. *Physiologia Plantarum* 176(1), e14126.

Jacques, C., Salon, C., Barnard, R.L., Vernoud, V. and Prudent, M. (2021) Drought stress memory at the plant cycle level: A review. *Plants* 10(9), 1873.

Jha, U.C., Nayyar, H., Roychowdhury, R., Prasad, P.V.V., Parida, S.K. *et al.* (2023) Non-coding RNAs (ncRNAs) in plants: Master regulators for adapting to extreme temperature conditions. *Plant Physiology and Biochemistry* 205, 108164.

Jiang, J., Ding, A.B., Liu, F. and Zhong, X. (2020) Linking signaling pathways to histone acetylation dynamics in plants. *Journal of Experimental Botany* 71(17), 5179–5190.

Kapoor, D., Bhardwaj, S., Landi, M., Sharma, A., Ramakrishnan, M. *et al.* (2020) The impact of drought in plant metabolism: How to exploit tolerance mechanisms to increase crop production. *Applied Sciences* 10(16), 5692.

Khan, A., Khan, V., Pandey, K., Sopory, S.K. and Sanan-Mishra, N. (2022) Thermo-priming mediated cellular networks for abiotic stress management in plants. *Frontiers in Plant Science* 13, 866409.

Kim, J.M., Sasaki, T., Ueda, M., Sako, K. and Seki, M. (2015) Chromatin changes in response to drought, salinity, heat, and cold stresses in plants. *Frontiers in Plant Science* 6, 124925.

Kim, Y.K., Chae, S., Oh, N.I., Nguyen, N.H. and Cheong, JJ. (2020) Recurrent drought conditions enhance the induction of drought stress memory genes in *Glycine max* L. *Frontiers in Genetics* 11, 576086.

Köhler, C. and Springer, N. (2017) Plant epigenomics-deciphering the mechanisms of epigenetic inheritance and plasticity in plants. *Genome Biology* 18(1), 3.

Kumar, M., Kesawat, M.S., Ali, A., Lee, S.C., Gill, S.S. *et al.* (2019) Integration of abscisic acid signaling with other signaling pathways in plant stress responses and development. *Plants* 8(12), 592.

Kumar, R., Chauhan, P.K. and Khurana, A. (2016) Identification and expression profiling of DNA methyltransferases during development and stress conditions in *Solanaceae*. *Functional & Integrative Genomics* 16(5), 513–528.

Kumar, S. and Singh, A. (2016) Epigenetic regulation of abiotic stress tolerance in plants. *Advances in Plants & Agriculture Research* 5(5), 00179.

Kumar, S. and Mohapatra, T. (2021) Dynamics of DNA methylation and its functions in plant growth and development. *Frontiers in Plant Science* 12, 596236.

Kumar, S., Seem, K. and Mohapatra, T. (2023) Biochemical and epigenetic modulations under drought: Remembering the stress tolerance mechanism in rice. *Life* 13(5), 1156.

Kumar, V., Thakur, J.K. and Prasad, M. (2021) Histone acetylation dynamics regulating plant development and stress responses. *Cellular and Molecular Life Sciences* 78, 4467–4486.

Lagiotis, G., Madesis, P. and Stavridou, E. (2023) Echoes of a stressful past: Abiotic stress memory in crop plants towards enhanced adaptation. *Agriculture* 13, 11.

Lata, C., Muthamilarasan, M. and Prasad, M. (2015) Drought stress responses and signal transduction in plants. In: Pandey, G.K. (ed.) *Elucidation of Abiotic Stress Signaling in Plants: Functional Genomics Perspectives*, Vol. 2. Springer, New York, pp. 195–225.

Lephatsi, M.M., Meyer, V., Piater, L.A., Dubery, I.A. and Tugizimana, F. (2021) Plant responses to abiotic stresses and rhizobacterial biostimulants: Metabolomics and epigenetics perspectives. *Metabolites* 11(7), 457.

Liu, H., Able, A.J. and Able, J.A. (2022) Priming crops for the future: Rewiring stress memory. *Trends in Plant Science* 27(7), 699–716.

Liu, J., Wu, Y., Dong, G., Zhu, G. and Zhou, G. (2023) Progress of research on the physiology and molecular regulation of sorghum growth under salt stress by gibberellin. *International Journal of Molecular Sciences* 24(7), 6777.

Marthandan, V., Geetha, R., Kumutha, K., Renganathan, V.G., Karthikeyan, A. *et al.* (2020) Seed priming: A feasible strategy to enhance drought tolerance in crop plants. *International Journal of Molecular Sciences* 21(21), 8258.

Mattick, J.S., Amaral, P.P., Carninci, P., Carpenter, S., Chang, H.Y. *et al.* (2023) Long non-coding RNAs: Definitions, functions, challenges and recommendations. *Nature Reviews. Molecular Cell Biology* 24(6), 430–447.

Mayer, A., Landry, H.M. and Churchman, L.S. (2017) Pause & go: From the discovery of RNA polymerase pausing to its functional implications. *Current Opinion in Cell Biology* 46, 72–80.

Negi, S., Singh, D., Pandey, H. and Singhal, R.K. (2021) Biofortification for high Fe and Zn in various poaceae crops by using different molecular breeding and biotechnological approaches. *Plant Physiology Reports* 26, 636–646.

Osnato, M., Cota, I., Nebhnani, P., Cereijo, U. and Pelaz, S. (2021) Photoperiod control of plant growth: Flowering time genes beyond flowering. *Frontiers in Plant Science* 12, 805635.

Pandian, TJ. (2022) *Evolution and Speciation in Plants*. CRC Press, Boca Raton, Florida.

Pirasteh-Anosheh, H., Saed-Moucheshi, A., Pakniyat, H. and Pessarakli, M. (2016) Stomatal responses to drought stress. In: Ahmad, P. (ed.) *Water Stress and Crop Plants: A Sustainable Approach*. John Wiley & Sons, Hoboken, New Jersey, pp. 24–40.

Pissolato, M.D., Martins, T.S., Fajardo, Y.C.G., Souza, G.M., Machado, E.C. *et al.* (2024) Stress memory in crops: What we have learned so far. *Theoretical and Experimental Plant Physiology* 1–31.

Raju, S.K.K., Shao, M.-R., Sanchez, R., Xu, Y.-Z., Sandhu, A. *et al.* (2018) An epigenetic breeding system in soybean for increased yield and stability. *Plant Biotechnology Journal* 16, 1836–1847.

Rathva, H., Pandey, A.K., Suthar, K., Suthar, H., Chakote, A. *et al.* (2024) Genetic relatedness analysis in sapota using SSR markers. *Ecological Genetics and Genomics* 100234.

Raza, A., Charagh, S., Karikari, B., Sharif, R., Yadav, V. *et al.* (2023) MiRNAs for crop improvement. *Plant Physiology and Biochemistry* 201, 107857.

Rissman, E.F. and Adli, M. (2014) Minireview: Transgenerational epigenetic inheritance: Focus on endocrine disrupting compounds. *Endocrinology* 155(8), 2770–2780.

Rossetto, D., Avvakumov, N. and Côté, J. (2012) Histone phosphorylation: A chromatin modification involved in diverse nuclear events. *Epigenetics* 7(10), 1098–1108.

Sadhukhan, A., Prasad, S.S., Mitra, J., Siddiqui, N., Sahoo, L. *et al.* (2022) How do plants remember drought? *Planta* 256(1), 7.

Saha, D., Choyal, P., Mishra, U.N., Dey, P., Bose, B. *et al.* (2022) Drought stress responses and inducing tolerance by seed priming approach in plants. *Plant Stress* 4, 100066.

Sahin, U., Thé, H. and Lallemand-Breitenbach, V. (2022) Sumoylation in physiology, pathology and therapy. *Cells* 11(5), 814.

Seleiman, M.F., Al-Suhaibani, N., Ali, N., Akmal, M., Alotaibi, M. *et al.* (2021) Drought stress impacts on plants and different approaches to alleviate its adverse effects. *Plants* 10(2), 259.

Senaldi, L. and Smith-Raska, M. (2020) Evidence for germline non-genetic inheritance of human phenotypes and diseases. *Clinical Epigenetics* 12(1), 136.

Sharma, S., Chauhan, A., Dobbal, S. and Kumar, R. (2022) Biology of plants coping stresses: Epigenetic modifications and genetic engineering. *South African Journal of Botany* 144, 270–283.

Singh, D., Pandey, H., Dhiman, V.K., Bang, N.T., Dhiman, V.K. *et al.* (2022a) Secondary metabolite engineering for plant immunity against various pathogens. In: Hakeem, K.R. and Aftab, T. (eds) *Metabolic Engineering in Plants*. Springer Nature Singapore, Singapore, pp. 123–143.

Singh, D., Pandey, H., Dhiman, V.K., Thakur, K., Sharma, I. *et al.* (2022b) Assessment of environmental impacts of metal/metalloid pollution on plants. In: Aftab, T. and Hakeem, K.R. (eds) *Metals and Metalloids in Soil-Plant-Water Systems*. Academic Press, pp. 217–232.

Singh, D., Pandey, H., Thakur, K., Sharma, I. and Naraian, R. (2022c) Aquaporins and their functions in water transportation in different plant species. *Environmental Sustainability* 5(4), 443–456.

Smolikova, G., Krylova, E., Petřík, I., Vilis, P., Vikhorev, A. *et al.* (2024) Involvement of abscisic acid in transition of pea (*Pisum sativum* L.) seeds from germination to post-germination stages. *Plants* 13(2), 206.

Sudarshna and Sharma, N. (2024) Endophytic bacteria associated with critically endangered medicinal plant *Trillium govanianum* (Wall ex. Royle) and their potential in soil nutrition alleviation. *Plant Stress* 11, 100349.

Sun, C., Liang, W., Yan, K., Xu, D., Qin, T. *et al.* (2022a) Expression of potato StDRO1 in *Arabidopsis* alters root architecture and drought tolerance. *Frontiers in Plant Science* 13, 836063.

Sun, M., Yang, Z., Liu, L. and Duan, L. (2022b) DNA methylation in plant responses and adaption to abiotic stresses. *International Journal of Molecular Sciences* 23(13), 6910.

Tabassum, T., Farooq, M., Ahmad, R., Zohaib, A., Wahid, A. *et al.* (2018) Terminal drought and seed priming improves drought tolerance in wheat. *Physiology and Molecular Biology of Plants* 24(5), 845–856.

Tyagi, S., Dhiman, V.K., Dhiman, V.K., Pandey, H., Singh, D. *et al.* (2024) Plant defense strategies and biomarkers against heavy metal-induced stress: A comprehensive review. *ACS Agricultural Science & Technology* 4(2), 129–143.

Vaughan, R.M., Kupai, A. and Rothbart, S.B. (2021) Chromatin regulation through ubiquitin and ubiquitin-like histone modifications. *Trends in Biochemical Sciences* 46(4), 258–269.

Venkat, A. and Muneer, S. (2022) Role of circadian rhythms in major plant metabolic and signaling pathways. *Frontiers in Plant Science* 13, 836244.

Verma, A., Kumar, S., Kumar, G., Saini, J.K., Agrawal, R. *et al.* (2018) Rhizosphere metabolite profiling: An opportunity to understand plant-microbe interactions for crop improvement. In: Prasad, R., Gill, S.S. and Tuteja, N. (eds) *Crop Improvement through Microbial Biotechnology*, Vol. 1. Elsevier, pp. 343–361.

Virdi, K.S., Laurie, J.D., Xu, Y.-Z., Yu, J., Shao, M.-R. *et al.* (2015) *Arabidopsis* MSH1 mutation alters the epigenome and produces heritable changes in plant growth. *Nature Communications* 6(1), 6386.

Vishwakarma, K., Upadhyay, N., Kumar, N., Yadav, G., Singh, J. *et al.* (2017) Abscisic acid signaling and abiotic stress tolerance in plants: A review on current knowledge and future prospects. *Frontiers in Plant Science* 08, 161.

Wang, K., Liu, H., Hu, Q., Wang, L., Liu, J. *et al.* (2022) Epigenetic regulation of aging: Implications for interventions of aging and diseases. *Signal Transduction and Targeted Therapy* 7(1), 374.

Williams, C.J., Dai, D., Tran, K.A., Monroe, J.G. and Williams, B.P. (2023) Dynamic DNA methylation turnover in gene bodies is associated with enhanced gene expression plasticity in plants. *Genome Biology* 24(1), 227.

Wojtyla, Ł., Paluch-Lubawa, E., Sobieszczuk-Nowicka, E. and Garnczarska, M. (2020) Drought stress memory and subsequent drought stress tolerance in plants. In: Hossain, M.A., Burritt, D.J., Fujita, M., Huang, B. and Liu, F. (eds) *Priming-Mediated Stress and Cross-Stress Tolerance in Crop Plants*. Academic Press, pp. 115–131.

Wu, J., Yamaguchi, N. and Ito, T. (2019) Histone demethylases control root elongation in response to stress-signaling hormone abscisic acid. *Plant Signaling & Behavior* 14(7), 1604019.

Yadav, A., Mathan, J., Dubey, A.K. and Singh, A. (2024) The emerging role of non-coding RNAs (ncRNAs) in plant growth, development, and stress response signaling. *Non-Coding RNA* 10(1), 13.

Zafar, S., Afzal, H., Ijaz, A., Mahmood, A., Ayub, A. *et al.* (2023) Cotton and drought stress: An updated overview for improving stress tolerance. *South African Journal of Botany* 161, 258–268.

Zaki, H.E.M. and Radwan, K.S.A. (2022) Response of potato (*Solanum tuberosum* L.) cultivars to drought stress under *in vitro* and field conditions. *Chemical and Biological Technologies in Agriculture* 9, 1–9.

Zhang, M., An, P., Li, H., Wang, X., Zhou, J. *et al.* (2019) The miRNA-mediated post-transcriptional regulation of maize in response to high temperature. *International Journal of Molecular Sciences* 20(7), 1754.

Zhao, X., Dou, L., Gong, Z., Wang, X. and Mao, T. (2019) BES1 hinders ABSCISIC ACID insensitive5 and promotes seed germination in *Arabidopsis*. *The New Phytologist* 221(2), 908–918.

Zuo, Z.F., He, W., Li, J., Mo, B. and Liu, L. (2021) Small RNAs: The essential regulators in plant thermotolerance. *Frontiers in Plant Science* 12, 726762.

10 Epigenetics for Combating Heat Stress in Plants: Updated Methods and Current Achievements

Khadija Benamar[1,2], Ilham Dehbi[1], Mohammed Radi[1,3], Rachid Ez-Zouggari[1,4], Kawtar Fikri Benbrahim[2], Yunfei Jiang[5] and Rachid Lahlali[1*]

[1]*Department of Plant Protection, École Nationale d'Agriculture de Meknès, Meknes, Morocco;* [2]*Department of Biology, Laboratory of Microbial Biotechnology and Bioactive Molecules, Sciences and Technologies Faculty, Sidi Mohamed Ben Abdellah University, Fez, Morocco;* [3]*Environment and Valorization of Microbial and Plant Resources Unit, Faculty of Sciences, Moulay Ismail University, Meknes, Morocco;* [4]*Laboratory of Biotechnology, Conservation and Valorisation of Natural Resources, Faculty of Sciences Dhar El Mehraz, Sidi Mohamed Ben Abdallah University, Fez, Morocco;* [5]*Department of Plant, Food and Environmental Sciences, Faculty of Agriculture, Dalhousie University, Truro, Nova Scotia, Canada*

Abstract

This chapter examines recent advances in epigenetic research aimed at understanding and improving plant responses to heat stress. Rising temperatures due to climate change pose a serious threat to global crop yields by causing heat stress, which can severely damage plants. This stress limits crop productivity, especially during critical growth phases. Epigenetic mechanisms, including DNA methylation, histone modifications, and non-coding RNA interference, play a crucial role in plants' ability to survive abiotic stress by creating a stress memory that is transmitted to subsequent generations. This chapter explores these mechanisms and their impact on plant resistance to heat stress. It also presents recent methods and current achievements in mitigating heat stress in plants, highlighting the importance of epigenetic regulation. Finally, it offers insights into developing heat-tolerant crops, which are essential for maintaining and improving agricultural productivity in the context of climate change.

Keywords: Heat stress, Climate change, Crop yields, Epigenetic mechanisms, DNA methylation, Histone modifications, Stress memory, Stress resistance

10.1 Introduction

Elevated temperatures due to climate change represent a serious threat to crop yields worldwide (Wang *et al.*, 2018), particularly when stress conditions occur during the most critical stage of plant growth. In general, heat stress in plants is defined as heat increases over

*Corresponding author: rlahlali@enameknes.ac.ma

 Epigenetics for Climate-Smart and Sustainable Agriculture (ed. Jen-Tsung Chen)
DOI: 10.1079/9781800626102.0010

a certain period of time, potentially causing permanent damage to plants, which usually happens with temperatures 10–15°C above the threshold (Willits and Peet, 1998). Elevated temperatures affect the activity of proteins and membrane lipids, impairing the activity of chloroplast- and mitochondria-based enzymes along with the integrity of membranes. Severe heat stress and prolonged exposure to mild elevated temperatures can result in cell damage and cell mortality. Heat stress is a major restricting factor in the production of field crops in tropical and subtropical climates (Hu *et al.*, 2020).

Plants can escape from heat stress or resist it by avoidance or tolerance (Bueckert and Clarke, 2013). Management and breeding approaches, such as shifting planting date and genetic manipulation of crop phenology, have been used to facilitate plants to escape from heat stress (Sadras and Dreccer, 2015). Heat avoidance is the ability of plants to avoid heat stress through some physiological mechanisms (Bueckert and Clarke, 2013). Heat tolerance is the ability of plants to survive under heat stress and usually produce lower economic yield compared with optimal conditions (Hasanuzzaman *et al.*, 2013). The terms heat tolerance and heat resistance are also used in the literature interchangeably to mean a trait with stress robustness compared with heat susceptibility. Plant responses to heat stress depend largely on temperature severity, the duration of heat stress, and the growth stage when plants are exposed to stress (Argosubekti, 2020). The rates of changing temperature and plants' degree of temperature resistance levels play an important role in remembering that plants have different degrees of heat resistance in order to survive periods of stress. Plants have evolved advanced mechanisms of their own, including regulating respiration, photosynthesis and water metabolism, to resist the impact of heat stress. For this reason, the scientific research community is attempting to clarify the epigenetically controlled mechanisms through which plants counteract the potential effects of severe heat stress (Verma *et al.*, 2022).

Coping with these extreme temperature fluctuations requires substantial changes at the molecular level. Over recent years, epigenetic regulation has emerged as one of the most successful controls on plant survival in the face of multiple abiotic stresses. Epigenetic processes engage changes in the chromatin structure of genes affected by diverse types of chemical modification along transcriptional and post-transcriptional events (Kim *et al.*, 2015; Lämke and Bäurle, 2017). Epigenetic marks, specifically DNA methylation, non-coding RNA interference (ncRNA), and histone modifications, play an extensive part in the ability of plants to survive abiotic stress. In addition, epigenetic processes can create a memory of stress that can be carried forward to the next generation of stressed plants (Friedrich *et al.*, 2019). Therefore, determining epigenetic processes may be essential for selecting stress-resistant plants. In recent years, considerable efforts have been focused on identifying molecular key variables that need to be genetically modified to produce tolerant lines capable of surviving wide variations in temperature extremes (Verma *et al.*, 2022).

This chapter aims to provide a comprehensive overview of recent advances in epigenetic research aimed at understanding and enhancing plant responses to heat stress. It explores updated methods and current achievements in epigenetics for mitigating heat stress in plants, emphasizing the role of epigenetic mechanisms like DNA methylation, histone modifications, and chromatin remodeling in plant stress memory. Additionally, it offers insights into developing heat-tolerant crops and enhancing crop productivity amidst changing environmental conditions.

10.2 Epigenetic Mechanisms in Response to Heat Stress

In response to unfavourable environmental conditions like heat stress, plants have exhibited different genetic and epigenetic modifications. The genetic regulatory system has been well studied, focusing on different plant responses under such conditions, like flowering acceleration, circadian clock modulation, and enhancement of plants' immunity and their thermotolerance to heat stress (Liu *et al.*, 2015). In contrast, epigenetic mechanisms are related to modifications in gene function without any changes in gene sequence. Indeed, these mechanisms involve several chemical and heritable changes in DNA occurring under different stresses, including

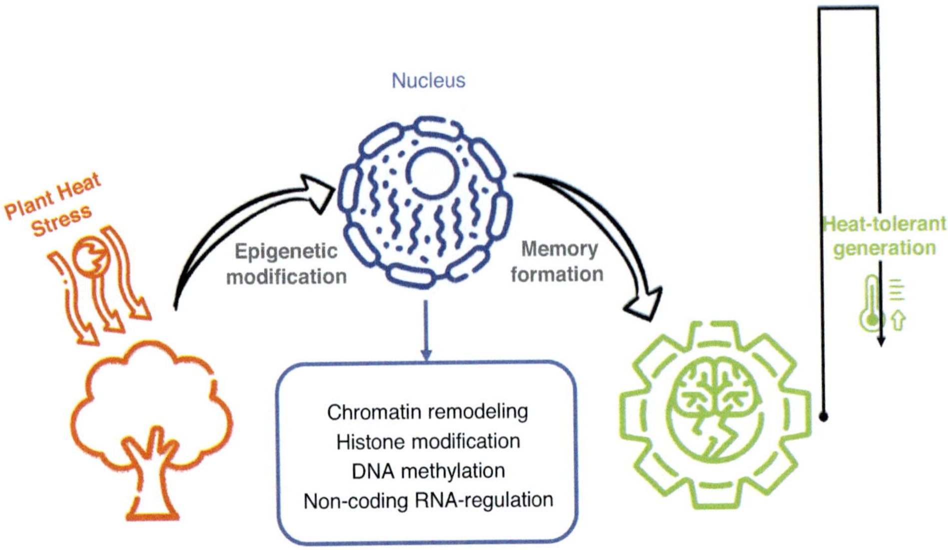

Fig. 10.1. Epigenetic modifications of plants under heat stress. Author's own image.

heat stress (Fig. 10.1), known as DNA methylation and demethylation, histone and chromatin modifications, and non-coding RNA regulation (Saraswat *et al.*, 2017). Different studies showing epigenetic modifications in several plant species under heat stress are presented in Table 10.1.

10.2.1 DNA methylation/demethylation

DNA methylation represents a covalent modification characterized by the insertion of methyl to the nucleic base cytosine at the site CpG of DNA molecule (Chang *et al.*, 2023). This modification ensures the regulation of chromatin architecture, the stability of DNA sequence, and its expression without leading to any modification of its nucleic bases. This process can occur either in the body region of genes and inducing their expression activation, or in gene promoters and leading to gene repression, or in transposable elements and contributing to genome integrity. Enzymes responsible for DNA methylation are known as DNA methyl transferases (Verma *et al.*, 2022).

It is necessary to note that DNA demethylation can also occur. Indeed, the lack of methyl transferase enzymes, or the lack of methyl donors after the replication of DNA sequence, can lead to passive DNA demethylation, whereas active DNA demethylation can take place when the plant involves several enzymes capable of DNA demethylation (Korotko *et al.*, 2021). Gao *et al.* (2014) studied the methylation responses of two *Brassica napus* subsp. *napus* (rapeseed) genotypes, one of which is tolerant of heat stress and the other sensitive to it, and they found that the tolerant genotype was prone to DNA demethylation epigenetic modifications, whereas the sensitive genotype was more characterized by methylation process. In *Oryza sativa* (rice), exposure to heat stress was associated with DNA methylation, and a reduction of OsFIE1 gene expression resulting in the modification of seed size, which is controlled by this gene (Folsom *et al.*, 2014). In another study, Min *et al.* (2014) observed a hypomethylation in the cotton line H05, resulting from heat stress which had induced a deficient level in S-ADENOSYL-L-HOMOCYSTEINE HYDROLASE1 and an inhibiting effect on methyltransferases. Moreover, in *Brassica napus* cv. Topas microspores, a DNA hypomethylation was found in CG and CHG sites under short-term heat treatment (Li *et al.*, 2016). Similarly, in *Glycine max* L. (soybean), a hypomethylation in the context CHH was observed (Hossain *et al.*, 2017). In *Setaria italica* L. (foxtail millet), it was revealed that tolerant cultivars

Table 10.1. Epigenetic modifications in different plants species under heat stress, according to some reported studies. Author's own table.

Plant species	Epigenetic modifications	References
Brassica napus subsp. *napus*	DNA demethylation in the tolerant genotype DNA methylation in the sensitive genotype	Gao *et al.*, 2014
Glycine max L.	Hypomethylation in CG and CHG sites	Hossain *et al.*, 2017
Zea mays L.	Methylation in CCGG and CCWGG sites	Qian *et al.*, 2019
Arabidopsis thaliana L.	DNA demethylation	Korotko *et al.*, 2021
Zea mays L.	A correlation between heat stressful conditions and histone acetylation	Zhao *et al.*, 2014
Arabidopsis thaliana	Histone acetylation activation	Weng *et al.*, 2014
Arabidopsis thaliana	A high induction of miR156	Stief *et al.*, 2014
Oryza sativa	A downregulation of the majority of miRNA	Sailaja *et al.*, 2014
Triticum aestivum	miRNA differentially expressed	Ravichandran *et al.*, 2019
Arabidopsis thaliana	Chromatin remodeling by the complex INO80	Xue *et al.*, 2021
Bemisia tabaci	Chromatin remodeling due to an upregulation of ISWI expression	Ji *et al.*, 2020
Cicer arietinum	Chromatin remodeling associated with an upregulation of PIE1 gene expression	Chidambaranathan *et al.*, 2016

showed a high expression of the SisHSP-27 gene which is responsible for the production of heat shock proteins, whereas in susceptible cultivars, a lower expression of these genes was observed and it was attributed to higher methylation in these genes in the susceptible cultivar (Singh *et al.*, 2016). Qian *et al.* (2019) demonstrated that in maize (*Zea mays* L.), a methylation process occurs under heat stress in CCGG and CCWGG sites. In addition, Pereira *et al.* (2021) studied the methylation process in *Pinus halepensis* under heat stress. They found that in response to heat conditions, DNA methylation was decreasing, and the DEHYDRATION INDUCED PROTEIN 19 gene was subjected to an upregulation. In addition, Malabarba *et al.* (2021) found that, under extreme heat stress, germinated seeds exhibited differentially methylated regions at the gene sequences, and in the promoter. In another study, Korotko *et al.* (2021) observed that DNA demethylation in *Arabidopsis thaliana* L. occurred under heat stress conditions. More recently, He *et al.* (2022) investigated the effect of short- and long-period heat stress on *Gossypium hirsutum* seedlings growth, and their results showed that methylation epigenetic modifications contribute to the thermotolerance in *Gossypium hirsutum*.

10.2.2 Histone modifications

Histones constitute a group of proteins that wrap around DNA by forming a compact structure called a nucleosome. This latter consists of histone octamers comprising H4, H3, H2A and H2B in two copies. The *N*-terminal tail of histones may undergo epigenetic modifications under heat stress, namely acetylation/deacetylation, and methylation (Singroha and Sharma, 2019).

Histone acetylation takes place at the *N*-terminal lysine and induces DNA transcription activation, whereas its deacetylation leads to gene transcription repression (Tang *et al.*,

2016). Indeed, in the acetylation process, a decrease in histones' positive charge occurs due to the insertion of acetyl group to the lysine. Therefore, the attractive force between DNA (negatively charged) and the histone proteins decreases. Hence, DNA loses its supercoiled structure and its transcription is therefore activated (Saraswat *et al.*, 2017). The acetylation is achieved by specific enzymes known as histone acetyl transferases. In rice, several histone acetyl transferases have been reported as belonging to four classes, namely GNAT, TAFII250, MYST and CBP (Liu *et al.*, 2012). The study of plant responses to heat stress showed the occurrence of histone acetylation/deacetylation epigenetic modifications in different plant species. For instance, in maize (*Zea mays* L.), a correlation between heat stressful conditions and roots acetylation level was obtained (Zhao *et al.*, 2014). In the study performed by Zheng *et al.* (2016), it was demonstrated that, in *Arabidopsis*, the enzymes histone deacetylases 9 (HDA9) negatively regulate the plant response to heat stress, by regulating different stress-resistant genes. This negative role of HDA9 in the regulation of heat stress in *Arabidopsis* was different from that of HDA19, HDA6, and HD1C, which were found to have a positive function in the control of different stress responses (Zheng *et al.*, 2016). Besides, it was demonstrated that HDA19 enzymes enhance tolerance in *Arabidopsis* against heat stress and other abiotic stresses (Ueda *et al.*, 2018). Moreover, Weng *et al.* (2014) showed that in *Arabidopsis*, under heat stress, histone H_3K_{56} acetylation activation is favored, and heat shock protein genes transcription is triggered. In addition, Hu *et al.* (2015) found that in *Arabidopsis*, acetyltransferase GCN5 function loss resulted in serious problems and defects regarding the transcription of heat stress responsive genes.

10.2.3 Regulation of non-coding RNA

Non-coding RNAs represent a class of RNA which do not encode proteins. They include long non-coding RNAs (lncRNAs), small interfering RNAs (siRNAs), microRNAs (miRNAs), and circular RNAs (circRNAs) (Zhao *et al.*, 2016). Many studies have shown the important role of non-coding RNAs in plant heat stress responses. Indeed, Stief *et al.* (2014) showed that in *Arabidopsis thaliana*, miR156 are characterized by a high induction after heat stress, and are responsible for downregulation of SQUAMOSA PROMOTOR BINDING PROTEIN-LIKE (SPL) genes which are encoding for specific transcription factors included in many processes, like flowering. Moreover, miR156 have also promoted the expression of heat stress-responsive genes (Stief *et al.*, 2014). For the same plant species, Wunderlich *et al.* (2014) found that long non-coding anti-sense RNA expression depends on the activity of heat shock transcription factors (HSFs) during heat stress, and has an important involvement in vegetative development. In addition, in rice (*Oryza sativa*), Sailaja *et al.* (2014) revealed that, after heat stress, the majority of miRNA were downregulated in the cultivar Nagina 22. They also found that miR160, miR1884, and miR169 were differentially expressed in rice shoot and root tissues. Ebrahimi Khaksefidi *et al.* (2015) showed differential regulation of several miRNA in *Helianthus annuus*, namely miR842, miR426, miR403, miR398, miR172, miR167, and miR160 under heat stress, with an overall downregulation of miRNA expression in the plant leaves except for miR403, not a significant modulation of miR172 expression in root tissues, and a similar expression pattern for the other miRNAs (miR160, miR426, miR842, miR167, and miR398) in both tissues. It is important to mention that after a long heat stress period, miR160 exhibited an opposite expression, which revealed the dependence of miRNA regulation not only on environmental stresses, but also on the period for which they are subjected to a specific stress. Ravichandran *et al.* (2019) found that in *Triticum aestivum* (wheat), miRNAs were differentially expressed after heat stress, and targeted an array of protein kinases, F-box proteins, and also superoxide dismutase.

10.2.4 Chromatin remodeling

ATP-dependent chromatin remodeling refers to chromatin conformation change by the alteration or destabilization of histone–DNA interactions through the use of the energy resulting from ATP hydrolysis. This process therefore induces chromatin disruption and remodeling, by the transfer of histone octamers to different points in the same DNA sequence. Hence, either transcription repression or activation occurs (Rehman and Tanti, 2020).

Chromatin remodeling factors (CRFs) are responsible for chromatin remodeling and include four subfamilies, namely chromodomain (CHD), imitation switch (ISWI), inositol 80 (INO80), and SWIching/sucrose non-fermentable (SWI/SNF) dependent factors. These CRFs are involved in the assessment of nucleosome spacing after replication (for CHD and ISWI), the exchange of H_2A.Z with the conical H_2A (for INO80/SWR1 complex), and nucleosome sliding, as well as histone–DNA interactions disintegration (for SWI/SNF) (Verma *et al.*, 2022). Under heat stress, ATP-dependent CRF induces chromatin remodeling and exhibits an important function in the control of such stresses. Buszewicz *et al.* (2016) showed that in *Arabidopsis*, an interaction between histone deacetylase HD2C and BRM-SWI/SNF remodeling complex occurs and induces a downregulation of heat-activated genes in the hd2c mutants. Further, Xue *et al.* (2021) demonstrated that INO80 chromatin remodeling complex promoted thermomorphogenesis, and activated thermoresponsive and auxin-related genes expression in *Arabidopsis*. Moreover, they revealed that an association between INO80 complex and phytochrome interacting factor 4 (PIF4) (which represents a potential thermomorphogenesis regulator) occurs and leads to the histone variant H_2AZ eviction at PIF4 targets, resulting in changes in the expression of thermoresponsive genes. In addition, H_2AZ removal was also promoted by the interaction between INO80 complex and transcription elongation factors SPT4 and SPT5. In another study, Chidambaranathan *et al.* (2016) identified in *Cicer arietinum* (chickpea), seven genes homologous to SWR1 chromatin remodeling complexes in *Arabidopsis*, namely photoperiod-independent early flowering 1 (PIE), actin-related protein (ARP6), two serrated leaf and early flowering (SEF) and three histone 2A variant Z (H_2AZs). Moreover, they analyzed their expression in chickpea during heat stress in different tissues. Their results showed that the PIE1 gene was upregulated after heat stress. In addition, the expression of PIE1 and SEF was higher in grain tissues, pod wall, flower, and in roots, in comparison with shoots. Overall, this study suggested the involvement of PIE1 gene in chromatin remodeling complexes under heat stress conditions, and the role of the three histone H_2AZ variants in the characterization of gene responsiveness basic processes in chickpea. Ji *et al.* (2020) assessed ISWI functions in invasive species of *Bemisia tabaci* under heat stress. Indeed, they noticed that *B. tabaci* ISWI expression was characterized by a significant upregulation after heat shock exposure. These findings indicate that *B. tabaci* ISWI expression could be induced by heat stress. Further, for the same species, Ji *et al.* (2022) identified 13 SWI/SNF and 10 ISWI, and revealed that their chromatin remodeling genes play an important role in thermal tolerance of *B. tabaci*.

10.3 Methods for Studying Epigenetic Responses to Heat Stress

10.3.1 Whole-genome bisulfite sequencing

Whole-genome bisulfite sequencing (WGBS) represents a method for determination of DNA methylation through treating DNA samples with sodium bisulfite. This method allows cytosine conversion to uracil, and keeps the residue 5 methyl-cytosine unaffected. Subsequently, a PCR reaction is required for the determination of methylated status by using specific methylation primers (Rehman and Tanti, 2020).

WGBS is an important tool for the study of plant responses to heat stress. It allows determination of the epigenetic variations regarding DNA methylation/demethylation in plants under such conditions. Li *et al.* (2016) showed through the use of WGBS that heat shock applied to *Brassica napus* cv. Topas microspores induced a hypomethylation in DNA at CG and CHG contexts. Moreover, thermal treatment at 32°C resulted in more variations of DNA methylation in the cultured microspores. Based on WGBS, Hossain *et al.* (2017) revealed that, under heat stress, soybean root hairs and stripped roots were subjected to hypomethylation, especially in the context CHH. Liu *et al.* (2018) demonstrated that, in *Brassica rapa*, DNA methylation occurred in the contexts CHH, CHG and CG near the transcription start and end sites, suggesting its role in genes transcription silencing. In another study, Zhang *et al.* (2019) relied on WGBS to map cytosine methylation in cotton (*Gossypium hirsutum* L.) anthers after heat stress. They revealed that heat stress resulted in a demethylation of transposable elements, and in methylation changes at the promoter site as well as in mitochondrial respiratory chain enzyme-related genes upregulation, namely GhATPBM, GhCX5B2, GhCOX6A and GhNDUS7, leading to an ultimate promotion of redox process to produce ATP, permitting the development of normal anthers under heat stress.

A study conducted to discover the impact of heat stress on DNA methylation regulation in *Arabidopsis* Col-0 (Malabarba *et al.*, 2021) exhibited a moderate effect of mild heat stress on DNA methylation, whereas under severe heat stress, a more important proportion of differentially methylated regions was observed in the promoters and in germination-related gene sequences. Korotko *et al.* (2021) found in *Arabidopsis thaliana* leaves a decrease of DNA methylation under heat stress in genes encoding heat shock proteins. The decrease of DNA methylation in such sites was suggested to be related to an active demethylation process, and not a passive demethylation which does not take place in a specific locus. Interestingly, Lu *et al.* (2021) investigated DNA methylation in *Arabidopsis* plants after multi-generational exposure to elevated temperatures. Their results displayed a significant increase of mutation rates in small indels and single-nucleotide, resulting from the multi-generational exposure of *Arabidopsis* to moderate and extreme heat conditions. Moreover, through WGBS, methylation was frequently observed in mutation sites. Furthermore, Yao *et al.* (2022) revealed that under a temperature of 22°C or 28°C, a hypermethylation occurs in the genome of broccoli (*Brassica oleracea*) head, especially in the promoter regions, resulting in the suppression of apex genes expression, leading to abnormal floral development. Overall, WGBS represents an important tool for the study of epigenetic modifications in plants under heat stress.

10.3.2 Chromatin immunoprecipitation sequencing (ChIP-seq)

Chromatin immunoprecipitation is a tool employed in epigenetic studies for detection of protein–DNA interactions in a cell. It allows the determination of a protein binding to a specific cellular DNA sequence, and it also permits the realization of transcription factors related to certain regions in the genome such as promoters. The ChIP method involves several steps; thus, during the ChIP process, cells are subjected to fixation using a specific agent like formaldehyde, which ensures the fixation of protein–DNA interactions in the cells. These are then lysed to release the chromatin, which is then fragmented using enzymatic digestion or sonication. Further, the obtained chromatin is immunoprecipitated by employing protein-specific antibodies. As a result, a co-precipitation of DNA sequences binding with the proteins occurs. The digestion of these proteins allows the release of DNA. Then, a PCR reaction is required for the amplification of DNA sequences to ensure their precipitation with the antibodies. ChIP is further associated with DNA micro-array assay and sequencing, and the resultant DNA is allowed to co-hybridize to the micro-array (Rehman and Tanti, 2020).

ChIP has been applied in several studies in order to investigate epigenetic modifications related to histones that occur in plants under heat stress. Buszewicz *et al.* (2016) used ChIP to study interactions between histone deacetylases (HDACs) and chromatin remodeling complexes (CRCs) during heat stress in *Arabidopsis*. Their results revealed the interaction between HD2C and BRM-containing SWI/SNF CRCs. The altered expression of HD2C regulated genes under heat stress coincided with an increase in histone acetylation at their loci. In addition, the ChIP technique has shown that, in *Arabidopsis* subjected to heat stress conditions, histone acetyltransferase GCN5 is enriched at the promoter regions of HSFA3 and UVH6 genes. Besides, GCN5 facilitated H3K9 and H3K14 acetylations, which are related to UVH6 and HSFA3 activation under heat stress (Hu *et al.*, 2015). Furthermore, a combination of ChIP-qPCR, RNA sequencing, and WGBS showed that mutations of histone H3K4 methyltransferases

SDG25 and ATX1 resulted in a decrease of histone H3K4me$_3$ levels, an increase of DNA cytosine methylation, and an inhibition of the expression of a subset of heat stress-responsive genes during the period of *Arabidopsis* stress recovery (Song *et al.*, 2021). In another study, Hou *et al.* (2019) demonstrated, through ChIP, the occurrence of alterations in H3K4me and H3K9 ac levels at the promoter regions of genes involved in the upregulation of maize seedlings heat stress factor and rRNA genes. More recently, He *et al.* (2022) revealed based on ChIP, the elevation of H3K4me$_2$ levels under the subjection of *Gossypium hirsutum* to short-term heat stress, and an increase of H4K5 ac levels during the whole heat stress process, especially under long-term exposure of *Gossypium hirsutum* to heat stress.

10.3.3 Small RNA sequencing

Various plant heat stress-responsive miRNAs were identified using RNA sequencing technology. For example, in switchgrass (an important biofuel crop species), RNA profiles were determined under heat and drought stresses using sequencing analysis. The study identified 62 novel and 29 conserved miRNA families. Moreover, differences and similarities regarding miRNA regulation were observed in plants when exposed to heat or drought stresses (Hivrale *et al.*, 2016). Similarly, Zhou *et al.* (2020) studied miRNA profiles in tomato plants subjected to drought and heat stress. They revealed 335 known miRNAs and 430 potential novel miRNAs in the studied plants. The comparison of combined heat and drought with the control plants, plants subjected to drought stress, and plants exposed to heat stress showed a differential regulation of 61, 74, and 37 miRNAs, respectively. In another study (Pokhrel and Meyers, 2022), following heat stress, 141 miRNAs were found in different tissues of *Linum usitatissimum* (flax), whereas 18 miRNAs were differentially accumulated in its reproductive tissues, including miR482/2118 and miR2275, which are known to trigger reproductive phased secondary small RNAs. Interestingly, Kushawaha *et al.* (2021) identified a wide number of miRNAs in rice plants under heat stress, after heat priming and without priming. Among the identified miRNAs, a differential expression of osa-miR167e-3p, osa-miR531b, osa-miR168a-5p, osa-miR5149, osa-miR1846d-5p, osa-miR5077, and osa-miR156b-3p was noticed between primed and non-primed plants. Furthermore, integrated analysis of small RNA and transcriptome sequencing in soybean revealed 21 differentially expressed miRNAs, and 15 differentially expressed target genes under heat stress, like gma-miR399a/inorganic phosphate transporter1-4, and gma-miR397a/laccase 2. To summarize, RNA sequencing is important for the study of epigenetic modifications regarding small RNA regulation in plants under heat stress (Ding *et al.*, 2021).

10.4 Current Achievements in Understanding Epigenetic Responses to Heat Stress

Currently, a highlight of plant heat priming and its association with several epigenetic modifications that induce heat memory formation and transgenerational inheritance of heat stress memory has been revealed. Indeed, when plants are exposed to moderately stressful conditions, they are subjected to a specific metabolite state known as priming. This state can be maintained for a certain period and leads to the creation of stress memory (Turgut-Kara *et al.*, 2020). It was noted that primed plants will respond effectively to subsequent stressful events (Oberkofler *et al.*, 2021). The study of chromatin context has revealed several epigenetic mechanisms that occur in memory genes (Fig. 10.2). These include genes showing transcription modifications persisting during the recovery period, and genes returning to the control of expression levels during recovery, but showing faster responses at stress recurrence. Moreover, it was suggested that short-term priming results in the activation and facilitation of memory genes transcription, whereas long-term priming may induce genes repression (Harris *et al.*, 2023). It was revealed that plants could exhibit stress memory even after short time exposure to stressful conditions. This is the case with *Arabidopsis thaliana*, for which 5 min of heat stress were found to be sufficient for the initiation of upregulation and transcription

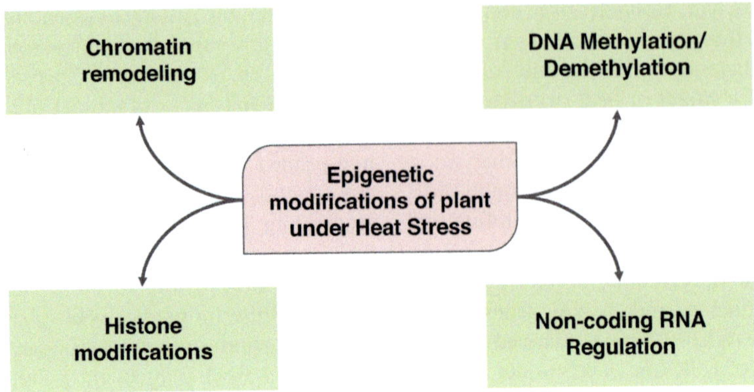

Fig. 10.2. Epigenetic modifications in plants under heat stress, and transgenerational memory formation. Author's own image.

of heat-responsive genes, including several heat shock factors, namely HsfA32, Hsp90, Hsp101, Hsp70, Hsp22, Hsp 21, and HSFA2 (Oyoshi *et al.*, 2020). Furthermore, it was mentioned that the intensity of priming heat stimulus affects the strength of further heat stress tolerance. Thus, for instance, it was reported that seedlings primed at 38°C showed a higher ability to overcome heat shock than those primed at lower temperatures. This differential response to heat shock was associated with the unequal level of heat shock proteins accumulation between both priming cases (Nair *et al.*, 2022). The epigenetic modifications resulting from plant exposure to a recurrent heat stress include histone modifications, chromatin remodeling, and also RNA regulation (Fig. 10.2).

10.4.1 Role of histone/chromatin modifications in plant heat stress memory formation

Several studies have shown the effect of heat priming on plant histone and chromatin modifications and the formation of transgenerational heat memory. Wang *et al.* (2016) investigated the effects of heat priming on wheat successive generation tolerance to post-anthesis high temperature. Their results revealed that heat-primed progeny possessed lower cell membrane damage and higher antioxidant enzymes activity, leaves photosynthesis and grain yield, in comparison with non-heat-primed plant progeny. This difference was attributed to the upregulation of the gene encoding for lysine-specific histone demethylase, involved in histone demethylation in heat-primed plants. Besides, it was claimed that the high accumulation of H_3K_4 methylation is associated with hyper-induction of genes expression under a recurring heat stress and the resulting heat stress memory was induced by the transcription factor HSFA2 (Lämke *et al.*, 2016). In this context, Migicovsky *et al.* (2014) showed that the progeny of heat-stressed *Arabidopsis thaliana* plants exhibited an enhanced expression of HSFA2. In addition, for the same plant species, Yamaguchi *et al.* (2021) found that JUMONJI proteins, which are demethylases involved in lysine 27 trimethylation at histone H3 (H3K27me3), play an important role in the response of *Arabidopsis thaliana* to a recurrent heat stress and are therefore necessary for its acclimation, through inducing the activation of heat shock protein genes. Pérez-Oliver *et al.* (2021) revealed that heat priming of *Pinus pinaster* embryogenic masses led to the upregulation of heat shock protein 70 (HSP 70) in old plants exposed to further heat stress. This acclimation to the heat stress was associated with a better osmotic adjustment, and higher levels of starch content, soluble sugars and chlorophyll. Similarly, Yadav *et al.* (2021) showed that the increased expression of heat shock proteins in primed chickpea (*Cicer arietinum* L.) plants was responsible for their further tolerance to heat stress. Brzezinka *et al.* (2016) illustrated the important role of FORGETTER (FGT1) locus in *Arabidopsis thaliana* heat stress memory. Indeed,

FGT1 interacts with the chromatin remodelers SWI/SNF and ISWI, and modulates nucleosome occupancy, resulting in heat stress memory formation.

10.4.2 Involvement of RNA regulation in plant heat stress memory

Apart from the epigenetic modifications affecting histones and chromatin, RNA regulation also plays a crucial function in plant heat stress memory. Stief *et al.* (2014) showed that the microRNA miR156 is highly induced after heat stress exposure, and the repression occurs of their target genes SQUAMOSA promoter binding protein like 2 and 11 (SPL11), resulting in the induction of heat stress memory genes transcription, including heat stress-associated 32 (HSA32), ascorbate peroxidase 2 (APX2), heat shock transcription factor A2 (HSFA2), heat shock protein 22 (HSP22), and heat shock protein 17.6A (HSP17.6A). The induction of miR156 in response to heat stress in other plant species, such as *Ginkgo biloba*, *Carthamus tinctorius* and *Medicago sativa*, was demonstrated in different studies suggesting their role in heat stress memory development, in addition to their involvement in heat stress management (Matthews *et al.*, 2019; Chang *et al.*, 2020; Kouhi *et al.*, 2020). It was revealed that miR824 induction after heat stress in Brassicaceae permits the overcoming of the impact of recurring heat stress, through its action as a post-transcriptional memory factor (Szaker *et al.*, 2019).

10.5 Applications of Epigenetics for Managing Heat Stress in Agriculture

As global temperatures continue to rise, heat stress has become a significant challenge in agriculture. Managing heat stress in crops may be improved by the application of epigenetics, the study of heritable variations in gene expression that do not include modifications to the underlying DNA sequence. This chapter will look into two key applications: (i) the use of epigenetic approaches for selecting heat-tolerant varieties; and (ii) the use of epigenetic engineering for enhancing heat stress resilience.

10.5.1 Epigenetic approaches for selecting heat-tolerant varieties

According to epigenetic research, variations in gene function that do not result in a change in DNA sequence are heritable during mitosis and/or meiosis. The chromatin state or epigenetic code supplies some of the information that determines a trait, with DNA sequence or genetic code supplying the remaining portion. The fundamental building block of chromatin is the nucleosome, which comprises roughly 150 bp (DNA), encircled by eight histone proteins: H2B, H2A, H3, and H4, with two copies of each (Dalakouras and Vlachostergios, 2021). Epigenetic modifications, like histone modifications and DNA methylation, can influence plant responses to heat stress. For studying these modifications, scientists can identify heat-responsive genes and select varieties that are more tolerant to heat stress. The process of DNA methylation involves adding a methyl group (CH_3) to a cytosine at the extremity 5', resulting in changes in the activity of a DNA segment without altering the sequence (Samantara *et al.*, 2021). Symmetric (CHG and CG) and asymmetric (CHH) methylation of cytosine bases has been observed in multiple plant species (Henderson and Jacobsen, 2007), like *Arabidopsis thaliana* (L.) (Korotko *et al.*, 2021), and *Triticum aestivum* (Gahlaut *et al.*, 2020), and other species already reported (Samantara *et al.*, 2021). For instance, a recent study on *Oryza sativa* revealed that heat stress induces specific changes in the patterns of DNA methylation, which in turn affects temperature-responsive gene expression (Rehman *et al.*, 2023). By selecting varieties with these specific methylation patterns, scientists were able to develop rice varieties with enhanced heat tolerance (Rehman *et al.*, 2023). Then, the methylation-sensitive amplified polymorphism (MSAP) method was employed to examine the variations in levels of DNA methylation during rice anthesis under both HT30 and typical circumstances. The MSAP findings indicated a considerable rise in DNA methylation levels in the vulnerable rice group and a decrease in the resilient rice group under HT treatment. In contrast, there was no discernible change in the moderate rice group. The results showed more instances of hypomethylation in the resilient rice group, while the susceptible rice group displayed more hypermethylation events (Li *et al.*, 2023). A study on wheat (*Triticum aestivum*) showed that heat stress leads to changes in histone modifications, particularly histone H3 lysine 9 acetylation (H3K9ac), linked to the activation of genes (Lamba *et al.*, 2023). Varieties with higher levels of H3K9ac for heat-responsive genes exhibited better heat tolerance, providing a potential marker for heat tolerance in wheat breeding programs (Lamba *et al.*, 2023). A study that used ChIP-qPCR, RNA-sequencing, and whole-genome bisulfite sequencing found that mutations in the SDG25 and ATX1 genes caused an increase in DNA cytosine methylation, a decrease in histone H3K4me3 levels, and the suppression of a subset of heat stress-responsive genes in *Arabidopsis* during the stress recovery phase. Epigenetic modifications such as H3K4me2 and H4K5ac may play a role in mediating HS signals, and consequently influence the transcription of several heat shock proteins, heat shock factors, senescence regulators, and genes that promote thermotolerance in *Gossypium hirsutum*. Additionally, GhHSFA1a and GhHSFA2 positively modulate cotton thermotolerance, while GhHXK1 and GhERF1A negatively modulate it (He *et al.*, 2022). Regarding the influence of elevated temperatures on the initial phase of *P. halepensis* SE and its consequences on the expression of genes linked to stress, an investigation demonstrated interesting and enduring changes in the differential expression of particular genes, including P444, DI19, and SOD, throughout different phases of SE (Pereira *et al.*, 2021). HS influences the germination of barley grains at the filling stage and modifies DNA methylation while acting on genes linked to GA biosynthesis (HvHA20ox, HvGA3ox), ABA catabolism (HvABA8'OH) and NADPH oxidase genes (HvRboh) responsible for seed germination (Sakai *et al.*, 2022).

10.5.2 Epigenetic engineering for heat stress resilience

Heat resilience and epigenetic engineering in plants is a fascinating area of research. It involves knowing the ways in which plants respond to hot and cold environmental conditions, including temperature variations on a daily and seasonal basis. These alterations, which include post-translational modifications of histone proteins and DNA methylation, affect how genes are expressed in plants both during their developmental phases and in stressful environments like cold and heat (Rehman *et al.*, 2023; Tiwari *et al.*, 2023). Even while temporary changes are frequent, some changes have the potential to last and cause stress memory that is handed on to the next generations. Comprehending the ways in which epigenomes react to stress and the elements that initiate stress memory is essential for creating climate-resilient farming practices (Rehman *et al.*, 2023). Heat resilience encompasses a vast range of adaptive traits and mechanisms that facilitate plants to cope with heat stress. This includes heat tolerance, which is the capacity of the plant to tolerate exposure to heat with variable degrees of harm. The less damage sustained, the greater is the heat tolerance. Heat avoidance, on the other hand, is the ability of the plant to maintain enzyme activity and tissue homeostasis to counter the effects of heat. As a result of its generally benign reaction to heat stimuli, the plant is exposed to heat without suffering any heat-related damage. Lastly, heat escape refers to the ability of the plant to adjust its phenology, growth pattern, and developmental plasticity, to avoid heat and ensure that critical physiological stages are not affected (Shanker *et al.*, 2020).

Epigenetic engineering, the targeted modification of epigenetic marks, offers another promising approach for enhancing heat stress resilience in crops. Several studies have been devoted to epigenetic engineering in plants in order to ensure their tolerance to heat stress (Table 10.2). DNA methylation is an epigenetic modification that can influence gene expression (Kakoulidou *et al.*, 2021); by targeting DNA methyltransferases or demethylases, researchers can modulate the methylation status of specific genes involved in heat stress responses. Then, by modifying DNA methylation patterns, a plant's resistance to high temperatures can be increased by activating or suppressing the production of genes linked to heat stress. In research by Saha *et al.* (2022), variations in the expression of the LuMET1 (DNA methyltransferase) gene were found to indicate that methylation and

Table 10.2. Examples of epigenetic plants involved in the control of heat stress. (Author's own table.)

Species	Epigenetic mechanisms	Genes /histones implicated	References
Cotton (*Gossypium hirsutum*)	DNA methylation and histone modifications	GhHSFA1a, GhHSFA2, GhHSP3, GhRBCS, GhERF1A and GhHXK1. H3K4me2, H4K5ac	He *et al.*, 2022
Arabidopsis thaliana (L.) Heynh.	DNA methylation	CpGs, CHGs, CHHs, CAB1, CAB2, CAB3 genes	Korotko *et al.*, 2021
Pinus halepensis	DNA methylation	ACT, P439, P444, DI19, SOD genes	Pereira *et al.*, 2021
Wild strawberry (*Fragaria vesca*)	DNA methylation	WRKY, AP2/EREBP, genes	López *et al.*, 2022
Zea mays	RNA-directed DNA methylation Histone modification (ChIP-qPCR analysis)	Mop1 mutants H3K9me2, H3K27me3	Guo *et al.*, 2021
Rice (*Oryza sativa*)	DNA methylation	TAEs genes	Li *et al.*, 2023
Triticum aestivum	DNA methylation	Cytosine5-DNA methyltransferases (C5-MTase) genes	Gahlaut *et al.*, 2020
Triticum aestivum	Histone modification	Histone H3 lysine 9 acetylation (H3K9ac)	Lamba *et al.*, 2023
Hordeum vulgare L.	DNA methylation	HvABA8'OH, HvHA20ox, HvGA3ox, and HvRboh genes	Sakai *et al.*, 2022
Arabidopsis	Histone modifications	H3K4me3	Song *et al.*, 2021
Brassica campestris L.	DE-siRNAs	Serine/threonine protein kinase-encoding genes	Ahmed *et al.*, 2021
Linum usitatissimum	DNA methylation et demethylation	the LuMET1 (DNA methyltransferase) gene	Saha *et al.*, 2022
Maize	Histone modification and chromatin immunoprecipitation (ChIP) analysis	Histone H3K4me2 and H3K9ac	Hou *et al.*, 2019
Cynodon dactylon (L.) Pers.	miRNAs (miRNA159a, miRNA160a, and miRNA164f)	GAMYB, ARF17, and NAC1 genes	Amini *et al.*, 2023
Tomato (*Solanum lycopersicum*)	RdDM and CRISPR/dCas9-based epigenome editing	Promoter of a heat-responsive gene	Rehman *et al.*, 2023
Arabidopsis thaliana	CRISPR/dCas9	Histone demethylase gene (JMJ14)	Rehman *et al.*, 2023

demethylation of DNA may have a significant impact on the heat stress response in flax (*Linum usitatissimum*). Histone modifications, such as methylation, and acetylation, play crucial roles in regulating gene expression and chromatin structure (Bhan *et al.*, 2017; Liu *et al.*, 2022). For instance, a short-term exposure to heat stress causes immediate changes in histone H3K4me2 and H3K9ac expression, which are linked to the activity of Hsf (heat shock transcription factor) and gene transcription of rRNA. During maize seedlings' acclimation period, this process is accompanied by disturbances in cell membranes and an increase in reactive oxygen species (ROS) (Hou *et al.*, 2019). Numerous methods can be used to accomplish this, including siRNAs and miRNAs, which are implicated in plant responses to heat stress and can control gene expression post-transcriptionally. The role of miRNAs and the genes they target in *Cynodon dactylon* (L.) Pers. response to heat stress was investigated. In the leaf and root tissues of the Gorgan accession under extreme heat stress, the expression levels of three miRNAs (miRNA159a, miRNA160a, and miRNA164f) and their target genes (GAMYB, ARF17, and NAC1, respectively) were examined (Amini *et al.*, 2023).

Serine/threonine protein kinase-encoding genes are another area of interest. Among the 795 DE-siRNA target genes that might be involved in controlling heat tolerance in Chinese cabbage (*Brassica campestris* L.) are SRK2I, CTR1-like, disease resistance protein RML1A-like, and RPP1 (Ahmed *et al.*, 2021). Also, RNA-directed DNA methylation (RdDM) and CRISPR/dCas9-based epigenome editing have been used to enhance heat tolerance in tomato (*Solanum lycopersicum*). By directing DNA methylation to the promoter of a heat-responsive gene, scientists were able to suppress its expression and enhance the heat tolerance of the tomato plants (Rehman *et al.*, 2023). In another study, scientists used the CRISPR/dCas9 system to target the histone demethylase gene JMJ14 in *Arabidopsis thaliana*. By inhibiting JMJ14, they were able to increase histone methylation levels and upregulate heat-responsive genes, thereby enhancing the heat tolerance of the plants (Rehman *et al.*, 2023). The preservation of epigenetic silencing at the border between transposable elements and neighboring genes is guaranteed by RdDM. The mediator of paramutation1 (Mop1) gene, which encodes a putative RNA-dependent ARN polymerase, is essential for RdDM in maize. A silent MuDR transposon requires RdDM to be methylated; nevertheless, the loss of this methylation does not result in the restoration of gene function. Even in the absence of DNA methylation, heritable silencing is mediated at this element's TIR (terminal inverted repeat) via H3 lysine 9 dimethylation (H3K9me2) and H3 lysine 27 dimethylation (H3K27me2). H3 lysine 27 trimethylation (H3K27me3), a signature often associated with somatically inherited gene silencing, mediates heritable silencing at the second TIR (Guo *et al.*, 2021).

10.6 Conclusion

High temperatures have emerged in large parts of the world as a major obstacle to plant growth and development. Research carried out in detail on the interaction and adaptability of cultivars to elevated temperatures has provided a greater insight into the responses of these organisms to heat stress. Through the application of epigenetic tools, we offer considerable prospectives for breeding crop varieties with respect to speeding plant adaptation, increasing yields and other agronomic parameters to the increasing global living conditions and food demands of our world today, through the development of new epiphytes, such as DNA methylation, histone modifications, regulation of transgenic expression and RNA interference (RNAi). Moreover, with the advent of chromatin profiling technology and advanced sequencing, research in epigenetics has moved to a different level. Nowadays, scientific research is increasingly concentrating on cultivated crop plants. To help meet current food security challenges by unraveling the intricate epigenetic pathways involved in heat stress responses, researchers can pave the way for innovative strategies in plant breeding and molecular biology to address the challenges posed by global warming and ensure sustainable crop production. Additionally, building stress-resistant cultures of high-yielding potential is a major focus of breeders' crop breeding programs. The fundamental mechanistic investigations of these changes will enable a better elucidation of the various genes and specific regions within them responsible for adaptations to various abiotic stresses, providing insight into which pathways to focus on for crop improvement.

Acknowledgements

This research was financially supported by the Phytopathology Unit at the Ecole Nationale d'Agriculture de Meknès, Morocco.

References

Ahmed, W., Xia, Y., Li, R., Zhang, H., Siddique, K.H.M. *et al.* (2021) Identification and analysis of small interfering RNAs associated with heat stress in flowering Chinese cabbage using high-throughput sequencing. *Frontiers in Genetics* 12, 746816. DOI: 0.3389/fgene.2021.746816.

Amini, Z., Salehi, H., Chehrazi, M., Etemadi, M. and Xiang, M. (2023) MiRNAs and their target genes play a critical role in response to heat stress in *Cynodon dactylon* (L.) pers. *Molecular Biotechnology* 65(12), 2004–2017. DOI: 10.1007/s12033-023-00713-2.

Argosubekti, N. (2020) A review of heat stress signaling in plants. *IOP Conference Series: Earth and Environmental Science* 484(1), 012041.

Bhan, A., Deb, P. and Mandal, S.S. (2017) Epigenetic code. In: Mandal, S.S. (ed.) *Gene Regulation, Epigenetics and Hormone Signaling*. Wiley VCH, Weinheim, Germany, pp. 29–58. DOI: 10.1002/9783527697274.

Brzezinka, K., Altmann, S., Czesnick, H., Nicolas, P., Gorka, M. *et al.* (2016) *Arabidopsis* FORGETTER1 mediates stress-induced chromatin memory through nucleosome remodeling. *eLife* 5, e17061. DOI: 10.7554/eLife.17061.

Bueckert, R.A. and Clarke, J.M. (2013) Review: Annual crop adaptation to abiotic stress on the Canadian prairies: Six case studies. *Canadian Journal of Plant Science* 93(3), 375–385. DOI: 10.4141/cjps2012-184.

Buszewicz, D., Archacki, R., Palusiński, A., Kotliński, M., Fogtman, A. *et al.* (2016) HD2C histone deacetylase and a SWI/SNF chromatin remodelling complex interact and both are involved in mediating the heat stress response in *Arabidopsis*. *Plant, Cell & Environment* 39(10), 2108–2122. DOI: 10.1111/pce.12756.

Chang, B., Ma, K., Lu, Z., Lu, J., Cui, J. *et al.* (2020) Physiological, transcriptomic, and metabolic responses of *Ginkgo biloba* L. to drought, salt, and heat stresses. *Biomolecules* 10(12), 1635. DOI: 10.3390/biom10121635.

Chang, M., Ge, J., Liao, M., Rong, X., Wang, Y. *et al.* (2023) Genome-wide DNA methylation and transcription analysis reveal the potential epigenetic mechanism of heat stress response in the sea cucumber *Apostichopus japonicus*. *Frontiers in Marine Science* 10. DOI: 10.3389/fmars.2023.1136926.

Chidambaranathan, P., Jagannadham, P.T.K., Satheesh, V., Jain, P.K. and Srinivasan, R. (2016) Expression analysis of six chromatin and remodeling complex genes (SWR1) in chickpea in different tissues and during heat stress. *Indian Journal of Genetics and Plant Breeding* 76(1), 47. DOI: 10.5958/0975-6906.2016.00007.9.

Dalakouras, A. and Vlachostergios, D. (2021) Epigenetic approaches to crop breeding: Current status and perspectives. *Journal of Experimental Botany* 72(15), 5356–5371. DOI: 10.1093/jxb/erab227.

Ding, X., Guo, J., Zhang, Q., Yu, L., Zhao, T. *et al.* (2021) Heat-responsive miRNAs participate in the regulation of male fertility stability in soybean CMS-based F_1 under high temperature stress. *International Journal of Molecular Sciences* 22(5), 2446. DOI: 10.3390/ijms22052446.

Ebrahimi Khaksefidi, R., Mirlohi, S., Khalaji, F., Fakhari, Z., Shiran, B. *et al.* (2015) Differential expression of seven conserved microRNAs in response to abiotic stress and their regulatory network in *Helianthus annuus*. *Frontiers in Plant Science* 6, 741. DOI: 10.3389/fpls.2015.00741.

Folsom, J.J., Begcy, K., Hao, X., Wang, D. and Walia, H. (2014) Rice fertilization-independent endosperm1 regulates seed size under heat stress by controlling early endosperm development. *Plant Physiology* 165(1), 238–248. DOI: 10.1104/pp.113.232413.

Friedrich, T., Faivre, L., Bäurle, I. and Schubert, D. (2019) Chromatin-based mechanisms of temperature memory in plants. *Plant, Cell & Environment* 42(3), 762–770. DOI: 10.1111/pce.13373.

Gahlaut, V., Samtani, H. and Khurana, P. (2020) Genome-wide identification and expression profiling of cytosine-5 DNA methyltransferases during drought and heat stress in wheat (*Triticum aestivum*). *Genomics* 112(6), 4796–4807. DOI: 10.1016/j.ygeno.2020.08.031.

Gao, G., Li, J., Li, H., Li, F., Xu, K. *et al.* (2014) Comparison of the heat stress induced variations in DNA methylation between heat-tolerant and heat-sensitive rapeseed seedlings. *Breeding Science* 64(2), 125–133. DOI: 10.1270/jsbbs.64.125.

Guo, W., Wang, D. and Lisch, D. (2021) RNA-directed DNA methylation prevents rapid and heritable reversal of transposon silencing under heat stress in *Zea mays*. *PLoS Genetics* 17(6), 1–28. DOI: 10.1371/journal.pgen.1009326.

Harris, C.J., Amtmann, A. and Ton, J. (2023) Epigenetic processes in plant stress priming: Open questions and new approaches. *Current Opinion in Plant Biology* 75, 102432. DOI: 10.1016/j.pbi.2023.102432.

Hasanuzzaman, M., Nahar, K., Alam, M.M., Roychowdhury, R. and Fujita, M. (2013) Physiological, biochemical, and molecular mechanisms of heat stress tolerance in plants. *International Journal of Molecular Sciences* 14(5), 9643–9684. DOI: 10.3390/ijms14059643.

He, S., Zhang, Y., Wang, J., Wang, Y., Ji, F. *et al.* (2022) H3K4me2, H4k5ac and DNA methylation function in short- and long-term heat stress responses through affecting the expression of the stress-related genes in *G. Hirsutum*. *Environmental and Experimental Botany* 194, 104699. DOI: 10.1016/j.envexpbot.2021.104699.

Henderson, I.R. and Jacobsen, S.E. (2007) Epigenetic inheritance in plants. *Nature* 447(7143), 418–424. DOI: 10.1038/nature05917.

Hivrale, V., Zheng, Y., Puli, C.O.R., Jagadeeswaran, G., Gowdu, K. *et al.* (2016) Characterization of drought- and heat-responsive microRNAs in switchgrass. *Plant Science* 242, 214–223. DOI: 10.1016/j.plantsci.2015.07.018.

Hossain, M.S., Kawakatsu, T., Kim, K.D., Zhang, N., Nguyen, C.T. *et al.* (2017) Divergent cytosine DNA methylation patterns in single-cell, soybean root hairs. *New Phytologist* 214(2), 808–819. DOI: 10.1111/nph.14421.

Hou, H., Zhao, L., Zheng, X., Gautam, M., Yue, M. *et al.* (2019) Dynamic changes in histone modification are associated with upregulation of HSF and RNAa genes during heat stress in maize seedlings. *Protoplasma* 256(5), 1245–1256. DOI: 10.1007/s00709-019-01364-4.

Hu, S., Ding, Y. and Zhu, C. (2020) Sensitivity and responses of chloroplasts to heat stress in plants. *Frontiers in Plant Science* 11, 375. DOI: 10.3389/fpls.2020.00375.

Hu, Z., Song, N., Zheng, M., Liu, X., Liu, Z. *et al.* (2015) Histone acetyltransferase GCN5 is essential for heat stress-responsive gene activation and thermotolerance in *Arabidopsis*. *The Plant Journal* 84(6), 1178–1191. DOI: 10.1111/tpj.13076.

Ji, S.-X., Wang, X.-D., Shen, X.-N., Liang, L., Liu, W.-X. *et al.* (2020) Using RNA interference to reveal the function of chromatin remodeling factor ISWI in temperature tolerance in *Bemisia tabaci* middle East-Asia minor 1 cryptic species. *Insects* 11(2), 113. DOI: 10.3390/insects11020113.

Ji, S.-X., Wang, X.-D., Lin, Z.-K., Wan, F.-H., Lü, Z.-C. *et al.* (2022) Characterization of chromatin remodeling genes involved in thermal tolerance of biologically invasive *Bemisia tabaci*. *Frontiers in Physiology* 13, 865172. DOI: 10.3389/fphys.2022.865172.

Kakoulidou, I., Avramidou, E.V., Baránek, M., Brunel-Muguet, S., Farrona, S. *et al.* (2021) Epigenetics for crop improvement in times of global change. *Biology* 10(8), 1–46. DOI: 10.3390/biology10080766.

Kim, J.-M., Sasaki, T., Ueda, M., Sako, K. and Seki, M. (2015) Chromatin changes in response to drought, salinity, heat, and cold stresses in plants. *Frontiers in Plant Science* 6, 114.

Korotko, U., Chwiałkowska, K., Sańko-Sawczenko, I. and Kwasniewski, M. (2021) DNA demethylation in response to heat stress in *Arabidopsis thaliana*. *International Journal of Molecular Sciences* 22(4), 1555. DOI: 10.3390/ijms22041555.

Kouhi, F., Sorkheh, K. and Ercisli, S. (2020) MicroRNA expression patterns unveil differential expression of conserved miRNAs and target genes against abiotic stress in safflower. *PloS ONE* 15(2), e0228850. DOI: 10.1371/journal.pone.0228850.

Kushawaha, A.K., Khan, A., Sopory, S.K. and Sanan-Mishra, N. (2021) Priming by high temperature stress induces microRNA regulated heat shock modules indicating their involvement in thermopriming response in rice. *Life* 11(4), 291. DOI: 0.3390/life11040291.

Lamba, K., Kumar, M., Singh, V., Chaudhary, L., Sharma, R. *et al.* (2023) Heat stress tolerance indices for identification of the heat tolerant wheat genotypes. *Scientific Reports* 13(1), 1–13. DOI: 0.1038/s41598-023-37634-8.

Lämke, J. and Bäurle, I. (2017) Epigenetic and chromatin-based mechanisms in environmental stress adaptation and stress memory in plants. *Genome Biology* 18, 1–11. DOI: 10.1186/s13059-017-1263-6.

Lämke, J., Brzezinka, K., Altmann, S. and Bäurle, I. (2016) A hit-and-run heat shock factor governs sustained histone methylation and transcriptional stress memory. *The EMBO Journal* 35(2), 162–175. DOI: 0.15252/embj.201592593.

Li, B., Cai, H., Liu, K., An, B., Wang, R. *et al.* (2023) DNA methylation alterations and their association with high temperature tolerance in rice anthesis. *Journal of Plant Growth Regulation* 42(2), 780–794. DOI: 0.1007/s00344-022-10586-5.

Li, J., Huang, Q., Sun, M., Zhang, T., Li, H. *et al.* (2016) Global DNA methylation variations after short-term heat shock treatment in cultured microspores of *Brassica napus* cv. Topas. *Scientific Reports* 6(1), 38401. DOI: 0.1038/srep38401.

Liu, G., Xia, Y., Liu, T., Dai, S. and Hou, X. (2018) The DNA methylome and association of differentially methylated regions with differential gene expression during heat stress in *Brassica rapa*. *International Journal of Molecular Sciences* 19(5), 1414. DOI: 0.3390/ijms19051414.

Liu, J., Feng, L., Li, J. and He, Z. (2015) Genetic and epigenetic control of plant heat responses. *Frontiers in Plant Science* 06. DOI: 0.3389/fpls.2015.00267.

Liu, X., Luo, M., Zhang, W., Zhao, J., Zhang, J. *et al.* (2012) Histone acetyltransferases in rice (*Oryza sativa* L.): Phylogenetic analysis, subcellular localization and expression. *BMC Plant Biology* 12(1), 145. DOI: 0.1186/1471-2229-12-145.

Liu, Y., Wang, J., Liu, B. and Xu, Z.Y. (2022) Dynamic regulation of DNA methylation and histone modifications in response to abiotic stresses in plants. *Journal of Integrative Plant Biology* 64(12), 2252–2274. DOI: 0.1111/jipb.13368.

López, M.-E., Roquis, D., Becker, C., Denoyes, B. and Bucher, E. (2022) DNA methylation dynamics during stress response in Woodland strawberry (*Fragaria vesca*). *Horticulture Research* 9. DOI: 0.1093/hr/uhac174.

Lu, Z., Cui, J., Wang, L., Teng, N., Zhang, S. *et al.* (2021) Genome-wide DNA mutations in *Arabidopsis* plants after multigenerational exposure to high temperatures. *Genome Biology* 22(1), 160. DOI: 0.1186/s13059-021-02381-4.

Malabarba, J., Windels, D., Xu, W. and Verdier, J. (2021) Regulation of DNA (de)methylation positively impacts seed germination during seed development under heat stress. *Genes* 12(3), 457. DOI: 0.3390/genes12030457.

Matthews, C., Arshad, M. and Hannoufa, A. (2019) Alfalfa response to heat stress is modulated by microRNA156. *Physiologia Plantarum* 165(4), 830–842. DOI: 0.1111/ppl.12787.

Migicovsky, Z., Yao, Y. and Kovalchuk, I. (2014) Transgenerational phenotypic and epigenetic changes in response to heat stress in *Arabidopsis thaliana*. *Plant Signaling & Behavior* 9(2), e27971. DOI: 0.4161/psb.27971.

Min, L., Li, Y., Hu, Q., Zhu, L., Gao, W. *et al.* (2014) Sugar and auxin signaling pathways respond to high-temperature stress during anther development as revealed by transcript profiling analysis in cotton. *Plant Physiology* 164(3), 1293–1308. DOI: 0.1104/pp.113.232314.

Nair, A.U., Bhukya, D.P.N., Sunkar, R., Chavali, S. and Allu, A.D. (2022) Molecular basis of priming-induced acquired tolerance to multiple abiotic stresses in plants. *Journal of Experimental Botany* 73(11), 3355–3371. DOI: 0.1093/jxb/erac089.

Oberkofler, V., Pratx, L. and Bäurle, I. (2021) Epigenetic regulation of abiotic stress memory: Maintaining the good things while they last. *Current Opinion in Plant Biology* 61, 102007. DOI: 0.1016/j.pbi.2021.102007.

Oyoshi, K., Katano, K., Yunose, M. and Suzuki, N. (2020) Memory of 5-min heat stress in *Arabidopsis thaliana*. *Plant Signaling & Behavior* 15(8), 1778919. DOI: 10.1080/15592324.2020.1778919.

Pereira, C., Castander-Olarieta, A., Sales, E., Montalbán, I.A., Canhoto, J. *et al.* (2021) Heat stress in *Pinus halepensis* somatic embryogenesis induction: Effect in DNA methylation and differential expression of stress-related genes. *Plants* 10(11), 2333. DOI: 0.3390/plants10112333.

Pérez-Oliver, M.A., Haro, J.G., Pavlović, I., Novák, O., Segura, J. *et al.* (2021) Priming maritime pine megagametophytes during somatic embryogenesis improved plant adaptation to heat stress. *Plants* 10(3), 446. DOI: 0.3390/plants10030446.

Pokhrel, S. and Meyers, B.C. (2022) Heat-responsive microRNAs and phased small interfering RNAs in reproductive development of flax. *Plant Direct* 6(2), e385. DOI: 0.1002/pld3.385.

Qian, Y., Hu, W., Liao, J., Zhang, J. and Ren, Q. (2019) The dynamics of DNA methylation in the maize (*Zea mays* L.) inbred line B73 response to heat stress at the seedling stage. *Biochemical and Biophysical Research Communications* 512(4), 742–749. DOI: 0.1016/j.bbrc.2019.03.150.

Ravichandran, S., Ragupathy, R., Edwards, T., Domaratzki, M. and Cloutier, S. (2019) MicroRNA-guided regulation of heat stress response in wheat. *BMC Genomics* 20(1), 488. DOI: 0.1186/s12864-019-5799-6.

Rehman, M. and Tanti, B. (2020) Understanding epigenetic modifications in response to abiotic stresses in plants. *Biocatalysis and Agricultural Biotechnology* 27, 101673. DOI: 0.1016/j.bcab.2020.101673.

Rehman, S., Ahmad, Z., Ramakrishnan, M., Kalendar, R. and Zhuge, Q. (2023) Regulation of plant epigenetic memory in response to cold and heat stress: Towards climate resilient agriculture. *Functional & Integrative Genomics* 23(4), 298. DOI: 0.1007/s10142-023-01219-5.

Sadras, V.O. and Dreccer, M.F. (2015) Adaptation of wheat, barley, canola, field pea and chickpea to the thermal environments of Australia. *Crop and Pasture Science* 66(1137), 1150. DOI: 0.1071/CP15129.

Saha, D., Shaw, A.K., Datta, S., Mitra, J. and Kar, G. (2022) DNA hypomethylation is the plausible driver of heat stress adaptation in *Linum usitatissimum*. *Physiologia Plantarum* 174(3), e13689. DOI: 0.1111/ppl.13689.

Sailaja, B., Voleti, S.R., Subrahmanyam, D., Sarla, N., Prasanth, V.V. *et al.* (2014) Prediction and expression analysis of miRNAs associated with heat stress in *Oryza sativa*. *Rice Science* 21(1), 3–12. DOI: 0.1016/S1672-6308(13)60164-X.

Sakai, Y., Suriyasak, C., Inoue, M., Hamaoka, N. and Ishibashi, Y. (2022) Heat stress during grain filling regulates seed germination through alterations of DNA methylation in barley (*Hordeum vulgare* L.). *Plant Molecular Biology* 110(4–5), 325–332. DOI: 0.1007/s11103-022-01278-5.

Samantara, K., Shiv, A., Sousa, L.L., Sandhu, K.S., Priyadarshini, P. *et al.* (2021) A comprehensive review on epigenetic mechanisms and application of epigenetic modifications for crop improvement. *Environmental and Experimental Botany* 188, 104479. DOI: 0.1016/j.envexpbot.2021.104479.

Saraswat, S., Yadav, A.K., Sirohi, P. and Singh, N.K. (2017) Role of epigenetics in crop improvement: Water and heat stress. *Journal of Plant Biology* 60(3), 231–240. DOI: 0.1007/s12374-017-0053-8.

Shanker, A.K., Bhanu, D. and Maheswari, M. (2020) Epigenetics and transgenerational memory in plants under heat stress. *Plant Physiology Reports* 25(4), 583–593. DOI: 0.1007/s40502-020-00557-x.

Singh, R.K., Jaishankar, J., Muthamilarasan, M., Shweta, S., Dangi, A. *et al.* (2016) Genome-wide analysis of heat shock proteins in C4 model, foxtail millet identifies potential candidates for crop improvement under abiotic stress. *Scientific Reports* 6(1), 32641. DOI: 0.1038/srep32641.

Singroha, G. and Sharma, P. (2019) Epigenetic modifications in plants under abiotic stress. In: *Epigenetics*. IntechOpen. DOI: 0.5772/intechopen.84455.

Song, Z.T., Zhang, L.L., Han, J.J., Zhou, M. and Liu, J.X. (2021) Histone H3K4 methyltransferases SDG25 and ATX1 maintain heat-stress gene expression during recovery in *Arabidopsis*. *The Plant Journal* 105(5), 1326–1338. DOI: 0.1111/tpj.15114.

Stief, A., Altmann, S., Hoffmann, K., Pant, B.D., Scheible, W.-R. *et al.* (2014) *Arabidopsis* miR156 regulates tolerance to recurring environmental stress through SPL transcription factors. *The Plant Cell* 26(4), 1792–1807. DOI: 0.1105/tpc.114.123851.

Szaker, H.M., Darkó, É., Medzihradszky, A., Janda, T., Liu, H.-C. *et al.* (2019) MiR824/AGAMOUS-LIKE16 module integrates recurring environmental heat stress changes to fine-tune poststress development. *Frontiers in Plant Science* 10, 1454. DOI: 0.3389/fpls.2019.01454.

Tang, N., Ma, S., Zong, W., Yang, N., Lv, Y. *et al.* (2016) MODD mediates deactivation and degradation of OsbZIP46 to negatively regulate ABA signaling and drought resistance in rice. *The Plant Cell* 28(9), 2161–2177. DOI: 0.1105/tpc.16.00171.

Tiwari, A., Pandey-Rai, S., Rai, K.K., Tiwari, A. and Pandey, N. (2023) Molecular and epigenetic basis of heat stress responses and acclimatization in plants. *The Nucleus* 66(1), 69–79. DOI: 0.1007/s13237-022-00400-8.

Turgut-Kara, N., Arikan, B. and Celik, H. (2020) Epigenetic memory and priming in plants. *Genetica* 148(2), 47–54. DOI: 0.1007/s10709-020-00093-4.

Ueda, M., Matsui, A., Nakamura, T., Abe, T., Sunaoshi, Y. *et al.* (2018) Versatility of HDA19-deficiency in increasing the tolerance of *Arabidopsis* to different environmental stresses. *Plant Signaling & Behavior* 1–4. DOI: 0.1080/15592324.2018.1475808.

Verma, N., Giri, S.K., Singh, G., Gill, R. and Kumar, A. (2022) Epigenetic regulation of heat and cold stress responses in crop plants. *Plant Gene* 29, 100351. DOI: 0.1016/j.plgene.2022.100351.

Wang, Q.-L., Chen, J.-H., He, N.-Y. and Guo, F.-Q. (2018) Metabolic reprogramming in chloroplasts under heat stress in plants. *International Journal of Molecular Sciences* 19(3), 849. DOI: 10.3390/ijms19030849.

Wang, X., Xin, C., Cai, J., Zhou, Q., Dai, T. *et al.* (2016) Heat priming induces trans-generational tolerance to high temperature stress in wheat. *Frontiers in Plant Science* 7. DOI: 0.3389/fpls.2016.00501.

Weng, M., Yang, Y., Feng, H., Pan, Z., Shen, W. *et al.* (2014) Histone chaperone ASF1 is involved in gene transcription activation in response to heat stress in *Arabidopsis thaliana*. *Plant, Cell & Environment* 37(9), 2128–2138. DOI: 0.1111/pce.12299.

Willits, D.H. and Peet, M.M. (1998) The effect of night temperature on greenhouse grown tomato yields in warm climates. *Agricultural and Forest Meteorology* 92, 191–202.

Wunderlich, M., Gross-Hardt, R. and Schöffl, F. (2014) Heat shock factor HSFB2a involved in gametophyte development of *Arabidopsis thaliana* and its expression is controlled by a heat-inducible long non-coding antisense RNA. *Plant Molecular Biology* 85(6), 541–550. DOI: 0.1007/s11103-014-0202-0.

Xue, M., Zhang, H., Zhao, F., Zhao, T., Li, H. *et al.* (2021) The INO80 chromatin remodeling complex promotes thermomorphogenesis by connecting H2A.Z eviction and active transcription in *Arabidopsis*. *Molecular Plant* 14(11), 1799–1813. DOI: 0.1016/j.molp.2021.07.001.

Yadav, R., Juneja, S. and Kumar, S. (2021) Cross priming with drought improves heat-tolerance in chickpea (*Cicer arietinum* L.) by stimulating small heat shock proteins and antioxidative defense. *Environmental Sustainability* 4(1), 171–182. DOI: 0.1007/s42398-020-00156-4.

Yamaguchi, N., Matsubara, S., Yoshimizu, K., Seki, M., Hamada, K. *et al.* (2021) H3K27me3 demethylases alter HSP22 and HSP17.6C expression in response to recurring heat in *Arabidopsis*. *Nature Communications* 12(1), 3480. DOI: 0.1038/s41467-021-23766-w.

Yao, Z., Yuan, L., Liu, K., Wang, T., Liu, B. *et al.* (2022) Warming-induced changes of broccoli head to cauliflower-like curd in *Brassica oleracea* are regulated by DNA methylation as revealed by methylome and transcriptome co-profiling. *Molecular Horticulture* 2(1), 26. DOI: 0.1186/s43897-022-00047-8.

Zhang, M., Zhang, X., Guo, L., Qi, T., Liu, G. *et al.* (2019) Single-base resolution methylomes of cotton CMS system reveal epigenomic changes in response to high-temperature stress during anther development. *Journal of Experimental Botany*. DOI: 0.1093/jxb/erz470.

Zhao, J., He, Q., Chen, G., Wang, L. and Jin, B. (2016) Regulation of non-coding RNAs in heat stress responses of plants. *Frontiers in Plant Science* 7. DOI: 0.3389/fpls.2016.01213.

Zhao, L., Wang, P., Hou, H., Zhang, H., Wang, Y. *et al.* (2014) Transcriptional regulation of cell cycle genes in response to abiotic stresses correlates with dynamic changes in histone modifications in maize. *PLoS ONE* 9(8), e106070. DOI: 0.1371/journal.pone.0106070.

Zheng, Y., Ding, Y., Sun, X., Xie, S., Wang, D. *et al.* (2016) Histone deacetylase HDA9 negatively regulates salt and drought stress responsiveness in *Arabidopsis*. *Journal of Experimental Botany* 67(6), 1703–1713. DOI: 0.1093/jxb/erv562.

Zhou, R., Yu, X., Ottosen, C.-O., Zhang, T., Wu, Z. *et al.* (2020) Unique miRNAs and their targets in tomato leaf responding to combined drought and heat stress. *BMC Plant Biology* 20(1), 107. DOI: 0.1186/s12870-020-2313-x.

11 Epigenetics for Combating Salinity Stress in Plants: Updated Methods and Current Achievements

Krishnasreya Mohandas, Elakkiya Maruthamuthu Rathinam and Chithra Devi Balasundaram Saraswathy*

Department of Botany, PSGR Krishnammal College for Women, Coimbatore, Tamil Nadu, India

Abstract

Unpredictable climate fluctuations expose plants to a variety of abiotic stresses, including high salinity, excessive temperature, dehydration, heavy metal toxicity, inadequate nutrition, and ultraviolet radiation, all of which reduce plant yield. Salinity stress has a significant detrimental impact on plant growth, primarily due to oxidative damage, sodium ion toxicity, and osmotic stress. Enhancing different agronomic characteristics, such as salt tolerance in crop plants, has been achieved primarily by harnessing genetic diversity present among crop plants, frequently accessing latent genetic diversity found in wild crop plant species. Epigenetic regulation is a fast-emerging field of research that affects the chromatin structure to enhance the expression of genes at particular positions within the genome. It has identified a stress memory in plants that enables the stress-exposed plants to better withstand future exposures to stress. DNA methylation is crucial in managing the expression of salt-responsive genes. The plant uses DNA methylation as an epigenetic regulatory mechanism to adapt to changing environmental stimuli. It happens as a result of a chemical change in which methyltransferases modify the cytosine base by adding a methyl group. Studies have demonstrated that changing DNA methylation patterns during stress is related to transcriptional control of genes. Since it appears promising in interpreting and imparting stress-adaptive potential in crops and other plant species, epigenetics has become a prominent field of research in the realm of plant functional genomics. This chapter reviews the epigenetic strategies for combating salinity stress in plants.

Keywords: Epigenetics, Salt tolerance, Histone modifications, DNA methylation, Chromatin remodeling, RNAi

11.1 Introduction

Soil salinity is a harmful environmental factor that inhibits plant growth, and diminishes crop quality, which ultimately decreases the agricultural output (Chen *et al.*, 2018; Yasmin *et al.*, 2021). It poses a significant challenge in the field of agriculture, leading to a notable decline in crop productivity. Its detrimental effects on plants primarily manifest through the inhibition of leaf expansion, restricted photosynthesis, reduction in chlorophyll levels, stunted growth, reduced leaf area index, seed germination, water absorption, and disrupted nutrient uptake (Gill *et al.*, 2024). Elevated levels of salt concentration in the soil can impede

*Corresponding author: bchitradevi@psgrkcw.ac.in

 Epigenetics for Climate-Smart and Sustainable Agriculture (ed. Jen-Tsung Chen)
DOI: 10.1079/9781800626102.0011

the ability of roots to absorb water, leading to water deficiencies in plant tissues. Insufficient water and excessive salt can induce osmotic stress, negatively impacting cell expansion and overall plant growth. As a result, plants undergo various physiological changes and experience decreased accumulation of biomass (Kravchik and Bernstein, 2013; Zafar *et al.*, 2024). To evaluate the plant's capacity to withstand salinity stress, the growth or survival of the plant is examined as it reflects the activation or suppression of numerous physiological processes that take place within the plant (Niknam and McComb, 2000; Ashraf, 2004).

Soil salinization occurs when soluble salts accumulate both on the soil surface and within the rhizosphere, resulting in the emergence of salt-affected soil (Siyal *et al.*, 2002). According to the Food and Agriculture Organization of the United Nations (FAO/UNESCO) soil map of the world from 1970 to 1980, the estimated area of saline soil covered an extensive area of 397 million hectares, while alkaline soils covered about 434 million hectares. The report also indicated that salt stress affected 19.49% of irrigated land and 2.13% of dry land (Chen *et al.*, 2018). Secondary salinization affects around 43 million hectares out of a total of 950 million hectares of salt-affected land globally (Pankova *et al.*, 2018). Irrigation using saline water has the potential to induce soil salinity, thereby creating negative impacts on crop development and yield, soil fertility, and groundwater quality. This alteration disrupts the typical operation of the soil–water–plant system, resulting in decreased nutrient absorption and ultimately culminating in a substantial decline in crop output (Solangi *et al.*, 2019).

Owing to these negative impacts, effective strategies to combat salinity stress have become a necessity (Fig. 11.1). Recent studies in plant stress biology have comprehended the mechanisms of regulation in the plants at the molecular levels that are subjected to diverse environmental conditions. Epigenetics is one such potent strategy to induce salt tolerance in plants. It refers to a semi-permanent alteration of DNA at the molecular level, without any changes to the genomic sequence. The resulting traits can be inherited by the following generations (Banerjee and Roychoudhury, 2017). According to the perceptions of epigenetics by Bird (2007), 'epigenetics is the structural

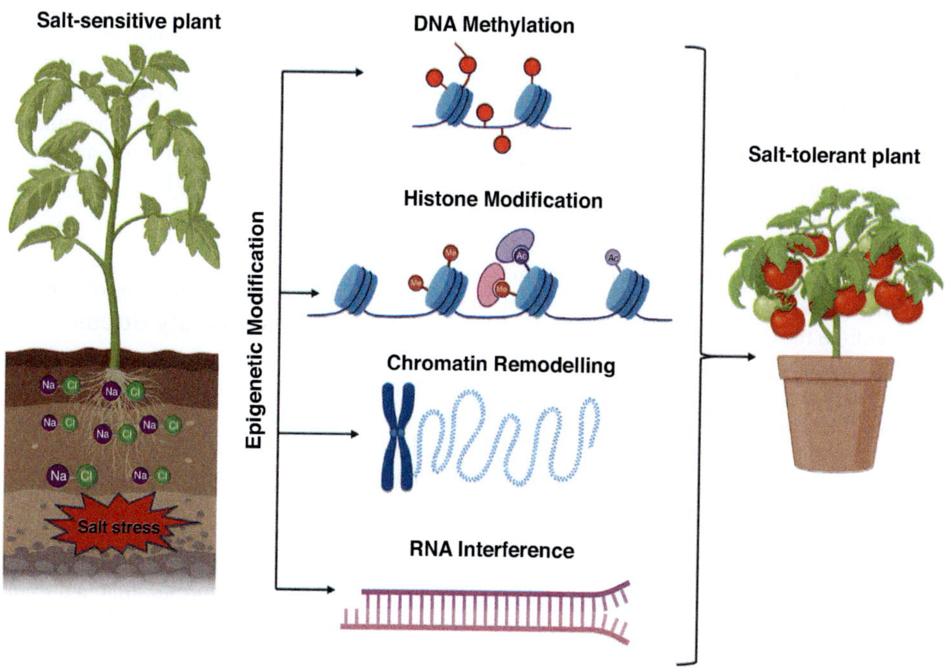

Fig. 11.1. Epigenetic mechanisms in tolerating salinity stress. Author's own image.

adaptation of chromosomal regions so as to register, signal, or perpetuate altered activity states.'

Salinity triggers an epigenetic mechanism that leads to phenotypic modifications through alterations in gene expression. The correlation between modifications in histones and changes in methylation of DNA, as well as the resulting differential gene expression, has been observed in many crops with salt sensitivity. Furthermore, epigenetics plays a role in creating a stress memory that helps plants effectively adapt and thrive in challenging environments (Rashid *et al.*, 2022).

11.2 Salinity Stress in Plants

Excessive salinity can hinder the development and growth of plants. Different plants are affected to varying degrees based on their ability to tolerate high salt levels (Blum, 2018). Halophytes are plants that can survive in soil with elevated salinity levels without a decrease in output, while glycophytes are plants that are sensitive to salt and perform poorly in high levels of salt concentration. Various types of plants exhibit different degrees of tolerance to salt stress. For example, barley has the ability to withstand salt concentrations up to 8.3 decisiemens per meter (dS/m) threshold level, onions up to 1.5 dS/m threshold level, soybeans up to a 6.4 dS/m threshold level, and carrots up to 1.8 dS/m threshold level. The salinity concentration is influenced by various factors such as the amount and standard of irrigation water, type of soil, evaporation rate of water used for irrigation, density and temperature of water, geologic surface, and even the arrival of salt water from the oceans. The soil in Pakistan is predominantly alkaline and lacks organic matter, resulting in unsuitable subterranean water for irrigation. This unfavorable condition leads to a rise in the salinity of the soil. When plants take in water from the soil, the elevated concentration of salt disrupts various physiological processes within the plant. These disruptions include water uptake, accumulation of certain solutes, water balance, potassium ions uptake, disruption in the maintenance of ionic hemostasis, respiration, photosynthesis and exchange of gases, transpiration, plant morphology, and anatomy, as well as growth patterns (Ismail and Horie, 2017).

Seedlings, vegetative, and reproductive growth stages are impacted by high concentrations of salt levels. Seedlings are particularly susceptible to salinity compared with fully grown plants (Jan *et al.*, 2017). Plant growth and development are hindered by salt stress due to three main factors: (i) osmotic stress, which results in a reduction in water potential in the root zone leading to the inability of plants to take up necessary water; (ii) specific ionic toxicity, caused by an excess of sodium and chloride ions being taken up by the plants; and (iii) lack of essential nutrients that plants struggle to take up, such as nitrates, calcium, and potassium (Shabala, 2000; Read *et al.*, 2006; Sevik and Cetin, 2015). High levels of chloride and sodium ions inhibit the uptake of potassium ions, which are essential for regulating stomatal function and activating cellular metabolic processes. The absence of K + negatively impacts plant growth by disrupting stomatal movements. Elevated internal chloride and sodium levels have been associated with hindered plant growth and reduced yields in various agricultural produce, such as mungbean (Saha *et al.*, 2010). The effect of salt stress on plant growth varies significantly based on their stages of growth. To illustrate, salt stress greatly affects the growth of wheat plants in the early vegetative stage, but its influence diminishes throughout the flowering and grain-filling stages (Carillo *et al.*, 2011). High levels of salt can also impact the growth of strawberries during the vegetative phase, leading to delays in flowering and fruit maturation (Tremblay *et al.*, 2010).

11.3 Epigenetic Mechanisms in Response to Salinity Stress

In nature, abiotic stresses often occur cyclically and may even worsen over time. In such cases, the plants are able to recall the previous events and use this stored information to cope with fresh obstacles. Plants develop stress memories that enable them to adjust more effectively in sequential encounters, which is known as priming. Plants acquire stress memories that assist in priming when they encounter mild stressful conditions beforehand (Crisp *et al.*, 2016). Numerous research studies have documented the emergence of stress memory in reaction to cold, abscisic acid, oxidative stress, drought and salinity in plants (Ding *et al.*, 2012). In some cases, the priming effect can be passed down to future generations, leading to intergenerational plasticity and providing acclimatization and evolutionary benefits to upcoming generations as well (Sharma *et al.*, 2022).

Different priming techniques have been discovered in a variety of plants, spanning from the build-up of compounds of an intermediate stage within compartments of the cells to the alteration of crucial regulatory proteins, as well as the involvement of epigenetic modifications (Hilker *et al.*, 2016).

Plants react to ecological pressure by making ever-changing alterations in gene expression and reconfiguring their physiology at the molecular level (Lämke and Bäurle, 2017; Luo and He, 2020). Over the past 20 years, researchers have extensively studied transcriptional responses to investigate the particular signaling pathways implicated in salinity stress reactions and to differentiate the distinct proteins involved in regulation and their corresponding targets. The structure of chromatin in eukaryotic cells is highly adaptable and undergoes modifications in response to ecological stimuli. To gain a deeper knowledge of the regulation of gene expression, it is crucial to uncover the underlying structural context (Zhang *et al.*, 2009).

Enhancing different agronomic characteristics, such as salt tolerance in crop plants, has historically relied on harnessing genetic diversity present among the agricultural produce. This often involves drawing from genetic diversity found in diverse species or varieties of wild crop plants. However, the decline of numerous natural ecosystems for crops, coupled with the continuous exploitation of genetic resources in nature, has steadily reduced the reservoirs of variation at the genetic level (; Govindaraj *et al.*, 2015).

Currently, the insufficient genetic diversity in different types of crops is a main obstacle in plant breeding efforts. To overcome this issue, a combination of radiation treatments and chemical mutagens has been utilized to introduce variation at the genetic level. These methods have resulted in a range of mutations, from small-scale point mutations to larger chromosomal abnormalities. Despite these efforts, the development of desirable alleles for certain traits has proven to be challenging at the desired rate. The efficient creation of genetic diversity is crucial to adapt to the rapidly changing climate and soil conditions. Traditional methods of crop enhancement are not just time-consuming, but they also face additional challenges. Firstly, it is exceptionally hard to distinguish favorable alleles from unfavorable ones that are closely linked genetically. Secondly, certain genomic regions suppress the occurrence of crossing over and further enhance the connection between various alleles even more. In many instances, the process of screening a significant number of individuals can be time-consuming and require a substantial amount of labor, making it impractical to employ this approach as part of current strategies (Agarwal *et al.*, 2020).

Epigenetics pertains to the control of gene expression via modulation of chromatin architecture. It serves as a crucial mechanism for enhancing or diminishing the activation of particular genes in reaction to environmental changes (Zhang *et al.*, 2009).

Epigenetic modifications alter the DNA structure without affecting the sequence, playing a significant role in the life cycle of plants (Duan *et al.*, 2018). DNA methylation, histone modification, histone variants, and certain non-coding RNAs are crucial epigenetic components. These modifications showcase a comprehensive influence on chromatin structure and influence its accessibility to the machinery responsible for transcribing genetic information, thus serving as a crucial factor in governing the expression of genes (Crisp *et al.*, 2016; Saroha *et al.*, 2017; Duan *et al.*, 2018; Singroha and Sharma, 2019). DNA methylation has been extensively researched as one of the well-studied epigenetic modifications. involving the introduction of the methyl group at the 6th position of adenine or the 5th position of cytosine (Liang *et al.*, 2018; Zhang *et al.*, 2018). DNA methylation is of utmost importance in various essential processes for plant growth and adaptation to stressful conditions (Zhang *et al.*, 2018).

To improve their odds of survival in demanding environmental circumstances, histone proteins undergo specific changes at their N-terminal regions to regulate gene expression. Recent research has affirmed the crucial function of methylation and histone acetylation as essential epigenetic markers in finely adjusting gene expression during unfavorable circumstances (Xie *et al.*, 2015). Epigenetics plays a crucial role in understanding the intricate mechanisms that underlie the response to adaptability and physical stress (Varotto *et al.*, 2020).

11.4 Current Achievements

11.4.1 DNA methylation

High salinity has detrimental effects on ionic balance, osmotic balance, energy and lipid metabolism, protein synthesis, and photosynthesis (Rahman *et al.*, 2021). In the context of salinity, DNA methylation in plants is commonly known as N6-methyladenine or 5-methylcytosine (Zhang *et al.*, 2018). However, it is worth noting that 6-methyladenine remains relatively unexplored, while the majority of studies focus on 5-methylcytosine under salt stress (Kumar, 2018). There is an increasing amount of evidence suggesting that DNA methylation is becoming more prominent in controlling the gene expression response to salinity (Al-Harrasi *et al.*, 2018).

Salinity triggers alterations in DNA methylation patterns across the genome, leading to varying impacts on the methylation of DNA in different species of plants or specific genes under salinity stress conditions (Mastan *et al.*, 2012). The content of methylome in alfalfa plants was found to be elevated under salinity stress conditions. However, when alfalfa seedlings were treated with 5-AzaC, an inhibitor of DNA methylation, there was a notable reduction in salt tolerance (Al-Lawati *et al.*, 2016). Conversely, CG methylation levels showed a notable decrease in the genomic areas studied in the epidermis when exposed to NaCl stress, with the reduction being more pronounced in *Arabidopsis* plants experiencing severe stress (Beyrne *et al.*, 2019). In the shoot of a salt-sensitive wheat genotype, an increase in 5-methylcytosine was observed in both CHH and CHG, whereas decreased 5-methylcytosine levels were identified in a salinity-tolerant wheat variety SR-3 (Miryeganeh, 2021). Tolerant genotypes of rice exhibited hypermethylation, while demethylation was observed in sensitive genotypes (Feng *et al.*, 2012). During salt stress in wheat and rice varieties, different levels of DNA methylation were observed in both the root and shoot systems (Ferreira *et al.*, 2015; Chang *et al.*, 2020).

Olive plants showed elevated levels of DNA methylation under salt stress conditions. The differences were more noticeable in salt-tolerant varieties, with Royal cultivars exhibiting more DNA methylation events compared with Koroneiki. This suggests that DNA methylation could potentially contribute to increasing tolerance to high salinity, varying across species and plant tissues (Mousavi *et al.*, 2019).

DNA methylation is essential for controlling gene expression under salt stress conditions in plants, impacting various transporter and secretion genes to control signals related to stress and trigger stress reactions. In wheat genotype Kharchia-65, a distinct rise in cytosine methylation occurs on potassium transporters under NaCl treatment, leading to their downregulation, ultimately enhancing salt tolerance ability in this genotype (Kumar *et al.*, 2017). Methylating salinity-responsive genes have been discovered to potentially increase the plant tolerance to salinity, including the wheat and barley genes *TaWRS15* and *TaFLS1*, which are involved in flavonol synthesis (Kong *et al.*, 2020).

DNA demethylation is also important in salt stress. When plants are exposed to salinity stress, it triggers the activation of genes that produce enzymes in the pathway of flavonoid biosynthesis and the antioxidant pathway. These gene expressions are linked to the methylation status and activity of the AtROS1 demethylase (Bharti *et al.*, 2015).

Nonabokra, a salt-tolerant rice variety, exhibited a notable rise in the levels of abscisic acid-responsive element binding factor (OsBZ8), compared with the IR64 variety when exposed to salinity stress. A decrease in the methylation of DNA at the OsBZ8 locus was also observed (Paul *et al.*, 2017).

Most gene promoters showed hypermethylation under salinity stress, in addition to the gene bodies in the descendents of stressed plants. Additionally, the majority of hypermethylated genes had reduced gene expression in *Arabidopsis* (Bilichak *et al.*, 2012).

Salt stress has the potential to impact the DNMTs, CMT, DRM, DML and DME gene expressions, leading to alterations in methylation patterns in plant DNA, which in turn allows plants to cope better with salt stress. In rapeseed, DNMT2 was discovered to be upregulated in reaction to salinity stress, whereas certain members of the MET and CMT gene families were notably downregulated (Fan *et al.*, 2020). In *Pyrus betulaefolia*, it has been documented that the expression of DME exhibits a tissue-specific response to salt, with a decrease in leaves and an increase in roots (Liu *et al.*, 2018). Boosting the expression of *AtROS1* (*Arabidopsis* demethylase gene) in tobacco plants enhances the demethylation rates of gene bodies and promoters associated with genes in the biosynthetic pathway of flavonoids and antioxidant pathways when exposed to salt stress. This results in elevated gene expression levels and improved salinity stress tolerance (Bilichak *et al.*, 2012).

Additional research is necessary to understand the connection between DNA methylation alterations and changes in gene expression patterns. Further investigation is necessary to uncover the mechanisms by which DNA methylation influences stress-responsive gene expression and the inheritance of these epigenetic modifications through generations (Hassan *et al.*, 2023).

11.4.2 Histone modification

Histones, composed primarily of lysine and arginine residues, play a fundamental role in establishing the structure of chromatin in the nucleosome (Zhou *et al.*, 2013). Histone tails, also referred to as the N-termini of histone proteins, undergo covalent modifications like phosphorylation, ubiquitination, acetylation, and methylation. These alterations have varying effects based on the specific amino acid residue that is modified, ultimately influencing gene activity (Banerjee and Roychoudhury, 2017). Tempering histone proteins has been proven to be an epigenetic mechanism that prepares plants for salt stress by regulating important salt-responsive genes throughout their growth cycle (Pikaard and Scheid, 2014).

Histone acetylation

An acetyl moiety with a negative charge on H4 and H3 histones decreases the binding strength between histone protein and DNA, allowing for easier access to DNA (Onufriev and Schiessel, 2019). Histone acetylation often correlates with elevated gene expression, whereas deacetylation is associated with the inhibition of transcription (Zheng *et al.*, 2016). Histone deacetylases and histone acetyltransferase are crucial enzymes that have an important role in regulating gene expression by adding or removing an acetyl group, respectively (Zhou *et al.*, 2017; Kim *et al.*, 2018).

Histone acetylation triggered by salt is linked with the activation of salinity stress-responsive genes in yeast (Magraner-Pardo *et al.*, 2014), *Zea mays* (Li *et al.*, 2014), and *Nicotiana tabacum* species (Sokol *et al.*, 2007). Active histone marks like H3K4ac and H3K9ac were found to be deposited on the peroxidases, leading to their activation in both *Beta vulgaris* and *B. maritime* (Yolcu *et al.*, 2016).

The activation of the abscisic acid pathway and antioxidant enzymes has been linked to the upregulation of the peroxidase gene, leading to reduced accumulation of ROS and higher levels of osmotic metabolites, ultimately enhancing tolerance towards salinity (Su *et al.*, 2020). The crucial role of enhanced histone acetylation in AtSOS1 and AtSOS3 is evident in salinity stress. The heightened acetylation aids in creating a more accessible chromatin structure, facilitating transcription (Sako *et al.*, 2016). It has been documented that the H3 acetylation by histone acetyltransferase in polyploid genes of wheat related to reactive oxygen species production can enhance the transcriptional alterations of these genes when exposed to salt stress. The manipulation of this gene in wheat and other crops has the potential to enhance salt tolerance (Zheng *et al.*, 2021).

Histone methylation

Histone methylation differs from acetylation as it does not impact the electrostatic properties of a histone protein. Rather, it enhances hydrophobic potential by altering the interactions at inter- or intramolecular levels, potentially leading to the formation of new binding sites for other proteins. Arginine methyltransferases are responsible for adding a methyl group to the arginine residue, while histone lysine methyltransferases catalyze the addition of a methyl group to the lysine amino acid residues. A total of two arginine amino acid methylation sites, such as H4R3 and H3R17, and five lysine amino acid methylation sites such as H4K20, H3K36, H3K27, H3K9, and H3K4 have been found in plants so far (Liu *et al.*, 2010).

Histone methylation in *Arabidopsis* serves as both active and repressive indicators (Liu *et al.*, 2016). The methylation status of arginine (Arg) and lysine (Lys) amino acids in histones plays a crucial role in modulating their binding affinity with reader proteins, thereby inducing changes in the structure of chromatin that may result in either transcriptional repression or activation (Teperino *et al.*, 2010). In castor and rice crops, the regulation of transcription for the critical salinity-response regulator is influenced by H3 DNA methylation at the 27th and 4th Lys positions (Karan *et al.*, 2012; Han *et al.*, 2020). Recently, it was discovered that the GmPHD6 protein, a reader of the H3K4me-0/1/2 code, plays a definite role in regulating the salt-tolerance gene transcription in *Glycine max* (Wei *et al.*, 2017). Methylation differences at H3K4me-3 and H3K27me-3 have been shown to impact the salt-responsive expression of the gene *OsBZ8* in the varieties of rice such as salt-resistant Nonabokra and salt-responsive IR64 (Paul *et al.*, 2017). The significant contribution of epigenetic features H3K4me-3 and H3K27me-3 in the regulation of genes responsive to salt stress and the enhancement of tolerance to salinity has been demonstrated by these observations. Additionally, the overexpression of the JMJ15 gene in *Arabidopsis*, which codes for H3K4 demethylase, under salinity stress conditions greatly increased the tolerance of plants to salinity (Shen *et al.*, 2014).

The impact of chemical changes on the structure of chromosomes and the accessibility of transcription factors, as well as the correlation between histone modifications and gene activity, remain incompletely understood despite their conservation across various organisms. Extensive research has been conducted on histone methylation and acetylation in various plant species exposed to salt stress. Exploring additional histone modifications could enhance our understanding of crucial epigenetic markers and their potential use in developing climate-resilient crops (Singroha *et al.*, 2022).

11.4.3 Chromatin remodeling

Chromatin remodeling complexes play a significant role in moving nucleosomes and modifying the core histone make-up of the nucleosome using ATP. This process helps control the genomic DNA accessibility, potentially in a specific site depending on the conditions (Han *et al.*, 2015). In *Arabidopsis thaliana*, research has shown that salinity stress leads to a reduction in chromatin accessibility, unlike the rise in accessibility of chromatin observed with cold or heat treatment (Raxwal *et al.*, 2020). Research on the phylogenetics of *A. thaliana* revealed that SNF2/BRM-type chromatin remodeling proteins are encoded by AtCHR-12, AtCHR-2/BRM, AtCHR23, and AtCHR-3/SPLAYED (Verbsky and Richards, 2001). The main role of AtCHR-12 is likely associated with development. However, the absence of AtCHR-12 eliminated the restriction on root growth caused by low sodium chloride concentration during salinity stress (Mlynárová *et al.*, 2007). Additionally, despite the lack of direct proof indicating BRM's involvement in salt resistance, the *A. thaliana brm* mutant with decreased function exhibited enhanced sensitivity to ABA and demonstrated greater resilience to dehydration (Han *et al.*, 2012). PKL (PICKLE) is a gene encoding a CHD3-type chromatin remodeling factor that was extensively studied in *A. thaliana*. Researchers have concluded that PICKLE plays a vital role in mediating the trimethylation of H3K27 at its target gene (Yang *et al.*, 2019).

Photoperiod-independent early flowering 1 (PIE1) and actin-related protein 6 (ARP6) are responsible for encoding constituents of the chromatin remodeling complex SWR1, regardless of photoperiod (Carter *et al.*, 2018). This complex facilitates the integration of H2A.Z into nucleosomes (Deal *et al.*, 2007; Kumar, 2018). ARP6, PIE1, and PICKLE have been recognized as key players in enhancing salt stress tolerance (Sura *et al.*, 2017; Yang *et al.*, 2019) and it has been observed that the mutation of any single gene among these would result in an *A. thaliana* phenotype that is sensitive to salinity conditions. An analysis of gene expression in *A. thaliana* mutants, pkl-1, pie1-5 and clf-28 (curly leaf), compared with the wild type revealed a shared control over a specific group of genes (Carter *et al.*, 2018).

The initial discovery of the *ein6 een* double mutant occurred during the examination of *A. thaliana* mutants that displayed modified responses to ethylene. The *ein6* gene is responsible for encoding the relative of early flowering 6 that functions as an H3K27me3 demethylase. On the other hand, the *een* gene, also known as an enhancer of *ein6*, encodes the INO80 subunit 6 (IES6), a component of the INO80 chromatin remodeling complex. Upon exposure to ethylene, EIN2 expression was upregulated in wild-type, *ein6-1*, and *een-1* single mutants. However, in the *ein6-1een-1* double mutant, EIN2 was significantly suppressed regardless of ethylene treatment. The inhibition of EIN2 might be due to the accumulation of H3K27me3, and H2A.Z, which are directly connected to the functions of REF6 and IES6. Despite the lack of direct evidence regarding the involvement of REF6 and IES6 in salt stress responses, considering EIN2's importance in salt tolerance, it is possible that REF6 and IES6 also have a role to play. Therefore, further investigation is warranted (Zander *et al.*, 2019).

11.4.4 RNA interference in response to salinity tolerance

Exposure to high salinity stress disrupts ion balance, induces imbalances in water potential at the cellular level, and affects the overall health of the plant. These cellular and molecular changes result in various morphological alterations like leaf yellowing, wilting, growth stunting, plant wilting, and even plant death. The physiological changes induced by salt stress also trigger ABA synthesis, which causes stomata closure in guard cells, ultimately reducing photosynthesis and causing oxidative damage (Mangrauthia *et al.*, 2013).

Numerous gene transcripts are downregulated and upregulated in response to salinity stress in plants. Consequently, post-transcriptional gene silencing is essential for enabling plants to adapt to salinity stress. MicroRNAs (miRNAs) are also significantly involved in the response to abiotic stress (Sunkar and Zhu, 2004; Navarro *et al.*, 2006).

Several research studies focusing on rice, *Arabidopsis* and a few other plants have highlighted the significance of miRNA in salinity stress. Notably, in *Arabidopsis*, a number of differentially regulated miRNAs have been pinpointed in tissues subjected to salt stress. Under salinity stress conditions, miR397, miR396, miR394, miR393, miR319, miR171, miR169, miR168, miR168, miR167, miR165, miR159, miR158, and miR156 were found to be upregulated in *Arabidopsis*, demonstrating the association of miRNAs in the salt stress response of plants (Liu *et al.*, 2008; Mangrauthia *et al.*, 2013).

In plants, it has been observed that the majority of miRNAs have the ability to target multiple genes within the same family. However, a study has revealed that, in certain circumstances, a few miRNAs are capable of regulating the expression of a particular target gene (Sunkar *et al.*, 2012).

Salinity-stressed maize plants have shown that certain miRNAs can control the target gene expression by targeting various transcription factors related to plant growth and organ formation. Specifically, zma-miR159a,b, zma-miR166l/m, and zma-miR164a,b,c,d could regulate transcription factors such as NAC1, myb, and homeo domain-leucine-zipper proteins. Similar findings have been observed in rice and *Arabidopsis* (Jones and Bartel, 2004). In addition to zinc-finger protein and MADS-box protein, researchers have identified a few other transcription factors as potential target genes for miRNA. There have been reports indicating that they also contribute to the salinity response of plants (Fang *et al.*, 2006; Xu *et al.*, 2008). Besides transcription factors, miRNA target genes encode proteins that are involved in different physiological processes or metabolic pathways in plants, either directly or indirectly (Mangrauthia *et al.*, 2013). Additionally, miRNAs that target the regulation of cytochrome oxidase and NADP-dependent malic enzymes have been discovered to be responsive to salt in plants (Yan *et al.*, 2005; Cheng and Long, 2007).

In *O. sativa*, miR393 targets the precursor of the phytosulfokine receptor, as well as the protein oxido-reductase or transport inhibitor response, which is crucial for auxin signaling in salt stress conditions (Liu *et al.*, 2008). miR396c selectively targets the heat shock protein (HSP 70 kDa), jasmonate o-ethyl transferase, leucine-rich repeat family protein, ubiquitin protein ligase caps, and TBP-associated 59 kDa subunit protein. These proteins are essential for plant growth and development, particularly in response to salinity stress (Gao *et al.*, 2010).

miR395 regulates the targets sulfurylase and ASP1 genes in a salt-stressed soybean variety experiencing sulfate deficiency conditions. This indicates that miR395 could be involved in a general salt-stress response pathway, potentially involving the maintenance of energy supply (Ding *et al.*, 2009). Artichoke's targets, cca-miR397 and 399, showed homology to laccase gene family members associated with salt-stress response (De Paola *et al.*, 2012). Laccase, a glycoprotein present in plants with several copper ions, contributes to lignin production. Research indicates that elevated NaCl concentrations can upregulate laccase gene expression in the roots of *Arabidopsis*, maize, and tomato (Liang *et al.*, 2006; Cai *et al.*, 2006). The expression of laccase in the root of artichoke could potentially be enhanced due to the reduction of miR397a levels under salinity conditions (De Paola *et al.*, 2012). In the context of salt stress experienced by *Arabidopsis*, there was a slight enhancement in the expression of miR398, culminating in a reduction of the expression level of the target enzyme known as superoxide dismutase. Moreover, it was noted that the regulation of aspartic proteinase APA1 was governed by a newly discovered miRNA called cca-novel-18, which specifically targets this protein (Attia *et al.*, 2011).

During salinity stress, *Populus trichocarpa* shows upregulation of miR482.2 and downregulation of miR1445. During stem development, these miRNAs play a crucial role by targeting the disease resistance protein and dihydropyrimidinase. This targeting mechanism contributes to the differentiation of cambium, which is essential for stem growth and development (Lu *et al.*, 2008). RNA interference is now widely recognized as a potent and efficient method in functional genomics for suppressing gene expression to enhance crop quality. The impact of miRNAs on plant regulation is an area that demands further exploration within the field of plant biology. More tactics must be utilized to explore the roles of miRNAs, as well as their linked signaling pathways and gene networks, during exposure to high salinity levels.

11.5 Conclusion

Epigenetic regulations have gained significant recognition as potent mechanisms for governing the impact of salt stress on plants, thereby establishing a solid foundation for the advancement of salt-tolerant crops. Epigenetic mechanisms play a pivotal role in tightly regulating gene expression in plants that are vulnerable to salt stress. The epigenetic modifications found in stress-responsive genes play a significant role in regulating chromatin accessibility and gene expression dynamically. Epigenetic modifications encompass a range of regulatory mechanisms that are important for the adaptation of plants to abiotic stress. These mechanisms include chromatin remodeling, methylation, histone variants, histone changes, and long non-coding RNAs (lncRNAs). By undergoing these modifications, several crops can adjust their genetic expression patterns and phenotypic response to cope with the adverse environmental conditions imposed by abiotic stressors. The phenotypic flexibility resulting from epigenetic variation, which is achieved by altering gene expression, directly impacts the fitness and overall survival prospects of plants through natural selection. In contrast to conventional DNA mutations, epimutations could occur rapidly and, despite their stability, they are mostly reversible. This makes them an ideal tool for promptly responding to unforeseen environmental pressures. It is important to emphasize that these modifications are usually influenced by the genetic variation present, and it is essential to consider both factors simultaneously. Extensive research is required in order to gain better knowledge about the epigenetic mechanisms responsible for chromatin modifications and how they affect the regulatory functions that influence the stress response of plants. Further investigation is necessary to comprehend the mechanism responsible for the inheritance of stress memory.

References

Agarwal, G., Kudapa, H., Ramalingam, A., Choudhary, D., Sinha, P. *et al.* (2020) Epigenetics and epigenomics: Underlying mechanisms, relevance, and implications in crop improvement. *Functional & Integrative Genomics* 20(6), 739–761. DOI: 10.1007/s10142-020-00756-7.

Al-Harrasi, I., Al-Yahyai, R. and Yaish, M.W. (2018) Differential DNA methylation and transcription profiles in date palm roots exposed to salinity. *PLoS ONE* 13(1), e0191492. DOI: 10.1371/journal.pone.0191492.

Al-Lawati, A., Al-Bahry, S., Victor, R., Al-Lawati, A.H. and Yaish, M.W. (2016) Salt stress alters DNA methylation levels in alfalfa (*Medicago* spp). *Genetics and Molecular Research* 15(1). DOI: 10.4238/gmr.15018299.

Ashraf, M. (2004) Some important physiological selection criteria for salt tolerance in plants. *Flora - Morphology, Distribution, Functional Ecology of Plants* 199(5), 361–376. DOI: 10.1078/0367-2530-00165.

Attia, H., Karray, N., Msilini, N. and Lachaâl, M. (2011) Effect of salt stress on gene expression of superoxide dismutases and copper chaperone in *Arabidopsis thaliana*. *Biologia Plantarum* 55(1), 159–163. DOI: 10.1007/s10535-011-0022-x.

Banerjee, A. and Roychoudhury, A. (2017) Epigenetic regulation during salinity and drought stress in plants: Histone modifications and DNA methylation. *Plant Gene* 11, 199–204.

Beyrne, C.C., Iusem, N.D. and González, R.M. (2019) Effect of salt stress on cytosine methylation within GL2, an *Arabidopsis thaliana* gene involved in root epidermal cell differentiation. Absence of inheritance in the unstressed progeny. *International Journal of Molecular Sciences* 20(18), 4446. DOI: 10.3390/ijms20184446.

Bharti, P., Mahajan, M., Vishwakarma, A.K., Bhardwaj, J. and Yadav, S.K. (2015) AtROS1 overexpression provides evidence for epigenetic regulation of genes encoding enzymes of flavonoid biosynthesis and antioxidant pathways during salt stress in transgenic tobacco. *Journal of Experimental Botany* 66(19), 5959–5969.

Bilichak, A., Ilnystkyy, Y., Hollunder, J. and Kovalchuk, I. (2012) The progeny of *Arabidopsis thaliana* plants exposed to salt exhibit changes in DNA methylation, histone modifications and gene expression. *PLoS ONE* 7(1), e30515. DOI: 10.1371/journal.pone.0030515.

Bird, A. (2007) Perceptions of epigenetics. *Nature* 447(7143), 396–398. DOI: 10.1038/nature05913.

Blum, A. (2018) *Plant Breeding for Stress Environments*. Taylor & Francis/CRC, Boca Raton, Florida.

Cai, X., Davis, E.J., Ballif, J., Liang, M., Bushman, E. *et al.* (2006) Mutant identification and characterization of the laccase gene family in *Arabidopsis*. *Journal of Experimental Botany* 57(11), 2563–2569. DOI: 10.1093/jxb/erl022.

Carillo, P., Annunziata, M.G., Pontecorvo, G., Fuggi, A. and Woodrow, P. (2011) Salinity stress and salt tolerance. In: Shanker, A. and Venkateswarlu, B. (eds) *Abiotic Stress in Plants – Mechanisms and Adaptations*. InTech Open, pp. 21–38. DOI: 10.5772/22331.

Carter, B., Bishop, B., Ho, K.K., Huang, R., Jia, W. *et al.* (2018) The chromatin remodelers PKL and PIE1 act in an epigenetic pathway that determines H3K27me3 homeostasis in *Arabidopsis*. *The Plant Cell* 30(6), 1337–1352. DOI: 10.1105/tpc.17.00867.

Chang, Y.N., Zhu, C., Jiang, J., Zhang, H., Zhu, J.K. *et al.* (2020) Epigenetic regulation in plant abiotic stress responses. *Journal of Integrative Plant Biology* 62(5), 563–580. DOI: 10.1111/jipb.12901.

Chen, H., Liu, X., Zhang, H., Yuan, X., Gu, H. *et al.* (2018) Advances in salinity tolerance of soybean: Genetic diversity, heredity, and gene identification contribute to improving salinity tolerance. *Journal of Integrative Agriculture* 17(10), 2215–2221. DOI: 10.1016/S2095-3119(17)61864-1.

Cheng, Y. and Long, M. (2007) A cytosolic NADP-malic enzyme gene from rice (*Oryza sativa* L.) confers salt tolerance in transgenic arabidopsis. *Biotechnology Letters* 29(7), 1129–1134. DOI: 10.1007/s10529-007-9347-0.

Crisp, P.A., Ganguly, D., Eichten, S.R., Borevitz, J.O. and Barry, J.P. (2016) Reconsidering plant memory: Intersections between stress recovery, RNA turnover, and epigenetics. *Science Advances* 2(2), e1501340. DOI: 10.1126/sciadv.1501340.

Deal, R.B., Topp, C.N., McKinney, E.C. and Meagher, R.B. (2007) Repression of Flowering *Arabidopsis* requires activation of *FLOWERING LOCUS C* expression by the histone variant H2A.Z. *The Plant Cell* 19(1), 74–83. DOI: 10.1105/tpc.106.048447.

De Paola, D., Cattonaro, F., Pignone, D. and Sonnante, G. (2012) The miRNAome of globe artichoke: Conserved and novel micro RNAs and target analysis. *BMC Genomics* 13(1), 41.

Ding, D., Zhang, L., Wang, H., Liu, Z., Zhang, Z. *et al.* (2009) Differential expression of miRNAs in response to salt stress in maize roots. *Annals of Botany* 103(1), 29–38. DOI: 10.1093/aob/mcn205.

Ding, Y., Fromm, M. and Avramova, Z. (2012) Multiple exposures to drought "train" transcriptional responses in *Arabidopsis*. *Nature Communications* 3(1), 740. DOI: 10.1038/ncomms1732.

Duan, C.-G., Zhu, J.-K. and Cao, X. (2018) Retrospective and perspective of plant epigenetics in China. *Journal of Genetics and Genomics* 45(11), 621–638. DOI: 10.1016/j.jgg.2018.09.004.

Fan, S., Liu, H., Liu, J., Hua, W., Xu, S. *et al.* (2020) Systematic analysis of the DNA methylase and demethylase gene families in rapeseed (*Brassica napus* L.) and their expression variations after salt and heat stresses. *International Journal of Molecular Sciences* 21(3), 953. DOI: 10.3390/ijms21030953.

Fang, Q., Xu, Z. and Song, R. (2006) Cloning, characterization and genetic engineering of *FLC* homolog in *Thellungiella halophila*. *Biochemical and Biophysical Research Communications* 347(3), 707–714. DOI: 10.1016/j.bbrc.2006.06.165.

Feng, Q., Yang, C., Lin, X., Wang, J., Ou, X. *et al.* (2012) Salt and alkaline stress induced transgenerational alteration in DNA methylation of rice ('*Oryza sativa*'). *Australian Journal of Crop Science* 6(5), 877–883.

Ferreira, L.J., Azevedo, V., Maroco, J., Oliveira, M.M. and Santos, A.P. (2015) Salt tolerant and sensitive rice varieties display differential methylome flexibility under salt stress. *PLoS ONE* 10(5), e0124060. DOI: 10.1371/journal.pone.0124060.

Gao, P., Bai, X., Yang, L., Lv, D., Li, Y. *et al.* (2010) Over-expression of osa-MIR396c decreases salt and alkali stress tolerance. *Planta* 231(5), 991–1001. DOI: 10.1007/s00425-010-1104-2.

Gill, S., Ramzan, M., Naz, G., Ali, L., Danish, S. *et al.* (2024) Effect of silicon nanoparticle-based biochar on wheat growth, antioxidants and nutrients concentration under salinity stress. *Scientific Reports* 14(1), 6380. DOI: 10.1038/s41598-024-55924-7.

Govindaraj, M., Vetriventhan, M. and Srinivasan, M. (2015) Importance of genetic diversity assessment in crop plants and its recent advances: An overview of its analytical perspectives. *Genetics Research International* 431487. DOI: 10.1155/2015/431487.

Han, B., Xu, W., Ahmed, N., Yu, A., Wang, Z. *et al.* (2020) Changes and associations of genomic transcription and histone methylation with salt stress in castor bean. *Plant and Cell Physiology* 61(6), 1120–1133.

Han, S.K., Sang, Y., Rodrigues, A., BIOL425, F., Wu, M.F. *et al.* (2012) SWI2/SNF2 chromatin remodeling ATPase BRAHMA represses abscisic acid responses in the absence of the stress stimulus in *Arabidopsis*. *The Plant Cell* 24(12), 4892–4906. DOI: 10.1105/tpc.112.105114.

Han, S.K., Wu, M.F., Cui, S. and Wagner, D. (2015) Roles and activities of chromatin remodeling ATPases in plants. *The Plant Journal* 83(1), 62–77. DOI: 10.1111/tpj.12877.

Hassan, A.H., Mokhtar, M.M. and El Allali, A. (2023) Transposable elements: multifunctional players in the plant genome. *Frontiers in Plant Science* 14, 1330127. DOI: 10.3389/fpls.2023.1330127.

Hilker, M., Schwachtje, J., Baier, M., Balazadeh, S., Bäurle, I. *et al.* (2016) Priming and memory of stress responses in organisms lacking a nervous system. *Biological Reviews* 91(4), 1118–1133. DOI: 10.1111/brv.12215.

Ismail, A.M. and Horie, T. (2017) Genomics, physiology, and molecular breeding approaches for improving salt tolerance. *Annual Review of Plant Biology* 68, 405–434. DOI: 10.1146/annurev-arplant-042916-040936.

Jan, A.U., Hadi, F., Midrarullah, Nawaz, M.A. and Rahman, K. (2017) Potassium and zinc increase tolerance to salt stress in wheat (*Triticum aestivum* L). *Plant Physiology and Biochemistry* 116, 139–149. DOI: 10.1016/j.plaphy.2017.05.008.

Jones-Rhoades, M.W. and Bartel, D.P. (2004) Computational identification of plant microRNAs and their targets, including a stress-induced miRNA. *Molecular Cell* 14(6), 787–799. DOI: 10.1016/j.molcel.2004.05.027.

Karan, R., DeLeon, T., Biradar, H. and Subudhi, P.K. (2012) Salt stress induced variation in DNA methylation pattern and its influence on gene expression in contrasting rice genotypes. *PLoS ONE* 7(6), e40203.

Kim, J., Yang, W., Forner, J., Lohmann, J.U., Noh, B. *et al.* (2018) Epigenetic reprogramming by histone acetyltransferase HAG1/atgcn5 is required for pluripotency acquisition in *Arabidopsis*. *The EMBO Journal* 37(20), e98726. DOI: 10.15252/embj.201798726.

Kong, L., Liu, Y., Wang, X. and Chang, C. (2020) Insight into the role of epigenetic processes in abiotic and biotic stress response in wheat and barley. *International Journal of Molecular Sciences* 21(4), Article 4. DOI: 10.3390/ijms21041480.

Kravchik, M. and Bernstein, N. (2013) Effects of salinity on the transcriptome of growing maize leaf cells point at cell-age specificity in the involvement of the antioxidative response in cell growth restriction. *BMC Genomics* 14(1), 24. DOI: 10.1186/1471-2164-14-24.

Kumar, S.V. (2018) H2A.Z at the core of transcriptional regulation in plants. *Molecular Plant* 11(9), 1112–1114. DOI: 10.1016/j.molp.2018.07.002.

Kumar, S., Beena, A.S., Awana, M. and Singh, A. (2017) Salt-induced tissue-specific cytosine methylation downregulates expression of HKT genes in contrasting wheat (*Triticum aestivum* L.) genotypes. *DNA and Cell Biology* 36(4), 283–294. DOI: 10.1089/dna.2016.3505.

Lämke, J. and Bäurle, I. (2017) Epigenetic and chromatin-based mechanisms in environmental stress adaptation and stress memory in plants. *Genome Biology* 18, 124. DOI: 10.1186/s13059-017-1263-6.

Li, H., Yan, S., Zhao, L., Tan, J., Zhang, Q. *et al.* (2014) Histone acetylation associated up-regulation of the cell wall related genes is involved in salt stress induced maize root swelling. *BMC Plant Biology* 14(1), 105. DOI: 10.1186/1471-2229-14-105.

Liang, M., Haroldsen, V., Cai, X. and Wu, Y. (2006) Expression of a putative laccase gene, ZmLAC1, in maize primary roots under stress. *Plant, Cell & Environment* 29(5), 746–753. DOI: 10.1111/j.1365-3040.2005.01435.x.

Liang, Z., Shen, L., Cui, X., Bao, S., Geng, Y. *et al.* (2018) DNA N-adenine methylation in *Arabidopsis thaliana*. *Developmental Cell* 45(3), 406–416. DOI: 10.1016/j.devcel.2018.03.012.

Liu, C., Lu, F., Cui, X. and Cao, X. (2010) Histone methylation in higher plants. *Annual Review of Plant Biology* 61, 395–420. DOI: 10.1146/annurev.arplant.043008.091939.

Liu, C., Li, H., Lin, J., Wang, Y., Xu, X. *et al.* (2018) Genome-wide characterization of DNA demethylase genes and their association with salt response in *Pyrus*. *Genes (Basel)* 9(8), 398. DOI: 10.3390/genes9080398.

Liu, C., Wang, C., Wang, G., Becker, C., Zaidem, M. *et al.* (2016) Genome-wide analysis of chromatin packing in *Arabidopsis thaliana* at single-gene resolution. *Genome Research* 26(8), 1057–1068. DOI: 10.1101/gr.204032.116.

Liu, H.H., Tian, X., Li, Y.J., Wu, C.A. and Zheng, C.C. (2008) Microarray-based analysis of stress-regulated microRNAs in *Arabidopsis thaliana*. *RNA* 1495, 836–843. DOI: 10.1261/rna.895308.

Lu, S., Sun, Y.H. and Chiang, V.L. (2008) Stress-responsive microRNAs in *Populus*. *The Plant Journal* 55(1), 131–151. DOI: 10.1111/j.1365-313X.2008.03497.x.

Luo, X. and He, Y. (2020) Experiencing winter for spring flowering: a molecular epigenetic perspective on vernalization. *Journal of Integrative Plant Biology* 62(1), 104–117. DOI: 10.1111/jipb.12896.

Magraner-Pardo, L., Pelechano, V., Coloma, M.D. and Tordera, V. (2014) Dynamic remodeling of histone modifications in response to osmotic stress in *Saccharomyces cerevisiae*. *BMC Genomics* 15(1), 247. DOI: 10.1186/1471-2164-15-247.

Mangrauthia, S.K., Agarwal, S., Sailaja, B., Madhav, M.S. and Voleti, S.R. (2013) MicroRNAs and their role in salt stress response in plants. In: Ahmad, P., Azooz, M.M. and Prasad, M.N.V. (eds) *Salt Stress in Plants: Signalling, Omics and Adaptations*. Springer, pp. 15–46. DOI: 10.1007/978-1-4614-6108-1_2.

Mastan, S.G., Rathore, M.S., Bhatt, V.D., Yadav, P. and Chikara, J. (2012) Assessment of changes in DNA methylation by methylation-sensitive amplification polymorphism in *Jatropha curcas* L. subjected to salinity stress. *Gene* 508(1), 125–129. DOI: 10.1016/j.gene.2012.07.063.

Miryeganeh, M. (2021) Plants' epigenetic mechanisms and abiotic stress. *Genes* 12(8), 1106. DOI: 10.3390/genes12081106.

Mlynárová, L., Nap, J. and Bisseling, T. (2007) The SWI/SNF chromatin-remodeling gene AtCHR12 mediates temporary growth arrest in *Arabidopsis thaliana* upon perceiving environmental stress. *The Plant Journal* 51(5), 874–885. DOI: 10.1111/j.1365-313X.2007.03185.x.

Mousavi, S., Regni, L., Bocchini, M., Mariotti, R., Cultrera, N.G.M. *et al.* (2019) Physiological, epigenetic and genetic regulation in some olive cultivars under salt stress. *Scientific Reports* 9(1), 1093. DOI: 10.1038/s41598-018-37496-5.

Navarro, L., Dunoyer, P., Jay, F., Arnold, B., Dharmasiri, N. *et al.* (2006) A plant miRNA contributes to antibacterial resistance by repressing auxin signaling. *Science* 312(5772), 436–439. DOI: 10.1126/science.1126088.

Niknam, S.R. and McComb, J. (2000) Salt tolerance screening of selected Australian woody species – A review. *Forest Ecology and Management* 139(1–3), 1–19. DOI: 10.1016/S0378-1127(99)00334-5.

Onufriev, A.V. and Schiessel, H. (2019) The nucleosome: From structure to function through physics. *Current Opinion in Structural Biology* 56, 119–130. DOI: 10.1016/j.sbi.2018.11.003.

Pankova, E.I., Gerasimova, M.I. and Korolyuk, T.V. (2018) Salt-affected soils in Russian, American, and international soil classification systems. *Eurasian Soil Science* 51(11), 1297–1308. DOI: 10.1134/S1064229318110078.

Paul, A., Dasgupta, P., Roy, D. and Chaudhuri, S. (2017) Comparative analysis of histone modifications and DNA methylation at OsBZ8 locus under salinity stress in IR64 and nonabokra rice varieties. *Plant Molecular Biology* 95(1–2), 63–88. .

Pikaard, C.S. and Scheid, O.M. (2014) Epigenetic regulation in plants. *Cold Spring Harbor Perspectives in Biology* 6(12), a019315. DOI: 10.1101/cshperspect.a019315.

Rahman, M.M., Mostofa, M.G., Keya, S.S., Siddiqui, M.N., Ansary, M.M.U. *et al.* (2021) Free full-text. adaptive mechanisms of halophytes and their potential in improving salinity tolerance in plants. *International Journal of Molecular Science* 22(19), 10733. DOI: 10.3390/ijms221910733.

Rashid, Md., Vaishnav, A., Verma, R.K., Sharma, P., Suprasanna, P. *et al.* (2022) Epigenetic regulation of salinity stress responses in cereals. *Molecular Biology Reports* 49(1), 761–772. DOI: 10.1007/s11033-021-06922-9.

Raxwal, V.K., Ghosh, S., Singh, S., Katiyar-Agarwal, S., Goel, S. *et al.* (2020) Abiotic stress-mediated modulation of the chromatin landscape in *Arabidopsis thaliana*. *Journal of Experimental Botany* 71(17), 5280–5293. DOI: 10.1093/jxb/eraa286.

Read, J.J., Reddy, K.R. and Jenkins, J.N. (2006) Yield and fiber quality of upland cotton as influenced by nitrogen and potassium nutrition. *European Journal of Agronomy* 24(3), 282–290. DOI: 10.1016/j.eja.2005.10.004.

Saha, P., Chatterjee, P. and Biswas, A.K. (2010) NaCl pretreatment alleviates salt stress by enhancement of antioxidant defense system and osmolyte accumulation in mungbean (*Vigna radiata* L. Wilczek). *Indian Journal of Experimental Biology* 48(6), 593–600.

Sako, K., Kim, J.-M., Matsui, A., Nakamura, K., Tanaka, M. *et al.* (2016) Ky-2, a histone deacetylase inhibitor, enhances high-salinity stress tolerance in *Arabidopsis thaliana*. *Plant & Cell Physiology* 57(4), 776–783. DOI: 10.1093/pcp/pcv199.

Saroha, M., Singroha, G., Sharma, M., Mehta, G., Gupta, O.P. *et al.* (2017) SRNA and epigenetic mediated abiotic stress tolerance in plants. *Indian Journal of Plant Physiology* 22(4), 458–469. DOI: 10.1007/s40502-017-0330-z.

Sevik, H. and Cetin, M. (2015) Effects of water stress on seed germination for select landscape plants. *Polish Journal of Environmental Studies* 24(2), 689–693. DOI: 10.15244/pjoes/30119.

Shabala, S. (2000) Ionic and osmotic components of salt stress specifically modulate net ion fluxes from bean leaf mesophyll. *Plant, Cell & Environment* 23(8), 825–837. DOI: 10.1046/j.1365-3040.2000.00606.x.

Sharma, M., Kumar, P., Verma, V., Sharma, R., Bhargava, B. *et al.* (2022) Understanding plant stress memory response for abiotic stress resilience: Molecular insights and prospects. *Plant Physiology and Biochemistry* 179, 10–24. DOI: 10.1016/j.plaphy.2022.03.004.

Shen, Y., Conde e silva, N., Audonnet, L., Servet, C., Wei, W. *et al.* (2014) Over-expression of histone H3K4 demethylase gene JMJ15 enhances salt tolerance in *Arabidopsis*. *Frontiers in Plant Science* 5. DOI: 10.3389/fpls.2014.00290.

Singroha, G. and Sharma, P. (2019) Epigenetic modifications in plants under abiotic stress. In: Meccariello, R. (ed.) *Epigenetics*. IntechOpen. DOI: 10.5772/intechopen.84455.

Singroha, G., Kumar, S., Gupta, O.P., Singh, G.P. and Sharma, P. (2022) Uncovering the epigenetic marks involved in mediating salt stress tolerance in plants. *Frontiers in Genetics* 13, 811732. DOI: 10.3389/fgene.2022.811732.

Siyal, A.A, Siyal, A.G and Abro, Z.A. (2002) Salt affected soils their identification and reclamation. *Journal of Applied Sciences* 2(5), 537–540. DOI: 10.3923/jas.2002.537.540.

Sokol, A., Kwiatkowska, A., Jerzmanowski, A. and Prymakowska-Bosak, M. (2007) Up-regulation of stress-inducible genes in tobacco and *Arabidopsis* cells in response to abiotic stresses and ABA treatment correlates with dynamic changes in histone H3 and H4 modifications. *Planta* 227(1), 245–254. DOI: 10.1007/s00425-007-0612-1.

Solangi, G.S., Siyal, A.A., Babar, M.M. and Siyal, P. (2019) Spatial analysis of soil salinity in the Indus River Delta, Pakistan. *Engineering, Technology & Applied Science Research* 9(3), 4271–4275. DOI: 10.48084/etasr.2818.

Su, P., Yan, J., Li, W., Wang, L., Zhao, J. *et al.* (2020) A member of wheat class III peroxidase gene family, TaPRX-2A, enhanced the tolerance of salt stress. *BMC Plant Biology* 20(1), 392. DOI: 10.1186/s12870-020-02602-1.

Sunkar, R. and Zhu, J.K. (2004) Novel and stress-regulated microRNAs and other small RNAs from *Arabidopsis*. *The Plant Cell* 16(8), 2001–2019. DOI: 10.1105/tpc.104.022830.

Sunkar, R., Li, Y.F. and Jagadeeswaran, G. (2012) Functions of microRNAs in plant stress responses. *Trends in Plant Science* 17(4), 196–203.

Sura, W., Kabza, M., Karlowski, W.M., Bieluszewski, T., Kus-Slowinska, M. *et al.* (2017) Dual role of the histone variant H2A.Z in transcriptional regulation of stress-response genes. *The Plant Cell* 29(4), 791–807. DOI: 10.1105/tpc.16.00573.

Teperino, R., Schoonjans, K. and Auwerx, J. (2010) Histone methyl transferases and demethylases; can they link metabolism and transcription? *Cell Metabolism* 12(4), 321–327. DOI: 10.1016/j.cmet.2010.09.004.

Tremblay, A.M., Dufour, C.R., Ghahremani, M. and Reudelhuber, T.L. (2010) Physiological genomics identifies estrogen-related receptor α as a regulator of renal sodium and potassium homeostasis and the renin-angiotensin pathway. *Molecular Endocrinology* 24(1), 22–32. DOI: 10.1210/me.2009-0254.

Varotto, S., Tani, E., Abraham, E., Krugman, T., Kapazoglou, A. *et al.* (2020) Epigenetics: Possible applications in climate-smart crop breeding. *Journal of Experimental Botany* 71(17), 5223–5236. DOI: 10.1093/jxb/eraa188.

Verbsky, M.L. and Richards, E.J. (2001) Chromatin remodeling in plants. *Current Opinion in Plant Biology* 4(6), 494–500. DOI: 10.1016/S1369-5266(00)00206-5.

Wei, W., Tao, J.J., Chen, H.W., Li, Q.T., Zhang, W.K. *et al.* (2017) A histone code reader and a transcriptional activator interact to regulate genes for salt tolerance. *Plant Physiology* 175(3), 1304–1320. DOI: 10.1104/pp.16.01764.

Xie, F., Wang, Q., Sun, R. and Zhang, B. (2015) Deep sequencing reveals important roles of microRNAs in response to drought and salinity stress in cotton. *Journal of Experimental Botany* 66(3), 789–804. DOI: 10.1093/jxb/eru437.

Xu, D.-Q., Huang, J., Guo, S.-Q., Yang, X., Bao, Y.-M. *et al.* (2008) Overexpression of a TFIIIA-type zinc finger protein gene ZFP252 enhances drought and salt tolerance in rice (*Oryza sativa* L). *FEBS Letters* 582(7), 1037–1043. DOI: 10.1016/j.febslet.2008.02.052.

Yan, S., Tang, Z., Su, W. and Sun, W. (2005) Proteomic analysis of salt stress-responsive proteins in rice root. *Proteomics* 5(1), 235–244. DOI: 10.1002/pmic.200400853.

Yang, R., Hong, Y., Ren, Z., Tang, K., Zhang, H. *et al.* (2019) A role for PICKLE in the regulation of cold and salt stress tolerance in *Arabidopsis*. *Frontiers in Plant Science* 10, 900. DOI: 10.3389/fpls.2019.0090.

Yasmin, H., Mazher, J., Azmat, A., Nosheen, A., Naz, R. *et al.* (2021) Combined application of zinc oxide nanoparticles and biofertilizer to induce salt resistance in safflower by regulating ion homeostasis and antioxidant defence responses. *Ecotoxicology and Environmental Safety* 218, 112262. DOI: 10.1016/j.ecoenv.2021.112262.

Yolcu, S., Ozdemir, F., Güler, A. and Bor, M. (2016) Histone acetylation influences the transcriptional activation of POX in *Beta vulgaris* L. and *Beta maritima* L. under salt stress. *Plant Physiology and Biochemistry* 100, 37–46. DOI: 10.1016/j.plaphy.2015.12.019.

Zafar, S., Hasnain, Z., Danish, S., Battaglia, M.L., Fahad, S. *et al.* (2024) Modulations of wheat growth by selenium nanoparticles under salinity stress. *BMC Plant Biology* 24(1), 35. DOI: 10.1186/s12870-024-04720-6.

Zander, M., Willige, B.C., He, Y., Nguyen, T.A., Langford, A.E. *et al.* (2019) Epigenetic silencing of a multifunctional plant stress regulator. *eLife* 8, e47835. DOI: 10.7554/eLife.47835.

Zhang, H., Lang, Z. and Zhu, J.-K. (2018) Dynamics and function of DNA methylation in plants. *Nature Review Molecular Cell Biology* 19(8), 489–506. DOI: 10.1038/s41580-018-0016-z.

Zhang, X., Bernatavichute, Y.V., Cokus, S., Pellegrini, M. and Jacobsen, S.E. (2009) Genome-wide analysis of mono-, di- and trimethylation of histone H3 lysine 4 in *Arabidopsis thaliana*. *Genome Biology* 10(6), R62. DOI: 10.1186/gb-2009-10-6-r62.

Zheng, M., Lin, J., Liu, X., Chu, W., Li, J. *et al.* (2021) Histone acetyltransferase tahag1 acts as a crucial regulator to strengthen salt tolerance of hexaploid wheat. *Plant Physiology* 186(4), 1951–1969. DOI: 10.1093/plphys/kiab187.

Zheng, Y., Ding, Y., Sun, X., Xie, S., Wang, D. *et al.* (2016) Histone deacetylase HDA9 negatively regulates salt and drought stress responsiveness in *Arabidopsis*. *Journal of Experimental Botany* 67(6), 1703–1713. DOI: 10.1093/jxb/erv562.

Zhou, B.-R., Feng, H., Kato, H., Dai, L., Yang, Y. *et al.* (2013) Structural insights into the histone H1-nucleosome complex. *Proceedings of the National Academy of Sciences* 110(48), 19390–19395. DOI: 10.1073/pnas.1314905110.

Zhou, S., Jiang, W., Long, F., Cheng, S., Yang, W. *et al.* (2017) Rice homeodomain protein WOX11 recruits a histone acetyltransferase complex to establish programs of cell proliferation of crown root meristem. *The Plant Cell* 29(5), 1088–1104. DOI: 10.1105/tpc.16.00908.

12 Epigenetics for Combating Chromium Stress in Plants

Sarah Jaison[1], Blassan George[2], P.M. Lakshmi[3] and Chithra Devi Balasundaram Saraswathy[1*]

[1]Department of Botany, PSGR Krishnammal College for Women, Coimbatore, Tamil Nadu, India; [2]Faculty of Health Sciences, University of Johannesburg, South Africa; [3]Directorate of Environment and Climate Change, Government of Kerala, India

Abstract

The past decades have witnessed rapid growth in industries and an accelerated increase in heavy metal (HM) pollution of the biosphere. Even though HMs are present in all environmental compartments in trace levels, their excess presence can be a serious cause for concern as it may lead to severe eco-toxicity resulting in loss of biodiversity and can cause hazardous effects on human health. The long-term persistence of these metals is another serious threat to the environment, especially plants, as they turn out to be conservative pollutants. The extent of concern over the adverse effects of HMs may differ, as some are toxic, while others may be essential nutrients at various magnitudes. Among the various HMs, chromium (Cr) is one of the major HMs that causes extensive environmental contamination, due to its widespread applications in textile, steel, electroplating, tanning, polishing, painting, and catalytic manufacturing industries. Chromium exists in different oxidation states, of which the most toxic one is Cr(VI). Although Cr is non-essential and possesses no known biological function in plants, all plant species can absorb Cr from the soil solution along with other essential micronutrients. The hexavalent Cr has a significant adverse effect on the growth and development of plants, due to accumulation of Cr(VI) in plant parts, inhibited nutrient uptake, chlorosis, reduction in photosynthesis, change in soil microbial flora, and oxidative stress. Epigenetics plays a very significant role in combating Cr(VI) stress in plants. DNA and histone acetylation, methylation, or phosphorylation have a profound effect on chromatin structure and gene expression, improving a plant's response to stress. These modifications might lead to the activation of stress-responsive genes. Epigenome editing and mRNAs also increase the stress tolerance in plants. These epigenetic variations are inherited across generations, helping plants to deal with the ill effects of long-term heavy metal stress. However, a more profound overview of epigenetic manipulation is required to understand and develop strategies to combat Cr stress.

Keywords: Abiotic stress, Epigenetic manipulation, Heavy metals, Stress-responsive genes

*Corresponding author: bschitradevi@psgrkcw.ac.in

 Epigenetics for Climate-Smart and Sustainable Agriculture (ed. Jen-Tsung Chen)
DOI: 10.1079/9781800626102.0012

12.1 Introduction

Exposure to heavy metals in the workplace and environment has long been acknowledged as a global public health hazard (Liu *et al.*, 2020). Contaminated food, water, and air are the primary sources of environmental exposure to heavy metals, impacting the health of millions of people worldwide. However, because of their unique physical and chemical characteristics, several heavy metals are widely used in modern industries; their widespread industrial use may expose workers to occupational risks. The air, soil, and water can become bioaccumulative sites for heavy metals such as nickel, cobalt, cadmium, copper, lead, chromium, and mercury. This can have detrimental effects on the health of all living species (Radwan and Salama, 2006). Numerous toxic consequences, including respiratory issues, gastrointestinal (GI) dysfunction, kidney toxicity, skin lesions, and vascular damage, can result from heavy metal exposure in different organs (Bao-yan *et al.*, 2012). Exposure to heavy metals can result in long-term health hazards because they are non-biodegradable and some of them exhibit effects that continue even after exposure is stopped.

Among the various HMs, chromium (Cr) is one of the major HMs that causes extensive environmental contamination, due to its widespread applications in textile, steel, electroplating, tanning, polishing, painting, and catalytic manufacturing industries (Peralta-Videa *et al.*, 2009). Chromium is a naturally occurring heavy metal found in the environment commonly in trivalent, Cr(III), and hexavalent, Cr(VI), forms. The reduction of Cr(VI) to Cr(III) results in the formation of reactive intermediates that contribute to the cytotoxicity, genotoxicity, and carcinogenicity of Cr(VI)-containing compounds. Although Cr is non-essential and possesses no known biological function in plants, all plant species can absorb Cr from the soil solution along with other essential micronutrients. The hexavalent Cr has a significant adverse effect on the growth and development of plants, due to accumulation of Cr(VI) in plant parts, inhibited nutrient uptake, chlorosis, reduction in photosynthesis, and change in soil microbial flora in addition to oxidative stress (Jaison and Muthukumar, 2017).

Epigenomics, the study of the complete set of epigenetic modifications on the genetic material of a cell, offers powerful strategies to combat cellular response (CR) stress in plants. By understanding and manipulating these epigenetic changes, researchers can enhance plant resilience to various stressors such as drought, salinity, extreme temperatures, and pathogen attacks. In this chapter, we will see a detailed exploration of how epigenomics can be harnessed to mitigate CR stress in plants

12.2 Effects of Chromium (VI) on Plants

Chromium poses significant environmental and agricultural challenges due to its toxicity and non-biodegradability. Particularly in its hexavalent form Cr(VI), chromium poses significant toxicity risks to plants, affecting various physiological and biochemical processes. Some of the key effects include the following.

12.2.1 Germination and seedling growth

Chromium VI can seriously impair seed germination and the growth of young seedlings. At high contamination levels, high concentrations of Cr(VI) in the soil can completely prevent the germination of some plants, such as sorghum and oats (López-Luna *et al.*, 2009; Hu *et al.*, 2016). Studies have shown that exposure to Cr can lead to reduced germination rates, lower seedling vigor, and impaired growth. Cr toxicity affects several physiological processes in plants, including water uptake and nutrient assimilation, leading to stunted growth and poor seedling establishment (Hu *et al.*, 2016). For instance, lettuce seedlings exposed to Cr exhibited inhibited root and shoot development, decreased biomass, and lower overall seedling vigor (Hou *et al.*, 2014). Cr significantly decreased the germination of seeds from a variety of plants, including crops like wheat (*Triticum aestivum* L.), barley (*Hordeum vulgare* L.), and maize (*Zea mays L.*), as well as vegetables like cauliflower (*Brassica oleracea* L.) and citrullus (*Citrullus vulgaris*) (Shahid *et al.*, 2017; Ao *et al.*, 2022). Higher soil Cr toxicity was found to decrease the germination rates of sugarcane (*Saccharum officinarum*), alfalfa (*Medicago sativa*), bush beans (*Phaseolus vulgaris*), and jungle rice (*Echinochloa colona*) by 25%, 48%, 23%, and 57%, respectively, in comparison with the control (Shanke *et al.*, 2005). Additionally, Cr exposure can cause oxidative stress, leading to increased levels of malondialdehyde and other markers of cellular damage, further impairing plant growth (Tang *et al.*, 2023).

12.2.2 Root and shoot development

Chromium stress significantly impacts the growth of plant roots and shoots, leading to various physiological and developmental challenges (Jing *et al.*, 2014). The primary effects of Cr toxicity include inhibited root elongation, reduced root hair growth, and overall root dieback, which impedes the plant's ability to absorb water and essential nutrients. These impacts on the root system lead to a cascade of negative effects on the entire plant, including stunted shoot growth, leaf chlorosis, and reduced biomass production. Roots are particularly sensitive to chromium toxicity. High concentrations of Cr, especially Cr(VI), can lead to the withering of root cells, detachment of the plasma membrane from the cell wall, and a reduction in root cell division and differentiation. This results in shorter root lengths, thicker primary roots, and an absence of lateral roots and root hairs. Chromium accumulates more in roots than in shoots, as roots are the first point of contact with the metal and exhibit a higher retention capacity due to their cell wall properties (Ao *et al.*, 2022). Chromium also adversely affects shoot growth, although to a lesser extent compared with roots. Singh and Sharma (2017) found that when Cr was present at higher concentrations (100 mg/L), the germination of green bean (*Phaseolus aureus*) and chickpea (*Cicer arietinum*) seeds was reduced by 42.60 and 53.53%, respectively. In a recent study by Hafiz and Ma (2021), more than 90% of the 45 tomato (*Solanum lycopersicum*) genotypes showed reduced and delayed germination within 14 days under 78 mg/L Cr(VI) stress. By interfering with the process of cell division in roots, higher concentrations of Cr dramatically reduced the number of bean roots (Zeid, 2001; Singh *et al.*, 2013). Proteins and starch that have accumulated as reserve resources during seed germination are digested to create precursors for the formation of the embryo axis as well as substrates for several metabolic processes (Ao *et al.*, 2022). Furthermore, the activity of the α- and β-amylases in the developing seeds diminished as the Cr content progressively rose, which could be the cause of the inhibition of seed germination (Oliveira, 2012). When Cr (VI) levels (300 μM) were present in the nutrient solution, black gram (*Vigna mungo*) seed germination decreased to 50.70% (Rath and Das, 2021).

12.2.3 Photosynthesis impairment

Chromium negatively impacts photosynthesis in plants through several mechanisms. One significant way Cr interferes with photosynthesis is by damaging the chloroplast structures and disrupting the chlorophyll biosynthesis process. This damage results in reduced chlorophyll content, which is crucial for capturing light energy during photosynthesis. As a result, the efficiency of the photosynthetic machinery declines, leading to impaired photosynthetic rates. Moreover, chromium exposure generates reactive oxygen species (ROS), which cause oxidative stress in plant cells. This oxidative stress further harms the photosynthetic apparatus, including photosystem II (PSII), reducing its quantum yield and overall efficiency. The impairment of PSII leads to decreased electron transport rates, thereby lowering the production of ATP and NADPH, essential molecules for the Calvin cycle and other biosynthetic processes in plants. Studies have shown that high concentrations of Cr can significantly reduce the net photosynthetic rate and stomatal conductance, limiting the plant's ability to perform gas exchange and assimilate carbon dioxide. These effects cumulatively hinder plant growth and productivity, as photosynthesis is the primary process driving biomass accumulation.

In microalgae, similar detrimental effects of Cr on photosynthesis have been observed. For instance, exposure to Cr(VI) reduces the quantum yield of photosystem II and inhibits photosynthetic activity by over 65% at higher concentrations, demonstrating a substantial decrease in photosynthetic efficiency and growth rates. High Cr concentrations disrupt water transport, reduce transpiration rates, and impair the uptake of essential minerals, leading to stunted shoot growth, yellowing, and dropping of leaves. This overall reduction in shoot biomass is a consequence of hindered photosynthetic processes and enzymatic activities within the plant (Bautista *et al.*, 2013).

12.2.4 Oxidative stress

Reactive oxygen species (ROS), which cause oxidative damage to biological components such as lipids, proteins, and nucleic acids, are induced by Cr(VI), which causes oxidative stress in plants, leading to detrimental effects on their physiology and development (Basit *et al.*, 2022). This oxidative stress arises primarily from the generation of ROS such as superoxide radicals and hydrogen peroxide when plants are exposed to Cr. Inhibition of CO_2 fixation and severe reduction of the electron transport chain are the main sources of ROS formation in plant organelles, such as mitochondria and chloroplasts (Ao *et al.*, 2022). These ROS can damage vital cellular components, including proteins, lipids, and DNA, thereby disrupting normal cellular functions and impairing metabolism. Moreover, Cr increases the generation of ROS such as H_2O_2 and O_2-, as well as MDA in plants like rice (Basit *et al.*, 2022), resulting in oxidative damage to DNA, RNA, proteins, and pigmentation (Tang *et al.*, 2023). The antioxidative defense system is similarly compromised in an environment high in Cr (Soliman *et al.*, 2022). However, excessive ROS production in plants causes damage to photosynthetic pigments, breakage in DNA strands, disruption of cell homeostasis, fragmentation of cell membranes or proteins, and deactivation and loss of genetic material (Srivastava *et al.*, 2021; Ao *et al.*, 2022). Ullah *et al.* (2019) observed similar results, reporting that enhanced ROS generation causes oxidative damage in plants with Cr toxicity, damaging proteins, lipids, DNA, and pigments while also inducing lipid peroxidation. According to Wakeel *et al.* (2020), these impacts reduce biomass output by blocking cell division or triggering cell death, which inhibits plant development. Shahid *et al.* (2014) stated that the length of exposure, Cr concentration, kind of plant, developmental stage, degree of stress, and specific organs all have an impact on how Cr toxicity manifests and affects plant physiology and biochemistry.

12.2.5 Phytotoxicity symptoms

Phytotoxicity, which is the toxic effect of certain substances on plant growth and development, manifests in several distinct symptoms. These symptoms vary depending on the type and concentration of the toxicant, as well as the plant species affected. When plants are exposed to toxic levels of chromium, a range of phytotoxic symptoms can be observed, including chlorosis or the yellowing of leaf tissue due to the loss of chlorophyll, a condition often seen as a primary symptom of heavy metal stress, including Cr exposure. Chlorosis typically starts at the leaf margins and progresses inward (Chow *et al.*, 2018). In severe cases of Cr toxicity, necrosis or the death of plant tissue occurs. This is often visible as brown or black spots on leaves, indicating areas where cells have died due to excessive stress and damage from ROS (Shahid *et al.*, 2017). Cr exposure leads to stunted growth, both in terms of plant height and root length. This is due to the interference with nutrient uptake and assimilation, as well as direct damage to cellular structures and enzymes involved in growth processes (Shahid *et al.*, 2017). Leaves may become deformed, curled, or display abnormal shapes due to the disruption of normal cell division and expansion processes caused by Cr toxicity (Zeid, 2001). The root system is particularly vulnerable to Cr, with symptoms including reduced root elongation, thickening of roots, and decreased formation of root hairs. This impairs the plant's ability to absorb water and nutrients, compounding the stress effects (Kumar *et al.*, 2016). Even in the presence of adequate water, plants exposed to Cr may wilt due to impaired root function and disrupted water transport within the plant. This symptom indicates severe internal physiological stress (Rath and Das, 2021). Cr toxicity can lead to reduced photosynthetic activity, decreased transpiration rates, and imbalances in hormonal levels such as increased abscisic acid (ABA), which promotes stomatal closure and limits gas exchange (Seneviratne *et al.*, 2019).

12.3 Enzymatic Activity

Chromium stress has a profound impact on the enzymatic activity in plants, affecting various enzymes involved in oxidative stress responses, detoxification processes, photosynthesis, and nitrogen metabolism. Cr stress has the potential to induce three different types of metabolic changes in plants: (i) changes in the synthesis of organic pigments, such as anthocyanin and chlorophyll, which help plants grow and develop (Shanke *et al.*, 2005; Shahid *et al.*, 2017); (ii) increased synthesis of metabolites, such as ascorbic acid and glutathione, in response to Cr stress, which will have an impact on the plants (Srivastava *et al.*, 2021); and (iii) changes in the metabolic pool to direct the synthesis of new biochemically associated metabolites, such as histidine and phytochelatins, which will confer tolerance or resistance to Cr stress (Shanke *et al.*, 2005; Ao *et al.*, 2022). According to Seneviratne *et al.* (2019), toxicity of Cr first caused seed imbibition and a decrease in germination rate by dramatically reducing the activity of gibberellin (GA) and increasing the activity of abscisic acid (ABA), a major component of seed dormancy. Enzymes such as acid phosphatases (ACPs), α-amylases, and proteases facilitate the efficient establishment and growth of seedlings by liberating food reserves from the endosperm. Under Cr stress, plants exhibit a multifaceted enzymatic response aimed at mitigating oxidative damage and maintaining metabolic functions. The upregulation of antioxidant and detoxification enzymes reflects the plant's effort to counteract the harmful effects of Cr-induced ROS. However, the inhibition of key metabolic enzymes, particularly those involved in photosynthesis and nitrogen metabolism, highlights the adverse impact of Cr on plant growth and productivity.

12.4 Epigenetics in Combating Cr(VI) in Plants

Epigenetics plays a significant role in the response of plants to various environmental stresses, including heavy metal toxicity such as Cr(VI) contamination. Chromium is a common environmental pollutant resulting from industrial activities such as leather tanning, electroplating, and metal finishing. Plants exposed to high levels of Cr(VI) can experience oxidative stress, DNA damage, and disruption of cellular functions. Epigenetic mechanisms, such as DNA methylation, histone modifications, and small RNA-mediated gene regulation, can modulate gene expression without altering the underlying DNA sequence (Fig. 12.1) (Table 12.1). These mechanisms can influence how plants respond to stressors like Cr(VI) by regulating the expression of stress-responsive genes, detoxification pathways, and antioxidant systems.

12.4.1 DNA methylation

Changes in DNA methylation, transfer of a methyl group onto the C5 position of the cytosine to form 5 methyl cytosine (5mC) patterns can regulate gene expression in response to any environmental stress. Studies have shown that exposure to Cr(VI) can alter DNA methylation patterns in plants, leading to changes in gene expression associated with stress response and detoxification pathways (Bartels *et al.*, 2018). Chromium contamination, especially in its hexavalent form Cr(VI), can induce alterations in DNA methylation patterns in plants. This phenomenon has been observed in various plant species exposed to Cr(VI) stress and is part of the plant's response to cope with the toxic effects of this heavy metal.

According to Labra *et al.* (2004), *Brassica napus* DNA methylation was dose-dependent and was stimulated by chromium stress. In contrast, *Trifolium repens* and *Cannabis sativa* showed decreased global DNA methylation in response to Cr stress in a different study (Aina *et al.*, 2004).

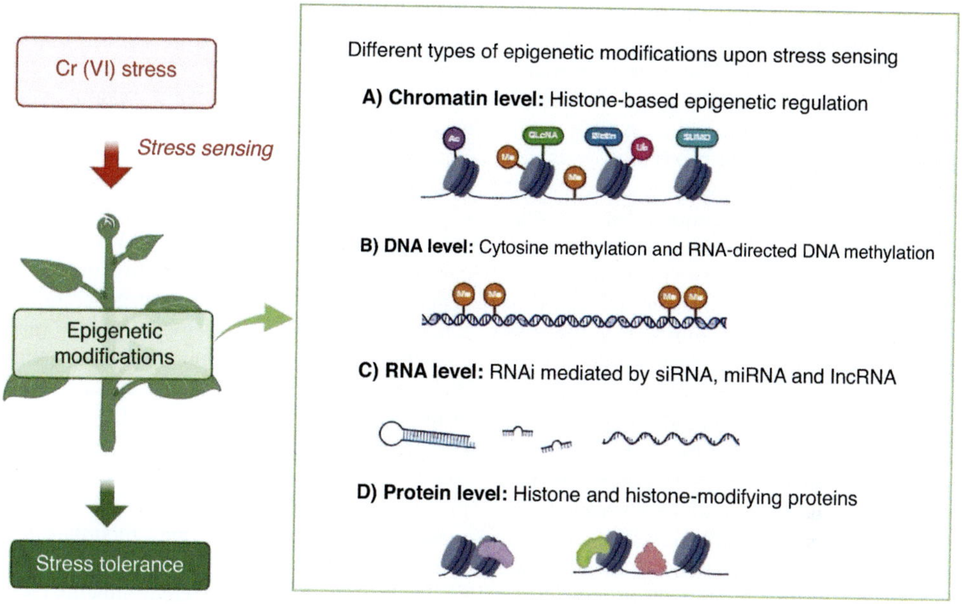

Fig. 12.1. Epigenetic modifications in plants in response to Cr (VI). Author's own image.

Table 12.1. Epigenetic mechanisms across various plant species in response to chromium stress. Author's own table.

Epigenetic mechanism	Plant species	Chromium stress impact	Reference
DNA methylation	*Brassica napus*	Altered methylation patterns in genes involved in stress response	Labra *et al.*, 2004
Histone modification	*Arabidopsis thaliana*	Changes in histone H3 acetylation and methylation associated with stress-responsive genes	Liu *et al.*, 2020
Non-coding RNAs	*Oryza sativa* (rice)	Upregulation of miRNAs targeting metal transport and detoxification genes	Yang *et al.*, 2022 Cong *et al.*, 2019
Chromatin remodeling	*Zea mays* (maize)	Enhanced chromatin accessibility in regions related to defense mechanisms	Ao *et al.*, 2022
DNA methylation	*Triticum ripens*	Global DNA hypomethylation observed under Cr exposure	Arora *et al.*, 2022
Histone modification	*Medicago sativa* (alfalfa)	Increased H3K4me3 and H3K27me3 marks at stress-related loci	
Non-coding RNAs	*Glycine max* (soybean)	Specific siRNAs and lncRNAs modulated in response to Cr toxicity	Yang *et al.*, 2022

These findings imply that various plant species' resistance to Cr may be attributed to distinct methylation processes. These investigations, however, mostly restricted the range of effects of Cr exposure on global alterations in DNA methylation levels in plants. The mechanism by which differentially methylated gene(s) control plant Cr tolerance is poorly understood. The increased DNA methylation level showed its essential role in kenaf seedlings growth Cr stress (Tang *et al.*, 2022). Techniques have looked at the targets of Cr-responsive microRNAs (miRNAs) in the roots of tobacco genotypes that were Cr-tolerant and Cr-sensitive, as well as roots that were Cr-free (control). Eleven miRNA families were found to be upregulated in Cr-tolerant species but not in Cr-sensitive species, whereas 17 miRNA families were found to be upregulated exclusively in Cr-sensitive species under Cr stress, according to comparative genomic analyses of 41 conserved Cr-responsive miRNA families. In this plant, just one family – miR6149 – was downregulated; in Cr-tolerant species, however, this family did not change. Fourteen of the 29 new miRNA families that were found to express differently in the two genotypes under Cr stress provide important insights into how miRNAs work in Cr tolerance (Bukhari *et al.*, 2015). In a different work, Cr used immunolabelling and MSAP techniques to change the methylation level of rape genomic DNA. Hypermethylation levels showed a positive correlation with the dosage of chromium stress, indicating *de novo* synthesis of methylated cytosine. On the other hand, other research has shown the opposite, showing that Cr stress decreased the amounts of cytosine methylation in hemp and clover by 20–40%, also proportionate to Cr concentrations (Aina *et al.*, 2004). These differences imply that various plant species have distinct methylation processes for chromium resistance (Peng and Zhang, 2009). Exposure to Cr(VI) can lead to changes in overall DNA methylation levels in plant genomes. Studies have shown both increases (hypermethylation) and decreases (hypomethylation) in global DNA methylation levels in response to Cr(VI) stress. Cr(VI) exposure can cause DNA methylation changes at specific genomic loci, including genes involved in stress response, detoxification, and defense mechanisms. These changes can influence the expression of these genes, modulating the plant's ability to withstand Cr(VI) toxicity. Cr-induced DNA methylation changes often occur at gene promoter regions, which can impact gene expression levels. Hypermethylation of promoter regions may lead to gene silencing, while hypomethylation can enhance gene expression. This modulation of gene expression can affect pathways related to stress response and metal detoxification. DNA methylation changes induced by Cr(VI) exposure can be heritable and passed on to subsequent generations of plants. This transgenerational inheritance of epigenetic modifications may contribute to the adaptation of plant populations to Cr-contaminated environments over multiple generations. The effects of Cr-induced DNA methylation changes can interact with those of other environmental stressors, such as drought, salinity, or exposure to other heavy metals. These interactions may modulate the overall stress response of plants and their ability to survive in contaminated environments (Sun *et al.*, 2022)

12.4.2 Histone modifications

Histone modifications, such as acetylation, methylation, and phosphorylation, can affect chromatin structure and gene expression. These modifications can be influenced by Cr(VI) exposure, and they may play a role in regulating the expression of genes involved in stress response and defence mechanisms. Small RNAs, including microRNAs (miRNAs) and small interfering RNAs (siRNAs), can regulate gene expression post-transcriptionally. Studies have identified specific miRNAs and siRNAs that are differentially expressed in response to Cr(VI) stress in plants. These small RNAs can target mRNAs encoding stress-related proteins, detoxification enzymes, and transporters, thereby modulating the plant's response to Cr(VI) toxicity (Bilichak *et al.*, 2012)

Histone acetylation

Cr(VI) exposure has been associated with alterations in histone acetylation levels in plants. Acetylation of histone proteins is generally associated with active gene transcription, and changes in histone acetylation patterns induced by Cr exposure can influence the expression of stress-responsive genes and detoxification pathways. Exposure to Cr(VI) can lead to global changes in histone acetylation levels in plant chromatin. Studies have shown that Cr exposure can result in both increases and decreases in histone acetylation, depending on the specific histone residues and the duration and concentration of Cr exposure. Changes in histone acetylation induced by Cr can influence the expression of stress-responsive genes and detoxification pathways in plants (Boycheva *et al.*, 2014). Histone acetylation is generally associated with transcriptional activation, so increased acetylation at specific gene loci may enhance the expression of stress-related genes involved in coping with Cr toxicity. Histone acetylation changes by Cr exposure can affect the chromatin structure of stress response genes and alter their accessibility to transcriptional machinery. This modulation of chromatin structure can regulate the expression of stress-responsive genes involved in antioxidant defense, metal detoxification, and other adaptive responses to Cr stress. Histone acetylation changes induced by Cr exposure may interact with other epigenetic modifications, such as DNA methylation and histone methylation, to fine-tune the plant's response to metal stress (Verdone *et al.*, 2006). Crosstalk between different epigenetic mechanisms can amplify or dampen the transcriptional response to Cr toxicity, influencing the overall stress tolerance of plants. Cr-induced changes in histone acetylation patterns may have transgenerational effects, impacting the phenotype and stress tolerance of subsequent plant generations. Epigenetic inheritance of histone modifications can contribute to the adaptation of plant populations to Cr-contaminated environments over multiple generations.

Histone methylation

Histone methylation can occur on different amino acid residues of histone proteins and can have varied effects on gene expression, depending on the specific residue and degree of methylation. Chromium exposure may lead to alterations in histone methylation marks associated with both transcriptional activation and repression, thereby modulating gene expression in response to stress (Wang *et al.*, 2021). Exposure to Cr(VI) can lead to changes in the levels of histone methylation marks in plant chromatin. Studies have demonstrated both increases and decreases in histone methylation at specific lysine residues in response to Cr stress. Histone methylation changes induced by Cr can modulate the expression of genes involved in stress response and detoxification pathways. Depending on the specific lysine residues and the degree of methylation (mono-, di-, or trimethylation), histone methylation can either activate or repress gene transcription. Cr-induced histone methylation alterations can affect the chromatin structure of stress-responsive genes, thereby influencing their accessibility to transcriptional regulators (Cui *et al.*, 2023). This regulation of chromatin structure can modulate the expression of genes involved in antioxidant defense, metal detoxification, and other adaptive responses to Cr toxicity. Histone methylation changes by Cr exposure may interact with other epigenetic modifications, such as histone acetylation and DNA methylation, to coordinate the plant's response to metal stress. Crosstalk between different epigenetic mechanisms can fine-tune the transcriptional response to Cr toxicity, shaping the overall stress tolerance of plants. Cr-induced alterations in histone methylation patterns may have transgenerational effects, impacting the phenotype and stress tolerance of subsequent plant generations. Epigenetic inheritance of histone modifications can contribute to the adaptation of plant populations to Cr-contaminated environments over multiple generations.

Histone phosphorylation

Histone phosphorylation is another histone modification that can be influenced by Cr exposure in plants. Phosphorylation of histone proteins is often associated with changes in chromatin structure and gene expression in response to environmental stimuli. Histone phosphorylation consists of the addition of a phosphate group, and thus of a negative charge, to serine, threonine, or tyrosine residues of the N-terminal tail of histones. Cr-induced histone phosphorylation changes may contribute to the activation or repression of stress-responsive genes involved in detoxification and defense mechanisms. Histone phosphorylation involves the addition of phosphate groups to specific serine, threonine, or tyrosine residues of histone proteins. This modification can affect chromatin structure by altering histone–DNA and histone–histone interactions. Depending on the location and context, histone phosphorylation can either promote or inhibit chromatin compaction, thus regulating gene accessibility and transcriptional activity (Wang *et al.*, 2015). Histone phosphorylation is often associated with transcriptional activation. Phosphorylation of histone H3 at serine 10 (H3S10ph) and serine 28 (H3S28ph), for example, has been linked to increased gene expression by facilitating the recruitment of transcriptional activators and chromatin remodeling complexes to target genes. Similarly, phosphorylation of histone H2AX (γH2AX) is involved in DNA damage response and repair mechanisms (Houben *et al.*, 2005).

Environmental stressors, including Cr contamination, can trigger changes in histone phosphorylation patterns in plants. Studies have shown that exposure to Cr can induce histone phosphorylation alterations, which may contribute to the plant's stress response mechanisms. For example, Cr-induced histone phosphorylation changes can modulate the expression of stress-responsive genes involved in antioxidant defense, detoxification pathways, and DNA repair processes. Histone phosphorylation can crosstalk with other histone modifications, such as acetylation, methylation, and ubiquitination, as well as with DNA methylation, to regulate chromatin structure and gene expression in response to environmental cues. The interplay between different epigenetic modifications adds complexity to the regulatory network controlling plant stress responses and adaptation to environmental challenges. Similar to other histone modifications, histone phosphorylation changes induced by Cr exposure may have transgenerational effects, influencing the phenotype and stress tolerance of subsequent plant generations. Epigenetic inheritance of histone phosphorylation marks can contribute to the long-term adaptation of plant populations to Cr-contaminated environments.

Histone ubiquitination and sumoylation

Chromium exposure may also impact histone ubiquitination and sumoylation, two other types of histone modifications involved in regulating gene expression and chromatin dynamics. These modifications can influence the accessibility of chromatin to transcription factors and RNA polymerase, thereby modulating gene transcription in response to Cr stress. Like histone ubiquitination, research on the effects of Cr exposure on histone sumoylation in plants is limited.

The complex interplay between ubiquitin code and chromatin remodeling dynamics is challenging to understand, yet the epigenetic regulatory networks of histone modification are essential for maintaining genomic stability, which in turn controls plants' responses to abiotic stress. When genotoxic stressors cause double-strand breaks (DSBs), which in turn cause DNA damage responses, lysine-rich histones – which are frequently ubiquitinated – play a crucial role (Mattiroli and Penengo, 2021). The most prevalent monoubiquitinated histone types, H2A and H2B, cause DNA damage to trigger an instant reaction. The H2A type histones K-63, K-27, and monoubiquitination of K13/15 are catalyzed by the recruitment of Ub ligases RNF-8 and RNF-168 at locations close to DSBs. After this, downstream repair machinery docks with ubiquitinated histones. *Ataxia telangiectasia* mutated (ATM)-mediated phosphorylation of H2AX (H2A variant) initiates RNF-168-mediated ubiquitination of H2A, which is then signaled by RNF-8 recruitment, which catalyzes ubiquitination of H1 linker histones (Aquila and Atanassov, 2020). Research reports that H2B monoubiquitination plays a role in rice and *Arabidopsis* ability to withstand abiotic stress. The *Arabidopsis* genome encodes three E2 conjugases, UBIQUITIN CARRIER PROTEIN1 (UBC1), UBC2, and UBC3, as well as two RING E3 ligases, HISTONE MONOUBIQUITINATION1 (HUB1) and HUB2. These are necessary for H2B ubiquitination (Cao *et al.*, 2008). Even studies in mammalian cells have demonstrated alterations in histone sumoylation patterns in response to environmental stressors, including heavy metals. Sumoylation of non-histone proteins has been implicated in plant responses to environmental stress, suggesting a potential role for histone sumoylation in mediating stress responses, including those induced by Cr exposure.

12.5 Epigenome Editing

Epigenome editing has shown promise in addressing heavy metal stress in plants by altering gene expression and enhancing stress tolerance without changing the underlying DNA sequence. Some key references explore this topic. Studies indicate that heavy metals like cadmium (Cd), lead (Pb), and zinc (Zn) can induce significant changes in the plant epigenome. For instance, in wheat, heavy metal stress has been linked to alterations in DNA methylation and histone modifications, which regulate the expression of metal detoxification transporters and other stress-related genes. This suggests that epigenetic modifications can play a crucial role in conferring heavy metal resistance in plants (Iqbal *et al.*, 2024). Advances in genome editing technologies, particularly CRISPR/Cas9, have facilitated precise epigenetic modifications. This approach can be used to target specific genes involved in stress responses, enhancing the plant's ability to tolerate heavy metals. Epigenetic editing has been used successfully in model plants like *Arabidopsis* to improve drought tolerance and could be applied similarly for heavy metal stress (Zhang *et al.*, 2019)

12.5.1 Applications and potential

The integration of whole-genome sequencing and epigenome editing can help identify and modify epigenetic loci associated with heavy metal stress tolerance. This can lead to the development of crops with improved resistance to environmental stresses. The potential of epigenome editing in creating desirable traits without genetic modification makes it a valuable tool for sustainable agriculture.

12.5.2 Challenges and considerations

Efficiently delivering epigenome editing tools to plant cells, particularly in diverse plant species, remains challenging. Methods include *Agrobacterium*-mediated transformation, particle bombardment, and protoplast transfection. Ensuring the specificity of epigenome editing to avoid off-target effects is crucial. Advances in guide RNA design and the use of high-fidelity dCas9 variants are improving specificity. Understanding the stability of induced epigenetic changes and their heritability across generations is essential for long-term agricultural applications. The use of epigenome editing in plants raises regulatory and ethical considerations, especially regarding genetically modified organisms (GMOs). Clear regulatory frameworks are needed to govern the use of these technologies.

12.6 Future Directions

The field of epigenome editing in plants is poised for significant advancements. Future research is likely to focus on improving delivery methods, enhancing the precision and efficiency of epigenetic modifications, and exploring the long-term effects and heritability of these changes. Combining epigenome editing with other biotechnological approaches, such as genomic selection and marker-assisted breeding, holds promise for developing crops with enhanced stress resilience and improved yields.

12.7 Conclusion

Chromium stress poses significant challenges to plant growth and development, primarily through the generation of reactive oxygen species that cause oxidative damage to cellular components. Epigenetic modifications, including DNA methylation, histone modifications, and non-coding RNAs, have emerged as critical regulators of plant responses to environmental stresses, including heavy metals like chromium. Understanding the epigenetic mechanisms involved in plant responses to Cr(VI) stress is essential for developing strategies to enhance plant tolerance and remediate contaminated environments. By manipulating epigenetic regulators or harnessing epigenetic variation in plant populations, researchers may be able to engineer crops with improved resilience to Cr(VI) toxicity and other environmental stresses.

References

Aina, R., Sgorbati, S., Santagostino, A., Labra, M., Ghiani, A. *et al.* (2004) Specific hypomethylation of DNA is induced by heavy metals in white clover and industrial hemp. *Physiologia Plantarum* 121(3), 472–480.

Ao, M., Chen, X., Deng, T., Sun, S., Tang, Y. *et al.* (2022) Chromium biogeochemical behaviour in soil-plant systems and remediation strategies: A critical review. *Journal of Hazardous Materials* 424(Pt A), 127233.

Aquila, L. and Atanassov, B.S. (2020) Regulation of histone ubiquitination in response to DNA double strand breaks. *Cells* 9(7), 1699.

Arora, H., Singh, R.K., Sharma, S., Sharma, N., Panchal, A. *et al.* (2022) DNA methylation dynamics in response to abiotic and pathogen stress in plants. *Plant Cell Reports* 41(10), 1931–1944.

Bao-yan, H., Yin, H., Feng, Y., Jin-shao, Y., Hai, P. *et al.* (2012) Improvement of chromium biosorption through protoplast electrofusion between *Candida tropicalis* and *Candida lipolytica*. *Journal of Central South University* 19(6), 1693–1701.

Bartels, A., Han, Q., Nair, P., Stacey, L., Gaynier, H. *et al.* (2018) Dynamic DNA methylation in plant growth and development. *International Journal of Molecular Sciences* 19(7), 2144.

Basit, F., Bhat, J.A., Dong, Z., Mou, Q., Zhu, X. *et al.* (2022) Chromium toxicity induced oxidative damage in two rice cultivars and its mitigation through external supplementation of brassinosteroids and spermine. *Chemosphere* 302, 134423.

Bautista, O.V., Fischer, G. and Cárdenas, J.F. (2013) Cadmium and chromium effects on seed germination and root elongation in lettuce, spinach and Swiss chard. *Agronomía Colombiana* 31, 48–57.

Bilichak, A., Ilnystkyy, Y., Hollunder, J. and Kovalchuk, I. (2012) The progeny of *Arabidopsis thaliana* plants exposed to salt exhibit changes in DNA methylation, histone modifications and gene expression. *PloS ONE* 7(1), e30515.

Boycheva, I., Vassileva, V. and Iantcheva, A. (2014) Histone acetyltransferases in plant development and plasticity. *Current Genomics* 15(1), 28–37. DOI: 10.2174/138920291501140306112742.

Bukhari, S.A., Shang, S., Zhang, M., Zheng, W., Zhang, G. *et al.* (2015) Genome-wide identification of chromium stress-responsive micro RNAs and their target genes in tobacco (*Nicotiana tabacum*) roots. *Environmental Toxicology and Chemistry* 34(11), 2573–2582.

Cao, Y., Dai, Y., Cui, S. and Ma, L. (2008) Histone H2B monoubiquitination in the chromatin of flowering locus C regulates flowering time in *Arabidopsis*. *The Plant Cell* 20(10), 2586–2602.

Chow, Y.N., Lee, L.K., Zakaria, N.A. and Foo, K.Y. (2018) Phytotoxic effects of trivalent chromium-enriched water irrigation in *Vigna unguiculata* seedling. *Journal of Cleaner Production* 202, 101–108.

Cong, W., Miao, Y., Xu, L., Zhang, Y., Yuan, C. *et al.* (2019) Transgenerational memory of gene expression changes induced by heavy metal stress in rice (*Oryza sativa* L.). *BMC Plant Biology* 19(1), 282.

Cui, G., Dong, Q., Gai, K. and Qi, S. (2023) Chromatin dynamics: Chromatin remodeler, epigenetic modification and diseases. In: Huang, T. (ed.) *Epigenetics – Regulation and New Perspectives*. Intech Open.

Hafiz, M.F. and Ma, L. (2021) Effect of chromium on seed germination, early seedling growth and chromium accumulation in tomato genotypes. *Acta Physiologiae Plantarum* 43(7), 100.

Houben, A., Demidov, D., Rutten, T. and Scheidtmann, K.H. (2005) Novel phosphorylation of histone H3 at threonine 11 that temporally correlates with condensation of mitotic and meiotic chromosomes in plant cells. *Cytogenetic and Genome Research* 109(1–3), 148–155.

Hou, J., Liu, G.-N., Xue, W., Fu, W.-J., Liang, B.-C. *et al.* (2014) Seed germination, root elongation, root-tip mitosis, and micronucleus induction of five crop plants exposed to chromium in fluvo-aquic soil. *Environmental Toxicology and Chemistry* 33(3), 671–676. DOI: 10.1002/etc.2489.

Hu, S., Gu, H., Cui, C. and Ji, R. (2016) Toxicity of combined chromium(VI) and phenanthrene pollution on the seed germination, stem lengths, and fresh weights of higher plants. *Environmental Science and Pollution Research* 23(15), 15227–15235.

Iqbal, B., Ahmad, N., Li, G., Jalal, A., Khan, A.R. *et al.* (2024) Unlocking plant resilience: advanced epigenetic strategies against heavy metal and metalloid stress. *Plant Science* 349, 112265. DOI: 10.1016/j.plantsci.2024.112265.

Jaison, S. and Muthukumar, T. (2017) Chromium accumulation in medicinal plants growing naturally on tannery contaminated and non-contaminated soils. *Biological Trace Element Research* 175(1), 223–235.

Jing, H., Liu, G.-N., Xue, W., Fu, W.-J., Liang, B.-C. *et al.* (2014) Seed germination, root elongation, root-tip mitosis, and micronucleus induction of five crop plants exposed to chromium in fluvo-aquic soil. *Environmental Toxicology and Chemistry* 57, 124–128.

Kumar, V., Suryakant, P., Kumar, S. and Kumar, N. (2016) Effect of chromium toxicity on plants: A review. *Agriways* 4, 107–120.

Labra, M., Grassi, F., Imazio, S., Di Fabio, T., Citterio, S. *et al.* (2004) Genetic and DNA-methylation changes induced by potassium dichromate in *Brassica napus* l. *Chemosphere* 54(8), 1049–1058. DOI: 10.1016/j.chemosphere.2003.10.024.

Liu, Y.-M., Liu, D.-Y., Zhang, W., Chen, X.-X., Zhao, Q.-Y. *et al.* (2020) Health risk assessment of heavy metals (Zn, Cu, Cd, Pb, As and Cr) in wheat grain receiving repeated Zn fertilizers. *Environmental Pollution* 257, 113581.

López-Luna, J., González-Chávez, M.C., Esparza-García, F.J. and Rodríguez-Vázquez, R. (2009) Toxicity assessment of soil amended with tannery sludge, trivalent chromium and hexavalent chromium, using wheat, oat and sorghum plants. *Journal of Hazardous Materials* 163(2–3), 829–834.

Mattiroli, F. and Penengo, L. (2021) Histone ubiquitination: An integrative signaling platform in genome stability. *Trends in Genetics* 37(6), 566–581.

Oliveira, H. (2012) Chromium as an environmental pollutant: Insights on induced plant toxicity. *Journal of Botany* 1–8.

Peng, H. and Zhang, J. (2009) Plant genomic DNA methylation in response to stresses: Potential applications and challenges in plant breeding. *Progress in Natural Science* 19(9), 1037–1045.

Peralta-Videa, J.R., Lopez, M.L., Narayan, M., Saupe, G. and Gardea-Torresdey, J. (2009) The biochemistry of environmental heavy metal uptake by plants: Implications for the food chain. *The International Journal of Biochemistry & Cell Biology* 41(8–9), 1665–1677.

Radwan, M.A. and Salama, A.K. (2006) Market basket survey for some heavy metals in Egyptian fruits and vegetables. *Food and Chemical Toxicology* 44(8), 1273–1278.

Rath, A. and Das, A.B. (2021) Chromium stress induced oxidative burst in *Vigna mungo* (L.) Hepper: Physio-molecular and antioxidative enzymes regulation in cellular homeostasis. *Physiology and Molecular Biology of Plants* 27(2), 265–279.

Seneviratne, M., Rajakaruna, N., Rizwan, M., Madawala, H.M.S.P., Ok, Y.S. *et al.* (2019) Heavy metal-induced oxidative stress on seed germination and seedling development: A critical review. *Environmental Geochemistry and Health* 41, 1813–1831.

Shahid, M., Pourrut, B., Dumat, C., Nadeem, M., Aslam, M. *et al.* (2014) Heavy-metal-induced reactive oxygen species: Phytotoxicity and physicochemical changes in plants. *Reviews of Environmental Contamination and Toxicology* 232, 1–44. DOI: 10.1007/978-3-319-06746-9_1.

Shahid, M., Shamshad, S., Rafiq, M., Khalid, S., Bibi, I. *et al.* (2017) Chromium speciation, bioavailability, uptake, toxicity and detoxification in soil-plant system: A review. *Chemosphere* 178, 513–533.

Shanke, A.K., Cervantes, C., Loza-Tavera, H. and Avudainayagam, S. (2005) Chromium toxicity in plants. *Environment International* 31(5), 739–753.

Singh, D. and Sharma, N. (2017) Effect of chromium on seed germination and seedling growth of green gram (*Phaseols aureus* L.) and chickpea (*Cicer arietinum* L). *International Journal of Applied and Natural Sciences* 6, 37–46.

Singh, H.P., Mahajan, P., Kaur, S., Batish, D.R. and Kohli, R.K. (2013) Chromium toxicity and tolerance in plants. *Environmental Chemistry Letters* 11(3), 229–254.

Soliman, M.M., Hesselberg, T., Mohamed, A.A. and Renault, D. (2022) Trophic transfer of heavy metals along a pollution gradient in a terrestrial agro-industrial food web. *Geoderma* 413, 115748.

Srivastava, D., Tiwari, M., Dutta, P., Singh, P., Chawda, K. *et al.* (2021) Chromium stress in plants: Toxicity, tolerance, and phytoremediation. *Sustainability* 13(9), 4629.

Sun, M., Yang, Z., Liu, L. and Duan, L. (2022) DNA methylation in plant responses and adaption to abiotic stresses. *International Journal of Molecular Sciences* 23(13), 6910.

Tang, M., Yue, J., Huang, Z., Hu, Y., Li, Z. *et al.* (2022) Physiological and DNA methylation analysis provides epigenetic insights into chromium tolerance in kenaf. *Environmental and Experimental Botany* 194, 104684.

Tang, M., Li, R. and Chen, P. (2023) Exogenous glutathione can alleviate chromium toxicity in kenaf by activating antioxidant system and regulating DNA methylation. *Chemosphere* 337, 139305.

Ullah, R., Hadi, F., Ahmad, S., Jan, A.U. and Rongliang, Q. (2019) Phytoremediation of lead and chromium contaminated soil improves with the endogenous phenolics and proline production in *Parthenium, Cannabis, Euphorbia, and Rumex* species. *Water, Air, & Soil Pollution* 230(2), 40.

Verdone, L., Agricola, E., Caserta, M. and Mauro, E.D. (2006) Histone acetylation in gene regulation. *Briefings in Functional Genomics and Proteomics* 5, 209–221.

Wakeel, A., Xu, M. and Gan, Y. (2020) Chromium-induced reactive oxygen species accumulation by altering the enzymatic antioxidant system and associated cytotoxic, genotoxic, ultrastructural, and photosynthetic changes in plants. *International Journal of Molecular Sciences* 21(3), 728.

Wang, Q., Liu, P., Jing, H., Zhou, X.F., Zhao, B. *et al.* (2021) JMJ27-mediated histone H3K9 demethylation positively regulates drought-stress responses in *Arabidopsis*. *The New Phytologist* 232(1), 221–236.

Wang, Z., Casas-Mollano, J.A., Xu, J., Riethoven, J.-J.M., Zhang, C. *et al.* (2015) Osmotic stress induces phosphorylation of histone H3 at threonine 3 in pericentromeric regions of *Arabidopsis thaliana*. *Proceedings of the National Academy of Sciences of the United States of America* 112(27), 8487–8492.

Yang, Y., Huang, J., Sun, Q., Wang, J., Huang, L. *et al.* (2022) MicroRNAs: Key players in plant response to metal toxicity. *International Journal of Molecular Sciences* 23(15), 8642.

Zeid, I.M. (2001) Responses of *Phaseolus vulgaris* to chromium and cobalt treatments. *Biologia Plantarum* 44(1), 111–115.

Zhang, P., Fan, Y., Sun, X., Chen, L., Terzaghi, W. *et al.* (2019) A large-scale circular RNA profiling reveals universal molecular mechanisms responsive to drought stress in maize and *Arabidopsis*. *The Plant Journal* 98(4), 697–713.

13 Plant Long Non-Coding RNAs (lncRNAs): An Updated Overview with Emphasis on the Regulation of Abiotic Stress Responses

Heena Gulia[1], Neha Verma[2], Shiv Kumar Giri[3], Gulab Singh[4] and Anil Kumar[1]*

[1]Centre for Medical Biotechnology, Maharshi Dayanand University, Rohtak, Haryana, India; [2]Department of Botany, Baba Mastnath University, Rohtak, Haryana, India; [3]Department of Biotechnology, Maharaja Agrasen University, Solan, Himachal Pradesh, India; [4]Department of Bio-Nanotechnology, Chaudhary Charan Singh Haryana Agricultural University, Hisar, Haryana, India

Abstract

Long non-coding RNAs (lncRNAs) in plants play a crucial role in mitigating various abiotic stresses, contributing to the better adaptation and survival of plants in challenging environments. These lncRNAs are involved in diverse mechanisms, including stress perception, epigenetic modifications, and gene regulation, that allow them to regulate various physiological processes, including stress signaling, osmotic regulation, ion homeostasis, and antioxidant defense mechanisms. Research indicates that lncRNAs can regulate stress-responsive genes, acting as molecular switches to modulate stress tolerance. Additionally, they participate in intricate regulatory networks, such as natural antisense transcripts (NATs), to fine-tune stress responses. Therefore, the characterization of stress-responsive lncRNAs and elucidation of their regulatory pathways hold immense potential for developing novel strategies to enhance plant resilience to environmental stresses. Furthermore, understanding the functions and regulatory mechanisms of plant lncRNAs provides insights into enhancing stress tolerance and crop productivity in the face of changing environmental conditions.

Keywords: Long non-coding RNAs, Abiotic stress, Gene regulation, Epigenetics

13.1 Introduction

Due to the constantly growing global population and changing environmental conditions, agriculture has struggled to ensure food security and sustainable crop production for decades. Various abiotic stresses, including drought, salt, temperature variations, and heavy metal toxicity, significantly affect plant health. Modifying biological, molecular, and physiological pathways helps plants tolerate such challenges, but the scientific community must find new strategies to make plants more resilient. In recent years, advancements in molecular biology and sequencing techniques, along with progress in *in-silico*-based approaches, have allowed the

*Corresponding author: anil.cmbt@mdurohtak.ac.in

(ed. Jen-Tsung Chen)
DOI: 10.1079/9781800626102.0013

identification of various forms of non-coding RNA (ncRNAs) in plants (Magar *et al.*, 2023). Based on their origin, biogenesis, and method of action, ncRNAs have been classified across organisms. Briefly, ncRNAs are characterized as short (< 200 nucleotides) or long (> 200 nucleotides) RNAs (Yang *et al.*, 2023). RNA polymerase II encodes major types of lncRNAs, which have open reading frames (ORFs) of less than 100 amino acids and do not code for proteins. Post-transcriptional processing, promoter epigenetic marks, and secondary and tertiary structure formation are similar to mRNA (Nejat and Mantri, 2018).

Over the years, various studies have deciphered how plants respond to the various elements of their agro-environment. Exposure of plants to one or more stressors may induce the expression of related genes and their intricate network of interactions. Plants can rebuild cell homeostasis via many gene regulatory systems to survive in harsh conditions. Significant efforts have been made to uncover stress-responsive genes and protein interaction networks. Scientific investigations have highlighted how gene expression and other mechanisms precisely regulate abiotic stress-associated responses at the transcriptional and post-transcriptional levels (Waititu *et al.*, 2020). Despite the focus on protein-coding gene regulation, ncRNAs, especially lncRNA, play crucial roles in modulating gene expression at transcriptional and post-transcriptional levels.

13.2 Non-Coding RNA Classification

The two main categories of ncRNA transcripts are regulatory and housekeeping ncRNAs. Ribosomal and cellular functions are largely mediated by housekeeping ncRNAs like tRNAs, snRNAs, rRNAs, and snoRNAs. In contrast, regulatory ncRNAs include lncRNAs, piwi-interacting RNAs (piRNAs), microRNAs (miRNAs), and short-interfering RNAs. Although transcribed from DNA, regulatory non-coding RNAs cannot be translated into proteins (Waititu *et al.*, 2020). Depending on their position and proximity to protein-coding genes, these can be categorized as intronic (incRNAs), long intergenic (lincRNAs), natural antisense (NATs), or circular lncRNAs (circRNAs) (Sun *et al.*, 2018; Jha *et al.*, 2020b). Long intergenic RNAs are produced from intergenic regions, while intronic regions transcribe intronic RNAs with weakly spliced, polyadenylated tissue-specific expression, and carry out *trans*-regulatory functions. NATs are produced from complementary sense-coding DNA strands and act as *cis*- and *trans*-regulators (Wu *et al.*, 2017).

13.3 Long Non-Coding RNA and Abiotic Stress

Abiotic stressors like drought, salinity, cold, heat, heavy metals, light intensity, etc. are the main negative constraints on agricultural productivity. These affect plant health physiologically, morphologically, and biochemically. Plants use a complex interaction network of miRNAs, protein-coding genes, and lncRNAs to combat the negative impacts of such stressors (Fig. 13.1) (Table 13.1).

13.3.1 Drought stress

Drought stands as a prominent ecological challenge worldwide, impacting various regions due to erratic rainfall patterns. This phenomenon exacerbates drought stress, affecting plant development and threatening global food security (Jha *et al.*, 2020a). Under drought stress, plants use a variety of physiological, biochemical, and molecular mechanisms, including sophisticated signaling pathways (Lamaoui *et al.*, 2018). Recent advances in molecular biology have unveiled the genetic and epigenetically regulated networks and molecular signaling cascades governing drought resistance in plants (Gupta *et al.*, 2017; Jha *et al.*, 2020b). Such responses to drought rely on a multitude of biochemical processes influenced by factors such as the nature and duration of stress, plant developmental stages, and diurnal rhythms. Studies have identified that, along with numerous functional candidate genes, certain lncRNA-mRNA pairs also exhibit *cis*- or *trans*-acting activities. Moreover, variable expression patterns of the genes related to drought stress have also revealed a significant correlation between plant

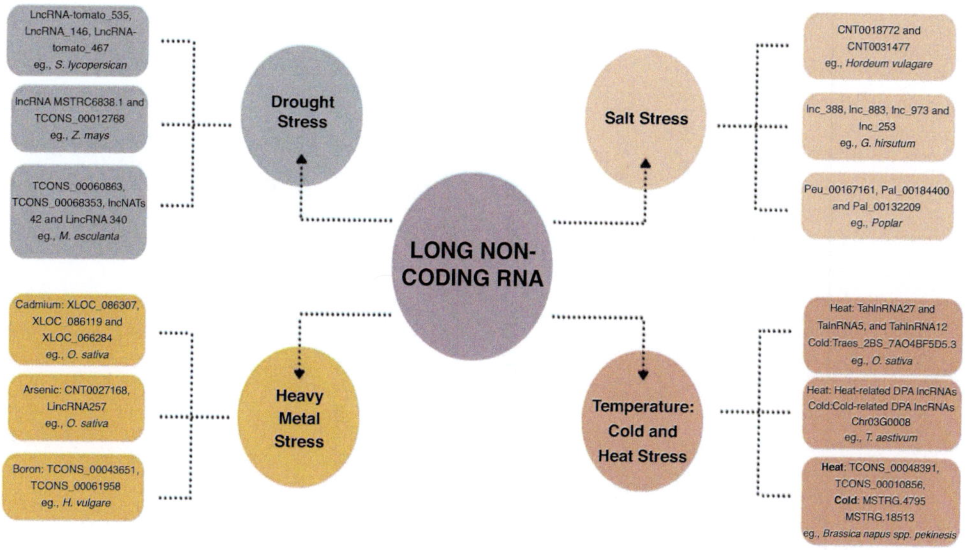

Fig. 13.1. Long non-coding RNAs (lncRNAs) under various abiotic stresses. Author's own image.

hormone signaling pathways. Comprehensive genome-wide transcriptome analyses have identified several stress-responsive lncRNA in *Arabidopsis*, *S. italica*, *Populus*, *Z. mays*, *O. sativa*, *Manihot esculenta*, and *Panicum virgatum*. A study has demonstrated that lncRNA acts in *cis*-manner on targeted hormone-related genes. For instance, lncRNAs (XLOC_115463 and XLOC-063105) may increase the expression of adjacent ethylene metabolic pathway genes (Lu *et al.*, 2016). Chung *et al.* (2016) found 98 lncRNAs and their antisense transcript in drought-stressed leaves. Deep sequencing analysis revealed that lncRNAs such as lincRNAs and lncNATs express differentially under drought stress in foxtail millet (Nejat and Mantri, 2018). Furthermore, a study has identified a total of 521 drought-responsive lncRNAs in tomato plants (Eom *et al.*, 2019). Melatonin and polyethylene glycol may influence the differential expression of drought-responsive lncRNA exerting control over abscisic acid (ABA) and ethylene metabolism, redox homeostasis, and calcium signaling, and modulate the expression network (Ding *et al.*, 2019a; b). Zhang *et al.* (2019a) revealed that *Arabidopsis* and maize harbor 1354 and 2174 high-confidence circular RNAs (circRNAs), respectively. Notably, two circRNAs were found to be intricately linked to specific genes: *GRMZM2G032852* encoding cytokinin oxidase/dehydrogenase (CKX) and *GRMZM5G817173* encoding calcium-dependent protein kinase (CPK). Both were generated via back-splicing between exons, resulting in allele insertion and deletion within the maize panel. Maize plants carrying insertion alleles exhibited significantly elevated circCPK expression levels and increased survival rates under drought and normal conditions. Dong *et al.* (2022) suggested that, under drought conditions, drought-induced intergenic lncRNA (DIR) and its interacting proteins target mRNA and genes involved in multiple metabolic pathways in cassava. Transcriptional analysis revealed that lncRNA (MSTRG.28732.3) interaction with Mir171 may increase rice drought resistance by modulating *Os06g0105350*, *Os02g0662700*, and *Os02g0663100* expression (Yang *et al.*, 2022). Drought stress upregulated Sly-miR169c and miR396b and probably downregulated lncRNA-tomato-503 and 535. A study revealed that sly-miR169c and miR396b may function as eTMs under drought influence (Patra *et al.*, 2023).

Zou *et al.* (2023) reported that 386 lncRNAs expressed differentially upon encountering drought stress in sugar beet cultivars. The lncRNAs TCONS_00055787 and TCONS_00038334 were upregulated (6000-fold) and downregulated (18,000-fold) during drought stress. Forty-two were projected

Table 13.1. Expression of long non-coding RNAs (lncRNAs) under various abiotic stresses. Author's own table.

Name of abiotic stress	Plant/crop	No. of lncRNAs identified	No. of lncRNAs expressed under stress	Platform and technique used for lncRNAs identification and their function	References
Drought	*Oryza sativa*	98	98	Illumina HiSeq 2500, qRT-PCR	Chung *et al*., 2016
	Oryza sativa *Cicer arietinum*	3714 3457	1010 1503	RT-qPCR, PLncPRO	Singh *et al*., 2017
	Triticum turgidum	63,773 (Kiziltan) 61,823 (TR39477) 43,932 (TTD-22)	59,110 57,944 40,858	Illumina HiSeq. 2000, qRT-PCR	Cagirici *et al*., 2017
	Pyrus betulifolia	14,478	251	Illumina HiSeq 4000, CNCI, CPC, qRT-PCR	Wang *et al*., 2018a
	Panicum virgatum L	16,551	441	HiSeq2500, qRT-PCR	Zhang *et al*., 2018
	Zea mays	3488	1535	Illumina HiSeq 2500, qRT-PCR	Pang *et al*., 2019
	Cleistogenes songorica	3397	468	HiSeq2500, CPC, CNCI, CPATqRT-PCR	Yan *et al*., 2019
	Manihot esculenta	833	124	Hiseq 4000, qRT-PCR, CNCI, CPC	Ding *et al*., 2019a
		1405	185	qRT-PCR	Ding *et al*., 2019b
		1379	194	qRT-PCR	Wu *et al*., 2019
	Brassica juncea	7613	1614	*In-silico*	Bhatia *et al*., 2020
	Oryza sativa	191	----	qRT-PCR	Yang *et al*., 2022
	Beta vulgaris	32,017	386	qRT-PCR	Zou *et al*., 2023

Cold	*Manihot esculenta*	682	318	HiSeq 2500, qRT-PCR, CNCI, CPC	Li *et al.*, 2017
	Musa itinerans	12,462	20	Illumina HiSeqTM 4000, qPCR, CPC	Liu *et al.*, 2018
	Arabidopsis thaliana	379	135	Illumina HiSeq 2500, RT-qPCR	Calixto *et al.*, 2019
	Brassica rapa	2088	549	Illumina HiSeqTM 2000, qPCR	Shea *et al.*, 2019
	Vitis vinifera	2088	466	HiSeq 2500, qRT-PCR, CNCI, CPC	Wang *et al.*, 2019b
	Oryza sativa	1485	566	Illumina HiSeq 2500 platform, qRT-PCR	Leng *et al.*, 2020
	Medicago truncatula	24,368	983/1288	Illumina HiSeq 4000, Q-PCR	Zhao *et al.*, 2020
	Sparassis latifolia	1003	495/483	Illumina HiseqX platform	Weng *et al.*, 2024
Heat	*Brassica rapa*	10,001	2236	Illumina HiSeq™ 2000 qRT-PCR, CPC	Song *et al.*, 2016
	Brassica rapa ssp. *chinensis*	4594	1686	Illumina Hiseq. 2500, qRT-PCR CPC, CNCI	Wang *et al.*, 2019a
	Brassica juncea	7613	1614	qRT-PCR	Bhatia *et al.*, 2020
	Cicer arietinum L.	894	----	qRT-PCR	Bhogireddy *et al.*, 2023
	Oryza sativa	2743	231	qRT-PCR, Illumina Hiseq. platform	Zhang *et al.*, 2022
	Triticum aestivum	11,054	5482	CPC, CNCI	Babaei *et al.*, 2024

Continued

Table 13.1. Continued

Name of abiotic stress	Plant/crop	No. of lncRNAs identified	No. of lncRNAs expressed under stress	Platform and technique used for lncRNAs identification and their function	References
Salinity	*Cicer arietinum* L.	3457	13	RT-qPCR, PLncPRO	Singh *et al*., 2017
	Gossypium hirsutum	1117	44	Illumina HiSeq 4000, RT-qPCR	Deng *et al*., 2018
	Camellia sinensis	16,452	172	qRT-PCR	Wan *et al*., 2020
	Cicer arietinum L.	3452	4446	RT-qPCR	Kumar *et al*., 2021
	Oryza sativa	1626/2208	30 (root) 81(leaves)	qRT-PCR	Jain *et al*., 2021
		75 l	75	qRT-PCR, CPC	Rehman *et al*., 2023
	Zea mays	12,817	742	Illumina HiSeq 4000	Liu *et al*., 2022b
	Camellia sinensis	16,452	172	qRT-PCR	Wan *et al*., 2020
Heavy metals	*Triticum aestivum*	44,698	19 (precursor) 1047 (target mimic)	qRT-PCR	Shumayla *et al*., 2017
Cadmium	*Oryza sativa*	3558	144	Illumina HiSeq 2000, CPC, RT-qPCR	Chen *et al*., 2018
	Betula platyphylla	30	30	Illumina HiSeq 2500	Wen *et al*., 2020
	Hordeum vulgare L.	9937	195	HiSeq RNA-sequencing	Zhou *et al*., 2023
	Oryza sativa	---	6	Illumina sequencing, RT-qPCR	Qiu *et al*., 2024
Boron	*Zea mays*	48,345	1710	Illumina MiSeq, RT-qPCR, AgriGO	Huanca-Mamani *et al*., 2018
Copper	*Citrus junos*	---	169	qRT-PCR	Fu *et al*., 2019
Lead	*Populus*	1171	226	qRT-PCR	Chen *et al*., 2022
Aluminium	*Medicago truncatula*	3284	515	Illumina NovaSeq 6000 platform, qRT-PCR	Gui *et al*., 2022

qRT-PCR, real-time quantitative polymerase chain reaction; RT-qPCR, Reverse transcription quantitative PCR; CPC, Coding Potential Calculator; CNCI, Coding Non-Coding Index; CPAT, Coding Potential Assessment Tool; PLncPRO, Plant Long Non-Coding RNA Prediction by Random forests;

target mimics, while the rest regulated secondary metabolite production and other metabolic processes. In *Arabidopsis*, lncRNA (DANA2) increases JMJ29 expression to positively modulate drought response (Zhang *et al.*, 2023). Drought-stressed and irrigated white mulberry plants were used to analyze their transcriptomes to determine drought resilience. The study identified 82 differently expressed mRNAs and 674 lncRNAs. Most lncRNAs target secondary metabolite production genes. It was also hypothesized that differentially expressed miRNA–lncRNA–mRNA regulatory networks interact with drought-stressed gene regulatory networks (Ackah *et al.*, 2024). The interaction between drought-induced long intergenic non-coding RNA (*DANA1*) and *DANA1*-Interacting Protein 1 (DIP1) enhances drought resistance in *Arabidopsis*. It was revealed that interaction between DNA1 and D1P1 stimulates the binding of histone deacetylase (HDA9) to *CYP707A1* and *CYP707A2* loci. Further, it was stated that mutations in DANA1 may elevate the expression of drought resistance-associated genes by promoting acetylation at CYP loci (Cai *et al.*, 2024). Transcriptome and gene expression analysis of drought stress-related ncRNAs were performed in stone apple. In total, 16,811 potential lncRNAs, 19 miRNAs, and six functional eTMs were identified. Further interaction between various ncRNAs and mRNA elucidated lncRNA/miRNA-mediated regulation of the expression of genes linked to drought stress (Tribhuvan *et al.*, 2024).

13.3.2 Salinity stress

Salinity stress is currently seen as a significant environmental challenge to agricultural productivity. Numerous studies have documented the regulatory role of lncRNAs against salt stress in a variety of plant species. Wang *et al.* (2015) identified two lncRNA, 7361 and 787, from the leaf and root samples of salt-stressed *Medicago truncatula*. Furthermore, results from expression profiling showed altered lncRNA expression in both roots and leaves experiencing salt stress. More specifically, osmotic (59.6%) and salt stress (63.7%) in leaves shared a sizeable fraction of stress-responsive lncRNAs. However, these percentages dropped to 47.0 and 33.2%, respectively, in roots. A study by Deng *et al.* (2018) revealed the presence of 1117 distinct lncRNAs in upland cotton under salt stress, 44 of which were differently expressed. Gene ontology analysis demonstrated that lincRNAs target stress-related protein-coding genes via *cis*-acting regulatory mechanisms. In addition, real-time quantitative PCR verified the expression of salt-responsive lincRNAs, suggesting their regulatory role in the expression of stress tolerance genes. Long intergenic RNAs may also operate as miRNA eTMs, influencing salt-induced stress responses. Thirteen protein-coding genes vital to signal transduction, transcription, and oxidation/reduction were regulated by lncRNAs (TCONS_00072494, TCONS_00074334, and TCONS_00075203). For instance, by controlling Medtr3g069280 expression, TCONS_00047650 may have an impact on signal transduction networks by increasing the expression of genes linked to salt stress. The expression pattern of the Pyran Aldehyde Reductase (*PLR*) gene was evaluated in barley cultivars exposed to 150 mM salt. The results demonstrated altered expression of genes, where AK372815 PLR was downregulated and AK370814 was upregulated during germination. About 2815 lncRNAs were identified in *S. polyrhiza*, of which 185 expressed differentially under salinity stress. Forty-two lncRNA–mRNA pairings were linked to *cis*-acting regulation, mainly in the cell wall and carbon metabolism, but co-expression analysis indicated that *trans*-acting lncRNAs regulate genes involved in different metabolic processes (Fu *et al.*, 2020).

Salt stress may pose serious threats to tea plants (*Camellia sinensis*) affecting their cultivation. High-throughput sequencing identified 16,452 distinct lncRNAs, of which 172 were differentially expressed (DE-lncRNAs) under salt stress. Further, DE-lncRNAs were engaged in pathways like calcium signaling and galactinol synthase (GOLS). The lncRNA (MSTRG.139242.1), by interacting with Ca^{2+}ATPase, may regulate responses to salt stress. LncRNAs are important regulators in tea plant stress responses, as evidenced by the prediction that 12 lncRNAs would resemble the targets of 17 miRNAs linked to salt stress (Wan *et al.*, 2020). In a further investigation, Sun *et al.* (2020a) identified various lncRNA types in salt-sensitive (Roma strains) and salt-tolerant (M-81E) varieties of sorghum. In M-81E, lncRNA2846, lncRNA11310, lncRNA13472, lncRNA14798, and lncRNA26929 may function as ceRNAs, regulating the target genes involved in ion transport, transcriptional regulation, and protein modification.

Liu *et al.* (2022b) studied the response of two maize-inbred lines to varying salt concentrations. Approximately 12,817 lncRNAs and 8320 new mRNAs were identified in the workflow analysis of maize seedling roots, of which 7835 mRNAs and 742 lncRNAs were expressed under salinity stress. Of these, 88 lncRNAs were linked to 41 *cis*- and 81 *trans*-target mRNAs, including transcripts connected to seven transcription factor families. Further, co-expression network analysis classified the transcripts into 28 modules, with the salt-tolerant module revealing an interaction network involving lncRNAs, TFs, and functional transcripts (FTs). The transgenic *Arabidopsis thaliana* plants overexpressing LNCRNA973 exhibited enhanced germination percentage, fresh weight, and root length under salt stress compared to the wild type. Conversely, silencing the *LNCRNA973* gene expression via viral induction resulted in wilted leaves under salt stress, indicating reduced salt stress tolerance compared to the wild type. LNCRNA973 is known to regulate the expression of genes involved in ROS scavenging, and salt-responsive genes (Li *et al.*, 2022c). When compared with the wild type, the transgenic *Arabidopsis thaliana* plants overexpressing LNCRNA973 showed improved germination percentage, fresh weight, and root length under salt stress whereas silencing the LNCRNA973 expression demonstrated a lower salt tolerance in the wild type (Liu *et al.*, 2022a).

Stress from alkaline salts can hinder plant growth and jeopardize crop harvests. Alkaline-salt responsive lncRNAs in two rice varieties – tolerant (WD20342) and sensitive (Caidao) – were identified by transcriptome analysis. The construction of a ceRNA network revealed that mRNAs, miRNAs, and lncRNAs were expressed differentially during stressful conditions. The variable expression was observed in 65 lncRNAs, 34 miRNAs, and 1549 mRNAs of the tolerant variety and 75 lncRNAs, 34 miRNAs, and 1725 mRNAs of the sensitive variety. Key lncRNA-miRNA-mRNA triplets (NONOSAT000455-osa_miR5809bLOC_Os11g01210) were found by the ceRNA network, resulting in salt stress tolerance (Rehman *et al.*, 2023). A study identified approximately 34,363 lncRNAs by analyzing the divergence of conserved lncRNAs in salt-tolerant and salt-sensitive poplar species. The conserved lncRNAs were expressed as differentially sensitive ones when compared with expression in salt-tolerant poplars. Elevated expression of antisense lncRNA (lncERF024) improved the ability of *P. alba* var. *pyramidalis* to withstand salt stress (Li *et al.*, 2023a). Yu *et al.* (2023) identified 172,986 mRNAs and 117,677 lncRNAs in two alfalfa cultivars exposed to 0 and 150 mM of NaCl. Differential expression analysis of several genes and lncRNAs related to osmotic and ionic stress was found to be linked to salinity tolerance, especially in the tolerant cultivar (GN5). Tobacco roots and leaves under salt stress express 5831 lncRNA, of which 2428 are differentially expressed. As per KEGG pathway analysis, the metabolism of starch and sucrose is linked to roots, whereas cysteine and methionine are linked to leaves. Four modules that were highly connected with salt stress were found by co-expression network analysis, and salt-associated miRNAs such as miR156, miR169, and miR396 were incorporated in a lncRNA-miRNA-mRNA network (Li *et al.*, 2023b).

An investigation was conducted to determine the importance of LAL, a lncRNA, for salt stress tolerance in *Medicago truncatula*. On encountering salinity stress, *MtLHCB* genes, lncRNA (LAL), exhibit a differential expression pattern. Further analysis revealed that LAL overexpression decreases salt tolerance, whereas mutants lacking LAL had increased tolerance. Transient co-expression in tobacco leaves indicates that LAL modulates MtLHCB1.4 expression (Zhao *et al.*, 2024). Transcriptome sequencing of cotton plants under normal and salt-stressed environments predicted 519 lncRNAs and 75,376 mRNAs. The two lncRNAs, XLOC_043671 and XLOC_034831, were key to the regulation of salinity stress (Feng *et al.*, 2024). Similarly, lncRNA (TRABA) was expressed significantly in a salt-tolerant cotton variety (DM37) and repressed the expression of *GhBGLU24-A*, which was highly expressive in a salt-sensitive variety (TM-1) but weakly expressed in DM37. GhBGLU24-A produces an ER-localized β-glucosidase that interacts with GhRUBL, to reduce salt and ER stress tolerance.

In one more study, Cui *et al.* (2024) reported that TRABA modulation of GhBGLU24-A offers valuable perspectives for enhancing crop tolerance to ER and salt stress. By increasing antioxidant activity and causing stomatal closure, the salt-responsive nuclear lncRNA (BplncSIR1) in *Betula platyphylla* (birch) promotes salt tolerance and plant growth. It was discovered that BplncSIR1 attaches itself to the BpNAC2 promoter, causing the gene to express. In birch plants, overexpression of BplncSIR1 and BpNAC2 improves growth and salt tolerance (Jia *et al.*, 2024).

13.3.3 Cold stress

Cold stress is considered a significant abiotic challenge to plant health. When encountering low temperatures, plants use sophisticated regulatory networks incorporating C-repeat binding factors (CBFs), cold-regulated genes (CORs), and many unique regulatory microRNAs. Recent genetic and genomic advancements have identified many candidate genes and quantitative trait loci (QTLs) associated with cold tolerance (Jha *et al.*, 2017). Various plant species, including *Arabidopsis*, cassava, *Brassica rapa*, banana, grapevine, and *Brachypodium*, have revealed that lncRNAs play an essential role in adaptation to cold stress. In *Arabidopsis*, RNAPII generates antisense-strand lncRNA (SVALKA) that transcriptionally represses CBF1 to facilitate cold adaptation. In cassava, 2127 siRNA reads aligned convergently to the overlapping area of two cold-responsive lncNAT pairs, viz., lncNAT14-Manes.05G207400 and lncNAT179-Manes.14G040500. Interestingly, siRNAs from the lncNAT14 gene have been expressed more during chilling than in control conditions (Li *et al.*, 2017). *In silico* and deep sequencing analysis identified 239 lncRNA in cold-stressed tomato plants, of which 138 were upregulated and 101 were downregulated. A gene ontology study identified 149 lncRNAs regulating 63 targets linked to low-temperature responses. In addition, 85 lncRNA were identified as ceRNAs. The findings revealed that plant miRNAs target lncRNAs and may act as ceRNAs, regulating cold stress responses (Wang *et al.*, 2018b).

A transcriptional regulatory cascade driven by CBF/DREB is the primary signaling mechanism for low-temperature responses. In grapevines, cold stress upregulated 487 and downregulated 326 lncRNAs, respectively, and targeted 326 stress-related genes involved in peroxisome biogenesis and late embryogenesis abundant proteins, etc. (Wang *et al.*, 2019b). A genome-wide investigation identified a potential ncRNAs in cassava. Among 2229 identified lncRNAs, 250 expressed differentially under cold and drought stress and regulated stress-related mechanisms post-transcriptionally (Suksamran *et al.*, 2020). A study by Lu *et al.* (2020) observed that low-temperature variation is inversely linked with tae-miR398 precursor expression in winter wheat, which cleaves CSD1. LincRNAs, viz. R9A, R117, and R616, interact with miR398. In *Arabidopsis* and *Brachypodium* transgenics, these lncRNAs indirectly downregulate CSD1 by binding to miR398, improving the tolerance to cold. In wheat, cold tolerance depends on miR398-lncRNA-target mRNA regulatory loop, resulting in stress-tolerant crops.

Virus-mediated gene silencing and cold treatments were used to study lincRNAs' regulatory roles in cold stress adaptation in cotton seedlings. Silencing lincRNA (XH123) increased cold sensitivity, resulted in chloroplast damage, and increased ROS production. Transcriptome analysis revealed that XH123-silenced seedlings resemble those with the silenced cold stress gene *PIF3*, indicating that XH123 plays an important role in cold stress regulation (Cao *et al.*, 2021). The FLC locus responsible for flowering in *Arabidopsis thaliana* is negatively regulated by lncRNAs like COLDAIR, COOLAIR, COLDWRAP, and ASL (Jampala *et al.*, 2021). In *Arabidopsis*, lncRNA-ARPL (Auxin-Regulated Promoter Loop) and WRKY42 interaction promoted the elongation of root hair cells under low temperatures (Moison *et al.*, 2021). Under low temperatures, pear buds were stimulated to various dormancy states. A study revealed that lncRNA-mediated dormancy release may involve interactions between mRNAs and miRNAs linked to the ABA degradation pathway (Li *et al.*, 2021b). In *Arabidopsis*, cold-induced lncRNA 1 (CIL1) may positively regulate cold stress-induced responses.

Transcriptome analysis predicted up- and down-regulation of various cold stress-related genes in *cil1* mutants, suggesting a potential role of CIL1 in cold stress responses (Liu *et al.*, 2022).

A systematic investigation identified 11,073 cold-responsive lncRNAs in cold-tolerant (C18) and sensitive (C6) varieties of *B. napus*. On applying the cold stress, C18 showed the expression of 2394 lncRNAs on the first day and 3508 after a week, while C6 expressed 3347 and 3872 lncRNAs. Fifteen lncRNAs were found as eTMs of 23 miRNAs, while 16 lncRNAs were projected as miRNA precursors. The cold stress response of *B. napus* cold stress response was better understood by gene coexpression analysis, which identified four modules associated with lncRNAs and protein-coding genes (Waseem *et al.*, 2022). Cold-responsive intergenic lncRNA (CRIR1) was identified as a new cold stress-positive regulator and strongly improved cold tolerance. The transcriptome analysis showed that lncRNA (CRIR1) in cassava regulates cold stress-related genes independently of the CBF pathway and interacts with cold shock-associated protein (MeCSP5) to enhance mRNA translation efficiency (Li *et al.*, 2022d). Similarly, cold-tolerant *Ammopiptanthus nanus* relies on lncRNAs for its cold stress response.

A study identified 4890 new lncRNAs, of which 1322 expressed differentially under cold stress (543 upregulated and 779 down-regulated). Approximately 421 lncRNA–mRNA modules regulated stress-related transcription factors and metabolic enzymes. Additionally, 31 lncRNAs functioned as miRNA precursors and eight as competitive miRNA targets, forming lncRNA–miRNA–mRNA regulatory modules. The TZP-regulating lncRNA (TCONS00065739), an eTMs of miR530, helped in adaptation to cold stress (Zhu *et al.*, 2023), although CBF/DREB1 transcription factors govern responses to cold stress. In *Medicago truncatula*, the cold-responsive lncRNA (MtCIR2) was found in a freezing tolerance QTL region. Cold-triggered MtCIR2 transcription is faster than MtCBF/DREB1s. Overexpression of MtCIR2 boosted freezing tolerance, resulting in enhanced cell survival and reduced cell membrane damage, while mutations increased cold sensitivity. Further, overexpressed MtCIR2 upregulated the DREB1s expression and elevated the soluble sugar concentration, suggesting that glycol metabolism and CBF/DREB1s may improve tolerance to cold stress (Zhao *et al.*, 2023). Antioxidant indices and whole-transcriptome data analysis have identified the gene modules linked to key antioxidant enzymes, viz., superoxide dismutase and peroxidase. Three ceRNA networks were constructed with seven lncRNAs, four microRNAs, and three mRNAs, suggesting a regulatory role in cold stress tolerance (Wang *et al.*, 2024).

13.3.4 Heat stress

Similar to cold stress, heat may affect plant health by interfering with essential physiological and metabolic processes. In *Arabidopsis*, HSFB2a governs the processes of vegetative and gametophyte development. The expression of HSFB2a has been selectively inhibited by the overexpression of NATlncRNA. The rise in temperature initiates antisense-mediated regulation, neutralizing the repressive effects of HSFB2a and promoting growth and subsequent development (Wunderlich *et al.*, 2014). Under heat stress, *Brassica rapa* differentially expressed lncRNAs that target the genes involved in the biosynthesis and signaling of plant hormones, including jasmonic acid, abscisic acid, salicylic acid, and auxin, also encoding heat shock proteins. The qRT-PCR analysis revealed three differentially expressed lncRNA (TCONS_00004594, TCONS_00010856 and, TCONS_00048391). The latter functioned as eTMs for bra-miR164a, regulating the heat stress response protein (Wang *et al.*, 2019a).

A study conducted in *Populus simonii* revealed that lncRNA TCONS_00260893 may regulate the Ca^{2+} flux and signaling by targeting the genes encoding the ion channel protein; i.e. Potri.017G089800 (Song *et al.*, 2021). In a similar study, Bhatia *et al.* (2020) revealed that heterologous expression of Potri.017G089800 may improve photosynthetic efficiency and reduce membrane peroxidation and DNA damage. They reported the involvement of co-expression networks consisting of mRNA and lncRNAs. The transcriptome and comparative phenotype analysis identified more than 1000 heat-responsive lncRNAs, expressed differentially in maize inbred lines. These lncRNAs were targeting the *cis*-regulated target genes and regulating their expression (Hu *et al.*, 2022). Strand-specific RNA sequencing identified 2743 lncRNAs in heat-sensitive (SYD2) and tolerant (ZS97B) rice varieties. Of these, 231 were expressed differentially, with 31 common to both varieties. Targeted genes were predicted using Kyoto Encyclopedia of Genes and Genomes (KEGG), and gene ontology (GO) annotations, and an interaction network was constructed (Zhang *et al.*, 2022). A study in chickpeas (*Cicer arietinum* L.) identified 894 lincRNAs and 61,110 mRNAs from three genotypes of stress (sensitive: ICC4567; tolerant: ICC1356, 15614) with varying responses to heat stress. Further analysis revealed significant co-expression of lincRNAs and mRNA in leaves. The targeted genes encoded heat shock-associated transcription factors and key proteins of sucrose and starch metabolism (Bhogireddy *et al.*, 2023). Using transcriptome analysis, 11,054 ncRNA were found in wheat cultivars, of which 5482 were expressed differentially under heat stress. By forming regulatory networks with mRNA and miRNA, lncRNAs targeted the genes regulating protein modification and folding, hormonal responses, heat-stress responses, and metabolic processes (Babaei *et al.*, 2024).

13.3.5 Heavy metals stress

Over the decades, environmental contamination with heavy metals has arisen at an alarming rate and is threatening the ecosystem. Heavy metals in soil interfere with plant viability, growth, and development, hence limiting agricultural productivity. They cause contamination by entering the food chain, posing serious health risks. Recently, plant lncRNAs have shown sensitivity to metals like copper (Cu), aluminum (Al), lead (Pb), iron (Fe), and manganese (Mn). Deep sequencing and genome-wide lncRNA analysis determined the lncRNA expression pattern in rice roots experiencing Cd stress. Specific lncRNAs were upregulated in Cd-treated roots of rice cultivars rather than untreated ones. Differentially expressed lncRNAs targeted the genes engaged in the metabolism of methionine and cysteine, carotenoid, and secondary metabolites (Chen *et al.*, 2018). A study by Tang *et al.* (2019) identified that, in rice, several short RNAs were susceptible to arsenic stress. Fifty-six lncRNAs were targeted by nine sRNAs, suggesting an important role in arsenic stress signaling.

Protein-coding sites as a part of regulatory networks control growth, reproduction, lipid metabolism, and stress responses. In *Malus domestica*, lncRNA MSTRG.85814 has complex structures and 13 projected RNA sequence variations, four of which are expressed more in iron-deficient roots. One splice variant of MSTRG.85814, such as MSTRG.85814.11, positively regulates the *cis*-targeted mRNA originating from the small auxin upregulated gene (*SAUR32*) (Sun *et al.*, 2020b). The Cd stress regulation of lncRNAs in birch trees was studied. Under Cd treatment, 16 lncRNAs were elevated and 14 were downregulated. Nine lncRNAs were chosen for further study; two showed higher Cd tolerance, and two were highly sensitive. LncRNAs also increased Cd tolerance by regulating target genes such as *HSP18.1* and lactate dehydrogenase A (Wen *et al.*, 2020). Transcriptional regulation of Pb^{2+} uptake in *Populus tomentosa* was studied by Chen *et al.* (2022). The study identified 226 Pb^{2+}-induced differentially expressed lncRNAs regulating key metabolic processes and signaling pathways. The epistatic interaction between the lncRNAP-MAT and the PtoMYB46 affects photosynthetic rate, leaf dry weight, and transketolase activity. Further genetic transformation analysis showed that PtoMYB46 decreased PtoMATE, affecting citric acid production and Pb^{2+} absorption. In Cd-exposed rapeseed, 5038 lncRNAs were identified, 2546 from the sense strand, and 2492 from native antisense transcripts. Cd stress modified the expression of roughly 301 lncRNA linked with nearby protein-coding genes, essential for Cd absorption, translocation, and stress responses (Patra *et al.*, 2023).

Zhou *et al.* (2023) examined the lncRNA transcriptome in the roots of W6nk2 (Cd-sensitive) and Zhenong8 (Cd-tolerant) barley genotypes. Among the 9937 newly identified lncRNAs, they projected 5758 as *cis*-acting and 4159 as *trans*-acting target pairings. Genotype comparison revealed 195 differentially expressed Cd-induced lncRNAs. Eight lncRNAs of barley were probable miRNA target mimics, with *HvGAMYB* as a target gene. Silencing *HvGAMYB* elevated Cd concentration in both roots and shoots, thereby triggering the damage as indicated by increased Cd sensitivity in ZN8. Melatonin may increase Cd tolerance in rice via lncRNA-mediated effects on photosynthesis and cell wall production. Qiu *et al.* (2024) reported variable expression of Cd tolerance and transport-associated genes in 2510 differentially expressed transcripts.

13.3.6 Light stress

Light is a key external trigger that regulates the growth and development of a plant. However, the role of plant lncRNAs in light response is still unknown. Many light-responsive NAT pairs have histone modification peaks linked to them. Yang *et al.* (2019) suggested that lncRNAs such as MLNC3.2 and MLNC4.6 could act as eTMs for miRNA156a. LncRNA upregulates the expression of transcription factors (SPL2 and SPL33) during the production of anthocyanin in response to light stimulus. A study by Sun *et al.* (2020c) discussed the role of lncRNA in response to mannitol and blue light-induced stress in *Arabidopsis* seedlings. These were treated with blue light for 2 and 8 hours, resulting in differential expression of 4197 mRNAs, 481 lncRNAs, and 375 miRNAs, or 5207, 545, and 286 miRNAs, respectively. A blue-light-induced lncRNA (BLIL1) mimics the endogenous target, and participates in blue light-mediated plant photomorphogenesis, inhibiting hypocotyl elongation and responses under mannitol stress. An appropriate amount of far-red light may enhance leaf area and stem elongation. Far-red light may affect the expression of lncRNAs linked to shade avoidance by *Dendrobium*. Research has highlighted the presence of multiple variably expressed lncRNAs that affect hormone signaling and DNA methylation (Li *et al.*, 2021a).

13.4 Mechanism of lncRNA-Mediated Regulation of Abiotic Stress

Many recent investigations have revealed plant lncRNA responses to abiotic stress. Studies have highlighted their role in ROS homeostasis, hormone signaling, and glucose metabolism as positive or negative regulators. The transcriptional effects of lncRNAs include DNA interaction and modulation of transcription factor expression, and they also affect the untranslated regions, promoters, exons, introns, and terminators. The *Arabidopsis* lncRNA (Alternative Splicing Competitor) regulates auxin-induced alternative splicing patterns, altering transcriptome reprogramming (Rigo *et al.*, 2020). lncRNAs may stabilize mRNA and regulate the expression of a gene by functioning as ceRNAs. As target mimics, they may affect miRNA activity. Some lncRNAs resemble miRNA targets in *Populus* under drought stress (Shuai *et al.*, 2014) while responding to salt stress in soybeans with miRNA–lncRNA interaction pairs (Li *et al.*, 2022a). lncRNAs can also generate miRNAs, with species-specific patterns in plants (Song *et al.*, 2021). miRNAs may target them to generate phased small interfering RNAs (phasiRNAs) that modulate abiotic stress responses (Chen *et al.*, 2021; Meng *et al.*, 2021).

Small open reading frames (smORFs) in lncRNAs help plants to develop and tolerate stress by regulating photosynthesis and oxidoreductase activity (Wierzbicki *et al.*, 2021). By interacting with its transcription factors, lncRNAs like APOLO may regulate the promoter activity of the target gene, affecting root hair growth under cold stress. In addition, nuclear lncRNAs directly affect transcription factors under abiotic stress (Baruah *et al.*, 2021). Moreover, complex lncRNA-splicing factor interactions may affect their stability and subcellular localization (Lucero *et al.*, 2021). The *cis*-regulatory element DglncTCP1 stimulates TCP1 expression to alter cold responses in *Chrysanthemum* (Li *et al.*, 2022b). Moreover, lncRNAs regulate histone modifications and DNA methylation under abiotic stress. Through RNA-mediated transcriptional gene silencing (TGS), they control gene expression with specialized RNA polymerases, DNA methyltransferases, and chromatin-modifying enzymes (Ding *et al.*, 2022; Sun *et al.*, 2022).

13.5 Conclusion

While significant advancements have been achieved in pinpointing lncRNAs responsive to stress, there is still a requirement for more in-depth explorations into the precise molecular mechanisms that facilitate lncRNA-mediated regulation of biological processes. Grasping the intricate web of interactions and cascades initiated by lncRNAs is crucial for leveraging their potential to bolster plant resilience against abiotic stresses. Additionally, individual lncRNAs must be functionally validated, and their effects on agricultural resilience in stress-prone regions need extensive examination. This may include improving nutrient usage efficiency, reducing heavy metal toxicity, and combating extreme temperature variations. Such endeavors should form the focal point of future research efforts. Systematic screening and comprehensive cross-referencing of lncRNAs present an opportunity to unearth essential lncRNAs pivotal for plant responses to recurring abiotic stresses. Furthermore, an in-depth understanding of lncRNA functions and metabolism across various plant groups remains a significant challenge. Therefore, functional assessments of plant lncRNA and species-wise comparisons may help in their better identification and characterization. Abiotic stress-associated functional candidate lncRNAs can be identified using GWAS-derived genomic studies to reveal single-nucleotide polymorphisms and functional variations. Additionally, transposable elements, known for their evolutionary significance and enrichment in the upstream regions of lncRNAs, may also enhance plant adaptation to abiotic stresses. Consequently, there is a need to develop new methodologies capable of mapping lncRNA structures, interactions, and evolutionary constraints accurately to unravel their underlying biology while acknowledging the potential underestimation of lncRNA conservation.

References

Ackah, M., Jin, X., Zhang, Q., Amoako, F.K., Wang, L. *et al.* (2024) Long noncoding RNA transcriptome analysis reveals novel lncRNAs in *Morus alba* "Yu-711" response to drought stress. *The Plant Genome* 17(1), e20273.

Babaei, S., Bhalla, P.L. and Singh, M.B. (2024) Identifying long non-coding RNAs involved in heat stress response during wheat pollen development. *Frontiers in Plant Science* 15, 1344928.

Baruah, P.M., Krishnatreya, D.B., Bordoloi, K.S., Gill, S.S. and Agarwala, N. (2021) Genome wide identification and characterization of abiotic stress responsive lncRNAs in *Capsicum annuum*. *Plant Physiology and Biochemistry* 162, 221–236.

Bhatia, G., Singh, A., Verma, D., Sharma, S. and Singh, K. (2020) Genome-wide investigation of regulatory roles of lncRNAs in response to heat and drought stress in *Brassica juncea* (Indian mustard). *Environmental and Experimental Botany* 171, 103922.

Bhogireddy, S., Kudapa, H., Bajaj, P., Garg, V., Chitikineni, A. *et al.* (2023) Transcriptome analysis of chickpea during heat stress unveils the signatures of long intergenic non-coding RNAs (lincRNAs) and mRNAs in the heat-QTL region. *Crop Design* 2(1), 100026.

Cagirici, H.B., Alptekin, B. and Budak, H. (2017) RNA sequencing and co-expressed long non-coding RNA in modern and wild wheats. *Scientific Reports* 7(1), 10670.

Cai, J., Zhang, Y., He, R., Jiang, L., Qu, Z. *et al.* (2024) LncRNA DANA1 promotes drought tolerance and histone deacetylation of drought responsive genes in *Arabidopsis*. *EMBO Reports* 25(2), 796–812.

Calixto, C.P., Tzioutziou, N.A., James, A.B., Hornyik, C., Guo, W. *et al.* (2019) Cold-dependent expression and alternative splicing of *Arabidopsis* long non-coding RNAs. *Frontiers in Plant Science* 10, 404796.

Cao, Z., Zhao, T., Wang, L., Han, J., Chen, J. *et al.* (2021) The lincRNA XH123 is involved in cotton cold-stress regulation. *Plant Molecular Biology* 106, 521–531.

Chen, J., Zhong, Y. and Qi, X. (2021) LncRNA TCONS_00021861 is functionally associated with drought tolerance in rice (*Oryza sativa* L.) via competing endogenous RNA regulation. *BMC Plant Biology* 21, 1–12.

Chen, L., Shi, S., Jiang, N., Khanzada, H., Wassan, G.M. *et al.* (2018) Genome-wide analysis of long non-coding RNAs affecting roots development at an early stage in the rice response to cadmium stress. *BMC Genomics* 19, 1–10.

Chen, P., Song, Y., Liu, X., Xiao, L., Bu, C. *et al.* (2022) LncRNA PMAT–PtoMYB46 module represses ptoMATE and PtoARF2 promoting Pb2+ uptake and plant growth in poplar. *Journal of Hazardous Materials* 433, 128769.

Chung, P.J., Jung, H., Jeong, D.H., Ha, S.H., Choi, Y.D. *et al.* (2016) Transcriptome profiling of drought responsive noncoding RNAs and their target genes in rice. *BMC Genomics* 17, 1–12.

Cui, C., Wan, H., Li, Z., Ai, N. and Zhou, B. (2024) Long noncoding RNA TRABA suppresses β-glucosidase-encoding BGLU24 to promote salt tolerance in cotton. *Plant Physiology* 194(2), 1120–1138.

Deng, F., Zhang, X., Wang, W., Yuan, R. and Shen, F. (2018) Identification of *Gossypium hirsutum* long non-coding RNAs (lncRNAs) under salt stress. *BMC Plant Biology* 18, 1–14.

Ding, Y., Zou, L.H., Wu, J., Ramakrishnan, M., Gao, Y. *et al.* (2022) The pattern of DNA methylation alteration, and its association with the expression changes of non-coding RNAs and mRNAs in Moso bamboo under abiotic stress. *Plant Science* 325, 111451.

Ding, Z., Tie, W., Fu, L., Yan, Y., Liu, G. *et al.* (2019a) Strand-specific RNA-seq based identification and functional prediction of drought-responsive lncRNAs in cassava. *BMC Genomics* 20, 1–13.

Ding, Z., Tie, W., Fu, L., Yan, Y., Liu, G. *et al.* (2019a) Strand-specific RNA-seq based identification and functional prediction of drought-responsive lncrnas in cassava. *BMC Genomics* 20(1), 214. DOI: 10.1186/s12864-019-5585-5.

Ding, Z., Wu, C., Tie, W., Yan, Y., He, G. *et al.* (2019b) Strand-specific RNA-seq based identification and functional prediction of lncRNAs in response to melatonin and simulated drought stresses in cassava. *Plant Physiology and Biochemistry* 140, 96–104.

Ding, Z., Wu, C., Tie, W., Yan, Y., He, G. *et al.* (2019b) Strand-specific RNA-seq based identification and functional prediction of lncrnas in response to melatonin and simulated drought stresses in cassava. *Plant Physiology and Biochemistry* 140, 96–104. DOI: 10.1016/j.plaphy.2019.05.008.

Dong, S.M., Xiao, L., Li, Z.B., Shen, J., Yan, H.B. *et al.* (2022) A novel long non-coding RNA, DIR, increases drought tolerance in cassava by modifying stress-related gene expression. *Journal of Integrative Agriculture* 21(9), 2588–2602.

Eom, S.H., Lee, H.J., Lee, J.H., Wi, S.H., Kim, S.K. *et al.* (2019) Identification and functional prediction of drought-responsive long non-coding RNA in tomato. *Agronomy* 9(10), 629.

Feng, W., Guo, L., Fang, H., Mehari, T.G., Gu, H. *et al.* (2024) Transcriptomic profiling reveals salt-responsive long non-coding RNAs and their putative target genes for improving salt tolerance in upland cotton (*Gossypium hirsutum*). *Industrial Crops and Products* 216, 118744.

Fu, L., Ding, Z., Tan, D., Han, B., Sun, X. *et al.* (2020) Genome-wide discovery and functional prediction of salt-responsive lncRNAs in duckweed. *BMC Genomics* 21, 1–14.

Fu, X.-Z., Zhang, X.-Y., Qiu, J.-Y., Zhou, X., Yuan, M. *et al.* (2019) Whole-transcriptome RNA sequencing reveals the global molecular responses and cerna regulatory network of mrnas, lncrnas, mirnas and circrnas in response to copper toxicity in ziyang xiangcheng (citrus junos sieb. ex tanaka). *BMC Plant Biology* 19(1), 509. DOI: 10.1186/s12870-019-2087-1.

Gui, Q., Yang, Z., Chen, C., Yang, F., Wang, S. *et al.* (2022) Identification and characterization of long noncoding RNAs involved in the aluminum stress response in *Medicago truncatula* via genome-wide analysis. *Frontiers in Plant Science* 13, 1017869.

Gupta, P.K., Balyan, H.S. and Gahlaut, V. (2017) QTL analysis for drought tolerance in wheat: Present status and future possibilities. *Agronomy* 7(1), 5.

Huanca-Mamani, W., Arias-Carrasco, R., Cárdenas-Ninasivincha, S., Rojas-Herrera, M., Sepúlveda-Hermosilla, G. *et al.* (2018) Long non-coding RNAs responsive to salt and boron stress in the hyper-arid Lluteno maize from Atacama Desert. *Genes* 9(3), 170.

Hu, X., Wei, Q., Wu, H., Huang, Y., Peng, X. *et al.* (2022) Identification and characterization of heat-responsive lncRNAs in maize inbred line CM1. *BMC Genomics* 23(1), 208.

Jain, P., Hussian, S., Nishad, J., Dubey, H., Bisht, D.S. *et al.* (2021) Identification and functional prediction of long non-coding RNAs of rice (*Oryza sativa* L.) at reproductive stage under salinity stress. *Molecular Biology Reports* 48, 2261–2271.

Jampala, P., Garhewal, A. and Lodha, M. (2021) Functions of long non-coding RNA in *Arabidopsis thaliana*. *Plant Signaling & Behavior* 16(9), 1925440.

Jha, U.C., Bohra, A. and Jha, R. (2017) Breeding approaches and genomics technologies to increase crop yield under low-temperature stress. *Plant Cell Reports* 36, 1–35.

Jha, U.C., Bohra, A. and Nayyar, H. (2020a) Advances in 'omics' approaches to tackle drought stress in grain legumes. *Plant Breeding* 139(1), 1–27.

Jha, U.C., Nayyar, H., Jha, R., Khurshid, M., Zhou, M. *et al.* (2020b) Long non-coding RNAs: Emerging players regulating plant abiotic stress response and adaptation. *BMC Plant Biology* 20, 1–20.

Jia, Y., Zhao, H., Niu, Y. and Wang, Y. (2024) Long noncoding RNA from *Betula platyphylla*, BplncSIR1, confers salt tolerance by regulating BpNAC2 to mediate reactive oxygen species scavenging and stomatal movement. *Plant Biotechnology Journal* 22(1), 48–65.

Kumar, N., Bharadwaj, C., Sahu, S., Shiv, A., Shrivastava, A.K. *et al.* (2021) Genome-wide identification and functional prediction of salt-stress related long non-coding RNAs (lncRNAs) in chickpea (*Cicer arietinum* L.). *Physiology and Molecular Biology of Plants* 27, 2605–2619.

Lamaoui, M., Jemo, M., Datla, R. and Bekkaoui, F. (2018) Heat and drought stresses in crops and approaches for their mitigation. *Frontiers in Chemistry* 6, 26.

Leng, Y., Sun, J., Wang, J., Liu, H., Zheng, H. *et al.* (2020) Genome-wide lncRNAs identification and association analysis for cold-responsive genes at the booting stage in rice (*Oryza sativa* L.). *The Plant Genome* 13(2), e20020.

Li, C., Nong, W., Zhao, S., Lin, X., Xie, Y. *et al.* (2022a) Differential microRNA expression, microRNA arm switching, and microRNA: Long noncoding RNA interaction in response to salinity stress in soybean. *BMC Genomics* 23(1), 65.

Li, G., Chen, Q., Bai, Q., Feng, Y., Mao, K. *et al.* (2023a) LncRNA expression analysis by comparative transcriptomics among closely related poplars and their regulatory roles in response to salt stress. *Tree Physiology* 43(7), 1233–1249.

Li, H., Ye, W., Wang, Y., Chen, X., Fang, Y. *et al.* (2021a) RNA sequencing-based exploration of the effects of far-red light on lncRNAs involved in the shade-avoidance response of *D. officinale*. *PeerJ* 9, e10769.

Li, J., He, H., Li, Y., Wang, A. and He, L.F. (2022b) Role of long non-coding RNA in plant responses to abiotic stresses. *Acta Physiologiae Plantarum* 44(5), 54.

Li, L., Liu, J., Liang, Q., Zhang, Y., Kang, K. *et al.* (2021b) Genome-wide analysis of long noncoding RNAs affecting floral bud dormancy in pears in response to cold stress. *Tree Physiology* 41(5), 771–790.

Li, N., Wang, Z., Wang, B., Wang, J., Xu, R. *et al.* (2022c) Identification and characterization of long non-coding RNA in tomato roots under salt stress. *Frontiers in Plant Science* 13, 834027.

Li, S., Yu, X., Lei, N., Cheng, Z., Zhao, P. *et al.* (2017) Genome-wide identification and functional prediction of cold and/or drought-responsive lncRNAs in cassava. *Scientific Reports* 7(1), 45981.

Li, S., Cheng, Z., Dong, S., Li, Z., Zou, L. *et al.* (2022d) Global identification of full-length cassava lncRNAs unveils the role of cold-responsive intergenic lncRNA 1 in cold stress response. *Plant, Cell & Environment* 45(2), 412–426.

Li, Z., Zhou, H., Xu, G., Zhang, P., Zhai, N. *et al.* (2023b) Genome-wide analysis of long noncoding RNAs in response to salt stress in *Nicotiana tabacum*. *BMC Plant Biology* 23(1), 646.

Liu, G., Liu, F., Wang, Y. and Liu, X. (2022a) A novel long noncoding RNA CIL1 enhances cold stress tolerance in *Arabidopsis*. *Plant Science* 323, 111370.

Liu, P., Zhang, Y., Zou, C., Yang, C., Pan, G. *et al.* (2022b) Integrated analysis of long non-coding RNAs and mRNAs reveals the regulatory network of maize seedling root responding to salt stress. *BMC Genomics* 23, 1–16.

Liu, P., Zhang, Y., Zou, C., Yang, C., Pan, G, *et al.* (2022) Integrated analysis of long non-coding RNAs and mRNAs reveals the regulatory network of maize seedling root responding to salt stress. *BMC Genomics* 23(1), 50. DOI: 10.1186/s12864-021-08286-7.

Liu, W., Cheng, C., Lin, Y., XuHan, X. and Lai, Z. (2018) Genome-wide identification and characterization of mRNAs and lncRNAs involved in cold stress in the wild banana (*Musa itinerans*). *PloS ONE* 13(7), e0200002.

Lu, Q., Guo, F., Xu, Q. and Cang, J. (2020) LncRNA improves cold resistance of winter wheat by interacting with miR398. *Functional Plant Biology* 47(6), 544–557.

Lu, X., Chen, X., Mu, M., Wang, J., Wang, X. *et al.* (2016) Genome-wide analysis of long noncoding RNAs and their responses to drought stress in cotton (*Gossypium hirsutum* L.). *PloS ONE* 11(6), e0156723.

Lucero, L., Ferrero, L., Fonouni-Farde, C. and Ariel, F. (2021) Functional classification of plant long noncoding RNAs: A transcript is known by the company it keeps. *The New Phytologist* 229(3), 1251–1260.

Magar, N.D., Shah, P., Barbadikar, K.M., Bosamia, T.C., Madhav, M.S. *et al.* (2023) Long non-coding RNA-mediated epigenetic response for abiotic stress tolerance in plants. *Plant Physiology and Biochemistry* 108165.

Meng, X., Li, A., Yu, B. and Li, S. (2021) Interplay between miRNAs and lncRNAs: Mode of action and biological roles in plant development and stress adaptation. *Computational and Structural Biotechnology Journal* 19, 2567–2574.

Moison, M., Pacheco, J.M., Lucero, L., Fonouni-Farde, C., Rodríguez-Melo, J. *et al.* (2021) The lncRNA APOLO interacts with the transcription factor WRKY42 to trigger root hair cell expansion in response to cold. *Molecular Plant* 14(6), 937–948.

Nejat, N. and Mantri, N. (2018) Emerging roles of long non-coding RNAs in plant response to biotic and abiotic stresses. *Critical Reviews in Biotechnology* 38(1), 93–105.

Pang, J., Zhang, X., Ma, X. and Zhao, J. (2019) Spatio-temporal transcriptional dynamics of maize long non-coding RNAs responsive to drought stress. *Genes* 10(2), 138.

Patra, G.K., Gupta, D., Rout, G.R. and Panda, S.K. (2023) Role of long non coding RNA in plants under abiotic and biotic stresses. *Plant Physiology and Biochemistry* 194, 96–110.

Qiu, C.W., Richmond, M., Ma, Y., Zhang, S., Liu, W. *et al.* (2024) Melatonin enhances cadmium tolerance in rice via long non-coding RNA-mediated modulation of cell wall and photosynthesis. *Journal of Hazardous Materials* 465, 133251.

Rehman, O.U., Uzair, M., Farooq, M.S., Saleem, B., Attacha, S. *et al.* (2023) Comprehensive insights into the regulatory mechanisms of lncRNA in alkaline-salt stress tolerance in rice. *Molecular Biology Reports* 50(9), 7381–7392.

Rigo, R., Bazin, J., Romero-Barrios, N., Moison, M., Lucero, L. *et al.* (2020) The *Arabidopsis* lnc RNA ASCO modulates the transcriptome through interaction with splicing factors. *EMBO Reports* 21(5), e48977.

Shea, D.J., Nishida, N., Takada, S., Itabashi, E., Takahashi, S. *et al.* (2019) Long noncoding RNAs in *Brassica rapa* L. following vernalization. *Scientific Reports* 9(1), 9302.

Shuai, P., Liang, D., Tang, S., Zhang, Z., Ye, C.Y. *et al.* (2014) Genome-wide identification and functional prediction of novel and drought-responsive lincRNAs in *Populus trichocarpa*. *Journal of Experimental Botany* 65(17), 4975–4983.

Shumayla, Sharma, S., Taneja, M., Tyagi, S., Singh, K. *et al.* (2017) Survey of high throughput RNA-seq data reveals potential roles for lncRNAs during development and stress response in bread wheat. *Frontiers in Plant Science* 8, 1019.

Singh, U., Khemka, N., Rajkumar, M.S., Garg, R. and Jain, M. (2017) PLncPRO for prediction of long non-coding RNAs (lncRNAs) in plants and its application for discovery of abiotic stress-responsive lncRNAs in rice and chickpea. *Nucleic Acids Research* 45(22), e183–e183.

Song, X., Liu, G., Huang, Z., Duan, W., Tan, H. *et al.* (2016) Temperature expression patterns of genes and their coexpression with lncRNAs revealed by RNA-seq in non-heading Chinese cabbage. *BMC Genomics* 17, 1–15.

Song, X., Hu, J., Wu, T., Yang, Q., Feng, X. *et al.* (2021) Comparative analysis of long noncoding RNAs in angiosperms and characterization of long noncoding RNAs in response to heat stress in Chinese cabbage. *Horticulture Research* 8.

Suksamran, R., Saithong, T., Thammarongtham, C. and Kalapanulak, S. (2020) Genomic and transcriptomic analysis identified novel putative cassava lncRNAs involved in cold and drought stress. *Genes* 11(4), 366.

Sun, M., Yang, Z., Liu, L. and Duan, L. (2022) DNA methylation in plant responses and adaption to abiotic stresses. *International Journal of Molecular Sciences* 23(13), 6910.

Sun, X., Zheng, H. and Sui, N. (2018) Regulation mechanism of long non-coding RNA in plant response to stress. *Biochemical and Biophysical Research Communications* 503(2), 402–407.

Sun, X., Zheng, H., Li, J., Liu, L., Zhang, X. *et al.* (2020a) Comparative transcriptome analysis reveals new lncRNAs responding to salt stress in sweet sorghum. *Frontiers in Bioengineering and Biotechnology* 8, 331.

Sun, Y., Hao, P., Lv, X., Tian, J., Wang, Y. *et al.* (2020b) A long non-coding apple RNA, MSTRG.85814.11, acts as a transcriptional enhancer of *SAUR32* and contributes to the Fe-deficiency response. *The Plant Journal* 103(1), 53–67.

Sun, Z., Huang, K., Han, Z., Wang, P. and Fang, Y. (2020c) Genome-wide identification of *Arabidopsis* long noncoding RNAs in response to the blue light. *Scientific Reports* 10(1), 6229.

Tang, Z., Xu, M., Ito, H., Cai, J., Ma, X. *et al.* (2019) Deciphering the non-coding RNA-level response to arsenic stress in rice (*Oryza sativa*). *Plant Signaling & Behavior* 14(9), 1629268.

Tribhuvan, K.U., Mishra, T., Kaur, S., Pandey, A., Choudhary, S.B. *et al.* (2024) Identification and characterization of novel drought-responsive lncRNAs in stone apple (*Aegle marmelos* L.) through whole-transcriptome analysis. *Current Plant Biology* 38, 100336.

Waititu, J.K., Zhang, C., Liu, J. and Wang, H. (2020) Plant non-coding RNAs: Origin, biogenesis, mode of action and their roles in abiotic stress. *International Journal of Molecular Sciences* 21(21), 8401.

Wan, S., Zhang, Y., Duan, M., Huang, L., Wang, W. *et al.* (2020) Integrated analysis of long non-coding RNAs (lncRNAs) and mRNAs reveals the regulatory role of lncRNAs associated with salt resistance in *Camellia sinensis*. *Frontiers in Plant Science* 11, 218.

Wang, A., Hu, J., Gao, C., Chen, G., Wang, B. *et al.* (2019a) Genome-wide analysis of long non-coding RNAs unveils the regulatory roles in the heat tolerance of Chinese cabbage (*Brassica rapa* ssp. *chinensis*). *Scientific Reports* 9(1), 5002.

Wang, H., Jia, Y., Bai, X., Gong, W., Liu, G. *et al.* (2024) Whole-transcriptome profiling and functional prediction of long non-coding RNAs associated with cold tolerance in *Japonica* rice varieties. *International Journal of Molecular Sciences* 25(4), 2310.

Wang, J., Lin, J., Kan, J., Wang, H., Li, X. *et al.* (2018a) Genome-wide identification and functional prediction of novel drought-responsive lncRNAs in *Pyrus betulifolia*. *Genes* 9(6), 311.

Wang, P., Dai, L., Ai, J., Wang, Y. and Ren, F. (2019b) Identification and functional prediction of cold-related long non-coding RNA (lncRNA) in grapevine. *Scientific Reports* 9(1), 6638.

Wang, T.-Z., Liu, M., Zhao, M.-G., Chen, R. and Zhang, W.-H. (2015) Identification and characterization of long non-coding RNAs involved in osmotic and salt stress in *Medicago truncatula* using genome-wide high-throughput sequencing. *BMC Plant Biology* 15, 131. DOI: 10.1186/s12870-015-0530-5.

Wang, Y., Gao, L., Zhu, B., Zhu, H., Luo, Y. *et al.* (2018b) Integrative analysis of long non-coding RNA acting as ceRNAs involved in chilling injury in tomato fruit. *Gene* 667, 25–33.

Wen, X., Ding, Y., Tan, Z., Wang, J., Zhang, D. *et al.* (2020) Identification and characterization of cadmium stress-related lncRNA from *Betula platyphylla*. *Plant Science* 299, 110601.

Waseem, M., Yang, X., Aslam, M.M., Li, M., Zhu, L. *et al.* (2022) Genome-wide identification of long non-coding RNAs in two contrasting rapeseed (*Brassica napus* L.) genotypes subjected to cold stress. *Environmental and Experimental Botany* 201, 104969.

Weng, M., Zhang, D., Wang, H., Yang, C., Lin, H. *et al.* (2024) Long non-coding RNAs and their potential function in response to postharvest senescence of *Sparassis latifolia* during cold storage. *Scientific Reports* 14(1), 747.

Wierzbicki, A.T., Blevins, T. and Swiezewski, S. (2021) Long noncoding RNAs in plants. *Annual Review of Plant Biology* 72, 245–271.

Wu, C., Ding, Z., Chen, M., Yang, G., Tie, W. *et al.* (2019) Identification and functional prediction of lncRNAs in response to PEG and ABA treatment in cassava. *Environmental and Experimental Botany* 166, 103809.

Wu, H., Yang, L. and Chen, L.L. (2017) The diversity of long noncoding RNAs and their generation. *Trends in Genetics* 33(8), 540–552.

Wunderlich, M., Groß-Hardt, R. and Schöffl, F. (2014) Heat shock factor HSFB2a involved in gametophyte development of *Arabidopsis thaliana* and its expression is controlled by a heat-inducible long non-coding antisense RNA. *Plant Molecular Biology* 85, 541–550.

Yang, H., Cui, Y., Feng, Y., Hu, Y., Liu, L. *et al.* (2023) Long non-coding RNAs of plants in response to abiotic stresses and their regulating roles in promoting environmental adaption. *Cells* 12(5), 729.

Yang, T., Ma, H., Zhang, J., Wu, T., Song, T. *et al.* (2019) Systematic identification of long noncoding RNAs expressed during light-induced anthocyanin accumulation in apple fruit. *The Plant Journal* 100(3), 572–590.

Yang, X., Liu, C., Niu, X., Wang, L., Li, L. *et al.* (2022) Research on lncRNA related to drought resistance of Shanlan upland rice. *BMC Genomics* 23(1), 336.

Yan, Q., Wu, F., Yan, Z., Li, J., Ma, T. *et al.* (2019) Differential co-expression networks of long non-coding RNAs and mRNAs in *Cleistogenes songorica* under water stress and during recovery. *BMC Plant Biology* 19(1), 19.

Yu, R., Li, Z., Rong, M., Yang, G., Xu, X. *et al.* (2023) Comparative physiology and transcriptome analysis to identify the important coding and non-coding RNAs imparting tolerance to salinity stress in alfalfa (*Medicago sativa* L.). *Preprint Research Square* 16. DOI: 10.21203/rs.3.rs-2428445/v1.

Zhang, C., Tang, G., Peng, X., Sun, F., Liu, S. *et al.* (2018) Long non-coding RNAs of switchgrass (*Panicum virgatum* L.) in multiple dehydration stresses. *BMC Plant Biology* 18, 1–15.

Zhang, P., Fan, Y., Sun, X., Chen, L., Terzaghi, W. *et al.* (2019a) A large-scale circular RNA profiling reveals universal molecular mechanisms responsive to drought stress in maize and *Arabidopsis*. *The Plant Journal* 98(4), 697–713.

Zhang, P., He, R., Yang, J., Cai, J., Qu, Z. *et al.* (2023) The long non-coding RNA DANA2 positively regulates drought tolerance by recruiting ERF84 to promote JMJ29-mediated histone demethylation. *Molecular Plant* 16(8), 1339–1353.

Zhang, Z., Zhong, H., Nan, B. and Xiao, B. (2022) Global identification and integrated analysis of heat-responsive long non-coding RNAs in contrasting rice cultivars. *Theoretical and Applied Genetics* 135(3), 833–852.

Zhao, M., Wang, T., Sun, T., Yu, X., Tian, R. *et al.* (2020) Identification of tissue-specific and cold-responsive lncRNAs in *Medicago truncatula* by high-throughput RNA sequencing. *BMC Plant Biology* 20, 1–16.

Zhao, M., Tian, R., Sun, X. and Zhang, W.H. (2023) lncRNA MtCIR2 positively regulates plant-freezing tolerance by modulating CBF/DREB1 gene clusters. *Plant, Cell & Environment* 46(8), 2450–2469.

Zhao, Y., Liu, Y., Zhang, F., Wang, Z.Y., Mysore, K.S. *et al.* (2024) The long noncoding RNA LAL contributes to salinity tolerance by modulating LHCB1s' expression in *Medicago truncatula*. *Communications Biology* 7(1), 289.

Zhou, X.Y., Wang, N.H., Qiu, C.W., Luo, L., Zhang, M. *et al.* (2023) Transcriptome profiling uncovers the lncRNA-mediated regulatory networks associated with tolerance to cadmium stress in barley. *Environmental and Experimental Botany* 206, 105156.

Zhu, M., Dong, Q., Bing, J., Songbuerbatu, L. and et al, A. (2023) Combined lncRNA and mRNA expression profiles identified the lncRNA–miRNA–mRNA modules regulating the cold stress response in *Ammopiptanthus nanus*. *International Journal of Molecular Sciences* 24(7), 6502.

Zou, C., Guo, Z., Zhao, S., Chen, J., Zhang, C. *et al.* (2023) Genome-wide analysis of long non-coding RNAs in sugar beet (*Beta vulgaris* L.) under drought stress. *Frontiers in Plant Science* 14, 1118011.

14 Drought Stress Memory in Plants

Anil Namdev Kale[1], Sachin Lalaso Abhang[2], Sanjay Kumar Sheshma[1], Mukul Kumar Gandhi[1], Sarita Kumari[1], Kumari Anjani[1], Karma Landup Bhutia[1], Bishun Deo Prasad[1] and Vinay Kumar Sharma[1]*

[1]Department of Agricultural Biotechnology and Molecular Biology, Dr. Rajendra Prasad Central Agricultural University, Pusa, Bihar, India; [2]Department of Agricultural Botany, Dr. Panjabrao Deshmukh Krishi Vidyapeeth, Akola, Maharashtra, India

Abstract

Plants acclimatize to drought stress by adjusting physiological, metabolic, and developmental responses to external environmental cues. The ability of plants to compete with repeated drought stresses more promptly and strongly illustrates the concept of drought stress memory. Epigenetic modifications are responsible for the activation of memory genes. A subset of the drought-responsive genes displays greatly elevated expression levels under recurrent drought conditions, due to epigenetic modifications. These memories are meiotically and mitotically heritable in nature and imprinted stress memory helps existing or upcoming generations of crop plants to adjust themselves more promptly during recurrent drought conditions. Repeated exposure to drought stress can lead to stress priming that helps plants' cross-tolerance to various abiotic stresses. In this chapter, the role of several physiological and epigenetic changes in the activation and development of drought stress memory in crops is explored.

Keywords: Drought stress, Memory genes, Epigenetic changes, Cross-tolerance

14.1 Introduction

Drought has been considered as the most significant abiotic stress impacting crop plants globally in terms of production and endurance (Palliotti *et al.*, 2014; Lee *et al.*, 2019; D'Odorico *et al.*, 2021). In response to drought stress, the synthesis of essential proteins and alterations in the expression levels of specific genes occur due to external signals during the initial occurrence (Natali *et al.*, 2007; Tosetti *et al.*, 2012; Balan *et al.*, 2017; Liu and He, 2020). During the subsequent drought conditions, these proteins are synthesized more quickly and alteration in the expression levels of particular genes is faster as compared with the first stress event (D'Odorico *et al.*, 2021). This leads to a decrease in the intensity of recurrent drought conditions that enables plants to function differently than naïve crops. Realization of such changes led to the development of the concept of memory, and genes activated by this mechanism are known as memory genes. Drought stress memory in crop plants refers to their ability to recall previous water scarcity conditions, enabling them to respond more effectively to subsequent drought stress compared with their initial encounter,

*Corresponding author: vksharma@rpcau.ac.in

 Epigenetics for Climate-Smart and Sustainable Agriculture (ed. Jen-Tsung Chen)
DOI: 10.1079/9781800626102.0014

thereby enhancing their survival during recurring drought conditions. The drought stress memories are mitotically and meiotically heritable in nature (Lämke and Bäurle, 2017; Mladenov *et al.*, 2021). Drought stress memory was discovered in *Arabidopsis thaliana* (Ding *et al.*, 2012; Ding *et al.*, 2013; Ding *et al.*, 2014).

Crop plants undergo a variety of physiological, metabolite and molecular alterations (Jacques *et al.*, 2021) to mitigate the adverse effects created by recurrent drought stress. In response to the physiological changes, the altered level of various osmatic accumulants such as proline, glycine betaine, and sucrose helps in the establishment of memory and enhances drought tolerance in crops (Li *et al.*, 2019a). Proline plays a vital role in encouraging the expression of specific genes that are required for crop plants to recover from drought stress (Szabados and Savouré, 2010). Plant hormones have an essential role in the growth and development of crop plants and variations in the concentrations of such hormones also help plants to perform differentially under recurrent drought stress conditions. Increased levels of ABA during drought stress play a vital role in the activation of reactive oxygen species (ROS) scavenger genes to help in the mitigation of photo-oxidative damage (Hasan *et al.*, 2021). Closing of stomata during drought stress also helps in reducing water loss from plants. Such stomata closing is induced by ABA during drought stress (Lee *et al.*, 2019).

Epigenetic changes modify the expression patterns of specific genes without altering the underlying DNA sequence. These changes encompass processes like DNA methylation, histone modifications, chromatin remodeling (Hilker *et al.*, 2016; Hilker and Schmülling, 2019) microRNAs associated with DNA methylation (Arshad *et al.*, 2017) and key protein synthesis (Sung and Amasino, 2004; Berry and Dean, 2015) perform a significant role in the activation of memory-related genes (Kinoshita and Seki, 2014). With the help of such epigenetic changes, the memory gene has a major effect on the ability of crop plants to function effectively under repeated drought stress. Long non-coding RNAs (lncRNAs), DNA methylation and indigenous phytohormones (particularly abscisic acid) play a role in the establishment of immediate dehydration memories in soybeans (Kim *et al.*, 2020).

Crop plants are impacted by more than one abiotic stress (for example. heat and drought stress) during the life cycle, which affects the yield of the crop more severely than single stress (Kumar *et al.*, 2024). Drought stress-induced priming, which enhances tolerance due to the establishment of drought stress memory during the first drought stress, plays a significant role in cross-tolerance by enhancing tolerance to another abiotic stress (Yadav *et al.*, 2022; Ru *et al.*, 2023). Researchers have identified certain kinds of cross-tolerance for drought combined with heat stress, drought combined with cold stress, and drought stress (Guo *et al.*, 2022; Sintaha *et al.*, 2022; Yadav *et al.*, 2022; Ru *et al.*, 2023). Such cross-tolerance has been observed in rice, wheat, maize, chickpeas, soybeans, potatoes, and *Arabidopsis thaliana*.

14.2 Types of Stress Memory in Crop Plants

Crop plants have three types of stress memory that are well understood (Table 14.1). The first type of stress memory is mitotically heritable (Lämke and Bäurle, 2017). It is memorized for several hours or days, limited to the lifespan of

Table 14.1. Comparative differences between different types of memory in crop plants,

Features of stress memory	Somatic memory	Intergenerational memory	Transgenerational memory
Nature of memory transfer	Mitotically	Meiotically	Meiotically
Offspring of the first stress-free condition	Absent	Present	Present
Offspring of the second stress-free condition and more	Absent	Absent	Present

plants and then erased when the plant appears to be in normal condition (Sako *et al.*, 2016, 2020). Long-term somatic stress memory has been shown to improve salt stress tolerance in rice salinity-tolerant genotypes (Do Amaral *et al.*, 2020). The second type of stress memory is known as intergenerational memory. This form of memory is scientifically recognized as meiotically heritable, meaning it is transmitted from parents to their first-generation offspring (Sani *et al.*, 2013; Mladenov *et al.*, 2021); it is also present in the first stress-free condition of the offspring, and it disappears during the second stress-free generation of the offspring. The maternal effect is responsible for memory transfer to the next generation. The last type is transgenerational memory, which is also meiotically heritable and transmitted through parents to first-generation offspring (Lämke and Bäurle, 2017; Mladenov *et al.*, 2021). In contrast to intergenerational memory, these memory genes can be detected in more than two stress-free conditions (Bhadouriya *et al.*, 2021). The maternal effect is responsible for the transmission of transgenerational memory to the following generation. Maternal genotypes transfer RNA, protein, lipids, carbohydrates, and other compounds into seeds, allowing the next generation of crop plants to protect themselves against stressful conditions (Crisp *et al.*, 2016; Lämke and Bäurle, 2017; Galviz *et al.*, 2020).

Plant performance declines due to the imposition of drought stress. During the first drought stress event, epigenetic modification in the genomic region occurs, impacting the expression patterns of certain genes (Natali *et al.*, 2007; Balan *et al.*, 2017; Liu and He, 2020) and triggering the production of vital proteins. This synthetic protein enables crop plants to exhibit different responses during recurrent drought stress events compared with their initial experience with drought stress. Fig. 14.1 depicts the flow of memory that allows crops to act differently in subsequent drought conditions.

14.3 Responses to Drought Stress Signals

Agricultural plants employ diverse physiological, metabolic, and developmental responses

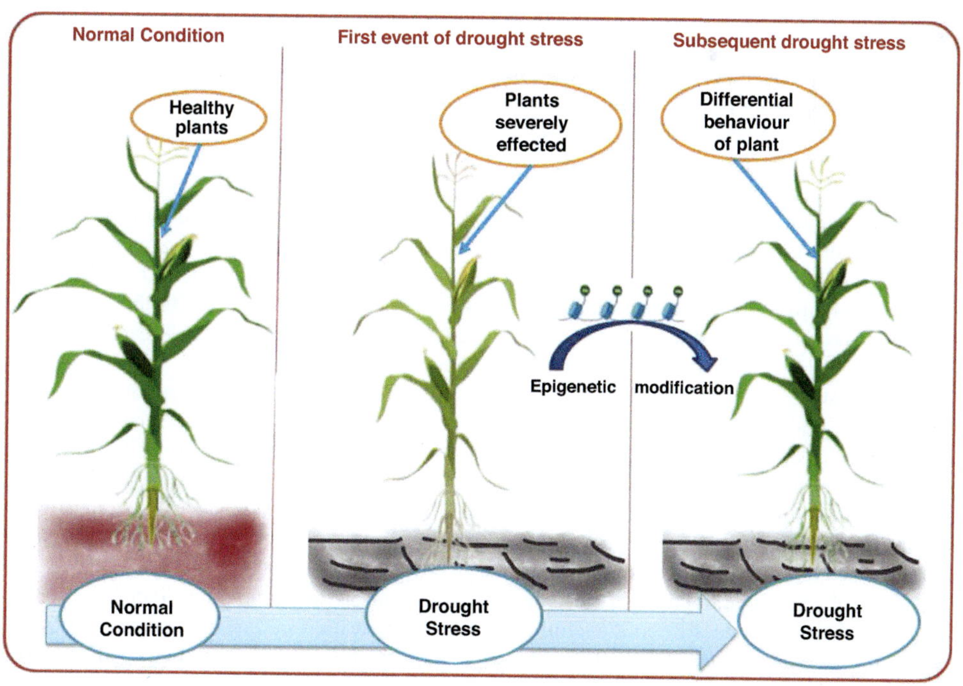

Fig. 14.1. Differential performance of plant after first drought stress event due to memory establishment. Author's own image.

to adapt to recurring drought stress conditions. These adaptive strategies enable them to effectively cope with the challenges posed by periods of water scarcity. Physiological alterations in chlorophyll content, osmotic accumulation, and hormone response play a crucial role during recurrent drought stress. Among them, total chlorophyll concentration is an important drought-tolerant indication in plants. Dehydration in the soil can affect both chlorophyll-a and chlorophyll-b (Farooq *et al.*, 2009). Photo-oxidation of pigments, oxidative stress, and chlorophyll breakdown all contribute to a decrease in chlorophyll levels during drought stress. Researchers have discovered that chlorophyll levels in plants decrease or remain stable, depending on the duration and severity of drought conditions. This discovery underscores the dynamic response of chlorophyll synthesis to varying degrees of water scarcity in plants (Kyparissis *et al.*, 1995; Zhang and Kirkham, 1996). According to Oñate *et al.* (2011), alteration in chlorophyll expression during the initial drought stress can improve drought resistance during subsequent incidences. Priming (i.e., exposure to first stress) in *Triticum aestivum* enhanced chlorophyll, ribulose 1,5 bisphosphate carboxylase, and photosynthetic efficiency during the second drought stress to help in drought stress memory establishment (Abid *et al.*, 2016). Priming keeps the photosynthetic apparatus functioning during stress.

The cytoplasm of plant cells contains a variety of organic and inorganic solutes in order to maintain cell turgidity and lower osmotic potential. During drought stress, plants increase the synthesis of osmo-regulators such as sucrose, proline, and glycine betaine. These compounds help maintain cellular osmotic balance and stabilize cellular structures under water-deficit conditions, soluble carbohydrates and other solutes, which are acquired in the cytoplasm, decrease water potential inside plant cells and lead to increases in water absorption from drying soil. Plants accumulate proline as an initial reaction to water stress in order to prevent cell damage. It acts as a signaling molecule, regulating mitochondrial processes, influencing cell growth and division, and promoting the expression of specific stress-responsive genes in plants to facilitate recovery from stress (Szabados and Savouré, 2010). Proline accumulation tends to be higher in stress-tolerant plant species compared with stress-sensitive ones, and it is often linked with enhanced stress tolerance across various plant species. Furthermore, it protects subcellular architecture, scavenges free radicals, and functions as a molecular chaperon that stabilizes protein structure (Szabados and Savouré, 2010). An increase in the expression level of the proline synthesis gene (*P5CS1*) was observed in rice during the first drought stress event; it reached peak value during the recovery period, and remained constant during subsequent drought stress conditions (Li *et al.*, 2019a). This indicates that proline plays a significant role in the formation of short-term memory for drought stress in plants. Plant growth hormones like auxin, cytokinin, ABA, and ethylene are crucial for regulating various aspects of plant development. Modifying the expression patterns of these hormones also contributes to enhancing drought resistance. During drought stress, there is a shift in the expression of ABA, which activates genes responsible for scavenging ROS. The altered expression of drought-responsive genes such as RD29B is dependent on the levels of ABA expressed during drought conditions (Ding *et al.*, 2012).

The synthesis of key metabolites (sugar, amino acids, amides, etc.) during drought stress is a crucial process for maintaining cellular homeostasis (Barnaby *et al.*, 2013). Production of ROS is one of the eukaryotic cell's earliest metabolic responses to abiotic stresses. Plants release ROS during an oxidative burst, which is the first stage of their defense mechanism in response to water stress. Plants use this reaction as a secondary messenger to initiate additional defense responses in the future. Reactive oxygen species are highly reactive molecules that can inflict substantial damage on plants by accelerating the degradation of proteins, lipids, and DNA, ultimately leading to cell death. This oxidative damage underscores the critical need for plants to maintain efficient ROS scavenging mechanisms to mitigate the harmful effects of ROS accumulation during stress conditions. The protection of the plants from ROS is conferred by the activation of ROS-scavenging genes. In rice, glutathione reductase (GR), APX1, and monodehydroascorbate reductase (MDAR) are the ROS scavenger genes involved in drought memory establishment (Li *et al.*, 2019a).

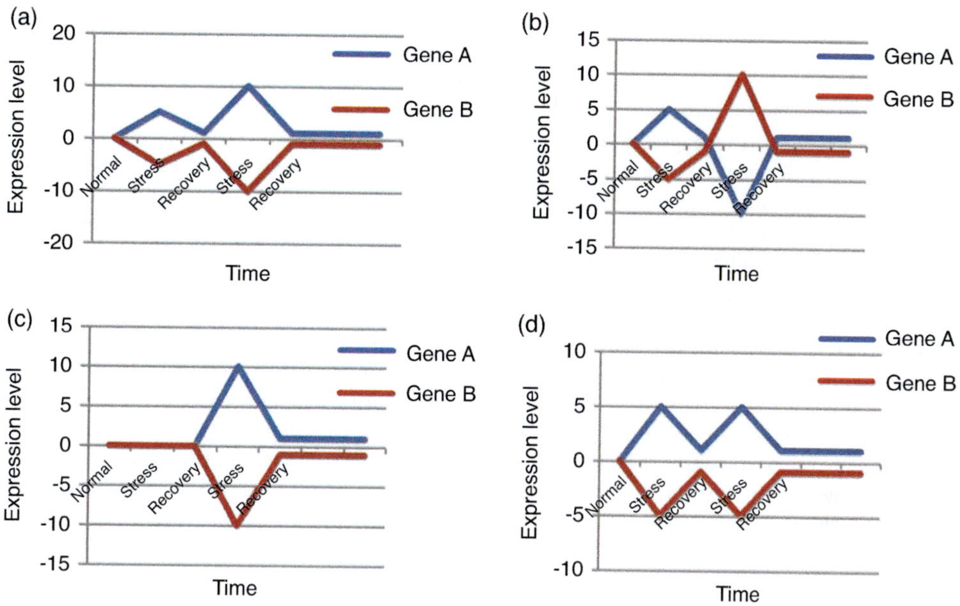

Fig. 14.2. Classification of memory and non-memory genes on the basis of differential expression pattern. Author's own image.

14.4 Classification of Drought Memory and Non-Memory Genes

Memory genes are identified based on their distinctive expression patterns during subsequent drought stress events compared with normal control conditions. This differential expression signifies their role in encoding molecular memory, enabling plants to respond more effectively to recurring incidents of drought stress. Drought stress memory genes are classified into four broad types (Ding *et al.*, 2013). The first is [+/+] or [−/−] type of gene expression patterns; the first plus sign represents the upregulation of a specific gene during the first drought stress event; the second plus sign also represents differential upregulation of the same gene in the second drought stress event as compared with the first drought stress event; and both negative signs represent differential downregulation of a specific gene during the first and second drought stress events (Fig. 14.2a). The second category depicts the opposite regulation in response to the first and second events. In the case of [−/+] memory genes (Fig. 14.2b), a specific gene's expression is decreased during the initial drought stress event and increased during the subsequent event. Conversely, for [+/−] memory genes, the gene expression is elevated during the first drought stress event but reduced during the subsequent stress event. Genes with equal transcript levels in response to each stress event are referred to as 'non-memory genes' and are labeled as [+/=] or [−/=] (Ding *et al.*, 2014) (Fig. 14.2c and Fig. 14.2d). The question arises with the function of non-memory genes. Non-memory genes with a [+/=] expression pattern contribute to nucleosome remodeling, while those with a [−/=] expression pattern contribute to DNA methylation in the DNA binding domain (de Freitas *et al.*, 2018). These differently expressed transcripts lead to physiological changes (Virlouvet *et al.*, 2018).

14.5 Molecular Basis of Drought Stress Memory Formation

Chromatin remodeling (Martienssen and Colot, 2001; Sahu *et al.*, 2013; Kinoshita and Seki, 2014) and important protein assimilation are two molecular bases for drought stress memory formation. Chromatin remodeling has been found responsible for changes in the transcription of stress-responsive genes (Oberkofler *et al.*, 2021). It activates genes undergoing stress and subsequently reduces the expression level of genes when stress disappears. Various epigenetic modifications, including DNA methylation, histone tail modification (Kinoshita and Seki, 2014), and RNA-directed DNA methylation (RdDM), are involved in chromatin remodeling. According to the position of methylation in the related genomic region, DNA methylation plays a role in epigenetic modifications in the genome (Singh *et al.*, 2014) by activating or silencing genes. Methylation in the gene body stimulates a particular gene expression, whereas methylation in the promoter region and transposable element suppresses a specific gene (Zilberman *et al.*, 2007; Li *et al.*, 2012). Histone modification is the subsequent most significant epigenetic modification. For a specific gene to be expressed, DNA must be unwound from histone proteins. Histone modification is responsible for the separation of packed DNA around the histone protein. Acetylation and methylation of histone proteins allow access to wounded DNA for expression and are hence responsible for the activation and silencing of specific genes. Gene activation and silencing are determined by the methylation and acetylation of lysine in H3 histone. Methylation and deacetylation of H3K9 and H3K27 enable the gene to be silenced, whilst acetylation and methylation of H3K4 and H3K36 enable it to be activated.

Non-coding RNA has an important function in drought stress memory development. RNA-directed DNA methylation (RdDM) is also responsible for chromatin remodeling in *Arabidopsis* (Matzke and Birchler, 2005). Double-stranded RNA (dsRNA) produced from the RNA-dependent RNA polymerase 2 (RDR2) from RNA polymerase II in the nucleus is used for DNA methylation. The same technique is used to process this dsRNA for RNA interference and is effected with the help of a Dicer-like family enzyme (DCL3). The processed dsRNA serves as a signal for direct DNA methylation. Drought stress causes the assimilation of crucial proteins or signaling molecules such as ABA, proline, and so on. Drought stress memory development relies heavily on ABA signaling. The endogenous level of ABA increases during drought stress conditions, whereas it decreases after re-watering and, after repeated rounds of treatment, the amount of ABA becomes constant, indicating the presence of drought stress memory (Virlouvet and Fromm, 2015).

Along with epigenetic change, some transcription factors are also responsible for activating drought stress-responsive genes in plants. Several transcription factors (TF) involved in drought stress resistance gene activation have been identified (Fang *et al.*, 2015; Sakuraba *et al.*, 2015). Among these, SNAC1 plays a significant role in the activation of drought resistance genes in rice. During drought stress, the transcription factor SNAC1 binds to the OsbZIP23 promoter, an essential ABA signaling regulator, allowing positive regulation of the expression of numerous ABA signaling genes, thereby increasing drought stress tolerance (Li *et al.*, 2019b).

14.6 Activation of Drought-Responsive Genes

Drought stress promotes the production of ROS, which affects the photosynthesis mechanism of plant cells. In order to protect plant cells from ROS damage, the expression levels of two genes, Responsive to Dessication 29B (RD29B) and Responsive to Abscisic Acid 18 (RAB18), have been found to increase in *Arabidopsis*, leading to the production of antioxidants and osmoprotectants (Kulcheski *et al.*, 2011; Kim *et al.*, 2012). In these two genes, methylation at H3K4 in histone proteins activates these genes, resulting in a greater number of transcripts that synthesize antioxidants and osmoprotectants, which contribute to ROS scavenging and reduced photochemical damage. Similarly, the JMJ27 gene plays an essential role in the activation of drought stress response genes (Wang *et al.*, 2021a). In *Arabidopsis*, demethylase at H3K9 activates this gene and enhanced transcription of this gene is responsible for the activation of drought stress response genes. Genes RD29A and RD20 produce Dehydrin and LEA proteins, which play a key role in protecting crop plants from drought stress, and they are activated by acetylation at the H3K9 site in the histone protein.

The elevated level of ABA during the drought also plays a vital function in activating the RD29B gene. In non-stressed crop plants, low concentrations of ABA hinder the activation of ABF transcription factors (TFs) by SnRK2 kinases. This inhibition prevents the binding of ABF TFs to the ABA response element, thereby reducing the expression of RD29B, a gene known to respond to ABA signaling. Under regulated environments, the level of ABA is low, resulting in a low level of SnRK2, which is unable to phosphorylate ABF proteins. During stress, the level of ABA increases, which likewise raises the level of SnRK2. A higher level of SnRK2 phosphorylates ABF proteins and activates related TFs, leading to the over-expression of RD29B genes. Furthermore, when exposed to the initial drought stress, the H3K4me3 mark level increases. Higher amounts of RD29B transcript, RNA Pol II, and H3K4me3 methylation marks have been observed as a result of repeated drought conditions. During the first and subsequent drought stress events, however, the H3K27me3 repressive mark level remains constant (Ding *et al.*, 2012; Virlouvet *et al.*, 2014; Virlouvet and Fromm, 2015; Alves *et al.*, 2019).

14.7 Drought Stress Priming and Cross-Tolerance to Various Abiotic Stresses

Cross-tolerance simply means that established memory (priming) for one abiotic stress helps crop plants minimize the intensity of the impact of other abiotic stresses during the next growth and development stages or generations (Table 14.2). Cross-tolerance reduces crop production loss and allows crop plants to function well during additional abiotic stress throughout the following crop life cycle.

14.7.1 Drought stress priming for cross-tolerance to heat stress

Rice is an important cereal crop that is exposed to a variety of abiotic stresses throughout its lifecycle. The effect of drought stress priming for cross-tolerance to heat stress was investigated in rice by applying drought stress at the vegetative stage, which leads to the establishment of drought stress priming, and such established priming helps enhance heat stress tolerance at the anthesis stage. The heat shock proteins (HSPs) operate as molecular chaperons to maintain the quaternary structure of proteins. The greater levels of HSPs, such as OsHSP16.9A, OsHSP70.1, and OsHSP70.6, in primed plants indicate improved cross-tolerance in rice (Feijó *et al.*, 2023). Wheat is another important grain crop in which productivity is influenced by heat stress. The effect of drought stress priming on heat stress was also investigated in wheat cultivars. Priming during drought conditions increases CO_2 absorption rates and yield results show that priming does not cause long-term damage to wheat (Mendanha *et al.*, 2020). Terminal heat stress in wheat is also responsible for the loss in yield of wheat. Drought priming in wheat crops was used to induce thermotolerance under terminal heat stress. It was induced by maintaining a mild drought situation for 8 days during wheat jointing. Drought priming induces the levels of heat shock factors, calmodulin, antioxidative genes, polyamines, and glutathione production genes, allowing wheat plants to withstand terminal heat stress (Kumar *et al.*, 2023). Drought stress priming in chickpeas induces cross-tolerance at 35°C, which helps in the identification of the best-performing and heat stress-tolerant genotype (Yadav *et al.*, 2022). Another study on temperate grass species was undertaken to find out how protein phosphorylation aids in cross-tolerance during drought and heat stress. The study confirmed that post-translational modification plays a significant role in cross-tolerance (Zhang *et al.*, 2020).

14.7.2 Drought stress priming for cross-tolerance to cold stress

Cold temperature is one of the most important abiotic elements influencing wheat development and yield. Drought priming is an important strategy for wheat during the vegetative stage to improve resilience to future cold stress. Drought stress memories are heritable in nature; therefore, it is possible to analyze the cross-tolerance for cold stress in the offspring of

Table 14.2. Drought stress priming in cross-tolerance to some abiotic stresses in plants

Primary stressor	Species	Cross-adaptation	Responsible factor	References
Water stress	Rice	Heat stress	Higher yield and number of panicles; increased expression of *OsHSP16.9A*, *OsHSP70.1*, and *OsHSP70.6*	Feijó *et al.*, 2023
Drought	Wheat	Heat stress	Higher CO_2 assimilation rate	Mendanha *et al.*, 2020
Drought	Wheat	Drought stress	More root growth rate and level of proline and glycine betaine; more transcription of genes involved in metabolism	Wang *et al.*, 2021b
Drought	Wheat	Cold stress	Higher ABA and lower osmotic potential; higher antioxidant activities and less H_2O_2; more ATP concentration and higher enzymatic activities	Guo *et al.*, 2022
Drought	Wheat	Heat stress	Lower membrane damage (ELI, MDA and LOX); more antioxidative activity; upregulation of HSFs, calmodulin, antioxidative genes, polyamine	Kumar *et al.*, 2023
Heat and drought	Maize	Heat and drought stress	Higher SOD, POD, CAT activities; more soluble sugar, protein, and proline content	Ru *et al.*, 2022
Heat and drought	Maize	Drought and heat stress	Reduced photosynthetic pigmentation, stomatal dimensions, and photosynthesis; elevated SOD, POD, CAT, and GR activity	Ru *et al.*, 2023
Drought	Soybean	Drought stress	Lower rate of transpiration and stomatal conductivity; more root-to-shoot ratio	Sintaha *et al.*, 2022

Continued

Table 14.2. Continued

Primary stressor	Species	Cross-adaptation	Responsible factor	References
Drought	Chickpea	Drought stress	More plant matter, fluorescence from chlorophyll, stomatal conductivity, total photosynthesis, and RWC	Makonya *et al.*, 2020
Drought	Chickpea	Heat stress	Higher SOD, CAT, GPX, APX, and GR activities; lower leaf water and chlorophyll content; higher electrolyte leakage, lipid peroxidation, proline, and total sugar	Yadav *et al.*, 2022
Drought	Chickpea	Chilling stress	Reduced membrane damage, chlorophyll b degradation, and cryoprotectant accumulation	Saini *et al.*, 2022
Drought	Chickpea	Heat stress	Upregulation of targeted ABFs, and GRFs; accumulation of 78 metabolites	Juneja *et al.*, 2023
High light (HL) and drought	*Arabidopsis thaliana*	Drought stress	Expression of specific stress response gene	Holness *et al.*, 2023
Water stress	*Coffea canephora*	Drought stress	Excelling the number of individuals	Sseremba *et al.*, 2023
Drought	Potato	Drought stress	More number of small vascular bundles; increasing water use efficiency and lowering transpiration; higher amount and rate of dry matter translocation	Lv *et al.*, 2024
Drought	*Festuca arundinacea*	Heat stress	Regulation of MAP kinases	Zhang *et al.*, 2020

parents with drought stress memories imposed during the grain filling stage. Drought stress-primed parents produce offspring with higher levels of ABA, decreased osmotic potential and improved antioxidant enzyme activity during cold stress, allowing offspring to withstand cold stress in wheat (Guo *et al.*, 2022). Chilling temperature is the most important abiotic element influencing chickpea growth and reproduction. Drought stress priming was applied to various chickpea cultivars during the vegetative phase and cold tolerance was examined throughout the reproductive cycle. Drought priming reduces membrane damage, improves water content, and increases cryoprotective solute accumulation, which benefits chickpeas during chilling stress and enhances crop production (Saini *et al.*, 2022).

14.7.3 Drought stress priming for cross-tolerance to drought stress

Drought stress priming has been used to improve drought resistance in wheat. The role of drought priming in wheat root apex, which plays an important role during drought stress, revealed that genes with differential expression patterns due to initial drought priming are mainly involved in hormonal pathways and play an important role in cell cycle regulation and cell wall loosening, which help roots to increase drought tolerance (Wang *et al.*, 2021b). The effect of drought stress priming on subsequent drought conditions of soybean revealed that the primed plant shows altered morpho-physiological traits, such as reduced drought stress index, increase in water use efficiency, increased root-to-shoot ratio, genes like late embryogenesis abundant (LEA) proteins and MYB, NAC and PP2C domain-containing transcription factors more in primed plants as compared with non-primed plants, which helps soybean to improve drought tolerance (Sintaha *et al.*, 2022). A study of drought priming in heat-tolerant chickpea genotypes demonstrated that drought priming at the vegetative stage improves heat-tolerant genotype acclimatization to drought stress at the flowering stage (Makonya *et al.*, 2020). The presence of high light and drought priming induces both ABA-dependent and ABA-independent responses, as well as decreased expression of stress-responsive genes in the aba-1 mutant in *Arabidopsis* (Holness *et al.*, 2023). Drought priming has been demonstrated with potatoes to boost photosynthesis and yield. Drought priming increased the amount and rate of dry matter transfer, minimizing detrimental effects on potato tubers during drought stress (Lv *et al.*, 2024).

14.7.4 Early heat and drought stress priming for cross-tolerance to later heat and drought stress

Researchers discovered that early drought priming improves heat and cold stress tolerance in crops. Ru *et al.* (2023) investigated the effects of combined heat and drought stress priming in respect of the impact of subsequent heat and drought stress in maize. The study revealed that primed maize plants had excellent antioxidant enzymes and osmoregulators, allowing them to perform better than maize plants non-primed under combined heat and drought stress (Ru *et al.*, 2023). Another study of such combined stress (heat and drought priming) established that heat, drought and combined heat and drought stress priming improved maize adaptation to subsequent heat, drought and combined heat and drought stress (Ru *et al.*, 2022).

14.8 Techniques Used for the Detection of Epigenetic Changes in Crop Plants

The detection of histone modifications is achieved by employing chromatin immunoprecipitation (ChIP) techniques. Basically, the approach relies on the principle of antigen and antibody precipitation (protein–protein interaction). Histone proteins are referred to as antigens and specially developed antibodies against histone proteins are used to precipitate antigens (Carey *et al.*, 2009; Xu *et al.*, 2021). Initially, proteins are cross-linked to DNA using formaldehyde. Subsequently, chromatin is broken down into minute fragments using ultrasound or micrococcal nuclease (MNase) digestion. The immuno-precipitation of antigen and a chosen antibody comprising precipitant (agarose beads) are used to precipitate antibodies bound to histone protein. PCR amplification with specific primers was employed to observe whether DNA sequences of interest had been precipitated or not. ChIP-seq and ChIP-qPCR methods are used to analyze the data using the ChIP methodology (Malone *et al.*, 2011; Zong *et al.*, 2020). ChIP-seq is a widely used approach; precipitated DNA is put into a DNA sequence machine, which generates data in the form of reads. The reads are then aligned using genome assembly methods. Finally, following peak calling, changes to the specific DNA sequence can be investigated. The second local method for data analysis is ChIP-qPCR. In this method, precipitated samples are utilized to perform qPCR on the desired gene. It indicates whether or not the gene of interest exists. Epigenetic modifications have the capacity to alter crop plant genotypes without affecting an organism's DNA sequence. This form of genotypic change is caused by DNA methylation at the promoter region. To detect this type of genomic alteration, a sodium bisulfite sequencing technique can be applied (Rajkumar *et al.*, 2020). It is an efficient method for detecting 5-methycytosine. This sequencing method uses sodium bisulfite as a deamination agent (Li and Tollefsbol, 2011). Cytosine in single-stranded DNA is transformed into uracil and identified as thymine. However, 5mCs are resistant to this conversion and remain cytosine, distinguishing them from unmethylated cytosines. The actual methylation status can be determined by either direct PCR product sequencing (detection of average methylation status) or sub-cloning sequencing (detection of single-molecule methylation pattern distribution). Furthermore, bisulfite sequencing analysis can identify DNA methylation status along a single DNA strand as well as the DNA methylation patterns of two DNA strands because the converted DNA strands are no longer self-complementary and the amplification products can be measured individually.

14.9 Conclusion

Plant drought stress causes photo-oxidative damage, production of ROS, reduced CO_2 uptake, stomatal closing, and other effects in crop plants, resulting in a loss in plant growth and productivity. Plants have the ability to recall drought conditions and perform well in recurring drought stress conditions, which contributes to the development of memory in crop plants. Crop plants exhibit various types of memory and these memories are heritable in nature. Drought stress memory is established by epigenetic changes and the synthesis of critical proteins. These memories enable crop plants to protect themselves against recurring drought stress conditions. Due to changed gene expression patterns and the manufacture of essential proteins or secondary metabolites, crop plants exhibit distinct responses compared with naïve plants. Drought stress priming can also assist crops in improving their resistance to another abiotic stress. Such cross-tolerance helps crops respond better to another abiotic stress during the subsequent growth, which leads to reduced crop production loss.

References

Abid, M., Tian, Z., Ata-Ul-Karim, S.T., Liu, Y., Cui, Y. *et al.* (2016) Improved tolerance to post-anthesis drought stress by pre-drought priming at vegetative stages in drought-tolerant and -sensitive wheat cultivars. *Plant Physiology and Biochemistry* 106, 218–227. DOI: 10.1016/j.plaphy.2016.05.003.

Alves de Freitas Guedes, F., Menezes-Silva, P.E., DaMatta, F.M. and Alves-Ferreira, M. (2019) Using transcriptomics to assess plant stress memory. *Theoretical and Experimental Plant Physiology* 31(1), 47–58. DOI: 10.1007/s40626-018-0135-0.

Arshad, M., Feyissa, B.A., Amyot, L., Aung, B. and Hannoufa, A. (2017) MicroRNA156 improves drought stress tolerance in alfalfa (*Medicago sativa*) by silencing SPL13. *Plant Science* 258, 122–136. DOI: 10.1016/j.plantsci.2017.01.018.

Balan, B., Caruso, T. and Martinelli, F. (2017) Gaining insight into exclusive and common transcriptomic features linked with biotic stress responses in Malus. *Frontiers in Plant Science* 8, 279625. DOI: 10.3389/fpls.2017.01569.

Barnaby, J.Y., Kim, M., Bauchan, G., Bunce, J. and Reddy, V. (2013) Drought responses of foliar metabolites in three maize hybrids differing in water stress tolerance. *PLoS ONE* 8(10), e77145. DOI: 10.1371/journal.pone.0077145.

Berry, S. and Dean, C. (2015) Environmental perception and epigenetic memory: Mechanistic insight through FLC. *The Plant Journal* 83(1), 133–148. DOI: 10.1111/tpj.12869.

Bhadouriya, S.L., Mehrotra, S., Basantani, M.K., Loake, G.J. and Mehrotra, R. (2021) Role of chromatin architecture in plant stress responses: An update. *Frontiers in Plant Science* 11, 603380. DOI: 10.3389/fpls.2020.603380.

Carey, M.F., Peterson, C.L. and Smale, S.T. (2009) Chromatin immunoprecipitation (chIP). *Cold Spring Harbor Protocols* 2009(9). DOI: 10.1101/pdb.prot5279.

Crisp, P.A., Ganguly, D., Eichten, S.R., Borevitz, J.O. and Pogson, B.J. (2016) Reconsidering plant memory: Intersections between stress recovery, RNA turnover, and epigenetics. *Science Advances* 2(2), e1501340. DOI: 10.1126/sciadv.1501340.

de Freitas Guedes, F.A., Nobres, P., Ferreira, D.C.R., Menezes-Silva, P.E., Ribeiro-Alves, M. *et al.* (2018) Transcriptional memory contributes to drought tolerance in coffee (*Coffea canephora*) plants. *Environmental and Experimental Botany* 147, 220–233. DOI: 10.1016/j.envexpbot.2017.12.004.

Ding, Y., Fromm, M. and Avramova, Z. (2012) Multiple exposures to drought "train" transcriptional responses in *Arabidopsis*. *Nature Communications* 3(1), 740. DOI: 10.1038/ncomms1732.

Ding, Y., Liu, N., Virlouvet, L., Riethoven, J.J., Fromm, M. *et al.* (2013) Four distinct types of dehydration stress memory genes in *Arabidopsis thaliana*. *BioMed Central Plant Biology* 13, 1–11. DOI: 10.1186/1471-2229-13-229.

Ding, Y., Virlouvet, L., Liu, N., Riethoven, J.J., Fromm, M. *et al.* (2014) Dehydration stress memory genes of *Zea mays*; comparison with *Arabidopsis thaliana*. *BioMed Central Plant Biology* 14, 1–15. DOI: 10.1186/1471-2229-14-141.

Do Amaral, M.N., Auler, P.A., Rossatto, T., Barros, P.M., Oliveira, M.M. *et al.* (2020) Long-term somatic memory of salinity unveiled from physiological, biochemical and epigenetic responses in two contrasting rice genotypes. *Physiologia Plantarum* 170(2), 248–268. DOI: 10.1111/ppl.13149.

D'Odorico, P., Schönbeck, L., Vitali, V., Meusburger, K., Schaub, M. *et al.* (2021) Drone-based physiological index reveals long-term acclimation and drought stress responses in trees. *Plant, Cell & Environment* 44(11), 3552–3570. DOI: 10.1111/pce.14177.

Fang, Z., Zhang, X., Gao, J., Wang, P., Xu, X. *et al.* (2015) A buckwheat (*Fagopyrum esculentum*) DRE-binding transcription factor gene, FeDREB1, enhances freezing and drought tolerance of transgenic *Arabidopsis*. *Plant Molecular Biology Reporter* 33, 1510–1525. DOI: 10.1007/s11105-015-0851-4.

Farooq, M., Wahid, A., Kobayashi, N.S.M.A., Fujita, D.B.S.M.A. and Basra, S.M. (2009) Plant drought stress: Effects, mechanisms and management. In: Lichtfouse, E., Navarrette, M., Debaeke, P., Souchere, V. and Alberola, C. (eds) *Sustainable Agriculture*. Springer, Dordrecht, pp. 153–188. DOI: 10.1007/978-90-481-2666-8_12.

Feijó, A.D.R., Viana, V.E., Balbinot, A., Fipke, M.V., Souza, G.M. *et al.* (2023) Water deficit at vegetative stage induces tolerance to high temperature during anthesis in rice. *Plants* 12(17), 3133. DOI: 10.3390/plants12173133.

Galviz, Y.C., Ribeiro, R.V. and Souza, G.M. (2020) Yes, plants do have memory. *Theoretical and Experimental Plant Physiology* 32(3), 195–202. DOI: 10.1007/s40626-020-00181-y.

Guo, J., Wang, H., Liu, S., Wang, Y., Liu, F. *et al.* (2022) Parental drought priming enhances tolerance to low temperature in wheat (*Triticum aestivum*) offspring. *Functional Plant Biology* 49(11), 946–957. DOI: 10.1071/FP22043.

Hasan, M.M., Skalicky, M., Jahan, M.S., Hossain, M.N., Anwar, Z. *et al.* (2021) Spermine: Its emerging role in regulating drought stress responses in plants. *Cells* 10(2), 261. DOI: 10.3390/cells10020261.

Hilker, M. and Schmülling, T. (2019) Stress priming, memory, and signalling in plants. *Plant, Cell & Environment* 42(3), 753–761. DOI: 10.1111/pce.13526.

Hilker, M., Schwachtje, J., Baier, M., Balazadeh, S., Bäurle, I. *et al.* (2016) Priming and memory of stress responses in organisms lacking a nervous system. *Biological Reviews* 91(4), 1118–1133. DOI: 10.1111/brv.12215.

Holness, S., Bechtold, U., Mullineaux, P., Serino, G. and Vittorioso, P. (2023) Highlight induced transcriptional priming against a subsequent drought stress in *Arabidopsis thaliana*. *International Journal of Molecular Sciences* 24(7), 6608. DOI: 10.3390/ijms24076608.

Jacques, C., Salon, C., Barnard, R.L., Vernoud, V. and Prudent, M. (2021) Drought stress memory at the plant cycle level: A review. *Plants* 10(9), 1873. DOI: 10.3390/plants10091873.

Juneja, S., Saini, R., Mukit, A. and Kumar, S. (2023) Drought priming modulates ABF, GRFs, related microRNAs and induce metabolic adjustment during heat stress in chickpea. *Plant Physiology and Biochemistry* 203, 108007. DOI: 10.1016/j.plaphy.2023.108007.

Kim, J.M., To, T.K., Ishida, J., Matsui, A., Kimura, H. *et al.* (2012) Transition of chromatin status during the process of recovery from drought stress in *Arabidopsis thaliana*. *Plant & Cell Physiology* 53(5), 847–856. DOI: 10.1093/pcp/pcs053.

Kim, Y.K., Chae, S., Oh, N.I., Nguyen, N.H. and Cheong, J.J. (2020) Recurrent drought conditions enhance the induction of drought stress memory genes in *Glycine max* L. *Frontiers in Genetics* 11, 576086. DOI: 10.3389/fgene.2020.576086.

Kinoshita, T. and Seki, M. (2014) Epigenetic memory for stress response and adaptation in plants. *Plant & Cell Physiology* 55(11), 1859–1863. DOI: 10.1093/pcp/pcu125.

Kulcheski, F.R., Oliveira, L.F., Molina, L.G., Almerão, M.P., Rodrigues, F.A. *et al.* (2011) Identification of novel soybean microRNAs involved in abiotic and biotic stresses. *BioMed Central Genomics* 12, 1–17. DOI: 10.1186/1471-2164-12-307.

Kumar, K.P., Pushpam, R., Manonmani, S., Raveendran, M., Santhiya, S. *et al.* (2024) Enhancing stress resilience in rice (*Oryza sativa* L.) through profiling early-stage morpho-physiological and molecular responses to multiple abiotic stress tolerance. *Frontiers in Plant Science* 15, 1342441. DOI: 10.3389/fpls.2024.1342441.

Kumar, R., Adhikary, A., Saini, R., Khan, S.A., Yadav, M. *et al.* (2023) Drought priming induced thermotolerance in wheat (*Triticum aestivum* L.) during reproductive stage; a multifaceted tolerance approach against terminal heat stress. Plant Physiology and Biochemistry 201, 107840. DOI: 10.1016/j.plaphy.2023.107840.

Kyparissis, A., Petropoulou, Y. and Manetas, Y. (1995) Summer survival of leaves in a soft-leaved shrub (*Phlomis fruticosa* L., *Labiatae*) under mediterranean field conditions: Avoidance of photoinhibitory damage through decreased chlorophyll contents. Journal of Experimental Botany 46(12), 1825–1831. DOI: 10.1093/jxb/46.12.1825.

Lämke, J. and Bäurle, I. (2017) Epigenetic and chromatin-based mechanisms in environmental stress adaptation and stress memory in plants. *Genome Biology* 18, 1–11. DOI: 10.1186/s13059-017-1263-6.

Lee, B.R., Islam, M.T., Park, S.H., Jung, H.I., Bae, D.W. *et al.* (2019) Characterization of salicylic acid-mediated modulation of the drought stress responses: Reactive oxygen species, proline, and redox state in *Brassica napus*. *Environmental and Experimental Botany* 157, 1–10. DOI: 10.1016/j.envexpbot.2018.09.013.

Li, Y. and Tollefsbol, T.O. (2011) DNA methylation detection: Bisulfite genomic sequencing analysis. *Epigenetics Protocols* 11–21. DOI: 10.1007/978-1-61779-316-5_2.

Li, P., Yang, H., Wang, L., Liu, H., Huo, H. *et al.* (2019b) Physiological and transcriptome analyses reveal short-term responses and formation of memory under drought stress in rice. *Frontiers in Genetics* 10, 55. DOI: 10.3389/fgene.2019.00055.

Li, X., Zhu, J., Hu, F., Ge, S., Ye, M. *et al.* (2012) Single-base resolution maps of cultivated and wild rice methylomes and regulatory roles of DNA methylation in plant gene expression. *BioMed Central Genomics* 13, 1–15. DOI: 10.1186/1471-2164-13-300.

Li, X., Chang, Y., Ma, S., Shen, J., Hu, H. *et al.* (2019a) Genome-wide identification of SNAC1-targeted genes involved in drought response in rice. *Frontiers in Plant Science* 10, 982. DOI: 10.3389/fpls.2019.00982.

Liu, J. and He, Z. (2020) Small DNA methylation, big player in plant abiotic stress responses and memory. *Frontiers in Plant Science* 11, 595603. DOI: 10.3389/fpls.2020.595603.

Lv, Z., Zhang, H., Huang, Y., Zhu, L., Yang, X. *et al.* (2024) Drought priming at seedling stage improves photosynthetic performance and yield of potato exposed to a short-term drought stress. *Journal of Plant Physiology* 292, 154157. DOI: 10.1016/j.jplph.2023.154157.

Makonya, G.M., Ogola, J.B., Muasya, A.M., Crespo, O., Maseko, S. *et al.* (2020) Intermittent moisture supply induces drought priming responses in some heat-tolerant chickpea genotypes. *Crop Science* 60(5), 2527–2542. DOI: 10.1002/csc2.20228.

Malone, B.M., Tan, F., Bridges, S.M. and Peng, Z. (2011) Comparison of four chIP-Seq analytical algorithms using rice endosperm H3K27 trimethylation profiling data. *PLoS ONE* 6(9), e25260. DOI: 10.1371/journal.pone.0025260.

Martienssen, R.A. and Colot, V. (2001) DNA methylation and epigenetic inheritance in plants and filamentous fungi. *Science* 293(5532), 1070–1074. DOI: 10.1126/science.293.5532.1070.

Matzke, M.A. and Birchler, J.A. (2005) RNAi-mediated pathways in the nucleus. *Nature Reviews Genetics* 6(1), 24–35. DOI: 10.1038/nrg1500.

Mendanha, T., Rosenqvist, E., Nordentoft Hyldgaard, B., Doonan, J.H. and Ottosen, C.O. (2020) Drought priming effects on alleviating the photosynthetic limitations of wheat cultivars (*Triticum aestivum* L.) with contrasting tolerance to abiotic stresses. *Journal of Agronomy and Crop Science* 206(6), 651–664. DOI: 10.1111/jac.12404.

Mladenov, V., Fotopoulos, V., Kaiserli, E., Karalija, E., Maury, S. *et al.* (2021) Deciphering the epigenetic alphabet involved in transgenerational stress memory in crops. *International Journal of Molecular Sciences* 22(13), 7118. DOI: 10.3390/ijms22137118.

Natali, L., Giordani, T., Lercari, B., Maestrini, P., Cozza, R. *et al.* (2007) Light induces expression of a dehydrin-encoding gene during seedling de-etiolation in sunflower (*Helianthus annuus* L.). *Journal of Plant Physiology* 164(3), 263–273. DOI: 10.1016/j.jplph.2006.01.015.

Oberkofler, V., Pratx, L. and Bäurle, I. (2021) Epigenetic regulation of abiotic stress memory: Maintaining the good things while they last. *Current Opinion in Plant Biology* 61, 102007. DOI: 10.1016/j.pbi.2021.102007.

Oñate, M., Blanc, J. and Munné-Bosch, S. (2011) Influence of stress history on the response of the dioecious plant *Urtica dioica* L. to abiotic stress. *Plant Ecology & Diversity* 4(1), 45–54. DOI: 10.1080/17550874.2011.557400.

Palliotti, A., Tombesi, S., Frioni, T., Famiani, F., Silvestroni, O. *et al.* (2014) Morpho-structural and physiological response of container-grown Sangiovese and Montepulciano cvv. (*Vitis vinifera*) to re-watering after a pre-veraison limiting water deficit. *Functional Plant Biology* 41(6), 634–647. DOI: 10.1071/FP13271.

Rajkumar, M.S., Shankar, R., Garg, R. and Jain, M. (2020) Bisulphite sequencing reveals dynamic DNA methylation under desiccation and salinity stresses in rice cultivars. *Genomics* 112(5), 3537–3548. DOI: 10.1016/j.ygeno.2020.04.005.

Ru, C., Hu, X., Chen, D., Wang, W. and Song, T. (2022) Heat and drought priming induce tolerance to subsequent heat and drought stress by regulating leaf photosynthesis, root morphology, and antioxidant defense in maize seedlings. *Environmental and Experimental Botany* 202, 105010. DOI: 10.1016/j.envexpbot.2022.105010.

Ru, C., Hu, X., Chen, D. and Wang, W. (2023) Droughts and thermo-priming enhance acclimation to later drought and heat stress in maize seedlings by improving leaf physiological activity. *Agronomy* 13(4), 1124. DOI: 10.3390/agronomy13041124.

Sahu, P.P., Pandey, G., Sharma, N., Puranik, S., Muthamilarasan, M. *et al.* (2013) Epigenetic mechanisms of plant stress responses and adaptation. *Plant Cell Reports* 32, 1151–1159. DOI: 10.1007/s00299-013-1462-x.

Saini, R., Das, R., Adhikary, A., Kumar, R., Singh, I. *et al.* (2022) Drought priming induces chilling tolerance and improves reproductive functioning in chickpea (*Cicer arietinum* L.). *Plant Cell Reports* 41(10), 2005–2022. DOI: 10.1007/s00299-022-02905-7.

Sako, K., Kim, J.M., Matsui, A., Nakamura, K., Tanaka, M. *et al.* (2016) Ky-2, a histone deacetylase inhibitor, enhances high-salinity stress tolerance in *Arabidopsis thaliana*. *Plant & Cell Physiology* 57(4), 776–783. DOI: 10.1093/pcp/pcv199.

Sako, K., Nguyen, H.M. and Seki, M. (2020) Advances in chemical priming to enhance abiotic stress tolerance in plants. *Plant and Cell Physiology* 61(12), 1995–2003. DOI: 10.1093/pcp/pcaa119.

Sakuraba, Y., Kim, Y.S., Han, S.H., Lee, B.D. and Paek, N.C. (2015) The *Arabidopsis* transcription factor NAC016 promotes drought stress responses by repressing AREB1 transcription through a trifurcate feed-forward regulatory loop involving NAP. *The Plant Cell* 27(6), 1771–1787. DOI: 10.1105/tpc.15.00222.

Sani, E., Herzyk, P., Perrella, G., Colot, V. and Amtmann, A. (2013) Hyperosmotic priming of *Arabidopsis* seedlings establishes a long-term somatic memory accompanied by specific changes of the epigenome. *Genome Biology* 14, 1–24. DOI: 10.1186/gb-2013-14-6-r59.

Singh, P., Yekondi, S., Chen, P.W., Tsai, C.H., Yu, C.W. *et al.* (2014) Environmental history modulates *Arabidopsis* pattern-triggered immunity in a HISTONE ACETYLTRANSFERASE1-dependent manner. *The Plant Cell* 26(6), 2676–2688. DOI: 10.1105/tpc.114.123356.

Sintaha, M., Man, C.K., Yung, W.S., Duan, S., Li, M.W. *et al.* (2022) Drought stress priming improved the drought tolerance of soybean. *Plants* 11(21), 2954. DOI: 10.3390/plants11212954.

Sseremba, G., Tongoona, P.B., Musoli, P., Eleblu, J.S.Y., Melomey, L.D. *et al.* (2023) Viability of deficit irrigation pre-exposure in adapting robusta coffee to drought stress. *Agronomy* 13(3), 674. DOI: 10.3390/agronomy13030674.

Sung, S. and Amasino, R.M. (2004) Vernalization in *Arabidopsis thaliana* is mediated by the PHD finger protein VIN3. *Nature* 427(6970), 159–164. DOI: 10.1038/nature02195.

Szabados, L. and Savouré, A. (2010) Proline: A multifunctional amino acid. *Trends in Plant Science* 15(2), 89–97. DOI: 10.1016/j.tplants.2009.11.009.

Tosetti, R., Martinelli, F., Tonutti, P. and Barupal, D.K. (2012) Metabolomics approach to studying minimally processed peach (*Prunus persica*) fruit. In: *XXVIII International Horticultural Congress on Science and Horticulture for People (IHC2010): International Symposium*, pp. 1017–1021. DOI: 10.17660/ActaHortic.2012.934.135.

Virlouvet, L. and Fromm, M. (2015) Physiological and transcriptional memory in guard cells during repetitive dehydration stress. *The New Phytologist* 205(2), 596–607. DOI: 10.1111/nph.13080.

Virlouvet, L., Ding, Y., Fujii, H., Avramova, Z. and Fromm, M. (2014) ABA signaling is necessary but not sufficient for RD 29 B transcriptional memory during successive dehydration stresses in *Arabidopsis thaliana*. *The Plant Journal* 79(1), 150–161. DOI: 10.1111/tpj.12548.

Virlouvet, L., Avenson, T.J., Du, Q., Zhang, C., Liu, N. *et al.* (2018) Dehydration stress memory: Gene networks linked to physiological responses during repeated stresses of *Zea mays*. *Frontiers in Plant Science* 9, 1058. DOI: 10.3389/fpls.2018.01058.

Wang, Q., Liu, P., Jing, H., Zhou, X.F., Zhao, B. *et al.* (2021a) JMJ27-mediated histone H3K9 demethylation positively regulates drought-stress responses in *Arabidopsis*. *The New Phytologist* 232(1), 221–236. DOI: 10.1111/nph.17593.

Wang, X., Chen, J., Ge, J., Huang, M., Cai, J. *et al.* (2021b) The different root apex zones contribute to drought priming induced tolerance to a reoccurring drought stress in wheat. *The Crop Journal* 9(5), 1088–1097. DOI: 10.1016/j.cj.2020.11.008.

Xu, S., Huang, J., Jin, J. and Huang, W. (2021) Global identification for targets of circadian transcription factors in *Arabidopsis* and rice using chromatin immunoprecipitation followed by sequencing (CHIP-seq). In: Yin, R., Li, L. and Zuo, K. (eds) *Plant Photomorphogenesis: Methods and Protocols*. Humana, New York, pp. 61–74. DOI: 10.1007/978-1-0716-1370-2_7.

Yadav, R., Juneja, S., Kumar, R., Saini, R. and Kumar, S. (2022) Understanding cross-tolerance mechanism and effect of drought priming on individual heat stress and combinatorial heat and drought stress in chickpea. *Journal of Crop Science and Biotechnology* 25(5), 515–533. DOI: 10.1007/s12892-022-00148-2.

Zhang, J. and Kirkham, M.B. (1996) Antioxidant responses to drought in sunflower and sorghum seedlings. *New Phytologist* 132(3), 361–373. DOI: 10.1111/j.1469-8137.1996.tb01856.x.

Zhang, X., Zhuang, L., Liu, Y., Yang, Z. and Huang, B. (2020) Protein phosphorylation associated with drought priming-enhanced heat tolerance in a temperate grass species. *Horticulture Research* 7. DOI: 10.1038/s41438-020-00440-8.

Zilberman, D., Gehring, M., Tran, R.K., Ballinger, T. and Henikoff, S. (2007) Genome-wide analysis of *Arabidopsis thaliana* DNA methylation uncovers an interdependence between methylation and transcription. *Nature Genetics* 39(1), 61–69. DOI: 10.1038/ng1929.

Zong, W., Yang, J., Fu, J. and Xiong, L. (2020) Synergistic regulation of drought-responsive genes by transcription factor OsbZIP23 and histone modification in rice. *Journal of Integrative Plant Biology* 62(6), 723–729. DOI: 10.1111/jipb.12850.

15 Plant Drought Resilience Based on Epigenetic Regulation

Muhammad Waseem[1,2,3,4]*, Sana Basharat[1,2,3,4], Iffat Shaheen[5] and Pingwu Liu[1,2,3,4]

[1]College of Tropical Agriculture and Forestry, Hainan University, Haikou, Hainan, China; [2]Key Laboratory of Tropical Horticultural Crop Quality Regulation, College of Horticulture, College of horticulture, Hainan University, Haikou, Hainan, China; [3]Hainan Yazhou Bay Seed Laboratory, Sanya Nanfan Research Institute of Hainan University, Sanya, China; [4]Fang Zhiyuan Academician Team Innovation Center of Hainan Province, China; [5]Department of Agriculture, Bahuddin Zakariya University, Multan, Pakistan

Abstract

Drought stress is one of the most devastating abiotic stresses in plant and agronomic crop plants and leads to lower yield and productivity. Plants encounter drought stress more than once in their whole life and evolve plenty of adaptive strategies to survive and acclimate to it. Among these adaptative strategies, epigenetic regulation of stress tolerance enables the plant to endure recurrent stress. In the past few decades, enormous progress has been made in unveiling the role of epigenetic variability to respond to drought stress. This chapter aims to comprehensively discuss the detailed studies into drought stress response in plants through epigenetic approaches, striving to identify drought resistance genes regulated by DNA methylation and histone modifications, specific signaling pathways, and their mechanisms of action. Lastly, this chapter discusses the possible consequences of comprehending the epigenetic pathways associated with plant drought stress memory for future implementation in enhancing crops and increasing resilience to drought.

Keywords: Drought, Epigenetics, Histone, Methylation, Mechanism, Tolerance

15.1 Introduction

Plants, being sessile organisms, face plenty of environmental challenges throughout their life and develop a variety of adaptive strategies by sensing the stress and survive (Sun *et al.*, 2021). In plants, abiotic stress responses are governed by complex epigenetic mechanisms and are heritable by the next generation (Latzel *et al.*, 2016). Such an ability of plants to remember specific environmental stress is designated as 'plant memory' or 'epigenetic memory' (Zheng *et al.*, 2017). In recent years epigenetic modifications have become popular, enabling plants with the ability to alter gene activity without modifying DNA sequences and adapt to challenging environments (Thiebaut *et al.*, 2019). Epigenetics is a dynamic process of gene expression regulation that affects chromatin stability and structure

*Corresponding author: m.waseem.botanist@gmail.com

 Epigenetics for Climate-Smart and Sustainable Agriculture (ed. Jen-Tsung Chen)
DOI: 10.1079/9781800626102.0015

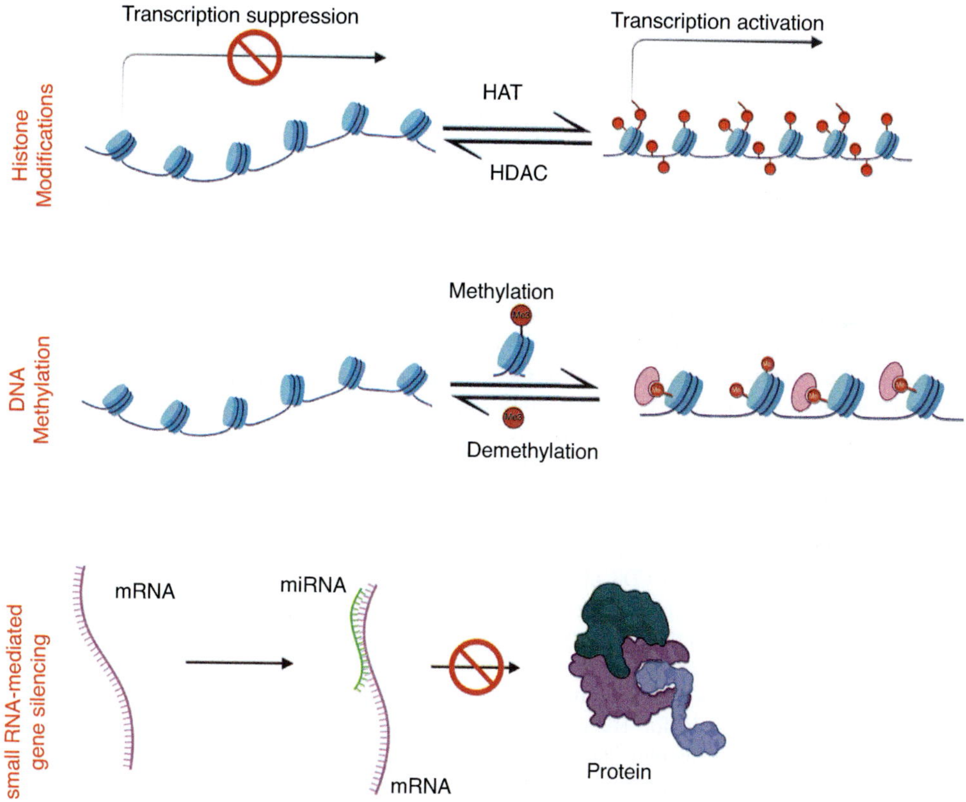

Fig. 15.1. Mechanism of epigenetic regulation in plants. These include histone modification, DNA methylation, and small RNA interference. Histone deacetylase (HDAC) and histone acetyltransferase (HAT) catalyze the interplay between acetylation and deacetylation mechanisms leading to activation and suppression of transcription of genes. DNA methylation includes histone methylation and demethylation catalyzed by methyltransferases and demethylases at lysine/arginine residues. In contrast, small RNA interference or (RNAi) also plays a pivotal role in gene expression regulation *viz* small non-coding RNAs (such as miRNA) or long non-coding RNAs. (Generated with Biorender.com). Author's own image.

through epigenetic markers such as histone modifications (HM), DNA methylation (DM), non-coding RNA (ncRNA), and chromatin remodeling (Fig. 15.1).

With the advancement of high-throughput technology, DM has become the most thoroughly investigated class of epigenetic modifications. DM is a highly conserved epigenetic modification in biological processes, which plays a pivotal role in regulating genome activity by altering DNA structure (Yuan *et al.*, 2020). Alterations in DM levels impact gene activity, subsequently affecting plant response to stress. DM is a form of biochemical process catalyzed by DNA methyltransferases (DNMT) that involve addition of methyl groups (CH3-) to cytosine C-5, guanine G-7 or adenine N-6, namely, 5-methyl cytosine (5mC), 7-methyl guanine (7mG), and N6-methyl purine (N6mA) from S-adenosylmethionine (Gao *et al.*, 2020). Among these, 5-mC is one of the most frequent methylation categories and regulates the transcription of plenty of stress response genes. It allows the plant to adapt to challenging environments through genome stability, gene expression regulation, and phenotypic alterations without DNA sequence modulation (Meiling *et al.*, 2021).

However, small RNAs (sRNAs) can guide RNA-directed DM (RdDM) and recruit DNMT to suppress its targets (Matzke and Mosher, 2014). The sRNAs include microRNA (miRNA) and small interfering RNAs (siRNAs) and play a pivotal role in transcriptional and post-transcriptional gene silencing against challenging environments (Gümüş *et al.*, 2023). Stress-induced sRNA interferes with stress-responsive genes, affecting their expression levels. For instance, downregulation of sRNA expression activates the expression of their target genes. Recent research showed the role of sRNA in epigenetic stress responses (Kambona *et al.*, 2023). Li *et al.* (2019b) uncovered long non-coding RNA as short-term epigenetic factors in rice drought tolerance.

The HMs such as H3K4me1, H3K4me2, H3K27me3, H3K4me3, and H3K27me3 are associated with promoters and CDS of stress-responsive genes in *Arabidopsis* (Zhang *et al.*, 2009). Multiprotein complexes Polycomb Repressive Complex 2 (PRC2), and PCR1 comprise H3K27-specific histone methyltransferase (HMT), which is responsible for H2A mono-ubiquitination. These proteins play key roles in HM-mediated epigenetic suppression of gene expression (Zhang *et al.*, 2007).

Drought is one of the most widely studied environmental constraints that negatively influence yield and production. Plants develop various strategies to tolerate drought stress, and epigenetic modification is one of such strategies (Ali *et al.*, 2020). For instance, drought stress induced differential methylation of a few drought-responsive genes (Liang *et al.*, 2014). Mao *et al.* (2015) discovered that RdDM-mediated DM and H3K9me2 deposition in transposable element, miniature inverted-repeat transposable element (MITE), make it unable to insert into ZmNAC1 associated with drought tolerance. The drought stress-mediated histone modifications were associated with transcriptional responses of drought-induced genes, for example several drought stress-responsive genes, RESPONSIVE to DEHYDRATION (RD29A, RD29B) and NCED3, critical integrators of multiple-epigenetic processes (Ali *et al.*, 2022). This chapter comprehensively summarizes the recent advances of epigenetic regulations in plant response to abiotic stress and insights into how plants survive under drought stress and their generational heritability.

15.2 Mechanism of Drought Stress Regulation in Plants

Plants face a continuously changing environment which influences plant survival and final productivity. At some point, drought stress affects plants and acts as one of the plant stress memories that endure recurrent droughts. Plants show a variety of responses to drought stress to mitigate its impact. To unveil drought resistance mechanisms in plants, the current progress and status quo of studies on the morphological, biochemical, physiological alterations, signal transduction pathways, and molecular regulation mechanisms have been examined in recent years. For instance, osmotic stress by drought stress inhibits plant development and growth (Ali *et al.*, 2020). Plants produce a variety of osmoprotectants and antioxidant enzymes such as proline, soluble sugar, glycine betaine, CAT, POD, SOD, MDA, and so on to alter the plant's state. Likewise, plants produce various signal molecules like abscisic acid (ABA), and IAA, and can directly or indirectly regulate the expression of downstream genes (Yang *et al.*, 2021). For example, ABA is the primer mediator of osmotic tolerance for regulating various drought-induced genes (Ding *et al.*, 2011; Waseem *et al.*, 2019). Various studies have demonstrated the plant molecular response related to transcriptional regulation of drought-induced genes and related pathways. These studies uncovered plenty of transcription factors, non-coding genes (miRNAs, lncRNAs), and drought-induced genes (Kudo *et al.*, 2017). A summary of the mechanisms of drought-induced stress responses is shown in Fig. 15.2.

15.3 Plant Epigenetic Regulators

15.3.1 Acetyltransferases and deacetyltransferases

Acetylation, the addition of acetyl groups to lysine, is associated with transcriptional activation of gene expression. Histone modifications are governed by two enzymes, including histone acetyltransferases (HATs) and histone deacetylases (HDACs) in reversible modification of lysine residues through the addition or removal

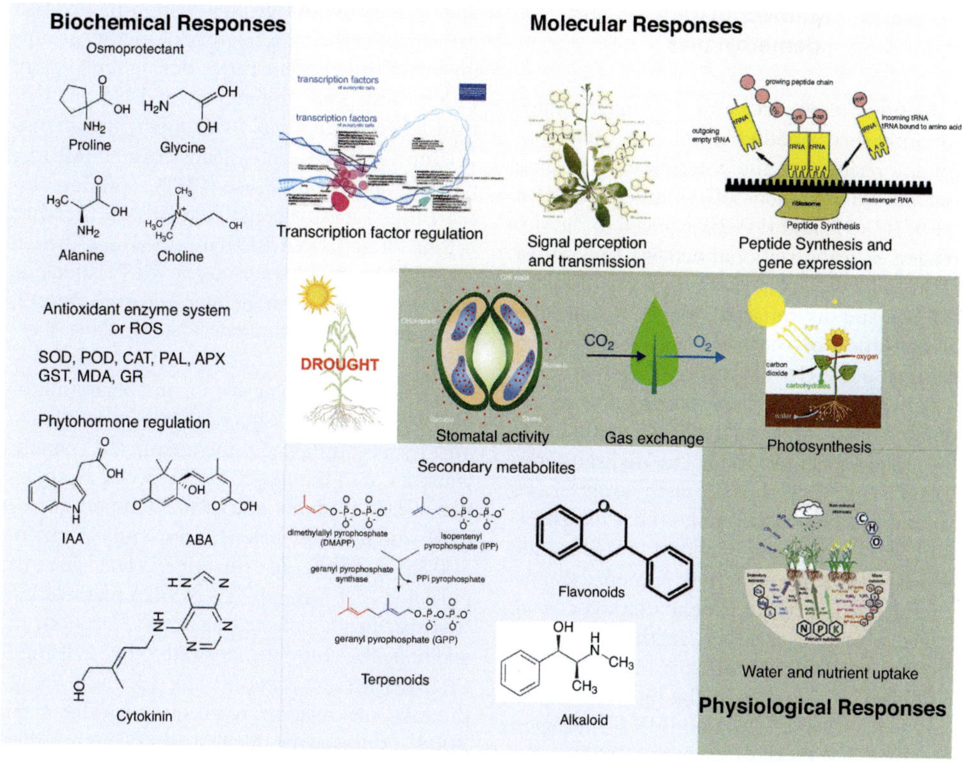

Fig. 15.2. Summary of various regulatory mechanisms of plant drought tolerance, including biochemical, molecular, and physiological responses. Among these processes, biochemical processes involve the production of osmoprotectants such as proline, glycine, alanine, and choline; the antioxidant system includes various protective enzymes like SOD, POD, CAT, PAL, GST, MDA, and so on; signaling pathways such as those involving ABA, IAA and cytokinin; secondary metabolites like terpenoids, alkaloids, and flavonoids. Molecular processes in response to drought stress include transcriptional modifications, alteration in stress signaling and transduction pathways, and alterations in gene expression and protein biosynthesis responsive to drought stress. Lastly, various physiological alterations in plants, including stomatal conductance, gas exchange, alterations in photosynthesis activity, and water and nutrient uptake, are categorized as physiological alterations in plants under drought stress. Author's own image.

of an acetyl group. In plants large numbers of HATs and HDACs have been identified. For instance, *Arabidopsis* contains a total of 12 HATs and 18 HDACs, and rice has 8 HATs and 19 HDACs (Hu *et al.*, 2009). Within the realm of plants, HATs have been categorized into four distinct groups, which are determined by their similarity in sequence and structure of domains. These groups include CREB-binding protein, TAFII-associated factors, GCN5-related N-terminal acetyltransferase (GNAT), and MOZ-YBF2/SAS3-SAS2/TIP60 (MYST) (Liu *et al.*, 2012). In the plant kingdom, HDACs can be subcategorized into three specific groups: type II HDAC (HD2), silent information regulator 2 (SIR2), and reduced potassium deficiency 3 (RPD3) (Luo *et al.*, 2017). The HDACs regulate non-inheritable epigenetic alterations, cause demethylation and function and cause transcriptional suppression of genes. These histone modification enzymes act specifically on lysine, such as H3K9/14/23/27 and H4K5/8/12/16 (Shen *et al.*, 2015). Lastly, acetylation and deacetylation are commonly linked to gene activation or repression, correspondingly (Liu *et al.*, 2016).

15.3.2 Methyltransferases and demethylases

Methyltransferases act as an enzymatic catalyst that introduces methyl groups to lysine and arginine residues at the N-terminus of histone proteins. Methylation of symmetric H4R3, H3K9, H3K27, and H4K20 is linked to the suppression of transcriptional activity. In contrast, the introduction of a methyl group to H3K4, H3K36, and asymmetric H4R3 is linked to the activation of the transcription machinery (Wang *et al.*, 2016). It has been demonstrated that all methyltransferases possess an intact SET domain which catalyzes the lysine methylation, also called histone lysine methyltransferase (HKMT). *Arabidopsis* has 32 methyltransferases, whereas rice has 43 SET genes (Lu *et al.*, 2013). Plant SET proteins are classified into four independent classes based on their sequence similarity and domain arrangement (Springer *et al.*, 2003). Protein arginine methyltransferases (PRMTs) particularly enhance arginine methylation on specific sites of the histone octamer, namely H3R2/8/17/26, and H4R3. *Arabidopsis* has a total of 9 PRMTs (Niu *et al.*, 2007).

Histone lysine demethylases (HDMs) have the ability to remove methyl groups from core histones, resulting in the removal of methylation marks (Liu *et al.*, 2010). There are two distinct types of plant histone demethylases that employ different mechanisms to remove methyl groups: the Jumonji C (JmjC) domain and lysine-specific demethylase1 (LSD1) (Shi *et al.*, 2004). *Arabidopsis* plants have 21 JmjC domain-containing (JMJs) and four KDM1/LSD1-like HDM proteins (Spedaletti *et al.*, 2008). In plants JMJ proteins are categorized into five subfamilies: KDM3/JHDM2, KDM5/JARID1, KDM4/JHDM3, JmjC, and JMJD6 (Liu *et al.*, 2010). JMJ proteins possess the capacity to eliminate methyl groups from mono-, di-, and tri-methylated lysine residues (Klose and Zhang, 2007).

DNA methyltransferase

DNA methyltransferase facilitates the catalytic transfer methylation of 5-cystine carbon from S-adenosylmethionine. In higher plants, two distinct types of DNA methylation processes exist. *De novo* methylation and methylation refer to the addition of methyl groups to non-methylated cytosine and the maintenance of methyl groups on partially methylated DNA during replication. Plants have four subfamilies of DNA methyltransferases, such as DNA methyltransferase 2 (Dnmt2), DNA methyltransferase 1 (MET1), chromomethyltransferase (CMT), and domain-rearranged methyltransferase (DRM). Rice has a total of ten DNA methyltransferases, which include two members from the MET1 subclass, three from CMT, a single member from Dnmt2, and four from DRM subclasses (Sharma *et al.*, 2009).

DRMs are responsible for the beginning of methylation in all sequence settings in plants. MET1 is accountable for the upkeep of CG methylation, CMT3 is engaged in the specific methylation of CHG in plants, and DRM2 is responsible for CHH methylation (Henderson and Jacobsen, 2007). Analysis of the *Arabidopsis* genome indicated that around 24% of DNA methylation was specifically detected at the CG location. In addition, the study demonstrated that CHH and CHG methylation accounted for 1.7 and 6.7% of the total methylation, respectively (Cokus *et al.*, 2008). Four enzymes facilitate the demethylation of methylated cytosines such as DEMETER-LIKE (DML2/3), DEMETER (DME), and REPRESSOR OF SILENCING 1 (ROS1) occurs through a base excision repair mechanism (Penterman *et al.*, 2007). Unlike mammals, cystine demethylation can be recognized and catalyzed by plant DNA glycosylases (Wu and Zhang, 2017). Blocking the process of newly synthesized DNA strands acquiring methyl groups or preventing the maintenance of methylated DNA imprints after DNA replication could result in passive demethylation (Sahu *et al.*, 2013).

15.3.3 Histone and DNA modifications

The complex chromatin structural changes such as DM and HMs induced differential expression of genes associated with environmental challenges, particularly drought stress (Lindermayr *et al.*, 2020). The plants develop future drought tolerance after being exposed to drought by a process called drought adaptation (Sani *et al.*, 2013). Histone proteins regulate gene expression through post-transcriptional modification like methylation, acetylation, and phosphorylation (Millán-Zambrano *et al.*, 2022) and are responsible for epigenetic memories in plants (Borg *et al.*, 2020). The HM causes transcriptional activation and suppression of genes depending on histone modification and the type of posttranscriptional modification. For instance, bivalent HMs like H3K4me3 or H3K27me3 are associated with gene activation or suppression (Zheng *et al.*, 2020).

Like histone modification, DM is a heritable genetic modification that impacts maintaining genome stability and gene expression. The nature of DM is based on the target genes and their function (Gao *et al.*, 2020). Alterations in DM negatively correlated with the transcription of genes (Lang *et al.*, 2017). The higher the level or frequency of DM, the lower the gene expression level, which subsequently corresponds to gene expression and final phenotype. The variation in agronomic traits among individuals of the same species with identical genomes can be attributed to the influence of challenging environmental conditions (Henderson and Jacobsen, 2007). In contrast, DM on the transposable elements moderately impacts corresponding gene expression with suppressing genes (Zilberman *et al.*, 2007).

15.4 Epigenetic Responses of Plants to Drought Stress

Plants are sessile organisms unable to escape environmental constraints. Drought is one of the most critical plant growth and development limiting factors and directly influences the plant's final productivity and quality of yield. Drought disrupts the molecular level, resulting in physiological and phenotypic disorders that stunt plant growth (Hu and Fu, 2018). Conversely, plants protect themselves against these damages by re-regulating cellular homeostasis. Recently, research revealed that plants develop an additional layer of protection against drought stress with epigenetic modification. These epigenetic alterations involve alterations in the magnitude of DM, activation of corresponding signaling pathways, and drought stress-responsive gene expression alteration which ultimately influence plant growth and development (Daccord *et al.*, 2017). Methylation and demethylation in the context of GC are believed to be major contributors to drought stress tolerance in plants and are heritable epigenetic alterations (Yuan *et al.*, 2020).

15.4.1 Drought stress-induced DNA methylation and histone modifications in plants

Epigenetic regulation is a complex multifaceted modulation process involved at transcriptional and post-transcriptional levels like DM, RNA interference, histone modifications, and so on (Santos *et al.*, 2011). Studies are focused on MD methylation correlated with phenotypic alterations (Springer and Schmitz, 2017). DM is an enzymatic reaction that alters DNA stability without modifying DNA sequence but a function of the corresponding gene is changed in response to external stimulus. At present, plenty of studies focus on DM response under drought stress and the impact of downstream gene expression regulation (Chinnusamy and Zhu, 2009). For instance, a few plant species, such as *Arabidopsis* and rice, display transgenerational DM with improved tolerance to drought stress in the offspring of multiple generations (Wang *et al.*, 2011; Zheng *et al.*, 2017). Epigenetic regulation encompasses both short-term and long-term processes that involve metabolic variations, such as ABA, flavonoids, and antioxidant enzymes, as well as changes in gene expression triggered by stress (Watkinson *et al.*, 2006; Ramírez *et al.*, 2015). However, it is unknown whether long-term memories can be transgenerational and short-term stress memories are regulated by physiological pathways.

DNA methylation has the ability to control the expression of genes that respond to stress. The study conducted by Verkest *et al.* (2015) involved subjecting canola plants to prolonged periods of drought stress. This was achieved by screening the plants across multiple generations, resulting in the selection of individuals with both strong drought tolerance and high nutrient usage efficiency, specifically in relation to nitrogen. Zhang *et al.* (2016) demonstrated that exposing winter wheat to dry conditions before flowering increases its ability to withstand high temperatures after flowering. The application of methylation inhibitor zebularine demethylase significantly alters overall plant biomass under drought stress through osmotic adjustment via alteration in antioxidant capabilities (Herman and Sultan, 2016). Likewise, the application of methylation inhibitor (5-azadC) in potatoes enhances drought tolerance via multiple metabolic alterations in soluble sugar, antioxidant pathways, and DM (Li *et al.*, 2019a). Wang *et al.* (2021) and Zheng *et al.* (2017) found 12% differences in DM levels among drought-sensitive or -tolerant rice genotypes. Among these, about 2/3 of DM/demethylation was reversed during a recovery phase. Liang *et al.* (2014) demonstrated a correlation between DM and expression of transcription factors induced under drought stress through single-base-resolution methylomes. A study in rice revealed DNA hypermethylation under drought stress at the grain-filling stage (Wang *et al.*, 2020).

Drought stress triggers DNA hypermethylation in rice's *SUGARS WILL EVENTUALLY BE EXPORTED TRANSPORTER 16* (*OsSWEET16b*), leading to an increase in its transcriptional activity (Wang *et al.*, 2020). Drought inhibits the DM status in the regulatory and intronic regions of *SlAsr1* and *SlAsr2* genes, leading to the activation of their expression. However, H3K9me2 suppresses the expression of *SlAsr2* (González *et al.*, 2013). Likewise, the expression of *ZmNAC111* is downregulated by DM and H3K9me2 alteration (Fig. 15.3). Additionally, the presence of an MITE insertion

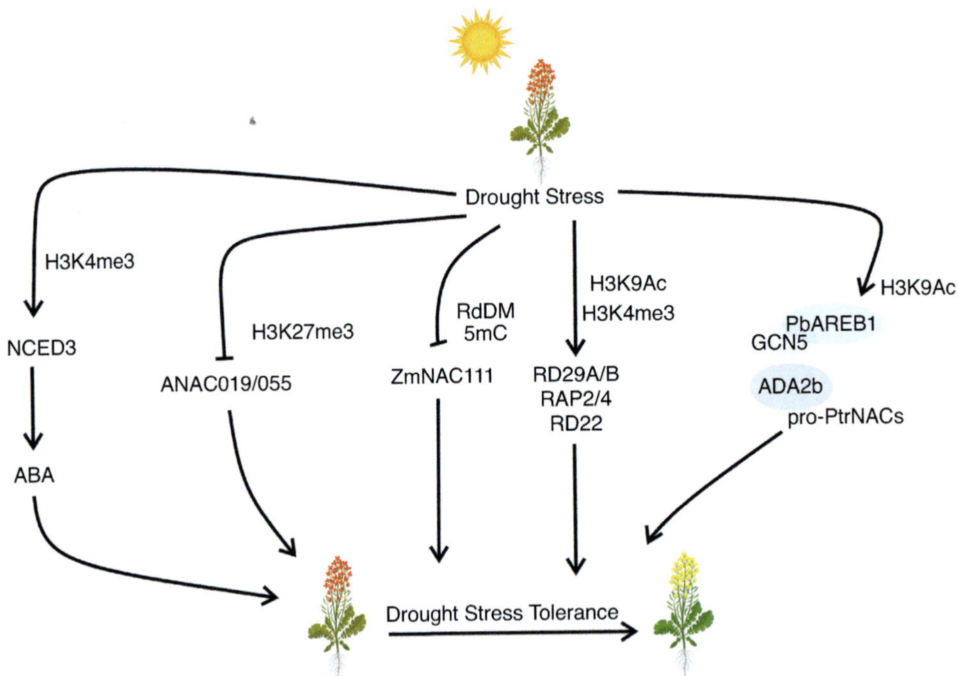

Fig. 15.3. Mechanism of action of epigenetic regulation in drought stress from a few popular studies. The promoter areas of the abscisic acid (ABA) biosynthesis enzyme-encoding gene *NINE CIS-EPOXYCAROTENOID DIOXYGENASE 3* (*NCED3*) and the drought-responsive genes *RD29A/B*, *RD22*, *RAP2.4* are upregulated in response to drought stress, leading to an increase in the levels of H3K4me3 and H3K9Ac. This activation of the promoter regions results in the enhanced expression of these genes in *Arabidopsis*. H3K27me3 deposition on the genes *ANAC019/055* represses the expression of AtNAC transcription factors. The expression of the NAC transcription factor gene ZmNAC111 in maize is inhibited by RNA-directed DNA methylation (RdDM)-dependent DNA methylation. *PtrAREB1* in *P. trichocarpa* interacts with the ABRE motif found in NAC genes and facilitates the recruitment of the ADA2b-GCN5 histone acetylation module. This recruitment leads to the deposition of H3K9Ac on the NAC genes. Author's own image.

in the promoter suppresses the transcription of *ZmNAC111* and diminishes its ability to withstand drought (Mao *et al.*, 2015). Barley plants respond to drought treatment by increasing the activity of certain enzymes like histone acetyltransferase. Other enzymes like HvELP3, HvMYST, histone demethylase HvPKDM7, and HvGCN5 play a favorable role in plant drought responses (Papaefthimiou and Tsaftaris, 2012). Drought stress induces a group of genes in rice that produce histone acetyltransferases, such as *OsHAC701*, *OsHAG702*, *OsHAF701*, etc. which leads to an increase in the acetylation levels of H4K5, H3K18, and H3K27 (Fang *et al.*, 2014). Certain drought-responsive genes in *Arabidopsis*, such as *AtCOR15A*, *AtNCED3*, *AtRD29A*, and *AtRD29B* activated by H3K4me3 (Fig. 15.3), are modified by the *Arabidopsis* histone transferase *trithorax-like factor*, *ATX1/4/5* and the histone demethylase *Jumonji C 17* (*AtJMJ17*) simultaneously (Gong *et al.*, 2002). Drought induces the expression of *ATX4/5*, but the transcription level of *AtJMJ17* is decreased. H3K9 acetylation (H3K9ac) also regulates genes that respond to drought (Huang *et al.*, 2019). The presence of drought stress leads to an increase in the levels of H3K9ac modifications and causes the upregulation of a few NAC genes such as *PtrNAC006*, *PtrNAC007*, and *PtrNAC120* (Fig. 15.3) (Li *et al.*, 2019c). The *PtrGCN5–PtrADA2* complex increases H3K9ac levels in the promoter of genes that respond to drought (Liu *et al.*, 2022). *ATX4/5* also induces the expression of *AtAHG3* by enhancing the H3K4me3 alteration in response to drought stress (Huang *et al.*, 2019). *AtHDA9*, *AtPWR*, and *AtABI4* in *Arabidopsis* work together to reduce the amount of histone acetylation in the gene regions responsible for ABA breakdown. This suppression of gene expression leads to an increase in ABA accumulation (Khan *et al.*, 2020). During periods of drought, the protein *AtHDA19* causes the removal of acetyl groups from histones in the gene *AtABI3*. This leads to the suppression of *AtABI3*, which in turn prevents *AtABI5* from responding to drought stress (Ryu *et al.*, 2014). Table 15.1 presents a few important studies on the epigenetic response of plants to drought stress and includes their underlying mechanisms.

15.5 Applications of Plant Epigenetic Regulation

Epigenetic regulation of plant stress response has great importance for environmental adaptability and agriculture. A few key potential implications are as follows.

1. Plenty of studies have revealed epigenetic memories in plants from physiological and biochemical perspectives in crop plants. However, knowledge related to the molecular perspective of epigenetic memories is limited. Therefore, it is desirable in modern crop breeding to utilize molecular mechanisms in addition to physiological and biochemical mechanisms for the selection of agronomically important phenotypic traits.
2. The application of modern epigenetic tools may also be fascinating for creating new epiallelic variants through the editing of DM and chromatin remodeling for crop improvement by the epigenetic engineering of plants.
3. Traditional breeding methods for introducing desirable agronomic traits have limitations in terms of time and precision. Researchers have identified plenty of plant stress-responsive markers for crop improvements. By integrating stress-associated epigenetic genes, the crop's ability to cope with stress could lead to the development of stress-resilient crops without compromising productivity and ensuing global food security.
4. Epigenetic studies should be considered at molecular levels, as it is essential to link plant stress responses to phenotypic traits. Epigenetic regulation of certain metabolites should be elucidated, as these molecules function as interspecies communication molecules.
5. Epigenetic regulations are inheritable characteristics, so targeted studies should be conducted to develop crops with improved stress memory.

Table 15.1. Epigenetic modification in response to drought stress in plants. Author's own table.

Plants	Processes	Mechanisms/Responses
Arabidopsis thaliana	DNA methylation	Hypomethylation (Luo and He, 2020)
Linum usitatissimum	DNA methylation	Alternative splicing and transcription (Wang *et al.*, 2023)
Zea mays	Transcriptome, miRNA, DM analysis	Promoting water retention (Min *et al.*, 2016)
Hordeum vulgare	DM at CKX2.1	Enhance DM (Surdonja *et al.*, 2017)
Populus trichocarpa	Single-base-resolution methylomes	Enhane methylated cytosines (Liang *et al.*, 2014)
Arabidopsis thaliana	Demethylation	Defense genes regulated in trans by DNA methylation (López Sánchez *et al.*, 2016)
Populus deltoides × P. nigra	DNA methylation	DNA hypermethylation correlated with phenotypic alterations under drought (Gourcilleau *et al.*, 2010)
Arabidopsis thaliana	Histone modification	Increased H3K9ac and negatively regulate drought-responsive genes (Zheng *et al.*, 2016)
Arabidopsis thaliana	Histone modification	Regulate ABA-mediated drought tolerance (Luo *et al.*, 2012)
Brachypodium distachyon	Histone modification	Certain HAT genes activation increase drought adaptation (Tan *et al.*, 2019)
Brassica rapa	Histone modification	Drought adaptations (Eom and Hyun, 2018)
Gossypium hirsutum	Histone modification	HDT4, a histone deacetylase, regulates the expression of a negative regulator of the drought response gene WRKY33 (Zhang *et al.*, 2020a)
Dendrobium officinale	Histone modification	Drought tolerance (Zhang *et al.*, 2020b)
Oryza sativa	Histone modification	Alteration in the expression of HAT genes through transcription regulation (Hou *et al.*, 2021)
Triticum aestivum	Histone modification	Upregulated in the expression of HAC2 genes under drought conditions (Li *et al.*, 2022)

15.6 Conclusion and Future Perspective

This chapter has reviewed recent progress in the epigenetic regulation of plant drought stress tolerance mechanisms, including DM dynamics and HMs. DNA methylation and HMs are heritable epigenetic alterations without introducing genetic or molecular modification, making it an effective approach to developing crops with desirable stress-tolerant traits. Therefore, epigenetics can be used as a molecular breeding tool to bypass genetics transformation to manipulate gene expression in agriculturally important crops. Drought stress occurring concurrently is a major challenge in field crop management systems. It is difficult to develop an efficient broad-spectrum drought stress response, due to limited knowledge of drought stress response in crops. However, the field of epigenetic regulation and drought stress is emerging among researchers, and the relationship between epigenetic regulation and drought at loci levels still needs exploration. Altogether, in a changing climatic era, it is pivotal to deal with unpredictable effects.

References

Ali, S., Hayat, K., Iqbal, A. and Xie, L. (2020) Implications of abscisic acid in the drought stress tolerance of plants. *Agronomy* 10(9), 1323.

Ali, S., Khan, N. and Tang, Y. (2022) Epigenetic marks for mitigating abiotic stresses in plants. *Journal of Plant Physiology* 275, 153740.

Borg, M., Jacob, Y., Susaki, D., LeBlanc, C., Buendía, D. *et al.* (2020) Targeted reprogramming of H3K27me3 resets epigenetic memory in plant paternal chromatin. *Nature Cell Biology* 22(6), 621–629.

Chinnusamy, V. and Zhu, J.-K. (2009) Epigenetic regulation of stress responses in plants. *Current Opinion in Plant Biology* 12(2), 133–139.

Cokus, S.J., Feng, S. and Zhang, X. (2008) Shotgun bisulphite sequencing of the *Arabidopsis* genome reveals DNA methylation patterning. *Nature* 452(7184), 215–219.

Daccord, N., Celton, J.-M., Linsmith, G., Becker, C., Choisne, N. *et al.* (2017) High-quality de novo assembly of the apple genome and methylome dynamics of early fruit development. *Nature Genetics* 49(7), 1099–1106.

Ding, Y., Avramova, Z. and Fromm, M. (2011) The *Arabidopsis* trithorax-like factor ATX1 functions in dehydration stress responses via ABA-dependent and ABA-independent pathways. *The Plant Journal* 66(5), 735–744.

Eom, S.H. and Hyun, T.K. (2018) Histone acetyltransferases (HATs) in Chinese cabbage: Insights from histone H3 acetylation and expression profiling of HATs in response to abiotic stresses. *Journal of the American Society for Horticultural Science* 143(4), 296–303.

Fang, H., Liu, X., Thorn, G., Duan, J. and Tian, L. (2014) Expression analysis of histone acetyltransferases in rice under drought stress. *Biochemical and Biophysical Research Communications* 443(2), 400–405.

Gao, H., Xu, D., Zhou, L., Luo, B. and Yang, Q. (2020) Advances in DNA methylation in plants. *Hubei Forestry Science and Technology* 49, 39–43.

Gong, Z., Morales-Ruiz, T., Ariza, R.R., Roldán-Arjona, T., David, L. *et al.* (2002) ROS1, a repressor of transcriptional gene silencing in *Arabidopsis*, encodes a DNA glycosylase/lyase. *Cell* 111(6), 803–814.

González, R.M., Ricardi, M.M. and Iusem, N.D. (2013) Epigenetic marks in an adaptive water stress-responsive gene in tomato roots under normal and drought conditions. *Epigenetics* 8(8), 864–872.

Gourcilleau, D., Bogeat-Triboulot, M.-B., Le Thiec, D., Lafon-Placette, C., Delaunay, A. *et al.* (2010) DNA methylation and histone acetylation: Genotypic variations in hybrid poplars, impact of water deficit and relationships with productivity. *Annals of Forest Science* 67(2), 208.

Gümüş, T., Aydın, T., Gündüz, B., Meriç, S., Ayan, A. *et al.* (2023) Stress memory and priming applications in plants: Potential for breeders. In: El-Esawi, M.A. (ed.) *Recent Trends in Plant Breeding and Genetic Improvement*. InTechOpen.

Henderson, I.R. and Jacobsen, S.E. (2007) Epigenetic inheritance in plants. *Nature* 447(7143), 418–424.

Herman, J.J. and Sultan, S.E. (2016) DNA methylation mediates genetic variation for adaptive transgenerational plasticity. *Proceedings of the Royal Society B: Biological Sciences* 283, 20160988.

Hou, J., Ren, R., Xiao, H., Chen, Z., Yu, J. *et al.* (2021) Characteristic and evolution of HAT and HDAC genes in Gramineae genomes and their expression analysis under diverse stress in *Oryza sativa*. *Planta* 253(3), 1–22.

Hu, H. and Fu, Y. (2018) Review of plant development and epigenetics. *Modern Agricultural Science and Technology* 47–50.

Hu, Y., Qin, F. and Huang, L. (2009) Rice histone deacetylase genes display specific expression patterns and developmental functions. *Biochemical and Biophysical Research Communications* 388(2), 266–271.

Huang, S., Zhang, A., Jin, J.B., Zhao, B., Wang, T.-J. *et al.* (2019) *Arabidopsis* histone H3K4 demethylase JMJ17 functions in dehydration stress response. *The New Phytologist* 223(3), 1372–1387.

Kambona, C.M., Koua, P.A., Léon, J. and Ballvora, A. (2023) Stress memory and its regulation in plants experiencing recurrent drought conditions. *TAG. Theoretical and Applied Genetics* 136(2), 26.

Khan, I.U., Ali, A., Khan, H.A., Baek, D., Park, J. *et al.* (2020) PWR/HDA9/ABI4 complex epigenetically regulates ABA dependent drought stress tolerance in *Arabidopsis*. *Frontiers in Plant Science* 11.

Klose, R.J. and Zhang, Y. (2007) Regulation of histone methylation by demethylimination and demethylation. *Nature Reviews Molecular Cell Biology* 8(4), 307–318.

Kudo, M., Kidokoro, S., Yoshida, T., Mizoi, J., Todaka, D. *et al.* (2017) Double overexpression of DREB and PIF transcription factors improves drought stress tolerance and cell elongation in transgenic plants. *Plant Biotechnology Journal* 15(4), 458–471.

Lang, Z., Wang, Y., Tang, K., Tang, D., Datsenka, T. *et al.* (2017) Critical roles of DNA demethylation in the activation of ripening-induced genes and inhibition of ripening-repressed genes in tomato fruit. *Proceedings of the National Academy of Sciences of the United States of America* 114(22), E4511–E4519.

Latzel, V., Rendina González, A.P. and Rosenthal, J. (2016) Epigenetic memory as a basis for intelligent behavior in clonal plants. *Frontiers in Plant Science* 7, 1354.

Li, H., Liu, H., Pei, X., Chen, H., Li, X. *et al.* (2022) Comparative genome-wide analysis and expression profiling of histone acetyltransferases and histone deacetylases involved in the response to drought in wheat. *Journal of Plant Growth Regulation* 41(3), 1065–1078.

Li, P., Yang, H., Wang, L., Liu, H., Huo, H. *et al.* (2019b) Physiological and transcriptome analyses reveal short-term responses and formation of memory under drought stress in rice. *Frontiers in Genetics* 10, 55.

Li, P.-C., Bi, Z.-Z., Liang, W.-J., Sun, C., Zhang, J.-L. *et al.* (2019a) DNA methylation involved in regulating drought stress response of potato. *Acta Agronomica Sinica* 45(1), 1595–1603.

Li, S., Lin, Y.-C.J., Wang, P., Zhang, B., Li, M. *et al.* (2019c) The AREB1 transcription factor influences histone acetylation to regulate drought responses and tolerance in *Populus trichocarpa*. *The Plant Cell* 31(3), 663–686.

Liang, D., Zhang, Z., Wu, H., Huang, C., Shuai, P. *et al.* (2014) Single-base-resolution methylomes of *Populus trichocarpa* reveal the association between DNA methylation and drought stress. *BMC Genetics* 15(Suppl.1), S9.

Lindermayr, C., Rudolf, E.E., Durner, J. and Groth, M. (2020) Interactions between metabolism and chromatin in plant models. *Molecular Metabolism* 38, 100951.

Liu, C., Lu, F., Cui, X. and Cao, X. (2010) Histone methylation in higher plants. *Annual Review of Plant Biology* 61(1), 395–420.

Liu, X., Luo, M., Zhang, W., Zhao, J., Zhang, J. *et al.* (2012) Histone acetyltransferases in rice (*Oryza sativa* L.): Phylogenetic analysis, subcellular localization and expression. *BMC Plant Biology* 12, 145.

Liu, X., Yang, S., Yu, C.-W., Chen, C.-Y. and Wu, K. (2016) Histone acetylation and plant development. *The Enzymes* 40, 173–199.

Liu, Y., Wang, J., Liu, B. and Xu, Z.-Y. (2022) Dynamic regulation of DNA methylation and histone modifications in response to abiotic stresses in plants. *Journal of Integrative Plant Biology* 64(12), 2252–2274.

López Sánchez, A., Stassen, J.H.M., Furci, L., Smith, L.M. and Ton, J. (2016) The role of DNA (de) methylation in immune responsiveness of *Arabidopsis*. *The Plant Journal* 88(3), 361–374.

Luo, M., Wang, Y.-Y., Liu, X., Yang, S., Lu, Q. *et al.* (2012) HD2C interacts with HDA6 and is involved in ABA and salt stress response in *Arabidopsis*. *Journal of Experimental Botany* 63(8), 3297–3306.

Luo, M., Cheng, K., Xu, Y., Yang, S. and Wu, K. (2017) Plant responses to abiotic stress regulated by histone deacetylases. *Frontiers in Plant Science* 8, 2147.

Luo, X. and He, Y. (2020) Experiencing winter for spring flowering: A molecular epigenetic perspective on vernalization. *Journal of Integrative Plant Biology* 62(1), 104–117.

Lu, Z., Huang, X., Ouyang, Y. and Yao, J. (2013) Genome-wide identification, phylogenetic and co-expression analysis of OsSET gene family in rice. *PloS ONE* 8(6), e65426.

Mao, H., Wang, H., Liu, S., Li, Z., Yang, X. *et al.* (2015) A transposable element in a NAC gene is associated with drought tolerance in maize seedlings. *Nature Communications* 6, 8326.

Matzke, M.A. and Mosher, R.A. (2014) RNA-directed DNA methylation: An epigenetic pathway of increasing complexity. *Nature Reviews Genetics* 15(6), 394–408.

Meiling, Z., Zhiqiang, X. and Wenquan, W. (2021) Research progress in DNA methylation in plant population. *Chinese Journal of Tropical Agriculture* 41, 49–59.

Millán-Zambrano, G., Burton, A., Bannister, A.J. and Schneider, R. (2022) Histone post-translational modifications – cause and consequence of genome function. *Nature Reviews Genetics* 23(9), 563–580.

Min, H., Chen, C., Wei, S., Shang, X., Sun, M. *et al.* (2016) Identification of drought tolerant mechanisms in maize seedlings based on transcriptome analysis of recombination inbred lines. *Frontiers in Plant Science* 7, 1080.

Niu, L., Lu, F. and Pei, Y. (2007) Regulation of flowering time by the protein arginine methyltransferase AtPRMT10. *EMBO Reports* 8(12), 1190–1195.

Papaefthimiou, D. and Tsaftaris, A.S. (2012) Significant induction by drought of *HvPKDM7-1,* a gene encoding a jumonji-like histone demethylase homologue in barley (*H. vulgare*). *Acta Physiologiae Plantarum* 34(3), 1187–1198.

Penterman, J., Zilberman, D. and Huh, J.H. (2007) DNA demethylation in the *Arabidopsis* genome. *Proceedings of the National Academy of Sciences of the United States of America* 104(16), 6752–6757.

Ramírez, D.A., Rolando, J.L., Yactayo, W., Monneveux, P., Mares, V. *et al.* (2015) Improving potato drought tolerance through the induction of long-term water stress memory. *Plant Science* 238, 26–32.

Ryu, H., Cho, H., Bae, W. and Hwang, I. (2014) Control of early seedling development by BES1/TPL/HDA19-mediated epigenetic regulation of ABI3. *Nature Communications* 5, 4138.

Sahu, P.P., Pandey, G., Sharma, N., Puranik, S., Muthamilarasan, M. *et al.* (2013) Epigenetic mechanisms of plant stress responses and adaptation. *Plant Cell Reports* 32(8), 1151–1159.

Sani, E., Herzyk, P., Perrella, G., Colot, V. and Amtmann, A. (2013) Hyperosmotic priming of *Arabidopsis* seedlings establishes a long-term somatic memory accompanied by specific changes of the epigenome. *Genome Biology* 14(6), 1–24.

Santos, A.P., Serra, T., Figueiredo, D.D., Barros, P., Lourenço, T. *et al.* (2011) Transcription regulation of abiotic stress responses in rice: A combined action of transcription factors and epigenetic mechanisms. *OMICS: A Journal of Integrative Biology* 15(12), 839–857.

Sharma, R., Mohan Singh, R.K., Malik, G., Deveshwar, P., Tyagi, A.K. *et al.* (2009) Rice cytosine DNA methyltransferases – gene expression profiling during reproductive development and abiotic stress. *The FEBS Journal* 276(21), 6301–6311.

Shen, Y., Wei, W. and Zhou, D.-X. (2015) Histone acetylation enzymes coordinate metabolism and gene expression. *Trends in Plant Science* 20(10), 614–621.

Shi, Y., Lan, F., Matson, C., Mulligan, P., Whetstine, J.R. *et al.* (2004) Histone demethylation mediated by the nuclear amine oxidase homolog LSD1. *Cell* 119(7), 941–953.

Spedaletti, V., Polticelli, F., Capodaglio, V., Schininà, M.E., Stano, P. *et al.* (2008) Characterization of a lysine-specific histone demethylase from *Arabidopsis thaliana*. *Biochemistry* 47(17), 4936–4947.

Springer, N.M. and Schmitz, R.J. (2017) Exploiting induced and natural epigenetic variation for crop improvement. *Nature Reviews Genetics* 18(9), 563–575.

Springer, N.M., Napoli, C.A., Selinger, D.A., Pandey, R., Cone, K.C. *et al.* (2003) Comparative analysis of SET domain proteins in maize and *Arabidopsis* reveals multiple duplications preceding the divergence of monocots and dicots. *Plant Physiology* 132(2), 907–925.

Sun, C., Ali, K., Yan, K., Fiaz, S., Dormatey, R. *et al.* (2021) Exploration of epigenetics for improvement of drought and other stress resistance in crops: A review. *Plants (Basel)* 10.

Surdonja, K., Eggert, K., Hajirezaei, M.-R., Harshavardhan, V., Seiler, C. *et al.* (2017) Increase of DNA methylation at the HvCKV2.1 promoter by terminal drought stress in barley. *Epigenomes* 1(2), 9.

Tan, S., Gao, L., Li, T. and Chen, L. (2019) Phylogenetic and expression analysis of histone acetyltransferases in *Brachypodium distachyon*. *Genomics* 111(6), 1966–1976.

Thiebaut, F., Hemerly, A.S. and Ferreira, P.C.G. (2019) A role for epigenetic regulation in the adaptation and stress responses of non-model plants. *Frontiers in Plant Science* 10, 246.

Verkest, A., Byzova, M., Martens, C., Willems, P., Verwulgen, T. *et al.* (2015) Selection for improved energy use efficiency and drought tolerance in canola results in distinct transcriptome and epigenome changes. *Plant Physiology* 168(4), 1338–1350.

Wang, G., Li, H., Meng, S., Yang, J., Ye, N. *et al.* (2020) Analysis of global methylome and gene expression during carbon reserve mobilization in stems under soil drying. *Plant Physiology* 183(4), 1809–1824.

Wang, J., Meng, X. and Yuan, C. (2016) The roles of cross-talk epigenetic patterns in *Arabidopsis thaliana*. *Briefings in Functional Genomics* 15(4), 278–287.

Wang, L., Wang, L., Tan, M., Wang, L., Zhao, W. *et al.* (2023) The pattern of alternative splicing and DNA methylation alteration and their interaction in linseed (*Linum usitatissimum* L.) response to repeated drought stresses. *Biological Research* 56(1), 12.

Wang, T.-J., Huang, S., Zhang, A., Guo, P., Liu, Y. *et al.* (2021) JMJ17–WRKY40 and HY5–ABI5 modules regulate the expression of ABA-responsive genes in *Arabidopsis*. *The New Phytologist* 230(2), 567–584.

Wang, W.-S., Pan, Y.-J., Zhao, X.-Q., Dwivedi, D., Zhu, L.-H. *et al.* (2011) Drought-induced site-specific DNA methylation and its association with drought tolerance in rice (*Oryza sativa* L.). *Journal of Experimental Botany* 62(6), 1951–1960.

Waseem, M., Rong, X. and Li, Z. (2019) Dissecting the role of a basic helix-loop-helix transcription factor, *SlbHLH22*, under salt and drought stresses in transgenic *Solanum lycopersicum* L. *Frontiers in Plant Science* 10, 734.

Watkinson, J.I., Hendricks, L., Sioson, A.A., Vasquez-Robinet, C., Stromberg, V. *et al.* (2006) Accessions of *Solanum tuberosum* ssp. *andigena* show differences in photosynthetic recovery after drought stress as reflected in gene expression profiles. *Plant Science* 171(6), 745–758.

Wu, X. and Zhang, Y. (2017) TET-mediated active DNA demethylation: Mechanism, function and beyond. *Nature Reviews Genetics* 18(9), 517–534.

Yang, X., Lu, M., Wang, Y., Wang, Y., Liu, Z. *et al.* (2021) Response mechanism of plants to drought stress. *Horticulturae* 7(3), 50.

Yuan, C., Zhang, S., Niu, Y., Tang, Q., Wei, D. *et al.* (2020) Advances in research on the mechanism of DNA methylation in plants. *Sheng Wu Gong Cheng Xue Bao (Chinese Journal of Biotechnology)* 36(5), 838–848.

Zhang, J.-B., He, S.-P., Luo, J.-W., Wang, X.-P., Li, D.-D. *et al.* (2020a) A histone deacetylase, GhHDt4D, is positively involved in cotton response to drought stress. *Plant Molecular Biology* 104(1–2), 67–79.

Zhang, M., Silva, J.A., Yu, Z., Wang, H., Si, C. *et al.* (2020b) Identification of histone deacetylase genes in *Dendrobium* officinale and their expression profiles under phytohormone and abiotic stress treatments. *PeerJ* 8, e10482.

Zhang, X., Clarenz, O., Cokus, S., Bernatavichute, Y.V., Pellegrini, M. *et al.* (2007) Whole-genome analysis of histone H3 lysine 27 trimethylation in *Arabidopsis*. *PLoS Biology* 5(5), e129.

Zhang, X., Bernatavichute, Y.V., Cokus, S., Pellegrini, M. and Jacobsen, S.E. (2009) Genome-wide analysis of mono-, di- and trimethylation of histone H3 lysine 4 in *Arabidopsis thaliana*. *Genome Biology* 10(6), 1–14.

Zhang, X., Wang, X., Zhong, J., Zhou, Q., Wang, X. *et al.* (2016) Drought priming induces thermo-tolerance to post-anthesis high-temperature in offspring of winter wheat. *Environmental and Experimental Botany* 127, 26–36.

Zheng, X., Chen, L., Xia, H., Wei, H., Lou, Q. *et al.* (2017) Transgenerational epimutations induced by multi-generation drought imposition mediate rice plant's adaptation to drought condition. *Scientific Reports* 7, 39843.

Zheng, Y., Ding, Y., Sun, X., Xie, S., Wang, D. *et al.* (2016) Histone deacetylase HDA9 negatively regulates salt and drought stress responsiveness in *Arabidopsis*. *Journal of Experimental Botany* 67(6), 1703–1713.

Zheng, Y., Ge, J., Bao, C., Chang, W., Liu, J. *et al.* (2020) Histone deacetylase HDA9 and WRKY53 transcription factor are mutual antagonists in regulation of plant stress response. *Molecular Plant* 13(4), 598–611.

Zilberman, D., Gehring, M., Tran, R.K., Ballinger, T. and Henikoff, S. (2007) Genome-wide analysis of *Arabidopsis thaliana* DNA methylation uncovers an interdependence between methylation and transcription. *Nature Genetics* 39(1), 61–69.

16 Plant Heat Resilience Based on Epigenetic Regulation

Aditya Pratap Singh[1,2]*, Naveen Aavula[3], H. A. Bhargavi[4], P. S. Basavaraj[5], Dibakar Das[1] and Kousik Atta[6]

[1]Department of Plant Breeding and Genetics, School of Agriculture, GIET University, Gunupur, Rayagada, Odisha, India; [2]Department of Genetics and Plant Breeding, Bidhan Chandra Krishi Viswavidyalaya, Mohanpur, Nadia, West Bengal; [3]Division of Genetics, ICAR – Indian Agricultural Research Institute, New Delhi, India; [4]Division of Vegetable Crops, ICAR – Indian Institute of Horticultural Research, Bengaluru, Karnataka, India; [5]School of Water Stress Management, ICAR – National Institute of Abiotic Stress Management, Baramati, Maharashtra, India; [6]Faculty of Agricultural Sciences, Institute of Applied Sciences and Humanities, GLA University, Mathura, Uttar Pradesh, India

Abstract

An increase in the mean global temperature and heat waves exert significant heat stress on plants, with varying and complex effects that range from changes in the levels of molecules to whole organisms and communities of plants. This is a threat to agricultural productivity and it is imperative to develop innovative strategies to enhance crops' heat stress resilience. Plants exposed to heat stress also exhibit modifications in the epigenome that vary with their level of tolerance and species. Lately, advances in epigenome mapping such as whole-genome bisulfite and chromatin immunoprecipitation sequencing have unraveled intricate epigenetic modifications associated with heat stress tolerance in diverse plant species, offering new opportunities to develop heat-tolerant varieties of crop plants. CRISPR-based epigenome editing provides precise control over marks linked to heat tolerance, thus modulating DNA methylation and histone modifications to enhance heat tolerance. This innovative approach holds promise for tailoring specific epigenetic changes in plants to confer increased thermotolerance. The identification of heat-responsive epigenetic marks, when combined with transgenerational epigenetic inheritance, may aid in developing new generations of heat-tolerant crop varieties.

Keywords: Climate resilience, Epigenetics, Sequencing, Thermotolerance

16.1 Introduction

Global climate change, driven largely by human activities such as greenhouse gas emissions, has become a pressing issue affecting agricultural systems worldwide (Shahzad *et al.*, 2021; Jägermeyr *et al.*, 2021). One of the most noticeable consequences of this phenomenon is the

*Corresponding author: adityapratapbckv@gmail.com

 Epigenetics for Climate-Smart and Sustainable Agriculture (ed. Jen-Tsung Chen)
DOI: 10.1079/9781800626102.0016

steady increase in global temperatures, which poses significant challenges to agricultural productivity and food security. In particular, heat stress has emerged as a critical environmental stressor, exerting detrimental effects on various crops and ecosystems (Malhi *et al.*, 2021). High temperatures not only disrupt physiological processes crucial for plant growth and development but also impede reproductive success, ultimately leading to reduced crop yields and economic losses for farmers (Saini and Bhatt, 2020).

While traditional breeding approaches have played a pivotal role in developing heat-tolerant crop varieties, they face inherent limitations in addressing the dynamic and unpredictable nature of climate change (Kamenya *et al.*, 2021; Xiong *et al.*, 2022). The long breeding cycles and genetic complexities associated with traditional breeding programs often hinder the timely development and deployment of resilient crop varieties capable of withstanding heat stress (Anderson *et al.*, 2020). Moreover, the narrow genetic diversity in popular crop species poses challenges in harnessing natural genetic variation for heat tolerance. In response to these challenges, there is a growing recognition of the need to explore alternative strategies to enhance heat tolerance in plants (Wahid *et al.*, 2007). Researchers and breeders are increasingly turning to innovative approaches that leverage advances in molecular biology, genomics, and biotechnology to accelerate the development of resilient crops capable of thriving in warmer climates (Massel *et al.*, 2021). Epigenetics, in particular, has emerged as a promising avenue for crop improvement, offering insights into the molecular mechanisms underlying stress responses and adaptation (Kong *et al.*, 2020; Chang *et al.*, 2020). By deciphering how plants modulate their gene expression in response to heat stress through epigenetic modifications, researchers hope to uncover new targets for breeding heat-tolerant crop varieties that can sustainably support global food production in a changing climate.

In recent years, epigenetics has emerged as a promising avenue for crop improvement, offering a deeper understanding of the molecular mechanisms underlying stress responses and adaptation (McCaw *et al.*, 2020). Epigenetics refers to heritable changes in gene expression that do not involve alterations in the DNA sequence itself (Holliday, 2006; Kumar *et al.*, 2020; Ashe *et al.*, 2021). These changes are mediated by modifications to DNA and histone proteins, as well as RNA-mediated mechanisms, which collectively regulate gene expression patterns in response to environmental cues (Jablonka and Lamb, 2002). The intersection of heat stress and epigenetics presents an intriguing area of research with profound implications for plant biology and agriculture. Understanding how plants perceive, respond to, and adapt to heat stress at the epigenetic level holds the potential to revolutionize crop breeding strategies and mitigate the adverse effects of climate change on food security. This chapter aims to provide an updated overview of the role of epigenetics in combating heat stress in plants. We delve into the fundamental principles of plant epigenetics, explore current methods for studying epigenetic modifications, and examine the specific epigenetic mechanisms implicated in heat stress responses. Furthermore, we discuss recent advancements in linking epigenetic variation with phenotypic variation and highlight strategies for harnessing epigenetics to build heat tolerance in plants. Finally, we outline future research directions and potential methodological advances to further unravel the complex interplay between epigenetics and heat stress adaptation in plants.

Through this chapter, we aim to consolidate existing knowledge, identify gaps in understanding, and stimulate further research efforts aimed at leveraging epigenetics for sustainable crop improvement in the face of global rising temperatures.

16.2 Plant Epigenetics

The study of epigenetics in plants has an extensive history, from early observations of traits not following Mendelian patterns to key discoveries of proteins and RNAs that control gene expression in plants and other organisms, such as humans (Inbar-Feigenberg *et al.*, 2013). To grow and reproduce successfully, plants must adjust to varying environmental stresses, prepare for potential future changes, and retain enough adaptability to respond to short-term fluctuations (Perrella *et al.*, 2020). It is known that

Fig. 16.1. Epigenetic manifestation in different parts or developmental stages of the plant. Author's own image.

epigenetic controls manage numerous activities across the plant life cycle, including genomic imprinting, stress reactions, and cell specialization (Takatsuka and Umeda, 2015). In recent years, small RNAs have become recognized as significant controllers that impact how genes are expressed across the plant genome. Since plants are stationary, they must precisely coordinate development with conditions. Epigenetics facilitates this synchronization, as seen in vernalization, a crucial process for certain accessions of *Arabidopsis* to flower after winter. In *Arabidopsis* varieties requiring winter cold, the FLOWERING LOCUS C (FLC) gene inhibiting flowering is transcriptionally silenced during cold weather and maintains epigenetic repression as the plant resumes growth when it is warmer (Henderson and Dean, 2004). This coordinated epigenetic regulation permits a timely onset of flowering. Additionally, epigenetic controls manage the root apex, critical for optimal growth under shifting environments, as studies in *Arabidopsis* show (Takatsuka and Umeda, 2015; Quadrana and Colot, 2016). Epigenetic changes at all the possible developmental stages of a plant are depicted in Fig. 16.1, while Table 16.1 describes the evolution of the epigenetic concept.

Unlike mammals, genomic imprinting in plants relies more on active DNA demethylation rather than methylation and often does not require reprogramming between generations (Quadrana and Colot, 2016). Heritable alterations in gene expression mediate numerous developmental processes in plants as well as animals (Rajewsky *et al.*, 2017). Indeed, research on epigenetic phenomena including DNA methylation has rapidly expanded in plants recently. DNA methylation primarily involves enzymatic transfer of methyl groups to the 5th carbon of cytosine pyrimidine rings, forming 5-methylcytosine (5mC), or the 6th nitrogen of adenine purine rings, forming N6-methyladenine (6mA) (Zhang *et al.*, 2018). These reversible base modifications can influence gene expression and chromatin architecture dynamics across plant growth, reproduction, and environmental responses. Advances in detecting precise methylation profiles genome-wide continue to reveal new complexities surrounding context-specific patterning and maintenance through cellular differentiation and transgenerational inheritance.

Developmental and transgenerational epigenetics are two linked aspects of a topic that investigate how epigenetic changes influence an organism's development and how these modifications might be passed down to future generations (Diez *et al.*, 2014; Gallusci *et al.*, 2017). Developmental epigenetics studies the role of epigenetic changes as an organism grows and differentiates from a single fertilized egg to a multicellular, functioning individual.

Table 16.1. Evolution of epigenetic concept and its manifestations. Author's own table.

Concepts	References	Summary of key findings
Genetics	Jablonka and Lamb, 2020	Initially, the term 'epigenetics' was linked to Bateson's 'genetics'. However, this concept evolved to incorporate Lederberg's distinctions and Waddington's focus on gene interactions and environmental influences.
Epigenetic concept introduction	Waddington, 1942	Proposed the concept of epigenetics in which gene expression is regulated by complex interactions between genes and environmental influences leading to effects on phenotype
5-methylcytosine discovery	Johnson and Coghill, 1925	Identified 5-methylcytosine as the first discovered DNA base modification using paper chromatography, laying the groundwork for research on the role of DNA methylation
5-methylcytosine distribution	Wyatt, 1951	Showed that 5-methylcytosine is extensively distributed across plant and animal genomes, implying conserved epigenetic regulatory functions
First plant methylase	Kalousek and Morris, 1969	Provided the first evidence of a methylase enzyme activity in plants that targets cytosine residues, expanding knowledge of epigenetic control
Additional DNA modification	Vanyushin and Ashapkin, 2011	Uncovered the presence of N6-methyladenine residues particularly in plant mitochondrial DNA, revealing additional dimensions of epigenetic modifications
Histone regulatory theory	Stedman and Stedman, 1951	Postulated that histones may serve as regulators of genetic activity and expression, foreshadowing later advances around histone modifications
Chromatin structure manipulations	Fucík *et al.*, 1970	Treatment with inhibitors caused uncoiling of chromatin and chromosomal aberrations, highlighting chromatin's dynamic structure governs access to gene expression
DNA methylation gene silencing	John and Amasino, 1989	Found DNA methylation to be involved in epigenetically silencing gene transcription, especially in plant environmental response pathways

Continued

Table 16.1. Continued

Concepts	References	Summary of key findings
Paramutation	Arteaga-Vazquez and Chandler, 2010	Described paramutation as an unusual epigenetic phenomenon allowing altered gene expression states to be inherited transgenerationally
Genomic imprinting	Alleman and Doctor, 2000	Revealed the occurrence of genomic imprinting in plants, in which gene expression depends on whether the allele is maternally or paternally inherited
Tissue culture epigenetic alteration	Miguel and Marum, 2011	Demonstrated extensive stress-induced epigenomic changes during plant tissue culture, enabling analysis of epigenomic plasticity and adaptation
DNA methylation developmental roles	Finnegan *et al.*, 1996	Emphasized DNA methylation's vital and widespread roles in regulating gene expression critical for proper growth and development
Gene silencing example	Jacobsen and Meyerowitz, 1997	Identified hypermethylation specifically silencing flower developmental gene SUPERMAN in *Arabidopsis* 'Clark Kent' mutants
De novo DNA methyltransferases	Cao and Jacobsen, 2002	Discovered plant enzymes capable of establishing new DNA methylation patterns, deepening mechanistic understanding of epigenome propagation
Diverse histone modifications	Wolffe, 1992	Highlighted the diverse repertoire of histone modifications beyond methylation/acetylation that combine to alter chromatin accessibility and integrity, influencing gene regulation
Evolutionary implications	Jablonka and Lamb, 2020	Epigenetic inheritance challenges traditional views of evolution by incorporating Lamarckian aspects. It broadens the concept of heredity and emphasizes the role of epigenetic variation in evolution.

The process of regulating gene expression and determining cell identity and function involves a precisely coordinated sequence of events that are controlled by epigenetic mechanisms. DNA methylation is an important component of developmental epigenetics (Zhang *et al.*, 2018; Liu and He, 2020). This involves the insertion of a methyl group onto the DNA molecule, which frequently occurs at cytosine residues in the setting of CpG dinucleotides. DNA methylation patterns alter dynamically throughout development, regulating gene expression and contributing to cell differentiation. The cytosine sequence context determines plant DNA methylation, which is catalyzed by DNA methyltransferases regulated by several processes. METHYLTRANSFERASE 1 (MET1) regulates CG cytosine methylation(Zhang *et al.*, 2018). A further component is the

Table 16.2. Examples of transgenerational inheritance in crops. Author's own table.

Phenotypic changes due to epigenetic changes	Plants	Epigenetic mechanisms	Explanations	References
Vernalization	*Arabidopsis thaliana*	Histone modifications, DNA methylation	*Arabidopsis* remembers cold exposure, which causes early flowering in subsequent generations.	Shirakawa *et al.*, 2021
Paramutation	Maize	DNA methylation, small RNA	Maize pigmentation variations are heritable due to methylation of the DNA and small RNAs.	Arteaga-Vazquez and Chandler, 2010
Drought stress memory	*Arabidopsis thaliana*	Histone modifications, RNA-based regulation	*Arabidopsis* plants subjected to drought stress exhibit heritable alterations in gene expression.	Ganguly *et al.*, 2017
Nutrient stress	*Brassica rapa*	DNA methylation, histone modifications	*Brassica rapa* plants experience transgenerational effects on flowering as well as seed production when exposed to nutritional stress.	Kellenberger *et al.*, 2018
Transposon silencing	Rice	DNA methylation, small RNA	Rice maintains the genome stable by epigenetically silencing transposable elements (TEs).	Cui *et al.*, 2013

alteration of histone proteins, which are essential for DNA packing into chromatin. Acetylation, methylation, phosphorylation, and other histone modifications affect the accessibility of DNA, influencing gene expression during development. Several non-coding RNAs, including microRNAs and long non-coding RNAs, also play a role in developmental epigenetic modifications. They can influence gene expression by interacting with messenger RNAs and affecting their stability or translation.

16.2.1 Transgenerational inheritance of epigenetic memory

Transgenerational epigenetics is a fascinating field that explores how epigenetic information is transmitted across generations, influencing the traits and characteristics of offspring (Table 16.2). This field challenges the traditional view that only genetic information encoded in DNA is inherited (Quadrana and Colot, 2016). Unlike developmental epigenetics, transgenerational epigenetics does not involve the resetting of epigenetic states between generations.

Plants, in contrast to mammals, seem to be more susceptible to this type of inheritance. For instance, in *Arabidopsis thaliana*, studies have shown that stress-induced epigenetic modifications can be inherited through several generations, with some phenotypic differences still present in the F2 generation (Xu *et al.*, 2021). In another study, it was found that an increase in growth temperature effectively inhibited transgene-induced post-transcriptional gene silencing (PTGS), and this warmth-induced PTGS release exhibited transgenerational epigenetic inheritance (Zhong *et al.*, 2013). Another study demonstrated that vernalization, a process by which prolonged cold exposure leads to flowering, can be inherited by subsequent generations. This epigenetic change, mediated by changes in histone methylation at the FLC locus, was observed up to four generations (Angel *et al.*, 2011)

In *Oryza sativa*, it has been observed that drought stress can induce genome-wide epigenetic segregation characterized by complete loss of heterozygosity (LOH) and nearly complete loss of an allele (LOA) in the F3 progenies of four rice populations in a single generation. Furthermore, about 25% of the stress-induced epigenetic segregation loci were transgenerational and inherited from their salt- and drought-selected parents (Li *et al.*, 2017).

In another example, flax (*Linum usitatissimum*) and tobacco (*Nicotiana rustica*) plants grown under nutrient-rich conditions exhibited heavier weight than those grown under normal conditions. This trait was inherited by the progeny, demonstrating the role of environmental factors in inducing epigenetic changes that can be passed on to subsequent generations (Durrant, 1962; Hill, 1965). In maize, a fascinating example of transgenerational epigenetic inheritance is the paramutation phenomenon, where an interaction between two alleles leads to a heritable change in the expression of one allele, without any change in the DNA sequence (Hollick, 2017).

Nonetheless, the extent to which this system of inheritance influences heritable alterations in the phenotype of natural populations remains not fully understood. The role of the environment in producing environmentally induced epimutations is also uncertain, and it appears to be less common than previously thought.

While the concept of the epigenome and epigenetic markers has a history spanning over 80 years, the majority of research on epigenetics has surged in the past 12 years, particularly following the identification of chromatin-modifying enzymes involved in histone protein modification. Notably, this enzyme discovery has played a crucial role in elucidating the complex mechanisms governing epigenetic regulation. Our understanding of epigenetics is still somewhat confined, with a primary focus on specific plant species like *Arabidopsis*, maize, and rice. Plants, known for their diversity, are believed to have evolved various epigenetic processes to facilitate gene activity in different genomic contexts, ensuring their adaptation to natural habitats (Granot *et al.*, 2009). Long-term adaptations and short-term responses to environmental factors like light and temperature often manifest through variations in gene expression (Franklin *et al.*, 2014; Wu *et al.*, 2014). Recent advancements in next-generation sequencing (NGS) technologies have revolutionized our ability to explore the epigenetic landscape of various plant species. NGS allows for the rapid and cost-effective acquisition of *de novo* genomic and transcriptome data for non-model plants, enabling the study of their unique epigenetic profiles. This technological progress has made gene discovery in wild plants feasible, opening new avenues for expanding our understanding of the diverse epigenetic processes operating in the plant kingdom

16.3 Methods to Study Epigenetic Modifications

Epigenetic variations in plants under heat stress are investigated by changes in DNA methylation, histone modifications, and small RNA expression (Li, 2021). Recent advances have resulted in fascinating findings, and the innovative nature of this field needs comprehensive methods and novel techniques to take epigenetics to the forefront of molecular biology. The methylation status of every single cytosine within genomes may now be determined by whole-genome sequencing after bisulfite treatment of DNA, which turns unmethylated cytosine into uracil (Rajewsky *et al.*, 2017). The following sections will cover the contemporary approaches available for detecting locus-specific and genome-wide alterations in all epigenetic codes.

16.3.1 DNA methylation analysis

DNA methylations, denoted by 5-methylcytosines (5mC), occur in three different sequence contexts such as CpG, CHH, and CHG (here H denotes A, C, and/or T). These contexts provide a variety of functions in controlling biological processes. A popular method for estimating total methylation is to degrade genomic DNA with endonucleases that have varying methylation sensitivity levels. Bisulfite sequencing is a commonly used method for detecting 5mC DNA methylation. This method involves treating DNA with sodium bisulfite, which converts unmethylated cytosines to uracils while retaining methylated cytosines.

After PCR amplification and sequencing, methylated regions can be detected (Li, 2021).

Despite its widespread use, bisulfite sequencing comes with limitations. Short-read sequencing impedes its ability to profile 5mCs in repetitive genome regions. Additionally, incomplete conversion and DNA degradation during bisulfite treatment can compromise specificity and sequencing diversity (Treangen and Salzberg, 2012). Another drawback is its inability to differentiate between 5mC and 5-hydroxymethylcytosine (5hmC). Techniques such as whole-genome bisulfite sequencing (WGBS) and reduced representation bisulfite sequencing (RRBS) provide high-resolution DNA methylation maps. Oxidative bisulfite sequencing (oxBS-seq) distinguishes between 5mC and 5hmC. In plant research, long nanapore reads are advantageous for detecting 5mCs compared with short-read bisulfite sequencing. Until 2021, existing methods could only detect 5mCs in the CpG context, limiting their application in plants. However, scientists recently introduced DeepSignal-plant, a deep learning tool capable of identifying genome-wide 5mCs in all three contexts from nanopore reads (Ni *et al.*, 2021).

16.3.2 Chromatin and histone modification analysis

Mass spectrometry enables large-scale quantitative study of histone post-translational modifications and combinations. DNaseI hypersensitivity assays, often positioned near promoter regions, facilitate the mapping of active and inactive chromatin by identifying hypersensitive sites. Chromatin immunoprecipitation (ChIP)-based methods, such as ChIP-seq, are commonly employed to examine histone modifications. Specific antibodies are utilized to precipitate DNA fragments associated with these modifications. ChIP variants, such as reverse chromatin immunoprecipitation (R-ChIP) and sequential ChIP (ChIP-re-ChIP), as well as high-throughput adaptations like ChIP-seq and ChIP-on-chip, have been developed (Saini *et al.*, 2023). Despite its utility, ChIP-seq has limitations, requiring a substantial amount of starting material and potentially introducing biases through cross-linking steps. Additionally, it offers information on modification presence but lacks the ability to distinguish between various histone variants or combinatorial patterns. Addressing these challenges, single-cell ChIP-seq allows the examination of histone modifications at the individual cell level. Advances in techniques such as ChIP-exo and ChIP-nexus enhance resolution and specificity, providing more precise mapping of histone modifications (Song *et al.*, 2014). To enhance our comprehension of the biological activities associated with the overall changes in histone modifications and variants in higher plants, some researchers investigated histone variants and post-translational modifications specifically in soybeans (Wu *et al.*, 2009).

16.3.3 Small RNA profiling and RNA interference

Tiny RNA molecules, known as small RNAs, play vital roles in essential processes like seed maturation, dormancy, and germination across different plant species. Particularly, a specific type called endogenous microRNAs (miRNAs) stands out as a group of evolutionarily conserved small non-coding RNA molecules. These miRNAs play a significant role in regulating plant responses to a variety of biotic and abiotic stresses by influencing both transcriptional and post-transcriptional mechanisms (Chiou, 2007).

The intricate interplay among DNA methylation, histone modification, and small interfering RNAs (siRNAs) plays a crucial role in orchestrating the transcriptional silencing of DNA. This silencing occurs at transposable elements (TEs), specific genes, or repetitive regions (Luján-Soto and Dinkova, 2021). Small RNA sequencing (sRNA-seq), a commonly used method, profiles various small RNAs, including miRNAs and siRNAs (Morgado and Johannes, 2019; Li, 2021). However, sRNA-seq may not effectively capture low-abundance small RNAs, and biases can be introduced during library preparation. Improved library preparation methods, such as adapter ligation-free approaches, enhance the detection of low-abundance small RNAs. To validate sequencing results, small RNA northern blotting and quantitative PCR remain valuable tools.

Under stress conditions, there is a notable upregulation of miRNAs, which typically leads to the downregulation of their target genes. These target genes frequently act as negative stress tolerance regulators, such as stress-responsive gene repressors (Sunkar *et al.*, 2007). Conversely, Sunkar *et al.* (2007) found that the downregulation of miRNAs during stress can lead to an increase in target gene mRNAs, which improves stress tolerance. These intricate molecular processes underscore the significance of small RNAs in governing essential aspects of plant developmental as well as stress responses.

16.3.4 Plant cell and tissue culture

In vitro culture is a well-studied approach that is commonly used to research various developmental processes, particularly in forestry species, due to their regular growth patterns and simplicity of manipulation (Kim *et al.*, 2021). There is evidence demonstrating that the stresses of plant tissue culture procedures substantially alter epigenetic marks across several key stages such as protoplast isolation and axillary shoot propagation (Smulders and Klerk, 2011). Culturing techniques provide opportunities to analyze rapid and dynamic shifts in regulatory factors like DNA methylation and histone modifications in response to induced stresses and dedifferentiation. Investigating these epigenetic changes can reveal influencing mechanisms that contribute to the emergence of new phenotypes subject to tissue culture. Further examination of specific epigenetic alterations at each phase may uncover regulatory roles related to transcriptional responses, hormonal signaling, and cellular reprogramming underlying plant developmental plasticity unlocked through *in vitro* manipulations. Examining the epigenome under different culture conditions, including temperature, light, and nutrient availability, is critical for understanding the impact of these factors on epigenetic changes. Examining how stress affects the epigenome during tissue culture can provide insights into adaptive responses and assist in breeding stress tolerance (Rodríguez *et al.*, 2012).

16.3.5 Advances in multi-omics integration

Integrative analysis combining epigenomic, transcriptomic, proteomic, and metabolomic data enables elucidating molecular mechanisms underlying plant heat stress responses. A common analytical approach, multi-omics integration (MOI), jointly evaluates methylome and transcriptome changes (Kim and Kim, 2021). Strategies like identifying overlapping patterns between differentially methylated and differentially expressed genes can provide insights. Advanced computational methods including machine learning and network analyses are also critical for detecting influential regulatory components and pathways (Gutschker *et al.*, 2022). Public databases such as Gene Ontology, KEGG and Reactome inform key analysis steps by contributing curated information on biological pathways. Overall, synergistic bioinformatic techniques allow harvesting of meaningful biological interpretations from the wealth of multi-omics data concerning epigenetic control of plant stress adaptation mechanisms.

16.3.6 Transcriptomic and epigenomic profiling

Technological progress has enabled more comprehensive and high-throughput analyses of gene expression and epigenetic patterns compared with previous techniques like polymerase chain reaction, pyrosequencing, and Sanger sequencing. While precise, those approaches are restricted to small target regions and sample sizes, due to long processing times (Gutschker *et al.*, 2022). In contrast, microarrays and RNA sequencing permit genome-wide characterization of transcript abundance. However, they generate extensive data requiring careful interpretation within relevant biological contexts. Ongoing advances aim to balance throughput, precision, accessibility, and physiological relevance to elucidate the relationships between complex gene expression and epigenetic changes. By integrating multiple lines of evidence, researchers can piece together how chromatin architecture and modifications coordinate plant responses to developmental and environmental cues.

16.3.7 Live imaging techniques

Recent advances have enabled new insights into chromatin architecture and dynamics regulating gene expression in plants. Super-resolution microscopy systems and computational image analysis now allow the visualization of tiny, transient chromatin structures. Moreover, fluorescence-based reporter systems permit real-time monitoring of heat stress-induced alterations in DNA methylation and histone modifications (Sato *et al.*, 2021). The ability to directly detect epigenetic changes in living cells represents a critical innovation (Table 16.3). For example, Kurita *et al.* (2017) demonstrated live imaging of histone H3 lysine 9 acetylation patterns in plant nuclei using newly developed probes. Ongoing microscopic and fluorescence-tagging advances continue expanding capabilities to uncover chromatin mechanisms underpinning gene regulation and environmental adaptation in plants.

Table 16.3. Various approaches for detecting epigenetic changes along with their advantages and limitations. Author's own table.

Methods	Applications	Advantages	Limitations
Restriction enzyme digestion	Estimates overall methylation	Simple, low-cost	Lacks nucleotide specificity, means cannot provide site-specific methylation
Bisulfite sequencing	Determines methylation at single nucleotides	High-resolution methylation quantitation	Risk of DNA degradation
ChIP	Identifies DNA–protein interactions	Genome-wide profiling in plant cells	Requires knowing the protein target
ChIP-sequencing	Maps DNA–protein binding genome-wide	High-throughput mapping	Lower binding site resolution
MethylC-sequencing	Provides nucleotide resolution methylation	Single nucleotide methylation maps	Expensive, computationally intensive
Whole-genome bisulfite sequencing	Comprehensive methylation analysis	Full genome methylation patterns	Costly, complex data analysis
Small RNA profiling	Quantifies regulatory small RNA expression	High-throughput small RNA identification	Limited to small RNA species
Mass spectrometry	Measures proteins and modifications	Quantitative proteomics data	Indirect DNA assessment
Plant cell/tissue culture	Manipulates plant cells and tissues	Controlled epigenetics experiments	May not reflect real physiology
Live cell imaging	Visualizes real-time epigenetic dynamics	Observe processes in real time	Resolution constraints
Long-read sequencing	Resolves complex genomic regions	Long DNA fragment sequencing	Higher error rates
DNase hypersensitivity	Identifies accessible chromatin	Maps open chromatin regions	Only denotes accessibility
CRISPR/Cas9 editing	Enables functional epigenetic studies	Precise genome editing	Off-target effects, ethical issues

16.3.8 Advanced functional genomics approaches

Researchers can use CRISPR/Cas9 technologies to modify individual genes involved in epigenetic regulation, allowing for a more direct assessment of their functions in heat stress responses (Pulecio *et al.*, 2017). Researchers could accurately target Cas9 nuclease to practically any site in the genome by modifying the guide RNA sequences in this system (Tingting *et al.*, 2023).

16.4 Epigenetic Modifications in Heat Stress

In reaction to heat stress, plants have developed dynamic responses across various levels, organs and whole plants, physiological, biochemical, and molecular – allowing them to either avoid or adapt to challenging environmental conditions. Avoidance is the primary response exhibited by plants to escape from elevated temperatures. In the face of challenging environmental conditions, plants adapt by reorganizing their genomic and molecular structures. One prevalent adaptation, known as epigenetic alterations, entails the transformation of transient changes in DNA or DNA-associated histone proteins into enduring modifications, ultimately shaping chromatin organization (Singh and Prasad, 2022). It significantly influences molecular adaptations to heat stress by establishing a lasting molecular imprint of previous environmental exposures. The regulation of gene expression in response to heat stress relies heavily on epigenetic mechanisms. These mechanisms include DNA methylation, ATP-dependent chromatin remodeling, modifications to histone proteins, and the involvement of long non-coding RNAs (Weng *et al.*, 2014). These modifications enable the precise control of when and where genes are expressed in reaction to external factors or specific developmental requirements. Recent advancements in next-generation sequencing technologies have greatly aided in identifying covalent modifications occurring at the chromatin level. Techniques such as chromatin immunoprecipitation (ChIP), along with its associated PCR and sequencing methods, as well as mass spectrometry, have been instrumental in uncovering changes associated with histone proteins in epigenomic investigations (Nakato and Sakata, 2021). Additionally, the study of genome-wide DNA methylation patterns in plants commonly employs methodologies such as methylated DNA immunoprecipitation (MeDIP) and bisulfite sequencing. Deciphering how epigenetic mechanisms control the expression of stress-related genes during heat stress opens numerous avenues for enhancing crops' stress resilience through molecular breeding or biotechnological strategies. In the following section, we discuss prominent epigenetic modifications, including DNA methylation, histone methylation/acetylation, and the impact of chromatin remodeling complexes. These processes collectively govern epigenetic modifications, exerting control over the expression of specific genes and the corresponding traits.

16.4.1 DNA methylation

DNA methylation refers to the addition of a methyl group (CH3) to cytosine bases positioned at the 5' position, leading to modifications in the activity of a DNA segment without altering its sequence (Singh *et al.*, 2019). In plant species, cytosine bases undergo both symmetric (CG and CHG) and asymmetric (CHH, where H = A, C, or T) methylation (Henderson and Jacobsen, 2007). Symmetric methylation denotes the presence of methylated cytosine in complementary DNA strands, whereas asymmetric methylation indicates methylation of cytosine in only one strand. The introduction of a methyl group to cytosine acts as a repressive chromatin mark, influencing chromatin structure and, subsequently, DNA accessibility. This epigenetic modification is linked to the regulation of gene activity, as methylated cytosines contribute to establishing a transcriptionally silent or repressive chromatin state. Thus, DNA methylation significantly contributes to the complex coordination of gene regulation and cellular functions, impacting a variety of biological processes and cellular outcomes (Wambui Mbichi *et al.*, 2020). Domains Rearranged Methyltransferase 2 (DRM2) is accountable for catalyzing *de novo* methylation across all sequence contexts. Conversely, the preservation of DNA methylation involves three distinct pathways: METHYLTRANSFERASE 1 (MET1) and DECREASE IN DNA METHYLATION 1 (DDM1) oversee CG methylation maintenance. CHROMOMETHYLASE 3 (CMT3), a plant-specific DNA methyltransferase, is responsible for preserving CHG methylation, and DRM2 ensures the perpetuation of asymmetric CHH methylation (Chan *et al.*, 2005). The functioning of DRM2 is under the regulatory influence of the RNA-directed DNA methylation (RdDM) pathway (Matzke and Mosher, 2014).

DNA methylation plays dual roles in plants, serving as a defense mechanism for the genome against selfish DNA elements and simultaneously functioning in the regulation of gene expression. This epigenetic modification triggers transcriptional gene silencing (TGS) of both transgenes and endogenous transposons and retrotransposons, thereby contributing to the preservation of genome stability (Chan *et al.*, 2005). Under conditions of heat stress, there is a noticeable upregulation of transcription observed in several transgenes previously subjected to TGS, including the 35S promoter of cauliflower mosaic virus and β-glucuronidase (GUS). Intriguingly, this upregulation occurs without any noticeable changes in DNA methylation levels (Tittel-Elmer *et al.*, 2010). While elevated temperatures lead to a slight increase in methylation levels in specific regions of GUS, they simultaneously result in a decrease in methylation in other regions (Zhong *et al.*, 2013). Thus, it seems that DNA methylation may not directly regulate heat responses. However, an examination of heat tolerance in mutants deficient in DNA methylation highlights the crucial role of the RdDM pathway in basal thermotolerance (Popova *et al.*, 2013).

Maintenance of methylation

Methylation is a critical cellular process characterized by the addition of methyl groups to a newly synthesized strand of DNA, occurring precisely opposite the methylated sites on the parental strand. This intricate task is orchestrated by the enzyme DNA methyltransferase 1 (DNMT1), which functions as a custodian of epigenetic information during DNA replication. As the DNA double helix undergoes the complex process of replication, DNMT1 actively recognizes and targets cytosine residues on the parent strand that are already methylated (Nishiyama *et al.*, 2016). In a precise and coordinated manner, DNMT1 then transfers methyl groups to the corresponding cytosines on the complementary, newly synthesized strand. This meticulous replication of methylation patterns ensures the faithful inheritance of epigenetic marks from one generation of cells to the next. The catalytic activity of DNMT1 in maintenance methylation is paramount for the stability and continuity of epigenetic information. By faithfully copying the methylation status of the parental DNA strand to the newly formed strand, DNMT1 contributes to the preservation of gene expression patterns, cellular identity, and overall genomic stability across successive rounds of cell division. Maintenance methylation is thus a fundamental mechanism in epigenetics that helps to perpetuate the epigenetic landscape and maintain cellular functionality (Yin *et al.*, 2012).

16.4.2 Histone modifications

First identified by Vincet Allfrey in the early 1960s, histone modification is an epigenetic mark that is extensively studied and plays a vital role in controlling gene expression. In the plant kingdom, a variety of histone variants and enzymes that modify them are key in directing epigenetic reactions to a wide range of environmental triggers. The expression of genes is closely linked with the structural organization of chromatin, where nucleosomes, the fundamental repeating units, wrap around 146 base pairs of DNA surrounding a histone core particle made up of eight units. The alteration of nucleosomes through histone modification is crucial in determining whether chromatin states are open or closed. The impact of histone modification on gene expression depends on the spatial arrangement of these modifications across a gene area, which outlines the landscape of histone modification, and the concurrent presence of other modifications. These modifications, which can be either reversible or irreversible, are attached to amino acids in the N-terminal tail of the histone. Notably, the complex patterns of post-translational modifications form the 'histone code' or signature, which is recognized by protein complexes known as effectors, leading to chromatin remodeling. A short exposure to heat stress triggers changes in histones, which is associated with increased transcription of heat shock factors (HSFs) and ribosomal RNA (rRNA) genes. This event is marked by disturbances to cell membranes and a rise in reactive oxygen species (ROS) levels (Hou *et al.*, 2019).

Histone acetylation

The importance of histone acetylation is recognized in numerous facets of plant growth, development, and adaptation to environmental changes. Enzymes known as histone acetyltransferases (HATs) and histone deacetylases (HDACs) are involved in the process of adding and removing acetyl groups on histones, respectively. The enzymatic function of HAT includes the attachment of an acetyl group to lysine residues. This acetylation neutralizes the electrostatic pull between the positively charged histones and the negatively charged DNA, resulting in a relaxation of the DNA coiling. This, in turn, enables the activation of transcription (Suto *et al.*, 2000).

Histone methylation

Histone methylations play a significant role in controlling heat stress responses and memory. Methyl groups attach to the side chains of lysines and arginines. Lysine residues can experience multiple methylation events, usually up to three, facilitated by histone lysine methyltransferase (HKMT). Most HKMTs, with the exception of the Dot1 enzyme that methylates H3K79, possess an active SET domain. On the other hand, arginine residues can undergo mono- and dimethylation, a process governed by protein arginine methyltransferase (PRMT). PRMTs are generally categorized into types I, II, III, or IV enzymes. Types I and II PRMTs, such as PRMT1, PRMT3, PRMT4, PRMT5, and PRMT7, are involved in the regulation of gene transcription through histone protein methylation. Type I PRMTs (PRMT1, PRMT3, PRMT4, and PRMT6) are associated with transcriptional activation and ribosomal biosynthesis, while Type II PRMTs (PRMT5 and PRMT7) are linked to transcriptional repression (Gonsalvez *et al.*, 2007).

Histone phosphorylation

Histone phosphorylation entails the action of histone kinases, which attach phosphate groups to serines, threonines, and tyrosines situated on N-terminal histone tails and linker histone H1. The function of these kinases is antagonized by histone phosphatases, which are accountable for removing phosphate groups (Chao *et al.*, 2010). Histone phosphorylation is a dynamic process that plays a crucial role in various cellular functions. For instance, it has been found to be involved in DNA damage repair, as phosphorylation of histone H2AX is a key early step in the response to DNA double-strand breaks (Vasko *et al.*, 2011). Additionally, histone phosphorylation is also implicated in the regulation of gene expression. Specifically, phosphorylation of histone H3 at serine 10 is associated with gene activation and is a hallmark of mitotic chromosomes (Park and Kim, 2012). Furthermore, researchers have shown that DNA damage induces histone depletion, which increases the accessibility and flexibility of the DNA fiber and enhances the rate of homology search during repair by homologous recombination (Kim *et al.*, 2019).

ADP ribosylation

Histones experience ribosylation at glutamate and arginine residues located on the N-terminal tail of the histone, which imparts a negative charge to the protein molecule. This ribosylation can take place as mono- or polyribosylation and is carried out by mono-ADP-ribosyltransferases and poly-ADP-ribose polymerase (PARP), which use nicotinamide adenine dinucleotide to transfer ADP-ribose to substrates. Importantly, the process of polyribosylation is reversible, facilitated by two classes of enzymes, namely ADP ribosylhydrolases (ARHs) and PAR glycohydrolases (PARGs) (Moss *et al.*, 1997). When activated, nuclear enzymes produce ADP-ribose polymers using NAD(+) as a substrate. In the context of plants, stressors such as drought, intense light, and heat trigger the activation of PARP, leading to the depletion of NAD(+) and the use of ATP. The reduction of PARP activity through the use of chemical inhibitors or gene silencing helps prevent cell death, thereby providing tolerance to various abiotic stresses like intense light, drought, and heat (Block *et al.*, 2005).

Histone ubiquitylation

The process of histone ubiquitylation (or ubiquitination) entails the addition of ubiquitin, a 76-amino-acid polypeptide, to lysine residues of histone proteins through the coordinated action of three enzymes: E1-activating, E2-conjugating, and E3-ligating enzymes (Saracco *et al.*, 2009).

Recent studies have shown that histone ubiquitylation plays a critical role in both active and repressed transcriptional states, and is central to DNA damage repair signaling (Chen *et al.*, 2022). The histone residues that are targeted for ubiquitylation are often highly conserved through evolution. Moreover, it has been found that K34-ubiquitylation of histone H2B can significantly enhance nucleosome dynamics, decrease nucleosome stability, and promote eviction of one histone H2A-H2B dimer (Chen *et al.*, 2022). This suggests that histone ubiquitylation can directly influence the dynamics and rearrangements of nucleosomes, thereby playing a crucial role in the regulation of gene expression and DNA repair.

Histone sumoylation

SUMO (small ubiquitin-related modifier) is a protein akin to ubiquitin, exhibiting an 18% similarity in its primary sequence to ubiquitin (Melchior, 2000). It attaches to lysine residues on histones through a coordinated process involving E1, E2, and E3 enzymes, mirroring the mechanism observed in ubiquitylation. Additionally, SUMO modification can be reversed by ULP/SUSP proteases through de-sumoylation. When exposed to heat stress, chromatin-associated SUMO signals exhibit an increase at promoter-transcriptional start site regions and a decrease within gene bodies (Han *et al.*, 2021).

Histone tail clipping

Relatively less explored irreversible post-translational modification is conserved across a spectrum from yeast to mammals. Specific proteases dedicated to histone tails facilitate the cleavage of both the N- and C-terminal ends of histones. This modification, initially identified in *Tetrahymena* in 1980, involves the removal of the first six amino acids from the H3 histone (David Allis *et al.*, 1980).

Other histone modifications

Histones undergo various post-translational modifications (PTMs), including β-N-glycosylation, carbonylation, citrullination (deimination), biotinylation, crotonylation, proline isomerization, aspartic acid isomerization, N-formylation, propionylation, and butyrylation. However, the precise roles of these modifications in the epigenetic control of plants remain uncertain and warrant further exploration.

16.4.3 Chromatin remodeling

ATP-dependent chromatin remodeling complexes are pivotal in the plant's response to heat stress (Liu *et al.*, 2019). The SWItch/sucrose non-fermentable (SWI/SNF) complex, first identified in *Arabidopsis*, plays a central role in temperature perception, with its crucial component ARP6 (SWI complex) being indispensable. When subjected to heat and drought stress, the overexpression of the SNF2/Brahma-type chromatin remodeling gene CHR12 (CHROMATIN REMODELING) caused the cessation of flower bud and primary stem growth in *A. thaliana*. Conversely, *Arabidopsis* plants with AtCHR12 knockout exhibited reduced growth arrest compared with wild-type plants. Furthermore, upon the return to normal temperatures, the H3-H4 chaperone CAF-1 (CHROMATIN ASSEMBLY FACTOR-1) participates in reloading nucleosomes onto chromosomes in *A. thaliana*, highlighting its importance in chromatin remodeling (Strenkert *et al.*, 2011). Additionally, heat stress induces the relaxation of transposon silencing through chromocenter decondensation mediated by HEAT-INTOLERANT 4 (HIT4). Meanwhile, MORPHEUS' MOLECULE 1 (MOM1) and DECREASE IN DNA METHYLATION 1 (DDM1)-mediated chromatin remodeling swiftly represses most of these transposons (Iwasaki and Paszkowski, 2014).

16.4.4 RNA-mediated modifications

The concept of RNA-directed chromatin modification was initially observed in plants, where the accumulation of double-stranded RNA (dsRNA) led to the targeted cytosine DNA methylation1. Small interfering RNAs (siRNAs) were identified as the key molecules guiding the deposition of DNA methylation and the methylation of histone H3 at lysine 9 (H3K9me). This process resulted in the formation of stable heterochromatin. While transcription factors are central regulators of transcription during heat stress, plant non-coding RNAs (ncRNAs) also play a significant role in responding to high temperatures. Non-coding RNAs, which include microRNAs (miRNAs), small interfering RNAs (siRNAs), long non-coding RNAs (lncRNAs), and circular RNAs (circRNAs), constitute a class of regulatory RNAs. They play crucial roles in the heat stress response by modulating the activity of transcription factors or genes (Zhao *et al.*, 2020). Both canonical and non-canonical forms of RdDM are investigated, with particular attention drawn to the interconnected relationship between post-transcriptional silencing and the establishment of chromatin marks.

16.5 Linking Epigenetic Variation with Phenotypic Variation

Phenotypic variation in every population can be caused by various mechanisms, each of which has a unique origin. This variation determines whether a trait can evolve through natural or artificial selection and adapt to environmental changes. Researchers have sought to understand the relative importance of genetic and environmental factors contributing to specific traits. Using this knowledge, scientists can predict the evolutionary trajectory of a population for a particular characteristic (Byers, 2008). Epigenetics contributes to phenotypic variability as it regulates the accessibility of genetic information. This results in heritable changes in gene expression and transposable element activity, which can be influenced by DNA methylation and histone modifications, such as acetylation, phosphorylation, and alterations in chromatin structure (Springer *et al.*, 2016).

The sources of phenotypic variation due to epigenetic variation vary and include epimutations and epigenetic inheritance, which can result in adaptive mechanisms or changes in plants, particularly during stress and environmental fluctuations (Bossdorf *et al.*, 2008). Epimutations, just as genetic mutations, can occur spontaneously or because of external factors and be passed down to subsequent generations, giving rise to new epigenetic variants (epialleles) that contribute to the diversity of phenotypes (Kakoulidou *et al.*, 2021). Epigenetic changes caused by DNA methylation, such as differentially methylated positions (DMPs) and differentially methylated regions (DMRs), can lead to epimutations and may be linked to the inheritance of stress responses (Zhang *et al.*, 2018). For instance, DMRs are associated with genes differentially expressed in regulating the immune system of *Arabidopsis thaliana* against *Pseudomonas syringae* (Dowen *et al.*, 2012). Additionally, proteins such as RNA Dicer-like 2 (DCL2) and DCL3, which usually play a role in the biogenesis of various types of siRNAs in *Arabidopsis*, are required for transgenerational changes in homologous recombination frequency in response to stresses such as heat, cold, drought, and UV exposure (Simon and Meyers, 2011; Migicovsky and Kovalchuk, 2014).

Recent research has highlighted transgenerationally heritable epigenetic changes, commonly called epimutations, which contribute to the formation of epigenetic stress memory amid stress conditions. This is believed to benefit the plant progeny by imparting a higher level of stress resistance (Ashapkin *et al.*, 2020). In a study of epigenetic mutants, Popova *et al.* (2013) discovered the role of the Rpd3-type histone deacetylase HDA6 in heat tolerance, emphasizing the significance of changes in histone modification and DNA methylation in both intergenerational and intragenerational stress memory. Recently, López *et al.* (2023) investigated the role of molecular modifications in high-temperature stress-induced DNA methylation and the transcriptome of *Fragaria vesca*. Their findings reveal that the precise nature of gene expression and heritability of DNA methylation and transcriptome is significantly higher during high temperatures, including the role of methylome in epigenetic stress memory, passed to the subsequent progeny.

16.6 Toward Building Heat Tolerance in Plants Using Epigenetics

Plants are inherently immobile, making it challenging for them to evade severe environmental changes. However, they have developed intricate stress-response systems through substantial levels of phenotypic plasticity and evolved multiple traits to withstand environmental stress conditions (Ashapkin *et al.*, 2020). The phenotypic plasticity response to environmental stress largely depends on the plant species and the underlying mechanisms at the physiological, genetic, and epigenetic levels, which vary based on the perceived stress's range, size, and duration. Heat stress, one of the various environmental stressors, causes irreversible cellular damage, such as protein denaturation, impairment of cell organelles, and accumulation of reactive oxygen species (ROS).

Various molecular breeding techniques are used to develop heat-tolerant varieties, including applying whole-genome sequencing (WGS) through various multi-omics approaches. These approaches involve the identification of signaling molecules and their gene expression mechanisms, as well as the identification of candidate genes for heat tolerance. For example, bHLH transcription factor PHYTOCHROME INTERACTING FACTOR 4 (PIF4) for thermomorphogenesis-signaling, genes encoding for heat shock transcription factors A1 (HsfA1s) such as DREB2A, HsfA2, HsfA7a, HsfBs, and MBF1C (Ashapkin *et al.*, 2020). These efforts have laid the foundation for studying the responses and adaptations of plants at the epigenetic level, intending to develop more heat-tolerant varieties using epi-breeding. Epigenetic variation, like the employment of genetic variation in plant breeding, is called epi-breeding. This method is utilized to improve crop resilience and is based on the molecular basis of improvement in both the plant genome and epigenome. Below are the main aspects related to epigenetics in plant breeding.

16.6.1 Epialleles breeding

Epialleles are heritable changes in DNA methylation patterns that can arise spontaneously or be induced by various stimuli. Natural epialleles, occurring without external intervention, can emerge in plant genomes, with spontaneous epimutations occurring at higher rates than genetic mutations (Bertozzi and Ferguson-Smith, 2020). However, these spontaneous changes rarely form stable and heritable epialleles, constituting only a small fraction of the methylome. Instead, the majority of methylated regions are faithfully inherited across generations, suggesting a strong association between DNA methylation and genetic variation. Epigenetic variation, which can result in phenotypic differences, is primarily driven by epialleles. These are alleles that modify the epigenetic status of the plant. Epialleles can be classified into three types based on their relationship with genetic variation: obligatory epialleles, which are associated with transposable element (TE) insertion; facilitated epialleles, which are highly stable and heritable due to their association with suitable genetic variants; and pure epialleles, which have no correlation with genome variability and are less stable (Richards, 2006; Achrem *et al.*, 2023).

Various steps can be used to incorporate epialleles into plant breeding, such as identifying sources of epigenetic variation, followed by pre-breeding selection or epigenome editing. There is also identified germplasm with epialleles that can be introduced into breeding programs (Crisp *et al.*, 2022). To identify the epialleles and epigenome modification various laboratory methods are employed to create novel phenotypes. Laboratory methodssuch as map-based gene cloning, genome-wide methylome analysis, tissue culture, speed breeding, grafting, chemical-induced products like mutants and transgenes, RNA interference (RNAi), transcription activator-like effectors (TALEs), and clustered regularly interspaced short palindromic repeats/CRISPR-associated protein 9 (CRISPR/Cas9) are employed (Dalakouras and Vlachostergios, 2021; Crisp *et al.*, 2022). The stability of the DNA methylome is maintained during environmental stress conditions, while epigenetic mechanisms such as DNA methylation and histone demethylation epialleles provide for stable Mendelian inheritance. For instance, when maize is subjected to short-term heat stress, it induces alterations in promoter regions, resulting in changes in H3K4me2 and H3K9ac levels. These changes are associated with the upregulation of heat stress factor (HSF) and rRNA genes, as well as perturbations in cell membranes and an increase in reactive oxygen species (Hou *et al.*, 2019). However, significant gaps must be addressed to utilize epialleles in plant breeding fully. First, a strong association between the trait and epialleles must be confirmed and validated. Second, tracking inheritance for natural or induced epialleles in various mapping populations is necessary. Finally, a better understanding of the heritability and stability of epialleles beyond *Arabidopsis* to various crop species needs to be explored.

Natural epialleles

Naturally occurring epialleles, arising from spontaneous epimutations, have been associated with significant plant traits in agriculture, including morphological, developmental, metabolic, and immunity characteristics (Srikant and Tri Wibowo, 2021). Examples include the hypermethylation of the COLORLESS NON-RIPENING (CNR) locus in tomatoes, inhibiting fruit maturation, and the hypomethylation of the FERTILIZATION-INDEPENDENT ENDOSPERM1 (FIE1) gene in rice, leading to reduced stature. These stable and meiotically independent epialleles, not reliant on genetic variation, offer potential targets for breeding or editing in agriculture. However, systematically harnessing them for crop improvement remains a challenge. Epigenetic editing, as an alternative to genetic modification, holds promise for introducing controlled variations, enhancing crop production, and adapting to environmental influences. It allows precise and reversible gene expression modifications without altering the DNA sequence, minimizing the risk of unintended consequences associated with traditional genome editing approaches.

Chemically induced epigenetic diversity

Chemicals capable of modifying epigenetic information are broadly categorized into two groups: inhibitors of DNA methyltransferases (DNMTs) and histone deacetylase inhibitors. DNMT inhibitors, such as cytidine analogues like 5-AzaC and Zebularine, operate by integrating into DNA during replication, entrapping DNMTs, and prompting their degradation. This leads to a passive reduction of DNA methylation in the cell. The second category comprises histone deacetylase inhibitors, which are further classified into hydroxamic acids, amino-benzamides, cyclic peptides, and short-chain fatty acids (Mondal *et al.*, 2022). Currently, DNA-demethylating compounds find wide-ranging applications, particularly in tissue cultures. For example, treatment with 5-AzaC has demonstrated benefits in inducing somatic embryogenesis, microspore embryogenesis, and shoot regeneration (Chen *et al.*, 2021). Another application involves the regulation of flowering, as 5-AzaC treatment has been shown to induce earlier or more frequent flowering compared with untreated plants. Given the role of DNA methylation in suppressing transposon activity, treatment with 5-AzaC or Zebularine could potentially increase transposon activity, as supported by various studies. In genetically transformed tissues where transgene activity is silenced by DNA methylation, treatment with 5-AzaC can reactivate silenced transgenes (Solís *et al.*, 2015). The involvement of DNA methylation in stress memory and priming is investigated through treatments with 5-AzaC or Zebularine. Studies have examined the effects of these treatments on the within-generation response to nutrient stress and phenotypic plasticity. Additionally, stress-induced memory effects on clonal offspring growth have been observed, with this transgenerational effect being attenuated in offspring of parents treated with 5-AzaC. The utilization of histone deacetylase inhibitors, the second category of epigenetically active substances, is less prevalent, primarily observed in plant tissue cultures (Yamagishi and Kikuta, 2021).

16.6.2 Epigenome-wide association studies (epi-GWAS)

Epigenome-wide association studies (Epi-GWAS) aim to uncover the relationship between genotype and phenotype by analyzing epigenetic genetic variations across the epigenomes. These studies are like GWAS in plants, as they concentrate on SNPs (single nucleotide polymorphisms (SNPs), recombination, and linkage disequilibrium (LD). The study methods are whole-genome bisulphite sequencing (WGBS), enzymatic equivalent, DNA methylation profiling approaches such as the whole genome, and base-pair resolution (Crisp *et al.*, 2022). The Epi-GWAS studies are reported in maize, soybean (Shen *et al.*, 2018), and wheat (Gardiner *et al.*, 2018). The major drawback in practical plant breeding is that epialleles might only be of value if SNP-based markers in linkage do not already capture them. For example, the study conducted by Xu *et al.* (2019) for maize SNP's identification and association for methylation polymorphism revealed no associations for SNPs. Moreover, the study methods are expensive, and approaches to capturing epialleles are largely unsuitable for discovering new ones.

16.6.3 Epigenetic recombinant inbred lines (epiRILs)

The production of epigenetic recombinant inbred lines (epiRILs) involves crossing recombinant inbred lines with strains that differ only in DNA methylation. These epiRILs are useful for studying the role of natural epialleles in transgenerational inheritance and how they activate transposable elements that lead to phenotypic changes. Several epigenetic mechanisms, such as DNA methylation, chromatin remodeling, and hypomethylation, have been reported to be passed from one generation to the next. For example, Quadrana and Colot (2016) found that a hypomethylation state in *Arabidopsis* could be inherited for eight generations. The research conducted by Raju *et al.* in 2018 involved creating hybrid soybean lines by crossing between wild-type and msh1-acquired epialleles, resulting in a diverse range of yield-related traits (Raju *et al.*, 2018).

Additionally, the obtained epitypes demonstrated low epitype–environment interactions, suggesting high yield stability and reduced susceptibility to environmental constraints. The study showed that novel epigenetic variation caused by MSH1 suppression could be inherited for up to three generations and utilized for crop improvement through a few rounds of selection to enhance and stabilize crop yield. EpiRILs can be developed using mutagenesis methods, such as T-DNA insertion, ethyl methanesulfonate, γ-rays, or more novel gene editing methods based on CRISPR. Developing epiRILs presents a significant challenge, due to the inherent instability of epigenetic mechanisms, which can result in genotypic variations (Varotto *et al.*, 2020). For instance, in rapeseed exposed to heat stress, significant differences in gene expression were observed for epigenetic methylation. Heat-tolerant genotypes exhibited higher DNA methylation levels than heat-susceptible genotypes (Gao *et al.*, 2014). Another challenge in generating crop-specific mutants is controlling unpredictable transposable element (TE) mobility, particularly in crops like maize rich in TEs (Dalakouras and Vlachostergios, 2021).

16.6.4 Targeting epigenetics for *in vitro* regeneration to improve and accelerate crop breeding

In biotechnology methods for *in vitro* plant regeneration, cell reprogramming is a key process, where epigenetic regulation plays a vital but not completely understood role. Somaclonal variations induced during *in vitro* culture, including genetic and epigenetic instabilities, are known challenges. Techniques such as somatic embryogenesis and microspore embryogenesis offer opportunities for clonal propagation, doubled-haploid plant production, and gene editing. Epigenetic memory, exemplified in *Picea abies*, shows that temperature-induced marks established during somatic embryogenesis influence adaptive traits in adult plants. Transcriptomic analyses reveal differential expression of specific epigenetic regulators, emphasizing the roles of DNA and histone methylation, as well as small RNAs (Zhong *et al.*, 2014). Despite the importance of *in vitro* plant regeneration, efficiency remains low in many crop and forest species due to incomplete understanding of molecular mechanisms. Evidence suggests that epigenetic mechanisms, in conjunction with hormones, play a crucial role in somatic embryogenesis induction and progression. Targeting epigenetics emerges as a potential biotech strategy to enhance *in vitro* plant regeneration and accelerate crop breeding. Chemically controlled mammalian cell reprogramming, including the use of synthetic small molecules and epigenetic inhibitors, has shown promise in controlling cellular processes

(Steward *et al.*, 2002). In plants, targeted chemical interventions with inhibitors of DNA methylation, H3K9 methylation, or histone deacetylases have shown improvements in *in vitro* plant cell reprogramming across various species (Song *et al.*, 2017). Leveraging knowledge in epigenetics extends to the production of secondary metabolites for industrial applications. The coordination between epigenetics and hormonal control can be utilized to enhance the production of secondary metabolites in plants for pharmaceutical purposes. Future challenges include identifying new components of regulatory epigenetic pathways, designing novel compounds for targeting epigenetic marks, and developing high-throughput screening assays for efficient epigenetic modulation. These strategies aim to enhance the efficiency of *in vitro* embryogenesis systems, especially in species facing challenges due to climate change.

Epigenetic factors play a regulatory role in various developmental and physiological processes by influencing chromatin organization and gene transcription. Epigenetic editing, akin to genetic editing, targets specific epigenetic marks to modulate transcription. While epigenetic editing has been implemented in tomatoes for genetic editing, its widespread application in crop species is still limited. Studies in *Arabidopsis* demonstrate the feasibility of epigenetic editing using programmable DNA-binding proteins fused to components of the RdDM machinery. This approach induces specific DNA methylation changes at predefined genomic sites, affecting traits such as flowering time. CRISPR/dCas9 technology has also been adapted for targeted methylation or demethylation, showing promise for crop enhancement (Guo *et al.*, 2019).

16.7 Looking Forward

As research in the intersection of epigenetics and heat stress tolerance in plants continues to evolve, there is a promising opportunity for synthesizing existing knowledge and uncovering novel insights into the molecular mechanisms underlying stress adaptation. Future studies are expected to shed light on the intricate regulatory networks governing epigenetic modifications in response to heat stress, providing a more comprehensive understanding of plant resilience in changing environments.

One area of focus for future research is the elucidation of epigenetic regulatory pathways that govern heat stress responses across different plant species and genotypes. By conducting comparative studies, researchers can identify conserved epigenetic mechanisms and key regulatory genes involved in heat stress tolerance, facilitating the development of broadly applicable strategies for crop improvement. Furthermore, methodological advances in high-throughput sequencing technologies and epigenome profiling techniques are anticipated to revolutionize our ability to study epigenetic modifications in plants under heat stress conditions. These advancements will enable researchers to comprehensively map epigenetic landscapes, identify stress-responsive epigenetic markers, and decipher the functional significance of specific epigenetic modifications in heat stress adaptation.

Despite the promise of epigenetics in enhancing heat tolerance in plants, several current limitations warrant attention. Challenges such as the complexity of epigenetic regulatory networks, the dynamic nature of epigenetic modifications, and the lack of standardized methodologies for studying epigenetic phenomena pose significant hurdles to progress in this field. To address these limitations, collaborative efforts among interdisciplinary research teams, including molecular biologists, geneticists, bioinformaticians, and agronomists, will be crucial. By integrating diverse expertise and leveraging emerging technologies, researchers can develop innovative approaches for overcoming current methodological challenges and unraveling the complexities of epigenetic regulation in heat stress responses. The convergence of epigenetics and heat stress biology presents a fertile ground for advancing our understanding of plant adaptation to environmental challenges. By embracing interdisciplinary collaborations, harnessing technological advancements, and addressing current limitations, future research endeavors hold the potential to unlock new avenues for breeding heat-tolerant crops and ensuring global food security in the face of climate change.

Epigenetic knowledge has been recognized as a promising toolbox for breeders, offering potential applications in crop improvement and adaptation to climate change. The use of epigenetic diversity, whether natural or induced, has been explored for creating new varieties acclimated to changing climates. Additionally, epigenetic markers have been proposed as molecular tools for selecting elite individuals and identifying new candidate genes (Rajkumar *et al.*, 2020).

16.8 Limitations

Some of the key challenges include the need to identify new epigenetically regulated traits, understand the stability of epigenetic modifications, and differentiate between transient and transgenerational changes. The identification of stable and heritable epigenetic changes is crucial for efficient selection during breeding. Moreover, integrating epigenetic data with other -omics information requires advanced statistical and modeling approaches to enhance crop improvement strategies. Efforts are also directed towards reducing the loss of genetic variability, as maintaining and increasing epigenetic variation can compensate for the genetic variability lost due to the use of a limited number of cultivars. Epigenetic analysis is considered valuable for discovering new epialleles involved in tolerance/resistance to biotic and abiotic stresses, aiding in stress-priming strategies for improved stress tolerance without yield reduction (Qian *et al.*, 2019). The standardization and normalization of epigenetic pipelines and methods are crucial for their adoption in molecular breeding. This includes the development of methodologies for epi-genotyping, epi-mutagenesis, and epigenome editing, as well as leveraging single-cell sequencing technology for dissecting cell heterogeneity. However, further research is needed to identify suitable strategies for different crops. There is a need for clarifying epigenetic mechanisms for public acceptance. Understanding the differences between transgenic and genome editing technologies and communicating the safety and health aspects of epigenetic approaches are essential for gaining public trust. Collaborative efforts, such as the EPI-CATCH consortium, bring together academic and private research to address the challenges of climate change at the epigenetic level. While there have been significant achievements, ongoing research is necessary to overcome limitations and fully exploit the potential of epigenetics in building heat tolerance in plants.

16.9 Conclusion

The emerging field of epigenetics offers promising avenues for enhancing heat stress tolerance in plants, thereby addressing the challenges posed by climate change to agricultural productivity and food security. Through a comprehensive exploration of epigenetic mechanisms underlying heat stress responses, researchers have gained valuable insights into the molecular basis of plant resilience and adaptation to environmental fluctuations. Leveraging these insights, innovative strategies for breeding heat-tolerant crop varieties are being developed, with the potential to mitigate the detrimental effects of rising temperatures on global food production. However, while significant progress has been made in elucidating the role of epigenetics in plant heat stress adaptation, several challenges remain to be addressed. Methodological limitations, the complexity of epigenetic regulatory networks, and the need for greater understanding of genotype–environment interactions pose ongoing hurdles to the advancement of this field. Nonetheless, with continued interdisciplinary collaboration, technological innovation, and concerted research efforts, these challenges can be overcome, paving the way for transformative breakthroughs in crop improvement and sustainable agriculture.

Looking ahead, future research endeavors hold the promise of unraveling new layers of epigenetic regulation in plant heat stress responses, uncovering novel targets for breeding heat-tolerant crops, and elucidating the interplay between epigenetics and other stress response pathways. By embracing a forward-looking approach, researchers can harness the power of epigenetics to develop resilient agricultural systems capable of withstanding the challenges of a changing climate, ensuring food security for future generations. The integration of epigenetics into crop breeding programs represents a paradigm shift in agricultural research, offering novel solutions to address the complex challenges posed by climate change. The insights gained from epigenetic studies hold immense potential to revolutionize crop improvement strategies and secure the future of global food production in a rapidly changing world.

References

Achrem, M., Stępień, E. and Kalinka, A. (2023) Epigenetic changes occurring in plant inbreeding. *International Journal of Molecular Sciences* 24(6), 5407.

Alleman, M. and Doctor, J. (2000) Genomic imprinting in plants: Observations and evolutionary implications. *Plant Molecular Biology* 43(2–3), 147–161.

Anderson, R., Bayer, P.E. and Edwards, D. (2020) Climate change and the need for agricultural adaptation. *Current Opinion in Plant Biology* 56, 197–202.

Angel, A., Song, J., Dean, C. and Howard, M. (2011) A polycomb-based switch underlying quantitative epigenetic memory. *Nature* 476(7358), 105–108.

Arteaga-Vazquez, M.A. and Chandler, V.L. (2010) Paramutation in maize: RNA mediated trans-generational gene silencing. *Current Opinion in Genetics & Development* 20(2), 156–163.

Ashapkin, V.V., Kutueva, L.I., Aleksandrushkina, N.I. and Vanyushin, B.F. (2020) Epigenetic mechanisms of plant adaptation to biotic and abiotic stresses. *International Journal of Molecular Sciences* 21(20), 7457.

Ashe, A., Colot, V. and Oldroyd, B.P. (2021) How does epigenetics influence the course of evolution? *Philosophical Transactions of the Royal Society of London. Series B, Biological Sciences* 376(1826), 20200111.

Bertozzi, T.M. and Ferguson-Smith, A.C. (2020) Metastable epialleles and their contribution to epigenetic inheritance in mammals. *Seminars in Cell & Developmental Biology* 97, 93–105.

Block, M.D., Verduyn, C., Brouwer, D.D. and Cornelissen, M. (2005) Poly(ADP-ribose) polymerase in plants affects energy homeostasis, cell death and stress tolerance. *The Plant Journal* 41(1), 95–106.

Bossdorf, O., Richards, C.L. and Pigliucci, M. (2008) Epigenetics for ecologists. *Ecology Letters* 11(2), 106–115.

Byers, D. (2008) Components of phenotypic variance. *Nature Education* 1(1), 161.

Cao, X. and Jacobsen, S.E. (2002) Role of the *Arabidopsis* DRM methyltransferases in de novo DNA methylation and gene silencing. *Current Biology* 12(13), 1138–1144.

Chan, S.W., Henderson, I.R. and Jacobsen, S.E. (2005) Gardening the genome: DNA methylation in *Arabidopsis thaliana*. *Nature Reviews. Genetics* 6, 351–360. DOI: 10.1038/nrg160.

Chang, Y.-N., Zhu, C., Jiang, J., Zhang, H., Zhu, J.-K. *et al.* (2020) Epigenetic regulation in plant abiotic stress responses. *Journal of Integrative Plant Biology* 62(5), 563–580.

Chao, S., Dubcovsky, J., Dvorak, J., Luo, M.-C., Baenziger, S.P. *et al.* (2010) Population- and genome-specific patterns of linkage disequilibrium and SNP variation in spring and winter wheat (*Triticum aestivum* L.). *BMC Genomics* 11, 727.

Chen, J.J., Stermer, D. and Tanny, J.C. (2022) Decoding histone ubiquitylation. *Frontiers in Cell and Developmental Biology* 10, 968398.

Chen, R., Chen, X., Huo, W., Zheng, S., Lin, Y. *et al.* (2021) Transcriptome analysis of azacitidine (5-AzaC)-treatment affecting the development of early somatic embryogenesis in longan. *The Journal of Horticultural Science and Biotechnology* 96(3), 311–323.

Chiou, T.-J. (2007) The role of microRNAs in sensing nutrient stress. *Plant, Cell & Environment* 30(3), 323–332. DOI: 10.1111/j.1365-3040.2007.01643.x.

Crisp, P.A., Bhatnagar-Mathur, P., Hundleby, P., Godwin, I.D., Waterhouse, P.M. *et al.* (2022) Beyond the gene: Epigenetic and cis-regulatory targets offer new breeding potential for the future. *Current Opinion in Biotechnology* 73, 88–94.

Cui, X., Jin, P., Cui, X., Gu, L., Lu, Z. *et al.* (2013) Control of transposon activity by a histone H3K4 demethylase in rice. *Proceedings of the National Academy of Sciences of the United States of America* 110(5), 1953–1958. DOI: 10.1073/pnas.1217020110.

Dalakouras, A. and Vlachostergios, D. (2021) Epigenetic approaches to crop breeding: Current status and perspectives. *Journal of Experimental Botany* 72(15), 5356–5371.

David Allis, C., Bowen, J.K., Abraham, G.N., Glover, C.V.C. and Gorovsky, M.A. (1980) Proteolytic processing of histone H3 in chromatin: A physiologically regulated event in *Tetrahymena micronuclei*. *Cell* 20(1), 55–64. DOI: 10.1016/0092-8674(80)90234-2.

Diez, C.M., Roessler, K. and Gaut, B.S. (2014) Epigenetics and plant genome evolution. *Current Opinion in Plant Biology* 18, 1–8.

Dowen, R.H., Pelizzola, M., Schmitz, R.J., Lister, R., Dowen, J.M. *et al.* (2012) Widespread dynamic DNA methylation in response to biotic stress. *Proceedings of the National Academy of Sciences of the United States of America* 109(32), E2183–2191. DOI: 10.1073/pnas.1217020110.

Durrant, A. (1962) The environmental induction of heritable change in linum. *Heredity* 17(1), 27–61.

Finnegan, E.J., Peacock, W.J. and Dennis, E.S. (1996) Reduced DNA methylation in *Arabidopsis thaliana* results in abnormal plant development. *Proceedings of the National Academy of Sciences of the United States of America* 93(16), 8449–8454.

Franklin, K.A., Toledo-Ortiz, G., Pyott, D.E. and Halliday, K.J. (2014) Interaction of light and temperature signalling. *Journal of Experimental Botany* 65(11), 2859–2871.

Fucík, V., Michaelis, A. and Rieger, R. (1970) On the induction of segment extension and chromatid structural changes in *Vicia faba* chromosomes after treatment with 5-azacytidine and 5-azadeoxycytidine. *Mutation Research/Fundamental and Molecular Mechanisms of Mutagenesis* 9(6), 599–606.

Gallusci, P., Dai, Z., Génard, M., Gauffretau, A., Leblanc-Fournier, N. *et al.* (2017) Epigenetics for plant improvement: Current knowledge and modeling avenues. *Trends in Plant Science* 22(7), 610–623.

Ganguly, D.R., Crisp, P.A., Eichten, S.R. and Pogson, B.J. (2017) The *Arabidopsis* DNA methylome is stable under transgenerational drought stress. *Plant Physiology* 175(4), 1893–1912. DOI: 10.1104/pp.17.00744.

Gao, G., Li, J., Li, H., Li, F., Xu, K. *et al.* (2014) Comparison of the heat stress induced variations in DNA methylation between heat-tolerant and heat-sensitive rapeseed seedlings. *Breeding Science* 64(2), 125–133.

Gardiner, L.-J., Joynson, R., Omony, J., Rusholme-Pilcher, R., Olohan, L. *et al.* (2018) Hidden variation in polyploid wheat drives local adaptation. *Genome Research* 28(9), 1319–1332.

Gonsalvez, G.B., Tian, L., Ospina, J.K., Boisvert, F.M., Lamond, A.I. *et al.* (2007) Two distinct arginine methyltransferases are required for biogenesis of Sm-class ribonucleoproteins. *The Journal of Cell Biology* 178(5), 733–740.

Granot, G., Sikron-Persi, N., Gaspan, O., Florentin, A., Talwara, S. *et al.* (2009) Histone modifications associated with drought tolerance in the desert plant *Zygophyllum dumosum* Boiss. *Planta* 231, 27–34.

Guo, H., Wu, T., Li, S., He, Q., Yang, Z. *et al.* (2019) The methylation patterns and transcriptional responses to chilling stress at the seedling stage in rice. *International Journal of Molecular Sciences* 20(20), 5089.

Gutschker, S., Corral, J.M., Schmiedl, A., Ludewig, F., Koch, W. *et al.* (2022) Multi-omics data integration reveals link between epigenetic modifications and gene expression in sugar beet (*Beta vulgaris* subsp. *vulgaris*) in response to cold. *BMC Genomics* 23(1), 144.

Han, D., Chen, C., Xia, S., Liu, J., Shu, J. *et al.* (2021) Chromatin-associated sumoylation controls the transcriptional switch between plant development and heat stress responses. *Plant Communications* 2(1), 100091.

Henderson, I.R. and Dean, C. (2004) Control of *Arabidopsis* flowering: The chill before the bloom 131(16), 3829–3838.

Henderson, I.R. and Jacobsen, S.E. (2007) Epigenetic inheritance in plants. *Nature* 447(7143), 418–424. DOI: 10.1038/nature05917.

Hill, J. (1965) Environmental induction of heritable change in *Nicotiana rustica*. *Genetics* 51(6), 797–812.

Hollick, J.B. (2017) Paramutation and related phenomena in diverse species. *Nature Reviews. Genetics* 18(1), 5–23.

Holliday, R. (2006) Epigenetics: A historical overview. *Epigenetics* 1(2), 76–80.

Hou, H., Zhao, L., Zheng, X., Gautam, M., Yue, M. *et al.* (2019) Dynamic changes in histone modification are associated with upregulation of HSF and rRNA genes during heat stress in maize seedlings. *Protoplasma* 256(5), 1245–1256.

Inbar-Feigenberg, M., Choufani, S., Butcher, D.T., Roifman, M. and Weksberg, R. (2013) Basic concepts of epigenetics. *Fertility and Sterility* 99(3), 607–615.

Iwasaki, M. and Paszkowski, J. (2014) Identification of genes preventing transgenerational transmission of stress-induced epigenetic states. *Proceedings of the National Academy of Sciences of the United States of America* 111(23), 8547–8552.

Jablonka, E. and Lamb, M.J. (2002) The changing concept of epigenetics. *Annals of the New York Academy of Sciences* 981(1), 82–96.

Jablonka, E. and Lamb, M.J. (2020) *Inheritance systems and the extended evolutionary synthesis*. Cambridge University Press.

Jacobsen, S.E. and Meyerowitz, E.M. (1997) Hypermethylated SUPERMAN epigenetic alleles in *Arabidopsis*. *Science* 277(5329), 1100–1103.

Jägermeyr, J., Müller, C., Ruane, A.C., Elliott, J., Balkovic, J. *et al.* (2021) Climate impacts on global agriculture emerge earlier in new generation of climate and crop models. *Nature Food* 2(11), 873–885.

John, M.C. and Amasino, R.M. (1989) Extensive changes in DNA methylation patterns accompany activation of a silent T-DNA *ipt* gene in *Agrobacterium tumefaciens*-transformed plant cells. *Molecular and Cellular Biology* 9(10), 4298–4303.

Johnson, T.B. and Coghill, R.D. (1925) Researches on pyrimidines. CIII. The discovery of 5-methylcytosine in tuberculinic acid, the nucleic acid of the *Tubercle bacillus*. *Journal of the American Chemical Society* 47(11), 2838–2844.

Kakoulidou, I., Avramidou, E.V., Baránek, M., Brunel-Muguet, S., Farrona, S. *et al.* (2021) Epigenetics for crop improvement in times of global change. *Biology* 10, 766.

Kalousek, F. and Morris, N.R. (1969) The purification and properties of deoxyribonucleic acid methylase from rat spleen. *The Journal of Biological Chemistry* 244(5), 1157–1163.

Kamenya, S.N., Mikwa, E.O., Song, B. and Odeny, D.A. (2021) Genetics and breeding for climate change in orphan crops. *TAG. Theoretical and Applied Genetics* 134(6), 1787–1815.

Kellenberger, R.T., Desurmont, G.A., Schlüter, P.M. and Schiestl, F.P. (2018) Trans-generational inheritance of herbivory-induced phenotypic changes in *Brassica rapa*. *Scientific Reports* 8(1), 3536.

Kim, D.Y. and Kim, J.M. (2021) Multi-omics integration strategies for animal epigenetic studies – A review. *Animal Bioscience* 34(8), 1271.

Kim, J.J., Lee, S.Y. and Miller, K.M. (2019) Preserving genome integrity and function: The DNA damage response and histone modifications. *Critical Reviews in Biochemistry and Molecular Biology* 54(3), 208–241.

Kim, T.D., Kim, N.H., Park, E.J. and Lee, N.N. (2021) High-frequency regeneration of plants *in vitro* from seedling-derived apical bud explants of *Tilia mandshurica* Rupr. & Maxim. *Journal of Plant Biotechnology* 48(1), 54–61.

Kong, L., Liu, Y., Wang, X. and Chang, C. (2020) Insight into the role of epigenetic processes in abiotic and biotic stress response in wheat and barley. *International Journal of Molecular Sciences* 21(4), 1480.

Kumar, H., Chaudhary, A., Singh, A., Sukhija, N., Panwar, A. *et al.* (2020) A review on epigenetics: Manifestations, modifications, methods & challenges. *Journal of Entomology and Zoology Studies* 8(4), 01–06.

Kurita, K., Sakamoto, T., Yagi, N., Sakamoto, Y., Ito, A. *et al.* (2017) Live imaging of H3K9 acetylation in plant cells. *Scientific Reports* 7(1), 45894.

Li, M., Wang, W.-S., Pang, Y.-L., Domingo, J.R., Ali, J. *et al.* (2017) Characterization of salt-induced epigenetic segregation by genome-wide loss of heterozygosity and its association with salt tolerance in rice (*Oryza sativa* L.). *Frontiers in Plant Science* 8, 977.

Li, Y. (2021) Modern epigenetics methods in biological research. *Methods (San Diego, Calif.)* 187, 104–113.

Liu, J. and He, Z. (2020) Small DNA methylation, big player in plant abiotic stress responses and memory. *Frontiers in Plant Science* 11, 595603.

Liu, J., Feng, L., Gu, X., Deng, X., Qiu, Q. *et al.* (2019) An H3k27me3 demethylase-HSFA2 regulatory loop orchestrates transgenerational thermomemory in *Arabidopsis*. *Cell Research* 29(5), 379–390.

López, M.E., Denoyes, B. and Bucher, E. (2023) Epigenetic and transcriptional landscape of heat-stress memory in woodland strawberry (*Fragaria vesca*). *Bio-Archive 2023-05*.

Luján-Soto, E. and Dinkova, T.D. (2021) Time to wake up: Epigenetic and small-RNA-mediated regulation during seed germination. *Plants* 10(2), 236. DOI: 10.3390/plants10020236.

Malhi, G.S., Kaur, M. and Kaushik, P. (2021) Impact of climate change on agriculture and its mitigation strategies: A review. *Sustainability* 13(3), 1318.

Massel, K., Lam, Y., Wong, A.C.S., Hickey, L.T., Borrell, A.K. *et al.* (2021) Hotter, drier, CRISPR: The latest edit on climate change. *TAG. Theoretical and Applied Genetics* 134(6), 1691–1709.

Matzke, M.A. and Mosher, R.A. (2014) RNA-directed DNA methylation: An epigenetic pathway of increasing complexity. *Nature Reviews Genetics* 15(6), 394–408.

McCaw, B.A., Stevenson, T.J. and Lancaster, L.T. (2020) Epigenetic responses to temperature and climate. *Integrative and Comparative Biology* 60(6), 1469–1480.

Melchior, F. (2000) SUMO – nonclassical ubiquitin. *Annual Review of Cell and Developmental Biology* 16(1), 591–626.

Migicovsky, Z. and Kovalchuk, I. (2014) Transgenerational changes in plant physiology and in transposon expression in response to UV-C stress in *Arabidopsis thaliana*. *Plant Signaling & Behavior* 9(11), e976490.

Miguel, C. and Marum, L. (2011) An epigenetic view of plant cells cultured *in vitro*: Somaclonal variation and beyond. *Journal of Experimental Botany* 62(11), 3713–3725.

Mondal, P., Natesh, J., Penta, D. and Meeran, S.M. (2022) Progress and promises of epigenetic drugs and epigenetic diets in cancer prevention and therapy: A clinical update. *Seminars in Cancer Biology* 83, 503–522.

Morgado, L. and Johannes, F. (2019) Computational tools for plant small RNA detection and categorization. *Briefings in Bioinformatics* 20(4), 1181–1192.

Moss, J., Zolkiewska, A. and Okazaki, I. (1997) ADP-ribosylarginine hydrolases and ADP-ribosyltransferases: Partners in ADP-ribosylation cycles. *Advances in Experimental Medicine and Biology* 419, 25–33. DOI: 10.1007/978-1-4419-8632-0_3.

Nakato, R. and Sakata, T. (2021) Methods for chIP-seq analysis: A practical workflow and advanced applications. *Methods* 187, 44–53.

Ni, P., Huang, N., Nie, F., Zhang, J., Zhang, Z. *et al.* (2021) Genome-wide detection of cytosine methylations in plant from nanopore data using deep learning. *Nature Communications* 12(1), 5976.

Nishiyama, A., Yamaguchi, L. and Nakanishi, M. (2016) Regulation of maintenance DNA methylation via histone ubiquitylation. *Journal of Biochemistry* 159(1), 9–15.

Park, C.H. and Kim, K.T. (2012) Apoptotic phosphorylation of histone H3 on ser-10 by protein kinase Cδ. *PLoS ONE* 7(9), e44307.

Perrella, G., Zioutopoulou, A., Headland, L.R. and Kaiserli, E. (2020) The impact of light and temperature on chromatin organization and plant adaptation. *Journal of Experimental Botany* 71(17), 5247–5255.

Popova, O.V., Dinh, H.Q., Aufsatz, W. and Jonak, C. (2013) The RdDM pathway is required for basal heat tolerance in *Arabidopsis*. *Molecular Plant* 6(2), 396–410.

Pulecio, J., Verma, N., Mejía-Ramírez, E., Huangfu, D. and Raya, A. (2017) CRISPR/Cas9-based engineering of the epigenome. *Cell Stem Cell* 21(4), 431–447.

Qian, Y., Hu, W., Liao, J., Zhang, J. and Ren, Q. (2019) The dynamics of DNA methylation in the maize (*Zea mays* L.) inbred line B73 response to heat stress at the seedling stage. *Biochemical and Biophysical Research Communications* 512(4), 742–749.

Quadrana, L. and Colot, V. (2016) Plant transgenerational epigenetics. *Annual Review of Genetics* 50(1), 467–491.

Rajewsky, N., Jurga, S. and Barciszewski, J. (eds) (2017) *Plant Epigenetics*. Springer International.

Rajkumar, M.S., Shankar, R., Garg, R. and Jain, M. (2020) Bisulphite sequencing reveals dynamic DNA methylation under desiccation and salinity stresses in rice cultivars. *Genomics* 112(5), 3537–3548.

Raju, S.K.K., Shao, M.-R., Sanchez, R., Xu, Y.-Z., Sandhu, A. *et al.* (2018) An epigenetic breeding system in soybean for increased yield and stability. *Plant Biotechnology Journal* 16(11), 1836–1847.

Richards, E.J. (2006) Inherited epigenetic variation – revisiting soft inheritance. *Nature Reviews. Genetics* 7(5), 395–401.

Rodríguez, J.L., Pascual, J., Viejo, M., Valledor, L., Meijón, M. *et al.* (2012) Basic procedures for epigenetic analysis in plant cell and tissue culture. *Methods in Molecular Biology (Clifton, N.J.)* 877, 325–341.

Saini, J. and Bhatt, R. (2020) Global warming – causes, impacts and mitigation strategies in agriculture. *Current Journal of Applied Science and Technology* 39(7), 93–107.

Saini, A., Rawat, Y., Jain, K. and Mani, I. (2023) State-of-the-art techniques to study epigenetics. In: Singh, V. and Mani, I. (eds) *Progress in Molecular Biology and Translational Science*, Vol. 197. Academic Press, pp. 23–50.

Saracco, S.A., Hansson, M., Scalf, M., Walker, J.M., Smith, L.M. *et al.* (2009) Tandem affinity purification and mass spectrometric analysis of ubiquitylated proteins in *Arabidopsis*. *The Plant Journal* 59(2), 344–358.

Sato, Y., Nakao, M. and Kimura, H. (2021) Live-cell imaging probes to track chromatin modification dynamics. *Microscopy* 70(5), 415–422.

Shahzad, A., Ullah, S., Dar, A.A., Sardar, M.F., Mehmood, T. *et al.* (2021) Nexus on climate change: Agriculture and possible solution to cope future climate change stresses. *Environmental Science and Pollution Research International* 28(12), 14211–14232.

Shen, Y., Zhang, J., Liu, Y., Liu, S., Liu, Z. *et al.* (2018) DNA methylation footprints during soybean domestication and improvement. *Genome Biology* 19(1), 128.

Shirakawa, M., Morisaki, Y., Gan, E.S., Sato, A. and Ito, T. (2021) Identification of a devernalization inducer by chemical screening approaches in *Arabidopsis thaliana*. *Frontiers in Plant Science* 12, 634068.

Simon, S.A. and Meyers, B.C. (2011) Small RNA-mediated epigenetic modifications in plants. *Current Opinion in Plant Biology* 14(2), 148–155.

Singh, R.K. and Prasad, M. (2022) Delineating the epigenetic regulation of heat and drought response in plants. *Critical Reviews in Biotechnology* 42(4), 548–561.

Singh, J., Mishra, V., Wang, F., Huang, H.Y. and Pikaard, C.S. (2019) Reaction mechanisms of pol IV, RDR2, and DCL3 drive RNA channeling in the siRNA-directed DNA methylation pathway. *Molecular Cell* 75(3), 576–589.

Smulders, M.J.M. and Klerk, G.J. (2011) Epigenetics in plant tissue culture. *Plant Growth Regulation* 63, 137–146.

Solís, M.-T., El-Tantawy, A.A., Cano, V., Risueño, M.C. and Testillano, P.S. (2015) 5-Azacytidine promotes microspore embryogenesis initiation by decreasing global DNA methylation, but prevents subsequent embryo development in rapeseed and barley. *Frontiers in Plant Science* 6, 472.

Song, J., Rutjens, B. and Dean, C. (2014) Detecting histone modifications in plants. *Methods in Molecular Biology (Clifton, N.J)* 165–175.

Song, Q., Zhang, T., Stelly, D.M. and Chen, Z.J. (2017) Epigenomic and functional analyses reveal roles of epialleles in the loss of photoperiod sensitivity during domestication of allotetraploid cottons. *Genome Biology* 18(1), 14.

Springer, N.M., Lisch, D. and Li, Q. (2016) Creating order from chaos: Epigenome dynamics in plants with complex genomes. *The Plant Cell* 28(2), 314–325.

Srikant, T. and Tri Wibowo, A. (2021) The underlying nature of epigenetic variation: Origin, establishment, and regulatory function of plant epialleles. *International Journal of Molecular Sciences* 22(16), 8618.

Stedman, E. and Stedman, E. (1951) The basic proteins of cell nuclei. *Philosophical Transactions of the Royal Society of London. Series B, Biological Sciences* 235(630), 565–595.

Steward, N., Ito, M., Yamaguchi, Y., Koizumi, N. and Sano, H. (2002) Periodic DNA methylation in maize nucleosomes and demethylation by environmental stress. *The Journal of Biological Chemistry* 277(40), 37741–37746.

Strenkert, D., Schmollinger, S., Sommer, F., Schulz-Raffelt, M. and Schroda, M. (2011) Transcription factor-dependent chromatin remodeling at heat shock and copper-responsive promoters in *Chlamydomonas reinhardtii*. *The Plant Cell* 23(6), 2285–2301.

Sunkar, R., Chinnusamy, V., Zhu, J. and Zhu, J.-K. (2007) Small RNAs as big players in plant abiotic stress responses and nutrient deprivation. *Trends in Plant Science* 12(7), 301–309.

Suto, R.K., Clarkson, M.J., Tremethick, D.J. and Luger, K. (2000) Crystal structure of a nucleosome core particle containing the variant histone H2A.Z. *Nature Structural Biology* 7(12), 1121–1124.

Takatsuka, H. and Umeda, M. (2015) Epigenetic control of cell division and cell differentiation in the root apex. *Frontiers in Plant Science* 6, 1178.

Tingting, L., Jinpeng, Z., Xi, Y., Kejian, W., Yuchun, R. *et al.* (2023) Development and application of prime editing in plants. *Rice Science* 30(6), 509–522.

Tittel-Elmer, M., Bucher, E., Broger, L., Mathieu, O., Paszkowski, J. *et al.* (2010) Stress-induced activation of heterochromatic transcription. *PLoS Genetics* 6(10), e1001175. DOI: 10.1371/journal.pgen.1001175.

Treangen, T.J. and Salzberg, S.L. (2012) Repetitive DNA and next-generation sequencing: Computational challenges and solutions. *Nature Reviews Genetics* 13(1), 36–46.

Vanyushin, B.F. and Ashapkin, V.V. (2011) DNA methylation in higher plants: Past, present and future. *Biochimica et Biophysica Acta (BBA) – Gene Regulatory Mechanisms* 1809(8), 360–368.

Varotto, S., Tani, E., Abraham, E., Krugman, T., Kapazoglou, A. *et al.* (2020) Epigenetics: Possible applications in climate-smart crop breeding. *Journal of Experimental Botany* 71(17), 5223–5236.

Vasko, M.R., Guo, C., Thompson, E.L. and Kelley, M.R. (2011) The repair function of the multifunctional DNA repair/redox protein APE1 is neuroprotective after ionizing radiation. *DNA Repair* 10(9), 942–952.

Waddington, C.H. (1942) The epigenotype. *Endeavour* 1, 18–20.

Wahid, A., Gelani, S., Ashraf, M. and Foolad, M.R. (2007) Heat tolerance in plants: An overview. *Environmental and Experimental Botany* 61(3), 199–223.

Wambui Mbichi, R., Wang, Q.-F. and Wan, T. (2020) RNA directed DNA methylation and seed plant genome evolution. *Plant Cell Reports* 39(8), 983–996.

Weng, M., Yang, Y., Feng, H., Pan, Z. and Shen, W.H. (2014) Histone chaperone ASF1 is involved in gene transcription activation in response to heat stress in *Arabidopsis thaliana*. *Plant, Cell & Environment* 37(9), 2128–2138.

Wolffe, A.P. (1992) *Chromatin: Structure and function*. Academic, San Diego, California.

Wu, H., Guang, X., Al-Fageeh, M.B., Cao, J., Pan, S. *et al.* (2014) Camelid genomes reveal evolution and adaptation to desert environments. *Nature Communications* 5(1), 5188.

Wu, T., Yuan, T., Tsai, S.-N., Wang, C., Sun, S.-M. *et al.* (2009) Mass spectrometry analysis of the variants of histone H3 and H4 of soybean and their post-translational modifications. *BMC Plant Biology* 9(1), 98.

Wyatt, G.R. (1951) Recognition and estimation of 5-methylcytosine in nucleic acids. *The Biochemical Journal* 48(5), 581–584.

Xiong, W., Reynolds, M. and Xu, Y. (2022) Climate change challenges plant breeding. *Current Opinion in Plant Biology* 70, 102308.

Xu, J., Chen, G., Hermanson, P.J., Xu, Q., Sun, C. *et al.* (2019) Population-level analysis reveals the widespread occurrence and phenotypic consequence of DNA methylation variation not tagged by genetic variation in maize. *Genome Biology* 20(1), 243.

Xu, P., Chen, H., Hu, J. and Cai, W. (2021) Potential evidence for transgenerational epigenetic memory in *Arabidopsis thaliana* following spaceflight. *Communications Biology* 4(1), 835.

Yamagishi, K. and Kikuta, Y. (2021) Nucleoside derivatives of 5-methylcytosine suppress 5-azacytidine-induced reactivation of a silent transgene in suspension-cultured tobacco cells. *Plant Biotechnology (Tokyo, Japan)* 38(1), 173–178.

Yin, L.-J., Zhang, Y., Lv, P.-P., He, W.-H., Wu, Y.-T. *et al.* (2012) Insufficient maintenance DNA methylation is associated with abnormal embryonic development. *BMC Medicine* 10, 1–10.

Zhang, H., Lang, Z. and Zhu, J.-K. (2018) Dynamics and function of DNA methylation in plants. *Nature Reviews Molecular Cell Biology* 19(8), 489–506. DOI: 10.1038/s41580-018-0016-z.

Zhao, J., Lu, Z., Wang, L. and Jin, B. (2020) Plant responses to heat stress: Physiology, transcription, noncoding RNAs, and epigenetics. *International Journal of Molecular Sciences* 22(1), 117.

Zhong, S.-H., Liu, J.-Z., Jin, H., Lin, L., Li, Q. *et al.* (2013) Warm temperatures induce transgenerational epigenetic release of RNA silencing by inhibiting siRNA biogenesis in *Arabidopsis*. *Proceedings of the National Academy of Sciences of the United States of America* 110(22), 9171–9176.

Zhong, X., Du, J., Hale, C.J., Gallego-Bartolome, J., Feng, S. *et al.* (2014) Molecular mechanism of action of plant DRM de novo DNA methyltransferases. *Cell* 157(5), 1050–1060.

17 Epigenetics for Crop Improvement and Breeding: Updated Methods and Current Achievements

Ambika [1] and Rajendra Kumar[2*]

[1]*Department of Genetics and Plant Breeding, UAS, GKVK, Bangalore, Karnataka, India;* [2]*Division of Genetics, ICAR-Indian Agricultural Research Institute, New Delhi, India*

Abstract

Maintaining global food security is a significant concern that necessitates cutting-edge crop improvement strategies. Some of the main obstacles to these strategies include the loss of agricultural lands as a result of urbanization in both developing and underdeveloped countries, extreme climate changes, plant diseases, and farmers who refuse to use modern agricultural techniques. The epigenetic regulation of plant responses to adversity provides a controllable mechanism for optimizing plant growth, flowering time, heterosis, and ultimately, yield. To design advantageous phenotypic qualities in target crops, epi-breeding uses agricultural techniques that depend on important epigenetic characteristics. DNA methylation, histone changes, and non-coding RNAs (ncRNAs) are the well-acknowledged epigenetic regulators. Recent advances that have resulted in intriguing discoveries, as well as the pioneering nature of this field, necessitate rigorous methodology and advanced technologies to propel epigenetics to the forefront of molecular biology. The methods currently in use to identify locus-specific and genome-wide modifications for all epigenetic codes will be covered in this chapter. Additionally, recent advances in epigenetic analysis as well as updated technology analysis are discussed. These techniques offer thorough recommendations and cutting-edge technologies in the field of epigenetics, facilitating further advancements in this exciting and quickly evolving subject.

Keywords: Epigenetic codes, Epigenetic regulation, Food security, Genetic engineering, Non-coding RNAs

17.1 Introduction

Epigenetics is the study of stable and heritable changes in gene expression that do not affect the underlying DNA sequence (Fujimoto *et al.*, 2012). Environmental or behavioral factors that may alter gene function can induce epigenetic changes (Parker *et al.*, 2021). Such changes are reversible in nature and affect how DNA sequences are read rather than the DNA itself. DNA methylation, histone modifications, and post-transcriptional silencing via small non-coding RNA mechanisms are the main mechanisms by which epigenetic regulation of gene expression is understood.

Epialleles are responsible for transmitting epigenetic changes. After being steadily maintained in the chromatin state, epialleles are transmitted to subsequent generations.

*Corresponding author: rajendrak64@yahoo.co.in

 Epigenetics for Climate-Smart and Sustainable Agriculture (ed. Jen-Tsung Chen)
DOI: 10.1079/9781800626102.0017

As additional sources of variation, epialleles control phenotypic diversity. Epialleles influence the morphology of flowers, flowering time, reproductive system, vigor of hybrids, resistance to diseases, and senescence of leaves in plant species (Brukhin and Albertini, 2021).

Since plants are sessile by nature, changes in the environment always have an impact on them. They can, however, modify their biological processes to take advantage of changing environmental circumstances. By changing the patterns of gene expression, they consistently adapt at the molecular level to their surroundings (Yaish *et al.*, 2011). Plants exhibit conservation of epigenetic modifications, which impact the chromatin structure and subsequently control the expression of genes. Any alteration associated with one of the epigenetic systems, which are crucial for controlling a variety of biological processes, leads to anomalies in plant development. Consequently, epigenetic modifications have a dynamic impact on plant growth and development (Zhang *et al.*, 2018).

17.2 Epigenetic Regulations

Epigenetic regulation is achieved through three mechanisms: DNA methylation, histone modifications, and the activity of non-coding RNA molecules, including small interfering RNAs (siRNAs) and microRNAs (miRNAs), as well as long non-coding RNAs (lncRNAs). Gene expression is impacted by chromatin architecture alterations brought about by the activation of these pathways. Gene activation or silencing is associated with either an accessible (active) or inaccessible (repressive) chromatin structure, respectively (Allis and Jenuwein, 2016). Under various environmental conditions and developmental stages, appropriate cellular function is dependent on the switch between active and repressive chromatin forms and interactions among the epigenetic mechanisms (Förderer *et al.*, 2016).

17.2.1 DNA methylation

In eukaryotes, DNA methylation is a well-conserved epigenetic mark that results from adding a methyl group to the 5' carbon of the cytosine base to generate 5-methylcytosine. DNA methylation takes place in three sequence contexts in plants: the asymmetric CHH context and the symmetric CG and CHG contexts (H = C, A, or T) (Feng *et al.*, 2010). DNA methylation is possible in two different ways: maintenance methylation, and *de novo* methylation. Retention of pre-existing methylation patterns follows DNA replication in maintenance methylation, whereas *de novo* methylation is the creation of new methylation patterns from the methylation of previously unmethylated cytosine residues (Chen and Li, 2004).

DNA methylation is an enzyme-mediated modification of bases that is reversible. Three different categories comprise the enzymes that catalyze cytosine methylation in plants. First, symmetric cytosine methylation (CpG) of the genome is maintained by DNA METHYLTRANSFERASE1 (MET1), a homolog of mammalian methyltransferase (Dnmt1). The second group consists of the plant-specific CHROMOMETHYLASE3 (CMT3), which adds a methyl group to sequences of the CHG type, particularly at transposons and centromeric repeats (Lindroth *et al.*, 2001; Tompa *et al.*, 2002). The aforementioned DNA methyltransferases cause symmetric cytosine methylation in the genome (Chan *et al.*, 2005), which is referred to as maintenance methylation. The third category includes the two DNA methyltransferases DOMAIN REARRANGED METHYLASE1 (DRM1) and DRM2, which are in charge of *de novo* methylation, catalyze asymmetrical cytosine methylation at the CpNpNp site (Ramsahoye *et al.*, 2000).

DNA methyltransferases and demethylases dynamically control various processes in plant development through their antagonistic actions. DNA glycosylases like REPRESSOR OF SILENCING 1 (ROS1), DEMETER (DME), and DEMETER LIKE 2 and 3 (DML2 and DML3) facilitate active demethylation in plants via a base-excision-repair pathway. Passive demethylation, on the other hand, refers to the loss of methylation as a result of inadequate maintenance during replication (Wu and Zhang, 2010).

The majority of the time, DNA methylation in promoter regions results in gene silencing; nevertheless, it is linked to gene expression in the gene body. It controls many different processes, including gene expression, genomic stability, gene imprinting, and the inactivation of transposable elements. Growth retardation, failure to reach the floral transition stage, aberrant embryo and endosperm development, reduced leaf size or total sterility, and inability to ripen fruit are among the developmental abnormalities that can arise when DNA methylation is affected (Baulcombe and Dean, 2014; Iwasaki and Paszkowski, 2014; Lang *et al.*, 2017; Zhang *et al.*, 2018; Gallego-Bartolomé, 2020). Methylation of DNA influences plant phenotypes and is heritable, which helps to increase crop productivity.

17.2.2 Histone modification

Histone modification is another important epigenetic process that affects many aspects of plant development (Förderer *et al.*, 2016). Histone modifications are chemical changes that occur on the N-tails of nucleosomal histones. These changes can be specific to lysine, arginine, serine, or threonine residues and include methylation, acetylation, phosphorylation, ubiquitination, biotinylation, and sumoylation (Engelhorn *et al.*, 2014). These chemical alterations in the chromatin architecture produce closed (repressive) or open (active) chromatin states, which regulate transcription and are carried on to progeny cells (Allis and Jenuwein, 2016).

One of the most commonly seen histone modifications is the methylation process, and plants have been shown to contain the proteins involved in both methylation and demethylation. Lysine and arginine residues are the sites of histone methylation. Histone lysine methyltransferase (HKMT) mediates the presence of mono-, di-, or trimethylation in lysines, which are typically found in lysines 4, 9, 27, and 36 of histone H3 (Cheng *et al.*, 2020). The catalytic activity of HKMT enzymes is attributed to their SET domain. Only mono- and dimethylation of arginine residues are found at the N-terminal tails of histone H3's arginine 2, 8, 7, 17, and 26 and histone H4's arginine 3 by the protein arginine methyltransferase (PRMT) (Bedford and Richard, 2005). PRMT enzymes can be categorized as type I, II, III, or IV depending on symmetrical or asymmetrical methylation forms. Furthermore, depending on the type of modified histone molecule and the amino acid residue, the number of methyl groups has varying biological implications that can either promote or inhibit the expression of a gene. For instance, trimethylations of H3K4 and H3K36 are linked to transcriptionally active genes, while transcriptional repression is linked to trimethylations of H3K27 and mono- and dimethylations of H3K9. It has been shown that arginine and lysine methylation contribute to nuclear transport, transcriptional regulation, signal transduction, RNA processing, and DNA damage repair (Liu *et al.*, 2010).

The dynamic process of histone demethylation demonstrates the reversibility of the histone methylation process. Lysine-specific demethylase 1 (LSD1), the first lysine demethylase, was discovered in 2004 and was found to use FAD as a co-factor to reverse mono- and dimethylated lysine substrates. Jumonji C domain-containing (JMJD) enzymes are the second class of lysine demethylases that were found to remove the trimethyl from lysine residues. Fe(II) and α-ketoglutarate are used as co-factors by this enzyme to remove the trimethyl group from H3K9me3 and H3K36me3 (Whetstine *et al.*, 2006). Thus far, PAD4 and JMJD6 have been identified as histone arginine demethylases (Chang *et al.*, 2007). JMJD6 directly converts methylarginine to arginine, while PAD4 transforms monomethylated arginine into citrulline.

Histone acetyltransferase (HAT) catalyzes histone acetylation, which adds an acetyl group to the lysine residue. Transcriptionally active genes are associated with the epigenetic mark of acetylation. Histone H3 residues K9, K14, K18, K23, and K27, as well as histone H4 residues K5, K8, K12, K16, and K20, are commonly acetylated in plants. There can be four acetylations on histones at most. In plants, acetylations are frequently associated with particular responses; for instance, H3K9/14ac or H3K27ac are correlated with the activation of genes induced by stress and developmental processes (Kim *et al.*, 2017). Similar to this, histones can also actively deacetylate via the action of histone deacetylases (HDACs), which represses transcription (Chen and Tian, 2007) and is associated with a wide range of processes such as photomorphogenesis (Jang *et al.*, 2011) or flowering (Yu *et al.*, 2011).

Histone phosphorylation is caused by histone kinases, which phosphorylate the serines, threonines, and tyrosines of N-terminal histone tails in addition to linker histone H1. Histone phosphatases remove the phosphate group from a molecule, reversing kinase activity. This histone mark is thought to affect a wide range of biological activities, including transcription, apoptosis, DNA repair, and chromatin compaction during cell division. The most well-documented signature for chromatin condensation linked to mitosis and meiosis in many eukaryotic organisms is likely the phosphorylation of H3 on S10, which is commonly used as a reference mark for these activities (de la Barre *et al.*, 2000). One significant histone modification that is crucial to the DNA damage response is the phosphorylation of H2A(X) (Rossetto *et al.*, 2012).

In histone ubiquitylation, three enzymes – E1-activating, E2-conjugating, and E3-ligating enzymes – add the 76-amino-acid polypeptide ubiquitin to the lysine residue of the histone protein (Saracco *et al.*, 2009). The degree of ubiquitylation (mono- or poly-) and substrate specificity are determined by the enzyme complexes. Two forms of histone mono-ubiquitylation – H2AK119ub1 for Polycomb pathway-mediated gene silencing and H2BK123ub1 for transcriptional elongation and initiation – have been thoroughly studied (Wang *et al.*, 2004). This mechanism modifies a histone by adding an exceptionally large molecule, in contrast to the other histone modifications previously stated. The overall shape of the nucleosome is probably altered by this extraordinarily big molecule, which influences intra-nucleosomal interactions (Kim *et al.*, 2009). Deubiquitinizing enzymes, a type of isopeptidase, dissolve the isopeptide link that binds ubiquitin to the histone molecule, thereby removing the modification.

17.2.3 RNA-directed DNA methylation

The term 'RNA-directed DNA methylation' (RdDM) describes the regulation of gene expression by small RNA-mediated methylation of particular DNA sequences. The RdDM pathway facilitates *de novo* methylation. It can be classified into canonical and non-canonical RdDM pathways according to the origin and biogenesis of the sRNAs and proteins involved. Canonical RdDM operates best by recruiting RNA polymerase IV (Pol IV) to the target region, where it transcribes roughly 30–45 nt of single-stranded RNAs (sRNAs). RNA-directed RNA polymerase 2 (RDR2) then converts the dsRNAs into 24 nt of sRNAs, which are then further cleaved by the DCL3 protein (Blevins *et al.*, 2015). They are stabilized by 3' end methylation in the cytoplasm through the methyltransferase protein HEN1. The sRNAs are inserted into the Argonaute (AGO4/AGO6) complex, which in turn employs DRM2 (DOMAINS REARRANGED METHYLTRANSFERASE 2) to methylate the desired DNA sequence.

Unlike the canonical RdDM pathway, the non-canonical RdDM pathway originates 21–22 nt sRNAs from transcripts generated by Pol II, transposable elements, viruses, transcripts containing inverted repeat sequences, and miRNA precursors (Nuthikattu *et al.*, 2013). Following their incorporation into AGO proteins, distinct DCL proteins cleave the dsRNAs into 21–24 nt sRNAs. After attaching itself to complementary RNA, the complex can cleave or repress translation, which silences genes post-transcriptionally and leads to post-transcriptional gene silencing (PTGS) (Martínez de Alba *et al.*, 2013).

17.3 Recent Advancements in Epigenetic Methodologies

DNA methylation is a transgenerationally inherited process that is dynamic, tissue-specific, and sequence-dependent (Yong *et al.*, 2016). In the past, studies based on DNA methylation have been centered on assessing the overall amount of 5mC and figuring out the methylation status in any gene of interest (Lisanti *et al.*, 2013). Fortunately, the use of microarray hybridization technology and the rapid advancement of high-throughput next-generation sequencing (NGS) technologies have made it feasible to explore the DNA methylation pattern all the way up to the genome. This allows for the construction of genomic maps of DNA methylation at the single-base resolution (Laird, 2010). In this context, many strategies are employed in the experimental approaches for DNA methylation characterization.

Bisulfite methylation analysis was one of the most significant advances in the field of epigenetics techniques. Cytosine is converted to uracil when denatured DNA is bisulfite treated, but 5-methylcytosine (5mC), a by-product of DNA methylation, is not changed. This concept has led to the development of several methods to determine 5mC in DNA, such as methylation-specific PCR (MSP), a widely used approach for evaluating methylation alterations in DNA. MSP is widely used for early disease diagnosis, despite the fact that it is not as sensitive as other techniques unless it is monitored in real-time, as in the case of the quantitative MS-PCR (qMS-PCR) approach.

While studies on DNA methylation specific to genes or loci have greatly advanced our understanding of the function of DNA methylation in numerous biological processes, it is frequently useful to assess the overall changes in DNA methylation within cells, tissues, or other samples. The technique known as enzyme-linked immunosorbent (ELISA) test is widely employed in the identification of global DNA methylation. It is capable of detecting 5-hydroxymethylcytosine (5hmC), which is also known to play a significant role in epigenetic-based gene control. Once the DNA mixes with anti-5mC or anti-5hmC, this approach compares DNA containing 5mC or 5hmC to a fully methylated or hydroxymethylated control. The percentage difference provides information about the global 5mC or 5hmC of the DNA under study.

There are numerous methods for mapping global DNA methylation. Reduced representation bisulfite sequencing (RRBS) is one of them. Meissner introduced it in 2005 as a tool for comparing and analyzing patterns of genomic methylation (Meissner *et al.*, 2005). Nonetheless, it is noteworthy for being an economical and sensitive technique that can be used extensively while still providing single-base resolution. After the genome is digested using restriction enzymes to identify CpG-dense regions, these regions are sequenced using next-generation sequencing technology to profile the methylome.

CRISPR/Cas systems are a new class of highly versatile programmable effectors. Relying on a short RNA guide (sgRNA) and an effector Cas protein, plant crops have shown the simplicity and versatility of this method. Plant breeding now has more opportunities as a result (Zaynab *et al.*, 2020). CRISPR systems can also be used to modify epigenetic modifications and function as efficient transcriptional regulators. Endonuclease-inactivated Cas9 protein, sometimes referred to as dead Cas9, or dCas9, is produced by mutating particular amino acids in the RuvC1 and HNH nuclease domains. Depending on the effector domain that is linked to the Cas9 protein, this protein can be readily converted into programmable activation (CRISPRa) or repression (CRISPRi) factors.

CRISPRi was first developed and used to selectively block target genes through transcription in *Escherichia coli* and human cells (Larson *et al.*, 2013). CRISPRi provides an efficient and targeted genomic targeting platform that does not alter the targeted DNA sequence, despite the possibility of transcriptional regulation and the fusion of the transcriptional repressor domain (TRD) to the dCas9 protein forming the chimeric dCas9-TRD protein to strengthen the transcriptional interference (Xu *et al.*, 2023). The expression of dCas9 fused 3 × SRDX in common wheat may efficiently suppress the expression of the target gene TaPDS, and this suppression can be inherited by the next generation (Zhou *et al.*, 2022).

Traditional gene overexpression techniques have limitations that can be overcome by the CRISPR/dCas9 activation (CRISPRa) system. Designating several guide RNAs in a single vector allows for the activation of several genes (Ding *et al.*, 2022). dCas9 can be coupled with different transcription activation domains (TADs) and other transcription activators to regulate the activation of target gene expression in this system (Gao *et al.*, 2014). Targeting four gene activation sites simultaneously is possible with this system. In *Arabidopsis*, it was successfully possible to activate three separate target genes simultaneously through transcription (Lowder *et al.*, 2018).

Selective epigenetic regulation such as DNA methylation/demethylation and histone acetylation of target sites can be achieved by recruiting distinct epigenetic effector domains using CRISPR/dCas. The human demethylase ten-eleven translocation 1 (TET1cd) catalytic domain combined with an improved SunTag system forms the basis of the targeted demethylation system, which is based on the CRISPR/dCas9 system. In *Arabidopsis*, SunTag-TET1cd can also target FWA or CACTA1, and demethylate and activate gene expression (Gallego-Bartolomé *et al.*, 2018). The tobacco DRM methyltransferase catalytic domain (NtDRMcd) was substituted for VP64 in the SunTag system based on the CRISPR/Cas9-SunTag targeting system. This effectively produced ectopic methylation of the *Arabidopsis* FWA promoter and an early flowering phenotype (Papikian *et al.*, 2019). *Arabidopsis* histone acetyltransferase 1 (AHT) was fused with dCas9, thereby improving the expression of the abscisic acid-responsive element-binding protein AREB1/ABF2 gene through chromatin remodeling and effectively increasing the drought tolerance of the plant (de Melo *et al.*, 2020). This suggests that the dCasHAT system has a promising future in plant epigenetic markers.

Plant growth and development are greatly influenced by epigenetic modification. With the rapid advancement of molecular biology methods like gene editing, epigenetics has demonstrated tremendous promise for the genetic enhancement of significant crop features. The application of epigenetic modification in crop improvement is also technically supported by the advent of epigenetic genome editing technology.

17.4 Epigenetics for Crop Improvement

In addition to crop development, epigenetic regulatory mechanisms also play a role in environmental adaptation. Maximizing the impact of epigenetics on crop development could be made possible by the application of epigenome analysis and engineering. This could allow agricultural development through epigenome engineering and the production of unique epiallelic variations by alterations in DNA methylation, chromatin remodeling, and RNA interference in conjunction with the new epigenome editing tools (Springer and Schmitz, 2017). Crops can be improved specifically and effectively by identifying the epigenetic mechanisms influencing key agronomic traits like yield, biomass, flowering time, resistance to biotic and abiotic stresses, fruit ripening, and heterosis (Table 17.1).

17.4.1 Flowering time

One of the most significant developmental shifts in a plant's life cycle is the transition from the vegetative to the floral stage. Plants need to flower in environments with favorable temperatures and sufficient day lengths to ensure successful reproduction. The combination of environmental inputs like temperature, light quality, and day length with endogenous factors like age and developmental stage determines flowering time genetically (Andrés and Coupland, 2012). Flowering can be controlled by varying the length of the day or the temperature through the photoperiodic or vernalization pathways, respectively (Andrés and Coupland, 2012; Song *et al.*, 2015). Many plants have their seasonal timing of flowering determined by the integration of these seasonal responses (Andrés and Coupland, 2012). FLOWERING LOCUS C (FLC) is the key gene involved in the vernalization response in *Arabidopsis* (Sheldon *et al.*, 1999; Michaels and Amasino, 1999). FLC is a MADS-box gene that results in a protein that continues to keep the plant's growing apex in a vegetative state.

The vernalization pathway, which is triggered by winter cold, enables Polycomb group (PcG)-mediated chromatin silencing at FLC (He and Li, 2018), which assemble into various chromatin-associated complexes, such as Polycomb repressive complex 1 (PRC1) and PRC2 that suppress target gene expression in plants and animals (Mozgova and Hennig, 2015). A PLANT HOMEODOMAIN (PHD) protein, encoded by VERNALIZATION INSENSITIVE 3 (VIN3), is expressed in response to prolonged cold exposure (Sung and Amasino, 2004) and, coupled with core PRC2 subunits, it forms a complex known as vernalization/cold-specific PHD (VIN3)-PRC2, which catalyzes the deposition of repressive H3K27me3 marks on FLC chromatin under cold temperatures (De Lucia *et al.*, 2008). The Cold Memory Element (CME) is the *cis*-regulatory element found in H3K27me3 that mediates FLC silencing both during and after winter cold exposure (Yuan *et al.*, 2016). H3K27me3 accumulates in the CME-bearing

Table 17.1. Epigenetic regulations influencing agronomic traits in different crops. Author's own table.

Crops	Traits	Epigenetic mechanism	References
Arabidopsis	Flowering time	DNA methylation	Zhang *et al.*, 2013
		Chromatin modifications	He, 2009
		Chromatin remodeling	Jing *et al.*, 2019
	Seed dormancy	DNA demethylation	Zhang *et al.*, 2018
	Mega-gametogenesis and endosperm development	Histone modification	Shen *et al.*, 2021
	Double-fertilization	DNA demethylation	Ibarra *et al.*, 2012
Rice	Frequency of crossover	DNA methylation and chromatin modifications	Yelina *et al.*, 2015
	Heterosis	DNA methylation	Feng *et al.*, 2021
	Flowering time	Histone modification	Sun *et al.*, 2014
	Flowering and reproduction	Histone modification, DNA methylation	Shi *et al.*, 2014
	Floral abnormalities	miRNA	Zhu *et al.*, 2009
	Leaf development	miRNA	Xie *et al.*, 2012
Maize	Heterosis	Small RNA	Seifert *et al.*, 2018
	Improved yield	DNA methylation	Tani *et al.*, 2005
	Controls cell division in leaves	DNA methylation	Candaele *et al.*, 2014
Tomato	Vitamin E	DNA methylation	Quadrana *et al.*, 2014
	Fruit ripening	DNA methylation	Zuo *et al.*, 2020
	Aberrant flowers and parthenocarpic fruits	Histone deacetylation	Ampomah-Dwamena *et al.*, 2002
	Enhanced plant vigor	Small RNA	Kundariya *et al.*, 2020
Soybean	Seed development	DNA methylation	An *et al.*, 2017
Sugar beet	Tolerance to bolting	DNA methylation	Hébrard *et al.*, 2016

nucleation region during vernalization as a result of the *trans*-acting VIVIPAROUS1/ABI3-LIKE 1/2 (VAL1/2) binding to the CME and physically interacting with PcG proteins to enrich the PHD (VIN3)-PRC2 complex at this location (Yuan *et al.*, 2016). This results in FLC silencing during vernalization by converting the active FLC chromatin produced by FRIsc to a silenced state with H3K27me3 marks (Yuan *et al.*, 2016).

As previously mentioned, CG, CHH, and CHG methylations are attributed to MET1, CMT2 and CMT3, and DRM1 and DRM2, respectively. Thus, in *Arabidopsis*, loss of function of DRM1/2 or MET1 results in delayed flowering (Kankel *et al.*, 2003). In contrast, quintuple mutants of met1 drm1 drm2 cmt3 cmt2 (mddcc) never flower under long-day conditions (He *et al.*, 2022), suggesting a critical role in DNA methylation in flowering.

17.4.2 Plant reproduction

During plant reproduction, epigenetic modifications are crucial for several molecular and physiological processes. Concerning male gametophytes, the haploid microspore undergoes mitosis to produce gametophytes. Even though there are just two mitotic divisions between the vegetative cell and the sperm, they are in different epigenetic states. Sperm DNA has low CHH methylation in retrotransposons but high methylation in the CG and CHG contexts. Conversely, the vegetative cell has low CG and CHG methylation and high CHH methylation in heterochromatic transposable elements (TE) (Slotkin *et al.*, 2009). The microspore mother cell, microspores, and sperm all have relatively similar levels of CG and CHG methylation (Walker *et al.*, 2018). While CHH methylation gradually rises, sperm still have lower CHH methylation than somatic cells (Walker *et al.*, 2018). In contrast, RdDM restores the levels of CHH methylation in the vegetative cell from the lower levels found in the microspore. The functional implications of these dynamics remain unknown.

With one to three mitoses separating them, the female gametes are cytologically distinct from the egg, showing less repressive histone modifications and decondensed chromatin in the secondary nucleus (Pillot *et al.*, 2010). *Arabidopsis* central cells show reduced expression of the maintenance and *de novo* methyltransferases (Jullien *et al.*, 2012) as a result of decreased CG methylation in genes and TEs (Park *et al.*, 2016). It is unclear, nevertheless, how female gametogenesis is related to epigenetic dynamics.

Despite the heritability of DNA methylation across generations, there is compelling evidence that chromatin-based epigenetic memory, which developed during vegetative growth and development, is erased during embryogenesis. It is well known that early transcription factors reprogrammed chromatin during embryonic development. For instance, in winter annuals or biennials, the memory of previous winter experiences is retained through mitotic cell divisions but is lost in the following generation (He and Li, 2018). Therefore, before flowering in the spring, each generation needs to overwinter. An embryonic regulatory pathway known as the LEC pathway, which is made up of three master TFs involved in embryogenesis – LEAF COTYLEDON 1 (LEC1), LEC2, and FUSCA 3 (FUS3) – erases or resets the cold memory in *Arabidopsis* during early embryogenesis (Tao *et al.*, 2019).

17.4.3 Fruit ripening

Fruit ripening is a critical stage in the development of fruit edible quality. It is a complicated developmental process that includes changes in fruit texture, pigment accumulation, and the synthesis of aroma and flavor compounds, and is regulated by a variety of internal and external stimuli (Ji and Wang, 2023). Among these, epigenetic regulation is one of the most significant. The discovery of the tomato variety known as colorless non-ripening (CNR) offered preliminary evidence that DNA methylation influences fruit ripening. It showed that fruit softening was inhibited and that flavor and carotenoid component accumulation were absent (Manning *et al.*, 2006). The gene SlDML2, which codes for the tomato DNA demethylase, is said to be responsible for a global decrease in DNA methylation in the tomato fruit's pericarp during ripening (Lang *et al.*, 2017). Ripeness is significantly delayed when SlDML2 is knocked down or deleted. It has been proposed that fruit development is regulated by the interaction of transcription factors and DNA methylation. The MADS-box transcription factor RIPENING INHIBITOR (RIN), which is hypomethylated in its promoter regions during ripening, is a crucial regulator of fruit ripening (Zhong *et al.*, 2013).

The control of fruit development and ripening appears to be mediated by histone deacetylases (HDACs), as evidenced by the delayed ripening of tomato fruits and reduced accumulation of carotenoids observed upon knockdown of tomato HDAC SlHDT3 (Guo *et al.*, 2017). In contrast, an increase in carotenoid accumulation and an earlier ripening were the outcomes of the knockdown of two additional HDACs, SlHDA1 and SlHDA3 (Guo *et al.*, 2018). Therefore, it is likely that HDACs play a role in the ripening kinetics as well as the fine-tuning of gene expression during this process.

17.4.4 Heterosis

Hybrid vigor or heterosis is defined as the phenomenon whereby a hybrid exhibits phenotypic superiority over its parents in traits such as growth rate, reproductive success, and yield. Numerous interactive characteristics, such as changes in metabolic activity, epigenetic regulation, and gene expression, have been linked to heterosis (Zhu *et al.*, 2017). DNA cytosine methylation may be a part of processes that lead to heterosis. Numerous investigations have revealed variations in the cytosine methylation patterns of heterotic F1 hybrids in maize relative to their respective parents (Zhao *et al.*, 2007). Similar to this, in rice, variations in transcript levels at certain differentially methylated regions (DMRs) in the F1 hybrids are correlated with cytosine methylation patterns that are seen between inbred lines (He *et al.*, 2010).

DNA methylation in the F1 hybrids was found to be higher overall than in the parents, as revealed by genome-wide methylation profiling of *Landsberg erecta* (Ler-0) and C24 parental inbred lines, as well as their reciprocal hybrid lines, Ler-0 ×C24 and C24 ×Ler-0 (Shen *et al.*, 2012). Two gene-activating marks in the *Arabidopsis* genome, H3K4me3 and H3K9ac, and two repressive marks, H3K27me3 and H3K9me2, were altered at certain locations in a Ler/C24 hybrid. A limited subset of genes showed altered expression in conjunction with these markers. It has been suggested that early developmental changes in gene activity and histone modifications play a crucial role in the later phases of development when seedling biomass increases (Zhu *et al.*, 2017).

17.4.5 Apomixis

Apomixis is reproduction via asexual seeds. It manifests as a result of the sexual reproduction-related genes' alternative expression. Meiosis in the nucellus of apomicts is either bypassed and replaced by apomeiosis, resulting in apospory, or blocked during the first meiotic division, in the case of diplospory (Albertini *et al.*, 2019). A diploid embryo sac, which morphologically resembles the sexual embryo sac, is formed in diploid apomicts, regardless of events. In both *Arabidopsis* and maize, epigenetic pathways seem to play a role in limiting the development of sexual female gametophytes to a single cell (Baulcombe and Dean, 2014). AGO genes from the ARGONAUTE family control the chromatin modifications caused by RNA interference that are linked to the apomixis epigenetic pathway.

Up to 70% of functional unreduced female gametes are produced by loss-of-function in ago104, a maize homolog of AGO9, which also causes apomixis-like features (Singh *et al.*, 2011). Ago104 mutations cause meiosis to resemble mitosis rather than affecting the number of germ cells in the ovule. Additional cytological data indicate that the mutant's meiosis begins identically to that of the wild type (Singh *et al.*, 2011). Ago9 mutants imitate the initial phases of apospory by inducing gametogenesis in somatic nucellar cells (without sporogenesis). Ago104 mutants are similar to diplospory in that they retain mitotic divisions in a distinct and appropriately specified archespore. Therefore, despite the phylogenetically related nature of both genes, their distinct functions are suggested by their phenotypes: AGO9 functions to suppress germ cell specification in the ovule beside the unique archespore, while AGO104 suppresses somatic (mitotic) fate within the archespore.

17.5 Conclusion

Since plants are sessile by nature, changes in the environment always have an impact on them. They can, however, modify their biological processes to take advantage of changing environmental circumstances. They consistently modify their molecular structure to interact with the environment. These modifications, which include DNA methylation, histone modification (acetylation/deacetylation, methylation/demethylation, ubiquitination, phosphorylation), and small RNA modifications, are part of the epigenetic machinery that regulates gene expression in plants in concert with external factors, transposon silencing, and hormone signaling. By enhancing phenotypic variation, these modifications aid in the survival of plants and increase their chances of sexual reproduction in adverse conditions. Additionally, it is necessary to detect and characterize the epiallelic variations that exist within and even between crop species' genome sequences. Numerous crops, such as rice, *Arabidopsis*, maize, soybeans, rapeseed, etc., have improved agronomic traits through the use of epigenetics. Utilizing epigenetic information could open up new possibilities for crop development since current breeding methods typically ignore epigenetic components in favor of genetics.

References

Albertini, E., Barcaccia, G., Carman, J.G. and Pupilli, F. (2019) Did apomixis evolve from sex or was it the other way around. *Journal of Experimental Botany* 70(11), 2951–2964. DOI: 10.1093/jxb/erz109.

Allis, C.D. and Jenuwein, T. (2016) The molecular hallmarks of epigenetic control. *Nature Reviews Genetics* 17(8), 487–500. DOI: 10.1038/nrg.2016.59.

Ampomah-Dwamena, C., Morris, B.A., Sutherland, P., Veit, B. and Yao, J.-L. (2002) Down-regulation of TM29, a tomato SEPALLATA homolog, causes parthenocarpic fruit development and floral reversion. *Plant Physiology* 130(2), 605–617. DOI: 10.1104/pp.005223.

An, Y.-Q.C., Goettel, W., Han, Q., Bartels, A., Liu, Z. *et al.* (2017) Dynamic changes of genome-wide DNA methylation during soybean seed development. *Scientific Reports* 7(1), 1–14. DOI: 10.1038/s41598-017-12510-4.

Andrés, F. and Coupland, G. (2012) The genetic basis of flowering responses to seasonal cues. *Nature Reviews Genetics* 13(9), 627–639. DOI: 10.1038/nrg3291.

Baulcombe, D.C. and Dean, C. (2014) Epigenetic regulation in plant responses to the environment. *Cold Spring Harbor Perspectives in Biology* 6(9), a019471. DOI: 10.1101/cshperspect.a019471.

Bedford, M.T. and Richard, S. (2005) Arginine methylation an emerging regulator of protein function. *Molecular Cell* 18(3), 263–272. DOI: 10.1016/j.molcel.2005.04.003.

Blevins, T., Podicheti, R., Mishra, V., Marasco, M., Wang, J. *et al.* (2015) Identification of pol IV and RDR2-dependent precursors of 24 nt siRNAs guiding de novo DNA methylation in *Arabidopsis*. *eLife* 4, e09591. DOI: 10.7554/eLife.09591.

Brukhin, V. and Albertini, E. (2021) Epigenetic modifications in plant development and reproduction. *Epigenomes* 5(4), 25. DOI: 10.3390/epigenomes5040025.

Candaele, J., Demuynck, K., Mosoti, D., Beemster, G.T.S., Inzé, D. *et al.* (2014) Differential methylation during maize leaf growth targets developmentally regulated genes. *Plant Physiology* 164(3), 1350–1364. DOI: 10.1104/pp.113.233312.

Chan, S.W.L., Henderson, I.R. and Jacobsen, S.E. (2005) Gardening the genome: DNA methylation in *Arabidopsis thaliana*. *Nature Reviews Genetics* 6(5), 351–360. DOI: 10.1038/nrg1601.

Chang, B., Chen, Y., Zhao, Y. and Bruick, R.K. (2007) JMJD6 is a histone arginine demethylase. *Science* 318(5849), 444–447. DOI: 10.1126/science.1145801.

Chen, T. and Li, E. (2004) Structure and function of eukaryotic DNA methyltransferases. *Current Topics in Developmental Biology* 60, 55–89. DOI: 10.1016/S0070-2153(04)60003-2.

Chen, Z.J. and Tian, L. (2007) Roles of dynamic and reversible histone acetylation in plant development and polyploidy. *Biochimica et Biophysica Acta* 1769(5–6), 295–307. DOI: 10.1016/j.bbaexp.2007.04.007.

Cheng, K., Xu, Y., Yang, C., Ouellette, L., Niu, L. *et al.* (2020) Histone tales: Lysine methylation, a protagonist in *Arabidopsis* development. *Journal of Experimental Botany* 71, 793–807. DOI: 10.1093/jxb/erz435.

de la Barre, A.E., Gerson, V., Gout, S., Creaven, M., Allis, C.D. *et al.* (2000) Core histone N-termini play an essential role in mitotic chromosome condensation. *The EMBO Journal* 19(3), 379–391. DOI: 10.1093/emboj/19.3.379.

De Lucia, F., Crevillen, P., Jones, A.M.E., Greb, T. and Dean, C. (2008) A PhD-polycomb repressive complex 2 triggers the epigenetic silencing of FLC during vernalization. *Proceedings of the National Academy of Sciences of the United States of America* 105(44), 16831–16836. DOI: 10.1073/pnas.0808687105.

de Melo, B.P., Lourenço-Tessutti, I.T., Paixão, J.F.R., Noriega, D.D., Silva, M.C.M. *et al.* (2020) Transcriptional modulation of AREB-1 by CRISPRa improves plant physiological performance under severe water deficit. *Scientific Reports* 10(1), 16231. DOI: 10.1038/s41598-020-72464-y.

Ding, X., Yu, L., Chen, L., Li, Y., Zhang, J. *et al.* (2022) Recent progress and future prospect of CRISPR/Cas-derived transcription activation (CRISPRa) system in plants. *Cells* 11(19), 3045. DOI: 10.3390/cells11193045.

Engelhorn, J., Blanvillain, R. and Carles, C.C. (2014) Gene activation and cell fate control in plants: A chromatin perspective. *Cellular and Molecular Life Sciences* 71(16), 3119–3137. DOI: 10.1007/s00018-014-1609-0.

Feng, J.-W., Lu, Y., Shao, L., Zhang, J., Li, H. *et al.* (2021) Phasing analysis of the transcriptome and epigenome in a rice hybrid reveals the inheritance and difference in DNA methylation and allelic transcription regulation. *Plant Communications* 2(4), 100185. DOI: 10.1016/j.xplc.2021.100185.

Feng, S., Jacobsen, S.E. and Reik, W. (2010) Epigenetic reprogramming in plant and animal development. *Science* 330(6004), 622–627. DOI: 10.1126/science.1190614.

Förderer, A., Zhou, Y. and Turck, F. (2016) The age of multiplexity: Recruitment and interactions of polycomb complexes in plants. *Current Opinion in Plant Biology* 29, 169–178. DOI: 10.1016/j.pbi.2015.11.010.

Fujimoto, R., Sasaki, T., Ishikawa, R., Osabe, K., Kawanabe, T. *et al.* (2012) Molecular mechanisms of epigenetic variation in plants. *International Journal of Molecular Sciences* 13(8), 9900–9922. DOI: 10.3390/ijms13089900.

Gallego-Bartolomé, J. (2020) DNA methylation in plants: Mechanisms and tools for targeted manipulation. *The New Phytologist* 227(1), 38–44. DOI: 10.1111/nph.16529.

Gallego-Bartolomé, J., Gardiner, J., Liu, W., Papikian, A., Ghoshal, B. *et al.* (2018) Targeted DNA demethylation of the *Arabidopsis* genome using the human TET1 catalytic domain. *Proceedings of the National Academy of Sciences of the United States of America* 115(9), E2125–E2134. DOI: 10.1073/pnas.1716945115.

Gao, X., Tsang, J.C.H., Gaba, F., Wu, D., Lu, L. *et al.* (2014) Comparison of TALE designer transcription factors and the CRISPR/dCas9 in regulation of gene expression by targeting enhancers. *Nucleic Acids Research* 42(20), e155. DOI: 10.1093/nar/gku836.

Guo, J.-E., Hu, Z., Li, F., Zhang, L., Yu, X. *et al.* (2017) Silencing of histone deacetylase SlHDT3 delays fruit ripening and suppresses carotenoid accumulation in tomato. *Plant Science* 265, 29–38. DOI: 10.1016/j.plantsci.2017.09.013.

Guo, J.-E., Hu, Z., Yu, X., Li, A., Li, F. *et al.* (2018) A histone deacetylase gene, SlHDA3, acts as a negative regulator of fruit ripening and carotenoid accumulation. *Plant Cell Reports* 37(1), 125–135. DOI: 10.1007/s00299-017-2211-3.

He, G., Zhu, X., Elling, A.A., Chen, L., Wang, X. *et al.* (2010) Global epigenetic and transcriptional trends among two rice subspecies and their reciprocal hybrids. *The Plant Cell* 22(1), 17–33. DOI: 10.1105/tpc.109.072041.

He, L., Huang, H., Bradai, M., Zhao, C., You, Y. *et al.* (2022) DNA methylation-free *Arabidopsis* reveals crucial roles of DNA methylation in regulating gene expression and development. *Nature Communications* 13(1), 1335. DOI: 10.1038/s41467-022-28940-2.

He, Y. (2009) Control of the transition to flowering by chromatin modifications. *Molecular Plant* 2(4), 554–564. DOI: 10.1093/mp/ssp005.

He, Y. and Li, Z. (2018) Epigenetic environmental memories in plants: Establishment, maintenance, and reprogramming. *Trends in Genetics* 34(11), 856–866. DOI: 10.1016/j.tig.2018.07.006.

Hébrard, C., Peterson, D.G., Willems, G., Delaunay, A., Jesson, B. *et al.* (2016) Epigenomics and bolting tolerance in sugar beet genotypes. *Journal of Experimental Botany* 67(1), 207–225. DOI: 10.1093/jxb/erv449.

Ibarra, C.A., Feng, X., Schoft, V.K., Hsieh, T.-F., Uzawa, R. *et al.* (2012) Active DNA demethylation in plant companion cells reinforces transposon methylation in gametes. *Science* 337(6100), 1360–1364. DOI: 10.1126/science.1224839.

Iwasaki, M. and Paszkowski, J. (2014) Epigenetic memory in plants. *The EMBO Journal* 33(18), 1987–1998. DOI: 10.15252/embj.201488883.

Jang, I.C., Chung, P.J., Hemmes, H., Jung, C. and Chua, N.H. (2011) Rapid and reversible light-mediated chromatin modifications of *Arabidopsis phytochrome* a locus. *The Plant Cell* 23(2), 459–470. DOI: 10.1105/tpc.110.080481.

Ji, Y. and Wang, A. (2023) Recent advances in epigenetic triggering of climacteric fruit ripening. *Plant Physiology* 192(3), 1711–1717. DOI: 10.1093/plphys/kiad206.

Jing, Y., Guo, Q., Zha, P. and Lin, R. (2019) The chromatin-remodelling factor PICKLE interacts with CONSTANS to promote flowering in *Arabidopsis*. *Plant, Cell & Environment* 42(8), 2495–2507. DOI: 10.1111/pce.13557.

Jullien, P.E., Susaki, D., Yelagandula, R., Higashiyama, T. and Berger, F. (2012) DNA methylation dynamics during sexual reproduction in *Arabidopsis thaliana*. *Current Biology* 22, 1825–1830.

Kankel, M.W., Ramsey, D.E., Stokes, T.L., Flowers, S.K., Haag, J.R. *et al.* (2003) *Arabidopsis* MET1 cytosine methyltransferase mutants. *Genetics* 163(3), 1109–1122. DOI: 10.1093/genetics/163.3.1109.

Kim, J., Guermah, M., McGinty, R.K., Lee, J.-S., Tan, Z. *et al.* (2009) RAD6-mediated transcription-coupled H2B ubiquitylation directly stimulates H3K4 methylation in human cells. *Cell* 137, 459–471. DOI: 10.1016/j.cell.2009.02.027).

Kim, J.-M., To, T.K., Matsui, A., Tanoi, K., Kobayashi, N.I. *et al.* (2017) Acetate-mediated novel survival strategy against drought in plants. *Nature Plants* 3, 17097. DOI: 10.1038/nplants.2017.97.

Kundariya, H., Yang, X., Morton, K., Sanchez, R., Axtell, M.J. *et al.* (2020) MSH1-induced heritable enhanced growth vigor through grafting is associated with the RdDM pathway in plants. *Nature Communications* 11(1), 5343. DOI: 10.1038/s41467-020-19140-x.

Laird, P.W. (2010) Principles and challenges of genomewide DNA methylation analysis. *Nature Reviews Genetics* 11(3), 191–203. DOI: 10.1038/nrg2732.

Lang, Z., Wang, Y., Tang, K., Tang, D., Datsenka, T. *et al.* (2017) Critical roles of DNA demethylation in the activation of ripening-induced genes and inhibition of ripening-repressed genes in tomato fruit. *Proceedings of the National Academy of Sciences of the United States of America* 114(22), E4511–E4519. DOI: 10.1073/pnas.1705233114.

Larson, M.H., Gilbert, L.A., Wang, X., Lim, W.A., Weissman, J.S. *et al.* (2013) CRISPR interference (CRISPRi) for sequence-specific control of gene expression. *Nature Protocols* 8(11), 2180–2196. DOI: 10.1038/nprot.2013.132.

Lindroth, A.M., Cao, X., Jackson, J.P., Zilberman, D., McCallum, C.M. *et al.* (2001) Requirement of CHROMOMETHYLASE3 for maintenance of CpXpG methylation. *Science* 292(5524), 2077–2080. DOI: 10.1126/science.1059745.

Lisanti, S., Omar, W.A.W., Tomaszewski, B., Prins, S., Jacobs, G. *et al.* (2013) Comparison of methods for quantification of global DNA methylation in human cells and tissues. *PLoS ONE* 8(11), e79044. DOI: 10.1371/journal.pone.0079044.

Liu, C., Lu, F., Cui, X. and Cao, X. (2010) Histone methylation in higher plants. *Annual Review of Plant Biology* 61(1), 395–420. DOI: 10.1146/annurev.arplant.043008.091939.

Lowder, L.G., Zhou, J., Zhang, Y., Malzahn, A., Zhong, Z. *et al.* (2018) Robust transcriptional activation in plants using multiplexed CRISPR-ACT2.0 and mTALE-act systems. *Molecular Plant* 11(2), 245–256. DOI: 10.1016/j.molp.2017.11.010.

Manning, K., Tör, M., Poole, M., Hong, Y., Thompson, A.J. *et al.* (2006) A naturally occurring epigenetic mutation in A gene encoding an SBP-box transcription factor inhibits tomato fruit ripening. *Nature Genetics* 38(8), 948–952. DOI: 10.1038/ng1841.

Martínez de Alba, A.E., Elvira-Matelot, E. and Vaucheret, H. (2013) Gene silencing in plants: A diversity of pathways. *Biochimica et Biophysica Acta* 1829(12), 1300–1308. DOI: 10.1016/j.bbagrm.2013.10.005.

Meissner, A., Gnirke, A., Bell, G.W., Ramsahoye, B., Lander, E.S. *et al.* (2005) Reduced representation bisulfite sequencing for comparative high-resolution DNA methylation analysis. *Nucleic Acids Research* 33, 5868–5877. DOI: 10.1093/nar/gki901.

Michaels, S.D. and Amasino, R.M. (1999) FLOWERING LOCUS C encodes a novel MADS domain protein that acts as a repressor of FLOWERING. *The Plant Cell* 11(5), 949–956. DOI: 10.1105/tpc.11.5.949.

Mozgova, I. and Hennig, L. (2015) The polycomb group protein regulatory network. *Annual Review of Plant Biology* 66(1), 269–296. DOI: 10.1146/annurev-arplant-043014-115627.

Nuthikattu, S., McCue, A.D., Panda, K., Fultz, D., DeFraia, C. *et al.* (2013) The initiation of epigenetic silencing of active transposable elements is triggered by RDR6 and 21-22 nucleotide small interfering RNAs. *Plant Physiology* 162(1), 116–131. DOI: 10.1104/pp.113.216481.

Papikian, A., Liu, W., Gallego-Bartolomé, J. and Jacobsen, S.E. (2019) Site-specific manipulation of *Arabidopsis loci* using CRISPR-Cas9 SunTag systems. *Nature Communications* 10(1), 729. DOI: 10.1038/s41467-019-08736-7.

Park, K., Kim, M.Y., Vickers, M., Park, J.-S., Hyun, Y. *et al.* (2016) DNA demethylation is initiated in the central cells of *Arabidopsis* and rice. *Proceedings of the National Academy of Sciences USA* 113, 15138–15143. DOI: 10.1073/pnas.1619047114.

Parker, A.H., Wilkinson, S.W. and Ton, J. (2021) Epigenetics: A catalyst of plant immunity against pathogens. *The New Phytologist* 2338–2429.

Pillot, M., Baroux, C., Vazquez, M.A., Autran, D., Leblanc, O. *et al.* (2010) Embryo and endosperm inherit distinct chromatin and transcriptional states from the female gamete in *Arabidopsis*. *Plant Cell* 22, 307–320. DOI: 10.1105/tpc.109.071647.

Quadrana, L., Almeida, J., Asís, R., Duffy, T., Dominguez, P.G. *et al.* (2014) Natural occurring epialleles determine vitamin E accumulation in tomato fruits. *Nature Communications* 5(1), 3027. DOI: 10.1038/ncomms5027.

Ramsahoye, B.H., Biniszkiewicz, D., Lyko, F., Clark, V., Bird, A.P. *et al.* (2000) Non-CpG methylation is prevalent in embryonic stem cells and may be mediated by DNA methyltransferase 3a. *Proceedings of the National Academy of Sciences of the United States of America* 97(10), 5237–5242. DOI: 10.1073/pnas.97.10.5237.

Rossetto, D., Avvakumov, N. and Côté, J. (2012) Histone phosphorylation: A chromatin modification involved in diverse nuclear events. *Epigenetics* 7(10), 1098–1108. DOI: 10.4161/epi.21975.

Saracco, S.A., Hansson, M., Scalf, M., Walker, J.M., Smith, L.M. *et al.* (2009) Tandem affinity purification and mass spectrometric analysis of ubiquitylated proteins in *Arabidopsis*. *Plant Journal* 59, 344–358.

Seifert, F., Thiemann, A., Schrag, T.A., Rybka, D., Melchinger, A.E. *et al.* (2018) Small RNA-based prediction of hybrid performance in maize. *BMC Genomics* 19(1), 371. DOI: 10.1186/s12864-018-4708-8.

Sheldon, C.C., Burn, J.E., Perez, P.P., Metzger, J., Edwards, J.A. *et al.* (1999) The FLF MADS box gene: A repressor of flowering in *Arabidopsis* regulated by vernalization and methylation. *The Plant Cell* 11(3), 445–458. DOI: 10.1105/tpc.11.3.445.

Shen, H., He, H., Li, J., Chen, W., Wang, X. *et al.* (2012) Genome-wide analysis of DNA methylation and gene expression changes in two *Arabidopsis* ecotypes and their reciprocal hybrids. *The Plant Cell* 24(3), 875–892. DOI: 10.1105/tpc.111.094870.

Shen, Q., Lin, Y., Li, Y. and Wang, G. (2021) Dynamics of H3K27me3 modification on plant adaptation to environmental cues. *Plants* 10(6), 1165. DOI: 10.3390/plants10061165.

Shi, J., Dong, A. and Shen, W.H. (2014) Epigenetic regulation of rice flowering and reproduction. *Frontiers in Plant Science* 5, 803. DOI: 10.3389/fpls.2014.00803.

Singh, M., Goel, S., Meeley, R.B., Dantec, C., Parrinello, H. *et al.* (2011) Production of viable gametes without meiosis in maize deficient for an ARGONAUTE protein. *The Plant Cell* 23(2), 443–458. DOI: 10.1105/tpc.110.079020.

Slotkin, R.K., Vaughn, M., Borges, F., Tanurdzic, M., Becker, J.D. *et al.* (2009) Epigenetic reprogramming and small RNA silencing of transposable elements in pollen. *Cell* 136, 461–472.

Song, Y.H., Shim, J.S., Kinmonth-Schultz, H.A. and Imaizumi, T. (2015) Photoperiodic flowering: Time measurement mechanisms in leaves. *Annual Review of Plant Biology* 66, 441–464. DOI: 10.1146/annurev-arplant-043014-115555.

Springer, N.M. and Schmitz, R.J. (2017) Exploiting induced and natural epigenetic variation for crop improvement. *Nature Reviews Genetics* 18(9), 563–575. DOI: 10.1038/nrg.2017.45.

Sun, C., Chen, D., Fang, J., Wang, P., Deng, X. *et al.* (2014) Understanding the genetic and epigenetic architecture in complex network of rice flowering pathways. *Protein & Cell* 5(12), 889–898. DOI: 10.1007/s13238-014-0068-6.

Sung, S. and Amasino, R.M. (2004) Vernalization in *Arabidopsis thaliana* is mediated by the PhD finger protein VIN3. *Nature* 427(6970), 159–164. DOI: 10.1038/nature02195.

Tani, E., Polidoros, A.N., Nianiou-Obeidat, I. and Tsaftaris, A.S. (2005) DNA methylation patterns are differentially affected by planting density in maize inbreds and their hybrids. *Maydica* 50, 19–23.

Tao, Z., Hu, H., Luo, X., Jia, B., Du, J. *et al.* (2019) Embryonic resetting of the parental vernalized state by two B3 domain transcription factors in *Arabidopsis*. *Nature Plants* 5(4), 424–435. DOI: 10.1038/s41477-019-0402-3.

Tompa, R., McCallum, C.M., Delrow, J., Henikoff, J.G., van Steensel, B. *et al.* (2002) Genome-wide profiling of DNA methylation reveals transposon targets of CHROMOMETHYLASE3. *Current Biology* 12(1), 65–68. DOI: 10.1016/s0960-9822(01)00622-4.

Walker, J., Gao, H., Zhang, J., Aldridge, B., Vickers, M. *et al.* (2018) Sexual-lineage-specific DNA methylation regulates meiosis in *Arabidopsis*. *Nature Genetics* 50, 130–137. DOI: 10.1038/s41588-017-0008-5.

Wang, H., Wang, L., Erdjument-Bromage, H., Vidal, M., Tempst, P. *et al.* (2004) Role of histone H2A ubiquitination in Polycomb silencing. *Nature* 431(7010), 873–878. DOI: 10.1038/nature02985.

Whetstine, J.R., Nottke, A., Lan, F., Huarte, M., Smolikov, S. *et al.* (2006) Reversal of histone lysine trimethylation by the JMJD2 family of histone demethylases. *Cell* 125(3), 467–481. DOI: 10.1016/j.cell.2006.03.028.

Wu, S.C. and Zhang, Y. (2010) Active DNA demethylation: Many roads lead to Rome. *Nature Reviews Molecular Cell Biology* 11(9), 607–620. DOI: 10.1038/nrm2950.

Xie, K., Shen, J., Hou, X., Yao, J., Li, X. *et al.* (2012) Gradual increase of mir156 regulates temporal expression changes of numerous genes during leaf development in rice. *Plant Physiology* 158(3), 1382–1394. DOI: 10.1104/pp.111.190488.

Xu, L., Sun, B., Liu, S., Gao, X., Zhou, H. *et al.* (2023) The evaluation of active transcriptional repressor domain for CRISPRi in plants. *Gene* 851, 146967. DOI: 10.1016/j.gene.2022.146967.

Yaish, M.W., Colasanti, J. and Rothstein, S.J. (2011) The role of epigenetic processes in controlling flowering time in plants exposed to stress. *Journal of Experimental Botany* 62(11), 3727–3735. DOI: 10.1093/jxb/err177.

Yelina, N.E., Lambing, C., Hardcastle, T.J., Zhao, X., Santos, B. *et al.* (2015) DNA methylation epigenetically silences crossover hot spots and controls chromosomal domains of meiotic recombination in *Arabidopsis*. *Genes & Development* 29(20), 2183–2202. DOI: 10.1101/gad.270876.115.

Yong, W.S., Hsu, F.M. and Chen, P.Y. (2016) Profiling genome-wide DNA methylation. *Epigenetics & Chromatin* 9, 26. DOI: 10.1186/s13072-016-0075-3.

Yu, C.-W., Liu, X., Luo, M., Chen, C., Lin, X. *et al.* (2011) HISTONE DEACETYLASE6 interacts with FLOWERING LOCUS D and regulates FLOWERING in *Arabidopsis*. *Plant Physiology* 156(1), 173–184. DOI: 10.1104/pp.111.174417.

Yuan, W., Luo, X., Li, Z., Yang, W., Wang, Y. *et al.* (2016) A cis cold memory element and A trans epigenome reader mediate Polycomb silencing of FLC by vernalization in *Arabidopsis*. *Nature Genetics* 48(12), 1527–1534. DOI: 10.1038/ng.3712.

Zaynab, M., Sharif, Y., Fatima, M., Afzal, M.Z., Aslam, M.M. *et al.* (2020) CRISPR/Cas9 to generate plant immunity against pathogen. *Microbial Pathogenesis* 141, 103996. DOI: 10.1016/j.micpath.2020.103996.

Zhang, H., Lang, Z. and Zhu, J.K. (2018) Dynamics and function of DNA methylation in plants. *Nature Reviews Molecular Cell Biology* 19(8), 489–506. DOI: 10.1038/s41580-018-0016-z.

Zhang, Y.Y., Fischer, M., Colot, V. and Bossdorf, O. (2013) Epigenetic variation creates potential for evolution of plant phenotypic plasticity. *New Phytologist* 197(1), 314–322. DOI: 10.1111/nph.12010.

Zhao, X., Chai, Y. and Liu, B. (2007) Epigenetic inheritance and variation of DNA methylation level and pattern in maize intra-specific hybrids. *Plant Science* 172(5), 930–938. DOI: 10.1016/j.plantsci.2007.01.002.

Zhong, S., Fei, Z., Chen, Y.-R., Zheng, Y., Huang, M. *et al.* (2013) Single-base resolution methylomes of tomato fruit development reveal epigenome modifications associated with ripening. *Nature Biotechnology* 31(2), 154–159. DOI: 10.1038/nbt.2462.

Zhou, H., Xu, L., Li, F. and Li, Y. (2022) Transcriptional regulation by CRISPR/dCas9 in common wheat. *Gene* 807, 145919. DOI: 10.1016/j.gene.2021.145919.

Zhu, A., Greaves, I.K., Dennis, E.S. and Peacock, W.J. (2017) Genome-wide analyses of four major histone modifications in *Arabidopsis* hybrids at the germinating seed stage. *BMC Genomics* 18(1), 137. DOI: 10.1186/s12864-017-3542-8.

Zhu, Q.-H., Upadhyaya, N.M., Gubler, F. and Helliwell, C.A. (2009) Over-expression of mir172 causes loss of spikelet determinacy and floral organ abnormalities in rice (*Oryza sativa*). *BMC Plant Biology* 9, 149. DOI: 10.1186/1471-2229-9-149.

Zuo, J., Grierson, D., Courtney, L.T., Wang, Y., Gao, L. *et al.* (2020) Relationships between genome methylation, levels of non-coding RNAs, mRNAs and metabolites in ripening tomato fruit. *The Plant Journal* 103(3), 980–994. DOI: 10.1111/tpj.14778.

18 Epigenetics for Combating Plant Pathogens: Updated Methods and Current Achievements

Mohamed Kouighat[1], Abdelaaziz Farhaoui[2], Amine Assouguem[3], Mohammaed Taoussi[2,4], Zineb Belabess[5] and Rachid Lahlali[2*]

[1]National Institute of Agricultural Research (INRA), Rabat, Morocco; [2]Phytopathology Unit, Department of Plant Protection, Ecole National of Agriculture Meknes, Meknes, Morocco; [3]Faculty of Sciences and Techniques, Sidi Mohammed Ben Abdellah University, Fez, Morocco; [4]Laboratory of Environment and Valorization of Microbial and Plant Resources, Faculty of Sciences, Moulay Ismail University, Zitoune, Meknès, Morocco; [5]Plant Protection Laboratory, Regional Center of Agricultural Research of Meknes, National Institute of Agricultural Research, Meknes, Morocco

Abstract

The field of epigenetics, which examines heritable modifications in gene expression that occur without modifications to the DNA sequence, presents promising opportunities for enhancing plant resilience against pathogens. By investigating how epigenetic changes control gene expression in response to pathogen challenges, scientists can develop novel approaches to bolster plant defense mechanisms. This chapter provides an overview of the most recent epigenetic research methods, including DNA methylation analysis, chromatin immunoprecipitation sequencing (ChIP-seq), and histone modification profiling. Using these techniques, researchers can identify epigenetic marks associated with defense genes, thereby shedding light on the molecular basis of plant disease resistance. Additionally, the chapter highlights current advances in the use of epigenetic mechanisms to enhance plant immunity, including manipulation of small RNAs, modification of chromatin structure, induction of epigenetic memory, and epigenetic editing techniques. Empirical evidence and case studies showcase the success of these methods in providing durable resistance to various plant pathogens. Ultimately, this chapter underscores the promise of epigenetics as a robust instrument for achieving sustainable crop protection, offering valuable perspectives on future research trajectories and their implications for agriculture.

Keywords: Epigenetic, Histone modification, Phytopathogen, DNA methylation, Epigenetic editing, Sustainability

*Corresponding author: rlahlali@enameknes.ac.ma

DOI: 10.1079/9781800626102.0018

18.1 Introduction

The dense monoculture of cultivated plants generally improves the productivity of farmland within intensive agricultural systems (Zhi and Chang, 2021; de Sousa *et al.*, 2024). Nevertheless, this agricultural approach renders crop plants highly susceptible to diseases caused by pathogens that have adapted to specific host plants. Thus, it is imperative to focus on the development of crop varieties resistant to diseases to ensure and enhance sustainable crop production (Silva *et al.*, 2018). To attain this objective, a deep comprehension of the molecular mechanisms governing plant–pathogen interactions is essential. Through continued coevolution with host-adapted pathogens, a complex defense system to combat these adversaries has been developed in plants (Zhi and Chang, 2021; Wang *et al.*, 2023). Cuticular barriers serve as physical obstacles to pathogens. Additionally, plant pathogens encounter a sophisticated set of specialized plant metabolites, called secondary metabolites, known for their ability to deter bacteria, fungi, and oomycetes (Ramos-Cruz *et al.*, 2021). Relying primarily on the detection of microbial effector proteins or pathogen-associated molecular patterns (PAMPs), plants have evolved a range of pathogen recognition strategies to trigger appropriate immune responses. Bigeard *et al.* (2015) reported that activation of various immune responses including generation of reactive oxygen species (ROS), nitric oxide, stomata closure, synthesis of antimicrobial substances, and activation of pathways hormonal defenses were triggered by the PAMP (pattern-triggered immunity, PTI). Some plant pathogens can circumvent PTI by secreting effector proteins that infiltrate plant cells. In a secondary defense strategy, immunity is activated by effector-triggered immunity (ETI) effectors after the recognition of PTI effectors by resistance proteins. ETI leads to programmed cell death to limit the pathogen and prevent its spread within plant tissues (Cui *et al.*, 2015). Although PTI and ETI vary in the intensity and duration of their defense responses, together they contribute to a comprehensive plant immune response (Ramos-Cruz *et al.*, 2021). Mounting evidence suggests that epigenetic mechanisms orchestrate the transcriptional reprogramming associated with defense responses and are crucial in modulating plant resistance to a diverse array of pathogens, encompassing fungi, oomycetes, bacteria, and viruses (Alonso *et al.*, 2019). According to Wilkinson *et al.* (2019), epigenetic mechanisms are important for ensuring the prolonged preservation of induced resistance in plants. Various abiotic and biotic factors are recognized to influence epigenetic processes in plants (Kotkar and Giri, 2020). There is considerable evidence to suggest that epigenetic mechanisms directly impact plant immune responses (Deleris *et al.*, 2016) and the formation of memory in response to environmental stress (He and Li, 2018). Thus, both the epigenetic status of the plant and interactions with pathogens have effects on the plant's immune response to biotic stress (Hannan Parker *et al.*, 2022). The epigenetic landscape of plants, like that of the majority of other eukaryotes, comprises histone modification, DNA methylation, RNA-associated silencing, and epigenetic editing (Iwasaki and Paszkowski, 2014) (Fig. 18.1). Whether acting alone or in combination, these epigenetic factors regulate the accessibility of local chromatin, ultimately impacting gene expression and potentially strengthening plant defense against pathogens and other biological processes (Ramos-Cruz *et al.*, 2021). DNA methylation stands out as one of the extensively researched epigenetic mechanisms governing gene expression in eukaryotes (Feng *et al.*, 2011). Within plant genomes, two forms of DNA methylation have been observed: 5-cytosine methylation and the less common methylation of N6-adenosine (Ramos-Cruz *et al.*, 2021). The prevailing pattern observed across numerous studies indicates that DNA methylation exerts a negative regulatory effect on plant defense (Wilkinson *et al.*, 2019). Histone post-translational modifications (PTMs), including methylation, ubiquitylation, and acetylation, are crucial epigenetic mechanisms. Typically, these modifications occur at the N-terminal tails of histones and play key roles in regulating chromatin functions and structures (Alonso *et al.*, 2019). RNA can silence or downregulate gene expression through a process known as RNA interference, also referred to as post-transcriptional gene silencing (PTGS). Gene expression can be affected by small interfering RNA (siRNA), which promotes DNA

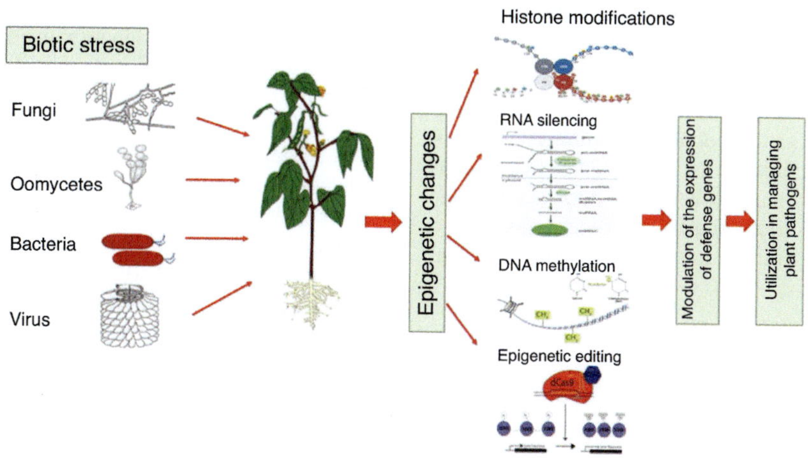

Fig. 18.1. Biotic stresses exerted on plants influence epigenetic processes. Plants can be affected by pathogens such as bacteria, fungi, oomycetes, and viruses. Epigenetic modifications occur through histone modifications, DNA methylation, RNA-associated silencing, and epigenetic editing, resulting in alterations in the expression of genes involved in plant defense. These mechanisms have been understood and utilized to combat phytopathogens and consequently enhance agricultural yields. Author's own image.

methylation, the formation of heterochromatin, or the induction of histone modifications. siRNAs exhibit high sensitivity to stress and are considered among the most responsive molecules to stress-induced changes. Notably, they possess the ability to retain a memory of stress encountered in previous instances and carry it forward to subsequent situations (Kotkar and Giri, 2020). Additionally, epigenomic investigations present promising avenues for genetic manipulation, including the modification of disease resistance gene expression through mechanisms such as transposon insertion, targeted DNA methylation, and demethylation systems utilizing dCas9, as well as the precise control of gene expression through techniques like adenine base editing (ABE) and cytosine base editing (CBE). Furthermore, translational enhancers can be introduced via the CRISPR/Cas system to finely regulate gene expression at the translation level (Xiong *et al.*, 2022).

Detection of epigenetic changes can be reached by several techniques, counting methylation-sensitive amplified polymorphism (MSAP), methylation capture sequencing (MC-seq), and methods based on bisulfite treatment such as enzymatic methyl sequencing (EM-seq), and methylated DNA immunoprecipitation sequencing (MeDIP-seq) (Corrêa *et al.*, 2024). Additionally, technological advances such as PacBio single-molecule sequencing in real time (SMRT) and Oxford Nanopore, along with immuno-based techniques like immunofluorescence and sequencing-based methods like chromatin immunoprecipitation-seq-chip (ChIP-chip) or ChIP-seq enable accurate detection of methylation (Wu *et al.*, 2020a). This integrated approach offers insights into the intricate relationship between epigenetics and plant immunity, advancing our understanding of molecular mechanisms underlying defense responses.

In this chapter, we delve into updated methods utilized in epigenetic research, encompassing DNA methylation analysis, histone modification, RNA-associated silencing, induction of epigenetic memory, and epigenetic editing. We explore current achievements in harnessing epigenetic mechanisms to bolster plant immunity. Through case studies and experimental insights, we unveil the efficacy of these strategies in imparting lasting resistance against diverse plant pathogens. Additionally, challenges, potentials, and strategies associated with exploiting epigenetic variations to enhance crop disease resistance will also be discussed.

18.2 Methods in Epigenetic Research

The field of epigenomics is rapidly advancing, resulting in the adoption of numerous established techniques to assess epigenetic status across various levels and dimensions. The choice of an appropriate method, tailored to a specific scientific question, is therefore of critical importance.

18.2.1 DNA methylation analysis

The DNA bisulfite treatment technique, invented in the early 1990s, is considered fundamental for detecting DNA methylation status. It is based on the use of sodium bisulfite as a treatment agent for genomic DNA; subsequently, uracil bases are obtained by the conversion of unmethylated cytosines without changing the methylated cytosines (5mC), thus allowing the detection of methylated cytosines (Kumar *et al.*, 2018). Bisulfite sequencing is the technique on which other methods of analyzing DNA methylation are based. Indeed, several methods of locus-specific analysis are available, such as subcloning sequencing, direct sequencing of polymerase chain reaction (PCR) products, methylation-specific PCR (MSP), and pyrosequencing (Monk and Kelsey, 2018). Whole-genome bisulfite sequencing (WGBS) stands out as a next-generation sequencing (NGS) technique that facilitates high-throughput, genome-wide examination of DNA methylation. Analysis of WGBS data requires significant computational resources and specific tools for alignment, quantification of methylation levels, and calling of methylation status (Xu and Corces, 2019). Targeted methylation capture sequencing (MC-seq) is a cost-effective method using bisulfite treatment to obtain methylation data at nucleotide resolution, involving targeted capture by hybridization with biotinylated oligonucleotide probes, followed by bisulfite treatment (Agius *et al.*, 2023). This method offers advantages such as low amount of DNA required, cost and time efficiency, and the ability to target specific regions. Alternatives to bisulfite processing include MeDIP-seq and EM-seq, which offer advantages such as better precision and low DNA requirements (Sun and Zhu, 2022). Third-generation sequencing (TGS) allows direct detection of base modifications, including DNA methylation, at nucleotide resolution. Technologies like PacBio SMRT and Oxford Nanopore have been used to map base modifications in different organisms, providing a promising alternative for methylation detection (Furuta, 2023).

Immunofluorescence (IF)-based techniques and enzyme-linked immunosorbent assay (ELISA) are also used to monitor DNA methylation patterns and overall levels in nuclei. IF allows investigation of the nuclear distribution of 5-methyl-deoxy-cytidine (5mdC), providing visualization of methylation patterns between different cell types (Agius *et al.*, 2023). ELISA tests, although developed for animal DNA, can be adapted for plant DNA, using 5mdC-specific antibodies to capture the methylated fraction of DNA, followed by colorimetric or fluorometric detection to quantify the methylation (Furuta, 2023). These techniques have been successfully applied to study DNA methylation in various plant contexts, highlighting their importance for understanding epigenetic regulation in plants.

18.2.2 Histone modification analysis

Histone modification constitutes one of the major epigenetic regulations, often influencing interactions with chromatin and gene expression by modulating transcriptional programs. Understanding the role of histone modifications hinges on two critical elements: the prevalence of modifications and the presence of interacting factors at specific loci, like transcription factors binding to regulatory regions. Conventional techniques for assessing histone modifications mainly rely on ChIP, which is commonly employed for discerning and measuring chromatin modifications and interaction patterns (Grzybowski *et al.*, 2019). When specific modification motifs are not defined, sequencing-based methods like ChIP-chip or ChIP-seq enable the analysis of protein–DNA binding events and the enrichment of histone modifications across multiple loci concurrently (Li, 2021). For examining the dynamic interplay between DNA methylation and chromatin interactions, approaches such as methylation sequencing after chromatin immunoprecipitation (BisChIP-seq) have been developed, allowing interpretation of the relationship between DNA methylation and chromatin (Tong and Wang, 2020). DNA, along with significant epigenetic regulators, operates at the genomic level. Additionally, advanced imaging methods like Hi-C can map out the three-dimensional chromatin structure extensively. Combining ChIP-seq and Hi-C data facilitates investigating how chromatin organization correlates with distinct histone modifications, unveiling intricate networks linking chromatin structure and the underlying epigenetic mechanisms that dynamically regulate gene expression (Li, 2021).

18.3 Current Achievements in Epigenetic Approaches

In recent years, notable advancements have emerged in epigenetics, especially concerning disease resilience and the augmentation of the immune system. Several review articles have addressed the modification of epigenetics in plants to improve their resistance to pathogens (Zhi and Chang, 2021; Hannan Parker *et al.*, 2022; Mierziak and Wojtasik, 2024).

18.3.1 DNA methylation

By adding the methyl group (CH_3) in the 5' position of the cytosine base, DNA methylation takes place, which plays an essential role in modulating the activity of DNA segments without modifying the genetic sequence adjacent (Khodadadi *et al.*, 2021). This modification can be symmetrical (CG and CHG) or asymmetrical (CHH), and its distribution varies depending on the plant species (Niederhuth and Schmitz, 2017). When methylation is not restored during DNA replication, demethylation occurs passively, while active methylation involves various enzymes such as DML2 and DML3 (Parrilla-Doblas *et al.*, 2019). Other proteins and biomolecules are involved in this demethylation process (Liu and Lang, 2020). ROS1/DME enzymes, with DNA glycosylase activity, remove CH_3 groups from DNA, followed by cleavage of the DNA chain, thereby generating a site filled with unmethylated cytosine through repair mechanisms DNA (Liu and Lang, 2020). Hannan Parker *et al.* (2022) signaled that the main purpose of DNA methylation is to suppress the activity of transposable elements, foreign DNA from pathogens, and other potentially damaging genetic elements, thus forming an integral part of the plant immune system. Therefore, resistance to biotic stress can be enhanced by mutations inducing DNA hypomethylation. Indeed, immune responses triggered by biotic stress are frequently associated with DNA hypomethylation in regions rich in transposable elements (Annacondia *et al.*, 2020; Atighi *et al.*, 2020). Transcriptional repression of genes encoding components of RNA-directed DNA methylation (RdDM) describes the link between decreased methylation levels of transposable elements and plant immunity. DNA hypomethylation at transposable elements stimulates defense gene expression through the *cis*-regulatory mechanism, where hypomethylation of transposable elements promotes more accessible chromatin structure, thereby increasing the accessibility of the transcriptional machinery to adjacent defense genes (Galindo-González *et al.*, 2018; Deniz *et al.*, 2019). Recent DNA methylation and genome-wide RNA profiling studies suggest that the beneficial effect of hypomethylation of transposable elements on resistance is not limited to mechanisms of action in *cis* (Gill *et al.*, 2021).

18.3.2 Histone modification

Chromatin, composed of nucleosomes, is the fundamental structure of DNA in eukaryotic cells. About 147 base pairs of DNA wrapped around a histone octamer form a nucleosome (McGinty and Tan, 2014). Histone H1 binds DNA between nucleosomes, thereby influencing chromatin structure at a higher level (Happel and Doenecke, 2009). Post-translational modifications such as methylation, acetylation, and ubiquitination of histone tails (which extend from nucleosomes) regulate DNA accessibility and gene expression by recruiting regulatory proteins. Histone acetylation promotes transcriptional activation, while methylation and ubiquitination may exert activating or repressive effects, contingent upon the circumstances (An, 2007). Chromatin structure can be remodeled by ATP-dependent chromatin remodeling enzymes, which reposition nucleosomes and improve DNA accessibility for the transcriptional machinery (Zhou *et al.*, 2016). Such chromatin modifications and structural rearrangements play a pivotal role in regulating gene expression associated with plant defense mechanisms. More recently, they have also been recognized as potential regulators of plant innate immunity, influencing the response to pathogens.

18.3.3 Small RNAs

In addition to the essential function of coding genes in the regulation of plant immunity, the importance of non-coding RNAs (ncRNAs) in the orchestration of plant immune responses is another avenue of research to improve plant resistance to environmental stresses. RNA interference (RNAi) is an important method for regulating gene expression in plants, used for almost 30 years by plant biotechnologists. RNAi approaches, such as PTGS and RNA-dependent DNA methylation (RdDM), are used to induce epigenetic modifications (Erdmann and Picard, 2020). However, the effectiveness of RdDM, particularly at endogenous promoters, remains a challenge due to intrinsic chromosomal characteristics and the time required for RdDM initiation. To improve this, researchers are exploring intronic RdDM systems based on hairpin dsRNAs (int-hpRNAs) (Dalakouras and Vlachostergios, 2021), which could be used in epi-breeding platforms. A transgene-free alternative is exo-RNAi, which requires strategies to ensure that the applied RNA is transported into the nucleus to enable RdDM. Plant genetics is regulated by two main classes of small RNAs: small interfering RNAs (siRNAs), which are 70 nt stem-loop RNAs produced from endogenous genomic loci, and microRNAs (miRNAs), which are long double-stranded RNAs of 21–22 nt, typically of exogenous origin (Yu *et al.*, 2022). These RNAs respectively induce perfect or imperfect cleavage of the target messenger RNA (mRNA), leading to PTGS. *Trans*-acting siRNAs (TasiRNAs), a subclass of siRNAs, are similar to miRNAs and are involved in the regulation of gene expression in plants (Yu *et al.*, 2022). Transgene-mediated RNA silencing involves transcriptional and post-transcriptional silencing of integrated transgenes, whereas pathogen RNA silencing is triggered by infection. Pathogens act as inducers and targets in siRNA pathways, natural defense mechanisms. Approaches such as hairpin RNA (HpRNA) transgene-mediated RNA silencing and artificial miRNAs offer valuable tools for research and crop improvement. HpRNA transgene-mediated RNA silencing has been widely used due to its simplicity. Transgenic constructs can generate RNAi by overexpressing sense and antisense sequences or by using self-complementary amplicon or hpRNA cassettes (Kumar *et al.*, 2022). Using spliced introns as spacers can improve the efficiency of RNA silencing. Another emerging technology is artificial miRNAs (amiRNAs), based on understanding the molecular mechanisms underlying the biogenesis of *Arabidopsis* miRNA precursors (Yu *et al.*, 2022). These approaches offer valuable tools to specifically silence target genes in plants, thereby opening new avenues for researching and improving crop resistance to pathogenic agents.

18.3.4 Epigenetic editing

Molecular biologists initially exploited two DNA-binding proteins, ZFPs (zinc finger proteins) and TALEs (transcription-activation artificial effectors), to engineer regions capable of binding selectively to a specified DNA sequence, which they then fused with specifically directed effectors (DEs), to manipulate epigenetic characters on DNA or histones. This field saw a significant leap thanks to the development of the CRISPR system. This facilitates the recognition of specific DNA sequences in the genome by engineered guide RNA (gRNA). Furthermore, the ongoing progress in unraveling epigenetic mechanisms, which includes the identification of novel DEs, has paved the way for more rational use of epigenome editing. In partnership with different structural proteins, these epigenetic factors alter the three-dimensional configuration of chromatin, facilitating either condensation, compression, or relaxation of the DNA helix (Cacabelos *et al.*, 2019). The most commonly studied modifications are DNA methylation and histone side chain modifications. CRISPR-based technologies have become popular for scientific research, most notably with the use of dCas9 for epigenome editing. dCas9 serves as a scaffold for engaging effector proteins like epigenetic effectors or chromatin modifiers, allowing for customizable gene regulation and epigenome editing (Brocken *et al.*, 2018). This makes it possible to specifically target DNA regions without inducing cutting, thus opening new perspectives for the precise manipulation of the epigenome without modifying the genome itself.

Epigenomic studies offer another promising perspective on genetic engineering technologies. The insertion of transposons into the promoter region via mechanisms such as non-homologous broken end joining (NHEJ) or homologous directed recombination (HDR) can alter the epigenetic status and, therefore, the level of expression of disease resistance genes (Salsman and Dellaire, 2017). This can be achieved through the use of inactive Cas9 (dCas9) systems, such as targeted DNA methylation and demethylation systems. Moreover, an advanced application of genetic engineering techniques, particularly cytosine base editors (CBE) and adenine base editors (ABE), involves the introduction of single base alterations into the miRNA binding site of the targeted gene, while preserving the amino acid sequence of the encoded protein (Xiong *et al.*, 2022). This can disrupt miRNA/mRNA base pairing, thereby influencing mRNA cleavage and, consequently, regulating the expression of target genes in a precise manner. Moreover, the CRISPR/Cas system can be utilized to introduce translational enhancers into open reading frames through HDR or NHEJ, enabling precise adjustment of gene expression at the translation level (Mushtaq *et al.*, 2021).

18.3.5 Induction of epigenetic memory

Immune memory is linked to cell priming, a process that reinforces the immune response of plants previously exposed to a stressor. Initiating changes at the phenotypic level through priming does not alter the DNA sequence and is potentially reversible. In response to stress, the plant undergoes alterations in gene expression, exhibiting accelerated or earlier expression, enhanced expression, or activation of genes typically dormant in a healthy state (Mierziak and Wojtasik, 2024). Epigenetic modifications equip plants with a form of immune memory (Lämke and Bäurle, 2017). In plants, epigenetic immune memory manifests in various forms: somatic memory, intergenerational memory, and transgenerational memory, persisting for at least two generations (Alonso *et al.*, 2019). Agents like chemical compounds, hormones, and microorganisms can induce epigenetic memory, enhancing resistance to pathogens (Pieterse *et al.*, 2014; Buswell *et al.*, 2018; Mierziak and Wojtasik, 2024).

18.4 The Efficacy of Epigenetic Strategies Against Various Plant Pathogens

Our understanding of the molecular underpinnings of epigenetics and its variations is advancing, yet the effect of epigenetics on host–pathogen interactions remains largely untapped (Dubey and Jeon, 2017). Epigenetic mechanisms play a significant role in the success, longevity, and interactions between pathogens and their hosts, suggesting that they could serve as the cornerstone for future research in epigenetic epidemiology within plant pathology (Dubey and Jeon, 2017). Plants employ epigenetic mechanisms to defend against fungal pathogens, with ATP-dependent chromatin remodeling, non-coding RNAs, covalent histone modifications, and DNA methylation playing significant roles (Mierziak and Wojtasik, 2024). These mechanisms can operate independently or in combination and are activated during both pathogenic infections and symbiotic relationships (Mierziak and Wojtasik, 2024). Plant priming and the establishment of stress-induced epigenetic memory also rely heavily on epigenetic mechanisms, permitting plants to remember past molecular experiences and adapt to new conditions (Mierziak and Wojtasik, 2024). Epigenetic modifications, such as variations in DNA methylation patterns known as differentially methylated regions (DMRs), linked to traits related to disease resistance, hold promise as molecular markers for enhancing crop resilience against diseases (Zhi and Chang, 2021). Epigenomic variations could also be used to predict crop resistance or susceptibility to disease, providing novel strategies to develop crop disease resistance over time (Zhi and Chang, 2021). Cases of scientific studies on the use of epigenetics to combat the interaction of plants by phytopathogens are listed in Table 18.1.

18.4.1 DNA methylation

The resistance of plants to diseases is influenced by DNA methylation and demethylation processes. López Sánchez *et al.* (2016) reported that the mutants deficient in DNA methylation (nrpe1, met1-3, and ddc) in *Arabidopsis* exhibited improved resistance to pathogens like *Hyaloperonospora arabidopsidis* and *Pseudomonas syringae*, but enhanced susceptibility to *Plectosphaerella cucumerina*. In contrast, the DNA hypermethylation mutant ros1-4 showed the opposite resistance patterns (Zhi and Chang, 2021; Mierziak and Wojtasik, 2024).

18.4.2 Histone modification

Histone modification disrupts interactions between nucleosomes, resulting in a more relaxed or compact chromatin state, thus making genes more accessible to transcription enzymes. Generally, a euchromatin state is achieved by acetylation of histone 3 (H3ac) or histone 4 (H4ac) and trimethylation of histone 3 at lysine 4 (H3K4me3). Whereas, heterochromatin formation is observed upon dimethylation of histone 3 at lysine 9 (H3K9me2), trimethylation of histone 3 at lysine 27 (H3K27me3), and deacetylation global (Zhou *et al.*, 2013). These dynamic histone modifications play crucial roles in various cellular processes and developmental events, as well as in the plant immune response against phytopathogens. Proteins like ATX1 and Atxr7 directly control the expression of defense genes, while SDG8 catalyzes the trimethylation of H3K36, thereby influencing resistance to some phytopathogens (Seni *et al.*, 2023). Additionally, ASH1 proteins participate in plant defense against pathogens by modulating the methylation state of histones (Ramirez-Prado *et al.*, 2018). These modifications mark chromosomal regions associated with plant defense, thus facilitating a rapid immune response when challenged. According to Li *et al.* (2013), the activation of defense genes in response to biotic and abiotic stress signals can be obtained by suppressing the repressive H3K27me3 state by stimulating the synthesis of the JmjC JMJ705 protein (Li *et al.*, 2013). Robust activation of salicylic acid signaling upon exposure of plants to virulent pathogens was achieved by histone modification at defense gene promoters using protein demethylases named FLD (Kang *et al.*, 2022). These findings highlight the importance of histone modifications in the fine regulation of the plant immune response against pathogens.

The regulation of gene expression heavily relies on histone acetylation, which is controlled by the interaction between histone deacetylases (HDACs) and histone acetyltransferases (HATs) (Lee and Grant, 2019). Among plant HDACs, HDA19 and HDA6 (members of the RPD3 family) have been studied for their involvement in plant defense against pathogens (Yu *et al.*, 2023). Work has shown that HDA19 is induced by stress signals and is involved in ET/JA defense pathways, regulating both JA-responsive gene transcription and basal defense (Berr *et al.*, 2012). Additionally, HDA19 interacts with WRKY transcription factors to modulate defense gene expression (Wani *et al.*, 2021). HDAC proteins of the Sir2 family, such as OsSRT1 and AtSRT2, are also implicated in the regulation of plant defense (Zheng, 2020). Furthermore, chromatin modifications, such as histone acetylation, are associated with the priming of innate immunity in plants, leading to increased resistance to microbial pathogens (Kang *et al.*, 2022).

Mono-ubiquitination of histones, particularly histone H2B, is also involved in transcriptional regulation and plant defense responses. Specific ubiquitin ligases have been identified to regulate mono-ubiquitination of histone H2B, which affects plant resistance to bacterial and fungal pathogens by modulating the expression of resistance genes and influencing chromatin dynamics (Ramirez-Prado *et al.*, 2018). In *Arabidopsis*, the RING E3 ligases HUB1 and HUB2 are crucial for orchestrating the mono-ubiquitination of histone H2B, which is essential for various developmental processes and plant defense against pathogens, particularly necrotrophic fungi (Sharma *et al.*, 2023). HUB1 interacts with MED21 to regulate RNA

Table 18.1. Examples of epigenetic modifications implicated in crop disease response

Method	Crop	Reference
Chromatin assembly and remodeling	Wheat	Wang *et al*., 2019
	Potato	Meller *et al*., 2018
	Potato	Yang *et al*., 2019
	Arabidopsis	Muñoz-Viana *et al*., 2017
DNA methylation	Maize	Li *et al*., 2015b
	Watermelon	Sun *et al*., 2019
	Rice	Akimoto *et al*., 2007; Li *et al*., 2011; Atighi *et al*., 2020
	Wheat	Geng *et al*., 2019
	Arabidopsis	Halter *et al*., 2021
Histone modification	Wheat	Chen *et al*., 2018; Jin *et al*., 2020; Kong *et al*., 2020
	Soy	Kong *et al*., 2017
	Maize	Walley *et al*., 2018
	Rice	Hou *et al*., 2015; He *et al*., 2018; Li *et al*., 2021
	Arabidopsis	Dhawan *et al*., 2009; Hu *et al*., 2014
Involvement of epigenetic memory	Common bean	Ramírez-Carrasco *et al*., 2017
	Flax	Wojtasik *et al*., 2024
	Arabidopsis	Flors *et al*., 2012; Nowicka *et al*., 2020
	Rice	Bertini *et al*., 2018
Small RNA	Rice	Zheng *et al*., 2017; Li *et al*., 2019
	Wheat	Feng *et al*., 2013; Panwar *et al*., 2018
	Cotton	Miao *et al*., 2022
	Tomato	Wu *et al*., 2020b
	Apple tree	Yu *et al*., 2022
	Brassica	Regmi *et al*., 2021
	Maize	Xu *et al*., 2022
	Tea plant	Wang *et al*., 2020a
	Sugarcane	Nandakumar *et al*., 2021
	Melon	Zhou *et al*., 2020
Epigenome editing	Tomato	García-Murillo *et al*., 2023
	Tobacco	Qi *et al*., 2018
	Soybean	Wang *et al*., 2020b; Chen *et al*., 2023
	Barley	Galli *et al*., 2022
	Rice	Chen *et al*., 2024
	Rapeseed	Gravot *et al*., 2024

polymerase II function, thereby influencing disease resistance (Dhawan *et al.*, 2009). While the response to the bacterial pathogen remains unchanged in 'hub1' mutant plants, the expression of the resistance genes *SNC1* and *Rpp4* is regulated by HUB1 and HUB2 (Kang *et al.*, 2022), thereby contributing to the fine-tuning of the immune response.

18.4.3 Chromatin remodelers

Plant defense mechanisms are influenced not only by covalent modifications of histones but also by chromatin remodeling factors, which regulate the expression of *R* genes and specific JA or *SA* pathways. More than 40 ATP-dependent chromatin remodeling factors have been identified in *Arabidopsis* (Gratkowska-Zmuda *et al.*, 2020), of which the DDM1 and MOS1 families are the best known for their antagonistic effect in the control of expression of the *R* gene *SNC1* (Yang *et al.*, 2022). The SWR1 complex, including *SWR1*, is involved in the substitution of histone H2A by H2A.Z, thereby maintaining SA-dependent repression of defense genes (Hannan Parker *et al.*, 2022). The SWI/SNF class ATPase SYD is recruited to the promoters of genes such as *PDF1.2a*, *VSP2*, and *MYC2* (Hannan Parker *et al.*, 2022), thereby regulating their expression and conferring resistance against necrotrophic pathogens. The rice SNF2/SWI ATPase gene *BRHIS1* is repressed in response to a phytopathogen, leading to the enhanced resistance of rice blast pathogens; this suggests the negative regulatory role of *BRHIS1* in plant immunity (Li *et al.*, 2015a). On the other hand, the expression of defense genes against pathogens can be controlled through the remodeling of nucleosome positions by chromatin remodelers such as BRM and SYD (Zhi and Chang, 2021; Mierziak and Wojtasik, 2024).

18.4.4 Non-coding RNA

Long non-coding RNA (lncRNAs) can exert their functions locally (in *cis*) or distantly (in *trans*), influencing gene regulation through the recruitment of transcription-activating factors or chromatin-modifying enzymes (Wojtasik *et al.*, 2024). For example, in *Vitis vinifera*, lncRNAs responsive to powdery mildew have been associated with proteins involved in calcium-binding, ROS metabolism, and phytoalexin production (Marsh *et al.*, 2010). In *Arabidopsis*, numerous lncRNAs induced in response to *Fusarium oxysporum* infection were found to be linked to disease expansion (Zhu *et al.*, 2014). miRNAs like miR393 in *Arabidopsis* are induced by PAMPs and silence auxin receptors to activate PTI (Li *et al.*, 2010). Another example is miR169 in rice, which exerts a negative regulatory effect on immunity against the blast fungus *Magnaporthe oryzae* by suppressing the expression of nuclear factor Y-A genes (Li *et al.*, 2017).

18.4.5 Epigenetic editing

The mentioned methodologies affect the plant epigenome in a global and stochastic manner, making their results changeable. However, approaches based on epigenetic editing, such as CRISPR, enable specific and targeted modifications (Johnson *et al.*, 2014). CRISPR can be used to deactivate genes coding to DNA methyltransferases, thereby altering the epigenome. By fusing dCas9 with DNA methyltransferases or histone acetyltransferases, CRISPR/dCas9 can impose precise epigenetic changes on genes in plants (Tadić *et al.*, 2019). A synthetic ZFP linked to *Arabidopsis* (DMS3) was utilized to induce methylation at the promoter region of the susceptibility gene *MeSWEET10a*, thereby decreasing fire disease symptoms (Veley *et al.*, 2023).

18.4.6 Involving epigenetic memory

A range of priming agents, encompassing chemical compounds, xenobiotic chemicals, microbes, hormones, growth-promoting rhizobacteria (PGPR), and arbuscular mycorrhizal fungi (AMF), can induce epigenetic memory, enhancing plant resistance to pathogens (Pieterse *et al.*, 2014; Buswell *et al.*, 2018; Mierziak and Wojtasik, 2024). Drugs, such as 5-azacytidine, decitabine, and zebularine, have been developed to alter the plant epigenome by inhibiting DNA methylation, leading to global hypomethylation Patnaik and Anupriya (2019). Additionally,

POL II inhibitors, such as α-amanitin, can also induce hypomethylation by interfering with transcription (Dalakouras and Vlachostergios, 2021). In addition to DNA methylation, certain chemicals can influence histone modifications, such as methylation and acetylation. These adjustments, although often transient, can induce stable and transmissible genetic changes, improving the ability of plants to adapt to their resistance to phytopathogens. Plant resistance to pathogens involves a complex interplay between SA and JA signals, influencing the transcriptome and epigenome. Studies show that JA induces histone deacetylases (HDACs) in *Arabidopsis*, affecting the epigenome (Luo *et al.*, 2019). Pathogens, such as fungi, can transfer RNA to the host plant to repress defense genes, inducing DNA hypomethylation near defense genes (He *et al.*, 2020). Although difficult to standardize, biotic stress results in DNA hypomethylation in various plants attacked by pathogens, such as *F. oxysporum* and *Magnaporthe grisea*, as well as in *Arabidopsis* infected by the nematode *Heterodera schachtii* (Çelik *et al.*, 2020).

18.5 Conclusion and Future Directions

Histone modifications, chromatin assembly and remodeling, and DNA methylation play a pivotal role in modulating plant defense mechanisms. These mechanisms offer promising avenues for developing breeding strategies to enhance crop disease resistance (Dubey and Jeon, 2017). During biotic stress responses, these epigenetic modifications orchestrate plant transcriptional reprogramming, which is crucial for both generational and transgenerational defense priming. Epigenetic variations, through modulation of defense-related transcription, respond to pathogen infections, enhancing plant phenotypic plasticity. This ultimately ensures that crop production remains secure in the face of pathogen challenges (Zhi and Chang, 2021).

However, additional research is necessary to comprehend the fundamental mechanisms behind epigenetic alterations in plant defense. This entails investigating the interplay and contribution of DNA methylation, chromatin remodeling, histone modifications, and lncRNAs toward the development of epi-breeding programs that harness both natural and induced epigenetic variations. These initiatives should prioritize the development of crop varieties possessing stable and heritable epigenetic traits that provide resistance against various pathogens. Additionally, they should incorporate epigenetic modeling to anticipate the impacts of particular epigenetic changes on plant characteristics.

Both inherent epigenetic diversity and experimentally induced epigenetic changes contribute to crop epi-breeding efforts aimed at enhancing disease resistance. The transgenerational inheritance of DNA methylation patterns extends the applicability of epi-breeding to a wide range of crops (Atighi *et al.*, 2020). In contrast, histone post-translational modifications, which are often erased during meiosis, are primarily relevant to clonally propagated crops. Epigenomic disparities, such as differentially methylated regions (DMRs) linked to disease resistance characters, can function as molecular markers assisting in the assessment and choice of crops (Birkenbihl *et al.*, 2017). Additionally, epigenetic modeling can predict how epigenetic variations affect crop disease resistance, guiding decisions on whether to induce/suppress these variations to enhance resistance in crop epi-breeding (Zhi and Chang, 2021). Further studies need to be conducted mainly to evaluate the stability and inheritability of epigenetic modifications across generations. Understanding how epigenetic marks are transmitted will be crucial for the long-term success of epi-breeding programs.

Acknowledgments

This research was financially supported by the Phytopathology Unit at the Ecole Nationale d'Agriculture de Meknès, Morocco.

References

Agius, D.R., Kapazoglou, A., Avramidou, E., Baranek, M., Carneros, E. *et al.* (2023) Exploring the crop epigenome: A comparison of DNA methylation profiling techniques. *Frontiers in Plant Science* 14, 1181039. DOI: 10.3389/fpls.2023.1181039.

Akimoto, K., Katakami, H., Kim, H.-J., Ogawa, E., Sano, C.M. *et al.* (2007) Epigenetic inheritance in rice plants. *Annals of Botany* 100(2), 205–217.

Alonso, C., Ramos-Cruz, D. and Becker, C. (2019) The role of plant epigenetics in biotic interactions. *New Phytologist* 221(2), 731–737. DOI: 10.1111/nph.15408.

An, W. (2007) Histone acetylation and methylation: Combinatorial players for transcriptional regulation. *Sub-Cellular Biochemistry* 41, 351–369.

Annacondia, M.L., Markovic, D., Reig-Valiente, J.L., Scaltsoyiannes, V., Pieterse, C.M. *et al.* (2020) Aphid feeding induces the relaxation of epigenetic control and the associated regulation of the defense response in *Arabidopsis*. *bioRxiv* 2020-01.

Atighi, M.R., Verstraeten, B., De Meyer, T. and Kyndt, T. (2020) Genome-wide DNA hypomethylation shapes nematode pattern-triggered immunity in plants. *New Phytologist* 227(2), 545–558. DOI: 10.1111/nph.16532.

Berr, A., Ménard, R., Heitz, T. and Shen, W.-H. (2012) Chromatin modification and remodelling: A regulatory landscape for the control of *Arabidopsis* defence responses upon pathogen attack: Chromatin regulation of plant defence. *Cellular Microbiology* 14(6), 829–839. DOI: 10.1111/j.1462-5822.2012.01785.x.

Bertini, L., Proietti, S., Focaracci, F., Sabatin, B. and Carusco, C. (2018) Epigenetic control of dense genes following MeJA-induced priming in rice (*O. sativa*). *Journal of Plant Physiology* 228, 166–177. DOI: 10.1016/j.jplph.2018.06.007.

Bigeard, J., Colcombet, J. and Hirt, H. (2015) Signaling mechanisms in pattern-triggered immunity (PTI). *Molecular Plant* 8(4), 521–539.

Birkenbihl, R.P., Liu, S. and Somssich, I.E. (2017) Transcriptional events defining plant immune responses. *Current Opinion in Plant Biology* 38, 1–9.

Brocken, D.J., Tark-Dame, M. and Dame, R.T. (2018) dCas9: A versatile tool for epigenome editing. *Current Issues in Molecular Biology* 26(1), 15–32.

Buswell, W., Schwarzenbacher, R.E., Luna, E., Sellwood, M., Chen, B. *et al.* (2018) Chemical priming of immunity without costs to plant growth. *New Phytologist* 218(3), 1205–1216. DOI: 10.1111/nph.15062.

Cacabelos, R., Tellado, I. and Cacabelos, P. (2019) The epigenetic machinery in the life cycle and pharmacoepigenetics. In: Cacabelos, R. (ed.) *Pharmacoepigenetics*. Academic Press, Cambridge, Massachusetts, pp. 1–100.

Çelik, Ö., Meriç, S., Ayan, A. and Atak, Ç. (2020) Biotic stress-tolerant plants through small RNA technology. In: Guleria, P. and Kumar, V. (eds) *Plant Small RNA: Biogenesis, Regulation and Application*. Academic Press, Cambridge, Massachusetts, pp. 435–468.

Chen, H., Fang, Y., Song, W., Shu, H., Li, X. *et al.* (2023) The SET domain protein PsKMT3 regulates histone H3K36 trimethylation and modulates effector gene expression in the soybean pathogen *Phytophthora sojae*. *Molecular Plant Pathology* 24(4), 346–358. DOI: 10.1111/mpp.13301.

Chen, X., Liu, C., Wang, H., Liu, Q., Yue, Y. *et al.* (2024) Ustilaginoidea virens-secreted effector Uv1809 suppresses rice immunity by enhancing OsSRT2-mediated histone deacetylation. *Plant Biotechnology Journal* 22(1), 148–164. DOI: 10.1111/pbi.14174.

Chen, Y., Wang, J., Yang, N., Wen, Z., Sun, X. *et al.* (2018) Wheat microbiome bacteria can reduce virulence of a plant pathogenic fungus by altering histone acetylation. *Nature Communications* 9(1), 3429.

Corrêa, R.L., Kutnjak, D., Ambrós, S., Bustos, M. and Elena, S.F. (2024) Identification of epigenetically regulated genes involved in plant-virus interaction and their role in virus-triggered induced resistance. *BMC Plant Biology* 24(1), 172. DOI: 10.1186/s12870-024-04866-3.

Cui, H., Tsuda, K. and Parker, J.E. (2015) Effector-triggered immunity: From pathogen perception to robust defense. *Annual Review of Plant Biology* 66(1), 487–511. DOI: 10.1146/annurev-arplant-050213-040012.

Dalakouras, A. and Vlachostergios, D. (2021) Epigenetic approaches to crop breeding: Current status and perspectives. *Journal of Experimental Botany* 72(15), 5356–5371.

de Sousa, T.R., deCarvalho, A.M., Ramos, M.L.G., de Oliveira, A.D., de Jesus, D.R. *et al.* (2024) Dynamics of carbon and soil enzyme activities under Arabica coffee intercropped with *Brachiaria decumbens* in the Brazilian cerrado. *Plants* 13(6), 835.

Deleris, A., Halter, T. and Navarro, L. (2016) DNA methylation and demethylation in plant immunity. *Annual Review of Phytopathology* 54(1), 579–603. DOI: 10.1146/annurev-phyto-080615-100308.

Deniz, Ö., Frost, J.M. and Branco, M.R. (2019) Regulation of transposable elements by DNA modifications. *Nature Reviews Genetics* 20(7), 417–431.

Dhawan, R., Luo, H., Foerster, A.M., AbuQamar, S., Du, H.-N. *et al.* (2009) HISTONE MONOUBIQUITINATION1 interacts with a subunit of the mediator complex and regulates defense against necrotrophic fungal pathogens in *Arabidopsis*. *The Plant Cell* 21(3), 1000–1019.

Dubey, A. and Jeon, J. (2017) Epigenetic regulation of development and pathogenesis in fungal plant pathogens. *Molecular Plant Pathology* 18(6), 887–898. DOI: 10.1111/mpp.12499.

Erdmann, R.M. and Picard, C.L. (2020) RNA-directed DNA methylation. *PLoS Genetics* 16(10), e1009034.

Feng, H., Zhang, Q., Wang, Q., Wang, X., Liu, J. *et al.* (2013) Target of tae-mir408, a chemocyanin-like protein gene (TaCLP1), plays positive roles in wheat response to high-salinity, heavy cupric stress and stripe rust. *Plant Molecular Biology* 83(4–5), 433–443. DOI: 10.1007/s11103-013-0101-9.

Feng, S., Rubbi, L., Jacobsen, S.E. and Pellegrini, M. (2011) Determining DNA methylation profiles using sequencing. In: Kwon, Y.M. and Ricke, S.C. (eds) *High-Throughput Next Generation Sequencing*, Vol. 733. Humana Press, pp. 223–238. DOI: 10.1007/978-1-61779-089-8_16.

Flors, V., Slaughter, A., Daniel, X., Luna Diez, E., Hohn, B. *et al.* (2012) Descendants of primed *Arabidopsis* plants exhibit resistance to biotic stress. *Plant Physiology* 158(2), 835–843. DOI: 10.1104/pp.111.191593.

Furuta, Y. (2023) Detection of DNA modification using nanopore sequencers. In: Arakawa, K. (ed.) *Nanopore Sequencing*, Vol. 2632. Springer US, pp. 79–90. DOI: 10.1007/978-1-0716-2996-3_6.

Galindo-González, L., Sarmiento, F. and Quimbaya, M.A. (2018) Shaping plant adaptability, genome structure and gene expression through transposable element epigenetic control: Focus on methylation. *Agronomy* 8(9), 180.

Galli, M., Hochstein, S., Iqbal, D., Claar, M., Imani, J. *et al.* (2022) CRISPR/SpCas9-mediated KO of epigenetically active MORC proteins increases barley resistance to bipolaris spot blotch and fusarium root rot. *Journal of Plant Diseases and Protection* 129(4), 1005–1011. DOI: 10.1007/s41348-022-00574-y.

García-Murillo, L., Valencia-Lozano, E., Priego-Ranero, N.A., Cabrera-Ponce, J.L., Duarte-Aké, F.P. *et al.* (2023) CRISPRa-mediated transcriptional activation of the *SlPR-1* gene in edited tomato plants. *Plant Science* 329, 111617. DOI: 10.1016/j.plantsci.2023.111617.

Geng, S., Kong, X., Song, G., Jia, M., Guan, J. *et al.* (2019) DNA methylation dynamics during the interaction of wheat progenitor *Aegilops tauschii* with the obligate biotrophic fungus *Blumeria graminis* f. sp. *tritici*. *The New Phytologist* 221(2), 1023–1035. DOI: 10.1111/nph.15432.

Gill, R.A., Scossa, F., King, G.J., Golicz, A.A., Tong, C. *et al.* (2021) On the role of transposable elements in the regulation of gene expression and subgenomic interactions in crop genomes. *Critical Reviews in Plant Sciences* 40(2), 157–189. DOI: 10.1080/07352689.2021.1920731.

Gratkowska-Zmuda, D.M., Kubala, S., Sarnowska, E., Cwiek, P., Oksinska, P. *et al.* (2020) The SWI/SNF ATP-dependent chromatin remodeling complex in *Arabidopsis* responds to environmental changes in temperature-dependent manner. *International Journal of Molecular Sciences* 21(3), 762.

Gravot, A., Liégard, B., Quadrana, L., Veillet, F., Aigu, Y. *et al.* (2024) Two adjacent NLR genes conferring quantitative resistance to clubroot disease in *Arabidopsis* are regulated by a stably inherited epiallelic variation. *Plant Communications* 5(5), 100824.

Grzybowski, A.T., Shah, R.N., Richter, W.F. and Ruthenburg, A.J. (2019) Native internally calibrated chromatin immunoprecipitation for quantitative studies of histone post-translational modifications. *Nature Protocols* 14(12), 3275–3302.

Halter, T., Wang, J., Amesefe, D., Lastrucci, E., Charvin, M. *et al.* (2021) The *Arabidopsis* active demethylase ROS1 cis-regulates defence genes by erasing DNA methylation at promoter-regulatory regions. *eLife* 10, e62994.

Hannan Parker, A., Wilkinson, S.W. and Ton, J. (2022) Epigenetics: A catalyst of plant immunity against pathogens. *New Phytologist* 233(1), 66–83. DOI: 10.1111/nph.17699.

Happel, N. and Doenecke, D. (2009) Histone H1 and its isoforms: Contribution to chromatin structure and function. *Gene* 431(1–2), 1–12.

He, C., Zhang, Z., Li, B. and Tian, S. (2020) The pattern and function of DNA methylation in fungal plant pathogens. *Microorganisms* 8(2), 227.

He, M., Xu, Y., Chen, J., Luo, Y., Lv, Y. *et al.* (2018) MoSnt2-dependent deacetylation of histone H3 mediates MoTor-dependent autophagy and plant infection by the rice blast fungus *Magnaporthe oryzae*. *Autophagy* 14(9), 1543–1561. DOI: 10.1080/15548627.2018.1458171.

He, Y. and Li, Z. (2018) Epigenetic environmental memories in plants: Establishment, maintenance, and reprogramming. *Trends in Genetics* 34(11), 856–866.

Hou, Y., Wang, L., Wang, L., Liu, L., Li, L. *et al.* (2015) JMJ704 positively regulates rice defense response against *Xanthomonas oryzae* pv. Oryzae infection via reducing H3K4me2/3 associated with negative disease resistance regulators. *BMC Plant Biology* 15(1), 286. DOI: 10.1186/s12870-015-0674-3.

Hu, M., Pei, B.-L., Zhang, L.-F. and Li, Y.-Z. (2014) Histone H2B monoubiquitination is involved in regulating the dynamics of microtubules during the defense response to verticillium dahliae toxins in *Arabidopsis*. *Plant Physiology* 164(4), 1857–1865.

Iwasaki, M. and Paszkowski, J. (2014) Epigenetic memory in plants. *The EMBO Journal* 33(18), 1987–1998. DOI: 10.15252/embj.201488883.

Jin, P., Gao, S., He, L., Xu, M., Zhang, T. *et al.* (2020) Genome-wide identification and expression analysis of the histone deacetylase gene family in wheat (*Triticum aestivum* L.). *Plants* 10(1), 19.

Johnson, L.M., Du, J., Hale, C.J., Bischof, S., Feng, S. *et al.* (2014) SRA- and SET-domain-containing proteins link RNA polymerase V occupancy to DNA methylation. *Nature* 507(7490), 124–128.

Kang, H., Fan, T., Wu, J., Zhu, Y. and Shen, W.-H. (2022) Histone modification and chromatin remodeling in plant response to pathogens. *Frontiers in Plant Science* 13, 986940.

Khodadadi, E., Fahmideh, L., Khodadadi, E., Dao, S., Yousefi, M. *et al.* (2021) Current advances in DNA methylation analysis methods. *BioMed Research International* 8827516. DOI: 10.1155/2021/8827516.

Kong, L., Qiu, X., Kang, J., Wang, Y., Chen, H. *et al.* (2017) A phytophthora effector manipulates host histone acetylation and reprograms defense gene expression to promote infection. *Current Biology* 27(7), 981–991.

Kong, L., Zhi, P., Liu, J., Li, H., Zhang, X. *et al.* (2020) Epigenetic activation of Enoyl-CoA reductase by an acetyltransferase complex triggers wheat wax biosynthesis. *Plant Physiology* 183(3), 1250–1267.

Kotkar, H. and Giri, A. (2020) Plant epigenetics and the 'intelligent' priming system to combat biotic stress. In: Kabelitz, D. and Bhat, J. (eds) *Epigenetics of the Immune System*. Academic Press, Cambridge, Massachusetts, pp. 25–38.

Kumar, M., Tripathi, P.K., Ayzenshtat, D., Marko, A., Forotan, Z. *et al.* (2022) Increased rates of gene-editing events using a simplified RNAi configuration designed to reduce gene silencing. *Plant Cell Reports* 41(10), 1987–2003.

Kumar, S., Chinnusamy, V. and Mohapatra, T. (2018) Epigenetics of modified DNA bases: 5-methylcytosine and beyond. *Frontiers in Genetics* 9, 429871.

Lämke, J. and Bäurle, I. (2017) Epigenetic and chromatin-based mechanisms in environmental stress adaptation and stress memory in plants. *Genome Biology* 18(1), 124. DOI: 10.1186/s13059-017-1263-6.

Lee, C.Y. and Grant, P.A. (2019) Role of histone acetylation and acetyltransferases in gene regulation. In: McCullough, S. and Dolinoy, D. (eds) *Toxicoepigenetics*. Elsevier, Amsterdam, pp. 3–30.

Li, Q., Song, J., West, P.T., Zynda, G., Eichten, S.R. *et al.* (2015a) Examining the causes and consequences of context-specific differential DNA methylation in maize. *Plant Physiology* 168(4), 1262–1274.

Li, T., Chen, X., Zhong, X., Zhao, Y., Liu, X. *et al.* (2013) Jumonji C domain protein JMJ705-mediated removal of histone H3 lysine 27 trimethylation is involved in defense-related gene activation in rice. *The Plant Cell* 25(11), 4725–4736.

Li, W., Xiong, Y., Lai, L.B., Zhang, K., Li, Z. *et al.* (2021) The rice RNase P protein subunit Rpp30 confers broad-spectrum resistance to fungal and bacterial pathogens. *Plant Biotechnology Journal* 19(10), 1988–1999. DOI: 10.1111/pbi.13612.

Li, X., Jiang, Y., Ji, Z., Liu, Y. and Zhang, Q. (2015b) BRHIS1 suppresses rice innate immunity through binding to monoubiquitinated H2A and H2B variants. *EMBO Reports* 16(9), 1192–1202.

Li, Y. (2021) Modern epigenetics methods in biological research. *Methods* 187, 104–113.

Li, Y., Zhang, Q., Zhang, J., Wu, L., Qi, Y. *et al.* (2010) Identification of microRNAs involved in pathogen-associated molecular pattern-triggered plant innate immunity. *Plant Physiology* 152(4), 2222–2231.

Li, Y., Xia, Q., Kou, H., Wang, D., Lin, X. *et al.* (2011) Induced *Pib* expression and resistance to *Magnaporthe grisea* are compromised by cytosine demethylation at critical promoter regions in rice. *Journal of Integrative Plant Biology* 53(10), 814–823. DOI: 10.1111/j.1744-7909.2011.01070.x.

Li, Y., Zhao, S.-L., Li, J.-L., Hu, X.-H., Wang, H. *et al.* (2017) Osa-miR169 negatively regulates rice immunity against the blast fungus *Magnaporthe oryzae*. *Frontiers in Plant Science* 8, 234434.

Li, Y., Cao, X., Zhu, Y., Yang, X., Zhang, K. *et al*. (2019) Osa-miR398b boosts H_2O_2 production and rice blast disease-resistance via multiple superoxide dismutases. *New Phytologist* 222(3), 1507–1522. DOI: 10.1111/nph.15678.

Liu, R. and Lang, Z. (2020) The mechanism and function of active DNA demethylation in plants. *Journal of Integrative Plant Biology* 62(1), 148–159. DOI: 10.1111/jipb.12879.

López Sánchez, A., Stassen, J.H.M., Furci, L., Smith, L.M. and Ton, J. (2016) The role of DNA (de)methylation in immune responsiveness of *Arabidopsis*. *The Plant Journal* 88(3), 361–374.

Luo, J., Xia, W., Cao, P., Xiao, Z., Zhang, Y. *et al*. (2019) Integrated transcriptome analysis reveals plant hormones jasmonic acid and salicylic acid coordinate growth and defense responses upon fungal infection in poplar. *Biomolecules* 9(1), 12.

Marsh, E., Alvarez, S., Hicks, L.M., Barbazuk, W.B., Qiu, W. *et al*. (2010) Changes in protein abundance during powdery mildew infection of leaf tissues of cabernet sauvignon grapevine (*Vitis vinifera* L.). *Proteomics* 10(10), 2057–2064.

McGinty, R.K. and Tan, S. (2014) Histone, nucleosome, and chromatin structure. In: Workman, J.L. and Abmayr, S.M. (eds) *Fundamentals of Chromatin*. Springer New York, pp. 1–28. DOI: 10.1007/978-1-4614-8624-4_1.

Meller, B., Kuźnicki, D., Arasimowicz-Jelonek, M., Deckert, J. and Floryszak-Wieczorek, J. (2018) BABA-primed histone modifications in potato for intergenerational resistance to *Phytophthora infestans*. *Frontiers in Plant Science* 9, 1228.

Miao, Y., Chen, K., Deng, J., Zhang, L., Wang, W. *et al*. (2022) MiR398b negatively regulates cotton immune responses to *Verticillium dahliae* via multiple targets. *The Crop Journal* 10(4), 1026–1036.

Mierziak, J. and Wojtasik, W. (2024) Epigenetic weapons of plants against fungal pathogens. *BMC Plant Biology* 24(1), 175. DOI: 10.1186/s12870-024-04829-8.

Monk, D. and Kelsey, G. (2018) Laboratory molecular methodologies to analyze DNA methylation. In: Camprubi, C. and Blanco, J. (eds) *Epigenetics and Assisted Reproduction*. CRC Press, Baton Rouge, pp. 69–97.

Muñoz-Viana, R., Wildhaber, T., Trejo-Arellano, M.S., Mozgová, I. and Hennig, L. (2017) *Arabidopsis* chromatin assembly factor 1 is required for occupancy and position of a subset of nucleosomes. *The Plant Journal* 92(3), 363–374. DOI: 10.1111/tpj.13658.

Mushtaq, M., Ahmad Dar, A., Skalicky, M., Tyagi, A., Bhagat, N. *et al*. (2021) CRISPR-based genome editing tools: Insights into technological breakthroughs and future challenges. *Genes* 12(6), 797.

Nandakumar, M., Malathi, P., Sundar, A.R., Rajadurai, C.P., Philip, M. *et al*. (2021) Role of miRNAs in the host-pathogen interaction between sugarcane and *Colletotrichum falcatum*, the red rot pathogen. *Plant Cell Reports* 40(5), 851–870. DOI: 10.1007/s00299-021-02682-9.

Niederhuth, C.E. and Schmitz, R.J. (2017) Putting DNA methylation in context: From genomes to gene expression in plants. *Biochimica et Biophysica Acta (BBA) – Gene Regulatory Mechanisms* 1860(1), 149–156.

Nowicka, A., Tokarz, B., Zwyrtková, J., Dvořák Tomaštíková, E., Procházková, K. *et al*. (2020) Comparative analysis of epigenetic inhibitors reveals different degrees of interference with transcriptional gene silencing and induction of DNA damage. *The Plant Journal* 102(1), 68–84. DOI: 10.1111/tpj.14612.

Panwar, V., Jordan, M., McCallum, B. and Bakkeren, G. (2018) Host-induced silencing of essential genes in *Puccinia triticina* through transgenic expression of RNA i sequences reduces severity of leaf rust infection in wheat. *Plant Biotechnology Journal* 16(5), 1013–1023. DOI: 10.1111/pbi.12845.

Parrilla-Doblas, J.T., Roldán-Arjona, T., Ariza, R.R. and Córdoba-Cañero, D. (2019) Active DNA demethylation in plants. *International Journal of Molecular Sciences* 20(19), 4683.

Patnaik, S. and Anupriya. (2019) Drugs targeting epigenetic modifications and plausible therapeutic strategies against colorectal cancer. *Frontiers in Pharmacology* 10, 588.

Pieterse, C.M.J., Zamioudis, C., Berendsen, R.L., Weller, D.M., Van Wees, S.C.M. *et al*. (2014) Induced systemic resistance by beneficial microbes. *Annual Review of Phytopathology* 52(1), 347–375. DOI: 10.1146/annurev-phyto-082712-102340.

Qi, T., Seong, K., Thomazella, D.P.T., Kim, J.R., Pham, J. *et al*. (2018) NRG1 functions downstream of EDS1 to regulate TIR-NLR-mediated plant immunity in *Nicotiana benthamiana*. *Proceedings of the National Academy of Sciences* 115, 46. DOI: 10.1073/pnas.1814856115.

Ramírez-Carrasco, G., Martínez-Aguilar, K. and Alvarez-Venegas, R. (2017) Transgenerational defense priming for crop protection against plant pathogens: A hypothesis. *Frontiers in Plant Science* 8, 263863.

Ramirez-Prado, J.S., Piquerez, S.J., Bendahmane, A., Hirt, H., Raynaud, C. *et al.* (2018) Modify the histone to win the battle: Chromatin dynamics in plant-pathogen interactions. *Frontiers in Plant Science* 9, 355.

Ramos-Cruz, D., Troyee, A.N. and Becker, C. (2021) Epigenetics in plant organismic interactions. *Current Opinion in Plant Biology* 61, 102060.

Regmi, R., Newman, T.E., Kamphuis, L.G. and Derbyshire, M.C. (2021) Identification of *Brassica napus* small RNAs responsive to infection by a necrotrophic pathogen. *BMC Plant Biology* 21(1), 366. DOI: 10.1186/s12870-021-03148-6.

Salsman, J. and Dellaire, G. (2017) Precision genome editing in the CRISPR era. *Biochemistry and Cell Biology* 95(2), 187–201. DOI: 10.1139/bcb-2016-0137.

Seni, S., Singh, R.K. and Prasad, M. (2023) Dynamics of epigenetic control in plants via SET domain containing proteins: Structural and functional insights. *Biochimica et Biophysica Acta (BBA) – Gene Regulatory Mechanisms* 194966.

Sharma, M., Sidhu, A.K., Samota, M.K., Gupta, M., Koli, P. *et al.* (2023) Post-translational modifications in histones and their role in abiotic stress tolerance in plants. *Proteomes* 11(4), 38.

Silva, M.S., Arraes, F.B.M., de Araújo Campos, M., Grossi-de-Sa, M., Fernandez, D. *et al.* (2018) Potential biotechnological assets related to plant immunity modulation applicable in engineering disease-resistant crops. *Plant Science* 270, 72–84.

Sun, R. and Zhu, P. (2022) Advances in measuring DNA methylation. *Blood Science* 4(1), 8–15. DOI: 10.1097/BS9.0000000000000098.

Sun, Y., Fan, M. and He, Y. (2019) DNA methylation analysis of the *Citrullus lanatus* response to *Cucumber green mottle mosaic virus* infection by whole-genome bisulfite sequencing. *Genes* 10(5), 344.

Tadić, V., Josipović, G., Zoldoš, V. and Vojta, A. (2019) CRISPR/Cas9-based epigenome editing: An overview of dCas9-based tools with special emphasis on off-target activity. *Methods* 164, 109–119.

Tong, J. and Wang, Z. (2020) Combinatorial ChIP and bisulfite sequencing (BisChIP-seq). In: Tollefsbol, T.O. (ed.) *Epigenetics Methods*. Academic Press, Cambridge, Massachusetts, pp. 387–395.

Veley, K.M., Elliott, K., Jensen, G., Zhong, Z., Feng, S. *et al.* (2023) Improving cassava bacterial blight resistance by editing the epigenome. *Nature Communications* 14(1), 85. DOI: 10.1038/s41467-022-35675-7.

Walley, J.W., Shen, Z., McReynolds, M.R., Schmelz, E.A. and Briggs, S.P. (2018) Fungal-induced protein hyperacetylation in maize identified by acetylome profiling. *Proceedings of the National Academy of Sciences* 115(1), 210–215. DOI: 10.1073/pnas.1717519115.

Wang, K., Miettinen, I., Jaber, E.H. and Asiegbu, F.O. (2023) Anatomical, chemical, molecular, and genetic basis for tree defenses. In: Asiegbu, F.O. and Kovalchuk, A. (eds) *Forest Microbiology*. Elsevier, pp. 33–57.

Wang, L., Chen, H., Li, J., Shu, H., Zhang, X. *et al.* (2020a) Effector gene silencing mediated by histone methylation underpins host adaptation in an oomycete plant pathogen. *Nucleic Acids Research* 48(4), 1790–1799. DOI: 10.1093/nar/gkz1160.

Wang, S., Liu, S., Liu, L., Li, R., Guo, R. *et al.* (2020b) miR477 targets the phenylalanine ammonia-lyase gene and enhances the susceptibility of the tea plant (*Camellia sinensis*) to disease during *Pseudopestalotiopsis* species infection. *Planta* 251(3), 59. DOI: 10.1007/s00425-020-03353-x.

Wang, X., Zhi, P., Fan, Q., Zhang, M. and Chang, C. (2019) Wheat CHD3 protein TaCHR729 regulates the cuticular wax biosynthesis required for stimulating germination of *Blumeria graminis* f. sp. *tritici*. *Journal of Experimental Botany* 70(2), sp.

Wani, S.H., Anand, S., Singh, B., Bohra, A. and Joshi, R. (2021) WRKY transcription factors and plant defense responses: Latest discoveries and future prospects. *Plant Cell Reports* 40(7), 1071–1085. DOI: 10.1007/s00299-021-02691-8.

Wilkinson, S.W., Magerøy, M.H., López Sánchez, A., Smith, L.M., Furci, L. *et al.* (2019) Surviving in a hostile world: Plant strategies to resist pests and diseases. *Annual Review of Phytopathology* 57(1), 505–529. DOI: 10.1146/annurev-phyto-082718-095959.

Wojtasik, W., Dymińska, L., Hanuza, J., Burgberger, M., Boba, A. *et al.* (2024) Endophytic non-pathogenic *Fusarium oxysporum* reorganizes the cell wall in flax seedlings. *Frontiers in Plant Science* 15, 1352105.

Wu, F., Qi, J., Meng, X. and Jin, W. (2020a) miR319c acts as a positive regulator of tomato against *Botrytis cinerea* infection by targeting TCP29. *Plant Science* 300, 110610.

Wu, J., Deng, Y., Hu, J., Jin, C., Zhu, X. *et al.* (2020b) Genome-wide analyses of direct target genes of an ERF11 transcription factor involved in plant defense against bacterial pathogens. *Biochemical and Biophysical Research Communications* 532(1), 76–81.

Xiong, X., Li, Z., Liang, J., Liu, K., Li, C. *et al.* (2022) A cytosine base editor toolkit with varying activity windows and target scopes for versatile gene manipulation in plants. *Nucleic Acids Research* 50(6), 3565–3580.

Xu, C. and Corces, V.G. (2019) Resolution of the DNA methylation state of single CpG dyads using in silico strand annealing and WGBS data. *Nature Protocols* 14(1), 202–216.

Xu, Y., Wang, R., Ma, P., Cao, J., Cao, Y. *et al.* (2022) A novel maize microRNA negatively regulates resistance to *Fusarium verticillioides*. *Molecular Plant Pathology* 23(10), 1446–1460. DOI: 10.1111/mpp.13240.

Yang, T., Wang, D., Tian, G., Sun, L., Yang, M. *et al.* (2022) Chromatin remodeling complexes regulate genome architecture in *Arabidopsis*. *The Plant Cell* 34(7), 2638–2651.

Yang, X., Lu, Y., Zhao, X., Jiang, L., Xu, S. *et al.* (2019) Downregulation of nuclear protein H2B induces salicylic acid mediated defense against PVX infection in *Nicotiana benthamiana*. *Frontiers in Microbiology* 10, 1000.

Yu, Q., Guo, Q., Li, B., Tan, X., Wang, L. *et al.* (2023) Identification of RPD3/HDA1 family genes in sugar beet and response to abiotic stresses. *Sugar Tech* 25(4), 834–845. DOI: 10.1007/s12355-023-01257-0.

Yu, X., Hou, Y., Cao, L., Zhou, T., Wang, S. *et al.* (2022) MicroRNA candidate miRcand137 in apple is induced by Botryosphaeria dothidea for impairing host defense. *Plant Physiology* 189(3), 1814–1832.

Zheng, M., Ding, H., Huang, L., Wang, Y., Yu, M. *et al.* (2017) Low-affinity iron transport protein Uvt3277 is important for pathogenesis in the rice false smut fungus *Ustilaginoidea virens*. *Current Genetics* 63(1), 131–144. DOI: 10.1007/s00294-016-0620-4.

Zheng, W. (2020) Review: The plant sirtuins. *Plant Science* 293, 110434.

Zhi, P. and Chang, C. (2021) Exploiting epigenetic variations for crop disease resistance improvement. *Frontiers in Plant Science* 12, 692328. DOI: 10.3389/fpls.2021.692328.

Zhou, C.Y., Johnson, S.L., Gamarra, N.I. and Narlikar, G.J. (2016) Mechanisms of ATP-dependent chromatin remodeling motors. *Annual Review of Biophysics* 45(1), 153–181. DOI: 10.1146/annurev-biophys-051013-022819.

Zhou, X., Cui, J., Cui, H., Jiang, N., Hou, X. *et al.* (2020) Identification of lncRNAs and their regulatory relationships with target genes and corresponding miRNAs in melon response to powdery mildew fungi. *Gene* 735, 144403.

Zhou, Y., Tan, B., Luo, M., Li, Y., Liu, C. *et al.* (2013) HISTONE DEACETYLASE19 interacts with HSL1 and participates in the repression of seed maturation genes in *Arabidopsis* seedlings. *The Plant Cell* 25(1), 134–148.

Zhu, Q., Stephen, S., Taylor, J., Helliwell, C.A. and Wang, M. (2014) Long noncoding RNAs responsive to *Fusarium oxysporum* infection in *Arabidopsis thaliana*. *The New Phytologist* 201(2), 574–584.

19 Plant Epigenomics: Technical Advances and Current Applications

Malin Paula Alf[1], Renate Horn[1], Ankica Kondić-Špika[2], Irene Luzzi[3], Dragana Miladinović[2*], Aleksandra Radanović[2] and Serena Varotto[3]

[1]Department of Plant Genetics, Institute of Biological Sciences, University of Rostock, Rostock, Germany; [2]Institute of Field and Vegetable Crops, National Institute of the Republic of Serbia, Maksima Gorkog, Novi Sad, Serbia; [3]Department of Agronomy, Food, Natural Resources, Animals and Environment, University of Padova, Legnaro (PD), Italy

Abstract

This chapter deals with the studied chemical modifications of DNA and histone proteins, and non-coding RNAs of plants that constitute the epigenome and contribute to the modulation of gene transcription and genome stability, independently of DNA sequence. The chapter also presents an overview of the current epigenomic profiling methods and data that can be produced and integrated into plant epigenomic maps. Case studies about applications of genome-wide profiling methods to the epigenomic investigations on abiotic stress response in plants are presented in the second part of the chapter.

Keywords: Data analysis, Epigenetics, Epigenome, Non-coding RNA

19.1 Introduction

Plant epigenomics is a multidisciplinary research field that aims to elucidate how plants adjust their phenotypes to external conditions through mechanisms other than DNA sequence variation, that can have biological, ecological, and evolutionary consequences (Quadrana and Colot, 2016). Epigenomic studies refer to the set of DNA and chromatin modifications that affect chromatin structure and consequently gene expression, the stability of plant genomes, and their phenotypic consequences and impacts on plant population dynamics and evolutionary processes (Becker and Weigel, 2012; Latzel *et al.*, 2013; Monroe *et al.*, 2022). Plant genomes have variable size, complexity, and structures and understanding genomes only based on comparative genomics and transcriptomics cannot provide all-standing biological answers. As demonstrated by many studies over the past years, epigenomic profiling can reveal global distribution of epigenetic modification and produce consistent information about chromosome organization and structure: integration of such information can help to better decode plant genomes (Ouyang *et al.*, 2020).

The capacity of plants to adapt to the changing environment largely depends on their levels of genetic diversity, phenotypic variation,

*Corresponding author: dragana.miladinovic@ifvcns.ns.ac.rs

 Epigenetics for Climate-Smart and Sustainable Agriculture (ed. Jen-Tsung Chen)
DOI: 10.1079/9781800626102.0019

and plasticity (Alonso-Blanco *et al.*, 2009; Lloyd and Lister, 2022). Particularly, plants as sessile organisms rely heavily on changes in gene expression to respond to their surrounding environment and in nature they interact with the environment through their phenotypes. The ability of plant cells to sense, respond to, and even remember environmental cues is essential to their survival. Genetic variation ascribable to allelic DNA sequence diversity is the source of heritable trait-based functional variation among genotypes within a species. At the genomic level, a major unresolved question is how much of the coding and non-coding genome diversity affects the twin evolutionary processes of plant adaptation and speciation. However, much evidence suggests that DNA sequence variation alone is not responsible for all standing phenotypic variation in plants, and epigenetic variation can contribute to phenotypic and functional diversity (Kilvitis *et al.*, 2014). Next-generation sequencing technologies (NGS) enable quick DNA and RNA sequencing and have broad applicability in plant genetic/genomic studies, with increasing resolution and precision. NGS can also be used to analyze different components of plant epigenomes. The rapid development of sequencing technologies makes it possible to map epigenomic changes in developmental and adaptative processes and to analyze their functional consequences. Novel high-throughput techniques for positioning DNA methylation, histone marks, and open chromatin take advantage of the ready availability of genomic data to make biological sense of the high-throughput data. Nowadays, epigenomic analysis has developed into big data science: comparative analysis of epigenome is an active research area in plants and crops with different objectives, such as comparisons across species, the study of time-course of a biological process for application in innovative breeding, and investigations of individual cell diversity.

In this chapter, we first briefly illustrate the most investigated epigenetic and chromatin marks and the mechanisms for their deposition in the plant genome, although for a more detailed description of epigenetic mechanisms, we suggest reading another chapter in this book. Then, we describe the recent technological advances that are revealing plant-specific epigenomic features. Finally, we review current applications for studying plant epigenome variations in response to stress and discuss the implications of potentially understanding the regulatory functions of the genomes.

19.2 Epigenome Marks and Mechanisms

Plant DNA can be methylated (with the addition of a methyl group CH3) to a cytosine nucleotide in all cytosine sequence contexts: CG, CHG and CHH (H represents A, T, or C), thereby distinguishing between symmetric and asymmetric DNA methylation (Quadrana and Colot, 2016). In all these sequence contexts, the RNA-directed DNA methylation (RdDM) pathway catalyzes DNA *de novo* methylation. RdDM involves small RNAs (sRNAs) produced by two RNA polymerases, Pol IV and Pol V, which are 24 nt sRNAs guide molecules for the DNA methylation writer DOMAINS REARRANGED METHYLTRANSFERASE 2 (DRM2) to specific loci (Matzke *et al.*, 2009; Law and Jacobsen, 2010). In plants, DNA methylation at different sequence contexts is also maintained by additional interfering pathways (Stroud *et al.*, 2014; To and Kakutani, 2022). METHYLTRANSFERASE 1 (MET1) is recruited by the VARIANT IN METHYLATION (VIM1) in hemi-methylated sequences, to catalyze CG methylation on the newly synthesized DNA strand by template copying. CHG DNA methylation depends primarily on two other methyltransferases writers, CHROMOMETHYLASE2 (CMT2) and 3 (CMT3), that act upon a reinforcing feedback loop between DNA methylation and the histone modifications, particularly histone H3 lysine nine dimethylation H3K9me2 (Zhong, 2016). Furthermore, CMT3 is recruited to H3K9me2-containing nucleosomes (Du *et al.*, 2012).

A family of plant-specific DNA glycosylase enzymes can remove DNA methylation within the genome (Zhang *et al.*, 2022). DEMETER (DME), REPRESSOR OF SILENCING 1 (ROS1) and DEMETER-LIKE2&3 (DML2/DML3) plant methylation erasers are essential for plant reproductive development, for maintaining epigenetic homeostasis in intergenic euchromatin for changing the levels of DNA methylation between tissues during somatic development (Zhang *et al.*, 2022).

DNA methylation patterns across the plant genome are shaped by methylation writers and erasers. Generally heavy methylation is present in heterochromatin, which is enriched with transposable elements and repetitive DNA sequences. Additionally, interspersed transposon-associated DNA methylation characterizes euchromatic chromosome arms. Gene body methylation characterizes the coding region of plant genes that are methylated specifically in the CG sequence context (Muyle *et al.*, 2022). Whereas the transcriptional silencing role of heterochromatic DNA methylation is well understood, the functional role of gene body methylation has not been fully elucidated until now (Zhang *et al.*, 2018).

DNA methylation is stable over multiple generations and has been associated with local adaptation (Dubin *et al.*, 2015; Meyer, 2015; He and Li, 2018). However, the extent to which DNA methylation variants are independent of changes in the underlying genotype is still unclear.

Histone proteins form the core of the nucleosome, which is the functional unit of chromatin around which the DNA is wrapped. In the plant genome multiple genes coding for the canonical histone forms are present, as well as variants containing minor differences in their amino acid sequences and often having distinct functions. Histones are highly conserved proteins, composed of a globular domain and an N-terminal tail, which can be modified at the post-translational level. Histones can be differently chemically modified, and their numerous modifications have potential effects on the underlying DNA, with a well-known additional role in gene expression regulation. In fact, histone modifications can be 'read' by specific 'reader' modules to translate the upstream histone mark signal to downstream effectors. Histone modifications may also mask the binding sites of other chromatin regulators. Canonical histones are mainly expressed during the S-phase of the cell cycle to ensure chromatin reorganization after DNA replication. By contrast, histone variants have introns and often show replication-independent expression and deposition (Deal and Henikoff, 2011). In plant genomes, several additional non-allelic histone sequence variants have introns, particularly of H2A and H3 (Simon and Probst, 2024), and can be incorporated into nucleosomes in a targeted and replication-independent manner. Due to their sequence divergence from canonical histones, histone variants give specific properties to the nucleosomes they occupy and can affect epigenetic states, transcription, and genome stability. Although most of the information about histone variants derives from a few model plants, their key role during the adaptation of plants to external stimuli, such as biotic and abiotic stresses, is emerging (Probst *et al.*, 2020). DNA accessibility and therefore transcription is affected by changes at the chromatin nucleosome level, through histone modifications, the incorporation of specific histone variants, or alterations in nucleosome occupancy. Histone lysine can be acetylated or deacetylated by histone acetyltransferase (HAT) and histone deacetylases (HDA), respectively. Histone acetylation is a well-known chromatin modification that facilitates chromatin relaxation by neutralizing the positive charges of histones. By weakening histone interactions with the negatively charged DNA and recruiting chromatin remodeling complexes, lysin acetylation can alter chromatin structure (Cui *et al.*, 2023). It is well documented that histone H3 lysine tri-methylation (H3K4me3) is crucial for and associated with gene activation in plants. Conversely, gene repression/silencing is obtained by H3K27me3 deposition by the Polycomb Repressive Complex 2 (PRC2; Mozgova and Hennig, 2015). The role of these histone modifications in plant development has been studied through numerous different genetic approaches (Simon and Probst, 2024). Other histone methylation marks, such as H3K9me2 and H3K27me1, are enriched at constitutively silenced heterochromatin and are required for the silencing of transposable elements (TEs) and other repetitive DNA. In particular, H3K9me2 is an evolutionary conserved mark of peripheral heterochromatin that is retained through mitosis (Poleshko *et al.*, 2019). In addition, as reported above, early genome-wide studies in plants have revealed a positive correlation between the distribution of H3K9me2 peaks and DNA methylation levels.

During the past few decades, numerous studies have indicated that chromatin organization in association with DNA methylation patterns is fundamental for gene expression control and genome stability. As specific mechanisms control chromatin remodeling to allow plasticity during the plant development transition phase, similarly chromatin states are maintained through DNA replication to properly regulate gene expression, determining what is defined as epigenetic regulation of gene transcription.

Non-coding RNAs (ncRNAs), including both small RNAs and long non-coding RNAs (lncRNAs), represent an additional layer of epigenetic regulation governing essential crop traits. By sequencing total RNA, additional information on lncRNAs can be retrieved. Long ncRNAs do not code for any polypeptides but are known to interact with DNA, RNA, and protein to affect transcriptional, post-transcriptional, and epigenetic modulation events. In plants, these regulatory elements might play a critical role in coping with environmental pressure during plant growth and development. Although plant lncRNA research is still in its initial stage, examples of gene expression regulation by non-coding RNAs have been reported for the *Arabidopsis* FLC locus (Swiezewski *et al.*, 2009; Kim and Sung, 2017) or APOLO (AUXIN-REGULATED PROMOTER LOOP) that regulates auxin-responsive genes (Ariel *et al.*, 2020).

19.3 Recent Technological Developments for Studying Plant Epigenomes

Recent advances in experimental methods have allowed the characterization of plant epigenomes and the identification of epi-targets that are part of plant development and/or answers to endogenous and exogenous signals. This was made possible by adopting both next-generation sequencing (NGS) methods and bioinformatic tools. Together these integrated approaches have permitted the study of plant epigenomes and epigenetic inheritance and the identification of numerous epiregulators and their respective patterns and pathways. Different methods have been developed for the detection of epigenetic modifications, variations, and distributions in plant genomes and for epigenomic mapping of different marks. The next sections will focus on the experimental methods for profiling modifications and analyzing data that concern genome sequences, such as methylation of DNA and post-translational modifications of histones.

19.3.1 Detection of DNA methylation states

For years, mostly 5-methylcytosine (5mC) has been investigated in plants, which in contrast to animals occurs not only at CG sites but also in CHG and CHH contexts. With advanced detection technologies N6-methyladenine (6mA) in plants is also becoming more and more interesting (Zhang *et al.*, 2023). In mammals, 5-hydroxymethyl-cytosine (5hmC) is also relevant, but the importance of 5hmC in plants is still being discussed (Shi *et al.*, 2017). In rye, 5hmC and other oxidized derivates were detected and associated with gene regions indicating a regulatory role, whereas in rice 5hmC was found in transposons and heterochromatin regions (Wang *et al.*, 2015; Kalinka *et al.*, 2023; Kaur *et al.*, 2023).

Recently, available techniques for profiling DNA methylation in crops have been extensively reviewed (Agius *et al.*, 2023). In addition, other publications have reviewed available epigenetic methods, but without focusing on plants (Li, 2021; Mehrmohamadi *et al.*, 2021).

Bisulfite sequencing and enzymatic methyl sequencing

Bisulfite sequencing (BS-seq) was invented in the early 1990s (Frommer *et al.*, 1992; Susan *et al.*, 1994) and is regarded as the gold standard for the detection of DNA methylation sites. Through the deamination of unmethylated cytosines to uracil under bisulfite treatment followed by gene-specific PCR or whole genome sequencing, unmethylated cytosines can be distinguished from methylated cytosines as they are sequenced as thymines. An alignment with the original sequence can show which cytosines were methylated. With this technique, it is possible to get information about the methylation status with a single-base resolution. This method can be applied as whole-genome bisulfite sequencing (WGBS) or alternatively as reduced representation bisulfite sequencing (RRBS) (Paun *et al.*, 2019). When performing bisulfide sequencing and library preparation there is a risk of biases. One disadvantage of bisulfite sequencing is uneven coverage of the genome due to extreme temperatures and pH needed for library preparation and resulting damage of unmethylated cytosines. This leads to an underrepresentation of dinucleotides containing G and C. The degradation of DNA also leads to problems and therefore bias in downstream steps such as DNA-polymerization and downstream analyses (Olova *et al.*, 2018).

An alternative approach proposed by New England Biolabs (NEB) is the enzymatic methyl-seq (EM-seq) (Hoppers *et al.*, 2020). This method tries to avoid the bias created by the bisulfite treatment. It uses APOBEC enzymes instead of bisulfite treatment to deaminate cytosines to uracil. For differentiation of 5mC and 5hmC, as they are both deaminated, two additional enzymes (TET2 and an oxidation enhancer) are used. The following steps are the same as in bisulfite sequencing (Hoppers *et al.*, 2020; Vaisvila *et al.*, 2021). This method showed more accurate results in *Arabidopsis*, while needing less starting material (Feng *et al.*, 2020).

For single-cell bisulfide sequencing, Li *et al.* (2019) developed a protocol using bisulfite-converted randomly integrated fragments sequencing (scBRIF-seq) in maize. This protocol achieved higher genome coverage in CG and CHG methylation-rich regions, providing a method that can cover the genome independently of regional differences in methylation states.

DATA ANALYSIS. There are different tools to analyze data coming from bisulfide sequencing (BS-seq) to find differential methylated cytosines (DMCs) and differential methylated regions (DMRs). Depending on the resources it is important to choose the right method to identify the epigenetic modifications (Fig. 19.1). Piao *et al.* (2021) tested eight different approaches and characterized them according to their sensitivity. RADmeth, methylKit, DSS, Fishers exact test, and Metilene were the most powerful tools, depending on the question and the available data (Fig. 19.1). For the DMRs, methylKit and Fishers exact test identified the highest number, whereas Metilene was the best tool for correctly identifying the boundaries of those DMRs.

Third-generation sequencing-based methods for epigenomics

With third-generation sequencing (Pacific Biosciences platform and Nanopore technology) producing longer reads directly from long DNA fragments, new opportunities for the detection of methylation states arise. With those new techniques, it is inevitable to have accurate methods for data analysis. There are different approaches for analyzing third-generation sequencing data using statistics and machine learning.

Compared with second-generation sequencing techniques, third-generation sequencing techniques have several advantages. First of all, the manipulation of samples and related errors can be avoided, as the DNA is directly sequenced. Because each strand stays in the nanopore it can be sequenced as often as necessary, creating the

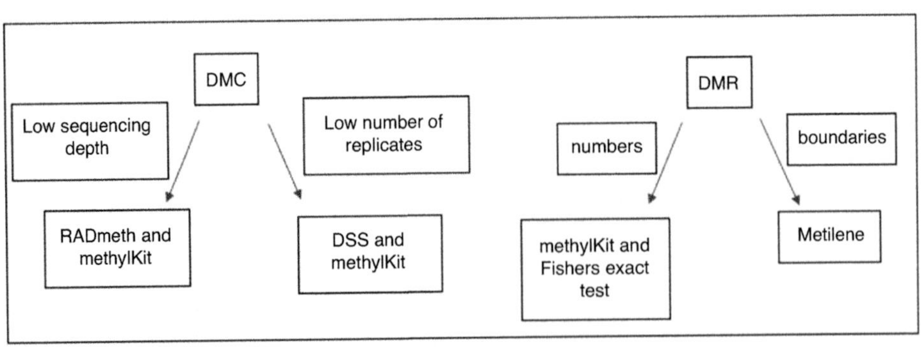

Fig. 19.1. Different tools to identify differential methylated cytosines/regions. Author's own image.

required sequencing depth resulting in high accuracy. Nanopore sequencing discriminates between methylated and unmethylated cytosine without the disruptive need of chemicals as in bisulfite treatment (Simpson *et al.*, 2017). Due to the long reads that are produced, a correlation of different methylation sites is possible (Schreiber *et al.*, 2013). Single-molecule real-time (SMRT) sequencing can also directly detect methylated DNA sites by variations in the kinetic of the DNA polymerase (Flusberg *et al.*, 2010).

OXFORD NANOPORE TECHNOLOGIES AND ITS DATA ANALYSIS. In nanopore sequencing the different methylation status of cytosine can be detected by changes in the electrical current produced by nanopore sequencing devices (Laszlo *et al.*, 2013). To achieve a high accuracy with low error rates it is important to have a well-annotated reference genome and a high amount of training data (Schreiber *et al.*, 2013). This might cause a problem in plant sciences, as most data will come from studies related to human health care.

DeepSignal is a tool developed by Ni *et al.* (2019). It is a deep learning method, using nanopore sequencing reads for the prediction of methylation states of DNA. It uses convolutional neural networks and bidirectional recurrent neural networks to implement sequence data with raw electric signals produced by the sequencing device. With this method, the researchers could predict 6mA and 5mC with more accuracy than with HMM-based methods like nanopolish (Simpson *et al.*, 2017). DeepSignal is a very robust software for detecting the DNA methylation status. This method was optimized for plants (DeepSignal-plants) detecting 5mC in the three methylation patterns CG, CHG, and CHH present in higher plants (Ni *et al.*, 2021).

19.3.2 Detection of histone modifications

Histones can be modified in a variety of ways. Histone acetylation, which leads to activation allowing responses to stress and developmental processes, has been well-studied in plants (Kumar *et al.*, 2021). Methylation is a lot more complex and can even lead to different responses in different contexts (Liu *et al.*, 2010). The first method to study histone modifications on a bigger scale was ChIP-chip (Zhu *et al.*, 2012), chromatin immunoprecipitation followed by microarray hybridization. This method has a few limitations, including low resolution, which has been overcome by combining ChIP with next-generation sequencing (ChIP-seq).

ChIP-seq and derived methods

The first protocol for chromatin immunoprecipitation (ChIP) in plants (*Arabidopsis* and rice) was established by Zhu *et al.* (2012). DNA-binding proteins, such as histones, are fixated through crosslinking. This step is followed by DNA extraction and fragmentation. Afterwards, it is essential to choose the right antibody for immunoprecipitation. This can be a limitation, even though a large number of histone-mark antibodies are commercially available. Antibodies against five marks of histone modifications (H3K4me3, H3K27ac, H3K4me1, H3K27me3, and H3K9me2) were used in rice (Zhao *et al.*, 2020) or more frequently only three antibodies or fewer are applied directed against, for example, H3K4me3, H3K27ac, and H3K27me3 (Forestan *et al.*, 2020; Xiao *et al.*, 2022). An enhanced chromatin immunoprecipitation (eChIP) method was developed for plants that directly sonicates the tissue fixed by formaldehyde and thereby reaches a highly improved recovery of chromatin, 72.1% versus 4.1% using the normal ChIP method. This allows starting with as low as 0.01g of plant material compared with the required 1–5g using the regular ChIP protocol (Zhao *et al.*, 2020).

From an epigenetic point of view, ChIP-seq can be used for the detection of transcription factor binding sites and the discovery of histone marks (Chen *et al.*, 2018). The combination of DNA methylation studies with eChiPseq, FAIRE-seq (formaldehyde-assisted isolation of regulatory elements) and transcriptome analyses resulted in a comprehensive epigenomic map for rice. This epigenomic landscape provides a picture of the involvement of chromatin modification in the gene regulation of different tissues (Zhao *et al.*, 2020).

In the past few years, single-cell ChIP-Seq has become more and more important (Mehrmohamadi *et al.*, 2021). In comparison with animals, this is especially difficult for plants because of the cell walls (Ouyang *et al.*, 2022; Zhang *et al.*, 2022).

Chromatin immunocleavage (ChIC) represents an alternative and a variant of the standard ChIP. In contrast to ChIP, where the DNA is fragmented randomly, ChIC uses restriction endonucleases to specifically cut the DNA (Schmid *et al.*, 2004). This enables a more specific resolution of protein–DNA interactions requiring less starting material. A reliable protocol for single-cell ChIC in plants is still needed (Rhaman *et al.*, 2024).

ATAC-seq

A hyperactive Tn5 transposase for labeling accessible DNA regions is employed in ATAC-seq (assay of transposase-accessible chromatin). In those regions, the cut DNA is immediately ligated to adapters for high-throughput sequencing so that these regions can be sequenced. With this method active, open DNA sections can be identified (Buenrostro *et al.*, 2013). An optimized ATAC protocol with an improved method for cell lysis, isolation of nuclei, and transposition includes the application of additional detergents such as digitonin and Tween-20 for animal cells (Corces *et al.*, 2017). In addition, three data pipelines for ATAC-seq (accessed 15 May 2024) were made available, which can be used for data processing: PEPATAC (https://github.com/databio/pepatac), ENCODE ATAC-seq (https://github.com/ENCODE-DCC/atac-seq-pipeline), and nf-core atac-seq (https://github.com/nf-core/atacseq) (Grandi *et al.*, 2022).

For plants, Debernardi *et al.* (2023) combined the ATAC-seq with the isolation of nuclei tagged in specific cell types (INTACT) in tetraploid wheat (*Triticum aestivum* subsp. *durum*). This approach of INTACT (Deal *et al.*, 2011), which requires transgenic lines, has also been successfully used for profiling accessible chromatin regions in *Arabidopsis*, *Medicago*, rice, and tomato to purify nuclei free of organellar DNA (plastids and mitochondria) (Maher *et al.*, 2018).

CUT&Tag and CUT&Run

CUT&Tag (cleavage under targets and tagmentation) and CUT&Run (cleavage under targets and release using nuclease) both represent advanced methods for chromatin profiling, based on the ChIC method developed by Schmid *et al.* (2004). Both methods represent alternatives for ChIP-seq that can be performed with native chromatin and do not require the fixation of chromatin with formaldehyde. For CUT&Run, a protein A/micrococcal nuclease (pA-MNase) complex is brought into the permeable cell and activated via the addition of Ca^{2+} (Skene and Henikoff, 2017). Limitations of CUT&Run are the time-consuming steps following the DNA extraction. CUT&Run is also not suited for single-cell applications because, after the cleavage of the DNA, the fragments are released into the supernatant. Therefore, CUT&Tag was established as a new and advanced protocol for chromatin profiling also usable for single-cell analyses (Kaya-Okur *et al.*, 2019). These techniques need considerably less starting material because. due to low background noise and high sensitivity, the coverage does not have to be so deep.

19.3.3 Simultaneous detection of histone modifications and DNA methylation state

The separate study of DNA methylation sites and histone modifications can present a problem due to the heterogeneity of cell populations and allele-specific chromatin markings when implementing the data (Yue *et al.*, 2022). The first approach to study histone modifications and DNA methylation simultaneously, and not only with correlation methods combining the different approaches, was performed by sequential ChIP-bisulfite sequencing (ChIP-BS-seq) (Brinkman *et al.*, 2012). A more recent study combined CUT&Tag with BS-seq (Li *et al.*, 2021). The combination of different approaches has also been used for single cells (Pott, 2017). All these approaches use BS-seq as one method and therefore have the biases discussed in the previous section ('Bisulfite sequencing and enzymatic methyl sequencing'). There are several publications using third-generation sequencing methods to investigate the different epigenetic variations simultaneously, but so far most of them have not been used for plants (Lee *et al.*, 2020; Altemose *et al.*, 2022; Yue *et al.*, 2022). In *Arabidopsis*, one study used single-molecule targeted accessibility and methylation sequencing (STAM-seq) to investigate chromatin states and methylation sites in centromere regions, rDNA and telomeres (Mo *et al.*, 2023).

19.3.4 Using epigenomic data from publicly available databases

Access to public databases for chromatin modifications in plants

The National Center for Biotechnology Information (NCBI) runs the database GEO (The Gene Expression Omnibus). NCBI GEO was published in 2002 (Edgar *et al.*, 2002). It is used to store and provide epigenetic and gene expression data. In 2023, it contained raw data, metadata and processed data representing more than 200,000 studies, including 6.5 million samples, from which about 20% were epigenomic data (Clough *et al.*, 2024). Alternatively, the plant chromatin state database (PCSD; http://systemsbiology.cau.edu.cn/chromstates, accessed 5 May 2024) provides access to the chromatin states of *Arabidopsis*, corn and rice (Liu *et al.*, 2018). More recently, the platform ChIP-Hub (https://biobigdata.nju.edu.cn/ChIPHub/, accessed 5 May 2024) has been established that provides integrative web-based access to datasets from >40 plant species for exploring plant regulatory elements (Fu *et al.*, 2022). AraENCODE (http://glab.hzau.edu.cn/AraENCODE/, accessed 5 May 2024) was specifically developed to access comprehensive knowledge about *Arabidopsis* epigenomics (Wang *et al.*, 2023), whereas PlantCADB (https://bioinfor.nefu.edu.cn/Plant-CADB/, accessed 5 May 2024) includes 649 samples derived from 37 species (Ding *et al.*, 2023).

Generating data for the public using FAIR principles

Submitting data according to FAIR (Findability, Accessibility, Interoperability and Reusability) standards is important to make data discoverable and available, both for humans and for machines. This includes contacts, but also authentication and authorization. For the interoperability and the reusability of data, it is important to use standardized vocabulary and richly describe the data with metadata, so that the integration and combination with other experiments are possible (https://www.go-fair.org/fair-principles/, accessed 5 May 2024).

Multidimensional profiling, which collects omics data from the whole plant, a single organ, or even a single cell, gives the possibility of integrating multiple types of data to generate more reliable overlapping information (Fig. 19.2). Moreover, integration of multiple omics data might facilitate the comprehension of the mechanisms which underlie plant complex traits, such as crop yield and biotic and abiotic stress tolerance in major crops.

19.4 Current Applications – Abiotic Stress Tolerance Studies

Recent advances in plant epigenomics enabled the identification of epigenetic changes involved in plant response to environmental stresses. In view of the global change of the climate, those newly discovered epigenetic marks and processes are gaining their place in crop improvement for resilience to environmental changes. Here, we summarize current research on epigenomics of plant responses to abiotic stresses, notably salinity, heat, and drought, that are increasingly having negative effects on plants and causing severe crop losses due to climate change.

19.4.1 Salinity stress studies

Increased soil salinity affects the plant's ability to uptake water and causes reduced growth by affecting plant metabolism and photosynthesis (Deinlein *et al.*, 2014). Similar to other environmental stresses, it was found that salinity also causes hyper- or hypomethylation of DNA, as well as alterations of histone post-translational modification patterns, hence modulating gene expression in plants affected by the stress (Yun *et al.*, 2024). This initiated a number of studies devoted to the identification of the epigenetic marks related to salt stress tolerance in plants.

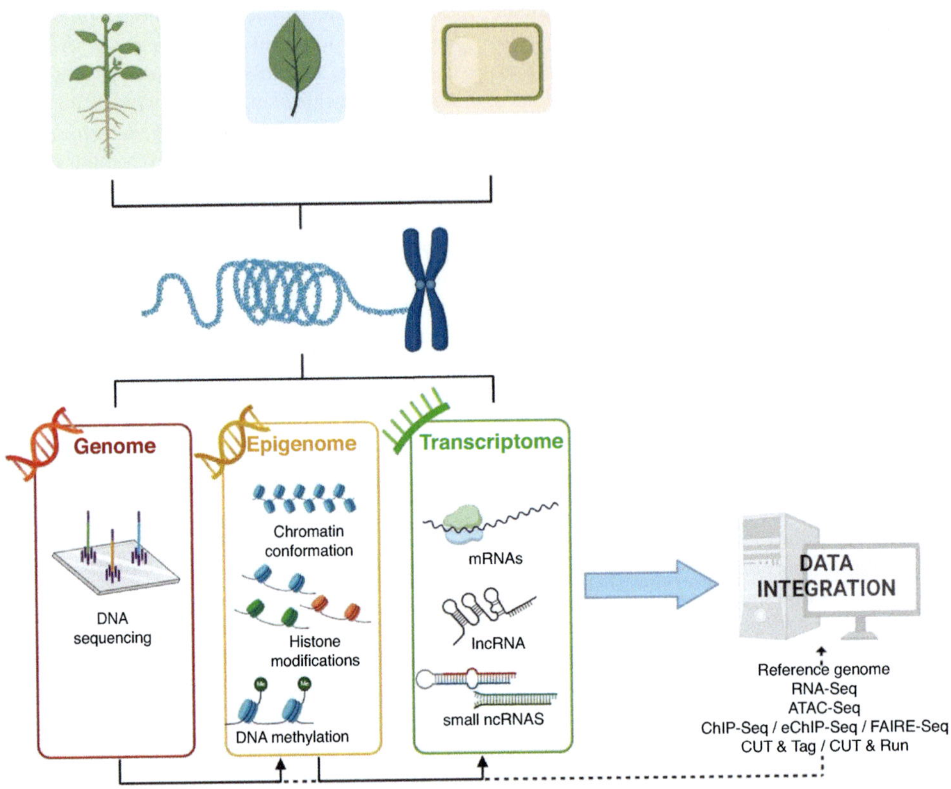

Fig. 19.2. Conceptual scheme of multi-omics approach. The combination of genomics, epigenomics, and transcriptomics datasets provides comprehensive information for translational research. Author's own image.

DNA methylation

Besides model plants, such as *Arabidopsis* and tobacco, most of the studies on plant reactions to salinity stress focus on 5mC (Singroha *et al.*, 2022), although DNA methylation in plants comprises N6-methyladenine (6mA). Studies of contrasting genotypes of wheat (*Triticum aestivum*) revealed that the coding region of TaHKT1;four genes showed only a minor variation in 5mC content that could not be correlated with both the differential expression of TaHKT1;four and the level of salt tolerance of tested genotypes. However, increased 5mC methylation in TaHKT2;one and TaHKT2;three due to salt stress was correlated with gene expression downregulation, indicating that epialleles for the differentially expressed genes can be identified from the salt-tolerant genotype (Kumar *et al.*, 2017). Similar results have also been reported in olive (*Olea europea*), where the level of DNA methylation increased in plants under salt stress, with more evident changes in salt-tolerant cultivar (Mousavi *et al.*, 2019). Furthermore, DNA methylation and some of the differentially methylated genes were expressed differently in susceptible compared with the tolerant cultivars under salt stress. Tolerant cultivars reacted to salt stress by reduce the expression of four (*OePIP1.1*, *OePetD*, *OePI4Kg4* and *OeXyla*) out of six identified genes that led to slower plant growth and confirming the significance of the identified genes for salt tolerance improvement.

Histone modifications

Recent studies have revealed the functional importance of chromatin modifiers in regulating salt tolerance in plants (Yung *et al.*, 2021). Although most of the findings are on the model plant *Arabidopsis thaliana*, progress has also been made in the identification of potential targets for improved salt tolerance in crops. Generally, acetylation and deacetylation of the histone proteins that occur at the N-terminal lysine residues of H4, H3, and H5 of H2A and H2B are the most studied modifications in the context of salt tolerance (Roy and Soni, 2022). In addition to these modifications, crosstalks between histone acetylation/deacetylation and methylation/demethylation could also affect gene expression in plants under salt stress, as observed by Chen *et al.* (2010) in *Arabidopsis*. In *Arabidopsis* HDA6 mutant, axe1-5, ABA- and salt-induced changes in H3K4me3 were halted or decreased, suggesting a synergistic interplay between histone demethylases and deacetylases during salt stress. Changes in histone methylation involved H3K27me3 modification levels under salt stress in soybean (*Glycine max*) in a comprehensive study in which the authors performed a whole-genome Chip-Seq analysis of this repressive epigenetic mark (Sun *et al.*, 2019). The same authors observed the changes in H3K27me3 level during soybean response to salt stress, indicating that complex chromatin regulation occurs at target loci to respond to salt stress. H3K27me3 modifications were also found to be involved in the regulation of salt stress response in castor bean, where depletion of H3K4me3 and increase of H3K27me3 affected the expression of salt-responsive gene, *RSM1*, providing a novel proof of the role of histone modifications in salt tolerance (Han *et al.*, 2020).

Epigenetic variability

Recent studies showed that the abiotic stresses can induce heritable epigenetic changes, thus creating natural epigenetic variability within the species (Rashid *et al.*, 2022). Hence, DNA methylation polymorphism observed between contrasting genotypes within the species may be the results of not only the stress application but also natural selection, as observed by Karan *et al.* (2012) who studied DNA methylation variation in rice cultivars with different salt tolerance. This observation further supports the significance of the identification and determination of epigenetic and chromatin states in crop germplasm collections and wild relatives and their further exploitation in epi-breeding programs (Varotto *et al.*, 2022). This approach has been used by Feng *et al.* (Feng *et al.*, 2022), who studied epigenetic regulation of tolerance to salt stress by histone acetyltransferase GsMYST1 identified in wild soybean (*Glycine soja*). The authors identified a GsSnRK1-GsMYST1-GsNAC83 module from wild soybean which takes part in epigenetic regulation of the COR gene and consequently enhances soybean tolerance to salt stress, providing a clue to uncover novel molecular mechanisms of plant adaptabilities to extreme environmental conditions.

19.4.2 Heat stress studies

Extreme temperatures, both heat and cold, significantly impact agricultural production by affecting plant productivity and causing morphological, physiological, and biochemical changes in plants (Duarte-Aké *et al.*, 2023; Mirosavljević *et al.*, 2021a, b). However, plants have different mechanisms to cope with such conditions, but the level of survival/tolerance varies among species and even genotypes belonging to the same species. One of these specific mechanisms is the ability to memorize a stressful experience using transcriptional, somatic, intergenerational, and transgenerational stress memory (Gallusci *et al.*, 2023). The plant stress memory is regulated by epigenetic changes, such as DNA methylation, histone modifications, and/or other regulator modules which can be inherited in the next generations (Ramakrishnan *et al.*, 2022).

DNA methylation

Different plant species react differently in terms of variation in DNA methylation when exposed to heat stress. In *Arabidopsis*, heat stress can cause an increased global DNA methylation and homologous recombination frequency (HRF) (Boyko *et al.*, 2010). Naydenov *et al.* (2015) reported that in response to heat stress, the upregulation of DRM2, NUCLEAR RNA POLYMERASE D 1(NRPD1) and NRPE1 occur in *Arabidopsis* and may result in increased methylation genome-wide. Similar results were observed in cork oak (*Quercus suber* L.) grown at 55°C (Correia *et al.*, 2013). In *Brassica napus* it was observed that the increase in DNA methylation was higher in heat-sensitive genotypes than in heat-tolerant genotypes (Gao *et al.*, 2014). In contrast, genome-wide hypomethylation was reported in cotton (*Gossypium hirsutum*) anthers, as a consequence of high temperatures, which significantly decreased the expression of S-ADENOSYL-LHOMOCYSTEINE HYDROLASE1 (SAHH1) and DNA methyltransferases (DRM1 and DRM3) (Min *et al.*, 2014). All these results suggest that the DNA methylation changes under heat stress are not consistent in different species and that DNA methylation may not always take part in the regulation of heat responses.

However, the study of heat tolerance in *Arabidopsis* mutants defective in DNA methylation revealed that the RNA-directed DNA methylation (RdDM) pathway is required for basic tolerance to temperature stress (Popova *et al.*, 2013). They reported that plants hypersensitive to heat stress are deficient in NRPD2, and while recovering from heat, in these mutants the misexpression of some protein-coding genes is connected with altered epigenetic regulation of nearby remnants transposons and retrotransposons. Naydenov *et al.* (2015) also investigated if the variations in DNA methylation of TEs have a causal role in the heat-induced activation of nearby genes. They found that the expression of the Calmodulin-like 41 (CML41) gene is upregulated by high temperature with reduced DNA methylation level in the TE insertion that is very close to the transcriptional start site. In a genome-wide association analysis of loci with an adaptive response to climate, CHROMOMETHYLASE 2 (CMT2) has been identified, among others (Shen *et al.*, 2014). It was found that the CMT2STOP allele, containing a premature stop codon, is associated with a changed genome-wide CHH-methylation pattern, increased leaf serration, and higher bacterial disease presence, while cmt2 mutant plants had an improved tolerance to heat stress and broader geographic distribution in relation to the wild-type (Shen *et al.*, 2014).

All the previously mentioned studies suggested the different roles of CMT2-dependent CHH methylation and the RdDM pathway in plant responses to heat stress, showing that cmt2 mutant plants were more tolerant to heat stress, while nrpd2 mutant plants are hypersensitive to heat stress. All these observations indicate that the role of DNA methylation in responses to heat needs more investigation in plants.

Histone modifications

Histone post-transcriptional modifications (PTMs) are added by so-called writer enzymes and removed by eraser enzymes, both having characteristic functional domains. In plants, as reported above, gene activation is associated with histone acetylation and the trimethylation of lysine 4 and lysine 36 on histone H3 (H3K4me3 and H3K36me3), while gene silencing is associated with the trimethylation of lysine 9 and lysine 27 (H3K27me3) and histone deacetylation. A recent comprehensive study of the organization and dynamics of chromatin and transcriptional changes under heat stress in tomato (*Solanum lycopersicum*) through Chip-seq revealed that enriched levels of H3K4me3, H3K9ac, H3K18ac, and H3K27ac were present in nucleosomes adjacent to accessible proximal regulatory elements, while moderate and low levels of H3K27ac and H3K4me3 were found in accessible distal regulatory elements, respectively (Huang *et al.*, 2023). The authors also investigated the role of heat shock factor (HSF) TFHSFA1a, a major transcriptional regulator of heat stress responses in plants, showing that it plays an essential role in the dynamic of formation of promoter–enhancer contacts in heat-stressed plants at several loci.

Strenkert *et al.* (2011) found higher levels of histone H3/4 acetylation and lower levels of H3K4me1 at promoter regions of active genes after heat stress in the unicellular green alga *Chlamydomonas reinhardtii*. However, Correia *et al.* (2013) found that, in the forest tree cork oak, temperature shift from 25°C to 45°C decreases H3Ac levels and they suggested that the deacetylated H3 enrichment might be responsible for repressive chromatin in gene promoters, thus negatively affecting gene transcription. It was also found that in rice (*Oryza sativa*) the H3K9me2 level at the *OsFIE1* locus is sensitive to moderate heat stress during seed development (Folsom *et al.*, 2014). Min *et al.* (2014) reported that in cotton (*Gossypium irsutum*) anthers, histone modification writers, namely one histone methyltransferase, one histone monoubiquitinase gene, and two jumonji C (jmjC) domain-containing genes, are downregulated upon high-temperature treatments. Furthermore, Singh *et al.* (2014) suggested that repetitive heat stress modulated *Arabidopsis* pattern-triggered immunity (PTI) in a HISTONE ACETYLTRANSFERASE1 (HAC1)-dependent manner.

Small RNAs

There are two main types of small RNAs in plants, microRNAs (miRNAs) and siRNAs, both considered key guide molecules in the regulation of gene expression (Liu *et al.*, 2015). Small interfering RNAs regulate gene expression by targeting mRNAs and by directing and guiding complexes modifying chromatin structure to specific locations in the genome. Small RNA-mediated gene silencing and chromatin remodeling complexes have a critical role in the adaption of plants to different environmental conditions.

Recently, sRNA deep-sequencing was used to identify conserved and non-conserved heat-responsive miRNAs in different plant species, but just a few of them have been experimentally validated (Liu *et al.*, 2015). The majority of the conserved heat-responsive miRNAs are differently regulated, depending on the plant species, except for miR159, 166 and 472 families. miR159 targets MYB transcription factors and it is downregulated by heat in *Arabidopsis* (Zong *et al.*, 2013), wheat (*Triticum aestivum*; Kumar *et al.*, 2017) and cassava (*Manihot esculenta*; Ballén-Taborda *et al.*, 2013). The study with tae-miR159 overexpression rice lines and *Arabidopsis* myb33myb65 double mutants showed that these lines and mutants are more susceptible to heat stress compared with the wild-types (Wang *et al.*, 2012). MiR166, which targets HD-Zip transcription factors, is upregulated by heat in *Arabidopsis* (Zhong *et al.*, 2013), wheat (Xin *et al.*, 2010), and barley (*Hordeum vulgare*; Kruszka *et al.*, 2014), while miR472 can be downregulated by heat in *Arabidopsis* (Zhong *et al.*, 2013) and Chinese white poplar (*Populus tomentosa*) (Chen *et al.*, 2012). All these findings need to be confirmed in further studies.

19.4.3 Drought stress studies

Drought is one of the most important abiotic stresses, gaining more and more importance as a direct consequence of climate change (Miladinović *et al.*, 2019). It induces complex responses in plants at all levels: phenotypic, morphological, psychological, biochemical, and molecular (Paul *et al.*, 2019). While there is an abundance of work performed on the identification of drought-resistant QTLs in different plant species (Andleeb *et al.*, 2020), research concerning the discovery of drought-related epi-QTLs is still scarce. Scrutinizing drought resistance in plants and discovering new epi-QTLs as well as gaining a comprehensive knowledge of the relationship between QTLs and epi-QTLs can be of help in translational research for enhancing plant tolerance to drought.

DNA methylation

DNA methylation under drought stress has been shown to have a variety of effects on gene expression. Plants that can better withstand drought have a more stable methylome compared with drought-sensitive plants. In a comprehensive study, it was shown that drought initiates an instructive mechanism guiding epigenetic machinery toward specific effector genes (Colaneri and Jones, 2013). Thus, in rice, upregulation of two genes involved in proline synthesis (*P5CS* and *IV-OAT*) was stably inherited and was associated with increased demethylation (Zhang *et al.*, 2013). Additionally, it was reported that methylation levels of chloride transporter (CLT1) and photosystem II polypeptide (PSBP) genes were lower in the tolerant genotype compared with the susceptible one (Ding *et al.*, 2022). In tomato, a decrease in DNA methylation levels in ABSCISIC ACID STRESS RIPENING 1 (Asr1) and Asr2 genes was observed. DNA methylation was also indicated as an epigenetic mechanism involved in priming for drought stress: in wheat, osmolyte accumulation was observed as a consequence of the demethylation of TaP5CS and TaBADH (Li *et al.*, 2023).

Histone modifications

In plants, H3K4me3 has been often associated with transcriptional gene activation in drought stress response. This association was initially observed in *Arabidopsis* subjected to dehydration (van Dijk *et al.*, 2010). Compared with H3K4me1 and H3K4me2, which were more widely distributed in nucleosomes, H3K4me3 was less distributed but showed weak and positive association with transcriptionally active genes. This positive correlation was later confirmed in rice and maize (*Zea mays*) subjected to drought stress (Zong *et al.*, 2013; Forestan *et al.*, 2018). In rice, with a stressed-related gene whose expression was associated with H3K4me3, this modification level was primarily enriched in low-expressed genes and depleted in highly expressed genes. In addition, Zong *et al.* (2020) demonstrated a collaboration between the transcription factor OsbZIP23 and H3K4me3 modification in the regulation of drought-responsive genes.

H3K36me3 has been also associated with gene transcription activation in plants. In rice, a promising epigenetic writer, SDG708, a H3K36 methyltransferase, was shown to be a drought tolerance regulator that activates two essential abscisic acid (ABA) biosynthetic genes boosting synthesis of ABA and improved drought resistance in rice (Chen *et al.*, 2021). This regulator also plays an important role in promoting stomatal closure as well as inducing hydrogen peroxide accumulation. In maize, in addition to H3K4me3, H3K27me3 may be a memory mark for both transcription factors and drought stress-responsive genes (Forestan *et al.*, 2018). As mentioned before, opposite to H3K4me3, H3K27me3 is considered a repressive mark, as genes marked with H3K27me3 are usually silenced in plants. Recently, investigations on drought tolerance in apple (*Malus domestica*) at chromatin and transcriptional levels revealed that MdRAD5B, a chromatin remodeling factor belonging to the SNF2 family, mediates the H3K27me3 deposition at drought-responsive gene loci through MdLHP1 (Song *et al.*, 2024). These authors showed that MdRAD5B can modulate gene expression in apple in response to drought, through the regulation of chromatin remodeling and H3K27me3.

Along with H3K4me3 and H3K27me3, the dynamics of H3K9ac are also associated with drought tolerance in plants. In barley, ChIP-Seq data showed that significantly more genes were enriched in H3K9ac compared with H3K4me3, with a higher flexibility and sensitivity of H3K9ac reaction to early stages of drought stress treatments (Ost *et al.*, 2023). H3K9ac appeared enriched in some crucial genes involved in ABA synthesis. In maize, natural antisense transcripts (NATs) genes associated with drought response have open chromatin configuration due to higher levels of H3K9ac, H3K4me3, and H3K36me3, and lower levels of H3K27me3 (Xu *et al.*, 2017). Forestan *et al.* (2018) indicated that H3K4me3 and H3K9ac enriched genes are different and emphasized the importance of the dynamics of H3K4me3 and H3K9ac for the expression of these genes in response to drought. Song *et al.* (2020) identified over 70 genes that were associated with either an increase or a decrease in H3K9ac under drought conditions and confirmed the positive correlation between H3K9ac levels and the level of drought tolerance in *Brachypodium distachyon*.

Non-coding RNA activity

Despite being identified in plants, the function and mode of action of lncRNAs remain little described and studied. Initial work in *Arabidopsis* showed an association between drought tolerance and overexpression of the DROUGHT INDUCED lncRNA (DRIR) which influences the expression patterns of stress-responsive genes, such as ABA response and aquaporin genes, and different genes coding for transcription factors (Qin *et al.*, 2017). Recently in *Arabidopsis*, Cai *et al.* (2024) characterized an intergenic non-coding RNA, denominated DANA1, and a DANA1-INTERACTING PROTEIN 1 (DIP1) which promoted drought resistance. DIP1 and DANA1 increase histone deacetylases HDA9 binding to CYP707A1 and CYP707A2 loci, affecting drought tolerance.

In maize, both Zhang *et al.* (2014) and Forestan *et al.* (2016) reported the important role of lncRNAs and siRNAs in regulating gene expression in drought conditions. Furthermore, Lunardon *et al.* (2016) found that long-term drought provokes changes in the accumulation of sRNAs and miRNAs, and identified novel MIRNA loci in drought stress experiments. In durum wheat, Liu *et al.* (2017) reported 2055 targets for 113 conserved miRNAs and 131 for four novel miRNAs associated with water deficit, while in barley, Surdonja *et al.* (2017) identified drought-specific 24mer-sized heterochromatic-siRNAs (hc-siRNAs) with a particular emphasis on hc-siRNA-mediated hypermethylation at the promoter region of CYTOKININ-OXIDASE 2.1.

In conclusion, there is an abundance of non-coding RNAs involved in drought tolerance. Still, the accumulation of miR408 transcript was found to be associated with drought stress in barley, rice and chickpea, while the opposite was reported in *Arabidopsis* (Kantar *et al.*, 2010; Hajyzadeh *et al.*, 2015; Ma *et al.*, 2015).

19.5 Conclusion and Perspectives

Epigenomic information in response to stress conditions has been produced in model and crop plants. Additionally, many studies focusing on plant stress response have elucidated the features of genes encoding various epigenetic/chromatin modification factors. However, comprehensive epigenome maps across different tissues and in response to stresses are available only for a few model and semi-model plant species (Zhao *et al.*, 2020). When available, epigenome maps are fundamental for elucidating the dynamics of chromatin changes, and their effects on expression profiles and for assessing both epigenetic and genetic variations in plants. It is expected that soon the adoption of advanced technologies for obtaining epigenomic datasets and performing integrative analyses will provide comprehensive maps of the epigenomic landscape in numerous plants and crops. Indeed, the exploitation of advanced multi-omics techniques and sequencing data has shown great potential in providing researchers with information ranging from the genome to the phenotype, as well as in assisting in mapping genetic loci and revealing high potentiality also in plant genetic breeding. Chromatin immunoprecipitation (ChIP) is considered the gold-standard method for the detection of epigenetic modifications and many adjusted protocols have been developed for different plant species. However, some plant-specific features still hinder the efficiency of immunoprecipitation, which requires a high amount of plant material to allow the ChIP-seq experiment to be successfully performed. The adoption of advanced alternative methods which require a small amount of starting plant materials, such as CUT&Tag and ATAC-seq, for studying chromatin marks and dynamics could accelerate the profiling of histone modifications in plant tissues. Similarly, the adoption of nanopore sequencing can make it easier to obtain the methylome of different plant tissues, coupled with chromatin accessibility information through STAM-seq. These advanced methods provide promising strategies for further application in the study of epigenomics in plants, particularly in response to a changing and challenging environment. Finally, very recently novel procedures have been developed and applied to study cell-type profiling, developmental trajectory, and epigenetic heterogeneity in plants (Ouyang *et al.*, 2022). Profiling plant histone modification and DNA methylation at single-cell resolution and integrating the epigenetic data with single-cell RNA sequencing might provide exceptional information in all fields of plant studies.

Acknowledgments

This work is supported by the European Commission through project CROPINNO, grant number 101,059,784. DM, AR and AKŠ were also supported by the Science Fund of the Republic of Serbia, through IDEAS project 'Creating climate smart sunflower for future challenges' (SMARTSUN) grant number 7732457, by the Ministry of Science, Technological Development and Innovations of Republic of Serbia, grant number 451-03-66/2024-03/200032, and by the Centre of Excellence for Innovations in Breeding of Climate-Resilient Crops – Climate Crops, Institute of Field and Vegetable Crops, Novi Sad, Serbia.

References

Agius, D.R., Kapazoglou, A., Avramidou, E., Baranek, M. and Carneros, E. (2023) Exploring the crop epigenome: A comparison of DNA methylation profiling techniques. *Frontiers in Plant Science* 14, 1–24. DOI: 10.3389/fpls.2023.1181039.

Alonso-Blanco, C., Aarts, M.G.M., Bentsink, L., Keurentjes, J.J.B., Reymond, M. *et al.* (2009) What has natural variation taught us about plant development physiology, and adaptation? *The Plant Cell* 21(7), 1877–1896. DOI: 10.1105/tpc.109.068114.

Altemose, N., Maslan, A., Smith, O.K., Sundararajan, K., Brown, R.R. *et al.* (2022) DiMeLo-seq: A long-read, single-molecule method for mapping protein-DNA interactions genome wide. *Nature Methods* 19(6), 711–723. DOI: 10.1038/s41592-022-01475-6.

Andleeb, T., Shah, T., Nawaz, R., Munir, I., Munsif, F. *et al.* (2020) QTL mapping for drought stress tolerance in plants. In: Hasanuzzaman, M. and Tanveer, M. (eds) *Salt and Drought Stress Tolerance in Plants: Signaling Networks and Adaptive Mechanisms*. Springer, Cham, pp. 383–403.

Ariel, F., Lucero, L., Christ, A., Mammarella, M.F., Jegu, T. *et al.* (2020) R-loop mediated trans action of the APOLO long noncoding RNA. *Molecular Cell* 77(5), 1055–1065. DOI: 10.1016/j.molcel.2019.12.015.

Ballén-Taborda, C., Plata, G., Ayling, S., Rodríguez-Zapata, F., Becerra Lopez-Lavalle, L.A. *et al.* (2013) Identification of cassava microRNAs under abiotic stress. *International Journal of Genomics* 857986. DOI: 10.1155/2013/857986.

Becker, C. and Weigel, D. (2012) Epigenetic variation: Origin and transgenerational inheritance. *Current Opinion in Plant Biology* 15(5), 562–567. DOI: 10.1016/j.pbi.2012.08.004.

Boyko, A., Blevins, T., Yao, Y., Golubov, A., Bilichak, A. *et al.* (2010) Transgenerational adaptation of *Arabidopsis* to stress requires DNA methylation and the function of dicer-like proteins. *PloS ONE* 5(3), e9514. DOI: 10.1371/journal.pone.0009514.

Brinkman, A.B., Gu, H., Bartels, S.J.J., Zhang, Y., Matarese, F. *et al.* (2012) Sequential XhIP-bisulfite sequencing enables direct genome-scale investigation of chromatin and DNA methylation cross-talk. *Genome Research* 22(6), 1128–1138. DOI: 10.1101/gr.133728.111.

Buenrostro, J.D., Giresi, P.G., Zaba, L.C., Chang, H.Y. and Greenleaf, W.J. (2013) Transposition of native chromatin for fast and sensitive epigenomic profiling of open chromatin, DNA-binding proteins, and nucleosome position. *Nature Methods* 10(12), 1213–1218. DOI: 10.1038/nmeth.2688.

Cai, J., Zhang, Y., He, R., Jiang, L., Qu, Z. *et al.* (2024) LncRNA DANA1 promotes drought tolerance and histone deacetylation of drought responsive genes in *Arabidopsis*. *EMBO Reports* 25(2), 796–812. DOI: 10.1038/s44319-023-00030-4.

Chen, K., Du, K., Shi, Y., Yin, L., Shen, W.-H. *et al.* (2021) H3K36 methyltransferase SDG708 enhances drought tolerance by promoting abscisic acid biosynthesis in rice. *The New Phytologist* 230(5), 1967–1984. DOI: 10.1111/nph.17290.

Chen, L., Ren, Y., Zhang, Y., Xu, J., Sun, F. *et al.* (2012) Genome-wide identification and expression analysis of heat-responsive and novel microRNAs in populus tomentosa. *Gene* 504(2), 160–165. DOI: 10.1016/j.gene.2012.05.034.

Chen, L.T., Luo, M., Wang, Y.Y. and Wu, K. (2010) Involvement of *Arabidopsis* histone deacetylase HDA6 in ABA and salt stress response. *Journal of Experimental Botany* 61(12), 3345–3353. DOI: 10.1093/jxb/erq154.

Chen, X., Bhadauria, V. and Ma, B. (2018) ChIP-seq: A powerful tool for studying protein-DNA interactions in plants. *Current Issues in Molecular Biology* 27, 171–180. DOI: 10.21775/CIMB.027.171.

Clough, E., Barrett, T., Wilhite, S.E., Ledoux, P., Evangelista, C. *et al.* (2024) NCBI GEO: Archive for gene expression and epigenomics data sets: 23-year update. *Nucleic Acids Research* 53, D138–D144.

Colaneri, A.C. and Jones, A.M. (2013) Genome-wide quantitative identification of DNA differentially methylated sites in *Arabidopsis* seedlings growing at different water potential. *PloS ONE* 8(4), e59878. DOI: 10.1371/journal.pone.0059878.

Corces, M.R., Trevino, A.E., Hamilton, E.G., Greenside, P.G., Sinnott-Armstrong, N.A. *et al.* (2017) An improved ATAC-seq protocol reduces background and enables interrogation of frozen tissues. *Nature Methods* 14(10), 959–962. DOI: 10.1038/nmeth.4396.

Correia, B., Valledor, L., Meijón, M., Rodriguez, J.L., Dias, M.C. *et al.* (2013) Is the interplay between epigenetic markers related to the acclimation of cork oak plants to high temperatures? *PloS ONE* 8(1), e53543. DOI: 10.1371/journal.pone.0053543.

Cui, X., Dard, A., Reichheld, J.P. and Zhou, D.X. (2023) Multifaceted functions of histone deacetylases in stress response. *Trends in Plant Science* 28(11), 1245–1256. DOI: 10.1016/j.tplants.2023.06.006.

Deal, R.B. and Henikoff, S. (2011) Histone variants and modifications in plant gene regulation. *Current Opinion in Plant Biology* 14(2), 116–122.

Deal, R.B., Henikoff, S. and Hutchinson, F. (2011) The INTACT method for cell type-specific gene expression and chromatin profiling in *Arabidopsis thaliana*. *Nature Protocols* 6(1), 56–68. DOI: 10.1038/nprot.2010.175.

Debernardi, J.M., Burguener, G., Bubb, K., Liu, Q., Queitsch, C. *et al.* (2023) Optimization of ATAC-seq in wheat seedling roots using INTACT-isolated nuclei. *BMC Plant Biology* 23(1), 1–15. DOI: 10.1186/s12870-023-04281-0.

Deinlein, U., Stephan, A.B., Horie, T., Luo, W., Xu, G. *et al.* (2014) Plant salt-tolerance mechanisms. *Trends in Plant Science* 19(6), 371–379. DOI: 10.1016/j.tplants.2014.02.001.

Ding, G., Cao, L., Zhou, J., Li, Z., Lai, Y. *et al.* (2022) DNA methylation correlates with the expression of drought-responsive genes and drought resistance in rice. *Agronomy* 12(6), 1445. DOI: 10.3390/agronomy12061445.

Ding, K., Sun, S., Luo, Y., Long, C., Zhai, J. *et al.* (2023) PlantCADB: A comprehensive plant chromatin accessibility database. *Genomics, Proteomics & Bioinformatics* 21(2), 311–323. DOI: 10.1016/j.gpb.2022.10.005.

Du, J., Zhong, X., Bernatavichute, Y.V., Stroud, H., Feng, S. *et al.* (2012) Dual binding of chromomethylase domains to H3K9me2-containing nucleosomes directs DNA methylation in plants. *Cell* 151(1), 167–180. DOI: 10.1016/J.CELL.2012.07.034.

Duarte-Aké, F., Us-Camas, R. and De-la-Peña, C. (2023) Epigenetic regulation in heterosis and environmental stress: The challenge of producing hybrid epigenomes to face climate change. *Epigenomes* 7(3), 14. DOI: 10.3390/epigenomes7030014.

Dubin, M.J., Zhang, P., Meng, D., Remigereau, M.-S., Osborne, E.J. *et al.* (2015) DNA methylation in *Arabidopsis* has a genetic basis and shows evidence of local adaptation. *eLife* 4, 1–23. DOI: 10.7554/eLife.05255.

Edgar, R., Domrachev, M. and Lash, A.E. (2002) Gene expression omnibus: NCBI gene expression and hybridization array data repository. *Nucleic Acids Research* 30(1), 207–210. DOI: 10.1093/nar/30.1.207.

Feng, P., Sun, X., Liu, X., Li, Y., Sun, Q. *et al.* (2022) Epigenetic regulation of plant tolerance to salt stress by histone acetyltransferase GsMYST1 from wild soybean. *Frontiers in Plant Science* 13, 1–19. DOI: 10.3389/fpls.2022.860056.

Feng, S., Zhong, Z., Wang, M. and Jacobsen, S.E. (2020) Efficient and accurate determination of genome-wide DNA methylation patterns in *Arabidopsis thaliana* with enzymatic methyl sequencing. *Epigenetics & Chromatin* 13(1), 1–17. DOI: 10.1186/s13072-020-00361-9.

Flusberg, B.A., Webster, D.R., Lee, J.H., Travers, K.J., Olivares, E.C. *et al.* (2010) Direct detection of DNA methylation during single-molecule, real-time sequencing. *Nature Methods* 7(6), 461–465. DOI: 10.1038/nmeth.1459.

Folsom, J.J., Begcy, K., Hao, X., Wang, D. and Walia, H. (2014) Rice fertilization-independent endosperm1 regulates seed size under heat stress by controlling early endosperm development. *Plant Physiology* 165(1), 238–248. DOI: 10.1104/pp.113.232413.

Forestan, C., Aiese Cigliano, R., Farinati, S., Lunardon, A., Sanseverino, W. *et al.* (2016) Stress-induced and epigenetic-mediated maize transcriptome regulation study by means of transcriptome reannotation and differential expression analysis. *Scientific Reports* 6, 1–20. DOI: 10.1038/srep30446.

Forestan, C., Farinati, S., Rouster, J., Lassagne, H., Lauria, M. *et al.* (2018) Control of maize vegetative and reproductive development, fertility, and rRNAs silencing by HISTONE DEACETYLASE 108. *Genetics* 208(4), 1443–1466. DOI: 10.1534/genetics.117.300625.

Forestan, C., Farinati, S., Zambelli, F., Pavesi, G., Rossi, V. *et al.* (2020) Epigenetic signatures of stress adaptation and flowering regulation in response to extended drought and recovery in *Zea mays*. *Plant, Cell & Environment* 43(1), 55–75. DOI: 10.1111/pce.13660.

Frommer, M., McDonald, L.E., Millar, D.S., Collis, C.M., Watt, F. *et al.* (1992) A genomic sequencing protocol that yields A positive display of 5-methylcytosine residues in individual DNA strands. *Proceedings of the National Academy of Sciences* 89(5), 1827–1831. DOI: 10.1073/pnas.89.5.1827.

Fu, L.Y., Zhu, T., Zhou, X., Yu, R., He, Z. *et al.* (2022) ChIP-hub provides an integrative platform for exploring plant regulome. *Nature Communications* 13(1). DOI: 10.1038/s41467-022-30770-1.

Gallusci, P., Agius, D.R., Moschou, P.N., Dobránszki, J., Kaiserli, E. *et al.* (2023) Deep inside the epigenetic memories of stressed plants. *Trends in Plant Science* 28(2), 142–153. DOI: 10.1016/j.tplants.2022.09.004.

Gao, G., Li, J., Li, H., Li, F., Xu, K. *et al.* (2014) Comparison of the heat stress induced variations in DNA methylation between heat-tolerant and heat-sensitive rapeseed seedlings. *Breeding Science* 64(2), 125–133. DOI: 10.1270/jsbbs.64.125.

Grandi, F.C., Modi, H., Kampman, L. and Corces, M.R. (2022) Chromatin accessibility profiling by ATAC-seq. *Nature Protocols* 17(6), 1518–1552. DOI: 10.1038/s41596-022-00692-9.

Hajyzadeh, M., Turktas, M., Khawar, K.M. and Unver, T. (2015) MiR408 overexpression causes increased drought tolerance in chickpea. *Gene* 555(2), 186–193. DOI: 10.1016/j.gene.2014.11.002.

Han, B., Xu, W., Ahmed, N., Yu, A., Wang, Z. *et al.* (2020) Changes and associations of genomic transcription and histone methylation with salt stress in castor bean. *Plant & Cell Physiology* 61(6), 1120–1133. DOI: 10.1093/pcp/pcaa037.

He, Y. and Li, Z. (2018) Epigenetic environmental memories in plants: Establishment, maintenance, and reprogramming. *Trends in Genetics* 34(11), 856–866. DOI: 10.1016/j.tig.2018.07.006.

Hoppers, A., Williams, L., Ponnaluri, V.K.C., Sexton, B. and Saleh, L. (2020) Enzymatic methyl-seq: Next generation methylomes. *Journal of Biomolecular Techniques* 31, S15.

Huang, Y., An, J., Sircar, S., Bergis, C., Lopes, C.D. *et al.* (2023) HSFA1a modulates plant heat stress responses and alters the 3D chromatin organization of enhancer-promoter interactions. *Nature Communications* 14(1), 469. DOI: 10.1038/s41467-023-36227-3.

Kalinka, A., Starczak, M., Gackowski, D., Stępień, E. and Achrem, M. (2023) Global DNA 5-hydroxymethylcytosine level and its chromosomal distribution in four rye species. *Journal of Experimental Botany* 74(12), 3488–3502. DOI: 10.1093/jxb/erad102.

Kantar, M., Unver, T. and Budak, H. (2010) Regulation of barley miRNAs upon dehydration stress correlated with target gene expression. *Functional and Integrative Genomics* 10(4), 493–507. DOI: 10.1007/s10142-010-0181-4.

Karan, R., DeLeon, T., Biradar, H. and Subudhi, P.K. (2012) Salt stress induced variation in DNA methylation pattern and its influence on gene expression in contrasting rice genotypes. *PloS ONE* 7(6), e40203. DOI: 10.1371/journal.pone.0040203.

Kaur, N., Nayakoti, S., Brock, N. and Halford, N.G. (2023) Uncovering plant epigenetics: New insights into cytosine methylation in rye genomes. *Journal of Experimental Botany* 74(12), 3395–3398. DOI: 10.1093/jxb/erad144.

Kaya-Okur, H.S., Wu, S.J., Codomo, C.A., Pledger, E.S., Bryson, T.D. *et al.* (2019) CUT&tag for efficient epigenomic profiling of small samples and single cells. *Nature Communications* 10(1), 1–10. DOI: 10.1038/s41467-019-09982-5.

Kilvitis, H.J., Alvarez, M., Foust, C.M., Schrey, A.W., Robertson, M. *et al.* (2014) Ecological epigenetics. *Advances in Experimental Medicine and Biology* 781, 191–210. DOI: 10.1007/978-94-7-7347-9_10.

Kim, D.H. and Sung, S. (2017) Vernalization-triggered intragenic chromatin loop formation by long non-coding RNAs. *Developmental Cell* 40(3), 302–312. DOI: 10.1016/j.devcel.2016.12.021.

Kruszka, K., Pacak, A., Swida-Barteczka, A., Nuc, P., Alaba, S. *et al.* (2014) Transcriptionally and post-transcriptionally regulated microRNAs in heat stress response in barley. *Journal of Experimental Botany* 65(20), 6123–6135. DOI: 10.1093/jxb/eru353.

Kumar, S., Beena, A.S., Awana, M. and Singh, A. (2017) Physiological, biochemical, epigenetic and molecular analyses of wheat (*Triticum aestivum*) genotypes with contrasting salt tolerance. *Frontiers in Plant Science* 8, 1–20. DOI: 10.3389/fpls.2017.01151.

Kumar, V., Thakur, J.K. and Prasad, M. (2021) Histone acetylation dynamics regulating plant development and stress responses. *Cellular and Molecular Life Sciences* 78(10), 4467–4486. DOI: 10.1007/s00018-021-03794-x.

Laszlo, A.H., Derrington, I.M., Brinkerhoff, H., Langford, K.W., Nova, I.C. *et al.* (2013) Detection and mapping of 5-methylcytosine and 5-hydroxymethylcytosine with nanopore MspA. *Proceedings of the National Academy of Sciences of the United States of America* 110(47), 18904–18909. DOI: 10.1073/pnas.1310240110.

Latzel, V., Allan, E., Bortolini Silveira, A., Colot, V., Fischer, M. *et al.* (2013) Epigenetic diversity increases the productivity and stability of plant populations. *Nature Communications* 4, 1–7. DOI: 10.1038/ncomms3875.

Law, J.A. and Jacobsen, S.E. (2010) Establishing, maintaining and modifying DNA methylation patterns in plants and animals. *Nature Reviews. Genetics* 11(3), 204–220. DOI: 10.1038/nrg2719.

Lee, I., Razaghi, R., Gilpatrick, T., Molnar, M., Gershman, A. *et al.* (2020) Simultaneous profiling of chromatin accessibility and methylation on human cell lines with nanopore sequencing. *Nature Methods* 17(12), 1191–1199. DOI: 10.1038/s41592-020-01000-7.

Li, Q., Wang, X., Sun, Z., Wu, Y., Malkodslo, M.M. *et al.* (2023) DNA methylation levels of TaP5CS and TaADH are associated with enhanced tolerance to PEG-induced drought stress triggered by drought priming in wheat. *Plant Physiology and Biochemistry* 200, 107769. DOI: 10.1016/J.PLAPHY.2023.107769.

Li, R., Grimm, S.A. and Wade, P.A. (2021) CUT&tag-BS for simultaneous profiling of histone modification and DNA methylation with high efficiency and low cost. *Cell Reports Methods* 1(8), 100118. DOI: 10.1016/j.crmeth.2021.100118.

Li, X., Chen, L., Zhang, Q., Sun, Y., Li, Q. *et al.* (2019) BRIF-seq: Bisulfite-converted randomly integrated fragments sequencing at the single-cell level. *Molecular Plant* 12(3), 438–446. DOI: 10.1016/j.molp.2019.01.004.

Li, Y. (2021) Modern epigenetics methods in biological research. *Methods* 187, 104–113. DOI: 10.1016/j.ymeth.2020.06.022.

Liu, C., Lu, F., Cui, X. and Cao, X. (2010) Histone methylation in higher plants. *Annual Review of Plant Biology* 61, 395–420. DOI: 10.1146/annurev.arplant.043008.091939.

Liu, H., Able, A.J. and Able, J.A. (2017) Water-deficit stress-responsive microRNAs and their targets in four durum wheat genotypes. *Functional & Integrative Genomics* 17(2–3), 237–251. DOI: 10.1007/s10142-016-0515-y.

Liu, J., Feng, L., Li, J. and He, Z. (2015) Genetic and epigenetic control of plant heat responses. *Frontiers in Plant Science* 6(APR), 1–21. DOI: 10.3389/fpls.2015.00267.

Liu, Y., Tian, T., Zhang, K., You, Q., Yan, H. *et al.* (2018) PCSD: A plant chromatin state database. *Nucleic Acids Research* 46(D1), D1157–D1167. DOI: 10.1093/nar/gkx919.

Lloyd, J.P.B. and Lister, R. (2022) Epigenome plasticity in plants. *Nature Reviews. Genetics* 23(1), 55–68. DOI: 10.1038/s41576-021-00407-y.

Lunardon, A., Forestan, C., Farinati, S., Axtell, M.J. and Varotto, S. (2016) Genome-wide characterization of maize small RNA loci and their regulation in the required to maintain repression6-1 (Rmr6-1) mutant and long-term abiotic stresses. *Plant Physiology* 170(3), 1535–1548. DOI: 10.1104/pp.15.01205.

Ma, C., Burd, S. and Lers, A. (2015) MiR408 is involved in abiotic stress responses in *Arabidopsis*. *The Plant Journal* 84(1), 169–187. DOI: 10.1111/tpj.12999.

Maher, K.A., Bajic, M., Kajala, K., Reynoso, M., Pauluzzi, G. *et al.* (2018) Profiling of accessible chromatin regions across multiple plant species and cell types reveals common gene regulatory principles and new control modules. *The Plant Cell* 30(1), 15–36. DOI: 10.1105/tpc.17.00581.

Matzke, M., Kanno, T., Daxinger, L., Huettel, B. and Matzke, A.J. (2009) RNA-mediated chromatin-based silencing in plants. *Current Opinion in Cell Biology* 21(3), 367–376. DOI: 10.1016/j.ceb.2009.01.025.

Mehrmohamadi, M., Sepehri, M.H., Nazer, N. and Norouzi, M.R. (2021) A comparative overview of epigenomic profiling methods. *Frontiers in Cell and Developmental Biology* 9, 1–14. DOI: 10.3389/fcell.2021.714687.

Meyer, P. (2015) Epigenetic variation and environmental change. *Journal of Experimental Botany* 66(12), 3541–3548. DOI: 10.1093/jxb/eru502.

Miladinović, D., Hladni, N., Radanović, A., Jocić, S. and Cvejić, S. (2019) Sunflower and climate change: Possibilities of adaptation through breeding and genomic selection. In: Kole, C. (ed.) *Genomic Designing of Climate-Smart Oilseed Crops*. Springer, Cham, pp. 173–238.

Min, L., Li, Y., Hu, Q., Zhu, L., Gao, W. *et al.* (2014) Sugar and auxin signaling pathways respond to high-temperature stress during anther development as revealed by transcript profiling analysis in cotton. *Plant Physiology* 164(3), 1293–1308. DOI: 10.1104/pp.113.232314.

Mirosavljević, M., Mikić, S., Špika, A.K., Župunski, V., Zhou, R. *et al.* (2021a) The effect of heat stress on some main spike traits in 12 wheat cultivars at anthesis and mid-grain filling stage. *Plant, Soil and Environment* 67(2), 71–76. DOI: 10.17221/457/2020-PSE.

Mirosavljević, M., Mikić, S., Župunski, V., Kondić Špika, A., Trkulja, D. *et al.* (2021b) Effects of high temperature during anthesis and grain filling on physiological characteristics of winter wheat cultivars. *Journal of Agronomy and Crop Science* 207(5), 823–832. DOI: 10.1111/jac.12546.

Mo, W., Shu, Y., Liu, B., Long, Y., Li, T. *et al.* (2023) Single-molecule targeted accessibility and methylation sequencing of centromeres, telomeres and rdnas in *Arabidopsis*. *Nature Plants* 9(9), 1439–1450. DOI: 10.1038/s41477-023-01498-7.

Monroe, J.G., Srikant, T., Carbonell-Bejerano, P., Becker, C., Lensink, M. *et al.* (2022) Mutation bias reflects natural selection in *Arabidopsis thaliana*. *Nature* 602(7895), 101–105. DOI: 10.1038/s41586-021-04269-6.

Mousavi, S., Regni, L., Bocchini, M., Mariotti, R., Cultrera, N.G.M. *et al.* (2019) Physiological, epigenetic and genetic regulation in some olive cultivars under salt stress. *Scientific Reports* 9(1), 1–17. DOI: 10.1038/s41598-018-37496-5.

Mozgova, I. and Hennig, L. (2015) The Polycomb group protein regulatory network. *Annual Review of Plant Biology* 66, 269–296. DOI: 10.1146/annurev-arplant-043014-115627.

Muyle, A.M., Seymour, D.K., Lv, Y., Huettel, B. and Gaut, B.S. (2022) Gene body methylation in plants: Mechanisms, functions, and important implications for understanding evolutionary processes. *Genome Biology and Evolution* 14(4), 1–18. DOI: 10.1093/gbe/evac038.

Naydenov, M., Baev, V., Apostolova, E., Gospodinova, N., Sablok, G. *et al.* (2015) High-temperature effect on genes engaged in DNA methylation and affected by DNA methylation in *Arabidopsis*. *Plant Physiology and Biochemistry* 87, 102–108. DOI: 10.1016/j.plaphy.2014.12.022.

Ni, P., Huang, N., Zhang, Z., Wang, D.-P., Liang, F. *et al.* (2019) DeepSignal: Detecting DNA methylation state from nanopore sequencing reads using deep-learning. *Bioinformatics* 35(22), 4586–4595. DOI: 10.1093/bioinformatics/btz276.

Ni, P., Huang, N., Nie, F., Zhang, J., Zhang, Z. *et al.* (2021) Genome-wide detection of cytosine methylations in plant from nanopore data using deep learning. *Nature Communications* 12(1), 1–11. DOI: 10.1038/s41467-021-26278-9.

Olova, N., Krueger, F., Andrews, S., Oxley, D., Berrens, R.V. *et al.* (2018) Comparison of whole-genome bisulfite sequencing library preparation strategies identifies sources of biases affecting DNA methylation data. *Genome Biology* 19(1), 1–19. DOI: 10.1186/s13059-018-1408-2.

Ost, C., Cao, H.X., Nguyen, T.L., Himmelbach, A., Mascher, M. *et al.* (2023) Drought-stress-related reprogramming of gene expression in barley involves differential histone modifications at ABA-related genes. *International Journal of Molecular Sciences* 24(15), 12065. DOI: 10.3390/ijms241512065.

Ouyang, W., Cao, Z., Xiong, D., Li, G. and Li, X. (2020) Decoding the plant genome: From epigenome to 3D organization. *Journal of Genetics and Genomics* 47(8), 425–435. DOI: 10.1016/j.jgg.2020.06.007.

Ouyang, W., Luan, S., Xiang, X., Guo, M., Zhang, Y. *et al.* (2022) Profiling plant histone modification at single-cell resolution using SnCUT&Tag. *Plant Biotechnology Journal* 20(3), 420–422. DOI: 10.1111/pbi.13768.

Paul, K., Pauk, J., Kondic-Spika, A., Grausgruber, H., Allahverdiyev, T. *et al.* (2019) Co-occurrence of mild salinity and drought synergistically enhances biomass and grain retardation in wheat. *Frontiers in Plant Science* 10, 1–15. DOI: 10.3389/fpls.2019.00501.

Paun, O., Verhoeven, K.J.F. and Richards, C.L. (2019) Opportunities and limitations of reduced representation bisulfite sequencing in plant ecological epigenomics. *The New Phytologist* 221(2), 738–742. DOI: 10.1111/nph.15388.

Piao, Y., Xu, W., Park, K.H., Ryu, K.H. and Xiang, R. (2021) Comprehensive evaluation of differential methylation analysis methods for bisulfite sequencing data. *International Journal of Environmental Research and Public Health* 18(15), 7975. DOI: 10.3390/ijerph18157975.

Poleshko, A., Smith, C.L., Nguyen, S.C., Sivaramakrishnan, P., Wong, K.G. *et al.* (2019) H3K9me2 orchestrates inheritance of spatial positioning of peripheral heterochromatin through mitosis. *eLife* 8, 1–24. DOI: 10.7554/eLife.49278.

Popova, O.V., Dinh, H.Q., Aufsatz, W. and Jonak, C. (2013) The RdDM pathway is required for basal heat tolerance in *Arabidopsis*. *Molecular Plant* 6(2), 396–410. DOI: 10.1093/mp/sst023.

Pott, S. (2017) Simultaneous measurement of chromatin accessibility, DNA methylation, and nucleosome phasing in single cells. *eLife* 6, 1–19. DOI: 10.7554/eLife.23203.

Probst, A.V., Desvoyes, B. and Gutierrez, C. (2020) Similar yet critically different: The distribution, dynamics and function of histone variants. *Journal of Experimental Botany* 71(17), 5191–5204. DOI: 10.1093/jxb/eraa230.

Qin, T., Zhao, H., Cui, P., Albesher, N. and Xiong, L. (2017) A nucleus-localized long non-coding RNA enhances drought and salt stress tolerance. *Plant Physiology* 175(3), 1321–1336. DOI: 10.1104/pp.17.00574.

Quadrana, L. and Colot, V. (2016) Plant transgenerational epigenetics. *Annual Review of Genetics* 50, 467–491. DOI: 10.1146/annurev-genet-120215-035254.

Ramakrishnan, M., Zhang, Z., Mullasseri, S., Kalendar, R., Ahmad, Z. *et al.* (2022) Epigenetic stress memory: A new approach to study cold and heat stress responses in plants. *Frontiers in Plant Science* 13, 1–17. DOI: 10.3389/fpls.2022.1075279.

Rashid, M.M., Vaishnav, A., Verma, R.K., Sharma, P., Suprasanna, P. *et al.* (2022) Epigenetic regulation of salinity stress responses in cereals. *Molecular Biology Reports* 49(1), 761–772. DOI: 10.1007/s11033-021-06922-9.

Rhaman, M.S., Ali, M., Ye, W. and Li, B. (2024) Opportunities and challenges in advancing plant research with single-cell omics. *Genomics, Proteomics & Bioinformatics* 22(2), qzae026. DOI: 10.1093/gpbjnl/qzae026.

Roy, S. and Soni, P. (2022) Unraveling the epigenetic landscape for salt tolerance in plants. *International Journal of Plant Biology* 13(4), 443–462. DOI: 10.3390/ijpb13040036.

Schmid, M., Durussel, T.R.S. and Laemmli, U.K. (2004) Technique ChIC and ChEC: Genomic mapping of chromatin proteins. *Molecular Cell* 16, 147–157.

Schreiber, J., Wescoe, Z.L., Abu-Shumays, R., Vivian, J.T., Baatar, B. *et al.* (2013) Error rates for nanopore discrimination among cytosine, methylcytosine, and hydroxymethylcytosine along individual DNA strands. *Proceedings of the National Academy of Sciences of the United States of America* 110(47), 18910–18915. DOI: 10.1073/pnas.1310615110.

Shen, X., De Jonge, J., Forsberg, S.K.G., Pettersson, M.E., Sheng, Z. *et al.* (2014) Natural CMT2 variation is associated with genome-wide methylation changes and temperature seasonality. *PLoS Genetics* 10, 12. DOI: 10.1371/journal.pgen.1004842.

Shi, D.Q., Ali, I., Tang, J. and Yang, W.C. (2017) New insights into 5hmc DNA modification: Generation, distribution and function. *Frontiers in Genetics* 8, 1–11. DOI: 10.3389/fgene.2017.00100.

Simon, L. and Probst, A.V. (2024) Maintenance and dynamic reprogramming of chromatin organization during development. *The Plant Journal* 118(3), 657–670. DOI: 10.1111/tpj.16119.

Simpson, J.T., Workman, R.E., Zuzarte, P.C., David, M., Dursi, L.J. *et al.* (2017) Detecting DNA cytosine methylation using nanopore sequencing. *Nature Methods* 14(4), 407–410. DOI: 10.1038/nmeth.4184.

Singh, P., Yekondi, S., Chen, P.W., Tsai, C.H., Yu, C.W. *et al.* (2014) Environmental history modulates *Arabidopsis* pattern-triggered immunity in a HISTONE ACETYLTRANSFERASE1-dependent manner. *Plant Cell* 26(6), 2676–2688. DOI: 10.1105/tpc.114.123356.

Singroha, G., Kumar, S., Gupta, O.P., Singh, G.P. and Sharma, P. (2022) Uncovering the epigenetic marks involved in mediating salt stress tolerance in plants. *Frontiers in Genetics* 13, 1–10. DOI: 10.3389/fgene.2022.811732.

Skene, P.J. and Henikoff, S. (2017) An efficient targeted nuclease strategy for high-resolution mapping of DNA binding sites. *eLife* 6, 1–35. DOI: 10.7554/eLife.21856.

Song, J., Henry, H. and Tian, L. (2020) Drought-inducible changes in the histone modification. *Plant Biology* 22(3), 433–440. DOI: 10.1111/plb.13057.

Song, Y., He, J., Guo, J., Xie, Y., Ma, Z. *et al.* (2024) The chromatin remodeller MdRAD5B enhances drought tolerance by coupling MdLHP11-mediated MdLhHP1 in apple. *Plant Biotechnology Journal* 22(3), 617–634. DOI: 10.1111/pbi.14210.

Strenkert, D., Schmollinger, S., Sommer, F., Schulz-Raffelt, M. and Schroda, M. (2011) Transcription factor-dependent chromatin remodeling at heat shock and copper-responsive promoters in *Chlamydomonas reinhardtii*. *The Plant Cell* 23(6), 2285–2301. DOI: 10.1105/tpc.111.085266.

Stroud, H., Do, T., Du, J., Zhong, X., Feng, S. *et al.* (2014) Non-CG methylation patterns shape the epigenetic landscape in *Arabidopsis*. *Nature Structural & Molecular Biology* 21(1), 64–72. DOI: 10.1038/nsmb.2735.

Sun, L., Song, G., Guo, W., Wang, W., Zhao, H. *et al.* (2019) Dynamic changes in genome-wide histone3 lysine27 trimethylation and gene expression of soybean roots in response to salt stress. *Frontiers in Plant Science* 10, 1–18. DOI: 10.3389/fpls.2019.01031.

Surdonja, K., Eggert, K., Hajirezaei, M.-R., Harshavardhan, V., Seiler, C. *et al.* (2017) Increase of DNA methylation at the Hvckx2.1 promoter by terminal drought stress in barley. *Epigenomes* 1(2), 9. DOI: 10.3390/epigenomes1020009.

Susan, J.C., Harrison, J., Paul, C.L. and Frommer, M. (1994) High sensitivity mapping of methylated cytosines. *Nucleic Acids Research* 22(15), 2990–2997. DOI: 10.1093/nar/22.15.2990.

Swiezewski, S., Liu, F., Magusin, A. and Dean, C. (2009) Cold-induced silencing by long antisense transcripts of an *Arabidopsis* Polycomb target. *Nature* 462(7274), 799–802. DOI: 10.1038/nature08618.

To, T.K. and Kakutani, T. (2022) Crosstalk among pathways to generate DNA methylome. *Current Opinion in Plant Biology* 68, 102248. DOI: 10.1016/j.pbi.2022.102248.

Vaisvila, R., Chaithany Ponnaluri, V.K., Sun, Z., Langhorst, B.W., Saleh, L. *et al.* (2021) Enzymatic methyl sequencing detects DNA methylation at single-base resolution from picograms of DNA. *Genome Research* 31(7), 1280–1289. DOI: 10.1101/gr.266551.120.

van Dijk, K., Ding, Y., Malkaram, S., Riethoven, J.-J.M., Liu, R. *et al.* (2010) Dynamic changes in genome-wide histone H3 lysine 4 methylation patterns in response to dehydration stress in *Arabidopsis thaliana*. *BMC Plant Biology* 10, 1–12. DOI: 10.1186/1471-2229-10-238.

Varotto, S., Krugman, T., Aiese Cigliano, R., Kashkush, K., Kondić-Špika, A. *et al.* (2022) Exploitation of epigenetic variation of crop wild relatives for crop improvement and agrobiodiversity preservation. *Theoretical and Applied Genetics* 135(11), 3987–4003. DOI: 10.1007/s00122-022-04122-y.

Wang, X., Song, S., Wu, Y.S., Li, Y.L., Chen, T. *et al.* (2015) Genome-wide mapping of 5-hydroxymethylcytosine in three rice cultivars reveals its preferential localization in transcriptionally silent transposable element genes. *Journal of Experimental Botany* 66(21), 6651–6663. DOI: 10.1093/jxb/erv372.

Wang, Y., Sun, F., Cao, H., Peng, H., Ni, Z. *et al.* (2012) TamiR159 directed wheat TaGAMYB cleavage and its involvement in anther development and heat response. *PloS ONE* 7(11), e48445. DOI: 10.1371/journal.pone.0048445.

Wang, Z., Liu, M., Lai, F., Fu, Q., Xie, L. *et al.* (2023) AraENCODE: A comprehensive epigenomic database of *Arabidopsis thaliana*. *Molecular Plant* 16(7), 1113–1116. DOI: 10.1016/j.molp.2023.06.005.

Xiao, M., Wang, J. and Xu, F. (2022) Methylation hallmarks on the histone tail as a linker of osmotic stress and gene transcription. *Frontiers in Plant Science* 13, 1–14. DOI: 10.3389/fpls.2022.967607.

Xin, M., Wang, Y., Yao, Y., Xie, C., Peng, H. *et al.* (2010) Diverse set of microRNAs are responsive to powdery mildew infection and heat stress in wheat (*Triticum aestivum* L.). *BMC Plant Biology* 10, 123. DOI: 10.1186/1471-2229-10-123.

Xu, J., Wang, Q., Freeling, M., Zhang, X., Xu, Y. *et al.* (2017) Natural antisense transcripts are significantly involved in regulation of drought stress in maize. *Nucleic Acids Research* 45(9), 5126–5141. DOI: 10.1093/nar/gkx085.

Yue, X., Xie, Z., Li, M., Wang, K., Li, X. *et al.* (2022) Simultaneous profiling of histone modifications and DNA methylation via nanopore sequencing. *Nature Communications* 13(1), 1–14. DOI: 10.1038/s41467-022-35650-2.

Yun, P., Kaya, C. and Shabala, S. (2024) Hormonal and epigenetic regulation of root responses to salinity stress. *The Crop Journal* 12(5), 1309–1320. DOI: 10.1016/j.cj.2024.02.007.

Yung, W.S., Li, M.W., Sze, C.C., Wang, Q. and Lam, H.M. (2021) Histone modifications and chromatin remodelling in plants in response to salt stress. *Physiologia Plantarum* 173(4), 1495–1513. DOI: 10.1111/ppl.13467.

Zhang, C.Y., Wang, N.N., Zhang, Y.H., Feng, Q.Z., Yang, C.W. *et al.* (2013) DNA methylation involved in proline accumulation in response to osmotic stress in rice (*Oryza sativa*). *Genetics and Molecular Research* 12(2), 1269–1277. DOI: 10.4238/2013.april.17.5.

Zhang, H., Lang, Z. and Zhu, J.K. (2018) Dynamics and function of DNA methylation in plants. *Nature Reviews Molecular Cell Biology* 19(8), 489–506. DOI: 10.1038/s41580-018-0016-z.

Zhang, H., Gong, Z. and Zhu, J.K. (2022) Active DNA demethylation in plants: 20 years of discovery and beyond. *Journal of Integrative Plant Biology* 64(12), 2217–2239. DOI: 10.1111/jipb.13423.

Zhang, W., Han, Z., Guo, Q., Liu, Y., Zheng, Y. *et al.* (2014) Identification of maize long non-coding RNAs responsive to drought stress. *PLoS ONE* 9(6), e98958. DOI: 10.1371/journal.pone.0098958.

Zhang, Y., Zhang, Q., Yang, X., Gu, X., Chen, J. *et al.* (2023) 6mA DNA methylation on genes in plants is associated with gene complexity, expression and duplication. *Plants* 12, 1949.

Zhao, L., Xie, L., Zhang, Q., Ouyang, W., Deng, L. *et al.* (2020) Integrative analysis of reference epigenomes in 20 rice varieties. *Nature Communications* 11(1), 1–16. DOI: 10.1038/s41467-020-16457-5.

Zhong, S.H., Liu, J.Z., Jin, H., Lin, L., Li, Q. *et al.* (2013) Warm temperatures induce transgenerational epigenetic release of RNA silencing by inhibiting sirna biogenesis in *Arabidopsis*. *Proceedings of the National Academy of Sciences of the United States of America* 110(22), 9171–9176. DOI: 10.1073/pnas.1219655110.

Zhong, X. (2016) Comparative epigenomics: A powerful tool to understand the evolution of DNA methylation. *The New Phytologist* 210(1), 76–80. DOI: 10.1111/nph.13540.

Zhu, J.Y., Sun, Y. and Wang, Z.Y. (2012) Genome-wide identification of transcription factor-binding sites in plants using chromatin immunoprecipitation followed by microarray (ChIP-chip) or sequencing (ChIP-Seq). *Methods in Molecular Biology* 876(1), 173–188. DOI: 10.1007/978-1-61779-809-2_14.

Zong, W., Zhong, X., You, J. and Xiong, L. (2013) Genome-wide profiling of histone H3K4-tri-methylation and gene expression in rice under drought stress. *Plant Molecular Biology* 81(1–2), 175–188. DOI: 10.1007/s11103-012-9990-2.

Zong, W., Yang, J., Fu, J. and Xiong, L. (2020) Synergistic regulation of drought-responsive genes by transcription factor OsbZIP23 and histone modification in rice. *Journal of Integrative Plant Biology* 62(6), 723–729. DOI: 10.1111/jipb.12850.

20 The Integration of Epigenetics and Epigenomics with Functional Genomics in Plants

Sitanshu Sekhar Satapathy, Sushil Barik, Manaswini Mahapatra, Madhuri Pradhan, Siddhartha Shankar Sharma and Jyoti Prakash Sahoo*
Faculty of Agriculture and Allied Sciences, C.V. Raman Global University, Bhubaneswar, India

Abstract

The integration of epigenetic and epigenomic data with functional genomics has become a pivotal approach to understanding the complex regulatory mechanisms governing plant development, responses to environmental stress, and overall genome dynamics. Epigenetics involves heritable changes in gene expression that do not alter the underlying DNA sequence, encompassing DNA methylation, histone modifications, and small RNA-mediated processes. Epigenomics, on the other hand, refers to the comprehensive analysis of epigenetic modifications on a genome-wide scale. The elucidation of epigenetic regulatory networks in plants has practical implications for crop improvement, as it enables the identification of epigenetic markers associated with desirable traits and the development of targeted breeding strategies. Additionally, the integration of epigenomic data with functional genomics enhances our ability to predict and manipulate gene expression patterns for desired outcomes in agriculture, conservation, and biotechnology. It is crucial to developing resilient crops and conservation of biodiversity. There are various tools and methods for interpreting and analyzing the data, such as whole-genome bisulfite sequencing (WGBS), chromatin immunoprecipitation followed by sequencing (ChIP-seq), small RNA sequencing, which is sequencing of RNA that provides insight into gene expression, and Cas9-fusion proteins, which is a CRISPR/Cas9-based epigenomic editing tool. There are also different tools for specific methods for analyzing and studying genetic expression and regulating data. This chapter is based on the study of epigenetic modifications in plants and their impacts on environments, changes in regulatory mechanisms of gene expression, and how their response to environmental stimuli works, and, according to this, finding the improvement in plants at the genetic level.

Keywords: Epigenetic changes, Epigenomic analysis, Crop improvement, CRISPR/Cas9 epigenomic editing, Gene expressions

20.1 Introduction

The main purpose of crop improvement in agriculture is to develop different varieties to prevent the loss of yields, stress resistance crops, disease resistance crop varieties, better quality crops, higher nutrition crops, etc. There are many reasons and purposes for performing crop improvement mechanisms. These occur at the genetic level or modifications in gene products

*Corresponding author: jyotiprakashsahoo2010@gmail.com

DOI: 10.1079/9781800626102.0020

and contents (Kumar, 2019). Epigenetics is the heritable changes in the gene expressions without changes in the DNA sequence but modifications of non-DNA sequence contents like DNA methylation, polyadenylations, histone modification, non-coding RNAs, etc. Epigenetic processes are essential for crop improvement because they may protect plants indirectly from biotic and abiotic stresses or any environmental changes (Rival *et al.*, 2010). Epigenetics includes meiotically and mitotically heritable changes in the expression of genes that are not directly coded with the DNA sequences (Tsaftaris *et al.*, 2005).

Epigenetic changes lead to changes in biochemical processes, modifications in structures, hormonal changes, and many more. Epigenetic changes may be inherited by the next generations but do not involve changes in DNA sequences. Epigenetic changes are acquired variations (Lacal and Ventura, 2018). Some major epigenetic changes include DNA methylation, with the addition of methyl group at the 5' cytosine bases in CpG island. This can repress or activate gene expression (Jones and Takai, 2001) and histone modification, as histone is a protein that compresses the long DNA and packaging to chromatin. Chemical modifications like phosphorylation, acetylation, and methylation can alter the chromatic structure and differ in gene expression (Bannister and Kouzarides, 2011). Non-coding RNAs, microRNAs, and long non-coding RNAs interact with the mRNA and affect gene expression and translation mechanism (Rinn and Chang, 2012). Chromatin remodeling complexes influence the DNA transcriptional machinery and may activate gene expression or repress gene expression (Narlikar *et al.*, 2013).

Some epigenetic modifications can be inherited in the next generations like transgenerational inheritance, in which epigenetic modifications like histone modification or DNA methylation in the germline can be inherited in offspring and also affect gene expressions (Bošković and Rando, 2018). With parental imprinting, imprinted genes are the genes that are copies of only one parent and get expressed, and it is present in the germline and maintained in offspring (Surani, 1994). Different environmental factors influence the epigenetic changes that will be inherited by the next generations. Epigenomics is the analysis and study of the effects of chromatin structures like DNA packaging, nuclear matrix, and covalent histone modifications like methylation, acetylation, ubiquitination, and phosphorylation on any non-DNA modifications (Fazzari and Greally, 2010). The epigenome plays a crucial role in cellular differentiation and gene regulation. It consists of the chemical modification of DNA molecules and their associated proteins, like histones (Bird, 2007).

Functional genomics is the study of genes and different intergenic regions of the genomes that contribute to different biological processes. It is typically the study of genes on wide scales and their regions in detail and their specific effects and functions in an organism. Genome sequencing is essential for the study of functional genomics (Fields *et al.*, 1999). It leads to the study of gene expressions, gene regulations, the role of genes, and gene interactions with biological processes. Functional genomics studies are also helpful for developing biotic and abiotic stress resistance crop improvement. It also can be helpful in the analysis of mutations in genomes and epigenetic changes in organisms (Haberkorn *et al.*, 2002). In this context, advances in high-throughput sequencing technologies have revolutionized the field, enabling researchers to explore the plant regulatory landscape at unprecedented resolutions. This chapter aims to provide an overview of the current state of research on the integration of epigenetic and epigenomic data with functional genomics in plants.

20.2 Importance of the Integration of Epigenetics and Epigenomics with Function-Genomics

Integration of epigenetics and epigenomics with functional genomics can lead to the study of the level of expression of genes in several environmental conditions and the interaction of genes with different biological processes in an organism to develop different stress-resistant crops. Some key aspects of these integrations are as follows.

20.2.1 Understanding gene regulation

Epigenetics and epigenomics highlight non-DNA modifications, and integration of these genomic data with the functional genomics to gain comprehensive research on different gene regulations (Allis and Jenuwein, 2016).

20.2.2 Identifying regulatory elements

Different techniques of epigenomics like DNA methylation profiling and ChIP-seq and ATAC-seq help in identifying the promoter, enhancer and regulatory elements. Integrating these data with functional genomics can connect the regulatory element with its genes (Kundaje *et al.*, 2015).

20.2.3 Phenotype association and disease

Due to many diseases, several epigenetic changes occur. Integrating epigenetic changes and functional genomic data can lead to analyzing the effect of gene expression and gene regulation and also to the study of the biochemical changes in an organism (Jones and Baylin, 2007).

20.2.4 Uncovering non-coding RNAs

In gene regulation, epigenetic changes also have a role in modifications of non-coding RNAs like microRNAs and long non-coding RNAs. Functional genomics can target these non-coding RNAs and make them functional (Brevini *et al.*, 2016; Wei *et al.*, 2017).

20.2.5 Epigenomics of environments

There should be analysis of how an environmental factor affects epigenetic modification and changes a gene expression; and how epigenetic changes interact with environmental factors and are critical for an environment, like modifications in DNA sequences and epi-mutation, and also the development of stress resistance changes in plants, which is also called acquired adaptive changes (Meyer, 2015).

20.3 Various Roles of Epigenetics in Plant Development and Stress Responses

In their lifespan, plants undergo various epigenetic changes that lead to survival and functioning in different types of environmental conditions, like temperature change, overcoming many diseases, scarcity conditions, and many others. These changes in gene expressions may be transferred to the next generation. Epigenetics plays several roles in the biological phenomena of plants, such as histone modification, silencing of transgenes and transposons, and regulations of gene expressions and their development (Ito, 2014). Epigenetic modifications include DNA methylation, where a methyl group becomes attached to a DNA molecule at a cytosine residue and plays key roles in gene silencing. It can be said that this modification directly regulates the gene expression (Zhang *et al.*, 2006).

20.3.1 Histone modification

Histone is a protein around which a DNA molecule is wrapped and helps in DNA packaging, and modification in this protein can affect gene accessibility. It plays a leading role in the activation or repression of genes (Fig. 20.1) (Hewezi, 2018).

20.3.2 Determination of cell fate

Epigenetic modification plays a crucial role in the determination of cell fate in plant development. The cell can undergo several types of differentiation into specific types of cells; in the meantime, epigenetic modifications have been induced, and then cell identity is established. In *Arabidopsis*, epidermal cells become either trichoblasts (hair-forming cells) or atrichoblasts (non-hair-forming); two types of trichoblast are formed into either trichomes or root hairs, both cell types being specified by the action of transcriptional regulators that define cell pattern (Guimil and Dunand, 2006).

20.3.3 Paramutation

Paramutation is the interaction of two alleles in a single locus where one allele induces

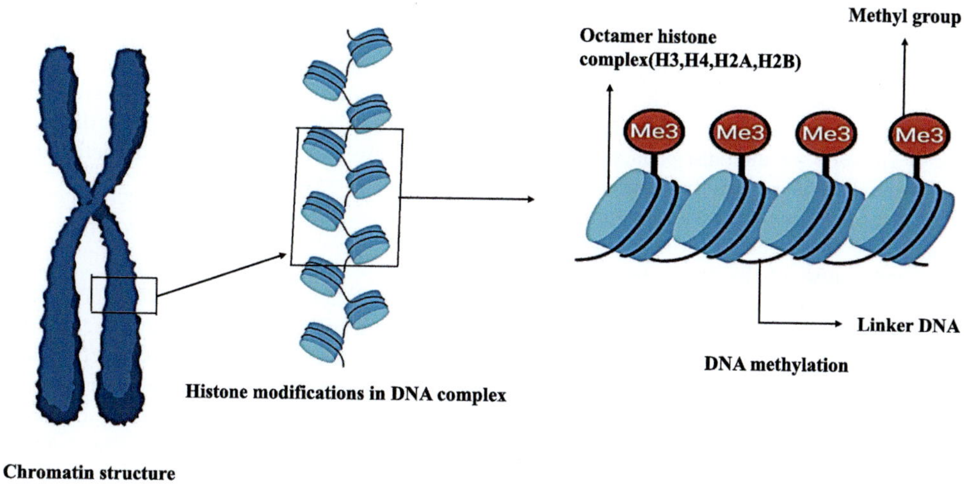

Fig. 20.1. Histone modification, which is an epigenetic change. Methylation and acetylation occur in histone proteins. Histone is an octamer structure. Author's own image.

heritable changes in the other allele (Hollick and Chandler, 2001). It is an epigenetic phenomenon involving modification of gene expression and transmission mitotically or meiotically. It was first observed in the case of plants (*Zea mays*) and later in mammals (Stam, 2009).

20.3.4 Response to environmental changes

Epigenetic changes in plants due to several changes in environmental conditions play a crucial role in the adaptation of plants, such as changes in acetylation histone modification in *Zea mays* in response to temperature fluctuation. *Arabidopsis* plants have epigenetic modification as DNA methylation in response to salt stresses and changes in light conditions (Turner, 2009; Meyer, 2015).

20.4 Major Epigenetic Modifications in Plants Due to Stresses

Some major epigenetic modifications are histone modification, paramutation, and hyper- and hypomethylation. In hypermethylation, stress conditions increase the methylation in the cytosine residue, which represses the gene expression; in hypomethylation, it increases the stress response gene expression (Klengel *et al.*, 2014). The SWI/SNF complexes are multi-subunit complexes involved in accessibility to remodeling of nucleosomes with the help of ATP, so they modulate the gene expression in response to various stresses or differences in any biological and biochemical process (Masliah-Planchon *et al.*, 2015). However, m6A is N6-methyladenosine that affects the stability of RNA and also changes translationally by methylation in RNA adenosine residue. It is a stress-induced change in the genome that modifies gene expression (Yue *et al.*, 2019). Paramutation, a non-Mendelian inheritance, is the stable heritable change in genetic expression on stress response in plants.

20.5 Epigenomics in Plants

Epigenomics is the study of epigenetic modifications, including various forms of growth and development according to different environmental impacts, responses to stimuli, physiological changes and adaptation of plants, which lead to changes in gene expressions. In various plant species, it was found that asexually reproducing clones by mitotic division can inherit epigenetic traits (Köhler and Springer, 2017). Epigenomic studies are used for multidisciplinary purposes. The epigenomic approach helps to study the adaptation to various stresses and environmental changes. For the development of climate-smart crops, epigenomes of the crops serve as a good source for studying the variations and modifications. Epigenomic engineering also has a crucial role in developing sustainable epi-markers linked with specific agronomic traits, making it easy to study the specific tolerance characters and then perform recombinant techniques (Singh *et al.*, 2021). Epigenomic natural variations occur by evolution, which is important in phenotypic diversity and also contributes to molecular change mechanisms like heterosis in plants. The genome stability and epigenomic regulation in chromatin structures are responsible for carrying genomic information and the information expression of various phenotypic characters (He *et al.*, 2011).

20.6 Tools and Methods for Studying Epigenomics

Plant genomic mapping can be obtained from the combination of chromatic technologies and genomic tiling microarrays and sequencing of chromatin to study genetic and epigenetic variations. The epigenetic modifications control the development of transcriptomics activities (He *et al.*, 2011). Epigenomics is the study of epigenetic changes that alter and modify gene expression. There are various tools and methods used to study specific epigenetic changes and modifications.

20.6.1 Techniques for DNA methylation analysis

Bisulfite sequencing

The bisulfite sequencing method is performed to analyze the methylation activities on the genome of interest. It is a high-throughput method of sequencing. When combined with restriction enzymes and bisulfite treatment, these techniques can deal with the genome sequence that has high CpG content. In this technique, unmethylated cytosines are converted into uracil and it is helpful in the differentiation between methylated and unmethylated cytosines (Wreczycka *et al.*, 2017).

MeDIP

MeDIP (methylated DNA immunoprecipitation) is a methylated DNA detecting method. It is helpful in the differentiation of methylated and non-methylated DNA and also extracts the methylated DNA from the sample of interest. Sonicated DNA is added with the monoclonal antibody which binds with methylated DNA. Non-methylated DNA remains unbound (Fig. 20.2) (Thu *et al.*, 2009).

20.6.2 Histone modification analysis

Chromatin immunoprecipitation (ChIP)

ChIP is an advanced method for the analysis of the protein–DNA interactions and density of a specific protein. It involves crosslinking, immunoprecipitation, and sequencing to identify the specific regions in the genome of histone modification (Fig. 20.3) (Nelson *et al.*, 2006).

Assay for transposase accessible chromatin sequencing (ATAC-seq)

ATAC-seq identifies the chromatin regions by the involvement of Tn5 transposases to tag accessible DNA. It is a method of chromatin mapping accessible genome-wide. It maps the transcription factors (TFs) and nucleosome positions in the genome (Fig. 20.4) (Buenrostro *et al.*, 2015).

20.6.3 Non-coding RNA analysis

Small RNA sequencing

Small RNA sequencing is a deep sequencing method which was discovered by Sanger

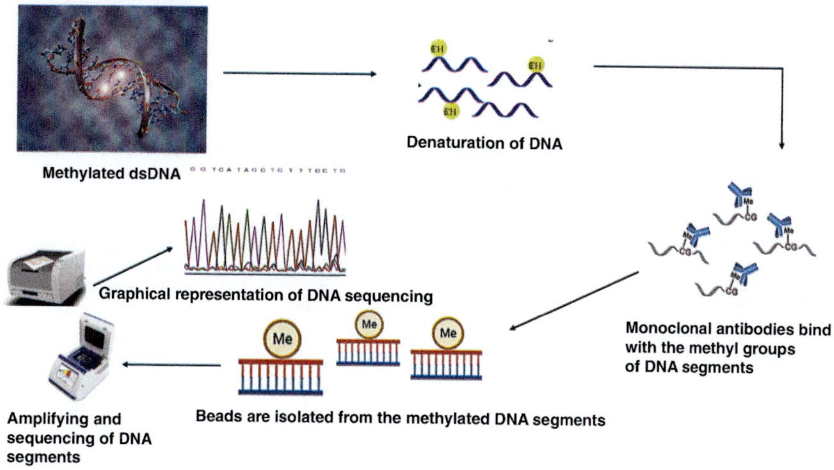

Fig. 20.2. The MeDIP method is performed to differentiate the methylated and non-methylated DNA sequences, which is an epigenetic modification. Author's own image.

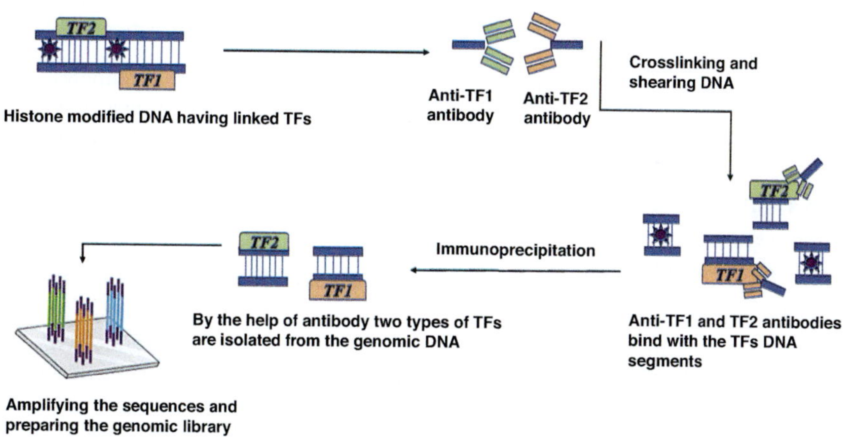

Fig. 20.3. The chromatin immunoprecipitation method, for isolating the TFs sequenced DNA regions, which is an epigenetic modification that amplifies the sequences and makes it a genomic library. Author's own image.

and also depends upon Sanger sequencing, like miRNAs, siRNAs and piRNAs. It is the quantification and sequencing of small RNA for analysis and modification. It is commonly used in microarray analysis (McCormick *et al.*, 2011).

Long non-coding RNAs (lncRNAs)

Long non-coding RNAs are crucial in the sequencing of non-coding RNAs and analysis of the activities of RNAs. They have various roles, like repressors, gene activators, or *cis* and *trans*

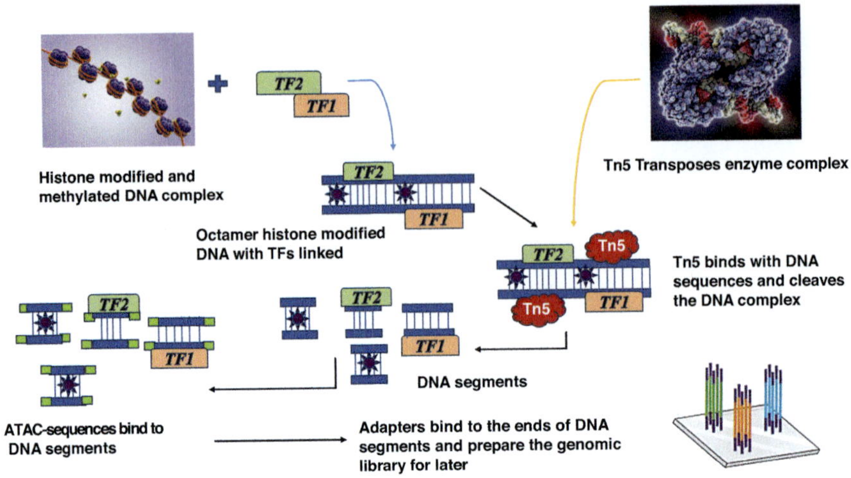

Fig. 20.4. ATAC-sequencing method performed for isolating the TFs sequences in DNA, which are the epigenetic modifications, and developing a genomic library for later use in crop improvements. Author's own image.

gene expression regulators. They have a crucial role in the modification of gene expression and epigenetic changes (Da Sacco *et al.*, 2012).

20.6.4 Integrated approaches

Epigenome-wide association studies (EWAS)

EWAS gives the concept of genetic epidemiology studies. This study is a modification of GWAS (Genome-Wide Association Studies). It specifically focuses on DNA methylation and helps in sample selection, confounding, and sample heterogeneity. It plays a major role in the analysis of specific genome disorders and traits (Tanić, 2020).

Multi-omics data integration

This is a high-throughput method, which is integrative to the analysis of a large sample of interest and provides omics data. It highlights the secondary metabolites and various biological processes, biomolecules and their functions. This integration provides data on transcriptomics, metabolomics, proteomics, and genomics as other cellular processes (Subramanian *et al.*, 2020).

Various bioinformatics tools

There are various tools for analysis and to provide mapping data of epigenomics, such as UCSC Genome Browser, Rnbeads, IGV, etc. (Angarica and Del Sol, 2017).

20.6.5 Single-cell epigenomics

Single-cell RNA sequencing and ATAC sequencing play a crucial role in studying transcriptomic analysis, and provide insight into cellular heterogeneity. In *Arabidopsis*, sNucRNA-seq and sNucATAC-seq technologies were performed to analyze the protoplast transcriptomes and cellular processes (Farmer *et al.*, 2021). CRISP/Cas9 genome editing technologies are also used for editing at the target epigenetic sites and modifications, along with isolating the specific trait or sequences to study the various consequences and activities (Kang *et al.*, 2019). There are various methods and protocols to analyze epigenetics or epigenomics. The integrations of these methods and analysis result from data in both epigenetic, epigenomics, and functional genomics can enable the development of an advanced plant variety for any specific stresses or ailments.

20.7 Tools and Technologies Used in Studying Functional Genomics in Plants

Functional genomics is the field of molecular biology where study and research about gene expression, functions and interaction within organisms take place. Various methods and tools are utilized in this field to analyze and interpret the genomic data of specific functions, regulations, and expressions of genes (Rai *et al.*, 2019). Various methods are included in the study of functional genomics.

20.7.1 Microarrays

In the functional genomics field of study, two types of microarrays are used: (i) DNA microarrays, which measure the expression level of thousands of DNA sequences or genes simultaneously; and (ii) RNA microarrays, which analyze the various RNA molecules' expressions and regulations, like mRNAs, miRNAs (Aharoni and Vorst, 2002).

20.7.2 Next-generation sequencing (NGS)

In these technologies, sequencer tools are used for studying or mapping a specific sequence of interest. Sequencers include Illumina sequencing, whole-genome sequencing, and many more. RNA sequencing is used to sequence the RNA molecules and study the gene expressions, transcripts and splicing (Kukurba and Montgomery, 2015). Sequencing of the entire genome and identifying the mutations, specific regulations and QTLs by DNA-seq. ChIP-sequencing is performed for the study and mapping of protein–DNA interactions and their regulation (Ben-Amar *et al.*, 2016).

20.7.3 RNA interference

Short hairpin RNA (shRNA) and small interference RNA (siRNA) are used to silence specific genes or target sequences and help to study the various consequences of the genes (Matthew, 2004).

20.7.4 Flow cytometry

This is a high-throughput method of analysis of the cells and their characters, based on cellular activities, regulations, gene expressions, and protein functions. These methods are used to understand the nuclear genomes and their relation with epigenetic functions (Zwyrtková *et al.*, 2021).

20.8 Integration of Epigenetics and Epigenomics with Functional Genomics

The combination of epigenetic studies and genomic studies within the organisms is a powerful approach. It is a multidisciplinary approach that plays a crucial role in advancements of cellular processes and analysis of epigenetic modification, then is applied to the enhancement of any phenotypic characteristics of plants. This field of study can examine the critical responses of various plants due to environmental impacts (de Los Reyes, 2019). There are lots of multidisciplinary approaches to this type of integration.

20.8.1 Genomics studies and bioinformatic tools

These contribute to the understanding of various genomic DNA sequences that interact with epigenetic changes and their expression. They are also crucial in the evolution of new bioinformatic tools that help create data and represent the findings of this type of integration (Angarica and Del Sol, 2017).

20.8.2 Molecular biology and epigenetic techniques

Various techniques such as bisulfite sequencing, ChIP sequencing and many more technologies have been used for the analysis and collection of data for the integration of epigenetics and epigenomics with functional genomic studies. This phenomenon also provides new ideas for developing new technologies for epigenetic studies and cellular activities (Cokus *et al.*, 2008).

20.8.3 Interaction of environmental studies with plant physiology

This also helps to research the connection between plant physiology and environmental studies and its impacts on plants. It includes understanding the physiological activities of plants in different environmental conditions and metabolism or studying the metabolomics of the plant (Chinnusamy *et al.*, 2007).

20.8.4 Computational biology and machine learning

This is the advancement of the study of epigenetic fields and genomics by evolutionary creations of technologies and machines with new ideas and ways. It entails advancing the machines and computational methods for getting the perfect omics data (Angermueller *et al.*, 2016).

20.8.5 Crop improvement and agronomy

Understanding the interaction of epigenetics and epigenomics and its integration gives high quality and quantity plant products. This integration can develop a stress tolerance plant variety, disease-free crops, enhanced nutritional content plants and many more (Varshney *et al.*, 2014).

20.8.6 Transcriptomics, proteomics, metabolomics, and functional genomics studies

Epigenetic changes and modification happen at transcription and translational levels. This integration mechanism can collaborate with transcriptomics, proteomics, and metabolomics studies. RNA molecules can be expressed into specific proteins that more or less play a role in metabolic functions and cellular activities within the organisms, and therefore, transcriptomic data collected from RNA profiling is likely to be valuable in uncovering crucial pathways. Analyzing these processes can perform the integration of epigenetics and epigenomics with functional genomics in plants (Wang *et al.*, 2009). There are many more multidisciplinary approaches to this integration mechanism, including advancing various methods for understanding the interaction of epigenetics, epigenomics, and functional genomics expression in plants. It includes sustainable agriculture and crop improvements.

20.9 Tools and Methodologies Performed for Integration of Epigenetics and Epigenomics with Functional Genomics in Plants

For integration, various bioinformatics tools and methodologies are used to generate and analyze data. Some methods include bisulfite sequencing; where methylated DNA is isolated and used for further advancement of plants, ChIP sequencing; identify the sequences associated with histone and TFs, ATAC sequencing, MeDIP sequencing, CRISPR/Cas9 genome editing technologies and many more methods were performed for isolating the modified sequences or epigenetic changes and utilizing them for next-generation increasing of crop productivity (Arora and Tollefsbol, 2021). There are many bioinformatics tools for collecting and interpreting the data, like Bismark and STAR for aligning the genomic sequencing data. HOMER and MACS tools are used in representing the data of ChIP and ATAC sequencing; Lemma and edgeR tools are used for representing the data of RNA sequencing and also define the rate of expressing; IGV (Integrative Genomic Viewer) and PyTorch computational tools classify and integrate the multiple omics data (Sheu *et al.*, 2024).

20.9.1 Challenges for the integration of epigenetics and epigenomics with functional genomics in plants

There are various challenges for these complex molecular mechanism and regulatory processes.

Various obstacles and challenges are formed during the integration of epigenetics and epigenomics with functional genomics with plants. There is a need for proper care for modified plants analysis of the phenotypical characters and many more (Tuberosa *et al.*, 2005).

Data integration and analysis

Understanding the omics data or integrated omics data is slightly complicated and so there are chances of failure in analysis and in processing and interpreting the large-scale genomic data sets (Han and He, 2016).

Dynamic nature of epigenetics modification

Dynamically epigenetic changes in various developmental changes and environmental impacts include epigenetic changes that depend upon the cell fate transitions (Lee and Seo, 2018).

Epigenetic variations and inheritance

This is studying and analyzing the epigenetic variations and their phenotypical nature in plants. The epigenetic modifications are also inherited by the next generations accurately or show variations. This slightly affects the integration mechanism (Jablonka and Lamb, 1989; Nikolayeva and Robinson, 2014).

Technological limitations

The limitations of tools and technologies also have an impact on data analysis or mapping. High-throughput sequencing and comprehensive analysis also differ, due to limitations in technologies (Meissner, 2010).

Cross-species comparison

After experimentation on a model plant and then applying that finding to crops, sometimes crops are unable to adapt or acquire the mechanism or modification. Due to species differences, the integral experiment may be unsuccessful (Leitch and Leitch, 2012).

20.9.2 Applications of integration of epigenetic and epigenomics with functional genomics in plants

Plant breeding and crop improvements

- **Precision breeding for improved traits:** This approach enhances the efficiency of breeding programs by providing molecular-level insights into trait variation (Zhang *et al.*, 2014).
- **Enhanced stress tolerance:** Understanding the epigenetic basis of stress responses in plants aids in developing crops with improved tolerance to environmental challenges. Epigenomic information guides the selection of genomic regions for manipulation to enhance stress resilience (Shen *et al.*, 2014).
- **Epigenetic markers for marker-assisted selection (MAS):** Identification of reliable epigenetic markers facilitates MAS in plant breeding. These markers, based on epigenomic data, enable breeders to select plants with desired traits more accurately.
- **Accelerated crop improvement:** Epigenomics expedites crop improvement by revealing epigenetic modifications linked to key agronomic traits. This knowledge streamlines the breeding process, reducing the time required to develop improved varieties (Kawakatsu *et al.*, 2016).
- **Unraveling complex traits through epigenomic insights:** Integration of epigenetic data with functional genomics aids in unraveling the complexity of traits controlled by multiple genes. This knowledge is crucial for developing crops with enhanced nutritional content or complex disease resistance (Li *et al.*, 2014).
- **Environmental adaptation strategies:** Epigenetic and epigenomic integration provides insights into how plants adapt to changing environments. This knowledge informs breeding strategies focused on developing crops capable of thriving under diverse environmental conditions (Richards *et al.*, 2017).
- **Transgenerational epigenetic inheritance:** Understanding transgenerational epigenetic effects allows for the development

of crops with improved heritability of desirable traits. This has implications for sustainable and resilient agriculture (Verhoeven *et al.*, 2016).

- **Disease resistance strategies:** Epigenomic insights into plant–pathogen interactions guide the development of crops with enhanced disease resistance. Targeting specific epigenetic mechanisms involved in defense responses improves crop protection (Dowen *et al.*, 2012).
- **Epigenetic engineering for biotechnological advances:** This approach allows for the precise manipulation of epigenetic marks to modulate gene expression patterns for specific agricultural or industrial purposes (Law and Jacobsen, 2010).

Response of plants to environmental changes

- **Environmental memory and response:** Epigenetic and epigenomic integration reveals how plants establish the memory of past environmental exposures, influencing their responses to subsequent changes (Lämke and Bäurle, 2017).
- **Adaptive epigenetic changes:** Plants exhibit adaptive epigenetic changes in response to environmental cues, influencing gene expression patterns for better adaptation (Byeon *et al.*, 2019).
- **Epigenetic regulation of stress-responsive genes:** Integration of epigenetic and functional genomic data elucidates the regulation of stress-responsive genes, providing insights into plant resilience mechanisms (Zhang *et al.*, 2018).
- **Transgenerational effects on stress tolerance:** Epigenetic changes induced by environmental stress can be inherited across generations, contributing to transgenerational stress tolerance (Luna *et al.*, 2012).
- **Epigenetic control of phenotypic plasticity:** Plants employ epigenetic mechanisms to regulate phenotypic plasticity, allowing them to adapt to diverse environmental conditions (Cortijo *et al.*, 2014).
- **DNA methylation dynamics under stress:** Epigenomic studies reveal dynamic changes in DNA methylation patterns under stress, influencing gene expression and plant responses (Zhang *et al.*, 2018).
- **Epigenetic regulation of abiotic stress responses:** Integration of epigenetic and functional genomic information highlights the role of epigenetic regulation in abiotic stress responses, such as drought and salinity (Kim *et al.*, 2012).
- **Epigenetic marks as environmental sensors:** Epigenetic marks serve as environmental sensors, allowing plants to interpret and respond to changing environmental conditions (Dowen *et al.*, 2012).

Disease and pest resistance plants

- **Identification of epigenetic marks associated with resistance:** Integration of epigenetic and functional genomic data aids in identifying specific epigenetic marks associated with disease and pest resistance in plants (Dowen *et al.*, 2012).
- **Epigenetic regulation of defence-related genes:** Understanding the epigenetic regulation of defense-related genes enables the targeted enhancement of plant defense mechanisms against diseases and pests (Luna *et al.*, 2012).
- **Epigenomic insights into plant–pathogen interactions:** Integration of epigenomic data provides insights into the dynamic changes in DNA methylation and histone modifications during plant–pathogen interactions, revealing potential targets for resistance breeding (Zhang *et al.*, 2018).
- **Transgenerational epigenetic inheritance of resistance traits:** Epigenetic modifications induced by pathogen exposure can be inherited across generations, contributing to transgenerational resistance in plants (Dowen *et al.*, 2012).
- **Epigenetic regulation of plant immune signaling pathways:** Integration of epigenomic and functional genomic data elucidates the epigenetic regulation of immune signaling pathways, providing avenues for manipulating plant defences responses (Yu *et al.*, 2013).

- **Epigenetic engineering for enhanced resistance:** Epigenetic engineering, guided by integrated epigenetic and functional genomic knowledge, allows for targeted modifications to enhance disease and pest resistance in plants (Lang *et al.*, 2017).
- **Epigenetic control of plant–microbe interactions:** Epigenetic and epigenomic integration sheds light on the complex regulatory networks controlling plant-microbe interactions, providing strategies for enhancing beneficial microbe-mediated resistance.
- **Marker-assisted selection for resistant traits:** Identification of epigenetic markers associated with disease and pest resistance facilitates marker-assisted selection in breeding programs, enabling the development of resistant plant varieties (Dowen *et al.*, 2012).

20.10 Future Aspects and Implications of Integration in Plants

The integration of epigenetics and epigenomics with functional genomics mainly focuses on epigenetic modifications due to environmental factors and other cellular processes. These implications can overcome any non-adaptive condition of the environment or any stress conditions (Agarwal *et al.*, 2020). There are many future aspect potentials of the integration of epigenetics and epigenomics with functional genomics in plants.

20.10.1 Crop improvement and precision breeding

Epigenetic modification analysis and gene expression change analysis give a major outcome in crop improvement. By focusing on the epigenetics modification gene expressions like nutrition content, stress tolerance, environmental impact on plants and many more, utilizing these data and records develops advanced plants and also performs precision breeding (Gahlaut *et al.*, 2020).

20.10.2 Adaptation to the environment

Epigenetics and epigenomics changes can adapt to environmental impacts like drought, chilling, and freezing, soil toxicity, etc. It involves modifications in plant responses, plant growth, and development. Identification of markers associated with epigenetic changes can help other plants; or, in the breeding process, plants gain the ability to adapt to extreme environments. Analyzing the metabolism, cellular processes, secondary metabolites, and gene expressions increases the adaptability of plants (Etchegaray and Mostoslavsky, 2016).

20.10.3 Epitranscriptomics changes

The modifications in mRNA and its regulation study are the epi-transcriptomics. Some epi-transcriptomics modifications are N6-methyl adenosine, 5-hydroxy methylcytidine, inosine, etc. These modifications increase the stability of mRNA or gene expression and also enhance the expression of a specific gene. Study and analysis of these modifications can develop bioinformatics tools for creating data and comparing with plant genomes or expressions; advancements in epi-transcriptomics technologies also investigate the functions of a specific modification (Xiong *et al.*, 2017)

20.10.4 Targeting the epigenetic changes

Targeting includes isolating the epigenetic sample of interest and applying it to other plants for the advancement of crop plant development; also focusing on target epigenetic modifications, removing any genetic disorders or abnormal cellular processes (Shin *et al.*, 2022). CRISPR genome editing technology is an advanced method of isolating or regulating the target sequences. This technology is used to modulate the epigenetics and transcriptional activities at specific target sequences (Lee *et al.*, 2019).

20.10.5 Transgenerational epigenetics and long-term effects

Concentration is on the analysis of how an epigenetic modification passes from one generation to the next generations and the way it affects the next-generation plant significantly and effectively. Studying these phenomena can develop a permanent epigenetic modification for long-term effects in plants (Youngson and Whitelaw, 2008)

20.10.6 Developing disease resistance and defence mechanism in plants

Epigenetic modifications and regulations can create defense mechanisms in response to the environment and specific pathogens. Future aspects of these modifications and regulations help to develop disease-resistant plant varieties, enhancing the quality of the fight against the pests and various pathogens in crops (Alvarez *et al.*, 2010). Epigenetic modifications analysis can study the various gene expressions during adverse conditions and also keep that adverse condition-expressed gene for future use or efficient breeding of advanced varieties of crops.

20.11 Conclusion

The integration of epigenetic and epigenomic approaches with functional genomics in plants has revolutionized the understanding of plant biology. This multidimensional perspective not only enriches basic scientific knowledge but also holds promise for addressing global challenges in agriculture and food security. Further research in this field will likely uncover additional layers of complexity in plant regulatory mechanisms, paving the way for innovative solutions in crop science and sustainable agriculture.

References

Agarwal, G., Kudapa, H., Ramalingam, A., Choudhary, D., Sinha, P. *et al.* (2020) Epigenetics and epigenomics: Underlying mechanisms, relevance, and implications in crop improvement. *Functional & Integrative Genomics* 20(6), 739–761.

Aharoni, A. and Vorst, O. (2002) DNA microarrays for functional plant genomics. *Plant Molecular Biology* 48, 99–118.

Allis, C.D. and Jenuwein, T. (2016) The molecular hallmarks of epigenetic control. *Nature Reviews. Genetics* 17(8), 487–500.

Alvarez, M.E., Nota, F. and Cambiagno, D.A. (2010) Epigenetic control of plant immunity. *Molecular Plant Pathology* 11(4), 563–576.

Angarica, V.E. and Del Sol, A. (2017) Bioinformatics tools for genome-wide epigenetic research. In: Delgado-Morales, R. (ed.) *Neuroepigenomics in Aging and Disease*. Springer International, pp. 489–512.

Angermueller, C., Pärnamaa, T., Parts, L. and Stegle, O. (2016) Deep learning for computational biology. *Molecular Systems Biology* 12(7), 878.

Arora, I. and Tollefsbol, T.O. (2021) Computational methods and next-generation sequencing approaches to analyze epigenetics data: Profiling of methods and applications. *Methods* 187, 92–103.

Bannister, A.J. and Kouzarides, T. (2011) Regulation of chromatin by histone modifications. *Cell Research* 21(3), 381–395.

Ben-Amar, A., Daldoul, S., Reustle, G.M., Krczal, G. and Mliki, A. (2016) Reverse genetics and high throughput sequencing methodologies for plant functional genomics. *Current Genomics* 17(6), 460–475.

Bird, A. (2007) Perceptions of epigenetics. *Nature* 447(7143), 396–398.

Bošković, A. and Rando, O.J. (2018) Transgenerational epigenetic inheritance. *Annual Review of Genetics* 52, 21–41.

Brevini, T.A.L., Pennarossa, G., Manzoni, E.F.M., Gandolfi, C.E., Zenobi, A. *et al.* (2016) The quest for an effective and safe personalized cell therapy using epigenetic tools. *Clinical Epigenetics* 8, 119.

Buenrostro, J.D., Wu, B., Chang, H.Y. and Greenleaf, W.J. (2015) ATAC-seq: A method for assaying chromatin accessibility genome-wide. *Current Protocols in Molecular Biology* 109(1), 21.
Byeon, B., Bilichak, A. and Kovalchuk, I. (2019) Transgenerational response to heat stress in the form of differential expression of noncoding RNA fragments in *Brassica rapa* plants. *The Plant Genome* 12(1), 180022.
Chinnusamy, V., Zhu, J. and Zhu, J.-K. (2007) Cold stress regulation of gene expression in plants. *Trends in Plant Science* 12(10), 444–451.
Cokus, S.J., Feng, S., Zhang, X., Chen, Z., Merriman, B. *et al.* (2008) Shotgun bisulphite sequencing of the *Arabidopsis* genome reveals DNA methylation patterning. *Nature* 452(7184), 215–219.
Cortijo, S., Wardenaar, R., Colomé-Tatché, M., Gilly, A., Etcheverry, M. *et al.* (2014) Mapping the epigenetic basis of complex traits. *Science (New York)* 343(6175), 1145–1148.
Da Sacco, L., Baldassarre, A. and Masotti, A. (2012) Bioinformatics tools and novel challenges in long non-coding RNAs (lncRNAs) functional analysis. *International Journal of Molecular Sciences* 13(1), 97–114.
de Los Reyes, B.G. (2019) Genomic and epigenomic bases of transgressive segregation – new breeding paradigm for novel plant phenotypes. *Plant Science* 288, 110213.
Dowen, R.H., Pelizzola, M., Schmitz, R.J., Lister, R., Dowen, J.M. *et al.* (2012) Widespread dynamic DNA methylation in response to biotic stress. *Proceedings of the National Academy of Sciences of the United States of America* 109(32), E2183–91.
Etchegaray, J.-P. and Mostoslavsky, R. (2016) Interplay between metabolism and epigenetics: A nuclear adaptation to environmental changes. *Molecular Cell* 62(5), 695–711.
Farmer, A., Thibivilliers, S., Ryu, K.H., Schiefelbein, J. and Libault, M. (2021) Single-nucleus RNA and ATAC sequencing reveals the impact of chromatin accessibility on gene expression in *Arabidopsis* roots at the single-cell level. *Molecular Plant* 14(3), 372–383.
Fazzari, M.J. and Greally, J.M. (2010) Introduction to epigenomics and epigenome-wide analysis. *Methods in Molecular Biology* 620, 243–265.
Fields, S., Kohara, Y. and Lockhart, D.J. (1999) Functional genomics. *Proceedings of the National Academy of Sciences of the United States of America* 96(16), 8825–8826.
Gahlaut, V., Zinta, G., Jaiswal, V. and Kumar, S. (2020) Quantitative epigenetics: A new avenue for crop improvement. *Epigenomes* 4(4), 25.
Guimil, S. and Dunand, C. (2006) Patterning of *Arabidopsis* epidermal cells: Epigenetic factors regulate the complex epidermal cell fate pathway. *Trends in Plant Science* 11(12), 601–609.
Haberkorn, U., Altmann, A. and Eisenhut, M. (2002) Functional genomics and proteomics--the role of nuclear medicine. *European Journal of Nuclear Medicine and Molecular Imaging* 29(1), 115–132.
Han, Y. and He, X. (2016) Integrating epigenomics into the understanding of biomedical insight. *Bioinformatics and Biology Insights* 10, 267–289.
He, G., Elling, A.A. and Deng, X.W. (2011) The epigenome and plant development. *Annual Review of Plant Biology* 62, 411–435.
Hewezi, T. (2018) Editorial: Epigenetic regulation of plant development and stress responses. *Plant Cell Reports* 37(1), 1–2.
Hollick, J.B. and Chandler, V.L. (2001) Genetic factors required to maintain repression of a paramutagenic maize pl1 allele. *Genetics* 157(1), 369–378.
Ito, H. (2014) Plant models of transgenerational epigenetic inheritance. In: Tollefsbol, T. (ed.) *Transgenerational Epigenetics*. Academic Press, pp. 147–161.
Jablonka, E. and Lamb, M.J. (1989) The inheritance of acquired epigenetic variations. *Journal of Theoretical Biology* 139(1), 69–83.
Jones, P.A. and Takai, D. (2001) The role of DNA methylation in mammalian epigenetics. *Science (New York)* 293(5532), 1068–1070.
Jones, P.A. and Baylin, S.B. (2007) The epigenomics of cancer. *Cell* 128(4), 683–692.
Kang, J.G., Park, J.S., Ko, J.-H. and Kim, Y.-S. (2019) Regulation of gene expression by altered promoter methylation using a CRISPR/Cas9-mediated epigenetic editing system. *Scientific Reports* 9(1), 11960.
Kawakatsu, T., Huang, S.-S.C., Jupe, F., Sasaki, E., Schmitz, R.J. *et al.* (2016) Epigenomic diversity in a global collection of *Arabidopsis thaliana* accessions. *Cell* 166(2), 492–505.
Kim, S.M., Kang, S.W., Kwon, O.N., Chung, D. and Pan, C.H. (2012) Fucoxanthin as a major carotenoid in *Isochrysis* aff. *galbana*: Characterization of extraction for commercial application. *Journal of the Korean Society for Applied Biological Chemistry* 55, 477–483.

Klengel, T., Pape, J., Binder, E.B. and Mehta, D. (2014) The role of DNA methylation in stress-related psychiatric disorders. *Neuropharmacology* 80, 115–132.

Köhler, C. and Springer, N. (2017) Plant epigenomics-deciphering the mechanisms of epigenetic inheritance and plasticity in plants. *Genome Biology* 18(1), 132.

Kukurba, K.R. and Montgomery, S.B. (2015) RNA sequencing and analysis. *Cold Spring Harbor Protocols* 2015(11), 951–969.

Kumar, S. (2019) Epigenetics and epigenomics for crop improvement: Current opinion. *Advances in Biotechnology & Microbiology* 14(1), 555879.

Lacal, I. and Ventura, R. (2018) Epigenetic inheritance: Concepts, mechanisms and perspectives. *Frontiers in Molecular Neuroscience* 11, 292.

Lämke, J. and Bäurle, I. (2017) Epigenetic and chromatin-based mechanisms in environmental stress adaptation and stress memory in plants. *Genome Biology* 18(1), 124.

Lang, M., Bischl, B. and Surmann, D. (2017) Batchtools: Tools for R to work on batch systems. *Journal of Open Source Software* 2(10), 135.

Law, J.A. and Jacobsen, S.E. (2010) Establishing, maintaining and modifying DNA methylation patterns in plants and animals. *Nature Reviews. Genetics* 11(3), 204–220.

Lee, K. and Seo, P.J. (2018) Dynamic epigenetic changes during plant regeneration. *Trends in Plant Science* 23(3), 235–247.

Lee, J.E., Neumann, M., Duro, D.I. and Schmid, M. (2019) CRISPR-based tools for targeted transcriptional and epigenetic regulation in plants. *PloS ONE* 14(9), e0222778.

Leitch, I.J. and Leitch, A.R. (2012) Genome size diversity and evolution in land plants. In: Greilhuber, J., Dolezel, J. and Wendel, J.F. (eds) *Plant Genome Diversity Volume 2: Physical Structure, Behaviour and Evolution of Plant Genomes*. Springer Vienna, Vienna, pp. 307–322.

Li, Q., Eichten, S.R., Hermanson, P.J., Zaunbrecher, V.M., Song, J. *et al.* (2014) Genetic perturbation of the maize methylome. *The Plant Cell* 26(12), 4602–4616.

Luna, E., Bruce, T.J.A., Roberts, M.R., Flors, V. and Ton, J. (2012) Next-generation systemic acquired resistance. *Plant Physiology* 158(2), 844–853.

Masliah-Planchon, J., Bièche, I., Guinebretière, J.M., Bourdeaut, F. and Delattre, O. (2015) SWI/SNF chromatin remodeling and human malignancies. *Annual Review of Pathology: Mechanisms of Disease* 10(1), 145–171.

Matthew, L. (2004) RNAi for plant functional genomics. *Comparative and Functional Genomics* 5(3), 240–244.

McCormick, K.P., Willmann, M.R. and Meyers, B.C. (2011) Experimental design, preprocessing, normalization and differential expression analysis of small RNA sequencing experiments. *Silence* 2(1), 1–19.

Meissner, A. (2010) Epigenetic modifications in pluripotent and differentiated cells. *Nature Biotechnology* 28(10), 1079–1088.

Meyer, P. (2015) Epigenetic variation and environmental change. *Journal of Experimental Botany* 66(12), 3541–3548.

Narlikar, G.J., Sundaramoorthy, R. and Owen-Hughes, T. (2013) Mechanisms and functions of ATP-dependent chromatin-remodeling enzymes. *Cell* 154(3), 490–503.

Nelson, J.D., Denisenko, O. and Bomsztyk, K. (2006) Protocol for the fast chromatin immunoprecipitation (chIP) method. *Nature Protocols* 1(1), 179–185.

Nikolayeva, O. and Robinson, M.D. (2014) *EdgeR* for differential RNA-seq and ChIP-seq analysis: An application to stem cell biology. In: Kidder, B.L. (ed.) *Stem Cell Transcriptional Networks: Methods and Protocols*. Humana New York, pp. 45–79.

Rai, A., Yamazaki, M. and Saito, K. (2019) A new era in plant functional genomics. *Current Opinion in Systems Biology* 15, 58–67.

Richards, C.L., Alonso, C., Becker, C., Bossdorf, O., Bucher, E. *et al.* (2017) Ecological plant epigenetics: Evidence from model and non-model species, and the way forward. *Ecology Letters* 20(12), 1576–1590.

Rinn, J.L. and Chang, H.Y. (2012) Genome regulation by long noncoding RNAs. *Annual Review of Biochemistry* 81, 145–166.

Rival, A., Beulé, T., Bertossi, F.A., Tregear, J. and Jaligot, E. (2010) Plant epigenetics: From genomes to epigenomes. *Notulae Botanicae Horti Agrobotanici Cluj-Napoca* 38(2), 09–15.

Roadmap Epigenomics Consortium, Kundaje, A., Meuleman, W., Ernst, J., Bilenky, M. *et al.* (2015) Integrative analysis of 111 reference human epigenomes. *Nature* 518, 317–330.

Shen, X., Jonge, J., Forsberg, S., Pettersson, M., Sheng, Z. *et al.* (2014) Natural CMT2 variation is associated with genome-wide methylation changes and temperature seasonality. *PLoS Genetics* 10(12), e100842.

Sheu, C.Y.L., Huang, Y.C., Lin, P.Y., Lin, G.J. and Chen, P.Y. (2024) Bioinformatics of epigenetic data generated from next-generation sequencing. In: Tollefsbol, T.O. (ed.) *Epigenetics in Human Disease*. Academic Press, pp. 37–82.

Shin, H., Choi, W.L., Lim, J.Y. and Huh, J.H. (2022) Epigenome editing: Targeted manipulation of epigenetic modifications in plants. *Genes & Genomics* 44(3), 1–19.

Singh, D., Chaudhary, P., Taunk, J., Kumar Singh, C., Sharma, S. *et al.* (2021) Plant epigenomics for extenuation of abiotic stresses: Challenges and future perspectives. *Journal of Experimental Botany* 72(20), 6836–6855.

Stam, M. (2009) Paramutation: A heritable change in gene expression by allelic interactions in *trans*. *Molecular Plant* 2(4), 578–588.

Subramanian, I., Verma, S., Kumar, S., Jere, A. and Anamika, K. (2020) Multi-omics data integration, interpretation, and its application. *Bioinformatics and Biology Insights* 14, 1177932219899051.

Surani, M.A. (1994) Genomic imprinting: Control of gene expression by epigenetic inheritance. *Current Opinion in Cell Biology* 6(3), 390–395.

Tanić, M. (2020) Epigenome-wide association study (EWAS): Methods and applications. In: Tollefsbol, T.O. (ed.) *Epigenetics Methods*. Academic Press, pp. 591–613.

Thu, K.L., Vucic, E.A., Kennett, J.Y., Heryet, C., Brown, C.J. *et al.* (2009) Methylated DNA immunoprecipitation. *Journal of Visualized Experiments* 23, e935.

Tsaftaris, A.S., Polidoros, A.N., Koumproglou, R., Tani, E., Kovacevic, N. *et al.* (2005) Epigenetic mechanisms in plants and their implications in plant breeding. In: Tuberosa, R., Phillips, R.L. and Gale, M. (eds) Proceedings of the International Congress *'In the Wake of the Double Helix: From the Green Revolution to the Gene Revolution'*, Avenue Media, Bologna, pp. 157–171.

Tuberosa, R., Phillips, R.L. and Gale, M. (eds) (2005) In: *Proceedings of the international congress "In the Wake of the Double Helix: From the Green Revolution to the Gene Revolution,"* Avenue media, Bologna, Italy, May 27–31, 2003, pp. 157–171.

Turner, B.M. (2009) Epigenetic responses to environmental change and their evolutionary implications. *Philosophical Transactions of the Royal Society of London. Series B, Biological Sciences* 364(1534), 3403–3418.

Varshney, R.K., Terauchi, R. and McCouch, S.R. (2014) Harvesting the promising fruits of genomics: Applying genome sequencing technologies to crop breeding. *PLoS Biology* 12(6), e1001883.

Verhoeven, K.J.F., vonHoldt, B.M. and Sork, V.L. (2016) Epigenetics in ecology and evolution: What we know and what we need to know. *Molecular Ecology* 25(8), 1631–1638.

Wang, Z., Gerstein, M. and Snyder, M. (2009) RNA-seq: A revolutionary tool for transcriptomics. *Nature Reviews. Genetics* 10(1), 57–63.

Wei, J.-W., Huang, K., Yang, C. and Kang, C.-S. (2017) Non-coding RNAs as regulators in epigenetics (review). *Oncology Reports* 37(1), 3–9.

Wreczycka, K., Gosdschan, A., Yusuf, D., Grüning, B., Assenov, Y. *et al.* (2017) Strategies for analyzing bisulfite sequencing data. *Journal of Biotechnology* 261, 105–115.

Xiong, X., Yi, C. and Peng, J. (2017) Epitranscriptomics: Toward a better understanding of RNA modifications. *Genomics, Proteomics & Bioinformatics* 15(3), 147–153.

Youngson, N.A. and Whitelaw, E. (2008) Transgenerational epigenetic effects. *Annual Review of Genomics and Human Genetics* 9, 233–257.

Yu, S., Zhang, G., Li, J., Zhao, Z. and Kang, X. (2013) Effect of endogenous hydrolytic enzymes pretreatment on the anaerobic digestion of sludge. *Bioresource Technology* 146, 758–761.

Yue, C.A., Men, L.R. and Ferguson, M.A. (2019) Bridging transformational leadership, transparent communication, and employee openness to change: The mediating role of trust. *Public Relations Review* 45(3), 101779.

Zhang, H., Lang, Z. and Zhu, J.-K. (2018) Dynamics and function of DNA methylation in plants. *Nature Reviews. Molecular Cell Biology* 19(8), 489–506.

Zhang, M., Xie, S., Dong, X., Zhao, X., Zeng, B. *et al.* (2014) Genome-wide high-resolution parental-specific DNA and histone methylation maps uncover patterns of imprinting regulation in maize. *Genome Research* 24(1), 167–176.

Zhang, X., Yazaki, J., Sundaresan, A., Cokus, S., Chan, S.W.-L. *et al.* (2006) Genome-wide high-resolution mapping and functional analysis of DNA methylation in *Arabidopsis*. *Cell* 126(6), 1189–1201.

Zwyrtková, J., Šimková, H. and Doležel, J. (2021) Chromosome genomics uncovers plant genome organization and function. *Biotechnology Advances* 46, 107659.

21 Epigenetics for Studying Plant Cell Signaling: Updated Methods and Current Achievements

Amarjeet Singh Bhogal[1], Dipen Nama Adhikari[2], Rahul Gogoi[2], Pratim Dutta[2], Praveen Kumar[3], Kishore Kumar Sharma[1], Nagendra Sarma Barua[1], Gargi Sharma[1] and Debojit Sarma[1]*

[1]Department of Plant Breeding and Genetics, Assam Agricultural University, Jorhat, Assam, India; [2]Department of Agricultural Biotechnology, Assam Agricultural University, Jorhat, Assam, India; [3]Department of Genetics and Plant Breeding, Agriculture University, Jodhpur, India

---

Abstract

Epigenetics, an enthralling field of scientific inquiry, instigates changes to gene function that are heritable at the mitotic or meiotic levels without altering the DNA sequence. These epigenetic pathways, now widely recognized, are pivotal to plant growth and their remarkable plasticity, enabling them to respond to subtle ecological signals. Many emerging signals regulate intricate epigenetic configurations, including transcription factors (TFs), ncRNAs, DNA methylation, and histone modifications. Transceptors, a term used to describe receptors that also function as transporters, play a crucial role in plant transcriptomic activity. For instance, NITRATE TRANSPORTER 1.1 (NRT1.1) or NIN-LIKE PROTEIN 7 (NLP7) acts as a transceptor, sensing nitrogen, leading to rapid changes in plant transcriptomic activity using second messengers such as Ca^{2+} and TFs, with changes occurring swiftly after nitrate exposure. The potential of our work in modifying these processes to enhance crop production is immense. Transgenic strategies that alter the gene expression levels of alleles involved in nitrogen transportation, assimilation, and signaling can significantly boost crop production or NUE. 5-aza deoxycytidine, a DNA-demethylating chemical, enhances bacterial defense against *Xanthomonas* in rice. This chapter delves into the intriguing concept of various plant cell signaling strategies through epigenetic processes, sparking curiosity about the adaptability of plants.

Keywords: DNA methylation, Defensive mechanisms, Epigenetics, Genes, Signaling

21.1 Introduction

The epigenetic gene expression regulation in vertebrates and plants is significantly influenced by the methylation of DNA bases at specific locations, such as the fifth carbon of cytosine (Kumar *et al.*, 2018). Plant functions, including growth, development, reproduction, and disease resistance, depend on epigenetic controls that help plants become more resilient to various environmental stresses. Constant modifications greatly aid evolution, offering an

*Corresponding author: debojit.sarma@aau.ac.in

 Epigenetics for Climate-Smart and Sustainable Agriculture (ed. Jen-Tsung Chen)
DOI: 10.1079/9781800626102.0021

established genomic foundation for phenotypic flexibility; short-term strategies play a critical role in stress survival. This flexibility facilitates the deliberate choice of offspring suited to continually changing surroundings, evidenced in naturally occurring populations with equivalent genetic backgrounds, implying that it is an epigenetic feature (Duarte-Aké *et al.*, 2023). During stress, plants utilize various epigenetic strategies to regulate gene expression. Controlling the expression of genes and ensuring genomic stability are critical activities, with each strategy serving a distinctive and significant purpose (Abdulraheem *et al.*, 2024). The N-terminus of histones, the fundamental element of nucleosomes, undergoes an array of post-translational alterations. Lysine (Lys) and arginine (Arg) remnants are capable of being methylated or acetylated, while serine (Ser) and threonine (Thr) remnants can be phosphorylated. Gene expression is hypothesized to be repressed by biotinylation and SUMOylation and activated by histone acetylation, phosphorylation, and ubiquitination (Kouzarides, 2007; Berger, 2007). Histone acetylation is a frequent epigenetic alteration that regulates a broad spectrum of life processes. Histone acetyltransferases (HATs) and histone deacetylases (HDACs) are primary enzymatic classes that conduct this dynamic change; histone tails are acetylated by HATs and deacetylated by HDACs (Liu *et al.*, 2010). RNA-directed DNA methylation (RdDM) in plants is accountable for the formation of DNA methylation. New DNA methylation is catalyzed by the DNA methyltransferase DOMAINS REARRANGED METHYLTRANSFERASE 2 (DRM2); entirely novel cytosine methylation is mediated via the RdDM pathway, and its maintenance across multiple sequence contexts is dependent on several DNA methyltransferases. (Zhang *et al.*, 2018; Li *et al.*, 2018).

As sessile organisms, plants constantly adjust to new surroundings, such as biotic and abiotic stressors brought on by climate change. Understanding how plants adapt to these changes can be gained by identifying gene sequences responsive to stress. Histone and DNA methylation are epigenetic alterations that regulate gene expression and impact stress memory and phenotypic plasticity. These alterations allow plants to adapt more successfully to recurring stress (Fig. 21.1) and better prepare progeny for future stressors. This chapter delves into the expansive concept of diverse plant cell signaling pathways via epigenetic systems.

21.2 Deciphering the Complexity of Epigenetics

The advancement of sophisticated tissues and systems like vascular structures, root systems, flowers, and seeds is vital for the prosperity of plants. To create tissues and organs, cells must specialize and keep their unique characteristics through mitotic divisions. A more profound comprehension of the genome is required to ensure proper development as the variety of cell types increase. (Bräutigam and Cronk, 2018). Hence, there is a sophisticated interaction between signals from within and outside the plant. Land plants possess a wide array of epigenetic pathways at the molecular level, likely contributing significantly to their complex development, diverse traits, and ability to adapt (Bossdorf *et al.*, 2008; Bräutigam *et al.*, 2013).

21.2.1 Epigenetic mechanisms: internal and external signals for developmental control

Normal development for an organism is characterized by constitutive (internal) developmental regulation, which relies on internal developmental cues. Constitutive epigenetic control in development has been well-documented in plants such as *Arabidopsis* and involves critical reproductive processes such as seed formation, the disruption of which may lead to deleterious consequences for the development of reproductive systems and embryos. For example, specific cells within the male and female reproductive structures undergo genome demethylation, which can help silence transposable elements in sperm, egg cells and embryos (Bräutigam and Cronk, 2018). Another instance of constitutive epigenetic control is exemplified by the CURLY LEAF (CLF) gene, which is pivotal in the normal development of leaves. CLF is categorized as a polycomb group (PcG) protein that remodels

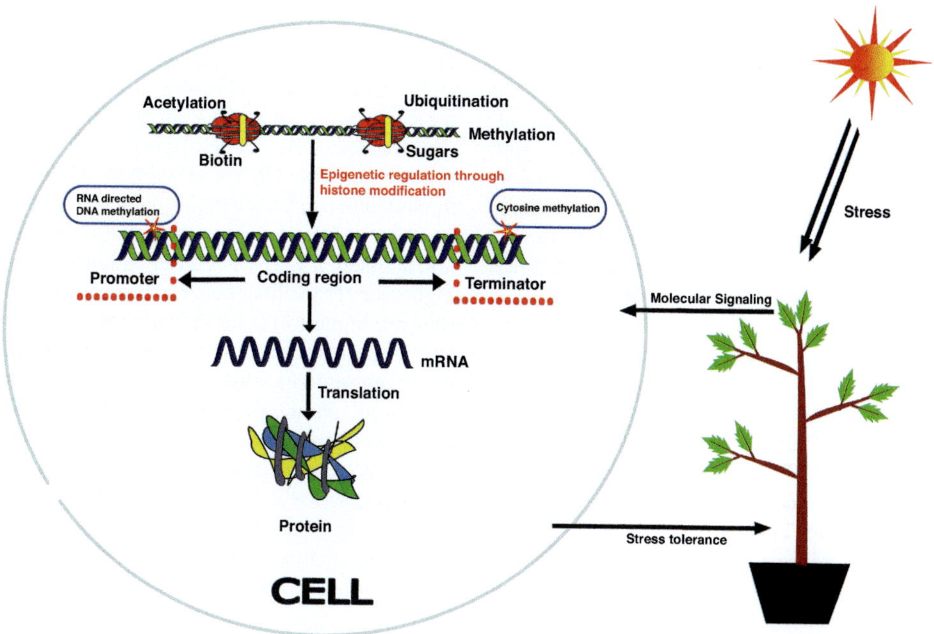

Fig. 21.1. Generalized diagram depicting epigenetic plant cell signaling. Author's own image.

chromatin, ensuring consistent gene repression across multiple cell divisions and lineage marking. PcG proteins form complex assemblies with diverse roles in development. For example, specific complexes containing the EMBRYONIC FLOWER 2 protein regulate vegetative growth and suppress flowering, while others, like the FIS-complex, prevent seed formation without fertilization in plants such as *Arabidopsis* (Hennig and Derkacheva, 2009).

The vernalization reaction is an example of facultative (external) developmental control in action, where a plant's ability to bloom is determined by its memory of exposure to winter conditions. He (2012) described the involvement of the VERNALIZATION 2 (VRN2) protein in this process and several epigenetic processes, including chromatin remodeling, histone changes, non-coding RNAs, and DNA methylation. Environmental conditions control whether a plant stays in a vegetative state or converts to flowering, a process known as vernalization, an adaptive epigenetic response (Bräutigam and Cronk, 2018).

It is necessary to understand light's influence throughout the critical developmental stages of seedlings, such as germination and switching to autotrophic development (photomorphogenesis). Environmental inputs, such as light, cause chromatin state alterations, resulting in considerable transcriptional regulation. When photoreceptors detect light, it reduces hypocotyl elongation, cotyledon opening, and chlorophyll production. Significant modifications in the organization of the genome are caused by this process, which also includes changes to histone proteins, enlargement of the nucleus, and modest increases in ploidy through endoreduplication and heterochromatin condensation (Bourbousse *et al.*, 2015). The chromatin landscape can be changed by environmental inputs, which can then fine-tune development and genomic activity. Subtle morphological changes can occur during post-embryonic stages by changing factors in their environment. Through hormone signaling, such signals incorporate within the chromatin of vegetal meristems, influencing the growth of leaves or the shape of roots (Xiao *et al.*, 2017). Hence, DNA methylation and other chromatin modifiers significantly shape plant characteristics without genetic alterations. This means that they can influence plant traits based on environmental conditions or developmental stages, even before genetic changes occur. They allow plants to create diversity in response to conditions, aiding short-term adaptive responses during growth and long-term evolutionary adaptation over generations (Jablonka, 2013).

21.2.2 Exploring DNA methylation patterns in green plants: insights into transposable element regulation

One common covalent alteration of DNA that can endure across mitotic and meiotic cell divisions is the methylation of cytosine (5-methylcytosine). DNA methylation significantly preserves and regulates the genome shape and function, even though it does not alter the genetic information or core DNA sequence. Plant DNA may include a range of non-canonical nucleotide alterations at lower levels, such as oxidized variants of 5-methylcytosine (5hmC, 5fC, and 5caC) or N6-methyladenine (6mA) (Liu *et al.*, 2013; Zhou *et al.*, 2018). These changes, albeit uncommon, were eventually discovered to have essential roles in gene regulation and plant growth. Greer *et al.* (2015) and Zhang *et al.* (2015) believed that 6mA may become a significant epigenetic mark in both plant and animal systems, albeit these base changes have yet to be adequately investigated regarding plant evolutionary processes. The methylation of DNA is an evolutionary regulatory trait common to all major types of green plants (Feng *et al.*, 2010; Zemach *et al.*, 2010). Because of its broad prevalence and importance, DNA methylation can significantly affect plant evolution. Scientists have recently been more interested in investigating the changes in methylation patterns of DNA found in distinct plant species.

The distribution of DNA methylation is not random in any plant genome. Centromeric repeats, rDNA, and transposable elements (TEs) are examples of repeating sections where it is frequently more prevalent (Slotkin and Martienssen, 2007). DNA segments known as TEs can migrate and disrupt genes and regulatory areas, thereby affecting the stability of the genome and perhaps leading to mutations. However, most TEs are dormant, silent, or otherwise unusable. DNA methylation is believed to contribute to TE silencing, which may have been one of its initial purposes (Slotkin and Martienssen, 2007). DNA methylation levels are higher for TEs in all sequence contexts, although they are highest in CG and CHG contexts. From gymnosperms and flowering plants to mosses like *Physcomitrella*, this trait has been seen in almost all plant epigenomes that have been investigated to date (Chan *et al.*, 2008). Silent TEs display a particular chromatin state marked by the restrictive histone mark H3K9me2, along with higher DNA methylation levels as well as other histone modifications in *Arabidopsis thaliana* (Cokus et *al.*, 2008).

21.2.3 Evolutionary shift: from control of TEs to mastery over genes through the methylation transition

In eukaryotes, TEs are regulated by the methylation of DNA. Organisms later evolved the capacity to target transcriptional start sites and promoters with DNA methylation to regulate genes. This characteristic is typically present in embryophytes. DNA methylation control is regulated by many genes in the moss *Physcomitrella patens* (Bräutigam and Cronk, 2018). These comprise the CHROMOMETHYLASE (CMT) gene, RdDM methylases, DNA METHYLTRANSFERASE 1 (MET1), and additional DNA methyltransferases (Malik *et al.*, 2012). Other genes in the moss genome express themselves more when these genes are deleted. This implies that these genes suppress the regulation of gene transcription. *Arabidopsis* variants exhibiting various DNA methylation levels exhibit various visible characteristics. A lack of function in the AtMET1 gene, which is in charge of CG methylation, results in several defects, including smaller plants, poorer fertility, shifted flowering times and changed floral structure. Increased production of FWA due to loss of DNA methylation at a transposable element near the gene is associated with delayed flowering (Soppe *et al.*, 2000). Furthermore, the development of many embryos in *Arabidopsis* is impacted by the lack of MET1.

21.2.4 Exploring complexity: the rising complexity of RNA-directed DNA methylation as an epigenetic pathway

Over the past 20 years, RdDM has undergone tremendous change, exposing a complex epigenetic process that affects many facets of plant life. The conventional RdDM pathway starts with short RNA molecules produced by Pol IV, RDR2, and DCL3. These siRNAs subsequently cause DNA methylation, mediated by Pol V and AGO4. However, other small RNA-guided DNA methylation forms exist in plants outside this classical mechanism. RdDM pathways exhibit specific variants that coincide with short RNA size classes and components commonly linked to post-transcriptional gene silencing (PTGS) pathways. This overlap likely explains the redundancy observed amongst DCLases within RdDM and early connections between PTGS and RdDM (Eun *et al.*, 2012). Examining potential differences between species is an exciting avenue for future RdDM research. For instance, compared with maize, the developmental consequences of individual RdDM component mutations in *Arabidopsis thaliana* are less noticeable. This shows that because of its more significant genome repeat concentration or outcrossing tendency, RdDM might be more critical in maize. Variations in RdDM within species contribute to hybrid vigour. It is also interesting to explore how variations in siRNA accumulation across genomes may affect the boundaries between species (Matzke and Mosher, 2014).

To completely understand RdDM pathways in plants, it is critical to study the emergence of RdDM constituents across the kingdom of plants and compare RdDM variants from different species. Even while our knowledge of the biological relevance and molecular processes of RdDM has advanced significantly, many unsolved concerns remain. This implies that there are exciting and interesting avenues for future RdDM research to pursue (Matzke and Mosher, 2014).

21.2.5 Unraveling the impact of DNA methylation in pest defense: Insights into plant–insect interactions

Research has indicated that DNA methylation is essential in plant defense mechanisms. Plant defense systems are generally weakened by DNA methylation. Put another way, plants grow more resistant to diseases when their DNA methylation levels drop (López Sánchez *et al.*, 2016). The bacterial pathogen *Pseudomonas syringae* (Pst) is more resilient to a range of *Arabidopsis* variations with decreased methylation of their DNA, notably met1-3 and the triple variants drm1 and drm2 cmt3. Compared with wild-type plants, they also exhibit higher expression of genes linked to defense (Dowen *et al.*, 2012; Espinas *et al.*, 2016).

López Sánchez *et al.* (2016) showed that *Arabidopsis thaliana* variants with varying degrees of methylation in the DNA had different effects when infected with *Hyaloperonospora arabidopsidis* (Hpa). They discovered that mutants with hypomethylated nuclear RNA polymerase E (nrpe), which had decreased RdDM activity due to faulty RNA polymerase V, have increased tolerance to Hpa. In contrast, the hypermethylated ros1 variant was more susceptible to infection. Their findings indicate a relationship between how infections are responded to and how genes that rely on SA are expressed. These genes were downregulated in Ros1 mutants but upregulated in NRPE mutants. In contrast, nrpe mutants showed increased sensitivity and downregulated JA-induced genes compared with ros1 variants when inoculated with *Plectosphaerella cucumerina*.

Arabidopsis is not the only plant species that links decreased DNA methylation to defense mechanisms. Researchers found higher resistance to the bacterial disease *Xanthomonas* in rice exposed to 5-azadeoxycytidine, which had lower DNA methylation levels (Akimoto *et al.*, 2007). Beyond interactions with bacteria or fungi, DNA methylation also affects interactions between plants and nematodes (Hewezi, 2020). *Heterodera glycines*, a soybean cyst nematode, has been studied to see how it affects soybean responses to infection (Rambani *et al.*, 2015). Infected plant roots were found to frequently have decreased DNA methylation or hypomethylation. According to a recent study, some of these methylation alterations also impacted the expression of genes known as plant microRNAs (Rambani *et al.*, 2020), at least one of which was involved in how the plant reacted to the insect.

21.3 Unraveling Molecular Pathways of Epigenetic Regulations

Plant gene expression is controlled by complex biochemical mechanisms known as epigenetic regulation, which go beyond the DNA sequence. These pathways include a variety of mechanisms, including chromatin remodeling, histone modifications, DNA methylations, small RNA-mediated gene silencing, non-coding RNAs, etc. In plant genomes, DNA methylation occurs in settings of the CG, CHG, and CHH sequences in patterns that show a balance between the activity of the enzymes that add, remove, or preserve methylation. Similar to other eukaryotes, plants' epigenetic states are influenced by histone-modifying enzymes. These enzymes are encoded by extensive gene families, which permits overlapping and diverse roles. Multiple routes are used in RNA-mediated gene silencing to fight viruses, control transposons, direct development, and aid in genome organization. These biochemical pathways interact to create a complex regulatory network in plants that controls epigenetic alterations essential to their growth, adaptability, and reaction to environmental cues. Furthermore, plants may flourish and reproduce in unpredictable conditions employing epigenetic regulatory systems, which enable metastable variations in gene activity and fine-tune gene expression patterns (Pikaard and Scheid, 2014). Knowledge of these mechanisms can help enhance crops, minimize environmental stress, and provide insights into plant biology.

21.3.1 Exploring methylations, demethylations, and their mechanistic underpinnings

Plants utilize the RNA-directed DNA methylation (RdDM) pathway to induce DNA methylation. The enzyme DRM2 sequences reorganized methyltransferase 2 contributes to new DNA methylation in various sequence circumstances (Zhang *et al.*, 2018). In this typical RdDM pathway, the enzyme RNA-DEPENDENT RNA POLYMERASE 2 (RDR2) converts single-stranded RNAs (ssRNAs) produced by the plant-specific RNA polymerase IV (Pol IV) into double-stranded RNAs (dsRNAs). DICER-LIKE 3 (DCL3) further processes these dsRNAs into 24-nucleotide small interfering RNAs (siRNAs), which are then incorporated into the proteins ARGONAUTE 4 (AGO4) and ARGONAUTE 6 (AGO6) (Zhang *et al.*, 2018). SAWADEE HOMEODOMAIN HOMOLOG 1 (SHH1), a histone reader, and the CLASSY family of potential chromatin remodelers are required for Pol IV to be recruited to chromatin (Zhang *et al.*, 2018; Zhou *et al.*, 2018). In the second leg of the RdDM pathway, RNA polymerase V (Pol V) binds to siRNA-loaded AGO4/AGO6 through sequence complementarity and transcribes non-coding RNAs (ncRNAs) (Liu *et al.*, 2018; Zhang *et al.*, 2018). DRM2 is recruited to target DNA methylation once the AGO–siRNA–ncRNA–Pol V ribonucleoprotein complex has formed (Zhang *et al.*, 2018). The DDR complex, which is also vital for Pol V's presence on chromatin, is recruited by the DNA methyl-readers SU (VAR)3–9 homologs SUVH2 and SUVH9, which subsequently recruit the DNA methyl-readers DEFECTIVE IN MERISTEM SILENCING 3 (DMS3), DEFECTIVE IN RNA-DIRECTED DNA METHYLATION 1 (DRD1), and RNA-DIRECTED DNA METHYLATION 1 (RDM1) (Zhang *et al.*, 2018; Gallego-Bartolomé *et al.*, 2019).

Non-canonical RdDM pathways, which allow short RNAs (sRNAs) from a variety of sources, including viruses and Pol II transcripts, to control the RdDM mechanism, have also been described alongside the canonical RdDM pathway (Cuerda-Gil and Slotkin, 2016). These sRNAs are produced from dsRNAs, which are then cut into 21–24 nt sRNAs by different DCL proteins. This process starts post-transcriptional gene silencing (PTGS) of complementary RNAs through breakage and/or translational inhibition. These sRNAs are then integrated into separate AGO proteins (de Alba *et al.*, 2013). By incorporating such sRNAs into AGO4 and/or AGO6, Pol V and DRM2 can methylate complementary DNA strands, potentially resulting in transcriptionally active TEs and gene TGS (Cuerda-Gil and Slotkin, 2016). Methyltransferase 1 (MET1) and CHROMOMETHYLASE 2 and 3 (CMT2 and 3), which operate in the CG, CHH, and CHG sequence contexts (where H is A, C, or T), respectively, sustain methylation after DNA replication following RdDM-mediated *de novo* establishment (Zhang *et al.*, 2018).

Methylation-causing enzymes are called writers as they methylate RNA; effectors are readers because they bind to methylated nucleotides; erasers reverse the operation by removing the methylation (Lim and Pawson, 2010). In eukaryotic organisms, alpha-ketoglutarate-dependent dioxygenase (AlkB) homolog (ALKBH) proteins are responsible for eliminating methylation marks, primarily via removing alkyl and methyl groups from proteins, RNAs, and DNA (Zaccara *et al.*, 2019). For example, when the alfalfa mosaic virus (AMV) infects *Arabidopsis*, AtALKBH9B functions as an eraser, removing the m6A tags from the RNAs to produce host resistance. By increasing the relative quantity of m6A marks on AMV RNAs, inhibition of AtALKBH9B hindered the plant's systemic invasion (Martínez-Pérez *et al.*, 2017). Hiring methylation loci binding proteins or readers is one way methylation influences RNA expression (Zaccara *et al.*, 2019). The processes of aberrant transcription termination and calcium signaling, respectively, are mediated by plant nuclear readers such as CPSF30 (Cleavage and Polyadenylation Specificity Factor 30) and ECT1 (EVOLUTIONARILY CONSERVED C-TERMINAL REGION 1). On the other hand, *Arabidopsis* ECT2, a cytoplasmic reader, can target several m6A-containing mRNAs, including TTG1 (TRANSPARENT TESTA GLABRA1), ITB1 (IRREGULAR TRICHOME BRANCH), whereas DIS2 (DISTORTED TRICHOME2) reader participates in trichome development. Furthermore, it comes to light that ECT2 readers are in charge of maintaining mRNA stability (Wei *et al.*, 2018). Writing the m5C methylation in *Arabidopsis* is done by two methyltransferase enzymes called TRM4A and TRM4B. However, no discernible phenotypes are altered by the TRM4A gene loss of function mutation. TRM4B's significance in root growth is implied by its absence of shortening roots (David *et al.*, 2017; Hu *et al.*, 2019).

Demethylation processes can be classified into two categories: active and passive. Agius *et al.* (2006) reported that passive demethylation is the result of freshly generated DNA strands failing to be methylated by methyltransferase during DNA replication, whereas active demethylation is the result of a sequence of enzymes and proteins functioning in concert. Certain DNA demethylase enzymes have been found in *Arabidopsis*, including DME (DNA glycosylase DEMETER), ROS1 (REPRESSOR OF SILENCING 1), DML2 (DEMETER-LIKE 2), and DML3 (DEMETER-LIKE 3) (Choi *et al.*, 2002; Morales-Ruiz *et al.*, 2006). In *Arabidopsis*, the MBD7-IDM protein complex regulates ROS1 targeting and recruitment to particular loci (Qian *et al.*, 2012). Comparably, the FACT complex (facilitates chromatin transactions) mediates the recruitment of DME to target loci (Liu and Lang, 2020). 5-methylcytosine (5mC) is removed from the DNA backbone by ROS1 or DME upon recruitment, with the help of their apyrimidinic/apurinic and glycosylase activities (Gong *et al.*, 2002). Enzymes having lyase activity break the DNA backbone, leaving a single nucleotide gap with a 3'-phosphor-α, β-unsaturated aldehyde (3'-PUA) or a 3' phosphate terminal (Agius *et al.*, 2006; Morales-Ruiz *et al.*, 2006). The 3'DNA phosphatase ZDP and the AP endonuclease-like protein APE1L then process this gap to form a 3'OH end (Morales-Ruiz *et al.*, 2006). DNA polymerase then fills the gap, which DNA ligase I seals using the base-excision repair route, replacing the unmethylated cytosine (Liu and Lang, 2020).

21.3.2 Epigenetic histone alterations

ADP-ribosylation, acetylation, methylation, phosphorylation, ubiquitination, and sumoylation are some changes that histone proteins undergo at their N-terminal tails. These modifications are essential for regulating chromatin structure and gene expression, and their distribution throughout the genome shows unique patterns linked to various genomic areas (Berger, 2007; Li *et al.*, 2007; Rando and Chang, 2009). Histone acetylation, mediated by histone acetyltransferases (HATs) enzymes, includes adding acetyl groups to lysine residues.

This mechanism neutralizes histones' positive charge, relaxing their association with DNA and increasing transcriptional activity (Suto *et al.*, 2000). In contrast, transcriptional repression results from removing those acetyl groups by histone deacetylases (HDACs) (Pandey, 2002). Histones H3 and H4 lysine residues are targeted explicitly by acetylation in plants such as *Arabidopsis* (Earley *et al.*, 2007). Protein arginine methyltransferases (PRMTs) and histone lysine methyltransferases (HKMTs) are two examples of enzymes that catalyze histone methylation, which adds methyl groups to lysine or arginine residues. Depending on the precise residue and level of methylation, adding methyl groups can either inhibit or increase the expression of a gene. For example, H3K4me3 methylation is linked to active transcription, but H3K9me2 and H3K27me2 methylation often results in gene suppression. According to Gonsalvez *et al.* (2007), Nishioka (2003), and Tollenaar and Lee (2002), most HKMTs have SET domains that are necessary for their enzyme activity, except Dot1, which methylates H3K79. Serine, threonine, or tyrosine residues can acquire phosphate groups by the process of histone phosphorylation, which is catalyzed by kinases. The alteration above plays a role in multiple physiological functions, including transcription, cell division, DNA repair, and apoptosis. According to Wei *et al.* (1998), Sauvé *et al.* (1999) de la Barre (2000), and others, phosphorylation typically promotes transcriptional activity by facilitating chromatin relaxation. H2A(X) phosphorylation is essential for DNA damage response pathways (Rossetto *et al.*, 2010). ADP-ribosylation is the process of binding ADP-ribose groups to glutamate and arginine residues on histones, which is mediated by enzymes such as mono-ADP-ribosyl transferases and poly-ADP ribose polymerases (PARPs). This alteration is essential for regulating chromatin structure and gene expression and is regulated by ADP ribosyl hydrolases (ARHs) and PAR glycohydrolases (PARGs) (Moss *et al.*, 1997; Hassa *et al.*, 2006; Koch-Nolte, 2008). Histone ubiquitination, promoted by E1 activating, E2 conjugating, and E3 ligating enzymes, involves attaching ubiquitin molecules to histone lysine residues. This alteration can cause histones to degrade or affect their function and location. Specific examples include H2AK119ub1 and H2BK123ub1, which are involved in gene silencing and transcriptional control, respectively (Wang *et al.*, 2004; Saracco *et al.*, 2009). To preserve the balance of regulation, deubiquitinating enzymes (DUBs) can reverse ubiquitination (Samantara *et al.*, 2021).

Sumoylation, like ubiquitination, involves the addition of SUMO proteins to histone lysine residues. This alteration is commonly connected with transcriptional repression and influences protein interactions and subcellular localization dynamics. According to Melchior (2000) and Bannister and Kouzarides (2011), SUMO-specific proteases (ULP/SUSP) are in charge of removing it from the process, which includes E1, E2, and E3 enzymes. These varied PTMs influence gene expression, chromatin structure, and cellular functions, demonstrating their critical roles in eukaryotic life.

21.3.3 Genomic chromatin reorganization

In a state of high condensation, the chromatin architecture is a barrier, hindering the access of transcription factors, polymerases, and other nuclear proteins to the DNA. However, in response to stress signals, modifications occur in the chromatin structure, facilitating accessibility to DNA. Such chromatin remodeling encompasses alterations such as repositioning or eviction of histones, incorporating histone variants, and post-translational modifications of pre-existing histones (Eberharter and Becker, 2002). Within the spectrum of various chromatin organization processes, two distinct strategies are discernible, each characterized by disparate enzymatic mechanisms. One strategy relies on chromatin remodelers, which induce alterations in DNA–histone interactions through the energy derived from ATP hydrolysis. Conversely, the alternative approach employs specialized enzymes tasked with methylating DNA or effecting modifications on histone residues through the covalent addition of chemical moieties (Cedar and Bergman, 2009).

Modifying chromatin structure requires the ATP-dependent chromatin remodeling system powered by ATP-dependent chromatin remodeling complexes. Genetic screens were utilized to identify the first of these, yeast SWI/SNF, which revealed mutants with impairments in their abilities to switch mating types (SWI) and not ferment sucrose (SNF) (Winston and Carlson, 1992). The genes found in these screens were subsequently identified as parts of the multi-subunit SWI/SNF complex and dubbed SWI or SNF genes (Roberts and Orkin, 2004). The SF2 subgroup of DNA helicases, which includes chromatin remodeling protein in the SNF2 family, has a helicase motif composed of seven conserved blocks (Flaus, 2006). The primary catalytic components of chromatin remodeling complexes are DNA-dependent ATPases from the SNF2 superfamily, which use ATP hydrolysis to break DNA–histone connections. This feature allows Snf2 ATPases to support a variety of chromatin modifications, such as changes in chromatin structure, histone composition, and nucleosomal DNA arrangement (Clapier and Cairns, 2009). Based on their catalytic Snf2 domain and other accessory domains, chromatin remodeling factors are classified into four main subfamilies: Inositol auxotrophy 80 (INO80), Imitation switch (ISWI), Switch/Sucrose non-fermentable (SWI/SNF), and Chromodomain helicase DNA-binding (CHD) (Clapier *et al.*, 2017). SNF2 proteins in *Arabidopsis* comprise 41 different groups: ISWI, CHD, LSH, RAD54, RAD5/16, SSO1653, SMARCAL1, INO80/SWR1, SWI/SNF, and so on (Knizewski *et al.*, 2008).

21.4 Recent Methods and Tools for Studying Plant Signaling

The use of epigenetic methods in studying plant signaling pathways has grown significantly. With the aid of these instruments, scientists may look into how alterations in histones, non-coding RNAs, and DNA methylation might affect gene expression, which in turn affects plant signaling pathways. Several epigenetic tools that are frequently employed in the study of plant signaling include small RNA sequencing, chromatin remodeling complexes, CRISPR/Cas9 technology, histone modification analysis, DNA methylation analysis, whole-genome bisulfite sequencing, and chromatin immunoprecipitation.

21.4.1 Small RNA sequencing

Plants rely heavily on short RNAs to control gene expression, particularly microRNAs (miRNAs) and small interfering RNAs (siRNAs). Advanced sequencing methods have improved our ability to detect and characterize these tiny RNAs, which play critical roles in plant signaling cascades. Processing these short RNAs by Dicer or Dicer-like (DCL) proteins and their integration into Argonaute (AGO) proteins allows for sequence-specific gene silence. Once incorporated into AGO proteins, they target specific genes for TGS or PTGS (Betti *et al.*, 2021). MicroRNAs are mainly generated from single-stranded primary RNA precursors with stem-loop structures, whereas siRNAs are formed from double-stranded RNAs (dsRNAs) derived from inverted repeats, sense–antisense transcripts pairs, or RNA-dependent RNA polymerases (RDRs). *Trans*-acting siRNAs, heterochromatic siRNAs (hc-siRNAs), natural antisense transcript-derived siRNAs (nat-siRNAs), long siRNAs (lsiRNAs), and secondary phased siRNAs (phasiRNAs) are some of the numerous types of plant siRNAs determined by their biogenesis processes. These diverse short RNA forms balance plant development and immune responses by controlling the expression of genes that signal plant defense and nucleotide-binding leucine-rich repeats (NLRs) (Lopez-Gomollon and Baulcombe, 2022). Transposable elements or genes that respond to stress can trigger the synthesis of short RNAs in response to environmental stress. These short RNAs help plants adapt to harsh environments by controlling stress-responsive genes. MiR398, for example, targets the transcripts of the copper superoxide dismutase (CSD) gene in *Arabidopsis thaliana* during oxidative stress, lowering CSD expression and increasing vulnerability to oxidative damage (Deneweth *et al.*, 2022).

Small RNAs also affect chromatin structure, influencing gene expression. Gene silencing via the RNA-directed DNA methylation (RdDM) pathway can occur when siRNAs produced from transposable elements or repetitive sequences target specific genomic areas for the DNA methyltransferase DRM2 to bind to. This route is critical for plant stress response, particularly in *Arabidopsis*, where it regulates heat shock protein production. Furthermore, to modify histone modifications and gene expression, short RNAs target specific genomic areas through interactions with the remodeling of chromatin complexes and histone modifiers (Gebrie, 2023).

21.4.2 Chromatin remodeling complexes

In plants, the remodeling of chromatin complexes is essential for the epigenetic control of gene expression, particularly under stressful circumstances. Stress-responsive gene expression is influenced by these complexes because they change the chromatin shape, which affects DNA accessibility for transcription machinery and regulatory proteins. Thanks to chromatin remodeling complexes, which induce structural changes that either activate or repress particular genes, plants can adapt and survive in harsh settings (Klemm *et al.*, 2019). This mechanism is essential for plant adaptation and survival. Plant stress-induced complexes modify gene expression by rearranging, removing, or altering nucleosome composition through ATP hydrolysis. Histones can be post-translationally modified by enzymes, which can either promote or prevent transcription. Transcription factors, polymerases, and other nuclear proteins cannot access dense chromatin structures, but stress signals cause particular chromatin structural alterations that increase DNA accessibility. Plants can rapidly adapt to various stressors without modifying their genetic code because chromatin remodeling complexes provide quick and reversible changes in gene expression in response to stress.

These complexes also cooperate with DNA methylation and short RNA-mediated silencing mechanisms to orchestrate a thorough response to stress. Plant adaptation, resilience, and stress response largely depend on how these many levels of epigenetic control interact. According to a recent study, the specificity and activity of chromatin remodeling complexes can also be influenced by environmental factors (Holoch and Moazed, 2015). This finding emphasizes the relationship between external stimuli and epigenetic alterations that control the expression of genes that respond to stress.

21.4.3 CRISPR/Cas9 technology

Plant genome editing can be effectively accomplished with the help of CRISPR/Cas9. This method allows researchers to alter particular genes involved in plant signaling pathways and investigate how those modifications affect signaling pathways.

CRISPR/Cas9 technology, derived from bacterial immune systems, is a flexible tool for gene editing that allows for large-scale DNA deletions, variations, and insertions. The target DNA sequence and a short-guide RNA (sgRNA) must have base-pairing complementarity for this method to function. The 20 target-specific nucleotides that make up the sgRNA direct the Cas9 endonuclease to the target site, where it is joined by the PAM (Protospacer-Adjacent Motif) repeat NGG, which is required for Cas9 recruitment. After Cas9 causes double-stranded breaks at the target region, the cell modifies the target sequence by either non-homologous end joining or homology-directed repair. CRISPR/Cas9 can detect methylated DNA, unlike ZFNs and TALENs. This RNA-guided gene editing method has improved our knowledge of several animal and yeast signaling networks, including the bone marrow survival strategy in mantle cell lymphoma cells and neural circuit abnormalities in induced pluripotent stem cells from humans (iPSCs) (Li *et al.*, 2020). Although CRISPR/Cas9 has made great strides in cell biology and *in vivo* studies, its application in plants is still early. The initial efforts have concentrated on proof-of-concept experiments to assess the tool's effectiveness and selectivity in plant cell cultures and live plants. Exhibited in model plants like *Nicotiana benthamiana* and *Arabidopsis*, as well as in crops like rice, tomato, and maize, is the effective target of the phytoene desaturase (PDS) gene, which causes chlorophyll oxidation and an albino phenotype. Recent successes with editing in woody plants have demonstrated how widely applicable the approach is (Wada *et al.*, 2020).

Studies on plant signaling networks have also been reported with CRISPR/Cas9. One example is the G1-T regulatory region upstream of the OsRAV2 promoter, which imparts NaCl sensitivity *in vivo*. This was discovered and verified by CRISPR-directed mutagenesis. This highlights CRISPR's efficiency in the rapid identification and functional investigation of regulatory elements in plants, which would take significantly longer using traditional approaches. Regarding CRISPR and TALE-based approaches, the editing events must be carefully chosen.

21.4.4 Histone modification analysis

Histone protein post-translational modifications (PTMs), which change the chromatin structure, impact the transcriptional control of gene expression. Writers add modifications, erasers erase them, and readers detect these changes and initiate downstream signaling pathways. Together, these three entities dynamically regulate these alterations. Different transcriptional states are correlated with specific histone PTMs: transcriptional activation is indicated by H3K4me, H3K36me, H3K9ac, and H3K27ac, whereas transcriptional silencing is linked to H3K27me3. Transposable elements (TEs) and repeats are highly concentrated in heterochromatic regions where H3K9me2 and H3K9me3 are abundant. These elements function as constitutive repressive elements there. Specifically, H3K9me2, which correlates with low expression levels, covers complete genes in euchromatic areas (Millán-Zambrano *et al.*, 2022). Plant histone and DNA methylation are linked and are essential for TE silencing, as demonstrated by the relationship between the histone methyltransferase Kryptonite/SUVH4 (KYP) and CMT3, which binds to KYP. According to Zhang *et al.* (2018), histone markers such as H3K9me2 affect neighboring NLR gene transcription and TE-regulated gene DNA methylation.

Research employing *Arabidopsis* variants has demonstrated the importance of DNA and H3K9 methylation in resistance to biotic stressors such as Pst infection and *Myzus persicae*, the green peach aphid, infestation. Phyp and nrpd1 variants showed increased resistance to aphids, suggesting that the relationship between H3K9 methylation and DNA plays a role in this resistance. However, the specific processes remain unclear (Züst and Agrawal, 2016).

21.4.5 DNA methylation analysis

Strategies like MethylC-seq and MeDIP-seq can be used to assess changes in DNA methylation that plants experience in response to signaling events. DNA methylation in plants occurs on cytosine residues at the symmetric CG, CHG, and asymmetric CHH (where H = A, C, or T) sites. The balance of methylation and demethylation mechanisms controls DNA methylation dynamics. DNA methylation is a significant epigenetic marker. Evidence indicates that methyl groups substantially impact gene expression regulation inside the genome. Most complicated diseases, including cancer and neurological disorders, have abnormal DNA methylation profiles. Methyl groups in CpG contexts can be written, removed, and read. DNA methylation occurs when DNMT1 and DNMT3a/3b add a methyl group to cytosine with SAM as a substrate (Halabian *et al.*, 2021). DNMT1 operates on hemimethylated DNA to preserve methylation patterns after replication, whereas DNMT3a/3b methylates DNA *de novo*. TET-1, TET-2, and TET-3 enzymes serve as oxygenases, removing methyl groups from 5-hydroxymethylcytosine (5hmC), 5-formyl cytosine (5fmC), and 5-carboxyl methylcytosine (5CamC). Methyl binding domain proteins (MBDs) act as readers by identifying and binding to methylated CpG sites and then translating these changes via interactions with other proteins, such as splicing-related proteins and transcription factors (David, 2012).

Since they cannot replicate methyl groups, conventional molecular methods like PCR amplification and cloning are inappropriate for detecting DNA methylation. Thus, the initial methylation DNA strand must be pretreated to discriminate between methylated and unmethylated sites independently or across the genome. Variations in phenotypes can result from dynamic DNA methylation alterations among species. Methylation of 5-methylcytosine in CpG dinucleotides influences X chromosome inactivation, genome imprinting, and gene regulation during human development (Deniz *et al.*, 2019). Transcription can be inhibited by increased DNA methylation density by methyl-binding protein recruitment, regulatory region obstruction, chromatin remodeling, and inhibiting transcription factor binding. Dynamic alterations in genome methylation patterns might cause changes in gene expression in response to varied stimuli (Kaluscha *et al.*, 2022). For example, cancer is associated with normal and abnormal methylation patterns. Choosing the best approach for measuring methylation necessitates considering several parameters, as each technique has unique advantages and limits.

The roles of normal and aberrant epigenetic alterations must be understood through investigations into targeted and genome-wide DNA methylation. The most frequent method combines bisulfite treatment with PCR, sequencing, or another technique. Furthermore, hybridization and PCR procedures are frequently associated with techniques that employ methylation-sensitive restriction enzymes (Nichols *et al.*, 2022). Methyl-binding proteins and antibodies can also bind methyl groups with great affinity. DNA methylation procedures are characterized as low, medium, or high throughput, depending on the data generated. Researchers can unravel the intricate regulatory processes that govern plant signaling pathways and responses to varied stimuli using epigenetic methods.

21.4.6 Whole-genome bisulfite sequencing (WGBS)

WGBS is extensively used for methylome profiling owing to its single-base resolution capability, enabling the determination of DNA methylation levels at 90–95% of cytosine locations in the genome (Urich *et al.*, 2015). The development process involved applying bisulfite conversion to whole genome sequencing before library amplification. Specifically, the MethylC-seq library preparation approach was utilized by splitting the genomic DNA, followed by processes involving end-prep, adapter ligature, bisulfite conversion, and library amplification (Lister *et al.*, 2008). Unmethylated cytosines are converted to uracil after bisulfite exposure to DNA, but methylated cytosines stay intact; sequencing reveals the unmethylated cytosines as thymines after PCR amplification. The amount of thymines at one specific location is employed to calculate the methylation status of each cytosine once the WGBS data have been aligned to a reference (Frommer *et al.*, 1992). WGBS typically requires greater precision than whole-genome sequencing (WGS) for identifying single nucleotide polymorphisms to reliably estimate methylation levels (Schmitz *et al.*, 2022). A whole-genome methylation study was performed in hydrogen peroxide overproducing transgenic tobacco resilient to biotic and abiotic stresses (Villagómez-Aranda *et al.*, 2021). A whole-genome bisulfite sequencing study found a barely noticeable effect on methylation's total amount and distribution. The tendency toward hypomethylation was noted at 9432 differentially methylated sites across various genomic regions, most of which were in the context of CHG.

21.4.7 Chromatin immunoprecipitation (ChIP)

ChIP is frequently utilized in plant science to investigate chromatin configurations and the enrichment of post-transcriptional alterations within histones during development and regulate the expression of genes under stress (Li *et al.*, 2017). Although it remains challenging to apply in other plant systems, it is commonly utilized for a few model species, such as *Arabidopsis*. It depends on the employment of a particular antibody produced against an undesirable transcription element and the histone modifications under evaluation (Vimont *et al.*, 2020). The function involving two chromatin marks in modifying gene expression in peach buds and mesocarp tissues was investigated by Canton *et al.* (2022); marks included trimethylation of histone H3 at lysine in position 4 (H3K4me3) and trimethylation of histone H3 at lysine 27 (H3K27me3). The techniques may prove relevant in determining the role of histone changes in regulating gene expression throughout bud dormancy and core ripening in fruits.

21.5 Diverse Signaling Pathways in Epigenetic Regulations

21.5.1 Sugar signaling in plants

Epigenetic changes in plants alter how genes are expressed, allowing them to adjust to different environments. This happens through chemical modifications on DNA or histones, which control which genes are turned on or off. Sugars are carbohydrates that are important for plants' energy. They also act as signaling molecules to regulate gene expression and are involved in their growth and development. Different chemical changes in the genetic material of plants help them detect and relay these sugar signals (Chen *et al.*, 2022a; Chen *et al.*, 2022b). In plants, sugars like sucrose are made in certain parts called source tissues, then moved to other parts, known as sink tissues, using sugar transporters (Wipf *et al.*, 2021. Sugar signaling controls various aspects of plant growth and development, including cell division, embryo development, seed germination, seedling growth, root and branch formation, stomata development, flowering, and ageing. Additionally, it helps plants survive in response to both living (biotic) and non-living (abiotic) stressors (Moore *et al.*, 2003). Glucose (Glc) is the primary sugar that sends signals in plants, although others like fructose, sucrose, and trehalose-6-phosphate also play roles in controlling gene expression.

Histone acetylation and deacetylation: masters of sugar signaling regulation

One typical epigenetic modification that is essential for regulating several biological processes is histone acetylation. Histone acetyltransferases (HATs), which add acetyl groups to histone tails, and histone deacetylases (HDACs), which remove these acetyl groups, are the two types of enzymes that are essential to this process (Shen *et al.*, 2015). When glucose levels are at normal physiological levels (15 mM), a protein called HAC1 (histone acetyl transferease1) interacts with two transcription factors: WRKY18 and WRKY53. Together, they add a unique chemical mark known as H3K27ac to genes that are activated by sugar, such as DFR (Dihydroflavonol 4-Reductase), Chalcone Synthase (CHS), and Glucose-6-phosphate/ Phosphate Translocator2 (GPT2). This gene activation prompts the movement of sugars to plastids for storing starch or to vacuoles for accumulating anthocyanin (Chen *et al.*, 2019). Additionally, HAC1 is crucial for triggering immune responses in *Arabidopsis* and collaborates with MEDIATOR 25 (MED25) to signal with jasmonic acid (Chen *et al.*, 2022b). HAC1 teams up with MED8/25 to jointly control flowering and the transition to flowering in plants (Guo *et al.*, 2021). Moreover, MED8/25 is critical for regulating gene expression in response to glucose treatment, highlighting the essential role of the Mediator complex in sugar signaling (Seguela-Arnaud *et al.*, 2015).

Plants control gene expression by adjusting the activity of HDACs, which decrease histone acetylation levels. The WRKY53 protein, known for its quick response to sugar, directly interacts with HDA9, belonging to the RPD3/ HDA1 family of histone deacetylases, and has been the subject of extensive research due to its function as a gene suppressor in various aspects of *Arabidopsis* growth and development, such as seed dormancy and germination (van Zanten *et al.*, 2014). This interaction reduces the levels of a specific histone acetylation mark, H3K27ac, on genes linked to leaf aging and responses to stress (Chen *et al.*, 2016a).

HIGH-LEVEL EXPRESSION OF SUGAR INDUCIBLE GENE2 (HSI2), also known as VAL1, is a B3 domain transcriptional repressor, works alongside its close relative HSI2-LIKE1 (HSL1) to suppress seed maturation after the seed has started to grow (Sharma *et al.*, 2013). Sucrose triggers the activation of gene HSI2 and its counterpart HSL1/VAL2, which work together to remove certain histone modifications by deploying histone-modifying enzymes. This process prevents the expression of genes associated with seed maturation in young seedlings. During the process of seed germination, a set of genes involved in seed maturation, known as LAFL (LEAFY COTYLEDON1, LEC2, ABA INSENSITIVE3, ABI3, and FUSCA3, FUS3), are turned off to facilitate the switch from seed maturation to the growth of vegetative tissues. HSI2 and HSL1 proteins join together to form pairs, either homodimers or heterodimers (Chhun *et al.*, 2016). These pairs physically interact with a protein called MED13. HSI2 interacts with HDA6, and HSL1 interacts with HDA19, enzymes responsible for removing acetyl groups from histones H3 or H4. This action leads to the suppression of LAFL gene expression. Altogether, HSI2 and HSL1 are crucial for the shift from seed to seedling growth triggered by sugar signaling (Chen *et al.*, 2020).

Sugar signaling and DNA methylation: harmonizing gene regulation in a symphony of cellular dynamics

DNA methylation is a crucial change in specific patterns of the DNA sequence, involving the chemical addition of a methyl group to cytosine bases. This process, occurring in particular sequence contexts known as CHH, CG, and CHG (where H stands for any base except guanine), helps control how genes are turned on or off and maintains the stability of the genome. DNA methylation does not alter the genetic code (Henderson and Jacobsen, 2007). In plants, different enzymes add methyl groups to specific types of cytosine sites in DNA. For CG sites, methyltransferase 1 (MET1) is usually in charge. Chromomethylase 3 (CMT3) takes care of CHG sites, and domains rearranged methyltransferase 1/2 (DRM1/2) along with chromomethylase 2 (CMT2) handle CHH sites (Kou *et al.*, 2021). The glucose (Glc)–Target of the Rapamycin (TOR) signaling pathway regulates *Arabidopsis*'s total DNA methylation levels, which affects seedling growth. Global DNA methylation decreases when TOR activity is suppressed, while methylation rises only in promoter regions near CHH sites. The changes in DNA methylation patterns impact numerous physiological systems, such as carbon metabolism, amino acid biosynthesis, and phytohormone signaling. The decrease in MET1 gene activity and the increase in DRM1 and DRM2 gene activity are linked to changes in DNA methylation levels. TOR signaling pathway plays a role in this process by controlling overall DNA methylation, affecting plant growth (Chen *et al.*, 2022a).

21.5.2 From defense signaling to epigenetics: unraveling the molecular complexity of plant immunity

Plants have developed advanced defense systems to cope with harmful microorganisms threatening their growth and survival in changing environments. To protect themselves from hazardous infections, plants have evolved natural defense systems. Two defense mechanisms that are part of these systems are known as effector-triggered immunity (ETI) and pathogen-associated molecular pattern (PAMP)-triggered immunity (PTI) (Chisholm *et al.*, 2006; Dodds and Rathjen, 2010). PTI is the initial line of defense in plants, activated by pattern recognition receptors (PRRs) that detect PAMPs from pathogens or endogenous signals known as damage-associated molecular patterns (DAMPs) released during pathogen attacks (Boller and Felix, 2009). On the flip side, ETI is usually controlled by receptors known as nucleotide-binding (NB)-leucine-rich repeat (LRR) receptors (Takken and Goverse, 2012). In rice, using a DNA demethylating agent called 5-azadeoxycytidine increases resistance to *Xanthomonas* bacteria, while boosting RdDM in ros1–4 plants reduces resistance to Pst DC3000 bacteria (Akimoto *et al.*, 2007).

In plant defense against biotic threats, HATs and HDACs play pivotal roles. An illustration of acetyltransferase activity is demonstrated by ELP2 and ELP3, components of the elongator complex, which facilitate the addition of acetyl groups to histones. These proteins are integral for effector-triggered immunity (ETI), a precise defense mechanism, and the fundamental defense response. Notably, mutations in the ELP3 gene of *Arabidopsis* plants render them more vulnerable to the effects of *Pseudomonas syringae* pv. *maculicola*, a specific bacterium (DeFraia *et al.*, 2013). Moreover, these mutated plants exhibit a delayed response in activating defense genes when exposed to pathogen attacks or salicylic acid (SA).

For instance, when specific genes like hda19 and hda6, mutants of *Arabidopsis* plants, they produce more of the defense-related genes PR1 and PR2 (Choi *et al.*, 2012; Wang *et al.*, 2017) and become more resistant to a bacterial pathogen called Pst. Similarly, when another gene called SRT2 is changed, resulting in mutants, these plants also produce more PR1 and show resistance to Pst infection (Wang *et al.*, 2010). In rice, when the HDT701 gene is turned off, it causes an increase in the activity of defense-related genes and makes the plants resistant to diseases caused by *Xanthomonas oryzae* pv. *oryzae* and *Magnaporthe oryzae*. Moreover, in *Arabidopsis*, a gene called HDA9 negatively influences plant defense by controlling the expression of another gene called NLR. When both HDA9 and its partner gene HOS15 are mutated, the plants have higher levels of expression of the

NLR gene SCN1 and can better resist Pst infection (Yang *et al.*, 2020).

21.5.3 Deciphering the epigenetic landscape: nitrogen signaling in focus

Nitrogen (N) is necessary for plant growth because it is a constituent of DNA, RNA, and proteins. Plants depend on effective nitrogen absorption and assimilation to get, retain, and metabolize nutrients for maximum growth (Krapp *et al.*, 2014). Through signal messengers (including calcium and transcription factors), receptors, including PROTEIN 7 (NLP7) and NITRATE TRANSPORTER 1.1 (NRT1.1), can rapidly modify plant transcriptional responses in response to nitrogen. Histone changes controlling nitrogen absorption in dicot plants include tri-methylation of lysine 36 or 27 on histone H3 (H3K36me3, H3K27me3). The H3K36me3 mark is added to genes in *Arabidopsis* by the protein SET DOMAIN GROUP8 (SDG8), which controls genes involved in energy metabolism, such as those for nitrogen metabolism (NMGs) and photosynthesis (Zhang *et al.*, 2023). SDG8 regulates the reprogramming of gene expression caused by nitrogen, influencing genes such as NRT2.1, NRT1.5, and GLUTAMINE SYNTHETASE 1;4. Mutant plants lacking SDG8 exhibit lower nitrate acquisition than wild-type plants cultivated under high nitrogen (Li *et al.*, 2019). Nitrogen supply influences root development and plant architecture, specifically tillering in monocots such as rice. Nitrogen fertilizer in rice produces genome-wide alterations in the histone modification H3K27me3 via NITROGEN-MEDIATED TILLER GROWTH RESPONSE 5 (NGR5). By altering H3K27me3, this process enlists the PRC2 complex, which represses genes that restrict branching, like Dwarf14 (D14) and SQUAMOSA-PROMOTER BINDING PROTEIN-LIKE 14 (SPL14) (Wu *et al.*, 2020). Recent research on monocots, notably wheat varieties, has revealed diverse techniques used by different varieties to adapt to low nitrogen (LN) conditions via histone changes. For example, elevated H3K27ac and reduced H3K27me3 levels in Kenong9204 (KN9204) promote the expression of genes involved in root formation in low-nitrogen environments. Nguyen *et al.* (2013) reported that the modification pattern of H3K27me3 is altered when SWINGER (SWN), a component of the PRC2 complex, is knocked down. This alteration affects root growth and nitrogen uptake. Nitrogen levels and auxin transport channels interact in dicot plants to affect the architecture of the root system. Histone alterations modulate auxin transport pathways, and histone deacetylase inhibitors (HDIs) promote PIN-FORMED 1 (PIN1) degradation by increasing the acetylation of nucleosome core histones in chromatin (Nguyen *et al.*, 2013). DNA methylation is essential for passing stress responses to offspring, especially in rice and other nitrogen-deficient crops. In rice, alterations in cytosine methylation patterns are critical for guaranteeing sustained inheritance of adaptive characteristics in response to nitrogen deprivation. The offspring of nitrogen-deficient rice plants exhibit enhanced tolerance to nitrogen scarcity, as shown by longer plant stature, increased plant mass, and higher nitrogen levels (Kou *et al.*, 2011). In maize, nitrogen deficit affects the overall methylation of CG sites in the root, with considerable regions displaying lower methylation levels than under normal nitrogen conditions (Mager and Ludewig, 2018). Conversely, nitrogen shortage in rice leaves has no significant effect on overall methylation levels (Kou *et al.*, 2011), which could be attributed to species-specific responses or discrepancies in experimental set-ups.

21.6 Recent Achievements in Plant Epigenetic Research

21.6.1 Understanding plant regeneration

Plant regeneration involves epigenetic processes that coordinate crucial changes in gene expression essential for tissue repair and growth. During plant regeneration, plant cells experience numerous swift transitions in cell fate. These transitions are accompanied by reprogramming the transcriptome and the chromatin landscape. Phytohormones induce cell identity genes, particularly transcription factor (TF) genes, to participate in plant regeneration (Sang *et al.*, 2018; Ikeuchi *et al.*, 2019; Sugimoto *et al.*,

2019; Mathew and Prasad, 2021). Histone modifications, including methylation and acetylation, are central to this composition and have a significant role in regulation. Histone methyltransferases (HMTs) catalyze the trimethylation of histones, such as H3K4me3, thereby activating genes associated with regeneration.

Conversely, HATs facilitate histone acetylation, exemplified by H3K9ac and H4K5ac, fostering an accessible chromatin configuration conducive to regeneration-related gene expression. Notably, dynamic alterations in DNA methylation patterns during regeneration involve the demethylation of specific genes, a prerequisite for their activation. Additionally, post-transcriptional fine-tuning of gene expression is achieved through the regulatory actions of miRNAs. These intricate epigenetic mechanisms empower plants to adapt, regenerate damaged tissues, and flourish amidst changing environmental conditions (Liu *et al.*, 2023).

21.6.2 Epigenetics in crop biotechnology

During the reproductive process in plants like *Arabidopsis*, rice, maize, and others, a specialized type of DNA modification known as DNA methylation occurs. This modification, which is specific to cells that will become reproductive cells, is guided by a process called RNA-directed DNA methylation (RdDM) (Chen *et al.*, 2016b).

21.6.3 Environmental adaptation and stress responses

Rapid plant adaptability and immunity depend on understanding epigenetic responses to biotic and abiotic environmental stress. Plant stress responses are influenced by various epigenetic regulatory mechanisms, particularly those involving transcription. When rice seedlings are submerged, they demonstrate higher H3K4 methylation and H3 acetylation levels on the submergence-inducible genes ADH1 and PDC1 (Tsuji *et al.*, 2006). An analogous phenomenon has been found in *Arabidopsis*, where drought stress induces histone changes at loci such as RD29A, RD29B, RD20, and At2g20880 (Kim *et al.*, 2008). Moreover, the histone deacetylase family members HDA6 and HDA19 exhibit elevated expression levels in response to environmental stressors such as low temperature, injury, or hormonal signals, indicating their potential involvement in the regulation of genes linked to stress (Zhou *et al.*, 2005). These changes resulting from epigenetic modifications underscore the significance of transcriptional control in plant adaptability and stress responses.

21.6.4 Breeding applications

DNA methylation effects on heterosis

The epistasis, dominant, and overdominant models each show areas for development; thus, even with their remarkable achievements, they fall short of a thorough explanation of heterosis (Khotyleva *et al.*, 2017). Moreover, these models frequently fail to provide a complete explanation for this occurrence (Groszmann *et al.*, 2013). In addition to genetic considerations, epigenetics may influence heterosis by contributing to non-Mendelian inheritance, cell destiny determination, and gene expression regulation (McKeown and Spillane 2014). One example of an epigenetic mechanism is DNA methylation, attaching a methyl group to the cytosine bases in DNA molecules. It can regulate gene expression by influencing how genes are turned on and off. Variations in DNA methylation patterns can affect the expression of specific genes in hybrids, affecting their phenotypes. For example, maize hybrids have DNA methylation profiles different from those of their inbred parents. Methylation levels link negatively with hybrid vigor for features like 'number of kernel rows' but positively with traits like 'kernel numbers'.

DNA methylation and inbreeding

When individuals with similar genetic make-up mate, a process known as inbreeding occurs, which can result in substantial genetic and epigenetic alterations. DNA methylation is an essential epigenetic mechanism caused by inbreeding (Labroo *et al.*, 2021). As shown by differences in DNA methylation between inbred parents and their F1 hybrid offspring, several studies emphasize the influence of inbreeding on forming epigenetic changes (Vergeer *et al.*,

2012; Lauss *et al.*, 2018; Pang *et al.*, 2019). These results clarify how important it is for inbreeding to drive epigenetic changes, which advances our knowledge of the dynamics of plant breeding.

21.6.5 Epigenetic memory

Epigenetics, now widely characterized, is concerned with heritable alterations to gene expression patterns that occur without mutations to the DNA sequence, including mitotic and/or meiotic events. Current research mainly focuses on chemical chromatin modifications, which affect transcriptional silence and activation by altering histone proteins and DNA. Methylation of DNA, a frequent covalent alteration of DNA, is essential in transcription regulation and has been extensively studied in plants and humans. In plants, specific short RNAs direct DNA methylation at homologous genomic locations in a process known as RNA-directed DNA methylation (RdDM).

According to Alvarez-Venegas *et al.*, 2003, trithorax (ATX1), a homolog of histone H3K4 methyltransferase, is a critical epigenetic regulator in beans. Rice plants have a short-term memory of rehydration due to endogenous processes such as DNA methylation and the plant hormone abscisic acid. These elements probably cause genes related to proline biosynthesis and photosynthesis to be transcriptionally activated, which improves the plant's capacity for flexible response to a succession of challenges (Li *et al.*, 2019a).

21.7 Conclusion and Future Prospects

Plant epigenetic regulation in response to signaling is a crucial field under investigation with promising potential implications. Epigenetic changes significantly impact gene expression and can alter plant responses to various stresses. Knowledge of the epigenetic processes underlying plant cell signaling can help increase crop resilience and boost agricultural and forest output. The reliance of plants on epigenetic control expresses their evolutionary history, way of existence, and development mode. Using the epigenetic regulatory systems, plants may effectively thrive and flourish in various uncertain situations by facilitating metastable alterations to gene activity and fine-tuning gene expression patterns. The molecular diversity, complicated development, and adaptive capacity of plants are probably greatly influenced by the vast array of epigenetic mechanisms that they possess. Wide arrays of tools frequently employed in plant cell signaling have been studied. When subjected to stress, the mechanisms underlying the control of plant epigenetics bear significant consequences. Scientists may improve stress tolerance in different crops and forest trees by figuring out the epigenetic regulatory networks involved in stress responses; changing DM patterns or histone alterations could help crops resist drought, salinity, and disease attacks. Moreover, discovering stress-responsive ncRNAs and their intended applications may open new paths for gene editing and breeding efforts to improve stress tolerance in various plants and crops.

References

Abdulraheem, M.I., Xiong, Y., Moshood, A.Y., Cadenas-Pliego, G., Zhang, H. *et al.* (2024) Mechanisms of plant epigenetic regulation in response to plant stress: Recent discoveries and implications. *Plants* 13(2), 163. DOI: 10.3390/plants13020163.

Agius, F., Kapoor, A. and Zhu, J.K. (2006) Role of the *Arabidopsis* DNA glycosylase/lyase ROS1 in active DNA demethylation. *Proceedings of the National Academy of Sciences of the United States of America* 103(31), 11796–11801. DOI: 10.1073/pnas.0603563103.

Akimoto, K., Katakami, H., Kim, H.-J., Ogawa, E., Sano, C.M. *et al.* (2007) Epigenetic inheritance in rice plants. *Annals of Botany* 100(2), 205–217. DOI: 10.1093/aob/mcm110.

Alvarez-Venegas, R., Pien, S., Sadder, M., Witmer, X., Grossniklaus, U. *et al.* (2003) ATX-1, an arabidopsis homolog of trithorax, activates flower homeotic genes. *Current Biology* 13(8), 627–637. DOI: 10.1016/S0960-9822(03)00243-4.

Bannister, A.J. and Kouzarides, T. (2011) Regulation of chromatin by histone modifications. *Cell Research* 21(3), 381–395. DOI: 10.1038/cr.2011.22.

Berger, S.L. (2007) The complex language of chromatin regulation during transcription. *Nature* 447(7143), 407–412. DOI: 10.1038/nature05915.

Betti, F., Ladera-Carmona, M.J., Weits, D.A., Ferri, G., Iacopino, S. *et al.* (2021) Exogenous miRNAs induce post-transcriptional gene silencing in plants. *Nature Plants* 7(10), 1379–1388. DOI: 10.1038/s41477-021-01005-w.

Boller, T. and Felix, G. (2009) A renaissance of elicitors: Perception of microbe-associated molecular patterns and danger signals by pattern-recognition receptors. *Annual Review of Plant Biology* 60(1), 379–406. DOI: 10.1146/annurev.arplant.57.032905.105346.

Bossdorf, O., Richards, C.L. and Pigliucci, M. (2008) Epigenetics for ecologists. *Ecology Letters* 11(2), 106–115. DOI: 10.1111/j.1461-0248.2007.01130.x.

Bourbousse, C., Mestiri, I., Zabulon, G., Bourge, M., Formiggini, F. *et al.* (2015) Light signaling controls nuclear architecture reorganization during seedling establishment. *Proceedings of the National Academy of Sciences of the United States of America* 112(21), E2836–44. DOI: 10.1073/pnas.1503512112.

Bräutigam, K. and Cronk, Q. (2018) DNA methylation and the evolution of developmental complexity in plants. *Frontiers in Plant Science* 9, 1447. DOI: 10.3389/fpls.2018.01447.

Bräutigam, K., Vining, K.J., Lafon-Placette, C., Fossdal, C.G., Mirouze, M. *et al.* (2013) Epigenetic regulation of adaptive responses of forest tree species to the environment. *Ecology and Evolution* 3(2), 399–415. DOI: 10.1002/ece3.461.

Canton, M., Farinati, S., Forestan, C., Joseph, J., Bonghi, C. *et al.* (2022) An efficient chromatin immunoprecipitation (chIP) protocol for studying histone modifications in peach reproductive tissues. *Plant Methods* 18(1), 43. DOI: 10.1186/s13007-022-00876-0.

Cedar, H. and Bergman, Y. (2009) Linking DNA methylation and histone modification: Patterns and paradigms. *Nature Reviews. Genetics* 10(5), 295–304. DOI: 10.1038/nrg2540.

Chan, A.P., Melake-Berhan, A., O'Brien, K., Buckley, S., Quan, H. *et al.* (2008) The highest-copy repeats are methylated in the small genome of the early divergent vascular plant *Selaginella moellendorffii*. *BMC Genomics* 9(1), 282. DOI: 10.1186/1471-2164-9-282.

Chen, J., Yang, S., Fan, B., Zhu, C. and Chen, Z. (2022b) The mediator complex: A central coordinator of plant adaptive responses to environmental stresses. *International Journal of Molecular Sciences* 23(11), 6170. DOI: 10.3390/ijms23116170.

Chen, N., Wang, H., Abdelmageed, H., Veerappan, V., Tadege, M. *et al.* (2020) HSI2/VAL1 and HSL1/VAL2 function redundantly to repress DOG1 expression in *Arabidopsis* seeds and seedlings. *New Phytologist* 227(3), 840–856. DOI: 10.1111/nph.16559.

Chen, Q., Zhang, J. and Li, G. (2022a). Dynamic epigenetic modifications in plant sugar signal transduction. *Trends in Plant Science*, 27(4), 379-390. DOI: 10.1016/j.tplants.2021.10.009.

Chen, Q., Xu, X., Xu, D., Zhang, H., Zhang, C. *et al.* (2019) WRKY18 and WRKY53 coordinate with HISTONE ACETYLTRANSFERASE1 to regulate rapid responses to sugar. *Plant Physiology* 180(4), 2212–2226. DOI: 10.1104/pp.19.00511.

Chen, X., Lu, L., Mayer, K.S., Scalf, M., Qian, S. *et al.* (2016a) POWERDRESS interacts with HISTONE DEACETYLASE 9 to promote aging in *Arabidopsis*. *eLife* 5, e17214. DOI: 10.7554/elife.17214.

Chen, Y., Müller, F., Rieu, I. and Winter, P. (2016b) Epigenetic events in plant male germ cell heat stress responses. *Plant Reproduction* 29(1–2), 21–29. DOI: 10.1007/s00497-015-0271-5.

Chhun, T., Chong, S.Y., Park, B.S., Wong, E.C.C., Yin, J.-L. *et al.* (2016) HSI2 repressor recruits MED13 and HDA6 to down-regulate seed maturation gene expression directly during *Arabidopsis* early seedling growth. *Plant and Cell Physiology* 57(8), 1689–1706. DOI: 10.1093/pcp/pcw095.

Chisholm, S.T., Coaker, G., Day, B. and Staskawicz, B.J. (2006) Host-microbe interactions: Shaping the evolution of the plant immune response. *Cell* 124(4), 803–814. DOI: 10.1016/j.cell.2006.02.008.

Choi, S.-M., Song, H.-R., Han, S.-K., Han, M., Kim, C.-Y. *et al.* (2012) HDA19 is required for the repression of salicylic acid biosynthesis and salicylic acid-mediated defense responses in *Arabidopsis*. *The Plant Journal* 71(1), 135–146. DOI: 10.1111/j.1365-313x.2012.04977.x.

Choi, Y., Gehring, M., Johnson, L., Hannon, M., Harada, J.J. *et al.* (2002) DEMETER, a DNA glycosylase domain protein, is required for endosperm gene imprinting and seed viability in *Arabidopsis*. *Cell* 110(1), 33–42. DOI: 10.1016/s0092-8674(02)00807-3.

Clapier, C.R. and Cairns, B.R. (2009) The biology of chromatin remodeling complexes. *Annual Review of Biochemistry* 78(1), 273–304. DOI: 10.1146/annurev.biochem.77.062706.153223.

Clapier, C.R., Iwasa, J., Cairns, B.R. and Peterson, C.L. (2017) Mechanisms of action and regulation of ATP-dependent chromatin-remodelling complexes. *Nature Reviews. Molecular Cell Biology* 18(7), 407–422. DOI: 10.1038/nrm.2017.26.

Cokus, S.J., Feng, S., Zhang, X., Chen, Z., Merriman, B. *et al.* (2008) Shotgun bisulphite sequencing of the *Arabidopsis* genome reveals DNA methylation patterning. *Nature* 452(7184), 215–219. DOI: 10.1038/nature06745.

Cuerda-Gil, D. and Slotkin, R.K. (2016) Non-canonical RNA-directed DNA methylation. *Nature Plants* 2(11), 16163. DOI: 10.1038/nplants.2016.163.

David, R. (2012) DNMT1 learns to be picky. *Nature Reviews Molecular Cell Biology* 13(3), 139–139. DOI: 10.1038/nrm3301.

David, R., Burgess, A., Parker, B., Li, J., Pulsford, K. *et al.* (2017) Transcriptome-wide mapping of RNA 5-methylcytosine in *Arabidopsis* mRNAs and noncoding RNAs. *The Plant Cell* 29(3), 445–460. DOI: 10.1105/tpc.16.00751.

de Alba, A.E.M., Elvira-Matelot, E. and Vaucheret, H. (2013) Gene silencing in plants: A diversity of pathways. *Biochimica Et Biophysica Acta. Gene Regulatory Mechanisms* 1829(12), 1300–1308. DOI: 10.1016/j.bbagrm.2013.10.005.

Defraia, C.T., Wang, Y., Yao, J. and Mou, Z. (2013) Elongator subunit 3 positively regulates plant immunity through its histone acetyltransferase and radical S-adenosylmethionine domains. *BMC Plant Biology* 13(1), 102. DOI: 10.1186/1471-2229-13-102.

de la Barre, A.E. (2000) Core histone N-termini play an essential role in mitotic chromosome condensation. *The EMBO Journal* 19(3), 379–391. DOI: 10.1093/emboj/19.3.379.

Deneweth, J., Peer, Y. and Vermeirssen, V. (2022) Nearby transposable elements impact plant stress gene regulatory networks: A meta-analysis in *A. Thaliana* and *S. Lycopersicum*. *BMC Genomics* 23(1), 18. DOI: 10.1186/s12864-021-08215-8.

Deniz, Ö., Frost, J.M. and Branco, M.R. (2019) Regulation of transposable elements by DNA modifications. *Nature Reviews Genetics* 20(7), 417–431. DOI: 10.1038/s41576-019-0106-6.

Dodds, P.N. and Rathjen, J.P. (2010) Plant immunity: Towards an integrated view of plant-pathogen interactions. *Nature Reviews. Genetics* 11(8), 539–548. DOI: 10.1038/nrg2812.

Dowen, R.H., Pelizzola, M., Schmitz, R.J., Lister, R., Dowen, J.M. *et al.* (2012) Widespread dynamic DNA methylation in response to biotic stress. *Proceedings of the National Academy of Sciences of the United States of America* 109(32), E2183–2191. DOI: 10.1073/pnas.1209329109.

Duarte-Aké, F., Us-Camas, R. and De-la-Peña, C. (2023) Epigenetic regulation in heterosis and environmental stress: The challenge of producing hybrid epigenomes to face climate change. *Epigenomes* 7(3), 14. DOI: 10.3390/epigenomes7030014.

Earley, K.W., Shook, M.S., Brower-Toland, B., Hicks, L. and Pikaard, C.S. (2007) In vitro specificities of *Arabidopsis* co-activator histone acetyltransferases: Implications for histone hyperacetylation in gene activation. *The Plant Journal* 52(4), 615–626. DOI: 10.1111/j.1365-313x.2007.03264.x.

Eberharter, A. and Becker, P.B. (2002) Histone acetylation: A switch between repressive and permissive chromatin. *EMBO Reports* 3(3), 224–229. DOI: 10.1093/embo-reports/kvf053.

Espinas, N.A., Saze, H. and Saijo, Y. (2016) Epigenetic control of defense signaling and priming in plants. *Frontiers in Plant Science* 7, 1201. DOI: 10.3389/fpls.2016.01201.

Eun, C., Lorkovic, Z.J., Sasaki, T., Naumann, U., Matzke, A.J.M. *et al.* (2012) Use of forward genetic screens to identify genes required for RNA-directed DNA methylation in *Arabidopsis thaliana*. *Cold Spring Harbor Symposia on Quantitative Biology* 77, 195–204. DOI: 10.1101/sqb.2012.77.015099.

Feng, S., Cokus, S.J., Zhang, X., Chen, P.-Y., Bostick, M. *et al.* (2010) Conservation and divergence of methylation patterning in plants and animals. *Proceedings of the National Academy of Sciences of the United States of America* 107(19), 8689–8694. DOI: 10.1073/pnas.1002720107.

Flaus, A. (2006) Identification of multiple distinct snf2 subfamilies with conserved structural motifs. *Nucleic Acids Research* 34(10), 2887–2905. DOI: 10.1093/nar/gkl295.

Frommer, M., McDonald, L.E., Millar, D.S., Collis, C.M., Watt, F. *et al.* (1992) A genomic sequencing protocol that yields A positive display of 5-methylcytosine residues in individual DNA strands.

Proceedings of the National Academy of Sciences of the United States of America 89(5), 1827–1831. DOI: 10.1073/pnas.89.5.1827.

Gallego-Bartolomé, J., Liu, W., Kuo, P.H., Feng, S., Ghoshal, B. *et al.* (2019) Co-targeting RNA polymerases IV and V promotes efficient de novo DNA methylation in *Arabidopsis*. *Cell* 176(5), 1068–1082. DOI: 10.1016/j.cell.2019.01.029.

Gebrie, A. (2023) Transposable elements as essential elements in the control of gene expression. *Mobile DNA* 14(1), 9. DOI: 10.1186/s13100-023-00297-3.

Gong, Z., Morales-Ruiz, T., Ariza, R.R., Roldán-Arjona, T., David, L. *et al.* (2002) ROS1, a repressor of transcriptional gene silencing in *Arabidopsis*, encodes a DNA glycosylase/lyase. *Cell* 111(6), 803–814.

Gonsalvez, G.B., Tian, L., Ospina, J.K., Boisvert, F.M., Lamond, A.I. *et al.* (2007) Two distinct arginine methyltransferases are required for biogenesis of Sm-class ribonucleoproteins. *Journal of Cell Biology* 733–740. DOI: 10.1083/jcb.200702147.

Greer, E.L., Blanco, M.A., Gu, L., Sendinc, E., Liu, J. *et al.* (2015) DNA methylation on N6-adenine in C. elegans. *Cell* 161(4), 868–878. DOI: 10.1016/j.cell.2015.04.005.

Groszmann, M., Greaves, I.K., Fujimoto, R., Peacock, W.J. and Dennis, E.S. (2013) The role of epigenetics in hybrid vigour. *Trends in Genetics* 29(12), 684–690. DOI: 10.1016/j.tig.2013.07.004.

Guo, J., Wei, L., Chen, S.-S., Cai, X.-W., Su, Y.-N. *et al.* (2021) The CBP/p300 histone acetyltransferases function as plant-specific MEDIATOR subunits in *Arabidopsis*. *Journal of Integrative Plant Biology* 63(4), 755–771. DOI: 10.1111/jipb.13052.

Halabian, R., Arshad, V., Ahmadi, A., Saeedi, P., Jamalkandi, S.A. *et al.* (2021) Laboratory methods to decipher epigenetic signatures: A comparative review. *Cellular & Molecular Biology Letters* 26(1), 46. DOI: 10.1186/s11658-021-00290-9.

Hassa, P.O., Haenni, S.S., Elser, M. and Hottiger, M.O. (2006) Nuclear ADP-ribosylation reactions in mammalian cells: Where are we today and where are we going? *Microbiology and Molecular Biology Reviews* 70(3), 789–829. DOI: 10.1128/mmbr.00040-05.

He, Y. (2012) Chromatin regulation of flowering. *Trends in Plant Science* 17(9), 556–562. DOI: 10.1016/j.tplants.2012.05.001.

Henderson, I.R. and Jacobsen, S.E. (2007) Epigenetic inheritance in plants. *Nature* 447(7143), 418–424. DOI: 10.1038/nature05917.

Hennig, L. and Derkacheva, M. (2009) Diversity of polycomb group complexes in plants: Same rules, different players. *Trends in Genetics* 25(9), 414–423. DOI: 10.1016/j.tig.2009.07.002.

Hewezi, T. (2020) Epigenetic mechanisms in nematode-plant interactions. *Annual Review of Phytopathology* 58(1), 119–138. DOI: 10.1146/annurev-phyto-010820-012805.

Holoch, D. and Moazed, D. (2015) RNA-mediated epigenetic regulation of gene expression. *Nature Reviews. Genetics* 16(2), 71–84. DOI: 10.1038/nrg3863.

Hu, J., Manduzio, S. and Kang, H. (2019) Epitranscriptomic RNA methylation in plant development and abiotic stress responses. *Frontiers in Plant Science* 10, 500. DOI: 10.3389/fpls.2019.00500.

Ikeuchi, M., Favero, D.S., Sakamoto, Y., Iwase, A., Coleman, D. *et al.* (2019) Molecular mechanisms of plant regeneration. *Annual Review of Plant Biology* 70(1), 377–406. DOI: 10.1146/annurev-arplant-050718-100434.

Jablonka, E. (2013) Epigenetic inheritance and plasticity: The responsive germline. *Progress in Biophysics and Molecular Biology* 111(2–3), 99–107. DOI: 10.1016/j.pbiomolbio.2012.08.014.

Kaluscha, S., Domcke, S., Wirbelauer, C., Stadler, M.B., Durdu, S. *et al.* (2022) Evidence that direct inhibition of transcription factor binding is the prevailing mode of gene and repeat repression by DNA methylation. *Nature Genetics* 54(12), 1895–1906. DOI: 10.1038/s41588-022-01241-6.

Khotyleva, L.V., Kilchevsky, A.V. and Shapturenko, M.N. (2017) Theoretical aspects of heterosis. *Russian Journal of Genetics* 7(4), 428–439. DOI: 10.1134/s2079059717040049.

Kim, J.-M., To, T.K., Ishida, J., Morosawa, T., Kawashima, M. *et al.* (2008) Alterations of lysine modifications on the histone H3 N-tail under drought stress conditions in *Arabidopsis thaliana*. *Plant and Cell Physiology* 49(10), 1580–1588. DOI: 10.1093/pcp/pcn133.

Klemm, S.L., Shipony, Z. and Greenleaf, W.J. (2019) Chromatin accessibility and the regulatory epigenome. *Nature Reviews. Genetics* 20(4), 207–220. DOI: 10.1038/s41576-018-0089-8.

Knizewski, L., Ginalski, K. and Jerzmanowski, A. (2008) Snf2 proteins in plants: Gene silencing and beyond. *Trends in Plant Science* 13(10), 557–565. DOI: 10.1016/j.tplants.2008.08.004.

Koch-Nolte, F. (2008) Mammalian ADP-ribosyltransferases and ADP-ribosylhydrolases. *Frontiers in Bioscience* 13, 6716. DOI: 10.2741/3184.

Kou, H.P., Li, Y., Song, X.X., Ou, X.F., Xing, S.C. *et al.* (2011) Heritable alteration in DNA methylation induced by nitrogen-deficiency stress accompanies enhanced tolerance by progenies to the stress in rice (*Oryza sativa* L.). *Journal of Plant Physiology* 168(14), 1685–1693. DOI: 10.1016/j.jplph.2011.03.017.

Kou, S., Gu, Q., Duan, L., Liu, G., Yuan, P. *et al.* (2021) Genome-wide bisulphite sequencing uncovered the contribution of DNA methylation to rice short-term drought memory formation. *Journal of Plant Growth Regulation* 41(7), 2903–2917. DOI: 10.1007/s00344-021-10483-3.

Kouzarides, T. (2007) Chromatin modifications and their function. *Cell* 128(4), 693–705. DOI: 10.1016/j.cell.2007.02.005.

Krapp, A., David, L.C., Chardin, C., Girin, T., Marmagne, A. *et al.* (2014) Nitrate transport and signalling in *Arabidopsis*. *Journal of Experimental Botany* 65(3), 789–798. DOI: 10.1093/jxb/eru001.

Kumar, S., Chinnusamy, V. and Mohapatra, T. (2018) Epigenetics of modified DNA bases: 5-methylcytosine and beyond. *Frontiers in Genetics* 9, 640. DOI: 10.3389/fgene.2018.00640.

Labroo, M.R., Studer, A.J. and Rutkoski, J.E. (2021) Heterosis and hybrid crop breeding: A multidisciplinary review. *Frontiers in Genetics* 12, 643761. DOI: 10.3389/fgene.2021.643761.

Lauss, K., Wardenaar, R., Oka, R., Hulten, M.H.A., Guryev, V. *et al.* (2018) Parental DNA methylation states are associated with heterosis in epigenetic hybrids. *Plant Physiology* 176(2), 1627–1645. DOI: 10.1104/pp.17.01054.

Li, B., Carey, M. and Workman, J.L. (2007) The role of chromatin during transcription. *Cell* 128(4), 707–719. DOI: 10.1016/j.cell.2007.01.015.

Li, G., Jagadeeswaran, G., Mort, A. and Sunkar, R. (2017) ChIP-seq analysis for identifying genome-wide histone modifications associated with stress-responsive genes in plants. In: *Methods in Molecular Biology*. pp. 139–149. DOI: 10.1007/978-1-4939-7136-7_8.

Li, H., Yang, Y., Hong, W., Huang, M., Wu, M. *et al.* (2020) Applications of genome editing technology in the targeted therapy of human diseases: Mechanisms, advances and prospects. *Signal Transduction and Targeted Therapy* 5(1), 1. DOI: 10.1038/s41392-019-0089-y.

Li, P., Yang, H., Wang, L., Liu, H., Huo, H. *et al.* (2019a) Physiological and transcriptome analyses reveal short-term responses and formation of memory under drought stress in rice. *Frontiers in Genetics* 10, 55. DOI: 10.3389/fgene.2019.00055.

Li, Y., Kumar, S. and Qian, W. (2018) Active DNA demethylation: Mechanism and role in plant development. *Plant Cell Reports* 37(1), 77–85. DOI: 10.1007/s00299-017-2215-z.

Li, Y., Brooks, M., Yeoh-Wang, J., McCoy, R.M., Rock, T.M. *et al.* (2019) SDG8-mediated histone methylation and RNA processing function in the response to nitrate signaling. *Plant Physiology* 182(1), 215–227. DOI: 10.1104/pp.19.00682.

Lim, W.A. and Pawson, T. (2010) Phosphotyrosine signaling: Evolving a new cellular communication system. *Cell* 142(5), 661–667. DOI: 10.1016/j.cell.2010.08.023.

Lister, R., O'Malley, R.C., Tonti-Filippini, J., Gregory, B.D., Berry, C.C. *et al.* (2008) Highly integrated single-base resolution maps of the epigenome in *Arabidopsis*. *Cell* 133(3), 523–536. DOI: 10.1016/j.cell.2008.03.029.

Liu, C., Lu, F., Cui, X. and Cao, X. (2010) Histone methylation in higher plants. *Annual Review of Plant Biology* 61(1), 395–420. DOI: 10.1146/annurev.arplant.043008.091939.

Liu, R. and Lang, Z. (2020) The mechanism and function of active DNA demethylation in plants. *Journal of Integrative Plant Biology* 62(1), 148–159. DOI: 10.1111/jipb.12879.

Liu, S., Dunwell, T.L., Pfeifer, G.P., Dunwell, J.M., Ullah, I. *et al.* (2013) Detection of oxidation products of 5-methyl-2'-deoxycytidine in *Arabidopsis* DNA. *PLoS ONE* 8(12), e84620. DOI: 10.1111/jipb.12879.

Liu, W., Duttke, S.H., Hetzel, J., Groth, M., Feng, S. *et al.* (2018) RNA-directed DNA methylation involves co-transcriptional small-RNA-guided slicing of polymerase V transcripts in *Arabidopsis*. *Nature Plants* 4(3), 181–188. DOI: 10.1038/s41477-017-0100-y.

Liu, X., Zhu, K. and Xiao, J. (2023) Recent advances in understanding of the epigenetic regulation of plant regeneration. *aBIOTECH* 4(1), 31–46. DOI: 10.1007/s42994-022-00093-2.

López Sánchez, A., Stassen, J.H.M., Furci, L., Smith, L.M. and Ton, J. (2016) The role of DNA (de)methylation in immune responsiveness of *Arabidopsis*. *The Plant Journal* 88(3), 361–374. DOI: 10.1111/tpj.13252.

Lopez-Gomollon, S. and Baulcombe, D.C. (2022) Roles of RNA silencing in viral and non-viral plant immunity and in the crosstalk between disease resistance systems. *Nature Reviews. Molecular Cell Biology* 23(10), 645–662. DOI: 10.1038/s41580-022-00496-5.

Mager, S. and Ludewig, U. (2018) Massive loss of DNA methylation in nitrogen-, but not in phosphorus-deficient *Zea mays* roots is poorly correlated with gene expression differences. *Frontiers in Plant Science* 9, 497. DOI: 10.3389/fpls.2018.00497.

Malik, G., Dangwal, M., Kapoor, S. and Kapoor, M. (2012) Role of DNA methylation in growth and differentiation in physcomitrella patens and characterization of cytosine DNA methyltransferases. *The FEBS Journal* 279(21), 4081–4094. DOI: 10.1111/febs.12002.

Martínez-Pérez, M., Aparicio, F., López-Gresa, M.P., Bellés, J.M., Sánchez-Navarro, J.A. *et al.* (2017) *Arabidopsis* m6A demethylase activity modulates viral infection of a plant virus and the m6a abundance in its genomic RNAs. *Proceedings of the National Academy of Sciences of the United States of America* 114(40), 10755–10760. DOI: 10.1073/pnas.1703139114.

Mathew, M.M. and Prasad, K. (2021) Model systems for regeneration: *Arabidopsis*. *Development* 148(6). DOI: 10.1242/dev.195347.

Matzke, M.A. and Mosher, R.A. (2014) RNA-directed DNA methylation: An epigenetic pathway of increasing complexity. *Nature Reviews. Genetics* 15(6), 394–408. DOI: 10.1038/nrg3683.

McKeown, P.C. and Spillane, C. (2014) Landscaping plant epigenetics. In: Spillane, C. and McKeown, P. (eds) *Plant Epigenetics and Epigenomics. Methods in Molecular Biology*, Vol. 1112. Humana Press, Totowa, New Jersey, pp. 1–24. DOI: 10.1007/978-1-62703-773-0_1.

Melchior, F. (2000) SUMO – nonclassical ubiquitin. *Annual Review of Cell and Developmental Biology* 16(1), 591–626. DOI: 10.1146/annurev.cellbio.16.1.591.

Millán-Zambrano, G., Burton, A., Bannister, A.J. and Schneider, R. (2022) Histone post-translational modifications - cause and consequence of genome function. *Nature Reviews. Genetics* 23(9), 563–580. DOI: 10.1038/s41576-022-00468-7.

Moore, B., Zhou, L., Rolland, F., Hall, Q. and Cheng, W.H. (2003) Role of the *Arabidopsis* glucose sensor HXK1 in nutrient, light, and hormonal signaling. *Science* 300(5617), 332–336. DOI: 10.1126/science.1080585.

Morales-Ruiz, T., Ortega-Galisteo, A.P., Ponferrada-Marín, M.I., Martínez-Macías, M.I., Ariza, R.R. *et al.* (2006) DEMETER and REPRESSOR OF SILENCING 1 encode 5-methylcytosine DNA glycosylases. *Proceedings of the National Academy of Sciences of the United States of America* 103(18), 6853–6858. DOI: 10.1073/pnas.0601109103.

Moss, J., Zolkiewska, A. and Okazaki, I. (1997) ADP-ribosylarginine hydrolases and ADP-ribosyltransferases. In: Haag, F. and Koch-Nolte, F. (eds) *ADP-Ribosylation in Animal Tissues. Advances in Experimental Medicine and Biology*, Vol. 419. Springer, Boston, Massachusetts, pp. 25–33. DOI: 10.1007/978-1-4419-8632-0_3.

Nguyen, H.N., Kim, J.H., Jeong, C.Y., Hong, S.W. and Lee, H. (2013) Inhibition of histone deacetylation alters *Arabidopsis* root growth in response to auxin via PIN1 degradation. *Plant Cell Reports* 32(10), 1625–1636. DOI: 10.1007/s00299-013-1474-6.

Nichols, R.V., O'Connell, B.L., Mulqueen, R.M., Thomas, J., Woodfin, A.R. *et al.* (2022) High-throughput robust single-cell DNA methylation profiling with sciMETv2. *Nature Communications* 13(1), 7627. DOI: 10.1038/s41467-022-35374-3.

Nishioka, K. (2003) Methods and tips for the purification of human histone methyltransferases. *Methods* 31(1), 49–58. DOI: 10.1016/s1046-2023(03)00087-2.

Pandey, R. (2002) Analysis of histone acetyltransferase and histone deacetylase families of *Arabidopsis thaliana* suggests functional diversification of chromatin modification among multicellular eukaryotes. *Nucleic Acids Research* 30(23), 5036–5055. DOI: 10.1093/nar/gkf660.

Pang, Y., Lu, R.J. and Chen, P. (2019) Behavioral epigenetics: Perspectives based on experience-dependent epigenetic inheritance. *Epigenomes* 3(3), 18. DOI: 10.3390/epigenomes3030018.

Pikaard, C.S. and Mittelsten Scheid, O. (2014) Epigenetic regulation in plants. *Cold Spring Harbor Perspectives in Biology* 6(12), a019315. DOI: 10.1101/cshperspect.a019315.

Qian, W., Miki, D., Zhang, H., Liu, Y., Zhang, X. *et al.* (2012) A histone acetyltransferase regulates active DNA demethylation in *Arabidopsis*. *Science* 336(6087), 1445–1448. DOI: 10.1126/science.1219416.

Rambani, A., Rice, J.H., Liu, J., Lane, T., Ranjan, P. *et al.* (2015) The methylome of soybean roots during the compatible interaction with the soybean cyst nematode. *Plant Physiology* 168(4), 1364–1377. DOI: 10.1104/pp.15.00826.

Rambani, A., Hu, Y., Piya, S., Long, M., Rice, J.H. *et al.* (2020) Identification of differentially methylated miRNA genes during compatible and incompatible interactions between soybean and soybean cyst nematode. *Molecular Plant-Microbe Interactions* 33(11), 1340–1352. DOI: 10.1094/mpmi-07-20-0196-r.

Rando, O.J. and Chang, H.Y. (2009) Genome-wide views of chromatin structure. *Annual Review of Biochemistry* 78(1), 245–271. DOI: 10.1146/annurev.biochem.78.071107.134639.

Roberts, C.W.M. and Orkin, S.H. (2004) The SWI/SNF complex—chromatin and cancer. *Nature Reviews. Cancer* 4(2), 133–142. DOI: 10.1038/nrc1273.

Rossetto, D., Truman, A.W., Kron, S.J. and Côté, J. (2010) Epigenetic modifications in double-strand break DNA damage signaling and repair. *Clinical Cancer Research* 16(18), 4543–4552. DOI: 10.1158/1078-0432.ccr-10-0513.

Samantara, K., Shiv, A., Sousa, L.L., Sandhu, K.S., Priyadarshini, P. *et al.* (2021) A comprehensive review on epigenetic mechanisms and application of epigenetic modifications for crop improvement. *Environmental and Experimental Botany* 188, 104479. DOI: 10.1016/j.envexpbot.2021.104479.

Sang, Y.L., Cheng, Z.J. and Zhang, X.S. (2018) Plant stem cells and de novo organogenesis. *The New Phytologist* 218(4), 1334–1339. DOI: 10.1111/nph.15106.

Saracco, S.A., Hansson, M., Scalf, M., Walker, J.M., Smith, L.M. *et al.* (2009) Tandem affinity purification and mass spectrometric analysis of ubiquitylated proteins in *Arabidopsis*. *The Plant Journal* 59(2), 344–358. DOI: 10.1111/j.1365-313x.2009.03862.x.

Sauvé, D.M., Anderson, H.J., Ray, J.M., James, W.M. and Roberge, M. (1999) Phosphorylation-induced rearrangement of the histone H3 NH2-terminal domain during mitotic chromosome condensation. *The Journal of Cell Biology* 145(2), 225–235. DOI: 10.1083/jcb.145.2.225.

Schmitz, R.J., Marand, A.P., Zhang, X., Mosher, R.A., Turck, F. *et al.* (2022) Quality control and evaluation of plant epigenomics data. *The Plant Cell* 34(1), 503–513. DOI: 10.1093/plcell/koab255.

Seguela-Arnaud, M., Smith, C., Uribe, M.C., May, S., Fischl, H. *et al.* (2015) The mediator complex subunits MED25/PFT1 and MED8 are required for transcriptional responses to changes in cell wall arabinose composition and glucose treatment in *Arabidopsis thaliana*. *BMC Plant Biology* 15(1), 215. DOI: 10.1186/s12870-015-0592-4.

Sharma, N., Bender, Y., Boyle, K. and Fobert, P.R. (2013) High-level expression of sugar inducible gene2 (HSI2) is a negative regulator of drought stress tolerance in *Arabidopsis*. *BMC Plant Biology* 13(1), 170. DOI: 10.1186/1471-2229-13-170.

Shen, Y., Wei, W. and Zhou, D.X. (2015) Histone acetylation enzymes coordinate metabolism and gene expression. *Trends in Plant Science* 20(10), 614–621. DOI: 10.1016/j.tplants.2015.07.005.

Slotkin, R.K. and Martienssen, R. (2007) Transposable elements and the epigenetic regulation of the genome. *Nature Reviews. Genetics* 8(4), 272–285. DOI: 10.1038/nrg2072.

Soppe, W.J., Jacobsen, S.E., Alonso-Blanco, C., Jackson, J.P., Kakutani, T. *et al.* (2000) The late flowering phenotype of *fwa* mutants is caused by gain-of-function epigenetic alleles of a homeodomain gene. *Molecular Cell* 6(4), 791–802. DOI: 10.1016/s1097-2765(05)00090-0.

Sugimoto, K., Temman, H., Kadokura, S. and Matsunaga, S. (2019) To regenerate or not to regenerate: Factors that drive plant regeneration. *Current Opinion in Plant Biology* 47, 138–150. DOI: 10.1016/j.pbi.2018.12.002.

Suto, R.K., Clarkson, M.J., Tremethick, D.J. and Luger, K. (2000) Crystal structure of a nucleosome core particle containing the variant histone H2A.Z. *Nature Structural Biology* 7(12), 1121–1124. DOI: 10.1038/81971.

Takken, F.L.W. and Goverse, A. (2012) How to build a pathogen detector: Structural basis of NB-LRR function. *Current Opinion in Plant Biology* 15(4), 375–384. DOI: 10.1016/j.pbi.2012.05.001.

Tollenaar, M. and Lee, E.A. (2002) Yield potential, yield stability and stress tolerance in maize. *Field Crops Research* 75(2–3), 161–169. DOI: 10.1016/s0378-4290(02)00024-2.

Tsuji, H., Saika, H., Tsutsumi, N., Hirai, A. and Nakazono, M. (2006) Dynamic and reversible changes in histone H3-lys4 methylation and H3 acetylation occurring at submergence-inducible genes in rice. *Plant & Cell Physiology/Plant and Cell Physiology* 47(7), 995–1003. DOI: 10.1093/pcp/pcj072.

Urich, M.A., Nery, J.R., Lister, R., Schmitz, R.J. and Ecker, J.R. (2015) MethylC-seq library preparation for base-resolution whole-genome bisulfite sequencing. *Nature Protocols* 10(3), 475–483. DOI: 10.1038/nprot.2014.114.

van Zanten, M., Zöll, C., Wang, Z., Philipp, C., Carles, A. *et al.* (2014) HISTONE DEACETYLASE 9 represses seedling traits in *Arabidopsis thaliana* dry seeds. *The Plant Journal* 80(3), 475–488. DOI: 10.1111/tpj.12646.

Vergeer, P., Wagemaker, N.C.A.M. and Ouborg, N.J. (2012) Evidence for an epigenetic role in inbreeding depression. *Biology Letters* 8(5), 798–801. DOI: 10.1098/rsbl.2012.0494.

Villagómez-Aranda, A.L., García-Ortega, L.F., Torres-Pacheco, I. and Guevara-González, R.G. (2021) Whole-genome DNA methylation analysis in hydrogen peroxide overproducing transgenic tobacco resistant to biotic and abiotic stresses 10(1), 178. DOI: 10.3390/plants10010178.

Vimont, N., Quah, F.X., Schöepfer, D.G., Roudier, F., Dirlewanger, E. *et al.* (2020) ChIP-seq and RNA-seq for complex and low-abundance tree buds reveal chromatin and expression co-dynamics during sweet cherry bud dormancy. *Tree Genetics & Genomes* 16(1). DOI: 10.1007/s11295-019-1395-9.

Wada, N., Ueta, R., Osakabe, Y. and Osakabe, K. (2020) Precision genome editing in plants: State-of-the-art in CRISPR/Cas9-based genome engineering. *BMC Plant Biology* 20(1), 234. DOI: 10.1186/s12870-020-02385-5.

Wang, C., Gao, F., Wu, J., Dai, J., Wei, C. *et al.* (2010) *Arabidopsis* putative deacetylase atsrt2 regulates basal defense by suppressing PAD4, EDS5 and SID2 expression. *Plant and Cell Physiology* 51(8), 1291–1299. DOI: 10.1093/pcp/pcq087.

Wang, H., Wang, L., Erdjument-Bromage, H., Vidal, M., Tempst, P. *et al.* (2004) Role of histone H2A ubiquitination in polycomb silencing. *Nature* 431(7010), 873–878. DOI: 10.1038/nature02985.

Wang, Y., Hu, Q., Wu, Z., Wang, H., Han, S. *et al.* (2017) HISTONE DEACETYLASE 6 represses pathogen defence responses in *Arabidopsis thaliana*. *Plant, Cell and Environment* 40(12), 2972–2986. DOI: 10.1111/pce.13047.

Wei, L.-H., Song, P., Wang, Y., Lu, Z., Tang, Q. *et al.* (2018) The m6A reader ECT2 controls trichome morphology by affecting mRNA stability in *Arabidopsis*. *The Plant Cell* 30(5), 968–985. DOI: 10.1073/pnas.95.13.7480.

Wei, Y., Mizzen, C.A., Cook, R.G., Gorovsky, M.A. and Allis, C.D. (1998) Phosphorylation of histone H3 at serine 10 is correlated with chromosome condensation during mitosis and meiosis in *Tetrahymena*. *Proceedings of the National Academy of Sciences of the United States of America* 95(13), 7480–7484. DOI: 10.1073/pnas.95.13.7480.

Winston, F. and Carlson, M. (1992) Yeast SNF/SWI transcriptional activators and the SPT/SIN chromatin connection. *Trends in Genetics* 8(11), 387–391. DOI: 10.1016/0168-9525(92)90300-s.

Wipf, D., Pfister, C., Mounier, A., Leborgne-Castel, N. *et al.* (2021) Identification of putative interactors of *Arabidopsis* sugar transporters. *Trends in Plant Science* 26(1), 13–22.

Wu, K., Wang, S., Song, W., Zhang, J., Wang, Y. *et al.* (2020) Enhanced sustainable green revolution yield via nitrogen-responsive chromatin modulation in rice. *Science* 367(6478). DOI: 10.1126/science.aaz2046.

Xiao, J., Jin, R. and Wagner, D. (2017) Developmental transitions: Integrating environmental cues with hormonal signaling in the chromatin landscape in plants. *Genome Biology* 18(1), 88. DOI: 10.1186/s13059-017-1228-9.

Yang, L., Chen, X., Wang, Z., Sun, Q., Hong, A. *et al.* (2020) HOS15 and HDA9 negatively regulate immunity through histone deacetylation of intracellular immune receptor NLR genes in *Arabidopsis*. *The New Phytologist* 226(2), 507–522. DOI: 10.1111/nph.16380.

Zaccara, S., Ries, R.J. and Jaffrey, S.R. (2019) Reading, writing and erasing mRNA methylation. *Nature Reviews. Molecular Cell Biology* 20(10), 608–624. DOI: 10.1038/s41580-019-0168-5.

Zemach, A., McDaniel, I.E., Silva, P. and Zilberman, D. (2010) Genome-wide evolutionary analysis of eukaryotic DNA methylation. *Science* 328(5980), 916–919. DOI: 10.1126/science.1186366.

Zhang, G., Huang, H., Liu, D., Cheng, Y., Liu, X. *et al.* (2015) N6-methyladenine DNA modification in drosophila. *Cell* 161(4), 893–906. DOI: 10.1016/j.cell.2015.04.018.

Zhang, H., Lang, Z. and Zhu, J.K. (2018) Dynamics and function of DNA methylation in plants. *Nature Reviews. Molecular Cell Biology* 19(8), 489–506. DOI: 10.1038/s41580-018-0016-z.

Zhang, H., Zhang, X. and Xiao, J. (2023) Epigenetic regulation of nitrogen signaling and adaptation in plants. *Plants (Basel, Switzerland)* 12(14), 2725. DOI: 10.3390/plants12142725.

Zhou, C., Zhang, L., Duan, J., Miki, B. and Wu, K. (2005) HISTONE DEACETYLASE19 is involved in jasmonic acid and ethylene signaling of pathogen response in *Arabidopsis*. *The Plant Cell* 17(4), 1196–1204. DOI: 10.1105/tpc.104.028514.

Zhou, M., Palanca, A.M.S. and Law, J.A. (2018) Locus-specific control of the de novo DNA methylation pathway in *Arabidopsis* by the CLASSY family. *Nature Genetics* 50(6), 865–873. DOI: 10.1038/s41588-018-0115-y.

Züst, T. and Agrawal, A.A. (2016) Mechanisms and evolution of plant resistance to aphids. *Nature Plants* 2(1), 15206. DOI: 10.1038/nplants.2015.206.

22 Epigenetic Strategies for Regulating Plant Growth and Development

Tharani Sureshkumar* **and Chithra Devi Balasundaram Saraswathy***
PSGR Krishnammal College for Women, Coimbatore, Tamil Nadu, India

Abstract
The challenge of achieving sustainable food security and crop adaptation to evolving climatic conditions is intrinsically linked to the need for broadening crop phenotypic diversity. Epigenetic regulatory mechanisms and chromatin marks play a fundamental role in controlling plant developmental processes and influencing plant phenotypic plasticity, especially in response to environmental stressors. Epigenetic strategies are methods and procedures designed to alter or comprehend the epigenetic processes that regulate gene expression without changing the underlying DNA sequence. The study of epigenetics focuses on variations in gene expression or cellular phenotype that do not arise from alterations in the DNA sequence. These changes are often mediated by modifications to DNA and histone proteins, which can influence how genes are turned on or off. Understanding the stability and heritability of epigenetic marks, along with insights into epigenetic regulatory mechanisms, holds great significance for applications in crop breeding. In the quest for sustainable agriculture and enhanced crop productivity, epigenetics has emerged as a promising frontier. Epigenetic modifications play a pivotal role in regulating gene expression, and understanding the underlying mechanisms provides novel opportunities for fine-tuning plant growth and development. This chapter delves into the current state of epigenetic strategies for plant growth regulation, shedding light on the importance of DNA methylation, histone modifications, and small RNAs in shaping plant phenotypes. Additionally, we explore the use of epigenome editing tools, such as CRISPR/Cas9-based technologies, to precisely modulate epigenetic marks for targeted growth control. Harnessing epigenetic insights promises not only to optimize crop yields and stress tolerance but also to revolutionize our approach to sustainable agriculture.

Keywords: Epigenetics, Plant growth, Plant development, DNA methylation, Histone modification, Non-coding RNAs, Chromatin remodeling, Gene regulation

22.1 Introduction

The challenge of achieving sustainable food security and crop adaptation to evolving climatic conditions is intrinsically linked to the need for broadening crop phenotypic diversity (Kakoulidou *et al.*, 2021). Epigenetic regulatory mechanisms and chromatin marks play a fundamental role in controlling plant developmental processes and influencing plant phenotypic plasticity, especially in response to environmental stressors (Kumar, 2018). Epigenetics, the study of heritable changes in gene expression that do not involve alterations to the underlying DNA sequence, has emerged as a crucial field in understanding the complex regulation of plant growth and development (Kumar and Mohapatra, 2021). Unlike genetic

*Corresponding authors: tharani@psgrkcw.ac.in and bchitradevi@psgrkcw.ac.in

 Epigenetics for Climate-Smart and Sustainable Agriculture (ed. Jen-Tsung Chen)
DOI: 10.1079/9781800626102.0022

modifications that alter the DNA sequence, epigenetic modifications influence gene activity through chemical changes to the DNA molecule itself or to histone proteins around which DNA is wound. These changes can be reversible and are often influenced by environmental factors, making epigenetics a dynamic interface between a plant's genetic code and its environment (Singroha and Sharma, 2019).

Understanding the stability and heritability of epigenetic marks, along with insights into epigenetic regulatory mechanisms, holds great significance for applications in crop breeding. Modifying epigenetic marks such as DNA methylation and histone modifications can be used to regulate gene expression and control plant growth (Li *et al.*, 2008). Epigenetic markers can be utilized to guide selective breeding efforts or genetic engineering to create plants with desired growth traits (Dalakouras and Vlachostergios, 2021). In plants, epigenetic regulation is vital for various physiological processes, including seed germination, flowering time, stress responses, and defense mechanisms against pathogens (Yaish *et al.*, 2011; Sahu *et al.*, 2013).

Key epigenetic mechanisms include DNA methylation, histone modification, and RNA-associated silencing, each contributing to the precise control of gene expression in response to developmental cues and environmental stimuli (Kalinina *et al.*, 2018). Similar to genetic editing technologies like CRISPR/Cas9, epigenome editing involves the targeted modification of epigenetic marks at specific genomic loci (Qi *et al.*, 2023a). This can be used to activate or repress genes that control growth and development. Understanding the roles of epigenetic regulatory proteins, such as DNA methyltransferases or histone-modifying enzymes, can provide insights into how they can be manipulated to control plant growth (Shin *et al.*, 2022).

Recent advances in epigenetic research have opened new avenues for enhancing plant traits through non-genetic approaches (Mounger *et al.*, 2021). By manipulating epigenetic marks, scientists can potentially improve crop yield, resilience to environmental stressors, and resistance to diseases without the controversies and regulatory hurdles associated with genetic modification (Guarino *et al.*, 2022). This epigenetic tuning holds promise for sustainable agriculture, particularly in the face of global challenges such as climate change and food security.

22.2 DNA Methylation

DNA methylation is a pivotal epigenetic mechanism that significantly influences plant growth and development. In plants, DNA methylation typically occurs at cytosine residues in CG, CHG, and CHH (where H represents A, T, or C) contexts (Kumar and Mohapatra, 2021). This process involves the addition of a methyl group to the 5' position of the cytosine ring, often leading to changes in gene expression without altering the underlying DNA sequence (Singal and Ginder, 1999). The regulation of gene expression through DNA methylation is crucial for various developmental processes and stress responses in plants (Liu *et al.*, 2022).

22.2.1 Mechanisms of DNA methylation

Establishment and maintenance of DNA methylation

De novo DNA methylation is established by the action of DNA methyltransferases such as DRM2 (Domains Rearranged Methyltransferase 2). These enzymes add methyl groups to previously unmethylated DNA, often guided by small interfering RNAs (siRNAs) through the RNA-directed DNA methylation (RdDM) pathway (Mahfouz, 2010; Zhong *et al.*, 2014). Once established, DNA methylation patterns are maintained through cell divisions by maintenance methyltransferases such as MET1 (Methyltransferase 1) in CG contexts and CMT3 (Chromomethylase 3) in CHG contexts. This ensures the stable inheritance of methylation patterns across generations (Ikeda *et al.*, 2018).

Demethylation

DNA demethylation, which removes methyl groups from cytosines, is mediated by DNA glycosylases such as ROS1 (Repressor of Silencing

1). This process is critical for reversing methylation marks and reactivating gene expression when needed (Zhu, 2009).

22.2.2 Functions of DNA methylation in plant development

Regulation of gene expression

DNA methylation often leads to the repression of gene expression (Newell-Price *et al.*, 2000). By methylating promoter regions or gene bodies, plants can silence genes that need to be turned off during specific developmental stages or in response to environmental stimuli (Zhang *et al.*, 2018).

MECHANISMS.

- **Gene silencing via promoter methylation:** DNA methylation within promoter regions can block the binding of transcription factors and RNA polymerase, thereby repressing gene expression (Boyes and Bird, 1991). Methylated promoters often recruit methyl-CpG-binding domain (MBD) proteins, which attract co-repressor complexes that modify chromatin to a condensed state, making it less accessible for transcription (Zlatanova *et al.*, 2004).
- **Developmental and environmental regulation:** During development, specific genes must be turned off at particular stages (Buchanan-Wollaston, 1997). For example, genes promoting juvenile traits must be silenced during the transition to adulthood (Wu and Poethig, 2006). Environmental stimuli, such as stress conditions, can trigger changes in promoter methylation, leading to the repression of genes that are not needed for immediate survival (Boyko and Kovalchuk, 2011).
- **Gene body methylation:** Gene body methylation occurs within the coding regions of genes and is generally associated with actively transcribed genes (Bewick and Schmitz, 2017). However, it can regulate gene expression by influencing transcriptional elongation, mRNA processing, and splicing (Brown *et al.*, 2012). This form of methylation prevents the initiation of transcription from cryptic promoters within gene bodies, ensuring accurate gene expression (Neri *et al.*, 2017). Gene body methylation can modulate the expression levels of genes, fine-tuning the production of proteins necessary for various developmental processes (Liang *et al.*, 2020).
- **Transcriptional repression through methylation:** Methylated DNA can recruit repressive protein complexes, including histone deacetylases (HDACs) and histone methyltransferases, which further modify histones to maintain a closed chromatin state (Bhadouriya *et al.*, 2021). This repressive chromatin environment prevents the assembly of transcriptional machinery, leading to stable gene silencing (Engelhorn *et al.*, 2014).

Advanced techniques such as bisulfite sequencing, MeDIP, ChIP, RNA-seq, and CRISPR-based epigenome editing provide powerful tools for studying and manipulating DNA methylation (Jordan, 2020). Understanding the regulation of gene expression through DNA methylation can lead to innovative strategies for crop improvement, enhancing plant growth, resilience, and productivity in a sustainable manner.

Transposon silencing

Transposons, also known as 'jumping genes', are DNA sequences capable of moving from one location to another within the genome (Bajus *et al.*, 2022). If not properly controlled, transposons can cause genomic instability by inserting themselves into or near genes, disrupting their function (Bennetzen and Wang, 2014). DNA methylation plays a crucial role in silencing transposons, thereby maintaining genomic integrity (Gill *et al.*, 2021).

MECHANISMS.

- **Types of transposons:**
  Class I transposons (tetrotransposons): These elements transpose via an RNA intermediate. The process involves transcription of the transposon, reverse transcription to produce a DNA copy, and

integration into a new genomic location (Finnegan, 2012).

Class II transposons (DNA transposons): These elements move directly as DNA through a 'cut-and-paste' mechanism, where the transposon is excised from one location and integrated into another (Feschotte and Pritham, 2007).

- **Role of DNA methylation in silencing transposons:**
  Promoter methylation: DNA methylation occurs predominantly at the promoter regions of transposons, preventing their transcription and subsequent mobilization (Slotkin and Martienssen, 2007).
  Histone modifications: DNA methylation often works in conjunction with histone modifications (e.g. H3K9me2) to establish a repressive chromatin environment, further preventing transposon activity (Walter, 2015).
- **RNA-directed DNA methylation (RdDM):**
  The RdDM pathway is a key mechanism for targeting DNA methylation to transposons. It involves the production of small interfering RNAs (siRNAs) derived from transposon sequences (Matzke and Mosher, 2014). Pol IV transcribes transposon sequences into RNA. The RNA is converted into double-stranded RNA (dsRNA) by RNA-dependent RNA polymerase (RDR2) (Gebert and Rosenkranz, 2015). The dsRNA is processed into siRNAs by the Dicer-like enzyme DCL3. The siRNAs are loaded into Argonaute (AGO) proteins, forming RNA-induced silencing complexes (RISCs) (Kryovrysanaki *et al.*, 2021). The siRNA-AGO complexes guide the *de novo* DNA methyltransferase DRM2 to the transposon sequences, where it methylates cytosines, leading to transcriptional silencing (Liu and Zhao, 2023).

Genomic imprinting

Genomic imprinting is an epigenetic phenomenon where only one allele of a gene is expressed depending on its parental origin (Bajrami and Spiroski, 2016). In plants, DNA methylation is a key regulator of this selective expression, which is crucial for seed development and viability (Vyskot, 2004).

MECHANISMS.

- **Genomic imprinting basics:**
  Genomic imprinting results in parent-of-origin-specific gene expression. Typically, either the maternal or paternal allele of a gene is epigenetically silenced, while the other is expressed. This phenomenon is essential for the proper development of seeds and endosperm in plants (Baroux *et al.*, 2002; Köhler *et al.*, 2012).
- **Role of DNA methylation:**
  Imprinting control regions (ICRs): Specific regions in the genome, known as imprinting control regions, are differentially methylated depending on their parental origin. These regions control the expression of imprinted genes (Adalsteinsson and Ferguson-Smith, 2014; Elhamamsy, 2017).
  Establishment of methylation patterns: During gametogenesis, DNA methylation marks are established differently in male and female gametes. These marks are maintained after fertilization, leading to the differential expression of imprinted genes in the offspring (Vyskot, 2004; Ben Maamar *et al.*, 2021). After fertilization, maintenance methyltransferases, such as MET1, ensure the propagation of these methylation patterns during cell division (FitzGerald *et al.*, 2008)
- **Regulatory mechanisms:**
  Methylation of promoters in the imprinted regions can prevent the binding of transcription factors, leading to allele-specific gene silencing. DNA methylation often interacts with histone modifications to maintain a repressive chromatin state, reinforcing the silencing of the imprinted allele (MacDonald, 2012).

Response to environmental stress

DNA methylation is a dynamic epigenetic modification that enables plants to rapidly adapt to environmental changes (Akhter *et al.*, 2021). Under stress conditions such as drought, salinity, and pathogen attack, DNA methylation patterns can be altered, leading to the activation or repression of stress-responsive genes

(Viggiano and de Pinto, 2017). This mechanism helps plants to survive and thrive under adverse conditions.

MECHANISMS.

- **Stress-induced DNA methylation changes:**
  Dynamic methylation: Environmental stress can induce both genome-wide and locus-specific changes in DNA methylation. These changes are often reversible, allowing plants to return to their baseline state once the stress is removed (Baker-Andresen *et al.*, 2013).
  Stress-responsive genes: Methylation changes typically occur in the promoter regions or gene bodies of stress-responsive genes, which can result in their activation or repression (Le *et al.*, 2014). For instance, demethylation of a promoter may activate a gene that helps the plant cope with drought. Stress conditions can lead to the *de novo* methylation of previously unmethylated DNA, mediated by enzymes such as DRM2 (Domains Rearranged Methyltransferase (Kumar and Mohapatra, 2021). This can silence genes that are not required during stress. DNA demethylation, facilitated by enzymes like ROS1 (Repressor of Silencing 1), can activate stress-responsive genes by removing methyl groups from their promoters or regulatory regions (Bhattarai and Bhattarai, 2021).
- **RNA-directed DNA methylation (RdDM) pathway:**
  Small interfering RNAs (siRNAs) play a crucial role in guiding the RdDM pathway to specific genomic regions under stress conditions. These siRNAs are derived from the transcription of stress-responsive genes or transposons (Glavincheska, 2022). The siRNA–AGO (Argonaute) complex directs the DNA methyltransferases to the target loci, resulting in methylation and gene silencing or activation (Rymen et al., 2020).

22.3 Gene Silencing

Gene silencing is a regulatory mechanism that reduces or prevents the expression of specific genes. This can occur at various stages of gene expression, including transcriptional and post-transcriptional levels (Chicas and Macino, 2001; Malecová and Morris, 2010). Understanding gene silencing is crucial for insights into how genes are regulated in different contexts, such as development, stress responses, and disease resistance in plants. The main types of gene silencing include transcriptional gene silencing (TGS) and post-transcriptional gene silencing (PTGS).

22.3.1 Transcriptional gene silencing (TGS)

Gene transcription is the process by which the genetic information in DNA is copied into messenger RNA (mRNA) by the enzyme RNA polymerase. This mRNA then serves as a template for protein synthesis during translation (Alberts *et al.*, 2002). TGS involves mechanisms that inhibit this transcription process, thereby preventing the production of mRNA from the DNA template (Fagard and Vaucheret, 2000). By repressing gene transcription, TGS effectively blocks the synthesis of mRNA, the intermediary molecule that conveys genetic information from DNA to the ribosomes, where proteins are synthesized (Pikaard and Mittelsten Scheid, 2014). This blockage can be crucial for various cellular processes, including development, differentiation, and response to environmental stimuli. TGS plays a significant role in regulating gene expression during development and differentiation, maintaining genome stability, and defending against viral infections and transposable elements. By silencing specific genes, cells can control which proteins are produced, ensuring that only the necessary genes are active at any given time (Lisch, 2009). This can occur through several mechanisms:

DNA methylation

DNA methylation primarily involves the addition of methyl groups to the cytosine bases of DNA, especially in CpG dinucleotides. Methylated DNA regions are often associated with repressed gene expression because the methyl groups can inhibit the binding of transcription factors and attract proteins that induce a closed chromatin structure. DNA methyltransferases (e.g., MET1, CMT3) are responsible for adding methyl groups to DNA (Attwood *et al.*, 2002).

Histone modifications

Histones, the protein components of chromatin, can undergo various chemical modifications, such as methylation, acetylation, phosphorylation, and ubiquitination. These modifications influence chromatin structure and gene accessibility. For example, trimethylation of histone H3 at lysine 9 (H3K9me3) is linked to gene repression. Histone methyltransferases (e.g., SUVH4/KYP for H3K9 methylation) and histone deacetylases (HDACs) play key roles in adding or removing these modifications (Bhan *et al.*, 2017).

RNA-directed DNA methylation (RdDM)

Involves small RNA molecules, particularly small interfering RNAs (siRNAs), that guide DNA methylation to specific genomic regions. siRNAs are generated from double-stranded RNA precursors and are incorporated into an RNA-induced silencing complex (RISC), which directs the methylation machinery to homologous DNA sequences (Castel and Martienssen, 2013). Key enzymes include RNA polymerase IV and V, Dicer-like proteins (DCL), Argonaute proteins (AGO), and DNA methyltransferases (Kapoor *et al.*, 2008).

22.3.2 Post-transcriptional gene silencing (PTGS)

PTGS occurs after transcription and involves mechanisms that degrade mRNA or inhibit its translation, preventing the synthesis of proteins from the mRNA transcript. PTGS can lead to the degradation of mRNA molecules, preventing them from being translated into proteins (Ashfaq *et al.*, 2020). This degradation can occur through the action of endonucleases or exonucleases, which cleave the mRNA into smaller fragments that are subsequently degraded by cellular machinery (Tourrière *et al.*, 2002). Alternatively, PTGS can inhibit the translation of mRNA into proteins without degrading the mRNA molecule itself. This inhibition can occur through various means, such as blocking the association of ribosomes with the mRNA, interfering with the initiation of translation, or promoting premature termination of translation (Chicas and Macino, 2001). PTGS is involved in diverse cellular processes, including development, differentiation, response to stress, and defense against pathogens. It allows cells to dynamically regulate gene expression by quickly adjusting protein levels in response to changing environmental conditions or internal signals.

RNA interference (RNAi)

RNAi is mediated by small RNA molecules, including siRNAs and microRNAs (miRNAs):

- **siRNAs:** Derived from long double-stranded RNA (dsRNA), siRNAs are processed by Dicer into short fragments that are incorporated into the RISC. The siRNA-RISC complex then targets complementary mRNA for degradation (Khraiwesh *et al.*, 2012).
- **miRNAs:** Endogenous non-coding RNAs transcribed from miRNA genes. They form hairpin structures that are processed by Dicer into mature miRNAs, which guide the RISC to partially complementary mRNA targets, leading to translational repression or mRNA degradation. (Allen and Howell, 2010; Khraiwesh *et al.*, 2012).

miRNA-mediated regulation

miRNAs are processed from primary miRNA transcripts (pri-miRNAs) into precursor miRNAs (pre-miRNAs) and then into mature miRNAs by the Dicer enzyme. Mature miRNAs are incorporated into the RISC, where they bind to target mRNAs based on sequence complementarity, leading to mRNA cleavage or translational repression (Ha and Kim, 2014; Michlewski and Cáceres, 2019).

22.3.3 Functional roles of gene silencing

Gene silencing plays a variety of crucial roles in plants:

- **Developmental regulation:** Controls the timing and expression of genes involved in developmental processes like flowering, seed germination, and leaf formation (El-Sappah *et al.*, 2021).
- **Stress responses:** Modulates the expression of genes in response to biotic and abiotic stresses, allowing plants to adapt to changing environmental conditions (Ashapkin *et al.*, 2020).
- **Defense mechanisms:** Silences transposable elements and pathogenic genes to maintain genome stability and protect against infections (Yu *et al.*, 2013).
- **Epigenetic memory:** Maintains long-term gene expression states across cell generations, ensuring that developmental and stress responses are consistent and heritable (Vineis *et al.*, 2017).

Gene silencing is a fundamental aspect of gene regulation in plants, involving complex and interrelated mechanisms at both the transcriptional and post-transcriptional levels. These processes ensure precise control of gene expression necessary for proper growth, development, and adaptation to the environment. Understanding these mechanisms provides valuable insights into plant biology and offers potential applications in agriculture and biotechnology.

22.4 Histone Modification

Histone modification is one of the primary mechanisms through which epigenetic regulation occurs. Histones are proteins around which DNA is wrapped to form chromatin, the complex structure that packages DNA within the cell nucleus. Post-translational modifications, such as methylation, acetylation, phosphorylation, and ubiquitination, can occur on histone proteins (Peterson and Laniel, 2004). These modifications influence the accessibility of DNA to transcription factors and the transcriptional machinery, thereby regulating gene expression (Albini *et al.*, 2019).

22.4.1 Regulation of gene expression

Histone modifications can either activate or repress gene expression, depending on the specific modification and its location within the chromatin. It includes acetylation, methylation, phosphorylation and ubiquitination.

Acetylation

Histone acetylation, catalyzed by histone acetyltransferases (HATs), typically leads to relaxed chromatin structure, allowing for increased accessibility of transcription factors and RNA polymerase to the DNA. This often results in enhanced gene expression (Lee and Grant, 2019).

Methylation

Histone methylation can either activate or repress gene expression, depending on which amino acid residue of the histone is methylated and the degree of methylation. For instance, trimethylation of histone H3 at lysine 4 (H3K4me3) is associated with active gene transcription, while trimethylation of histone H3 at lysine 9 (H3K9me3) is associated with gene silencing (Liu *et al.*, 2010; Zhou *et al.*, 2020).

Phosphorylation

Phosphorylation of histone proteins, typically at serine, threonine, or tyrosine residues, can influence chromatin structure and gene expression. For example, phosphorylation of histone H3 at serine 10 (H3S10ph) is associated with the activation of gene transcription during mitosis and meiosis (Wang *et al.*, 2015).

Ubiquitination

Ubiquitination involves the addition of ubiquitin molecules to histone proteins, which can lead to changes in chromatin structure and gene expression (Shilatifard, 2006). For example, monoubiquitination of histone H2B at lysine 120 (H2BK120ub) is associated with transcriptional activation, while polyubiquitination can target histones for degradation (Vaughan *et al.*, 2021).

22.4.2 Role in plant growth and development

In plants, histone modifications play critical roles in regulating various aspects of growth and development, including the following.

- **Developmental transitions:** Histone modifications are involved in regulating transitions between different developmental stages, such as germination, flowering, and senescence (Fig. 22.1) (Xu *et al.*, 2018).
- **Stress responses:** Plants employ histone modifications to modulate their responses to environmental stresses, such as drought, salinity, and pathogens. These modifications enable plants to activate stress-responsive genes and adapt to changing environmental conditions (Ueda and Seki, 2020).
- **Organogenesis:** Histone modifications contribute to the regulation of organogenesis, including the formation of roots, shoots, leaves, and flowers, by controlling the expression of genes involved in tissue differentiation and morphogenesis (Chen *et al.*, 2018a, b).

The role of histone modifications in plant growth and development has significant implications for agricultural biotechnology. Researchers can manipulate histone modification patterns to enhance crop yield, improve stress tolerance, and modulate plant architecture.

22.5 Chromatin Remodeling

Chromatin remodeling is a fundamental process within epigenetics that plays a crucial role in regulating plant growth and development (Jarillo *et al.*, 2009). Chromatin is the complex of DNA and proteins, primarily histones that package the genome within the nucleus of eukaryotic cells (Wolffe, 1998). The structure of chromatin can be dynamic, transitioning between condensed and open configurations (Rosa and Shaw, 2013). This dynamic nature of chromatin is essential for regulating gene expression.

Chromatin remodeling involves the action of specialized protein complexes known as chromatin remodeling complexes (Clapier and Cairns, 2009). These complexes utilize the energy derived from ATP hydrolysis to alter the structure of nucleosomes, the basic repeating units of chromatin, thereby modulating the accessibility of DNA to transcription factors and

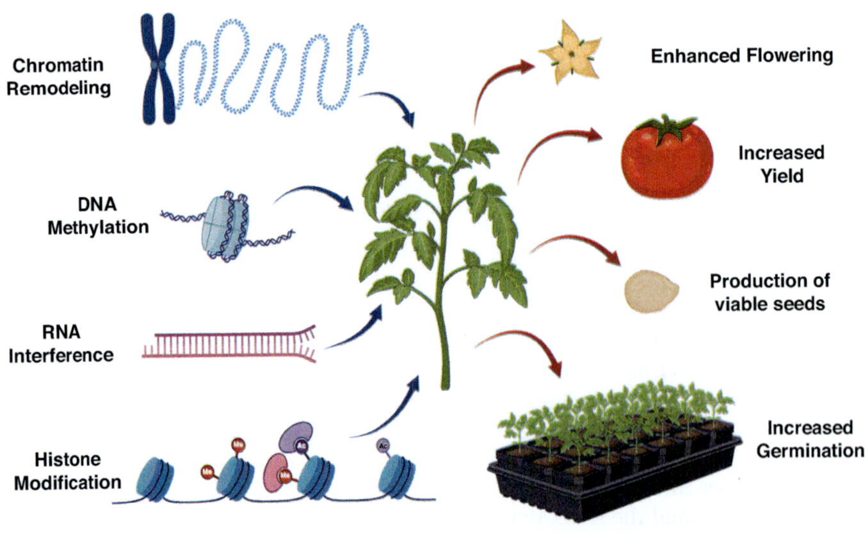

Fig. 22.1. The epigenetic landscape of plant growth and development. Author's own image.

the transcriptional machinery (Becker and Hörz, 2002).

22.5.1 Mechanisms of chromatin remodeling

Chromatin remodeling complexes can perform various functions, including nucleosome sliding, nucleosome ejection, and histone variant exchange.

Nucleosome sliding

Chromatin remodelers can slide nucleosomes along the DNA, repositioning them to expose or conceal specific DNA sequences. This process can regulate the accessibility of DNA to transcription factors and other regulatory proteins (Narlikar *et al.*, 2002; Clapier and Cairns, 2009).

Nucleosome ejection

Chromatin remodelers can evict histone proteins from the nucleosome, leading to the formation of nucleosome-free regions. This facilitates the binding of transcription factors and promotes transcriptional activation (Lorch *et al.*, 2006; Venkatesh and Workman, 2015).

Histone variant exchange

Chromatin remodelers can exchange canonical histones with histone variants, altering the properties of the nucleosome and influencing gene expression. Histone variants may have distinct functions compared with canonical histones and can contribute to the regulation of specific genomic regions (Szenker *et al.*, 2011; Henikoff and Smith, 2015).

22.5.2 Role in plant growth and development

Chromatin remodeling plays critical roles in various aspects of plant growth and development, including developmental transitions, cell differentiation and specialization, and environmental responses.

- **Developmental transitions:** Chromatin remodeling is involved in regulating transitions between different developmental stages, such as germination, vegetative growth, flowering, and senescence. It controls the expression of genes associated with each developmental phase, ensuring proper timing and coordination (Table 22.1). (Gu *et al.*, 2013).
- **Cell differentiation and specialization:** Chromatin remodeling regulates the expression of genes involved in cell differentiation and specialization, leading to the formation of different cell types and tissues during plant development. This process is essential for the development of roots, shoots, leaves, flowers, and other structures (Reyes, 2006; Kang *et al.*, 2020).
- **Environmental responses:** Plants utilize chromatin remodeling to respond to environmental cues and stresses, such as light, temperature, water availability, and pathogen attack. Chromatin modifications allow plants to activate stress-responsive genes and adapt their growth and development to changing environmental conditions (Kim *et al.*, 2012; Sani *et al.*, 2013).

Chromatin remodeling in plant growth and development has significant implications for agricultural biotechnology (Table 22.1). Chromatin remodeling processes can be manipulated to enhance crop yield, improve stress tolerance, and modulate plant architecture. Techniques such as genome editing, epigenome editing, and the use of chromatin-modifying compounds offer promising avenues for targeted manipulation of chromatin structure and gene expression in crops.

22.6 Epigenome Editing

Epigenome editing is an emerging field within molecular biology and biotechnology that focuses on precisely modifying epigenetic marks and chromatin structure at specific genomic loci (Kumar, 2019). Unlike traditional genome editing techniques such as CRISPR/Cas9, which target DNA sequence alterations, epigenome editing enables the direct manipulation of epigenetic modifications, such as DNA methylation,

Table 22.1. Different aspects of plant growth and development by epigenetic strategies. Author's own table.

Plant aspect	Description	Epigenetic strategy	References
Seed germination	Initiation of growth from a seed, including water uptake, metabolism activation, and emergence of the radicle	DNA methylation, histone modification	Qi *et al.*, 2023b
Vegetative growth	Growth of stems, leaves, and roots during the vegetative phase, involves cell division and elongation	Chromatin remodeling, non-coding RNAs (ncRNAs)	Ojolo *et al.*, 2018; Wang *et al.*, 2023
Flowering time	Transition from vegetative to reproductive phase, regulated by environmental cues and internal signals	Histone modification, RNA-directed DNA methylation (RdDM)	Jeong *et al.*, 2015; Matzke *et al.*, 2015
Floral organogenesis	Development of floral organs (sepals, petals, stamens, carpels) from floral meristems. The transition from the vegetative to the reproductive phase is regulated by environmental cues and internal signals	Histone modification, non-coding RNAs (ncRNAs)	Berr *et al.*, 2011; Wambui *et al.*, 2020
Pollination	Transfer of pollen grains from the anther to the stigma, facilitating fertilization	DNA methylation, non-coding RNAs (ncRNAs)	Hu *et al.*, 2021; Nie *et al.*, 2023
Fruit development	Maturation of ovary following successful pollination, involving fruit enlargement and seed development	Histone modification, chromatin remodeling	Tang *et al.*, 2020; Castañeda *et al.*, 2022
Senescence	Programmed degradation and death of plant tissues and organs	DNA methylation, histone modification	Brusslan *et al.*, 2015; Miryeganeh, 2022

histone modifications, and chromatin accessibility, without altering the underlying DNA sequence (Miglani *et al.*, 2020). Epigenome editing has emerged as a promising tool within the realm of epigenetics for precisely modulating gene expression and chromatin structure to regulate plant growth and development.

22.6.1 Mechanisms

CRISPR-based epigenome editing

CRISPR/Cas9 technology, which is widely used for genome editing, can be adapted for epigenome editing by fusing the Cas9 enzyme with epigenetic effector domains. These effector domains include DNA methyltransferases (DNMTs) for DNA methylation, histone acetyltransferases (HATs) or histone deacetylases (HDACs) for histone modifications, and chromatin remodeling complexes for altering chromatin structure. By targeting specific genomic loci with guide RNAs (gRNAs), the fused effector domains can be recruited to modify epigenetic marks at precise genomic locations (Hilton *et al.*, 2015).

Transcriptional regulation

Epigenome editing can also involve the modulation of gene expression through the targeted modification of regulatory regions, such as promoters and enhancers. By recruiting transcriptional activators or repressors to these regions using CRISPR-based systems, researchers can manipulate gene expression levels without altering the underlying DNA sequence (Thakore *et al.*, 2016; Vojta *et al.*, 2016).

22.6.2 Targeting epigenetic modifications in plants

DNA methylation

DNA methylation plays a crucial role in regulating gene expression in plants. Epigenome editing can be utilized to target DNA methyltransferases (DNMTs) to specific genomic loci using CRISPR/Cas9 technology (Thakore *et al.*, 2015). By guiding DNMTs to promoter regions, enhancers, or other regulatory elements, researchers can induce targeted DNA methylation changes, leading to altered gene expression patterns (Vojta *et al.*, 2016).

Histone modifications

Histone modifications, such as acetylation, methylation, phosphorylation, and ubiquitination, influence chromatin structure and gene expression in plants. Epigenome editing approaches can be employed to modulate histone-modifying enzymes, such as histone acetyltransferases (HATs), histone deacetylases (HDACs), histone methyltransferases (HMTs), and histone demethylases (HDMs), at specific genomic loci. This enables precise control over histone modification patterns and gene expression profiles associated with plant growth and development (Bannister and Kouzarides, 2011; Zentner and Henikoff, 2013).

Chromatin accessibility

Chromatin accessibility is another critical aspect of epigenetic regulation in plants. Epigenome editing techniques can be utilized to remodel chromatin structure by targeting chromatin remodeling complexes to specific genomic regions. This can result in changes in nucleosome positioning and accessibility of DNA to transcription factors, thereby influencing gene expression and developmental processes (Li *et al.*, 2007; Clapier and Cairns, 2009).

22.6.3 Applications in regulating plant growth and development

Enhanced stress tolerance

Epigenome editing can be employed to modulate the expression of stress-responsive genes in plants, thereby enhancing their ability to withstand various environmental stresses, such as drought, salinity, heat, and pathogens (Ali *et al.*, 2022). By inducing targeted epigenetic modifications associated with stress tolerance pathways, researchers can engineer crops with improved resilience to adverse environmental conditions.

Optimization of yield and quality traits

Epigenome editing offers a means to fine-tune the expression of genes involved in plant growth, flowering, fruit development, and nutrient metabolism (Shi *et al.*, 2024). By precisely modulating gene expression patterns through epigenetic modifications, researchers can optimize agronomic traits related to yield, quality, and nutritional content in crop plants (Kakoulidou *et al.*, 2021).

Controlled developmental transitions

Epigenome editing can be used to manipulate the timing and coordination of developmental transitions in plants, such as germination, vegetative growth, flowering, and senescence. By targeting key regulatory genes involved in these processes, researchers can exert precise control over developmental timing and phenotypic outcomes in crop plants (Ay *et al.*, 2014; Kakoulidou *et al.*, 2021).

22.6.4 Future directions and challenges

While epigenome editing holds tremendous potential for regulating plant growth and development, several challenges remain to be addressed. These include optimizing the efficiency and specificity of epigenome editing tools, developing robust delivery methods for *in planta* applications, and ensuring the stability and heritability of epigenetic modifications across generations. Additionally, ethical considerations surrounding the use of epigenome editing in agriculture, such as potential unintended consequences and regulatory frameworks, need to be carefully addressed (Kuzma and Kokotovich, 2011).

Epigenome editing represents a powerful approach for harnessing epigenetic mechanisms to regulate plant growth and development. By precisely modulating gene expression and chromatin structure, researchers can engineer plants with enhanced stress tolerance, optimized agronomic traits, and controlled developmental processes, thereby contributing to sustainable agriculture and food security in a changing climate (Raza *et al.*, 2023).

22.7 Bisulphite Sequencing

Bisulfite sequencing is a powerful epigenetic technique used to analyze DNA methylation patterns in plants and other organisms (Foerster and Scheid, 2010). It provides valuable insights into the role of DNA methylation in regulating gene expression, chromatin structure, and various biological processes, including plant growth and development (Sun *et al.*, 2018).

22.7.1 Principle of bisulfite sequencing

Bisulfite sequencing is based on the chemical conversion of unmethylated cytosine residues in genomic DNA to uracil (and subsequent thymine after PCR amplification), while methylated cytosines remain unaffected (Hayatsu, 2008). This conversion is achieved by treating genomic DNA with sodium bisulfite under denaturing conditions, where cytosine residues are deaminated to uracil, except for those protected by methylation (Hajkova *et al.*, 2002). After bisulfite treatment, DNA is subjected to PCR amplification, and the resulting PCR products are sequenced to determine the methylation status of individual cytosine residues (Li and Tollefsbol, 2011).

22.7.2 Role of bisulphite sequencing in plant growth and development

DNA methylation profiling

Bisulfite sequencing allows genome-wide profiling of DNA methylation patterns in plants, revealing the distribution and dynamics of methylated cytosines across the genome (Chen *et al.*, 2018b). By mapping DNA methylation landscapes in different tissues, developmental stages, and environmental conditions, researchers can identify methylated regions associated with gene regulatory elements, transposable elements, and other functional genomics regions (Law and Jacobsen, 2010).

Identification of differentially methylated regions (DMRs)

Comparative bisulfite sequencing analyses enable the identification of differentially methylated regions (DMRs) between samples, such as wild type versus mutant plants, or control versus stress-treated plants (Kenchanmane Raju *et al.*, 2018). DMRs represent genomic regions with altered DNA methylation levels in response to genetic or environmental cues, providing insights into the epigenetic regulation of gene expression and stress responses (Secco *et al.*, 2015).

Epigenome-wide association studies (EWAS)

Bisulfite sequencing can be used in epigenome-wide association studies (EWAS) to investigate the relationship between DNA methylation patterns and phenotypic traits, such as plant morphology, yield, and stress tolerance (Can *et al.*, 2021). By correlating DNA methylation variation with phenotypic variation, researchers can identify candidate genes and regulatory elements underlying trait variation and potential targets for crop improvement (Yadav *et al.*, 2018).

Functional characterization of epigenetic regulators

Bisulfite sequencing can be coupled with genetic and functional genomic approaches to study the role of specific epigenetic regulators, such as DNA methyltransferases and demethylases, in plant growth and development (Liu and Zhong, 2024). By analyzing DNA methylation changes in mutant or transgenic plants lacking or overexpressing these regulators, researchers can elucidate their functions and regulatory mechanisms in shaping the plant epigenome (Ashapkin *et al.*, 2020).

22.7.3 Challenges and considerations

- **Bioinformatics analysis** Bisulfite sequencing generates large datasets that require sophisticated bioinformatics tools and pipelines for data processing, alignment, and methylation calling (Rauluseviciute *et al.*, 2019).
- **PCR bias** PCR amplification bias and errors can introduce artifacts in bisulfite sequencing data, leading to inaccurate methylation estimates. Quality control measures and appropriate normalization techniques are essential to minimize PCR bias and ensure data reliability (Hernández *et al.*, 2013).
- **Single-base resolution** Bisulfite sequencing provides single-base resolution of DNA methylation patterns but may have limitations in regions with repetitive sequences or low DNA methylation levels (Li and Tollefsbol, 2021). Complementary techniques, such as whole-genome bisulfite sequencing and nanopore sequencing, can overcome these limitations and provide comprehensive methylation profiles at higher resolution (Gouil and Keniry, 2019).

Bisulfite sequencing is a valuable tool in plant epigenetics for deciphering DNA methylation dynamics and its role in regulating gene expression and plant development. By unraveling the epigenetic mechanisms underlying stress responses, growth processes, and trait variation in plants, bisulfite sequencing contributes to the development of epigenetic strategies for crop improvement and sustainable agriculture.

22.8 Small RNA Technology

Small RNA technology refers to the manipulation and utilization of small RNA molecules, typically ranging from 20 to 30 nucleotides in length, for various biological applications (Kamthan *et al.*, 2015). Small RNA technology has emerged as a powerful tool within the realm of epigenetics for regulating gene expression and controlling various aspects of plant growth and development.

22.8.1 Types of small RNAs

MicroRNAs (miRNAs)

miRNAs are endogenous small RNAs that play crucial roles in post-transcriptional gene regulation. They typically target messenger RNAs (mRNAs) for degradation or translational repression by base-pairing with complementary sequences (Jones-Rhoades *et al.*, 2006). miRNAs are involved in diverse biological processes, including development, differentiation, stress responses, and disease.

Small interfering RNAs (siRNAs)

siRNAs are double-stranded RNA molecules that trigger RNA interference (RNAi) pathways. They guide sequence-specific gene silencing by targeting complementary RNA molecules for degradation. siRNAs are widely used in research and biotechnology for gene knockdown and functional genomics studies (Sanan-Mishra *et al.*, 2021).

Piwi-interacting RNAs (piRNAs)

piRNAs are a class of small RNAs that interact with Piwi proteins and play roles in transposon silencing and germ cell development. They are predominantly found in animal cells and are involved in maintaining genome stability and epigenetic inheritance (Ozata *et al.*, 2019).

22.8.2 Techniques involved in small RNA technology

RNA interference (RNAi)

RNA interference (RNAi) is a highly conserved cellular mechanism for gene silencing that involves the use of small RNA molecules, such as siRNAs (small interfering RNAs) or miRNAs (microRNAs), to regulate gene expression (Younis *et al.*, 2014). In RNAi, double-stranded RNA molecules are processed into small RNA duplexes, which are then incorporated into an RNA-induced silencing complex (RISC). Within the RISC, one of the strands, known as the guide strand, directs the complex to complementary sequences in target mRNAs. This leads to mRNA degradation or translational repression, resulting in gene silencing. Synthetic siRNAs or miRNA mimics can be designed and introduced into cells to exploit the RNAi pathway for sequence-specific gene silencing (Iwakawa and Tomari, 2022). RNAi technology has revolutionized biological research by providing a powerful tool for studying gene function, pathway analysis, and therapeutic intervention in various biological systems (Bharathi *et al.*, 2023).

Antisense oligonucleotides

Antisense oligonucleotides are short single-stranded nucleic acids designed to bind to target RNA molecules and modulate gene expression. They can be used to inhibit miRNA function or to target specific mRNA transcripts for degradation (Bajan and Hutvagner, 2020). Antisense oligonucleotides (ASOs) are short single-stranded nucleic acid sequences, typically composed of DNA or modified RNA bases, designed to complementarily bind to specific target RNA molecules (Wdowikowska *et al.*, 2021). By forming base-pairing interactions with the target RNA, ASOs can modulate gene expression through various mechanisms. One approach involves inhibiting the function of microRNAs (miRNAs) by binding to them and preventing their interaction with their target mRNAs, thereby blocking their regulatory activity (Meng *et al.*, 2011). Additionally, ASOs can be designed to target specific messenger RNA (mRNA) transcripts for degradation through mechanisms such as RNase H-mediated cleavage or steric hindrance, leading to reduced levels of the targeted protein (Quemener *et al.*, 2020). This targeted modulation of gene expression makes ASOs valuable tools for studying gene function, elucidating regulatory pathways, and potentially developing therapeutic interventions for various diseases in plants (Krasnodębski *et al.*, 2023).

CRISPR/Cas systems

CRISPR/Cas systems, originally discovered as adaptive immune systems in bacteria and archaea, have been repurposed as powerful tools for genome engineering and gene regulation across diverse organisms, including plants (Kirchner and Schneider, 2015). In addition to their well-known role in DNA editing, CRISPR/Cas systems can also be harnessed for manipulating small RNA pathways and gene regulation. One approach involves engineering Cas proteins, such as Cas9 or Cas13, by fusing them with RNA-binding domains or other effector domains. These engineered Cas proteins can then be programmed to target specific small RNA molecules, such as miRNAs or siRNAs, by recognizing complementary sequences. By binding to these small RNAs, the engineered Cas proteins can either sequester them, interfere with their function, or modulate their processing or stability (Bhaya *et al.*, 2011). Furthermore, CRISPR/Cas systems can also be used to target endogenous small RNA pathways for regulation. For instance, by designing guide RNAs (gRNAs) to target components of the RNA interference (RNAi) pathway or the microRNA (miRNA) biogenesis machinery, researchers can manipulate the activity of small RNA pathways and study their roles in gene regulation and cellular processes (Sharma *et al.*, 2022; Ding *et al.*, 2022). Overall, the adaptability of CRISPR/Cas systems for small RNA engineering and gene regulation offers versatile tools for dissecting the functions of small RNAs and developing novel strategies for therapeutic intervention and biotechnological applications (Nidhi *et al.*, 2021).

22.8.3 Role of small RNA technology in plant growth and development

Regulation of developmental transitions

miRNAs are key regulators of developmental transitions in plants, including germination, vegetative growth, flowering, and senescence. They modulate the expression of genes involved in developmental processes such as leaf morphogenesis, shoot branching, root architecture, and floral organ identity (Jones-Rhoades *et al.*, 2006; Axtell and Bowman, 2008).

Maintenance of stem cell niches

miRNAs play a critical role in maintaining stem cell niches in plant meristems, which are essential for continuous growth and organ formation. They regulate the balance between stem cell proliferation and differentiation by controlling the expression of key transcription factors and signaling molecules (Schauer *et al.*, 2002; Yang *et al.*, 2006).

Environmental responses

miRNAs mediate plant responses to environmental cues and stresses, such as drought, salinity, temperature fluctuations, and pathogen attack (Sunkar and Zhu, 2004). They regulate the expression of stress-responsive genes and modulate physiological and biochemical processes involved in stress tolerance and adaptation (Khraiwesh *et al.*, 2012).

Transposon silencing and genome stability

siRNAs play a critical role in silencing transposable elements (TEs) and maintaining genome stability in plants. They target TE transcripts for degradation or transcriptional silencing, preventing their mobilization and potential harmful effects on genome integrity (Lisch, 2009).

Epigenetic regulation

siRNAs are involved in epigenetic regulation through RNA-directed DNA methylation (RdDM) pathways. They guide DNA methylation machinery to specific genomic loci, leading to transcriptional gene silencing and establishment of epigenetic memory (Mahfouz, 2010).

Antiviral defense

Plants use siRNAs as part of their antiviral defense mechanisms to combat viral infections. siRNAs derived from viral RNA genomes trigger RNAi-mediated degradation of viral RNAs, thereby restricting viral replication and spread within the plant host (Leonetti *et al.*, 2021).

Small RNA technology is a powerful tool for understanding and manipulating gene expression and epigenetic regulation in plants. Its applications in crop improvement, functional genomics, and epigenome editing hold promise for addressing global challenges in agriculture and food security. Continued research efforts and technological innovations in small RNA technology are expected to drive advancements in plant biotechnology and sustainable agriculture.

22.9 Pharmacological Inhibitors and Activators

Pharmacological inhibitors and activators for plants are chemical compounds or molecules that can modulate specific biological processes, pathways, or molecular targets within plant cells. These compounds are valuable tools for studying plant biology, elucidating signaling pathways, and manipulating plant traits for agricultural and biotechnological purposes.

22.9.1 Pharmacological inhibitors

Pharmacological inhibitors are compounds that bind to specific proteins or enzymes, blocking their activity and interfering with biological processes. They are designed to target key components of signaling pathways, metabolic pathways, or enzymatic reactions involved in disease pathogenesis or cellular functions.

DNA methylation inhibitors

DNA methylation inhibitors, such as 5-azacytidine and zebularine, block the activity of DNA methyltransferases (DNMTs) in plants. By inhibiting DNA methylation, these compounds induce DNA demethylation, leading to changes in gene expression patterns and potentially altering plant growth, development, and stress responses. These inhibitors can induce DNA demethylation and reactivate silenced genes, providing insights into the role of DNA methylation in regulating plant growth and development (Lopez *et al.*, 2022).

Histone modification inhibitors

Histone modification inhibitors target enzymes involved in histone post-translational modifications, such as histone acetylation and methylation. For example, histone deacetylase inhibitors (HDACis) like trichostatin A inhibit histone deacetylases (HDACs), leading to increased histone acetylation levels and altered gene expression in plants. HDACis, such as trichostatin A and suberoylanilide hydroxamic acid (SAHA), inhibit the activity of histone deacetylases (HDACs), leading to hyperacetylation of histones and relaxation of chromatin structure. This results in the activation of transcriptionally silenced genes involved in plant development, stress responses, and hormone signaling (Delcuve *et al.*, 2012).

Protein kinase inhibitors

Protein kinase inhibitors, such as staurosporine and genistein, target protein kinases involved in various signaling pathways regulating plant growth, development, and stress responses. By inhibiting specific protein kinases, these compounds can modulate signal transduction and downstream cellular responses in plants (Baier and Szyszka, 2020).

Histone methyltransferase inhibitors

Pharmacological inhibitors targeting histone methyltransferases (HMTs), such as chaetocin and BIX-01294, disrupt histone methylation patterns and alter gene expression profiles. By inhibiting specific HMTs, these compounds can modulate chromatin structure and regulate gene expression associated with plant growth and development (Greiner *et al.*, 2005; Kubicek *et al.*, 2007).

22.9.2 Pharmacological activators

Pharmacological activators are compounds that enhance the activity of specific proteins or enzymes, promoting their biological functions. These compounds are designed to bind to target proteins or enzymes, inducing conformational changes that increase their catalytic activity, substrate affinity, or stability. Pharmacological activators act as molecular switches, amplifying signaling pathways, metabolic reactions, or cellular processes involved in physiological functions or therapeutic interventions.

Histone acetyltransferase activators

Pharmacological activators of histone acetyltransferases (HATs), such as anacardic acid and curcumin, enhance histone acetylation levels and promote transcriptional activation of target genes. These activators can stimulate plant growth and development by modulating the expression of genes involved in cell differentiation, organ development, and stress responses (Balasubramanyam *et al.*, 2003; Hossain *et al.*, 2012).

DNA demethylase activators

Emerging pharmacological compounds, such as vitamin C and alpha-ketoglutarate analogs, have been shown to enhance the activity of DNA demethylases, leading to active DNA demethylation and gene reactivation. These activators play a role in reshaping the epigenome and regulating gene expression dynamics during plant development and environmental responses (Laing *et al.*, 2007; Yin *et al.*, 2013).

Chromatin remodeling complex activators

Pharmacological compounds targeting chromatin remodeling complexes, such as SWI/SNF and ISWI complexes, can activate their enzymatic activities and promote chromatin remodeling events. By facilitating nucleosome repositioning and altering chromatin accessibility, these activators contribute to the regulation of gene expression programs underlying plant growth and development (Vincent *et al.*, 2008; Clapier and Cairns, 2009).

Phytohormone analogs

Phytohormone analogs mimic the activity of endogenous plant hormones, such as auxins, cytokinins, gibberellins, abscisic acid, and ethylene. These compounds can act as pharmacological activators by stimulating hormone-responsive pathways, promoting specific developmental processes, or enhancing stress tolerance in plants (Bleecker and Kende, 2000; Fleet and Sun, 2005; Sakakibara, 2006; Cutler *et al.*, 2010; Zhao, 2010).

Stress signaling activators

Certain compounds, such as hydrogen peroxide (H_2O_2) and jasmonic acid (JA) analogs, can activate stress signaling pathways in plants. By mimicking endogenous stress signals, these activators can induce stress-responsive gene expression, trigger defense mechanisms, and enhance plant resilience to biotic and abiotic stresses (Wasternack and Hause, 2013; Baxter *et al.*, 2014). Mittler and Blumwald, 2015).

Photosynthesis enhancers

Compounds like 3-(3,4-dichlorophenyl)−1,1-di methylurea (DCMU) can act as pharmacological activators by enhancing photosynthetic efficiency in plants. DCMU inhibits electron flow in photosystem II (PSII), leading to increased ATP and NADPH production, which can promote plant growth and biomass accumulation under optimal conditions (Kyle *et al.*, 1984).

22.9.3 Applications and considerations

Pharmacological inhibitors and activators provide valuable tools for dissecting the roles of epigenetic modifications in plant growth and development (Law and Jacobsen, 2010). These compounds can be used in conjunction with genetic approaches to validate epigenetic regulatory pathways and identify potential targets for crop improvement and biotechnological applications.

However, it is essential to consider potential off-target effects, dose-dependent responses, and specificity of pharmacological compounds when using them in epigenetic studies and agricultural applications. Further research is needed to develop more selective and potent inhibitors and activators targeting specific epigenetic regulators, thereby advancing our understanding of epigenetic mechanisms and their applications in plant biology (Balasubramanian *et al.*, 2009).

22.10 Environmental Stress

Environmental stressors, such as drought, salinity, temperature extremes, and pathogen attacks, exert significant impacts on plant growth and development. Epigenetic mechanisms play a crucial role in mediating plant responses to environmental stress by regulating gene expression and modulating physiological and biochemical pathways (Ashapkin *et al.*, 2020).

22.10.1 Epigenetic responses to environmental stress

DNA methylation dynamics

Environmental stress can induce changes in DNA methylation patterns, leading to alterations in gene expression profiles. Stress-responsive genes may undergo dynamic DNA methylation changes in their regulatory regions, influencing their transcriptional activity and contributing to stress adaptation (Viggiano and de Pinto, 2017). DNA methylation is an epigenetic modification involving the addition of a methyl group to the 5' position of cytosine residues in DNA, typically occurring in the context of CpG dinucleotides. DNA methylation patterns are dynamic and can change in response to environmental stimuli, developmental cues, and stress conditions (Singal and Ginder, 1999).

Changes in DNA methylation patterns at regulatory regions can have profound effects on gene expression profiles. Stress-induced alterations in DNA methylation may influence the accessibility of transcription factor binding sites, chromatin structure, and histone modifications, thereby modulating the transcriptional activity of stress-responsive genes (Liu *et al.*, 2022). Stress-responsive genes that undergo dynamic DNA methylation changes may exhibit altered transcriptional activity in response to environmental stress. Hypermethylation of gene promoters may lead to transcriptional silencing, while hypomethylation may result in transcriptional activation, depending on the context and functional consequences for stress adaptation (Mifsud *et al.*, 2011).

Histone modifications

Environmental stressors can affect the abundance and distribution of histone modifications, such as acetylation, methylation, phosphorylation, and ubiquitination, on chromatin (Czajka *et al.*, 2021). These histone modifications influence chromatin structure and accessibility, thereby regulating the expression of stress-responsive genes and signaling pathways.

Histone modifications, collectively known as the 'histone code', serve as signals for chromatin remodeling complexes to alter nucleosome positioning and chromatin accessibility (Vanzan *et al.*, 2023). Environmental stress-induced histone modifications can remodel chromatin structure, allowing stress-responsive transcription factors to access their target genes and modulate their expression in response to stress stimuli (Kim, 2021).

Histone modifications directly influence the transcriptional activity of stress-responsive genes by promoting or inhibiting the recruitment of transcriptional machinery and regulatory factors. Changes in histone modification patterns at stress-responsive gene loci govern their transcriptional activation or repression, orchestrating adaptive responses to environmental stressors (Ueda and Seki, 2020). Environmental stress-induced histone modifications can confer a form of 'epigenetic memory' in plants, enabling rapid and robust transcriptional responses to recurring stress events. These histone marks may persist through mitotic divisions and influence gene expression patterns in subsequent generations, contributing to transgenerational stress adaptation and memory (Kinoshita and Seki, 2014).

Small RNA-mediated regulation

Small RNAs, including microRNAs (miRNAs) and small interfering RNAs (siRNAs), play roles in mediating plant responses to environmental stress. Stress-responsive small RNAs can target and regulate the expression of genes involved in stress signaling, defense responses, and adaptive pathways (Wang and Chekanova, 2016).

Stress-responsive miRNAs are induced or repressed in response to environmental cues, such as drought, salinity, temperature fluctuations, and pathogen attacks. These stress-responsive miRNAs target and regulate the expression of genes involved in stress signaling, defense responses, and adaptive pathways, thereby modulating plant stress tolerance and resilience (Pagano *et al.*, 2021). For example, miR393 has been implicated in regulating plant immunity by targeting auxin signaling components, while miR398 targets copper superoxide dismutase (CSD1 and CSD2) genes involved in oxidative stress responses (Singh *et al.*, 2017).

Stress-responsive siRNAs are involved in plant defense mechanisms against biotic and abiotic stressors. They target and regulate the expression of genes encoding stress-related proteins, transcription factors, and signaling components involved in stress responses. For instance, siRNAs generated from the processing of viral or transposable element-derived double-stranded RNA precursors mediate antiviral defense and transposon silencing, respectively, thereby protecting plants from viral infections and genomic instability under stress conditions (Ramakrishnan *et al.*, 2021).

22.10.2 Crosstalk and integration

Integration with hormone signaling

Small RNAs often intersect with hormone signaling pathways, such as those involving abscisic acid (ABA), salicylic acid (SA), jasmonic acid (JA), and ethylene (ET), to coordinate plant responses to environmental stress. Crosstalk between small RNAs and hormone signaling networks enables plants to fine-tune stress responses and optimize stress tolerance mechanisms (Parwez *et al.*, 2022).

Epigenetic regulation

Small RNAs can also influence epigenetic regulation by directing DNA methylation and histone modification activities to target loci. Through RNA-directed DNA methylation (RdDM) and RNA-directed histone modification pathways, small RNAs contribute to the establishment of stress-responsive epigenetic marks and transcriptional reprogramming under stress conditions (Matzke and Mosher, 2014).

22.10.3 Epigenetic memory and priming

Stress memory

Epigenetic modifications induced by environmental stress can confer a form of 'stress memory' in plants, allowing them to respond more effectively to recurring stress events (Lämke and Bäurle, 2017). These epigenetic changes may persist through mitotic divisions and influence gene expression patterns in subsequent generations, contributing to transgenerational stress adaptation (Turner, 2009).

Epigenetic priming

Environmental stress can induce epigenetic priming mechanisms that prepare plants for future stress encounters. Priming involves the establishment of epigenetic marks, such as histone modifications and DNA methylation, that enhance the responsiveness of stress-responsive genes upon re-exposure to stress stimuli, thereby promoting faster and more robust stress responses (Nair *et al.*, 2022).

22.10.4 Epigenetic strategies for stress tolerance

Epigenome editing

Epigenome editing technologies, such as CRISPR-based systems, can be used to engineer plants with altered epigenetic profiles that confer enhanced stress tolerance. Targeted modifications of DNA methylation patterns or histone modifications at stress-responsive genes can improve plant resilience to environmental stressors (Qi *et al.*, 2023a).

Epigenetic modulators

Pharmacological agents targeting epigenetic regulators, such as DNA methyltransferases, histone deacetylases, or histone methyltransferases, can be applied to modulate epigenetic responses to environmental stress. These epigenetic modulators can enhance stress tolerance by manipulating gene expression and stress-responsive pathways (Choudhuri *et al.*, 2010).

Breeding for epigenetic variation

Breeding programs can leverage natural epigenetic variation within plant populations to select individuals with enhanced stress tolerance traits. Screening for epigenetic markers associated with stress adaptation and incorporating them into breeding schemes can facilitate the development of stress-resistant crop varieties (Sun *et al.*, 2021).

22.10.5 Future directions and challenges

Further research is needed to elucidate the mechanisms underlying epigenetic plasticity in response to environmental stress and its implications for plant adaptation and evolution. Strategies for harnessing natural epigenetic variation and engineering targeted epigenetic modifications in crops require optimization and validation to ensure their effectiveness and stability under field conditions (Gahlaut *et al.*, 2020). Investigating the transgenerational inheritance of stress-induced epigenetic changes and their ecological significance in natural populations is an area of growing interest and importance (Weinhold, 2018).

Environmental stress profoundly influences plant growth and development by eliciting epigenetic responses that regulate gene expression and stress adaptation pathways (Rossnerova *et al.*, 2020). Understanding the interplay between environmental stress and epigenetic regulation holds promise for developing innovative strategies to enhance crop resilience and ensure food security in a changing climate.

22.11 Conclusion

In conclusion, epigenetics offers a rich and dynamic framework for understanding the intricate regulatory mechanisms governing plant growth and development in response to environmental cues. Through epigenetic modifications such as DNA methylation, histone modifications, and small RNA-mediated regulation, plants can fine-tune gene expression patterns, chromatin structure, and stress responses to optimize their fitness and adaptability in changing environments.

Epigenetic strategies hold immense potential for advancing agricultural practices and crop improvement efforts. By deciphering the epigenetic code underlying plant traits and stress tolerance mechanisms, researchers can develop innovative approaches for enhancing crop productivity, resilience, and sustainability. Epigenome editing technologies, pharmacological interventions targeting epigenetic regulators, and precision breeding strategies based on epigenetic variation offer promising avenues for engineering crops with desired traits and environmental adaptability.

Furthermore, the integration of epigenetic with other disciplines such as genomics, transcriptomics, and systems biology provides a holistic understanding of plant biology and ecosystem dynamics. Leveraging multi-omics approaches and computational modeling enables the elucidation of complex gene regulatory networks and the prediction of genotype–phenotype associations in diverse environmental contexts.

Overall, epigenetics represents a frontier in agricultural research and biotechnology, offering transformative opportunities for addressing global challenges such as food security, climate change, and sustainable agriculture. By harnessing the power of epigenetic regulation, we can unlock the full potential of plants to thrive in a rapidly changing world and meet the growing demands of a burgeoning population.

References

Adalsteinsson, B.T. and Ferguson-Smith, A.C. (2014) Epigenetic control of the genome-lessons from genomic imprinting. *Genes* 5(3), 635–655.

Akhter, Z., Bi, Z., Ali, K., Sun, C., Fiaz, S. *et al.* (2021) In response to abiotic stress, DNA methylation confers epigenetic changes in plants. *Plants* 10(6), 1096.

Alberts, B., Johnson, A., Lewis, J., Raff, M., Roberts, K. *et al.* (2002) From DNA to RNA. In: Alberts, B. (ed.) *Molecular Biology of the Cell*, 4th edn. Garland Science, New York.

Albini, S., Zakharova, V. and Ait-Si-Ali, S. (2019) Histone modifications. In: Palacios, D. (ed.) *Epigenetics and Regeneration*. Academic Press, pp. 47–72.

Ali, S., Khan, N. and Tang, Y. (2022) Epigenetic marks for mitigating abiotic stresses in plants. *Journal of Plant Physiology* 275, 153740.

Allen, E. and Howell, M.D. (2010) MiRNAs in the biogenesis of trans-acting siRNAs in higher plants. In: *Seminars in Cell & Developmental Biology*, Vol. 21, No.8. Academic Press, pp. 798–804.

Ashapkin, V.V., Kutueva, L.I., Aleksandrushkina, N.I. and Vanyushin, B.F. (2020) Epigenetic mechanisms of plant adaptation to biotic and abiotic stresses. *International Journal of Molecular Sciences* 21(20), 7457.

Ashfaq, M.A., Kumar, V.D., Reddy, P.S.S., Kumar, C.H.A., Kumar, K.S. *et al.* (2020) Post-transcriptional gene silencing: Basic concepts and applications. *Journal of Biosciences* 45(1), 128.

Attwood, J.T., Yung, R.L. and Richardson, B.C. (2002) DNA methylation and the regulation of gene transcription. *Cellular and Molecular Life Sciences* 59(2), 241–257.

Axtell, M.J. and Bowman, J.L. (2008) Evolution of plant microRNAs and their targets. *Trends in Plant Science* 13(7), 343–349.

Ay, N., Janack, B. and Humbeck, K. (2014) Epigenetic control of plant senescence and linked processes. *Journal of Experimental Botany* 65(14), 3875–3887.

Baier, A. and Szyszka, R. (2020) Compounds from natural sources as protein kinase inhibitors. *Biomolecules* 10(11), 1546.

Bajan, S. and Hutvagner, G. (2020) RNA-based therapeutics: From antisense oligonucleotides to miRNAs. *Cells* 9(1), 137.

Bajrami, E. and Spiroski, M. (2016) Genomic imprinting. *Open Access Macedonian Journal of Medical Sciences* 4(1), 181–184.

Bajus, M., Macko-Podgórni, A., Grzebelus, D. and Baránek, M. (2022) A review of strategies used to identify transposition events in plant genomes. *Frontiers in Plant Science* 13, 1080993.

Baker-Andresen, D., Ratnu, V.S. and Bredy, T.W. (2013) Dynamic DNA methylation: A prime candidate for genomic metaplasticity and behavioral adaptation. *Trends in Neurosciences* 36(1), 3–13.

Balasubramanian, S., Verner, E. and Buggy, J.J. (2009) Isoform-specific histone deacetylase inhibitors: The next step. *Cancer Letters* 280(2), 211–221.

Balasubramanyam, K., Swaminathan, V., Ranganathan, A. and Kundu, T.K. (2003) Small molecule modulators of histone acetyltransferase p300. *The Journal of Biological Chemistry* 278(21), 19134–19140.

Bannister, A.J. and Kouzarides, T. (2011) Regulation of chromatin by histone modifications. *Cell Research* 21(3), 381–395.

Baroux, C., Spillane, C. and Grossniklaus, U. (2002) Genomic imprinting during seed development. *Advances in Genetics* 46, 165–214.

Baxter, A., Mittler, R. and Suzuki, N. (2014) ROS as key players in plant stress signalling. *Journal of Experimental Botany* 65(5), 1229–1240.

Becker, P.B. and Hörz, W. (2002) ATP-dependent nucleosome remodeling. *Annual Review of Biochemistry* 71(1), 247–273.

Ben Maamar, M., Nilsson, E.E. and Skinner, M.K. (2021) Epigenetic transgenerational inheritance, gametogenesis and germline development. *Biology of Reproduction* 105(3), 570–592.

Bennetzen, J.L. and Wang, H. (2014) The contributions of transposable elements to the structure, function, and evolution of plant genomes. *Annual Review of Plant Biology* 65, 505–530.

Berr, A., Shafiq, S. and Shen, W.H. (2011) Histone modifications in transcriptional activation during plant development. *Biochimica et Biophysica Acta (BBA) – Gene Regulatory Mechanisms* 1809(10), 567–576.

Bewick, A.J. and Schmitz, R.J. (2017) Gene body DNA methylation in plants. *Current Opinion in Plant Biology* 36, 103–110.

Bhadouriya, S.L., Mehrotra, S., Basantani, M.K., Loake, G.J. and Mehrotra, R. (2021) Role of chromatin architecture in plant stress responses: An update. *Frontiers in Plant Science* 11, 603380.

Bhan, A., Deb, P. and Mandal, S.S. (2017) Epigenetic code: Histone modification, gene regulation, and chromatin dynamics. In: Mandal, S.S. (ed.) *Gene Regulation, Epigenetics and Hormone Signaling*. Wiley-VCH, pp. 29–58.

Bharathi, J.K., Anandan, R., Benjamin, L.K., Muneer, S. and Prakash, M.A.S. (2023) Recent trends and advances of RNA interference (RNAi) to improve agricultural crops and enhance their resilience to biotic and abiotic stresses. *Plant Physiology and Biochemistry* 194, 600–618.

Bhattarai, K. and Bhattarai, B. (2021) Mechanism of DNA methylation and its role in biotic and abiotic stress response in plants: A review. *Farming and Management* 6(1), 39–46.

Bhaya, D., Davison, M. and Barrangou, R. (2011) CRISPR-cas systems in bacteria and archaea: Versatile small RNAs for adaptive defense and regulation. *Annual Review of Genetics* 45, 273–297.

Bleecker, A.B. and Kende, H. (2000) Ethylene: A gaseous signal molecule in plants. *Annual Review of Cell and Developmental Biology* 16(1), 1–18.

Boyes, J. and Bird, A. (1991) DNA methylation inhibits transcription indirectly via a methyl-CpG binding protein. *Cell* 64(6), 1123–1134.

Boyko, A. and Kovalchuk, I. (2011) Genome instability and epigenetic modification – heritable responses to environmental stress. *Current Opinion in Plant Biology* 14(3), 260–266.

Brown, S.J., Stoilov, P. and Xing, Y. (2012) Chromatin and epigenetic regulation of pre-mRNA processing. *Human Molecular Genetics* 21(R1), R90–R96.

Brusslan, J.A., Bonora, G., Rus-Canterbury, A.M., Tariq, F., Jaroszewicz, A. *et al.* (2015) A genome-wide chronological study of gene expression and two histone modifications, H3k4me3 and H3k9ac, during developmental leaf senescence. *Plant Physiology* 168(4), 1246–1261.

Buchanan-Wollaston, V. (1997) The molecular biology of leaf senescence. *Journal of Experimental Botany* 48(2), 181–199.

Can, S.N., Nunn, A., Galanti, D., Langenberger, D., Becker, C. *et al.* (2021) The epiDiverse plant epigenome-wide association studies (EWAS) pipeline. *Epigenomes* 5(2), 12.

Castañeda, L., Giménez, E., Pineda, B., García-Sogo, B., Ortiz-Atienza, A. *et al.* (2022) Tomato CRABS CLAW paralogues interact with chromatin remodelling factors to mediate carpel development and floral determinacy. *The New Phytologist* 234(3), 1059–1074.

Castel, S.E. and Martienssen, R.A. (2013) RNA interference in the nucleus: Roles for small RNAs in transcription, epigenetics and beyond. *Nature Reviews Genetics* 14(2), 100–112.

Chen, D.H., Huang, Y., Jiang, C. and Si, J.P. (2018a) Chromatin-based regulation of plant root development. *Frontiers in Plant Science* 9, 1509.

Chen, Y.R., Yu, S. and Zhong, S. (2018b) Profiling DNA methylation using bisulfite sequencing (BS-seq). In: Berner, M. and Baroux, C. (eds) *Plant Chromatin Dynamics. Methods in Molecular Biology*, Vol. 1675. Humana Press, New York, pp. 31–43.

Chicas, A. and Macino, G. (2001) Characteristics of post-transcriptional gene silencing. *EMBO Reports* 2(11), 992–996.

Choudhuri, S., Cui, Y. and Klaassen, C.D. (2010) Molecular targets of epigenetic regulation and effectors of environmental influences. *Toxicology and Applied Pharmacology* 245(3), 378–393.

Clapier, C.R. and Cairns, B.R. (2009) The biology of chromatin remodeling complexes. *Annual Review of Biochemistry* 78, 273–304.

Cutler, S.R., Rodriguez, P.L., Finkelstein, R.R. and Abrams, S.R. (2010) Abscisic acid: Emergence of a core signaling network. *Annual Review of Plant Biology* 61, 651–679.

Czajka, K., Mehes-Smith, M. and Nkongolo, K. (2021) DNA methylation and histone modifications induced by abiotic stressors in plants. *Genes & Genomics* 1–19.

Dalakouras, A. and Vlachostergios, D. (2021) Epigenetic approaches to crop breeding: Current status and perspectives. *Journal of Experimental Botany* 72(15), 5356–5371.

Delcuve, G.P., Khan, D.H. and Davie, J.R. (2012) Roles of histone deacetylases in epigenetic regulation: Emerging paradigms from studies with inhibitors. *Clinical Epigenetics* 4(1), 1–13.

Ding, X., Yu, L., Chen, L., Li, Y., Zhang, J. *et al.* (2022) Recent progress and future prospect of CRISPR/Cas-derived transcription activation (CRISPRa) system in plants. *Cells* 11(19), 3045.

Elhamamsy, A.R. (2017) Role of DNA methylation in imprinting disorders: An updated review. *Journal of Assisted Reproduction and Genetics* 34(5), 549–562.

El-Sappah, A.H., Yan, K., Huang, Q., Islam, M.M., Li, Q. *et al.* (2021) Comprehensive mechanism of gene silencing and its role in plant growth and development. *Frontiers in Plant Science* 12, 705249.

Engelhorn, J., Blanvillain, R. and Carles, C.C. (2014) Gene activation and cell fate control in plants: A chromatin perspective. *Cellular and Molecular Life Sciences* 71(16), 3119–3137.

Fagard, M. and Vaucheret, H. (2000) Trans gene silencing in plants: How many mechanisms. *Annual Review of Plant Biology* 51(1), 167–194.

Feschotte, C. and Pritham, E.J. (2007) DNA transposons and the evolution of eukaryotic genomes. *Annual Review of Genetics* 41, 331–368.

Finnegan, D.J. (2012) Retrotransposons. *Current Biology* 22(11), R432–R437.

FitzGerald, J., Luo, M., Chaudhury, A. and Berger, F. (2008) DNA methylation causes predominant maternal controls of plant embryo growth. *PloS ONE* 3(5), e2298.

Fleet, C.M. and Sun, T. (2005) A DELLAcate balance: The role of gibberellin in plant morphogenesis. *Current Opinion in Plant Biology* 8(1), 77–85.

Foerster, A.M. and Mittelsten Scheid, O. (2010) Analysis of DNA methylation in plants by bisulfite sequencing. *Plant Epigenetics: Methods in Molecular Biology* 631, 1–11.

Gahlaut, V., Zinta, G., Jaiswal, V. and Kumar, S. (2020) Quantitative epigenetics: A new avenue for crop improvement. *Epigenomes* 4(4), 25.

Gebert, D. and Rosenkranz, D. (2015) RNA-based regulation of transposon expression. *Wiley Interdisciplinary Reviews: RNA* 6(6), 687–708.

Gill, R.A., Scossa, F., King, G.J., Golicz, A.A., Tong, C. *et al.* (2021) On the role of transposable elements in the regulation of gene expression and subgenomic interactions in crop genomes. *Critical Reviews in Plant Sciences* 40(2), 157–189.

Glavincheska, I. (2022) The regulatory potential of transposable elements in plant defence pathways. Master thesis, Utrecht University.

Gouil, Q. and Keniry, A. (2019) Latest techniques to study DNA methylation. *Essays in Biochemistry* 63(6), 639–648.

Greiner, D., Bonaldi, T., Eskeland, R., Roemer, E. and Imhof, A. (2005) Identification of a specific inhibitor of the histone methyltransferase SU (VAR)3-9. *Nature Chemical Biology* 1(3), 143–145.

Gu, X., Le, C., Wang, Y., Li, Z., Jiang, D. *et al.* (2013) *Arabidopsis* FLC clade members form flowering-repressor complexes coordinating responses to endogenous and environmental cues. *Nature Communications* 4(1).

Guarino, F., Cicatelli, A., Castiglione, S., Agius, D.R., Orhun, G.E. *et al.* (2022) An epigenetic alphabet of crop adaptation to climate change. *Frontiers in Genetics* 13, 818727.

Ha, M. and Kim, V.N. (2014) Regulation of microRNA biogenesis. *Nature Reviews. Molecular Cell Biology* 15(8), 509–524.

Hajkova, P., El-Maarri, O., Engemann, S., Oswald, J., Olek, A. *et al.* (2002) DNA-methylation analysis by the bisulfite-assisted genomic sequencing method. *DNA Methylation Protocols* 143–154.

Hayatsu, H. (2008) The bisulfite genomic sequencing used in the analysis of epigenetic states, a technique in the emerging environmental genotoxicology research. *Mutation Research/Reviews in Mutation Research* 659(1–2), 77–82.

Henikoff, S. and Smith, M.M. (2015) Histone variants and epigenetics. *Cold Spring Harbor Perspectives in Biology* 7(1), a019364.

Hernández, H.G., Tse, M.Y., Pang, S.C., Arboleda, H. and Forero, D.A. (2013) Optimizing methodologies for PCR-based DNA methylation analysis. *BioTechniques* 55(4), 181–197.

Hilton, I.B., D'Ippolito, A.M., Vockley, C.M., Thakore, P.I., Crawford, G.E. *et al.* (2015) Epigenome editing by a CRISPR-Cas9-based acetyltransferase activates genes from promoters and enhancers. *Nature Biotechnology* 33(5), 510–517.

Hossain, M.S., Soejima, T. and Tase, K. (2012) Curcumin: Novel plant-derived HAT activator modulating histone acetylation and gene expression in plants. *Plant Biotechnology Reports* 6(1), 81–89.

Hu, G., Huang, B., Wang, K., Frasse, P., Maza, E. *et al.* (2021) Histone posttranslational modifications rather than DNA methylation underlie gene reprogramming in pollination-dependent and pollination-independent fruit set in tomato. *The New Phytologist* 229(2), 902–919.

Ikeda, Y., Nishihama, R., Yamaoka, S., Arteaga-Vazquez, M.A., Aguilar-Cruz, A. *et al.* (2018) Loss of CG methylation in *Marchantia polymorpha* causes disorganization of cell division and reveals unique DNA methylation regulatory mechanisms of non-CG methylation. *Plant and Cell Physiology* 59(12), 2421–2431.

Iwakawa, H.O. and Tomari, Y. (2022) Life of RISC: Formation, action, and degradation of RNA-induced silencing complex. *Molecular Cell* 82(1), 30–43.

Jarillo, J.A., Piñeiro, M., Cubas, P. and Martínez-Zapater, J.M. (2009) Chromatin remodeling in plant development. *The International Journal of Developmental Biology* 53(8–10), 1581–1596.

Jeong, H.J., Yang, J., Yi, J. and An, G. (2015) Controlling flowering time by histone methylation and acetylation in *Arabidopsis* and rice. *Journal of Plant Biology* 58(4), 203–210.

Jones-Rhoades, M.W., Bartel, D.P. and Bartel, B. (2006) MicroRNAS and their regulatory roles in plants. *Annual Review of Plant Biology* 57, 19–53.

Jordan, W.T. (2020) *Methods for plant-based genome and epigenome editing.* Doctoral dissertation, University of Georgia.

Kakoulidou, I., Avramidou, E.V., Baránek, M., Brunel-Muguet, S., Farrona, S. *et al.* (2021) Epigenetics for crop improvement in times of global change. *Biology* 10(8), 766.

Kalinina, N.O., Makarova, S., Makhotenko, A., Love, A.J. and Taliansky, M. (2018) The multiple functions of the nucleolus in plant development, disease and stress responses. *Frontiers in Plant Science* 9, 132.

Kamthan, A., Chaudhuri, A., Kamthan, M. and Datta, A. (2015) Small RNAs in plants: Recent development and application for crop improvement. *Frontiers in Plant Science* 6, 208.

Kang, H., Wu, D., Fan, T. and Zhu, Y. (2020) Activities of chromatin remodeling factors and histone chaperones and their effects in root apical meristem development. *International Journal of Molecular Sciences* 21(3), 771.

Kapoor, M., Arora, R., Lama, T., Nijhawan, A., Khurana, J.P. *et al.* (2008) Genome-wide identification, organization and phylogenetic analysis of dicer-like, argonaute and RNA-dependent RNA polymerase gene families and their expression analysis during reproductive development and stress in rice. *BMC Genomics* 9, 1–17.

Kenchanmane Raju, S.K., Shao, M.R., Wamboldt, Y. and Mackenzie, S. (2018) Epigenomic plasticity of *Arabidopsis* msh1 mutants under prolonged cold stress. *Plant Direct* 2(8), e00079.

Khraiwesh, B., Zhu, J.K. and Zhu, J. (2012) Role of miRNAs and siRNAs in biotic and abiotic stress responses of plants. *Biochimica et Biophysica Acta (BBA)-Gene Regulatory Mechanisms* 1819(2), 137–148.

Kim, J.H. (2021) Multifaceted chromatin structure and transcription changes in plant stress response. *International Journal of Molecular Sciences* 22(4), 2013.

Kim, J.M., To, T.K. and Seki, M. (2012) An epigenetic integrator: New insights into genome regulation, environmental stress responses and developmental controls by histone deacetylase 6. *Plant & Cell Physiology* 53(5), 794–800.

Kinoshita, T. and Seki, M. (2014) Epigenetic memory for stress response and adaptation in plants. *Plant and Cell Physiology* 55(11), 1859–1863.

Kirchner, M. and Schneider, S. (2015) CRISPR-Cas: From the bacterial adaptive immune system to a versatile tool for genome engineering. *Angewandte Chemie International Edition* 54(46), 13508–13514.

Köhler, C., Wolff, P. and Spillane, C. (2012) Epigenetic mechanisms underlying genomic imprinting in plants. *Annual Review of Plant Biology* 63, 331–352.

Krasnodębski, C., Sawuła, A., Kaźmierczak, U. and Żuk, M. (2023) Oligo – not only for silencing: Overlooked potential for multidirectional action in plants. *International Journal of Molecular Sciences* 24(5), 4466.

Kryovrysanaki, N., James, A., Tselika, M., Bardani, E. and Kalantidis, K. (2021) RNA silencing pathways in plant development and defense. *The International Journal of Developmental Biology* 66(1–2–3), 163–175.

Kubicek, S., O'Sullivan, R.J., August, E.M., Hickey, E.R., Zhang, Q. *et al.* (2007) Reversal of H3k9me2 by a small-molecule inhibitor for the G9a histone methyltransferase. *Molecular Cell* 25(3), 473–481.

Kumar, S. (2018) Epigenomics of plant responses to environmental stress. *Epigenomes* 2(1), 6.

Kumar, S. (2019) Genome editing to epigenome editing: Towards unravelling the enigmas in developmental biology. *Trends in Developmental Biology* 12, 31–38.

Kumar, S. and Mohapatra, T. (2021) Dynamics of DNA methylation and its functions in plant growth and development. *Frontiers in Plant Science* 12, 596236.

Kuzma, J. and Kokotovich, A. (2011) Renegotiating GM crop regulation: Targeted gene-modification technology raises new issues for the oversight of genetically modified crops. *EMBO Reports* 12(9), 883–888.

Kyle, D.J., Ohad, I. and Arntzen, C.J. (1984) Membrane protein damage and repair: Selective loss of a quinone-protein function in chloroplast membranes. *Proceedings of the National Academy of Sciences* 81(13), 4070–4074.

Laing, W.A., Wright, M.A., Cooney, J. and Bulley, S.M. (2007) The missing step of the L-galactose pathway of ascorbate biosynthesis in plants, an L-galactose guanyltransferase, increases leaf ascorbate content. *Proceedings of the National Academy of Sciences* 104(22), 9534–9539.

Lämke, J. and Bäurle, I. (2017) Epigenetic and chromatin-based mechanisms in environmental stress adaptation and stress memory in plants. *Genome Biology* 18(1), 1–11.

Law, J.A. and Jacobsen, S.E. (2010) Establishing, maintaining and modifying DNA methylation patterns in plants and animals. *Nature Reviews Genetics* 11(3), 204–220.

Le, T.N., Schumann, U., Smith, N.A., Tiwari, S., Au, P.C.K. *et al.* (2014) DNA demethylases target promoter transposable elements to positively regulate stress responsive genes in *Arabidopsis*. *Genome Biology* 15(9), 1–18.

Lee, C.Y. and Grant, P.A. (2019) Role of histone acetylation and acetyltransferases in gene regulation. In: McCullough, S.D. and Dolinoy, D.C. (eds) *Toxicoepigenetics: Core Principles and Applications*. Academic Press, pp. 3–30.

Leonetti, P., Stuttmann, J. and Pantaleo, V. (2021) Regulation of plant antiviral defense genes via host RNA-silencing mechanisms. *Virology Journal* 18(1), 1–10.

Li, B., Carey, M. and Workman, J.L. (2007) The role of chromatin during transcription. *Cell* 128(4), 707–719.

Li, S. and Tollefsbol, T.O. (2021) DNA methylation methods: Global DNA methylation and methylomic analyses. *Methods* 187, 28–43.

Li, Y. and Tollefsbol, T.O. (2011) DNA methylation detection: Bisulfite genomic sequencing analysis. *Methods in Molecular Biology* 791, 11–21.

Li, X., Wang, X., He, K., Ma, Y., Su, N. *et al.* (2008) High-resolution mapping of epigenetic modifications of the rice genome uncovers interplay between DNA methylation, histone methylation, and gene expression. *The Plant Cell* 20(2), 259–276.

Liang, Z., Riaz, A., Chachar, S., Ding, Y., Du, H. *et al.* (2020) Epigenetic modifications of mRNA and DNA in plants. *Molecular Plant* 13(1), 14–30.

Lisch, D. (2009) Epigenetic regulation of transposable elements in plants. *Annual Review of Plant Biology* 60, 43–66.

Liu, B. and Zhao, M. (2023) How transposable elements are recognized and epigenetically silenced in plants. *Current Opinion in Plant Biology* 75, 102428.

Liu, J. and Zhong, X. (2024) Population epigenetics: DNA methylation in the plant omics era. *Plant Physiology* 194(4), 2039–2048.

Liu, C., Lu, F., Cui, X. and Cao, X. (2010) Histone methylation in higher plants. *Annual Review of Plant Biology* 61, 395–420.

Liu, Y., Wang, J., Liu, B. and Xu, Z.Y. (2022) Dynamic regulation of DNA methylation and histone modifications in response to abiotic stresses in plants. *Journal of Integrative Plant Biology* 64(12), 2252–2274.

Lopez, M., Gilbert, J., Contreras, J., Halby, L. and Arimondo, P.B. (2022) Inhibitors of DNA methylation. *Advances in Experimental Medicine and Biology* 471–513.

Lorch, Y., Maier-Davis, B. and Kornberg, R.D. (2006) Chromatin remodeling by nucleosome disassembly *in vitro*. *Proceedings of the National Academy of Sciences* 103(9), 3090–3093.

MacDonald, W.A. (2012) Epigenetic mechanisms of genomic imprinting: Common themes in the regulation of imprinted regions in mammals, plants, and insects. *Genetics Research International* 585024.

Mahfouz, M.M. (2010) RNA-directed DNA methylation: Mechanisms and functions. *Plant Signaling & Behavior* 5(7), 806–816.

Malecová, B. and Morris, K.V. (2010) Transcriptional gene silencing mediated by non-coding RNAs. *Current Opinion in Molecular Therapeutics* 12(2), 214.

Matzke, M.A. and Mosher, R.A. (2014) RNA-directed DNA methylation: An epigenetic pathway of increasing complexity. *Nature Reviews Genetics* 15(6), 394–408.

Matzke, M.A., Kanno, T. and Matzke, A.J.M. (2015) RNA-directed DNA methylation: The evolution of a complex epigenetic pathway in flowering plants. *Annual Review of Plant Biology* 66(1), 243–267.

Meng, Y., Shao, C., Wang, H. and Chen, M. (2011) The regulatory activities of plant microRNAs: A more dynamic perspective. *Plant Physiology* 157(4), 1583–1595.

Michlewski, G. and Cáceres, J.F. (2019) Post-transcriptional control of miRNA biogenesis. *RNA* 25(1), 1–16.

Mifsud, K.R., Gutièrrez-Mecinas, M., Trollope, A.F., Collins, A., Saunderson, E.A. *et al.* (2011) Epigenetic mechanisms in stress and adaptation. *Brain, Behavior, and Immunity* 25(7), 1305–1315.

Miglani, G.S., Kaur, A. and Kaur, L. (2020) Plant gene expression control using genome- and epigenome-editing technologies. *Journal of Crop Improvement* 34(1), 1–63.

Miryeganeh, M. (2022) Epigenetic mechanisms of senescence in plants. *Cells* 11(2), 251.

Mittler, R. and Blumwald, E. (2015) The roles of ROS and ABA in systemic acquired acclimation. *The Plant Cell* 27(1), 64–70.

Mounger, J., Ainouche, M.L., Bossdorf, O., Cavé-Radet, A., Li, B. *et al.* (2021) Epigenetics and the success of invasive plants. *Philosophical Transactions of the Royal Society B* 376(1826), 20200117.

Nair, A.U., Bhukya, D.P.N., Sunkar, R., Chavali, S. and Allu, A.D. (2022) Molecular basis of priming-induced acquired tolerance to multiple abiotic stresses in plants. *Journal of Experimental Botany* 73(11), 3355–3371.

Narlikar, G.J., Fan, H.Y. and Kingston, R.E. (2002) Cooperation between complexes that regulate chromatin structure and transcription. *Cell* 108(4), 475–487.

Neri, F., Rapelli, S., Krepelova, A., Incarnato, D., Parlato, C. *et al.* (2017) Intragenic DNA methylation prevents spurious transcription initiation. *Nature* 543(7643), 72–77.

Newell-Price, J., Clark, A.J.L. and King, P. (2000) DNA methylation and silencing of gene expression. *Trends in Endocrinology & Metabolism* 11(4), 142–148.

Nidhi, S., Anand, U., Oleksak, P., Tripathi, P., Lal, J.A. *et al.* (2021) Novel CRISPR-Cas systems: An updated review of the current achievements, applications, and future research perspectives. *International Journal of Molecular Sciences* 22(7), 3327.

Nie, H., Cheng, C., Kong, J., Li, H. and Hua, J. (2023) Plant non-coding RNAs function in pollen development and male sterility. *Frontiers in Plant Science* 14, 1109941.

Ojolo, S.P., Cao, S., Priyadarshani, S.V.G.N., Li, W., Yan, M. *et al.* (2018) Regulation of plant growth and development: A review from a chromatin remodeling perspective. *Frontiers in Plant Science* 9, 372977.

Ozata, D.M., Gainetdinov, I., Zoch, A., O'Carroll, D. and Zamore, P.D. (2019) PIWI-interacting RNAs: Small RNAs with big functions. *Nature Reviews Genetics* 20(2), 89–108.

Pagano, L., Rossi, R., Paesano, L., Marmiroli, N. and Marmiroli, M. (2021) MiRNA regulation and stress adaptation in plants. *Environmental and Experimental Botany* 184, 104369.

Parwez, R., Aftab, T., Gill, S.S. and Naeem, M. (2022) Abscisic acid signaling and crosstalk with phytohormones in regulation of environmental stress responses. *Environmental and Experimental Botany* 199, 104885.

Peterson, C.L. and Laniel, M.A. (2004) Histones and histone modifications. *Current Biology* 14(14), R546–R551.

Pikaard, C.S. and Mittelsten Scheid, O. (2014) Epigenetic regulation in plants. *Cold Spring Harbor Perspectives in Biology* 6(12), a019315.

Qi, Q., Hu, B., Jiang, W., Wang, Y., Yan, J. *et al.* (2023a) Advances in plant epigenome editing research and its application in plants. *International Journal of Molecular Sciences* 24(4), 3442.

Qi, X., Wan, C., Zhang, X., Sun, W., Liu, R. *et al.* (2023b) Effects of histone methylation modification on low temperature seed germination and growth of maize. *Scientific Reports* 13(1), 5196.

Quemener, A.M., Bachelot, L., Forestier, A., Donnou-Fournet, E., Gilot, D. *et al.* (2020) The powerful world of antisense oligonucleotides: From bench to bedside. *Wiley Interdisciplinary Reviews: RNA* 11(5), e1594.

Ramakrishnan, M., Satish, L., Kalendar, R., Narayanan, M., Kandasamy, S. *et al.* (2021) The dynamism of transposon methylation for plant development and stress adaptation. *International Journal of Molecular Sciences* 22(21), 11387.

Rauluseviciute, I., Drabløs, F. and Rye, M.B. (2019) DNA methylation data by sequencing: Experimental approaches and recommendations for tools and pipelines for data analysis. *Clinical Epigenetics* 11(1), 1-13.

Raza, A., Tabassum, J., Fakhar, A.Z., Sharif, R., Chen, H. *et al.* (2023) Smart reprograming of plants against salinity stress using modern biotechnological tools. *Critical Reviews in Biotechnology* 43(7), 1035–1062.

Reyes, J.C. (2006) Chromatin modifiers that control plant development. *Current Opinion in Plant Biology* 9(1), 21–27.

Rosa, S. and Shaw, P. (2013) Insights into chromatin structure and dynamics in plants. *Biology* 2(4), 1378–1410.

Rossnerova, A., Izzotti, A., Pulliero, A., Bast, A., Rattan, S.I.S. *et al.* (2020) The molecular mechanisms of adaptive response related to environmental stress. *International Journal of Molecular Sciences* 21(19), 7053.

Rymen, B., Ferrafiat, L. and Blevins, T. (2020) Non-coding RNA polymerases that silence transposable elements and reprogram gene expression in plants. *Transcription* 11(3–4), 172–191.

Sahu, P.P., Pandey, G., Sharma, N., Puranik, S., Muthamilarasan, M. *et al.* (2013) Epigenetic mechanisms of plant stress responses and adaptation. *Plant Cell Reports* 32, 1151–1159.

Sakakibara, H. (2006) Cytokinins: Activity, biosynthesis, and translocation. *Annual Review of Plant Biology* 57, 431–449.

Sanan-Mishra, N., Abdul Kader Jailani, A., Mandal, B. and Mukherjee, S.K. (2021) Secondary siRNAs in plants: Biosynthesis, various functions, and applications in virology. *Frontiers in Plant Science* 12, 610283.

Sani, E., Herzyk, P., Perrella, G., Colot, V. and Amtmann, A. (2013) Hyperosmotic priming of *Arabidopsis* seedlings establishes a long-term somatic memory accompanied by specific changes of the epigenome. *Genome Biology* 14(6), 1–24.

Schauer, S.E., Jacobsen, S.E., Meinke, D.W. and Ray, A. (2002) DICER-LIKE1: Blind men and elephants in *Arabidopsis* development. *Trends in Plant Science* 7(11), 487–491.

Secco, D., Wang, C., Shou, H., Schultz, M.D., Chiarenza, S. *et al.* (2015) Stress induced gene expression drives transient DNA methylation changes at adjacent repetitive elements. *eLife* 4, e09343.

Sharma, V.K., Marla, S., Zheng, W., Mishra, D., Huang, J. *et al.* (2022) CRISPR guides induce gene silencing in plants in the absence of Cas. *Genome Biology* 23(1), 1–24.

Shi, M., Wei, Z., Zhang, P., Guan, C., Chachar, S. *et al.* (2024) Epigenetic modifications and breeding applications in horticultural plants. *Horticulturae* 10(2), 143.

Shilatifard, A. (2006) Chromatin modifications by methylation and ubiquitination: Implications in the regulation of gene expression. *Annual Review of Biochemistry* 75, 243–269.

Shin, H., Choi, W.L., Lim, J.Y. and Huh, J.H. (2022) Epigenome editing: Targeted manipulation of epigenetic modifications in plants. *Genes & Genomics* 44, 1–9.

Singal, R. and Ginder, G.D. (1999) DNA methylation. *Blood, The Journal of the American Society of Hematology* 93(12), 4059–4070.

Singh, R., Parihar, P., Singh, S., Singh, M.P.V.V.B., Singh, V.P. *et al.* (2017) Micro RNAs and nitric oxide cross talk in stress tolerance in plants. *Plant Growth Regulation* 83(2), 199–205.

Singroha, G. and Sharma, P. (2019) Epigenetic modifications in plants under abiotic stress. In: Meccariello, R. (ed.) *Epigenetics* (10). Intech Open.

Slotkin, R.K. and Martienssen, R. (2007) Transposable elements and the epigenetic regulation of the genome. *Nature Reviews Genetics* 8(4), 272–285.

Sun, C., Ali, K., Yan, K., Fiaz, S., Dormatey, R. *et al.* (2021) Exploration of epigenetics for improvement of drought and other stress resistance in crops: A review. *Plants* 10(6), 1226.

Sun, Q., Qiao, J., Zhang, S., He, S., Shi, Y. *et al.* (2018) Changes in DNA methylation assessed by genomic bisulfite sequencing suggest a role for DNA methylation in cotton fruiting branch development. *PeerJ* 6, e4945.

Sunkar, R. and Zhu, J.K. (2004) Novel and stress-regulated microRNAs and other small RNAs from *Arabidopsis*. *The Plant Cell* 16(8), 2001–2019.

Szenker, E., Ray-Gallet, D. and Almouzni, G. (2011) The double face of the histone variant H3.3. *Cell Research* 21(3), 421–434.

Tang, D., Gallusci, P. and Lang, Z. (2020) Fruit development and epigenetic modifications. *New Phytologist* 228(3), 839–844.

Thakore, P.I., D'Ippolito, A.M., Song, L., Safi, A., Shivakumar, N.K. *et al.* (2015) Highly specific epigenome editing by CRISPR-Cas9 repressors for silencing of distal regulatory elements. *Nature Methods* 12(12), 1143–1149.

Thakore, P.I., Black, J.B., Hilton, I.B. and Gersbach, C.A. (2016) Editing the epigenome: Technologies for programmable transcription and epigenetic modulation. *Nature Methods* 13(2), 127–137.

Tourrière, H., Chebli, K. and Tazi, J. (2002) MRNA degradation machines in eukaryotic cells. *Biochimie* 84(8), 821–837.

Turner, B.M. (2009) Epigenetic responses to environmental change and their evolutionary implications. *Philosophical Transactions of the Royal Society Series B: Biological Sciences* 364(1534), 3403–3418.

Ueda, M. and Seki, M. (2020) Histone modifications form epigenetic regulatory networks to regulate abiotic stress response. *Plant Physiology* 182(1), 15–26.

Vanzan, L., Sklias, A., Boskovic, M., Herceg, Z., Murr, R. *et al.* (2023) Mechanisms of histone modifications. In: Tollefsbol, T.O. (ed.) *Handbook of Epigenetics*. Academic Press, pp. 27–54.

Vaughan, R.M., Kupai, A. and Rothbart, S.B. (2021) Chromatin regulation through ubiquitin and ubiquitin-like histone modifications. *Trends in Biochemical Sciences* 46(4), 258–269.

Venkatesh, S. and Workman, J.L. (2015) Histone exchange, chromatin structure and the regulation of transcription. *Nature Reviews Molecular Cell Biology* 16(3), 178–189.

Viggiano, L. and de Pinto, M.C. (2017) Dynamic DNA methylation patterns in stress response. *Plant Epigenetics* 281–302.

Vincent, J.A., Kwong, T.J. and Tsukiyama, T. (2008) ATP-dependent chromatin remodeling shapes the DNA replication landscape. *Nature Structural & Molecular Biology* 15(5), 477–484.

Vineis, P., Chatziioannou, A., Cunliffe, V.T., Flanagan, J.M., Hanson, M. *et al.* (2017) Epigenetic memory in response to environmental stressors. *FASEB Journal* 31(6), 2241–2251.

Vojta, A., Dobrinić, P., Tadić, V., Bočkor, L., Korać, P. *et al.* (2016) Repurposing the CRISPR-Cas9 system for targeted DNA methylation. *Nucleic Acids Research* 44(12), 5615–5628.

Vyskot, B. (2004) The role of DNA methylation in plant reproductive development. In: Ainsworth, C.C. (ed.) *Sex Determination in Plants*. Garland Science, New York and London, pp. 101–121.

Walter, M. (2015) *Transposon regulation upon dynamic loss of DNA methylation*. Doctoral dissertation, Université Pierre et Marie Curie-Paris VI.

Wambui Mbichi, R., Wang, Q.F. and Wan, T. (2020) RNA directed DNA methylation and seed plant genome evolution. *Plant Cell Reports* 39(8), 983–996.

Wang, H.V. and Chekanova, J.A. (2016) Small RNAs: Essential regulators of gene expression and defenses against environmental stresses in plants. *Wiley Interdisciplinary Reviews: RNA* 7(3), 356–381.

Wang, X., Fan, H., Wang, B. and Yuan, F. (2023) Research progress on the roles of lncRNAs in plant development and stress responses. *Frontiers in Plant Science* 14, 1138901.

Wang, Z., Casas-Mollano, J.A., Xu, J., Riethoven, J.J.M., Zhang, C. *et al.* (2015) Osmotic stress induces phosphorylation of histone H3 at threonine 3 in pericentromeric regions of *Arabidopsis thaliana*. *Proceedings of the National Academy of Sciences* 112(27), 8487–8492.

Wasternack, C. and Hause, B. (2013) Jasmonates: Biosynthesis, perception, signal transduction and action in plant stress response, growth and development. An update to the 2007 review in annals of botany. *Annals of Botany* 111(6), 1021–1058.

Wdowikowska, A., Janicka, M. and Bozhkov, P. (2021) Antisense oligonucleotide technology as a research tool in plant biology. *Functional Plant Biology* 49(1), 1–12.

Weinhold, A. (2018) Transgenerational stress-adaption: An opportunity for ecological epigenetics. *Plant Cell Reports* 37(1), 3–9.

Wolffe, A. (1998) *Chromatin: Structure and Function*, 3rd edn. Academic Press, San Diego, California.

Wu, G. and Poethig, R.S. (2006) Temporal regulation of shoot development in *Arabidopsis thaliana* by miR156 and its target SPL3. *Development* 133, 3539–3547.

Xu, Y., Zhang, L. and Wu, G. (2018) Epigenetic regulation of juvenile-to-adult transition in plants. *Frontiers in Plant Science* 9, 390420.

Yadav, C.B., Pandey, G., Muthamilarasan, M. and Prasad, M. (2018) Epigenetics and epigenomics of plants. In: Varshney, R.K., Pandey, M.K. and Chitikineni, A. (eds) *Advances in Biochemical Engineering/ Biotechnology: Plant Genetics and Molecular Biology*. Springer International, pp. 237–261.

Yaish, M.W., Colasanti, J. and Rothstein, S.J. (2011) The role of epigenetic processes in controlling flowering time in plants exposed to stress. *Journal of Experimental Botany* 62(11), 3727–3735.

Yang, L., Liu, Z., Lu, F., Dong, A. and Huang, H. (2006) SERRATE is a novel nuclear regulator in primary microRNA processing in *Arabidopsis*. *The Plant Journal* 47(6), 841–850.

Yin, R., Mao, S.-Q., Zhao, B., Chong, Z., Yang, Y. *et al.* (2013) Ascorbic acid enhances tet-mediated 5-methylcytosine oxidation and promotes DNA demethylation in mammals. *Journal of the American Chemical Society* 135(28), 10396–10403.

Younis, A., Siddique, M.I., Kim, C.K. and Lim, K.B. (2014) RNA interference (RNAi) induced gene silencing: A promising approach of hi-tech plant breeding. *International Journal of Biological Sciences* 10(10), 1150.

Yu, A., Lepère, G., Jay, F., Wang, J., Bapaume, L. *et al.* (2013) Dynamics and biological relevance of DNA demethylation in *Arabidopsis* antibacterial defense. *Proceedings of the National Academy of Sciences* 110(6), 2389–2394.

Zentner, G.E. and Henikoff, S. (2013) Regulation of nucleosome dynamics by histone modifications. *Nature Structural & Molecular Biology* 20(3), 259–266.

Zhang, H., Lang, Z. and Zhu, J.K. (2018) Dynamics and function of DNA methylation in plants. *Nature Reviews Molecular Cell Biology* 19(8), 489–506.

Zhao, Y. (2010) Auxin biosynthesis and its role in plant development. *Annual Review of Plant Biology* 61(1), 49–64.

Zhong, X., Du, J., Hale, C.J., Gallego-Bartolome, J., Feng, S. *et al.* (2014) Molecular mechanism of action of plant DRM de novo DNA methyltransferases. *Cell* 157(5), 1050–1060.

Zhou, H., Liu, Y., Liang, Y., Zhou, D., Li, S. *et al.* (2020) The function of histone lysine methylation related SET domain group proteins in plants. *Protein Science* 29(5), 1120–1137.

Zhu, J.K. (2009) Active DNA demethylation mediated by DNA glycosylases. *Annual Review of Genetics* 43, 143–166.

Zlatanova, J., Stancheva, I. and Caiafa, P. (2004) DNA methylation and chromatin structure. *New Comprehensive Biochemistry* 39, 309–341.

23 Epigenetic Strategies for Plant Protection: Updated Methods and Current Achievements

Noru Rajasekhar Reddy[1], Bandela Chandana[1], Nalam Sindhu Rani[1], Abigna Narla[1] and Lellapalli Rithesh[2]*

[1]Department of Genetics and Plant Breeding, Kerala Agricultural University, Thiruvananthapuram, Kerala, India; [2]Department of Plant Pathology, Kerala Agricultural University, Thiruvananthapuram, Kerala, India

Abstract

Plant stresses pose a significant threat to global agriculture, resulting in considerable crop losses and economic repercussions. Historically, plant protection strategies have heavily relied on chemical approaches, raising environmental concerns and fostering resistance development. In recent times, epigenetic modifications have emerged as promising avenues for enhancing plant resilience. Scientists have begun leveraging these modifications, such as DNA methylation, histone modifications, and histone acetylation/deacetylation, to bolster plants against diverse stressors. The chapter explores advanced methods for studying plant epigenetics, revealing regulatory networks governing plant immunity and leveraging epigenetic diversity to develop resilient crops.

Keywords: Plant stress, Epigenetic regulation, Disease resistance, DNA methylation, Histone modifications, Plant immunity

23.1 Introduction

Food security and sustainable agriculture are paramount challenges facing humanity in the 21st century, compounded by a rapidly growing global population and escalating environmental pressures. Plants face numerous threats from biotic stresses without effective plant protection measures; these threats can devastate crops, leading to yield losses, reduced quality, and compromised food supplies. Moreover, spreading plant pests and diseases can have extensive consequences, causing ecological imbalances and disrupting ecosystems. By implementing integrated pest management (IPM) strategies, employing resistant crop varieties, and utilizing sustainable agricultural practices, we can mitigate these risks and safeguard our farming systems.

Genetics and plant breeding have been fundamental to agriculture, allowing breeders to propagate desirable crop traits selectively for thousands of years. Through traditional breeding methods, significant strides have been made in improving crop yields, disease resistance, and adaptation to diverse environmental conditions. However, the pace of conventional breeding is often constrained by the long generation times

*Corresponding author: rithesh132@gmail.com

 Epigenetics for Climate-Smart and Sustainable Agriculture (ed. Jen-Tsung Chen)
DOI: 10.1079/9781800626102.0023

of many crop species and the limited genetic diversity available within breeding populations. Plant breeding is a core strategy of plant protection to reinforce plants' immunity to biotic and abiotic stresses and increase crop productivity. Through selective breeding, breeders endeavor to develop cultivars endowed with disease resistance, pest tolerance, and abiotic stress resilience. These efforts are indispensable for mitigating the detrimental impacts of pathogens, pests, and adverse environmental conditions on crop production. As agricultural systems confront mounting challenges posed by climate change, evolving pathogens, and emerging pests, the imperative for robust plant protection measures becomes increasingly pronounced.

Epigenetics, a burgeoning field at the intersection of genetics and molecular biology, offers profound insights into regulating gene expression and its impact on cellular identity, development, and adaptation. In contrast to conventional genetics, which centers on the sequence of DNA bases, epigenetics investigates the ever-changing modifications to DNA and its related proteins, impacting gene expression without changing the genetic sequence itself. Unlike transcription factors and RNA interference, which primarily affect gene expression at the transcriptional or post-transcriptional stages, epigenetic mechanisms bring about enduring changes to chromatin structure, thus influencing gene accessibility and activity. Moreover, epigenetic modifications exhibit transgenerational heritability, allowing gene behavior patterns to be passed on to subsequent generations, thereby contributing to the persistence of phenotypic traits beyond the initial environmental stimuli. The various epigenetic alterations, such as DNA methylation, alterations to histones, and processes mediated by non-coding RNA, create a sophisticated regulatory system governing gene expression patterns across various biological scenarios. In this context, epigenetics offers a powerful toolkit for addressing modern agriculture's complex challenges. Integrating these disciplines, from developing climate-resilient crop varieties to enhancing resource use efficiency and resistance to pests and diseases, is critical to ensuring a resilient and sustainable food system for future generations.

This chapter delves into the intricate nature of epigenetics and its significance in plant protection strategies. It examines the mechanisms governing epigenetic regulation and their role in fortifying plant defenses against biotic stressors. Additionally, the chapter investigates the significance of epigenetic variations in fostering plant immunity, along with the potential applications of epigenome editing technologies. Emphasizing the vital role of epigenetic strategies in preserving crop health and resilience, its objective is to consolidate current knowledge and spotlight emerging trends in epigenetics and plant protection.

23.2 Fundamentals of Epigenetics in Plants

The fundamentals of epigenetics in plants revolve around the intricate interplay between genetic inheritance and environmental influences, shaping plant phenotypes and responses. In 1942, a well-known embryologist, Conrad Waddington, introduced the term epigenetics in modern biology, linking it to the traditional concept of epigenesis. Waddington described epigenetics as covering all developmental processes from the 'genotype to phenotype' (Deichmann, 2016). The term epigenetic refers to heritable patterns of gene expression influenced by how a gene's DNA is packaged rather than just its underlying DNA sequence. When tightly packed, DNA makes genes less accessible to transcription machinery, resulting in reduced expression (Bender, 2002).

Understanding the fundamentals of plant epigenetics provides insights into diverse biological processes, including development, growth, and stress responses. These epigenetic mechanisms enable plants to adjust their gene expression profiles dynamically in response to environmental cues, allowing for phenotypic plasticity and adaptation. Elucidating the molecular basis of epigenetic regulation in plants holds promise for applications in crop improvement, environmental conservation, and sustainable agriculture.

23.2.1 Epigenetic mechanisms in plants

Epigenetic mechanisms in plants are pivotal regulators of gene expression and development, encompassing a diverse array of processes that modulate heritable changes in gene function without alterations to the underlying DNA sequence. These mechanisms, including DNA methylation, involving the addition of methyl groups to cytosine residues, influence transcriptional activity and genome stability, while histone modifications, including acetylation, methylation, and phosphorylation, modulate chromatin accessibility and gene silencing or activation (Samantara *et al.*, 2021). Additionally, small RNAs , such as microRNAs (miRNAs) and small interfering RNAs (siRNAs), guide gene regulation through post-transcriptional gene silencing or chromatin modification; these processes play crucial roles in plant growth, stress responses, and adaptation to changing environments, offering insights into both fundamental biology and potential applications in agriculture and ecosystem management (Arnholdt-Schmitt, 2004).

DNA methylation

DNA methylation adds a methyl group to cytosine's fifth carbon, a heritable modification crucial in eukaryotes for regulating genome elements like transposable elements (Holliday, 1996; Nasrullah *et al.*, 2022). Methylation levels vary in insects (0–3%), vertebrates (2–7%), fish/amphibians (~10%), and certain plants (> 30%) (Adams, 1996). Plants use DNA methyltransferases for symmetrical (CG, CHG) and asymmetrical (CHH) methylation (Maeji and Nishimura, 2018) (Fig. 23.1).

DNA methylation, once considered irreversible, is now known to be reversible. Demethylation occurs passively during DNA replication or actively through processes like base excision repair (BER) in plants. Enzymes like DEMETER (DME) and REPRESSOR OF SILENCING 1 (ROS1) remove methylation marks. In plants, the siRNA-directed RNA-directed DNA methylation (RdDM) pathway can influence

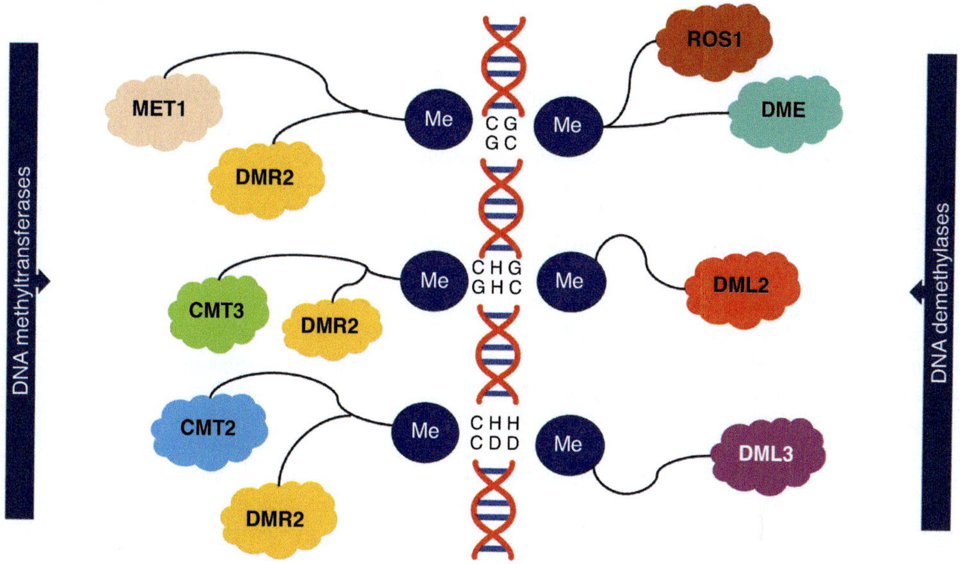

Fig. 23.1. The enzymes, including DNA methyltransferases and demethylases, play roles in adding (depicted as blue circles) and eliminating methyl groups from cytosine bases within different sequence environments. MET1, CMT3, and CMT2 are accountable for CG, CHG, and CHH methylation, respectively, while DRM2, a component of the RdDM pathway, regulates methylation across all sequence contexts. Demethylation processes involve enzymes like ROS1, DME, DML2, and DML3. Author's own image.

CHH methylation initiation. Small, non-coding RNAs (sRNAs) play vital roles in biological functions in mammals and plants (Chen, 2012). DNA methylation regulates plant growth, development, leaf morphology, flowering time, floral organization, reproductive capacity, and embryo growth, suppressing transposable elements and repetitive DNA sequences while enhancing gene activation. Mutations in DNA methyltransferase MET1 and chromatin-remodeling factor DDM1, an ATP-dependent SWI2/SNF2 family member, alter seed development (Kim *et al.*, 2008; Pikaard and Scheid, 2014; Kellenberger *et al.*, 2016).

New studies indicate dynamic DNA methylation shifts during plant growth stages, notably in soybeans. Variations occur in gene promoter methylation levels across seed maturation phases (An *et al.*, 2017). DNA methylation diversity is studied extensively in plants, algae, cereal crops, vegetables, and trees. Methylation levels and patterns vary significantly among species. For instance, *Chlamydomonas reinhardtii* algae have low methylation levels, while terrestrial plants generally exhibit higher methylation, particularly in CG and CHG contexts. *Arabidopsis* leaves show moderate methylation, rice leaves demonstrate intermediate levels, and beet leaves exhibit the highest methylation levels (Niederhuth *et al.*, 2016).

Different plant species exhibit diverse genome-wide DNA methylation patterns, with CG methylation being the most common, followed by CHG and CHH methylation. CG methylation dominates in over 50% of flowering plants, but CHG and CHH methylation levels vary greatly among species (Díaz-Manzano *et al.*, 2018). For instance, *Beta vulgaris* has high CHG and CHH methylation levels, while *Eutrema salsugineum* shows low CHG and moderate CG and CHH methylation. *Vitis vinifera* displays the lowest CHH methylation and moderate CG and CHG methylation. Brassicaceae species typically have reduced CG and CHG methylation, while some Poaceae species exhibit decreased CHH methylation, illustrating diverse methylation patterns in plants (Niederhuth *et al.*, 2016).

DNA methylation is crucial for plant environmental responses. Soybean under heat stress showed reduced methylation levels. CG and CHG methylation varied minimally (< 10%) between control and stressed plants, while CHH methylation decreased notably by 25% within genes and 37% within transposable elements (TEs) in stressed samples (Hossain *et al.*, 2017). DNA methylation in eukaryotes safeguards genomic integrity, suppressing TEs and repetitive sequences to control plant gene expression. Plant germlines uniquely reset and restore DNA methylation patterns, prompting ongoing debates about their establishment during development (Lanfear, 2018). Unlike animals, the prevailing belief in plants is that they retain germline cells throughout development rather than completely erasing and re-establishing the genome during gametogenesis. This suggests that epigenetic changes initiated in parents can be inherited by offspring. This capability allows plants to maintain stable epigenetic patterns while adapting quickly to environmental changes. These methylation patterns are regulated by specific methyltransferases and demethylases unique to plants (Lira-Medeiros *et al.*, 2010).

MEDEA (MEA), a Polycomb group gene, was the endosperm's first recognized imprinted plant gene. MET1 maintains methylation at the MEA promoter, while DEMETER (DME) removes 5-methylcytosine from the maternal MEA allele via the BER pathway (Choi *et al.*, 2002). Mutations in ROS1 inhibit transgene transcription (Gong *et al.*, 2002). DME and ROS1 actively prevent 5-methylcytosine accumulation by demethylating genes. DNA methylation enhances ROS1 expression but decreases upon demethylation. Methylation near ROS1 can restore its expression in an RdDM mutant, demonstrating ROS1's role in regulating epigenome stability through methylation adjustments (Williams *et al.*, 2015).

Histone modifications

Histones are conserved spherical proteins that expose N-terminal tails on nucleosome octamers, allowing chemical alterations crucial for eukaryotic gene expression regulation. They compact DNA by wrapping 146 base pairs, forming nucleosomes (H2A, H2B, H3, and H4). Chemical changes to histone tails create a 'histone code', shaping chromatin structure and gene expression (Jenuwein and Allis, 2001). Acetyl group additions or removals on histones modify chromatin, influencing gene expression patterns (Zhou, 2009) (Fig. 23.2).

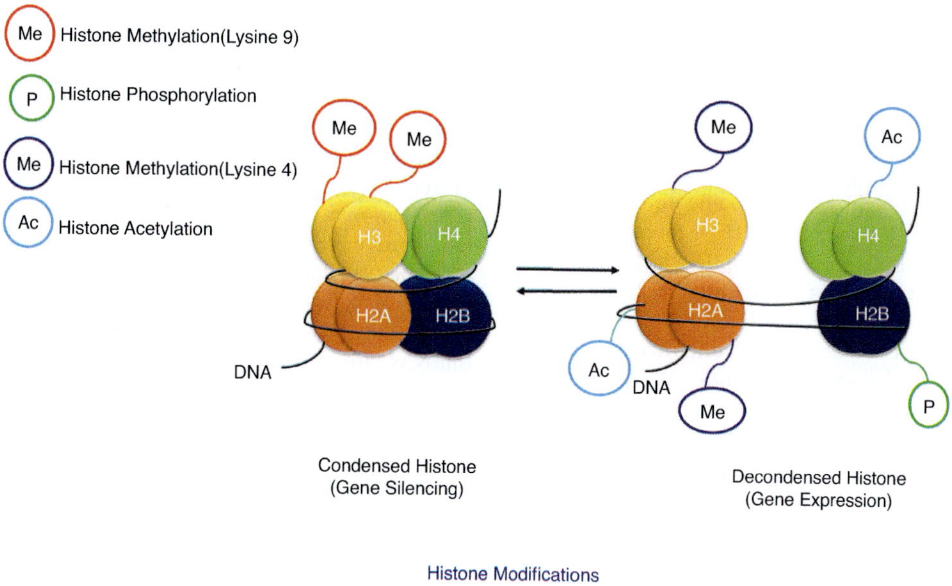

Fig. 23.2. Histone modifications for gene expression. Author's own image.

Histone lysine methylation, a pivotal epigenetic change, plays dual roles in gene regulation through mono-, di-, or trimethylation states, each with distinct biological implications. Modifications like H3K9me2, H3K9me3, H3K27me3, and H4K20me3 repress genes, while H3K4me1, H3K4me2, H3K4me3, H3K36me2, and H3K36me3 activate transcription. Recent findings challenge irreversible histone methylation beliefs, particularly through demethylase enzymes. Investigations into plant histone modifications, including H3K9 and H3K27 acetylation and methylation, reveal unique facets of plant epigenetic regulation (Zhou, 2009). Histone modifications like methylation, acetylation, phosphorylation, sumoylation, carbonylation, glycosylation, and ubiquitination target specific amino acid sites, notably lysine and arginine in histone N-terminal segments, regulating nearby gene expression (Dutta *et al.*, 2017; Gambino and Pantaleo, 2017; Canton *et al.*, 2022).

Histone acetylation and deacetylation

In eukaryotes, acetylating lysine residues on histones are essential for regulating gene expression. Acetylating core histones enhances the accessibility of chromatin, thereby activating genes. Conversely, deacetylating histones through histone deacetylation reduces accessibility, resulting in gene repression (Shahbazian and Grunstein, 2007). Both histone acetyltransferases (HATs) and histone deacetylases (HDACs) are involved in the reversible modification of histones through acetylation.

Histone acetylation

The enzymatic acetylation of core histone N-terminal tails is crucial for epigenetic signaling (Turner, 2000). Recent studies suggest that histone acetyltransferases (HATs) and complexes respond to environmental stresses. GCN5-related N-acetyl transferases (GNAT) interact with phosphatase 2C protein (AtPP2C-6–6). Mutants lacking GCN5 show increased stress-related gene expression (Hewezi *et al.*, 2008; Servet *et al.*, 2008). *Arabidopsis* mutants lacking SGF29a, a GCN5 complex component, resist salt stress better than wild-type plants (Kaldis *et al.*, 2011). Studies emphasize the significance of the GCN5 complex in *Arabidopsis* response to stress, particularly under salt stress conditions. Mutations in

ADA2b, another complex component, reduce histone acetylation (H3 and H4) and suppress stress-responsive gene transcription (e.g. COR6.6, RAB18, RD29b) during salt stress. This highlights GCN5 complex involvement, including ADA2b, in gene expression regulation and chromatin modification under plant stress (Kaldis *et al.*, 2011). Moreover, in cold conditions, the transcription factor CBF1 can physically interact with ADA2 and GCN5 (Mao *et al.*, 2006). Recent evidence suggests that GCN5 and ADA2 enhance the activity of cold-responsive genes as temperatures decrease in *Arabidopsis* (Pavangadkar *et al.*, 2010). These findings suggest that activator complexes containing GCN5 may be recruited by the transcription factor CBF1 to initiate the activation of genes responding to cold environments.

Histone deacetylation

Histone deacetylases (HDACs) remove acetyl groups from lysine residues in histone and non-histone proteins (Sengupta and Seto, 2004). In *Arabidopsis*, histone deacetylase proteins, particularly HDA19, crucially regulate responses to environmental stresses. HDA19 enhances *Arabidopsis* defense against pathogens by activating JA and ET signaling pathways. Elevated HDA19 levels improve resistance to *Alternaria brassicicola* by boosting ERF1 expression (Zhou *et al.*, 2005). Recent research shows that HDA19 suppresses *Arabidopsis* defense mechanisms involving salicylic acid (SA). Loss of HDA19 function increases SA levels and PR gene expression, enhancing resistance to *Pseudomonas syringae* (Choi *et al.*, 2012). HDA19 interacts with ERF and WRKY transcription factors, critical for responding to abiotic and biotic stresses (Kim *et al.*, 2008). ERF7, a transcriptional repressor, interacts with an *Arabidopsis* SIN3 counterpart, inhibiting gene responses to abiotic stress through HDA19. HDA19 also interacts with WRKY38 and WRKY62, suppressing their roles in basal defense mechanisms (Kim *et al.*, 2008). These studies underscore HDA19's significant role in hormonal signaling pathways during various environmental stresses.

RNA-based modifications: microRNAa and small interfering RNAs

In plants, small RNAs are categorized into microRNAs (miRNAs) and small interfering RNAs (siRNAs), each playing distinct roles in regulatory pathways. miRNAs, originally identified in *Caenorhabditis elegans*, regulate eukaryote developmental processes, including leaf polarity, floral organogenesis timing, and various developmental pathways (Lee *et al.*, 1993; Kidner and Martienssen, 2005). The miRNAs are transcribed and processed into pre-miRNA transcripts of 80–100 base pairs, forming hairpin loops recognized by Dicer-like proteins such as DCL1 (Ketting *et al.*, 2001). These loops are cleaved to produce mature miRNAs, typically 21–24 nucleotides long, which bind to ARGONAUTE proteins through their PAZ domain (Carmell *et al.*, 2002; Lingel *et al.*, 2003). MiRNA–ARGONAUTE complexes target complementary transcripts for degradation.

The other type of small RNA, siRNAs, derive from double-stranded RNAs, silencing 'foreign' DNA like retroviruses and retrotransposons. They form exact pairs with target mRNAs to suppress gene expression. RNA-dependent RNA polymerases generate siRNA duplexes, which are then loaded onto ARGONAUTE proteins within the RNA-induced silencing complex (RISC) to degrade mRNA targets. This process increases siRNA levels, leading to targeted methylation of genomic sites regulating their transcription (Gendrel and Colot, 2005).

23.2.2 Role of epigenetics in plant development and stress response

Plants remember stress by adjusting to it, but the breakdown of stress-related molecules can restrict this. Yet, stress memory lasts longer if it changes the plant's growth and structure. Epigenetic processes like DNA methylation and histone modifications help maintain this activity. Modifying cell specialization due to environmental pressure triggers adaptive changes in both timing and formative processes, known as phenological and developmental flexibility.

These adaptations serve as crucial defense mechanisms against stress. Phenotypic flexibility enables plants to regulate the timing of different growth stages, thereby shielding critical phases, notably reproductive development, from stressors. Moreover, fine-tuning growth and developmental patterns becomes essential for optimizing resource utilization amidst stressful conditions (Chinnusamy and Zhu, 2009).

Epigenetic regulation during the extreme temperature stress response

Temperature significantly influences the growth and development of plants, and with ongoing climate changes, extreme heat and cold pose significant challenges to crop productivity. Cold stress mainly affects plant biochemistry and genetic activity, leading to the direct inhibition of essential metabolic enzymes and alterations in gene regulation (Zhu, 2016). In recent years, extensive research efforts have elucidated the mechanisms underlying plant responses to heat and cold stress. With increasing research findings, there is a growing body of literature exploring the involvement of epigenetic regulation in plants' responses to heat or cold stress conditions (Chang *et al.*, 2020).

Salt-induced epigenetic changes in crop plants

Epigenetic mechanisms are critical in regulating plant gene expression under salt stress. Research focuses on understanding these crop changes, essential for their adaptation to environmental fluctuations like climate-induced salt stress, which affects productivity (Kumar *et al.*, 2017). Promoter and gene-body methylation regulate genotype-specific gene expression under stress. However, genetic variations in salt tolerance do not always correlate with stress-induced DNA methylation changes, indicating distinct adaptation pathways. Methylation alterations may occur randomly, suggesting stochastic processes. Wheat varieties respond differently to salt stress due to variations in high-affinity potassium transporter (HKT) expression, influenced by genetic and epigenetic factors (Kumar *et al.*, 2017).

In wheat, TaHKT2;1 and TaHKT2;3 coding regions show varied 5-methylcytosine (5mC) levels among genotypes, increasing under salt stress. Salt-sensitive genotypes exhibit full methylation of cytosine in the CG context, whereas salt-tolerant ones show higher 5-mC content, especially in CG contexts, under stress. Differences in 5mC content and differentially methylated regions (DMRs) correlate with reduced HKT2 gene expression in response to salt stress. In contrast, minor 5mC variations were found in TaHKT1;4 coding regions, with no significant effect of salt stress on methylation or gene expression in salt-tolerant genotypes. The precise impact of cytosine methylation on gene expression remains unclear, influenced by both abiotic and biotic factors, aiding plants in adapting to challenging environments (Kumar, 2018).

Epigenetic regulation during biotic stress

Intensive monoculture farming and fertilization increase plant disease incidence, exacerbated by climate change-induced extreme weather events (Yan *et al.*, 2019; Zytynska *et al.*, 2020). Despite traditional and sustainable disease management methods, plant diseases still cause a significant global yield loss of about 20% (Oliva and Quibod, 2017). Traditional breeding relying on R genes for disease resistance faces challenges due to rapid pathogen shifts and reduced genetic diversity from heavy artificial selection (Rodriguez-Moreno *et al.*, 2017). Integrating genetic material from wild relatives and GMO engineering offers innovative approaches to enhance crop resilience. Exploring epigenetic processes like DNA methylation and histone modifications in response to pathogen infections presents new avenues for bolstering crop resilience through epi-breeding (Varotto *et al.*, 2020). Natural epigenetic variations and intentional modifications hold promise for enhancing plant disease resistance, with inheritable epialleles identified in various model and crop plants like *Arabidopsis*, rice, and tomato. Stress-induced DNA methylation changes suggest potential epigenetic alterations linked to crop disease resistance (Latutrie *et al.*, 2019).

23.3 Epigenetic Regulation of Plant Defence Mechanisms

Despite appearing inactive, plants have evolved intricate defense structures to protect against threats. These include physical, chemical, and molecular defenses (Kramer *et al.*, 2023). Physical defenses, such as thorns, trichomes, and toughened surfaces, deter herbivores. Some plants use rapid leaf movements or shedding to evade attacks. Chemical defenses involve producing secondary metabolites like alkaloids and terpenoids to deter herbivores and inhibit pathogens. These compounds can also serve as signaling molecules to attract herbivore predators.

At the molecular level, plants activate specific defense pathways triggered by recognizing pathogen-associated or damage-associated molecular patterns (Jones and Dangl, 2006; Ay *et al.*, 2009). This response includes inducing defense-related genes, producing antimicrobial proteins, and reinforcing cell walls (Malinovsky *et al.*, 2014). Additionally, plants can develop induced systemic resistance (ISR) to pathogens, producing signaling molecules that prepare distant plant parts for a more robust response to subsequent attacks.

Hormonal regulation via jasmonic acid (JA) and salicylic acid (SA) coordinates plant defenses against pests and diseases (Pieterse *et al.*, 2009; Alvarez *et al.*, 2010). Symbiotic relationships with microbes like mycorrhizal fungi and rhizobacteria enhance nutrient uptake, stress tolerance, and pathogen defense. Epigenetic mechanisms, including DNA methylation and nucleoprotein modifications, enable plants to adapt and refine defense responses to ecological challenges (Fig. 23.3). Exploring plant defenses unveils their resilience strategies, vital for survival and ecological balance (Dubey and Jeon, 2017).

Epigenetic modifications are pivotal in plant defense, dynamically regulating gene expression without altering DNA sequences (Gijzen *et al.*, 2014; Aerts *et al.*, 2021). This exploration highlights their critical role in adapting plant responses to environmental challenges through mechanisms like DNA methylation, histone modifications, and small RNA-mediated silencing (An *et al.*, 2022; Chen *et al.*, 2022). Understanding these processes is crucial for advancing plant biology and agricultural research, offering insights into how plants fine-tune their defense responses and enabling the potential engineering of crops with enhanced

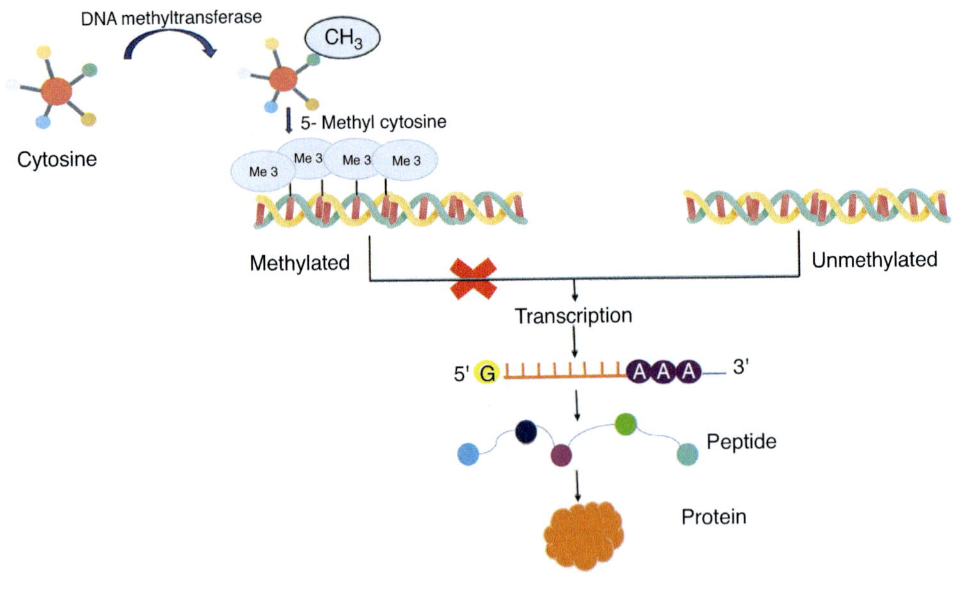

Fig. 23.3. Epigenetic regulation during gene expression. Author's own image.

resistance. Beyond agriculture, insights into epigenetic control promise to enhance ecological balance by elucidating how plants interact with their environment, thus advancing fundamental knowledge and practical solutions in both agriculture and conservation efforts.

23.3.1 Epigenetic regulation of specific defense genes

Hoang *et al.* (2018) discovered that rice employs epigenetic mechanisms to defend itself against pathogens. One such mechanism involves DNA methylation, where methyl (CH3) groups are added to DNA bases, causing alterations in chromatin structure. These modifications create barriers that make it more challenging for transcription factors to reach the promoter region of specific genes (Lim and Maher, 2010). In plants, DNA methylation primarily impacts two nucleotide bases: cytosine, known as 5-methylcytosine (5mC), found in the genetic material, and adenine, referred to as N6-methyladenine (6mA), which is present in mitochondrial DNA (Ratel *et al.*, 2006).

Cytosine methylation at a promoter typically shuts down gene expression. Song *et al.* (1995) discovered that *Xa21* plays a critical role as a rice R gene in defending against various *Xanthomonas oryzae* pv. *oryzae* (*Xoo*) races. Akimoto *et al.* (2007) isolated the *Xa21G* protein, similar to *Xa21*, using methylation-sensitive amplification polymorphisms. They observed these polymorphisms in a mutated lineage produced by exposing germinating seeds to the DNA methylation inhibitor 5-aza-2′-deoxycytidine. In conventional plants, the promoter area of *Xa21G* was methylated. The *Xa21G* transcripts were not detectable in the original type, but they significantly increased in the Line-2 mutant with reduced cytosine methylation. This indicated that *Xa21G* was effectively silenced in the original type. Moreover, Line-2 showed resistance to *Xoo* infection, while the conventional type was highly susceptible, probably because of the suppression of *Xa21G* expression (Deng *et al.*, 2017).

He *et al.* (2020) discovered that DNA methylation in fungi, an epigenetic change linked to plant diseases, involves various types, patterns, and functions. They identified distinct families of DNA methyltransferases (MTases) responsible for this modification, targeting transposons, gene promoters, and repeat sequences. These patterns vary with cell type, developmental stage, and environmental stress, influencing critical processes like repeat-induced point mutation, fungal growth, disease virulence, and secondary metabolism pathways. Luo *et al.* (2016) studied DNA methylation changes in banana leaves responding to Fusarium wilt disease caused by *Fusarium oxysporum* f. sp. *cubense* tropical strain 4 (Foc TR4). They used methylation-sensitive amplification polymorphism (MSAP) to analyze DNA methylation at specific sites in infected leaves compared with mock-treated leaves at different infection stages. Through sequencing and RT-PCR, they identified increased and decreased DNA methylation induced by Foc TR4 in the banana genome. Some methylated genes were related to plant–pathogen interactions, including resistance genes and transcription factors.

Luo *et al.* (2016) explored the role of DNA methyltransferases (MTases) in *Botrytis cinerea*, a fungal pathogen causing gray mold rot in horticultural crops. They found that two MTases, BcDIM2 and BcRID2, collaborate to control the pathogenicity of *B. cinerea*. Their study revealed that a mutant lacking BcDIM2 and BcRID2 loses its ability to cause disease, displaying deficiencies in growth, spore germination, oxidative tolerance, and gene expression. The researchers also noted a link between low methylation levels and increased expression of RID genes, higher AT content, and frequent repeat-induced point mutations (RIP) within the *B. cinerea* genome.

In plant pathogenic fungi, DNA methylation mainly happens in transposable elements (TEs), gene promoter regions, and repetitive DNA sequences. This important and heritable epigenetic change is associated with repressing gene expression and transposons, influencing various biological processes in fungi.

Recent research has focused on Jumonji C (JmjC) domain-containing proteins, a group of conserved histone lysine demethylases found in organisms with a cell nucleus (Chen *et al.*, 2011). In rice, 20 proteins (JMJ701 to JMJ720) specialize in removing di- and tri-methylations from histone lysines using ferrous ion- and α-ketoglutaric acid-dependent oxidative reactions. These proteins play crucial roles in rice's developmental processes and responses to biotic stress. For instance, the jmj704 mutant, with elevated levels of trimethylation of H3 lysine 4 (H3K4me3), showed heightened susceptibility to *Xoo*, which positively impacts resistance against bacterial blight.

Further investigations revealed that JMJ704 suppresses defense-negative regulator genes by reducing activation marks. JMJ705, a histone lysine demethylase reversing H3K27me3, also demonstrated induced expression under stress and pathogen infection (Li *et al.*, 2013). Overexpressing JMJ705 enhanced resistance to *Xoo*, whereas loss-of-function reduced resistance, indicating its potential role in defense-related gene expression through demethylation of H3K27me3. Additional studies indicated that infection with *Xoo* induces the transcription of 15 JmjCs (Hou *et al.*, 2015), highlighting JmjCs as potential elements involved in rice's defensive response against this pathogen.

Polycomb group proteins (PcGs) are crucial epigenetic regulators in plants, modifying chromatin to suppress gene expression through histone modifications. They play vital roles in developmental processes like cell differentiation, epigenetic imprinting, and stem cell maintenance. PcG proteins assemble into two main complexes: Polycomb Repressive Complex 1 (PRC1) and Polycomb Repressive Complex 2 (PRC2) (Roy *et al.*, 2018). PRC2 methylates histone H3 at lysine 27 (H3K27me), marking genes for silencing. PRC1 recognizes this modification and further enhances gene repression by promoting histone H2A ubiquitination at lysine 119 (H2AK119ub), thereby maintaining a repressive chromatin environment. The fundamental elements of PRC1/2 are conserved in both the animal and plant kingdoms. *Arabidopsis* harbors three H3K27 methyltransferases within the PRC2 complexes, specifically MEDEA (MEA), CURLY LEAF (CLF), and SWINGER (SWN). Lately, MEA has been identified as an inhibitory factor in plant immunity. Elevated MEA expression heightens susceptibility to *Botrytis cinerea*, *Pseudomonas syringae* pv. *Tomato* DC3000 and *Pseudomonas syringae* pv. *Tomato Avirulence* (Pst-AvrRpt2).

Additionally, MEA associates with the transcription factor DROUGHT-INDUCED 19 (DIL9) and is engaged to methylate the histone H3 at lysine 27 (H3K27me3), MEA and DIL9 repress the expression of RESISTANT TO *P. SYRINGAE* 2 (RPS2) RPS2 and reduce the plant's defense response (Ramirez *et al.*, 2019). Recent research suggests that the wheat histone deacetylase TaHDT701, which belongs to the histone deacetylase 2 (HD2) family, regulates wheat's resistance to *Blumeria graminis* f.sp. *tritici* (Bgt) infection (Geng *et al.*, 2019). It has been discovered that TaHDT701 collaborates with TaHDA6 and TaHOS15 to form a histone deacetylase complex. This complex, named TaHDT701-TaHDA6-TaHOS15, negatively affects wheat's resistance to powdery mildew by altering the chromatin structure of defense-related genes such as WRKY 45 (TaWRKY45), *Triticum aestivum* Pathogenesis-Related 1, 2 and 3 (TaPR1, TaPR2, and TaPR5) (Mapuranga *et al.*, 2022).

23.3.2 Impact of epigenetic modifications on plant immunity

Nucleotide-binding domain and leucine-rich repeat (NBS and LRR) immune receptors, a key group of R proteins, are vital in triggering effector-triggered immunity (ETI) against pathogens. These NLR genes are often clustered in the genome, containing repetitive sequences and transposable elements (TEs), which enable rapid gene expansion through unequal recombination (Friedman and Baker, 2007). In *Arabidopsis*, intronic COPIA-type retrotransposons, including TEs, regulate NLR gene expressions like RPP7, crucial for resisting downy mildew. EDM2 is essential for proper RPP7 transcription and splicing. In EDM2 mutants, intronic COPIA TEs show repressive DNA and H3K9 methylation marks, impairing RPP7 expression and immunity. Despite TE sequence variation, their selective insertion suggests a role in finely tuning RPP7 expression during biotic stress. This dynamic TE epigenetic regulation underscores their regulatory role in NLR gene expression in plant–pathogen interactions.

23.4 Current Achievements in Epigenetic Plant Protection

Plants are continually exposed to many biotic and abiotic stressors, such as fluctuations in temperature and light, a shortage of water, the availability of nutrients and minerals in the soil, and invasions by significant and minor insect pests and diseases. As a result, plants have evolved in many ways to alter their physiology and development. Gene-to-environment or gene-to-gene interactions result in epigenetic changes in genomic expression patterns that activate or deactivate certain genes and modify cellular and physiological characteristics without affecting the DNA sequence (Moore, 2015; Vaiserman and Koliada, 2017; Talarowska, 2020).

These alterations have been created by rapid and coordinated changes at the transcriptional and post-transcriptional levels and epigenetic regulation of genes to adapt to severe environments. As a result, it is critical to concentrate on epigenetic gene expressions to illustrate the epigenome's potential as a source of variation in desirable agronomic characteristics. This will lay a solid foundation for using the epigenome in crop development methods (Álvarez-Venegas and De-la-Peña, 2016). Understanding the epigenetic regulation of plant growth and development may result in novel genetic variants for improved agricultural output and stress adaptation, which has sparked much interest in plant epigenetics in recent years for basic and applied research (Mirouze and Paszkowski, 2011; Kotkar and Giri, 2020).

23.4.1 Successful applications of epigenetic strategies in crop protection

Plants face challenging environmental conditions that hinder their growth and development. In response to pathogen attacks, they employ defense mechanisms and protective barriers, while pathogens evolve strategies to overcome these defenses. Epigenetics involves heritable changes in gene expression without altering the DNA sequence, crucial for plant growth and stress responses through mechanisms like small RNA, DNA methylation, and histone modifications.

Studies show that epigenetic modifications play a significant role in plants under biotic stress. For instance, in *Arabidopsis* mutants, disruptions in small RNA biogenesis reduce susceptibility to nematodes. DNA methylation influences nucleosome location, histone modifications, and gene expression. It also plays a role in transgenerational responses to herbivory, as seen in tomatoes and *Arabidopsis* (Medina *et al.*, 2017). Kuć (1982) demonstrated that small amounts of pathogen exposure in plants can induce immunity. In *Arabidopsis*, pathogen attacks alter the DNA methylome and defense gene expression, with reduced DNA methylation increasing disease tolerance (López Sánchez *et al.*, 2016). This suggests that epigenetic processes can convert environmental cues into heritable changes affecting natural selection. Mukherjee *et al.* (2017) showed that epigenetic processes play a role in developing pathogen resistance. Exposure to *Bacillus thuringiensis* increased resistance in wax moth larvae, with variations in miRNA expression linked to resistance.

Pre-treating potato plants with BABA improved resistance to *Phytophthora infestans*, linked to changes in DNA methylation and demethylation, indicating long-lasting priming memory. BABA treatment altered gene expression patterns, suggesting a complex interaction between promoter methylation and gene expression (Kuźnicki *et al.*, 2019).

Ashapkin *et al.* (2020) found significant differences in methylation patterns in soybean roots infected with soybean cyst nematode (SCN). Genes with altered methylation patterns also showed changes in expression, highlighting the role of DNA methylation in response to SCN infection. Epigenetic modifications are crucial for plant defense and stress responses, influencing gene expression and providing a mechanism for heritable changes in response to environmental conditions (Table 23.1).

23.5 Updated Methods in Epigenetic Research for Plant Protection

The dideoxy end termination method (Sanger *et al.*, 1977) marked the inception of first-generation sequencing technology, pivotal in genetic research. Since then, DNA sequencing has revolutionized biological inquiry, leading to the complete sequencing of the human genome. At present, with genomics and proteomics driving research, high-throughput sequencing

Table 23.1. Case studies on improved resistance to pests and diseases. Author's own table.

Epigenetic process	Pathogen	Crop	Gene involved	Finding	Reference
Histone modifications	*Pseudomonas syringae* pv. *tomato*	*Arabidopsis*	*met1* and *DDC*	In response to salicylic acid stress, transposon-associated differentially methylated areas were found, accompanied by up-regulation of 21-nt siRNAs. This indicates dynamic DNA methylation changes inside repetitive sequences can affect neighboring genes.	Dowen *et al.* (2012)
	Uromyces appendiculatus.	French bean	R proteins and stress-related transcription factors.	Changes in acetylation levels affect the transcription of stress-responsive genes, such as R proteins and stress-related transcription factors.	Ayyappan *et al.* (2015)
	Phytophthora infestans	Potato cv. Sarpo Mira	*NPR1* and the *SNI1*	Increased resistance to *P. infestans* due to the upregulation of SA-responsive genes.	Meller *et al.* (2018)
	Botrytis cinerea	Tomato	*SlyWRKY75*	Histone modifications on the gene encoding Sly-miR1127-3p in response to *B. cinerea or P. syringae*.	Crespo-Salvador *et al.* (2020)
	Pseudomonas syringae	*Arabidopsis*	*NLR*	Inactivation of either HOS15 or HDA9 results in heightened resistance against pathogen invasion, coupled with elevated expression of approximately one-third of the 207 NLR genes.	Yang *et al.* (2020)
	Phytophthora infestans	*Nicotiana benthamiana*	*StPUB17*	Through degrading the negative regulator KH17, *StPUB17* positively modulates immunity.	McLellan *et al.* (2020)
	Pseudomonas syringae and *Botrytis cinerea*	*Arabidopsis*		H1 histone mutations displayed altered epigenetic control of defense priming.	Sheikh *et al.* (2023)

DNA (Hypo) methylation	*Meloidogyne graminicola*	Rice	*PTI* genes	DNA hypomethylation results in basal pattern-triggered immune response and confers lower sensitivity in rice to root-parasitic nematodes.	Atighi *et al*. (2020)
DNA methylation	*Magnaporthe oryzae*	Rice	*DNMTase*	DNA methylation limits fungal cells' ability to proliferate, and their genomes experience a significant loss of their current mC sites.	Jeon *et al*. (2015)
	Xanthomonas oryzae pv. *oryzae*.	Rice	OsWRKY62, and Os-11N3	Rice enzyme JMJ704 removes histone lysine methyl groups, regulating defense against bacterial blight by inhibiting harmful regulator genes NRR, OsWRKY62, and Os-11N3 through reducing activating histone marks H3K4me2/3.	Hou *et al*. (2015)
	Heterodera glycine Ichinohe	Soybean	Homologs of *Arabidopsis* O-acetyltransferases, S-adenosyl-L-Met-dependent methyltransferase protein, dicer-like4, DEFECTIVE IN MERISTEM SILENCING2, Suppressor of variegation 3–9 homolog9, chromatin remodeling complex subunit B, chromatin remodeling factor17, and homeodomain finger family protein	Differentially methylated genes altered expression in the soybean syncytium, indicating a distinct role of SCN-induced DNA methylation in controlling gene expression differences during parasitism.	Rambani *et al*. (2015)

Continued

Table 23.1. Continued

Epigenetic process	Pathogen	Crop	Gene involved	Finding	Reference
	Botrytis cinerea	Tomato	SDG8 and SDG25	Regulate the expression of defense genes globally, affecting the methylation of histone H3 at lysine 4 (H3K4) and lysine 36 (H3K36). Plants with mutations in either one or both of these genes show increased susceptibility to both *Botrytis cinerea*.	Lee *et al*. (2016)
	Phytophthora infestans and *Phytophthora sojae*	Soybean	*PsDAMT3*	Adenine N6-methylation is linked to adaptive evolution in *Phytophthora* sp.	Chen *et al*. (2018)
Chromatin assembly and remodeling	*Magnaporthe oryzae*	Rice	*OsHLP1*	Endoplasmic reticulum homeostasis is provided by overexpression of *OsHLP1*, promoting blast resistance.	Meng *et al*. (2022)
Chromatin assembly and remodeling	*Xanthomonas oryzae* pv. oryzae	Rice	*OsCHR11*	*OsCHR11* showed differential expression of numerous defense response genes and conferred enhanced resistance to bacterial leaf blight.	Liu *et al*. (2023)

methods are essential for studying epigenetic regulations. These techniques promise insights into gene expression, fostering a deeper understanding of epigenetics.

23.5.1 Advanced tools for studying epigenetic regulation in plants

Single-cell epigenomics

Recent advances in single-cell sequencing have revolutionized plant epigenetic research by enabling detailed individual cell-level investigation. Techniques like single-cell ATAC-seq and bisulfite sequencing uncover epigenetic variability among cells, offering insights into plant growth and adaptation (Madlung and Comai, 2004; Clark *et al.*, 2018; Luo *et al.*, 2020). Despite similar genomes, different cell types perform distinct functions due to epigenomic differences that reflect genome functionality (Schwartzman and Tanay, 2015). Single-cell whole-genome amplification characterizes epigenomic features using various biochemical methods. The plant epigenome responds to environmental changes, biological signals, and developmental cues. Techniques like single-cell reduced representation bisulfite sequencing and improved single-nucleus methylcytosine sequencing enhance DNA methylation analysis with broader genomic coverage and mapping precision.

CRISPR/Cas9-based epigenome editing

Clustered regularly interspaced short palindromic repeats (CRISPR)/Cas9 technology has been adapted for targeted epigenome editing, allowing for the precise manipulation of epigenetic marks at specific plant genomic loci. By fusing the Cas9 nuclease with epigenetic modifiers such as DNA methyltransferases or histone-modifying enzymes, researchers can investigate the causal roles of these marks in plant development and physiology. This approach offers unprecedented opportunities to engineer epigenetic modifications and study their functional consequences in plants. Hilton *et al.* (2015) demonstrated that targeting dCas9-KRAB (dead Cas9 with Krüppel-associated box domain) to the HS2 enhancer induced specific H3K9 tri-methylation and reduced chromatin accessibility (HS2 refers to Hypersensitive Site 2, which is a genomic region that exhibits hypersensitivity to cleavage by certain enzymes, often associated with regulatory elements such as enhancers or promoters, while H3K9 refers to Histone H3 Lysine 9, which is a specific modification of the histone protein H3 at the ninth lysine residue). This resulted in the silencing of numerous globin genes' expression with limited off-target effects. These findings show that dCas9-KRAB can disrupt individual enhancer activity via local epigenetic modifications.

Long-read sequencing technologies

Long-read sequencing platforms, like PacBio and Oxford Nanopore, have revolutionized plant genomics by providing the ability to sequence long DNA fragments (Michael *et al.*, 2018). These technologies are particularly useful for resolving complex sections of the plant genome, for example repetitive sequences and structural variations, which may influence epigenetic regulation. By generating high-quality, long-range sequencing data, researchers can better characterize the epigenomic landscape of plants and elucidate its role in gene regulation and genome evolution.

23.5.2 Multi-omics integration: integrating epigenomic data and other omics

Datasets, like transcriptomics and metabolomics, have emerged as a powerful approach for obtaining comprehensive insights into the functional consequences of epigenetic regulation in plants (Kawakatsu *et al.*, 2016). Advanced computational methods are employed to analyze and integrate these multi-omics datasets, enabling researchers to uncover complex regulatory networks and pathways underlying plant phenotypes. By combining multiple layers of omics data, researchers can elucidate the interplay between epigenetic modifications, gene expression, and metabolic pathways, providing a holistic understanding of plant biology.

High-throughput sequencing techniques

High-throughput sequencing, next-generation, deep, or second-generation sequencing presents notable advantages over traditional Sanger sequencing. It facilitates efficient parallel sequencing at a large scale through microarray technology. Moreover, it eliminates the need for electrophoretic separation technology, leading to cost savings and efficiencies in time and manpower.

Next-generation sequencing (NGS) technologies have transformed epigenetic research by enabling the analysis of DNA methylation patterns, histone modifications, and chromatin accessibility throughout the entire genome. These methods empower researchers to pinpoint epigenetic alterations linked to particular genomic regions or biological functions, offering valuable insights into gene regulation and the dynamics of plant chromatin.

SEQUENCING BY SYNTHESIS AFTER AMPLIFICATION.

Pyrosequencing and Roche 454 (gs flx titanium system). The Roche 454 sequencing technology is a pioneering second-generation sequencing system recognized for using a cyclic microarray approach and sequencing by synthesis (SBS) technology (Margulies *et al.*, 2005). It uses bioluminescence to detect DNA sequences, starting with PCR amplification of the template post-library construction. DNA polymerase, ATP sulfurylase, bisphosphatase, and luciferase work together to pair the template with deoxyribonucleoside triphosphates (dNTPs), releasing pyrophosphate groups (Li *et al.*, 2016). This release generates a fluorescence signal, which the system detects to determine the DNA sequence in real-time. Each run can produce 1 million sequences with an average length of 400 base pairs, totaling about 500 million base pairs of data. Despite challenges in measuring homopolymer lengths accurately, leading to insertion and omission errors, modern machines consistently produce reads of 400–500 base pairs and up to 600 million raw sequence data in a 10 hour run, with recent advancements enabling reads of around 1000 base pairs. The study shows that pyrosequencing effectively analyzes DNA methylation in plant gene promoter regions, particularly in tomatoes. It confirms a significant decrease in DNA methylation during early fruit development, consistent with previous whole-genome bisulfite sequencing findings (How-Kit *et al.*, 2015).

Illumina Solexa. Illumina's Solexa technology uses single-molecule array sequencing featuring 'bridge PCR' and 'reversible terminator sequencing' (Turcatti *et al.*, 2008). Bridge PCR amplifies DNA to create clusters detectable by fluorescence. DNA fragments are attached to a solid-phase carrier and amplified to form identical clusters for sequencing. During sequencing, paired ddNTPs are added, unpaired ones are washed away, and fluorescently labeled nucleotides are imaged. Solexa adds one dNTP per cycle to address homopolymer issues. Despite advantages like high accuracy and cost-effectiveness, it has shorter read lengths than other methods. Deep resequencing of reduced representation libraries using the Illumina Genome Analyzer discovered 2618 SNPs in rice and 1682 SNPs in soybean. This extensive SNP discovery for two genotypes per species highlights the Illumina platform's efficacy in rapidly characterizing genetic variation in plant genomes (Deschamps *et al.*, 2010).

Solid. Applied Biosystems SOLiD sequencing is an advanced ligase-based technique using emulsion PCR for high-throughput sequencing. It features constant ligation synthesis with 8-base oligonucleotides labeled in four colors, and a 'double base correction' method enhances accuracy (Smith *et al.*, 2008). SOLiD supports two library types: fragment (RNA-seq, ChIP-seq, methylation analysis) and mate-paired (for whole-genome sequencing, SNP analysis, structural rearrangements, and CNVs). Despite its high accuracy, SOLiD's short read lengths and slower speed are notable drawbacks.

SEQUENCING FROM SINGLE DNA MOLECULES. In contrast to earlier methods, this group of techniques eliminates the need to amplify DNA fragment libraries and allows direct sequencing of any DNA molecule. These methods are occasionally labeled as third-generation sequencing approaches.

Helicos technology. The Helicos platform revolutionized DNA sequencing by directly sequencing individual DNA molecules using highly fluorescent, reversible terminator nucleotides. DNA fragments tagged with poly(A) tails attach to an oligo(dT)-coated array surface. Sequencing occurs in a flow cell where fluorescent nucleotides enable single-nucleotide extensions of primers. After imaging, excess nucleotides are washed away, and the fluorescent tags are removed. This cycle repeats until sequences of around 30 nucleotides are read. The system can generate 30–40 gigabases in an 8 day run and was initially used to sequence the genomes of M13 bacteriophage and *Escherichia coli* (Harris *et al.*, 2008; Eid *et al.*, 2009).

Real-time sequencing with pacific biosciences technology. Pacific Biosciences SMRT sequencing technology uses a chip with tiny wells containing a single Φ 29-phage DNA polymerase and a DNA template with primers. Hexaphosphate nucleotides labeled with different fluorochromes are introduced, and their brief presence at the polymerase's active site allows real-time fluorescent signal recording. The nano well design ensures high sensitivity and minimal background noise. After incorporation, the fluorochrome is released, resulting in a dark phase until the next nucleotide arrives. The incorporation rate is three to five nucleotides per second, with standard reads around 1 kb in 15 min runs and possible reads up to 10 kb.

Other emerging technologies

OXFORD NANOPORE TECHNOLOGIES. Oxford Nanopore Technologies devised a unique approach in which the DNA molecule slated for sequencing is broken down by an exonuclease within a nanopore. As each base is sequentially released, it disrupts an electric current passing through the nanopore in a manner characteristic of the base type (Jain *et al.*, 2018). Therefore, it eliminates the need for DNA labeling or intricate optical devices, marking a significant advancement.

Oxford Nanopore Technologies has indeed commercialized its sequencing technology, offering products such as the MinION, GridION, and PromethION sequencing platforms to researchers and institutions worldwide, which are versatile, scalable, and portable solutions for DNA sequencing applications across various fields, including plant genomics and molecular biology research (Zhang *et al.*, 2020).

ION TORRENT SYSTEMS. The Ion Torrent systems employ a dense array of microwells packed within a semiconductor-based framework. Here, each and every nucleotide incorporation by DNA polymerase is precisely captured by detecting the release of hydrogen ions, a by-product of DNA synthesis. Presently, Ion Torrent systems boast reads of 100–200 base pairs for 1–2 hour runs. The Ion Torrent systems developed by Thermo Fisher Scientific are commercialized and available for research.

GENOME-WIDE ANALYSIS OF EPIGENETIC MODIFICATIONS. The introduction of genome-wide analysis of epigenetic modifications in plants holds significant promise for enhancing plant protection strategies. By comprehensively profiling epigenetic marks, such as DNA methylation, histone modifications, and chromatin remodeling, researchers can uncover novel mechanisms underlying plant responses to biotic stresses, including pathogen attacks and herbivore infestations (Hu *et al.*, 2015; Zhang *et al.*, 2020). Understanding the epigenetic regulation of defense-related genes and pathways enables the development of targeted approaches for enhancing plant immunity and resilience against pathogens and pests.

Sequencing technology enables genome-wide epigenetic analysis, with techniques such as bisulfite conversion, methylation-sensitive restriction enzyme digestions, ChIP-seq (chromatin immunoprecipitation sequencing) etc. Typically, promoters, which are CpG-rich regions in the genome, undergo methylation, with a hypermethylated state observed in promoter regions containing CpG islands, often influenced by various environmental factors, thereby leading to the inactivation of specific genes (Moore *et al.*, 2013). Therefore, quantifying the differentially methylated regions (DMRs) in the plant genome during diverse stress responses becomes crucial for understanding the regulatory mechanisms involved.

BISULFITE SEQUENCING. Bisulfite sequencing is a comprehensive technique for studying genome-wide DNA methylation patterns (Cokus *et al.*, 2008). It involves treating DNA with sodium bisulfite and converting unmethylated cytosine to uracil while preserving methylated cytosines. PCR amplification and subsequent analysis reveal DNA methylation status at single-nucleotide resolution. This technique is crucial for understanding methylation in untranslated regions, promoters, and coding regions and the relationship between DNA methylation and transcription (Parent *et al.*, 2021).

Current research uses whole-genome bisulfite sequencing to generate methylome maps for various plant species, aided by bioinformatic tools like methylKit, Methy-Pipe, and Bismark (Krueger and Andrews, 2011; Akalin *et al.*, 2012; Jiang *et al.*, 2014; Lou *et al.*, 2014). These tools facilitate the analysis of bisulfite sequencing data, enhancing our understanding of epigenetic interactions.

METHYLATION-SENSITIVE ENZYME RESTRICTION DIGESTION (MSAP). MSAP studies DNA methylation patterns without prior sequence knowledge, revealing cytosine methylation insights in genomes. It uses methylation-sensitive enzymes HpaII and MspI, which cleave at specific 5-CCGG recognition sequences. HpaII cuts hemimethylated 5mCCGG sequences but not fully methylated ones, while MspI cleaves fully or hemimethylated C5mCGG but not 5mCCGG sequences. MSAP distinguishes methylated from unmethylated DNA by analyzing restriction site presence (Fulneček and Kovařík, 2014). Recent studies used MSAP to analyze soybean cultivar methylation, quantitatively assessing differentially methylated regions linked to soybean agronomic traits (Shen *et al.*, 2018).

CHROMATIN IMMUNOPRECIPITATION SEQUENCING (CHIP-SEQ). Epigenomic changes in chromatin structure vary across developmental stages, disease states, tissue types, and environmental stresses. High-throughput technologies, such as ChIP-seq, analyze these dynamic chromatin states genome-wide (Kidder *et al.*, 2011). ChIP-seq identifies DNA–protein interactions, transcription factor binding sites, and patterns of histone modifications (Johnson *et al.*, 2007; Mahony and Pugh, 2015). The process involves cross-linking chromatin, fragmenting it, and using antibodies to capture protein–DNA complexes. These complexes are then reversed, purified, and analyzed via high-throughput sequencing (Mundade *et al.*, 2014).

During ChIP-seq, chromatin is extracted, fragmented, and immunoprecipitated with modification-specific antibodies. The DNA is then PCR-amplified, denatured, and fluorescently labeled. The labeled single-stranded DNA hybridizes with target sequences on a DNA microarray, forming double-stranded DNA that fluoresces. The fluorescent signals are analyzed to identify methylated regions and mapped to reference genomes to determine probe positions.

MEDIP-CHIP. MeDIP-Chip, or methylated DNA immunoprecipitation followed by microarray analysis, is a method for examining DNA methylation patterns in genomes. It has been utilized in plants to explore changes in DNA methylation across different species and under various conditions. By profiling all DNA methylation patterns across the genome of plants subjected to different environmental conditions, identification of differentially methylated regions (DMRs) associated with specific genes or genomic regions for stress tolerance, growth, and development is possible (Zhang *et al.*, 2008).

MeDIP-Chip is also used in plants to explore how transposable elements (TEs), mobile genetic elements, are regulated epigenetically. TEs can affect genome stability and gene expression. MeDIP-Chip helps identify shifts in DNA methylation levels at various TE sites in response to environmental or developmental signals, offering insights into the epigenetic controls governing TE activity and plant genome protection.

SMALL RNA-MEDIATED METHYLATION. Small RNA sequencing studies small RNA populations, such as microRNAs and small interfering RNAs (siRNAs), involved in epigenetic regulation (Rajagopalan *et al.*, 2006). siRNAs guide chromatin to specific genomic regions. The process begins with isolating RNA from samples, followed by size fractionation. DNA adaptors are then ligated to the small RNA molecules, providing primer binding sites for PCR amplification and reverse transcription.

Next-generation sequencing (NGS) is performed, enabling large-scale reads across the entire genome.

ASSAY FOR TRANSPOSASE-ACCESSIBLE CHROMATIN USING SEQUENCING (ATAC-SEQ). Assay for transposase-accessible chromatin using sequencing is a technique that assesses chromatin accessibility by sequencing DNA fragments accessible to a transposase enzyme. These genome-wide approaches provide comprehensive maps of epigenetic modifications and chromatin states, facilitating the identification of regulatory elements and functional elements in the plant genome (Buenrostro *et al.*, 2013).

23.6 Environmental and Economic Benefits of Epigenetics

Plant epigenetics studies how internal, environmental, and organism-related factors influence plant development without altering the genetic code. These heritable epigenetic changes regulate growth and enhance crop resilience to environmental stressors. Epigenetic methods offer sustainable crop protection, as factors like nutrition, temperature, and pollutants can induce inherited epigenetic changes and help plants tolerate stress (Skinner, 2014; Tiwari *et al.*, 2023). Epigenetic modifications can improve disease resistance and control gene expression related to defense, such as AGO1 and H2A.Z (Zhi and Chang, 2021). These changes can prime plants for long-lasting defense responses (Roberts and López Sánchez, 2019).

Epigenetics also impacts insect resistance, with DNA methylation regulating gene expression and potentially aiding early adaptation to insecticides and climate change (Rasmann *et al.*, 2012; Oppold and Müller, 2017; Mogilicherla and Roy, 2023). Epigenetic markers in arthropods can influence pesticide resistance and climate resilience (Bras *et al.*, 2022).

In agriculture, epigenetic insights improve crop management, enhance productivity, and create stress-resilient plants, promoting sustainability and economic benefits for farmers (Abdulraheem *et al.*, 2024).

23.7 Challenges and Future Directions

Challenges persist in epigenetics strategies for plant protection despite current achievements and updated methods. One major hurdle involves elucidating the complex interplay between epigenetic modifications and environmental factors, which can vary widely across plant species and conditions. Additionally, scaling up epigenetic interventions for agricultural applications remains challenging, requiring cost-effective methods and strategies that ensure broad-spectrum efficacy without adverse effects on plant growth and development. Future directions may involve harnessing advances in computational biology and high-throughput technologies to understand plants' epigenetic landscape better and develop targeted epigenetic manipulation techniques for sustainable crop protection against biotic and abiotic stresses. Integration with other omics approaches and interdisciplinary collaborations will be important for realizing the full capacity of epigenetics in enhancing plant resilience and food security in the face of global environmental challenges.

23.8 Conclusion

In conclusion, this chapter underscores the pivotal role of epigenetics in revolutionizing plant protection strategies. Through cutting-edge methodologies and recent advancements, it becomes evident that epigenetic approaches offer immense potential for enhancing plant resilience against various stresses, pathogens, and environmental challenges. By elucidating the intricate interplay between epigenetic mechanisms and plant defense responses, researchers can harness this knowledge to develop innovative solutions for sustainable agriculture. As we continue to unravel the complexities of epigenetic regulation in plants, the prospects for leveraging epigenetics in crop improvement and protection appear increasingly promising, suggesting a future where the power of epigenetic manipulation bolsters agricultural sustainability and food security.

References

Abdulraheem, M.I., Xiong, Y., Moshood, A.Y., Cadenas-Pliego, G., Zhang, H. *et al.* (2024) Mechanisms of plant epigenetic regulation in response to plant stress: Recent discoveries and implications. *Plants* 13(2), 163.

Adams, R.L. (1996) DNA methylation. In: Bittar, E. (ed.) *Molecular and Cellular Genetics. Principles of Medical Biology*, Vol. 5. Elsevier, pp. 33–66.

Aerts, N., Pereira Mendes, M. and Wees, S.C. (2021) Multiple levels of crosstalk in hormone networks regulating plant defense. *The Plant Journal* 105(2), 489–504.

Akalin, A., Kormaksson, M., Li, S., Garrett-Bakelman, F.E., Figueroa, M.E. *et al.* (2012) MethylKit: A comprehensive R package for the analysis of genome-wide DNA methylation profiles. *Genome Biology* 13(10), R87.

Akimoto, K., Katakami, H., Kim, H.J., Ogawa, E., Sano, C.M. *et al.* (2007) Epigenetic inheritance in rice plants. *Annals of Botany* 100(2), 205–217.

Alvarez, M.E., Nota, F. and Cambiagno, D.A. (2010) Epigenetic control of plant immunity. *Molecular Plant Pathology* 11(4), 563–576.

Álvarez-Venegas, R. and De-la-Peña, C. (2016) Recent advances of epigenetics in crop biotechnology. *Frontiers in Plant Science* 7, 196572.

An, C., Deng, L., Zhai, H., You, Y., Wu, F. *et al.* (2022) Regulation of jasmonate signaling by reversible acetylation of TOPLESS in *Arabidopsis*. *Molecular Plant* 15(8), 1329–1346.

An, Y.Q.C., Goettel, W., Han, Q., Bartels, A., Liu, Z. *et al.* (2017) Dynamic changes of genome-wide DNA methylation during soybean seed development. *Scientific Reports* 7(1), 12263.

Arnholdt-Schmitt, B. (2004) Stress-induced cell reprogramming. A role for global genome regulation? *Plant Physiology* 136(1), 2579–2586.

Ashapkin, V.V., Kutueva, L.I., Aleksandrushkina, N.I. and Vanyushin, B.F. (2020) Epigenetic mechanisms of plant adaptation to biotic and abiotic stresses. *International Journal of Molecular Sciences* 21(20), 7457.

Atighi, M.R., Verstraeten, B., Meyer, T. and Kyndt, T. (2020) Genome-wide DNA hypomethylation shapes nematode pattern-triggered immunity in plants. *New Phytologist* 227(2), 545–558.

Ay, N., Irmler, K., Fischer, A., Uhlemann, R., Reuter, G. *et al.* (2009) Epigenetic programming via histone methylation at WRKY53 controls leaf senescence in *Arabidopsis thaliana*. *The Plant Journal* 58(2), 333–346.

Ayyappan, V., Kalavacharla, V., Thimmapuram, J., Bhide, K.P., Sripathi, V.R. *et al.* (2015) Genome-wide profiling of histone modifications (H3K9me2 and H4K12ac) and gene expression in rust (*Uromyces appendiculatus*) inoculated common bean (*Phaseolus vulgaris* L). *PloS ONE* 10(7), e0132176.

Bender, J. (2002) Plant epigenetics. *Current Biology* 12(12), R412–R414.

Bras, A., Roy, A., Heckel, D.G., Anderson, P. and Karlsson Green, K. (2022) Pesticide resistance in arthropods: Ecology matters too. *Ecology Letters* 25(8), 1746–1759.

Buenrostro, J.D., Giresi, P.G., Zaba, L.C., Chang, H.Y. and Greenleaf, W.J. (2013) Transposition of native chromatin for fast and sensitive epigenomic profiling of open chromatin, DNA-binding proteins and nucleosome position. *Nature Methods* 10(12), 1213–1218.

Canton, M., Farinati, S., Forestan, C., Joseph, J., Bonghi, C. *et al.* (2022) An efficient chromatin immunoprecipitation (CHIP) protocol for studying histone modifications in peach reproductive tissues. *Plant Methods* 18(1), 43.

Carmell, M.A., Xuan, Z., Zhang, M.Q. and Hannon, G.J. (2002) The argonaute family: Tentacles that reach into RNAi, developmental control, stem cell maintenance, and tumorigenesis. *Genes & Development* 16(21), 2733–2742.

Chang, Y.N., Zhu, C., Jiang, J., Zhang, H., Zhu, J.K. *et al.* (2020) Epigenetic regulation in plant abiotic stress responses. *Journal of Integrative Plant Biology* 62(5), 563–580.

Chen, H., Shu, H., Wang, L., Zhang, F., Li, X. *et al.* (2018) *Phytophthora* methylomes are modulated by 6mA methyltransferases and associated with adaptive genome regions. *Genome Biology* 19, 1–16.

Chen, X. (2012) Small RNAs in development – insights from plants. *Current Opinion in Genetics & Development* 22(4), 361–367.

Chen, X., Hu, Y. and Zhou, D.X. (2011) Epigenetic gene regulation by plant jumonji group of histone demethylase. *Biochimica et Biophysica Acta (BBA)-Gene Regulatory Mechanisms* 1809(8), 421–426.

Chen, X., Duan, Y., Qiao, F., Liu, H., Huang, J. *et al.* (2022) A secreted fungal effector suppresses rice immunity through host histone hypoacetylation. *New Phytologist* 235(5), 1977–1994.

Chinnusamy, V. and Zhu, J.K. (2009) Epigenetic regulation of stress responses in plants. *Current Opinion in Plant Biology* 12(2), 133–139.

Choi, S.M., Song, H.R., Han, S.K., Han, M., Kim, C.Y. *et al.* (2012) HDA19 is required for the repression of salicylic acid biosynthesis and salicylic acid-mediated defense responses in *Arabidopsis*. *The Plant Journal* 71(1), 135–146.

Choi, Y., Gehring, M., Johnson, L., Hannon, M., Harada, J.J. *et al.* (2002) DEMETER, a DNA glycosylase domain protein, is required for endosperm gene imprinting and seed viability in *Arabidopsis*. *Cell* 110(1), 33–42.

Clark, S.J., Argelaguet, R., Kapourani, C.A., Stubbs, T.M., Lee, H.J. *et al.* (2018) ScNMT-seq enables joint profiling of chromatin accessibility DNA methylation and transcription in single cells. *Nature Communications* 9(1), 1–10.

Cokus, S.J., Feng, S., Zhang, X., Chen, Z., Merriman, B. *et al.* (2008) Shotgun bisulphite sequencing of the *Arabidopsis* genome reveals DNA methylation patterning. *Nature* 452(7184), 215–219.

Crespo-Salvador, Ó., Sánchez-Giménez, L., López-Galiano, M.J., Fernández-Crespo, E., Schalschi, L. *et al.* (2020) The histone marks signature in exonic and intronic regions is relevant in early response of tomato genes to *Botrytis cinerea* and in miRNA regulation. *Plants Basel* 9, 300.

Deichmann, U. (2016) Epigenetics: The origins and evolution of a fashionable topic. *Developmental Biology* 416(1), 249–254.

Deng, Y., Zhai, K., Xie, Z., Yang, D., Zhu, X. *et al.* (2017) Epigenetic regulation of antagonistic receptors confers rice blast resistance with yield balance. *Science* 355(6328), 962–965.

Deschamps, S., Llaca, V. and May, G.D. (2012) Genotyping-by-Sequencing in Plants. *Biology* 25(1), 460-83. DOI: 10.3390/biology1030460. PMID: 24832503; PMCID: PMC4009820.

Díaz-Manzano, F.E., Cabrera, J., Ripoll, J.-J., Del Olmo, I., Andrés, M.F. *et al.* (2018) A role for the gene regulatory module microRNA172/TARGET OF EARLY ACTIVATION TAGGED 1/FLOWERING LOCUS T (mi RNA 172/TOE 1/FT) in the feeding sites induced by *Meloidogyne javanica* in *Arabidopsis thaliana*. *New Phytologist* 217(2), 813–827.

Dowen, R.H., Pelizzola, M., Schmitz, R.J., Lister, R., Dowen, J.M. *et al.* (2012) Widespread dynamic DNA methylation in response to biotic stress. *Proceedings of the National Academy of Sciences of the United States of America* 109(32), E2183–91.

Dubey, A. and Jeon, J. (2017) Epigenetic regulation of development and pathogenesis in fungal plant pathogens. *Molecular Plant Pathology* 18(6), 887–898.

Dutta, A., Choudhary, P., Caruana, J. and Raina, R. (2017) JMJ27, an *Arabidopsis* H3K9 histone demethylase, modulates defense against *Pseudomonas syringae* and flowering time. *The Plant Journal* 91(6), 1015–1028.

Eid, J., Fehr, A., Gray, J., Luong, K., Lyle, J. *et al.* (2009) Real-time DNA sequencing from single polymerase molecules. *Science* 323(5910), 133–138.

Friedman, A.R. and Baker, B.J. (2007) The evolution of resistance genes in multi-protein plant resistance systems. *Current Opinion in Genetics & Development* 17(6), 493–499.

Fulneček, J. and Kovařík, A. (2014) How to interpret methylation sensitive amplified polymorphism (MSAP) profiles? *BMC Genetics* 15, 2.

Gambino, G. and Pantaleo, V. (2017) Epigenetics in plant–pathogen interactions. In: *Plant Epigenetics*. pp. 385–404.

Gendrel, A.V. and Colot, V. (2005) *Arabidopsis* epigenetics: When RNA meets chromatin. *Current Opinion in Plant Biology* 8(2), 142–147.

Geng, S., Kong, X., Song, G., Jia, M., Guan, J. *et al.* (2019) DNA methylation dynamics during the interaction of wheat progenitor *Aegilops tauschii* with the obligate biotrophic fungus *Blumeria graminis* f. sp. *tritici*. *New Phytologist* 221(2), 1023–1035.

Gijzen, M., Ishmael, C. and Shrestha, S.D. (2014) Epigenetic control of effectors in plant pathogens. *Frontiers in Plant Science* 5, 638.

Gong, Z., Morales-Ruiz, T., Ariza, R.R., Roldán-Arjona, T., David, L. *et al.* (2002) ROS1, a repressor of transcriptional gene silencing in *Arabidopsis*, encodes a DNA glycosylase/lyase. *Cell* 111(6), 803–814.

Harris, T.D., Buzby, P.R., Babcock, H., Beer, E., Bowers, J. *et al.* (2008) Single-molecule DNA sequencing of a viral genome. *Science* 320, 106–109.

He, C., Zhang, Z., Li, B. and Tian, S. (2020) The pattern and function of DNA methylation in fungal plant pathogens. *Microorganisms* 8(2), 227.

Hewezi, T., Howe, P., Maier, T.R. and Baum, T.J. (2008) *Arabidopsis* small RNAs and their targets during cyst nematode parasitism. *Molecular Plant-Microbe Interactions* 21(12), 1622–1634.

Hilton, I.B., D'Ippolito, A.M., Vockley, C.M., Thakore, P.I., Crawford, G.E. *et al.* (2015) Epigenome editing by a CRISPR-Cas9-based acetyltransferase activates genes from promoters and enhancers. *Nature Biotechnology* 33(5), 510–517.

Hoang, T.V., Vo, K.T.X., Hong, W.J., Jung, K.H. and Jeon, J.S. (2018) Defense response to pathogens through Epigenetic regulation in rice. *Journal of Plant Biology* 61, 1–10.

Holliday, R. (1996) DNA methylation in eukaryotes: 20 years on. *Cold Spring Harbor Monograph Series* 32, 5–28.

Hossain, M.S., Kawakatsu, T., Kim, K.D., Zhang, N., Nguyen, C.T. *et al.* (2017) Divergent cytosine DNA methylation patterns in single-cell, soybean root hairs. *The New Phytologist* 214(2), 808–819.

Hou, Y., Wang, L., Wang, L., Liu, L., Li, L. *et al.* (2015) JMJ704 positively regulates rice defense response against *Xanthomonas oryzae* pv. *oryzae* infection via reducing H3K4me2/3 associated with negative disease resistance regulators. *BMC Plant Biology* 15, 1–13.

How-Kit, A., Dejeux, E., Dousset, B., Renault, V., Baudry, M. *et al.* (2015) DNA methylation profiles distinguish different subtypes of gastroenteropancreatic neuroendocrine tumors. *Epigenomics* 7(8), 1245–1258.

Hu, J., Chen, X., Zhang, H. and Ding, Y. (2015) Genome-wide analysis of DNA methylation in photoperiod- and thermo-sensitive male sterile rice Peiai 64S. *BMC Genomics* 16(1), 14.

Jain, M., Koren, S., Miga, K.H., Quick, J., Rand, A.C. *et al.* (2018) Nanopore sequencing and assembly of a human genome with ultra-long reads. *Nature Biotechnology* 36(4), 338–345.

Jenuwein, T. and Allis, C.D. (2001) Translating the histone code. *Science* 293(5532), 1074–1080.

Jeon, J., Choi, J., Lee, G.W., Park, S.Y., Huh, A. *et al.* (2015) Genome-wide profiling of DNA methylation provides insights into epigenetic regulation of fungal development in a plant pathogenic fungus, *Magnaporthe oryzae*. *Scientific Reports* 5(1), 8567.

Jiang, P., Sun, K., Lun, F.M.F., Guo, A.M., Wang, H. *et al.* (2014) Methy-pipe: An integrated bioinformatics pipeline for whole genome bisulfite sequencing data analysis. *PLoS ONE* 9(6), e100360.

Johnson, D.S., Mortazavi, A., Myers, R.M. and Wold, B. (2007) Genome-wide mapping of *in vivo* protein–DNA interactions. *Science* 316(5830), 1497–1502.

Jones, J.D. and Dangl, J.L. (2006) The plant immune system. *Nature* 444(7117), 323–329.

Kaldis, A., Tsementzi, D., Tanriverdi, O. and Vlachonasios, K.E. (2011) *Arabidopsis thaliana* transcriptional co-activators ADA2b and SGF29a are implicated in salt stress responses. *Planta* 233(4), 749–762.

Kawakatsu, T., Huang, S.S.C., Jupe, F., Sasaki, E., Schmitz, R.J. *et al.* (2016) Epigenomic diversity in a global collection of *Arabidopsis thaliana* accessions. *Cell* 166(2), 492–505.

Kellenberger, R.T., Schlüter, P.M. and Schiestl, F.P. (2016) Herbivore-induced DNA demethylation changes floral signalling and attractiveness to pollinators in *Brassica rapa*. *PloS ONE* 11(11), e0166646.

Ketting, R.F., Fischer, S.E.J., Bernstein, E., Sijen, T., Hannon, G.J. *et al.* (2001) Dicer functions in RNA interference and in synthesis of small RNA involved in developmental timing in *C. elegans*. *Genes & Development* 15(20), 2654–2659.

Kidder, B.L., Hu, G. and Zhao, K. (2011) ChIP-seq: Technical considerations for obtaining high-quality data. *Nature Immunology* 12(10), 918–922.

Kidner, C.A. and Martienssen, R.A. (2005) The developmental role of microRNA in plants. *Current Opinion in Plant Biology* 8(1), 38–44.

Kim, M., Ohr, H., Lee, J.W., Hyun, Y., Fischer, R.L. *et al.* (2008) Temporal and spatial downregulation of *Arabidopsis* MET1 activity results in global DNA hypomethylation and developmental defects. *Molecules and Cells* 26(6), 611–615.

Kotkar, H. and Giri, A. (2020) Plant epigenetics and the 'intelligent' priming system to combat biotic stress. In: *Epigenetics of the Immune System*. Academic Press, pp. 25–38.

Kramer, H.M., Cook, D.E., Seidl, M.F. and Thomma, B.P. (2023) Epigenetic regulation of nuclear processes in fungal plant pathogens. *PLoS Pathogens* 19(8), e1011525.

Krueger, F. and Andrews, S.R. (2011) Bismark: A flexible aligner and methylation caller for bisulfite-seq applications. *Bioinformatics* 27(11), 1571–1572.

Kuć, J. (1982) Induced immunity to plant disease. *BioScience* 32(11), 854–860.

Kumar, S. (2018) Epigenomics of plant responses to environmental stress. *Epigenomes* 2(1), 6.

Kumar, S., Beena, A.S., Awana, M. and Singh, A. (2017) Physiological, biochemical, epigenetic and molecular analyses of wheat (*Triticum aestivum*) genotypes with contrasting salt tolerance. *Frontiers in Plant Science* 8, 280121.

Kuźnicki, D., Meller, B., Arasimowicz-Jelonek, M., Braszewska-Zalewska, A., Drozda, A. *et al.* (2019) BABA-induced DNA methylome adjustment to intergenerational defense priming in potato to *Phytophthora infestans*. *Frontiers in Plant Science* 10, 650. DOI: 10.3389/fpls.2019.00650.

Lanfear, R. (2018) Do plants have a segregated germline? *PLoS Biology* 16(5), e2005439.

Latutrie, M., Gourcilleau, D. and Pujol, B. (2019) Epigenetic variation for agronomic improvement: An opportunity for vegetatively propagated crops. *American Journal of Botany* 106(10), 1281.

Lee, R.C., Feinbaum, R.L. and Ambros, V. (1993) The *C. elegans* heterochronic gene lin-4 encodes small RNAs with antisense complementarity to lin-14. *Cell* 75(5), 843–854.

Lee, S., Fu, F., Xu, S., Lee, S.Y., Yun, D.J. *et al.* (2016) Global regulation of plant immunity by histone lysine methyl transferases. *The Plant Cell* 28(7), 1640–1661.

Li, J., Gong, Y. and Zhang, Y. (2016) The identification ability of pyrosequencing technology was evaluated by VITEK2Compact system for rare clinical pathogens with low accuracy. *Biomedical Engineering and Clinical* 20(6), 626–630.

Li, T., Chen, X., Zhong, X., Zhao, Y., Liu, X. *et al.* (2013) Jumonji C domain protein JMJ705-mediated removal of histone H3 lysine 27 trimethylation is involved in defense-related gene activation in rice. *The Plant Cell* 25(11), 4725–4736.

Lim, D.H. and Maher, E.R. (2010) DNA methylation: A form of epigenetic control of gene expression. *The Obstetrician & Gynaecologist* 12(1), 37–42.

Lingel, A., Simon, B., Izaurralde, E. and Sattler, M. (2003) Structure and nucleic-acid binding of the drosophila argonaute 2 PAZ domain. *Nature* 426(6965), 465–469.

Lira-Medeiros, C.F., Parisod, C., Fernandes, R.A., Mata, C.S., Cardoso, M.A. *et al.* (2010) Epigenetic variation in mangrove plants occurring in contrasting natural environment. *PloS ONE* 5(4), e10326.

Liu, H., Li, J., Wang, S., Hua, J. and Zou, B. (2023) CHROMATIN REMODELING 11-dependent nucleosome occupancy affects disease resistance in rice. *Plant Physiology* 193(2), 1635–1651.

López Sánchez, A., Stassen, J.H., Furci, L., Smith, L.M. and Ton, J. (2016) The role of DNA (de) methylation in immune responsiveness of *Arabidopsis*. *The Plant Journal* 88(3), 361–374.

Lou, S., Lee, H.M., Qin, H., Li, J.W., Gao, Z. *et al.* (2014) Whole-genome bisulfite sequencing of multiple individuals reveals complementary roles of promoter and gene body methylation in transcriptional regulation. *Genome Biology* 15(7), 408.

Luo, C., Fernie, A.R. and Yan, J. (2020) Single-cell genomics and epigenomics: Technologies and applications in plants. *Trends in Plant Science* 25(10), 1030–1040.

Luo, J., Pan, X.L., Peng, T.C., Chen, Y.Y., Hui, Z. *et al.* (2016) DNA methylation patterns of banana leaves in response to *Fusarium oxysporum* f. sp. *cubense* tropical race 4. *Journal of Integrative Agriculture* 15(12), 2736–2744.

Madlung, A. and Comai, L. (2004) The effect of stress on genome regulation and structure. *Annals of Botany* 94(4), 481–495.

Maeji, H. and Nishimura, T. (2018) Epigenetic mechanisms in plants. In: Gallusci, P., Bucher, E. and Mirouze, M. (eds) *Plant Epigenetics Coming of Age for Breeding Application. (Advances in Botanical Research*, Vol. 88). Academic Press, pp. 21–47.

Mahony, S. and Pugh, B.F. (2015) Protein-DNA binding in high-resolution. *Critical Reviews in Biochemistry and Molecular Biology* 50(4), 269–283.

Malinovsky, F.G., Fangel, J.U. and Willats, W.G. (2014) The role of the cell wall in plant immunity. *Frontiers in Plant Science* 5, 86833.

Mao, Y., Pavangadkar, K.A., Thomashow, M.F. and Triezenberg, S.J. (2006) Physical and functional interactions of *Arabidopsis* ADA2 transcriptional coactivator proteins with the acetyltransferase GCN5 and with the cold-induced transcription factor CBF1. *Biochimica et Biophysica Acta (BBA) – Gene Structure and Expression* 1759(1–2), 69–79.

Mapuranga, J., Chang, J. and Yang, W. (2022) Combating powdery mildew: Advances in molecular interactions between *Blumeria graminis* f. sp. *tritici* and wheat. *Frontiers in Plant Science* 13, 1102908.

Margulies, M., Egholm, M., Altman, W.E., Attiya, S., Bader, J.S. *et al.* (2005) Genome sequencing in microfabricated high-density picolitre reactors. *Nature* 437, 376–380.

McLellan, H., Chen, K., He, Q., Wu, X., Boevink, P.C. *et al.* (2020) The ubiquitin E3 ligase PUB17 positively regulates immunity by targeting a negative regulator, KH17, for degradation. *Plant Communications* 1(4).

Medina, C., Rocha, M., Magliano, M., Ratpopoulo, A., Revel, B. *et al.* (2017) Characterization of microRNAs from *Arabidopsis* galls highlights a role for miR159 in the plant response to the root-knot nematode *Meloidogyne incognita*. *New Phytologist* 216(3), 882–896.

Meller, B., Kuźnicki, D., Arasimowicz-Jelonek, M., Deckert, J. and Floryszak-Wieczorek, J. (2018) BABA-primed histone modifications in potato for intergenerational resistance to *Phytophthora infestans*. *Frontiers in Plant Science* 9, 1228.

Meng, F., Zhao, Q., Zhao, X., Yang, C., Liu, R. *et al.* (2022) A rice protein modulates endoplasmic reticulum homeostasis and coordinates with A transcription factor to initiate blast disease resistance. *Cell Reports* 39, 11.

Michael, T.P., Jupe, F., Bemm, F., Motley, S.T., Sandoval, J.P. *et al.* (2018) High contiguity *Arabidopsis thaliana* genome assembly with a single nanopore flow cell. *Nature Communications* 9(1), 1–11.

Mirouze, M. and Paszkowski, J. (2011) Epigenetic contribution to stress adaptation in plants. *Current Opinion in Plant Biology* 14(3), 267–274.

Mogilicherla, K. and Roy, A. (2023) Epigenetic regulations as drivers of insecticide resistance and resilience to climate change in arthropod pests. *Frontiers in Genetics* 13, 1044980.

Moore, D.S. (2015) *The Developing Genome: An Introduction to Behavioral Epigenetics*. Oxford University Press, Oxford.

Moore, L.D., Le, T. and Fan, G. (2013) DNA methylation and its basic function. *Neuropsychopharmacology* 38(1), 23–38.

Mukherjee, K., Grizanova, E., Chertkova, E., Lehmann, R., Dubovskiy, I. *et al.* (2017) Experimental evolution of resistance against *Bacillus thuringiensis* in the insect model host *Galleria mellonella* results in epigenetic modifications. *Virulence* 8(8), 1618–1630.

Mundade, R., Ozer, H.G., Wei, H., Prabhu, L. and Lu, T. (2014) Role of CHIP-seq in the discovery of transcription factor binding sites, differential gene regulation mechanism, epigenetic marks and beyond. *Cell Cycle* 13(18), 2847–2852.

Nasrullah Hussain, A., Ahmed, S., Rasool, M. and Shah, A.J. (2022) DNA methylation across the tree of life, from micro to macro-organism. *Bioengineered* 13(1), 1666–1685.

Niederhuth, C.E., Bewick, A.J., Ji, L., Alabady, M.S., Kim, K.D. *et al.* (2016) Widespread natural variation of DNA methylation within angiosperms. *Genome Biology* 17(1), 1–19.

Oliva, R. and Quibod, I.L. (2017) Immunity and starvation: New opportunities to elevate disease resistance in crops. *Current Opinion in Plant Biology* 38, 84–91.

Oppold, A.M. and Müller, R. (2017) Epigenetics: A hidden target of insecticides. In: Verlinden, H. (ed.) *Insect Epigenetics, Advances in Insect Physiology*. Academic Press, Cambridge, pp. 313–324.

Parent, J.S., Cahn, J., Herridge, R.P., Grimanelli, D. and Martienssen, R.A. (2021) Small RNAs guide histone methylation in *Arabidopsis* embryos. *Genes & Development* 35(11–12), 841–846.

Pavangadkar, K., Thomashow, M.F. and Triezenberg, S.J. (2010) Histone dynamics and roles of histone acetyltransferases during cold-induced gene regulation in *Arabidopsis*. *Plant Molecular Biology* 74, 183–200.

Pieterse, C.M., Leon-Reyes, A., Ent, S. and Wees, S.C. (2009) Networking by small-molecule hormones in plant immunity. *Nature Chemical Biology* 5(5), 308–316.

Pikaard, C.S. and Scheid, O.M. (2014) Epigenetic regulation in plants. *Cold Spring Harbor Perspectives in Biology* 6(12), a019315.

Rajagopalan, R., Vaucheret, H., Trejo, J. and Bartel, D.P. (2006) A diverse and evolutionarily fluid set of microRNAs in *Arabidopsis thaliana*. *Genes & Development* 20(24), 3407–3425.

Rambani, A., Rice, J.H., Liu, J., Lane, T., Ranjan, P. *et al.* (2015) The methylome of soybean roots during the compatible interaction with the soybean cyst nematode. *Plant Physiology* 168(4), 1364–1377.

Ramirez-Prado, J.S., Latrasse, D., Rodriguez-Granados, N.Y., Huang, Y., Manza-Mianza, D. *et al.* (2019) The Polycomb protein LHP 1 regulates *Arabidopsis thaliana* stress responses through the repression of the MYC 2-dependent branch of immunity. *The Plant Journal* 100(6), 1118–1131.

Rasmann, S., Vos, M., Casteel, C.L., Tian, D., Halitschke, R. *et al.* (2012) Herbivory in the previous generation primes plants for enhanced insect resistance. *Plant Physiology* 158(2), 854–863.

Ratel, D., Ravanat, J.L., Berger, F. and Wion, D. (2006) N6-methyladenine: The other methylated base of DNA. *BioEssays* 28(3), 309–315.

Roberts, M.R. and López Sánchez, A. (2019) Plant epigenetic mechanisms in response to biotic stress. In: Alvarez-Venega, R., De-la-Peña, C. and Casas-Mollano, J.A. (eds) *Epigenetics in Plants of*

Agronomic Importance: Fundamentals and Applications. Transcriptional Regulation and Chromatin Remodelling in Plants. Springer, Cham, pp. 65–113.
Rodriguez-Moreno, L., Song, Y. and Thomma, B.P. (2017) Transfer and engineering of immune receptors to improve recognition capacities in crops. *Current Opinion in Plant Biology* 38, 42–49.
Roy, S., Gupta, P., Rajabhoj, M.P., Maruthachalam, R. and Nandi, A.K. (2018) The polycomb-group repressor MEDEA attenuates pathogen defense. *Plant Physiology* 177(4), 1728–1742.
Samantara, K., Shiv, A., Sousa, L.L., Sandhu, K.S., Priyadarshini, P. *et al*. (2021) A comprehensive review on epigenetic mechanisms and application of epigenetic modifications for crop improvement. *Environmental and Experimental Botany* 188, 104479.
Sanger, F., Air, G.M., Barrell, B.G., Brown, N.L., Coulson, A.R. *et al*. (1977) Nucleotide sequence of bacteriophage φX174 DNA. *Nature* 265(5596), 687–695.
Schwartzman, O. and Tanay, A. (2015) Single-cell epigenomics: techniques and emerging applications. *Nature Reviews Genetics* 16(12), 716–726.
Sengupta, N. and Seto, E. (2004) Regulation of histone deacetylase activities. *Journal of Cellular Biochemistry* 93(1), 57–67.
Servet, C., Benhamed, M., Latrasse, D., Kim, W., Delarue, M. *et al*. (2008) Characterization of a phosphatase 2C protein as an interacting partner of the histone acetyltransferase GCN5 in *Arabidopsis*. *Biochimica et Biophysica Acta (BBA) – Gene Regulatory Mechanisms* 1779(6–7), 376–382.
Shahbazian, M.D. and Grunstein, M. (2007) Functions of site-specific histone acetylation and deacetylation. *Annual Review of Biochemistry* 76(1), 75–100.
Sheikh, A.H., Nawaz, K., Tabassum, N., Almeida-Trapp, M., Mariappan, K.G. *et al*. (2023) Linker histone H1 modulates defense priming and immunity in plants. *Nucleic Acids Research* 51(9), 4252–4265.
Shen, Y., Zhang, J., Liu, Y., Liu, S., Liu, Z. *et al*. (2018) DNA methylation footprints during soybean domestication and improvement. *Genome Biology* 19(1), 128.
Skinner, M.K. (2014) Endocrine disruptor induction of epigenetic transgenerational inheritance of disease. *Molecular and Cellular Endocrinology* 398(1–2), 4–12.
Smith, D.R., Quinlan, A.R., Peckham, H.E., Makowsky, K., Tao, W. *et al*. (2008) Rapid whole-genome mutational profiling using next-generation sequencing technologies. *Genome Research* 18(10), 1638–1642.
Song, W.Y., Wang, G.L., Chen, L.L., Kim, H.S., Pi, L.Y. *et al*. (1995) A receptor kinase-like protein encoded by the rice disease resistance gene, Xa21. *Science* 270(5243), 1804–1806.
Talarowska, M. (2020) Epigenetic mechanisms in the neurodevelopmental theory of depression. *Depression Research and Treatment* 2020(1–2), 1–9.
Tiwari, A., Pandey-Rai, S., Rai, K.K., Tiwari, A. and Pandey, N. (2023) Molecular and epigenetic basis of heat stress responses and acclimatization in plants. *The Nucleus* 66(1), 69–79.
Turcatti, G., Romieu, A., Fedurco, M. and Tairi, A.P. (2008) A new class of cleavable fluorescent nucleotides: synthesis and optimization as reversible terminators for DNA sequencing by synthesis. *Nucleic Acids Research* 36(4), e25.
Turner, B.M. (2000) Histone acetylation and an epigenetic code. *BioEssays* 22(9), 836–845.
Vaiserman, A.M. and Koliada, A.K. (2017) Early-life adversity and long-term neurobehavioral outcomes: Epigenome as a bridge? *Human Genomics* 11, 1–15.
Varotto, S., Tani, E., Abraham, E., Krugman, T., Kapazoglou, A. *et al*. (2020) Epigenetics: possible applications in climate-smart crop breeding. *Journal of Experimental Botany* 71(17), 5223–5236.
Williams, B.P., Pignatta, D., Henikoff, S. and Gehring, M. (2015) Methylation-sensitive expression of a DNA demethylase gene serves as an epigenetic rheostat. *PLoS Genetics* 11(3), e1005142.
Yan, L., Fan, G. and Li, X. (2019) Genome-wide analysis of three histone marks and gene expression in *Paulownia fortunei* with phytoplasma infection. *BMC Genomics* 20(1), 14.
Yang, L., Chen, X., Wang, Z., Sun, Q., Hong, A. *et al*. (2020) HOS15 and HDA9 negatively regulate immunity through histone deacetylation of intracellular immune receptor NLR genes in *Arabidopsis*. *New Phytologist* 226(2), 507–522.
Zhang, L., Lu, Q. and Chang, C. (2020) Epigenetics in health and disease. In: Chang, C. and Lu, Q. (eds) *Epigenetics in Allergy and Autoimmunity. Advances in Experimental Medicine and Biology*, Vol. 1253. Springer, Singapore, pp. 3–55. DOI: 10.1007/978-981-15-3449-2_1.
Zhang, X., Yazaki, J., Sundaresan, A., Cokus, S., Chan, S.W.L. *et al*. (2008) Genome-wide high-resolution mapping and functional analysis of DNA methylation in *Arabidopsis*. *Cell* 126(6), 1189–1201.

Zhi, P. and Chang, C. (2021) Exploiting epigenetic variations for crop disease resistance improvement. *Frontiers in Plant Science* 12, 692328.

Zhou, D.X. (2009) Regulatory mechanism of histone epigenetic modifications in plants. *Epigenetics* 4(1), 15–18.

Zhou, C., Zhang, L., Duan, J., Miki, B. and Wu, K. (2005) HISTONE DEACETYLASE19 is involved in jasmonic acid and ethylene signaling of pathogen response in *Arabidopsis*. *The Plant Cell* 17(4), 1196–1204.

Zhu, J.K. (2016) Abiotic stress signaling and responses in plants. *Cell* 167(2), 313–324.

Zytynska, S.E., Eicher, M., Rothballer, M. and Weisser, W.W. (2020) Microbial-mediated plant growth promotion and pest suppression varies under climate change. *Frontiers in Plant Science* 11, 573578.

24 Plant Epigenetic Regulation Under Stress Conditions: Machinery and Applications

C. A. Cedillo-Jiménez, A. L. Villagómez-Aranda and R. G. Guevara-González*

Center of Applied Research in Biosystems (CARB-CIAB), Autonomous University of Querétaro, Querétaro, México

Abstract

Plants continuously cope with biotic and abiotic adverse conditions. It is currently known that changes regarding chromatin environment and gene expression are of high importance in this phenomenon. Epigenetic modifications involving changes at DNA methylation level, covalent modifications in histones, and even specific expression of small non-coding RNAs are used by plants as responses to biotic and abiotic stresses. In addition, these epigenetic changes are inherited in a somatic (within the same generation), inter- and transgenerational level (stress memory), that is used for plant adaptation to the environment. This chapter aims to display current knowledge, as well as examples, regarding epigenetic mechanisms used by plants to cope with biotic and abiotic stresses and stress memory. The possibility of using epigenetic stress memory in plant breeding is also dealt with. Finally, some perspectives about the future trends in plant epigenetic studies and applications of this knowledge are mentioned and discussed.

Keywords: Epigenetic, Plant breeding, Stress memory, Plant stress

24.1 Introduction: Current Knowledge on the Role of Epigenetic Regulation in Plant Stress

Plants are sessile organisms that have developed enormous flexibility and plasticity to cope with predation and stress conditions to adapt to the environment. This adaptation, in part, is regulated via epigenetic mechanisms, involving DNA methylation, chromatin modifications, and production of specific microRNAs (Villagómez-Aranda *et al.*, 2022). Overall, it has been found that epigenetic diversity in plants is related to environmental changes. Several recent studies have shown that transcriptional rewiring occurs associated with various epigenetic signals in plants exposed to biotic and abiotic stresses, such as exogenous DNA applications, 2,6-dichloro isonicotinic acid (INA), extreme temperatures, drought, and salinity (Vega-Muñoz *et al.*, 2018; Cao *et al.*, 2023).

The majority of the studies have focused on abiotic stress and its memory on DNA methylation and histone modifications (Villagómez-Aranda *et al.*, 2022). By contrast, studies of biotic interactions and epigenetic and phenotypic variation have been scarce, likely because suitable strategies to study

*Corresponding author: ramon.guevara@uaq.mx

 Epigenetics for Climate-Smart and Sustainable Agriculture (ed. Jen-Tsung Chen)
DOI: 10.1079/9781800626102.0024

these models need to be more developed. Interestingly it is noteworthy that some plant stress studies have shown interplay between specific genetic and epigenetic changes, making plant epigenetic knowledge more complex than initially might be expected (Huang *et al.*, 2019).

On the one hand, in some recent studies on plant epigenetic changes associated with drought stress in three Brassicaceae species, an overall signal upregulating epigenetic reprogramming transcription associated with stress response genes in the most drought-adapted species has been reported (Marín-de la Rosa *et al.*, 2019). Studies using a transgenic tobacco (*Nicotiana tabacum* cv. Xanthi nc) model constitutively expressing a gene encoding to a superoxide dismutase (SOD) activity from *Capsicum chinense* Jacq. tolerant to several biotic and abiotic stresses (Guevara-Olvera *et al.*, 2012; Cardenas-Manríquez *et al.*, 2016; Sáenz-de la O *et al.*, 2020, 2021, 2023), showed a clear correlation between stress responses mediated by hydrogen peroxide endogen levels (produced by SOD activity in these plants) and epigenetic changes related to cytosine DNA methylation, especially in CHG context (Villagómez-Aranda *et al.*, 2021). In this latter study, cytosine DNA methylation changes were associated with genes related to respiration, energy, DNA repair, and calcium signaling. As commonly found in other plant models, this study reported main changes in intergenic regions, transposable elements (TEs), exons, introns, and promoters, depending on the cytosine context (Villagómez-Aranda *et al.*, 2021). The case of TEs has recently shown interesting results regarding their role in epigenetic responses to stress likely helping in plant adaptation to new environmental conditions (Wang *et al.*, 2018). In this latter study, it was shown that stress-induced reactivations of TEs become epigenetically controlled by siRNA-mediated CHH methylation in *Rhizophora apiculata*.

On the other hand, in general, plant epigenetic responses to biotic stresses seem to be related to similar pathways and epigenetic changes found in abiotic responses. It is clear that, currently, plant responses to biotic stress have mainly shown the epigenetic changes produced and not the role of these epigenetic changes in the interactions, thus a wide field of opportunity to unravel these aspects is opened for current and future research. These epigenetic studies mainly report changes in non-coding RNAs and histone marks associated with plant defense against pathogenic fungi (Wang *et al.*, 2017). Moreover, some studies have shown the role of DNA methylation changes in the recognition of some damage-associated molecular patterns (DAMPs) based on self- or non-self DNA in plants (Vega-Muñoz *et al.*, 2018). It is well known that plant priming is a defense response that recalls a previous interaction with biotic or abiotic stress, to further quickly mount a defense in case of subsequent contacts (Godínez-Mendoza *et al.*, 2023). Recent evidence has shown that there is an epigenetic component in priming defense response, especially in long-term defense that might even be inherited over several generations, strongly related to the intensity of the initial interaction (Godínez-Mendoza *et al.*, 2023). This aspect, related to the role of epigenetic mechanisms in plant stress memory, will be discussed in more detail in another section of this chapter. Additionally, further sections of this chapter will deal with some specific examples of epigenetic strategies used by plants under stress conditions, as well as showing and discussing possibilities of epigenetic approaches in the design of plant breeding strategies. Finally, some perspectives and concluding remarks, including current approaches using plant genome editing associated with epigenetic changes, will be discussed.

24.2 Insights Into Plant Epigenetic Responses to Environmental Stress: Mechanisms for Plant Survival

Epigenetic responses to stress conditions manifest at the molecular level through mechanisms such as DNA methylation, histone modifications, and RNA-directed DNA methylation (RdDM), significantly affecting plant physiology (Kumar and Rani, 2023). This section explores cases in which plants are exposed to stressors that endanger their survival, as well as the epigenetic events that not only contribute to their survival but may also facilitate the inheritance of stress adaptations.

24.2.1 RdDM and salt stress tolerance: Dynamic DNA methylation regulates MYB transcription factor in *Arabidopsis thaliana*

A study conducted by Xu *et al.* (2015) focused on a putative R2R3-MYB member in *Arabidopsis thaliana*, named AtMYB74, particularly its gene expression and DNA methylation under salt stress conditions. Additionally, the methylation status related to 24 nt small RNA (sRNA) was examined to explore the possibility of RdDM affecting the AtMYB74 promoter, aiming to elucidate the epigenetic modifications involved. AtMYB74 is consistently expressed at low levels across various tissues, with a specific localization within the nucleus. Exposure to salt stress induces the expression of AtMYB74.

Researchers observed that transgenic plants overexpressing AtMYB74 display hypersensitivity to salt stress during seed germination and a decreased survival rate. Additionally, analysis of the nucleotide sequence for RdDM prediction using starPRO DB by Baev *et al.* (2010, no longer available) identified five putative small interfering RNA (siRNA) candidates targeting the AtMYB74 promoter region. Transgenic lines overexpressing these siRNA candidates exhibited CHH *de novo* methylations in the AtMYB74 promoter regions, which strongly suggests the RdDM regulatory pathway.

In the absence of salt stress, plants accumulate siRNAs that induce CHH hypermethylation at the AtMYB74 promoter, leading to a decrease in AtMYB74 transcript levels. Conversely, under stress treatment, siRNA levels decrease, resulting in demethylation of the promoter region and an increase in AtMYB74 expression. Given that constitutive overexpression of AtMYB74 renders plants hypersensitive to salt stress, it is suggested that dynamic DNA methylation plays a crucial role in stress tolerance (Fig. 24.1).

24.2.2 Histone methylation and drought stress tolerance: SDG708's regulatory influence on ABA-related gene activation in rice

The investigation by Liu *et al.* (2016) was centered on the SET DOMAIN GROUP 708 (SDG708), recognized as a histone H3 lysine 36 (H3K36) methyltransferase, which plays a pivotal role in catalyzing the epigenetic response. Chen *et al.* (2021) identified that SDG708 is crucial for drought tolerance in rice, distinguishing its function from responses to cold or heat stress. It was revealed that SDG708 directly interacts with abscisic acid-related genes through H3K36 methylation, initiating a mechanism for drought stress tolerance. It was discovered that SDG708 increases the levels of chlorophyll, and stress biomarkers such as malondialdehyde and proline, along with the accumulation of reactive oxygen species (ROS) in guard cells, accompanied by changes in stomatal aperture. Given the role of ROS in modulating stomatal movements via abscisic acid (ABA) regulation (Qi *et al.*, 2018), the influence of SDG708 under drought stress and after ABA application was examined, demonstrating an increased expression in the ABA-related genes, *OsNCED3* and *OsNCED5*.

Evidence was presented that delineates the mechanism for drought tolerance in rice: SDG708 targets the *OsNCED3/5* DNA locus for H3K36 methylation, which in turn activates the transcription of ABA-related genes. This activation leads to the accumulation of ROS in guard cells, inducing stomatal closure to conserve water. This process signifies that SDG708, as a histone methyltransferase, not only responds to drought stress but also instigates a tolerance mechanism compared with other histone methyltransferases, highlighting SDG708's role in controlling mechanisms that reduce water loss (Fig. 24.2). The findings suggest that histone methylation may activate as well as inhibit transcription.

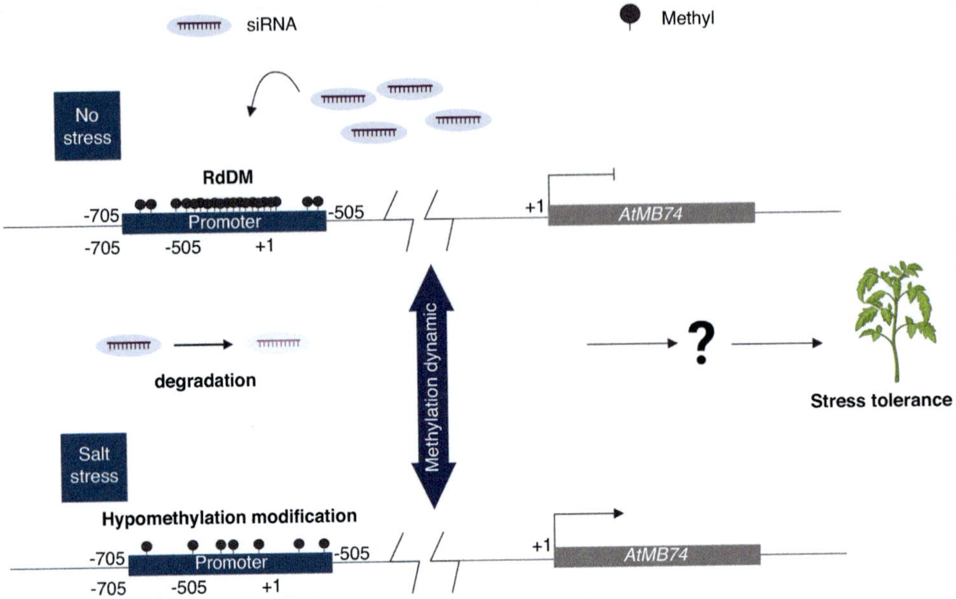

Fig. 24.1. RdDM and salt stress tolerance. A schematic representation of the accumulation of siRNA, its role in the reduction of methylation under salt stress, and the expression of AtMB74 under said stress are presented. The subsequent molecular and biochemical mechanisms are unknown but result in stress. The R2R3-MYB transcription factors, AtMYB2, AtMYB20, and AtMYB73, have been implicated in regulating gene expression under salt stress, and it is suggested that their roles may exhibit redundancy (Abe *et al*., 2003; Cui *et al*., 2013; Kim *et al*., 2013). Xu *et al*. (2015) shed light on the role of AtMYB74 in modulating transcriptional responses to salt stress, thus adding a new layer of complexity to our understanding of epigenetic regulation by RdDM. However, the molecular and biochemical mechanisms underlying salt stress tolerance remain to be fully elucidated. Created with BioRender.com. Author's own image.

24.2.3 Histone acetylation and cold stress tolerance: The role of HOS15 in regulating COR gene expression in *Arabidopsis* and *Cannabis sativa*

It was reported by Park *et al.* (2018) that cold tolerance in *Arabidopsis* emerges through gene de-repression, facilitated by chromatin modification. In their research, epigenetic mechanisms were demonstrated to enhance the expression of COLD RESPONSIVE (COR) genes, essential for cold tolerance. Absent stress, the WD40-repeat protein HOS15, which does not possess a DNA-binding domain, is assembled by C-REPEAT BINDING FACTORS (CBFs) along with HISTONE DEACETYLASE 2C (HD2C) to suppress COR gene expression via hypoacetylation of COR chromatin. In this scenario, CBF might perform a constitutive role in recruiting HOS15, serving as part of a repressor complex for COR genes under non-stress conditions, yet assumes a divergent role upon the accumulation of CBF under stress conditions.

Under conditions of cold stress, the mechanism through which cold tolerance is switched by HOS15 is found to require the regulation of COR genes. In this process, HOS15 ubiquitinates HD2C, signaling it for proteasomal degradation, and prompts histone H3 hyperacetylation, which eases chromatin access. Contrary to normal conditions, CBFs accumulate and interact with the promoter region, resulting in an increase in the expression of COR genes (Fig. 24.3). Moreover, Mayer *et al.* (2015) observed that *Cannabis sativa* plants undergoing cold acclimation exhibited an increase in COR genes and an accumulation of soluble sugar. Furthermore, during deacclimation, an increase in methylcytosine at COR gene loci was noted, which might represent a locus-specific DNA methylation

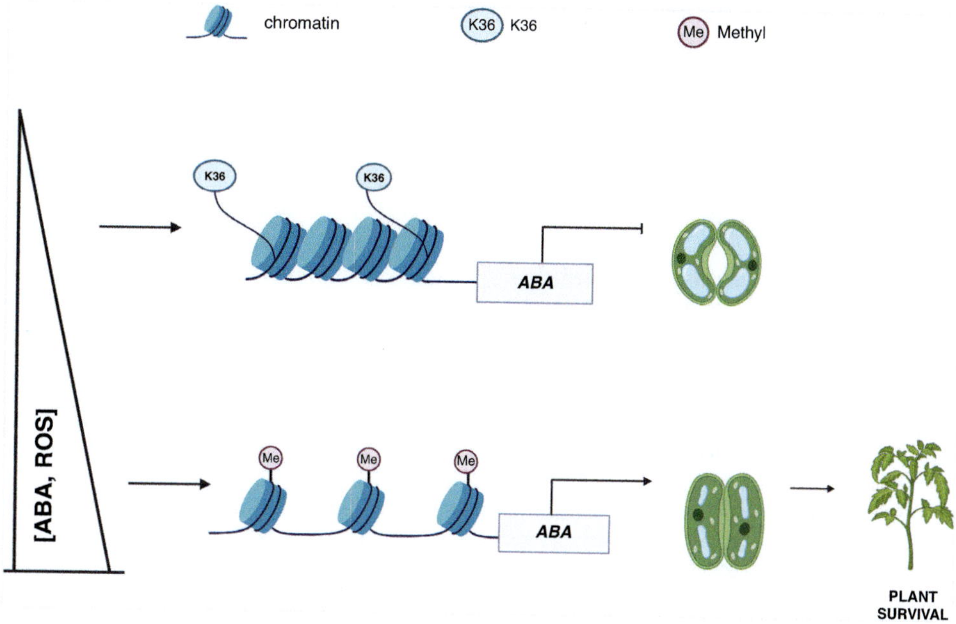

Fig. 24.2. Histone methylation and drought stress tolerance. A schematic representation of the chromatin array status under drought stress, where SDG708 induces its methylation, leading to the activation of genes associated with abscisic acid. The accumulation of ABA and ROS induces stomatal closure. Created with BioRender.com. Author's own image.

strategy following cold stress. No stress memory was reported.

24.2.4 Heat-stress memory: the role of generational stress memory and methylation status in *Arabidopsis*

The impact of sustained heat stress over 25 generations of *Arabidopsis thaliana* was investigated by Yadav *et al.* (2022). The lineage known as F25H, which had endured multigenerational heat stress, displayed enhanced heat tolerance. This study also revealed that *Arabidopsis* plants can develop a stress memory when exposed to repeated heat stress over 25 consecutive generations. A single *Arabidopsis thaliana* plant from the Col-0 line was utilized as the seed origin for the study, and across 25 generations the plants were either subjected to heat stress at 42°C or maintained in a controlled environment at 22°C. At the 25th generation, both the heat-stressed progeny (F25H) and the control group (F25C) underwent a 6 hour heat treatment at 42°C followed by a 10 day recovery period. This treatment resulted in a higher survival rate and greater fresh weight in the F25H group compared with the control. Genetic and epigenetic modifications were observed in the multigenerational heat-stressed progeny. The F25H heat-stressed progeny exhibited a higher somatic homologous recombination, while over three times more single nucleotide polymorphisms, insertions, and deletions were detected. It was predicted that these genetic variants were mostly located in the 5′ and 3′ untranslated regions. In the CG context, hypermethylation was predominantly observed in the gene body, with the order being F25H>F25C > F2C, and in all cases, exon >promoter > intron >intergenic regions. F25H plants displayed global hypomethylation in the CHH DNA methylation context. Additionally, in both stress and control groups, hyper- and hypomethylation in the CHH context were more pronounced in promoter >intergenic > exon > intron regions.

Furthermore, there was a 35.9-fold increase in the epigenetic/genetic ratio observed for F25H compared with F25C. It is unclear why Yadav *et al.* (2022) conducted their analysis specifically at F25 rather than at other generational stages. However, it was indicated that the epigenetic changes observed between the stressed progeny and the control were more substantial than the genetic modifications. These epigenetic modifications were associated with responses to abiotic and biotic stimuli, cellular organization and biogenesis, and were predominantly observed in processes related to DNA or RNA metabolism. Nonetheless, the functions of the genes affected and their roles in the mechanism of thermoresistance remain to be fully elucidated (Fig. 24.3).

In this section, cases in which plants epigenetically respond to stress resulting in survival have been discussed. Although a substantial amount of information on epigenetic phenomena is available, particularly for the plant *Arabidopsis thaliana*, it is noted that crop plants exhibit greater complexity, and their survival holds practical significance (Zhang *et al.*, 2018). For an enhanced understanding of assessing epigenetic changes, it is advised that both the combo protocols provided by Zhang *et al.* (2014) and Yamaguchi *et al.* (2014) be consulted.

24.3 Epigenetic Strategies in Plant Breeding

Plant breeding has existed since the beginning of agriculture, with the selection of desirable traits in crops. According to scientific and technological advances, the techniques and methods have progressed to the present. However, the exponential population growth and challenging climatic conditions demand the unlocking of new alternatives for optimizing plant characteristics (Springer and Schmitz, 2017; Shi *et al.*, 2024). There is no doubt that traditional breeding approaches will remain as a pillar in crop improvement strategies, but they can be enhanced with novel technologies to meet the

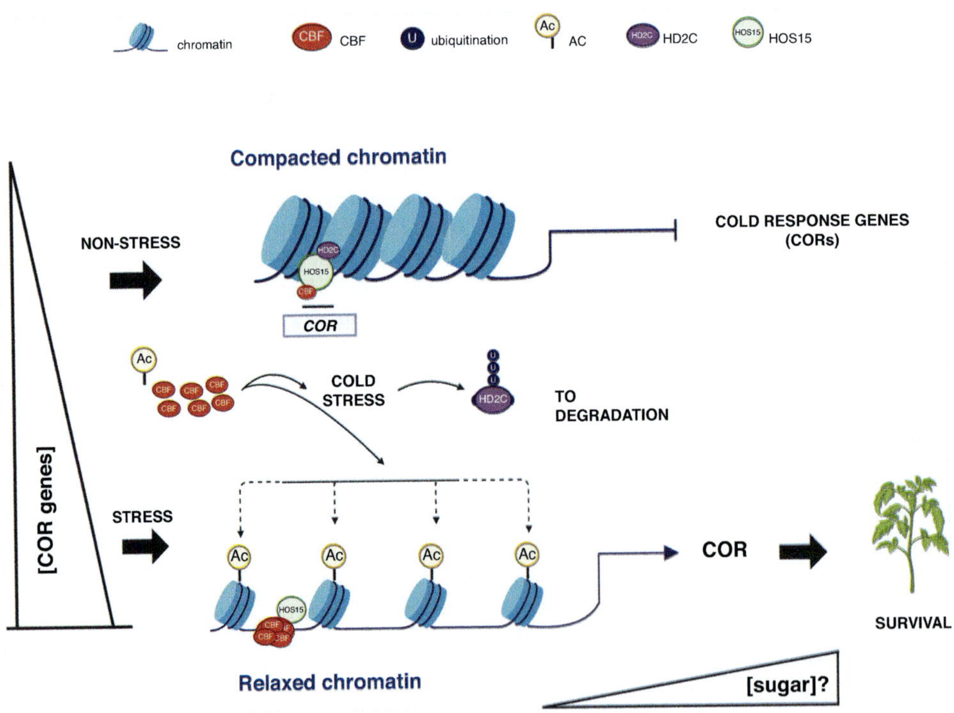

Fig. 24.3. Heat-stress memory. Heat-stressed progeny resulted in a higher somatic homologous recombination, nucleotide polymorphisms, insertions, and deletions and the epigenetic modifications were related to DNA or RNA metabolism. Created with BioRender.com. Author's own image.

demands of food security and sustainable agriculture (Crisp *et al.*, 2022). Among these, given the key role of epigenetic mechanisms in gene regulation and the phenotypic diversity that they can generate, the new perspectives focus on epigenetic variation and epigenome editing to enhance crop resilience, yield, and nutritional quality (Varotto *et al.*, 2020; Tonosaki *et al.*, 2022).

24.3.1 Epigenetic explorations and profiling

Considering that epigenetic states can be unstable concerning maintenance during the life cycle and the transfer of information throughout the progeny, it is necessary to accumulate knowledge on the role of epigenetic regulation in response to stress and identify specific epigenetic marks associated with desirable traits, where genotyping assays and sequencing are a key point to the marker-assisted selection for achieving epigenetic variation (Varotto *et al.*, 2020; Crisp *et al.*, 2022). Sequencing techniques at whole-genome base-pair resolution have been the gold standard for profiling and there are multiple specific techniques for profiling the epigenetic landscape and exploring the key regulatory elements and epialleles that influence important agronomic traits. There are several techniques for characterization and analysis with different degrees of depth and specificity in the genome that are used in studies for the identification of epialleles. Targeting DNA methylation, there could be mentioned whole-genome bisulfite sequencing (WGBS), reduced representation bisulfite sequencing (RRBS), methylated DNA immunoprecipitation sequencing (MeDIP-seq), methylation-sensitive amplification polymorphism (MSAP), and bisulfite PCR and sequencing. WGBS consists of a high-resolution single-based mapping of DNA methylation patterns by the treatment of DNA with sodium bisulfite, which converts unmethylated cytosines to uracil while leaving methylated cytosines unchanged. MeDIP-seq uses antibodies to precipitate DNA fragments with methylated cytocines. RRBS uses restriction enzymes to digest the DNA and select preferential methylation-sensitive fragments for sequencing. Similarly, MSAP involves the digestion of DNA with restriction enzymes but uses specific PCR amplification to detect DNA methylation changes at specific loci. In another way, bisulfite PCR and sequencing approaches utilize bisulfite treatment, and PCR for the amplification of target genes or regulatory regions (Baubec and Akalin, 2016; Kapazoglou *et al.*, 2018; Wang *et al.*, 2022).

Histone modifications are accessed through chromatin immunoprecipitation (ChIP) techniques, which use an antibody specific to the protein of interest to precipitate the protein–DNA assembly and enrich the DNA fragments associated with the protein. Targeting specific regions by PCR (ChIP-PCR) or by direct genome-wide sequencing (ChIP-seq) can be approached. Another technology is cleavage under targets and segmentation (CUT&Tag), by the use of an A-Tn5 transposase fusion protein to characterize various histone modifications in small samples and single cells (Baubec and Akalin, 2016; Kapazoglou *et al.*, 2018; Wang *et al.*, 2022). For small RNA studies, there is small RNA sequencing (sRNA-seq), which is utilized for profiling and quantifying small RNA molecules, including microRNAs (miRNAs) and small interfering RNAs (siRNAs) by sequencing through the genome.

In addition to these profiling techniques, advanced epigenomic analysis methods such as epigenome-wide association studies (epi-GWAS), epi-genotype by sequencing (epiGBS), epigenetic recombinant inbred line (epiRIL) populations, and epigenetic quantitative trait loci (epiQTL) mapping have emerged as powerful tools for dissecting the genetic basis of epigenetic variation and its phenotypic consequences in plants. Also, methods such as the Plant MutS homolog1 (MSH1) are used to exploit epigenetic variations in epigenetic-assisted breeding to develop superior crops. These integrated approaches hold great promise for uncovering epigenetic regulators associated with the control of phenotypic variability to facilitate the selection of epigenetic markers for targeting the development of epigenetically tailored crop varieties with improved performance and resilience (Gogolev *et al.*, 2021; Chandana *et al.*, 2022).

24.3.2 Chemical treatments

Among the epigenetic mechanisms, DNA methylation epialleles occupy a unique and useful position, by and large exhibiting highly stable Mendelian inheritance (Crisp *et al.*, 2022). However, engineering DNA methylation states is impractical, due to the unlimited number of possibilities in crop genomes, but one possibility is the widespread random perturbation of DNA methylomes through chemical treatment (Springer and Schmitz, 2017).

For DNA methylation approaches, the more frequent compounds are 5-AzaC (5-azacytidine, 5-aza-2′-deoxy-cytidine) and zebularine, which are non-methylate cytidine analogs that integrated into the DNA and disrupt the function of the DNA methyltransferases during the replication, reducing DNA methylation levels at genome-wide manner inducing a hypomethylation state (Shaikh *et al.*, 2022; Kumar *et al.*, 2024). Another variant is polymerase inhibitors such as α-amanitin and actinomycin D, which reduce the POL II activity by disrupting the interaction with the enzyme and active site, leading to DNA hypomethylation and, thus, variations in phenotype (Kumar *et al.*, 2024). Histone modification is also targeted with chemicals. In the first group, the histone deacetylation inhibitors such as trichostatin A, nicotinamide, diallyl disulfide, and HC toxin, bind to the histone deacetylase (HDAC) in the catalytic pocket, blocking their activity. Also in this group, sodium butyrate triggers acetylation by activating rRNA gene promoters (Kumar *et al.*, 2024). Other groups are the histone methylation inhibitors, such as sulfamethazine and 3-deazaneplanocin A (DZNep), which impede the conversion of SAH into adenosine and homocysteine, thus reducing histone methylation (Kumar *et al.*, 2024). Although chemical treatments are valuable sources of epigenetic variation, they have some weaknesses as non-specific targeting which leads to a broad effect due to the entire genome being influenced, such that it may lead to undesired phenotypes such as developmental abnormality (Sampson *et al.*, 2024). Also, there is a low survival rate of plants exposed to this type of treatment, which limits their application in breeding (Shaikh *et al.*, 2022).

24.3.3 Epigenome editing

Recently, with the boom of genomic tools, there are epigenome engineering techniques that allow manipulation and editing in a fine-tuning manner due to site-specific targeting by the fusion of enzymes of the epigenetic machinery with proteins that are guided to specific DNA sequences. These approaches include: the clustered regularly interspaced short palindromic repeats (CRISPR) and CRISPR-associated (Cas) system (CRISPR/Cas); transcription activator-like effector (TALE) system; and zinc finger protein (ZNF) systems (Springer and Schmitz, 2017; Varotto *et al.*, 2020; Sampson *et al.*, 2024), as is schematized in Fig. 24.4.

Another important consideration is that this type of technology involves the introduction of a transgene into the plant to direct the epigenome modification. This implies that its use must be under GMO regulations, and therefore its use in open-field plant breeding is problematic (Tonosaki *et al.*, 2022). However, these techniques provide a promising tool for genome editing in a site-specific manner, allowing the creation of desirable germplasm in different crops and manipulating key genes responsible for specific agronomic traits – such that some countries, such as China, the United States, Japan, Brazil, Argentina, Israel, Canada, Australia, and the United Kingdom, are loosening regulatory policies on gene editing crops with precision techniques (Tang *et al.*, 2023).

24.3.4 CRISPR/Cas9 systems

CRISPR/Cas systems are adaptative antiviral immunity systems found in bacteria and archaea, which are classified into several types and subtypes, but the most frequently used in plant genome editing is a variation of the type CRISPR/Cas system of *Streptococcus pyogenes* (El-Mounadi *et al.*, 2020). CRISPR/Cas technology consists of three components: the single guide RNA transcript (sgRNA) that is direct to the target DNA sequence, the Cas9 protein

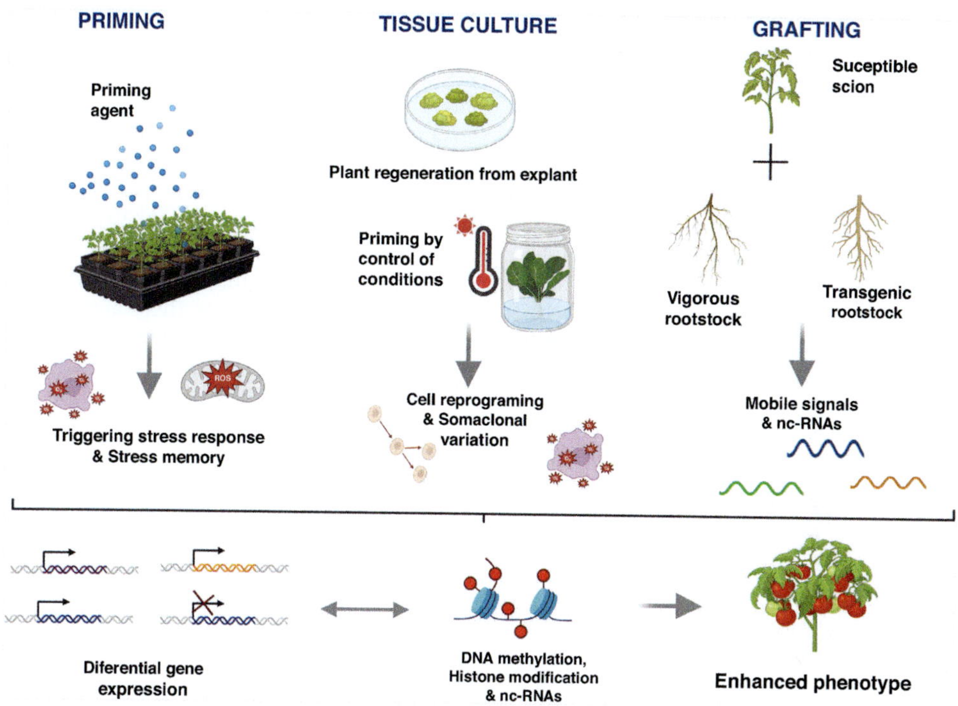

Fig. 24.4. Epigenome editing strategies. Once a specific gene of the genome segment is targeted, there are several platforms for specific epigenome editing. The most used are CRISPR/Cas technology, the ZF and TALE systems, and RNAi technology. The first two differ in the structure of the system on which they are based, but all involve fusion with an effector protein. In this case, the effector protein consists of a DNA methylase or demethylase (DNMT and DMT) or histone modifiers such as acetylase (HAT), deacetylase (HDAC), methylase (HMT), or demethylase (HDMT). Created using Biorender. Author's own image.

with DNA endonuclease activity, and a protospacer adjacent motif (PAM) that is required for binding to the target (Hussain *et al.*, 2018). CRISPR/Cas9 technology creates double-strand breaks (DSBs) at specific locations in the DNA, which can then be repaired by mechanisms such as non-homologous end joining (NHEJ) or homology-directed repair (HDR), allowing insertions, deletions, or substitutions at the target site, leading to genome editing. Thus CRISPR technology has multiple functions as gene knockout, gene deletion, base editing, gene insertion or replacement, targeted random mutagenesis, epigenetic modulation, transcriptional regulation, and RNA editing (El-Mounadi *et al.*, 2020; Tang *et al.*, 2023).

There are several types and variations of the CRISPR/Cas systems developed to enhance and improve its performance in epigenome editing. A popular variant used for transcriptional regulation is the endonuclease-deactivated Cas9, also known as dead Cas9 (dCas9). This has a point mutation in the nuclease domain that inactivates nuclease activity but retains the binding activity, so that dCas9 is fused with other effector protein domains as epigenetic modifiers for the epigenetic edition without altering the underlying DNA sequence (Xu and Li, 2020). Another variant is the SunTag system, which involves the use of tandem GCN4 peptide repeats fused to the dCas9 and a single-chain variable fragment (scFv) GCN4 antibody fused to multiple effector molecules. The SunTag system can be used to activate targeted multiple loci like transposable elements and protein-coding genes (Brezgin *et al.*, 2019; Jogam *et al.*, 2022).

The Scaffold technique utilizes RNA-binding motifs and synthetic RNA scaffolds containing aptamer-specific protein fused to transcriptional activators, enabling transcriptional regulation both by transcriptional domains linked to dCas9 and domains recruited by gRNA aptamers (Brezgin *et al.*, 2019; Zhang *et al.*, 2019).

Epigenetic effectors used commonly in plants with CRISPR/Cas systems include the RdDM methyltransferase DRM2 catalytic domain (DNMT), which adds DNA methylation at the target DNA sequence. Also, the demethylase TET1 is used as an effector to remove methyl residues from the DNA target (Brezgin *et al.*, 2019; Jogam *et al.*, 2022). For acetylation of histones there is the p900 histone acetyltransferase fused with dCas9 to favor DNA accessibility and transcription activation, and for histone methylation the lysine-specific demethylase 1 (LSD1) and Jumonji C (JmjC) can be fused with sCas9 to remove methyl groups from the target site, leading to the repression of the targeted gene (Jogam *et al.*, 2022).

In plant cells, the CRISPR/Cas system can be introduced through conventional plant transformation techniques such as biobalistic, nanoparticles, *Agrobacterium*-mediated system, and viral vectors such as the tobacco rattle virus (TRV), as well as plasmids dsDNA, RNA, and RNPs through PEG-mediated transformation (Hussain *et al.*, 2018; Zhang *et al.*, 2019). However, the genetic transformation process limits the widespread application of CRISPR/Cas technology in different crop, due to the low efficiency and high genotype depending on the transformation methods (Tang *et al.*, 2023).

24.3.5 ZF- and TALE-based systems

Other engineered genome editing platforms are zinc finger (ZF) and transcription activator-like effector (TALE)-based systems. Both of them possess DNA-binding domains with programmable specificity fused with different catalytic or effector domains such as epigenetic modifiers for DNA methylation (DNMT) or histone modification (HAT, HDAC, HMT, HDM), and both systems can incorporate domains that regulate gene expression and nucleases for targeted gene editing. TALE proteins are derived from *Xanthomonas*; their tandem repeat segments allow the single-base recognition of DNA targets. By contrast, ZFs are present in transcriptional factors in eukaryotes; their characteristic finger-like structures involved in their DNA binding allow the recognition of 3 base pair segments (Sanjana *et al.*, 2012; Jankele and Svoboda, 2014; Cano-Rodriguez and Rots, 2016). Both systems are potential tools for epigenetic engineering in crop improvement, but their complex production limits their application compared with the CRISPR/Cas system (Gogolev *et al.*, 2021).

24.3.6 Transgenic RNA Interference (RNAi) Techniques

RNAi techniques aim for gene expression regulation through post-transcriptional modifications by the presence of small RNA molecules, primarily to protect plants from pathogens (Chandana *et al.*, 2022). Recently, the challenges for GMO adoption have led to the exogenous use of RNA molecules such as double-stranded RNAs (dsRNAs), short (or small) interfering RNAs (siRNAs), or hairpin RNAs (hpRNAs), which are capable of triggering RNAi. Generally, the RNA is processed by Dicer-like endonucleases (DCLs) into siRNAs and loaded onto AGO1, which leads to the formation of secondary siRNAs that recognize cognate DNA and recruit PolV to form RdDM. It also can stimulate histone modifications for transcriptional gene silencing and heterochromatin formation (Das and Sherif, 2020; Koeppe *et al.*, 2023; Kumar *et al.*, 2024).

A crucial step in the use of RNAi machinery is the application methods as well as the sensitivity of plant organs to the RNA molecules applications. The exogenous methods are approaches to developing eco-friendly techniques for plant protection and crop improvement. Among these, exogenous applications include spraying, infiltration, injection, spreading, mechanical inoculation, and root/seed soaking (Das and Sherif, 2020; Koeppe *et al.*, 2023; Kumar *et al.*, 2024). The major limitation is the short-term stability of the naked RNA, such that to improve the stability peptide complex-based delivery systems are used: layered double hydroxide (LDH) clay nanosheets-based delivery, and cationic fluorescent nanoparticles. However, considering the expensive cost of nanotechnology, the application of naked-dsRNAs is efficient to induce plant RNAi for the desired traits in crops (Das and Sherif, 2020; Koeppe *et al.*, 2023).

24.3.7 Induced natural epigenetic variation

As mentioned above, epigenetic mechanisms are present for physiologic, biochemical, and stress response processes during the plant life cycle, including memory mechanisms by the plant's adaptation to abiotic and biotic stresses. The possibility to trigger and mediate epigenetic remodeling in plants can increase phenotypic diversity in crops, offering a promising avenue to enhance breeding efficiency and potentially improve crop yields. For the induced natural epigenetic variation methods such as tissue culture, imprinting by stress treatments, and grafting are used as nuanced strategies to modulate epigenetic mechanisms and shape the phenotypic landscape of crops (Fig. 24.5).

24.3.8 Tissue culture

Tissue culture is a technique for micropropagation and propagation by vegetative

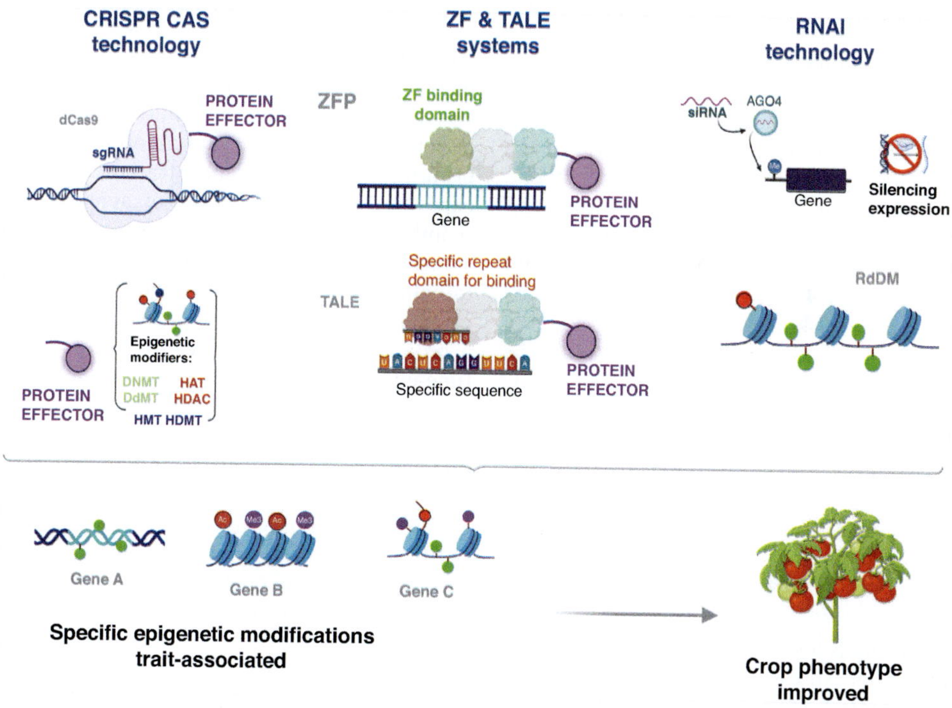

Fig. 24.5. Strategies for natural epigenetic induced variation for plant breeding. The principal strategies are priming, tissue culture, and grafting. In priming, seedlings are exposed to an agent (compound of mild stress) that triggers stress response and metabolism changes. In tissue culture, the regeneration of a plant from the explant is highly stressful and implies cell reprogramming. It also can include an alteration of the overgrowth conditions for a major or specific stimulus. In grafting, the scion is enhanced by the vigor of the rootstock through the transport of signal molecules as nc-RNAs. In all three, the final objective is differential gene expression by change in epigenetic modifications, leading to a plant with enhanced phenotype and epigenetic variability to endure climatic challenges. Created using Biorender. Author's own image.

reproduction. Here, a fragment of tissue from the mother plant known as an explant grows in a nutrient medium until the plant regenerates. It is considered that these techniques produce clone-identical individuals, but they also can create variants at a relatively high rate, named somaclonal variation (Alfalahi *et al.*, 2022; Tonosaki *et al.*, 2022). Tissue culture has been reported to induce genetic and epigenetic changes due to the process itself being highly stressful for the cells. Among the epigenetic changes, the most common is the hypomethylation on promoter associated with active gene expression (Chandana *et al.*, 2022; Tonosaki *et al.*, 2022), as well as histone alteration at methylation and acetylation and an increase in siRNAs. The interplay of these elements is critical for the restructuring of the plant and induces epiallelic variants that are transmitted to the plant and fixed in the genome. Factors contributing to the accuracy of the process and the feasibility of reprogramming epigenetics marks are the cell type used, and the developmental and physical age of the explants (Smulders and Klerk, 2011). Another factor implied in tissue culture and with great potential is imprinting through the manipulation of culture parameters and precise control of overgrowth conditions and hormonal stimuli, which facilitates the generation of phenotypic diversity and the screening and selection of desirable epigenetic variants with enhanced agronomic traits (Springer and Schmitz, 2017; Wijerathna-Yapa and Hiti-Bandaralage, 2023).

24.3.9 Genome imprinting

Imprinting is a phenomenon where there is a differential expression of alleles depending on the parent-of-origin-specific manner, which involves epigenetic mechanisms (Springer and Schmitz, 2017). The imprinting is achieved through priming, meaning the exposure to stimuli of stressors, which triggers dynamic changes in stress responses and, consequently, alterations in gene expression and epigenetic marks. These changes can enable the plant to retain information about the challenges to respond against stress events in the future in a better, more efficient, and faster way, to prevail and even be heritable over generations, known as stress memory. This is enhanced by the presence of sRNAs as the main regulators of plant hormones, reactive oxygen species (ROS), and transcriptional factors; as well as the establishment of other epigenetic marks as DNA methylation and histone modification for their prolonged continuance (Crisp *et al.*, 2016; He and Li, 2018; Chandana *et al.*, 2022). The priming agent can be mild abiotic stresses (temperature, drought, salinity, etc.), biotic stresses (pathogen infection or beneficial microbe colonization), or chemical treatments, which include natural compounds, volatile organic chemicals, salts, synthetic stimulants, plant hormones, vitamins and antioxidants, and even signaling molecules as sRNAs (Martinez-Medina *et al.*, 2016; Turgut-Kara *et al.*, 2020). The priming effect depends on factors such as the plant's physiological stage in which it is stimulated (seeds, seedlings, adult plant in vegetative or reproductive phase), and according to the priming agent used, concentration, time of exposure, doses, and periods of applications (Lämke and Bäurle, 2017). Priming offers several advantages for crop breeding as a sustainable alternative to potentiate the innate capability of adaptation of plants to generate a broad-spectrum resistance, as a cost- and time-effective strategy. However, the inheritance of priming-induced epigenetic changes is variable, and their stability and heritability on phenotypic effects across generations is not fully understood yet. More research is needed to integrate priming into breeding programs with the potential to rapidly enhance crop yields in the face of environmental challenges (Villagómez-Aranda *et al.*, 2022).

24.3.10 Grafting

Plant grafting is a technique commonly used to increase production and reduce losses due to environmental stress, by grafting scions (usually perennial horticulture species) with vigorous rootstock, especially from different species (heterografting), to reduce the stress in the shoot (Fuentes-Merlos *et al.*, 2023).

Grafting induces phenotypic variants in plants by the changes in several plant characteristics associated with plant performance through the trafficking of information between the scion and the rootstock and vice versa, which are critical for the growth, interaction, and adaptation of grafted plants (Crisp *et al.*, 2022). The mobile signals are transported by non-coding (nc) RNAs (siRNAs and miRNAs) across graft segments, which directly impacts the gene expression of plants. Also, these lead to other epigenetic alterations such as changes in DNA methylation and transposon silencing via the RdDM (Kapazoglou *et al.*, 2018; Tsaballa *et al.*, 2021), and some histone modifications such as acetylation and methylation. All of these mechanisms induce an acquired stress memory (Fuentes-Merlos *et al.*, 2023), where some changes can be retained for five generations after grafting (Tsaballa *et al.*, 2021). Recently, with the boom of the genomic era, a new method of grafting has arisen: trans-grafting, which combines the traditional grafting practice with transgenic plants. Here a naïve scion is combined with a genetically modified rootstock, such that the scion acquires benefits and traits from the transgenes in the rootstock without the insertion of the transgene, hence considering the end products as non-genetically modified, and consequently without biosafety concerns (Lusser *et al.*, 2012; Limera *et al.*, 2017; Kapazoglou *et al.*, 2021). However, the benefits of grafting are neglected due to the high production costs, high labor intensity, the need for skilled staff, and the narrow time frame for planting (Tsaballa *et al.*, 2021).

24.4 Concluding Remarks and Perspectives

Based on the above-mentioned in this chapter, it is clear that, although the basic knowledge and application of plant epigenetic mechanisms to cope with stress have shown a significant advance in recent years, there is still a wide range of possibilities to be tested in order to better understand the phenomenon and thus use this knowledge to cope with plant stress physiology and breeding strategies in the current climate change scenario. Future studies using omics and machine learning approaches trying to find relationships among gene expression and epigenetic marks, as well as the stability of their inheritance during several generations, must be carried out (Rico-Chávez *et al.*, 2022). Interestingly, approaches involving controlled elicitation using different types of eustressors (i.e. positive stressors) under a hormetic scheme should be evaluated regarding 'stress memory' induction and stability in different plant species (Vázquez-Hernández *et al.*, 2019; Godínez-Mendoza *et al.*, 2023). Molecular tools based on epigenetic changes, such as those related to dead Cas9 proteins used in a CRISPR coupled with DNA methylases (Pavithra *et al.*, 2024), or gene silencing strategies using artificial double-strand small non-coding RNAs (Cedillo-Jimenez *et al.*, 2024), among others, must also be explored in order to become an actual technology that can be used in crop protection from stress, either for research or for industrial purposes.

Acknowledgments

The authors thank FONFIVE 2024 and FOPER 2023 from Universidad Autónoma de Queretaro (UAQ), for funding research relevant to some of the information discussed in this chapter.

References

Abe, H., Urao, T., Ito, T., Seki, M., Shinozaki, K. *et al.* (2003) *Arabidopsis* AtMYC2 (BHLH) and AtMYB2 (MYB) function as transcriptional activators in abscisic acid signaling. *The Plant Cell* 15(1), 63–78. DOI: 10.1105/tpc.006130.

Alfalahi, A.O., Hussein, Z.T., Khalofah, A., Sadder, M.T., Qasem, J.R. *et al.* (2022) Epigenetic variation as A new plant breeding tool: A review. *Journal of King Saud University – Science* 34(8), 102302. DOI: 10.1016/j.jksus.2022.102302.

Baev, V., Naydenov, M., Apostolova, E., Ivanova, D., Doncheva, S. *et al.* (2010) Identification of RNA-dependent DNA-methylation regulated promoters in *Arabidopsis*. *Plant Physiology and Biochemistry* 48(6), 393–400. DOI: 10.1016/j.plaphy.2010.03.013.

Baubec, T. and Akalin, A. (2016) Genome-wide analysis of DNA methylation patterns by high-throughput sequencing. In: Aransay, A.M. and Trueba, J.L.L. (eds) *Field Guidelines for Genetic Experimental Designs in High-Throughput Sequencing*. Springer, Cham, pp. 197–221. DOI: 10.1007/978-3-319-31350-4.

Brezgin, S., Kostyusheva, A., Kostyushev, D. and Chulanov, V. (2019) Dead cas systems: Types, principles, and applications. *International Journal of Molecular Sciences* 20(23), 1–26. DOI: 10.3390/ijms20236041.

Cano-Rodriguez, D. and Rots, M.G. (2016) Epigenetic editing: On the verge of reprogramming gene expression at will. *Current Genetic Medicine Reports* 4(4), 170–179. DOI: 10.1007/s40142-016-0104-3.

Cao, H., Ding, R., Kang, S., Du, T., Tong, L. *et al.* (2023) Drought, salt, and combined stresses in plants: Effects, tolerance mechanisms, and strategies. *Advances in Agronomy* 178, 107–163.

Cardenas-Manríquez, G., Vega-Muñoz, I., Villagómez-Aranda, A.L., León-Galvan, M.F., Cruz-Hernandez, A. *et al.* (2016) Proteomic and metabolomic profiles in transgenic tobacco (N. *tabacum xanthi* nc) to CchGLP from *Capsicum chinense* BG-3821 resistant to biotic and abiotic stresses. *Environmental and Experimental Botany* 130, 33–41.

Cedillo-Jimenez, C.A., Guevara-Gonzalez, R.G. and Cruz-Hernandez, A. (2024) Exogenous dsRNA sequence based on miR1917 downregulates its target gene related to ethylene signaling in tomato seedlings and fruit. *Scientia Horticulturae* 331, 113090.

Chandana, B.S., Mahto, R.K., Singh, R.K., Ford, R., Vaghefi, N. *et al.* (2022) Epigenomics as potential tools for enhancing magnitude of breeding approaches for developing climate resilient chickpea. *Frontiers in Genetics* 13, 1–25. DOI: 10.3389/fgene.2022.900253.

Chen, K., Du, K., Shi, Y., Yin, L., Shen, W.H. *et al.* (2021) H3K36 methyltransferase SDG708 enhances drought tolerance by promoting abscisic acid biosynthesis in rice. *New Phytologist* 230(5), 1967–1984. DOI: 10.1111/nph.17290.

Crisp, P.A., Ganguly, D., Eichten, S.R., Borevitz, J.O. and Pogson, B.J. (2016) Reconsidering plant memory: Intersections between stress recovery, RNA turnover, and epigenetics. *Science Advances* 2(2), e1501340. DOI: 10.1126/sciadv.1501340.

Crisp, P.A., Bhatnagar-Mathur, P., Hundleby, P., Godwin, I.D., Waterhouse, P.M. *et al.* (2022) Beyond the gene: Epigenetic and cis-regulatory targets offer new breeding potential for the future. *Current Opinion in Biotechnology* 73, 88–94. DOI: 10.1016/j.copbio.2021.07.008.

Cui, M.H., Yoo, K.S., Hyoung, S., Nguyen, H.T., Kim, Y.Y. *et al.* (2013) An *Arabidopsis* R2R3-MYB transcription factor, AtMYB20, negatively regulates type 2C serine/threonine protein phosphatases to enhance salt tolerance. *FEBS Letters* 587, 1773–1778. DOI: 10.1016/j.febslet.2013.04.028.

Das, P.R. and Sherif, S.M. (2020) Application of exogenous dsRNAs-induced RNAi in agriculture: Challenges and triumphs. *Frontiers in Plant Science* 11. DOI: 10.3389/fpls.2020.00946.

El-Mounadi, K., Morales-Floriano, M.L. and Garcia-Ruiz, H. (2020) Principles, applications, and biosafety of plant genome editing using CRISPR-Cas9. *Frontiers in Plant Science* 11, 1–16. DOI: 10.3389/fpls.2020.00056.

Fuentes-Merlos, M.I., Bamba, M., Sato, S. and Higashitani, A. (2023) Self-grafting-induced epigenetic changes leading to drought stress tolerance in tomato plants. *DNA Research* 30(4), 1–10. DOI: 10.1093/dnares/dsad016.

Godínez-Mendoza, P.L., Rico-Chávez, A.K., Ferrusquía-Jimenez, N.I., Carbajal-Valenzuela, I.A., Villagómez-Aranda, A.L. *et al.* (2023) Plant hormesis: Revising of the concepts of biostimulation, elicitation and their application in a sustainable agricultural production. *Science of the Total Environment* 894, 164883.

Gogolev, Y.V., Ahmar, S., Akpinar, B.A., Budak, H., Kiryushkin, A.S. *et al.* (2021) Omics, epigenetics, and genome editing techniques for food and nutritional security. *Plants* 10(7), 1–44. DOI: 10.3390/plants10071423.

Guevara-Olvera, L., Ruíz-Nito, M.L., Rangel-Cano, R.M., Torres-Pacheco, I., Rivera-Bustamante, R.F. *et al.* (2012) Expression of a germin-like protein gene (CchGLP) from a geminivirus-resistant pepper

(*Capsicum chinense* jacq.) enhances tolerance to geminivirus infection in transgenic tobacco. *Physiological and Molecular Plant Pathology* 78, 45–50.

He, Y. and Li, Z. (2018) Epigenetic environmental memories in plants: Establishment, maintenance, and reprogramming. *Trends in Genetics* 34(11), 856–866. DOI: 10.1016/j.tig.2018.07.006.

Huang, S., Zhang, A., Jin, J.B., Zhao, B., Wang, T.J. *et al.* (2019) *Arabidopsis* histone H3K4 demethylase jmj17 functions in dehydration stress response. *New Phytologist* 223, 1372–1387.

Hussain, B., Lucas, S.J. and Budak, H. (2018) CRISPR/Cas9 in plants: At play in the genome and at work for crop improvement. *Briefings in Functional Genomics* 17, 319–328. DOI: 10.1093/bfgp/ely016.

Jankele, R. and Svoboda, P. (2014) TAL effectors: Tools for DNATargeting. *Briefings in Functional Genomics* 13(5), 409–419. DOI: 10.1093/bfgp/elu013.

Jogam, P., Sandhya, D., Alok, A., Peddaboina, V., Rao, V. *et al.* (2022) A review on CRISPR/Cas-based epigenetic regulation in plants. *International Journal of Biological Macromolecules* 219, 1261–1271. DOI: 10.1016/j.ijbiomac.2022.08.182.

Kapazoglou, A., Ganopoulos, I., Tani, E. and Tsaftaris, A. (2018) Epigenetics, epigenomics and crop improvement. *Advances in Botanical Research* 86, 287–324. DOI: 10.1016/bs.abr.2017.11.007.

Kapazoglou, A., Tani, E., Avramidou, E.V., Abraham, E.M., Gerakari, M. *et al.* (2021) Epigenetic changes and transcriptional reprogramming upon woody plant grafting for crop sustainability in a changing environment. *Frontiers in Plant Science* 11, 1–16. DOI: 10.3389/fpls.2020.613004.

Kim, J.H., Nguyen, N.H., Jeong, C.Y., Nguyen, N.T., Hong, S.W. *et al.* (2013) Loss of the R2R3 MYB, atmyb73, causes hyper-induction of the SOS1 and SOS3 genes in response to high salinity in *Arabidopsis*. *Journal of Plant Physiology* 170, 1461–1465. DOI: 10.1016/j.jplph.2013.05.011.

Koeppe, S., Kawchuk, L. and Kalischuk, M. (2023) RNA interference past and future applications in plants. *International Journal of Molecular Sciences* 24(11), 9755. DOI: 10.3390/ijms24119755.

Kumar, M. and Rani, K. (2023) Epigenomics in stress tolerance of plants under the climate change. *Molecular Biology Reports* 50(7), 6201–6216. DOI: 10.1007/s11033-023-08539-6.

Kumar, V., Ahmed, S., Kumar, D. and Gahlaut, V. (2024) Manipulating epigenetic diversity in crop plants: Techniques, challenges and opportunities. *(BBA) – General Subjects* 1868(2), 130544. DOI: 10.1016/j.bbagen.2023.130544.

Lämke, J. and Bäurle, I. (2017) Epigenetic and chromatin-based mechanisms in environmental stress adaptation and stress memory in plants. *Genome Biology* 18(1), 1–11. DOI: 10.1186/s13059-017-1263-6.

Limera, C., Sabbadini, S., Sweet, J.B. and Mezzetti, B. (2017) New biotechnological tools for the genetic improvement of major woody fruit species. *Frontiers in Plant Science* 8, 1–16. DOI: 10.3389/fpls.2017.01418.

Liu, B., Wei, G., Shi, J., Jin, J., Shen, T. *et al.* (2016) SET DOMAIN GROUP 708, a histone H3 lysine 36-specific methyltransferase, controls flowering time in rice (*Oryza sativa*). *New Phytologist* 210(2), 577–588. DOI: 10.1111/nph.13768.

Lusser, M., Parisi, C., Plan, D. and Rodríguez-Cerezo, E. (2012) Deployment of new biotechnologies in plant breeding. *Nature Biotechnology* 30(3), 231–239. DOI: 10.1038/nbt.2142.

Marín-de la Rosa, N., Lin, C.W., Kang, Y.J., Dhondt, S., Gonzalez, N. *et al.* (2019) Drought resistance is mediated by divergent strategies in closely related Brassicaceae. *New Phytologist* 223, 783–797.

Martinez-Medina, A., Flors, V., Heil, M., Mauch-Mani, B., Pieterse, C.M.J. *et al.* (2016) Recognizing plant defense priming. *Trends in Plant Science* 21(10), 818–822. DOI: 10.1016/j.tplants.2016.07.009.

Mayer, B.F., Ali-Benali, M.A., Demone, J., Bertrand, A. and Charron, J.-B. (2015) Cold acclimation induces distinctive changes in the chromatin state and transcript levels of COR genes in *Cannabis sativa* varieties with contrasting cold acclimation capacities. *Physiologia Plantarum* 155(3), 281–295. DOI: 10.1111/ppl.12318.

Park, J., Lim, C.J., Shen, M., Park, H.J., Cha, J.Y. *et al.* (2018) Epigenetic switch from repressive to permissive chromatin in response to cold stress. *Proceedings of the National Academy of Sciences of the United States of America* 115, 23. DOI: 10.1073/pnas.1721241115.

Pavithra, A., Sashtika, C., Anand, A.V., Kalaiselvi, S., Krishnapriya, S. *et al.* (2024) Characterization of CRISPR-associated protein for epigenetic manipulation in plants. In: Chen, J.-T. (ed.) *CRISPR and Plant Functional Genomics*. CRC Press, pp. 223–238.

Qi, J., Song, C.P., Wang, B., Zhou, J., Kangasjärvi, J. *et al.* (2018) Reactive oxygen species signaling and stomatal movement in plant responses to drought stress and pathogen attack. *Journal of Integrative Plant Biology* 60(9), 805–826. DOI: 10.1111/jipb.12654.

Rico-Chávez, A.K., Franco, J.A., Fernandez-Jaramillo, A.A., Contreras-Medina, L.M., Guevara-González, R.G. *et al.* (2022) Machine learning for plant stress modeling: A perspective towards hormesis management. *Plants* 11(7), 970.

Sáenz-de la O, D., Cedillo-Jimenez, C.A., García-Ortega, L.F., Martínez-Reséndiz, M., Arné-Robles, D. *et al.* (2020) Response of transgenic tobacco overexpressing the Cchglp gene to cadmium and aluminium: Phenotypic and microRNAs expression changes. *Physiology and Molecular Biology of Plants* 26, 3–13.

Sáenz-de la O, D., Morales, L.O., Strid, Å., Torres-Pacheco, I. and Guevara-González, R.G. (2021) Ultraviolet-B exposure and exogenous hydrogen peroxide application lead to cross-tolerance toward drought in *Nicotiana tabacum* L. *Physiologia Plantarum* 173(3), 666–679.

Sáenz-de la O, D., Morales, L.O., Strid, Å., Feregrino-Perez, A.A., Torres-Pacheco, I. *et al.* (2023) Antioxidant and drought-acclimation responses in UV-B-exposed transgenic *Nicotiana tabacum* displaying constitutive overproduction of H2O2. *Photochemical & Photobiological Sciences* 22(10), 2373–2387.

Sampson, C., Holyword, T., Innocent, I., Okagu, U. and Inuwa, I. (2024) Epigenetics: Toward improving crop disease resistance and agronomic characteristics. *Plant Biotechnology Reports* 18(1), 1–20. DOI: 10.1007/s11816-023-00876-z.

Sanjana, N.E., Cong, L., Zhou, Y., Cunniff, M.M., Feng, G. *et al.* (2012) A transcription activator-like effector toolbox for genome engineering. *Nature Protocols* 7(1), 171–192. DOI: 10.1038/nprot.2011.431.

Shaikh, A.A., Chachar, S., Chachar, M., Ahmed, N., Guan, C. *et al.* (2022) Recent advances in DNA methylation and their potential breeding applications in plants. *Horticulturae* 8(562), 1–28.

Shi, M., Wei, Z., Zhang, P., Guan, C., Chachar, S. *et al.* (2024) Epigenetic modifications and breeding applications in horticultural plants. *Horticulturae* 10(2), 4–9.

Smulders, M.J.M. and Klerk, G.J. (2011) Epigenetics in plant tissue culture. *Plant Growth Regulation* 63(2), 137–146. DOI: 10.1007/s10725-010-9531-4.

Springer, N.M. and Schmitz, R.J. (2017) Exploiting induced and natural epigenetic variation for crop improvement. *Nature Reviews Genetics* 18(9), 563–575. DOI: 10.1038/nrg.2017.45.

Tang, Q., Wang, X., Jin, X., Peng, J., Zhang, H. *et al.* (2023) CRISPR/Cas technology revolutionizes crop breeding. *Plants* 12(3119), 1–13.

Tonosaki, K., Fujimoto, R., Dennis, E.S., Raboy, V. and Osabe, K. (2022) Will epigenetics be a key player in crop breeding? *Frontiers in Plant Science* 13, 958350. DOI: 10.3389/fpls.2022.958350.

Tsaballa, A., Xanthopoulou, A., Madesis, P., Tsaftaris, A. and Nianiou-Obeidat, I. (2021) Vegetable grafting from a molecular point of view: The involvement of epigenetics in rootstock-scion interactions. *Frontiers in Plant Science* 11. DOI: 10.3389/fpls.2020.621999.

Turgut-Kara, N., Arikan, B. and Celik, H. (2020) Epigenetic memory and priming in plants. *Genetica* 148(2), 47–54. DOI: 10.1007/s10709-020-00093-4.

Varotto, S., Tani, E., Abraham, E., Krugman, T., Kapazoglou, A. *et al.* (2020) Epigenetics: Possible applications in climate-smart crop breeding. *Journal of Experimental Botany* 71(17), 5223–5236. DOI: 10.1093/jxb/eraa188.

Vázquez-Hernández, M.C., Parola-Contreras, I., Montoya-Gómez, L.M., Torres-Pacheco, I., Schwarz, D. *et al.* (2019) Eustressors: Chemical and physical stress factors used to enhance vegetables production. *Scientia Horticulturae* 250, 223–229.

Vega-Muñoz, I., Feregrino-Pérez, A.A., Torres-Pacheco, I. and Guevara-González, R.G. (2018) Exogenous fragmented DNA acts as a damage-associated molecular pattern (DAMP) inducing changes in CPG DNA methylation and defence-related responses in *Lactuca sativa*. *Functional Plant Biology* 45, 1065–1072.

Villagómez-Aranda, A.L., García-Ortega, L.F., Torres-Pacheco, I. and Guevara-González, R.G. (2021) Whole-genome DNA methylation analysis in hydrogen peroxide overproducing transgenic tobacco resistant to biotic and abiotic stresses. *Plants (Basel, Switzerland)* 10(1), 178. DOI: 10.1007/s00299-022-02858-x.

Villagómez-Aranda, A.L., Feregrino-Pérez, A.A., García-Ortega, L.F., González-Chavira, M.M., Torres-Pacheco, I. *et al.* (2022) Activating stress memory: Eustressors as potential tools for plant breeding. *Plant Cell Reports* 41, 1481–1498.

Wang, C., Yang, Q., Wang, W., Li, Y., Guo, Y. *et al.* (2017) A transposon-directed epigenetic change in ZmCCT underlies quantitative resistance to Gibberella stalk rot in maize. *New Phytologist* 215, 1503–1515.

Wang, S., Yan, J., Hu, B., Wang, R. and Xu, J. (2022) Advanced epigenomic engineering in crop quality improvement. *Current Opinion in Food Science* 46, 100843. DOI: 10.1016/j.cofs.2022.100843.

Wang, Y., Liang, W. and Tang, T. (2018) Constant conflict between Gypsy LTR retrotransposons and CHH methylation within a stress-adapted mangrove genome. *New Phytologist* 220, 922–935.

Wijerathna-Yapa, A. and Hiti-Bandaralage, J. (2023) Tissue culture-A sustainable approach to explore plant stresses. *Life* 13(3), 1–16. DOI: 10.3390/life13030780.

Xu, R., Wang, Y., Zheng, H., Lu, W., Wu, C. *et al.* (2015) Salt-induced transcription factor MYB74 is regulated by the RNA-directed DNA methylation pathway in *Arabidopsis*. *Journal of Experimental Botany* 66(19), 5997–6008. DOI: 10.1093/jxb/erv312.

Xu, Y. and Li, Z. (2020) CRISPR-Cas systems: Overview, innovations and applications in human disease research and gene therapy. *Computational and Structural Biotechnology Journal* 18, 2401–2415. DOI: 10.1016/j.csbj.2020.08.031.

Yadav, N.S., Titov, V., Ayemere, I., Byeon, B., Ilnytskyy, Y. *et al.* (2022) Multigenerational exposure to heat stress induces phenotypic resilience, and genetic and epigenetic variations in *Arabidopsis thaliana* offspring. *Frontiers in Plant Science* 13. DOI: 10.3389/fpls.2022.728167.

Yamaguchi, N., Winter, C.M., Wu, M.-F., Kwon, C.S., William, D.A. *et al.* (2014) PROTOCOLS: Chromatin immunoprecipitation from *Arabidopsis* tissues. *The Arabidopsis Book* 12, e0170. DOI: 10.1199/tab.0170.

Zhang, H., Tang, K., Wang, B., Duan, C.-G., Lang, Z. *et al.* (2014) Protocol: A beginner's guide to the analysis of RNA-directed DNA methylation in plants. *Plant Methods* 10(1). DOI: 10.1186/1746-4811-10-18.

Zhang, Y., Malzahn, A.A., Sretenovic, S. and Qi, Y. (2019) The emerging and uncultivated potential of CRISPR technology in plant science. *Nature Plants* 5(8), 778–794. DOI: 10.1038/s41477-019-0461-5.

Zhang, Y.Y., Latzel, V., Fischer, M. and Bossdorf, O. (2018) Understanding the evolutionary potential of epigenetic variation: A comparison of heritable phenotypic variation in epiRILs, RILs, and natural ecotypes of *Arabidopsis thaliana*. *Heredity (Edinb)* 121(3), 257–265. DOI: 10.1038/s41437-018-0095-9.

25 Plant MicroRNAs: Recent Advances in Their Roles Under Stress Conditions

Jagyaswani Nayak, Subharshree Pattnaik, Tuward J. Dweh, Salma Kayastha, Jyoti Prakash Sahoo*, Manaswini Mahapatra and Madhuri Pradhan
Faculty of Agriculture and Allied Sciences, C.V. Raman Global University, Bhubaneswar, Odisha, India

Abstract
MicroRNAs (miRNAs) are small non-coding RNA molecules that play a significant role in post-transcriptional gene regulation. In recent years, research in this field has expanded rapidly, shedding light on the crucial functions of miRNAs in plant stress responses. The chapter highlights the versatility of miRNAs in regulating plant responses to a wide range of stressors, including abiotic stresses (e.g., drought, salinity, temperature extremes) and biotic stresses (e.g., pathogen infections). It underscores how miRNAs are central players in the plant's adaptive mechanisms. The chapter delves into the intricate molecular mechanisms through which miRNAs modulate gene expression during stress conditions. It explains how miRNAs can target stress-responsive genes, transcription factors, and signaling components, thereby fine-tuning the plant's stress-related responses. The chapter also discusses the interconnected nature of miRNA-mediated regulation and the signaling pathways involved in stress responses. It elucidates how miRNAs participate in crosstalk between different stress signaling cascades, contributing to the plant's ability to integrate and prioritize responses to multiple stresses. The chapter also touches upon the emerging concept of epigenetic regulation by miRNAs, where miRNA-mediated changes in gene expression can be inherited by subsequent plant generations, influencing their stress tolerance and adaptability. Additionally, it provides insights into the potential applications of miRNA research in crop improvement and stress tolerance enhancement, highlighting the promising avenues for genetic engineering to develop stress-resistant plants.

Keywords: Crop improvement, Epigenetic regulation, Plant microRNAs, Signaling, Stress conditions

25.1 Introduction

Plant microRNAs (miRNAs) are small, non-coding RNA molecules that regulate gene expression in response to various stresses, such as nutrient deficiency, salinity, drought, and heavy metals. They play critical roles in sensing the abundance of nutrients, controlling nutrient uptake and transport, and maintaining nutrient homeostasis (Ma and Hu, 2023). They also promote plant growth, development, and adaptation to extreme adverse environmental conditions by targeting stress-responsive genes. Recent advances in sequencing technologies and bioinformatics tools have enabled the identification and functional characterization of plant miRNAs under different stress conditions. Several plant miRNAs have been shown to improve agronomic properties and confer stress

*Corresponding author: jyotiprakashsahoo2010@gmail.com

 Epigenetics for Climate-Smart and Sustainable Agriculture (ed. Jen-Tsung Chen)
DOI: 10.1079/9781800626102.0025

resistance in plants, such as miR156, miR159, miR160, miR164, miR167, miR169, miR171, miR172, miR393, miR395, miR396, miR397, miR398, miR399, miR408, and miR528. These miRNAs target various genes involved in stress signaling, hormone biosynthesis and signaling, nutrient transport and metabolism, and antioxidant defense (Xu *et al.*, 2021). Plant miRNAs are promising tools for crop improvement and sustainable agriculture, as they can be manipulated to enhance plant performance and stress tolerance. However, there are still many challenges and knowledge gaps in understanding the complex regulatory networks of plant miRNAs and their targets under stress conditions. Further research is needed to elucidate the molecular mechanisms, spatiotemporal expression patterns, and crosstalk of plant miRNAs and their targets, as well as to develop efficient and precise methods for miRNA delivery and expression in plants.

The intricate world of plant biology has continuously captivated researchers, especially in understanding the molecular mechanisms governing plant responses to stress. In recent years, miRNAs have emerged as pivotal regulators, orchestrating a complex network of genetic expression and fine-tuning responses under various stressors. miRNAs, typically composed of 20–24 nucleotides, are well-known for their post-transcriptional gene regulation prowess. In plants, they act as master modulators, regulating a myriad of biological processes, including development, growth, and importantly, stress responses. Recent advancements have shed light on their remarkable versatility in modulating plant stress reactions, showcasing a multilayered regulatory system.

Under stress conditions like drought, salinity, extreme temperatures, or pathogen attacks, plants trigger intricate molecular mechanisms to survive. This is where miRNAs exhibit their prowess by targeting specific genes involved in stress response pathways. For instance, miR398 has been found to regulate the copper superoxide dismutase (CSD) gene family, essential for scavenging reactive oxygen species (ROS) generated during stress, thereby safeguarding plant cells from oxidative damage. Furthermore, the versatility of miRNAs is exemplified by their crosstalk and synergistic action. They form regulatory cascades, where one miRNA might regulate the expression of another miRNA or transcription factor, creating a domino effect in stress signaling pathways. This intricate web of interactions highlights their multifaceted role in fine-tuning stress adaptations. In this comprehensive chapter, we embark on a detailed exploration of the pivotal role played by miRNAs in orchestrating the gene regulatory networks essential for plant survival under diverse stress conditions. We will dissect the intricate molecular mechanisms through which miRNAs modulate key pathways.

25.2 Molecular Mechanisms of miRNA-Mediated Gene Regulation

MicroRNAs are a class of endogenous small non-coding RNAs that participate in a majority of biological processes via regulating target gene expression. As the canonical way of miRNA-mediated gene regulation, post-transcriptional repression via miRNA seed region binding to 3′ UTR (untranslated region) of target mRNA is thought to occur. New research, nevertheless, points to the possibility of other regulatory mechanisms outside of the standard process. Specifically, evidence increasingly reveals the role of intranuclear miRNA in the transcriptional control of genes, indicating their contribution to either activating or silencing genes. As a result, the discovery of miRNA-mediated control of gene transcription contributes to our knowledge of the molecular mechanism underpinning miRNA regulatory function and offers fresh proof of the mechanism's applicability to the intricate regulation of several bioprocesses (Pu *et al.*, 2019).

As explained by Zhao and Liu (2009), it had been discovered that a novel family of non-protein regulators called microRNAs (miRNAs) provides extensive control over gene expression in eukaryotes. Though there were still some unresolved issues, significant progress had been made in recent years in understanding the roles and mechanisms of miRNA. The mechanisms of miRNA repression had been explained by a number of theories and models, such as translational inhibition at the initiation or elongation level, quick peptide degradation, mRNA degradation, and mRNA sequestration into stress granules (SGs) and P bodies (processing bodies) for storage or degradation. Some non-canonical miRNA regulation had been discovered recently, including de-repression of miRNA inhibition and miRNA activation. The review covered some new developments about the several ways that miRNAs act and how they regulate their targets.

25.2.1 Target genes of miRNAs during stress conditions

Abiotic stress-associated miRNAs in plants

Environmental factors such as temperature, light, water, and mineral elements are essential for the growth and development of plants. The response to abiotic stress, facilitated by miRNA, is a crucial mechanism in how plants react to various environmental challenges. Abiotic stresses, including extreme temperature, drought, high salinity, and heavy metals, are primary contributors to global crop losses, often leading to oxidative stress. This oxidative stress hinders plant growth and reduces yields in both model and crop plants (Kerchev *et al.*, 2020). The involvement of miRNAs in the response to abiotic stress, particularly in managing oxidative stress and influencing DNA effects in different plant species, has been observed (Pagano *et al.*, 2021). The diverse roles of miRNAs in plant responses to growth, development, and environmental stimuli highlight their significance. miRNAs actively participate in mitigating unfavourable conditions by regulating the expression of related target genes in plants.

RESPONSE OF MIRNAS TO EXTREME TEMPERATURE STRESS IN PLANTS

Low temperature. Challenging conditions like low temperatures and frost compromise the survival of plants, resulting in slowed growth and reduced yields (Das *et al.*, 2021). When exposed to chilling or freezing, plants exhibit various phenotypic symptoms, including poor germination, stunted seedlings, yellowing of leaves, diminished leaf expansion, wilting, and potential tissue death (Ding *et al.*, 2019). In response to low-temperature stress, miRNAs employ three distinct tactics based on their targets: (i) a direct response to abiotic stress; (ii) an indirect response to external stimuli by regulating transcription factors associated with stress responses; and (iii) miRNAs can react to multiple stresses, with their target genes potentially coding for specific hydrolases or oxidoreductases (Yang *et al.*, 2017). Apart from its involvement in miR168-regulated immunity, miR1320 has been found to have an additional role in reacting to abiotic stress. Rice that had miR1320 overexpressed had a higher tolerance to cold. By suppressing the jasmonic acid (JA)-mediated cold signaling pathway, AP2/ERF TF OsERF096 – a target of miR1320 – co-regulates cold tolerance (Sun *et al.*, 2022). Likewise, rice's ability to withstand cold can also be enhanced by overexpressing Osa-miR156, Osa-miR319, and Osa-miR528 (Huo *et al.*, 2022).

High temperature. Heat stress gradually becomes one of the elements that impede plant growth and development as a result of global warming and climatic changes. Stress from high temperatures is a danger to the sustainability of agricultural productivity. Heat-responsive miRNA has been thoroughly researched to mitigate the negative impacts, which primarily show up in photosynthesis, respiration, nitrogen metabolism, protein turnover, nucleic acid metabolism, and lipid metabolism (Grover *et al.*, 2013; Zuo *et al.*, 2021; Begum, 2022). Generally, species-specific miRNAs tend to be expressed at lower levels (Fahlgren *et al.*, 2007). Over the course of increasing research efforts, numerous conserved and novel species-specific miRNAs have been identified in significant crops, and these miRNAs with species-specific attributes have developed stable and specialized functions (Pan *et al.*, 2017; Chen *et al.*, 2020). In *Arabidopsis*, examples of species-specific miRNAs include miR824, miR858, miR856, and miR857. The target gene of miR824 is AGL16 (At3g57230, MADSbox); miR858 targets MYB12 (At2g47460, R2R3-MYB family); and miR856 targets an unrelated transcript encoding the efflux protein ZAT1, though the functions of these targets remain unknown. Additionally, miR857 has been validated to target a laccase family gene, LAC7 (At3g09220), with a predicted multicopper oxidase function. miRNAs play a role in responding to and mitigating the detrimental effects associated with high-temperature stress. They achieve this by regulating the abundance of transcription factors and proteins linked to reactive oxygen species (ROS) metabolism, as well as various mitogen-activated protein kinases (MAPK) signaling pathways (Ahmed *et al.*, 2019; Zhang *et al.*, 2019).

RESPONSE OF MIRNAS AND THEIR TARGETS TO SALT AND DROUGHT STRESSES. Salinity and drought stresses are among the most severe and widespread factors that negatively impact the growth and development of plants, resulting in a significant reduction in agricultural production. Furthermore, extensive areas remain unutilized due to the adverse environmental effects of high salinity or drought on plants. Fortunately, plants have evolved various strategies to cope with such challenging conditions, demonstrating varying degrees of tolerance. Substantial efforts have been directed toward unraveling the molecular mechanisms that underlie plant tolerance. It has been acknowledged that plant adaptation to stressful environments involves, in part, the regulation of gene expression at both the transcriptional and post-transcriptional levels. Nevertheless, the precise molecular details of the gene expression regulation mechanism remain unclear.

MicroRNAs play crucial regulatory roles in post-transcriptional gene expression through mRNA cleavage or translation repression (Bartel, 2004; Zhang *et al.*, 2006). These miRNAs have been reported to participate in various biological processes in plants by regulating the expression of their target protein-coding genes. These processes encompass organ development (Aukerman and Sakai, 2003), hormone signaling (Mallory *et al.*, 2005), defense against pathogens (Navarro *et al.*, 2006), and responses to abiotic stress (Sunkar *et al.*, 2007; Lu and Huang, 2008; Eldem *et al.*, 2013). The abiotic stresses to which plants respond include salinity (Li *et al.*, 2011), drought (Zhao *et al.*, 2007; Covarrubias and Reyes, 2010; Eldem *et al.*, 2012), cold (Zhou *et al.*, 2008), heavy metals (Huang *et al.*, 2009), and nutritional stresses (Lu and Huang, 2008). Currently, more than 40 miRNA families have been associated with plant response to abiotic stress (Sunkar, 2010), with many implicated in salt and drought stress responses (Covarrubias and Reyes, 2010). Some miRNAs have been identified in multiple plant species (Covarrubias and Reyes, 2010; Sunkar, 2010), suggesting that their function in stress response may be conserved across different plants.

RESPONSE OF MIRNAS TO HEAVY METAL STRESS IN PLANTS. Heavy metal (HM) stress stands as a prominent factor limiting plant yields (Ding *et al.*, 2019). Certain HMs, including manganese (Mn), copper (Cu), boron (B), zinc (Zn), and iron (Fe), play a beneficial role at lower concentrations, serving as essential components of plant cells. Conversely, elevated concentrations of these HMs induce toxic reactions in plants, significantly impeding their growth and development. Metals such as cadmium (Cd), aluminum (Al), arsenic (As), lead (Pb), and mercury (Hg) are potentially highly toxic to plants (Gupta *et al.*, 2014). Previous research indicates that miRNAs, as part of plants' survival strategies, are potential targets for regulating heavy metal tolerance in plants (Noman and Aqeel, 2017).

In *Arabidopsis thaliana*, miRNAs responding to excessive Cd and Cu were identified using a spl7 knockout mutant. The transcription factor SPL7, a key regulator of Cu homeostasis, was affected by Cd, leading to Cd-induced Cu deficiency. This, in turn, caused opposite effects on miRNAs such as miR857, miR397a, and miR398b/c under Cu and Cd exposures (Gielen *et al.*, 2016). In *Brassica napus*, 84 miRNAs and 802 target genes were identified through small RNA high-throughput sequencing and degradome sequencing. Among them, the gene encoding the NRAMP metal transporter was recognized as the target gene of miR167 (Ding *et al.*, 2018). BnNRAMP1b was further confirmed as a target of miR167 in *B. napus* (Meng *et al.*, 2017). Moreover, a set of miRNAs responding to cadmium stress was identified in *B. napus* (Huang *et al.*, 2010), with miR395 reported to be involved in the detoxification of cadmium in *B. napus* (Zhang *et al.*, 2013). Interestingly, the same metal stress elicits contrasting miRNA responses in different species; for instance, miR167 was upregulated in *Arabidopsis* but suppressed in maize under salt stress (Khraiwesh *et al.*, 2012).

Biotic stress-associated miRNAs in plants

ROLE OF MIRNAS AGAINST BACTERIAL PATHOGENS. The initial identification of miRNA's significant role in plant–bacterial interactions occurred in *Arabidopsis* exposed to the bacterial pathogen-associated molecular pattern (PAMP) peptide flg22, resulting in the upregulation of miR393 expression (Navarro *et al.*, 2006). Subsequent high-throughput sequencing of small RNAs from *Arabidopsis* leaves infected with the non-pathogenic *Pseudomonas syringae* pv. tomato strain DC3000 hrc (mutated in the type III secretion system) at 1 hour and 3 hours post-inoculation (hpi) unveiled the significant upregulation of three miRNAs (miR160, miR167, and miR393), while miR825 was downregulated at 3 hpi (Fahlgren *et al.*, 2007). Transgenic lines overexpressing miR393 exhibited inhibited Pst DC3000 growth by suppressing auxin signaling pathways through the regulation of TIR1 and related proteins.

Additionally, miR160 and miR167 were induced at 3 hpi in *Arabidopsis* leaves during bacterial infection, targeting mRNAs of the auxin-response factor (ARF) family genes involved in auxin signaling (Yang *et al.*, 2013). Notably, miR162 and miR168 were previously reported to be downregulated upon bacterial infection, targeting DCL1 and AGO1, respectively (Fahlgren *et al.*, 2007). The modulation of the miRNA pathway by these genes was deemed crucial for bacterial defense. Similarly, miR825 was downregulated at 3 hpi during bacterial infection, predicted to target mRNAs of remorin, frataxin-related, and zinc finger homeobox family encoding genes (Fahlgren *et al.*, 2007). Expression profiling of *Arabidopsis* miRNAs in response to *Pseudomonas syringae* pv. tomato (Pst) infection revealed distinctive patterns, with 15, 27, and 20 miRNA families exhibiting differential expression upon infection with Pst DC3000 hrcC (non-pathogenic), Pst DC3000 EV (virulent), and Pst DC3000 avrRpt2 (avirulent), respectively (Zhang *et al.*, 2011). Notably, the majority of miRNAs regulated by bacterial infection targeted genes involved in hormone biosynthesis and signaling pathways, including auxin, jasmonic acid (JA), and abscisic acid (ABA) pathways, indicating their crucial role in plant defense signaling.

In a similar context, RNA sequencing was employed to analyse miRNA expression in cassava (*Manihot esculenta*) tissue during infection by *Xanthomonas axonopodis* pv. *manihotis* (Xam) (Pérez-Quintero *et al.*, 2012). The study reported a total of 56 conserved and 12 novel cassava-specific miRNA families exhibiting differential expression upon Xam infection. Among the upregulated miRNAs, some played a role in mediating defense by regulating auxin-responsive factors. Conversely, downregulated miRNAs were implicated in the regulation of disease resistance-associated genes. These findings underscore the intricate involvement of miRNAs in shaping the plant's response to bacterial infections, with implications for both defense mechanisms and hormonal signaling pathways.

ROLE OF MIRNAS AGAINST FUNGAL PATHOGENS. miRNAs are pivotal in plant defense against not only bacterial infections but also fungal ones. In loblolly pine (*Pinus taeda*) infected with the rust fungus *Cronartium quercuum* f. sp. Fusiforme, the identified 26 miRNAs exhibited differential expression in both spindle-shaped galls and the surrounding tissues. Notably, these miRNAs targeted defense-related genes, including nucleotide-binding site leucine-rich repeat (NB-LRR) receptors, laccases, ubiquitin ligases, and peroxidases (Balmer and Mauch-Mani, 2013). In wheat (*Triticum aestivum* L.) infected with *Erysiphe graminis* f. sp. *tritici* (Egt), the expression profiles of miRNAs differed between susceptible and near-isogenic resistant lines, with miR156, miR164, and miR393 showing distinct downregulation patterns in response to the pathogen. Particularly, miR393, known to target auxin response factors, suggested a modulation of the auxin signaling pathway in the defense mechanism against *E. graminis* infection.

Soybean (*Glycine max*) and cotton (*Gossypium barbadense* and *G. hirsutum*) exhibited differential miRNA expression in response to *Phakopsora pachyrhizi* and *Verticillium dahliae* infections, respectively. The suppression of miR1917 and miR2118 in cotton roots upon *V. dahliae* infection indicated their involvement in defense responses. Furthermore, small RNA profiling of eggplant (*Solanum melongena* L.) and upland cotton (KV-1 variety) revealed conserved and novel miRNAs in response to *V. dahliae* strains, suggesting their regulatory roles in plant–pathogen interactions. In garlic (*Allium sativum* L.) infected with *Fusarium oxysporum* f. sp. *cepae* (FOC), miRNAs, including miR164a, miR168a, and miR393, exhibited enhanced resistance. Transgenic garlic lines overexpressing these miRNAs showed reduced fungal growth and increased expression of defense-responsive genes, providing further evidence of the crucial role of miRNAs in conferring resistance to FOC infection (Chand *et al.*, 2017). Overall, these studies emphasize the complex regulatory networks orchestrated by miRNAs in plant–fungal interactions, contributing significantly to our understanding of plant defense strategies against diverse pathogens.

ROLE OF MIRNAS AGAINST VIRAL DISEASES. Prior investigations have demonstrated the involvement of small RNAs, including miRNAs, in plant antiviral defense during viral infections (Niu *et al.*, 2006; Ding and Voinnet, 2007; Qu *et al.*, 2007; Bester *et al.*, 2017; Kundu *et al.*, 2017; Bao *et al.*, 2018). Viral RNA silencing is orchestrated by small RNAs processed from virus-derived double-stranded RNAs (dsRNAs) through the small interfering RNA (siRNA) pathways (Balmer and Mauch-Mani, 2013). In *Brassica rapa* infected with Turnip mosaic virus (TuMV), the upregulation of bramiR158 and bra-miR1885 was observed, with bra-miR1885 targeting the mRNA of the Toll/interleukin-1, nucleotide-binding site leucine-rich repeat (TIR-NBS-LRR) disease-resistant gene (He *et al.*, 2008). Deep sequencing analysis of miRNA expression in rice infected with Rice dwarf virus (RDV) and Rice stripe virus (RSV) revealed that RSV induced the abundance of miRNAs instead of the corresponding miRNAs, while RDV infection upregulated the expression levels of RNA-dependent RNA polymerase (RDR) genes in rice. The elevated expression of Dicer-like (DCL), Argonaute (AGO), or RDR genes in defense mechanisms, however, remains unclear (Du *et al.*, 2011).

In *Nicotiana benthamiana* infected with Tobacco mosaic virus (TMV) and TMV-43A, a mutated TMV variant, deep sequencing analysis demonstrated a differential expression pattern of miRNAs. This differential expression impacted the corresponding putative mRNA targets, influencing the moderate host response to viral infection (Guo *et al.*, 2017). Overall, these studies highlight the intricate role of miRNAs in plant defense against viral infections, shedding light on the regulatory mechanisms involved in antiviral responses.

Recently, a comprehensive study identified a total of 422 miRNAs in *Phaseolus vulgaris* subjected to infection by Mungbean yellow mosaic India virus (MYMIV) compared with mock-treated samples (Patwa *et al.*, 2019). The downregulation of miR482 was noted during MYMIV infection in the resistant genotype, leading to the upregulation of the NB-LRR encoding resistance gene. This upregulation is associated with enhanced resistance to MYMIV (Patwa *et al.*, 2019). In general, transcriptome analysis of miRNAs in various plants undergoing viral infections has uncovered alterations in miRNA expression. Despite numerous recent studies aiming to unravel the role of miRNAs in antiviral resistance, the precise mechanisms governing antiviral defense mechanisms remain to be fully elucidated.

25.3 Regulation of Transcription Factors by miRNAs

The identification of intricate regulatory networks in higher organisms, such as plants and animals, has become increasingly acknowledged in the realm of plant biology (Morris and Mattick, 2014). These intricate networks, encompassing chromatin modification at the epigenetic level, mRNA splicing, cell signaling, polyadenylation, and mechanisms governing protein activation and degradation, have garnered significant attention. Understanding the comprehensive regulation of plant systems necessitates an in-depth exploration of these multifaceted processes (Boucas *et al.*, 2012; Holoch and Moazed, 2015). This review aims to concentrate on gene regulation at both the transcriptional and post-transcriptional levels, with a focus on miRNAs and transcription factors (TFs) as pivotal regulatory entities.

On the transcriptional front, the well-established interaction of TFs with enhancers to orchestrate gene expression has been a focal point in the past decade (Yu and Gerstein, 2006; Osório, 2016). This is evident in the discovery of various TF families that play diverse roles in plant systems (Duval *et al.*, 2014). On the post-transcriptional level, an intriguing mechanism of gene regulation emerged decades ago, involving a substantial class known as miRNAs. These miRNAs serve as gene regulators in both plants and animals, exerting negative control over mRNAs (Bartel, 2004; Thomson and Dinger, 2016). As the number of regulators involved in gene networks continues to grow, it becomes compelling to observe and comprehend the dynamic relationships among miRNAs, TFs, and mRNAs.

Presently, there exist 58 families of TFs, comprising 320,370 members across 165 plant species (Jin *et al.*, 2017). While miRNAs exclusively downregulate their targets, TFs play a dual role by either activating or repressing the transcription of their targets. This dual functionality ultimately dictates the fate of specific genes, determining whether they are switched 'on' or 'off' (Istrail *et al.*, 2007; Chow *et al.*, 2016). Intriguingly, a significant proportion of miRNA targets are TFs (Rhoades *et al.*, 2002; Mitsuda and Ohme-Takagi, 2009; Kamthan *et al.*, 2015). Given the substantial impact both regulators have on the plant genetic system, the interplay between miRNAs and TFs facilitates the coordination of numerous biological processes with a high degree of reliability.

It is essential to illustrate an integrated picture of the regulatory relationships between miRNAs, TFs, and target genes. However, it is quite difficult to develop a clear-cut regulatory relationship between miRNAs and TFs, because, in addition to the interaction of these regulators with their target genes, they sometimes interact with each other; leading to some different results. Here, we have summarized the similarities (Table 25.1) and differences (Table 25.2) between miRNAs-mediated and TFs-mediated regulatory systems. We also propose a model to relate these two regulators with their target genes and the consequences of this model to the plant regulatory network under normal and stress conditions. The existence of both miRNAs and TFs in gene regulatory networks will reveal the regulatory role involving both direct and indirect regulatory relationships. In this review, we have tried to bring together previous findings, related to the interaction between miRNAs and TFs, mostly in model plants and some non-model plants. Rigorous and time-consuming web lab/experimental work was a big hurdle in developing interactions between miRNAs and TFs in plants (Le *et al.*, 2013). But now, with the advancement in public databases and bioinformatics tools, developing an interaction between the above-mentioned regulators is relatively easy.

25.3.1 Interconnected nature of miRNA-mediated regulation

MicroRNAs are said to be endogenous. It is the small non-coding RNA sequences that play an important role in the regulation of gene expression. The interconnected nature of miRNA-mediated regulation refers to complex networks of interactions between the miRNAs and the target mRNA sequence, the post-transcriptional repression activities (Bartel, 2009).

Table 25.1. Similarities between miRNA and TF. Author's own table.

Factors	miRNA and TF	Reference
Gene regulator	Both are gene regulators	Lai *et al.*, 2016
Stimulus response	Both are inducible toward external stimuli	Nazarov *et al.*, 2013
Number of targets	Both can regulate from single to multiple targets at a time	Lindemose *et al.*, 2014

Table 25.2. Differences between miRNA and transcription factor. Author's own table.

Factors	miRNA	Transcription factor	Reference
Biogenesis	Produced by a sequence of cleavages that DCL facilitated	Directly synthesized from genes and go through a folding process	Rogers and Chen, 2013; Boeva, 2016
Molecular composition	Short non-coding RNA	Proteins	Rogers and Chen, 2013; Boeva, 2016
Level of regulation	Post-transcription	Transcription	Rogers and Chen, 2013; Boeva, 2016
Functional requirement	Argonaute protein is required for proper function.	No extra protein is required for full functionality.	Boeva, 2016; Miyoshi *et al.*, 2016
Mode of action	Use translational inhibition or cleavage to repress the target gene	Bind to the target gene's promoter region to either repress or activate it.	Brodersen *et al.*, 2008; Rogers and Chen, 2013; Boeva, 2016
Target region	Bind to coding region or UTR	*Cis* region of the promoter	Rogers and Chen, 2013; Boeva, 2016
Family classification	Based on sequence conservation	Based on DNA-binding domain	Kozomara and Griffiths-Jones, 2014; Salih *et al.*, 2016

25.3.2 Crosstalk between different stress signaling cascades

This refers to the complicated interactions and communications of various cellular pathways which are activated by different stresses. Various molecules are involved, including TFs and various enzymes and kinases in the crosstalk between stress-signaling pathways. Evidence includes hormonal signaling pathways during stress such as jasmonic acid, abscisic acid, etc., and ROS signaling pathways, which also play an important role in the crosstalk between stress-signaling cascades (Fujita *et al.*, 2006). There are complexes for a network of signaling pathways that play key roles, such as ER complex, nucleic acids, and different adaptations of oxidative stresses, DNA damage, etc. MicroRNAs emerge as the key regulators of phytohormones in response to various stress conditions and by disturbing the metabolism process (Curaba *et al.*, 2014). The modification in miRNAs enhances the abiotic and biotic stress tolerances. It functions by focusing on the target gene network and by controlling the signal pathway. It performs through inducing and inhibiting the various pathways for adaptation in plants during stresses. MicroRNA modification in biological processes facilitates the epigenetic changes by regulation of various transcription factors.

Whole-genome sequencing method advancement of miRNA modifications and sequencing promotes the comprehensive view of miRNA-mediated gene regulatory networks in plants (Patel *et al.*, 2019). The mitogen-activated protein linase (MAPK) cascade converts developmental signals and environmental signals to adaptation responses of plants. It regulates various processes like hormonal signaling in response to stresses, developmental programs, and modification in response to stresses. Signaling occurs through protein–protein interaction or kinase signaling and mechanisms like feedback controls, cross-inhibition, and scaffolding. It enables the integration of signals from various pathways (Cristina *et al.*, 2010). Salicylic acid is an important molecule that defines against parasitic and symbiont or some other host-dependent pathogens, and ethylene response to various biotic and abiotic stresses: the ethylene and SA signaling pathway interaction response to disease-causing pathogens (Zhang *et al.*, 2014).

25.3.3 Integration and prioritization of responses to multiple stresses

The crucial part of the capability or ability of any organism, biological system, or entity to effectively carry out various reactions against stresses and different challenges is the capability of organisms to prioritize and integrate responses to various complex environmental stresses (Sokolova, 2013). This response capacity maintains balance, resilience, and proper functioning in complex environments. In biological systems, organisms face various stressors like temperature changes, extreme climatic conditions, nutrient availability, pathogenic activity, etc. The response integration interacts with different physiological processes to maintain the internal stability of any organisms under different stresses. Prioritization makes sure that critical activities like energy production and immunity response occur when resources are limited (Feder and Hofmann, 1999).

Ecosystems are also subjected to various stressors, such as climate change, toxicity of soils and air, habitat loss, and many more. The integration response is the interaction between different species and their adaptation according to different environmental changes and conditions. Prioritization is in identifying the important species or ecosystems that are needed to protect the overall ecosystem's health (Sala *et al.*, 2000). Stress management needs to focus on the various stressors that show more impact on the organisms. In this case, response integration is for the organisms unable to overcome the stress, which may die or remain non-functional and unable to adapt to the environmental changes. It involves removing stress conditions in various environmental fields and levels (Bohnert *et al.*, 1995).

25.3.4 Signaling pathways involved in stress response

The stress response is the physiological activities that drive so many signaling pathways in various organisms for adaptations, development, modifications, etc. Various stress signaling pathways perform the biological or cellular process response to stresses. Stress response signaling includes MAPK, protein kinase A signaling, heat shock protein signaling, etc. Stress signaling pathways play a major role in maintaining homeostasis (Hotamisligil and Davis, 2016). MAPK signaling pathways are present in the eukaryotic cells and are involved in transmitting signals from the plasma membrane to the nucleus and regulation of cellular processes during stresses. They comprise three modules: (i) extracellular signal-regulated kinase (ERK); (ii) c-JUN N-terminal kinase (JNK); and (iii) p38 MAPK (Davis, 2000). ERK is activated by growth factors and tyrosine molecules. JNK and p38MAPK are induced by various stresses like UV radiation, osmotic shock, and cytokinin. MAPKs get phosphorylated and change in gene expression and cellular processes. There may be integration of MAPK pathways with various network signals like PI3K/AKT, Wnt signaling, and other cellular processes (Chang and Karin, 2001).

Heat shock proteins (HSPs) are activated during stress conditions or cellular stress. During stress conditions, they protect the cells. HSPs are also called molecular chaperons and are controlled by heat shock transcription factor (Bingxiang *et al.*, 1997). HSPs play a crucial role in maintaining cellular homeostasis. Complex mechanisms of the HSPs include HSP70/90, ATPase-coupled confirmational modification, and interaction cochaperones (Divya *et al.*, 2019). The important function of HSPs is to maintain the folding and unfolding of proteins, protein aggregations, and degrading the damaged protein. Thus they are activated during the stress conditions, particularly by HSFs (heat shock factors). This leads to the synthesis of different HSPs, like HSP70 and HSP90 (Hartl *et al.*, 2011). It contributes to cellular proteostasis and the proper folding and unfolding of the protein. It works against stresses like climatic changes, various environmental conditions, both biotic and abiotic stresses, etc. (Kampinga and Craig, 2010). The P53 signaling pathway is involved in response to different stresses and DNA damage. Mi-34a is a miRNA which is linked with the P53 pathway and contributes to cell cycle arrest and apoptosis. The miRNA controls and regulates the expression of P53 and its activities (Hermeking, 2012).

25.4 Epigenetic Regulation by miRNAs

Epigenetic regulations are the modifications in gene expression not in DNA sequences, but in histone modification and other molecules that are associated with the DNA. MicroRNA acts on post-transcriptional regulation of gene expression. It binds at 3′ of untranslated region of mRNA for degradation of target mRNA, so it leads to changes in gene expression and other epigenetic changes (Yao *et al.*, 2019). Some epigenetic modifications and regulations are DNA methylation, polyadenylation, histone modifications, RNA methylation, etc., and also some other cellular modifications.

25.4.1 Inheritance of miRNA-mediated changes in gene expression

This is a dynamic process and complex modification in both genetics and epigenetics. MicroRNA cannot be passed to or inherited by the next generation but its action or effects on various cellular processes and genetics can be inherited by the next generations (Rassoulzadegan *et al.*, 2006). Some key inheritances of miRNA-mediated changes are epigenetic inheritance, maternal inheritance, cell lineage specificity, and environmental influences, as follows.

Epigenetic inheritance

MicroRNA can contribute to epigenetic and cellular modification which can be passed to the next generations or also passed during cell division. For example, histone modification or DNA methylation can pass to daughter cells and also from generation to generation (Martin and Zhang, 2007).

Maternal inheritance

From the egg, the maternal role of miRNAs during early embryonic development influences the expression of genes in the developing embryo. Maternal miRNA plays a role in regulating the developmental processes. Regulating various developmental processes shows an impact on the phenotype of offspring (Hutchison *et al.*, 1974).

Cell lineage specificity

MicroRNA-mediated gene expression modification can be cell lineage-specific. Cells can differentiate and develop into various cell types; miRNA-mediated changes in gene expression can pass to daughter cells of specific lineage. This cell lineage specificity contributes to maintaining and developing the identity of cells (Li *et al.*, 2018).

Environmental influences

The functions and expressions of miRNA are also affected by various environmental factors. Due to changes in environmental factors or conditions, changes in miRNA activities according to these epigenetic regulations also get modified from one generation to the next generations. Research on this interaction of environment and epigenetics is known as environmental epigenetics (Hudder and Novak, 2008).

25.4.2 Influence on subsequent plant generations' stress tolerance and adaptability

Stress tolerance capability and adaptability in plants are affected or impacted by various biotic and abiotic factors. Due to the influence of various factors, there are many changes in cellular processes, epigenetic changes, genetic factors, genomic modifications, and many more. Some major influences are epigenetic modification, variation in genomic level, transgenerational plasticity, and also the influence of microbes on the plants (Bláha and Středa, 2016).

Genetic variations

In plant populations, genetic variations play an important role in adaptability and stress tolerance. Survival of the fittest and natural selection act on these genetic variations. In different environmental stress conditions due to the adaptability and survival ability of plants, variations happen and, in this way, evolutionary changes also occur (Hamrick and Godt, 1996).

Parental stress priming

Stress tolerance activities in parent plants can also be induced in their offspring; this phenomenon is known as stress priming. In this case, epigenetic regulations also go through from parents to their offspring or the next generation. It results in the enhancement of stress tolerance capability and adaptability in the next generations (Rasmann and Agrawal, 2011).

Epigenetic modification

Epigenetic changes like DNA methylation, histone modifications, and other epigenetic changes can also influence gene expression against stress tolerances. It also can be inherited by the next generations and affect stress responses in subsequent generations (Becker and Weigel, 2012).

Transgenerational plasticity

This is known as the phenotypic changes due to various environmental and ecological factors or stresses and it is also inherited by subsequent generations. Its major contribution is the stress tolerance of plants (Herman *et al.*, 2012).

Endophytic microorganisms

Endophytic microbes inhabit internal tissues of other organisms without causing any damage or diseases to them. Endophytic microbes and plant microbiomes also can influence the stress tolerance in plants. It is also inherited by subsequent generations and contributes to adaptability against various stresses. Examples include *Xanthomonas*, *Variovorax*, *Inquilinus*, etc. (Lau and Lennon, 2012).

25.5 Potential Applications of miRNA Research

25.5.1 Advancements in developing stress-resistant plants through miRNA

The exploration of miRNAs in crops like rice and maize illustrates their significant impact on key yield factors. For instance, the miR156/miR529/SPL complex in rice is pivotal for tillering and panicle branching, essential for grain yield. Similarly, in maize, the miR172-IDS1 pathway and miR156a-l's interaction with SPL genes are crucial for meristem identity and ear development (Lauter *et al.*, 2005; Chuck *et al.*, 2007; Salvi *et al.*, 2007). The role of miRNAs extends to direct yield enhancement, as seen in the case of OsmiR397 in rice, which downregulates OsLAC, leading to increased grain size and yield (Zhang *et al.*, 2013).

The discovery that miRNA gene promoters react to environmental and hormonal signals indicates their role in stress adaptation. AGO10's regulation by auxin, brassinolide, and light is a prime example of this potential. The approach of overexpressing stress-positive miRNAs has demonstrated effectiveness in improving plant tolerance (Fujii *et al.*, 2005; Chuck *et al.*, 2011). However, the complexity of miRNA pathways necessitates refined strategies such as RNAi for specific targeting (Yang *et al.*, 2013) and the use of target mimics to counteract miRNAs' negative regulation in stress responses.

25.5.2 Promising avenues in genetic engineering through miRNAs

The development of artificial miRNAs (amiRNAs) and techniques like Short Tandem Target Mimic (STTM) mark significant advancements in genetic engineering (Warthmann *et al.*, 2008). These innovations enable targeted gene regulation, opening up possibilities for stress-tolerant and nutritionally enhanced crops. The use of stress-inducible and tissue-specific promoters facilitates fine-tuned gene expression and can also minimize the constitutive expression of genes that have negative effects (Coca *et al.*, 1996; Kasuga *et al.*, 2004; Rai *et al.*, 2009). The successes in nutritional enhancement through RNAi technology, as seen in high-lysine maize and enriched tomatoes, demonstrate the potential for miRNA applications in creating genetically modified crops with improved nutritional value.

25.6 Conclusion

The burgeoning field of plant miRNAs has unveiled pivotal roles in mediating stress responses and epigenetic regulation, marking a significant advancement in our understanding of plant biology. These small but mighty regulators are central to the ability of plants to withstand environmental challenges, from extreme climatic conditions to pathogenic attacks. Through intricate interactions with stress signaling pathways like MAPK, HSPs, and P53, and the nuanced regulation of gene expression, miRNAs emerge as critical players in maintaining plant homeostasis and resilience. With the development of innovative tools like artificial miRNAs and Short Tandem Target Mimicry, the precise manipulation of miRNA pathways is within reach. This could lead to the creation of crops that are not only more resistant to environmental stresses but also nutritionally enhanced. As we confront the challenges of climate change and a growing global population, the strategic exploitation of miRNA-mediated pathways stands out as a promising avenue for securing food production and promoting resilient agricultural systems.

References

Ahmed, W., Xia, Y., Zhang, H., Li, R., Bai, G. *et al.* (2019) Identification of conserved and novel mirnas responsive to heat stress in flowering Chinese cabbage using high-throughput sequencing. *Scientific Reports* 9(1), 14922.

Aukerman, M.J. and Sakai, H. (2003) Regulation of flowering time and floral organ identity by a microRNA and its APETALA2-like target genes. *The Plant Cell* 15(11), 2730–2741.

Balmer, D. and Mauch-Mani, B. (2013) Small yet mighty – microRNAs in plant-microbe interactions. *MicroRNA* 2(1), 73–80.

Bao, D., Ganbaatar, O., Cui, X., Yu, R., Bao, W. *et al.* (2018) Down-regulation of genes coding for core RNAi components and disease resistance proteins via corresponding microRNAs might be correlated with successful soybean mosaic virus infection in soybean. *Molecular Plant Pathology* 19(4), 948–960.

Bartel, D.P. (2004) MicroRNAs: Genomics, biogenesis, mechanism, and function. *Cell* 116(2), 281–297.

Bartel, D.P. (2009) MicroRNAs: Target recognition and regulatory functions. *Cell* 136(2), 215–233.

Becker, C. and Weigel, D. (2012) Epigenetic variation: Origin and transgenerational inheritance. *Current Opinion in Plant Biology* 15(5), 562–567.

Begum, Y. (2022) Regulatory role of microRNAs (miRNAs) in the recent development of abiotic stress tolerance of plants. *Gene* 821, 146283.

Bester, R., Burger, J.T. and Maree, H.J. (2017) Differential expression of miRNAs and associated gene targets in grapevine leafroll-associated virus 3-infected plants. *Archives of Virology* 162(4), 987–996.

Bingxiang, L., Huiluo, C. and Yonglin, C. (1997) Insect heat shock response and heat shock proteins. *Kun chong xue bao. Acta Entomologica Sinica* 40(4), 417–427.

Bláha, L. and Středa, T. (2016) Plant integrity – the important factor of adaptability to stress conditions. In: Shanker, A.K. and Chanker, C. (eds) *Abiotic and Biotic Stress in Plants-Recent Advances and Future Perspectives. IntechOpen*. IntechOpen.

Boeva, V. (2016) Analysis of genomic sequence motifs for deciphering transcription factor binding and transcriptional regulation in eukaryotic cells. *Frontiers in Genetics* 7, 24.

Bohnert, H.J., Nelson, D.E. and Jensen, R.G. (1995) Adaptations to environmental stresses. *The Plant Cell* 7(7), 1099–1111.

Boucas, J., Höpker, K., Chen, S., Herter-Sprie, G., Jokic, M. *et al.* (2012) Posttranscriptional regulation of gene expression-adding another layer of complexity to the DNA damage response. *Frontiers in Genetics* 3, 159.

Brodersen, P., Sakvarelidze-Achard, L., Bruun-Rasmussen, M., Dunoyer, P., Yamamoto, Y.Y. *et al.* (2008) Widespread translational inhibition by plant miRNAs and siRNAs. *Science* 320(5880), 1185–1190.

Chand, S.K., Nanda, S., Mishra, R. and Joshi, R.K. (2017) Multiple garlic (*Allium sativum* L.) microRNAs regulate the immunity against the basal rot fungus *Dusarium oxysporum* f. sp. cepae. *Plant Science: An International Journal of Experimental Plant Biology* 257, 9–21.

Chang, L. and Karin, M. (2001) Mammalian MAP kinase signalling cascades. *Nature* 410(6824), 37–40.

Chen, J., Pan, A., He, S., Su, P., Yuan, X. *et al.* (2020) Different microRNA families involved in regulating high temperature stress response during cotton (*Gossypium hirsutum* L.) anther development. *International Journal of Molecular Sciences* 21(4), 1280.

Chow, C.-N., Zheng, H.-Q., Wu, N.-Y., Chien, C.-H., Huang, H.-D. *et al.* (2016) PlantPAN 2.0: An update of plant promoter analysis navigator for reconstructing transcriptional regulatory networks in plants. *Nucleic Acids Research* 44, D1.

Chuck, G., Cigan, A.M., Saeteurn, K. and Hake, S. (2007) The heterochronic maize mutant corngrass1 results from overexpression of a tandem microRNA. *Nature Genetics* 39(4), 544–549.

Chuck, G.S., Tobias, C., Sun, L., Kraemer, F., Li, C. *et al.* (2011) Overexpression of the maize corngrass1 microRNA prevents flowering, improves digestibility, and increases starch content of switchgrass. *Proceedings of the National Academy of Sciences* 108(42), 17550–17555.

Covarrubias, A.A. and Reyes, J.L. (2010) Post-transcriptional gene regulation of salinity and drought responses by plant microRNAs. *Plant, Cell & Environment* 33(4), 481–489.

Cristina, M.S., Petersen, M. and Mundy, J. (2010) Mitogen-activated protein kinase signalling in plants. *Annual Review of Plant Biology* 61, 621–649.

Curaba, J., Singh, M.B. and Bhalla, P.L. (2014) miRNAs in the crosstalk between phytohormone signalling pathways. *Journal of Experimental Botany* 65(6), 1425–1438.

Das, R., Mukherjee, A., Basak, S. and Kundu, P. (2021) Plant miRNA responses under temperature stress. *Plant Gene* 28(3), 100317.

Davis, R.J. (2000) Signal transduction by the JNK group of MAP kinases. *Cell* 103(2), 239–252.

Ding, S.-W. and Voinnet, O. (2007) Antiviral immunity directed by small RNAs. *Cell* 130(3), 413–426.

Ding, Y., Gong, S., Wang, Y., Wang, F., Bao, H. *et al.* (2018) MicroRNA166 modulates cadmium tolerance and accumulation in rice. *Plant Physiology* 177(4), 1691–1703.

Ding, Y., Shi, Y. and Yang, S. (2019) Advances and challenges in uncovering cold tolerance regulatory mechanisms in plants. *New Phytologist* 222(4), 1690–1704.

Divya, K., Bhatnagar-Mathur, P., Sharma, K.K. and Reddy, P.S. (2019) Heat shock proteins (Hsps) mediated signalling pathways during abiotic stress conditions. In: Khan, M.R.K., Reddy, P.S., Ferrante, A. and Khan, N.A. (eds) *Plant Signaling Molecules*. Woodhead Publishing, Sawston, Cambridge, UK, pp. 499–516.

Du, P., Wu, J., Zhang, J., Zhao, S., Zheng, H. *et al.* (2011) Viral infection induces expression of novel phased microRNAs from conserved cellular microRNA precursors. *PLOS Pathogens* 7(8), e1002176.

Duval, I., Lachance, D., Giguère, I., Bomal, C., Morency, M.-J. *et al.* (2014) Large-scale screening of transcription factor–promoter interactions in spruce reveals a transcriptional network involved in vascular development. *Journal of Experimental Botany* 65(9), 2319–2333.

Eldem, V., Akçay, U.Ç., Ozhuner, E., Bakır, Y., Uranbey, S. *et al.* (2012) Genome-wide identification of miRNAs responsive to drought in peach (*Prunus persica*) by high-throughput deep sequencing. *PLoS ONE* 7(12), e50298.

Eldem, V., Okay, S. and Ünver, T. (2013) Plant microRNAs: New players in functional genomics. *Turkish Journal of Agriculture and Forestry* 37(1), 1–21.

Fahlgren, N., Howell, M.D., Kasschau, K.D., Chapman, E.J., Sullivan, C.M. *et al.* (2007) High-throughput sequencing of *Arabidopsis* microRNAs: Evidence for frequent birth and death of miRNA genes. *PLoS ONE* 2(2), e219.

Feder, M.E. and Hofmann, G.E. (1999) Heat-shock proteins, molecular chaperones, and the stress response: Evolutionary and ecological physiology. *Annual Review of Physiology* 61(1), 243–282.

Fujii, H., Chiou, T.J., Lin, S.I., Aung, K. and Zhu, J.K. (2005) A miRNA involved in phosphate-starvation response in *Arabidopsis*. *Current Biology* 15(22), 2038–2043.

Fujita, M., Fujita, Y., Noutoshi, Y., Takahashi, F., Narusaka, Y. *et al.* (2006) Crosstalk between abiotic and biotic stress responses: A current view from the points of convergence in the stress signaling networks. *Current Opinion in Plant Biology* 9(4), 436–442.

Gielen, H., Remans, T., Vangronsveld, J. and Cuypers, A. (2016) Toxicity responses of Cu and Cd: The involvement of miRNAs and the transcription factor SPL7. *BMC Plant Biology* 16(1), 145.

Grover, A., Mittal, D., Negi, M. and Lavania, D. (2013) Generating high temperature tolerant transgenic plants: Achievements and challenges. *Plant Science* 38–47.

Guo, L., Qiu, J., Ye, C., Jin, G., Mao, L. *et al.* (2017) *Echinochloa crus-galli* genome analysis provides insight into its adaptation and invasiveness as a weed. *Nature Communications* 8(1), 1031.

Gupta, O.P., Sharma, P., Gupta, R.K. and Sharma, I. (2014) MicroRNA mediated regulation of metal toxicity in plants: Present status and future perspectives. *Plant Molecular Biology* 84(1–2), 1–18.

Hamrick, J.L. and Godt, M.W. (1996) Effects of life history traits on genetic diversity in plant species. *Philosophical Transactions of the Royal Society of London. Series B: Biological Sciences* 351(1345), 1291–1298.

Hartl, F.U., Bracher, A. and Hayer-Hartl, M. (2011) Molecular chaperones in protein folding and proteostasis. *Nature* 475(7356), 324–332.

He, X.-F., Fang, Y.-Y., Feng, L. and Guo, H.-S. (2008) Characterization of conserved and novel microRNAs and their targets, including a tumv-induced TIR-NBS-LRR class R gene-derived novel miRNA in brassica. *FEBS Letters* 582(16), 2445–2452.

Herman, J.J., Sultan, S.E., Horgan-Kobelski, T. and Riggs, C. (2012) Adaptive transgenerational plasticity in an annual plant: Grandparental and parental drought stress enhance performance of seedlings in dry soil. *Integrative and Comparative Biology* 52(1), 77–88.

Hermeking, H. (2012) MicroRNAs in the p53 network: Micromanagement of tumour suppression. *Nature Reviews. Cancer* 12(9), 613–626.

Holoch, D. and Moazed, D. (2015) RNA-mediated epigenetic regulation of gene expression. *Nature Reviews. Genetics* 16(2), 71–84.

Hotamisligil, G.S. and Davis, R.J. (2016) Cell signaling and stress responses. *Cold Spring Harbor Perspectives in Biology* 8(10), a006072.

Huang, S.Q., Peng, J., Qiu, C.X. and Yang, Z.M. (2009) Heavy metal-regulated new microRNAs from rice. *Journal of Inorganic Biochemistry* 103(2), 282–287.

Huang, S.Q., Xiang, A.L., Che, L.L., Chen, S., Li, H. *et al.* (2010) A set of miRNAs from *Brassica napus* in response to sulphate deficiency and cadmium stress. *Plant Biotechnology Journal* 8(8), 887–899.

Hudder, A. and Novak, R.F. (2008) MiRNAs: Effectors of environmental influences on gene expression and disease. *Toxicological Sciences* 103(2), 228–240.

Huo, C., Zhang, B. and Wang, R. (2022) Research progress on plant noncoding RNAs in response to low-temperature stress. *Plant Signaling & Behavior* 17(1), 2004035.

Hutchison, C.A. III., Newbold, J.E., Potter, S.S. and Edgell, M.H. (1974) Maternal inheritance of mammalian mitochondrial DNA. *Nature* 251(5475), 536–538.

Istrail, S., De-Leon, S.B.-T. and Davidson, E.H. (2007) The regulatory genome and the computer. *Developmental Biology* 310(2), 187–195.

Jin, J., Tian, F., Yang, D.-C., Meng, Y.-Q., Kong, L. *et al.* (2017) PlantTFDB 4.0: Toward a central hub for transcription factors and regulatory interactions in plants. *Nucleic Acids Research* 45(D1), D1040–D1045.

Kampinga, H.H. and Craig, E.A. (2010) The HSP70 chaperone machinery: J proteins as drivers of functional specificity. *Nature Reviews. Molecular Cell Biology* 11(8), 579–592.

Kamthan, A., Chaudhuri, A., Kamthan, M. and Datta, A. (2015) Small RNAs in plants: Recent development and application for crop improvement. *Frontiers in Plant Science* 6, 208.

Kerchev, P., Meer, T., Sujeeth, N., Verlee, A., Stevens, C.V. *et al.* (2020) Molecular priming as an approach to induce tolerance against abiotic and oxidative stresses in crop plants. *Biotechnology Advances* 40, 107503.

Khraiwesh, B., Zhu, J.-K. and Zhu, J. (2012) Role of miRNAs and siRNAs in biotic and abiotic stress responses of plants. *Biochimica et Biophysica Acta* 1819(2), 137–148.

Kozomara, A. and Griffiths-Jones, S. (2014) MiRBase: Annotating high confidence microRNAs using deep sequencing data. *Nucleic Acids Research* (Database Issue) 42.

Kundu, A., Paul, S., Dey, A. and Pal, A. (2017) High throughput sequencing reveals modulation of microRNAs in *Vigna mungo* upon Mungbean Yellow Mosaic India virus inoculation highlighting stress regulation. *Plant Science* 257, 96–105.

Lai, X., Wolkenhauer, O. and Vera, J. (2016) Understanding microRNA-mediated gene regulatory networks through mathematical modelling. *Nucleic Acids Research* 44(13), 6019–6035.

Lau, J.A. and Lennon, J.T. (2012) Rapid responses of soil microorganisms improve plant fitness in novel environments. *Proceedings of the National Academy of Sciences* 109(35), 14058–14062.

Lauter, N., Kampani, A., Carlson, S., Goebel, M. and Moose, S.P. (2005) MicroRNA172 down-regulates glossy15 to promote vegetative phase change in maize. *Proceedings of the National Academy of Sciences* 102(26), 9412–9417.

Le, T.D., Liu, L., Liu, B., Tsykin, A., Goodall, G.J. *et al.* (2013) Inferring microRNA and transcription factor regulatory networks in heterogeneous data. *BMC Bioinformatics* 14(1), 92.

Li, L., Miu, K.K., Gu, S., Cheung, H.H. and Chan, W.Y. (2018) Comparison of multi-lineage differentiation of hiPSCs reveals novel miRNAs that regulate lineage specification. *Scientific Reports* 8(1), 9630.

Li, Y., Zhou, W., Hu, B., Min, M., Chen, P. *et al.* (2011) Integration of algae cultivation as biodiesel production feedstock with municipal wastewater treatment: Strains screening and significance evaluation of environmental factors. *Bioresource Technology* 102(23), 10861–10867.

Lindemose, S., Jensen, M.K., Velde, J., O'Shea, C., Heyndrickx, K.S. *et al.* (2014) A DNA-binding-site landscape and regulatory network analysis for NAC transcription factors in *Arabidopsis thaliana*. *Nucleic Acids Research* 42(12), 7681–7693.

Lu, X.-Y. and Huang, X.-L. (2008) Plant miRNAs and abiotic stress responses. *Biochemical and Biophysical Research Communications* 368(3), 458–462.

Ma, Z. and Hu, L. (2023) MicroRNA: A dynamic player from signalling to abiotic tolerance in plants. *International Journal of Molecular Sciences* 24(14), 11364.

Mallory, A.C., Bartel, D.P. and Bartel, B. (2005) MicroRNA-directed regulation of *Arabidopsis* AUXIN RESPONSE FACTOR17 is essential for proper development and modulates expression of early AUXIN RESPONSE genes. *The Plant Cell* 17(5), 1360–1375.

Martin, C. and Zhang, Y. (2007) Mechanisms of epigenetic inheritance. *Current Opinion in Cell Biology* 19(3), 266–272.

Meng, J.G., Zhang, X.D., Tan, S.K., Zhao, K.X. and Yang, Z.M. (2017) Genome-wide identification of cd-responsive NRAMP transporter genes and analyzing expression of NRAMP 1 mediated by miR167 in *Brassica napus*. *Biometals* 30(6), 917–931.

Mitsuda, N. and Ohme-Takagi, M. (2009) Functional analysis of transcription factors in *Arabidopsis*. *Plant & Cell Physiology* 50(7), 1232–1248. DOI: 10.1038/ncomms11846.

Miyoshi, T., Ito, K., Murakami, R. and Uchiumi, T. (2016) Structural basis for the recognition of guide RNA and target DNA heteroduplex by Argonaute. *Nature Communications* 7(1), 11846. DOI: 10.1038/ncomms11846.

Morris, K.V. and Mattick, J.S. (2014) The rise of regulatory RNA. *Nature Reviews. Genetics* 15(6), 423–437.

Navarro, L., Dunoyer, P., Jay, F., Arnold, B., Dharmasiri, N. *et al.* (2006) A plant miRNA contributes to antibacterial resistance by repressing auxin signaling. *Science* 312(5772), 436–439.

Nazarov, P.V., Reinsbach, S.E., Muller, A., Nicot, N., Philippidou, D. *et al.* (2013) Interplay of microRNAs, transcription factors and target genes: Linking dynamic expression changes to function. *Nucleic Acids Research* 41(5), 2817–2831.

Niu, Q.-W., Lin, S.-S., Reyes, J.L., Chen, K.-C., Wu, H.-W. *et al.* (2006) Expression of artificial microRNAs in transgenic *Arabidopsis thaliana* confers virus resistance. *Nature Biotechnology* 24(11), 1420–1428.

Noman, A. and Aqeel, M. (2017) MiRNA-based heavy metal homeostasis and plant growth. *Environmental Science and Pollution Research International* 24(11), 10068–10082.

Osório, J. (2016) Gene regulation: Landscape and mechanisms of transcription factor cooperativity. *Nature Reviews. Genetics* 17(1), 5. DOI: 10.1038/nrg.2015.11.

Pagano, L., Rossi, R., Paesano, L., Marmiroli, N. and Marmiroli, M. (2021) MiRNA regulation and stress adaptation in plants. *Environmental and Experimental Botany* 184, 104369.

Pan, C., Ye, L., Zheng, Y., Wang, Y., Yang, D. *et al.* (2017) Identification and expression profiling of microRNAs involved in the stigma exsertion under high-temperature stress in tomato. *BMC Genomics* 18(1), 843.

Patel, P., Yadav, K., Ganapathi, T.R. and Penna, S. (2019) Plant miRNAome: Cross talk in abiotic stressful times. In: *Genetic Enhancement of Crops for Tolerance to Abiotic Stress: Mechanisms and Approaches*, Vol. I. Springer, pp. 25–52.

Patwa, N., Nithin, C., Bahadur, R.P. and Basak, J. (2019) Identification and characterization of differentially expressed *Phaseolus vulgaris* miRNAs and their targets during mungbean yellow mosaic india virus infection reveals new insight into phaseolus-MYMIV interaction. *Genomics* 111(6), 1333–1342.

Pérez-Quintero, Á.L., Quintero, A., Urrego, O., Vanegas, P. and López, C. (2012) Bioinformatic identification of cassava miRNAs differentially expressed in response to infection by *Xanthomonas axonopodis* pv.Manihotis. *BMC Plant Biology* 12(1), 29.

Pu, M., Chen, J., Tao, Z., Miao, L., Qi, X. *et al.* (2019) Regulatory network of miRNA on its target: Co-ordination between transcriptional and post-transcriptional regulation of gene expression. *Cellular and Molecular Life Sciences* 76(3), 441–451.

Qu, J., Ye, J. and Fang, R. (2007) Artificial microRNA-mediated virus resistance in plants. *Journal of Virology* 81(12), 6690–6699. DOI: 10.1128/JVI.02457-06.

Rasmann, S. and Agrawal, A.A. (2011) Latitudinal patterns in plant defense: Evolution of cardenolides, their toxicity and induction following herbivory. *Ecology Letters* 14(5), 476–483.

Rassoulzadegan, M., Grandjean, V., Gounon, P., Vincent, S., Gillot, I. *et al.* (2006) RNA-mediated non-mendelian inheritance of an epigenetic change in the mouse. *Nature* 441(7092), 469–474.

Rhoades, M.W., Reinhart, B.J., Lim, L.P., Burge, C.B., Bartel, B. *et al.* (2002) Prediction of plant microRNA targets. *Cell* 110(4), 513–520.

Rogers, K. and Chen, X. (2013) Biogenesis, turnover, and mode of action of plant microRNAs. *The Plant Cell* 25(7), 2383–2399.

Sala, O.E., Stuart Chapin, F., Armesto, J.J., Berlow, E., Bloomfield, J. *et al.* (2000) Global biodiversity scenarios for the year 2100. *Science* 287(5459), 1770–1774.

Salih, H., Gong, W., He, S., Sun, G., Sun, J. *et al.* (2016) Genome-wide characterization and expression analysis of MYB transcription factors in *Gossypium hirsutum*. *BMC Genetics* 17(1), 129.

Salvi, S., Sponza, G., Morgante, M., Tomes, D., Niu, X. *et al.* (2007) Conserved noncoding genomic sequences associated with a flowering-time quantitative trait locus in maize. *Proceedings of the National Academy of Sciences* 104(27), 11376–11381.

Sokolova, I.M. (2013) Energy-limited tolerance to stress as a conceptual framework to integrate the effects of multiple stressors. *Integrative and Comparative Biology* 53(4), 597–608.

Sun, M., Shen, Y., Chen, Y., Wang, Y., Cai, X. *et al.* (2022) Osa-miR1320 targets the ERF transcription factor oserf096 to regulate cold tolerance via JA-mediated signaling. *Plant Physiology* 189(4), 2500–2516.

Sunkar, R. (2010) MicroRNAs with macro-effects on plant stress responses. *Seminars in Cell & Developmental Biology* 21(8), 805–811.

Sunkar, R., Chinnusamy, V., Zhu, J. and Zhu, J.-K. (2007) Small RNAs as big players in plant abiotic stress responses and nutrient deprivation. *Trends in Plant Science* 12(7), 301–309.

Thomson, D.W. and Dinger, M.E. (2016) Endogenous microRNA sponges: Evidence and controversy. *Nature Reviews. Genetics* 17(5), 272–283.

Warthmann, N., Chen, H., Ossowski, S., Weigel, D. and Hervé, P. (2008) Highly specific gene silencing by artificial miRNAs in rice. *PLoS ONE* 3(3), e1829.

Xu, T., Zhang, L., Yang, Z., Wei, Y. and Dong, T. (2021) Identification and functional characterization of plant miRNA under salt stress shed light on salinity resistance improvement through mirRNA manipulation in crops. *Frontiers in Plant Science* 12, 665439. DOI: 10.3389/fpls.2021.665439.

Yang, C., Li, D., Mao, D., Liu, X., Ji, C. *et al.* (2013) Overexpression of microRNA319 impacts leaf morphogenesis and leads to enhanced cold tolerance in rice (*Oryza sativa* L.). *Plant, Cell and Environment* 36(12), 2207–2218.

Yang, X., Liu, F., Zhang, Y., Wang, L. and Cheng, Y. (2017) Cold-responsive miRNAs and their target genes in the wild eggplant species *Solanum aculeatissimum*. *BMC Genomics* 18(1), 1000.

Yao, Q., Chen, Y. and Zhou, X. (2019) The roles of microRNAs in epigenetic regulation. *Current Opinion in Chemical Biology* 51, 11–17.

Yu, H. and Gerstein, M. (2006) Genomic analysis of the hierarchical structure of regulatory networks. *Proceedings of the National Academy of Sciences* 103(40), 14724–14731.

Zhang, B., Pan, X., Cobb, G.P. and Anderson, T.A. (2006) Plant microRNA: A small regulatory molecule with big impact. *Developmental Biology* 289(1), 3–16.

Zhang, L.W., Song, J.B., Shu, X.X., Zhang, Y. and Yang, Z.M. (2013) MiR395 is involved in detoxification of cadmium in *Brassica napus*. *Journal of Hazardous Materials* 250–251, 204–211.

Zhang, M., An, P., Li, H., Wang, X., Zhou, J. *et al.* (2019) The miRNA-mediated post-transcriptional regulation of maize in response to high temperature. *International Journal of Molecular Sciences* 20(7), 1754.

Zhang, W., Gao, S., Zhou, X., Chellappan, P., Chen, Z. *et al.* (2011) Bacteria-responsive microrRNAs regulate plant innate immunity by modulating plant hormone networks. *Plant Molecular Biology* 75(1), 93–105.

Zhang, Z., Shrestha, J., Tateda, C. and Greenberg, J.T. (2014) Salicylic acid signaling controls the maturation and localization of the *Arabidopsis* defense protein ACCELERATED CELL DEATH6. *Molecular Plant* 7(8), 1365–1383.

Zhao, B., Liang, R., Ge, L., Li, W., Xiao, H. *et al.* (2007) Identification of drought-induced microRNAs in rice. *Biochemical and Biophysical Research Communications* 354(2), 585–590.

Zhao, S. and Liu, M.-F. (2009) Mechanisms of microRNA-mediated gene regulation. *Science in China Series C: Life Sciences* 52(12), 1111–1116.

Zhou, X., Wang, G., Sutoh, K., Zhu, J.-K. and Zhang, W. (2008) Identification of cold-inducible microRNAs in plants by transcriptome analysis. *Biochimica et Biophysica Acta (BBA) – Gene Regulatory Mechanisms* 1779(11), 780–788.

Zuo, Z.-F., He, W., Li, J., Mo, B. and Liu, L. (2021) Small RNAs: The essential regulators in plant thermotolerance. *Frontiers in Plant Science* 12, 726762.

26 Short Interfering RNAs-based Immunity Against Biotic Stressors: An Updated Overview

Reham I. Abdelhamid[1]*, Ahmed G. Soliman[2], Mariam A. Abd El-Samea[3], Nourhan M. Battor[4], Nesma A. Zakari[3], Yasmeen A. Roby[3], Mayar A. Elnagar[3] and Sozan E. El-Abeid[4]

[1]Sugar Crops Research Institute, Agricultural Research Center (ARC), Giza, Egypt; [2]Faculty of Agriculture, Department of Biotechnology, Ain Shams University; [3]Faculty of Science, Department of Biotechnology, Cairo University; [4]Nanotechnology & Advanced Nano-Materials Laboratory (NANML), Plant Pathology Research Institute, Agricultural Research Center, Giza, Egypt

Abstract

Short interfering RNAs (siRNAs) have emerged as integral components of plant immunity against biotic stressors. Upon perception of pathogens or pests, the RNA interference pathway is activated, leading to the biogenesis of siRNAs that mediate targeted gene silencing in the host and pathogen genes. This post-transcriptional regulation aids in plant defense through the inhibition of pathogen virulence factors, systemic signaling, and crosstalk with other defense mechanisms. Bioinformatics plays a pivotal role in deciphering the complex functions of siRNAs in plant biotic stress response. Computational tools enable the profiling of stress-induced siRNAs, prediction of their gene targets, analysis of pathogen genetic variation, and functional annotation via integration with genomic datasets. Advances in the understanding of siRNA-mediated plant immunity through bioinformatics can ultimately inform resilient crop development and enhance biotic stress management.

Keywords: siRNAs, Biotic stress, Plant immunity, RNA interference, Bioinformatics

26.1 Role of siRNAs in Plant Immunity

Short interfering RNAs (siRNAs) are small RNA molecules that play pivotal roles in plant immunity against various biotic stressors, including pathogens and pests. These stressors trigger a complex network of molecular responses in plants, which produces specific siRNAs (Islam *et al.*, 2018). One of their varying roles is participating in host defense mechanisms against pathogens. RNA contributes to the host defense mechanism in two significant ways (Muthamilarasan and Prasad, 2013). The first is pattern recognition of non-self-motifs of RNA by sequence-specific RNA binding proteins (RNA helicases and ribonucleases) that activate host defense response to

*Corresponding author: rehamezz78@gmail.com

 Epigenetics for Climate-Smart and Sustainable Agriculture (ed. Jen-Tsung Chen)
DOI: 10.1079/9781800626102.0026

destroy foreign RNA. Sucrose non-fermenting protein kinase and protein kinase R (PKR) phosphorylate eIF2α prevent initiation of translation. The second is the production of small RNA species that mediate sequence-specific RNA degradation to silence gene expression of foreign viruses, plasmids, and transposable genetic elements (Casas-Mollano *et al.*, 2016; Hung and Slotkin, 2021). RNA-mediated gene silencing is now appreciated as a fundamental mode of gene regulation in eukaryotes. However, RNA interference (RNAi) was only described in 1998 when sequence-specific degradation of endogenous mRNA was observed in *Caenorhabditis elegans* with its homologous dsRNA (Sugimoto, 2005). This phenomenon was a reverse from the researchers' expectation, and they were concerned about off-targeting gene silencing. Studies have shown that off-targeting is dose- and sequence-dependent and makes an invaluable contribution to cancer therapy and the treatment of genetic disorders. RNA silencing also occurs in plants through the degradation of transcripts from transgenes, viruses, and viroids through the post-transcriptional gene silencing (PTGS) pathway, considered to be the plant equivalent of RNAi. Another mode of RNA-mediated defense in plants is through the cleavage of foreign RNA by RNA ribonuclease proteins to initiate the RNA degradation and mRNA degradation-dependent defense mechanisms. RNA-mediated gene silencing is now considered a common strategy for genetic modifications to protect plants against viral pathogens and quality traits for desired phenotypes (Bartok and Hartmann, 2020; Uehata and Takeuchi, 2020; Guillemin *et al.*, 2022).

26.2 Biogenesis of siRNA

The mechanisms of siRNA generation can vary among different organisms and tissues according to the unique requirements for the RNA silencing pathways in those cell types. The size and structural characteristics of the RNA substrate are likely primary determinants of which Dicer enzyme gets loaded and how extensively Dicer will process that RNA into small RNAs of diverse sizes and functions (Sanan-Mishra *et al.*, 2021). The differential association of Dicer and its dsRNA binding partner proteins with distinct RNA substrates may be key for specifying the siRNA pathway vs other small RNA pathways, RNA-induced silencing complex (RISC) loading, and the selection between silencing and target RNA regulation by the miRNA pathway, the specificity between related gene silencing and repression of translation, and potentially influencing the RNA-induced transcriptional silencing (RITS) pathway involved in heterochromatin formation (Vergani-Junior *et al.*, 2021; Fujimoto and Iwakawa, 2023).

Small RNA molecules play a crucial role in DNA and RNA silencing as guided by homology between the small RNA and the target nucleic acid. In the yeast *Schizosaccharomyces pombe*, RNA-directed RNA polymerases (Rdp1 and Rdp2) are responsible for producing double-stranded RNA (dsRNA) (Wagnon, 2009). Dicer then cleaves this dsRNA to generate siRNAs specific for heterochromatic regions. siRNAs enhance the silencing signal by directing the degradation of primary transcripts or other homologous RNAs. In addition, endogenous hairpin RNA transcripts and convergent genes serve as substrates for Dicer, leading to homology-dependent gene silencing and the creation of heterochromatin (Jose and Hunter, 2007; Djupedal and Ekwall, 2009).

26.2.1 Generation of siRNA

RNA polymerase, the enzyme responsible for the transcription of RNA from DNA, generates a single-stranded precursor to siRNA, termed long hairpin dsRNA. In model organisms where RNAi is well characterized, the long hairpin dsRNA can be introduced into the cell, or it can be generated from an inverted repeat transgene (Guo *et al.*, 2016). Biochemical and genetic analyses have shown that Drosha, a member of the RNase III family of enzymes, functions in the nucleus and cleaves the long hairpin dsRNA to produce a short RNA duplex with 2-nt 3' overhangs. This first step in the RNAi pathway is now referred to as step 0, since this RNA duplex is an siRNA precursor (Ciechanowska *et al.*, 2021). Sense shuttles between the nucleus and cytoplasm.

This shuttling activity requires specific protein motifs and an intact ATPase motif, and when these motifs are disrupted, the enzyme remains in the cytoplasm, resulting in no RNAi activity (Haward *et al.*, 2021).

26.2.2 Dicer enzyme

Drosophila melanogaster Dicer 1 (DmDcr 1) is crucial for siRNA biogenesis (Fig. 26.1). The 1836 amino acid polypeptide is essential for producing siRNAs, with N-terminal helicase and two catalytic RNase III domains necessary for its activity. Deletion of these domains in *Drosophila* cells impairs DmDcr 1's ability to process double-stranded RNA into siRNAs, vital in the RNA interference pathway for gene regulation. Overall, DmDcr 1's role in siRNA production aligns with established molecular biology understanding (Soleimani *et al.*, 2020; Du and Patel, 2022).

In vitro assays using bacterially expressed and refolded recombinant proteins have demonstrated that the minimal short dsRNA processing activity of the enzyme is located in the two

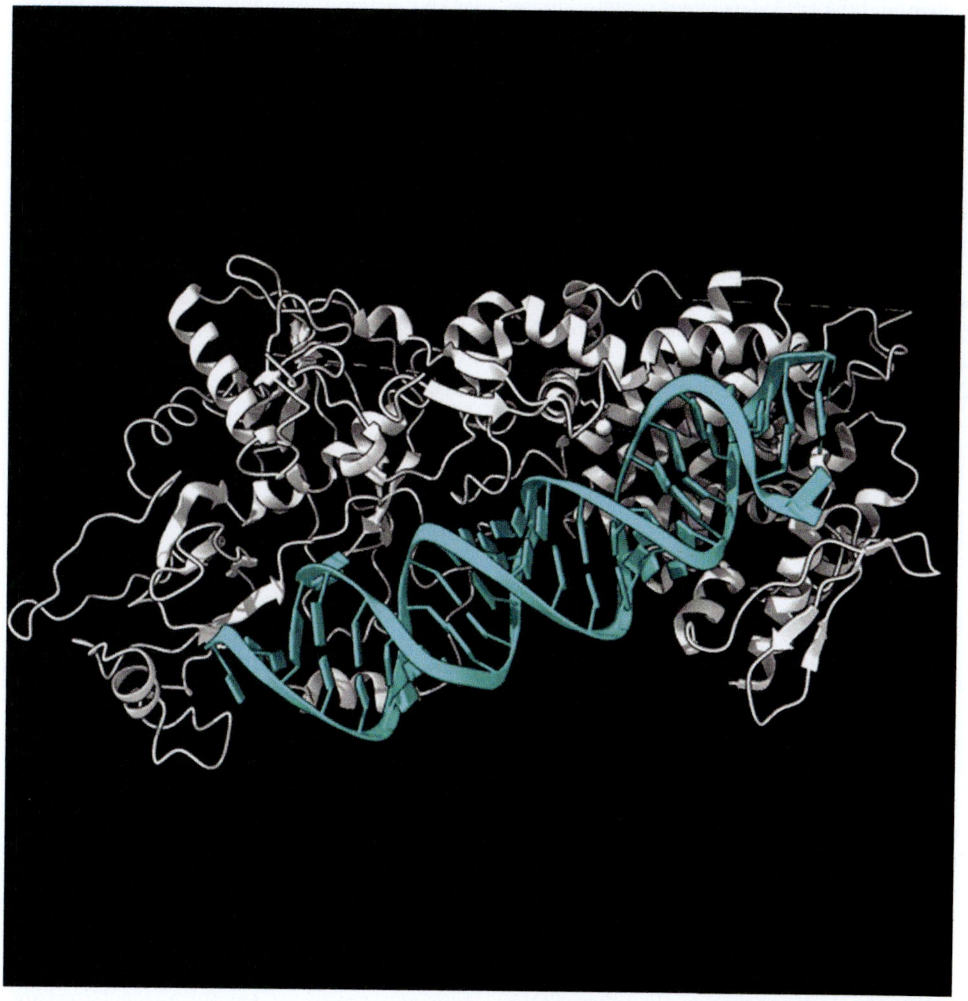

Fig. 26.1. Dicer enzyme. White: Dicer enzyme; cyan: RNA (73-MER). Presented from Chimera software retrieved from PDB ID: 7XW2. Author's own image.

RNase III domains. This has been demonstrated for humans for Dicer as well as the *Arabidopsis* and *Drosophila* enzymes (Su *et al.*, 2022). Full-length recombinant proteins also demonstrate robust ATP-independent unwinding of long dsRNA concurrent with the single dsRNA cleavage event. Given that the helicase domain is dispensable for *in vitro* processing activity, it has been proposed that Dicer may interact with other cellular helicases to facilitate dsRNA unwinding *in vivo*. It has been suggested from genetic studies in yeast and other organisms that RNA helicases are required for RNAi, and from the identification of Dicer protein interacting factors such as the drsh mRNA cap-binding protein and the DNA-dependent protein kinase (Su *et al.*, 2022; Zhang *et al.*, 2023).

The resultant product of the Dicer enzyme is known as short interfering RNA or siRNA and is the universal initiator of RNA interference in any organism. It has been shown that siRNAs can be synthesized using bulky integrated protein BIP labeling, as well as using *in vitro* transcribed RNA, some of which have a 5' cap and 3' polyA tail. Although the observation of a variety of siRNAs migrating at different positions in denaturing PAGE is likely due to RNA structure inter-heterogeneity, it is unclear if Dicer processing is somehow selectively targeted to release distinct groups of siRNAs (Vergani-Junior *et al.*, 2021).

26.2.3 RISC complex

The RNA-induced silencing complex (RISC) is composed of Argonaute, the protein with endonuclease activity, and the siRNA which is cleaved between the 10th and 11th nucleotide once it has been incorporated into the RISC. There are several different Argonaute proteins in *Arabidopsis*, three of which (AGO2, AGO7, and ZLL/AGO10) have been shown to be involved in antiviral defense and the silencing of transgenes with varying levels of importance (Jones-Rhoades *et al.*, 2006). AGO2 is the primary protein involved in antiviral defense and is the one that directly incorporates the siRNA. The exact function of the other siRNA-binding Argonaute proteins in antiviral defense and transgene silencing is unknown. Once the siRNA has been incorporated, the RISC complex scans the mRNA for complementary sequences (Odokonyero, 2012). This is where there is currently a gap in knowledge between the understanding of siRNA action in PTGS and antiviral defense. It is not known whether the siRNA anneals to the viral mRNA and then cleaves it, or the RISC cleaves mRNA that has the complementary sequence to the siRNA regardless of where it had come from. Once the siRNA has guided the RISC to viral mRNA, there is evidence in *Arabidopsis* and other organisms that there is systemic movement of the signal to initiate silencing. This is a potential future area of research in plant antiviral defense and PeaDSV action (Shamandi *et al.*, 2015; Sanan-Mishra *et al.*, 2021).

Studies in *Arabidopsis* using Potato Virus X and Tobacco mosaic virus have revealed a link between RNA silencing of different genes and the ability to resist symptoms caused by viruses. Using transgenes for PVX coat protein and inducible clones of GFP silencing from TMV, it was noticed that symptoms worsened in transgenic lines lacking siRNA-specific AGO2 or with mutations in the AGO2 protein. The same outcome was observed when a line of *Arabidopsis* with reduced expression of the AGO7 gene was infected with TMV. This research holds great potential in enhancing comprehension of RISC–siRNA roles in antiviral defense and potentially in creating transgenic lines that are resistant in the future (Harvey *et al.*, 2011; Garcia-Ruiz *et al.*, 2015; Jagram and Dasgupta, 2023).

26.2.4 Stability of siRNA

The high G/C content of siRNA is another factor that contributes to the stability of siRNA. This is again due to the increased stability of the double-stranded region of siRNA within RISC. In contrast to this, it is suggested that low-stability siRNA is recognized by RISC and causes immediate degradation of the siRNA, effectively clearing it from the gene silencing pool (Tuttolomondo and Ditzel, 2021). siRNA with 3' overhangs (the typical product of Dicer activity) displays great resistance to various denaturing conditions and to basic conditions. This stability is due to the protection of the 3' overhang within a siRNA by the Argonaute2 protein. It is believed that this stabilized siRNA is sequestered within RISC for prolonged periods of time, causing increased effectiveness and duration of gene silencing (Wojnicka *et al.*, 2020). siRNA stability is an important factor in understanding the duration and effectiveness of siRNA-mediated gene silencing. It is interesting to note that various studies have provided evidence that the stability of siRNA may differ depending on the mode of introduction of siRNA. For example, chemically synthesized siRNA might have different stability compared with the *in vitro* transcribed siRNA (Safari *et al.*, 2017; Wojnicka *et al.*, 2020).

26.2.5 Transport of siRNA

The significance of gene regulation in plants during disease is underscored by siRNA transport. It is necessary to analyze the difference between natural and induced therapy methods in gene regulation. If gene silencing is induced, it needs to target the specific pathogen, and there is proof that viroid or virus-induced diseases can also be controlled by their natural RNA silencing mechanisms. This indicates that pathogens have ways to counteract the host's RNA-silencing mechanisms (Mann *et al.*, 2023). When there is no specific gene regulation process involved in fighting off a pathogen, there can be cytotoxicity – a non-selective killing of cells linked to a powerful defense reaction. This could potentially be dangerous for the host and has been proposed as a possibility in defense mechanisms involving induced gene silencing, resulting in the suppression of a resistant gene's expression. In order to successfully alter the defense mechanism for a particular pathway, it is important to evaluate the duration of effect and off-target inhibition for gene regulation, as it plays a crucial role in maintaining homeostasis for other defense responses. Through comprehending the different methods and routes for RNA silencing processes and gene regulation in plants, novel approaches for antiviral treatment can be developed, with a focus on targeting particular diseases and minimizing impacts on the host (Guzman-Aranguez *et al.*, 2013; Backer *et al.*, 2019).

siRNAs are known for their specific gene regulation by assisting target mRNA degradation in plants, fungi, and animals (Neumeier and Meister, 2021). Such regulation is important in the ability to ward off virus attacks or mutation. It is suggested that the gene from which siRNA is derived is a defense mechanism and can be identified as the infection of the virus itself. Plants have an apparent need for gene regulation (Tan and Yin, 2004). Microarray experiments have examined the changes in mRNA expression levels of *Arabidopsis* genes upon an attackable pathogen. It was more often than not that genes regulated by siRNAs were inactivated upon attack from a pathogen. In other experiments done with *Arabidopsis* infected with the TMV virus, a virus-resistant transgenic plant containing an altered tobacco gene which made it resistant to the virus was found to not develop disease symptoms, and it maintained high levels of the altered tobacco gene mRNA (Girke *et al.*, 2000). Further study would be to examine the particular mechanism from which it is taking place, but the evidence infers that the resistance to the virus was due to siRNA derived from the altered tobacco gene, effectively regulating the pathogen to a specific gene, and will endure in the plant's memory for the experience of a later attack (Kryovrysanaki *et al.*, 2021).

26.2.6 RNA interference mechanism

The siRNAs and microRNAs (miRNAs) from endogenous genes function in several gene-silencing pathways to either degrade or inhibit translation of mRNA or to direct chromatin alterations at the DNA level. The si/miRNAs involved determine the specific pathway and gene targets are determined by sequence homology (Zeng *et al.*, 2003).

At the heart of the RNAi system is a nuclease that degrades RNA, called RNAseIII. It acts to process long double-stranded RNA molecules into small 21–23 nt RNA fragments with 2-nt 3' overhanging ends, which are the hallmarks of RNA silencing. In plants, these small RNA molecules are termed small interfering RNAs (siRNAs) as they are homologous to the RNA from which they were processed and to other RNAs that have similar sequences (Elbashir *et al.*, 2001). The RNaseIII catalytic component is called Dicer. It is a protein and because it acts on long double-stranded RNAs, it can generate many siRNAs from a single RNA molecule. Dicers have been shown to function in antiviral defense mechanisms in *Arabidopsis* through the specific cleavage of viral RNA. This process turns on when a plant protein called RDRC recruits the Dicer to viral RNA to create siRNA, thus signaling the onset of RNA silencing against the virus (Elbashir *et al.*, 2001; Song and Rossi, 2017).

Small RNA molecules have emerged as major regulators of gene expression. They have been shown to directly interfere with gene transcription at the DNA level and with the synthesis of mRNA or its use in translation into protein. The main components of RNA interference (RNAi) engage in a series of steps. The result, which is a molecule of double-stranded RNA (dsRNA), may be designed to trigger degradation of a particular mRNA, degrade and/or inhibit translation of an mRNA with a complementary sequence, or through a complex network of base pairing that involves siRNA and miRNA will force alterations to chromatin structure at the DNA level (Portnoy *et al.*, 2011).

26.3 Targeted Gene Silencing

Short interfering RNAs (siRNAs) are essential components of plant defense mechanisms and play a crucial role in targeted gene silencing to combat pathogen virulence and enhance host defense pathways. When plants detect the presence of pathogens or pests, they activate the RNA interference (RNAi) pathway, initiating the production of siRNAs that serve as molecular tools for precise gene regulation in both the host plant and invading pathogens (Kong *et al.*, 2022).

26.3.1 Specificity of siRNA-mediated silencing

The specificity of siRNA-mediated gene silencing is fundamental to effective plant immunity. For instance, when a plant is infected by a pathogen carrying a gene responsible for producing a toxin that weakens the plant's defenses, siRNAs can be generated to specifically target and silence the toxin-producing gene. Through base-pairing interactions, siRNAs guide the RNA-induced silencing complex (RISC) to recognize and degrade mRNA molecules containing sequences complementary to the siRNA. This targeted approach ensures that only genes critical for pathogen virulence and host susceptibility are silenced, allowing the plant to mount a tailored defense response against the invading threat (Katiyar-Agarwal *et al.*, 2006)

26.3.2 Regulation of gene expression

By regulating gene expression through targeted silencing, siRNAs modulate the activity of key genes involved in pathogen virulence and host defenses. For example, in response to a pest that targets a specific defense-related gene in the plant, siRNAs can be deployed to silence the gene responsible for its virulence. Silencing disrupts essential biological processes in pests, weakening their ability to infect plants. Concurrently, plants can adjust their gene expression to bolster their defense mechanisms, enhancing their overall immunity against biotic stressors. The precise regulation of gene expression by siRNAs contributes to the ability of plants to effectively fight pathogens and pests (Kong *et al.*, 2022).

26.3.3 Role in host–pathogen interaction

In the intricate interplay between hosts and pathogens, siRNAs act as molecular tools that target specific genes critical for pathogen survival and host susceptibility. For instance, when a plant encounters a pathogen that produces virulence factors that compromise the defense of the plant, siRNAs can be employed to silence the genes responsible for these virulence factors. By disrupting the pathogen's ability to cause harm, siRNAs weaken its infectivity and enhance plant defense responses. This targeted gene-silencing mechanism exemplifies how plants utilize siRNAs to strategically combat pathogens and pests and safeguard their health and productivity in the face of biotic stressors.

Through the precise targeting, regulation, and role of siRNA-mediated gene silencing in host–pathogen interactions, plants can effectively defend themselves against pathogens and pests, highlighting the importance of siRNAs in plant immunity and biotic stress management.

26.4 Systemic Defense Signaling

Systemic defense signals are a vital factor in plant immunity. They include transmitting shielding alerts from the site of infection to different parts of the plant, preparing it for ability assault via pathogens. Here are some key factors.

- **Systemin:** This is an 18-amino acid peptide plant hormone that coordinates nearby and systemic immune responses (Zhang *et al.*, 2020). Systemin-mediated activation of the canonical systemic signaling pathway includes the release of systemin from its precursor prosystemin, binding of systemin to its membrane receptor SYSTEMIN RECEPTOR1 (SYR1), and long-distance delivery of signaling molecules (Zhang *et al.*, 2020).
- **Salicylic acid (SA):** SA performs a vital role in systemic defense signaling. In chickpea flowers, for instance, SA biogenesis takes place through the interest of phenylalanine ammonia-lyase (PAL) that turns on other SA-responsive genes and TGA transcription elements to induce energetic defense in opposition to *Fusarium oxysporum* f. sp. Ciceris Race 1 (Foc1) (Bhar *et al.*, 2018).
- **Reactive oxygen species (ROS):** ROS act as chemical indicators in stimulating a downstream systemic resistance response (Bhar *et al.*, 2018).
- **Other molecules:** Other molecules such as jasmonic acid, procystemin mRNA, VOCs, and perhaps systemin itself also are involved in long-distance signaling (Zhang *et al.*, 2020).

These mechanisms assist vegetation in integrating more than one long-range indicator into their systemic defense response. They are a part of a complex signaling device concerning various molecules (Zhang *et al.*, 2020). This is a captivating region of research with many ongoing studies aimed at increasing knowledge of these strategies and their effects on plant health and productivity.

26.4.1 Regulation of gene expression

An interesting example of targeted gene silencing in plants comes from the work on *Medicago truncatula* resistance to the oomycete pathogen *Aphanomyces euteiches*. A loss-of-function mutation in the *Medicago truncatula* Dicer (MtDcl1) gene leads to increased susceptibility to *Aphanomyces euteiches* (Shamandi *et al.*, 2015). This gene encodes a protein involved in the first step of the RNA-dependent DNA methylation pathway, so one would expect that the Dicer gene would be involved in gene silencing through siRNA production. This was indeed the case, and gene expression analysis of wild-type and Dicer mutant plants revealed that Dicer was required for the gene silencing of an R gene that imparted resistance to *Aphanomyces euteiches*. This absence of gene silencing resulted in constitutive R gene expression and led to the autoimmune response mentioned earlier. This example pairs an R gene and a known disease resistance response, so the effects of the Dicer mutation on overall plant resistance were relatively easy to assess. The fact that only Dicer and not AGO proteins were required to prevent autoimmunity suggests that siRNAs were being directly produced from the R gene transcript without its translation into protein. Although a more detailed analysis of MtDcl1 and siRNA in R gene silencing was not possible, this scenario likely involved primary siRNAs and phased siRNAs that were acting on a specific mRNA transcript (Shamandi *et al.*, 2015).

Regulation of gene expression in plants is a complex process that can involve proteins and/or small RNAs. It can either increase or decrease gene expression. It is known that increased expression of a gene involved in defense can be regulated by a TF to coordinate a better defense response. Many R genes can increase susceptibility to certain pathogens, so this increased expression of an R gene will need to be specifically inhibited for constitutive host defense to work properly. The most well-understood examples of gene expression regulation in plant defense are related to the targeted silencing of R genes and other defense-related genes (Rehem *et al.*, 2012; Lee *et al.*, 2021).

26.5 Crosstalk with Other Defense Mechanisms

Plant immunity involves essential systemic defense signals in response to biotic stresses. When plants come across pathogens or pests, they trigger various defense mechanisms. One key systemic reaction involves the creation of signaling molecules like salicylic acid, jasmonic acid, and ethylene, which are able to travel through the plant to trigger defense reactions in other parts of the plant (Rejeb *et al.*, 2014). These pathways of communication assist the plant in organizing a holistic defense plan, frequently leading to the production of proteins related to disease development and the fortification of cell walls to inhibit pathogen transmission. This systemic acquired resistance (SAR) is a crucial component of the plant's defense mechanism against biotic stressors, such as bacteria, viruses, fungi, nematodes, and insects. Understanding these defense mechanisms is crucial for improving sustainable agriculture and preventing economic crop losses from biotic stress (Gimenez *et al.*, 2018; Ghosh *et al.*, 2019).

26.5.1 Epigenetic modifications (EM)

Though it is not completely understood at this time, gene body DNA methylation seems to be promoting gene silencing through siRNA production, which leads to RNA silencing and PTGS. Transposable elements are also silenced through DNA methylation, histone modification, and siRNA production and it has been shown in several instances that transposon siRNAs can move systemically and have secondary effects on other transposable elements or gene body regions. The result of these EM mechanisms is nucleotide identity between the silenced gene and the siRNA that is produced, considerable homology at transgene loci, and, in the case of transposon silencing, homology between transposon siRNAs and siRNAs with secondary effects. Thus, the link between EM mechanisms and siRNA production has widespread implications for gene silencing in plants (Mann *et al.*, 2023).

Recent data in *Arabidopsis* has shown that targeting the *de novo* DNA methyltransferase, DRM2, to a gene body region results in an increase in siRNA production from that gene and leads to an increased probability that the DNA will be histone modified and then silenced through histone modification mechanisms. It is not known whether DNA methylation at gene body regions recruits siRNA production by the normal RNA-dependent RNA polymerase 2 (RDR2) pathway or if it recruits RDR6-mediated siRNA production from a transcript of the gene. Consumer information for siRNA is also not known. These mechanistic questions are currently under investigation in *Arabidopsis* (Naumann *et al.*, 2011; Kumar and Mohapatra, 2024).

Epigenetic modifications have been recognized as a mechanism of gene silencing in plants. In plants, data is accumulating that it also occurs in gene body regions and is associated with active gene expression. The difference in gene body DNA methylation as compared with mammals suggests a somewhat different role for DNA methylation in gene expression (Naumann *et al.*, 2011; Kumar and Mohapatra, 2024).

26.5.2 Recent developments in CRISPR/Cas9 and epigenetics

CRISPR/Cas9 is an immune defense mechanism in bacteria and archaea that originated from Nakata's research at Osaka University in 1987 (Fig. 26.2). The system consists of genes flanking CRISPR loci (Cas1 to Cas4) linked to bacterial immune systems (Barrangou *et al.*, 2007). The locus includes a Cas protein gene, a tracrRNA gene, and a CRISPR locus with lead, repeat, and spacer sequences (Liu *et al.*, 2022). CRISPR/Cas9 system modifications have improved efficiency, stability, and specificity. These modifications have enabled gene editing for crop quality and medication development while addressing safety concerns related to transgenes. CRISPR/Cas9 technology can eliminate transgenes inserted into crops, producing transgene-free lines with minimal risk of off-target mutations

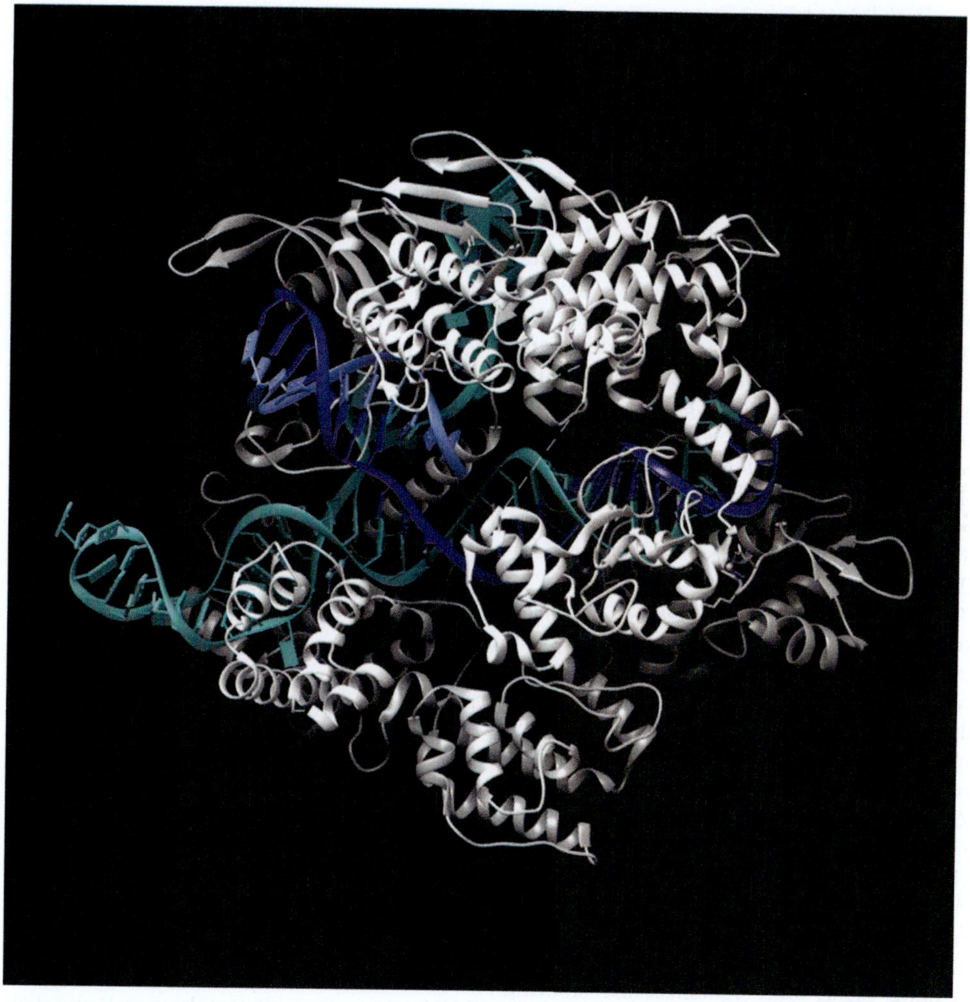

Fig. 26.2. CRISPR/Cas9. White: Cas9 protein; cyan: gRNA; blue: targeted DNA; cornflower blue: mono-targeted DNA, represented by chimera. From PDB ID: 4un3. Author's own image.

(Manghwar *et al.*, 2020; Zheng *et al.*, 2020). The mechanism involves the complex formation of tracrRNA and crRNA to guide Cas9 in cutting specific DNA sequences. Scientists have developed single-stranded sgRNA to enhance targeting. Mutations in the RuvC domain of Cas9 nickase have successfully reduced off-target effects, while modifications to the catalytic domains have allowed for transcriptional control in *Escherichia coli*. The development of CRISPR/Cas9 technology has opened up new possibilities for genetic editing in various fields (Liu *et al.*, 2022).

CRISPR/Cas9 technology is a revolutionary genome editing tool in plants. It works by unwinding the DNA strands, recognizing the protospacer adjacent motif (PAM), and creating a single guide RNA (sgRNA) to target specific genes. This sgRNA, when coupled with the sequence upstream of the PAM, creates a double-stranded break (DSB) that can be repaired using non-homologous end joining (NHEJ) or homology-directed repair (HDR). Gene mutations can occur during NHEJ, while HDR accurately inserts or replaces bases. The technology is widely used in plant research for precision molecular breeding and functional evaluation of crop genes. It enhances stress tolerance, disease resistance, and agronomic performance, improving plant growth characteristics (Liu *et al.*, 2022).

By targeting specific genes, CRISPR/Cas9 enables the development of improved crop varieties with enhanced traits and increased resilience, contributing to global food security. Ongoing research aims to address challenges such as off-target effects and delivery optimization. The system consists of the Cas9 endonuclease and a guide RNA (gRNA) for precise gene targeting and alterations. The sgRNA binds to the target DNA sequence and creates a DSB, which the cell's DNA repair machinery can repair through HDR or NHEJ. Overall, CRISPR/Cas9 holds great promise for sustainable agriculture by providing a tool for precise plant genetic modifications (Wang *et al.*, 2021).

CRISPR/Cas9 plant multiplex genome editing is a new technique that can replace existing transformation methods. Researchers can edit multiple genes simultaneously by introducing Cas9 and sgRNA genes into transgenic plants, particularly those related to plant cell wall construction (Zhang and Showalter, 2020; Zlobin *et al.*, 2020). The planta transformation technique, such as the floral dip method, eliminates the need for *in vitro* regeneration from single transformed cells, saving time and effort, especially for slow-growing plant species. This approach allows for editing redundant genes, creating higher-order mutants, and studying complex gene networks. The CRISPR/Cas9 system has been proven effective in various crops, offering precise genetic engineering capabilities. In rice, it outperforms other methods like TILLING, while in maize, wheat, and tomato, it has shown heritable inheritance of site-specific mutations. Epigenetics, a popular research area, involves histone modification, non-coding RNAs, and DNA methylation, which regulate gene expression (Su *et al.*, 2022). CRISPR/Cas9 technology opens up new possibilities for studying gene relationships and gene families, providing a valuable tool for researchers in the field (Wang *et al.*, 2021). Overall, CRISPR/Cas9 plant multiplex genome editing offers a faster, more efficient way to edit multiple genes simultaneously and explore the complexities of gene regulation in plants (Mushtaq *et al.*, 2021; Zhang and Showalter, 2020).

26.5.3 Examples of successful application: CRISPR/Cas9 enhances crop tolerance

CRISPR/Cas9 technology has demonstrated considerable potential in improving crop resilience to various abiotic and biotic stresses that threaten global food security. Precise gene editing facilitated by this tool has led to advancements in enhancing tolerance to environmental factors, such as drought, salinity, temperature extremes, and heavy-metal pollution. Additionally, CRISPR/Cas9 has been effectively utilized to enhance resistance against fungal pathogens by targeting specific genes such as the wheat MLO gene (Paul *et al.*, 2021). Targeted mutagenesis using this technique has improved plant growth characteristics, including increased iron and zinc concentrations, drought resistance, larger grain size, disease resistance, and salt tolerance.

In tomato breeding, CRISPR/Cas9 has successfully optimized fruit characteristics such as size, yield, color, and texture, while inhibiting ripening inhibitors to extend shelf-life without compromising taste or nutritional value. Targeted genes such as HyPRP1 have enhanced tomato resistance to high salinity levels (Paul *et al.*, 2021). These studies highlight the efficacy of CRISPR/Cas9 in improving the resilience and quality of crops.

Furthermore, soybean research has identified a family of transcription factors called AITRs that play a crucial role in abscisic acid (ABA) response and stress tolerance. Wang *et al.* (2021) utilized CRISPR/Cas9 to edit GmAITR genes, resulting in enhanced salt tolerance in soybean. The GmAITR mutants exhibited altered ABA responses during germination, but increased salt tolerance in various laboratory and field tests, presenting a potentially effective strategy for improving abiotic stress tolerance.

In grapevines, CRISPR/Cas9 has been employed to investigate the impact of the VvPR4b gene on downy mildew resistance. Li *et al.* (2020) designed specific single-guide RNAs (sgRNAs) to target VvPR4b based on its genomic location and potential off-target effects. A CRISPR/Cas9 knockout vector was constructed by recombining the pCACRISPR/Cas9 vector with the VvPR4b-AtU6-sgRNA cassette. Through CRISPR/Cas9-mediated editing of

VvPR4b, a decrease in downy mildew resistance was observed in edited grapevine lines (Li *et al.*, 2020; Zlobin *et al.*, 2020).

These studies demonstrate the versatility and precision of CRISPR/Cas9 in improving crop resilience by targeting specific genes associated with stress tolerance and disease resistance, paving the way for sustainable agriculture and enhancing food security.

26.6 Bioinformatics in siRNA-based Immunity Against Biotic Stressors

As with other biomolecules, the ability to identify coding loci and elucidate the structure of a gene remains the first step in gleaning knowledge about a gene's expression, underlying transcriptional and translational regulation, and functional information about potential biomolecular activity (e.g. catalytic, structural, or recognition function during stress response). Depending on the extent of biotic stress in need of the research objective, nucleotide sequence data generated can readily be analyzed to characterize a few primary protein-coding gene members or whole large-scale genetic datasets that include numerous miRNAs, siRNAs, and well-supported genes (mRNAs) known to quell the stress response caused by unicellular or larger biotic organisms (e.g. bacteria, virus, viroids), organisms of the same species, or other biotic organisms within the same or different ecological communities. Thus, the knowledge obtained about individual genes can be leveraged for partially high-impact experimentation – either by knockout or inhibiting expression of genes; likewise, annotation of genetic up- and down-regulatory (e.g. miRNA, siRNA) expression can permit research on how these regulators manipulate abiotic and biotic stress responses mediated by RNA.

The advent of genome sequencing and assembly has impacted the way we conduct research on every living organism known to humanity. From PhD students conducting simple polymerase chain reaction (PCR) experiments in the lab to global initiatives that require supercomputing, every instance of nucleic acid research often touches on the general bioinformatic conundrum: sequence alignment. By aligning the sequenced nucleic acid against biologically meaningful reference sequences (e.g. RNA, transcriptome, and related gene families), one can obtain a wealth of information; for instance, a gene's location in the genome where it resides, the expressed RNA (i.e. copy number and spliced variants due to alternative splicing events), and fundamentally, the unknown: functional insight about the gene as pertains to biotic and abiotic events (i.e. responses to stimuli and stress due to exposure to environment and biota). Thus, access to gene expression datasets and available functional annotations translate into a rapid pace of biological discovery, as evidenced by an exponential growth in database works and publications centered on meta-genomic, meta-transcriptomic, and meta-proteomic data analysis. The improved access and availability of bioinformatic tools and the sequence data generated by concurrent sequencing and sequencing/alignment pipelines have, inter alia, permitted machine learning and data analytics dissemination (e.g. meta-omics, genetic variant discovery) with Bloom and other initiatives, while also matching the computational capacity required for the bioinformatically heavy workflow.

26.7 Roles of Bioinformatics in siRNA-based Immunity

Bioinformatics plays a crucial role in the design and optimization of siRNA-based immunity strategies. siRNA is a powerful tool for gene silencing and has shown great potential in developing targeted therapies for various diseases. Bioinformatics approaches are employed at various stages of siRNA-based immunity research and application.

26.7.1 Profiling siRNA

Designing effective siRNA sequences is a critical step in achieving successful gene silencing. Bioinformatics tools are instrumental in this process. Computational algorithms are used to analyze the target gene sequence and identify potential siRNA candidates based on specific criteria, such as thermodynamic stability, target accessibility, and off-target effects (Naito *et al.*, 2006; Vert *et al.*, 2006). These tools also consider factors like sequence specificity and potential cross-reactivity with other genes to minimize undesired effects.

Moreover, bioinformatics approaches can be used to predict the efficacy of siRNA candidates and rank them based on various parameters (Shabalina *et al.*, 2006). This helps researchers prioritize the most promising siRNA sequences for further experimental validation, saving time and resources.

Role of molecular docking in siRNA design

Designing effective siRNAs for gene silencing applications requires not only potent binding to the target mRNA but also minimizing potential off-target effects. Two computational approaches, free energy calculations and molecular docking, can aid in this process by predicting the thermodynamic favorability of siRNA–target interactions and identifying potential off-target binding sites.

Using IntRNA for Δg predictions

The first step is to evaluate the thermodynamic favorability of siRNA binding to the intended target mRNA. This can be achieved through free energy (ΔG) calculations using tools like IntRNA. IntRNA is a computational program that predicts the ΔG of RNA–RNA interactions based on a combination of energy terms, including hydrogen bonding, base stacking, and conformational entropy changes.

By providing the sequences of the siRNA and target mRNA, IntRNA can compute the ΔG of their hybridization. More negative ΔG values indicate stronger, more favorable binding. This allows for screening and ranking of siRNA candidates based on their predicted binding affinities to the target (Mann *et al.*, 2017). However, ΔG calculations alone do not account for potential off-target interactions or the accessibility of the target site within the mRNA structure.

Molecular docking with HNADOCK server

To address these limitations, molecular docking can be employed as a complementary approach. The HNADOCK server is a web-based tool specifically designed for docking RNA–RNA interactions, including siRNA–mRNA docking. The workflow typically involves the following steps:

- **Prepare the siRNA and mRNA structures:** The siRNA sequence is converted into a 3D structure, while the mRNA structure can be obtained from databases or predicted using RNA structure prediction tools.
- **Define the docking site:** The user specifies the target region on the mRNA where the siRNA is expected to bind. This can be guided by the ΔG calculations or other sequence-based predictions.
- **Perform the docking:** HNADOCK uses a rigid-body docking algorithm to generate numerous poses of the siRNA bound to the target site on the mRNA. These poses are then scored and ranked based on their estimated binding affinities (He *et al.*, 2019).
- **Analyze the docking results:** The top-ranked docking poses provide insights into the predicted binding mode and specific interactions between the siRNA and the target mRNA.

This information can be used to assess the likelihood of off-target interactions and guide the selection of the most promising siRNA candidates. By combining ΔG calculations and molecular docking, researchers can not only predict the thermodynamic favorability of siRNA-target binding but also gain insights into the structural aspects of these interactions and potential off-target effects.

Advantages and limitations

The strengths of this approach lie in its ability to rapidly screen and rank siRNA candidates based on thermodynamic and structural considerations, reducing the need for extensive experimental testing. Additionally, molecular docking can identify potential off-target interactions that may be overlooked by sequence-based methods alone.

However, it is important to note that these computational predictions are based on approximations and simplified models, and their accuracy can be influenced by factors such as the quality of the input structures and the limitations of the scoring functions used.

Experimental validation through techniques like pull-down assays, gene silencing experiments, and off-target profiling is still necessary to confirm the computational predictions (Mosher and Melnyk, 2010).

26.7.2 Ribozyme design

Ribozymes (ribonucleic acid enzymes) are catalytic RNA molecules that can cleave specific RNA sequences. They have been explored as potential therapeutic agents for gene regulation and disease treatment. Bioinformatics plays a vital role in the design and engineering of ribozymes.

Computational tools are employed to analyze the target RNA sequence and identify suitable cleavage sites for ribozyme activity (Zuker, 2003). Algorithms are used to predict the secondary and tertiary structures of the target RNA, which is crucial for designing efficient ribozymes (Ruan *et al.*, 2004). Additionally, bioinformatics approaches are used to optimize ribozyme sequences for improved stability, catalytic activity, and specificity (Kazakov *et al.*, 2001).

26.7.3 Epigenetics, siRNA and ribozymes

Epigenetics, the study of heritable changes in gene expression without alterations in the DNA sequence, has emerged as a critical aspect of siRNA-based immunity and ribozyme-mediated therapies. Bioinformatics plays a significant role in understanding the epigenetic mechanisms involved in gene regulation and their implications for siRNA and ribozyme design.

Computational tools are used to analyze epigenetic patterns, such as DNA methylation, histone modifications, and non-coding RNA expression, which can influence the accessibility and efficacy of siRNA and ribozyme targets (Liang, 2013; Huang *et al.*, 2015). Bioinformatics approaches are also employed to identify potential epigenetic modulators that could enhance or inhibit the activity of siRNA or ribozymes (Muhonen and Holthofer, 2010).

Furthermore, bioinformatics is instrumental in integrating epigenetic data with other omics data sources, such as transcriptomics and proteomics, to gain a comprehensive understanding of the complex regulatory networks involved in siRNA-based immunity and ribozyme-mediated therapies (Sanches *et al.*, 2024).

26.8 Ribozymes: Structure, Function, and Applications in Agricultural Pest Control

Ribozymes (ribonucleic acid enzymes) are RNA molecules that have the ability to catalyze chemical reactions like enzymes. They were first identified in the 1980s and have been found in various organisms, including animals, plants, and bacteria. Ribozymes are exceptional as they can edit genetic information and catalyze chemical reactions, making them useful for a range of biological processes such as genome editing, gene knockdown, and biological control for insects and nematodes. Ribozymes have wide range of structures and types; the main mechanism is that ribozymes identify the specific part of mRNA substrate then use a ‘hammerhead’ or machinery to cleave the cutting site (Fig. 26.3). Here we delve into the various kinds of ribozymes, their structure, how they function, and agriculture applications in root-knot nematode control.

26.8.1 Structure of ribozymes

Ribozymes have a complex three-dimensional structure (Fig. 26.4) that allows them to catalyze chemical reactions. The structure of ribozymes is determined by the sequence of nucleotides that make up the RNA molecule. The shape of the ribozyme is critical to its function, as it allows the ribozyme to interact with other molecules in the cell and catalyze specific chemical reactions.

The researchers obtained a crystal structure of a hammerhead ribozyme and observed the binding of Na(+) ions in the ribozyme's active site. They found at least two Na(+) ions, with one binding to the N7 of G10.1 and the

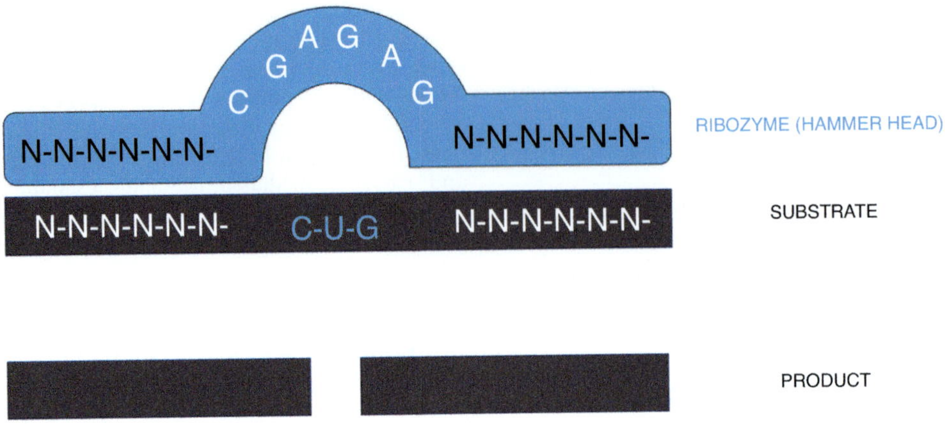

Fig. 26.3. Ribozyme mode of action. Ribozyme (blue) identifies mRNA sequence substrates using flanks, then uses its hammerhead to cleave the cutting site. Author's own image.

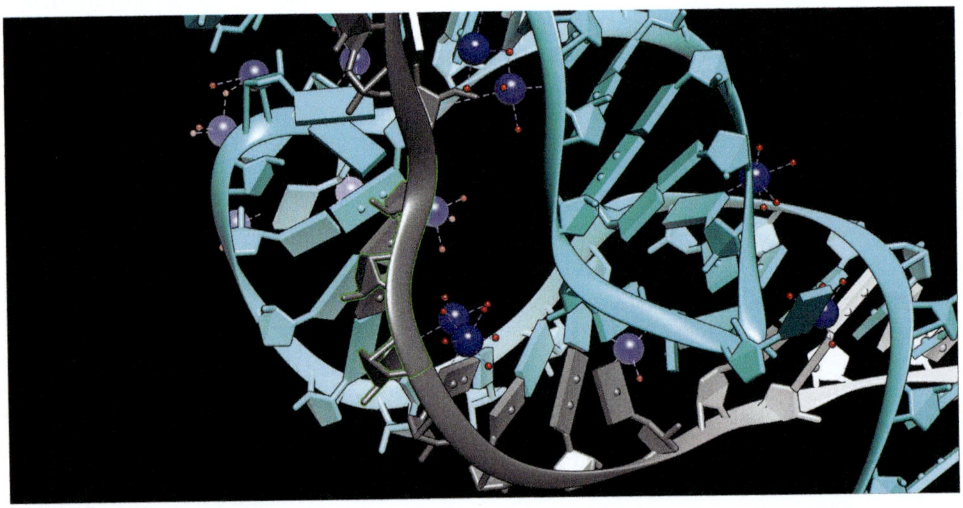

Fig. 26.4. 3D schematic presentation for ribozyme hammerhead from pdb record 3zp8 ribozyme (sky blue) and substrate strand (gray), and targeted nucleic acid (dark blue), and H_2O (red) taken from Chimera ucsf.

adjacent A9 phosphate, and the other binding to the Hoogsteen face of G12. The second Na(+) ion may neutralize the negative charge of the catalytically active enolate form of the nucleotide base. A potential third Na(+) binding site was observed between the A9 and scissile phosphates. The results suggest that Na(+) can substitute for divalent cations in the hammerhead active site. The high-resolution structure allows for detailed observation of the pre-catalytic active-site complex.

26.8.2 Types of ribozymes

There are several types of ribozymes, each with its own specific function. The most well-known type of ribozyme is the self-cleaving ribozyme, which is able to break itself apart into two separate molecules; then this type of ribozyme was modified to break a specific substrate sequence. Other types of ribozymes include the hammerhead ribozyme, the hairpin ribozyme, and the hepatitis delta virus ribozyme. Each of these ribozymes has a unique structure and mechanism of action, making them useful for a variety of applications.

GIR1 branching ribozyme

The GIR1 ribozyme is found in the plant pathogenic fungus, *Ustilago maydis*. It catalyzes the formation of a 2',5'-phosphodiester bond and has a unique branched structure (Nielsen *et al.*, 2005).

glmS ribozyme

The glmS ribozyme is found in bacteria and regulates the biosynthesis of glucosamine-6-phosphate. It is activated by the binding of small molecules, such as glucosamine-6-phosphate. It can be used as a ribo-switch to activate or inhibit some pathways in plants to induce anti-nematode mechanisms; in addition, it can be used as a target for antibacterial drugs (Traykovska *et al.*, 2021).

Group I self-splicing intron

Group I introns are found in bacteria, bacteriophages, and some eukaryotic organelles. They are capable of self-splicing and have a unique secondary structure that includes a catalytic core (Puttaraju and Been, 1996).

Hairpin ribozyme

The hairpin ribozyme is a small catalytic RNA molecule that is able to cleave RNA molecules at specific sites. It has a unique structure that includes a hairpin loop and stem regions (Symons, 1997).

Hammerhead ribozyme

The hammerhead ribozyme is another small catalytic RNA molecule that is able to cleave RNA molecules at specific sites. It has a unique structure that includes a catalytic core and three stem-loop regions.

HDV ribozyme

The HDV ribozyme is found in the hepatitis delta virus and is critical for its replication. It is able to cleave RNA molecules at specific sites and has a unique structure that includes a catalytic core and several stem-loop regions (Been, 2006).

RNase P

RNase P is a ribonucleoprotein complex that is involved in the processing of transfer RNA (tRNA) molecules. It has a catalytic RNA component and several protein components.

Twister ribozyme

The twister ribozyme is a small catalytic RNA molecule that is able to cleave RNA molecules at specific sites. It has a unique structure that includes a large internal loop (Gebetsberger and Micura, 2017).

Twister sister ribozyme

The twister sister ribozyme is a recently discovered ribozyme that is structurally similar to the twister ribozyme.

VS ribozyme

The vs ribozyme is found in some plant viruses and is involved in the processing of viral RNA. It has a unique structure that includes a catalytic core and several stem-loop regions. (DasGupta and Piccirilli, 2021).

26.8.3 Ribozymes in agriculture

One promising application of ribozymes is in agriculture, particularly in the control of plant parasitic nematodes such as root-knot nematodes. Root-knot nematodes are a major problem for farmers, causing significant damage to crops and reducing yields. Ribozymes have been shown to be effective in downregulating genes in root-knot nematodes, effectively reducing their ability to infect and damage plants. This has the potential to significantly improve crop yields and reduce the use of chemical pesticides.

Research on this point is low, but Caswell-Chen *et al.* (1993) and Kawahara and Ohshima (1992) made nine different hammerhead ribozymes to cut a specific part of unc-22 mRNA in a tiny worm, *C. elegans*. These ribozymes have a common core and some extra parts that help them match up with the mRNA. They were then tested in a lab dish and it was found that they were all excellent at cleaving a short piece of RNA at body temperature. The researchers also tested them in the worm and found that they could be expressed using different promoters.

26.8.4 Mechanism of action

The mechanism of action of ribozymes involves a complex series of chemical reactions that result in the catalysis of a specific chemical reaction. The ribozyme is able to selectively bind to a specific molecule, such as an mRNA molecule, and then catalyze the cleavage of that molecule. This results in the downregulation of the gene that is encoded by that mRNA molecule, effectively silencing its expression.

Ribozymes are RNA molecules that have catalytic activity, meaning they can accelerate chemical reactions. The catalytic activity of ribozymes is based on their ability to fold into complex three-dimensional structures that create specific pockets or active sites where chemical reactions can occur.

There are different types of ribozymes, but one of the most well-known is the hammerhead ribozyme. The hammerhead ribozyme is a small, self-cleaving RNA molecule that can cleave its own phosphodiester bonds. Phosphodiester bonds are the covalent bonds that link nucleotides together in RNA and DNA molecules.

When a hammerhead ribozyme encounters its target RNA molecule, it binds to a specific sequence of nucleotides within the target RNA, called the recognition sequence. This binding triggers a conformational change in the ribozyme that brings its catalytic site into close proximity with the phosphodiester bond to be cleaved.

The catalytic site of the hammerhead ribozyme contains three nucleotides, which act as a catalytic triad. The first nucleotide in the triad acts as a general base, abstracting a proton from the 2'-hydroxyl group of the nucleotide to be cleaved. This creates a nucleophile, which attacks the phosphorus atom of the adjacent nucleotide, breaking the phosphodiester bond and releasing the cleaved RNA fragment.

The second nucleotide in the catalytic triad acts as an acid, donating a proton to the leaving group, which stabilizes the negative charge that develops on the oxygen atom of the cleaved RNA fragment. The third nucleotide in the triad stabilizes the transition state of the reaction, lowering the activation energy required for the reaction to occur.

After the cleavage reaction, the hammerhead ribozyme releases the cleaved RNA fragment and can bind to another target RNA molecule to repeat the process.

Overall, the cleavage reaction catalyzed by ribozymes involves a complex interplay of molecular interactions and conformational changes, which create a specific microenvironment that promotes and accelerates the chemical reaction (Lilley, 2011).

26.8.5 Bioinformatic ribozyme design with RIBOSOFT

The RIBOSOFT (Awad, 2016) webserver was used to design and filter many types of ribozymes, by adding mRNA sequence target and then using an open reading frame (ORF) finder (NCBI, 2021) to get the ORF index for the sequence used, then define if the experiment was used *in vivo* or *in vitro*. The main defect for RIBOSOFT is low number of organisms used in *in vivo* experiments. This scheme can be used for many types of application due to the purpose of use ribozyme.

26.8.6 Future directions

The use of ribozymes in agriculture has the potential to revolutionize the way that crops are grown and protected from pests. As research in this area continues, we can expect to see new and innovative applications of ribozymes in the field of agriculture. This may include the development of new ribozymes that are able to target specific genes in nematodes or other pests, as well as the use of ribozymes in combination with other biological and chemical controls (James, 1994).

Ribozymes have the potential to be used in agriculture for the detection of plant pathogens and other contaminants. Researchers have already demonstrated the use of ribozymes in detecting viruses and other plant pathogens in the laboratory. By designing ribozymes that target specific RNA sequences unique to these pathogens, they can be used to provide rapid and accurate detection of these harmful agents. This could have significant benefits for agriculture, allowing for the early detection and prevention of plant diseases, and ultimately improving crop yields and food security. Additionally, ribozymes are relatively easy to design and produce, making them an attractive option for use in the field. With further research and development, ribozyme-based detection methods could become an important tool in ensuring the safety and security of our food supply (Klauser *et al.*, 2015).

Ribozymes also have the potential to be used in agriculture for gene editing purposes. By using RNA-cleaving ribozymes, specific RNA sequences can be precisely cut and removed in plants to introduce new traits or modify existing ones. This could be used to enhance crop productivity, improve disease resistance, and create plants with desirable traits. Ribozyme-based gene editing approaches could be particularly useful in cases where precision is crucial, as they can target specific RNA sequences with high accuracy. However, as with any new technology, there are still challenges that need to be addressed, such as improving the efficiency of ribozyme-based gene editing and ensuring safety. Nevertheless, the potential benefits of ribozyme-based gene editing in agriculture are significant and could help to address some of the challenges facing the agricultural industry today (Majid *et al.*, 2017).

26.9 Target Prediction

Bioinformatics and target prediction play a crucial role in advancing our understanding of siRNA-based immunity against biotic stressors, such as pathogens, pests, and parasites. RNA interference (RNAi) mediated by siRNAs has emerged as a powerful tool for gene silencing and holds great potential for developing novel defense strategies against these biotic stressors. Bioinformatics approaches and target prediction techniques contribute significantly to the effective design and implementation of siRNA-based immunity (Kang *et al.*, 2023).

26.9.1 Target identification and selection

Bioinformatics tools like BLAST (Basic Local Alignment Search Tool), ClustalW, and MAFFT are used for sequence analysis and comparative genomics to identify potential target genes within the genomes or transcriptomes of biotic stressors (Muhonen and Holthofer, 2010). Functional annotation databases like Gene Ontology (GO) and Kyoto Encyclopedia of Genes and Genomes (KEGG) can help prioritize targets based on their biological roles (Muhonen and Holthofer, 2010). Tools like siRNA Wizard and siDESIGN Center employ machine learning or rule-based algorithms for target prediction and ranking (Muhonen and Holthofer, 2010).

26.9.2 siRNA design and off-target analysis

Tools like siRNA Scan, siRNA Design, and DSIR (Design of siRNA) are used for designing potent siRNA molecules based on factors like thermodynamic stability, specificity, and accessibility within the target mRNA (Muhonen and Holthofer, 2010). Off-target prediction algorithms like siRNA Scales and CSTU (Computational Selection of Effective siRNA) can assess the potential for unintended gene silencing by analyzing sequence complementarity between the siRNA and non-target transcripts (Bai *et al.*, 2024).

26.9.3 siRNA efficacy and specificity evaluation

Computational methods like RNAstructure, Mfold, and IntRNA are used for free energy calculations and structural modeling of siRNA–target interactions (Muhonen and Holthofer, 2010). Molecular docking tools like HNADOCK and rDock can provide insights into the binding affinities and accessibility of siRNA candidates to target sites, aiding in the evaluation of their predicted efficacy and specificity.

26.9.4 Functional genomics and pathway analysis

Tools like DAVID (Database for Annotation, Visualization and Integrated Discovery), Ingenuity Pathway Analysis (IPA), and Reactome enable the mapping of predicted target genes to biological pathways, regulatory networks, or cellular processes critical for the survival or virulence of the biotic stressors (Muhonen and Holthofer, 2010).

26.9.5 Comparative and evolutionary analysis

Multiple sequence alignment tools like MUSCLE, MAFFT, and ClustalW, along with phylogenetic analysis tools like MEGA (Molecular Evolutionary Genetics Analysis) and PAML (Phylogenetic Analysis by Maximum Likelihood), can be used for comparative and evolutionary analyses of target genes and siRNA sequences across different strains, species, or populations of biotic stressors (Muhonen and Holthofer, 2010). By integrating these bioinformatics tools and target prediction techniques, researchers can gain valuable insights into the molecular mechanisms underlying siRNA-based immunity against biotic stressors. This knowledge can guide the rational design of potent and specific siRNA molecules, minimize off-target effects, and ultimately contribute to the development of effective and sustainable defense strategies against pathogens, pests, and parasites.

26.10 Pathogen Variation Analysis

26.10.1 Sequence alignment and variant identification

Tools like MUSCLE, MAFFT, and ClustalW are utilized for aligning nucleotide or amino acid sequences from different strains or isolates of a pathogen. This alignment aids in identifying conserved regions, variable sites, or mutations. Tools such as SNPdetector, VariBench, and VarScan are employed to detect single nucleotide polymorphisms (SNPs) and other variants within the aligned sequences (Kang *et al.*, 2023).

26.10.2 Phylogenetic analysis

Phylogenetic analysis tools like MEGA, PAML, and RAxML are used to reconstruct evolutionary relationships among pathogen strains or species. This analysis guides the selection of target regions or siRNA sequences that are conserved across closely related pathogens, facilitating the development of broad-spectrum siRNA-based strategies (Kang *et al.*, 2023).

26.10.3 Structural analysis

RNA structure prediction tools like Mfold, RNA structure, and RNA fold are employed to analyze the secondary and tertiary structures of target regions within pathogen genomes or transcripts. This analysis helps identify structurally accessible regions suitable for siRNA targeting, even in the presence of sequence variations (Kang *et al.*, 2023). By leveraging these bioinformatics tools and analytical techniques, researchers can gain insights into pathogen variation, evolutionary relationships, and structural features that are crucial for designing effective siRNA-based strategies against biotic stressors. This comprehensive analysis aids in the identification of conserved target regions and the development of tailored siRNA approaches for combating pathogen diversity and enhancing defense mechanisms.

26.11 Functional Annotation

Functional annotation of pathogen genomes or transcriptomes is crucial for understanding the biological roles of potential siRNA targets and their impact on pathogenicity or virulence. Bioinformatics tools and databases provide valuable resources for functional annotation:

26.11.1 Sequence homology searches

BLAST (Basic Local Alignment Search Tool) is widely used for searching sequence similarities between pathogen genes or proteins and those with known functions in other organisms (Mahmood *et al.*, 2020). This approach can provide insights into the putative functions of target genes based on their homology to well-characterized sequences.

26.11.2 Protein domain analysis

Databases like Pfam, InterPro, and SMART are employed to identify conserved protein domains or functional motifs within pathogen proteins (Miryeganeh and Saze, 2021). This information can help predict the potential roles of target genes based on the presence of known functional domains.

26.11.3 Gene Ontology (GO) annotation

GO annotation tools like Blast2GO, InterProScan, and PANNZER assign Gene Ontology terms to pathogen genes, providing information about their molecular functions, biological processes, and cellular components (Miryeganeh and Saze, 2021). This annotation can guide the selection of targets with roles in critical processes for pathogen survival or virulence.

26.11.4 Pathway and network analysis

Tools like KEGG (Kyoto Encyclopedia of Genes and Genomes), Reactome, and PathwayTools map pathogen genes to metabolic pathways, regulatory networks, or signaling cascades (Mahmood *et al.*, 2020). This can help identify potential targets that are part of essential pathways or networks crucial for pathogen biology or host-pathogen interactions.

References

Awad, I.M.A. (2016) The biological function of hammerhead RANA ribozyme in human genome. MSc thesis, United Arab Emirates University, Abu Dhabi.

Backer, R., Naidoo, S. and van den Berg, N. (2019) The NONEXPRESSOR OF PATHOGENESIS-RELATED GENES 1 (NPR1) and RELATED family: Mechanistic insights in plant disease resistance. *Frontiers in Plant Science* 10, 430215.

Bai, Y., Zhong, H., Wang, T. and Lu, Z.J. (2024) OligoFormer: An accurate and robust prediction method for siRNA design. *Bioinformatics* 10, btae577. 578533. DOI: 10.1093/bioinformatics/btae577.

Barrangou, R., Fremaux, C., Deveau, H., Richards, M., Boyaval, P. *et al.* (2007) CRISPR provides acquired resistance against viruses in prokaryotes. *Science* 315(5819), 1709–1712.

Bartok, E. and Hartmann, G. (2020) Immune sensing mechanisms that discriminate self from altered self and foreign nucleic acids. *Immunity* 53(1), 54–77.

Been, M.D. (2006) HDV ribozymes. *Current Topics in Microbiology and Immunology* 307, 47–65. DOI: 10.1007/3-540-29802-9_3.

Bhar, A., Chatterjee, M., Gupta, S. and Das, S. (2018) Salicylic acid regulates systemic defense signaling in chickpea during *Fusarium oxysporum* f. sp. *ciceri* race 1 infection. *Plant Molecular Biology Reporter* 36(2), 162–175.

Casas-Mollano, J.A., Zacarias, E., Ma, X., Kim, E.J. and Cerutti, H. (2016) RNA-mediated silencing in eukaryotes: Evolution of protein components and biological roles. In: Jagus, R. and Hernández, G. (eds) *Evolution of the Protein Synthesis Machinery and Its Regulation*. Springer Nature, Cham, pp. 513–529.

Caswell-Chen, E.P., Williamson, V.M. and Westerdahl, B.B. (1993) Applied biotechnology in nematology. *Journal of Nematology* 25(4 Suppl), 719–730.

Ciechanowska, K., Pokornowska, M. and Kurzyńska-Kokorniak, A. (2021) Genetic insight into the domain structure and functions of dicer-type ribonucleases. *International Journal of Molecular Sciences* 22(2), 616.

DasGupta, S. and Piccirilli, J.A. (2021) The Varkud satellite ribozyme: A thirty-year journey through biochemistry, crystallography, and computation. *Accounts of Chemical Research* 54(11), 2591–2602. DOI: 10.1021/acs.accounts.1c00052.

Djupedal, I. and Ekwall, K. (2009) Epigenetics: Heterochromatin meets RNAi. *Cell Research* 19(3), 282–295.

Du, J. and Patel, D.J. (2022) Cofactor-assisted dicing: Insights from structural snapshots. *Cell Research* 32(11), 965–966.

Elbashir, S.M., Lendeckel, W. and Tuschl, T. (2001) RNA interference is mediated by 21- and 22-nucleotide RNAs. *Genes & Development* 15(2), 188–200.

Fujimoto, Y. and Iwakawa, H.O. (2023) Mechanisms that regulate the production of secondary siRNAs in plants. *Journal of Biochemistry* 174(6), 491–499.

Garcia-Ruiz, H., Carbonell, A., Hoyer, J.S., Fahlgren, N., Gilbert, K.B. *et al.* (2015) Roles and programming of *Arabidopsis* ARGONAUTE proteins during turnip mosaic virus infection. *PLoS Pathogens* 11(3), e1004755.

Gebetsberger, J. and Micura, R. (2017) Unwinding the twister ribozyme: From structure to mechanism. *Wiley Interdisciplinary Reviews. RNA* 8(3), e1402. DOI: 10.1002/wrna.1402.

Ghosh, S., Malukani, K.K., Chandan, R.K., Sonti, R.V. and Jha, G. (2019) How plants respond to pathogen attack: Interaction and communication. In: Sopory, S. (ed.) *Sensory Biology of Plants*. Springer Singapore, pp. 537–568.

Gimenez, E., Salinas, M. and Manzano-Agugliaro, F. (2018) Worldwide research on plant defense against biotic stresses as improvement for sustainable agriculture. *Sustainability* 10(2), 391.

Girke, T., Todd, J., Ruuska, S., White, J., Benning, C. *et al.* (2000) Microarray analysis of developing *Arabidopsis* seeds. *Plant Physiology* 124(4), 1570–1581.

Guillemin, A., Kumar, A., Wencker, M. and Ricci, E.P. (2022) Shaping the innate immune response through post-transcriptional regulation of gene expression mediated by RNA-binding proteins. *Frontiers in Immunology* 12, 796012.

Guo, Q., Liu, Q., A. Smith, N., Liang, G. and Wang, M.-B. (2016) RNA silencing in plants: Mechanisms, technologies and applications in horticultural crops. *Current Genomics* 17(6), 476–489.

Guzman-Aranguez, A., Loma, P. and Pintor, J. (2013) Small-interfering RNAs (siRNAs) as a promising tool for ocular therapy. *British Journal of Pharmacology* 170(4), 730–747.

Harvey, J.J.W., Lewsey, M.G., Patel, K., Westwood, J., Heimstädt, S. *et al.* (2011) An antiviral defense role of AGO2 in plants. *PloS ONE* 6(1), e14639.

Haward, F., Maslon, M.M., Yeyati, P.L., Bellora, N., Hansen, J.N. *et al.* (2021) Nucleo-cytoplasmic shuttling of splicing factor SRSF1 is required for development and cilia function. *eLife* 10, e65104.

He, J., Wang, J., Tao, H., Xiao, Y. and Huang, S.Y. (2019) HNADOCK: A nucleic acid docking server for modeling RNA/DNA-RNA/DNA 3D complex structures. *Nucleic Acids Research* 47(W1), W35–W42. DOI: 10.1093/nar/gkz412.

Huang, Y., Kendall, T., Forsythe, E.S., Dorantes-Acosta, A., Li, S. *et al.* (2015) Ancient origin and recent innovations of RNA polymerase IV and V. *Molecular Bioloogy and Evolution* 32(7), 1788–1799.

Hung, Y.H. and Slotkin, R.K. (2021) The initiation of RNA interference (RNAi) in plants. *Current Opinion in Plant Biology* 61, 102014.

Islam, W., Noman, A., Qasim, M. and Wang, L. (2018) Plant responses to pathogen attack: Small RNAs in focus. *International Journal of Molecular Sciences* 19(2), 515.

Jagram, N. and Dasgupta, I. (2023) Principles and practice of virus induced gene silencing for functional genomics in plants. *Virus Genes* 59(2), 173–187.

James, H.A. (1994) Development of hammerhead ribozymes and modified snRNAs to down-regulate gene expression. Thesis, University of Liverpool, UK.

Jones-Rhoades, M.W., Bartel, D.P. and Bartel, B. (2006) MicroRNAs and their regulatory roles in plants. *Annual Review of Plant Biology* 57, 19–53.

Jose, A.M. and Hunter, C.P. (2007) Transport of sequence-specific RNA interference information between cells. *Annual Review of Genetics* 41, 305–330.

Kang, H., Ga, Y.J., Kim, S.H., Cho, Y.H., Kim, J.W. *et al.* (2023) Small interfering RNA (siRNA)-based therapeutic applications against viruses: Principles, potential, and challenges. *Journal of Biomedical Science* 30, 88. DOI: 10.1186/s12929-023-00981-9.

Katiyar-Agarwal, S., Morgan, R., Dahlbeck, D., Borsani, O., Villegas, A. Jr, Jr *et al.* (2006) A pathogen-inducible endogenous siRNA in plant immunity. *Proceedings of the National Academy of Sciences of the United States of America* 103(47), 18002–18007. DOI: 10.1073/pnas.0608258103.

Kawahara, T. and Ohshima, Y. (1992) Ribozymes for specific inhibition of mRNA function in the nematode *Caenorhabditis elegans*. *Nucleic Acids Symposium Series* 27(27), 45–46.

Kazakov, S.A., Litvinyuk, V.V. and Baratova, L.A. (2001) *Biochemistry (Moscow)* 66(10), 1093–1099.

Klauser, B., Atanasov, J., Siewert, L.K. and Hartig, J.S. (2015) Ribozyme-based aminoglycoside switches of gene expression engineered by genetic selection in *S. cerevisiae*. *ACS Synthetic Biology* 4(5), 516–525. DOI: 10.1021/sb500062p.

Kong, X., Yang, M., Le, B.H., He, W. and Hou, Y. (2022) The master role of siRNAs in plant immunity. *Molecular Plant Pathology* 23(10), 1565–1574. DOI: 10.1111/mpp.13250.

Kryovrysanaki, N., James, A., Tselika, M., Bardani, E. and Kalantidis, K. (2021) RNA silencing pathways in plant development and defense. *The International Journal of Developmental Biology* 66(1–2–3), 163–175.

Kumar, S. and Mohapatra, T. (2024) Epigenetic modifications in genome help remembering the stress tolerance strategy adopted by the plant. *Frontiers in Bioscience-Landmark* 29(3), 126.

Lee, D., Suza, W., Hanneman, M. and Hain, P. (2021) Gene expression: Applied example (part 1). In: Suza, W. and Lee, D. (eds) *Genetics, Agriculture, and Biotechnology*. Iowa State University Digital Press, Chapter 7.

Li, M.-Y., Jiao, Y.-T., Wang, Y.-T., Zhang, N., Wang, B.-B. *et al.* (2020) CRISPR/Cas9-mediated VvPR4b editing decreases downy mildew resistance in grapevine (*Vitis vinifera* L.). *Horticulture Research* 7(1).

Liang, K.H. (2013) *Bioinformatics for biomedical science and clinical applications*. Elsevier.

Lilley, D.M.J. (2011) Mechanisms of RNA catalysis. *Philosophical Transactions of the Royal Society of London. Series B, Biological Sciences* 366(1580), 2910–2917. DOI: 10.1098/rstb.2011.0132.

Liu, H., Chen, W., Li, Y., Sun, L., Chai, Y. *et al.* (2022) CRISPR/Cas9 technology and its utility for crop improvement. *International Journal of Molecular Sciences* 23(18), 10442.

Mahmood, K., Orabi, J., Kristensen, P.S., Sarup, P., Jørgensen, L.N. *et al.* (2020) De novo transcriptome assembly, functional annotation, and expression profiling of rye (*Secale cereale* L.) hybrids inoculated with ergot (*Claviceps purpurea*). *Scientific Reports* 10(1), 13475. DOI: 10.1038/s41598-020-70406-2.

Majid, A., Parray, G.A., H. Wani, S., Kordostami, M., Sofi, N.R. *et al.* (2017) Genome editing and its necessity in agriculture. *International Journal of Current Microbiology and Applied Sciences* 6(11), 5435–5443.

Manghwar, H., Li, B., Ding, X., Hussain, A., Lindsey, K. *et al.* (2020) CRISPR/Cas systems in genome editing: Methodologies and tools for sgRNA design, off-target evaluation, and strategies to mitigate off-target effects. *Advanced Science (Weinheim, Baden-Wurttemberg, Germany)* 7(6), 1902312.

Mann, C.W., Sawyer, A., Gardiner, D.M., Mitter, N., Carroll, B.J. *et al.* (2023) RNA-based control of fungal pathogens in plants. *International Journal of Molecular Sciences* 24(15), 12391.

Mann, M., Wright, P.R. and Backofen, R. (2017) IntaRNA 2.0: Enhanced and customizable prediction of RNA-RNA interactions. *Nucleic Acids Research* 45(W1), W435–W439.

Miryeganeh, M. and Saze, H. (2021) De novo transcriptome assembly, functional annotation, and transcriptome dynamics analyses reveal stress tolerance genes in Mangrove tree (*Bruguiera gymnorhiza*). *International Journal of Molecular Sciences* 22(18), 9874. DOI: 10.3390/ijms22189874.

Mosher, R.A. and Melnyk, C.W. (2010) SiRNAs and DNA methylation: Seedy epigenetics. *Trends in Plant Science* 15(4), 204–210. DOI: 10.1016/j.tplants.2010.01.002.

Muhonen, P. and Holthofer, H. (2010) Bioinformatic approaches to siRNA selection and optimization. *Methods in Molecular Biology (Clifton, N.J)* 623, 93–107.

Muthamilarasan, M. and Prasad, M. (2013) Plant innate immunity: An updated insight into defense mechanism. *Journal of Biosciences* 38(2), 433–449.

Mushtaq, M., Ahmad Dar, A., Skalicky, M., Tyagi, A., Bhagat, N. *et al.* (2021) CRISPR-based genome editing tools: insights intotechnological breakthroughs and future challenges. *Genes* 12(6), 797.

Naito, Y., Ui-Tei, K., Nishikawa, T., Takebe, Y. and Saigo, K. (2006) SiVirus: Web-based antiviral siRNA design software for highly divergent viral sequences. *Nucleic Acids Research* 34, W448–W450.

National center for biotechnology information (NCBI) (2021) ORF finder. Available at: www.ncbi.nlm.nih.gov/orffinde

Naumann, U., Daxinger, L., Kanno, T., Eun, C., Long, Q. *et al.* (2011) Genetic evidence that DNA methyltransferase DRM2 has a direct catalytic role in RNA-directed DNA methylation in *Arabidopsis thaliana*. *Genetics* 187(3), 977–979.

Neumeier, J. and Meister, G. (2021) SiRNA specificity: RNAi mechanisms and strategies to reduce off-target effects. *Frontiers in Plant Science* 11, 526455.

Nielsen, H., Westhof, E. and Johansen, S. (2005) An mRNA is capped by a 2', 5' lariat catalyzed by a group I-like ribozyme. *Science (New York)* 309(5740), 1584–1587. DOI: 10.1126/science.1113645.

Odokonyero, D. (2012) *Identification of argonautes involved in antiviral RNA silencing in Nicotiana benthamiana*. [Master's thesis, Texas A&M University]. Department of Plant Pathology, College of Agriculture and Life Sciences, Texas A&M.

Paul, N.C., Park, S.-W., Liu, H., Choi, S., Ma, J. *et al.* (2021) Plant and fungal genome editing to enhance plant disease resistance using the CRISPR/Cas9 system. *Frontiers in Plant Science* 12, 700925.

Portnoy, V., Huang, V., Place, R.F. and Li, L.C. (2011) Small RNA and transcriptional upregulation. *Wiley Interdisciplinary Reviews. RNA* 2(5), 748–760.

Puttaraju, M. and Been, M.D. (1996) Circular ribozymes generated in *Escherichia coli* using group I self-splicing permuted intron-exon sequences. *The Journal of Biological Chemistry* 271(42), 26081–26087. DOI: 10.1074/jbc.271.42.26081.

Rehem, B.C., Bertolde, F.Z. and Almeida, A.A.F. (2012) Regulation of gene expression in response to abiotic stress in plants. In: Bubulya, P. (ed.) *Cell Metabolism – Cell Homeostasis and Stress Response*. IntechOpen, pp. 13–38.

Rejeb, I.B., Pastor, V. and Mauch-Mani, B. (2014) Plant responses to simultaneous biotic and abiotic stress: Molecular mechanisms. *Plants* 3(4), 458–475.

Ruan, J., Stormo, G.D. and Zhang, W. (2004) An iterated loop matching approach to the prediction of RNA secondary structures with pseudoknots. *Bioinformatics* 20(1), 58–60.

Safari, F., Rahmani Barouji, S. and Tamaddon, A.M. (2017) Strategies for improving siRNA-induced gene silencing efficiency. *Advanced Pharmaceutical Bulletin* 7(4), 603–609.

Sanan-Mishra, N., Abdul Kader Jailani, A., Mandal, B. and Mukherjee, S.K. (2021) Secondary siRNAs in plants: Biosynthesis, various functions, and applications in virology. *Frontiers in Plant Science* 12, 610283.

Sanches, P.H.G., de Melo, N.C., Porcari, A.M. and de Carvalho, L.M. (2024) Integrating molecular perspectives: strategies for comprehensive multi-omics integrative data analysis and machine learning applications in transcriptomics, proteomics, and metabolomics. *Biology* 13(11), 848.

Shabalina, S.A., Spiridonov, A.N. and Ogurtsov, A.Y. (2006) Computational models with thermodynamic and composition features improve siRNA design. *BMC Bioinformatics* 7(1), 65.

Shamandi, N., Zytnicki, M., Charbonnel, C., Elvira-Matelot, E., Bochnakian, A. *et al.* (2015) Plants encode a general siRNA suppressor that is induced and suppressed by viruses. *PLoS Biology* 13(12), e1002326.

Soleimani, S., Valizadeh Arshad, Z., Moradi, S., Ahmadi, A., Davarpanah, S.J. *et al.* (2020) Small regulatory noncoding rnas in *Drosophila melanogaster*: Biogenesis and biological functions. *Briefings in Functional Genomics* 19(4), 309–323.

Song, M.S. and Rossi, J.J. (2017) Molecular mechanisms of dicer: Endonuclease and enzymatic activity. *The Biochemical Journal* 474(10), 1603–1618.

Su, S., Wang, J., Deng, T., Yuan, X., He, J. *et al.* (2022) Structural insights into dsRNA processing by drosophila dicer-2-loqs-PD. *Nature* 607(7918), 399–406.

Sugimoto, A. (2005) High-throughput RNAi by soaking in *Caenorhabditis elegans*. In: Appasani, K. (ed.) *RNA Interference Technology: From Basic Science to Drug Development*. Cambridge University Press, pp. 419–432.

Symons, R.H. (1997) Plant pathogenic RNAs and RNA catalysis. *Nucleic Acids Research* 25(14), 2683–2689. DOI: 10.1093/nar/25.14.2683.

Tan, F.L. and Yin, J.Q. (2004) RNAi, a new therapeutic strategy against viral infection. *Cell Research* 14(6), 460–466.

Traykovska, M., Popova, K.B. and Penchovsky, R. (2021) Targeting glmS ribozyme with chimeric antisense oligonucleotides for antibacterial drug development. *ACS Synthetic Biology* 10(11), 3167–3176. DOI: 10.1021/acssynbio.1c00443.

Tuttolomondo, M. and Ditzel, H.J. (2021) Evaluation of siRNA stability and interaction with serum components using an agarose gel-based single-molecule FRET labeling method. In: Ditzel, H.J., Tuttolomondo, M. and Kauppinen, S. (eds) *Design and Delivery of SiRNA Therapeutics*. Humana, New York, pp. 43–56.

Uehata, T. and Takeuchi, O. (2020) RNA recognition and immunity-innate immune sensing and its post-transcriptional regulation mechanisms. *Cells* 9(7), 1701.

Vergani-Junior, C.A., Tonon-da-Silva, G., Inan, M.D. and Mori, M.A. (2021) DICER: Structure, function, and regulation. *Biophysical Reviews* 13(6), 1081–1090.

Vert, J.P., Foveau, N., Lajaunie, C. and Vandenbrouck, Y. (2006) An accurate and interpretable model for siRNA efficacy prediction. *Bioinformatics* 7, 520.

Wagnon, J.L. (2009) *Characterization of Ade-M26 Hotspot RNA and Its Role in the Activation of Meiotic Homologous Recombination in the Fission Yeast Schizosaccharomyces Pombe*. Doctoral dissertation, University of Arkansas for Medical Sciences.

Wang, J.-L., Liang, L.-Y., Lang, N., Li, Y.-Y., Guo, W. *et al.* (2021) An improved plant CRISPR-Cas9 system for generating cas9 -free multiplex mutants. *Biotechnology & Biotechnological Equipment* 35(1), 1850–1857.

Wojnicka, M., Szczepanska, A. and Kurzynska-Kokorniak, A. (2020) Unknown areas of activity of human ribonuclease dicer: A putative deoxyribonuclease activity. *Molecules (Basel, Switzerland)* 25(6), 1414.

Zeng, Y., Yi, R. and Cullen, B.R. (2003) MicroRNAs and small interfering RNAs can inhibit mRNA expression by similar mechanisms. *Proceedings of the National Academy of Sciences* 100(17), 9779–9784.

Zhang, H., Zhang, H. and Lin, J. (2020) Systemin-mediated long-distance systemic defense responses. *The New Phytologist* 226(6), 1573–1582.

Zhang, J., Liu, Y., Li, C., Xiao, Q., Zhang, D. *et al.* (2023) Recent advances and innovations in the preparation and purification of in vitro-transcribed-mRNA-based molecules. *Pharmaceutics* 15(9), 2182.

Zhang, Y. and Showalter, A.M. (2020) CRISPR/cas9 genome editing technology: A valuable tool for understanding plant cell wall biosynthesis and function. *Frontiers in Plant Science* 11, 589517.

Zheng, M., Zhang, L., Tang, M., Liu, J., Liu, H. *et al.* (2020) Knockout of two BnaMAX1 homologs by CRISPR/Cas9-targeted mutagenesis improves plant architecture and increases yield in rapeseed (*Brassica napus* L.). *Plant Biotechnology Journal* 18(3), 644–654.

Zlobin, N.E., Lebedeva, M.V. and Taranov, V.V. (2020) CRISPR/Cas9 genome editing through *in planta* transformation. *Critical Reviews in Biotechnology* 40(2), 153–168.

Zuker, M. (2003) Mfold web server for nucleic acid folding and hybridization prediction. *Nucleic Acids Research* 31(13), 3406–3415.

27 Genomic Imprinting in Plants: Molecular Mechanisms and Future Directions

Varucha Misra*
ICAR-Indian Institute of Sugarcane Research, Lucknow (UP), India

Abstract
Genomic imprinting, a phenomenon observed and studied extensively in both mammals and plants, involves gene expression which is determined by the parent-of-origin. Understanding the developmental consequences and mechanisms of genomic imprinting has been a significant focus in research involving both kingdoms. This parent-of-origin effect on gene expression is a shared epigenetic characteristic and is a notable feature in species showing this phenomenon. In plants, imprinting is closely intertwined with other epigenetic phenomena such as paramutation and transgene silencing. Hundreds of imprinted genes in plants have been identified through genome-scale surveys. Important regulators governing plant genome imprinting, including DNA and histone methylation, have been revealed, throwing light on the mechanisms behind the varied expression of imprinted genes. This chapter comprehensively explores the multifaceted nature of genome imprinting in plants, shedding light on the mechanisms that govern genomic imprinting in plants, and considering evolutionary theories in light of recent discoveries.

Keywords: Genome, Imprinting, Mechanisms, Plant, Theory, Locus, Parent-of-origin

27.1 Introduction

Genomic imprinting manifests as the favored expression of one allele over the other and this expression is influenced by the parent's sex, contributing to allele expression. Intriguingly, this one-sided expression can even occur when both alleles possess identical DNA sequences. A gene expressing from the mother transmissible allele is termed a maternally expressed imprinted gene, while a gene from the paternally inherited allele is termed a paternally expressed imprinted gene (Rodrigues and Zilberman, 2023). The significance of both types of imprinted genes stems from the differentiation undergone by maternally and paternally inherited alleles during post-fertilization. This is attributed to a pre-existing epigenetic mark rather than DNA sequence variations. This epigenetic mark, known as the imprint, is the basis of the distinction within imprinted genes showing the derivation from maternally and paternally inherited alleles (Rodrigues and Zilberman, 2023). Imprinted genes are known to exhibit imprinting in the endosperm in particular, but at times it may influence endosperm and seed size. This concept is in line with predictions from the parental conflict theory (McKeown *et al.*, 2011).

*Corresponding author: misra.varucha@gmail.com

DOI: 10.1079/9781800626102.0027

The concept of genomic imprinting originated in maize from its genetic studies, mainly focusing on the R pigmentation gene. This gene exhibited preferential expression from the maternal allele (Kermicle, 1970). Furthermore, in *Arabidopsis thaliana*, the first imprinted gene, *MEDEA* (*MEA*, formerly known as *FIS1*), was identified at the molecular level. This gene was crucial for the seed development of the plant (Kinoshita *et al.*, 1999; Vielle-Calzada *et al.*, 1999). The discovery of imprinting at the MEA locus in 1999 initiated a continuous exploration and has led to the foundation of identifying additional imprinted genes in *Arabidopsis thaliana* up to 2009. Among these findings were notable genes such as *FWA* (Kinoshita *et al.*, 2004), *PHERES1* (PHE1) (Köhler *et al.*, 2005), *FIS2* (Jullien *et al.*, 2006b, *MPC* (Tiwari *et al.*, 2008), *AtFH5* (Fitzgerald *et al.*, 2009), *AtMYB3R2*, *HDG3*, *HDG8*, *HDG9*, and *At5G62110*, encoding an unexplored protein (Gehring *et al.*, 2009).

During plant meiosis, haploid products undergo mitotic divisions, ultimately forming multicellular gametophytes. These gametophytes have pollen containing two sperm cells and a vegetative cell responsible for forming a pollen tube. In flowering plants, the zygote forms through the fusion of one sperm with an egg, while the other sperm contributes to the production of triploid endosperm by fusing with the central cell (Rodrigues and Zilberman, 2023). Despite the differentiation in two-sperm cell morphology in plants, these appear functionally interchangeable, revealing limited differential imprinting between them (Russell, 1991; Faure *et al.*, 2003). Consequently, primary imprints can be established in flowering plants with potential maternal imprints. Most of the studies have shown that imprinted expression occurs mostly in endosperm and few in embryos (Waters *et al.*, 2013; Pignatta *et al.*, 2014; Xu *et al.*, 2014; Zhang *et al.*, 2014).

27.2 Exploring Allelic and Locus Imprinting

Two distinct types of imprinting have been identified, namely allelic imprinting, and locus imprinting. Allelic imprinting involves only alleles from a specific genetic background, while locus imprinting involves alleles from diverse backgrounds.

27.2.1 Allelic imprinting

Allelic imprinting involves the imprinting of genes within alleles, a phenomenon confirmed in plant seeds where ten genes have been identified to demonstrate this behaviour. Of these, six were in maize and the remainder in *Arabidopsis thaliana* (Table 27.1). The allelic imprinting is explained through a few examples. The maize *R* gene displays the allelic imprinting in aleurone layer coloration due to its crucial role in anthocyanin accumulation in angiosperms. The

Table 27.1. Allelic imprinting in plants and their allele expression. Author's own table.

Imprinted gene	Plant species	Expression of allele	References
MEDEA	*Arabidopsis thaliana*	Maternal	Verona *et al.* (2003); Guitton and Berger (2005)
FWA	*Arabidopsis thaliana*	Maternal	Kakutani *et al.* (2004)
PHE 1	*Arabidopsis thaliana*	Paternal	Köhler *et al.* (2003)
FIS2	*Arabidopsis thaliana*	Maternal	Gehring *et al.* (2006)
fie 1	*Zea mays*	Maternal	Makarevich *et al.* (2006)
fie 2	*Zea mays*	Maternal	Danilevskaya *et al.* (2003)
meg1	*Zea mays*	Maternal	Danilevskaya *et al.* (2003)
Nrp	*Zea mays*	Maternal	Gutierrez-Marcos *et al.* (2004)
peg1	*Zea mays*	Paternal	Guo *et al.* (2003)
Mez1	*Zea mays*	Maternal	Gutierrez-Marcos *et al.* (2003)

dominance of specific alleles governs the coloration of the aleurone layer in the progeny in crosses between male and female traits, resulting in either consistent coloration or mottled pigmentation (Fig. 27.1). For instance, progeny with consistently colored aleurone is obtained from crosses between RR females (exhibiting red coloration) and rr males (exhibiting no color). Conversely, mottled aleurone pigmentation due to the irregular distribution of anthocyanin in progenies is the result of reciprocal crosses (Kermicle, 1970). This phenomenon is confined to the endosperm, with no observable reciprocal distinctions in embryos or seedlings (Brink *et al.*, 1970). Furthermore, Kermicle (1970) demonstrated that the R-mottled phenotype is not influenced by dosage effects but rather by the inheritance pattern of the *R* allele. Irrespective of the number of paternally inherited *R* alleles, mottling occurs in kernels, while consistent solid coloration is observed when the *R* allele is inherited maternally. Additionally, it is important to note that the imprinting effect is not solely dependent on *R* alleles; for example, Rst responds in a dosage-dependent manner independent of sex.

In addition to the *R* gene, alleles of *Zea mays* genes, such as *dzr1* and *α-zein*, also exhibit imprinting in the endosperm. The *dzr1* locus is responsible for regulating the post-transcriptional quantity of 10 kD α-zeins in the endosperm (Chaudhuri and Messing, 1994). Zeins, the main reserve proteins in grain kernel tissue, remain dormant within the embryonic structure (Lopes and Larkins, 1993). The *dzr1* gene influences fluctuating amounts of zein storage across different genetic lineages, with increased amounts in BSSS53 and decreased amounts in MO17 (Chaudhuri and Messing, 1994). Progeny resulting from a cross between a BSSS53 female and a MO17 male displayed a prominent rise in zein RNA and high protein accumulation. Conversely, in reciprocal crosses between a MO17 female and a BSSS53 male, low content of zein RNA and protein accumulation was observed in the endosperm, thus showing a dosage effect. Despite the number of paternal copies of BSSS53 dzr1, the zein accumulation

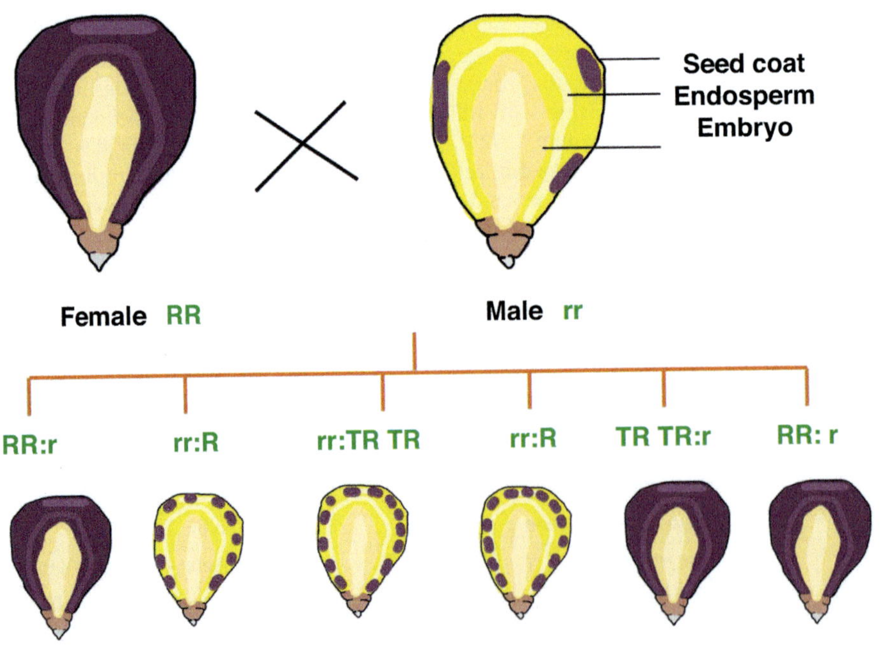

Fig. 27.1. Gene imprinting at r1 locus in corn showing crosses between R, r alleles and TR translocation. Female parent having kernel with maternal traits displaying full pigmentation and male parent having paternal inheritance of the R allele displaying mottled pigmentation. The male parent is obtained after crossing with the recessive r allele having no coloration. Author's own image.

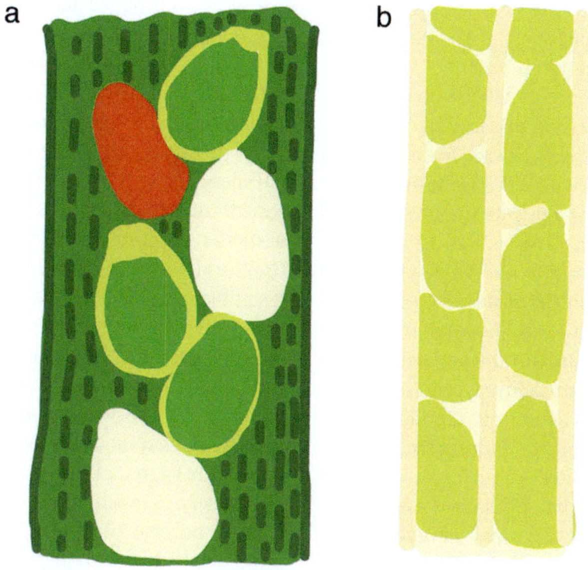

Fig. 27.2. Maternal effect seed abortion in the *mea* mutant. (a) Cross resultant of male wild type with Female mea. Opened siliques show developing seeds as a result of the reciprocal cross. If the mutant mea allele is derived from the female, seed abortion occurs (red and brown seeds), while if the allele is from the male, seed development remains unaffected. (b) Cross resultant of male mea with female wild type. In the reciprocal cross, if the wild-type allele is inherited from the female and the mutant allele from the female then the development of all seeds occurs normally. Author's own image.

was low. Conversely, maternally inherited BSSS53 dzr1 resulted in elevated zein accumulation in the endosperm (Chaudhuri and Messing, 1994). On the other hand, when the *BSSS53 dzr1* gene was introduced into another background, a dosage-dependent behaviour was seen, contrasting with *MO17 dzr1*. Additionally, specific α-zein alleles in maize endosperm have been identified to undergo imprinting. Through RNase protection assays, it was shown that α-zeins belonging to the SF2 subfamily exhibit maternal expression, with this imprinting phenomenon restricted to the W64A genetic lineage (Lund *et al.*, 1995).

27.2.2 Locus imprinting

Locus imprinting refers to genes within a MEA that were identified through screening processes. Two distinct phenotypes emerged, namely, elongation of siliques without fertilization during reproductive development, and seed abortion (Fig. 27.2) (Chaudhury *et al.*, 1997; Grossniklaus *et al.*, 1998; Kiyosue *et al.*, 1999). In the absence of fertilization, mutations in *MEA* lead to the division of the diploid nucleus within the central region of the female gametophyte, resulting in a central cell with multiple nuclei resembling syncytial endosperm, with a particular genetic site undergoing imprinting. In many cases, imprinted loci regulate multiple genes through coordinated epigenetic mechanisms (Eggermann *et al.*, 2015). The *MEA* gene in *Arabidopsis* has been extensively researched on its imprinting. However, evidence has shown that despite reaching the cellularization stage, there is no proliferation of the central cell, and subsequently, no formation of endosperm tissue capable of nourishing the embryo (Chaudhury *et al.*, 1997; Kiyosue *et al.*, 1999). Instead, sporophytic fertilization programs are stimulated, leading to the development of maternal seed coat, elongation of siliques, and degeneration in seed-like structures. This phenotype displays partial penetrance (Kiyosue *et al.*,

1999), revealing that one critical role of *MEA* is to restrain the duplication of the central cell nucleus when fertilization is absent.

Additional functions of *MEA* become apparent in the post-fertilization phase, where distinct phenotypes emerge. Seeds arising from fertilized female gametophytes with *MEA* mutations show an overabundance of endosperm growth and developmental arrest of the embryo. This leads to abortion. In contrast to the wild type, *MEA* mutations cause endosperm nuclei to continue proliferating beyond the normal cessation point, which results in the formation of large balloon-like seeds with delayed cellularization. Sørensen *et al.* (2001) demonstrated that *MEA* enlarges the central zone of endosperm, expanding more toward anterior regions compared with wild type, thereby regulating endosperm proliferation post-fertilization. Furthermore, each morphogenetic stage in the embryo is prolonged, eventually halting at the heart stage, indicating MEA's role in orchestrating the timing of embryonic development events. *MEA* encodes a SET-domain Polycomb group protein homologous to Drosophila Enhancer of Zeste [E(z)], which is known for forming complexes capable of modifying histones to maintain repressed states of gene expression (Grossniklaus *et al.*, 1998; Kiyosue *et al.*, 1999; Luo *et al.*, 1999; Orlando, 2003). Köhler *et al.* (2003) illustrated *MEA's* crucial role in suppressing genes linked to cellular growth, with the Type I MADS-box gene PHERES1 pinpointed as a direct target. The mutation in the *mea* gene demonstrates parent-specific impacts on seed development, with observable effects appearing solely when passed down maternally. Roughly half of the seeds fail to develop when an *MEA/mea* mother is paired with a wild-type father, whereas all seeds progress normally when the pairing involves a wild-type mother and an *MEA/mea* father. This highlights the exclusive reliance of seed viability on the maternal *MEA* variant, with the paternal variant being unnecessary. The recessive mea allele in males can sporadically pass to the succeeding generation via the maternal line. This resulted in a significant elevation in seed abortion rates. Moreover, the expression pattern of the *MEA* gene indicates maternal allele expression and paternal allele silencing, consistent with findings by Vielle-Calzada *et al.* (1999). Vielle-Calzada *et al.* (2000) later uncovered a broader paternal genome silencing phenomenon persisting until 80 hours after pollination (HAP), suggesting that the absence of paternal *MEA* expression at 52 HAP may reflect a genome-wide effect rather than a gene-specific one. The imprinting of *mea* genes showed maternal silencing by *MEA*, while paternal expression contributed to endosperm growth. This aligns with the parental conflict theory proposed by Haig and Westoby (1989). The *mea* phenotype resembles an abundance of paternal genetic material, marked by extended mitotic activity in the endosperm, delayed formation of individual cells, and an enlarged chalazal endosperm. Additionally, *MEA* aids as a potential chromatin-modifying enzyme accountable for maintaining gene repression (Schubert and Goodrich, 2003). Conversely, Kinoshita *et al.* (1999) exploited ecotype polymorphisms within the *MEA* coding sequence to differentiate the expression patterns of maternal and paternal alleles. These distinct expression patterns of maternal and paternal alleles were observable in the embryo during various developmental stages such as torpedo, walking-stick, and early maturation stages. The biallelic expression pattern of MEA was evident in various vegetative tissues, including seedlings, rosette leaves, stems, and roots (Kinoshita *et al.*, 1999).

Furthermore, mutations in the *Fertilization-Independent Endosperm* (FIE) and *Fertilization-Independent Seed2* (FIS2) genes display similar phenotypic expressions to those observed in mutations of the *mea* gene, as reported by Ohad *et al.* (1996) and Chaudhury *et al.* (1997), lending further support to the parent-of-origin theory. These mutations profoundly affect seed development. Interestingly, a maize homolog of *Arabidopsis* FI was identified, demonstrating locus-specific imprinting (Danilevskaya *et al.*, 2003). This marked the first instance of such a phenomenon in maize where two FIE homologs, *fie1* and *fie2* were found. These homologs share 78% similarity in exonic sequences but differ in introns or promoter sequences homology. Fie1 displays specific expression in the endosperm from 6 days after pollination (DAP), while fie2 exhibits biallelic expression in the embryo, endosperm, and various vegetative tissues. Moreover, consistent expression of Fie1 is solely attributed to the maternal allele, underscoring the crucial role of imprinting in the repressive function of the Polycomb group complex compared with imprinting a specific component. Similar to mutations in the *mea* gene, mutations in FIE also exhibited similarities with differences. In normal seed development, the wild-type maternal allele of *FIE* plays a significant function compared with the paternal allele (Yadegari *et al.*, 2000). This showed that the *FIE* locus is not subject to imprinting. Contrary to the *mea* mutation, Ohad *et al.* (1999) demonstrated that a mutated maternal fie allele is never passed down through the maternal reproductive structure. This emphasizes the essential role of FIE presence within the female reproductive unit before or shortly following fertilization. This necessity for FIE likely contributes to the observed parent-specific effects.

Furthermore, regarding *FIS2* gene, Luo *et al.* (2000) utilized the FIS2 promoter for delving into β-glucuronidase (GUS) transgene inheritance patterns through maternal or paternal lines. Phenotypic expression originating from the maternal parent was evident in unfused polar nuclei and the central cell nucleus before fertilization, as well as in dividing endosperm nuclei during post-fertilization. However, GUS expression is markedly absent in ovules at 24 HAP when derived from the paternal parent. Despite these findings, the imprinting status of the endogenous *FIS2* gene remains unknown.

27.3 Development and Challenges of Imprinted Genes in Plants and the Complexities of Their Functionality

Imprinted genes in plants can exhibit either binary or preferential imprinting, as discussed by Dilkes and Comai (2004) and Khatib (2007). Binary imprinting involves complete repression or activation of the imprinted allele, whereas preferential imprinting entails consistently higher expression of the maternally or paternally derived allele compared with its counterpart. Establishing a cut-off for the expression ratio qualifying as an imprinted locus poses challenges, particularly in cases of preferential imprinting where no obvious parent-of-origin effect on phenotype is observed (Kohler *et al.*, 2012). The tissue specificity of imprinting in plants further complicates research efforts. In angiosperms, genomic imprinting predominantly occurs in the developing F1 seed, particularly in the products of endosperm and embryo fertilization. Although there were exceptions like the *mee1* gene in maize, Hsieh *et al.* (2011) and Luo *et al.* (2011) highlighted imprinting in both developing endosperm and embryos. This suggests unknown erasure and resetting mechanisms of genomic imprinting.

Imprinting studies encounter several challenges, partly stemming from the genetically chimeric nature of seed tissues. While the seed coat originates from the maternal parent, the embryo and endosperm are obtained from both parents. The variation was seen in proportions across species and in developmental stages within a single species (Jahnke and Scholten, 2009). The interpretation of allele-specific expression levels is additionally complicated by these variations, particularly in tissue composition. Despite these challenges, research on 12,311 genes expressed in sporophytic tissues of *Arabidopsis thaliana* suggested restricted imprinting effects beyond the embryonic stage (Zhang *et al.*, 2006). Moreover, the chimeric nature of seed tissues adds complexity to interpreting allele-specific expression levels. These expression levels are exacerbated by tissue-specific expression variations during seed development and across plant species. Another hurdle involves discerning whether the detection of a maternally expressed gene in the seed endosperm arises from *de novo* post-fertilization transcription or the presence of long-lived transcripts deposited in the central cell during female gametogenesis. McKeown *et al.* (2013) and Vielle-Calzada *et al.* (1999) emphasized these challenges as crucial for advancing comprehension of genomic imprinting in plants. Furthermore, before the occurrence of the fertilization process, imprinted genes exhibit distinct markings in gametes, establishing functional disparities between maternal and paternal chromosomes.

27.4 Genome Imprinting Mechanisms

Recent genome-wide analyses of imprinted genes in plants have significantly advanced our understanding of the mechanisms behind genomic imprinting (Gehring *et al.*, 2009; Hsieh *et al.*, 2009).

27.4.1 Epigenetic modifications and imprinted genes

Epigenetic regulation of the genome plays a significant function in development, enabling heritable changes in gene expression without altering the DNA sequence. This regulation involves recruiting molecular processes that either encourage/inhibit transcription or degrade existing transcripts. Genomic imprinting, an epigenetic process, involves the marking of DNA in a sex-dependent manner. This results in differential gene expression which is based on parental origin. The formation of stable male and female imprints during gametogenesis is essential for imprinting. Imprint erasure occurs in the germline of the preceding generation, followed by re-establishment based on the organism's sex. Macdonald (2012) elucidated that each stage of this imprinting process relies on epigenetic marks being understood by the genome and action.

DNA methylation and patterns associated with imprinted genes

DNA methylation serves as a vital mechanism in suppressing harmful sequences like transposable elements and repeats, while simultaneously controlling the activity of genes. Within plant systems, DNA methylation takes place in three different sequence contexts, each managed by unique mechanisms for creation and upkeep. The small RNA pathway, particularly RNA-dependent DNA methylation, is responsible for *de novo* methylation in all sequence contexts and is essential for CHH methylation maintenance. CG methylation is maintained by *MET1*, *DDM1*, and *VIM* in *Arabidopsis*. Conversely, *CMT3* activity is required for CHG methylation maintenance (Feng *et al.*, 2010; Law and Jacobsen, 2010). *MET1* serves a crucial role in imprinting processes (Law and Jacobsen, 2010).

Seminal studies demonstrated that this process is catalyzed by DNA methyltransferase enzymes (Sapienza *et al.*, 1987). These enzymes possess a highly conserved group of enzymes responsible for methylating cytosine-C5 (Xiao *et al.*, 2006; Spada *et al.*, 2007). Numerous imprint control regions display differentially methylated regions in plants, which serve as guides for the epigenetic regulation of imprinted domains. Methylation within these regions is often established during gametogenesis and subsequently maintained throughout development. This emphasizes the pivotal role of this process in both the establishment and preservation of numerous imprinted domains (Macdonald, 2012).

Furthermore, enzymes belonging to the Dnmt1 family, known as MET1 in plants, methylate cytosines within symmetric CG dinucleotides, perpetuating methylation patterns (Feng *et al.*, 2010; Law and Jacobsen, 2010; Zemach *et al.*, 2010). This is observed when a cytosine on one DNA strand is already methylated, particularly following DNA replication. Traditionally linked to gene and transposable element silencing, DNA methylation can also contribute to gene activation (Makarevich *et al.*, 2008; Shibuya *et al.*, 2009; Jones, 2012). Consequently, differential methylation of maternal and paternal alleles can result in imprinted gene expression. The intricate interplay between DNA methylation patterns and their impact on gene regulation introduces a layer of complexity to the epigenetic control of genetic information (Rigal *et al.*, 2012; Deng and Chua, 2015; Lei *et al.*, 2015; Williams *et al.*, 2015).

Histone modifications and chromatin structure in imprinted regions

Histone modification is the second mechanism of genomic imprinting in plants, involving trimethylation of lysine 27 on histone H3, particularly in flowering plants. This has emerged as an important imprinting mark. The histone variants have not been directly implicated in plant imprinting; however, the incorporation of a specific histone H3 variant in *Arabidopsis* sperm cells could potentially differentiate the paternal allele in the endosperm or zygote (Ingouff *et al.*, 2007). Talbert and Henikoff (2010) categorized the core histones into subtypes or variants with distinct structures and functions. The istone modification catalyzed by Polycomb-repressive complex 2 (PRC2) has been linked to the imprinting of many genes. These genes are mostly paternally expressed (Hsieh *et al.*, 2011; Wolff *et al.*, 2011). Mutations in PRC2 disrupt these gene imprintings with H3K27me3 (maternally inherited alleles), correlating with DNA hypomethylation (Makarevich *et al.*, 2008; Weinhofer *et al.*, 2010). Perturbations in DNA methylation at some imprinted genes have even led to a loss of imprinting. This implies that DNA methylation is the primary imprint causing differential recognition by PRC2. PRC2-mediated regulation of maternally expressed *Arabidopsis* genes, such as *MEDEA* and *AtFH5*, involves silencing the paternal allele without DNA methylation involvement (Baroux *et al.*, 2006; Gehring *et al.*, 2006; Jullien *et al.*, 2006a; Fitzgerald *et al.*, 2009). While PRC2-catalyzed H3K27me3 is inherited after DNA replication and recruits PRC2, creating a self-reinforcing loop, it remains unclear whether *H3K27me3* serves as the primary imprint for some genes. PRC2-mediated silencing, with its ability to establish an allele-specific silent state, can be inherited by the endosperm. This could explain imprinted gene expression in the embryo (Raissig *et al.*, 2013). Furthermore, the epigenetic modifications undergone in plants can help in the alteration of histones, thus impacting chromatin conformation. Histone acetylation typically results in an open chromatin conformation, facilitating accessibility. However, histone deacetylation, frequently coupled with histone methylation, induces a condensed chromatin conformation. This is conducive to gene silencing and heterochromatin formation (Berger, 2002).

27.4.2 Imprinting control regions (ICRs)

Imprinted genes are often organized into clusters that are governed by ICRs. The ICRs are the specific chromosomal segments that exert methylation-sensitive control over the expression of imprinted genes based on their parental origin. Acting as regulated transcriptional insulators, these regions interact with CTCF (CCCTC-binding factor) to influence whether imprinted genes derived from a particular parent are activated or silenced. Imprinting in flowering plants extends to both the embryo and the supporting tissues responsible for nourishment, namely the placenta and endosperm. The imprinted genes serve in the regulation of nutrient transfer from the maternal source to the developing offspring (Kohler and Weinhofer-Molisch, 2010). The kinship theory, put forth by Haig and Westoby in 1989, serves as a conceptual framework aimed at elucidating the adaptive advantages of imprinting in facilitating parental regulation of resource allocation to offspring. However, it does not explicitly elucidate the mechanisms underlying imprinted gene expression (Haig and Westoby, 1989). According to the kinship theory, the persistence of imprinting is justified by the scenario where the gene expression level in one individual affects the other expression. Both gene expressions vary in their likelihood of carrying maternal and paternal alleles (Haig, 2004).

Identification and characterization of ICRs

ICRs play a crucial role in orchestrating preferential gene transcription based on the parent of origin (Li *et al.*, 1993; Constância *et al.*, 2004). These comprise DNA segments abundant in CpG dinucleotides, where cytosine nucleotides exhibit distinct methylation patterns contingent on parental alleles. The resultant outcome, whether the region is silenced or active, hinges on the default imprint state (i.e. methylated or unmethylated) of the region. Disruptions in DNA methylation within ICRs have been shown to result in aberrant expression of imprinted genes. The discovery of the first imprinted gene in any species stemmed from observing a reciprocal kernel-color phenotype in maize (Kermicle, 1970). As stated earlier in this chapter, the imprinted genes are regulated by DNA methylation. The CG methylation in the *Arabidopsis* endosperm is particularly noteworthy for the genome-wide reduction (Hsieh *et al.*, 2009). Gehring *et al.* (2009) further showed that, for the novel imprinted genes, identification preference is given to the presence of DMRs and expression in endosperm. The occurrence of DMRs in the imprinted genes at either 5', 3', or gene body regions between the endosperm and embryo is essentially required in imprinted genes of *Arabidopsis*. Genetic advancements further showed that numerous candidates imprinted genes in plants, including *Oyza sativa*, *Arabidopsis*, and *Zea mays*. There is enough potential for gene inclusion which might have occurred due to the presence of maternal tissues that surround the embryo and endosperm. This is evident by the results of the studies by Gehring *et al.* (2011) and Hsieh *et al.* (2011).

Furthermore, imprinting genes identification is dependent on the p-values. Several candidate genes have shown partial imprinting which is characterized by biased expression rather than complete monoallelic expression. Hsieh *et al.* (2011) identified 116 maternally expressed genes (MEGs) and ten paternally expressed genes (PEGs) in the endosperm. Simultaneously, Gehring *et al.* (2011) reported 165 MEGs and 43 PEGs, which included some of the overlapping gene sets identified by Hsieh *et al.* (2011). Concurrently, Wolff *et al.* (2011) recognized 39 MEGs and 27 PEGs with potential imprinting with limited overlap genes as revealed by the other two studies. Differences in accession (SNP variations) and/or experimental conditions may cause discrepancies to arise. On the other hand, Luo *et al.* (2011) found 93 MEGs and 72 PEGs as potential imprinted candidates in the endosperm in plants of *Oryza sativa* in six DAP embryos and five DAP endosperm of reciprocal hybrid seeds crossed between Nipponbare and the 93–11 subspecies. Besides, Zhang *et al.* (2011) revealed 68 and 111 MEGs and PEGs, respectively, in *Zea mays* from backcrosses between B73 and Mo17 inbreds at 10 DAP. Among the identified imprinted genes across the three mentioned plant species, the number of conserved genes varied where 27 were in the cross between *Oryza sativa* and *Arabidopsis*, 15 in *Oryza sativa* and *Zea mays*, and six in *Zea mays* and *Arabidopsis*. The studies highlighted that genomic imprinting predominantly occurs in rice (Gehring *et al.*, 2011; Hsieh *et al.*, 2011; Luo *et al.*, 2011). Furthermore, variation in the site of genomic imprinting was also seen (endosperm in both *Arabidopsis* and *Oryza sativa* while embryo and the endosperm in *Zea mays*) (Jahnke and Scholten, 2009).

Genomic analysis of imprinted genes also has certain limitations. The availability of single nucleotide polymorphisms (SNPs) on the desired genes in the designated accessions and subspecies is the first limitation. SNPs are essential in the finding of allele-specific expression. Another challenge is the low-level expression by imprinted genes at certain levels. These expressions can be detected by cDNA-AFLP. New imprinted genes were recognized using this method in *Zea mays* and *Arabidopsis* (Guo *et al.*, 2003; McKeown *et al.*, 2011). One interesting result obtained by this method was the identification of candidate *Arabidopsis* imprinted genes which varied from other research groups. Furthermore, several nuclear proteins have been identified by genome-wide analyses (Table 27.2).

Precise identification of imprinted genes has included the following approaches: (i) the detection of genes exhibiting genetic patterns generated by endosperm distinct genes reminiscent of those seen in established imprinted genes (Kinoshita *et al.*, 2004; Bratzel *et al.*, 2012); (ii) unveiling targets of imprinting regulators (Köhler *et al.*, 2003; Köhler *et al.*, 2005); (iii) identifying genes whose expression is dysregulated in endosperm with modified ratios of maternal and paternal genomes (Tiwari *et al.*, 2008; Shirzadi *et al.*, 2011); and (iv) identifying genes linked to variances in DNA methylation between the embryo and endosperm (Gehring *et al.*, 2009). These endeavors have led to the identification of many imprinted genes in plants, more prominently in *Zea mays* and *Arabidopsis*.

An effective approach entails directly examining the parental origins of transcripts (McKeown *et al.*, 2011). The advent of high efficiency in mRNA sequencing technologies has facilitated the swift discovery of additional potential imprinted genes in flowering plants (DeVeale *et al.*, 2012; Luo *et al.*, 2022). These technologies utilized the crossing of strains having known polymorphisms for revealing the allele-specific gene expression in F1 progeny. Subsequently, the backcross was performed to manage the *cis* effects on transcription to assess the genomic imprinting. Several candidate imprinted genes have been unveiled by using this technology and have significantly expanded our understanding of the imprinted gene landscape (Waters *et al.*, 2011). It is important to note that mRNA-seq monitors steady-state RNA populations and does not directly measure transcription, and there is no information about the proportion of transcripts that undergo translation. The use of global run-on sequencing or *in situ* hybridization of specific loci could offer more precise insights into the variation at the transcription step which distinguishes maternal and paternal alleles (Vielle-Calzada *et al.*, 1999; Core *et al.*, 2008).

Table 27.2. Imprinted genes identified in plants. Author's own table.

Imprinted genes	Encoding	Maternal/ paternal expression	Plant	References
Homeodomain Glabrous 8 (HDG8)	Class IV homeodomain-leucine zipper (HD-ZIP) transcription factor	MEG	*Arabidopsis*	Gehring *et al.* (2009)
Homeodomain Glabrous 9 (HDG9)		MEG		
FWA		MEG		Kinoshita *et al.* (2004)
HDG3		PEG		Gehring *et al.* (2009)
PEG PHE1	MADS-box genes	PEG and MEG		Wolff *et al.* (2011)
VARIATION IN METHYLATION 5 (VIM5)	Putative 5-methylcytosine-binding protein	PEG	*Arabidopsis*, rice, and maize	Gehring *et al.* (2011); Hsieh *et al.* (2011); Luo *et al.* (2011); Zhang *et al.* (2011)
Homologs of VIM5, VIM1 and VIM6		PEG	*Arabidopsis*	Gehring *et al.* (2011)
SU(VAR)3–9 Homologue 7 (SUVH7) and SUVH9	Histone methyltransferase genes	PEG		Kraft *et al.* (2008)
Jumonji C Domain 15 (JMJ15)	SUVH8 and the putative histone demethylase	MEG		Gehring *et al.* (2011); Hsieh *et al.* (2011)
Cytidine deaminase genes	–	MEG		Hsieh *et al.* (2011); Wolff *et al.* (2011)
E3 SCF complexes Arabidopsis Skp1-like 8 (ASK8) and ASK10	E3 ubiquitin degradation pathway	MEG	*Arabidopsis*	Hsieh *et al.* (2011)
DRM1 DRM2 CMT3 (SDC)	F-box protein suppressor	MEG		
YUCCA10 (YUC10)	Auxin transport during embryogenesis	PEG	*Arabidopsis*, rice, and maize	Gehring *et al.* (2011); Hsieh *et al.* (2011); Luo *et al.* (2011); Zhang *et al.* (2011)

27.4.3 Parent-of-origin effects

Parent-of-origin effects refer to situations where the phenotypic gene/allele expression is dependent on inheritance lineage. These effects are often associated with genomic imprinting (Lawson *et al.*, 2013). Gene expression relying on parent-of-origin operates independently and is governed by genome imprinting mechanisms (Satyaki and Gehring, 2017). The gene activity regulation by Polycomb repressive complex 2 (PRC2) occurs through methylation of Lys-27 on histone H3 (H3K27me3). PRC2 consists of the histone methyltransferase MEDEA (MEA) and the

FIS2, a homolog of the Suppressor of zeste12, both of which exhibit a maternal bias in expression in *Arabidopsis* endosperm (Thorstensen *et al.*, 2011). Berger *et al.* (2006) recognized these genes for independent endosperm growth. This also showed that these gene complexes aid in the regulation of sporogamy. Achieving parent-of-origin expression in genetically similar alleles relies on specific epigenetic modifications, constituting the imprint. A significant distinction in angiosperms is that imprinting is chiefly confined to the transient endosperm (Hsieh *et al.*, 2011; Luo *et al.*, 2011). However, this perspective is challenged by the finding of *mee1* imprinting in embryos, prompting a reconsideration of mechanisms during plant embryogenesis (Jahnke and Scholten, 2009).

The study of entire genomes or specific portions of chromosomes has revealed the significant influence of the parental origin of genetic information on its functionality. Imprinted genes were characterized by varying expression patterns based on parental inheritance. McKeown *et al.* (2011) revealed that only 18 imprinted genes were identified across all flowering plant species, with 11 in *Arabidopsis thaliana*. Similarly, Wolff *et al.* (2011) also identified candidate imprinted genes in F1 hybrid seeds resulting from reciprocal crosses between Bur-0 and Col-0.

27.5 Theories for the Evolution of Genomic Imprinting

There are several theories for the evolution of genomic imprinting and some of these are briefly summarized here. Among these, the parental conflict theory has garnered significant attention.

27.5.1 Maternal–offspring coadaptation theory

According to the Wolf and Hager (2006) theory of maternal-offspring coadaptation, the suitability of a genotype is determined by its straight result on the individual bearing it and interrelations between the observable traits of different individuals, particularly mothers, and offspring. The theory posits that when interactions between maternal and offspring traits significantly influence suitability, then genetic correlations between these traits are likely to occur (Wolf and Brodie, 1998; Sinervo and Svensson, 2002). The theory explores two scenarios, namely the single-locus model and the two-locus model. As the names indicate, the single-locus model shows the influence of the same locus influences on traits of maternal and offspring alleles, while the two-locus model involves different loci (Wolf and Hager, 2006). In the framework of the single-locus model, the alleles at a diploid locus within an offspring are deemed non-equivalent in terms of fitness, owing to the varying likelihood of the maternal allele being present in the mother compared with the paternal allele. This theoretical framework posits that if there exists a stronger genetic covariance – akin to phenotypic matching – between the mother and offspring, selection favours the expression of the maternal allele while suppressing the paternal allele (Patten *et al.*, 2014). In contrast, within the two-locus model, historical correlational selection acting on the maternal–offspring interaction engenders genetic correlations, thereby establishing linkage disequilibrium between the loci. Consequently, the segment of the offspring's genotype originating from the maternal lineage becomes positively correlated with any maternal effects that interact with the offspring's phenotype. As a result, the theory predicts that confining expression to the matrigenic allele is more likely to yield a high-fitness interaction, given that this allele would have undergone some degree of coadaptation with the alleles present in the mother. Ultimately, the theory proposes that whether the matrigenic or patrigenic allele is expressed hinges on the nature of the interaction, with evolutionary history shaping the genetic correlations and determining the allele expression pattern that maximizes fitness (Patten *et al.*, 2014).

27.5.2 Kinship theory

The kinship theory, rooted in kin selection models, delves into the differential relatedness patterns experienced by matrigenic and patrigenic alleles in a social environment. This variation arises because individuals are more likely to encounter matrilineal relatives than patrilineal ones, leading to distinct consequences for their inclusive fitness (Haig, 2002). The theory focuses on genes influencing physiological or behavioural interactions between individuals. Haig (1997) introduced the concept of parental antagonism, analogous to sexual antagonism, where a trait favoured in one parent is disfavoured in the other.

In scenarios characterized by parental conflict, it is advantageous for an allele to modulate its expression level to align with the optimal level associated with its parental origin. If parental origin-specific expression is feasible, the kinship theory aims to establish an evolutionarily stable strategy (ESS) for gene expression originating from both parental sources. This ESS involves the preferential expression of the allele from the parental origin favouring a higher expression level, while the allele from the other parental origin is suppressed, a phenomenon recognized as genomic imprinting. At this evolutionarily stable point, the growth factor gene is expressed exclusively from the patrigenic allele at the optimal level dictated by the patrigene (Patten *et al.*, 2014).

The kinship theory encompasses both strong and weak versions, with the strong version proposing that imprinted expression arises from parental antagonism, while the weak version suggests that once imprinted expression is established, the gene evolves based on the inclusive fitness consequences for the remaining expressed copy (matrigene or patrigene). Originally centered on multiple mating as a source of relatedness asymmetry, subsequent research extends the kinship theory to encompass other sources such as haplodiploidy and sex-biased dispersal. Furthermore, the theory can be applied to any behavior impacting kin fitness, not exclusively limited to offspring solicitation behaviors. For instance, the provision of shared resources by adults demonstrates parentally antagonistic inclusive fitness effects when the alleles of the provider are asymmetrically related to the recipients of the resource (Patten *et al.*, 2014).

27.5.3 Sexual antagonism theory

This theory is entirely reliant on sex-specific selection pressures (Day and Bonduriansky, 2004; Bonduriansky, 2007). In cases where there is variation in selection among parents, whether in magnitude or orientation, the proportions of alleles in eggs and sperm become distinct in the next generation. This is responsible for two scenarios. If a gene experiences sexually antagonistic selection, it implies that alleles benefiting males will be favoured in patrigenes, while alleles benefiting females will be favoured in matrigenes. This fosters a conducive environment for a novel regulator that selectively modulates gene activity in an imprinted and gender-specific fashion. Specifically, the gene should exhibit paternal expression in males and maternal expression in females. Furthermore, there exists an overall selective advantage in altering gene expression in the classical imprinted manner, irrespective of the individual's sex. This benefit arises when the allele inherited from the parent experiencing more pronounced selection in the previous cycle is activated. According to the theory of sexual antagonism, imprinted expression is favoured because, on average, the two alleles at a diploid locus carry disparate information or instructions, with one copy usually containing more advantageous information than the other. The existence of differences among groups of individuals in certain genetic traits, along with the lack of alternative epigenetic processes regulating their gender-specific effects on overall health and reproductive success, leads to conflict within the same genetic loci between the sexes. This conflict emerges when one gender faces limitations in attaining its ideal characteristics because of varying selective forces acting on the identical genetic site in the other gender (Bonduriansky and Chenoweth, 2009). In scenarios marked by contrasting selection pressures among genders, the parent sharing the offspring's gender typically provides genetic material that, on average, bestows higher fitness upon the offspring compared with the genetic contributions from the parent of the opposite gender. As a result, imprinted suppression acts as a direct mechanism allowing individuals to manifest the advantageous array of directives customized to their gender. Within the context of sexual conflict theory, imprinting works to alleviate conflict within the same genetic loci by helping each gender to better approach its specific ideal phenotype (Patten *et al.*, 2014).

27.5.4 Parental conflict theory (PCT)

PCT, presented by Haig and Westoby in 1989 and refined by Moore and Haig in 1991, has emerged as a prominent framework for comprehending the evolution of imprinting. However, this theory is not without its share of controversy, with critiques and alternative perspectives challenging its dominance (de Jong and Scott, 2007; Moore and Mills, 2008). Brandvain and Haig (2005) extended the PCT to accommodate variations in breeding systems, adding a layer of nuance to the original concept. Beyond the PCT and kinship models, diverse hypotheses have been proposed to elucidate the evolution of genomic imprinting in plants (Garnier *et al.*, 2008a). With the advancement of next-generation sequencing, the understanding of genomic imprinting continues to expand, offering a wealth of data for critical testing of these evolutionary theories. Importantly, these models are not mutually exclusive, as genomic imprinting could evolve in response to diverse evolutionary pressures at different genomic loci (Hurst and McVean, 1998). While this theory generates predictions consistent with observed phenotypes in imprinted mutants, discrepancies challenge its universal applicability. Recent findings, such as the involvement of the imprinted gene *ZmMeg1* in enhancing offspring size, deviate from expectations from the models based on this theory (Costa *et al.*, 2012). Additionally, the prevalence of identified MEGs surpassing that of PEGs in species like *A. thaliana* contradicts the PCT's prediction of PEG predominance (Wolff *et al.*, 2011). The evolving landscape of imprinted gene discoveries promises ongoing exploration and refinement of theories surrounding their evolutionary origins.

The disparity in the prevalence of MEGs and PEGs has been reported. The outnumbering of PEGs by MEGs has often shown a spark in intriguing exploration. While the possibility of false positives in MEG identification due to seed coat contamination is acknowledged, the current data align more convincingly with theories rooted in maternal coevolution or care, as proposed by Wolf and Hager in 2006. This pattern is not unique to *A. thaliana* but extends to *Oryza sativa* (Luo *et al.*, 2011) and aligns with predictions for *Zea mays* (Guo *et al.*, 2003), where the risk of maternal seed coat contamination is minimal. Insights derived from studies on plant imprinting in interploidy crosses provide evidence supporting theories centered on the management of dosage (Jullien and Berger, 2010). However, the investigations remain ambiguous, as no single theory definitively elucidates the development of genomic imprinting in plants at this stage. Within extended PCT frameworks, the interaction between male and female germplasm hinges on their average degree of connectivity. This dynamic gives rise to the 'WISO' (weak inbreeder, strong outbreeder) hypothesis, suggesting that predominantly outcrossing species have a competitive advantage in reproductive success compared with predominantly inbreeding ones (Brandvain and Haig, 2005). When applied to *A. thaliana*, this concept predicts that accessions with a stronger tendency towards outcrossing should demonstrate superior capabilities in producing interspecific hybrids. This hypothesis may align with natural populations where outcrossing is prevalent (Bomblies and Weigel, 2010). Conversely, disparities in imprinting levels among *A. thaliana* accessions might be linked to factors associated with the extent of selfing and crossing. The theory suggests that different mating systems serve as mechanisms for plant divergence (Kondoh and Higashi, 2000), implying that there are potentially more phenotypes of imprinted gene mutants yet to be discovered. The ongoing endeavor to unravel the complexities of imprinting in plants holds the promise of uncovering additional layers of this evolutionary puzzle.

This theory provides a conceptual framework for explaining growth-related defects observed in genetic variations impacting imprinted genes across both plant and mammalian species. The similarity in the manifestation of certain mutant traits and those found in numerous mouse mutants, especially regarding cell proliferation and growth, suggests that akin selective pressures might have prompted the enlistment of growth-regulating genes governed by genomic imprinting (Vielle-Calzada *et al.*, 1999). These parallels might be rooted in the shared characteristic of mammals and plants having a placental habit, where the mother provides all nutrients supporting the developing embryo. In polygamous organisms, this scenario is anticipated to create a conflict between maternal and paternal interests over the allocation of nutrients from the mother to the offspring, aligning with David Haig's parental conflict theory.

While the parental discord concept aligns with the attributes of some imprinted genes, most imprinted and potentially imprinted genes in plants either have non-vital roles in development or their phenotypes remain undisclosed (Hurst and McVean, 1998). Challenges to the idea that inter-ploidy mating and relocation investigations provide support for genomic imprinting in plants have arisen, with increased focus placed on dosage effects, especially for gene products sensitive to dosage within regulatory complexes comprising multiple subunits (Birchler, 1993; Birchler *et al.*, 2001; von Wangenheim and Peterson, 2004). In this context, both the maternal-to-paternal genome ratio and the gametophytic-to-zygotic genome ratio contribute to dosage sensitivity. Imprinting functions as a mechanism to precisely adjust gene expression levels, enabling imprinted genes to be expressed at relative levels of 1 or 2, contingent upon the inactivation of either the maternal or paternal gene variant, unlike non-imprinted genes whose expression is either 3 (active) or 0 (inactive) in the nutritive tissue (Kermicle and Alleman, 1990).

While precise fine-tuning of gene expression levels is crucial for normal development, it does not preclude the possibility of parental conflict as the underlying basis for the evolution of imprinted loci. The spectrum of imprinting processes and the varied traits displayed by genes subject to imprinting suggest that a range of environmental pressures have influenced the development of these genetic sites. Moreover, in plants featuring a haploid, post-meiotic phase and the production of distinct cell types, imprinting may contribute to the epigenetic divergence of egg and central cells, thereby influencing their distinct developmental trajectories post-fertilization (Messing and Grossniklaus, 1999). Imprinting configurations could vary among these cellular types, potentially influencing the formation of fertilization outcomes. Although the hypothesis regarding the distinct epigenomes of egg and central cells has not been extensively verified, due to challenges in these cells, the presence of numerous imprinted genes precise to the endosperm supports such a differentiation.

27.6 Conclusion and Future Directions

The intricate landscape of genomic imprinting in plants continues to unfold as researchers delve deeper into the molecular intricacies and functional implications of imprinted genes. The beginning of next-generation sequencing technology has substantially broadened our understanding, unveiling a multitude of imprinted loci in plants like *A. thaliana*, rice, and maize. The cross-species imprinting patterns observed in both monocots and dicots suggest a potential selective benefit, particularly in the crucial process of seed development.

The evolutionary significance of genomic imprinting is being unraveled. There remains a pivotal need for comprehensive studies exploring imprinted genes in outcrossing relatives of *A. thaliana*. Genomic analyses in diverse plant species and the comparative examination of imprinted genes across these taxa hold promise for identifying conserved imprinting elements. While genome-wide studies underscore the involvement of transposable elements in imprinting evolution, the complete elucidation of the biological processes underpinning imprinted gene expression remains an ongoing challenge.

The trajectory of future research is poised to address these gaps by leveraging modern sequencing technologies and emerging tools across various plant species. This proactive approach will not only refine our understanding of the differentiation of offspring regulation by parents and seed size through genomic imprinting but also facilitate the formulation and testing of new hypotheses. In essence, the journey to unravel the complexities of genomic imprinting in plants is an evolving narrative, where continued exploration promises to yield deeper insights into the fundamental roles and evolutionary nuances of this intriguing epigenetic phenomenon.

References

Baroux, C., Gagliardini, V., Page, D.R. and Grossniklaus, U. (2006) Dynamic regulatory interactions of polycomb group genes: MEDEA autoregulation is required for imprinted gene expression in *Arabidopsis*. *Genes & Development* 20(9), 1081–1086.

Berger, F., Grini, P.E. and Schnittger, A. (2006) Endosperm: An integrator of seed growth and development. *Current Opinion in Plant Biology* 9, 664–670.

Berger, S.L. (2002) Histone modifications in transcriptional regulation. *Current Opinion in Genetics & Development* 12(2), 142–148.

Birchler, J.A. (1993) Dosage analysis of maize endosperm development. *Annual Review of Genetics* 27, 181–204.

Birchler, J.A., Bhadra, U., Bhadra, M.P. and Auger, D.L. (2001) Dosage-dependent gene regulation in multicellular eukaryotes: Implications for dosage compensation, aneuploid syndromes, and quantitative traits. *Developmental Biology* 234(2), 275–288.

Bomblies, K. and Weigel, D. (2010) *Arabidopsis* and relatives as models for the study of genetic and genomic incompatibilities. *Philosophical Transactions of the Royal Society B* 365, 811815–811823.

Bonduriansky, R. (2007) The genetic architecture of sexual dimorphism: The potential roles of genomic imprinting and condition dependence. In: Fairbairn, D.J., Blanckenhorn, W.U. and Székely, T. (eds) *Sex, Size, and Gender Roles: Evolutionary Studies of Sexual Dimorphism*. Oxford University Press, Oxford, UK, pp. 176–184.

Bonduriansky, R. and Chenoweth, S.F. (2009) Intralocus sexual conflict. *Trends in Ecology and Evolution* 24(5), 280–288.

Brandvain, Y. and Haig, D. (2005) Divergent mating systems and parental conflict as a barrier to hybridization in flowering plants. *The American Naturalist* 166(3), 330–338.

Bratzel, F., Yang, C., Angelova, A., López-Torrejón, G., Koch, M. *et al.* (2012) Regulation of the new *Arabidopsis* imprinted gene AtBMI1C requires the interplay of different epigenetic mechanisms. *Molecular Plant* 5(1), 260–269.

Brink, R.A., Kermicle, J.L. and Ziebur, N.K. (1970) Derepression in the female gametophyte in relation to paramutant R expression in maize endosperms, embryos, and seedlings. *Genetics* 66(1), 87–96.

Chaudhuri, S. and Messing, J. (1994) Allele-specific parental imprinting of dzr1, a posttranscriptional regulator of zein accumulation. *Proceedings of the National Academy of Sciences USA* 91, 4867–4871.

Chaudhury, A.M., Ming, L., Miller, C., Craig, S., Dennis, E.S. *et al.* (1997) Fertilization-independent seed development in *Arabidopsis thaliana*. *Proceedings of the National Academy of Sciences USA* 94, 4223–4228.

Constância, M., Kelsey, G. and Reik, W. (2004) Resourceful imprinting. *Nature* 432, 53–57.

Core, L.J., Waterfall, J.J. and Lis, J.T. (2008) Nascent RNA sequencing reveals widespread pausing and divergent initiation at human promoters. *Science* 322(5909), 1845–1848.

Costa, L.M., Yuan, J., Rouster, J., Paul, W., Dickinson, H. *et al.* (2012) Maternal control of nutrient allocation in plant seeds by genomic imprinting. *Current Biology* 22(2), 160–165.

Danilevskaya, O.N., Hermon, P., Hantke, S., Muszynski, M.G., Kollipara, K. *et al.* (2003) Duplicated fie genes in maize: Expression pattern and imprinting suggest distinct functions. *The Plant Cell* 15(2), 425–438.

Day, T. and Bonduriansky, R. (2004) Intralocus sexual conflict can drive the evolution of genomic imprinting. *Genetics* 167(4), 1537–1546.

de Jong, T.J. and Scott, R.J. (2007) Parental conflict does not necessarily lead to the evolution of imprinting. *Trends in Plant Science* 12(10), 439–443.

Deng, S. and Chua, N.H. (2015) Inverted-repeat RNAs targeting FT intronic regions promote FT expression in *Arabidopsis*. *Plant & Cell Physiology* 56(8), 1667–1678.

DeVeale, B., van der Kooy, D. and Babak, T. (2012) Critical evaluation of imprinted gene expression by RNA-Seq: A new perspective. *PLoS Genetics* 8(3), e1002600.

Dilkes, B.P. and Comai, L. (2004) A differential dosage hypothesis for parental effects in seed development. *The Plant Cell* 16(12), 3174–3180.

Eggermann, T., Perez de Nanclares, G., Maher, E.R., Temple, I.K., Tümer, Z. *et al.* (2015) Imprinting disorders: A group of congenital disorders with overlapping patterns of molecular changes affecting imprinted loci. *Clinical Epigenetics* 7(1), 123. DOI: 10.1186/s13148-015-0143-8.

Faure, J.E., Rusche, M.L., Thomas, A., Keim, P., Dumas, C. *et al.* (2003) Double fertilization in maize: The two male gametes from a pollen grain have the ability to fuse with egg cells. *The Plant Journal* 33(6), 1051–1062.

Feng, S., Cokus, S.J., Zhang, X., Chen, P.Y., Bostick, M. *et al.* (2010) Conservation and divergence of methylation patterning in plants and animals. *Proceedings of the National Academy of Sciences USA* 107, 8689–8694.

Fitzgerald, J.N., Hui, P.S. and Berger, F. (2009) Polycomb group-dependent imprinting of the actin regulator AtFH5 regulates morphogenesis in *Arabidopsis thaliana. Development* 136, 3399–3404.

Garnier, O., Laoueille-Duprat, S. and Spillane, C. (2008a) Genomic imprinting in plants. *Epigenetics* 3, 14–20.

Gehring, M., Huh, J.H., Hsieh, T.F., Penterman, J., Choi, Y. *et al.* (2006) DEMETER DNA glycosylase establishes MEDEA polycomb gene self-imprinting by allele-specific demethylation. *Cell* 124, 495–506.

Gehring, M., Bubb, K.L. and Henikoff, S. (2009) Extensive demethylation of repetitive elements during seed development underlies gene imprinting. *Science* 324, 1447–1451.

Gehring, M., Missirian, V. and Henikoff, S. (2011) Genomic analysis of parent-of-origin allelic expression in *Arabidopsis thaliana* seeds. *PloS ONE* 6(8), e23687.

Grossniklaus, U., Vielle-Calzada, J.-P., Hoeppner, M.A. and Gagliano, W.B. (1998) Maternal control of embryogenesis by MEDEA, a polycomb group gene in *Arabidopsis. Science* 280(5362), 446–450.

Guitton, A.E. and Berger, F. (2005) Control of reproduction by polycomb group complexes in animals and plants. *The International Journal of Developmental Biology* 49, 707–716.

Guo, M., Rupe, M.A., Danilevskaya, O.N., Yang, X. and Hu, Z. (2003) Genome-wide mRNA profiling reveals heterochronic allelic variation and a new imprinted gene in hybrid maize endosperm. *The Plant Journal* 36(1), 30–44.

Gutierrez-Marcos, J.F., Costa, L.M., Biderre-Petie, C. and Khbaya, B. (2004) Maternally expressed gene1 is a novel maize endosperm transfer cell-specific gene with a maternal parent-of-origin pattern of expression. *The Plant Cell* 16(5), 1288–1301.

Gutierrez-Marcos, J.F., Pennington, P.D., Costa, L.M. and Dickinson, H.G. (eds) (2003) Imprinting in the endosperm: A possible role in preventing wide hybridization. *Philosophical Transactions of the Royal Society of London. Series B, Biological Sciences* 358(1434), 1105–1111.

Haig, D. (1997) Parental antagonism, relatedness asymmetries, and genomic imprinting. *Proceedings in Biological Sciences* 264(1388), 1657–1662.

Haig, D. (2002) *Genomic Imprinting and Kinship*. Rutgers University Press, New Brunswick, New Jersey.

Haig, D. (2004) Genomic imprinting and kinship: How good is the evidence? *Annual Review of Genetics* 38(1), 553–585.

Haig, D. and Westoby, M. (1989) Parent-specific gene expression and the triploid endosperm. *The American Naturalist* 134(1), 147–155.

Hsieh, T.F., Ibarra, C.A., Silva, P., Zemach, A., Eshed-Williams, L. *et al.* (2009) Genome-wide demethylation of *Arabidopsis* endosperm. *Science* 324, 1451–1454.

Hsieh, T.F., Shin, J., Uzawa, R., Silva, P., Cohen, S. *et al.* (2011) Regulation of imprinted gene expression in *Arabidopsis* endosperm. *Proceedings of the National Academy of Sciences USA* 108(5), 1755–1762.

Hurst, L.D. and McVean, G.T. (1998) Do we understand the evolution of genomic imprinting? *Current Opinion in Genetics & Development* 8(6), 701–708.

Ingouff, M., Hamamura, Y., Gourgues, M., Higashiyama, T. and Berger, F. (2007) Distinct dynamics of HISTONE3 variants between the two fertilization products in plants. *Current Biology* 17(12), 1032–1037.

Jahnke, S. and Scholten, S. (2009) Epigenetic resetting of a gene imprinted in plant embryos. *Current Biology* 19(19), 1677–1681.

Jones, P.A. (2012) Functions of DNA methylation: Islands, start sites, gene bodies and beyond. *Nature Reviews Genetics* 13(7), 484–492.

Jullien, P.E. and Berger, F. (2010) Parental genome dosage imbalance deregulates imprinting in *Arabidopsis*. *PLoS Genetics* 6(3), e1000885. DOI: 10.1371/journal.pgen.1000885.

Jullien, P.E., Katz, A., Oliva, M., Ohad, N. and Berger, F. (2006a) Polycomb group complexes self-regulate imprinting of the polycomb group gene MEDEA in *Arabidopsis*. *Current Biology* 16(5), 486–492.

Jullien, P.E., Kinoshita, T., Ohad, N. and Berger, F. (2006b) Maintenance of DNA methylation during the *Arabidopsis* life cycle is essential for parental imprinting. *The Plant Cell* 18(6), 1360–1372.

Kakutani, T., Kato, M., Kinoshita, T. and Miura, A. (2004) Control of development and transposon movement by DNA methylation in *Arabidopsis thaliana*. *Cold Spring Harbor Symposia on Quantitative Biology* 69, 139–144.

Kermicle, J.L. (1970) Dependence of the R-mottled aleurone phenotype in maize on mode of sexual transmission. *Genetics* 66(1), 69–85.

Kermicle, J.L. and Alleman, M. (1990) Gametic imprinting in maize in relation to the angiosperm life cycle. *Development* 108(Supplement), 9–14.

Khatib, H. (2007) Is it genomic imprinting or preferential expression? *BioEssays* 29(10), 1022–1028.

Kinoshita, T., Yadegari, R., Harada, J.J., Goldberg, R.B. and Fischer, R.L. (1999) Imprinting of the MEDEA polycomb gene in the *Arabidopsis endosperm*. *The Plant Cell* 11(10), 1945.

Kinoshita, T., Miura, A., Choi, Y., Kinoshita, Y., Cao, X. *et al.* (2004) One-way control of FWA imprinting in *Arabidopsis* endosperm by DNA methylation. *Science (New York)* 303(5657), 521–523.

Kiyosue, T., Ohad, N., Yadegari, R., Hannon, M., Dinneny, J. *et al.* (1999) Control of fertilization-independent endosperm development by the MEDEA polycomb gene in *Arabidopsis*. *Proceedings of the National Academy of Sciences USA* 96, 4186–4191.

Kohler, C. and Weinhofer-Molisch, I. (2010) Mechanisms and evolution of genomic imprinting in plants. *Heredity* 105, 57–63.

Köhler, C., Hennig, L., Spillane, C., Pien, S., Gruissem, W. *et al.* (2003) The polycomb-group protein MEDEA regulates seed development by controlling expression of the MADSbox gene PHERES1. *Genes & Development* 17(12), 1540–1553.

Köhler, C., Page, D.R., Gagliardini, V. and Grossniklaus, U. (2005) The *Arabidopsis thaliana* MEDEA polycomb group protein controls expression of PHERES1 by parental imprinting. *Nature Genetics* 37(1), 28–30.

Kohler, C., Wolff, P. and Spillane, C. (2012) Epigenetic mechanisms underlying genomic imprinting in plants. *Annual Review of Plant Biology* 63, 331–352.

Kondoh, M. and Higashi, M. (2000) Reproductive isolation mechanism resulting from resolution of intragenomic conflict. *The American Naturalist* 156, 511–518.

Kraft, E., Bostick, M., Jacobsen, S.E. and Callis, J. (2008) ORTH/VIM proteins that regulate DNA methylation are functional ubiquitin E3 ligases. *The Plant Journal* 56(5), 704–715.

Law, J.A. and Jacobsen, S.E. (2010) Establishing, maintaining and modifying DNA methylation patterns in plants and animals. *Nature Reviews Genetics* 11, 204–220.

Lawson, H.A., Cheverud, J.M. and Wolf, J.B. (2013) Genomic imprinting and parent-of-origin effects on complex traits. *Nature Reviews. Genetics* 14(9), 609–617.

Lei, M., Zhang, H., Julian, R., Tang, K., Xie, S. *et al.* (2015) Regulatory link between DNA methylation and active demethylation in *Arabidopsis*. *Proceedings of the National Academy of Sciences USA* 112(11), 3553–3557.

Li, E., Beard, C. and Jaenisch, R. (1993) Role for DNA methylation in genomic imprinting. *Nature* 366(6453), 362–365.

Lopes, M.A. and Larkins, B.A. (1993) Endosperm origin, development, and function. *Plant Cell* 5, 1383–1399.

Lund, G., Ciceri, P. and Viotti, A. (1995) Maternal-specific demethylation and expression of specific alleles of zein genes in the endosperm of *Zea mays* L. *The Plant Journal* 8(4), 571–581.

Luo, H., Mipam, T.D., Wu, S., Xu, C., Yi, C. *et al.* (2022) DNA methylome of primary spermatocyte reveals epigenetic dysregulation associated with male sterility of cattleyak. *Theriogenology* 191, 153–167.

Luo, M., Bilodeau, P., Koltunow, A., Dennis, E.S., Peacock, W.J. *et al.* (1999) Genes controlling fertilization-independent seed development in *Arabidopsis thaliana*. *Proceedings of the National Academy of Sciences of the United States of America* 96(1), 296–301.

Luo, M., Bilodeau, P., Dennis, E.S., Peacock, W.J. and Chaudhury, A.M. (2000) Expression and parent-of-origin effects for FIS2, MEA, and FIE in the endosperm and embryo of developing *Arabidopsis* seeds. *Proceedings of the National Academy of Sciences of the United States of America* 97(19), 10637–10642.

Luo, M., Taylor, J.M., Spriggs, A., Zhang, H., Wu, X. *et al.* (2011) A genome-wide survey of imprinted genes in rice seeds reveals imprinting primarily occurs in the endosperm. *PLoS Genetics* 7(6), e1002125.

Macdonald, W.A. (2012) Epigenetic mechanisms of genomic imprinting: Common themes in the regulation of imprinted regions in mammals, plants, and insects. *Genetics Research International* 585024. DOI: 10.1155/2012/585024.

Makarevich, G., Leroy, O., Akinci, U., Schubert, D., Clarenz, O. *et al.* (2006) Different polycomb group complexes regulate common target genes in *Arabidopsis*. *EMBO Reports* 7(9), 947–952.

Makarevich, G., Villar, C.B.R., Erilova, A. and Köhler, C. (2008) Mechanism of PHERES1 imprinting in *Arabidopsis*. *Journal of Cell Science* 121, 906–912.

McKeown, P.C., Laouielle-Duprat, S., Prins, P., Wolff, P., Schmid, M.W. *et al.* (2011) Identification of imprinted genes subject to parent-of-origin specific expression in *Arabidopsis thaliana* seeds. *BMC Plant Biology* 11(1), 113. DOI: 10.1186/1471-2229-11-113.

McKeown, P.C., Fort, A. and Spillane, C. (2013) Genomic imprinting: Parental control of gene expression in higher plants. In: Chen, Z.J. and Birchler, J.A. (eds) *Polyploid and Hybrid Genomics. Wiley-Blackwell.* pp. 257–270. DOI: 10.1002/9781118552872.ch16.

Messing, J. and Grossniklaus, U. (1999) Genomic imprinting in plants. *Results and Problems in Cell Differentiation* 25, 23–40.

Moore, T. and Mills, W. (2008) Evolutionary theories of imprinting – enough already!. *Advances in Experimental Medicine and Biology* 626, 116–122.

Ohad, N., Margossian, L., Hsu, Y.C., Williams, C., Repetti, P. *et al.* (1996) A mutation that allows endosperm development without fertilization. *Proceedings of the National Academy of Sciences USA* 93(11), 5319–5324. DOI: 10.1073/pnas.93.11.5319.

Ohad, N., Yadegari, R., Margossian, L., Hannon, M., Michaeli, D. *et al.* (1999) Mutations in FIE, a WD polycomb group gene allow endosperm development without fertilization. *The Plant Cell* 117, 407–415.

Orlando, V. (2003) Polycomb, epigenomes and control of cell identity. *Cell* 112, 599–606.

Patten, M.M., Ross, L., Curley, J.P., Queller, D.C., Bonduriansky, R. *et al.* (2014) The evolution of genomic imprinting: Theories, predictions and empirical tests. *Heredity* 113(2), 119–128.

Pignatta, D., Erdmann, R.M., Scheer, E., Picard, C.L., Bell, G.W. *et al.* (2014) Natural epigenetic polymorphisms lead to intraspecific variation in *Arabidopsis* gene imprinting. *eLife* 3, e03198.

Raissig, M.T., Bemer, M., Baroux, C. and Grossniklaus, U. (2013) Genomic imprinting in the *Arabidopsis* embryo is partly regulated by PRC2. *PLoS Genetics* 9(12), e1003862.

Rigal, M., Kevei, Z., Pélissier, T. and Mathieu, O. (2012) DNA methylation in an intron of the IBM1 histone demethylase gene stabilizes chromatin modification patterns. *The EMBO Journal* 31(13), 2981–2993.

Rodrigues, J.A. and Zilberman, D. (2023) Evolution and function of genomic imprinting in plants. *Genes and Development* 29, 2517–2531.

Russell, S.D. (1991) Isolation and characterization of sperm cells in flowering plants. *Annual Review of Plant Physiology and Plant Molecular Biology* 42(1), 189–204.

Sapienza, C., Peterson, A.C., Rossant, J. and Balling, R. (1987) Degree of methylation of transgenes is dependent on gamete of origin. *Nature* 328(6127), 251–254.

Satyaki, P.R.V. and Gehring, M. (2017) DNA methylation and imprinting in plants: Machinery and mechanisms. *Critical Reviews in Biochemistry and Molecular Biology* 52(2), 163–175.

Schubert, D. and Goodrich, J. (2003) Plant epigenetics: MEDEA's children take centre stage. *Current Biology* 13(16), R638–R640.

Shibuya, K., Fukushima, S. and Takatsuji, H. (2009) RNA-directed DNA methylation induces transcriptional activation in plants. *Proceedings of the National Academy of Sciences USA* 106(5), 1660–1665.

Shirzadi, R., Andersen, E.D., Bjerkan, K.N., Gloeckle, B.M., Heese, M. *et al.* (2011) Genome-wide transcript profiling of endosperm without paternal contribution identifies parent-of-origin-dependent regulation of AGAMOUS-LIKE36. *PLoS Genetics* 7(2), e1001303.

Sinervo, B. and Svensson, E. (2002) Correlational selection and the evolution of genomic architecture. *Heredity* 89(5), 329–338.

Sørensen, M.B., Chaudhury, A.M., Robert, H., Bancharel, E. and Berger, F. (2001) Polycomb group genes control pattern formation in plant seed. *Current Biology* 11(4), 277–281.

Spada, F., Haemmer, A., Kuch, D., Rothbauer, U., Schermelleh, L. *et al.* (2007) DNMT1 but not its interaction with the replication machinery is required for maintenance of DNA methylation in human cells. *The Journal of Cell Biology* 176(5), 565–571.

Talbert, P.B. and Henikoff, S. (2010) Histone variants--ancient wrap artists of the epigenome. *Nature Reviews Molecular Cell Biology* 11(4), 264–275.

Thorstensen, T., Grini, P.E. and Aalen, R.B. (2011) SET domain proteins in plant development. *Biochimica et Biophysica Acta (BBA) – Gene Regulatory Mechanisms* 1809(8), 407–420.

Tiwari, S., Schulz, R., Ikeda, Y., Dytham, L., Bravo, J. *et al.* (2008) MATERNALLY expressed PAB C-TERMINAL, a novel imprinted gene in *Arabidopsis*, encodes the conserved C-TERMINAL domain of polyadenylate binding proteins. *The Plant Cell* 20(9), 2387–2398.

Verona, R.I., Mann, M.R.W. and Bartolomei, M.S. (2003) Genomic imprinting: Intricacies of epigenetic regulation in clusters. *Annual Review of Cell and Developmental Biology* 19(1), 237–259.

Vielle-Calzada, J.P., Thomas, J., Spillane, C., Coluccio, A., Hoeppner, M.A. *et al.* (1999) Maintenance of genomic imprinting at the *Arabidopsis* medea locus requires zygotic DDM1 activity. *Genes & Development* 13(22), 2971–2982.

Vielle-Calzada, J.-P., Baskar, R. and Grossniklaus, U. (2000) Delayed activation of the paternal genome during seed development. *Nature* 404(6773), 91–94.

von Wangenheim, K.-H. and Peterson, H.-P. (2004) Aberrant endosperm development in interploidy crosses reveals a timer of differentiation. *Developmental Biology* 270(2), 277–289.

Waters, A.J., Makarevitch, I., Eichten, S.R., Swanson-Wagner, R.A., Yeh, C.T. *et al.* (2011) Parent-of-origin effects on gene expression and DNA methylation in the maize endosperm. *Plant Cell* 23(12), 4221–4233.

Waters, A.J., Bilinski, P., Eichten, S.R., Vaughn, M.W., Ross-Ibarra, J. *et al.* (2013) Comprehensive analysis of imprinted genes in maize reveals allelic variation for imprinting and limited conservation with other species. *Proceedings of the National Academy of Sciences USA* 110(48), 19639–19644.

Weinhofer, I., Hehenberger, E., Roszak, P., Hennig, L. and Köhler, C. (2010) H3K27me3 profiling of the endosperm implies exclusion of polycomb group protein targeting by DNA methylation. *PLoS Genetics* 6(10), e1001152.

Williams, B.P., Pignatta, D., Henikoff, S. and Gehring, M. (2015) Methylation-sensitive expression of a DNA demethylase gene serves as an epigenetic rheostat. *PLoS Genetics* 11(3), e1005142.

Wolf, J.B. and Hager, R. (2006) A maternal-offspring coadaptation theory for the evolution of genomic imprinting. *PLoS Biology* 4(12), e380.

Wolf, J.B. and Brodie, E.D. (eds) (1998) The coadaptation of parental and offspring characters. *Evolution* 52(2), 299–308.

Wolff, P., Weinhofer, I., Seguin, J., Roszak, P., Beisel, C. *et al.* (2011) High-resolution analysis of parent-of-origin allelic expression in the *Arabidopsis* endosperm. *PLoS Genetics* 7(6), e1002126.

Xiao, W., Custard, K.D., Brown, R.C., Lemmon, B.E., Harada, J.J. *et al.* (2006) DNA methylation is critical for *Arabidopsis* embryogenesis and seed viability. *The Plant Cell* 18(4), 805–814.

Xu, W., Dai, M., Li, F. and Liu, A. (2014) Genomic imprinting, methylation and parent-of-origin effects in reciprocal hybrid endosperm of castor bean. *Nucleic Acids Research* 42(11), 6987–6996.

Yadegari, R., Kinoshita, T., Lotan, O., Cohen, G., Katz, A. *et al.* (2000) Mutations in the FIE and MEA genes that encode interacting polycomb proteins cause parent-of-origin effects on seed development by distinct mechanisms. *The Plant Cell* 12(12), 2367–2382.

Zemach, A., McDaniel, I.E., Silva, P. and Zilberman, D. (2010) Genome-wide evolutionary analysis of eukaryotic DNA methylation. *Science* 328(5980), 916–919.

Zhang, M., Zhao, H., Xie, S., Chen, J., Xu, Y. *et al.* (2011) Extensive, clustered parental imprinting of protein-coding and noncoding rnas in developing maize endosperm. *Proceedings of the National Academy of Sciences USA* 108(50), 20042–20047.

Zhang, M., Xie, S., Dong, X., Zhao, X., Zeng, B. *et al.* (2014) Genome-wide high resolution parental-specific DNA and histone methylation maps uncover patterns of imprinting regulation in maize. *Genome Research* 24(1), 167–176.

Zhang, X., Yazaki, J., Sundaresan, A., Cokus, S., Chan, S.W.-L. *et al.* (2006) Genome-wide high-resolution mapping and functional analysis of DNA methylation in *Arabidopsis*. *Cell* 126(6), 1189–1201.

28 Plant Stress Priming: The Role of Epigenetic Regulation

Burcu Arıkan[1]*, Elif Pulat[2,3], Özgür Çakır[1] and Neslihan Turgut Kara[1]
[1]Istanbul University, Faculty of Science, Department of Molecular Biology and Genetics, Vezneciler, Istanbul, Türkiye; [2]Acibadem University, Faculty of Engineering and Natural Sciences, Department of Molecular Biology and Genetics, Istanbul, Türkiye; [3]Istanbul University, Institute of Science, Program of Molecular Biology and Genetics, Istanbul, Türkiye

Abstract

Plant stress priming, characterized by brief exposure to mild stress conditions, induces the formation of stress memory in plants. This memory equips them to respond more effectively when faced with recurrent stressors, as if the plant had learned from previous experiences. The effects of this priming process are far-reaching and often visible at the genetic and epigenetic levels. Epigenetic mechanisms lead to shifts in gene expression and the appearance of variances in transcript levels, enabling the plant to fine-tune its response to future stressors. Crucially, stress memory is linked to epigenetic modifications, and these changes can sometimes be limited to the recovery process, while in other instances, they are inherited across generations. The formation of stress memory gives plants plasticity and increases their chances of adaptation to environmental stimuli. This chapter discusses the relationship between plant stress priming and epigenetic mechanisms, with diverse examples. However, the molecular mechanisms of priming are not fully elucidated and are waiting to be discovered, since it is important in terms of developing new strategies for the sustainability of agriculture and food safety in the coming years.

Keywords: Priming, Epigenetics, Stress response, Epigenetic regulation, Stress priming

28.1 Introduction

The human population continues to grow daily, with an increasing need for food sources. However, due to global warming and climate change, the area of arable lands and crop productivity are decreasing, endangering food security and the sustainability of agriculture. With the increasing demand for food, the world is getting closer to facing hunger daily. To circumvent this situation, researchers have been studying and developing plant breeding techniques to combat global warming (Aswathi *et al.*, 2022).

Both conventional and modern techniques play crucial roles in the continuous improvement of crops, ensuring food supply and agricultural sustainability. While conventional breeding strategies consisting of selection, hybridization, crossing, mutation breeding, and inducing polyploidy rely on phenotypic selection and natural genetic variation, modern breeding strategies including marker-assisted selection (MAS), genomic selection, genetic engineering,

*Corresponding author: burcuarikan@istanbul.edu.tr

 Epigenetics for Climate-Smart and Sustainable Agriculture (ed. Jen-Tsung Chen)
DOI: 10.1079/9781800626102.0028

genome editing via TALENs, ZFNs, CRISPR/Cas9 and omic technologies utilize advanced biotechnological tools for more precise, faster, and more efficient breeding (Lamichhane and Thapa, 2022). Even though these methods have positive aspects, they also present disadvantages, particularly in terms of biosafety and ethics.

Plants develop sophisticated mechanisms in order to adapt to fluctuating environmental cues through morphological, physiological, and molecular pathways. These include generating signal molecules, expressing defense-related genes, ion homeostasis, hardening cell walls, producing primary and secondary metabolites, and regulating osmotic pressure (Haak *et al.*, 2017). After stress conditions subside, some changes return to their baseline state, while others may contribute to memory formation and increased resilience in plants. This memory can be formed as somatic stress memory, or transgenerational or intergenerational stress memory (Do Amaral *et al.*, 2020). The formation of stress memory may be regulated by epigenetic mechanisms such as DNA methylation, post-translational modifications of histone tails, chromatin remodeling, and RNA-directed regulation (Ding *et al.*, 2020; Do Amaral *et al.*, 2020; Srivastava *et al.*, 2021).

Stress priming helps plants develop a heightened ability to tolerate diverse environmental stressors such as drought, salinity, extreme temperatures, and pathogen attacks. By exposing plants to controlled, sub-lethal levels of stress, researchers can activate the plant's defense mechanisms and help them build a 'memory' of the stress, enabling the plants to respond more effectively when exposed to similar or more severe stress conditions later on. Primed plants are more likely to maintain their productivity under adverse conditions, thereby ensuring a more reliable food supply. Priming is a low-cost technique that can be easily adapted to plant-breeding strategies in addition to conventional breeding techniques.

It is promising to develop stress-tolerant plants with high yield and productivity to fight the loss of arable lands and ensure the sustainability of agriculture (Martinez-Medina *et al.*, 2016; Dhanya Thomas *et al.*, 2020). Current advances are focused on developing agroecological and agro-friendly breeding strategies to combat the harnessing effects of stress factors that limit crop yield and productivity. This chapter explores the importance of priming studies in the development of stress-tolerant plants, highlighting the role of epigenetic regulation in plant stress responses.

28.2 Stress Response of Plants

In nature, plants encounter diverse abiotic and biotic stress conditions, including salinity, cold, high temperature, drought, UV radiation, waterlogging, nutrient deficiency, pathogen attacks, and insects throughout their lifespan. These challenging circumstances limit their growth, productivity, and yield, consequently threatening crop yield and food security in a world where the human population is increasing day by day (Pereira, 2016; Thiebaut *et al.*, 2019; Chang *et al.*, 2020; Haider *et al.*, 2021). To cope with these environmental fluctuations, breeding strategies have been developed for many years. Breeding technologies, started with the search for new characteristics by crossing crops, continue today with the use of modern technologies.

Plant stress response mechanisms consist of the perception of stress clues, signaling cascades, generation of transcription factors, changes in hormonal levels, and gene expression. For the initiation of these, some triggering is needed. This triggering is called priming; also, in some cases, it is described as conditioning, hardening, or acclimation. The term priming is also used for the milder applications of stress conditions in order to trigger stress response mechanisms (Conrath *et al.*, 2006). After priming, signaling cascades are activated, plant hormone levels are changed, transcription factors are generated, and some secondary messengers like Ca^{2+} are produced in addition to the generation of epigenetic regulators (Boulc'h *et al.*, 2020; Leuendorf *et al.*, 2020; Harris *et al.*, 2023). But these parameters depend on the duration, type, and severity of the stress conditions or triggering. According to the studies, the type of priming and the duration between priming and stress exposure are important. Studies have shown that gradual exposure to stress or priming results in better adaptation. In addition, the time interval between each stress exposure is also important, because the effect of triggering will diminish after each cell division (Bruce *et al.*, 2007).

Unlike animals, plants do not have nervous systems, so they need to develop other ways to remember and adapt to past encounters. After several exposures to stress, plants may become resilient to stress conditions by forming a stress memory controlled by metabolomic, proteomic, transcriptomic, and epigenetic mechanisms (Ding *et al.*, 2014). The formed memory patterns can be limited to the same generation or transmitted to the new generations (Harris *et al.*, 2023). This stress memory enables plants to adapt to fluctuating environmental stimuli. During somatic stress memory, changes can be passed on mitotically to daughter cells. However, if these changes occur in germline cells, they can be transmitted to the next generations. When this memory response is seen after one stress-free generation, it is described as transgenerational memory, while if it is seen after two stress-free generations, it is described as transgenerational stress memory (Lämke and Bäurle, 2017).

The molecular mechanisms underlying this memory formation are not fully elucidated. Among these mechanisms, epigenetic mechanisms, consisting of DNA methylation, histone modifications, chromatin remodeling, and non-coding RNAs (ncRNAs), play significant roles in forming memory and its transmission to new generations. This stress memory enables plants to survive in recurring stress conditions. The stored information may be inherited by the next generations, thus enhancing the resilience of plants. Therefore, elucidating the molecular mechanisms underlying epigenetic regulations related to plant stress response and adaptation seems promising.

28.3 Priming

Plants are constantly exposed to environmental stimuli and may become resistant to stressors after several exposures, creating a stress memory that can be described as priming, hardening, acclimation, or conditioning (Bruce *et al.*, 2007; Conrath, 2011; Herman and Sultan, 2011).

Plant stress priming is a critical technique in modern agriculture and horticulture. It involves exposing plants to mild stressors to enhance their ability to withstand more severe stress in the future. This practice has gained significant importance due to its potential to improve plant resilience, productivity, and sustainability. It triggers a stress memory in plants, resulting in a faster and stronger stress response (Bruce *et al.*, 2007; Ganie *et al.*, 2024). This triggering is achieved by natural or synthetic exposure of plants to biotic and abiotic stress conditions. In addition, the stress response is stimulated by priming with signaling metabolites such as jasmonic acid (JA) and salicylic acid (SA) (Conrath *et al.*, 2015). Moreover, it is reported that β-aminobutyric acid (BABA) induces a stress response in some cases, making plants primed. The researchers discovered that this perception is mediated by an enzyme called aspartyl-tRNA synthetase, which acts as a receptor for BABA. Upon binding with BABA, the aspartyl-tRNA synthetase initiates a signaling cascade that boosts the plant's defense mechanisms, thereby improving its stress tolerance (Luna *et al.*, 2014). Similarly, another study (Luna *et al.*, 2016) explored the use of BABA and JA to induce long-lasting resistance in tomatoes against the fungal pathogen *Botrytis cinerea*. In this study, diverse application methods such as seed coating, seed soaking, and root treatments were analyzed in order to develop strategies to enhance the resilience of tomato plants.

Priming treatments can be made at different stages of plant development. According to the growth stage or duration of the application, the outcomes can differ. Each priming event could result differently and influence diverse mechanisms. For instance, seed priming affects germination rates and induces plant growth. Seed priming techniques are based on UV-B priming, osmopriming, hydropriming, halopriming and chemical priming. These mild stress factors activate the stress response of plants by transcribing stress-response genes and also creating an epigenetic memory (Bruce *et al.*, 2007; Hossain *et al.*, 2018; Aswathi *et al.*, 2022).

Increasing data in recent years show that plant stress priming has a major role in memory formation. It enables plants to activate their metabolism and accumulate hormones, signaling molecules, and transcription factors (TFs), resulting in faster and stronger responses to recurring stress conditions (Conrath *et al.*, 2006). However, the lasting effect of memory is not fully understood. On the other hand, duration, intensity, or time interval between recurring stress conditions vary in nature, and sometimes, plants have to face more than one stress condition simultaneously. Thus, their sensitivity, plasticity and adaptation to environmental cues have great importance.

After priming, plants go from the so-called naïve phase to the primed phase, where transcriptional, post-translational, and epigenetic changes occur (Luna *et al.*, 2012). Priming application stimulates several molecular mechanisms, including the accumulation of regulatory proteins, defense-related substances, transcription factors, and epigenetic modifications (Chinnusamy and Zhu, 2009; Conrath, 2011). It encourages plants to store information from previous stress exposure, which results in stress memory formation during a recovery phase. This stress memory can stay at the somatic level or be transmitted to the next generations, creating intergenerational or transgenerational stress memory. During somatic stress memory, priming can last for days or weeks. In intergenerational memory, the priming effect can be observed in at least two consecutive generations. In transgenerational stress memory, the effects can persist across multiple generations (Lämke and Bäurle, 2017; Tardieu *et al.*, 2018). Thus, it is an important perspective for combating stress in nature.

A study by Luna *et al.* (2012) explored the link between priming and epigenetic mechanisms. Researchers have shown that infection of *Arabidopsis* plants with bacteria initiated defense priming. According to the study, plants were triggered to activate defense genes related to salicylic acid (SA), resulting in systemic acquired resistance (SAR). Subsequently, plant progeny demonstrated resilience to infection by oomycetes, in contrast to unprimed plants. The research highlights the role of chromatin remodeling and DNA methylation in regulating SAR, suggesting that these epigenetic changes can be inherited, providing a transgenerational defense mechanism.

To date, several strategies have been developed to produce stress-tolerant plants. Conventional breeding technologies involve crossing and selection to develop new traits such as inbreeding, crossbreeding, backcrossing, hybridization, marker-assisted selection (MAS), mass selection, pure line selection, and segregation. These conventional breeding strategies have been used for centuries and still continue to be part of breeding technologies. They are often integrated into modern breeding strategies that involve genome editing, marker-assisted breeding (MAB), metabolomics, and bioinformatics. These modern breeding strategies leverage advancements in genetics, genomics, and biotechnology to address challenges in agriculture and develop crop varieties with improved yield, quality, and resilience (Anand *et al.*, 2022; Lamichhane and Thapa, 2022). However, they have limitations in food safety and toxicity. Moreover, in some cases, applying these strategies is not very economical and increases the cost significantly. On the other hand, priming has several advantages. It can be adapted to each plant according to the need, and, due to its low cost and toxicity, it can be considered agro-friendly. By enhancing the natural stress resistance of plants, stress priming can reduce the need for chemical fertilizers, pesticides, and herbicides. This not only lowers production costs for farmers but also mitigates the environmental impact associated with the overuse of agricultural chemicals, such as soil degradation, water contamination, and loss of biodiversity. In addition, priming can be combined with other crop improvement technologies to enhance the 'overall' effectiveness (Kerchev *et al.*, 2020; Gaude *et al.*, 2024).

In recent years, the importance of epigenetic regulation in plant stress response has been revealed. Many studies explain the effects of priming on plant stress response and memory formation related to epigenetic regulation, but the underlying molecular mechanisms are still not fully understood. That is why research on plant stress priming has become important.

28.4 Epigenetic Regulation in Stress Priming

The term epigenetics describes variations in the genome that do not affect the DNA sequence

(Eichten *et al.*, 2014). Epigenetic changes can be inherited through mitotic or meiotic division, either in the same generation or over generations. DNA is wrapped with histone proteins into nucleosomes and each nucleosome core is approximately 147 bp, consisting of two units of each of H2A, H2B, H3, and H4 histone protein and connected with the H1 linker histone (Cutter and Hayes, 2015). Due to their dynamic structure, epigenetic modifications can increase chromatin accessibility, leading to changes in gene expression. Key mechanisms of epigenetic regulation include DNA methylation, histone modification, and the action of non-coding RNAs. Collectively, these mechanisms enable plants to swiftly and efficiently respond to various stressors such as drought, salinity, extreme temperatures, and pathogen attacks (Bertini *et al.*, 2018; Miryeganeh, 2021; Oberkofler *et al.*, 2021; Harris *et al.*, 2023).

Epigenetics has been widely used in crop breeding to improve desired traits. For example, many studies with the *MSH1* gene are related to epigenetic memory formation and the development of agronomic traits in different plants, including *Arabidopsis*, soybean, and tomato (Raju *et al.*, 2018; Kundariya *et al.*, 2020; Yang *et al.*, 2020). In a study with seagrasses, the relationship between thermopriming and epigenetic modifications was analyzed (Nguyen *et al.*, 2020). According to their findings, methylation-related genes have been regulated significantly, resulting in enhanced response to reoccurring stress. In a study conducted with *Oryza sativa*, Bertini *et al.* (2018) investigated the role of methyl jasmonate (MeJA) priming on wounding response. They found that after priming plants with MeJA, the expression of defense-related genes *OsBBPI* and *OsPOX* increased. Moreover, the results demonstrated histone modifications in the promoter region of the *OsBBPI* gene and genome-wide changes in DNA methylation levels. Similarly, treating the seeds of *Arabidopsis* plants with menadione sodium bisulfite (MSB) affected the promoter regions of genes involved in proline metabolism, resulting in changes in DNA methylation that contribute to salt stress tolerance. MSB induced hypomethylation in the promoters of *P5CS1* and *ERD5* genes, which are involved in the biosynthesis and degradation of proline, respectively. While hypomethylation led to an increase in the expression of *P5CS1*, it resulted in a decrease in the expression of *ERD5*, which correlated with enhanced resistance (Jiménez-Arias *et al.*, 2015).

Priming not only affects one stress condition but may also influence other stress response mechanisms. Researchers examined the effect of high light (HL) and drought stress response following priming. According to their findings, HL induced transcriptional priming. They discovered that after HL exposure, the expression of genes such as *RD29A*, *AREB1*, *NCED3*, *ABA1*, and *DREB2* increased; however, after 5 days, this increase returned to the basal state. Interestingly, when drought stress was applied after HL exposure, these genes demonstrated a higher increase compared with non-exposed plants. Additionally, histone modifications were investigated, leading to the conclusion that priming affected the expression of stress response genes, with the enrichment of the H3K4me3 histone mark being associated with upregulated genes (Holness *et al.*, 2023).

After exposure to high temperatures, plants can acclimatize and acquire thermotolerance, referred to as heat shock (HS) priming (Perrella *et al.*, 2022). It has been demonstrated that epigenetic regulation plays a major role in adapting to HS (Bäurle, 2016; Ohama *et al.*, 2017). HS priming influences nuclear organization, altering transcriptomic activity in heterochromatic regions and causing changes in the transcription. Moreover, prolonged HS affects the organization of repetitive elements, disrupting their silencing mechanisms, and resulting in changes in genome stability (Pecinka *et al.*, 2010; Tittel-Elmer *et al.*, 2010). In addition, hypermethylation was seen at the histone H3K4 mark in HS memory formation (Bäurle, 2016; Liu *et al.*, 2018). In a study with *Arabidopsis*, it was revealed that recurring heat stress triggers changes in histone methylation patterns, specifically the removal of the H3K27me3 mark, through the action of demethylases. This alteration in histone modifications leads to the upregulation of *HSP22* and *HSP17.6C*, which are important for the plant's response to heat stress, preparing the related genes for faster activation (Yamaguchi *et al.*, 2021). In another study involving heat stress priming, transgenerational tolerance was observed in wheat (Wang *et al.*, 2016). This tolerance was induced by epigenetic modifications affecting defense metabolite production. The study revealed that the expression of the lysine-specific histone demethylase 1 (*LSD1*), involved in histone demethylation, was increased, along with genes associated with signal transduction and defense pathways. Interestingly, primed plants exhibited higher grain weight and increased yield.

Understanding epigenetic regulation in plant stress response is pivotal, focusing on how plants utilize reversible alterations in gene expression, which are heritable but not encoded in the DNA sequence, to adapt to environmental stresses. Recent advancements in epigenetics provide fresh insights into deciphering plant stress response mechanisms. Typically, epigenetic changes induced by stress return to their basal state over time. However, some persist, establishing a priming stress memory (Bruce *et al.*, 2007; Perrone and Martinelli, 2020). Table 28.1 presents a summary of epigenetic regulation in stress priming events, elucidating the mechanisms and types of priming applications.

28.4.1 DNA methylation

DNA methylation is one of the most studied epigenetic mechanisms in plant stress biology. It includes the addition of a methyl group at the 5′ position of the cytosine base or 6′ position of the adenine base, creating 5-methylcytosine (5mC) and N6-methyladenine (6mA), respectively (Liang *et al.*, 2018; Zhang *et al.*, 2018). This modification affects gene expression by altering DNA accessibility and influencing transcription factor binding (Kribelbauer *et al.*, 2020). Additionally, DNA methylation is important for transposon silencing and the inheritance of traits (Zhang *et al.*, 2018).

DNA methylation can occur symmetrically on CG and CHG sequences or asymmetrically as in CHH (H = A, T, or C). In plants, DNA methylation can occur through different processes, such as *de novo* DNA methylation or the maintenance of methylation. *De novo* DNA methylation is executed through the RNA-directed DNA methylation (RdDM) pathway, involving plant-specific RNA polymerase enzymes Pol IV and Pol V, and DOMAINS REARRANGED METHYLASE 2 (DRM2). The maintenance of methylation is controlled by METHYLTRANSFERASE 1 (MET1) and CHROMOMETHYLASE 3 (CMT3) enzymes (Tirnaz and Batley, 2019).

In Tartary buckwheat (*Fagopyrum tataricum*), researchers conducted methylome and transcriptome analyses in order to investigate the epigenetic mechanisms underlying the cold response, with a particular focus on DNA methylation (Song *et al.*, 2020). Through comprehensive phenome, methylome, and transcriptome analyses, the researchers examined how cold priming affects the plant. The findings revealed a global decrease in DNA methylation during the cold response, particularly in intergenic regions. Several genes, including *CuAO*, *RPB1*, and *DHE1*, exhibited DNA methylation changes during cold treatments. Overall, the study suggests that DNA methylation plays a crucial role in regulating the cold response and shaping agronomic traits in Tartary buckwheat. In another study, it was observed that abiotic stress factors including salt, cold, and water deficit caused demethylation in the promoter regions of the genes *ACD6*, *ACO3* and *GSTF14* in *Arabidopsis thaliana*, resulting in activation of the gene expression (Yang *et al.*, 2022).

Researchers investigated the effect of priming potato plants with β-aminobutyric acid (BABA), a chemical known to enhance plant defenses (Kuźnicki *et al.*, 2019). The study explored how potato plants respond to the presence of *Phytophthora infestans*, a pathogen causing late blight disease. Priming with BABA affected the DNA methylation dynamics throughout the genome, particularly in regions associated with defense-related genes. Through hypermethylation, the expression of the genes *CMT3*, *DRM2*, and *ROS1* was upregulated. Moreover, the expression of the genes *NPR1*, *StWRKY1*, and *PR1* was also upregulated, but not directly associated with DNA methylation; rather, it resulted from alterations related to methylation. These changes were inherited by subsequent generations, suggesting an intergenerational transmission of epigenetic defense mechanisms. In another study, researchers investigated the effect of artificial salicylic acid (SA) on *Arabidopsis* and apple to understand how plants respond to biotic stress through priming and de-priming of transcriptional memory (Gully *et al.*, 2019). Their findings concluded that SA priming induced DNA

Table 28.1. Summary of type of acquired tolerance, epigenetic regulation and/or memory and related genes affected by priming applications. Author's own table.

Plant	Priming method	Tolerance acquired	Priming intervals	Type of memory	Associated genes	Reference
Arabidopsis thaliana	NaCl	PGPR	2 × PGPR, 48 hour intervals	Induced systemic tolerance	RD29A-RD29B HSA32	Liu *et al.*, 2023
Arabidopsis thaliana	Heat	Heat	37°C, 90 min at RT, 45 min at 44°C	Chromatin remodeling	BRUSHY1/TONSOKU/ MGOUN3	Brzezinka *et al.*, 2019
Arabidopsis thaliana	Jasmonic acid (JA)	Biotic factors	1 day – short 3 weeks – long	Induced resistance Secondary metabolism RdDM, sRNA regulation	MYC transcription factors, ROS1, AGO1, ATREP2	Wilkinson *et al.*, 2023
Tritucum durum	Drought Heat	N starvation	Transgenerational	Non-coding RNA-transcriptional regulation	miR169 – leucine-rich repeat receptor-like protein kinase family	Liu *et al.*, 2021
Arabidopsis thaliana	Heat	Heat	37°C – 20 min 43.5°C – 1 hour	Histone demethylation	HSP22, HSP17.6C, JMJ (JUMAJI)	Yamaguchi *et al.*, 2021
Solanum lycopersicum	BABA (β-aminobutyric acid)	Pathogen resistance (Fungus-*Botrytis cinerea*)	7 days after stress	Induced resistance CHH hypomethylation of tolerance genes and Class II TEs.	CMT3, DRM2, DML2, ROS1	Catoni *et al.*, 2022
Arabidopsis thaliana	Cold NaCl Drought	Cold NaCl Drought	4°C – 24 hour NaCl – 3 days NW – 7 days	DNA demethylation, triggering SA defense pathway.	ACD6, ACO3, GSTF14	Yang *et al.*, 2022
Arabidopsis thaliana	flg22 (bacterial PAMP)	*Pseudomonas syringae* *Botrytis cinerea*	24 hour – flg22 priming / Early 6 hour Late 24 hour infection	H1-linker-associated DNA methylation and histone acetylation	HAC2, HDA18	Sheikh *et al.*, 2023
Festuca arundinacea Schreb.	Heat	Heat	2 hour/4 hour – 34°C 44 hour/acc	Transcriptional memory due to low ROS and strong PSII activity	FaHSP17.8-CII	Bi *et al.*, 2021
Arabidopsis thaliana	Cold	Cold	4°C – 3 days 22°C – 3 days acc	Memory acquired in the true leaves	CBFs	Leuendorf *et al.*, 2020

Continued

Table 28.1. Continued

Plant	Priming method	Tolerance acquired	Priming intervals	Type of memory	Associated genes	Reference
Brachypodium distachyon	Cold	Cold	4°C – 14 days 22°C – 3 days acc	Transcriptional memory caused by DNA methylation	VRN1 – permanent Others are short-term and transient (CBFs, etc.)	Mayer and Charron, 2021
Arabidopsis thaliana	Heat	Heat	44°C – 2 h/day for 7 days 21°C – acc	Transcription and H3K4me3 resetting after stress	Constitutive photomorphogenesis 5A (CSN5A) resets memory genes APX2, HSP22	Singh *et al.*, 2021
Brassica napus	Gibberellic acid Melatonin	Drought	6 hour – priming Overnight, dark, until dry	Higher ROS scavenging, proline accumulation	Unknown	Khan *et al.*, 2020
Oryza sativa	UVB	NaCl Drought	four kJm^2 UVB light	Metabolite accumulation and transcriptional regulation	HSP90, Group3 LEA	Sen *et al.*, 2021
Solanum lycopersicum cv. Qianxi	Heat High light	Heat	38°C – day 33°C – night High light intensity 2-day recovery	Somatic stress memory Stress avoidance Leaf cooling	Unknown	Zhou *et al.*, 2020
Cucumis sativus L.	Flooding	Hypoxia	X1 – 7 days water, 7 days water, 14 days rec, 7 days water, 14 days rec, water (multiple days)	Post-transcriptional regulation debated	Uncharacterized DEGs	Kęska *et al.*, 2021

methylation, which is related to transcriptional stress memory.

28.4.2 Histone modifications

Histone proteins come together to form the nucleosome structure. Each nucleosome consists of four different histones, including H2A, H2B, H3, and H4, composing a nucleosome octamer. These histone proteins are highly conserved among eukaryotes. The N-terminal tails of histone proteins emerge and are open to modifications. These modifications frequently occur at arginine (R) or lysine (K) residues of the histones H3 or H4, including acetylation, methylation, phosphorylation, ubiquitination, biotinylation, and sumoylation. While some histone modifications cause an increase in gene expression, including acetylation, phosphorylation, and ubiquitination, others may cause a decrease, like sumoylation and biotinylation (Camporeale *et al.*, 2007; Kouzarides, 2007; Zhang *et al.*, 2007; Perrella and Kaiserli, 2016).

Histone modifications are catalyzed by enzymes categorized as writers and erasers. Writer enzymes consist of methyltransferases, ubiquitinates, kinases, and histone acetyltransferases (HATs), whilst eraser enzymes comprise histone demethylases (HDMs), deubiquitinases, phosphatases, and histone deacetylases (HDACs) (Xu *et al.*, 2017). Histone methylation is catalyzed with histone methyltransferases (HTMs), while removed by HDMs (Ng *et al.*, 2007; Rando, 2012). Histone methylation can be related to the activation or the repression of gene expression, depending on the number of methyl residues and the sites. For instance, the repression of the gene expression is often related to the H3K27me3, H3K27me2, H3K9me3, and H3K9me2 histone marks whereas the activation of the gene expression is often related to the H3K4me2, H3K4me3, H3K36me3, and H3K36me2 histone marks (Hu and Du, 2022). On the other hand, histone acetylation seems to be increasing the accessibility of the DNA sequence by means of the neutralization effect on basically charged histones (Allis and Jenuwein, 2016; Onufriev and Schiessel, 2019).

Histone modifications are important for plant stress response via increasing or decreasing the transcription of stress-responsive and/or defense-related genes in addition to the propagation of somatic stress memory (Sani *et al.*, 2013; Lämke and Bäurle, 2017). H3K4me3 has been proven to be related to the dehydration stress memory genes and priming of plants (Jaskiewicz *et al.*, 2011; Ding *et al.*, 2012). In a study conducted with *Arabidopsis*, researchers investigated the effect of recurring dehydration stress on stress memory (Ding *et al.*, 2012). According to the findings, recurring stress induced transcriptional stress memory by increasing the expression of defense-related genes. However, during the recovery phase, the expression of these genes returned to basal levels, while the levels of H3K4me3 and Ser5P polymerase II remained elevated. In heat shock (HS) priming, it has been proven that histone H3K4me2 and H3K4me3 were increased compared with non-primed plants. The accumulation of histone mark H3K4me3 increased the transcriptional activity of memory-related genes, and afterwards HS-primed plants showed higher expression rates (Lämke *et al.*, 2016; Liu *et al.*, 2018).

The H3K27me3 histone mark is known as a repressive mark in general. In a study with *Arabidopsis* plants, researchers investigated the effects of salt priming on memory formation (Sani *et al.*, 2013). After salt priming, plants were exposed to stress, and changes in the epigenome were analyzed. After ChIP-seq and RNA-seq analysis, they found that priming induced changes in H3K27me3 histone marks, and this effect lasted more than 10 days. In addition, they discovered that after priming, some genes displayed altered transcription activities during the next stress exposure. A study conducted with soybeans found that priming induced histone modifications, which led to the modification of transcriptional responses and, consequently, to tolerance to salt stress (Yung *et al.*, 2022). After priming, the study analyzed the genome-wide H3K4me2, H3K4me3, and H3K9ac histone modifications and found alterations in these histone marks. These modifications were correlated with stress response mechanisms such as cell wall modification, ion homeostasis, and transcriptional changes. Moreover, it has been shown that priming *Arabidopsis* plants with azelaic acid resulted in the accumulation of histone H2A related to salicylic acid (SA) production for defense response (Jung *et al.*, 2009). Similarly, acibenzolar S-methyl treatment in *Arabidopsis* plants induced histone modifications in genes related to plant stress response, creating a memory (Jaskiewicz *et al.*, 2011). Additionally, BABA treatment has been shown to prime plants against different stressors, including beans, necrotrophic bacteria *Pectobacterium carotovorum*, and oomycete *Phytophthora infestans*. During these priming events, alterations in histone modifications, including H3K9K14ac, H3K4me2, H3K4me3, and H3K36me3, have been observed (Po-Wen *et al.*, 2013; Martínez-Aguilar *et al.*, 2016; Meller *et al.*, 2018).

28.4.3 Chromatin changes/remodeling

In eukaryotes, DNA is packed with histone proteins in nucleosomes, which affects the accessibility of DNA (Kouzarides, 2007). The density and position of nucleosomes have great importance for gene regulation. Tightly packed chromatin regions are referred to as heterochromatin, while loosely packed regions are called euchromatin. Generally, genes located in heterochromatin regions are not expressed, whereas gene expression is active in euchromatic regions (Yadav *et al.*, 2021). In this aspect, chromatin remodeling plays a crucial role in changes in transcription (Teves *et al.*, 2014). ATP-dependent chromatin remodeling protein complexes (CRCs) can alter the relationship between DNA and histone proteins (Ho and Crabtree, 2010). Besides, many non-histone proteins such as chromatin remodeling factors (CRFs) affect the chromatin structure, resulting in changes in transcription (Song *et al.*, 2021).

Chromatin modeling exerts a significant influence on the stress response of plants. Under unfavorable environmental stimuli, chromatin dynamics change (Kim, 2021). In many studies, the importance of chromatin remodeling has been emphasized. For instance, it has been demonstrated that nucleosome repositioning resulting in heterochromatin decondensation, affecting repetitive elements, has occurred through histone modifications. In addition, when the chromatin assembly factor 1 (CAF-1) mutants were analyzed for stress recovery, the delay of chromatin organization was found after heat stress. It has been suggested that CAF-1 may act in the regulation of priming stress memory (Pecinka *et al.*, 2010; Mozgová *et al.*, 2015; Brzezinka *et al.*, 2016; Wang *et al.*, 2022). The chromatin modification related to priming was also observed in systemic acquired resistance (SAR) (Jaskiewicz *et al.*, 2011). In the study, priming was induced by bacterial pathogens and a salicylic acid analog, benzothiadiazole (BTH). The research revealed that histone methylation and acetylation act as forms of memory for SAR, enabling plants to mount a faster and more effective defense response upon subsequent pathogen encounters. The results confirmed the relationship between priming and transcriptional stress memory through the H3K3me3 histone mark. Additionally, this study identified specific proteins and enzymes involved in histone modification in response to pathogen infection, shedding light on the molecular mechanisms underlying chromatin modification-mediated memory formation in SAR.

Drought significantly limits crop productivity by reducing water availability, which hampers plant growth, development, and yield. In this aspect, deciphering drought stress response-related genes is important. In a study with *Arabidopsis thaliana*, after *Paenibacillus polymyxa* CR1 priming, it was observed that it initiated a drought response enhancing tolerance (Liu *et al.*, 2023). The researchers discovered two drought-responsive genes, *Response to Desiccation (RD)29A* and *RD29B*. These genes play a role in triggering the drought response through the remodeling of histone complexes, resulting in a stronger and faster stress response.

The role of linker histone H1 is crucial for modulating defense priming and overall immunity in plants. Through a series of experiments, researchers demonstrated that histone H1 influences the expression of key defense genes and the plant's ability to mount an effective immune response (Sheikh *et al.*, 2023). The findings highlight the significance of chromatin organization in regulating defense mechanisms, with histone H1 playing a central role in enhancing the plant's readiness to respond to pathogen attacks.

28.4.4 Non-coding RNAs (ncRNAs)

Non-coding RNAs (ncRNAs) in plants play critical roles in regulating gene expression and maintaining cellular functions without being translated into proteins. These RNAs are involved in many biological processes, including development, stress responses, and adaptation to environmental changes. They ensure precise control of gene expression, enabling plants to develop properly, respond to environmental challenges, and maintain genome stability. Understanding the roles of ncRNAs in plants opens up possibilities for improving crop resilience and productivity through biotechnological interventions (Zhang *et al.*, 2022; Yadav *et al.*, 2024).

Non-coding RNAs can be classified based on their size and function, such as microRNAs (miRNAs), small interfering RNAs (siRNAs), long non-coding RNAs (lncRNAs), natural antisense transcripts (NATs), and piwi-interacting RNAs (piRNAs). Many studies have shown the relationship between these ncRNAs and the plant stress response. They exhibit diverse activating and silencing effects upon stress exposure (Verma *et al.*, 2022; Yadav *et al.*, 2024).

siRNAs are involved in RNA interference (RNAi) pathways, mediating mRNA degradation, and are crucial for defense against viruses and transposon silencing. In addition, siRNAs take place in the RdDM pathway. In a study investigating the long-term effects of the stress hormone jasmonic acid (JA) on *Arabidopsis* plants, transcriptome analysis showed that long-term JA-induced resistance involved changes in glucosinolate composition and required MYC2/3/4 transcription factors, RdDM, the DNA demethylase ROS1, and the small RNA-binding protein AGO1. Wilkinson *et al.* (2023) suggested that AGO1-related small RNAs from hypomethylated ATREP2 TEs trans-regulate long-term memory of JA-dependent immunity.

MicroRNAs participate in plant stress response mechanisms by inducing chromatin-based silencing. They help increase the plant's plasticity to stress conditions (Matzke *et al.*, 2009). One study has shown the relationship between miRNA156, priming, and stress tolerance (Stief *et al.*, 2014). miR156 regulates the expression of SQUAMOSA PROMOTER BINDING PROTEIN-LIKE (SPL) transcription factors. It was found that miR156 levels increase under stress conditions, which in turn decreases SPL levels. This adjustment helps enhance the plant's tolerance to repeated stress by modulating developmental processes and stress response pathways. Liu *et al.* (2021) investigated the influence of nitrogen deficiency in durum wheat under water deficit and heat stress. After small RNA sequencing and bioinformatic analysis, they discovered that miR169 functions as a regulator in the expression of leucine-rich repeat receptor genes. Wibowo *et al.* (2016) investigated the environmentally induced priming responses in plants involving DNA (de) methylation pathways, which control the inheritance and maintenance of acquired epigenetic states across generations. The study provided evidence that stress can specifically alter methylation near key stress-response regulators, mediated by the RdDM and DNA demethylation pathways. Additionally, DNA methylation and demethylation activities are regulated by adjacent upstream TEs acting as epigenetic sensors. The study also found that epigenetic regulation of hyperosmotic stress-differentially methylated regions (HS-DMRs) is associated with the transcription of antisense lncRNAs, which influence chromatin changes, transcription, splicing, and transcript stability. Methylation changes in these regions can affect transcription factor binding or chromatin regulatory loops. Young rice seedlings that undergo alternating cycles of mild drought and rewatering exhibit short-term stress memory due to changes in DNA methylation, levels of lncRNAs, and phytohormones (Li *et al.*, 2019). Hence, the degree of priming and the frequency of recurrence can affect subsequent stress tolerance by activating various regulatory pathways (Nair *et al.*, 2022).

28.5 Concluding Remarks and Future Perspectives

Due to climate change, arable land is decreasing daily, endangering food security and the sustainability of agriculture. Since the loss of crop yield due to climate change is inevitable, acquiring plants resilient to stress conditions is crucial to mitigate potential risks in the future (Lämke and Bäurle, 2017). Although conventional breeding techniques (hybridization and selection) and genetic engineering strategies (mutations, transgenes, etc.) have promising results, they also have limitations, including biosafety, time, and workload. Thus, alternative breeding technologies are needed. Studies focused on epigenetics seem to be yielding promising results. Recent studies have pointed out the importance of epigenetic regulation in gene expression, memory formation, and plant stress response. To date, some data has been gathered about the molecular mechanisms of this regulation. However, future studies are crucial in detailing each aspect of the process to understand the basis (Khan *et al.*, 2022).

Stress priming is a promising and essential approach in the context of modern agriculture. It leverages the natural adaptive capabilities of plants to build resilience against environmental challenges, thereby supporting food security, reducing chemical dependency, and promoting sustainable farming practices. As research advances, stress priming is likely to become an integral part of crop management strategies to meet the demands of a growing global population under changing climatic conditions. Mounting evidence has demonstrated the importance of priming in stress response mechanisms on the epigenetic level to create effective and sustainable technologies for the future. Although the helpful effect of priming on plant stress response has been demonstrated in many studies, the molecular mechanisms underlying this phenomenon are still waiting to be discovered. By deciphering the mechanisms behind priming and epigenetics, researchers could improve stress tolerance, enhance memory, and produce higher-yielding crops for the future of sustainable agriculture.

Consequently, this chapter has discussed the importance of epigenetic regulation in stress priming. Research on adapting priming techniques to stress is steadily growing, offering promising prospects for the future. Yet, many points still need to be clarified.

References

Allis, C.D. and Jenuwein, T. (2016) The molecular hallmarks of epigenetic control. *Nature Reviews Genetics* 17, 487–500. DOI: 10.1038/nrg.2016.59.

Anand, A., Subramanian, M. and Kar, D. (2022) Breeding techniques to dispense higher genetic gains. *Frontiers in Plant Science* 13, 1076094. DOI: 10.3389/fpls.2022.1076094.

Aswathi, K.P.R., Kalaji, H.M. and Puthur, J.T. (2022) Seed priming of plants aiding in drought stress tolerance and faster recovery: A review. *Plant Growth Regulation* 97, 235–253. DOI: 10.1007/s10725-021-00755-z.

Bäurle, I. (2016) Plant heat adaptation: Priming in response to heat stress. *F1000Research* 5, 694. DOI: 10.12688/f1000research.7526.1.

Bertini, L., Proietti, S., Focaracci, F., Sabatini, B. and Caruso, C. (2018) Epigenetic control of defense genes following MeJA-induced priming in rice (*O. sativa*). *Journal of Plant Physiology* 228, 166–177. DOI: 10.1016/j.jplph.2018.06.007.

Bi, A., Wang, T., Wang, G., Zhang, L., Wassie, M. *et al.* (2021) Stress memory gene FaHSP17.8-CII controls thermotolerance via remodeling PSII and ROS signaling in tall fescue. *Plant Physiology* 187, 1163–1176. DOI: 10.1093/plphys/kiab205.

Boulc'h, P.-N., Caullireau, E., Faucher, E., Gouerou, M., Guérin, A. *et al.* (2020) Abiotic stress signalling in extremophile land plants. *Journal of Experimental Botany* 71, 5771–5785. DOI: 10.1093/jxb/eraa336.

Bruce, T.J.A., Matthes, M.C., Napier, J.A. and Pickett, J.A. (2007) Stressful "memories" of plants: Evidence and possible mechanisms. *Plant Science* 173, 603–608. DOI: 10.1016/j.plantsci.2007.09.002.

Brzezinka, K., Altmann, S., Czesnick, H., Nicolas, P., Gorka, M. *et al.* (2016) *Arabidopsis* FORGETTER1 mediates stress-induced chromatin memory through nucleosome remodeling. *eLife* 5, e17061. DOI: 10.7554/eLife.17061.

Brzezinka, K., Altmann, S. and Bäurle, I. (2019) BRUSHY1/TONSOKU/MGOUN3 is required for heat stress memory. *Plant, Cell & Environment* 42, 771–781. DOI: 10.1111/pce.13365.
Camporeale, G., Oommen, A.M., Griffin, J.B., Sarath, G. and Zempleni, J. (2007) K12-biotinylated histone H4 marks heterochromatin in human lymphoblastoma cells. *The Journal of Nutritional Biochemistry* 18, 760–768. DOI: 10.1016/j.jnutbio.2006.12.014.
Catoni, M., Alvarez-Venegas, R., Worrall, D., Holroyd, G., Barraza, A. *et al.* (2022) Long-lasting defence priming by β-aminobutyric acid in tomato is marked by genome-wide changes in DNA methylation. *Frontiers in Plant Science* 13, 836326. DOI: 10.3389/fpls.2022.836326.
Chang, Y.N., Zhu, C., Jiang, J., Zhang, H., Zhu, J.K. *et al.* (2020) Epigenetic regulation in plant abiotic stress responses. *Journal of Integrative Plant Biology* 62(5), 563–580.
Chinnusamy, V. and Zhu, J.-K. (2009) Epigenetic regulation of stress responses in plants. *Current Opinion in Plant Biology* 12, 133–139. DOI: 10.1016/j.pbi.2008.12.006.
Conrath, U. (2011) Molecular aspects of defence priming. *Trends in Plant Science* 16, 524–531. DOI: 10.1016/j.tplants.2011.06.004.
Conrath, U., Beckers, G.J., Flors, V., García-Agustín, P., Jakab, G. *et al.* (2006) Priming: Getting ready for battle. *Molecular Plant–Microbe Interactions* 19, 1062–1071. DOI: 10.1094/MPMI-19-1062.
Conrath, U., Beckers, G.J., Langenbach, C.J. and Jaskiewicz, M.R. (2015) Priming for enhanced defense. *Annual Review of Phytopathology* 53, 97–119. DOI: 10.1146/annurev-phyto-080614-120132.
Cutter, A.R. and Hayes, J.J. (2015) A brief review of nucleosome structure. *FEBS Letters* 589, 2914–2922. DOI: 10.1016/j.febslet.2015.05.016.
Dhanya Thomas, T.T., Dinakar, C. and Puthur, J.T. (2020) Effect of UV-B priming on the abiotic stress tolerance of stress-sensitive rice seedlings: Priming imprints and cross-tolerance. *Plant Physiology and Biochemistry* 147, 21–30. DOI: 10.1016/j.plaphy.2019.12.002.
Ding, Y., Fromm, M. and Avramova, Z. (2012) Multiple exposures to drought “train” transcriptional responses in *Arabidopsis*. *Nature Communications* 3, 740. DOI: 10.1038/ncomms1732.
Ding, Y., Virlouvet, L., Liu, N., Riethoven, J.-J., Fromm, M. *et al.* (2014) Dehydration stress memory genes of *Zea mays*; comparison with *Arabidopsis thaliana*. *BMC Plant Biology* 14, 141. DOI: 10.1186/1471-2229-14-141.
Ding, Y., Shi, Y. and Yang, S. (2020) Molecular regulation of plant responses to environmental temperatures. *Molecular Plant* 13, 544–564. DOI: 10.1016/j.molp.2020.02.004.
Do Amaral, M.N., Auler, P.A., Rossatto, T., Barros, P.M., Oliveira, M.M. *et al.* (2020) Long-term somatic memory of salinity unveiled from physiological, biochemical and epigenetic responses in two contrasting rice genotypes. *Physiologia Plantarum* 170, 248–268. DOI: 10.1111/ppl.13149.
Eichten, S.R., Schmitz, R.J. and Springer, N.M. (2014) Epigenetics: Beyond chromatin modifications and complex genetic regulation. *Plant Physiology* 165(3), 933–947.
Ganie, S.A., McMulkin, N. and Devoto, A. (2024) The role of priming and memory in rice environmental stress adaptation: Current knowledge and perspectives. *Plant, Cell & Environment* 47(5), 1895–1915. DOI: 10.1111/pce.14855.
Gaude, A.A., Siqueira, R.H., Botelho, S.B. and Jalmi, S.K. (2024) Epigenetic arsenal for stress mitigation in plants. *Biochimica et Biophysica Acta (BBA) – General Subjects* 1868 130620. DOI: 10.1016/j.bbagen.2024.130620.
Gully, K., Celton, J.M., Degrave, A., Pelletier, S., Brisset, M.N. *et al.* (2019) Biotic stress-induced priming and de-priming of transcriptional memory in *Arabidopsis* and apple. *Epigenomes 14* 3(1), 3. DOI: 10.3390/epigenomes3010003.
Haak, D.C., Fukao, T., Grene, R., Hua, Z., Ivanov, R. *et al.* (2017) Multilevel regulation of abiotic stress responses in plants. *Frontiers in Plant Science* 8, 1564. DOI: 10.3389/fpls.2017.01564.
Haider, S., Iqbal, J., Naseer, S., Yaseen, T., Shaukat, M. *et al.* (2021) Molecular mechanisms of plant tolerance to heat stress: Current landscape and future perspectives. *Plant Cell Reports* 1–25.
Harris, C.J., Amtmann, A. and Ton, J. (2023) Epigenetic processes in plant stress priming: Open questions and new approaches. *Current Opinion in Plant Biology* 75, 102432. DOI: 10.1016/j.pbi.2023.102432.
Herman, J.J. and Sultan, S.E. (2011) Adaptive transgenerational plasticity in plants: Case studies, mechanisms, and implications for natural populations. *Frontiers in Plant Science* 2, 102. DOI: 10.3389/fpls.2011.00102.
Ho, L. and Crabtree, G.R. (2010) Chromatin remodelling during development. *Nature* 463, 474–484. DOI: 10.1038/nature08911.
Holness, S., Bechtold, U., Mullineaux, P., Serino, G. and Vittorioso, P. (2023) Highlight induced transcriptional priming against a subsequent drought stress in *Arabidopsis thaliana*. *International Journal of Molecular Sciences 1* 24(7), 6608. DOI: 10.3390/ijms24076608.

Hossain, M.A., Li, Z.-G., Hoque, T.S., Burritt, D.J., Fujita, M. *et al.* (2018) Heat or cold priming-induced cross-tolerance to abiotic stresses in plants: Key regulators and possible mechanisms. *Protoplasma* 255, 399–412. DOI: 10.1007/s00709-017-1150-8.

Hu, H. and Du, J. (2022) Structure and mechanism of histone methylation dynamics in *Arabidopsis*. *Current Opinion in Plant Biology* 67, 102211. DOI: 10.1016/j.pbi.2022.102211.

Jaskiewicz, M., Conrath, U. and Peterhänsel, C. (2011) Chromatin modification acts as a memory for systemic acquired resistance in the plant stress response. *EMBO Reports* 12, 50–55. DOI: 10.1038/embor.2010.186.

Jiménez-Arias, D., Borges, A.A., Luis, J.C., Valdés, F., Sandalio, L.M. *et al.* (2015) Priming effect of menadione sodium bisulphite against salinity stress in *Arabidopsis* involves epigenetic changes in genes controlling proline metabolism. *Environmental and Experimental Botany* 120, 23–30. DOI: 10.1016/j.envexpbot.2015.07.003.

Jung, H.W., Tschaplinski, T.J., Wang, L., Glazebrook, J. and Greenberg, J.T. (2009) Priming in systemic plant immunity. *Science* 324, 89–91. DOI: 10.1126/science.1170025.

Kerchev, P., van der Meer, T., Sujeeth, N., Verlee, A., Stevens, C.V. *et al.* (2020) Molecular priming as an approach to induce tolerance against abiotic and oxidative stresses in crop plants. *Biotechnology Advances* 40, 107503. DOI: 10.1016/j.biotechadv.2019.107503.

Kęska, K., Szcześniak, M.W., Makałowska, I. and Czernicka, M. (2021) Long-term waterlogging as factor contributing to hypoxia stress tolerance enhancement in cucumber: Comparative transcriptome analysis of waterlogging sensitive and tolerant accessions. *Genes* 12, 189. DOI: 10.3390/genes12020189.

Khan, A., Khan, V., Pandey, K., Sopory, S.K. and Sanan-Mishra, N. (2022) Thermo-priming mediated cellular networks for abiotic stress management in plants. *Frontiers in Plant Science* 13, 866409. DOI: 10.3389/fpls.2022.866409.

Khan, M.N., Khan, Z., Luo, T., Liu, J., Rizwan, M. *et al.* (2020) Seed priming with gibberellic acid and melatonin in rapeseed: Consequences for improving yield and seed quality under drought and non-stress conditions. *Industrial Crops and Products* 156, 112850. DOI: 10.1016/j.indcrop.2020.112850.

Kim, J.H. (2021) Multifaceted chromatin structure and transcription changes in plant stress response. *International Journal of Molecular Sciences 18* 22(4), 2013. DOI: 10.3390/ijms22042013.

Kouzarides, T. (2007) Chromatin modifications and their function. *Cell* 128, 693–705. DOI: 10.1016/j.cell.2007.02.005.

Kribelbauer, J.F., Lu, X.-J., Rohs, R., Mann, R.S. and Bussemaker, H.J. (2020) Toward a mechanistic understanding of DNA methylation readout by transcription factors. *Journal of Molecular Biology* 432, 1801–1815. DOI: 10.1016/j.jmb.2019.10.021.

Kundariya, H., Yang, X., Morton, K., Sanchez, R., Axtell, M.J. *et al.* (2020) MSH1-induced heritable enhanced growth vigor through grafting is associated with the RdDM pathway in plants. *Nature Communications* 11, 5343. DOI: 10.1038/s41467-020-19140-x.

Kuźnicki, D., Meller, B., Arasimowicz-Jelonek, M., Braszewska-Zalewska, A., Drozda, A. *et al.* (2019) BABA-induced DNA methylome adjustment to intergenerational defense priming in potato to *Phytophthora infestans*. *Frontiers in Plant Science 31* 10, 650. DOI: 10.3389/fpls.2019.00650.

Lamichhane, S. and Thapa, S. (2022) Advances from conventional to modern plant breeding methodologies. *Plant Breeding and Biotechnology* 10, 1–14. DOI: 10.9787/PBB.2022.10.1.1.

Lämke, J. and Bäurle, I. (2017) Epigenetic and chromatin-based mechanisms in environmental stress adaptation and stress memory in plants. *Genome Biology* 18, 124. DOI: 10.1186/s13059-017-1263-6.

Lämke, J., Brzezinka, K., Altmann, S. and Bäurle, I. (2016) A hit-and-run heat shock factor governs sustained histone methylation and transcriptional stress memory. *The EMBO Journal* 35, 162–175. DOI: 10.15252/embj.201592593.

Leuendorf, J.E., Frank, M. and Schmülling, T. (2020) Acclimation, priming and memory in the response of *Arabidopsis thaliana* seedlings to cold stress. *Scientific Reports* 10, 689. DOI: 10.1038/s41598-019-56797-x.

Li, P., Yang, H., Wang, L., Liu, H., Huo, H. *et al.* (2019) Physiological and transcriptome analyses reveal short-term responses and formation of memory under drought stress in rice. *Frontiers in Genetics* 10, 55. DOI: 10.3389/fgene.2019.00055.

Liang, Z., Shen, L., Cui, X., Bao, S., Geng, Y. *et al.* (2018) DNA N6-adenine methylation in *Arabidopsis thaliana*. *Developmental Cell* 45, 406–416. DOI: 10.1016/j.devcel.2018.03.012.

Liu, H., Able, A.J. and Able, J.A. (2021) Nitrogen starvation-responsive micrornas are affected by transgenerational stress in durum wheat seedlings. *Plants* 10, 826. DOI: 10.3390/plants10050826.

Liu, H.-C., Lämke, J., Lin, S.-Y., Hung, M.-J., Liu, K.-M. *et al.* (2018) Distinct heat shock factors and chromatin modifications mediate the organ-autonomous transcriptional memory of heat stress. *The Plant Journal* 95, 401–413. DOI: 10.1111/tpj.13958.

Liu, W., Thapa, P. and Park, S.W. (2023) RD29A and RD29B rearrange genetic and epigenetic markers in priming systemic defense responses against drought and salinity. *Plant Science* 337, 111895. DOI: 10.1016/j.plantsci.2023.111895.

Luna, E., Bruce, T.J., Roberts, M.R., Flors, V. and Ton, J. (2012) Next-generation systemic acquired resistance. *Plant Physiology* 158, 844–853. DOI: 10.1104/pp.111.187468.

Luna, E., van Hulten, M., Zhang, Y., Berkowitz, O., López, A. *et al.* (2014) Plant perception of β-aminobutyric acid is mediated by an aspartyl-tRNA synthetase. *Nature Chemical Biology* 10(6), 450–456. DOI: 10.1038/nchembio.1520.

Luna, E., Beardon, E., Ravnskov, S., Scholes, J. and Ton, J. (2016) Optimizing chemically induced resistance in tomato against *Botrytis cinerea*. *Plant Disease* 100, 704–710. DOI: 10.1094/PDIS-03-15-0347-RE.

Martínez-Aguilar, K., Ramírez-Carrasco, G., Hernández-Chávez, J.L., Barraza, A. and Alvarez-Venegas, R. (2016) Use of BABA and INA as activators of a primed state in the common bean (*Phaseolus vulgaris* L.). *Frontiers in Plant Science* 18 7, 653. DOI: 10.3389/fpls.2016.00653.

Martinez-Medina, A., Flors, V., Heil, M., Mauch-Mani, B., Pieterse, C.M.J. *et al.* (2016) Recognizing plant defense priming. *Trends in Plant Science* 21, 818–822. DOI: 10.1016/j.tplants.2016.07.009.

Matzke, M., Kanno, T., Daxinger, L., Huettel, B. and Matzke, A.J. (2009) RNA-mediated chromatin-based silencing in plants. *Current Opinion in Cell Biology* 21(3), 367–376.

Mayer, B.F. and Charron, J.B. (2021) Transcriptional memories mediate the plasticity of cold stress responses to enable morphological acclimation in *Brachypodium distachyon*. *The New Phytologist* 229(3), 1615–1634. DOI: 10.1111/nph.16945.

Meller, B., Kuźnicki, D., Arasimowicz-Jelonek, M., Deckert, J. and Floryszak-Wieczorek, J. (2018) BABA-primed histone modifications in potato for intergenerational resistance to *Phytophthora infestans*. *Frontiers in Plant Science* 29 9, 1228. DOI: 10.3389/fpls.2018.01228.

Miryeganeh, M. (2021) Plants' epigenetic mechanisms and abiotic stress. *Genes* 12, 1106. DOI: 10.3390/genes12081106.

Mozgová, I., Wildhaber, T., Liu, Q., Abou-Mansour, E., L'Haridon, F. *et al.* (2015) Chromatin assembly factor CAF-1 represses priming of plant defence response genes. *Nature Plants* 1, 15127. DOI: 10.1038/nplants.2015.127.

Nair, A.U., Bhukya, D.P.N., Sunkar, R., Chavali, S. and Allu, A.D. (2022) Molecular basis of priming-induced acquired tolerance to multiple abiotic stresses in plants. *Journal of Experimental Botany 2* 73(11), 3355–3371. DOI: 10.1093/jxb/erac089.

Ng, D.W., Wang, T., Chandrasekharan, M.B., Aramayo, R., Kertbundit, S. *et al.* (2007) Plant SET domain-containing proteins: Structure, function and regulation. *Biochimica et Biophysica Acta* 1769, 316–329. DOI: 10.1016/j.bbaexp.2007.04.003.

Nguyen, H.M., Kim, M., Ralph, P.J., Marín-Guirao, L., Pernice, M. *et al.* (2020) Stress memory in seagrasses: First insight into the effects of thermal priming and the role of epigenetic modifications. *Frontiers in Plant Science 28* 11, 494. DOI: 10.3389/fpls.2020.00494.

Oberkofler, V., Pratx, L. and Bäurle, I. (2021) Epigenetic regulation of abiotic stress memory: Maintaining the good things while they last. *Current Opinion in Plant Biology* 61, 102007. DOI: 10.1016/j.pbi.2021.102007.

Ohama, N., Sato, H., Shinozaki, K. and Yamaguchi-Shinozaki, K. (2017) Transcriptional regulatory network of plant heat stress response. *Trends in Plant Science* 22(1), 53–65. DOI: 10.1016/j.tplants.2016.08.015.

Onufriev, A.V. and Schiessel, H. (2019) The nucleosome: From structure to function through physics. *Current Opinion in Structural Biology* 56, 119–130. DOI: 10.1016/j.sbi.2018.11.003.

Pecinka, A., Dinh, H.Q., Baubec, T., Rosa, M., Lettner, N. *et al.* (2010) Epigenetic regulation of repetitive elements is attenuated by prolonged heat stress in *Arabidopsis*. *The Plant Cell* 22(9), 3118–3129. DOI: 10.1105/tpc.110.078493.

Pereira, A. (2016) Plant abiotic stress challenges from the changing environment. *Frontiers in Plant Science* 7, 1123.

Perrella, G. and Kaiserli, E. (2016) Light behind the curtain: Photoregulation of nuclear architecture and chromatin dynamics in plants. *New Phytologist* 212, 908–919. DOI: 10.1111/nph.14269.

Perrella, G., Bäurle, I. and van Zanten, M. (2022) Epigenetic regulation of thermomorphogenesis and heat stress tolerance. *New Phytologist* 234(4), 1144–1160. DOI: 10.1111/nph.17970.

Perrone, A. and Martinelli, F. (2020) Plant stress biology in epigenomic era. *Plant Science* 294, 110376. DOI: 10.1016/j.plantsci.2019.110376.

Po-Wen, C., Singh, P. and Zimmerli, L. (2013) Priming of the *Arabidopsis* pattern-triggered immunity response upon infection by necrotrophic *Pectobacterium carotovorum* bacteria. *Molecular Plant Pathology* 14, 58–70. DOI: 10.1111/j.1364-3703.2012.00827.x.

Raju, S.K.K., Shao, M.-R., Sanchez, R., Xu, Y.-Z., Sandhu, A. *et al.* (2018) An epigenetic breeding system in soybean for increased yield and stability. *Plant Biotechnology Journal* 16, 1836–1847. DOI: 10.1111/pbi.12919.

Rando, O.J. (2012) Combinatorial complexity in chromatin structure and function: Revisiting the histone code. *Current Opinion in Genetics and Development* 22, 148–155. DOI: 10.1016/j.gde.2012.02.013.

Sani, E., Herzyk, P., Perrella, G., Colot, V. and Amtmann, A. (2013) Hyperosmotic priming of *Arabidopsis* seedlings establishes a long-term somatic memory accompanied by specific changes of the epigenome. *Genome Biology* 14(6), R59. DOI: 10.1186/gb-2013-14-6-r59.

Sen, A., Challabathula, D. and Puthur, J.T. (2021) UV-B priming of *Oryza sativa* seeds augments the innate tolerance potential in a tolerant variety more effectively toward NaCl and PEG stressors. *Journal of Plant Growth Regulation* 40, 1166–1180. DOI: 10.1007/s13205-019-1903-5.

Sheikh, A.H., Nawaz, K., Tabassum, N., Almeida-Trapp, M., Mariappan, K.G. *et al.* (2023) Linker histone H1 modulates defense priming and immunity in plants. *Nucleic Acids Research* 51(9), 4252–4265. DOI: 10.1093/nar/gkad106.

Singh, A.K., Dhanapal, S., Finkelshtein, A. and Chamovitz, D.A. (2021) CSN5A subunit of cop9 signalosome is required for resetting transcriptional stress memory after recurrent heat stress in *Arabidopsis*. *Biomolecules* 11(5), 668. DOI: 10.3390/biom11050668.

Song, Y., Jia, Z., Hou, Y., Ma, X., Li, L. *et al.* (2020) Roles of DNA methylation in cold priming in tartary buckwheat. *Frontiers in Plant Science 7* 11, 608540. DOI: 10.3389/fpls.2020.608540.

Song, Z.-T., Liu, J.-X. and Han, J.-J. (2021) Chromatin remodeling factors regulate environmental stress responses in plants. *Journal of Integrative Plant Biology* 63(3), 438–450. DOI: 10.1111/jipb.13064.

Srivastava, A.K., Suresh Kumar, J. and Suprasanna, P. (2021) Seed "primeomics": Plants memorize their germination under stress. *Biological Reviews* 96(5), 1723–1743. DOI: 10.1111/brv.12722.

Stief, A., Altmann, S., Hoffmann, K., Pant, B.D., Scheible, W.R. *et al.* (2014) *Arabidopsis* miR156 regulates tolerance to recurring environmental stress through SPL transcription factors. *Plant Cell* 26(4), 1792–1807. DOI: 10.1105/tpc.114.123851.

Tardieu, F., Simonneau, T. and Muller, B. (2018) The physiological basis of drought tolerance in crop plants: A scenario-dependent probabilistic approach. *Annual Review of Plant Biology* 69, 733–759. DOI: 10.1146/annurev-arplant-042817-040218.

Teves, S.S., Weber, C.M. and Henikoff, S. (2014) Transcribing through the nucleosome. *Trends in Biochemical Sciences* 39(12), 577–586. DOI: 10.1016/j.tibs.2014.10.004.

Thiebaut, F., Hemerly, A.S. and Ferreira, P.C.G. (2019) A role for epigenetic regulation in the adaptation and stress responses of non-model plants. *Frontiers in Plant Science* 10, 246.

Tirnaz, S. and Batley, J. (2019) DNA methylation: Toward crop disease resistance improvement. *Trends in Plant Science* 24(12), 1137–1150. DOI: 10.1016/j.tplants.2019.08.007.

Tittel-Elmer, M., Bucher, E., Broger, L., Mathieu, O., Paszkowski, J. *et al.* (2010) Stress-induced activation of heterochromatic transcription. *PLoS Genetics* 6(10), e1001175. DOI: 10.1371/journal.pgen.1001175.

Verma, N., Giri, S.K., Singh, G., Gill, R. and Kumar, A. (2022) Epigenetic regulation of heat and cold stress responses in crop plants. *Plant Gene* 29, 100351. DOI: 10.3389/fpls.2022.1075279.

Wang, X., Xin, C., Cai, J., Zhou, Q., Dai, T. *et al.* (2016) Heat priming induces trans-generational tolerance to high temperature stress in wheat. *Frontiers in Plant Science* 7, 501. DOI: 10.3389/fpls.2016.00501.

Wang, Y., Liu, Y., Qu, S., Liang, W., Sun, L. *et al.* (2022) Nitrogen starvation induces genome-wide activation of transposable elements in *Arabidopsis*. *Journal of Integrative Plant Biology* 64(12), 2374–2384. DOI: 10.1111/jipb.13376.

Wibowo, A., Becker, C., Marconi, G., Durr, J., Price, J. *et al.* (2016) Hyperosmotic stress memory in *Arabidopsis* is mediated by distinct epigenetically labile sites in the genome and is restricted in the male germline by DNA glycosylase activity. *eLife* 5, e13546. DOI: 10.7554/eLife.13546.

Wilkinson, S.W., Hannan Parker, A., Muench, A., Wilson, R.S., Hooshmand, K. *et al.* (2023) Long-lasting memory of jasmonic acid-dependent immunity requires DNA demethylation and ARGONAUTE1. *Nature Plants* 9(1), 81–95. DOI: 10.1038/s41477-022-01313-9.

Xu, Y., Zhang, S., Lin, S., Guo, Y., Deng, W. *et al.* (2017) WERAM: A database of writers, erasers and readers of histone acetylation and methylation in eukaryotes. *Nucleic Acids Research* 4 45(D1), D264–D270. DOI: 10.1093/nar/gkw1011.

Yadav, A., Mathan, J., Dubey, A.K. and Singh, A. (2024) The emerging role of non-coding RNAs (ncRNAs) in plant growth, development, and stress response signaling. *Non-Coding RNA* 10(1), 13. DOI: 10.3390/ncrna10010013.

Yadav, V.K., Santos-González, J. and Köhler, C. (2021) INT-Hi-C reveals distinct chromatin architecture in endosperm and leaf tissues of *Arabidopsis*. *Nucleic Acids Research* 49(8), 4371–4385. DOI: 10.1093/nar/gkab191.

Yamaguchi, N., Matsubara, S., Yoshimizu, K., Seki, M., Hamada, K. *et al.* (2021) H3K27me3 demethylases alter HSP22 and HSP17.6C expression in response to recurring heat in *Arabidopsis*. *Nature Communications* 12(1), 3480. DOI: 10.1038/s41467-021-23766-w.

Yang, L., Lang, C., Wu, Y., Meng, D., Yang, T. *et al.* (2022) ROS1-mediated decrease in DNA methylation and increase in expression of defense genes and stress response genes in *Arabidopsis thaliana* due to abiotic stresses. *BMC Plant Biology* 22(1), 104. DOI: 10.1186/s12870-022-03473-4.

Yang, X., Sanchez, R., Kundariya, H., Maher, T., Dopp, I. *et al.* (2020) Segregation of an MSH1 RNAi transgene produces heritable non-genetic memory in association with methylome reprogramming. *Nature Communications* 11(1), 2214. DOI: 10.1038/s41467-020-16036-8.

Yung, W.S., Wang, Q., Huang, M., Wong, F.L., Liu, A. *et al.* (2022) Priming-induced alterations in histone modifications modulate transcriptional responses in soybean under salt stress. *The Plant Journal* 109(6), 1575–1590. DOI: 10.1111/tpj.15652.

Zhang, H., Lang, Z. and Zhu, J.K. (2018) Dynamics and function of DNA methylation in plants. *Nature Reviews Molecular Cell Biology* 19(8), 489–506. DOI: 10.1038/s41580-018-0016-z.

Zhang, K., Sridhar, V.V., Zhu, J., Kapoor, A. and Zhu, J.-K. (2007) Distinctive core histone post-translational modification patterns in *Arabidopsis thaliana*. *PloS ONE* 2(11), e1210. DOI: 10.1371/journal.pone.0001210.

Zhang, Y., Zhou, Y., Zhu, W., Liu, J. and Cheng, F. (2022) Non-coding RNAs fine-tune the balance between plant growth and abiotic stress tolerance. *Frontiers in Plant Science* 13, 965745. DOI: 10.3389/fpls.2022.965745.

Zhou, R., Yu, X., Li, X., Mendanha Dos Santos, T., Rosenqvist, E. *et al.* (2020) Combined high light and heat stress induced complex response in tomato with better leaf cooling after heat priming. *Plant Physiology and Biochemistry* 151, 1–9. DOI: 10.1016/j.plaphy.2020.03.011.

29 Epigenetics for Combating Pests in Plants: Updated Methods and Current Achievements

Varucha Misra*

ICAR-Indian Institute of Sugarcane Research, Lucknow (U.P.), India

Abstract

In the face of escalating challenges posed by degenerating soil health, climate change, and evolving pest dynamics, modern agriculture seeks innovative strategies to safeguard plant yields sustainably. Traditional high-input agricultural practices are proving insufficient, prompting a paradigm shift towards novel approaches such as harnessing epigenetics to bolster plant defenses against pests. Epigenetics offers a promising avenue for enhancing plant resilience through modulation of gene expression without altering the underlying DNA sequence. Plants deploy a sophisticated defense mechanism against herbivore attacks, comprising direct and indirect defenses. Direct defenses involve physiological and behavioral modifications to deter herbivores, while indirect defenses utilize volatile organic compounds to attract predators and parasites of pests. Herbivore-induced elicitors trigger the production of volatile compounds in plants, initiating a cascade of defensive responses. These elicitors encompass a diverse array of compounds including fatty acid–amino acid conjugates, cell wall fragments, peptides, enzymes, etc. Epigenetic priming, an adaptive strategy observed in plants, enhances the activation of induced defense mechanisms, fortifying plants against impending herbivore threats. By elucidating the epigenetic mechanisms underlying plant defense responses, researchers can unlock new avenues for developing pest-resistant crop varieties. Understanding how epigenetic modifications influence the expression of defense-related genes enables targeted manipulation to bolster plant immunity. Leveraging epigenetics holds immense potential for sustainable agriculture by reducing reliance on chemical pesticides while fostering resilient crop systems capable of withstanding evolving pest pressures.

Keywords: Insects, Systemic acquired Resistance, Plant defense system, Defense priming

29.1 Introduction

Plants captivate scientists with their extraordinary capacity to adjust developmental processes and adapt to environmental cues through epigenetic alterations. These modifications, surpassing genetic predispositions, introduce an additional layer of regulation (Lu and Hyun, 2021; Hemenway and Gehring, 2023). Within the plant realm, epigenetic inheritance manifests in two ways: transmitting information not embedded in DNA across generations, and maintaining epigenetic changes within an individual while resetting between generations (Chachar *et al.*, 2022; Lloyd and Lister, 2022). Understanding the nuances distinguishing these two forms of epigenetic inheritance carries immense importance. This research domain

*Corresponding author: misra.varucha@gmail.com

DOI: 10.1079/9781800626102.0029

is particularly captivating because stress can trigger stress-signaling pathways, amplifying stress-responsive genes, which can inform strategies for enhancing yield, quality, and stress resilience in plants, thus benefiting agriculture and ecosystems (Akhter *et al.*, 2021; Balažová *et al.*, 2022). Plants employ epigenetic regulation to bolster immunity and foster diversity during interactions with pathogens and pests (Jones *et al.*, 2016; Arnold *et al.*, 2019; Bakhtiari *et al.*, 2019; Zhou and Zhang, 2020; Huang and Jin, 2022).

29.2 Epigenetic Responses to Pest Infestations in Plants

Epigenetic reactions to pests in plants encompass shifts in gene activation not prompted by changes in the DNA sequence itself but by adjustments in chromatin organization or DNA methylation configurations. Epigenetic alterations serve pivotal functions in plants' defense strategies against pests, enabling them to adjust and counter environmental adversities. Take, for instance, the soybean cyst nematode (SCN) (*Heterodera glycines*), a notable threat to soybean crops globally. This nematode is among the most economically impactful pests affecting soybean cultivation worldwide. Upon infesting soybean roots, the nematode triggers the syncytium. Researchers conducted a comparative whole-genome bisulphite sequencing analysis to examine changes in genome methylation between infected and uninfected soybean roots (Rambani *et al.*, 2015). On average, methylation levels displayed no notable contrast between the infected and uninfected soybean roots. However, within the cytosine guanine context, increased methylation was seen in 718 regions, while decreased methylation was noted in 1408 regions of infected roots compared with their uninfected counterparts. Similarly, 1142 and 605 were hypermethylated regions and 2074 and 1210 were hypomethylated regions within the CHG context and CHH context, respectively. Thus, a predominant trend of hypomethylation emerged across all sequence contexts. In total, 703 genes displayed heightened methylation, while 1346 genes showcased reduced methylation. Only 25 genes displayed heightened methylation across multiple contexts, whereas 45 genes demonstrated both increased and decreased methylation in distinct regions. Among a selection of 24 genes chosen at random and harboring differentially methylated regions (DMRs), 21 genes exhibited modified expression levels in response to nematode infection in soybean. The diverse effects of this nematode-induced differential methylation were evident in gene expression patterns. Among the genes experiencing unique methylation changes in response to this infection, 93 genes with hyperDMRs and 193 genes with hypoDMRs coincided with the 6903 genes undergoing altered expression levels due to the same infection. In a subsequent investigation, researchers delved into how soybean cyst nematode infection influences genome-wide DNA methylation patterns in soybean near-isogenic lines (NILs) with differing levels of resistance to soybean cyst nematode (Rambani *et al.*, 2020). Although the highly susceptible and highly resistant soybean lines share close genetic similarities, their DNA methylation profiles displayed notable differences. Employing stringent criteria (≥ 50% methylation difference, FDR = 0.01), scientists pinpointed 21,852 distinct DMRs between susceptible and resistant plants.

These DMRs comprised 4180 DMRs overlapping with 3666 protein-coding genes and 11,211 DMRs overlapping with 6033 transposable elements. CG-DMRs were primarily situated within gene bodies, with fewer instances in gene promoters and untranslated regions.

Approximately 70% of CHG-DMRs were located within gene bodies, while DMRs associated with transposable elements were mainly found in LTR retrotransposons. A total of 948 differentially expressed genes (DEGs) were identified, with 587 upregulated and 361 downregulated in susceptible plants compared with resistant ones. Gene ontology analysis uncovered significant enrichment in functions linked to defense responses, membrane disassembly, and intracellular signal transduction, suggesting that differences in gene expression between NILs under non-infected conditions might impact their responses to SCN infection. Intriguingly, in susceptible plants, SCN infection led to diminished DNA methylation levels across protein-coding genes in all sequence contexts, whereas in resistant plants, the trend was reversed. Following SCN infection, susceptible plants exhibited 50,040 DMRs relative to non-infected controls, with 7584 overlapping with 6252 protein-coding genes and 28,100 with transposable elements. Conversely, the count of DMRs in resistant plants was considerably lower, with 5080 DMRs identified in infected resistant samples compared with resistant controls, encompassing 1296 overlapping with 1293 protein-coding genes and 2356 with transposable elements. Significant alterations in DNA methylation patterns were evident exclusively during vulnerable interactions with the pathogen. Susceptible plants exhibited a total of 1668 DEGs, whereas only 112 DEGs were detected in resistant plants 5 days after SCN infection compared with non-infected controls. Analysis of these DEG sets using gene ontology revealed notable enrichment in resistant plants for responses to oxidative stress, chemical cues, and oxidation-reduction processes. Conversely, susceptible plants showed enrichment in responses to diverse stimuli and phytohormone signaling pathways, including ethylene, salicylic acid (SA), jasmonic acid (JA), and abscisic acid (ABA). Remarkably, only a single gene overlapped between the 1293 differentially methylated genes and the 112 DEGs identified in resistant plants post-SCN infection. When comparing the same differentially methylated genes (DMGs) with the DEGs associated with syncytium formation, a total of 188 genes exhibited overlap. In a similar analysis conducted on susceptible plants, 123 genes were found to be shared between DMGs and DEGs identified following SCN infection. Furthermore, 50 DEGs were discovered to contain differentially methylated transposable elements (TEs) within their gene bodies or promoters, resulting in a unique set of 147 differentially expressed DMGs. Upon comparison with syncytium DEGs, an overlap of 756 genes was observed. These results collectively suggest that alterations in DNA methylation during SCN infection play a role in modulating gene expression. When analyzing the DNA methylomes of non-infected susceptible and resistant plants alongside those of their parental lines, researchers discovered 58 DMRs within near-isogenic lines (NILs) derived from parental lines. Among these, 38 DMRs displayed hypomethylation in the NIL-resistant line and its SCN-resistant parental line, Fowler, while being hypermethylated in the NIL-susceptible line and its nematode-susceptible parental line, Anand. Conversely, 20 DMRs exhibited hypermethylation in Fowler and NIL-resistant, but hypomethylation in Anand and NIL-susceptible. These 58 DMRs were associated with 57 protein-coding genes, including four that demonstrated differential expression in SCN-induced soybean syncytium. Remarkably, researchers identified 56 DMRs unique to NIL-susceptible, indicating a gain or loss of methylation contrary to that observed in the parental lines and NIL-resistant. These 56 DMRs overlapped with 55 protein-coding genes, nine of which were previously identified as syncytium DEGs. Thus, the sustained modifications in methylation levels may wield significant influence over the interaction between soybean and soybean cyst nematode (Ashapkin *et al.*, 2020).

29.3 Epigenetic Regulation in Plants for Pest Resistance

Epigenetic modifications are pivotal in overseeing gene expression by reconfiguring the architecture and operations of chromatin, all while preserving the DNA sequence intact. In recent times, considerable attention has been directed towards uncovering the intricacies of epigenetic regulation in the genetic systems of both animals and plants. Unlike animals, plants are stationary organisms, incapable of fleeing environmental alterations or enduring diverse biotic pressures such as herbivore attacks and pathogen invasions. As a reaction, plants have developed intricate defense mechanisms, extensively examined in numerous research endeavours (Jones *et al.*, 2016; Arnold *et al.*, 2019; Bakhtiari *et al.*, 2019; Zhou and Zhang, 2020). Moreover, epigenetic control is of paramount importance in influencing plant defense mechanisms and facilitating diverse traits during the interplay between plants and microorganisms (Huang and Jin, 2022). The epigenetic regulation in plants for pest resistance occurs through induced or acquired resistance, particularly systemic acquired resistance and induced resistance.

29.3.1 Acquired resistance in plants against pests

The phenomenon of plants exhibiting heightened resistance to pests after exposure to specific stimuli is termed induced or acquired resistance (AR). AR represents a form of phenotypic plasticity, as it enables plants with identical genotypes to display varying resistance traits. One classic instance of AR is systemic acquired resistance (SAR), initially documented in tobacco in 1961 (Ross, 1961). It was soon evident that SAR constitutes a durable resistance response, detectable even 42 days post-induction (Bozarth and Ross, 1964). SAR relies on the involvement of the SA hormone (Loake and Grant, 2007), and the downstream signaling protein NPR1 (Durrant and Dong, 2004), both of which are pivotal in initiating SAR. Throughout SAR, there is an induction of transcription in SA-responsive genes, including those responsible for encoding antimicrobial pathogenesis-related proteins (PR) (Ryals *et al.*, 1996). Increased SA levels instigate modifications in chromatin structure at targeted genes. For example, there is an elevation in H3 acetylation, H4 acetylation, and H3K4 methylation levels at the PR-1 promoter (Butterbrodt *et al.*, 2006). The exact impact of these alterations on SAR stability, particularly in bolstering the memory of the initial pest or pathogen assault, remains uncertain. Nonetheless, it has been proposed that histone modification and/or substitution with histone variants might prepare pathogen-responsive genes for rapid activation when facing subsequent pathogen challenges. Histone acetyltransferases (HATs) and deacetylases (HDACs) play essential roles in the defense response of plants (Ramos-Cruz *et al.*, 2021). Throughout subsequent years, a myriad of both biotic and abiotic cues has been identified to trigger AR across diverse plant species and against a broad spectrum of pests. Additionally, various natural and synthetic compounds have been found to induce AR (Walters *et al.*, 2013). AR typically operates via two complementary mechanisms, namely sustained upregulation of inducible defenses and defense priming.

29.3.2 Induced resistance in plants against pests

Induced resistance (IR) in plants refers to a phenomenon where exposure to certain stimuli primes the plant's defense mechanisms, making them more effective in warding off pests and pathogens. This heightened defense response is a crucial aspect of plant immunity and represents a sophisticated strategy for plant survival in challenging environments. Various triggers can induce resistance in plants against pests, namely oviposition, chemical elicitors, and pathogen attack. For instance, IR against herbivores may be triggered by oviposition, granting plants an opportunity to develop an appropriate defense reaction upon the hatching of herbivore eggs (Reymond, 2013). Stahl *et al.* (2020) discovered that phosphatidylcholines found in eggs laid by the specialized herbivore *Pieris brassicae* can activate immune responses associated with induced resistance. Through a comparative transcriptomic analysis of *Arabidopsis* plants, Ojeda-Martinez *et al.* (2022) revealed that oviposition by spider mites (*T. urticae*) and lepidopterans (*P. brassicae*) elicits distinct responses, displaying a time-dependent sequence of unique processes. Notably, the regulation of transcription involves defense-related mechanisms such as SA, JA, reactive oxygen species (ROS), and glucosinolates, along with modifications in cell wall structure, responses to abiotic stress, and alterations in energy metabolism. This suggests the activation of specific pathways to optimize the induced resistance response tailored to the invading herbivore (Ojeda-Martinez *et al.*, 2022).

Cues associated with herbivores can arise from either biological or physical origins. Biological cues consist of oral secretions, microbes carried by insects, insect-associated molecular patterns (IAMPs), and signals emitted during oviposition (Howe and Jander, 2008; Hilker and Meiners, 2010; Hilker and Fatouros,

2015). Physical cues involve repetitive spatiotemporal patterns and the detection of insects by trichomes as they move across leaf surfaces (Howe and Jander, 2008; Hilker and Meiners, 2010). Additionally, herbivore-induced plant volatiles have been identified as triggers for priming, as they act as signals for nearby plants (Frost *et al.*, 2008). While these stimuli naturally occur during interactions with arthropods, their experimental application can prompt a quicker and/or more robust defensive response in targeted plants.

Oral secretions play a crucial role in enabling plants to discern between mechanical damage and insect feeding. In caterpillar secretions, fatty-acid–amino-acid conjugates play a crucial role in triggering specific responses from plants during attacks (Howe and Jander, 2008). Volicitin is among the first fatty-acid–amino-acid conjugates identified in lepidopteran larvae (Alborn *et al.*, 1997). Additionally, insect oral peptides and sulphated fatty acids act as key signals in facilitating interactions between insects and plants (Alborn *et al.*, 2007; Hilker and Meiners, 2010).

Aside from the aforementioned stimuli, trichomes play a significant role in detecting insect contact, prompting the plant to ready itself for defense against herbivores (Peiffer *et al.*, 2009). Moreover, particular entomophytophagous beneficial insects can act as stimuli by inserting their stylets into the plant's stem, thus initiating indirect defense mechanisms and antixenosis (Pérez-Hedo *et al.*, 2015). A high density of trichomes negatively impacts the egg-laying, feeding behavior, and larval development of insect pests (Handley *et al.*, 2005). Moreover, dense trichomes mechanically hinder herbivory and restrict the movement of insects and other arthropods on plant surfaces, thus limiting their access to the leaf epidermis (Agrawal *et al.*, 2009). Trichomes exhibit various forms, such as straight, spiral, hooked, branched, or unbranched, and they may be glandular or non-glandular (Hanley *et al.*, 2007). Glandular trichomes secrete secondary compounds such as flavonoids, terpenoids, and alkaloids, which can be toxic, repellent, or ensnare insects and other organisms, creating a dual structural and chemical defense mechanism (Hanley *et al.*, 2007; Sharma *et al.*, 2009). Trichome induction as a response to insect damage is observed across various plant species (Agrawal, 1999). This augmentation in trichome density post-damage is evident solely in leaves that develop during or after insect attacks, as the density of trichomes on existing leaves remains unaltered (Agrawal *et al.*, 2009). Research indicates that damage inflicted by adult leaf beetles like *Phratora vulgatissima* L. triggers increased trichome density in subsequent new leaves of plants such as *Salix cinerea* L. (Dalin and Björkman, 2003). Similarly, an upsurge in trichome density following damage by coleopterans has been observed in *S. cinerea* (Björkman and Ahrné, 2005). Instances of heightened trichome density subsequent to insect damage have also been documented in *Lepidium virginicum* L. and *Raphanus raphanistrum* L. (Agrawal, 1999). For instance, feeding by *Pieris rapae* (L.) on black mustard led to increased trichome density and glucosinolate levels (Traw, 2002). Similarly, damage by beetles resulted in elevated trichome density in *Alnus incana* Moench (Baur *et al.*, 1991). Typically, the rise in trichome density due to herbivory falls within the range of 25–100%, although instances of a 500–1000% increase have been documented. Changes in trichome density manifest within days or weeks following insect damage (Agrawal, 1999; Dalin and Björkman, 2003; Björkman and Ahrné, 2005). Additionally, herbivory induces a shift in the relative proportion of glandular and non-glandular trichomes (Agrawal, 1999). A positive correlation exists between the abundance of natural enemies and trichome density. Trichome exudates also function as extra-floral nectar (EFN) for the scelonid egg parasitoid *Gryon pennsylvanicum* Ashmead, which preys on squash bugs (Olson and Nechols, 1995).

Among these compounds are β-aminobutyric acid (BABA), probenazole, benzothiadiazole (BTH), and SA, all of which possess the capability to confer resistance in plants by providing protection against a broad spectrum of pathogens (Oostendorp *et al.*, 2001). SA acts as a hormone that triggers various direct responses in plants; however, it has been noted that at low concentrations, it can enhance flg22-induced activation of MITOGEN-ACTIVATED PROTEIN KINASE 3 (MPK3) and MPK6 (Yi *et al.*, 2015). Extensive investigations have been conducted on BTH and BABA as priming agents against pathogens and insects (Pastor *et al.*, 2013;

Balmer *et al.*, 2015; Baccelli and Mauch-Mani, 2016). Similar to SA, these compounds may also directly stimulate defense mechanisms when applied in high concentrations (van Hulten *et al.*, 2006). Dempsey and Klessig (2012) provided a comprehensive review of natural secondary metabolites implicated in mediating systemic acquired resistance, which include JA, azelaic acid, dehydroabietinal, glycerol-3-phosphate, methyl salicylate, and pipecolic acid. However, it has been confirmed, for instance, that azelaic acid and pipecolic acid trigger priming effects (Jung *et al.*, 2009; Návarová *et al.*, 2012), though the precise molecular mechanisms underlying induced resistance by these chemicals remain incompletely understood, making their classification as priming stimuli not always straightforward.

Given the enhancement of basal defense mechanisms by induced resistance, exploring genetic diversity in the efficacy of basal defense responses could aid in selecting cultivars with improved induced resistance potential. A thorough analysis of the transcriptome and metabolome of different quinoa varieties (*Chenopodium quinoa* Willd.), ranging from resistant to susceptible, has pinpointed crucial metabolites (like alkaloids and phenolic acids) and genes (such as bZIP transcription factors) that govern specific pathways against insects like *Spodoptera exigua* (Liu *et al.*, 2022). Additional investigations will be necessary to ascertain whether these biochemical indicators of innate defense reactions can reliably signify the potential for induced resistance.

29.4 Mechanisms of Acquired Plant Resistance Against Pests

29.4.1 Inducible defenses in plants against insects

Inducible defenses in plants manifest as toxins, antifeedants, or antinutrients. For instance, certain plants produce proteinase inhibitor proteins in response to herbivory, which hinders insect digestive enzymes and exhibits direct toxicity (Ryan, 1990). Furthermore, oxidative enzymes such as polyphenol oxidase (PPO), peroxidase, and lipoxygenase contribute to anti-nutritive defenses by modifying essential amino acids and fatty acids (Duffey *et al.*, 1991). Certain species enhance their defensive capabilities with secondary metabolites like alkaloids, terpenoids, or phenolics (Walling, 2000; Kessler and Baldwin, 2002). Consumption by herbivores triggers the production and emission of terpenoid, aromatic, and aliphatic volatile compounds, serving as potential cues to entice predators and parasitoids of the herbivore. Consequently, this assists in bolstering defense mechanisms (Arimura *et al.*, 2005). Significantly, herbivore defenses are frequently activated systemically beyond the direct feeding area, enhancing resilience in unaffected foliage of impacted plants (Havill and Raffa, 1999). This systemic defensive reaction and its control are comprehensively understood in tomato plants, where the hormone JA prominently participates in systemic signaling (Howe, 2004; Schilmiller and Howe, 2005).

In plants, two types of inducible defenses are recognized, namely, direct defenses and indirect defenses (Fig. 29.1). Direct defenses refer to plant characteristics that independently affect the susceptibility of host plants to insect attacks (Kessler and Baldwin, 2002). These defenses can be categorized into two main groups: anti-nutrition and toxicity. Anti-nutrition strategies aim to limit food availability both before ingestion, by reducing the food supply, and after ingestion, by reducing nutrient value for attacking insects. Toxicity involves causing physical harm and chemical disruptions to attacking insects through specific plant traits. During the pre-ingestion phase, host plants employ various mechanisms to limit food availability to insects. This includes inducing physical barriers, reinforcing cell walls, triggering hypersensitive reactions, employing anti-manipulation tactics, and producing insect repellents. Plants strengthen cell walls to create mechanical barriers against insect feeding, particularly for insects with piercing mouthparts (Goussain *et al.*, 2005; Divol *et al.*, 2007). Enhanced cell walls also increase the indigestibility of plant tissues to insects with chewing mouthparts (Clissold *et al.*, 2004). Upon insect attack, a multitude of genes involved in strengthening the cell wall are upregulated, including pectin esterases, expansins, cellulose synthases, and xyloglucan endotransglycosylases/hydrolases (XTH) (Goggin, 2007; Liu

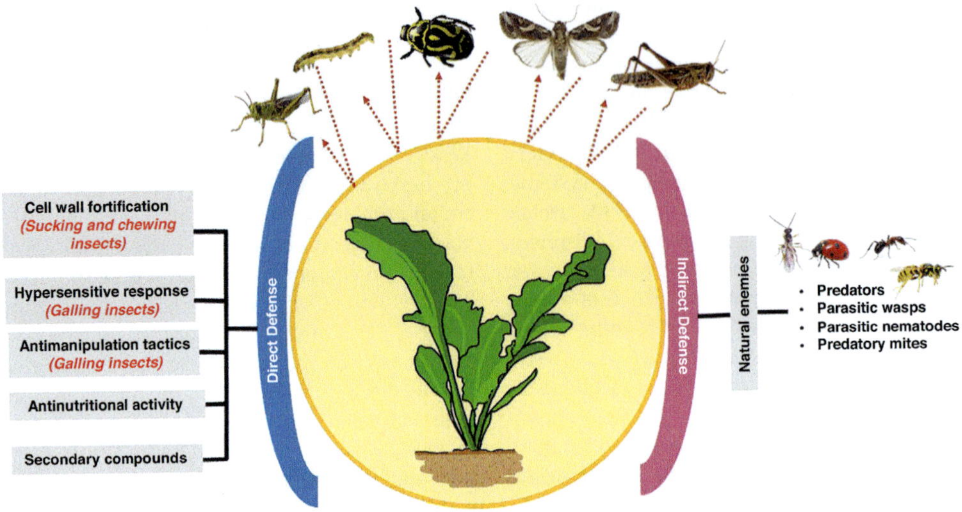

Fig. 29.1. Plant defense system against insects/pests. Author's own image.

et al., 2007). The importance of reinforcing the cell wall in plant defense is highlighted by studies such as the mutation in an XTH gene in *Arabidopsis* (xth33), which increases the vulnerability of plants to pests like *Myzus persicae* (Divol *et al.*,). Another direct defense strategy in plants against certain insect herbivores, particularly those with a fixed feeding site like galling insects, is the hypersensitive response (HR) (Fernandes and Negreiros, 2001; Cornelissen *et al.*, 2002; Larsson and Wingsle, 2005). HR has been observed in numerous varieties of willow trees (*Salix viminalis*) resistant to *Dasineura marginemtorquens* (Höglund *et al.*, 2005; Larsson and Wingsle, 2005). HR is also linked to wheat's resistance against *Mayetiola destructor* (Grover, 1995), rice's resistance against *Orseolia oryzae* (Bentur and Kalode, 1996), peach's resistance against aphids (Sauge *et al.*, 1998), and bean's resistance against *Apion godmani* (Garza *et al.*, 2001). Insects employ anti-manipulation tactics as another defense strategy against host plants, representing a significant mechanism of anti-nutrition, especially for galling insects and other herbivores that modify local environments to acquire nutrients (Agrios, 1997; Danks, 2002; Rohfritsch, 2005). Invasive galling insects infiltrate vulnerable plants by triggering a localized alteration in metabolic habitat within plant tissues, often termed a gall (Goethals *et al.*, 2001). Within this gall formation, the insect secures a competitive edge by accessing augmented nutritional supplies (arising from the establishment of a nutrient sink) and diminished plant defensive capabilities (Stone and Schönrogge, 2003; Harris *et al.*, 2006). Resistant plants possess mechanisms to hinder gall formation, thus preventing the insect's access to nutrition (Harris *et al.*, 2006). Moreover, specific plants react to insect herbivory by releasing intricate mixtures of volatile organic compounds (VOCs), which lure the natural predators of the herbivores in strategies of indirect defense (Kessler and Baldwin, 2001; Kost and Heil, 2006). Nonetheless, certain VOCs may also function as deterrents against the invading insects, acting as direct defensive mechanisms (Kessler and Baldwin, 2001). For example, tobacco plants emit volatile blends in response to herbivore attacks, which exhibit temporal fluctuations. Certain VOCs are exclusively released during nighttime. These nocturnal VOCs serve as a deterrent to female moths of *Heliothis virescens*, preventing them from depositing eggs on plants that have been previously damaged or on neighboring undamaged ones (De Moraes *et al.*, 2001). The production of VOCs is triggered by elicitors found in the oral secretions of invading insect herbivores (Truitt *et al.*, 2004; Schmelz *et al.*, 2006).

Furthermore, anti-nutritional activity in the post-ingestion phase works to diminish the nutritional quality by either stripping away essential nutrients or hindering their digestion. Certain plant proteins exhibit considerable stability within the digestive systems of insect herbivores after being consumed (Chen *et al.*, 2007). Within the insect gut, some plant enzymes retain their activity and effectively break down nutrients that would otherwise be available for the insect's use. Similar to humans, insects rely on dietary intake for essential amino acids like arginine, lysine, methionine, and threonine, which they cannot produce internally. When crucial amino acids undergo degradation by plant enzymes within the insect gut, it hinders the growth and development of the insect. Numerous enzymes involved in amino acid degradation are activated by insect feeding, remaining active even after being ingested by the insect. For example, enzymes like arginase and threonine deaminase, responsible for breaking down arginine and threonine, respectively, are significantly stimulated in response to insect feeding (Chen *et al.*, 2005). These enzymes continue to function within the insect gut, contributing to the removal of essential amino acids arginine and threonine, hence being considered anti-nutritional (Felton, 2005; Chen *et al.*, 2007). Asparaginase, another enzyme involved in amino acid degradation, is notably upregulated in resistant plants during insect attacks (Liu *et al.*, 2007). Another method to diminish nutrient value is by inhibiting the herbivore's digestive enzymes. When subjected to attacks by insect herbivores, plants produce various protein inhibitors capable of blocking insect digestive enzymes such as proteases and amylases (Koiwa *et al.*, 1997). Furthermore, diverse plant secondary compounds also influence the consumption and absorption of food, potentially by obstructing digestive enzymes in insects (Akhtar and Isman, 2004). By inhibiting insect digestive enzymes, the ingestion and assimilation of nutrients are compromised, ultimately impeding insect growth and development.

In contrast, indirect defenses involve plant characteristics that, while not directly affecting the vulnerability of host plants, act as lures to natural enemies of invading insects (Table 29.1). These natural enemies subsequently reduce the insect population, thereby lessening the damage inflicted on plants by the insects (Chen, 2008). Predators commonly attracted by volatile compounds encompass lady beetles, predatory mites, and various other predatory species (van Den Boom *et al.*, 2004; Hagenbucher *et al.*, 2013). It was discovered that 34 species of parasitic wasps belonging to ten distinct families have been lured by volatile emissions from plants infested with insect herbivores. Even attracted parasitic wasps belonging to different families like Braconidae, Pteromalidae, Scelionidae, Aphelinidae, Encyrtidae, and Aphidiidae play a role in indirect defense to plants. Among parasites belonging to the Braconidae family are the prevalent beneficiaries enticed by volatiles released from plants targeted by lepidopterans, coleopterans, aphids, and select Dipterans (Poelman *et al.*, 2009; Bruinsma *et al.*, 2009; Hanyu *et al.*, 2009, 2011; Girling *et al.*, 2011). Other Hymenopteran parasites also perform similar functions in plant defense against pests (Lou and Cheng, 2003; Colazza, 2004; Snoeren *et al.*, 2010). Future investigations ought to prioritize examining the effectiveness of particular parasites, ideal distances, and concentrations of plant volatile mixtures to attract parasites. Non-insect invertebrates are similarly enticed by plant volatiles. Two types of parasitic nematodes, *Heterorhabditis megidis* and *H. bacteriophora*, have been observed to diminish the population of *Diabrotica virgifera virgifera* in reaction to emissions from corn (Rasmann and Turlings, 2008). *H. bacteriophora* assists in defending *Asclepias syriaca* underground against attacks from the root-boring beetle *Tetraopes tetraophthalmus* (Rasmann *et al.*, 2011). Certain predatory mites, such as those belonging to the Tydeidae and Phytoseiidae families, display the capacity to consume various types of sustenance depending on circumstances. When prey numbers are low, these predators can sustain themselves by consuming alternative food sources, such as foliar pathogenic fungi, and occasionally pollen (Duso *et al.*, 2005; Pozzebon *et al.*, 2009). Plants damaged by insect herbivores are prone to secondary fungal infections. Predatory mites, which are generalists drawn to injured plants, play a beneficial function in tritrophic relationships (Norton *et al.*, 2000; English-Loeb *et al.*, 2005, 2007; Tempfli *et al.*, 2015).

Table 29.1. A few natural enemies targeting insects as indirect plant defense. Author's own table.

Natural enemies group	Specific natural enemies	Targeted insects	References
Lady beetles	*Coccinella septempunctata*	Aphid eaters	Dicke and Sabelis (1988); Dicke (1999); Krips *et al.* (2001); Horiuchi *et al.* (2003); van Den Boom *et al.* (2004)
	Chilocorus kuwanae	Scale insects, mealy bugs, aphids	
	Aiolocaria hexaspilota	Aphids, mealy bugs, scale insects, whiteflies, mites	
Predatory mites	*Amblyseius andersoni*	Whitefly, tarsonemid mites, rust mites	
	Phytoseiulus persimilis	Eggs and adults of spider mites	
	A. finlandicus	Spider mites, whiteflies, thrips	
	A. womerleyi	Thrips, whiteflies, spider mites	
Predators	Minute pirate bug	Spider mites, aphids, thrips, psyllids, whiteflies, insect eggs, small caterpillars	Venzon *et al.* (1999); James (2003); Hagenbucher *et al.* (2013)
	Two-spotted stink bug	Eggs and larvae of Colorado potato beetle, and smaller caterpillars and larvae	
	Bigeyed bug	Bug nymphs, flea beetles, insect eggs, small caterpillars, and all stages of aphids, mites, and whiteflies	
	Green lacewings	Aphids, thrips, whiteflies, spider mites, caterpillars, insect eggs	
	Predatory thrips	Mites, lace bugs, whiteflies, scale insects	
	Predatory bug	Aphids, mites, caterpillars, insect nymphs, larvae, and eggs	

29.4.2 Plant defense priming against insects

Plant defense priming represents a type of plant immunological 'memory' whereby the plant interprets an external stimulus as a warning signal, prompting a slight activation of induced defense mechanisms (Fig. 29.2). This encounter is stored as memory, resulting in a swifter and/or more robust defense response upon subsequent encounters with pests or pathogens. The external stimulus may encompass chemical, biotic (such as pests, pathogens, or beneficial microbes), or abiotic factors (like environmental stressors) (Desmedt *et al.*, 2021). Defense priming involves the activation of defense-related genes or the increase in other metabolites, which can trigger biochemical signal transduction

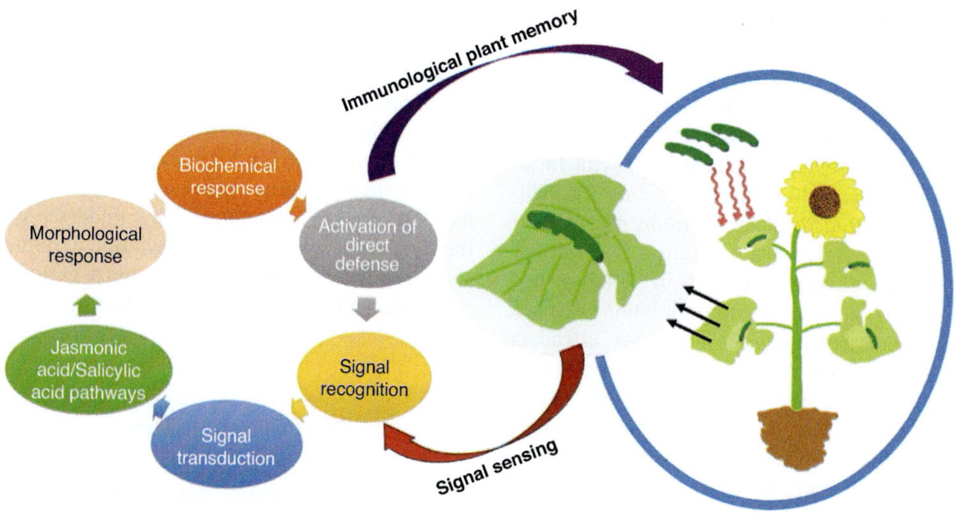

Fig. 29.2. Plant defense priming against pests involving signaling pathways. Author's own image.

pathways leading to a primed state. This process prepares the organism to mount a more rapid and effective response to subsequent stressors or threats (Frost *et al.*, 2008). The process of priming notably includes the upregulation of genes associated with defense mechanisms or signaling pathways, as evidenced by studies such as those conducted by van Hulten *et al.* (2006), Engelberth *et al.* (2007), and Ton *et al.* (2007). This phenomenon suggests that organisms enhance their readiness to respond to subsequent stressors or challenges through the activation of these genetic elements. Upon exposure to a resistance-triggering stimulus, inducible defenses may remain elevated, providing ongoing AR against subsequent attacks. Alternatively, the stimulus may prime the plant's immune system, resulting in quicker and/or more robust upregulation of inducible defenses upon subsequent pathogen or herbivore assault. The optimal strategy is contingent upon the cost–benefit equilibrium of resistance within a particular environment (Burow and Halkier, 2017).

Another epigenetic mechanism possibly involved in memory formation in SAR is the histone variant H2A.Z. Being highly preserved among eukaryotes, H2A.Z, one of the most evolutionarily conserved histone variants, is notably concentrated at the initiation sites of gene transcription. It has been proposed that its integration contributes to gene activation, the retention of transcriptional information, the suppression of heterochromatin, and the detection of thermal stimuli (Dhillon *et al.*, 2006; Brickner *et al.*, 2007; Zlatanova and Thakar, 2008; Kumar and Wigge, 2010; Light *et al.*, 2010). In *Arabidopsis* mutants devoid of the SWR1 complex, essential for depositing H2A.Z, a considerable proportion of genes activated in SAR exhibit persistent expression (March-Diaz *et al.*, 2008). Given that H2A.Z incorporation is linked to the retention of transcriptional information and swift reinstatement of genes, it suggests the potential importance of H2A.Z in preparing genes engaged in SAR initiation.

29.5 Methods for Developing Plant Resistance Against Pests Through Epigenetics

29.5.1 Epigenome editing

Epigenome editing involves modifying specific epigenetic marks, such as DNA methylation or histone modifications, at targeted loci within the plant genome. By altering the epigenetic landscape, researchers can activate or suppress the gene expression involved in plant defense against pests (Qi *et al.*, 2023). The introduction of genome editing has revealed extensive opportunities for engineering a range of agronomic characteristics, particularly long-lasting resistance against insect pests. Novel methods for developing insect-resistant crops via genome editing encompass altering effector-target engagements, disabling host susceptibility genes, crafting artificial immune receptors to induce comprehensive resistance, and mitigating contradictory functions of defense hormones. In parallel, the manipulation of insect genomes is being explored to either introduce gene drives or combat resistance to diverse insecticides. Notably, genome editing systems offer a distinct advantage by enabling precise targeting of specific regions in the genome without retaining unwanted vector backbones (Bisht *et al.*, 2019). Utilizing genome editing to control insects through gene drive mechanisms necessitates the disruption of specific sites within the insect genome. However, such disruptions may inadvertently lead to lethality or negatively impact factors such as sex ratio or fecundity. Alternatively, targeting key genes involved in plant immunity, including susceptibility (S-genes), resistance (R-genes), and those governing interactions between the plant and insect effectors, presents an alternative approach. While current research predominantly focuses on optimizing CRISPR/Cas methods within individual crops, insights from fundamental investigations into plant–insect interactions offer immense potential for developing insect resistance through CRISPR/Cas9-mediated genome editing (Bisht *et al.*, 2019).

29.5.2 Epigenetic modifiers

Chemical compounds or natural substances that influence epigenetic marks can be applied to plants to modulate their defense responses against pests. These modifiers can alter DNA methylation patterns, histone modifications, or microRNA expression profiles, thereby enhancing the plant's resistance to pest attacks. Emerging findings suggest that DNA methylation plays a crucial role in modulating plant defense responses to pests and nematodes, as indicated by Leonetti and Molinari (2020). Notably, *Arabidopsis* mutants deficient in DNA and H3K9 methylation, such as kyp, have demonstrated heightened resistance to infestations by *Myzus persicae* (Annacondia *et al.*, 2021). Likewise, mutations in *Oryza sativa* associated with RNA-directed DNA methylation (RdDM) and decrease in DNA methylation (DDM1) have been linked to reduced susceptibility to infections by the nematode *Meloidogyne graminicola*. This underscores the significant role of DNA methylation dynamics in regulating plant defense against various pests and nematodes (Atighi *et al.*, 2020).

29.6 Conclusion

Epigenetic mechanisms, encompassing DNA methylation, histone modifications, and chromatin remodeling, play pivotal roles in coordinating plant responses to insect pests. These mechanisms not only govern the expression of defense-related genes for immediate and transgenerational defense priming but also react to insect attacks, potentially manipulated by elements to augment plant adaptability and ensure crop yield amid insect/pest challenges. Both innate epigenetic diversity and stress-induced epigenetic alterations, facilitated by factors such as chemical treatments, mutations in epigenetic machinery, induced changes in DNA methylation, and epigenome editing, offer prospects for improving pest resistance in crops through epigenetic breeding strategies. Although DNA methylation patterns can be inherited across generations and are universally applicable to all crop species, information concerning histone modifications may be lost during meiosis and hence is solely pertinent to epi-breeding in crops propagated clonally. Epigenomic variances, exemplified by differentially methylated regions linked to pest-resistance traits, could function as molecular indicators for assessing and selecting disease-resistant crop cultivars. Furthermore, anticipatory modeling of the impacts of epigenetic variances on crop disease resistance could steer epigenetic engineering endeavors in resistance breeding initiatives. As epigenetic methodologies evolve and comprehension of plant–pathogen interactions progresses, harnessing epigenetic variability stands poised to elevate crop pest resistance in the years ahead.

References

Agrawal, A.A. (1999) Induced responses to herbivory in wild radish: Effects on several herbivores and plant fitness. *Ecology* 80, 1713–1723. DOI: 10.1890/0012-9658(1999)080[1713:IRTHIW]2.0.CO;2.

Agrawal, A.A., Fishbein, M., Jetter, R., Salminen, J.P., Goldstein, J.B. *et al.* (2009) Phylogenetic ecology of leaf surface traits in the milkweeds (*Asclepias* spp.): Chemistry, ecophysiology, and insect behavior. *New Phytologist* 183, 848–867. DOI: 10.1111/j.1469-8137.2009.02897.x.

Agrios, G.N. (1997) Plant diseases caused by mollicutes: Phytoplasmas and spiroplasmas. In: Agrios, G.N. (ed.) *Plant Pathology*, 4th edn. Academic Press, New York, pp. 457–470.

Akhtar, Y. and Isman, M.B. (2004) Comparative growth inhibitory and antifeedant effects of plant extracts and pure allelochemicals on four phytophagous insect species. *Journal of Applied Entomology* 128, 32–38.

Akhter, Z., Bi, Z., Ali, K., Sun, C., Fiaz, S. *et al.* (2021) In response to abiotic stress, DNA methylation confers epigenetic changes in plants. *Plants* 10(6), 1096.

Alborn, H.T., Turlings, T.C.J., Jones, T.H., Stenhagen, G., Loughrin, J.H. *et al.* (1997) An elicitor of plant volatiles from beet armyworm oral secretion. *Science* 276, 945–949.

Alborn, H.T., Hansen, T.V., Jones, T.H., Bennett, D.C., Tumlinson, J.H. *et al.* (2007) Disulfooxy fatty acids from the American bird grasshopper *Schistocerca americana*, elicitors of plant volatiles. *Proceedings of the National Academy of Sciences of the United States of America* 104, 12976–12981.

Annacondia, M.L., Markovic, D., Reig-Valiente, J.L., Scaltsoyianned, V., Pieterse, C.M.J. *et al.* (2021) Aphid feeding induced the relaxation of epigenetic control and the associated regulation of the defense response in *Arabidopsis*. *New Phytologist* 230(3), 1185–1200. DOI: 10.1111/nph.17226.

Arimura, G., Kost, C. and Boland, W. (2005) Herbivore-induced, indirect plant defences. *Biochimica et Biophysica Acta* 1734(2), 91–111.

Arnold, P.A., Kruuk, L.E.B. and Nicotra, A.B. (2019) How to analyse plant phenotypic plasticity in response to a changing climate. *New Phytologist* 222(3), 1235–1241. DOI: 10.1111/nph.15656.

Ashapkin, V.V., Kutueva, L.I., Aleksandrushkina, N.I. and Vanyushin, B.F. (2020) Epigenetic mechanisms of plant adaptation to biotic and abiotic stresses. *International Journal of Molecular Sciences* 21(20), 7457.

Atighi, M.R., Verstraeten, B., Meyer, T. and Kyndt, T. (2020) Genome-wide DNA hypomethylation shapes nematode pattern-triggered immunity in plants. *New Phytologist* 227(2), 545–558. DOI: 10.1111/nph.16532.

Baccelli, I. and Mauch-Mani, B. (2016) Beta-aminobutyric acid priming of plant defense: The role of ABA and other hormones. *Plant Molecular Biology* 91(6), 703–711.

Bakhtiari, M., Formenti, L., Caggìa, V., Glauser, G. and Rasmann, S. (2019) Variable effects on growth and defense traits for plant ecotypic differentiation and phenotypic plasticity along elevation gradients. *Ecology and Evolution* 9(7), 3740–3755. DOI: 10.1002/ece3.4999.

Balažová, E., Balažová, A. and Obložinský, M. (2022) Epigenetic modifications in plants – impact on phospholipid signaling and secondary metabolism. *Chemické Listy* 116(7), 416–422.

Balmer, A., Pastor, V., Gamir, J., Flors, V. and Mauch-Mani, B. (2015) The "prime-ome": Towards a holistic approach to priming. *Trends in Plant Science* 20(7), 443–452.

Baur, R., Binder, S. and Benz, G. (1991) Nonglandular leaf trichomes as short-term inducible defense of the grey alder, *Alnus incana* (L.), against the chrysomelid beetle, *Agelastica alni* L. *Oecologia* 87(2), 219–226.

Bentur, J.S. and Kalode, M.B. (1996) Hypersensitive reaction and induced resistance in rice against the Asian rice gall midge *Orseolia oryzae*. *Entomologia Experimentalis et Applicata* 78(1), 77–81.

Bisht, D.S., Bhatia, V. and Bhattacharya, R. (2019) Improving plant-resistance to insect-pests and pathogens: The new opportunities through targeted genome editing. *Seminars in Cell & Developmental Biology* 96, 65–76.

Björkman, C. and Ahrné, K. (2005) Influence of leaf trichome density on the efficiency of two polyphagous insect predators. *Entomologia Experimentalis et Applicata* 115(1), 179–186. DOI: 10.1111/j.1570-7458.2005.00284.x.

Bozarth, R.F. and Ross, A.F. (1964) Systemic resistance induced by localized virus infections: Extent of changes in uninfected plant parts. *Virology* 24(3), 446–455.

Brickner, D.G., Cajigas, I., Fondufe-Mittendorf, Y., Ahmed, S., Lee, P.C. *et al.* (2007) H2A.Z-mediated localization of genes at the nuclear periphery confers epigenetic memory of previous transcriptional state. *PLoS Biology* 5(4), e81.

Bruinsma, M., Posthumus, M.A., Mumm, R., Mueller, M.J., van Loon, J.J.A. *et al.* (2009) Jasmonic acid-induced volatiles of *Brassica oleracea* attract parasitoids: Effects of time and dose, and comparison with induction by herbivores. *Journal of Experimental Botany* 60(9), 2575–2587.

Burow, M. and Halkier, B.A. (2017) How does a plant orchestrate defense in time and space? Using glucosinolates in *Arabidopsis* as case study. *Current Opinion in Plant Biology* 38, 142–147.

Butterbrodt, T., Thurow, C. and Gatz, C. (2006) Chromatin immunoprecipitation analysis of the tobacco PR-1a- and the truncated CaMV 35S promoter reveals differences in salicylic acid-dependent TGA factor binding and histone acetylation. *Plant Molecular Biology* 61(4–5), 665–674.

Chachar, S., Chachar, M., Riaz, A., Shaikh, A.A., Li, X. *et al.* (2022) Epigenetic modification for horticultural plant improvement comes of age. *Scientia Horticulturae* 292, 110633.

Chen, H., Wilkerson, C.G., Kuchar, J.A., Phinney, B.S. and Howe, G.A. (2005) Jasmonate-inducible plant enzymes degrade essential amino acids in the herbivore midgut. *Proceedings of the National Academy of Sciences of the United States of America* 102(52), 19237–19242.

Chen, H., Gonzales-Vigil, E., Wilkerson, C.G. and Howe, G.A. (2007) Stability of plant defense proteins in the gut of insect herbivores. *Plant Physiology* 143(4), 1954–1967.

Chen, M.S. (2008) Inducible direct defense against insect herbivores: A review. *Insect Science* 15(2), 101–114.

Clissold, F.J., Sanson, G.D. and Read, J. (2004) Indigestibility of plant cell wall by the Australian plague locust, *Chortoicetes terminifera*. *Entomologia Experimentalis et Applicata* 112(3), 159–168.

Colazza, S. (2004) Insect oviposition induces volatile emission in herbaceous plants that attracts egg parasitoids. *Journal of Experimental Biology* 207(1), 47–53.

Cornelissen, T., Negreiros, D. and Fernandes, G. (2002) Plant resistance against gall-forming insects: The role of hypersensitivity. In: Wagner, M.R., Clancy, K.M., Lieutier, F. and Paine, T.D. (eds) *Mechanisms and Deployment of Resistance in Trees to Insects*. Kluwer Academic, Dordrecht, Netherlands, pp. 137–152.

Dalin, P. and Björkman, C. (2003) Adult beetle grazing induces willow trichome defence against subsequent larval feeding. *Oecologia* 134(1), 112–118. DOI: 10.1007/s00442-002-1093-3.

Danks, H.V. (2002) Modification of adverse conditions by insects. *Oikos* 99(1), 10–24.

De Moraes, C.M., Mescher, M.C. and Tumlinson, J.H. (2001) Caterpillar-induced nocturnal plant volatiles repel conspecific females. *Nature* 410(6828), 577–580.

Dempsey, D.A. and Klessig, D.F. (2012) SOS – too many signals for systemic acquired resistance? *Trends in Plant Science* 17(9), 538–545.

Desmedt, W., Vanholme, B. and Kyndt, T. (2021) Plant defense priming in the field: A review. In: Maienfisch, P. and Mangelinckx, S. (eds) *Recent Highlights in the Discovery and Optimization of Crop Protection Products*. Academic Press, pp. 87–124.

Dhillon, N., Oki, M., Szyjka, S.J., Aparicio, O.M. and Kamakaka, R.T. (2006) H2A.Z functions to regulate progression through the cell cycle. *Molecular and Cellular Biology* 26(2), 489–501.

Dicke, M. (1999) Evolution of induced indirect defense of plants. In: Tollrian, R. and Harvell, C.D. (eds) *The Ecology and Evolution of Inducible Defenses*. Princeton University Press, New Jersey, pp. 62–88.

Dicke, M. and Sabelis, M.W. (1988) How plants obtain predatory mites as bodyguards. *Netherlands Journal of Zoology* 38(2–4), 148–165.

Divol, F., Vilaine, F., Thibivilliers, S., Kusiak, C., Sauge, M.H. *et al.* (2007) Involvement of the xyloglucan endotransglycosylase/hydrolases encoded by celery XTH1 and *Arabidopsis* XTH33 in the phloem response to aphids. *Plant, Cell & Environment* 30(2), 187–201.

Duffey, S.S., Felton, G.W. and Hedin, P. (1991) Enzymatic antinutritive defenses of the tomato plant against insects. In: Hedin, P. (ed.) *Naturally Occurring Pest Bioregulators*. ACS Press, Washington, DC, pp. 167–197.

Durrant, W.E. and Dong, X. (2004) Systemic acquired resistance. *Annual Review of Phytopathology* 42, 185–209.

Duso, C., Pozzebon, A., Capuzzo, C., Malagnini, V., Otto, S. *et al.* (2005) Grape downy mildew spread and mite seasonal abundance in vineyards: Effects on *Tydeus caudatus* and its predators. *Biological Control* 32(1), 143–154.

Engelberth, J., Seidl-Adams, I., Schultz, J.C. and Tumlinson, J.H. (2007) Insect elicitors and exposure to green leafy volatiles differentially upregulate major octadecanoids and transcripts of

12-oxophytodienoic acid reductases in *Zea mays*. *Molecular Plant-Microbe Interactions* 20(6), 707–716.

English-Loeb, G., Norton, A.P., Gadoury, D., Seem, R. and Wilcox, W. (2005) TRI-trophic interactions among grapevines, a fungal pathogen, and a mycophagous mite. *Ecological Applications* 15(5), 1679–1688.

English-Loeb, G., Norton, A.P., Gadoury, D., Seem, R. and Wilcox, W. (2007) Biological control of grape powdery mildew using mycophagous mites. *Plant Disease* 91(4), 421–429.

Felton, G.W. (2005) Indigestion is a plant's best defense. *Proceedings of the National Academy of Sciences of the United States of America* 102(52), 18771–18772.

Fernandes, G.W. and Negreiros, D. (2001) The occurrence and effectiveness of hypersensitive reaction against galling herbivores across host taxa. *Ecological Entomology* 26(1), 46–55.

Frost, C.J., Mescher, M.C., Carlson, J.E. and Moraes, C.M. (2008) Plant defense priming against herbivores: Getting ready for a different battle. *Plant Physiology* 146(3), 818–824.

Garza, R., Vera, J., Cardona, C., Barcenas, N. and Singh, S.P. (2001) Hypersensitive response of beans to *Apion godmani* (Coleoptera: Curculionidae). *Journal of Economic Entomology* 94(4), 958–962.

Girling, R.D., Stewart-Jones, A., Dherbecourt, J., Staley, J.T., Wright, D.J. *et al*. (2011) Parasitoids select plants more heavily infested with their caterpillar hosts: A new approach to aid interpretation of plant headspace volatiles. *Proceedings of Biological Sciences* 278(1718), 2646–2653.

Goethals, K., Vereecke, D., Jaziri, M., Montagu, M. and Holsters, M. (2001) Leafy gall formation by *Rhodococcus fascians*. *Annual Review of Phytopathology* 39, 27–52.

Goggin, F.L. (2007) Plant–aphid interactions: Molecular and ecological perspectives. *Current Opinion in Plant Biology* 10(4), 399–408.

Goussain, M.M., Prado, E. and Moraes, J.C. (2005) Effect of silicon applied to wheat plants on the biology and probing behaviour of the green bug *Schizaphis graminum* (Rond.) (Hemiptera: Aphididae). *Neotropical Entomology* 34(5), 807–813.

Grover, P.B. (1995) Hypersensitive response of wheat to the hessian fly. *Entomologia Experimentalis et Applicata* 74(3), 283–294.

Hagenbucher, S., Olson, D.M., Ruberson, J.R., Wäckers, F.L. and Romeis, J. (2013) Resistance mechanisms against arthropod herbivores in cotton and their interactions with natural enemies. *Critical Reviews in Plant Sciences* 32(6), 458–482.

Handley, R., Ekbom, B. and Ågren, J. (2005) Variation in trichome density and resistance against a specialist insect herbivore in natural populations of *Arabidopsis thaliana*. *Ecological Entomology* 30(3), 284–292. DOI: 10.1016/j.ppees.2007.01.001.

Hanley, M.E., Lamont, B.B., Fairbanks, M.M. and Rafferty, C.M. (2007) Plant structural traits and their role in anti-herbivore defence. *Perspectives in Plant Ecology, Evolution and Systematics* 8(4), 157–178. DOI: 10.1016/j.ppees.2007.01.001.

Hanyu, K., Ichiki, R.T., Nakamura, S. and Kainoh, Y. (2009) Duration and location of attraction to herbivore damaged plants in the tachinid parasitoid *Exorista japonica*. *Applied Entomology and Zoology* 44(3), 371–378.

Hanyu, K., Ichiki, R.T., Nakamura, S. and Kainoh, Y. (2011) Behavior of the tachinid parasitoid *Exorista japonica* (Diptera:Tachinidae) on herbivore-infested plants. *Applied Entomology and Zoology* 46(4), 565–571.

Harris, M.O., Freeman, T.P., Rohfritsch, O., Anderson, K.G., Payne, S.A. *et al*. (2006) Virulent hessian fly (Diptera: Cecidomyiidae) larvae induce a nutritive tissue during compatible interactions with wheat. *Annals of the Entomological Society of America* 99(2), 305–316.

Havill, N.P. and Raffa, K.F. (1999) Effects of elicitation treatment and genotypic variation on induced resistance in *Populus*: Impacts on gypsy moth (Lepidoptera: Lymantriidae) development and feeding behavior. *Oecologia* 120(2), 295–303.

Hemenway, E.A. and Gehring, M. (2023) Epigenetic regulation during plant development and the capacity for epigenetic memory. *Annual Review of Plant Biology* 74, 87–109.

Hilker, M. and Fatouros, N.E. (2015) Plant responses to insect egg deposition. *Annual Review of Entomology* 60, 493–515.

Hilker, M. and Meiners, T. (2010) How do plants "notice" attack by herbivorous arthropods? *Biological Reviews of the Cambridge Philosophical Society* 85(2), 267–280.

Höglund, S., Larsson, S. and Wingsle, G. (2005) Both hypersensitive and non-hypersensitive responses are associated with resistance in *Salix viminalis* against the gall midge *Dasineura marginemtor-quens*. *Journal of Experimental Botany* 56(422), 3215–3222.

Horiuchi, J.-I., Arimura, G.-I., Ozawa, R., Shimoda, T., Takabayashi, J. *et al*. (2003) A comparison of the responses of *Tetranychus urticae* (Acari: Tetranychidae) and *Phytoseiulus persimilis* (Acari: Phytoseiidae) to volatiles emitted from lima bean leaves with different levels of damage made by *T. urticae* or *Spodoptera exigua* (Lepidoptera: Noctuidae). *Applied Entomology and Zoology* 38(1), 109–116.

Howe, G.A. (2004) Jasmonates as signals in the wound response. *Journal of Plant Growth Regulation* 23(3), 223–237.

Howe, G.A. and Jander, G. (2008) Plant immunity to insect herbivores. *Annual Review of Plant Biology* 59, 41–66.

Huang, C.-Y. and Jin, H. (2022) Coordinated epigenetic regulation in plants: A potent managerial tool to conquer biotic stress. *Frontiers in Plant Science* 12, 795274.

James, D.G. (2003) Synthetic herbivore-induced plant volatiles as field attractants for beneficial insects. *Environmental Entomology* 32(5), 977–982.

Jones, J.D.G., Vance, R.E. and Dangl, J.L. (2016) Intracellular innate immune surveillance devices in plants and animals. *Science* 354(6316), aaf6395. DOI: 10.1126/science.aaf6395.

Jung, H.W., Tschaplinski, T.J., Wang, L., Glazebrook, J. and Greenberg, J.T. (2009) Priming in systemic plant immunity. *Science* 324(5923), 89–91.

Kessler, A. and Baldwin, I.T. (2001) Defensive function of herbivore: Induced plant volatile emissions in nature. *Science (New York)* 291(5511), 2141–2144.

Kessler, A. and Baldwin, I.T. (2002) Plant responses to insect herbivory: The emerging molecular analysis. *Annual Review of Plant Biology* 53, 299–328.

Koiwa, H., Bressan, R.A. and Hasegawa, P.M. (1997) Regulation of protease inhibitors and plant defense. *Trends in Plant Science* 2(10), 379–384.

Kost, C. and Heil, M. (2006) Herbivore-induced plant volatiles induce an indirect defence in neighbouring plants. *Journal of Ecology* 94(3), 619–628.

Krips, O.E., Willems, P.E.L., Gols, R., Posthumus, M.A., Gort, G. *et al*. (2001) Comparison of cultivars of ornamental crop *Gerbera jamesonii* on production of spider mite-induced volatiles, and their attractiveness to the predator *Phytoseiulus persimilis*. *Journal of Chemical Ecology* 27(7), 1355–1372.

Kumar, S.V. and Wigge, P.A. (2010) H2A.Z-containing nucleosomes mediate the thermosensory response in *Arabidopsis*. *Cell* 140(1), 136–147.

Larsson, H.S.S. and Wingsle, G. (2005) Both hypersensitive and non-hypersensitive responses are associated with resistance in *Salix viminalis* against the gall midge *Dasineura marginemtorquens*. *Journal of Experimental Botany* 56, 3215–3222.

Leonetti, P. and Molinari, S. (2020) Epigenetic and metabolic changes in root-knot nematode-plant interactions. *International Journal of Molecular Science* 21, 7759. DOI: 10.3390/ijms21207759.

Light, W.H., Brickner, D.G., Brand, V.R. and Brickner, J.H. (2010) Interaction of a DNA zip code with the nuclear pore complex promotes H2A.Z incorporation and INO1 transcriptional memory. *Molecular Cell* 40, 112–125.

Liu, J., Li, L., Liu, Y. and Kong, Z. (2022) Combined transcriptome and metabolome analysis of the resistance mechanism of quinoa seedlings to *Spodoptera exigua*. *Frontiers in Plant Science* 13. DOI: 10.3389/fpls.2022.931145.

Liu, X.-M., Bai, J., Huang, L., Zhu, L., Liu, X. *et al*. (2007) Gene expression of different wheat genotypes during attack by virulent and avirulent Hessian fly (*Mayetiola destructor*) larvae. *Journal of Chemical Ecology* 33, 2171–2194.

Lloyd, J.P.B. and Lister, R. (2022) Epigenome plasticity in plants. *Nature Reviews. Genetics* 23(1), 55–68.

Loake, G. and Grant, M. (2007) Salicylic acid in plant defence–the players and protagonists. *Current Opinion in Plant Biology* 10(5), 466–472.

Lou, Y. and Cheng, J. (2003) Role of rice volatiles in the foraging behaviour of the predator *Cyrtorhinus lividipennis* for the rice brown plant hopper *Nilaparvata lugens*. *Biological Control* 48, 73–86.

Lu, X. and Hyun, T.K. (2021) The role of epigenetic modifications in plant responses to stress. *Botanica Serbica* 45(1), 3–12.

March-Díaz, R., García-Domínguez, M., Lozano-Juste, J., León, J., Florencio, F.J. *et al*. (2008) Histone H2A.Z and homologues of components of the SWR1 complex are required to control immunity in *Arabidopsis*. *The Plant Journal* 53(3), 475–487.

Návarová, H., Bernsdorff, F., Döring, A.-C. and Zeier, J. (2012) Pipecolic acid, an endogenous mediator of defense amplification and priming, is a critical regulator of inducible plant immunity. *The Plant Cell* 24(12), 5123–5141.

Norton, A.P., English-Loeb, G., Gadoury, D. and Seem, R.C. (2000) MYCOPHAGOUS mites and foliar pathogens: Leaf domatia mediate tritrophic interactions in grapes. *Ecology* 81(2), 490–499.

Ojeda-Martinez, D., Diaz, I. and Santamaria, M.E. (2022) Transcriptomic landscape of herbivore oviposition in *Arabidopsis*: A systematic review. *Frontiers in Plant Science* 12, 772492. DOI: 10.3389/fpls.2021.772492.

Olson, D.L. and Nechols, J.R. (1995) Effects of squash leaf trichome exudates and honey on adult feeding, survival, and fecundity of the squash bug (Heteroptera: Coreidae) egg parasitoid *Gryon pennsylvanicum* (Hymenoptera: Scelionidae). *Environmental Entomology* 24(2), 454–458.

Oostendorp, M., Kunz, W., Dietrich, B. and Staub, T. (2001) Induced disease resistance in plants by chemicals. *European Journal of Plant Pathology* 107(1), 19–28.

Pastor, V., Luna, E., Mauch-Mani, B., Ton, J. and Flors, V. (2013) Primed plants do not forget. *Environmental and Experimental Botany* 94, 46–56.

Peiffer, M., Tooker, J.F. and Felton, G.W. (2009) Plants on early alert: Glandular trichomes as sensors for insect herbivores. *The New Phytologist* 184(3), 644–656.

Pérez-Hedo, M., Bouagga, S., Jaques, J.A., Flors, V. and Urbaneja, A. (2015) Tomato plant responses to feeding behavior of three zoophytophagous predators (Hemiptera: Miridae). *Biological Control* 86, 46–51.

Poelman, E.H., Oduor, A.M.O., Broekgaarden, C., Hordijk, C.A., Jansen, J.J. *et al.* (2009) Field parasitism rates of caterpillars on *Brassica oleracea* plants are reliably predicted by differential attraction of *Cotesia* parasitoids. *Functional Ecology* 23(5), 951–962.

Pozzebon, A., Loeb, G.M. and Duso, C. (2009) Grape powdery mildew as a food source for generalist predatory mites occurring in vineyards: Effects on life-history traits. *Annals of Applied Biology* 155(1), 81–89.

Qi, Q., Hu, B., Jiang, W., Wang, Y., Yan, J. *et al.* (2023) Advances in plant epigenome editing research and its application in plants. *International Journal of Molecular Sciences* 24(4), 3442.

Rambani, A., Rice, J.H., Liu, J., Lane, T., Ranjan, P. *et al.* (2015) The methylome of soybean roots during the compatible interaction with the soybean cyst nematode. *Plant Physiology* 168(4), 1364–1377. DOI: 10.1104/pp.15.00826.

Rambani, A., Pantalone, V., Yang, S., Rice, J.H., Song, Q. *et al.* (2020) Identification of introduced and stably inherited DNA methylation variants in soyabean associated with soyabean cyst nematode parasitism. *New Phytologist* 227(1), 168–184.

Ramos-Cruz, D., Troyee, A.N. and Becker, C. (2021) Epigenetics in plant organismic interactions. *Current Opinion in Plant Biology* 61, 102060.

Rasmann, S. and Turlings, T.C.J. (2008) First insights into specificity of belowground tritrophic interactions. *Oikos* 117(3), 362–369.

Rasmann, S., Erwin, A.C., Halitschke, R. and Agrawal, A.A. (2011) Direct and indirect root defences of milkweed (*Asclepias syriaca*): Trophic cascades, trade-offs and novel methods for studying subterranean herbivory. *Journal of Ecology* 99(1), 16–25.

Reymond, P. (2013) Perception, signaling and molecular basis of oviposition-mediated plant responses. *Planta* 238(2), 247–258. DOI: 10.1007/s00425-013-1908-y.

Rohfritsch, O. (2005) Gall making. In: Goodman, R.M. (ed.) *Encyclopedia of Plant and Crop Science*. Marcel Dekker, New York, pp. 1021–1022.

Ross, A.F. (1961) Systemic acquired resistance induced by localized virus infections in plants. *Virology* 14(3), 340–358.

Ryals, J.A., Neuenschwander, U.H., Willits, M.G., Molina, A., Steiner, H.Y. *et al.* (1996) Systemic acquired resistance. *The Plant Cell* 8(10), 1809–1819.

Ryan, C.A. (1990) Protease inhibitors in plants: Genes for improving defenses against insects and pathogens. *Annual Review of Phytopathology* 28(1), 425–449.

Sauge, M.H., Kervella, J. and Pascal, T. (1998) Settling behaviour and reproductive potential of the green peach aphid *Myzus persicae* on peach varieties and a related wild *Prunus*. *Entomologia Experimentalis et Applicata* 89(3), 233–242.

Schilmiller, A.L. and Howe, G.A. (2005) Systemic signaling in the wound response. *Current Opinion in Plant Biology* 8(4), 369–377.

Schmelz, E.A., Carroll, M.J., LeClere, S., Phipps, S.M., Meredith, J. *et al.* (2006) Fragments of ATP synthase mediate plant perception of insect attack. *Proceedings of the National Academy of Sciences of the United States of America* 103(23), 8894–8899.

Sharma, H.C., Sujana, G. and Manohar Rao, D. (2009) Morphological and chemical components of resistance to pod borer, *Helicoverpa armigera* in wild relatives of pigeon pea. *Arthropod–Plant Interactions* 3(3), 151–161. DOI: 10.1007/s11829-009-9068-5.

Snoeren, T.A., Mumm, R., Poelman, E.H., Yang, Y., Pichersky, E. *et al.* (2010) The herbivore-induced plant volatile methyl salicylate negatively affects attraction of the parasitoid *Diadegma semiclausum*. *Journal of Chemical Ecology* 36, 479–489.

Stahl, E., Brillatz, T., Ferreira-Queiroz, E., Marcourt, L., Schmiesing, A. *et al.* (2020) Phosphatidylcholines from *Pieris brassicae* eggs activate an immune response in *Arabidopsis*. *Elife* 28(9), e60293. DOI: 10.7554/eLife.60293.

Stone, G.N. and Schönrogge, K. (2003) The adaptive significance of insect gall morphology. *Trends in Ecology & Evolution* 18, 512–522.

Tempfli, B., Pénzes, B., Fail, J. and Szabó, Á. (2015) The occurrence of tydeoid mites (Acari: Tydeoidea) in Hungarian vineyards. *Systematic and Applied Acarology* 20, 937–954.

Ton, J., D'Alessandro, M., Jourdie, V., Jakab, G., Karlen, D. *et al.* (2007) Priming by airborne signals boosts direct and indirect resistance in maize. *Plant Journal* 49, 16–26.

Traw, M.B. (2002) Is induction response negatively correlated with constitutive resistance in black mustard? *Evolution; International Journal of Organic Evolution* 56(11), 2196–2205.

Truitt, C.L., Wei, H.-X. and Paré, P.W. (2004) A plasma membrane protein from *Zea mays* binds with the herbivore elicitor volicitin. *The Plant Cell* 16(2), 523–532.

van Den Boom, C.E.M., Beek, T.A., Posthumus, M.A., Groot, A. and Dicke, M. (2004) Qualitative and quantitative variation among volatile profiles induced by *Tetranychus urticae* feeding on plants from various families. *Journal of Chemical Ecology* 30(1), 69–89.

van Hulten, M., Pelser, M., Loon, L.C., Pieterse, C.M.J. and Ton, J. (2006) Costs and benefits of priming for defense in *Arabidopsis*. *Proceedings of the National Academy of Sciences of the United States of America* 103(14), 5602–5607.

Venzon, M., Janssen, A. and Sabelis, M.W. (1999) Attraction of a generalist predator towards herbivore-infested plants. *Entomologia Experimentalis et Applicata* 93(3), 303–312.

Walling, L.L. (2000) The myriad plant responses to herbivores. *Journal of Plant Growth Regulation* 19(2), 195–216.

Walters, D.R., Ratsep, J. and Havis, N.D. (2013) Controlling crop diseases using induced resistance: Challenges for the future. *Journal of Experimental Botany* 64(5), 1263–1280.

Yi, S.Y., Min, S.R. and Kwon, S.Y. (2015) NPR1 is instrumental in priming for the enhanced flg22-induced MPK3 and MPK6 activation. *The Plant Pathology Journal* 31(2), 192–194.

Zhou, J.M. and Zhang, Y. (2020) Plant immunity: Danger perception and signaling. *Cell* 181(5), 978–989. DOI: 10.1016/j.cell.2020.04.028.

Zlatanova, J. and Thakar, A. (2008) H2A.Z: View from the top. *Structure (London)* 16(2), 166–179.

30 Epigenetics for Combating Bacterial Diseases in Plants: Updated Methods and Current Achievements

Ramkumar Govindarajan[1], K. Mohammed Rafiq Khan[2], P. Servin Wesley[3], Sarmad Moin[4] and Chithra Devi Balasundaram Saraswathy[5]*

[1]BIRAC E-YUVA Centre, PSGR Krishnammal College for Women, Coimbatore, Tamil Nadu, India; [2]Department of Biotechnology, Indian Academy Degree College - Autonomous, Kalyan Nagar, Bangalore, Karnataka, India; [3]Department of Biotechnology, SAFI Institute of Advanced Study, Malappuram, Kerala, India; [4]School of Applied Sciences, Suresh Gyan Vihar University, Gyan Vihar Marg, Jagatpura, Jaipur, Rajasthan, India; [5]Department of Botany, PSGR Krishnammal College for Women, Coimbatore-, Tamil Nadu, India

Abstract

Plant diseases produced by pathogenic bacteria are a persistent challenge to world agriculture, causing major economic losses and raising worries about food security. Traditional approaches to bacterial illness control frequently use genetic alteration and pharmacological therapies. Epigenetics has emerged as a unique and promising route for increasing plant resistance to various diseases in recent years. This chapter highlights the latest methods and current achievements in harnessing epigenetic mechanisms to mitigate bacterial diseases in plants. The review explores how numerous epigenetic alterations, such as DNA methylation, histone modifications, small RNA regulation, and chromatin remodeling, affect plant immunity. Epigenetic editing technologies, including CRISPR/Cas-based techniques for targeted changes, are being investigated for their ability to provide long-term resistance to infectious bacteria. Case studies of successful applications of epigenetic strategies for disease resistance are presented, showcasing the remarkable achievements in controlling bacterial infections. These advancements not only bolster our understanding of plant–pathogen interactions but also open up new avenues for sustainable disease management in agriculture. The review also addresses the current challenges and limitations in the field, including ethical and regulatory considerations. As we delve deeper into the realm of epigenetics, it is imperative to evaluate its potential impact on plant health, crop yields, and ecosystem sustainability. This chapter provides a glimpse into the growing body of research and innovation at the intersection of epigenetics and plant pathology. The ongoing pursuit of epigenetic solutions for bacterial diseases in plants promises to transform the future of agriculture and foster global food security.

Keywords: Epigenetics, Bacterial diseases, Plants, Crop protection

*Corresponding author: bchitradevi@psgrkcw.ac.in

 Epigenetics for Climate-Smart and Sustainable Agriculture (ed. Jen-Tsung Chen)
DOI: 10.1079/9781800626102.0030

30.1 Introduction

Epigenetics refers to the study of heritable changes in gene expression or cellular phenotype that do not involve alterations to the underlying DNA sequence (Deans and Maggert, 2015). This field has garnered significant attention due to its profound implications across various biological processes, including development, ageing, and disease (Pagiatakis *et al.*, 2021). Significantly, epigenetic modifications play a crucial role in regulating gene expression by influencing how genes are activated or silenced without changing the DNA sequence itself. These modifications include DNA methylation, histone modifications, and non-coding RNAs, each of which can affect chromatin structure and accessibility, thereby modulating gene expression patterns in response to environmental cues and developmental signals (Kumar *et al.*, 2020).

Bacterial diseases pose a significant threat to global agriculture and food security, affecting crop yield, quality, and profitability (Lindsey *et al.*, 2020). These diseases are caused by a diverse array of bacterial pathogens that can infect various plant tissues, leading to symptoms such as wilting, leaf spots, necrosis, and ultimately crop loss (Saxena *et al.*, 2020). The economic impact of bacterial diseases on agriculture is substantial, with billions of dollars lost annually due to decreased productivity and increased management costs (Vurro *et al.*, 2010). Furthermore, the widespread use of monocultures and globalization has facilitated the rapid spread of bacterial pathogens, exacerbating the challenge of disease management.

Harnessing epigenetic mechanisms presents a promising approach for enhancing plant immunity and mitigating the impact of bacterial diseases (Mushtaq *et al.*, 2019; Saradadevi *et al.*, 2021). Epigenetic regulation offers a dynamic and adaptable means of modulating gene expression in response to pathogen invasion, thereby priming plants for more effective defense responses. Recent research has unveiled the potential of epigenetic modifications, such as DNA methylation and histone acetylation, in regulating key genes involved in plant defense pathways (Zhu *et al.*, 2016; Kang *et al.*, 2022). Additionally, studying epigenetic changes during plant–pathogen interactions has provided novel insights into the molecular mechanisms underlying disease resistance, offering new targets for crop improvement and disease management strategies (Zhi and Chang, 2021).

The significance of epigenetics in plant biology has become increasingly apparent in recent years, with advancements in technology enabling researchers to unravel the complexity of epigenetic regulatory networks (Grant-Downton and Dickinson, 2005). Techniques such as bisulfite sequencing, chromatin immunoprecipitation followed by sequencing (ChIP-seq), and small RNA sequencing have provided unprecedented insights into the epigenomic landscape of plants, facilitating the identification of epigenetic marks associated with specific biological processes, including immune responses (Furci *et al.*, 2023). Furthermore, the development of genome editing tools, such as CRISPR/Cas9, has revolutionized the field by allowing precise manipulation of epigenetic marks at targeted genomic loci, offering new avenues for functional validation and crop improvement (Arora and Narula, 2017).

Against this backdrop, the exploration of epigenetic mechanisms in combating bacterial diseases in plants represents a critical frontier in agricultural research. Bacterial pathogens pose significant threats to crop productivity and food security worldwide, causing devastating losses in yield and quality across a wide range of agricultural systems (Strange and Scott, 2005). Traditional methods of disease management, including chemical pesticides and genetic resistance, have been met with challenges such as pesticide resistance and limited genetic diversity in crop germplasm (Van Esse *et al.*, 2020). In this context, harnessing the power of epigenetics to bolster plant immunity and enhance disease resistance offers a promising alternative approach.

30.2 Epigenetic Mechanisms in Plant Immunity

Epigenetic mechanisms orchestrate plant immune responses against microbial pathogens, including bacteria, viruses, fungi, and oomycetes (Wang *et al.*, 2022). Understanding the interplay between epigenetic modifications and defense signaling pathways is crucial for elucidating the molecular basis of disease resistance in plants.

30.2.1 DNA methylation in plant immunity

DNA methylation is a fundamental epigenetic mechanism that plays a crucial role in regulating gene expression and chromatin structure in plants. It involves the covalent addition of a methyl group to the 5' position of cytosine residues in DNA, primarily occurring in the context of CpG dinucleotides (Bommarito and Fry, 2019). DNA methylation can occur in three main sequence contexts: CpG, CpHpG (where H is any nucleotide except guanine), and CpHpH (Klupczyńska and Ratajczak, 2021).

The process of DNA methylation is catalyzed by a family of enzymes called DNA methyltransferases (DNMTs). In plants, DNMTs include several members, such as MET1 (METHYLTRANSFERASE 1), CMT3 (CHROMOMETHYLASE 3), and DRM (DOMAINS REARRANGED METHYLTRANSFERASE) (Wada, 2005). Each DNMT has specific functions in maintaining DNA methylation patterns across the genome, including maintenance methylation during DNA replication and *de novo* methylation at specific genomic loci (Jeltsch and Jurkowska, 2014). In plants, DNA methylation occurs at cytosine residues in three different sequence contexts: CG, CHG, and CHH (where H represents A, T, or C) (Vanyushin and Ashapkin, 2011). Each context is associated with different regulatory mechanisms and effects on gene expression.

1. **CG methylation:** This is the most common form and is often associated with stable gene silencing. It is maintained through cell divisions by the maintenance of methyltransferase MET1 (Vaillant and Paszkowski, 2007).
2. **CHG methylation:** CHG methylation is maintained by the chromomethylase CMT3 and is also linked to gene silencing (Wendte *et al.*, 2019).
3. **CHH methylation:** This type of methylation is often found in transposable elements and is maintained by the RNA-directed DNA methylation (RdDM) pathway involving the DRM2 methyltransferase (Lee *et al.*, 2023).

DNA methylation has profound effects on gene expression and chromatin structure by influencing the accessibility of DNA to transcription factors and other regulatory proteins (Loaeza-Loaeza *et al.*, 2020). Methylated DNA can recruit proteins that modify histones, leading to changes in chromatin conformation and gene silencing (Ehrlich and Lacey, 2013). Conversely, DNA demethylation processes mediated by DNA glycosylases and DNA repair mechanisms can activate gene expression by removing methyl groups from cytosine residues (Gehring *et al.*, 2009) (Fig. 30.1).

In the context of plant immunity, changes in DNA methylation patterns have been associated with the regulation of defense-related genes and the modulation of immune responses against microbial pathogens (Deleris *et al.*, 2016). Upon pathogen recognition, plants undergo dynamic changes in DNA methylation patterns, resulting in the activation or suppression of immune-related genes (Hannan Parker *et al.*, 2022). For example, hypomethylation of defense genes can lead to their increased expression, priming plants for more robust defense responses against pathogens (Espinas *et al.*, 2016). To study DNA methylation in plant immunity, there are a variety of techniques and approaches, which include bisulfite sequencing and chromatin immunoprecipitation.

Bisulfite sequencing

Bisulfite sequencing is a widely used method for analyzing DNA methylation patterns at single-base resolution (Peng *et al.*, 2016). Bisulfite treatment converts unmethylated cytosines to uracils, while methylated cytosines remain unchanged (Clark *et al.*, 2006). By sequencing bisulfite-treated DNA and comparing it with the reference genome, researchers can map DNA methylation patterns across the genome and identify differentially methylated regions associated with immune responses (Agius *et al.*, 2023).

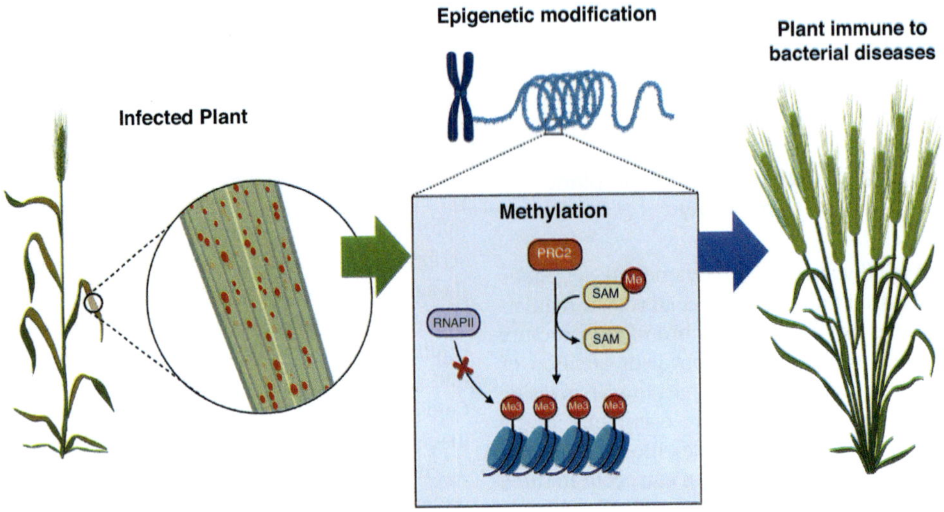

Fig. 30.1. DNA methylation to enhance plant immunity.

Chromatin immunoprecipitation

Chromatin immunoprecipitation followed by sequencing (ChIP-seq) is another powerful technique used to investigate the association of histone modifications with DNA methylation patterns in plant immunity (Chen *et al.*, 2018). By immunoprecipitating chromatin fragments bound to specific histone modifications, such as H3K4me3 (trimethylation of histone H3 at lysine 4) or H3K9me2 (dimethylation of histone H3 at lysine 9), it is possible to identify genomic regions where DNA methylation and histone modifications intersect to regulate immune gene expression (Du *et al.*, 2013). Furthermore, genetic and molecular approaches, such as mutant analysis and transgenic plant studies, are essential for elucidating the functions of DNA methyltransferases and other epigenetic regulators in plant immunity (Alvarez *et al.*, 2010). Loss-of-function mutants or RNA interference (RNAi)-mediated knockdown lines targeting specific DNMTs can be used to assess the impact of altered DNA methylation patterns on plant defense responses (Miglani *et al.*, 2020). Conversely, overexpression of DNMTs or other epigenetic modifiers can be employed to investigate their roles in immune gene regulation and defense priming (Ashapkin *et al.*, 2020).

DNA methylation is a dynamic epigenetic mechanism that modulates gene expression and chromatin structure in response to pathogen infection in plants (Arora *et al.*, 2022). Changes in DNA methylation patterns influence the expression of defense-related genes and shape immune responses against microbial pathogens (Arora *et al.*, 2022). By employing a combination of techniques and approaches, the regulatory roles of DNA methylation in plant immunity can be deciphered, leading to the development of strategies for enhancing crop resistance to diseases.

30.2.2 Histone modifications and plant defense responses

Histone modifications play a critical role in regulating chromatin accessibility and gene expression in plants, including in response to pathogen infection (Kang *et al.*, 2022). These modifications, which include acetylation, methylation, phosphorylation, and ubiquitination, alter the chemical properties of histone proteins and thereby affect the accessibility of DNA to transcriptional machinery (Berr *et al.*, 2011).

Acetylation

Acetylation of histone proteins typically occurs at lysine residues and is associated with transcriptional activation. This modification neutralizes the positive charge of lysine, thereby loosening the interaction between histones and DNA, making the chromatin more accessible to transcription factors and RNA polymerase (Liu *et al.*, 2016). Histone acetyltransferases (HATs) add acetyl groups to histones, while histone deacetylases (HDACs) remove them. In plant defense responses, histone acetylation often correlates with the activation of immune-related genes, promoting robust defense responses against pathogens (Kim *et al.*, 2020).

Methylation

Methylation of histone proteins can occur at lysine or arginine residues and can have either activating or repressive effects on gene expression, depending on the specific site and degree of methylation (Prakash *et al.*, 2014). For example, methylation of histone H3 at lysine 4 (H3K4me3) is associated with transcriptional activation, while methylation of histone H3 at lysine 9 (H3K9me2/3) or lysine 27 (H3K27me3) is linked to gene repression. Histone methyltransferases (HMTs) catalyze the addition of methyl groups, while histone demethylases (HDMs) remove them (Dambacher *et al.*, 2010). Dynamic changes in histone methylation patterns are involved in the regulation of immune-related genes and defense responses in plants (Kang *et al.*, 2022).

Phosphorylation

Phosphorylation of histone proteins occurs primarily at serine or threonine residues and can influence chromatin structure and gene expression (Bhan *et al.*, 2017). Histone phosphorylation is often associated with transcriptional activation and is mediated by various protein kinases in response to signaling pathways activated during stress or pathogen infection (Dong and Hamon, 2020). Although less studied compared with acetylation and methylation, histone phosphorylation has been implicated in plant defense responses, potentially by regulating the recruitment of chromatin-modifying complexes or transcription factors to target genes (Ding and Wang, 2015).

Ubiquitination

Ubiquitination involves the covalent attachment of ubiquitin molecules to histone proteins and can affect chromatin structure and gene expression (Zhang and Zeng, 2020). Histone ubiquitination has diverse effects on transcriptional regulation, depending on the specific site and type of ubiquitin modification (Zhang, 2003). It can serve as a signal for gene activation or repression, as well as for DNA repair or degradation (Hammond-Martel *et al.*, 2012). While the role of histone ubiquitination in plant immunity remains less understood compared with other histone modifications, emerging evidence suggests its involvement in regulating immune-related genes and defense signaling pathways.

30.2.3 Techniques and methodologies

Dynamic changes in histone modifications contribute to the activation or repression of immune-related genes, shaping plant defense responses against microbial pathogens. By employing a combination of techniques and approaches, researchers can decipher the regulatory roles of histone modifications in plant immunity and develop strategies for enhancing crop resistance to diseases (Jenuwein and Allis, 2001; Li *et al.*, 2007).

Chromatin immunoprecipitation (ChIP)

ChIP is a powerful technique used to investigate the association of histone modifications with specific genomic loci. By crosslinking histones to DNA in intact chromatin, immunoprecipitating chromatin fragments bound to specific histone modifications using antibodies, and analyzing the precipitated DNA by PCR or sequencing, it is possible to map the genome-wide distribution of histone modifications and identify target genes regulated by these modifications (Johnson *et al.*, 2007).

ChIP-seq (chromatin immunoprecipitation followed by sequencing)

ChIP-seq combines chromatin immunoprecipitation with high-throughput sequencing technology to provide genome-wide maps of histone modifications at single-nucleotide resolution. By sequencing the DNA fragments immunoprecipitated with histone modification-specific antibodies, genomic regions enriched for specific histone modifications can be identified and correlated with gene expression profiles and biological processes, including plant defense responses (Yadav *et al.*, 2018).

Mutant analysis and transgenic plant studies

Genetic and molecular approaches, such as mutant analysis and transgenic plant studies, are essential for elucidating the functions of histone-modifying enzymes and their impact on plant defense (Ramirez-Prado *et al.*, 2018b). Loss-of-function mutants or RNA interference (RNAi)-mediated knockdown lines targeting specific histone modifiers can be used to assess the effects of altered histone modifications on immune gene regulation and defense responses (Dhawan *et al.*, 2009). Conversely, overexpression of histone modifiers or histone-modifying enzymes can be employed to investigate their roles in immune gene activation or repression (Kouzarides, 2007).

30.2.4 Small RNAs in plant immune regulation

Small RNAs (sRNAs), including microRNAs (miRNAs) and small interfering RNAs (siRNAs), are key regulators of gene expression and play essential roles in post-transcriptional gene silencing and plant immune gene regulation in plants (Waheed *et al.*, 2021). These small RNA molecules, typically 20–24 nucleotides (nt) in length, function primarily through post-transcriptional gene silencing mechanisms to fine-tune gene expression and regulate various biological processes, including plant defense responses against microbial pathogens (Weiberg *et al.*, 2014). They modulate plant defense responses by targeting mRNAs encoding immune receptors, transcription factors, and pathogen effectors, thereby fine-tuning immune gene expression and defense signaling pathways. By employing a combination of experimental approaches and techniques, researchers can decipher the regulatory roles of small RNAs in plant immunity and develop strategies for enhancing crop resistance to diseases.

MicroRNAs (miRNAs)

miRNAs are endogenous small RNAs that regulate gene expression by guiding the RNA-induced silencing complex (RISC) to target mRNAs for cleavage or translational repression (Jungers and Djuranovic, 2022). In plants, miRNAs are processed from longer precursor transcripts by the DICER-LIKE 1 (DCL1) enzyme and loaded onto ARGONAUTE (AGO) proteins to form the miRNA-induced silencing complex (miRISC) (Li and Yu, 2021). miRNAs typically bind to complementary sequences in the 3' untranslated regions (UTRs) of target mRNAs, leading to their degradation or translational inhibition (Dalmay, 2013). Numerous miRNAs have been implicated in plant immune regulation by targeting mRNAs encoding immune receptors, transcription factors, and defense-related proteins.

Small interfering RNAs (siRNAs)

siRNAs are another class of endogenous small RNAs that mediate RNA interference (RNAi) pathways to regulate gene expression (Pattanayak *et al.*, 2005). In plants, siRNAs are generated from double-stranded RNA precursors, such as viral RNAs or transposable elements, by DICER-LIKE (DCL) proteins and are incorporated into RNA-induced silencing complexes (RISCs) to guide sequence-specific gene silencing (Sanan-Mishra *et al.*, 2021). siRNAs can target mRNAs for degradation or inhibit their translation, leading to post-transcriptional gene silencing. Plant defense responses often involve the production of siRNAs targeting pathogen-derived RNAs, such as viral genomes or transposon transcripts, to restrict pathogen proliferation and spread (Lopez-Gomollon and Baulcombe, 2022).

Small RNA-mediated pathways modulate plant defense responses by regulating the expression of genes involved in immune signaling, pathogen recognition, and defense gene activation (Seo *et al.*, 2013). By targeting mRNAs encoding immune receptors, such as nucleotide-binding leucine-rich repeat (NB-LRR) proteins, or transcription factors, small RNAs can fine-tune immune gene expression and modulate defense signaling pathways (Park and Shin, 2015). Additionally, small RNAs can target pathogen effectors or virulence factors, leading to their degradation and attenuating pathogen virulence (Diallo and Provost, 2020).

To understand small RNA-mediated regulation of plant immunity, a variety of experimental approaches and techniques are utilized.

Small RNA sequencing (sRNA-seq)

sRNA-seq is a high-throughput sequencing method used to profile small RNA populations in plants. By sequencing small RNA libraries prepared from plant tissues or cells under different experimental conditions, researchers can identify and quantify miRNAs, siRNAs, and other small RNA species involved in immune regulation (Katiyar-Agarwal and Jin, 2007). sRNA-seq data analysis enables the identification of small RNA–target interactions and the characterization of small RNA-mediated regulatory networks in plant immunity (Guleria and Kumar, 2020).

Gene expression analysis

Quantitative real-time PCR (qRT-PCR) and RNA sequencing (RNA-seq) are commonly used techniques to analyze the expression levels of small RNA target genes in response to pathogen infection or small RNA manipulation (Sun *et al.*, 2016). By measuring mRNA abundance and translational efficiency, researchers can assess the impact of small RNA-mediated gene regulation on plant defense responses and immune signaling pathways (Peláez and Sanchez, 2013).

Small RNA manipulation

Genetic and molecular approaches, such as small RNA overexpression, artificial miRNA/siRNA expression, or small RNA mutant analysis, are used to manipulate small RNA levels and activity in plants (Kamthan *et al.*, 2015). Transgenic plants expressing small RNA precursors or artificial miRNAs/siRNAs targeting specific immune-related genes can be generated to investigate the functional roles of small RNAs in plant defense responses. Conversely, mutants defective in small RNA biogenesis or small RNA pathway components can be analyzed to assess the contribution of small RNA-mediated regulation to plant immunity (Seo *et al.*, 2013).

30.2.5 Integration of epigenetic mechanisms with defense signaling pathways

The integration of epigenetic mechanisms with defense signaling pathways is a complex and dynamic process that plays a crucial role in coordinating plant defense responses against pathogens (Espinas *et al.*, 2016). Plant immune signaling pathways, including the salicylic acid (SA), jasmonic acid (JA), and ethylene (ET) pathways, interact with epigenetic regulators to modulate immune gene expression, chromatin structure, and systemic acquired resistance (SAR) in plants (Ramirez-Prado *et al.*, 2018a).

Salicylic acid (SA) pathway

The SA pathway is a central signaling pathway involved in plant defense against biotrophic pathogens, which depend on living host cells for their nutrition (Kou *et al.*, 2021). SA signaling activates defense responses, such as the production of antimicrobial compounds, pathogenesis-related (PR) proteins, and the induction of systemic acquired resistance (SAR) (Kamle *et al.*, 2020). Epigenetic mechanisms, including DNA methylation and histone modifications, intersect with the SA pathway to regulate the expression of SA-responsive genes and modulate plant immune responses (Roberts and López Sánchez, 2019). Changes in DNA methylation patterns and histone modifications have been associated with the activation or repression of SA-inducible genes, contributing to the fine-tuning of SA-mediated defense responses (Caarls *et al.*, 2015).

Jasmonic acid (JA) pathway

The JA pathway is another essential signaling pathway involved in plant defense against necrotrophic pathogens, which kill host cells to obtain nutrients (Pandey *et al.*, 2016). JA signaling activates defense responses, such as the production of defense-related proteins, phytoalexins, and the induction of wound responses (Mostafa *et al.*, 2022). Crosstalk between JA signaling and epigenetic mechanisms has been reported to regulate the expression of JA-responsive genes and modulate plant defense against necrotrophic pathogens. Histone modifications, such as histone acetylation and methylation, have been implicated in the regulation of JA-inducible genes involved in defense against necrotrophs (Li *et al.*, 2019).

Ethylene (ET) pathway

The ET pathway is a hormone signaling pathway involved in plant defense responses to various biotic and abiotic stresses, including pathogen infection. ET signaling activates defense responses, such as the production of defense-related proteins, ethylene-responsive transcription factors, and the induction of senescence and programmed cell death (Checker *et al.*, 2018). Epigenetic mechanisms, including DNA methylation and histone modifications, have been shown to interact with the ET pathway to modulate immune gene expression and defense responses (Alvarez *et al.*, 2010). Changes in DNA methylation patterns and histone modifications can influence the expression of ET-responsive genes and contribute to the regulation of ET-mediated defense signaling (Xie and Duan, 2023).

Systemic acquired resistance (SAR)

SAR is a systemic immune response in plants induced by localized pathogen infection, leading to enhanced resistance against subsequent pathogen attacks in distal parts of the plant. SAR is mediated by long-distance signaling molecules, such as SA, and involves the activation of defense genes and the establishment of a primed state for enhanced defense responses (Shah and Zeier, 2013). Epigenetic mechanisms play a critical role in regulating SAR by mediating changes in chromatin structure and gene expression at distal loci in response to local pathogen infection (Xie and Duan, 2023). DNA methylation and histone modifications have been implicated in the establishment and maintenance of SAR-associated epigenetic memory, enabling plants to respond more rapidly and effectively to subsequent pathogen challenges.

30.2.6 Experimental approaches and techniques

The integration of epigenetic mechanisms with defense signaling pathways plays a critical role in coordinating plant immune responses against pathogens. Crosstalk between epigenetic regulators and immune signaling components modulates immune gene expression, chromatin structure, and systemic acquired resistance (SAR) in plants (Xie and Duan, 2023). By employing a combination of experimental approaches and techniques, researchers can elucidate the molecular mechanisms underlying the integration of epigenetic mechanisms with defense signaling pathways and develop strategies for enhancing crop resistance to diseases.

Genetic and molecular analysis

Mutant analysis, transgenic plant studies, and RNA interference (RNAi)-mediated knockdown approaches are used to dissect the roles of epigenetic regulators and defense signaling components in plant immunity (Muhammad *et al.*, 2019). Loss-of-function mutants or transgenic plants deficient in specific epigenetic factors or signaling proteins can be generated to assess their effects on immune gene expression, defense responses, and SAR induction (Spoel and Dong, 2012).

Epigenomic profiling

High-throughput sequencing techniques, such as chromatin immunoprecipitation followed by sequencing (ChIP-seq) and bisulfite sequencing (BS-seq), are used to profile epigenetic modifications, including histone modifications and DNA methylation, at genome-wide scales (Mehrmohamadi *et al.*, 2021). Epigenomic profiling allows researchers to identify genomic regions associated with immune gene regulation and characterize changes in chromatin structure and DNA methylation patterns in response to pathogen infection or signaling pathway activation (Hannan Parker *et al.*, 2022).

Functional genomics and bioinformatics

Functional genomics approaches, such as genome-wide association studies (GWAS), forward and reverse genetics screens, and bioinformatics analysis of omics data, are used to identify candidate genes, regulatory elements, and signaling networks involved in the integration of epigenetic mechanisms with defense signaling pathways (Uffelmann *et al.*, 2021). Computational modeling and network analysis provide insights into the regulatory relationships between epigenetic regulators, signaling components, and immune-responsive genes in plants (Aerts *et al.*, 2022).

30.2.7 Epigenetic regulation of immune receptor genes

Immune receptor genes, particularly nucleotide-binding leucine-rich repeat (NB-LRR) proteins, are pivotal components of plant defense mechanisms against pathogens. These receptors recognize specific pathogen-derived molecules, such as effector proteins, or pathogen-associated molecular patterns (PAMPs), triggering downstream defense responses (Noman *et al.*, 2019). The regulation of NB-LRR gene expression and activity is subject to epigenetic control, wherein modifications to chromatin structure and DNA methylation at NB-LRR gene loci influence their transcriptional regulation and subsequent impact on plant susceptibility to pathogen infection (Garner *et al.*, 2016).

Chromatin structure and NB-LRR gene regulation

Epigenetic modifications, such as histone modifications, play a crucial role in modulating chromatin accessibility and gene expression at NB-LRR gene loci. Histone modifications, including acetylation, methylation, and phosphorylation, can either promote or repress NB-LRR gene expression by altering the recruitment of transcriptional regulators and RNA polymerase complexes to NB-LRR gene promoters (Ma *et al.*, 2011). For example, histone acetylation at NB-LRR gene loci is associated with transcriptional activation, while histone methylation can lead to gene repression. Dynamic changes in histone modifications at NB-LRR gene loci in response to pathogen infection or signaling pathway activation modulate the expression and activity of NB-LRR proteins, influencing plant defense responses (Kang *et al.*, 2022).

DNA methylation and NB-LRR gene regulation

DNA methylation is another epigenetic mechanism that regulates NB-LRR gene expression by modulating chromatin structure and transcriptional activity. Changes in DNA methylation patterns, particularly at CpG dinucleotides within NB-LRR gene promoters or regulatory regions, can affect the binding of transcription factors and chromatin-modifying complexes, thereby influencing NB-LRR gene expression (Gambino and Pantaleo, 2017). Hypomethylation of NB-LRR gene loci is often associated with enhanced expression and activity of NB-LRR proteins, leading to heightened plant resistance against pathogens. Conversely, hypermethylation of NB-LRR gene promoters can result in transcriptional silencing and increased susceptibility to pathogen infection (Halter *et al.*, 2021).

Transcriptional regulation of NB-LRR genes by small RNAs

Small RNAs, such as microRNAs (miRNAs) and small interfering RNAs (siRNAs), can also regulate NB-LRR gene expression through post-transcriptional gene silencing mechanisms. miRNAs or siRNAs targeting NB-LRR transcripts can lead to mRNA degradation or translational repression, thereby modulating the abundance and activity of NB-LRR proteins (Zhai *et al.*, 2011). Small RNA-mediated regulation of NB-LRR genes contributes to the fine-tuning of plant defense responses and the establishment of immune priming against pathogens.

Epigenetic regulation of NB-LRR genes plays a critical role in modulating plant defense responses against pathogens. Changes in chromatin structure, histone modifications, DNA methylation, and small RNA-mediated gene silencing at NB-LRR gene loci influence their expression and activity, thereby shaping plant susceptibility to pathogen infection. By employing a combination of experimental approaches and techniques, researchers can elucidate the molecular mechanisms underlying the epigenetic regulation of NB-LRR genes and develop strategies for enhancing crop resistance to diseases.

30.2.8 Epigenetic memory and defense priming

Epigenetic memory refers to the heritable changes in gene expression without altering the underlying DNA sequence, allowing plants to 'remember' previous pathogen encounters and respond more effectively to subsequent attacks. This phenomenon is particularly crucial in defense priming, where primary exposure to a pathogen or stressor primes the plant for enhanced defense responses upon subsequent challenges (Lephatsi *et al.*, 2021). The mechanisms underlying epigenetic memory and defense priming involve histone modifications, DNA methylation, and the regulation of non-coding RNAs, which collectively contribute to a molecular basis for immune memory and adaptation (Roberts and López Sánchez, 2019).

Histone modifications in epigenetic memory

Histone modifications, such as acetylation, methylation, phosphorylation, and ubiquitination, play pivotal roles in establishing and maintaining epigenetic memory. These modifications alter chromatin structure, making it either more accessible or more condensed, thereby influencing gene expression. In defense priming, specific histone marks, like H3K4me3 (associated with active transcription) or H3K27me3 (associated with gene repression), are established at defense-related gene loci. For instance, after an initial pathogen attack, certain defense genes might acquire active histone marks that persist over time, enabling a faster and stronger transcriptional response upon re-exposure to the pathogen (Bruce *et al.*, 2007; Conrath *et al.*, 2015).

DNA methylation in defense priming

DNA methylation, typically occurring at cytosine residues within CpG dinucleotides, is another crucial component of epigenetic memory. Methylation patterns can be stably maintained through cell divisions and even across generations, providing a long-term memory of stress responses (Agius *et al.*, 2023). In the context of plant immunity, changes in DNA methylation at defense gene promoters can either suppress or enhance gene expression. For example, the demethylation of promoters of key defense genes can lead to their sustained activation, thus priming the plant for improved defense responses against subsequent pathogen attacks (Mauch-Mani *et al.*, 2017).

Non-coding RNAs and epigenetic memory

Small RNAs, including miRNAs and siRNAs, also contribute to epigenetic memory and defense priming. These non-coding RNAs can guide DNA methylation and histone modification complexes to specific genomic loci, reinforcing the transcriptional states established during the initial stress response (Liu and He, 2020).

Mechanisms of defense priming

Defense priming involves several steps: recognition of the initial pathogen attack, establishment of epigenetic marks, and the maintenance of these marks through cell divisions (Conrath *et al.*, 2015). During initial infection, plants recognize pathogen-associated molecular patterns (PAMPs) or specific effectors through pattern recognition receptors (PRRs) or NB-LRR proteins. This recognition triggers a cascade of signaling events leading to the establishment of epigenetic marks at defense gene loci. These marks, such as histone modifications and DNA methylation changes, are maintained and can be reactivated upon subsequent pathogen encounters, resulting in a more robust and rapid defense response (Noman *et al.*, 2019).

30.2.9 Advances in genomic and epigenomic technologies

Recent advancements in genomic and epigenomic technologies have significantly enhanced our understanding of the complex regulatory networks underlying plant immunity. High-throughput sequencing, chromatin immunoprecipitation (ChIP), and other omics technologies have revolutionized the field, providing comprehensive datasets and novel insights into epigenetic mechanisms involved in plant defense.

High-throughput sequencing

Next-generation sequencing (NGS) technologies have enabled the genome-wide profiling of DNA methylation, histone modifications, and RNA expression (Ku *et al.*, 2011). Techniques such as whole-genome bisulfite sequencing (WGBS) allow for the precise mapping of DNA methylation patterns at single-base resolution, while RNA sequencing (RNA-seq) provides insights into the transcriptomic changes associated with defense responses (Hagmann, 2017). These high-throughput approaches have uncovered numerous differentially methylated regions and histone modification sites linked to immune regulation.

Chromatin immunoprecipitation (ChIP)

ChIP followed by sequencing (ChIP-seq) is a powerful technique for studying protein–DNA interactions and histone modifications. ChIP-seq enables the identification of genomic regions bound by specific transcription factors or modified histones, providing insights into the regulatory elements and epigenetic marks associated with immune gene expression (Barski and Zhao, 2009). This technique has been instrumental in mapping histone modifications across the genome and understanding their roles in defense priming and epigenetic memory.

Omics datasets and integrative analysis

The integration of multiple omics datasets, including genomics, transcriptomics, proteomics, and metabolomics, offers a holistic view of the molecular networks regulating plant immunity. Integrative analysis of these datasets helps identify key epigenetic regulators and target genes involved in immune responses. For instance, combining RNA-seq data with ChIP-seq and WGBS can reveal correlations between gene expression, histone modifications, and DNA methylation, providing a comprehensive understanding of epigenetic regulation (Wang *et al.*, 2009; Jamil *et al.*, 2020).

CRISPR/Cas9 and epigenome editing

The advent of CRISPR/Cas9 technology has enabled precise genome and epigenome editing in plants. CRISPR/Cas9 can be used to manipulate specific genes or regulatory elements, allowing researchers to study the functional roles of epigenetic modifications in plant immunity (Khan *et al.*, 2018). Additionally, CRISPR-based epigenome editing tools, such as dCas9 fused with histone modifiers or DNA methyltransferases, enable targeted modification of epigenetic marks, providing a powerful approach to dissecting the causal relationships between epigenetic changes and immune responses (Goell and Hilton, 2021).

Single-cell epigenomics

Advances in single-cell sequencing technologies have opened new avenues for studying epigenetic regulation at the single-cell level (Luo *et al.*, 2020). Single-cell epigenomics techniques, such as single-cell ATAC-seq (assay for transposase-accessible chromatin using sequencing) and single-cell RNA-seq, allow for the characterization of chromatin accessibility and gene expression in individual cells (Farmer *et al.*, 2021). These approaches provide insights into cell-specific epigenetic states and their contributions to plant immune responses, revealing heterogeneity in cell populations and uncovering novel regulatory mechanisms.

Integration of epigenetic mechanisms with defense signaling pathways plays a crucial role in plant immunity, with histone modifications, DNA methylation, and small RNAs contributing to the regulation of immune receptor genes and defense responses. Recent advancements in genomic and epigenomic technologies have provided valuable insights into these complex regulatory networks, enabling the identification of key epigenetic regulators and target genes involved in plant defense. These technological advancements have revolutionized our understanding of epigenetic regulation in plant immunity, offering new strategies for enhancing crop resistance to diseases.

30.2.10 Functional characterization of epigenetic regulators in plant immunity

Functional characterization of epigenetic regulators in plant immunity involves understanding how specific proteins and pathways influence the expression of immune-related genes (Panigrahi *et al.*, 2021). This is achieved through genetic and biochemical studies that focus on key components such as DNA methyltransferases, histone-modifying enzymes, and RNA interference (RNAi) factors. Loss-of-function mutants, transgenic plants, and various molecular techniques are instrumental in dissecting the roles of these epigenetic regulators in immune gene regulation and defense priming.

DNA methyltransferases

These enzymes are responsible for adding methyl groups to cytosine residues in DNA, a modification that can lead to gene silencing or activation (Finnegan and Kovac, 2000). In plants, DNA methyltransferases such as MET1 (maintenance methyltransferase 1), CMT3 (chromomethylase 3), and DRM2 (domains rearranged methyltransferase 2) are critical for maintaining DNA methylation patterns (Du, 2016). Studies using mutants deficient in these methyltransferases reveal their roles in plant immunity. For instance, MET1 mutants show altered resistance to pathogens due to changes in the methylation status of immune genes, leading to their dysregulation. These findings highlight the importance of DNA methylation in modulating plant defense responses (Tirnaz and Batley, 2019).

Histone modifying enzymes

These enzymes, including histone acetyltransferases (HATs), histone deacetylases (HDACs), histone methyltransferases (HMTs), and histone demethylases (HDMs), modify histones to regulate chromatin structure and gene expression (Lu, 2013). Genetic studies using mutants and transgenic plants have elucidated the roles of these enzymes in plant immunity. For example, mutants of the HAT gene HAG4 display reduced resistance to pathogens, linked to decreased acetylation of histones at defense gene loci. In contrast, mutants of the HDAC gene HDA6 exhibit enhanced pathogen resistance, indicating that histone deacetylation represses defense gene expression. These studies demonstrate how histone modifications can either promote or repress immune responses (Ding and Wang, 2015; Tornesaki, 2023).

RNA interference (RNAi) factors

The RNAi pathway, involving small RNAs such as miRNAs and siRNAs, plays a crucial role in post-transcriptional gene silencing (Tan *et al.*, 2020). Key components include DICER-LIKE (DCL) proteins, which process precursor RNAs into small RNAs, and ARGONAUTE (AGO) proteins, which incorporate small RNAs to target complementary mRNAs for degradation or translational repression (Chapman and Carrington, 2007). Mutants of DCL1, essential for miRNA biogenesis, exhibit altered pathogen resistance due to disrupted miRNA regulation of immune genes. Similarly, AGO1 mutants show compromised defense responses, underscoring the importance of RNAi in regulating immune gene expression (Song *et al.*, 2021).

Transgenic plants and functional analysis

Creating transgenic plants that overexpress or suppress specific epigenetic regulators provides a powerful approach to functional analysis (Pikaard and Scheid, 2014). Overexpression of genes encoding HATs can lead to increased histone acetylation and enhanced expression of defense genes, improving resistance to pathogens (Liu *et al.*, 2012). Conversely, silencing HDAC genes can elevate histone acetylation levels and activate immune responses (Licciardi and Karagiannis, 2012). These transgenic models help to elucidate the roles of epigenetic modifications in plant defense and identify potential targets for enhancing disease resistance.

30.2.11 Manipulating epigenetic pathways for enhanced disease resistance

Manipulating epigenetic pathways offers promising strategies for improving plant immunity and disease resistance. Chemical genetics approaches using small molecule inhibitors and inducers, alongside genetic engineering techniques, provide tools for targeting specific epigenetic regulators to enhance defense responses in crops.

Small molecule inhibitors and inducers

Chemical agents that modulate the activity of epigenetic enzymes can be used to enhance plant immunity. For example, HDAC inhibitors like trichostatin A (TSA) increase histone acetylation levels, thereby activating the expression of defense genes and improving pathogen resistance (Xu *et al.*, 2022). Similarly, DNA methylation inhibitors such as 5-azacytidine reduce DNA methylation, activating immune-related genes and enhancing defense responses (Wrangle *et al.*, 2013). These approaches offer a rapid and reversible means of modulating plant immunity through epigenetic modifications.

Genetic engineering of epigenetic regulators

Advanced genetic engineering techniques, including CRISPR/Cas9, allow precise manipulation of epigenetic regulators in plants (Farooq *et al.*, 2018). By editing genes encoding DNA methyltransferases, histone modifying enzymes, or RNAi components, researchers can create plants with improved disease resistance. For instance, CRISPR/Cas9-mediated knockout of an HDAC gene can increase histone acetylation and activate defense genes, leading to enhanced pathogen resistance (Bajpai *et al.*, 2024). Overexpression of HAT genes through genetic engineering can similarly boost immune responses by promoting chromatin accessibility and gene activation. These strategies offer durable and precise solutions for developing disease-resistant crops (Smale and Fisher, 2002).

Epigenome editing

Epigenome editing tools, such as CRISPR/dCas9 fused with epigenetic modifiers, enable targeted modification of epigenetic marks at specific genomic loci (Shin *et al.*, 2022). This approach allows for the precise regulation of gene expression by altering DNA methylation or histone modification patterns. For example, targeting dCas9 fused with a DNA demethylase to an immune receptor gene promoter can reduce methylation levels and enhance gene expression, improving disease resistance (Sung and Yim, 2020). Similarly, directing dCas9 fused with a HAT to a defense gene enhancer can increase histone acetylation and activate gene expression. Epigenome editing provides a sophisticated and reversible method for modulating plant immunity and developing resilient crops.

Agricultural applications and crop protection

The manipulation of epigenetic pathways has significant implications for crop protection and sustainable agriculture (Tiwari *et al.*, 2022). Enhancing the innate immune responses of plants through epigenetic modifications can reduce the reliance on chemical pesticides and improve crop yields. Developing crop varieties with enhanced resistance to a broad spectrum of pathogens can contribute to sustainable agricultural practices and food security (Krishnaswami *et al.*, 2022). Understanding the epigenetic regulation of plant immunity can also lead to the identification of biomarkers for disease resistance, facilitating the selection of resistant varieties through marker-assisted breeding.

Functional characterization of epigenetic regulators has advanced our understanding of their roles in plant immunity. Genetic and biochemical studies using loss-of-function mutants and transgenic plants have elucidated the mechanisms by which DNA methyltransferases, histone-modifying enzymes, and RNAi factors regulate immune gene expression and defense priming. Manipulating epigenetic pathways through chemical genetics and genetic engineering offers innovative strategies for enhancing

Table 30.1. Epigenetic strategies and their applications for combating bacterial diseases in plants. Author's own table.

Epigenetic strategy	Application	Plants	References
DNA methylation	Modifying DNA methylation to enhance disease resistance	Rice, tomato	Zhang *et al.*, 2018
Histone modification	Altering histone acetylation/ methylation to change gene expression	Barley, maize	Rossi *et al.*, 2007; Demetriou *et al.*, 2009
RNA interference (RNAi)	Using siRNA/miRNA to silence bacterial genes or plant susceptibility genes	Potato, tobacco	Akbar *et al.*, 2022.
Chromatin remodeling	Restructuring chromatin to make resistant genes more accessible	Wheat, soybean	Yung *et al.*, 2021
Non-coding RNAs (ncRNAs)	Employing lncRNAs or other ncRNAs to regulate gene expression related to disease resistance	Rice, wheat	Sharma *et al.*, 2022
Epigenetic priming	Priming plants with stress to induce long-lasting resistance against bacterial pathogens	*Arabidopsis*, cucumber	Conrath *et al.*, 2015
CRISPR/dCas9 Epigenome Editing	Targeting specific genomic regions to modify epigenetic marks and boost immune response	Tomato, *Arabidopsis*	Miglani *et al.*, 2020.

disease resistance in crops, with significant implications for agricultural sustainability and food security (Table 30.1).

30.3 Bacterial Diseases in Plants

Bacterial diseases in plants pose a significant threat to agriculture, impacting crop yield, quality, and food security worldwide (Strange and Scott, 2005). These diseases are caused by a diverse range of bacterial pathogens that infect various parts of the plant, including leaves, stems, roots, and fruits. Some of the most notorious bacterial pathogens include *Xanthomonas*, *Pseudomonas*, *Erwinia*, and *Agrobacterium* (Mansfield *et al.*, 2012). These pathogens can cause a variety of symptoms, such as leaf spots, blights, wilts, galls, and cankers, leading to severe economic losses (Sawicka *et al.*, 2020). The mechanisms through which these bacteria infect plants, evade their defense systems, and proliferate involve complex interactions between bacterial virulence factors and plant immune responses (Pfeilmeier *et al.*, 2016).

30.3.1 Impact on agriculture and food security

Bacterial plant diseases can have devastating effects on agriculture, reducing both the quantity and quality of produce. For instance, *Xanthomonas oryzae* pv. *oryzae*, the causal agent of bacterial blight in rice, is a major concern in Asia, where rice is a staple food (Sharma *et al.*, 2017). Similarly, *Pseudomonas syringae* causes bacterial specks in tomatoes and peppers, leading to significant losses in vegetable crops (Griffin *et al.*, 2017). The impact extends beyond direct crop damage; bacterial infections can also increase susceptibility to secondary infections by other pathogens, further exacerbating yield losses (Savary and Willocquet, 2020). Additionally, the presence of bacterial diseases can limit the marketability of produce, affecting farmers' income and overall economic stability in agricultural regions (Vurro *et al.*, 2010).

30.3.2 Challenges in disease management

Managing bacterial diseases in plants is particularly challenging due to several factors. First, bacterial pathogens can spread rapidly through water, soil, seed, and contaminated equipment, making containment difficult (Hayes *et al.*, 2022). Second, many bacteria can survive in plant debris, soil, or alternative hosts for extended periods, complicating eradication efforts (Leben, 1981). Traditional control measures, such as chemical bactericides and antibiotics, often have limited efficacy and can lead to the development of resistant bacterial strains (Gill *et al.*, 2015). Furthermore, these chemicals can have negative environmental impacts, harming beneficial microorganisms and contributing to soil and water pollution (Kumar *et al.*, 2019). Cultural practices, such as crop rotation, sanitation, and resistant cultivars, are essential components of integrated disease management but may not always provide complete control (Palti, 2012).

30.3.3 Mechanisms of bacterial pathogenesis

Bacterial pathogens deploy a variety of strategies to infect plants and evade their defense mechanisms. Key among these is the secretion systems that bacteria use to deliver effector proteins directly into plant cells (Pfeilmeier *et al.*, 2016; Wang *et al.*, 2022). The Type III secretion system (T3SS) is particularly well-studied; it allows bacteria to inject effector proteins that can suppress plant immune responses, manipulate host cellular processes, and facilitate nutrient acquisition (Alfano and Collmer, 2004). For example, *Pseudomonas syringae* uses T3SS to deliver a suite of effector proteins that inhibit plant defense signaling pathways and promote bacterial proliferation (Lee *et al.*, 2012). Other secretion systems, such as Type II and Type IV, are also involved in the secretion of enzymes and toxins that degrade plant cell walls and disrupt cellular integrity (Cascales and Christie, 2003; Deng *et al.*, 2017).

30.3.4 Plant immune responses to bacterial infection

Plants have evolved sophisticated immune systems to detect and respond to bacterial pathogens (Nimchuk *et al.*, 2003). The primary layer of defense involves pattern recognition receptors (PRRs) that detect conserved microbial signatures known as pathogen-associated molecular patterns (PAMPs), such as flagellin and lipopolysaccharides (Boller and Felix, 2009). Recognition of PAMPs triggers PAMP-triggered immunity (PTI), which includes the production of reactive oxygen species, cell wall reinforcement, and the activation of defense-related genes (Yang *et al.*, 2022). However, bacterial pathogens can overcome PTI by deploying effectors that interfere with these responses (Hann *et al.*, 2010). In turn, plants have developed a secondary layer of defense, known as effector-triggered immunity (ETI), which involves the recognition of specific bacterial effectors by intracellular resistance (R) proteins (Wu *et al.*, 2014). ETI often leads to a stronger defense response, including localized cell death known as the hypersensitive response (HR), which restricts pathogen spread (Alhoraibi *et al.*, 2019).

30.3.5 Integrated disease management

Effective management of bacterial diseases in plants requires an integrated approach that combines multiple strategies. Cultural practices, such as crop rotation, the use of disease-free planting material, and proper irrigation management, are fundamental to reducing disease incidence (Palti, 2012). Resistant cultivars play a crucial role, and ongoing breeding programs aim to develop varieties with improved resistance to bacterial pathogens (Khan *et al.*, 2014). Chemical control, including bactericides and antibiotics, should be used judiciously to avoid the development of resistance and minimize environmental impact (Manyi-Loh *et al.*, 2018). Biological control, involving the use of beneficial microbes to suppress bacterial pathogens, offers a promising alternative to chemical treatments (Heydari and Pessarakli, 2010). Advances in molecular biology and biotechnology are also paving the way for innovative approaches, such as the use of CRISPR/Cas9 for targeted resistance gene editing and the application of RNA interference (RNAi) to silence essential genes in bacterial pathogens (Bisht *et al.*, 2019).

30.3.6 Future directions and challenges

The ongoing battle against bacterial diseases in plants faces several challenges and opportunities. Climate change is expected to alter the distribution and severity of bacterial diseases, necessitating adaptive management strategies (Velásquez *et al.*, 2018). Emerging bacterial pathogens and the evolution of new virulent strains pose continuous threats to crop production (McDonald and Stukenbrock, 2016). Understanding the molecular basis of plant–bacterial interactions, including the role of the plant microbiome in disease resistance, is crucial for developing effective control measures. Advances in genomic and epigenomic technologies are providing new tools for studying these interactions and identifying key regulatory mechanisms (Montesinos *et al.*, 2002). Ultimately, a holistic approach that integrates traditional practices with cutting-edge scientific innovations will be essential for safeguarding global food security and ensuring sustainable agricultural production (National Academies of Sciences *et al.*, 2019).

Bacterial diseases in plants represent a significant challenge to global agriculture, impacting crop yield and quality. The complexity of bacterial pathogenesis and plant immune responses underscores the need for integrated disease management strategies that combine cultural practices, resistant cultivars, chemical control, biological control, and advanced biotechnological approaches. Continued research into the molecular mechanisms of plant–bacterial interactions, including the role of epigenetics, will be vital for developing innovative solutions to combat these diseases and ensure sustainable agricultural productivity.

30.4 Epigenetic Strategies for Disease Resistance

30.4.1 Stable DNA methylation patterns

Stable DNA methylation patterns involve long-term changes to the DNA methylation status of specific genomic regions, which can be maintained through cell divisions and passed on to progeny (Cedar and Bergman, 2012). This mechanism allows for the inheritance of beneficial traits, such as disease resistance, without altering the underlying DNA sequence. These stable methylation marks can influence gene expression by modifying the chromatin structure, making certain genes more or less accessible to the transcriptional machinery, thereby enhancing the plant's ability to resist pathogens (Bird, 2002). DNA methylation in plants primarily involves the addition of a methyl group to the cytosine base of DNA, typically at CpG, CHG, and CHH sequences (where H represents A, T, or C). This modification is mediated by a family of enzymes known as DNA methyltransferases (DNMTs), which include MET1 (maintains CG methylation), CMT3 (maintains CHG methylation), and DRM2 (responsible for *de novo* methylation in all sequence contexts through the RNA-directed DNA methylation pathway) (Wada, 2005; Klupczyńska and Ratajczak, 2021). These methylation marks can be stably inherited through somatic cell divisions and, in some cases, through sexual reproduction, leading to transgenerational inheritance of the modified epigenetic state (Brukhin and Albertini, 2021). During DNA replication, maintenance methyltransferases like MET1 recognize hemi-methylated DNA (where one strand is methylated) and add methyl groups to the newly synthesized strand, ensuring that methylation patterns are faithfully copied to daughter cells (Du, 2016). For non-CG methylation, proteins like CMT3 and chromomethylase 2 (CMT2) recognize specific histone marks and maintain methylation at CHG and CHH sites (Kenchanmane Raju *et al.*, 2019).

Environmental stimuli, such as pathogen attack, can induce *de novo* DNA methylation. RNA-directed DNA methylation (RdDM) involves small interfering RNAs (siRNAs) guiding the DRM2 enzyme to specific genomic locations to establish new methylation marks (Zhang and Zhu, 2011). Once established, these marks can create an epigenetic memory that persists across generations. During gametogenesis and embryogenesis, a portion of the DNA methylation marks are retained, allowing the transmission of the epigenetic state to the next generation (Ben Maamar *et al.*, 2021). This inheritance mechanism provides offspring with pre-adapted responses to environmental challenges faced by the parent, such as enhanced disease resistance.

Techniques for studying stable DNA methylation include whole-genome bisulfite sequencing (WGBS), methylated DNA immunoprecipitation sequencing (MeDIP-seq), and reduced representation bisulfite sequencing (RRBS) (Beck *et al.*, 2022). WGBS involves converting unmethylated cytosines to uracil, allowing researchers to identify methylated regions across the genome, while MeDIP-seq uses antibodies to pull down specific histone modifications, enabling the identification of associated DNA regions (Yong *et al.*, 2016). RRBS focuses on CpG-rich regions of the genome, providing a cost-effective approach to analyzing DNA methylation (Fouse *et al.*, 2010).

The inheritance of epigenetic memory allows plants to pass on these modifications to their progeny, ensuring that future generations are better prepared to respond to similar pathogen challenges (Kakoulidou *et al.*, 2021). For instance, plants exposed to pathogens may develop DNA methylation patterns that enhance the expression of defense genes (Mladenov *et al.*, 2021). These patterns can be inherited, providing offspring with a heightened ability to resist pathogens. Applications of epigenetic memory in agriculture include the development of disease-resistant crop varieties through epigenetic breeding, where plants with desirable epigenetic traits are selected and propagated (Krishnaswami *et al.*, 2022). This approach can lead to more resilient crops that require fewer chemical inputs, promoting sustainable agriculture. Additionally, understanding epigenetic memory can help in developing strategies to enhance plant stress tolerance and adaptability, contributing to improved crop performance under various environmental conditions (Sun *et al.*, 2021). Challenges in leveraging epigenetic memory include fully understanding the mechanisms that govern the stable inheritance of epigenetic marks and their context-specific effects. More research is needed to elucidate how these marks are established, maintained, and transmitted across generations.

Advances in epigenomic technologies and single-cell analysis will be crucial in addressing these challenges and unlocking the potential of epigenetic memory for crop improvement and sustainable agricultural practices. Stable DNA methylation patterns play a critical role in the long-term adaptation of plants to environmental challenges and pathogen attacks (Viggiano and Pinto, 2017). Techniques like WGBS, MeDIP-seq, and RRBS enable detailed mapping and analysis of these patterns, offering valuable insights for crop improvement and sustainable agriculture. Understanding and harnessing these stable epigenetic marks can lead to the development of more resilient and productive crop varieties, contributing to global food security.

30.4.2 Transgenerational inheritance

Transgenerational inheritance refers to the phenomenon where epigenetic modifications, such as DNA methylation and histone modifications, are passed from one generation to the next, leading to inherited traits such as disease resistance (Xavier *et al.*, 2019). Unlike genetic changes, which involve alterations to the DNA sequence, epigenetic modifications influence gene expression without changing the underlying DNA. This form of inheritance can provide long-term adaptation to environmental challenges, including pathogen attacks, by maintaining beneficial epigenetic states across generations (Holeski *et al.*, 2012). The mechanisms underlying this inheritance involve the establishment of DNA methylation patterns in response to environmental factors, including pathogen exposure, which can alter gene expression to enhance disease resistance (Tirnaz and Batley, 2019). Maintenance methylation processes, facilitated by DNA methyltransferases like MET1, CMT3, and DRM2, ensure that these patterns are faithfully copied during cell division (Hsieh, 2016). Some of these methylation patterns can escape the reprogramming processes during gametogenesis and embryogenesis, thus being passed to progeny. Similarly, histone proteins, around which DNA is wrapped, can undergo chemical modifications that affect chromatin structure and gene expression, with some modifications being maintained during DNA replication and cell division (Ehrenhofer-Murray, 2004). Non-coding RNAs, such as siRNAs and miRNAs, also play roles in guiding DNA methylation and histone modification, establishing epigenetic states that can be inherited (Wei *et al.*, 2017).

Techniques for studying transgenerational inheritance include whole-genome bisulfite sequencing (WGBS), chromatin immunoprecipitation sequencing (ChIP-seq), and RNA sequencing (RNA-seq). WGBS, by enabling the detection of methylated cytosines across the entire genome, provides a comprehensive map of DNA methylation (Lister *et al.*, 2008). ChIP-seq identifies DNA regions associated with specific histone modifications using antibodies, allowing for the study of histone modifications on a genome-wide scale (Barski *et al.*, 2007). RNA-seq analyzes the expression levels of non-coding RNAs involved in epigenetic regulation, providing insights into the regulatory roles of these RNAs (Wang *et al.*, 2009).

Transgenerational studies involve cross-generational analysis to examine methylation and histone modification patterns, as well as small RNA profiles in parental and offspring generations. These studies include treating parental plants with specific environmental stimuli, such as pathogens, and monitoring the epigenetic changes to analyze progeny for the inheritance of these modifications and associated phenotypes (Boyko *et al.*, 2010). The applications of transgenerational inheritance in disease resistance are significant, including the use of inherited epigenetic marks in breeding programs to develop resistant plant varieties and ensure long-term adaptation to pathogens. Plants can develop and pass on resistance traits to their offspring, ensuring continued resilience (Luna *et al.*, 2012).

In the context of environmental stress adaptation, stress memory developed by plants exposed to environmental stresses like drought and salinity can enhance tolerance, and these beneficial modifications can be inherited by future generations, providing stress resilience and promoting sustainable agriculture (Nguyen *et al.*, 2022). For crop improvement, identifying epigenetic marks for desirable traits such as yield, stress tolerance, and resistance to pests and diseases is crucial. Genome editing tools like CRISPR/dCas9 are being used to specifically modify epigenetic marks to induce beneficial traits without altering the DNA sequence (Papikian *et al.*, 2019).

However, challenges remain in understanding the detailed mechanisms governing the stable inheritance of epigenetic marks and the context-specific effects of these modifications. Further research is needed to fully comprehend how genomic context and environmental factors influence the inheritance of epigenetic modifications (Heard and Martienssen, 2014). Technological advancements are also required, including the development of more precise and high-throughput techniques for studying epigenetic inheritance and single-cell analysis to uncover cellular heterogeneity (Luo *et al.*, 2020). Integration with conventional breeding approaches is essential, combining epigenetic strategies with traditional breeding and genetic modification to develop superior plant varieties while addressing ethical and regulatory concerns regarding the use of epigenetic modifications in agriculture (Springer and Schmitz, 2017). Transgenerational inheritance of epigenetic modifications offers a powerful mechanism for plants to adapt to environmental challenges and pathogens, providing long-term benefits such as disease resistance. Techniques like WGBS, ChIP-seq, and RNA-seq enable detailed study of these processes and pave the way for innovative applications in crop improvement and sustainable agriculture (Baulcombe and Dean, 2014).

30.4.3 Epigenetic memory

Epigenetic memory in plants refers to the concept that plants can maintain a 'memory' of pathogen exposure through stable epigenetic marks, which prepare future generations for similar challenges. This form of memory does not involve changes to the DNA sequence but rather relies on heritable changes in gene expression facilitated by mechanisms such as DNA methylation, histone modifications, and the involvement of non-coding RNAs (Crisp *et al.*, 2016). When plants are exposed to pathogens, they can undergo specific epigenetic changes that alter the expression of defense-related genes. These changes often involve the addition of methyl groups to cytosines in DNA, which can silence or activate genes, depending on their location and context. Key enzymes, such as DNA methyltransferases (e.g. MET1, CMT3, and DRM2), play crucial roles in establishing and maintaining these methylation patterns (Matzke and Mosher, 2014).

Histone modifications, including methylation and acetylation, also contribute to epigenetic memory by altering chromatin structure and thus regulating gene accessibility. Specific histone marks can make chromatin more open or closed, thereby facilitating or hindering the transcription of defense genes (Kouzarides, 2007). For instance, histone acetylation, generally associated with gene activation, loosens chromatin structure, making DNA more accessible to transcription factors. In contrast, histone methylation can either activate or repress gene expression, depending on the specific amino acid residues that are modified and the number of methyl groups added (Jenuwein and Allis, 2001).

Non-coding RNAs, such as small interfering RNAs (siRNAs) and microRNAs (miRNAs), play crucial roles in guiding these histone modifications and participate in the regulation of gene expression in response to pathogen attack. siRNAs can direct the RNA-induced silencing complex (RISC) to target and degrade complementary mRNA transcripts, thereby reducing gene expression. Additionally, siRNAs can guide DNA methylation and histone modification complexes to specific genomic loci, reinforcing silencing marks (Baulcombe, 2004). miRNAs, on the other hand, primarily function by base-pairing with complementary sequences in target mRNAs, leading to their degradation or inhibition of translation, thereby fine-tuning gene expression in response to environmental stimuli, including pathogen attacks (Jones-Rhoades *et al.*, 2006).

Techniques to study epigenetic memory in plants include whole-genome bisulfite sequencing (WGBS) for mapping DNA methylation patterns, chromatin immunoprecipitation sequencing (ChIP-seq) for identifying histone modifications, and RNA sequencing (RNA-seq) for analyzing non-coding RNAs. WGBS allows researchers to identify methylated regions across the genome (Lister *et al.*, 2008). ChIP-seq enables the identification of associated DNA regions (Barski *et al.*, 2007). RNA-seq helps in quantifying and characterizing non-coding RNAs involved in the epigenetic regulation of defense responses (Wang *et al.*, 2009).

The inheritance of epigenetic memory allows plants to pass on these modifications to their progeny, ensuring that future generations are better prepared to respond to similar pathogen challenges. For instance, plants exposed to pathogens may develop DNA methylation patterns that enhance the expression of defense genes. These patterns can be inherited, providing offspring with a heightened ability to resist pathogens (Heard and Martienssen, 2014).

Applications of epigenetic memory in agriculture include the development of disease-resistant crop varieties through epigenetic breeding, where plants with desirable epigenetic traits are selected and propagated. This approach can lead to more resilient crops that require fewer chemical inputs, promoting sustainable agriculture (Springer and Schmitz, 2017). Epigenetic breeding harnesses the natural ability of plants to adapt to environmental stresses and pass on beneficial traits to their offspring without altering the underlying DNA sequence.

Additionally, understanding epigenetic memory can help in developing strategies to enhance plant stress tolerance and adaptability, contributing to improved crop performance under various environmental conditions. Challenges in leveraging epigenetic memory include fully understanding the mechanisms that govern the stable inheritance of epigenetic marks and their context-specific effects. More research is needed to elucidate how these marks are established, maintained, and transmitted across generations. Advances in epigenomic technologies and single-cell analysis will be crucial in addressing these challenges and unlocking the potential of epigenetic memory for crop improvement and sustainable agricultural practices (Luo *et al.*, 2020).

30.5 Case Studies

These case studies collectively highlight the significant progress made in understanding and utilizing epigenetic mechanisms to combat bacterial diseases in plants. From DNA methylation and histone modifications to small RNA regulation and advanced epigenome editing technologies, each study underscores the potential of epigenetics in enhancing plant immunity and developing resilient crop varieties. Future research and application of these techniques promise to contribute significantly to sustainable agriculture and global food security.

Case study 1: Epigenetic regulation of tomato resistance to *Ralstonia solanacearum*

Ralstonia solanacearum is a bacterial pathogen causing bacterial wilt in a variety of crops, including tomatoes. This pathogen invades plant roots, colonizes the xylem, and leads to wilting and plant death. Traditional breeding for resistance has had limited success, necessitating alternative approaches such as epigenetic modifications. Bisulfite sequencing was used to analyze DNA methylation patterns in resistant and susceptible tomato cultivars before and after infection with *R. solanacearum*. Nair (2020) identified differentially methylated regions (DMRs) associated with defense-related genes. The study found that resistant tomato cultivars exhibited significant hypomethylation at promoter regions of key defense genes, leading to their upregulation upon infection. In contrast, susceptible cultivars showed hypermethylation and suppressed expression of these genes. This suggests epigenetic priming of defense genes via DNA. The findings highlight the potential of manipulating DNA methylation patterns to enhance disease resistance in crops. By targeting specific DMRs, it may be possible to breed or engineer tomato plants with improved resistance to bacterial wilt.

Case study 2: Histone modifications in *Arabidopsis* resistance to *Pseudomonas syringae*

Pseudomonas syringae is a well-studied bacterial pathogen that infects *Arabidopsis thaliana*, a model plant species. The interaction between *Arabidopsis* and *P. syringae* has provided valuable insights into plant immunity. Chromatin immunoprecipitation followed by sequencing (ChIP-seq) was employed to examine histone modifications at defense gene loci in *Arabidopsis* upon infection with *P. syringae* (De-la-Peña *et al.*, 2012). The focus was on histone acetylation and methylation marks known to correlate with gene activation or repression. The research revealed that histone acetylation at the promoters of several key defense genes increased significantly in response to *P. syringae* infection, correlating with heightened gene expression. Conversely, histone methylation marks associated with gene repression were reduced at these loci. Mutant plants lacking specific histone acetyltransferases showed reduced resistance, underscoring the role of histone acetylation in activating defense responses. This study demonstrates that histone modifications are crucial regulators of plant immune responses. Targeting histone acetylation could be a viable strategy to enhance disease resistance in crops.

Case study 3: Small RNA-mediated regulation of plant immunity in rice against *Xanthomonas oryzae*

Xanthomonas oryzae pv. *oryzae* causes bacterial blight, a major disease affecting rice production. Understanding the role of small RNAs in regulating rice immunity can provide new avenues for disease control. Small RNA sequencing (sRNA-seq) was conducted on rice plants infected with *X. oryzae*. blight (Sharma *et al.*, 2017; Reshetnyak *et al.*, 2021). The researchers identified miRNAs and siRNAs differentially expressed during infection and predicted their target genes involved in immune responses. Several miRNAs were found to be downregulated in response to infection, leading to the upregulation of their target genes, which included key components of the immune signaling pathways. Transgenic rice plants overexpressing these miRNAs showed increased susceptibility to *X. oryzae*, confirming their regulatory roles in immunity. The studies illustrate that small RNAs are integral to the fine-tuning of plant immune responses. Manipulating the levels of specific miRNAs could potentially bolster rice resistance to bacterial blight.

Case study 4: Epigenetic memory and systemic acquired resistance in tobacco

Systemic acquired resistance (SAR) is a 'primed' state of enhanced defense that occurs in plants following an initial localized exposure to a pathogen. Epigenetic changes are believed to play a role in the establishment and maintenance of SAR. Luna *et al.* (2012) used tobacco plants, a model for studying SAR, and employed both bisulfite sequencing and ChIP-seq to investigate DNA methylation and histone modifications, respectively, following SAR induction with an avirulent strain of *Pseudomonas syringae*. The study found that SAR was associated with sustained hypomethylation and increased histone acetylation at defense gene promoters. These epigenetic marks were maintained long after the initial pathogen exposure, suggesting an epigenetic memory that primes the plant for faster and stronger responses to subsequent infections. This research highlights the potential for leveraging epigenetic memory in crop protection strategies. Inducing a primed state through controlled exposure to avirulent pathogens or chemical elicitors could enhance resistance to future infections.

30.6 Achievements and Outcomes

The application of epigenetic mechanisms in combating bacterial diseases in plants has led to several notable achievements and outcomes. These advancements are not only enhancing our understanding of plant–pathogen interactions but also paving the way for innovative strategies in crop protection and sustainable agriculture. Through detailed studies of DNA methylation, histone modifications, and small RNA pathways, researchers have elucidated the complex regulatory networks governing plant immunity, providing insights into how specific epigenetic changes correlate with the activation or repression of defense-related genes (Deleris *et al.*, 2016). These insights have led to the identification of key epigenetic marks associated with disease resistance and susceptibility, such as hypomethylation at promoter regions of defense genes in resistant plants and hypermethylation in susceptible cultivars (Tirnaz and Batley, 2019). Furthermore, the development of novel crop protection strategies has been facilitated by epigenetic priming, which involves inducing a 'memory' of pathogen attack through specific epigenetic modifications. This approach has shown promise in enhancing disease resistance, as evidenced by sustained hypomethylation and increased histone acetylation associated with systemic acquired resistance (SAR) (Durrant and Dong, 2004). Additionally, the use of CRISPR/dCas9 technology to precisely edit epigenetic marks at specific genomic loci represents a groundbreaking achievement, offering a highly specific tool for crop improvement (Miglani *et al.*, 2020). These advancements have practical applications in crop improvement through both transgenic and non-transgenic approaches, as well as in breeding programs and marker-assisted selection, accelerating the development of resistant crop varieties (Ragimekula *et al.*, 2013). Furthermore, by leveraging epigenetic mechanisms to enhance natural plant immunity, there is potential to reduce the reliance on chemical pesticides, contributing to more sustainable agricultural practices and promoting ecological balance. Ultimately, these achievements and outcomes collectively underscore the significant progress made in utilizing epigenetic mechanisms to combat bacterial diseases in plants, promising a brighter future for crop protection and global food security.

30.7 Challenges and Considerations

While the application of epigenetics in combating bacterial diseases in plants holds great promise, several challenges and considerations must be addressed to realize its full potential and ensure its successful implementation.

30.7.1 Epigenetic complexity and specificity

The intricate interplay between different epigenetic modifications and their coordination with other regulatory mechanisms poses a significant challenge (Pikaard and Scheid, 2014). Deciphering the complex networks of epigenetic regulation in plant immunity requires comprehensive studies integrating multiple omics datasets and advanced computational analyses (Murmu *et al.*, 2024). Achieving precise and targeted epigenetic modifications without unintended off-target effects remains a significant challenge. Ensuring the specificity and accuracy of epigenome editing tools, such as CRISPR/dCas9 systems, is essential to minimize potential side effects and maximize efficacy (Mushtaq *et al.*, 2021).

30.7.2 Regulatory approval and public acceptance

The regulatory landscape surrounding genetically modified organisms (GMOs) and novel biotechnologies presents regulatory hurdles and public acceptance challenges. Addressing safety concerns, ethical considerations, and regulatory requirements is crucial for gaining approval and public trust in epigenetic-based crop protection strategies (Rozas *et al.*, 2022).

30.7.3 Technical and methodological limitations

Continued advancements in genomic and epigenomic technologies are necessary to overcome technical limitations and expand our capabilities in studying and manipulating epigenetic mechanisms (Ku *et al.*, 2011). Improving accessibility to state-of-the-art equipment and bioinformatics resources is essential for fostering innovation and collaboration in the field. Ensuring the stability and heritability of desired epigenetic modifications across generations presents a challenge. Understanding the factors influencing epigenetic memory and transgenerational inheritance is essential for developing durable and sustainable crop protection strategies (Tiwari *et al.*, 2022).

30.7.4 Ethical and societal implications

Ethical considerations and equity

Addressing ethical considerations, such as intellectual property rights, equitable access to technologies, and socioeconomic impacts, is essential for ensuring the responsible and equitable deployment of epigenetics-based crop protection strategies. Engaging stakeholders and fostering dialogue with diverse communities is crucial for navigating complex ethical and societal implications (Pickersgill *et al.*, 2013).

Environmental and ecological impact

Assessing the potential environmental and ecological impact of epigenetic interventions on ecosystems and biodiversity is paramount. Conducting comprehensive risk assessments and environmental monitoring studies is necessary to mitigate potential risks and safeguard environmental integrity (Segelbacher *et al.*, 2022).

Addressing the challenges and considerations surrounding the application of epigenetics in combating bacterial diseases in plants requires collaborative efforts across scientific disciplines, regulatory bodies, industry stakeholders, and societal actors. By overcoming technical limitations, navigating regulatory hurdles, and addressing ethical and societal concerns, we can harness the power of epigenetics to develop sustainable and resilient agricultural systems that ensure food security, promote environmental stewardship, and advance global well-being (Kuzma and Kokotovich, 2011).

30.8 Future Prospects

The future prospects of utilizing epigenetics in combating bacterial diseases in plants are promising, offering innovative solutions for enhancing crop resilience, improving agricultural sustainability, and ensuring global food security. Further refinement of CRISPR-based epigenome editing technologies, such as CRISPR/dCas9 systems, holds promise for achieving precise and targeted modifications of epigenetic marks associated with disease resistance genes (Wang *et al.*, 2018; Gallego-Bartolomé *et al.*, 2019). Advances in delivery methods and optimization of editing efficiencies will enhance the applicability and scalability of these tools for crop improvement. Multiplex editing approaches and synthetic epigenetic circuits offer innovative strategies for engineering complex epigenetic modifications in plants, allowing researchers to tailor defense responses and enhance crop resilience against diverse pathogens (Li *et al.*, 2016). Integration of multi-omics datasets, including genomics, epigenomics, transcriptomics, and metabolomics, will enable a systems-level understanding of epigenetic regulation in plant immunity (Meng *et al.*, 2016; Zhang *et al.*, 2018). Computational modeling and network analysis approaches will elucidate the complex interplay between epigenetic mechanisms and signaling pathways, facilitating the prediction of novel targets for crop protection (Belhaj *et al.*, 2015). Single-cell epigenomic profiling techniques offer insights into cell-type-specific epigenetic dynamics and heterogeneity within plant tissues, unraveling the epigenetic landscapes of individual cells during pathogen infection (Li *et al.*, 2024). Exploiting epigenetic memory mechanisms, such as transgenerational inheritance and induced priming, holds promise for developing durable and sustainable disease-resistance traits in crops (Slaughter *et al.*, 2012). Understanding the molecular basis of epigenetic memory and its transmission across generations will enable the development of crop varieties with enhanced resilience to recurring pathogen threats. Epigenome-wide association studies (EWAS) offer a powerful approach for identifying epigenetic variants associated with natural variation in disease resistance among crop germplasm, uncovering novel epialleles and epigenetic regulatory networks underlying plant immunity traits (Cortijo *et al.*, 2014). Establishing transparent and inclusive ethical frameworks and engaging stakeholders in decision-making processes is essential for navigating the ethical, regulatory, and societal implications of epigenetic-based crop protection strategies. Harmonizing regulatory frameworks across jurisdictions and fostering public acceptance through science communication and education are critical for facilitating the adoption of epigenetic technologies in agriculture, ensuring responsible innovation, and equitable deployment (Kuzma and Kokotovich, 2011; Jasanoff *et al.*, 2015).

30.9 Conclusion

In conclusion, the intricate field of epigenetics has emerged as a promising frontier in the ongoing battle against bacterial diseases in plants. Through updated methods and relentless scientific pursuit, researchers have uncovered a wealth of knowledge regarding the epigenetic mechanisms governing plant immunity and bacterial pathogenesis. This newfound understanding has paved the way for innovative strategies aimed at bolstering plant defenses and mitigating the impact of devastating bacterial infections. The future of utilizing epigenetics in combating bacterial diseases in plants holds immense promise for transforming agriculture and addressing global food security challenges.

Advances in precision epigenome editing, systems-level understanding of epigenetic regulation, and exploitation of epigenetic memory mechanisms offer innovative strategies for enhancing crop resilience and sustainability. By harnessing technologies like CRISPR-based epigenome editing and integrating multi-omics datasets, researchers can develop targeted interventions to boost plant immunity and mitigate the impact of bacterial pathogens. Furthermore, the integration of epigenetic insights into breeding programs and crop management practices holds the promise of more resilient and disease-resistant plant varieties. Through strategic deployment of epigenetic modifiers and tailored interventions, farmers can better safeguard their crops against bacterial diseases while minimizing the reliance on conventional pesticides and antibiotics.

However, it is crucial to acknowledge the challenges and complexities inherent in harnessing epigenetics for agricultural benefit. Continued research efforts are needed to unravel the intricate interplay between epigenetic regulation, host–pathogen interactions, and environmental cues. Moreover, ethical considerations and potential unintended consequences must be carefully evaluated as epigenetic-based interventions are translated into real-world applications. In the quest for sustainable agriculture and global food security, epigenetics stands as a beacon of hope, offering innovative solutions to age-old challenges. Through collaboration, innovation, and a steadfast commitment to scientific inquiry, we can harness the power of epigenetics to combat bacterial diseases in plants and pave the way toward a more resilient and sustainable agricultural future.

References

Aerts, N., Chhillar, H., Ding, P. and Wees, S.C. (2022) Transcriptional regulation of plant innate immunity. *Essays in Biochemistry* 66(5), 607–620.

Agius, D.R., Kapazoglou, A., Avramidou, E., Baranek, M., Carneros, E. *et al.* (2023) Exploring the crop epigenome: A comparison of DNA methylation profiling techniques. *Frontiers in Plant Science* 14, 1181039.

Akbar, S., Wei, Y. and Zhang, M.Q. (2022) RNA interference: Promising approach to combat plant viruses. *International Journal of Molecular Sciences* 23(10), 5312.

Alfano, J.R. and Collmer, A. (2004) Type III secretion system effector proteins: Double agents in bacterial disease and plant defense. *Annual Review of Phytopathology* 42, 385–414.

Alhoraibi, H., Bigeard, J., Rayapuram, N., Colcombet, J. and Hirt, H. (2019) Plant immunity: The MTI-ETI model and beyond. *Current Issues in Molecular Biology* 30(1), 39–58.

Alvarez, M.E., Nota, F. and Cambiagno, D.A. (2010) Epigenetic control of plant immunity. *Molecular Plant Pathology* 11(4), 563–576.

Arora, L. and Narula, A. (2017) Gene editing and crop improvement using CRISPR-Cas9 system. *Frontiers in Plant Science* 8, 296675.

Arora, H., Singh, R.K., Sharma, S., Sharma, N., Panchal, A. *et al.* (2022) DNA methylation dynamics in response to abiotic and pathogen stress in plants. *Plant Cell Reports* 41(10), 1931–1944.

Ashapkin, V.V., Kutueva, L.I., Aleksandrushkina, N.I. and Vanyushin, B.F. (2020) Epigenetic mechanisms of plant adaptation to biotic and abiotic stresses. *International Journal of Molecular Sciences* 21(20), 7457.

Bajpai, S.K., NishaPandita, S., Bahadur, A. and Verma, P.C. (2024) Recent advancements in the role of histone acetylation dynamics to improve stress responses in plants. *Molecular Biology Reports* 51(1), 413.

Barski, A. and Zhao, K. (2009) Genomic location analysis by ChIP-seq. *Journal of Cellular Biochemistry* 107(1), 11–18.

Barski, A., Cuddapah, S., Cui, K., Roh, T.-Y., Schones, D.E. *et al.* (2007) High-resolution profiling of histone methylations in the human genome. *Cell* 129(4), 823–837.

Baulcombe, D. (2004) RNA silencing in plants. *Nature* 431(7006), 356–363.

Baulcombe, D.C. and Dean, C. (2014) Epigenetic regulation in plant responses to the environment. *Cold Spring Harbor Perspectives in Biology* 6(9), a019471–a019471.

Beck, D., Ben Maamar, M. and Skinner, M.K. (2022) Genome-wide cpg density and DNA methylation analysis method (MeDIP, RRBS, and WGBS) comparisons. *Epigenetics* 17(5), 518–530.

Belhaj, K., Chaparro-Garcia, A., Kamoun, S., Patron, N.J. and Nekrasov, V. (2015) Editing plant genomes with CRISPR/Cas9. *Current Opinion in Biotechnology* 32, 76–84.

Ben Maamar, M., Nilsson, E.E. and Skinner, M.K. (2021) Epigenetic transgenerational inheritance, gametogenesis and germline development. *Biology of Reproduction* 105(3), 570–592.

Berr, A., Shafiq, S. and Shen, W.H. (2011) Histone modifications in transcriptional activation during plant development. *Biochimica et Biophysica Acta (BBA) – Gene Regulatory Mechanisms* 1809(10), 567–576.

Bhan, A., Deb, P. and Mandal, S.S. (2017) Epigenetic code: Histone modification, gene regulation, and chromatin dynamics. In: Mandal, S.S. (ed.) *Gene Regulation, Epigenetics and Hormone Signaling*. Blackwell / Wiley, pp. 29–58.

Bird, A. (2002) DNA methylation patterns and epigenetic memory. *Genes & Development* 16(1), 6–21.

Bisht, D.S., Bhatia, V. and Bhattacharya, R. (2019) Improving plant-resistance to insect-pests and pathogens: The new opportunities through targeted genome editing. In: *Seminars in Cell & Developmental Biologym*, Vol. 96. Academic Press, pp. 65–76.

Boller, T. and Felix, G. (2009) A renaissance of elicitors: Perception of microbe-associated molecular patterns and danger signals by pattern-recognition receptors. *Annual Review of Plant Biology* 60, 379–406.

Bommarito, P.A. and Fry, R.C. (2019) The role of DNA methylation in gene regulation. In: McCullough, S.D. and Dolinoy, D.C. (eds) *Toxicoepigenetics*. Academic Press, pp. 127–151.

Boyko, A., Blevins, T., Yao, Y., Golubov, A., Bilichak, A. *et al.* (2010) Transgenerational adaptation of *Arabidopsis* to stress requires DNA methylation and the function of dicer-like proteins. *PloS ONE* 5(3), e9514.

Bruce, T.J.A., Matthes, M.C., Napier, J.A. and Pickett, J.A. (2007) Stressful "memories" of plants: Evidence and possible mechanisms. *Plant Science* 173(6), 603–608.

Brukhin, V. and Albertini, E. (2021) Epigenetic modifications in plant development and reproduction. *Epigenomes* 5(4), 25.

Caarls, L., Pieterse, C.M. and Wees, S.C. (2015) How salicylic acid takes transcriptional control over jasmonic acid signaling. *Frontiers in Plant Science* 6, 170.

Cascales, E. and Christie, P.J. (2003) The versatile bacterial type IV secretion systems. *Nature Reviews Microbiology* 1(2), 137–149.

Cedar, H. and Bergman, Y. (2012) Programming of DNA methylation patterns. *Annual Review of Biochemistry* 81, 97–117.

Chapman, E.J. and Carrington, J.C. (2007) Specialization and evolution of endogenous small RNA pathways. *Nature Reviews Genetics* 8(11), 884–896.

Checker, V.G., Kushwaha, H.R., Kumari, P. and Yadav, S. (2018) Role of phytohormones in plant defense: Signaling and cross talk. In: Singh, A. and Singh, I.K. (eds) *Molecular Aspects of Plant–Pathogen Interaction*. Springer Singapore, pp. 159–184.

Chen, X., Bhadauria, V. and Ma, B. (2018) ChIP-seq: A powerful tool for studying protein–DNA interactions in plants. *Current Issues in Molecular Biology* 27(1), 171–180.

Clark, S.J., Statham, A., Stirzaker, C., Molloy, P.L. and Frommer, M. (2006) DNA methylation: Bisulphite modification and analysis. *Nature Protocols* 1(5), 2353–2364.

Conrath, U., Beckers, G.J.M., Langenbach, C.J.G. and Jaskiewicz, M.R. (2015) Priming for enhanced defense. *Annual Review of Phytopathology* 53, 97–119.

Cortijo, S., Wardenaar, R., Colomé-Tatché, M., Gilly, A., Etcheverry, M. *et al.* (2014) Mapping the epigenetic basis of complex traits. *Science* 343(6175), 1145–1148.

Crisp, P.A., Ganguly, D., Eichten, S.R., Borevitz, J.O. and Pogson, B.J. (2016) Reconsidering plant memory: Intersections between stress recovery, RNA turnover, and epigenetics. *Science Advances* 2(2), e1501340.

Dalmay, T. (2013) Mechanism of miRNA-mediated repression of mRNA translation. *Essays in Biochemistry* 54, 29–38.

Dambacher, S., Hahn, M. and Schotta, G. (2010) Epigenetic regulation of development by histone lysine methylation. *Heredity* 105(1), 24–37.

De-la-Peña, C., Rangel-Cano, A. and Alvarez-Venegas, R. (2012) Regulation of disease-responsive genes mediated by epigenetic factors: Interaction of *Arabidopsis*–pseudomonas. *Molecular Plant Pathology* 13(4), 388–398.

Deans, C. and Maggert, K.A. (2015) What do you mean, "epigenetic"? *Genetics* 199(4), 887–896.

Deleris, A., Halter, T. and Navarro, L.D.N.A. (2016) DNA methylation and demethylation in plant immunity. *Annual Review of Phytopathology* 54(1), 579–603.

Demetriou, K., Kapazoglou, A., Tondelli, A., Francia, E., Stanca, M.A. *et al.* (2009) Epigenetic chromatin modifiers in barley: I. cloning, mapping and expression analysis of the plant specific HD2 family of histone deacetylases from barley, during seed development and after hormonal treatment. *Physiologia Plantarum* 136(3), 358–368.

Deng, W., Marshall, N.C., Rowland, J.L., McCoy, J.M., Worrall, L.J. *et al.* (2017) Assembly, structure, function and regulation of type III secretion systems. *Nature Reviews Microbiology* 15(6), 323–337.

Dhawan, R., Luo, H., Foerster, A.M., Abuqamar, S., Du, H.N. *et al.* (2009) HISTONE MONOUBIQUITINATION1 interacts with a subunit of the mediator complex and regulates defense against necrotrophic fungal pathogens in *Arabidopsis*. *The Plant Cell* 21(3), 1000–1019.

Diallo, I. and Provost, P. (2020) RNA-sequencing analyses of small bacterial RNAs and their emergence as virulence factors in host-pathogen interactions. *International Journal of Molecular Sciences* 21(5), 1627.

Ding, B. and Wang, G.L. (2015) Chromatin versus pathogens: The function of epigenetics in plant immunity. *Frontiers in Plant Science* 6, 152822.

Dong, W. and Hamon, M.A. (2020) Revealing eukaryotic histone-modifying mechanisms through bacterial infection. In: *Seminars in Immunopathology*, Vol. 42, No. 2. Springer Berlin Heidelberg, pp. 201–213.

Du, J. (2016) Structure and mechanism of plant DNA methyltransferases. In: Jeltsch, A. and Jurkowska, R. (eds) *DNA Methyltransferases – Role and Function*, Vol. 945. Advances in Experimental Medicine and Biology, Springer, Cham, pp. 173–192.

Du, Z., Li, H., Wei, Q., Zhao, X., Wang, C. *et al.* (2013) Genome-wide analysis of histone modifications: H3k4me2, H3k4me3, H3k9ac, and H3k27ac in *Oryza sativa* L. *Japonica. Molecular Plant* 6(5), 1463–1472.

Durrant, W.E. and Dong, X. (2004) Systemic acquired resistance. *Annual Review of Phytopathology* 42, 185–209.

Ehrenhofer-Murray, A.E. (2004) Chromatin dynamics at DNA replication, transcription and repair. *European Journal of Biochemistry* 271(12), 2335–2349.

Ehrlich, M. and Lacey, M. (2013) DNA methylation and differentiation: Silencing, upregulation and modulation of gene expression. *Epigenomics* 5(5), 553–568.

Espinas, N.A., Saze, H. and Saijo, Y. (2016) Epigenetic control of defense signaling and priming in plants. *Frontiers in Plant Science* 7, 202758.

Farmer, A., Thibivilliers, S., Ryu, K.H., Schiefelbein, J. and Libault, M. (2021) Single-nucleus RNA and ATAC sequencing reveals the impact of chromatin accessibility on gene expression in *Arabidopsis* roots at the single-cell level. *Molecular Plant* 14(3), 372–383.

Farooq, R., Hussain, K., Nazir, S., Javed, M.R. and Masood, N. (2018) CRISPR/Cas9; A robust technology for producing genetically engineered plants. *Cellular and Molecular Biology* 64(14), 31–38.

Finnegan, E.J. and Kovac, K.A. (2000) Plant DNA methyltransferases. *Plant Molecular Biology* 43, 189–201.

Fouse, S.D., Nagarajan, R.P. and Costello, J.F. (2010) Genome-scale DNA methylation analysis. *Epigenomics* 2(1), 105–117.

Furci, L., Berthelier, J., Juez, O., Miryeganeh, M. and Saze, H. (2023) Plant epigenomics. In: Tollefsbol, T.O. (ed.) *Handbook of Epigenetics*. Academic Press, pp. 263–286.

Gallego-Bartolomé, J., Liu, W., Kuo, P.H., Feng, S., Ghoshal, B. *et al.* (2019) Co-targeting RNA polymerases IV and V promotes efficient de novo DNA methylation in *Arabidopsis*. *Cell* 176(5), 1068–1082.

Gambino, G. and Pantaleo, V. (2017) Epigenetics in plant–pathogen interactions. *Plant Epigenetics* 385–404.

Garner, C.M., Kim, S.H., Spears, B.J. and Gassmann, W. (2016) Express yourself: Transcriptional regulation of plant innate immunity. In: Cerase, A. (ed.) *Plant Immunity. Seminars in Cell & Developmental Biology*, Vol. 56. Academic Press, pp. 150–162.

Gehring, M., Reik, W. and Henikoff, S. (2009) DNA demethylation by DNA repair. *Trends in Genetics* 25(2), 82–90.

Gill, E.E., Franco, O.L. and Hancock, R.E.W. (2015) Antibiotic adjuvants: Diverse strategies for controlling drug-resistant pathogens. *Chemical Biology & Drug Design* 85(1), 56–78.

Goell, J.H. and Hilton, I.B. (2021) CRISPR/Cas-based epigenome editing: Advances, applications, and clinical utility. *Trends in Biotechnology* 39(7), 678–691.

Grant-Downton, R.T. and Dickinson, H.G. (eds) (2005) Epigenetics and its implications for plant biology. 1. The epigenetic network in plants. *Annals of Botany* 96(7), 1143–1164.

Griffin, K., Gambley, C., Brown, P. and Li, Y. (2017) Copper-tolerance in *Pseudomonas syringae* pv. tomato and *Xanthomonas* spp. and the control of diseases associated with these pathogens in tomato and pepper. A systematic literature review. *Crop Protection* 96, 144–150.

Guleria, P. and Kumar, V. (eds) (2020) *Plant small RNA: Biogenesis, Regulation and Application*. Academic Press.

Hagmann, J. (2017) Characterizing short-term evolution of DNA methylation in *A. thaliana* using next-generation sequencing. Doctoral dissertation, Universität Tübingen.

Halter, T., Wang, J., Amesefe, D., Lastrucci, E., Charvin, M. *et al.* (2021) The *Arabidopsis* active demethylase ROS1 *cis*-regulates defence genes by erasing DNA methylation at promoter-regulatory regions. *eLife* 10, e62994.

Hammond-Martel, I., Yu, H. and Affar, E.B. (2012) Roles of ubiquitin signaling in transcription regulation. *Cellular Signalling* 24(2), 410–421.

Hann, D.R., Gimenez-Ibanez, S. and Rathjen, J.P. (2010) Bacterial virulence effectors and their activities. *Current Opinion in Plant Biology* 13(4), 388–393.

Hannan Parker, A., Wilkinson, S.W. and Ton, J. (2022) Epigenetics: A catalyst of plant immunity against pathogens. *New Phytologist* 233(1), 66–83.

Hayes, M.M., Dewberry, R.J., Babujee, L., Moritz, R. and Allen, C. (2022) Validating methods to eradicate plant-pathogenic *Ralstonia* strains reveals that growth *in planta* increases bacterial stress tolerance. *Microbiology Spectrum* 10(6), e02270–22.

Heard, E. and Martienssen, R.A. (2014) Transgenerational epigenetic inheritance: Myths and mechanisms. *Cell* 157(1), 95–109.

Heydari, A. and Pessarakli, M. (2010) A review on biological control of fungal plant pathogens using microbial antagonists. *Journal of Biological Sciences* 10(4), 273–290.

Holeski, L.M., Jander, G. and Agrawal, A.A. (2012) Transgenerational defense induction and epigenetic inheritance in plants. *Trends in Ecology & Evolution* 27(11), 618–626.

Hsieh, P.H. (2016) *Maintenance and inheritance of DNA methylation in Arabidopsis*. Doctoral dissertation, UC Berkeley.

Jamil, I.N., Remali, J., Azizan, K.A., Nor Muhammad, N.A., Arita, M. *et al.* (2020) Systematic multi-omics integration (MOI) approach in plant systems biology. *Frontiers in Plant Science* 11, 540561.

Jasanoff, S., Hurlbut, J.B. and Saha, K. (2015) CRISPR democracy: Gene editing and the need for inclusive deliberation. *Issues in Science and Technology* 32(1), 25–32.

Jeltsch, A. and Jurkowska, R.Z. (2014) New concepts in DNA methylation. *Trends in Biochemical Sciences* 39(7), 310–318.

Jenuwein, T. and Allis, C.D. (2001) Translating the histone code. *Science* 293(5532), 1074–1080.

Johnson, D.S., Mortazavi, A., Myers, R.M. and Wold, B. (2007) Genome-wide mapping of *in vivo* protein-DNA interactions. *Science* 316(5830), 1497–1502.

Jones-Rhoades, M.W., Bartel, D.P. and Bartel, B. (2006) MicroRNAS and their regulatory roles in plants. *Annual Review of Plant Biology* 57, 19–53.

Jungers, C.F. and Djuranovic, S. (2022) Modulation of miRISC-mediated gene silencing in eukaryotes. *Frontiers in Molecular Biosciences* 9, 832916.

Kakoulidou, I., Avramidou, E.V., Baránek, M., Brunel-Muguet, S., Farrona, S. *et al.* (2021) Epigenetics for crop improvement in times of global change. *Biology* 10(8), 766.

Kamle, M., Borah, R., Bora, H., Jaiswal, A.K., Singh, R.K. *et al.* (2020) Systemic acquired resistance (SAR) and induced systemic resistance (ISR): Role and mechanism of action against phytopathogens. In: Hesham, A.L., Upadhyay, R., Sharma, G., Manoharachary, C. and Gupta, V. (eds) *Fungal Biotechnology and Bioengineering*. Springer, Cham, pp. 457–470.

Kamthan, A., Chaudhuri, A., Kamthan, M. and Datta, A. (2015) Small RNAs in plants: Recent development and application for crop improvement. *Frontiers in Plant Science* 6, 124569.

Kang, H., Fan, T., Wu, J., Zhu, Y. and Shen, W.H. (2022) Histone modification and chromatin remodeling in plant response to pathogens. *Frontiers in Plant Science* 13, 986940.

Katiyar-Agarwal, S. and Jin, H. (2007) Discovery of pathogen-regulated small RNAs in plants. *Methods in Enzymology* 427, 215–227.

Kenchanmane Raju, S.K., Ritter, E.J. and Niederhuth, C.E. (2019) Establishment, maintenance, and biological roles of non-CG methylation in plants. *Essays in Biochemistry* 63(6), 743–755.

Khan, M.A., Naeem, M. and Iqbal, M. (2014) Breeding approaches for bacterial leaf blight resistance in rice (*Oryza sativa* L.), current status and future directions. *European Journal of Plant Pathology* 139(1), 27–37.

Khan, M.H.U., Khan, S.U., Muhammad, A., Hu, L., Yang, Y. *et al.* (2018) Induced mutation and epigenetics modification in plants for crop improvement by targeting CRISPR/Cas9 technology. *Journal of Cellular Physiology* 233(6), 4578–4594.

Kim, S., Piquerez, S.J.M., Ramirez-Prado, J.S., Mastorakis, E., Veluchamy, A. *et al.* (2020) GCN5 modulates salicylic acid homeostasis by regulating H3k14ac levels at the 5' and 3' ends of its target genes. *Nucleic Acids Research* 48(11), 5953–5966.

Klupczyńska, E.A. and Ratajczak, E. (2021) Can forest trees cope with climate change? – effects of DNA methylation on gene expression and adaptation to environmental change. *International Journal of Molecular Sciences* 22(24), 13524.

Kou, M.Z., Bastías, D.A., Christensen, M.J., Zhong, R., Nan, Z.B. *et al.* (2021) The plant salicylic acid signalling pathway regulates the infection of a biotrophic pathogen in grasses associated with an *Epichloë* endophyte. *Journal of Fungi* 7(8), 633.

Kouzarides, T. (2007) Chromatin modifications and their function. *Cell* 128(4), 693–705.

Krishnaswami, V., Kumar, M. and Vijayaraghavalu, S. (2022) Advances in epigenetics for crop improvement and sustainable agriculture. In: Singh, R.L., Mondal, S., Parthar, A. and Singh, P.K. (eds) *Plant Genomics for Sustainable Agriculture*. Springer Nature Singapore, Singapore, pp. 351–370.

Ku, C.S., Naidoo, N., Wu, M. and Soong, R. (2011) Studying the epigenome using next generation sequencing. *Journal of Medical Genetics* 48(11), 721–730.

Kumar, H., Chaudhary, A., Singh, A., Sukhija, N., Panwar, A. *et al.* (2020) A review on epigenetics: Manifestations, modifications, methods & challenges. *Journal of Entomology and Zoology Studies* 8(4), 01–06.

Kumar, R., Kumar, R. and Prakash, O. (2019) The impact of chemical fertilizers on our environment and ecosystem. In: Sharma, P. (ed.) *Research Trends in Environmental Sciences*. AkiNik Publications, New Delhi, pp. 60–86.

Kuzma, J. and Kokotovich, A. (2011) Renegotiating GM crop regulation: Targeted gene-modification technology raises new issues for the oversight of genetically modified crops. *EMBO Reports* 12(9), 883–888.

Leben, C. (1981) How plant-pathogenic bacteria survive. *Plant Disease* 65(8), 633–637.

Lee, J., Teitzel, G.M., Munkvold, K., Pozo, O., Martin, G.B. *et al.* (2012) Type III secretion and effectors shape the survival and growth pattern of *Pseudomonas syringae* on leaf surfaces. *Plant Physiology* 158(4), 1803–1818.

Lee, J., Lee, S., Park, K., Shin, S.Y., Frost, J.M. *et al.* (2023) Distinct regulatory pathways contribute to dynamic CHH methylation patterns in transposable elements throughout *Arabidopsis* embryogenesis. *Frontiers in Plant Science* 14, 1204279.

Lephatsi, M.M., Meyer, V., Piater, L.A., Dubery, I.A. and Tugizimana, F. (2021) Plant responses to abiotic stresses and rhizobacterial biostimulants: Metabolomics and epigenetics perspectives. *Metabolites* 11(7), 457.

Li, B., Carey, M. and Workman, J.L. (2007) The role of chromatin during transcription. *Cell* 128(4), 707–719.

Li, J., Meng, X., Zong, Y., Chen, K., Zhang, H. *et al.* (2016) Gene replacements and insertions in rice by intron targeting using CRISPR-Cas9. *Nature Plants* 2(10), 1–6.

Li, M. and Yu, B. (2021) Recent advances in the regulation of plant miRNA biogenesis. *RNA Biology* 18(12), 2087–2096.

Li, N., Han, X., Feng, D., Yuan, D. and Huang, L.J. (2019) Signaling crosstalk between salicylic acid and ethylene/jasmonate in plant defense: Do we understand what they are whispering? *International Journal of Molecular Sciences* 20(3), 671.

Li, P., Liu, Q., Wei, Y., Xing, C., Xu, Z. *et al.* (2024) Transcriptional landscape of cotton roots in response to salt stress at single-cell resolution. *Plant Communications* 5(2), 100740.

Licciardi, P.V. and Karagiannis, T.C. (2012) Regulation of immune responses by histone deacetylase inhibitors. *International Scholarly Research Notices* 690901.

Lindsey, A.P.J., Murugan, S. and Renitta, R.E. (2020) Microbial disease management in agriculture: Current status and future prospects. *Biocatalysis and Agricultural Biotechnology* 23, 101468.

Lister, R., O'Malley, R.C., Tonti-Filippini, J., Gregory, B.D., Berry, C.C. *et al.* (2008) Highly integrated single-base resolution maps of the epigenome in *Arabidopsis*. *Cell* 133(3), 523–536.

Liu, J. and He, Z. (2020) Small DNA methylation, big player in plant abiotic stress responses and memory. *Frontiers in Plant Science* 11, 595603.

Liu, X., Luo, M., Zhang, W., Zhao, J., Zhang, J. *et al.* (2012) Histone acetyltransferases in rice (*Oryza sativa* L.): Phylogenetic analysis, subcellular localization and expression. *BMC Plant Biology* 12, 1–17.

Liu, X., Yang, S., Yu, C.W., Chen, C.Y. and Wu, K. (2016) Histone acetylation and plant development. *The Enzymes* 40, 173–199.

Loaeza-Loaeza, J., Beltran, A.S. and Hernández-Sotelo, D. (2020) DNMTs and impact of cpg content, transcription factors, consensus motifs, lncRNAs, and histone marks on DNA methylation. *Genes* 11(11), 1336.

Lopez-Gomollon, S. and Baulcombe, D.C. (2022) Roles of RNA silencing in viral and non-viral plant immunity and in the crosstalk between disease resistance systems. *Nature Reviews Molecular Cell Biology* 23(10), 645–662.

Lu, D. (2013) Epigenetic modification enzymes: Catalytic mechanisms and inhibitors. *Acta Pharmaceutica Sinica B* 3(3), 141–149.
Luna, E., Bruce, T.J., Roberts, M.R., Flors, V. and Ton, J. (2012) Next-generation systemic acquired resistance. *Plant Physiology* 158(2), 844–853.
Luo, C., Fernie, A.R. and Yan, J. (2020) Single-cell genomics and epigenomics: Technologies and applications in plants. *Trends in Plant Science* 25(10), 1030–1040.
Ma, K.W., Flores, C. and Ma, W. (2011) Chromatin configuration as a battlefield in plant-bacteria interactions. *Plant Physiology* 157(2), 535–543.
Mansfield, J., Genin, S., Magori, S., Citovsky, V., Sriariyanum, M. *et al.* (2012) Top 10 plant pathogenic bacteria in molecular plant pathology. *Molecular Plant Pathology* 13(6), 614–629.
Manyi-Loh, C., Mamphweli, S., Meyer, E. and Okoh, A. (2018) Antibiotic use in agriculture and its consequential resistance in environmental sources: Potential public health implications. *Molecules* 23(4), 795.
Matzke, M.A. and Mosher, R.A. (2014) RNA-directed DNA methylation: An epigenetic pathway of increasing complexity. *Nature Reviews Genetics* 15(6), 394–408.
Mauch-Mani, B., Baccelli, I., Luna, E. and Flors, V. (2017) Defense priming: An adaptive part of induced resistance. *Annual Review of Plant Biology* 68(1), 485–512.
McDonald, B.A. and Stukenbrock, E.H. (2016) Rapid emergence of pathogens in agro-ecosystems: Global threats to agricultural sustainability and food security. *Philosophical Transactions of the Royal Society of London. Series B: Biological Sciences* 371(1709), 20160026.
Mehrmohamadi, M., Sepehri, M.H., Nazer, N. and Norouzi, M.R. (2021) A comparative overview of epigenomic profiling methods. *Frontiers in Cell and Developmental Biology* 9, 714687.
Meng, D., Dubin, M., Zhang, P., Osborne, E.J., Stegle, O. *et al.* (2016) Limited contribution of DNA methylation variation to expression regulation in *Arabidopsis thaliana*. *PLoS Genetics* 12(7), e1006141.
Miglani, G.S., Kaur, A. and Kaur, L. (2020) Plant gene expression control using genome- and epigenome-editing technologies. *Journal of Crop Improvement* 34(1), 1–63.
Mladenov, V., Fotopoulos, V., Kaiserli, E., Karalija, E., Maury, S. *et al.* (2021) Deciphering the epigenetic alphabet involved in transgenerational stress memory in crops. *International Journal of Molecular Sciences* 22(13), 7118.
Montesinos, E., Bonaterra, A., Badosa, E., Francés, J., Alemany, J. *et al.* (2002) Plant–microbe interactions and the new biotechnological methods of plant disease control. *International Microbiology* 5(4), 169–175.
Mostafa, S., Wang, Y., Zeng, W. and Jin, B. (2022) Plant responses to herbivory, wounding, and infection. *International Journal of Molecular Sciences* 23(13), 7031.
Muhammad, T., Zhang, F., Zhang, Y. and Liang, Y. (2019) RNA interference: A natural immune system of plants to counteract biotic stressors. *Cells* 8(1), 38.
Murmu, S., Sinha, D., Chaurasia, H., Sharma, S., Das, R. *et al.* (2024) A review of artificial intelligence-assisted omics techniques in plant defense: Current trends and future directions. *Frontiers in Plant Science* 15, 1292054.
Mushtaq, M., Sakina, A., Wani, S.H., Shikari, A.B., Tripathi, P. *et al.* (2019) Harnessing genome editing techniques to engineer disease resistance in plants. *Frontiers in Plant Science* 10, 550.
Mushtaq, M., Ahmad Dar, A., Skalicky, M., Tyagi, A., Bhagat, N. *et al.* (2021) CRISPR-based genome editing tools: Insights into technological breakthroughs and future challenges. *Genes* 12(6), 797.
Nair, R.G. (2020) Molecular determinants of host adaptation in the plant pathogen *Ralstonia solanacearum*. Doctoral dissertation, Université Paul Sabatier-Toulouse III.
National Academies of Sciences, Engineering, and Medicine (2019) *Science Breakthroughs to Advance Food and Agricultural Research by 2030*. National Academies Press, Washington, DC.
Nguyen, N.H., Vu, N.T. and Cheong, J.J. (2022) Transcriptional stress memory and transgenerational inheritance of drought tolerance in plants. *International Journal of Molecular Sciences* 23(21), 12918.
Nimchuk, Z., Eulgem, T., Holt Iii, B.F. and Dangl, J.L. (2003) Recognition and response in the plant immune system. *Annual Review of Genetics* 37(1), 579–609.
Noman, A., Aqeel, M. and Lou, Y. (2019) PRRs and NB-LRRs: From signal perception to activation of plant innate immunity. *International Journal of Molecular Sciences* 20(8), 1882.
Pagiatakis, C., Musolino, E., Gornati, R., Bernardini, G. and Papait, R. (2021) Epigenetics of aging and disease: A brief overview. *Aging Clinical and Experimental Research* 33(4), 737–745.
Palti, J. (2012) *Cultural Practices and Infectious Crop Diseases*. Springer, Berlin, Heidelberg.

Pandey, D., Rajendran, S.R.C.K., Gaur, M., Sajeesh, P.K. and Kumar, A. (2016) Plant defense signaling and responses against necrotrophic fungal pathogens. *Journal of Plant Growth Regulation* 35, 1159–1174.

Panigrahi, G.K., Sahoo, A. and Satapathy, K.B. (2021) Insights to plant immunity: Defense signaling to epigenetics. *Physiological and Molecular Plant Pathology* 113, 101568.

Papikian, A., Liu, W., Gallego-Bartolomé, J. and Jacobsen, S.E. (2019) Site-specific manipulation of *Arabidopsis* loci using CRISPR-Cas9 suntag systems. *Nature Communications* 10(1), 729.

Park, J.H. and Shin, C. (2015) The role of plant small RNAs in NB-LRR regulation. *Briefings in Functional Genomics* 14(4), 268–274.

Pattanayak, D., Agarwal, S., Sumathi, S., Chakrabarti, S.K., Naik, P.S. *et al.* (2005) Small but mighty RNA-mediated interference in plants. *Indian Journal of Experimental Biology* 43(1), 7–24.

Peláez, P. and Sanchez, F. (2013) Small RNAs in plant defense responses during viral and bacterial interactions: Similarities and differences. *Frontiers in Plant Science* 4, 62127.

Peng, J., Xia, B. and Yi, C. (2016) Single-base resolution analysis of DNA epigenome via high-throughput sequencing. *Science China Life Sciences* 59(3), 219–226.

Pfeilmeier, S., Caly, D.L. and Malone, J.G. (2016) Bacterial pathogenesis of plants: Future challenges from a microbial perspective: Challenges in bacterial molecular plant pathology. *Molecular Plant Pathology* 17(8), 1298–1313.

Pickersgill, M., Niewöhner, J., Müller, R., Martin, P. and Cunningham-Burley, S. (2013) Mapping the new molecular landscape: Social dimensions of epigenetics. *New Genetics and Society* 32(4), 429–447.

Pikaard, C.S. and Scheid, O.M. (2014) Epigenetic regulation in plants. *Cold Spring Harbor Perspectives in Biology* 6(12), a019315.

Prakash, S., Singh, R. and Lodhi, N. (2014) Histone demethylases and control of gene expression in plants. *Cellular and Molecular Biology* 60(5), 97–105.

Ragimekula, N., Varadarajula, N.N., Mallapuram, S.P., Gangimeni, G., Reddy, R.K. *et al.* (2013) Marker assisted selection in disease resistance breeding. *Journal of Plant Breeding and Genetics* 1(2), 90–109.

Ramirez-Prado, J.S., Abulfaraj, A.A., Rayapuram, N., Benhamed, M. and Hirt, H. (2018a) Plant immunity: From signaling to epigenetic control of defense. *Trends in Plant Science* 23(9), 833–844.

Ramirez-Prado, J.S., Piquerez, S.J., Bendahmane, A., Hirt, H., Raynaud, C. *et al.* (2018b) Modify the histone to win the battle: Chromatin dynamics in plant-pathogen interactions. *Frontiers in Plant Science* 9, 355.

Reshetnyak, G., Jacobs, J.M., Auguy, F., Sciallano, C., Claude, L. *et al.* (2021) An atypical class of non-coding small RNAs is produced in rice leaves upon bacterial infection. *Scientific Reports* 11(1), 24141.

Roberts, M.R. and López Sánchez, A. (2019) Plant epigenetic mechanisms in response to biotic stress. In: Alvarez-Venegas, R., De-la-Peña, C. and Casas-Mollano, J.A. (eds) *Epigenetics in Plants of Agronomic Importance: Fundamentals and Applications: Transcriptional Regulation and Chromatin Remodelling in Plants*. Springer Cham, pp. 65–113.

Rossi, V., Locatelli, S., Varotto, S., Donn, G., Pirona, R. *et al.* (2007) Maize histone deacetylase *hda101* is involved in plant development, gene transcription, and sequence-specific modulation of histone modification of genes and repeats. *The Plant Cell* 19(4), 1145–1162.

Rozas, P., Kessi-Pérez, E.I. and Martínez, C. (2022) Genetically modified organisms: Adapting regulatory frameworks for evolving genome editing technologies. *Biological Research* 55(1).

Sanan-Mishra, N., Abdul Kader Jailani, A., Mandal, B. and Mukherjee, S.K. (2021) Secondary siRNAs in plants: Biosynthesis, various functions, and applications in virology. *Frontiers in Plant Science* 12, 610283.

Saradadevi, G.P., Das, D., Mangrauthia, S.K., Mohapatra, S., Chikkaputtaiah, C. *et al.* (2021) Genetic, epigenetic, genomic and microbial approaches to enhance salt tolerance of plants: A comprehensive review. *Biology* 10(12), 1255.

Savary, S. and Willocquet, L. (2020) Modeling the impact of crop diseases on global food security. *Annual Review of Phytopathology* 58, 313–341.

Sawicka, B., Egbuna, C., Nayak, A.K. and Kala, S. (2020) Plant diseases, pathogens and diagnosis. In: Egbuna, C. and Sawicka, B. (eds) *Natural Remedies for Pest, Disease and Weed Control*. Academic Press, pp. 17–28.

Saxena, A., Rai, D., Kalra, A. and Saxena, M.J. (2020) Bacterial diseases of crops: Epidemiology symptoms and management. *International Journal of Current Microbiology and Applied Sciences* 9(6), 1483–1499.

Segelbacher, G., Bosse, M., Burger, P., Galbusera, P., Godoy, J.A. *et al.* (2022) New developments in the field of genomic technologies and their relevance to conservation management. *Conservation Genetics* 23(2), 217–242.

Seo, J.K., Wu, J., Lii, Y., Li, Y. and Jin, H. (2013) Contribution of small RNA pathway components in plant immunity. *Molecular Plant-Microbe Interactions* 26(6), 617–625.

Shah, J. and Zeier, J. (2013) Long-distance communication and signal amplification in systemic acquired resistance. *Frontiers in Plant Science* 4, 30.

Sharma, P., Bora, L.C., Puzari, K.C., Baruah, A.M. and Baruah, R. (2017) Review on bacterial blight of rice caused by *Xanthomonas oryzae* pv. *Oryzae*: Different management approaches and role of *Pseudomonas fluorescens* as a potential biocontrol agent. *International Journal of Current Microbiology and Applied Sciences* 6(3), 982–1005.

Sharma, Y., Sharma, A., Madhu, S., Singh, K. and Upadhyay, S.K. (2022) Long non-coding RNAs as emerging regulators of pathogen response in plants. *Non-Coding RNA* 8(1), 4.

Shin, H., Choi, W.L., Lim, J.Y. and Huh, J.H. (2022) Epigenome editing: targeted manipulation of epigenetic modifications in plants. *Genes & Genomics* 44, 307–315. DOI: 10.1007/s13258-021-01199-5.

Slaughter, A., Daniel, X., Flors, V., Luna, E., Hohn, B. *et al.* (2012) Descendants of primed *Arabidopsis* plants exhibit resistance to biotic stress. *Plant Physiology* 158(2), 835–843.

Smale, S.T. and Fisher, A.G. (2002) Chromatin structure and gene regulation in the immune system. *Annual Review of Immunology* 20(1), 427–462.

Song, L., Fang, Y., Chen, L., Wang, J. and Chen, X. (2021) Role of non-coding RNAs in plant immunity. *Plant Communications* 2(3), 100180.

Spoel, S.H. and Dong, X. (2012) How do plants achieve immunity? Defence without specialized immune cells. *Nature Reviews Immunology* 12(2), 89–100.

Springer, N.M. and Schmitz, R.J. (2017) Exploiting induced and natural epigenetic variation for crop improvement. *Nature Reviews Genetics* 18(9), 563–575.

Strange, R.N. and Scott, P.R. (2005) Plant disease: A threat to global food security. *Annual Review of Phytopathology* 43, 83–116.

Sun, C., Ali, K., Yan, K., Fiaz, S., Dormatey, R. *et al.* (2021) Exploration of epigenetics for improvement of drought and other stress resistance in crops: A review. *Plants* 10(6), 1226.

Sun, F., Fang, P., Li, J., Du, L., Lan, Y. *et al.* (2016) RNA-seq-based digital gene expression analysis reveals modification of host defense responses by rice stripe virus during disease symptom development in *Arabidopsis*. *Virology Journal* 13, 1–13.

Sung, C.K. and Yim, H. (2020) CRISPR-mediated promoter de/methylation technologies for gene regulation. *Archives of Pharmacal Research* 43(7), 705–713.

Tan, H., Li, B. and Guo, H. (2020) The diversity of post-transcriptional gene silencing mediated by small silencing RNAs in plants. *Essays in Biochemistry* 64(6), 919–930.

Tirnaz, S. and Batley, J. (2019) DNA methylation: Toward crop disease resistance improvement. *Trends in Plant Science* 24(12), 1137–1150.

Tiwari, M., Pati, D., Mohapatra, R., Sahu, B.B. and Singh, P. (2022) The impact of microbes in plant immunity and priming induced inheritance: A sustainable approach for crop protection. *Plant Stress* 4, 100072.

Tornesaki, A. (2023) The *Arabidopsis* MYST histone acetyltransferase, HAG4, modulates root immunity responses. Doctoral dissertation, University of Warwick.

Uffelmann, E., Huang, Q.Q., Munung, N.S., Vries, J., Okada, Y. *et al.* (2021) Genome-wide association studies. *Nature Reviews Methods Primers* 1(1), 59.

Vaillant, I. and Paszkowski, J. (2007) Role of histone and DNA methylation in gene regulation. *Current Opinion in Plant Biology* 10(5), 528–533.

Van Esse, H.P., Reuber, T.L. and Does, D. (2020) Genetic modification to improve disease resistance in crops. *The New Phytologist* 225(1), 70–86.

Vanyushin, B.F. and Ashapkin, V.V. (2011) DNA methylation in higher plants: Past, present and future. *Biochimica et Biophysica Acta (BBA) – Gene Regulatory Mechanisms* 1809(8), 360–368.

Velásquez, A.C., Castroverde, C.D.M. and He, S.Y. (2018) Plant-pathogen warfare under changing climate conditions. *Current Biology* 28(10), R619–R634.

Viggiano, L. and Pinto, M.C. (2017) Dynamic DNA methylation patterns in stress response. *Plant Epigenetics* 281–302.

Vurro, M., Bonciani, B. and Vannacci, G. (2010) Emerging infectious diseases of crop plants in developing countries: Impact on agriculture and socio-economic consequences. *Food Security* 2(2), 113–132.

Wada, Y. (2005) Physiological functions of plant DNA methyltransferases. *Plant Biotechnology* 22(2), 71–80.

Waheed, S., Anwar, M., Saleem, M.A., Wu, J., Tayyab, M. *et al*. (2021) The critical role of small RNAs in regulating plant innate immunity. *Biomolecules* 11(2), 184.

Wang, X., Tu, M., Li, Z., Wang, Y. and Wang, X. (2018) Current progress and future prospects for the clustered regularly interspaced short palindromic repeats (CRISPR) genome editing technology in fruit tree breeding. *Critical Reviews in Plant Sciences* 37(4), 233–258.

Wang, Y., Pruitt, R.N., Nürnberger, T. and Wang, Y. (2022) Evasion of plant immunity by microbial pathogens. *Nature Reviews Microbiology* 20(8), 449–464.

Wang, Z., Gerstein, M. and Snyder, M. (2009) RNA-seq: A revolutionary tool for transcriptomics. *Nature Reviews Genetics* 10(1), 57–63.

Wei, J.W., Huang, K., Yang, C. and Kang, C.S. (2017) Non-coding RNAs as regulators in epigenetics. *Oncology Reports* 37(1), 3–9.

Weiberg, A., Wang, M., Bellinger, M. and Jin, H. (2014) Small RNAs: A new paradigm in plant-microbe interactions. *Annual Review of Phytopathology* 52, 495–516.

Wendte, J.M., Zhang, Y., Ji, L., Shi, X., Hazarika, R.R. *et al*. (2019) Epimutations are associated with CHROMOMETHYLASE 3-induced de novo DNA methylation. *eLife* 8, e47891.

Wrangle, J., Wang, W., Koch, A., Easwaran, H., Mohammad, H.P. *et al*. (2013) Alterations of immune response of non-small cell lung cancer with azacytidine. *Oncotarget* 4(11), 2067–2079.

Wu, L., Chen, H., Curtis, C. and Fu, Z.Q. (2014) Go in for the kill: How plants deploy effector-triggered immunity to combat pathogens. [corrected], *Virulence* 5(7), 710–721.

Xavier, M.J., Roman, S.D., Aitken, R.J. and Nixon, B. (2019) Transgenerational inheritance: How impacts to the epigenetic and genetic information of parents affect offspring health. *Human Reproduction Update* 25(5), 518–540.

Xie, S.S. and Duan, C.G. (2023) Epigenetic regulation of plant immunity: From chromatin codes to plant disease resistance. *aBIOTECH* 4(2), 124–139.

Xu, Y., Miao, Y., Cai, B., Yi, Q., Tian, X. *et al*. (2022) A histone deacetylase inhibitor enhances rice immunity by derepressing the expression of defense-related genes. *Frontiers in Plant Science* 13, 1041095.

Yadav, C.B., Pandey, G., Muthamilarasan, M. and Prasad, M. (2018) Epigenetics and epigenomics of plants. *Plant Genetics and Molecular Biology* 237–261.

Yang, C., Dolatabadian, A. and Fernando, W.G.D. (2022) The wonderful world of intrinsic and intricate immunity responses in plants against pathogens. *Canadian Journal of Plant Pathology* 44(1), 1–20.

Yong, W.S., Hsu, F.M. and Chen, P.Y. (2016) Profiling genome-wide DNA methylation. *Epigenetics & Chromatin* 9, 26.

Yung, W.S., Li, M.W., Sze, C.C., Wang, Q. and Lam, H.M. (2021) Histone modifications and chromatin remodelling in plants in response to salt stress. *Physiologia Plantarum* 173(4), 1495–1513.

Zhai, J., Jeong, D.-H., Paoli, E., Park, S., Rosen, B.D. *et al*. (2011) MicroRNAs as master regulators of the plant NB-LRR defense gene family via the production of phased, trans-acting siRNAs. *Genes & Development* 25(23), 2540–2553.

Zhang, H. and Zhu, J.K. (2011) RNA-directed DNA methylation. *Current Opinion in Plant Biology* 14(2), 142–147.

Zhang, H., Lang, Z. and Zhu, J.K. (2018) Dynamics and function of DNA methylation in plants. *Nature Reviews Molecular Cell Biology* 19(8), 489–506.

Zhang, Y. (2003) Transcriptional regulation by histone ubiquitination and deubiquitination. *Genes & Development* 17(22), 2733–2740.

Zhang, Y. and Zeng, L. (2020) Crosstalk between ubiquitination and other post-translational protein modifications in plant immunity. *Plant Communications* 1(4), 100041.

Zhi, P. and Chang, C. (2021) Exploiting epigenetic variations for crop disease resistance improvement. *Frontiers in Plant Science* 12, 692328.

Zhu, Q.H., Shan, W.X., Ayliffe, M.A. and Wang, M.B. (2016) Epigenetic mechanisms: An emerging player in plant-microbe interactions. *Molecular Plant-Microbe Interactions* 29(3), 187–196.

31 Epigenetics for Combating Viral Diseases in Plants: Updated Methods and Current Achievements

Ajay Samuel Siva Daniel[1], Sarmad Moin[2], S. Basker[3], Sangilimuthu Alagar Yadav[4], Sahaya Shibu Babu[5] and Chithra Devi Balasundaram Saraswathy[6]*

[1]GRG BIRAC EYUVA Centre, PSGR Krishnammal College for Women, Coimbatore, Tamil Nadu, India; [2]School of Applied Sciences, Suresh Gyan Vihar University, Gyan Vihar Marg, Jagatpura, Jaipur, Rajasthan, India; [3]Department of Botany, Government Arts College, Salem, Tamil Nadu, India; [4]Department of Biotechnology, Karpagam Academy of Higher Education, Coimbatore, Tamil Nadu, India; [5]Department of Biotechnology, Safi Institute of Advanced Study (SIAS), Via Ramanattukara, Malappuram Dist, Kerala, India; [6]Department of Botany, PSGR Krishnammal College for Women, Coimbatore, Tamil Nadu, India

Abstract

In recent years, advancements in biotechnology and genomics have improved our understanding of plant–virus interactions, enabling the development of more targeted and sustainable approaches to managing viral diseases in plants. One approach is epigenetics, the study of heritable changes in gene expression that do not involve alterations to the underlying DNA sequence, which is an emerging field with promising applications in combating viral diseases in plants. Epigenetic modifications, such as DNA methylation and histone modifications, can influence the activation of genes involved in plant defense mechanisms. By targeting specific genes and pathways through epigenetic modifications, plants can enhance their resistance to viral infections. The battle against plant viruses remains an ongoing challenge, as these pathogens continue to evolve and adapt to new environmental conditions and host plants. In conclusion, epigenetics offers an innovative approach to enhancing plant resilience against viral diseases, and its potential for sustainable agriculture is increasingly evident. This chapter provides a comprehensive update on the methods and current achievements in the field, underlining the promise of epigenetics as a valuable tool in addressing the ongoing challenges posed by plant viral diseases.

Keywords: Epigenetics, Plant viral disease, DNA methylation, Histone modification, CRISPR/Cas9

31.1 Introduction

In recent years, the intersection of epigenetics and plant virology has opened new and promising avenues for combating viral diseases in plants. Traditional approaches to plant disease management often rely on genetic modifications or external treatments, but the emerging field of epigenetics offers a unique and sophisticated strategy. Epigenetics involves the modulation of

*Corresponding author: bchitradevi@psgrkcw.ac.in

 Epigenetics for Climate-Smart and Sustainable Agriculture (ed. Jen-Tsung Chen)
DOI: 10.1079/9781800626102.0031

gene expression without altering the underlying DNA sequence, providing a means to enhance plant immunity against viral invaders. This chapter explores the cutting-edge methods and remarkable achievements in the realm of epigenetics for combating viral diseases in plants, shedding light on how harnessing the power of epigenetic modifications is revolutionizing our ability to protect crops and ensure food security in an ever-evolving agricultural landscape. As global populations continue to grow, the importance of sustainable agriculture and food security becomes increasingly paramount. Plant viral diseases pose a significant threat to crop yields, making the development of innovative strategies for disease management imperative. One of the remarkable aspects of epigenetics is its potential to reduce the reliance on traditional interventions like chemical pesticides. By promoting the plant's innate resistance through epigenetic modifications, it is possible to achieve robust viral disease protection while minimizing the environmental and health-related risks associated with chemical treatments.

Viral diseases in plants, caused by ultra-microscopic organisms, are a significant concern for agriculture (Brooks, 1923). These diseases can affect a wide range of crops, including fruit trees and grapevines, leading to symptoms such as leaf discoloration, fruit malformation, and plant decline (Martelli and Uyemoto, 2008). Some notable examples of viral diseases in crops include *tobacco mosaic virus*, *tomato spotted wilt virus*, and *papaya ringspot virus*, which can cause significant yield losses (Sumida, 2020). Emerging and re-emerging viral diseases, such as *pepino mosaic virus* and *cucumber vein yellowing virus*, pose new threats to crops and require preventive measures such as sanitation and use of certified propagation material (Martelli and Gallitelli, 2008).

In the ever-evolving landscape of epigenetics, a host of advanced techniques and tools have emerged to dissect the intricate world of gene regulation, chromatin structure, and epigenetic modifications. These innovative methods have provided researchers with unprecedented insights into the dynamic and reversible changes that orchestrate gene expression, offering new possibilities for understanding and manipulating the epigenome. The advent of high-throughput sequencing technologies, such as ChIP-seq (chromatin immunoprecipitation sequencing) and RNA-seq, has revolutionized our ability to map epigenetic marks and gene expression on a genome-wide scale. These methods allow researchers to explore the intricate relationship between epigenetic modifications, chromatin structure, and gene activity, offering a comprehensive view of the epigenome.

Single-cell epigenomics is another cutting-edge area of research that permits the analysis of epigenetic modifications at the individual cell level. By examining epigenetic heterogeneity among cells within a population, researchers can gain a deeper understanding of complex biological processes, such as development, disease progression, and tissue regeneration. CRISPR/Cas9, originally known for its genome-editing capabilities, has been adapted for epigenome editing. Researchers can now precisely target specific genes or regions of the epigenome to modify DNA methylation patterns, histone marks, and other epigenetic features. This method provides a powerful tool for investigating the functional consequences of specific epigenetic changes and their role in gene regulation.

The three-dimensional organization of the genome within the nucleus plays a crucial role in gene regulation. Techniques like Hi-C (High-throughput Chromosome Conformation Capture) and 4C (Circular Chromosome Conformation Capture) enable the mapping of long-range chromatin interactions. These methods have shed light on how epigenetic marks can influence the spatial organization of genes and regulatory elements.

Recent advancements in plant virology have focused on the use of epigenetic modifications to combat viral diseases. Wani and Sanghera (2010) and Wang *et al.* (2019) both highlighted the potential of transgene-based approaches and epigenetic modifications, such as DNA methylation, in enhancing plant resistance to viruses. These methods have shown promise in improving the understanding of plant host methylation regulation in response to viral infection. Furthermore, Mehetre *et al.* (2021) discussed the development of new diagnostic techniques, including next-generation sequencing, which can aid in the identification and management of plant viral diseases. Lastly, Fondong *et al.* (2016) emphasized the potential of RNA-based technologies and genome editing approaches, such as CRISPR/Cas9, in generating virus-resistant plants. These studies collectively underscore the importance of integrating various methods, including epigenetic modifications, in the fight against plant viral diseases. Epigenetic modifications, such as DNA methylation, play a crucial role in plant–virus interactions, potentially providing a new avenue for antiviral defense (Wang *et al.*, 2019). This is further supported by the use of RNA-based technologies, including artificial microRNA and transacting small interfering RNA, in generating virus-resistant plants (Fondong *et al.*, 2016). Genetic engineering, particularly the development of transgene-based approaches, has also shown promise in managing viral diseases in plants (Wani and Sanghera, 2010). Additionally, the use of viral expression vectors for epigenetic modification of plants with systemic RNA viruses has been successful in identifying new gene functions (Fitzmaurice *et al.*, 2002). These studies collectively highlight the potential of epigenetics and genetic engineering in combating viral diseases in plants.

31.2 Plant Viral Diseases: A Growing Agricultural Challenge

Plant viral diseases pose a significant threat to agriculture, with a range of viruses affecting crops worldwide (Jones and Naidu, 2019). In Bangladesh, these diseases are particularly problematic, with vegetables and pulses being the most affected (Akhter *et al.*, 2019). In Europe, plant viruses cause substantial economic damage, yet they have received less attention compared with other pathogens (Vlugt, 2006). The impact of agriculture on the distribution and prevalence of plant viruses has been highlighted, with a significant influence observed in cultivated areas (Bernardo *et al.*, 2017). These studies underscore the need for a comprehensive understanding of plant viral diseases and the development of effective management strategies. Plant viruses are becoming potent to economically important crops, causing yield and quality reductions (Manjunatha *et al.*, 2022). In the Yemen Arab Republic, viruses such as *alfalfa mosaic virus*, watermelon chlorotic stunt disease, and *turnip mosaic virus* have been found to affect crops (Walkey *et al.*, 1990). The emergence of new viral diseases, such as *pepino mosaic virus* in tomatoes, presents ongoing challenges (Hanssen *et al.*, 2010). Prevention and control strategies, including genetic resistance and hygienic measures, are crucial in managing these diseases (Singh and Srivastava, 2020).

The rise of new viruses or strains due to genetic changes and transmission from reservoir host species, alterations in farming methods, co-infections leading to disease synergy, and the effects of climate change present ongoing obstacles for controlling epidemics caused by emerging plant virus diseases. This analysis outlines some of the most destructive virus diseases as well as specific virus diseases that have regional significance in key crops, resulting in substantial crop losses. The absence of effective treatments for plant virus infections necessitates the implementation of risk-reducing strategies to manage plant virus diseases. These strategies encompass exclusion, avoidance, and eradication methods, in addition to vector control practices. The introduction of next-generation sequencing technologies holds promise for identifying previously unknown viruses in quarantine samples. Utilizing genetic resistance in crop plants is a valuable and effective approach to managing virus diseases. Various dominant and recessive resistance genes have been utilized to combat virus diseases in crops.

The agricultural system is undergoing rapid changes due to climate change, the rapid increase in population, and food insecurity. These changes are creating favorable conditions for the outbreak of destructive viral diseases. When plants are infected by viruses, their physiology is altered, leading to the development of symptoms that ultimately result in significant yield losses. It is estimated that these losses amount to more than 30 billion dollars annually (Jones and Naidu, 2019). An effective strategy commonly utilized in agriculture involves utilizing crop genetic resistance, which relies on the mechanisms that control plant–virus interactions. Therefore, understanding the molecular foundations of viral infections and crop resistance is crucial for combating viral threats in agricultural settings (Nicaise, 2014). Numerous strategies have been attempted to reduce the negative impact of these diseases. These strategies primarily revolve around avoiding sources of infection, managing or eliminating vectors, altering cultural practices, utilizing resistant varieties obtained through traditional breeding methods, implementing cross-protection, promoting systemic acquired resistance, and employing transgenic plants with foreign genes that confer virus resistance (Varma, 2007). The *begomoviruses*, *tospoviruses*, and *potyviruses* are part of three major plant virus groups that pose a threat to food security by causing destructive diseases in tropical and subtropical food crops. These viruses are spread by whiteflies, thrips, and aphids, respectively (Jones, 2020). Despite the fact that these viruses continue to cause significant economic losses in various crops, their value as scientific tools has elevated their importance among researchers. It is worth mentioning that the situation has changed for viruses like *potato virus X* (PVX): although they initially posed a major problem resulting in substantial losses, certification schemes and breeding programs have successfully mitigated their impact. Furthermore, the initial interest from scientists has transformed PVX into an exceptional model system, not only for studying its virology but also for understanding plant–virus interactions (Scholthof *et al*., 2011). Apart from the commonly known virus family, most of the mosaic viruses play a major role in the spreading of viral disease among economically important plants.

31.3 Epigenetics: An Innovative Approach to Viral Disease Control

Conrad H. Waddington, a British developmental biologist, is credited with coining the term 'epigenetics'. Epigenetics focuses on the study of gene interactions and their products, which ultimately influence the phenotype of a system. Throughout an organism's development, cell fate is shaped not only by genes but also by other epigenetic factors, giving rise to the concept of 'epigenesis'. In modern biology, epigenetics is now understood as a process in which a gene's activity is regulated by modifications to nucleic acids or the physical structure of the chromatin in which it is housed (Rajeevkumar *et al*., 2015). Epigenetics has emerged as a promising field of study in the control of plant viral diseases. It offers potential applications in modifying both host and viral genomes, as highlighted by Fondong *et al*. (2016). DNA methylation and histone post-translational modifications are essential epigenetic factors that interact with regulatory proteins, non-coding RNAs, and each other to restructure chromatin into various domains (Delcuve *et al*., 2009). In addition to epigenetics, other genetic engineering strategies such as pathogen-derived resistance and RNA-mediated virus gene silencing, as discussed by Wani and Sanghera (2010), can complement this approach. Chauhan *et al*. (2019) conducted a systematic review of conventional and advanced methods for plant virus control, emphasizing the potential of nucleic acid-based techniques like RNA silencing and transgenic plants. Burdon *et al*. (2016) suggested that the use of non-pathogen-derived resistance could further enhance these methods, potentially providing more durable resistance. Epigenetic modifications in DNA and proteins are remarkable in the antiviral defense mechanism of plants. In response to viral challenges, methylation changes occur, leading to a reprogramming of the plant's defense system. This process holds the potential to enhance our comprehension of how plants regulate methylation in response to viral infections (Wang *et al*., 2019). Plants utilize a variety of molecular mechanisms to combat viruses, with RNA silencing being a key player in their defense system. Recent research has linked the plant's antiviral RNA silencing to changes in epigenetic modifications in both DNA and proteins, ultimately affecting the expression of coding genes (Samantara *et al*., 2021). When faced with a viral threat, the plant's methylation levels are altered as a response. Additionally, certain viruses have been found to impact epigenetic modifications through small RNA-mediated processes, leading to gene silencing at both transcriptional and post-transcriptional levels (Wang *et al*., 2019). Epigenetic regulation in plants, such as vernalization and virus-induced silencing, plays a crucial role in adapting to different environments and influencing fitness through natural selection (Baulcombe and Dean, 2014). Growing evidence indicates that the regulation of gene expression is most prominent in the epigenetic control of stress-induced phenotypic response in plants. It is becoming increasingly clear that DNA methylation, triggered by stress, contributes to the observed variations in phenotype (Sahu *et al.*, 2013). Small interfering RNAs (siRNAs) and methylation-based epigenetic defensive mechanisms can effectively protect plants against diverse plant viruses by inhibiting viral gene expression and promoting resistance through RNA silencing pathways (Sharma *et al.*, 2013).

31.4 Methods for Studying Epigenetics in Plant–Virus Interactions

Various molecular defense systems have been evolved by plants to fend against viruses. An important part of antiviral defense is RNA silencing. Through altering the expression levels of coding genes, recent research has linked plant antiviral suppression to epigenetic modifications in genomic DNA and protein. The methylation level of the plant host is altered in reaction to viral challenge. Certain virus genomes have been linked to epigenetic modification through post-transcriptional and transcriptional gene silencing mediated by short RNAs. Through methylation modifications for antiviral defense, these pathways can be primed before a virus attack (Wang *et al.*, 2019).

Methods for studying epigenetics in plant–virus interactions can be analyzed through; next-generation sequencing, bio-informatics tools, and experimental approaches, as follows.

31.4.1 Next-generation sequencing techniques (NGS)

NGS offers fast, cost-effective sequencing of plant viruses and viroids genomes, as well as specific small RNAs produced during infection, with high efficiency. These small RNAs, which cover frequently the whole genome of the infectious agent, are 21–24 nt long and are known as vsRNAs for viruses and vd-sRNAs for viroids (Hadidi *et al.*, 2016). Within the field of epigenetics, NGS is essential for interpreting the intricate regulatory webs that control gene expression in a range of biological processes, such as interactions between plants and viruses. This section aims to explore the concepts, procedures, uses, and developments of NGS methods in the investigation of epigenetics in plant–virus interactions. It emphasizes the importance of these methods in deciphering the molecular mechanisms that underlie plant defense responses against viruses (Meaburn and Schulz, 2012).

The fundamental idea behind NGS is the massive parallel sequencing of millions of DNA fragments (Behjati and Tarpey, 2013). This approach is used in a wide range of high-throughput sequencing methods. These methods make use of several sequencing-by-synthesis or sequencing-by-ligation strategies to produce enormous volumes of sequencing data, which allow for the highly accurate and efficient simultaneous analysis of several genomic regions.

Methodologies of next-generation sequencing

Three primary categories can be used to broadly classify next-generation sequencing methodologies: (i) transcriptome sequencing; (ii) targeted sequencing; and (iii) whole-genome sequencing (WGS).

A complete picture of an organism's genetic composition is obtained using whole-genome sequencing (WGS). Selected sequencing concentrates on certain genomic regions of interest, enabling high-coverage deep sequencing of selected regions. The process of capturing and sequencing the transcriptome, commonly referred to as RNA sequencing (RNA-seq), yields information about alternative splicing events and gene expression levels (Chaitankar *et al.*, 2016).

Next-generation sequencing techniques have revolutionized the study of epigenetics by facilitating the comprehensive analysis of various epigenetic modifications, including DNA methylation, histone modifications, and small RNAs. Bisulfite sequencing is widely used to profile DNA methylation patterns by converting unmethylated cytosines to uracil through bisulfite treatment, followed by sequencing.

Hirst and Marra (2010) showed that Chromatin immunoprecipitation followed by sequencing (ChIP-seq) enables the genome-wide mapping of histone modifications and transcription factor binding sites, providing insights into chromatin structure and gene regulation. Small RNA sequencing allows the characterization of small RNAs, including microRNAs (miRNAs) and small interfering RNAs (siRNAs), involved in post-transcriptional gene regulation and RNA silencing pathways.

Advancements in next-generation sequencing technologies

Read length, accuracy, cost-effectiveness, and sequencing throughput have all improved as a result of developments in NGS technologies. With the help of long-read sequencing capabilities provided by third-generation sequencing technologies like single-molecule real-time sequencing (SMRT) and nanopore sequencing, raw DNA or RNA molecules can be directly sequenced without the need for fragment cloning or PCR amplification (Cheng *et al.*, 2023). The ability to characterize complicated genomic areas, such as repeated sequences, structural changes, and epigenetic modifications in individual cells or tissues, has broadened the field of epigenetic study.

With the noteworthy advancements in NGS technology, a number of obstacles still need to be addressed. These include the need for data interpretation and analysis, protocol standardization, and scalable sequencing platforms (Tatineni and Hein, 2023). Compiling multi-omics datasets and creating data mining and visualization computational tools are crucial for understanding the intricate regulatory networks that underlie epigenetic processes. Furthermore, advances in spatial transcriptomics and single-cell sequencing technology show promise for deciphering the variability of epigenetic states within tissues and clarifying cell-specific reactions to viral infection.

In summary, the study of epigenetics in plant–virus interactions has been transformed by NGS tools, which allow researchers to understand the intricate relationship between plant defense mechanisms and epigenetic alterations. Next-generation sequencing has advanced our understanding of the molecular mechanisms underlying plant–virus interactions and holds promise for the development of novel strategies to improve crop resilience against viral diseases by offering comprehensive insights into DNA methylation, histone modifications, and small RNA populations.

31.4.2 Bioinformatics tools for epigenetic analysis

Bioinformatics tools play a crucial role in the analysis and interpretation of vast amounts of sequencing data generated from NGS techniques in epigenetic research (Pereira *et al.*, 2020). These tools enable researchers to identify and characterize epigenetic modifications, analyze their functional implications, and integrate multi-omics datasets to unravel the complex regulatory networks governing gene expression. In this section, we delve into the principles, methodologies, applications, and advancements of bioinformatics tools for epigenetic analysis, highlighting their significance in elucidating the molecular mechanisms underlying plant–virus interactions.

The processing, analysis, and interpretation of sequencing data from several epigenetic assays, including short RNA sequencing, chromatin immunoprecipitation followed by sequencing (ChIP-seq), and bisulfite sequencing, are the functions of bioinformatics tools for epigenetic analysis (He and Song, 2017). These tools use computational frameworks, statistical techniques, and algorithms to infer regulatory networks from sequencing data, annotate genomic areas, and discover epigenetic alterations.

Bioinformatics tool methodologies for epigenetic analysis

Numerous features are available in bioinformatics tools for epigenetic studies, such as route enrichment analysis, read alignment, peak calling, differential analysis, motif finding, and data visualization. Sequencing reads are mapped to a reference genome or transcriptome using alignment techniques like Bowtie, BWA, and HISAT. Peak calling methods, which extract regions of enrichment for epigenetic markers from ChIP-seq data, include MACS2 and SICER. Differential analysis methods that compare epigenetic profiles between various experimental circumstances are DESeq2 and edgeR. Algorithms for motif discovery, such as HOMER and MEME Suite, find DNA sequence motifs that are abundant in regulatory areas. Tools for pathway enrichment analysis include DAVID and Metascape.

Bioinformatics tools for epigenetic analysis are widely used in various research areas, including genome-wide DNA methylation profiling, histone modification mapping, transcription factor binding site identification, and small RNA characterization (Chenarani *et al.*, 2021). These tools enable researchers to identify epigenetic changes associated with a viral infection, elucidate the regulatory mechanisms underlying antiviral defense responses, and identify potential targets for crop improvement. In addition, bioinformatics tools facilitate the integration of multi-omics datasets, such as genomics, transcriptomics, proteomics, and metabolomics, to provide a holistic understanding of the molecular interactions between plants and viruses.

Advancements in bioinformatics tools for epigenetic analysis

Advancements in bioinformatics tools for epigenetic analysis have led to the development of user-friendly interfaces, cloud-based platforms, and integrated pipelines for streamlined data analysis and interpretation. Tools like Galaxy, Bioconductor, and ENCODE provide comprehensive suites of bioinformatics tools and resources for epigenetic research. Additionally, machine-learning algorithms and deep-learning models are being increasingly utilized to predict epigenetic modifications, infer regulatory networks, and classify epigenetic states based on sequencing data (Kagohara *et al.*, 2017). These advancements have expanded the scope of epigenetic research, enabling researchers to address complex biological questions and accelerate discoveries in plant–virus interactions.

Despite the progress in bioinformatics tools for epigenetic analysis, several challenges remain, including data integration, reproducibility, scalability, and standardization of analysis pipelines. Integrating multi-omics datasets and developing computational frameworks for data integration is essential for deciphering the complex regulatory networks underlying epigenetic processes. Future directions in bioinformatics tools for epigenetic analysis include the development of advanced algorithms for single-cell epigenomics, spatial transcriptomics, and multi-omics data integration, as well as the integration of artificial intelligence (Mani and Singh, 2023).

By leveraging algorithms, statistical methods, and computational frameworks, bioinformatics tools provide insights into the molecular mechanisms underlying antiviral defense responses, identify potential targets for crop improvement, and accelerate discoveries in plant biology and agriculture. Continued advancements in bioinformatics tools and methodologies will further enhance our understanding of epigenetic regulation and pave the way for innovative strategies to combat viral diseases and ensure global food security.

31.4.3 Experimental approaches

Experimental approaches in the study of epigenetics in plant–virus interactions encompass a wide array of techniques aimed at deciphering the molecular mechanisms underlying gene regulation in response to viral infection. These approaches enable researchers to elucidate the dynamic interplay between epigenetic modifications and antiviral defense responses in plants, ultimately contributing to the development of strategies to enhance crop resilience against viral diseases (Wang *et al.*, 2019).

Experimental approaches in epigenetics are grounded in the principles of molecular biology, genetics, and biochemistry. These approaches involve the manipulation of biological samples, the application of specific assays to interrogate epigenetic modifications, and the analysis of experimental data to infer regulatory networks and functional implications. By combining various experimental techniques, researchers can gain insights into the mechanisms underlying epigenetic regulation and its impact on gene expression in response to viral infection (Thompson *et al.*, 2010).

Methodologies of experimental approaches in epigenetics

Experimental approaches in epigenetics encompass a diverse range of techniques, including bisulfite sequencing, chromatin immunoprecipitation followed by sequencing (ChIP-seq), quantitative PCR (qPCR), transgenic and mutant analysis, and biochemical assays. Bisulfite sequencing enables the genome-wide profiling of DNA methylation patterns by treating DNA with bisulfite, which converts unmethylated cytosines to uracil while leaving methylated cytosines unchanged. ChIP-seq allows the genome-wide mapping of histone modifications and transcription factor binding sites by immunoprecipitating chromatin fragments bound to specific epigenetic marks (Sadeh *et al.*, 2021). qPCR enables the quantification of gene expression changes associated with epigenetic modifications, providing insights into the regulatory roles of these modifications in plant–virus interactions (Statham *et al.*, 2012). Biochemical assays, such as chromatin immunoprecipitation followed by qPCR (ChIP-qPCR) and enzyme activity assays, provide complementary information on the epigenetic landscape and regulatory mechanisms underlying plant–virus interactions.

Experimental approaches in epigenetics have diverse applications in studying plant–virus interactions, including the identification of epigenetically regulated genes involved in antiviral defense pathways, the characterization of epigenetic modifications induced by viral infection, and the validation of candidate genes for crop improvement (Kakoulidou *et al.*, 2021). Bisulfite sequencing has been instrumental in elucidating dynamic changes in DNA methylation induced by viral infection and identifying epigenetically regulated genes involved in antiviral defense mechanisms. ChIP-seq has enabled the genome-wide mapping of histone modifications associated with viral infection, providing insights into chromatin remodeling and gene regulation. qPCR has facilitated the quantification of gene expression changes associated with epigenetic modifications, allowing researchers to dissect the regulatory roles of these modifications in plant–virus interactions. Transgenic and mutant analysis has elucidated the roles of specific genes or epigenetic regulators in antiviral defense pathways, providing valuable insights into host–pathogen interactions. Biochemical assays have provided mechanistic insights into the epigenetic regulation of gene expression in response to viral infection, paving the way for the development of targeted interventions to enhance crop resistance against viral diseases.

Advancements in experimental approaches in epigenetics

Advancements in experimental approaches in epigenetics have led to improvements in assay sensitivity, throughput, and scalability, as well as the development of novel techniques for single-cell epigenomics, spatial transcriptomics, and multi-omics integration. Spatial transcriptomics techniques allow the visualization of gene expression patterns within tissue microenvironments, providing insights into the spatial organization of host–pathogen interactions. Multi-omics integration approaches enable the integration of epigenomic, transcriptomic, proteomic, and metabolomic datasets to unravel the complex regulatory networks governing plant–virus interactions (Corrêa *et al.*, 2024).

31.5 Epigenetic Modifications in Plant Responses to Viral Infections

Although epigenetic variables can affect how defense genes are expressed in plants, little is known about how they function in antiviral responses and how virus evolution and adaptation have shaped these relationships. To build sufficient antiviral defenses, RNA silencing and constitutive immune systems, DNA methylation, and histone modification pathways may also be necessary. The degree to which the virus has adapted to its host greatly influences how well these types of control work. Part of a wider, conserved system including gene regulatory pathways is plant RNA-based defense against viruses. Transposable elements (TEs) can be expressed and controlled by altering DNA or histone proteins linked to DNA with epigenetic markers. Double-stranded RNA (dsRNA) molecules are recognized and degraded to activate the RNA-based immune system. The mechanism mostly relates to antivirus defense. In order to inhibit related single-stranded RNAs, Dicer-like (DCL) proteins break viral dsRNA into small RNAs (sRNAs), which are then loaded onto Argonaute (AGO) proteins (Brosseau *et al.*, 2016). The RNA silencing response can also be improved by producing fresh dsRNAs from targets using RNA-dependent RNA polymerase (RDR) proteins. In particular, quickly changing RNA viruses can subdue host defenses through the quickly producing wildly varied swarms of mutants with escape variants uncontrollably encoding particular proteins that actively interact and suppress host defenses as viral suppressors of RNA silencing (Corrêa *et al.*, 2020).

31.5.1 DNA methylation

DNA methylation is a crucial epigenetic modification that plays a significant role in plant defense against viral diseases. This process involves the addition of a methyl group to the cytosine residues in DNA (Fig. 31.1), often leading to gene silencing. In the context of viral infections, DNA methylation can modulate the expression of genes involved in immune responses and antiviral defense mechanisms. Studies have shown that plants can methylate viral DNA, thereby restricting viral replication and spread. For instance, in *Arabidopsis*, the RNA-directed DNA methylation (RdDM) pathway has been found to target and silence viral genomes, enhancing resistance to viruses like the *cucumber mosaic* (Diezma-Navas *et al.*, 2019). Additionally, changes in DNA methylation patterns can activate plant defense genes, enhancing the overall immune response. Advances in next-generation sequencing have enabled detailed mapping of methylation changes during viral infections, providing insights into the dynamics of host–virus interactions (Zhang *et al.*, 2018). These findings highlight the potential of manipulating DNA methylation as a strategy for developing virus-resistant plants. However, further research is needed to fully understand the mechanisms and to apply this knowledge effectively in agricultural practices. The integration of epigenetic tools with traditional breeding could revolutionize the management of plant viral diseases.

31.5.2 Histone modifications

Histone modifications in plant defense against viral diseases through epigenetic regulation of gene expression, including methylation, acetylation, phosphorylation, and ubiquitination of histone proteins, can alter chromatin structure and subsequently influence the accessibility of DNA to transcriptional machinery. In plant antiviral defense, histone modifications are critical in

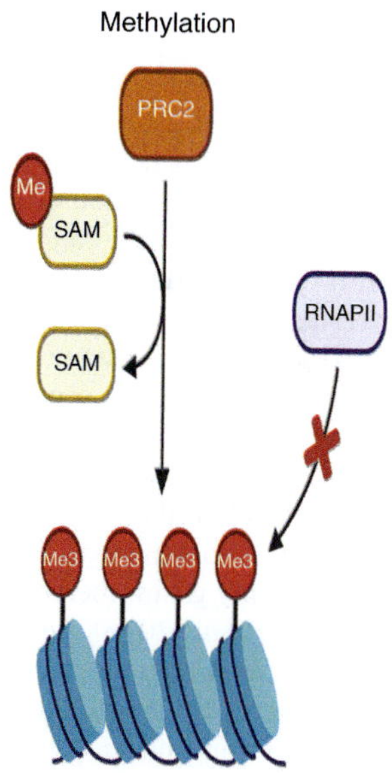

Fig. 31.1. DNA methylation. This diagram illustrates how PRC2 uses SAM to methylate histones (shown as Me3), which leads to chromatin compaction and transcriptional repression by preventing RNAPII activity. Author's own image. PRC2 – Polycomb Repressive Complex 2; SAM – S-AdenosylMethionine (methyl group donor in methylation reactions); Me – Methyl Group; Me3 – Trimethylation (three methyl groups added, typically to lysine residues on histones); RNAPII – RNA Polymerase II.

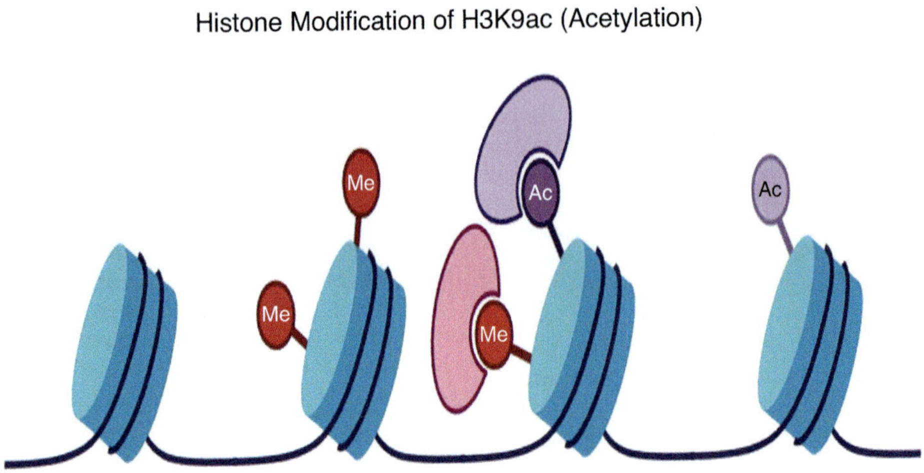

Fig. 31.2. Histone modification of H3K9ac. Author's own image.

regulating genes involved in immune responses. For example, histone acetylation is often associated with the activation of defense genes. Studies have shown that the acetylation of histone H3 lysine 9 (H3K9ac) (Fig. 31.2) can enhance the expression of pathogenesis-related (PR) genes, which are crucial for antiviral responses (Wang *et al.*, 2019). Conversely, histone methylation can either activate or repress gene expression, depending on the specific residues modified. For instance, the trimethylation of histone H3 lysine 27 (H3K27me3) is linked to gene silencing and has been observed to suppress the expression of viral susceptibility genes, thereby conferring resistance (Deleris *et al.*, 2016). Advanced techniques like chromatin immunoprecipitation followed by sequencing (ChIP-seq) have facilitated the mapping of histone modifications across the genome during viral infections, offering insights into host–virus interactions. Understanding these modifications provides a pathway to developing new strategies for enhancing plant resistance to viruses through targeted epigenetic interventions (Zhi and Chang, 2021).

31.5.3 Small RNA-mediated silencing

Small RNA-mediated silencing is a crucial epigenetic mechanism in plant defense against viral diseases. This process involves small RNAs (sRNAs), such as small interfering RNAs (siRNAs) and microRNAs (miRNAs), which regulate gene expression and maintain genome integrity by targeting specific RNA molecules for degradation or translational repression. In the context of viral infections, sRNAs play a significant role in antiviral defense by targeting viral RNA directly and silencing genes that facilitate viral replication.

One well-studied pathway is RNA interference (RNAi), where virus-derived siRNAs (vsiRNAs) are produced by the plant's Dicer-like (DCL) enzymes. These vsiRNAs are incorporated into the RNA-induced silencing complex (RISC), guiding it to complementary viral RNA sequences for cleavage and degradation, thereby limiting viral proliferation (Llave, 2010) (Fig. 31.3). For instance, in tobacco plants, vsiRNAs derived from the *tobacco mosaic virus* (TMV) have been shown to effectively target and degrade viral RNA, reducing the viral load (Ding and Voinnet, 2007).

Moreover, miRNAs also contribute to antiviral defense by regulating host gene expression. Some miRNAs modulate the expression of genes involved in immune responses, thereby enhancing the plant's overall defense mechanisms. For example, miR482/2118 family members in tomato plants target and silence

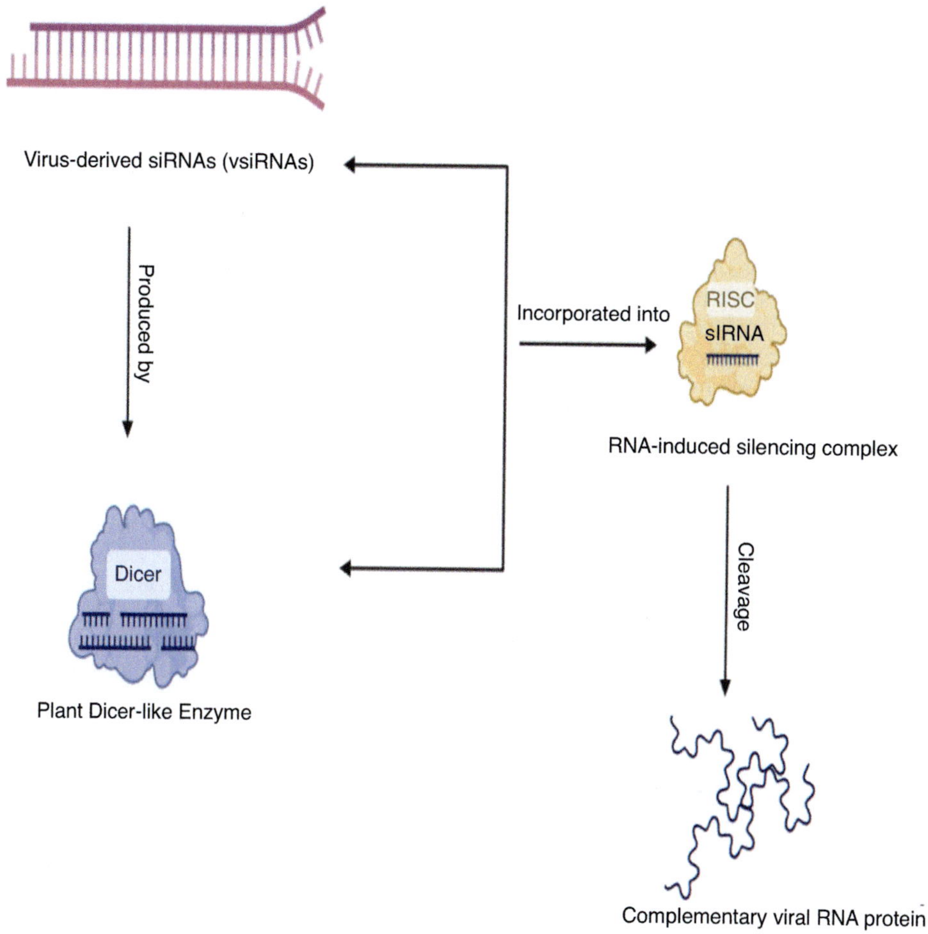

Fig. 31.3. RNAi Pathway involving vsiRNA. Author's own image.

NBS-LRR disease resistance genes to fine-tune the immune response, balancing defense and growth (Shivaprasad *et al.*, 2012).

Advances in high-throughput sequencing and bioinformatics have enabled comprehensive profiling of sRNAs during viral infections, revealing complex regulatory networks and identifying novel antiviral sRNAs. These discoveries highlight the potential of harnessing sRNA pathways for developing virus-resistant crops. For example, engineering plants to over-express specific antiviral siRNAs or miRNAs could provide a robust and sustainable strategy for controlling viral diseases.

DNA (de)methylation-deficient mutants showed altered viral susceptibility/resistance, indicating that DNA methylation is a key regulatory mechanism regulating TRV proliferation. When RdDM targets plant DNA viruses, the result is altered methylation of the viral minichromosomes in plants infected with geminiviruses or caulimoviruses, as well as modifications to a histone modification and chromatin compacting. *De novo* DNA cytosine methylation, maintenance of methylation, and demethylation are all components of RNA-mediated epigenetic silencing. Transposable elements (TEs) and repetitive DNA sequences are the primary targets of DNA methylation in plants, which is crucial for preserving genome integrity and controlling gene expression (Diezma-Navas *et al.*, 2019).

In order to control immunity, histone modifications must also work in tandem with DNA methylation markers. For instance, histone H3 lysine 4 trimethylation (H3K4m3) is frequently linked to the transcriptional memory of several plant defense genes and the activation of gene expression. There have been few studies on the epigenetic control of immunity genes against pathogens, and even fewer on RNA viral infections. It has been shown that plants with specific RdDM and histone modification functions have different viral susceptibilities (Corrêa *et al.*, 2024).

Different plant–microbe interactions have been documented to involve transcriptional modifications in host gene expression that are associated with modifications in epigenetic marks. RNA-directed DNA methylation (RdDM), a plant-specific mechanism, mediates the *de novo* establishment and strengthening of DNA methylation in plants. The majority of DNA methylation in eukaryotes is associated with cytosines, especially those that come after guanine (CG). However, in plants, cytosine methylation can also happen outside of the CG context, as in CHG and CHH (where H stands for A, C, or T) (Pérez-Cañamás *et al.*, 2020).

Plants face ongoing challenges from a variety of phytopathogens. Plants have developed multilevel surveillance systems to defend themselves against all diseases. Plants have evolved defense mechanisms against intracellular parasite viruses, including innate immunity, RNA silencing, translation repression, autophagy- and ubiquitination-mediated protein degradation, and other dominant resistance gene-mediated responses. Additionally, plant viruses have developed a variety of defense mechanisms to inhibit and even take advantage of host defenses in order to survive. The development of effective and broad-spectrum virus resistance for sustainable agriculture would undoubtedly aid in a better understanding of the defense and counter-defense mechanisms between plants and viruses (Wu *et al.*, 2019).

Plants have developed a variety of defense mechanisms to fend off viral infection and pathology as a result of their coevolution with harmful viruses. Viruses, however, coevolve to evade these resistance reactions. By blocking the production of virus genes and halting the systemic spread of viruses, plants regulate viral populations. The mainline defence against geminiviruses is essential in fighting geminivirus infections, even if one of the basic principles of protecting the host from infection is to limit virus gene expression (Gupta *et al.*, 2021).

When viruses and plants interact, defense and counter-defense processes are set off, which often results in coevolution and reciprocal adaptation. Numerous immunological responses in plants are involved in the defense against virus. Generally speaking, these antiviral factors can be categorized into two groups: (i) basal, which are pre-existing and limit cell-to-cell and within-cell propagation; and (ii) inducible, which are activated after infection and impede replication and systemic migration. This may be either directly, if epigenetic regulation affects how plant resistance genes are expressed, which in turn affects how viral factors interact with them, or indirectly, if it causes a significant physiological dysregulation that mimics a different cellular environment for the virus (Ambrós *et al.*, 2024).

The genomes are frequently broken down into siRNAs by the immunological response that RNA viruses elicit. The genomes of ssRNA viruses are first transformed into dsRNA molecules by RNA-dependent RNA polymerases, as DCL family endoribonucleases act on dsRNA. When plants are infected with RNA viruses, they identify the viral double-strand RNA molecules and cause DCL2 and DCL4 to break them down into siRNAs. Through RdDM, a different mechanism called TGS (transcriptional gene silencing) offers a more robust protection against DNA viruses. In addition to providing defense against viral infections, post-transcriptional gene silencing (PTGS) and TGS also control the expression of genes within plants (Ashapkin *et al.*, 2020).

Virus-induced gene silencing (VIGS) is a successful gene-silencing technique that inhibits the expression of particular invasive viral transcripts by using a plant's antiviral defense system. Van Kammen first used the term VIGS to describe the phenomenon known as 'recovery from viral infection'. DNA methylation is connected to epigenetic markers and the heredity of virus-induced silence. A viral vector (DNA or RNA) containing a sequence matching the targeted gene is used to inoculate plants. After the inoculation, endogenous RNA-directed RNA polymerase (RdRp) is activated, replicating and generating viral dsRNA. The Dicer enzyme homolog, also known as the mother of proteins, recognizes these dsRNAs and cleaves them into siRNA duplexes that are roughly 21–24 nt in length. The viral vector insert must match the transgenic promoter rather than the coding sequence in order to accomplish epigenetic silencing. An epigenetic mark known as early DNA methylation is reinforced through the PolIV route of RNA-directed DNA methylation (RdDM), resulting in the formation of a heritable epigenome. DNA methylation results in genomic imprinting and gene silence, as well as temporary and/or genetically inherited changes in chromatin structure and gene expression that do not significantly affect nucleotide sequence (Zulfiqar *et al.*, 2023).

31.6 Conclusion

In conclusion, the intersection of epigenetics and plant virology represents a paradigm shift in our approach to combating viral diseases in plants. Traditional methods of disease management are being complemented and, in some cases, supplanted by innovative strategies rooted in the manipulation of epigenetic mechanisms. Epigenetics offers a sophisticated means of enhancing plant immunity against viral invaders without altering the underlying DNA sequence, presenting a promising avenue for sustainable agriculture and food security.

The advancement of cutting-edge techniques and tools in epigenetic research has provided unprecedented insights into gene regulation, chromatin structure, and epigenetic modifications. High-throughput sequencing technologies, single-cell epigenomics, and CRISPR-based epigenome editing have revolutionized our ability to map epigenetic marks, understand their functional consequences, and manipulate them with precision. Moreover, the elucidation of the three-dimensional organization of the genome within the nucleus has shed light on how epigenetic marks influence gene expression through spatial organization.

Recent studies underscore the potential of epigenetic modifications, such as DNA methylation and RNA-based technologies, in conferring resistance to viral diseases in plants. The integration of various methods, including genetic engineering and viral expression vectors, highlights the multifaceted approach required to tackle the complex challenges posed by plant viral diseases.

As we continue to unravel the intricate interactions between plants and viruses at the epigenetic level, it is imperative to harness this knowledge to develop sustainable solutions for disease management. By leveraging the power of epigenetics, we can enhance crop resilience, minimize environmental impacts, and ensure food security in the face of evolving agricultural landscapes and emerging viral threats. Thus, the future of plant virology lies in embracing the potential of epigenetics to revolutionize disease management and safeguard global food production.

References

Akhter, M., Akanda, A., Kobayashi, K., Jain, R. and Mandal, B. (2019) Plant virus diseases and their management in Bangladesh. *Crop Protection* 118, 57–65.

Ambrós, S., Olmo-Uceda, M.J., Corrêa, R.L. and Elena, S.F. (2024) Phenotypic and genomic changes during turnip mosaic virus adaptation to *Arabidopsis thaliana* mutants lacking epigenetic regulatory factors. *Evolution* 78(1), 69–85.

Ashapkin, V.V., Kutueva, L.I., Aleksandrushkina, N.I. and Vanyushin, B.F. (2020) Epigenetic mechanisms of plant adaptation to biotic and abiotic stresses. *International Journal of Molecular Sciences* 21(20), 7457.

Baulcombe, D.C. and Dean, C. (2014) Epigenetic regulation in plant responses to the environment. *Cold Spring Harbor Perspectives in Biology* 6(9), a019471.

Behjati, S. and Tarpey, P.S. (2013) What is next generation sequencing? Archives of disease in childhood. *Education & Practice Edition* 98(6), 236–238.

Bernardo, P., Charles-Dominique, T., Barakat, M., Ortet, P., Fernandez, E. *et al.* (2017) Geometagenomics illuminates the impact of agriculture on the distribution and prevalence of plant viruses at the ecosystem scale. *The ISME Journal* 12, 173–184.

Brooks, F.T. (1923) Virus diseases of plants. *Nature* 112(2826), 955.

Brosseau, C., El Oirdi, M., Adurogbangba, A., Ma, X. and Moffett, P. (2016) Antiviral defense involves AGO4 in an *Arabidopsis*-potexvirus interaction. *Molecular Plant-Microbe Interactions* 29(11), 878–888.

Burdon, J.J., Zhan, J., Barrett, L.G., Papaïx, J. and Thrall, P.H. (2016) Addressing the challenges of pathogen evolution on the world's arable crops. *Phytopathology* 106(10), 1117–1127.

Chaitankar, V., Karakülah, G., Ratnapriya, R., Giuste, F.O., Brooks, M.J. *et al.* (2016) Next generation sequencing technology and genomewide data analysis: Perspectives for retinal research. *Progress in Retinal and Eye Research* 55, 1–31.

Chauhan, P., Singla, K., Rajbhar, M., Singh, A., Das, N. *et al.* (2019) A systematic review of conventional and advanced approaches for the control of plant viruses. *Journal of Applied Biology & Biotechnology* 7(4), 89–98.

Chenarani, N., Emamjomeh, A., Allahverdi, A., Mirmostafa, S., Afsharinia, M.H. *et al.* (2021) Bioinformatic tools for DNA methylation and histone modification: A survey. *Genomics* 113(3), 1098–1113.

Cheng, C., Fei, Z. and Xiao, P. (2023) Methods to improve the accuracy of next-generation sequencing. *Frontiers in Bioengineering and Biotechnology* 11, 982111.

Corrêa, R.L., Sanz-Carbonell, A., Kogej, Z., Müller, S.Y., Ambrós, S. *et al.* (2020) Viral fitness determines the magnitude of transcriptomic and epigenomic reprograming of defense responses in plants. *Molecular Biology and Evolution* 37(7), 1866–1881.

Corrêa, R.L., Kutnjak, D., Ambrós, S., Bustos, M. and Elena, S.F. (2024) Identification of epigenetically regulated genes involved in plant-virus interaction and their role in virus-triggered induced resistance. *BMC Plant Biology* 24(1), 172.

Delcuve, G.P., Rastegar, M. and Davie, J.R. (2009) Epigenetic control. *Journal of Cellular Physiology* 219(2), 243–250.

Deleris, A., Halter, T. and Navarro, L.D.N.A. (2016) DNA methylation and demethylation in plant immunity. *Annual Review of Phytopathology* 54, 579–603.

Diezma-Navas, L., Pérez-González, A., Artaza, H., Alonso, L., Caro, E. *et al.* (2019) Crosstalk between epigenetic silencing and infection by tobacco rattle virus in *Arabidopsis*. *Molecular Plant Pathology* 20, 1439–1452.

Ding, S.W. and Voinnet, O. (2007) Antiviral immunity directed by small RNAs. *Cell* 130(3), 413–426.

Fitzmaurice, W., Holzberg, S., Lindbo, J., Padgett, H., Palmer, K. *et al.* (2002) Epigenetic modification of plants with systemic RNA viruses. *Omics* 6(2), 137–151.

Fondong, V.N., Nagalakshmi, U. and Dinesh-Kumar, S.P. (2016) Novel functional genomics approaches: A promising future in the combat against plant viruses. *Phytopathology* 106(10), 1231–1239.

Gupta, N., Reddy, K., Bhattacharyya, D. and Chakraborty, S. (2021) Plant responses to geminivirus infection: Guardians of the plant immunity. *Virology Journal* 18(1), 143.

Hadidi, A., Flores, R., Candresse, T. and Barba, M. (2016) Next-generation sequencing and genome editing in plant virology. *Frontiers in Microbiology* 7.

Hanssen, I.M., Lapidot, M. and Thomma, B.P. (2010) Emerging viral diseases of tomato crops. *Molecular Plant–Microbe Interactions* 23(5), 539–548.

He, Y. and Song, J. (2017) Bioinformatics analysis of epigenetics. In: Liu, Z.J. (ed.) *Bioinformatics in Aquaculture: Principles and Methods*. John Wiley & Sons, pp. 247–273.

Hirst, M. and Marra, M.A. (2010) Next generation sequencing based approaches to epigenomics. *Briefings in Functional Genomics* 9(5–6), 455–465.

Jones, R.A. (2020) Disease pandemics and major epidemics arising from new encounters between indigenous viruses and introduced crops. *Viruses* 12.

Jones, R. and Naidu, R. (2019) Global dimensions of plant virus diseases: Current status and future perspectives. *Annual Review of Virology* 6, 387–409.

Kagohara, L., Stein-O'Brien, G., Kelley, D., Flam, E., Wick, H. *et al.* (2017) Epigenetic regulation of gene expression in cancer: Techniques, resources and analysis. *Briefings in Functional Genomics* 17, 49–63.

Kakoulidou, I., Avramidou, E.V., Baránek, M., Brunel-Muguet, S., Farrona, S. *et al.* (2021) Epigenetics for crop improvement in times of global change. *Biology* 10(8), 766.

Llave, C. (2010) Virus-derived small interfering RNAs at the core of plant-virus interactions. *Trends in Plant Science* 15(12), 701–707.

Mani, I. and Singh, V. (2023) Applications of bioinformatics in epigenetics. *Progress in Molecular Biology and Translational Science* 198, 1–13.

Manjunatha, L., Rajashekara, H., Uppala, L., Ambika, D., Patil, B. *et al.* (2022) Mechanisms of microbial plant protection and control of plant viruses. *Plants (Basel)* 11(24), 3449.

Martelli, G.P. and Gallitelli, D. (2008) Emerging and re-emerging viruses of plants. In: Bamford, D.H. and Zuckerman, M. (eds) *Encyclopedia of Virology*. Elsevier, pp. 86–92.

Martelli, G.P. and Uyemoto, J.K. (2008) Plant virus diseases: Fruit trees and grapevine. In: Bamford, D.H. and Zuckerman, M. (eds) *Encyclopedia of Virology*. Elsevier, pp. 201–207.

Meaburn, E. and Schulz, R. (2012) Next generation sequencing in epigenetics: Insights and challenges. *Seminars in Cell & Developmental Biology* 23(2), 192–199.

Mehetre, G.T., Leo, V.V., Singh, G., Sorokan, A.V., Maksimov, I.V. *et al.* (2021) Current developments and challenges in plant viral diagnostics: A systematic review. *Viruses* 13.

Nicaise, V. (2014) Crop immunity against viruses: Outcomes and future challenges. *Frontiers in Plant Science* 5, 118474.

Pereira, R., Oliveira, J. and Sousa, M. (2020) Bioinformatics and computational tools for next-generation sequencing analysis in clinical genetics. *Journal of Clinical Medicine* 9(1), 132.

Pérez-Cañamás, M., Hevia, E. and Hernández, C. (2020) Epigenetic changes in host ribosomal DNA promoter induced by an asymptomatic plant virus infection. *Biology* 9(5), 91.

Rajeevkumar, S., Anunanthini, P. and Sathishkumar, R. (2015) Epigenetic silencing in transgenic plants. *Frontiers in Plant Science* 6, 693.

Sadeh, R., Sharkia, I., Fialkoff, G., Rahat, A., Gutin, J. *et al.* (2021) ChIP-seq of plasma cell-free nucleosomes identifies gene expression programs of the cells of origin. *Nature Biotechnology* 39(5), 586–598.

Sahu, P.P., Pandey, G., Sharma, N., Puranik, S., Muthamilarasan, M. *et al.* (2013) Epigenetic mechanisms of plant stress responses and adaptation. *Plant Cell Reports* 32(8), 1151–1159.

Samantara, K., Shiv, A., Sousa, L.L., Sandhu, K.S., Priyadarshini, P. *et al.* (2021) A comprehensive review on epigenetic mechanisms and application of epigenetic modifications for crop improvement. *Environmental and Experimental Botany* 188, 104479.

Scholthof, K.-B.G., Adkins, S., Czosnek, H., Palukaitis, P., Jacquot, E. *et al.* (2011) Top 10 plant viruses in molecular plant pathology. *Molecular Plant Pathology* 12(9), 938–954.

Sharma, N., Sahu, P.P., Puranik, S. and Prasad, M. (2013) Recent advances in plant-virus interaction with emphasis on small interfering RNAs (siRNAs). *Molecular Biotechnology* 55(1), 63–77.

Shivaprasad, P.V., Chen, H.-M., Patel, K., Bond, D.M., Santos, B.A.C.M. *et al.* (2012) A microRNA superfamily regulates nucleotide binding site-leucine-rich repeats and other mRNAs. *The Plant Cell* 24(3), 859–874.

Singh, R. and Srivastava, A. (2020) Prevention and control of viral diseases of crops. In: Aswathi, L.P. (ed.) *Applied Plant Virology*. Academic Press, pp. 593–599.

Statham, A.L., Robinson, M.D., Song, J.Z., Coolen, M.W., Stirzaker, C. *et al.* (2012) Bisulfite sequencing of chromatin immunoprecipitated DNA (Bischip-seq) directly informs methylation status of histone-modified DNA. *Genome Research* 22(6), 1120–1127.

Sumida, C. (2020) Viral diseases of crops: A critical review. In: Awasthi, L.P. (ed.) *Applied Plant Virology*. Academic Press, pp. 471–474.

Tatineni, S. and Hein, G.L. (2023) Plant viruses of agricultural importance: Current and future perspectives of virus disease management strategies. *Phytopathology* 113(2), 117–141.

Thompson, R.F., Fazzari, M.J. and Greally, J.M. (2010) Experimental approaches to the study of epigenomic dysregulation in ageing. *Experimental Gerontology* 45(4), 255–268.

Varma, A. (2007) Integrated management of plant viral diseases. In: *Ciba Foundation Symposium 177-Crop Protection and Sustainable Agriculture: Crop Protection and Sustainable Agriculture: Ciba Foundation Symposium 177*. John Wiley & Sons, pp. 140–157.

Vlugt, R. van der (2006) Plant viruses in European Agriculture: Current problems and future aspects. In: Cooper, I., Kuhne, T. and Polishchuk, V.P. (eds) *Virus Diseases and Crop Biosecurity*. NATO Security through Science series, Springer, Dordrecht, pp. 33–44.

Walkey, D.G.A., Alhubaishi, A.A. and Webb, M.J.W. (1990) Plant virus diseases in the Yemen Arab Republic. *Tropical Pest Management* 36(3), 195–206.

Wang, C., Wang, C., Zou, J., Yang, Y., Li, Z. *et al.* (2019) Epigenetics in the plant-virus interaction. *Plant Cell Reports* 38(9), 1031–1038.

Wani, S.H. and Sanghera, G.S. (2010) Genetic engineering for viral disease management in plants. *Notulae Botanicae Horti Agrobotanici Cluj-Napoca* 2, 20–28.

Wu, X., Valli, A., García, J.A., Zhou, X. and Cheng, X. (2019) The tug-of-war between plants and viruses: Great progress and many remaining questions. *Viruses* 11(3), 203.

Zhang, H., Lang, Z. and Zhu, J.K. (2018) Dynamics and function of DNA methylation in plants. *Nature Reviews Molecular Cell Biology* 19(8), 489–506.

Zhi, P. and Chang, C. (2021) Exploiting epigenetic variations for crop disease resistance improvement. *Frontiers in Plant Science* 12, 692328.

Zulfiqar, S., Farooq, M.A., Zhao, T., Wang, P., Tabusam, J. *et al.* (2023) Virus-induced gene silencing (VIGS): A powerful tool for crop improvement and its advancement towards epigenetics. *International Journal of Molecular Sciences* 24(6), 5608.

32 Current Status of Epigenetic Research in Plant–Nematode Interactions

Tushar K. Dutta*
Division of Nematology, ICAR-Indian Agricultural Research Institute, New Delhi

Abstract

Plant-parasitic nematodes (PPNs) continue to wreak havoc on crops all over the world. The relationships between plants and nematodes are quite diverse, which warrants investigating the regulatory networks and coordinated functional features of both organisms. To feed and grow, the nematodes bore into host-plant roots and construct feeding structures, which re-differentiate inside plant cells to create giant cells (induced by root-knot nematodes) and syncytia (induced by cyst and reniform nematodes), which are enormous multi-nucleate cells that provide continual supplies of nutrition for PPN development. Various epigenetic, proteomic, metabolomic, and transcriptome alterations regulate the intimate interrelationship between plants and nematodes. In this chapter, the progress made so far on epigenetic modulations during plant–nematode interaction is discussed. A vast network has been uncovered by transcriptomics that aims to decipher the feeding mechanism of PPNs by means of gene expression programming of certain genes. Plant epigenetic profiling is also well-explained by regulatory circuits of epigenetic changes via DNA methylation, non-coding RNAs, and histone modifications. The potential function of microRNAs (miRNAs) and short RNAs is also involved. Through the post-transcriptional silencing of different effector proteins and transcriptional factors, these RNAs mediate the expressional changes in plants, serving as major genetic elements. The pragmatic role of miRNAs is also explored, which influences gene expression during PPN infection and feeding site formation. We are now moving closer to a fundamental molecular understanding of plant–nematode relationships by studying epigenetic regulations.

Keywords: DNA methylation, Histone acetylation, Root-knot nematode, Cyst nematode, Giant cell, Syncytium

32.1 Introduction

Due to global climate change, global agriculture faces a serious threat from plant-parasitic nematodes (PPNs), which are predominantly soil-dwelling obligatory parasites (Dutta and Phani, 2023). Global food security is negatively impacted by the incidence of two major groups of PPNs such as cyst (*Heterodera* spp., *Globodera* spp.) and root-knot (*Meloidogyne* spp.) nematodes. Additionally, reniform (*Rotylenchulus reniformis*), citrus (*Tylenchulus semipenetrans*), lesion (*Pratylenchus* spp.) and burrowing (*Radopholus* spp.) nematodes cause significant crop losses (Phani *et al.*, 2023). PPNs drastically alter the host root morphology. Root-knot and cyst nematodes form feeding sites in the root vasculature by inducing hypermetabolic, multinucleate giant cells and syncytia, respectively. These nematodes secrete a repertoire of effector proteins (secreted from the nematode pharyngeal glands via the stylet orifice into the host tissue) that interact with specific host proteins to completely hijack the host metabolic pathway

*Corresponding author: nemaiari@gmail.com

DOI: 10.1079/9781800626102.0032

to their advantage. Both giant cells and syncytia serve as the continual nutrition source for the feeding of nematodes (Dutta *et al.*, 2023). In the case of root-knot nematode infestation, cortical tissues around giant cells undergo hypertrophy and hyperplasia to form macroscopic root galls (Dutta and Akhil, 2023).

The PPN-secreted effector proteins change the expression of several host genes that code for regulatory functions to modulate a number of cellular processes (Hewezi, 2015). In addition, several epigenetic and regulatory circuits have been found to play an important role. However, the exact molecular mechanisms underlying these activities related to the modification of gene regulation associated with feeding mechanisms and developmental processes are still not completely understood (Hewezi and Baum, 2017). The epigenetic toolbox and post-transcriptional and translational alterations are crucial for recognizing the invading nematodes, organ regeneration processes (involved in giant cell and syncytium initiation), and host defense gene expression or suppression (Hewezi *et al.*, 2017; Hewezi, 2020).

Epigenetics-based gene regulation is dependent on specific regulatory triggers, including DNA methylation, histone acetylation, messenger RNAs (mRNAs), and non-coding small RNAs (sRNAs) that act as a coordinated circuitry (Dalakouras and Vlachostergios, 2021). sRNAs are short RNA molecules (21-nucleotide (nt)) microRNAs (miRNAs) and 21 or 25 nt epigenetic small-interfering RNAs (siRNAs) that are processed from long double-stranded RNA (dsRNA) molecules by RNaseIII or Dicer enzymes (Simon and Meyers, 2011). Base pairing with target genes that have complementary or nearly identical sequences is the mechanism by which miRNAs cause gene silencing. The binding of miRNAs to complementary base pairs triggers the target messenger RNA (mRNA) degradation or post-translational suppression (Tirnaz and Batley, 2019). Additionally, 21 or 25 nt siRNAs lead to transcriptional gene silencing (TGS) by altering the transposable elements by DNA methylation or modulating the RNA-directed DNA methylation (RdDM) pathway (Simon and Meyers, 2011; Tirnaz and Batley, 2019). DNA methylation most commonly occurs in CG, CHG and CHH sequence contexts (H = thymine (T) or adenine (A) bases). Different enzymes involved in the process are DNA methyl transferase 1 (MET1), chromomethylase 3 (CMT3), and domain rearrangement methyl transferase 2 (DRM2). Methylation of cysteine (C) base establishes a reversible epigenetic signature, which can be reverted to its original state by DNA-glycosylase-mediated demethylation process (Dalakouras and Vlachostergios, 2021).

The post-translational modification of histone proteins constitutes an additional regulatory modulator in the epigenetic toolbox for plant–PPN relationships (Hewezi, 2020). A significant biochemical change is imposed on histones, where methylation and acetylation occur at lysine (K) residues. These changes lead to major shifts in chromosomal structure and transcriptional and translational patterns of gene expression. Acetylation of H3 and H4 histone proteins favors euchromatin formation *vis-à-vis* activation of gene transcription. Conversely, H3K9 and H3K27 methylation cause euchromatin formation *vis-à-vis* suppression of gene transcription (Dalakouras and Vlachostergios, 2021). Similar to reciprocal changes in DNA methylation, histone demethylases and deacetylases can remove the epigenetic marks. The acetylation and methylation states impact the entire chromatin structure as well as transcription factors to modulate the gene expression pathways (Lawrence *et al.*, 2016).

Research suggests that these epigenetic components can detect PPNs during the different parasitic stages and initiate defense responses by regulating genes at the transcriptional and post-transcriptional levels. Intriguingly, PPN parasitism can stimulate the synthesis of sRNAs, histone modifications, histone acetylation, DNA hypomethylation, and DNA hypermethylation. Epigenetic alterations allow plants to establish an immediate and effective cell defense response to harmfully assault the PPNs (Medina *et al.*, 2018; Jaubert-Possamai *et al.*, 2019; Han *et al.*, 2022; Bennett *et al.*, 2023; Ye *et al.*, 2024).

32.2 Regulatory Roles of Small RNAs (SRNAs) in Plant–PPN Interactions

The regulatory role of sRNAs in plant–PPN interactions was initially unraveled in the model plant *Arabidopsis thaliana*. *A. thaliana* mutants for impaired sRNA production (more specifically, mutated RNA-directed RNA-polymerases (RdRP) and Dicer protein-encoding genes) exhibited reduced susceptibility to root-knot (*M. incognita*) and cyst-nematode (*H. schachtii*) infection (Ruiz-Ferrer *et al.*, 2018). Likewise, *A. thaliana* mutants for Argonaute (AGO) protein-encoding genes (*ago1-25*, *ago1-27*) were less vulnerable to *M. incognita* infection (Medina *et al.*, 2017). Additionally, Medina and co-workers Medina *et al.* (2017) found that the widespread inhibition of sRNA biogenesis in host plants is correlated with reduced *M. incognita* parasitism. A high-throughput sRNA sequencing of *M. incognita*-infected *A. thaliana* galls revealed a substantial number of siRNA clusters, which were discretely transcribed in infected and uninfected roots at 7 and 14 days after inoculation (Medina *et al.*, 2018). The siRNA clusters (protein-coding genes) were categorized into two major types: those with 20–22 nt and those with 23–25 nt. Among these, 23–25 nt heterochromatic siRNAs were plentiful and had preferential accumulation in the gene promoter regions and lower abundance in the gene body regions. Further, these siRNAs were predominantly expressed in the infected roots at 7 and 14 days after inoculation, compared with uninfected roots, exemplifying the activation of siRNA biosynthesis and siRNA silencing upon nematode parasitism (Medina *et al.*, 2018). The 23–25 nt heterochromatic siRNAs putatively regulate gene expression in the galls via the canonical RNA-directed DNA methylation (RdDM) pathway. Unlike 23–25 nt siRNAs, 20–22 nt siRNAs are abundant in the gene body regions and weakly expressed in the galled roots compared with the uninfected ones. An identical finding on sRNA biosynthesis and abundance was recorded in *A. thaliana–M. javanica* pathosystem when root galls of 3 days old were examined (Cabrera *et al.*, 2016).

The post-transcriptional changes in the plant upon PPN infection are mediated by a repertoire of miRNA genes. Hewezi and co-workers Hewezi *et al.* (2008) reported the differential expression of 30 miRNA transcripts in *A. thaliana* upon beet cyst nematode (*H. schachtii*) infection. In another study, 40 miRNAs were specifically transcribed in the soybean (*Glycine max*) roots upon soybean cyst nematode (*H. glycines*) infection (Li *et al.*, 2012). Additionally, 60 *hour. H. glycines*-inducible miRNA genes were identified from *G. max*, implicating the regulatory role of miRNAs in host plant–*Heterodera* spp. interactions (Tian *et al.*, 2017). In the case of root-knot nematode (*M. javanica*)-infected *A. thaliana* galls, 62 miRNA transcripts were variably expressed (11 and 51 were up- and downregulated, respectively) (Cabrera *et al.*, 2016). Albeit the specific roles of downregulated miRNAs are yet to be determined, attenuated miRNA expression was also documented in *A. thaliana–H. schachtii* interaction (Hewezi *et al.*, 2008). Incidentally, in *A. thaliana–M. incognita* pathosystem, 24 differentially expressed miRNAs could be identified, and their spatial and temporal expression dynamics in the galled roots were investigated by fusing the miRNA promoter genes with a reporter gene such as β-glucuronidase (GUS) (Medina *et al.*, 2017). A multitude of miRNAs were detected during the different parasitic phases of tomato (*Solanum lycopersicum*)–*M. incognita* interaction (Kaur *et al.*, 2017). Notably, a greater proportion of these detected miRNAs are extremely conserved across the plant species. Considering the broader host status of *Melodogyne* spp., these miRNAs are putatively involved in an identical gene pathway or network that ultimately regulates the nematode-induced gall or knot formation. However, miRNAs are inducible to both *Heterodera* spp. (narrow host range), and *Meloidogyne* spp. may confer identical or contrasting modes of regulatory reprogramming during the parasitism process. These investigations shed some light on the commonalities and differences in the infection patterns (including giant cell and syncytium establishment) of two different groups of PPNs.

32.3 Specific Roles of microRNAs (MiRNAs) in Plant–PPN Interactions

The non-coding plant miRNAs are typically 20–24 nt long that regulate the post-transcriptional gene silencing (PTGS) mechanism (Li and Mao, 2007). The complete pathway of miRNA biosynthesis is schematically represented in Fig. 32.1. Briefly, miRNA genes (*MIR*) are transcribed via RNA polymerase II to produce primary miRNAs (pri-miRNA), which

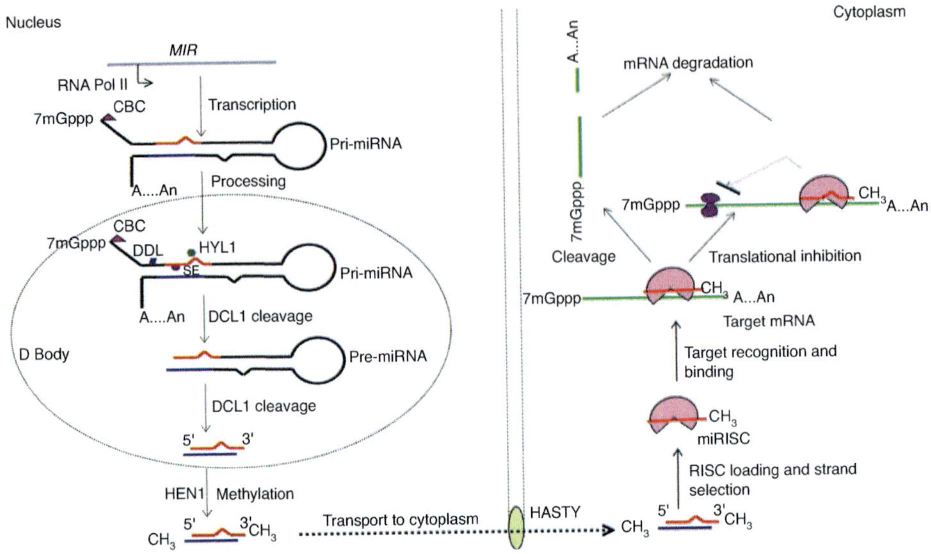

Fig. 32.1. Schematic representation of the plant miRNA biosynthesis and the putative mode of action during host–PPN interactions. miRNA genes (*MIR*) are transcribed via RNA polymerase II to produce primary miRNAs (pri-miRNA), which are single-stranded transcripts that contain hairpin loops. Pri-miRNAs are cleaved by Dicer (DCL1) in the nuclear region of the cell (facilitated by the HYL1 RNA-binding protein and SE1 zinc finger protein) to generate precursor miRNAs (pre-miRNA). Pre-miRNAs are sliced via DCL1 and its cofactors to generate miRNA duplexes, which contain mature miRNAs and their complementary strands. A methyl (CH_3) group is added to each of the 3′ hydroxyl group via HEN1 protein to inhibit the degradation of duplexes. An EXPORTIN protein ortholog HASTY mediates the transport of miRNA duplex from nucleus to the cytoplasm. Any of the two strands of the duplex are loaded onto the AGO1 protein, which is the major effector in the multiprotein RNA-induced silencing complex (RISC). Based on the sequence complementarity, guide strand (AGO1-associated) then guides the RISC to the target mRNA. A partial complementarity between guide strand and target mRNA leads to translational inhibition and no protein synthesis, whereas absolute complementarity causes target mRNA cleavage. Author's own image.

are single-stranded transcripts that contain hairpin loops. Pri-miRNAs are cleaved by Dicer (DCL1) in the nuclear region of the cell (facilitated by the HYL1 RNA-binding protein and SE1 zinc finger protein) to generate precursor miRNAs (pre-miRNA). Pre-miRNAs are sliced via DCL1 and its cofactors to generate miRNA duplexes, which contain mature miRNAs and their complementary strands. A methyl (CH_3) group is added to each of the 3′ hydroxyl groups via HEN1 protein to inhibit the degradation of duplexes. An EXPORTIN protein ortholog HASTY mediates the transport of miRNA duplex from the nucleus to the cytoplasm. Any of the two strands of the duplex is loaded onto the AGO1 protein, which is the major effector in the multiprotein RNA-induced silencing complex (RISC). Based on the sequence complementarity, the guide strand (AGO1-associated) then guides the RISC to the target mRNA. A partial complementarity between the guide strand and target mRNA leads to translational inhibition and no protein synthesis, whereas perfect complementarity causes target mRNA cleavage (Yu *et al.*, 2017). In the case of absolute complementarity, mRNA cleavage occurs at the 10th or 11th nucleotide position (Bartel, 2009). In *A. thaliana*, miR172 inhibits the translation of *APETALA2* gene due to partial sequence complementarity (Zhang and Li, 2013). Since miRNA targets can also elicit the activity of miRNA regulators such as CUC2 and MIR164a, gene regulation via miRNAs does not necessarily indicate a negative correlation between the expression of mature miRNA and its target transcript (Nikovics *et al.*, 2006). Intriguingly, *MIR* genes across different plants can have common target mRNAs because *MIR* belongs to multigene families, for which although precursor miRNA sequences may vary, mature miRNA sequences are nearly identical (Palatnik *et al.*, 2007). Apart from that, several *MIR* gene families share commonalities (target conserved genes) across the evolutionarily distant plant species (Jones-Rhoades, 2012). Various *MIR* gene families play key roles during host–PPN interactions (Table 32.1). PPNs induce ontogenesis of giant cells and syncytia inside host tissue (and suppress host defense) that involves extensive reprogramming of gene expression in target root cells (Jaubert-Possamai *et al.*, 2019). Genome and transcriptome sequencing of host plant mutants (corresponding to disruption of specific miRNA/siRNA pathway) led to identifying the function of specific miRNAs in different plant–PPN interactions. Those miRNAs are discussed in greater detail in the following paragraphs.

32.3.1 Role of plant miR396 during cyst nematode infection

A. thaliana miR396 is the first miRNA that has been investigated in greater depth (Hewezi *et al.*, 2012). *A. thaliana* miR396 played a prominent role during the transition of syncytium (*H. schachtii*-induced) initiation to the maintenance phase. During the initiation phase, miR396 was downregulated, while during the maintenance phase, miR396 was strongly induced (Hewezi *et al.*, 2012). Most specifically, miR396 targets two growth-regulating transcription factors (GRF), i.e. GRF1 and GRF3, in the syncytium and regulates them via post-transcriptional modification. Upon interference of miR396 or GRF1/3 expression, an impeded syncytium development was observed along with disruption in *H. schachtii* life cycle progression. Therefore, a concerted expression of miR396 and its *GRF* targets is necessary for syncytium neoformation that ultimately defines host susceptibility to nematodes (Fig. 32.2). The expression equilibrium (miR396 and GRF1/3 regulate the transcription of each other) of this particular regulatory module is maintained via a reciprocal feedback loop (Hewezi and Baum, 2012). The critical regulatory function of GRFs in establishing the syncytium transcriptome has been unraveled (Hewezi *et al.*, 2012). Additionally, GRFs modulate the synchronization between plant development and defense response induction in the PPN-induced feeding sites (Liu *et al.*, 2014). This finding was further validated using chromatin immunoprecipitation (ChIP) sequencing (Piya *et al.*, 2020). Out of 1510 tested, 345 GRF1 and GRF3 transcripts were found to be direct targets of miR396 in the nematode-induced syncytium. Intriguingly, miR396 is recruited by *H. schachtii* to reprogram host defense and syncytium development via GRF1 and GRF3 switching on or off. A similar mechanism was reported in *Glycine max–H. glycines* pathosystem (Noon *et al.*, 2019). The miR396–GRF regulatory circuit can be exploited in molecular breeding programs for developing cyst nematode resistance.

32.3.2 Role of plant miR858 during cyst nematode infection

As explained in the previous section, miRNA-targeted transcription factors (TFs) can remarkably alter the cell metabolic and physiological pathways because these TFs can directly modulate the transcription of a multitude of target genes. *A. thaliana* miR858 is one such kind of miRNA, which can target a number of TFs. In *H. schachtii*-induced syncytium, MYB83 has been detected as the major MYB TF, which is targeted by miR858 (Piya *et al.*, 2017). MYB83 was repressed post-transcriptionally by miR858 in the early developing syncytium, implicating the significance of miR858–MYB83 regulon during the initial phase of *A. thaliana–H. schachtii* interaction (Fig. 32.2B). Using forward- and reverse-genetics screening of MYB83 and miR858 function, it was clearly demonstrated that MYB83 can positively regulate *H. schachtii* parasitic success in *A. thaliana*. Transcriptome analyses showed that 1286 genes are regulated by MYB83 in the syncytium and among them 471 genes are the direct targets, since they harbor a MYB83 *cis*-regulatory motif in their

Table 32.1. List of plant miRNAs differentially expressed during PPN infection. Author's own table.

S. No.	Host plant	PPN species	Infection site examined	miRNA class	Reference
1	*A. thaliana*	*H. schachtii*	Syncytium	miR396	Hewezi *et al.*, 2012
2	*Glycine max*	*H. glycines*	Root	miR393, miR1507, miR1510, miR171, miR2118	Xu *et al.*, 2014
3	*S. lycopersicum*	*M. incognita*	Root	miR319, miR396	Zhao *et al.*, 2015
4	*S. lycopersicum, Gossypium hirsutum*	*M. incognita*	Root, gall, giant cell	miR100, miR124, miR228, miR71, miR92, miR34, miR50, miR279, miR87	Zhang *et al.*, 2016
5	*Nicotiana benthamiana*	*H. schachtii*	Syncytium	miR827	Hewezi *et al.*, 2016
6	*Glycine max*	*H. glycines*	Root	miR166, miR1510, miR3522, miR482, miR159, miR396, miR2109, miR2118, miR4996, miR1509	Tian *et al.*, 2017
7	*Pinus massoniana*	*Bursaphelenchus xylophilus*	Needle	miR5054, miR384, miR6196, miR5059	Xie *et al.*, 2017
8	*S. lycopersicum*	*M. incognita*	Root	miR156, miR159, miR164, miR396	Kaur *et al.*, 2017
9	*A. thaliana*	*M. incognita*	Giant cell, gall	miR159	Medina *et al.*, 2017
10	*S. lycopersicum*	*G. rostochiensis*	Root	miR408.22/ Solyc07g043390, miR156.64/ Solyc03g11936	Koter *et al.*, 2018
11	*Glycine max*	*H. glycines*	Syncytium	miR10186, miR10187, miR10193	Lei *et al.*, 2019
12	*G. hirsutum*	*M. incognita*	Root, egg mass	miR252, miRN19, miR1, miR124, miR2-3p, miR279, miR57-5p, miR7904, miR87, miR92, miR4738, miRN3, miRN5	Liu *et al.*, 2019
13	*G. hirsutum*	*M. incognita*	Root	miR159, miR319, miR167	Pan *et al.*, 2019
14	*Glycine max*	*H. glycines*	Root	miR5032, miR5043, miR1520b, miR5770a, miR5770b	Rambani *et al.*, 2020

Continued

Table 32.1. Continued

S. No.	Host plant	PPN species	Infection site examined	miRNA class	Reference
15	*Cucumis metuliferus*	*M. incognita*	Root	miR156, miR390, miR159a, miR827	Ye *et al.*, 2020
16	*S. lycopersicum*	*M. incognita*	Giant cell	miR167	Noureddine *et al.*, 2023
17	*C. metuliferus*	*M. incognita*	Root	miR396	Ye *et al.*, 2024

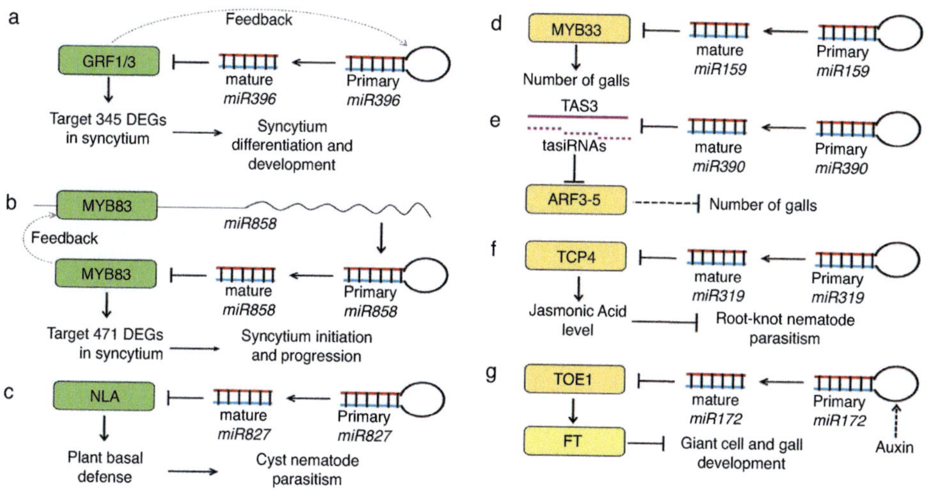

Fig. 32.2. A summary of the plant miRNA gene function in different host–PPN interactions. (A) During *H. schachtii* infection, miR396 suppresses the expression of GRF1 and GRF3 TFs, which positively regulate the transcription of 345 genes in the syncytium. A feedback regulatory mechanism adjusts the expression of GRF1/3 and miR396. (B) During *H. schachtii* infection, miR858 suppresses the expression of *MYB83* TF, which positively regulates the transcription of 471 genes in the syncytium. A feedback regulatory mechanism adjusts the expression of MYB83 and miR858. (C) During *H. schachtii* parasitism, miR827 downregulates NLA expression in the syncytium that suppresses host basal defense to expedite infection progression. (D) During *M. incognita* infection, miR159 suppresses MYB33 TF expression in giant cells leading to enhanced host susceptibility (unknown mechanism). (E) During *M. javanica* infection, miR390-induced biosynthesis of TAS3-derived tasiRNA negatively regulates ARF3/4/5 expression in giant cell to confer host susceptibility. (F) During *M. incognita* infection, miR319 suppresses *TCP4* (a positive regulator of JA biogenesis) expression. Elevated miR319 levels cause attenuated JA production to facilitate host susceptibility. (G) During *M. javanica* parasitism, auxin-recruited miR172 downregulates TOE1 TF in the giant cell that alleviates the TOE1 suppressive effect on *FT* gene, which modulates giant cell ontogenesis. DEG – differentially expressed genes. Author's own image.

promoters (Piya *et al.*, 2017). According to the gene ontology analysis, MYB83-regulated genes are associated with various cellular processes, including syncytial expansion, sugar transport, glucosinolate biogenesis, phytohormone signaling, cell wall modification, and defense response. Like the miR396–GRF1/3 regulation mechanism, miR858–MYB83 also maintains a feedback regulatory module that balances out the transcript abundance of miR858 and MYB83, as well as downstream transcriptional oscillation of various target genes at the syncytium (Hewezi, 2020). The *de novo* molecular and functional characterization of miR396 and miR858 led to our enhanced understanding of how miRNAs and their TFs fine-tune the expressions of each other, which tightly controls gene regulation in the PPN feeding sites.

32.3.3 Role of plant miR827 during cyst nematode infection

Endoparastic sedentary PPNs spend a considerable duration of their life cycle in direct contact with the host roots, which warrants active suppression of plant defense responses. PPNs recruit miR827 to perpetually suppress the host defense signaling in their feeding sites that aid in PPN disease progression in the infected root (Hewezi *et al.*, 2016). *A. thaliana* miR827 (encoded by a single gene) targets a ubiquitin E3 ligase gene *NLA*. Upon *H. schachtii* infection, miR827 was greatly induced both in early and fully developed syncytium. Activation of miR827 is linked with post-transcriptional silencing of *NITROGEN LIMITATION ADAPTATION* (*NLA*) gene in the syncytium throughout the nematode developmental stages. At the steady-state level, *NLA* can activate host basal defense by eliciting the upregulation of a number of pathogenesis-related (PR) genes manifold (Fig. 32.2C). As revealed by overexpression analysis, *A. thaliana* plants overexpressing miR827 transcripts downregulated *NLA* upon *H. schachtii* infection, eventually leading to increased nematode susceptibility of host plants. On the contrary, elevated *NLA* transcription via inactivation of the NLA negative regulator or overexpression of its non-degradable transcript in *A. thaliana* substantially reduced *H. schachtii* susceptibility. The interacting protein substrates of NLA were identified via yeast two-hybrid assays. A number of those substrates are involved in defense signaling, such as mitogen-activated protein (MAP) kinase 2 and 17, PR4, thaumatin, papain cysteine proteinase, and β−1,3-endoglucanase (Hewezi *et al.*, 2016). Additionally, miR827-NLA regulon also controls plant response to PO_4^{3-} and NO_3^- deficiency (Hewezi, 2020). Accordingly, it is speculated that miR827-NLA regulon may aid in acquiring and utilizing the phosphorus (P) and nitrogen (N) elements in the PPN feeding sites.

32.3.4 Role of plant miRNAs during root-knot nematode infection

Differentially expressed miRNAs were identified from the *Meloidogyne* species-induced galled roots in several instances. From the *A. thaliana*–*M. incognita* compatible interaction, differential expression of four miRNAs, namely miR159, miR319, miR398, and miR408, were documented (Medina *et al.*, 2017). Functional analysis including overexpression and T-DNA mutation indicated that miR159 plays a key role in modulating plant susceptibility to nematodes. Additionally, the mature miR159 molecule exhibited a strong hybridization signal in the giant cell as revealed by the *in situ* expression localization study. The higher expression of miR159 at the feeding site was correlated with weaker expression of the target gene *MYB33* TF (Medina *et al.*, 2017). Nevertheless, it is still unclear how miR159-mediated *MYB33* downregulation in the giant cell increases plant susceptibility to nematodes (Fig. 32.2D).

miR390 was abundantly expressed in the *M. javanica*-induced galled root of *A. thaliana* (Cabrera *et al.*, 2016). miR390 can mediate the biosynthesis of TAS3-derived tasiRNAs (*trans*-acting siRNAs) that interact with ARF3/4/5 (auxin response factor) transcripts to regulate their degradation (Marin *et al.*, 2010). Both miR390 and its TAS3 target were abundantly expressed in the early developing giant cell as investigated via promoter: GUS reporter fusion. On the contrary, T-DNA insertional mutagenesis of miR390 and TAS3 caused reduced plant susceptibility to nematodes. The regulatory role of the miR390/TAS3 module was linked to the TAS3-derived tasiRNA-guided transcriptional alteration of ARF3. Association of miR390/TAS3 regulon in mediating the repression of ARF3, ARF4, and ARF5 (involved in auxin signaling) led to the giant cell ontogenesis (Cabrera *et al.*, 2016). However, specific downstream factors that aid the feeding cell induction process are yet to be determined (Fig. 32.2E).

The systematic signaling molecule jasmonic acid (JA) modulates the compatible or incompatible interaction between *Meloidogyne* spp. and their plant hosts (Gheysen and Mitchum, 2019). miR319 is a JA-responsive miRNA that regulates *Solanum lycopersicum* vulnerability to *M. incognita* infection by targeting the *TCP4* gene (*TEOSINTE BRANCHED/CYCLOIDEA/PROLIFERATING CELL FACTOR 4*). While overexpressed miR319 increased tomato susceptibility to *M. incognita*, overexpressed *TCP4* decreased tomato susceptibility to the root-knot nematode, suggesting a negative feedback loop (Zhao *et al.*, 2015). Endogenous JA levels were markedly increased in *TCP4* overexpressed plants (that exhibited decreased nematode susceptibility), suggesting the putative role of *TCP4* in JA biogenesis (Schommer *et al.*, 2008; Zhao *et al.*, 2015) (Fig. 32.2F).

miR172 is considered one of the most greatly conserved miRNA genes across the plant kingdom. It was abundantly expressed in the *M. javanica*-induced giant cells of *A. thaliana*, *S. lycopersicum*, and *Pisum sativum* (Díaz-Manzano *et al.*, 2018). In *A. thaliana–M. javanica* compatible interaction, miR172 negatively regulates its AP2-like TF target *TOE1* (*TARGET OF EARLY ACTIVATION TAGGED 1*). When miR172 was inactivated and *TOE1* was overexpressed, a prominent decline in giant cell size as well as decreased gall numbers per root system was documented. Thus, miR172-mediated suppression of *TOE1* is necessary for giant cell differentiation and gall formation (Díaz-Manzano *et al.*, 2018). Given that the *FLOWERING LOCUS T* (*FT*) gene is the direct target of *TOE1* (Zhang *et al.*, 2015), suppression of *TOE1* alternatively circumvents its inhibitory effect on *FT*, causing *FT* abundance in feeding cells (Fig. 32.2G). Additionally, T-DNA insertional mutagenesis of *FT* indicated its association in morphogenesis (like *FT* regulation during the flowering process) of pseudo organs such as giant cells followed by gall induction (Díaz-Manzano *et al.*, 2018). The members of miR172 gene family harbor auxin response *cis*-regulatory elements in their promoter regions and, because of that, miR172 expression is modulated by auxin during giant cell development. Incidentally, ARF-mediated miR172 regulation has been described in other plant organogenesis processes (Ripoll *et al.*, 2015). Therefore, the miR172-TOE1-FT regulatory module is tightly controlled by the auxin signaling pathway in the feeding cells. Overall, the putative role of miRNAs in modulating the crosstalk between auxin and JA signaling pathway cannot be ruled out, because of the functional conservation between miR390-TAS3-ARF3, miR319-TCP4, and miR172-TOE1-FT regulatory cascades.

32.4 Role of Long Non-Coding RNAs in Plant–PPN Interaction

Long non-coding RNAs (lncRNAs), which are more than 200 nt long, also function in gene regulation. Characterized in numerous plant species, lncRNAs play a pivotal role during plant development and plant defense responses (Kim and Sung, 2012; Liu *et al.*, 2015; Nejat and Mantri, 2018; Rai *et al.*, 2019). lncRNAs are able to act as *cis*-regulatory elements, sRNA precursors, protein complex scaffolds, RdDM pathway mediators, and miRNA target mimic (Rai *et al.*, 2019). Currently, the literature is quite scanty regarding the role of lncRNAs in plant–nematode interactions. In an isolated study, 565 lncRNAs were differentially expressed in *Nicotiana tabacum* upon *M. incognita* invasion (Li *et al.*, 2018). Nevertheless, the exact regulatory role of these induced lncRNAs is yet to be unraveled. Functional analysis of specific lncRNA molecules in nematode-induced feeding cells will enrich our understanding of the regulatory mechanism of non-coding RNAs in host–PPN interactions.

32.5 DNA Methylation as the Epigenome Regulatory Mechanism

Epigenome marks in terms of DNA hypomethylation and hypermethylation (which contribute to transcriptome programming and reprogramming) have been studied quite extensively in host–PPN interactions. DNA methylation patterns are very discrete during compatible and incompatible plant–nematode interactions. Using bisulfite genomic sequencing, a high-resolution methylome was generated from soybean (*Glycine max*) root infected with the cyst nematode *Heterodera glycines* (Rambani *et al.*, 2015). *H. glycines* infection considerably changed the root methylation patterns towards prominent hypomethylation marks rather than hypermethylation (Rambani *et al.*, 2015). The findings are in accordance with previous reports, which showed active DNA demethylation during a plant's susceptible response to various pathogens (Dowen *et al.*, 2012; Yu *et al.*, 2013; Deleris *et al.*, 2016; López Sánchez *et al.*, 2016; Zhu *et al.*, 2016; Hewezi *et al.*, 2018). Methylation of CG, CHG, and CHH sequence contexts occurred independently of each other, and concurrent cytosine methylation of the targeted region (approximately 6% of the methylated sites) indicated the biological significance of DNA methylation during *H. glycines* parasitism of soybean (Rambani *et al.*, 2015). Further, DNA methylation preferably occurred in the duplicated genes rather than the ancient genes, suggesting that *H. glycines*-elicited differential methylation might have contributed to genetic equilibrium between ancient and duplicated (presumably evolved for defense function) copies of the target gene (Hewezi, 2020). Additionally, *H. glycines*-elicited hypomethylation and hypermethylation impacted a substantial amount of genes that were earlier identified as differentially expressed genes in the syncytium. Those genes are associated with different physiological functions such as syncytia formation, defense responses, cell wall biosynthesis, cytoskeletal dynamics, phytohormone signaling, signal transduction, epigenetic alterations, ubiquitination, and gene regulation (Rambani *et al.*, 2015). Therefore, DNA methylation plays a critical role in transcriptional reprogramming during host–PPN interactions.

A correlation between DNA methylation marks and differential gene expression in *A. thaliana* during different phases of *H. schachtii* parasitism (syncytium initiation and maintenance) was observed (Hewezi *et al.*, 2017). Like *H. glycines*, *H. schachtii* considerably changed root methylome during both the syncytium initiation and maintenance phases. More than 90% of the differentially methylated regions were of hypomethylated types (Hewezi *et al.*, 2017). Moreover, DNA methylation preferably occurs in certain locations of the protein-coding genes and transposable elements. Hypermethylation of transposable elements and gene promoters was accompanied by the greater accumulation of siRNAs 24 nt long, as displayed by methylomes at the nematode-infected sites (Hewezi *et al.*, 2017). Conversely, DNA hypomethylation predominantly occurred in specific transposons in microspores and sperm cells in *A. thaliana* (Calarco *et al.*, 2012). Nevertheless, a substantial proportion of *H. schachtii*-induced hypermethylated sites were not linked with the greater accumulation of siRNAs (Hewezi *et al.*, 2017). Notably, DNA methylation and siRNA abundance are not always correlated, because each of these processes can occur independently (Lister *et al.*, 2008).

DNA methylation alterations elicited by the parasitizing nematode in protein-coding host genes were ascertained upon observing the overlap between differentially methylated genes and differentially expressed genes in the feeding cell. Generating loss of function mutants via T-DNA insertional or CRISPR/Cas9 mutagenesis of promoter-methylated and gene-body-methylated genes can advance our fundamental understanding of DNA methylation roles in host–PPN communications. It is well known that altered methylation marks in gene promoters can ultimately modify gene transcription. However, the impact of gene body methylation in gene regulation is not well understood. It is hypothesized that gene body methylation of PPN-inducible genes suppresses the transcription initiation in intragenic regions or suppresses the incorporation of transposable elements in actively transcribing genes. In addition, gene body methylation enhances splicing efficiency (Hewezi, 2020).

Gene expressions are also controlled via transposable elements that are located proximally to the protein-coding genes (Wang *et al.*, 2013). Methylome signatures of *A. thaliana* roots upon *H. schachtii* infection displayed transposon methylation in CHH sequence contexts that primed the expression of proximal genes. For example, a hypomethylated (CHH-type) transposable element was positioned at 3 kb distance from the neighboring gene. However, *H. schachtii*-elicited hypomethylation of transposons was associated with lower expression of closely located genes (Hewezi *et al.*, 2017). An identical scenario was reported from *A. thaliana* exposed to various abiotic and biotic stressors

(Le *et al.*, 2014; Secco *et al.*, 2015). Elimination of methylation marks or demethylation in these transposons might have caused activation of transposon-derived cryptic transcripts that might have disrupted the transcription of nearby genes. It is also possible that demethylation of the transposon might have facilitated the induction of PPN-regulated genes during the late parasitic phase, because DNA methylation occasionally precedes the transcriptional alterations (Hewezi, 2020).

Transposon transcriptional activity during host–PPN interaction and the impending effect on the expression of nearby genes have been studied in *A. thaliana–H. schachtii* pathosystems using transcriptome sequencing (Piya *et al.*, 2017). Intriguingly, 192 differentially expressed transposable elements (belonging to rolling-circle/Helitron and MuDR families) were detected from *H. schachtii*-infected roots. Most interestingly, > 70% of the differentially expressed transposons were located within 5 kb of candidate genes, which had differential expression in the feeding cell. The differentially expressed candidate genes code for various functions, including control of gene transcription, primary metabolic pathway, cell wall modification, phytohormone signaling, and defense response (Piya *et al.*, 2017). Therefore, in line with the gene regulation in host plants upon exposure to various abiotic and biotic stresses (Mirouze and Paszkowski, 2011; Feng *et al.*, 2013; Makarevitch *et al.*, 2015; Hossain *et al.*, 2017), PPNs can modulate transcriptional activity of plant transposons as well as proximal PPN-inducible genes to transcriptionally reprogram the feeding cells for enhanced host susceptibility. However, this hypothesis is derived from RNA-seq data only. Generation of loss-of-function mutants via forward or reverse genetics screen would be necessary to obtain a concrete validation.

A number of siRNAs (linked with transposable elements) were differentially accumulated in the galled roots of *A. thaliana* upon *M. javanica* infection (Ruiz-Ferrer *et al.*, 2018). Predominant were siRNAs that were 22 and 24 nt long and were associated with Gypsy and Copia family of retrotransposon elements. siRNA abundance was associated with a notable decline in the transcript levels of different retrotransposons (Ripoll *et al.*, 2015). Gene expression analysis and mutation studies indicated the existence of canonical and non-canonical RdDM pathways while retrotransposons were repressed in the galls (Ruiz-Ferrer *et al.*, 2018). Nevertheless, the DNA methylation profile of the retrotransposons remains to be understood. Collectively, the findings implied that epigenome modification is modulated via differential expression of siRNAs, and DNA methylome is repurposed towards tighter command on transposon movement and augmentation. This type of regulatory control is a prerequisite to conserve genome integrity and steadiness while host plants carry out genome-wide reprogramming upon PPN infestation.

The putative role of Copia retrotransposon in soybeans has been reported to regulate host resistance to *H. glycines* via activating the *rhg1-a R* gene (Bayless *et al.*, 2019). Interestingly, Copia elements were detected within *rhg1-a* gene in several *Glycine max* accessions. However, the expression of *rhg1-a* gene and the existence of Copia elements were not correlated. It was hypothesized that *H. glycines*-elicited epigenetic alteration in Copia element might regulate the maturation of *rhg1-a* gene and production of its splicing variants, considering that retrotransposons are quite vulnerable to DNA methylation-related alterations compared with other families of transposable elements (Hewezi *et al.*, 2017). Speculatively, DNA methylation and other epigenome marks might trigger plant machinery for recruiting splicing factors to the *rhg1-a* precursor mRNA followed by altering the RNA polymerase II elongation rate and controlling exon insertion or exclusion in mature mRNA (Hewezi, 2020).

32.6 Effect of *R* Genes in Epigenome Modification

R gene loci *Rhg1* (which constitutes three genes within 31 kb repeat region at chromosome 18) and *Rhg4* (constitutes solo gene that codes for serine hydroxymethyltransferase or *GmSHMT08* at chromosome 8) contribute to *H. glycines* resistance in soybean (Cook *et al.*, 2012; Liu *et al.*, 2012). Methylome analyses of *Rhg1* locus across various *Glycine max* accessions displayed elevated DNA methylation in resistant accessions (contained a triple copy of *Rhg1*) compared with susceptible ones (single-copy containing). Hypermethylation (in CG, CHG, and CHH sequence contexts) was evident specifically in the upstream and downstream positions of three *Rhg1*-encoded genes. Nevertheless, these findings could not correlate elevated expression of hypermethylated genes with resistance to *H. glycines*. Probably, *H. glycines* invasion induced hypomethylation of specific genes in the syncytium, and to counteract that transcriptional activation occurred in resistant lines (Hewezi, 2020).

The definitive role of *Rhg4/GmSHMT08* in reprograming the DNA methylation profile in soybean during *H. glycines* infection has been implicated, as revealed by the genome-wide methylation study in two homozygous near-isogenic lines (NILs) that differed by only *GmSHMT08* locus. The susceptible NIL (functional *GmSHMT08* allele was absent) displayed decreased global DNA methylation levels in protein-coding genes and transposable elements. On the contrary, resistant NIL displayed a sharp increase in global DNA methylation patterns in CG, CHG, and CHH sequence contexts (Hewezi, 2020). *Rhg4* as a DNA methylation modulator must be investigated further for probable utility in molecular breeding programs intended for PPN resistance.

32.7 PPN Effectors Can Modulate Plant Epigenetic Modifications

PPNs secrete a multitude of parasitism proteins (from their pharyngeal glands) that function as effector molecules to suppress host immunity and aid the parasitic process (Eves-van den Akker, 2021). Advancements made in next-generation sequencing technology as well as precise extraction of PPN pharyngeal glands (Maier *et al.*, 2013) led to the identification and expression localization of numerous effector genes from different groups of PPNs. A significant proportion of these effectors show negligible sequence homology to proteins in public databases, and thus their role in PPN parasitism is as yet underexplored. Nonetheless, a number of these effectors displayed notable sequence similarity to candidate proteins that are involved in epigenome modifications (Jones *et al.*, 2009; Noon *et al.*, 2015; Eves-van den Akker *et al.*, 2016; Noon and Baum, 2016; Gardner *et al.*, 2018). For instance, the GLAND1 effector of *H. glycines* exhibited significant sequence conservation with GCN5-related *N*-acetyltransferase (GNAT) of *Streptomyces violaceusniger* (Noon *et al.*, 2015). Phylogeny studies suggested possible acquisition of GLAND1 in nematodes was from bacteria via the lateral gene transfer event (Noon and Baum, 2016). Additionally, phylogenetic analyses showed the presence of GLAND1 homologues in *H. avenae*, *G. pallida*, *G. rostochiensis*, and *R. reniformis* (Noon and Baum, 2016). Although GNATs are one of the most common acyltransferases encountered across biological systems, a single GNAT effector is reported from *Mycobacterium tuberculosis* that is involved in the suppression of host immune responses via acetylation (Kim *et al.*, 2012). Although effector-mediated suppression of plant defense is central to nematode infection progression (Goverse and Smant, 2014; Holbein *et al.*, 2016), GLAND1 was not expressed during the early stage of *H. glycines* parasitism (Noon *et al.*, 2015). Intriguingly, PPN GNATs are capable of mimicking the histone acyltransferase function of host GCN5, thus modifying acetylation levels at particular loci that lead to alteration in gene expression and host susceptibility. In line with the GNAT function during PPN infection progression, *Phytophthora sojae* effector PsAvh23 suppressed GCN5-mediated H3K9 acetylation to increase *Glycine max* susceptibility (Kong *et al.*, 2017). Identification of plant-interacting proteins of PPN GNATs and their target loci would shed some light on the PPN-induced histone acetylation for reprograming plant epigenomes toward nematode susceptibility.

PPN effectors showed considerable homology to the epigenome-regulating proteins of *Ascaris suum*, including histone H4 TF (activates H4 gene transcription), histone deacetylase 1 (deacetylate lysine residue of core histone), Set1/Ash2 histone methyltransferase (methylate fourth lysine residue of H3 histone), peregrin (acetylate H3 histone at Lys23 site), histone-lysine *N*-methyltransferase SUV39H2 (trimethylate Lys9 of H3 histone), etc. (Gardner *et al.*, 2018). PPN effectors also showed similarity to *Aedes aegypti* nucleosome binding factor SPN, associated with chromatin structure and dynamics (Gardner *et al.*, 2018).

A *H. schachtii* effector 32E03 interacted with *A. thaliana* tuin-type histone deacetylase HDT one and histone chaperone FKBP53, as revealed by the co-localization assay and yeast two-hybrid screening (Vijayapalani *et al.*, 2018). Since HDT1 and FKBP53 suppress the ribosomal RNA (rRNA) gene expression (Lawrence *et al.*, 2004; Li and Luan, 2010), 32E03 might regulate the transcriptional activity of rRNA genes to facilitate *H. schachtii* infection progression in *A. thaliana*. This was validated using 32E03-expressing transgenic *A. thaliana* and *in vitro* assay with recombinant 32E03 protein. These functional studies indicated that 32E03 acts as the inhibitor of histone deacetylase resulting in elevated levels of histone H3 acetylation at the ribosomal DNA (rDNA) chromatin region (Vijayapalani *et al.*, 2018). Since elevated levels of histone acetylation are largely associated with elevated gene expression levels, 45S pre-rRNA transcript levels were estimated in transgenic *A. thaliana* expressing high and low levels of 32E03 coding sequence. rRNA expression was dramatically upregulated in 32E03 low-expressing *A. thaliana*, whereas rRNA expression was attenuated in 32E03 high-expressing *A. thaliana*. This was correlated with *H. schachtii* infection levels, because 32E03 high-expressing *A. thaliana* exhibited reduced nematode susceptibility, whilst 32E03 low-expressing *A. thaliana* exhibited increased nematode susceptibility. Intriguingly, 32E03 upregulation in *A. thaliana* induced RdDM of rDNA that led to gene silencing (Vijayapalani *et al.*, 2018). Collectively, these interesting findings suggest that PPNs can deploy nucleus-targeting effectors to elicit epigenetic alterations in rDNA chromatin in a dose-dependent manner to facilitate PPN parasitism. Nevertheless, how increased rRNA accumulation contributes to greater host susceptibility (to PPNs) is yet to be elucidated.

32.8 Conclusion and Future Directions

Significant advancements have been made in delineating the regulatory function of different epigenetic components such as sRNAs, miRNAs, DNA methylation, and histone modification in host–PPN interactions. However, the probable interplay between these components is yet to be elucidated. The interconnectedness of various epigenetic changes is becoming increasingly evident. Elucidating the regulatory and functional linkages between distinct layers of epigenetic changes requires integration among different genomic strategies that include multi-omics analysis. It is still not entirely known how PPNs trigger epigenetic alterations in their host plants. The involvement of PPN effectors in initiating these epigenetic reactions to infection is highly probable. Robust characterization of PPN effector along with identification of target plant protein should provide light on the biochemical and molecular mechanisms underpinning nematode-elicited epigenetic reprogramming in hosts. To potentially deorphanize epigenetic signals unique to each PPN species, further research into the epigenome alterations caused by specific PPN in specific host plants is required. To generate these types of epigenetic markers, it would be necessary to conduct a thorough temporal and spatial investigation of epigenetic modifications, particularly in the feeding cells. With the advent of single-cell sequencing technology, it is now possible to generate data with pinpoint accuracy. Numerous epigenome alterations are stably inherited across many generations of plant species. By deploying novel strategies such as genome editing, newly created meiotically heritable epialleles can be identified that significantly impact host responses to PPN parasitism. This ushered in an entirely novel way to include an epigenetic selection system into breeding programs, allowing for the selection of individuals with the predicted epialleles.

References

Bartel, D.P. (2009) MicroRNAs: Target recognition and regulatory functions. *Cell* 136(2), 215–233.

Bayless, A.M., Zapotocny, R.W., Han, S., Grunwald, D.J., Amundson, K.K. *et al.* (2019) The rhg1-a (Rhg1 low-copy) nematode resistance source harbors a copia-family retrotransposon within the Rhg1-encoded α-SNAP gene. *Plant Direct* 3(8), e00164.

Bennett, M., Hawk, T.E., Lopes-Caitar, V.S., Adams, N., Rice, J.H. *et al.* (2023) Establishment and maintenance of DNA methylation in nematode feeding sites. *Frontiers in Plant Science* 13, 1111623.

Cabrera, J., Barcala, M., García, A., Rio-Machín, A., Medina, C. *et al.* (2016) Differentially expressed small RNAs in *Arabidopsis* galls formed by *Meloidogyne javanica*: A functional role for miR390 and its TAS 3-derived tasi RNAs. *The New Phytologist* 209(4), 1625–1640.

Calarco, J.P., Borges, F., Donoghue, M.T.A., Van Ex, F., Jullien, P.E. *et al.* (2012) Reprogramming of DNA methylation in pollen guides epigenetic inheritance via small RNA. *Cell* 151(1), 194–205.

Cook, D.E., Lee, T.G., Guo, X., Melito, S., Wang, K. *et al.* (2012) Copy number variation of multiple genes at rhg1 mediates nematode resistance in soybean. *Science* 338(6111), 1206–1209.

Dalakouras, A. and Vlachostergios, D. (2021) Epigenetic approaches to crop breeding: Current status and perspectives. *Journal of Experimental Botany* 72(15), 5356–5371.

Deleris, A., Halter, T. and Navarro, L. (2016) DNA methylation and demethylation in plant immunity. *Annual Review of Phytopathology* 54, 579–603.

Díaz-Manzano, F.E., Cabrera, J., Ripoll, J.-J., Del Olmo, I., Andrés, M.F. *et al.* (2018) A role for the gene regulatory module microrna172/TARGET OF EARLY ACTIVATION TAGGED 1/FLOWERING LOCUS T (miRNA172/TOE1/FT) in the feeding sites induced by *Meloidogyne javanica* in *Arabidopsis thaliana*. *The New Phytologist* 217(2), 813–827.

Dowen, R.H., Pelizzola, M., Schmitz, R.J., Lister, R., Dowen, J.M. *et al.* (2012) Widespread dynamic DNA methylation in response to biotic stress. *Proceedings of the National Academy of Sciences of the United States of America* 109(32), E2183–E2191.

Dutta, T.K. and Akhil, V.S. (2023) The relative infection potential of *Meloidogyne incognita* and *M. graminicola* in a basmati rice cultivar PB1121. *Plant Pathology* 73(2), 262–271.

Dutta, T.K. and Phani, V. (2023) Transgenics, application in plant nematode management. In: Khan, M.R. (ed.) *Novel Biological and Biotechnological Applications in Plant Nematode Management*. Springer, Singapore. DOI: 10.1007/978-981-99-2893-4_9.

Dutta, T.K., Ray, S. and Phani, V. (2023) The status of the CRISPR/Cas9 research in plant–nematode interactions. *Planta* 258(6), 103.

Eves-van den Akker, S. (2021) Plant–nematode interactions. *Current Opinion in Plant Biology* 62, 102035.

Eves-van den Akker, S., Laetsch, D.R., Thorpe, P., Lilley, C.J., Danchin, E.G.J. *et al.* (2016) The genome of the yellow potato cyst nematode, *Globodera rostochiensis*, reveals insights into the basis of parasitism and virulence. *Genome Biology* 17(1), 124.

Feng, G., Leem, Y.E. and Levin, H.L. (2013) Transposon integration enhances expression of stress response genes. *Nucleic Acids Research* 41(2), 775–789.

Gardner, M., Dhroso, A., Johnson, N., Davis, E.L., Baum, T.J. *et al.* (2018) Novel global effector mining from the transcriptome of early life stages of the soybean cyst nematode *Heterodera glycines*. *Scientific Reports* 8(1), 2505.

Gheysen, G. and Mitchum, M.G. (2019) Phytoparasitic nematode control of plant hormone pathways. *Plant Physiology* 179(4), 1212–1226.

Goverse, A. and Smant, G. (2014) The activation and suppression of plant innate immunity by parasitic nematodes. *Annual Review of Phytopathology* 52, 243–265. DOI: 10.1146/annurev-phyto-102313-050118.

Han, X., Shi, Q., He, Z., Song, W., Chen, Q. *et al.* (2022) Transcriptome-wide N6-methyladenosine (m6A) methylation in soybean under *Meloidogyne incognita* infection. *aBIOTECH* 3(3), 197–211.

Hewezi, T. (2015) Cellular signaling pathways and posttranslational modifications mediated by nematode effector proteins. *Plant Physiology* 169(2), 1018–1026.

Hewezi, T. (2020) Epigenetic mechanisms in nematode-plant interactions. *Annual Review of Phytopathology* 58, 119–138.

Hewezi, T. and Baum, T.J. (2012) Complex feedback regulations govern the expression of miRNA396 and its GRF target genes. *Plant Signaling & Behavior* 7(7), 749–751.

Hewezi, T. and Baum, T.J. (2017) Communication of sedentary plant-parasitic nematodes with their host plants. *Advances in Botanical Research* 82, 305–324.

Hewezi, T., Howe, P., Maier, T.R. and Baum, T.J. (2008) *Arabidopsis* small RNAs and their targets during cyst nematode parasitism. *Molecular Plant–Microbe Interactions* 21(12), 1622–1634.

Hewezi, T., Maier, T.R., Nettleton, D. and Baum, T.J. (2012) The *Arabidopsis* microRNA396-GRF1/GRF3 regulatory module acts as a developmental regulator in the reprogramming of root cells during cyst nematode infection. *Plant Physiology* 159(1), 321–335.

Hewezi, T., Piya, S., Qi, M., Balasubramaniam, M., Rice, J.H. *et al.* (2016) *Arabidopsis* miR827 mediates post-transcriptional gene silencing of its ubiquitin E3 ligase target gene in the syncytium of the cyst nematode *Heterodera schachtii* to enhance susceptibility. *The Plant Journal* 88(2), 179–192.

Hewezi, T., Lane, T., Piya, S., Rambani, A., Rice, J.H. *et al.* (2017) Cyst nematode parasitism induces dynamic changes in the root epigenome. *Plant Physiology* 174(1), 405–420.

Hewezi, T., Pantalone, V., Bennett, M., Neal Stewart, C., Jr and Burch-Smith, T.M. (2018) Phytopathogen-induced changes to plant methylomes. *Plant Cell Reports* 37(1), 17–23.

Holbein, J., Grundler, F.M.W. and Siddique, S. (2016) Plant basal resistance to nematodes: An update. *Journal of Experimental Botany* 67(7), 2049–2061.

Hossain, M.S., Kawakatsu, T., Kim, K.D., Zhang, N., Nguyen, C.T. *et al.* (2017) Divergent cytosine DNA methylation patterns in single-cell, soybean root hairs. *The New Phytologist* 214(2), 808–819.

Jaubert-Possamai, S., Noureddine, Y. and Favery, B. (2019) MicroRNAs, new players in the plant–nematode interaction. *Frontiers in Plant Science* 10, 1180.

Jones, J.T., Kumar, A., Pylypenko, L.A., Thirugnanasambandam, A., Castelli, L. *et al.* (2009) Identification and functional characterization of effectors in expressed sequence tags from various life cycle stages of the potato cyst nematode *Globodera pallida*. *Molecular Plant Pathology* 10(6), 815–828.

Jones-Rhoades, M.W. (2012) Conservation and divergence in plant microRNAs. *Plant Molecular Biology* 80(1), 3–16.

Kaur, P., Shukla, N., Joshi, G., VijayaKumar, C., Jagannath, A. *et al.* (2017) Genome-wide identification and characterization of miRNAome from tomato (*Solanum lycopersicum*) roots and root-knot nematode (*Meloidogyne incognita*) during susceptible interaction. *PloS ONE* 12(4), e0175178.

Kim, E.D. and Sung, S. (2012) Long noncoding RNA: Unveiling hidden layer of gene regulatory networks. *Trends in Plant Science* 17(1), 16–21.

Kim, K.H., An, D.R., Song, J., Yoon Ji, Y., Kim, H.S. *et al.* (2012) Mycobacterium tuberculosis eis protein initiates suppression of host immune responses by acetylation of DUSP16/MKP-7. *Proc Natl. Acad. Sci. U S A* 109(20), 7729–7734. DOI: 10.1073/pnas.1120251109.

Kong, L., Qiu, X., Kang, J., Wang, Y., Chen, H. *et al.* (2017) A phytophthora effector manipulates host histone acetylation and reprograms defense gene expression to promote infection. *Current Biology* 27(7), 981–991.

Koter, M.D., Święcicka, M., Matuszkiewicz, M., Pacak, A., Derebecka, N. *et al.* (2018) The miRNAome dynamics during developmental and metabolic reprogramming of tomato root infected with potato cyst nematode. *Plant Science* 268, 18–29.

Lawrence, M., Daujat, S. and Schneider, R. (2016) Lateral thinking: How histone modifications regulate gene expression. *Trends in Genetics* 32(1), 42–56.

Lawrence, R.J., Earley, K., Pontes, O., Silva, M., Chen, Z.J. *et al.* (2004) A concerted DNA methylation/histone methylation switch regulates rRNA gene dosage control and nucleolar dominance. *Molecular Cell* 13(4), 599–609.

Le, T.-N., Schumann, U., Smith, N.A., Tiwari, S., Au, P.C.K. *et al.* (2014) DNA demethylases target promoter transposable elements to positively regulate stress responsive genes in *Arabidopsis*. *Genome Biology* 15(9), 458.

Lei, P., Han, B., Wang, Y., Zhu, X., Xuan, Y. *et al.* (2019) Identification of microRNAs that respond to soybean cyst nematode infection in early stages in resistant and susceptible soybean cultivars. *International Journal of Molecular Sciences* 20(22), 5634.

Li, A. and Mao, L. (2007) Evolution of plant microRNA gene families. *Cell Research* 17(3), 212–218.

Li, H. and Luan, S. (2010) AtFKBP53 is a histone chaperone required for repression of ribosomal RNA gene expression in *Arabidopsis*. *Cell Research* 20(3), 357–366.

Li, X., Wang, X., Zhang, S., Liu, D., Duan, Y. *et al.* (2012) Identification of soybean microRNAs involved in soybean cyst nematode infection by deep sequencing. *PLoS ONE* 7(6), e39650.

Li, X., Xing, X., Xu, S., Zhang, M., Wang, Y. *et al.* (2018) Genome-wide identification and functional prediction of tobacco lncRNAs responsive to root-knot nematode stress. *PLoS ONE* 13(11), e0204506.

Lister, R., O'Malley, R.C., Tonti-Filippini, J., Gregory, B.D., Berry, C.C. *et al.* (2008) Highly integrated single-base resolution maps of the epigenome in *Arabidopsis*. *Cell* 133(3), 523–536.

Liu, H., Nichols, R.L., Qiu, L., Sun, R., Zhang, B. *et al.* (2019) Small RNA sequencing reveals regulatory roles of microRNAs in the development of *Meloidogyne incognita*. *International Journal of Molecular Sciences* 20(21), 5466.

Liu, J., Rice, J.H., Chen, N., Baum, T.J. and Hewezi, T. (2014) Synchronization of developmental processes and defense signaling by growth regulating transcription factors. *PLoS ONE* 9(5), e98477.

Liu, S., Kandoth, P.K., Warren, S.D., Yeckel, G., Heinz, R. *et al.* (2012) A soybean cyst nematode resistance gene points to a new mechanism of plant resistance to pathogens. *Nature* 492(7428), 256–260.

Liu, X., Hao, L., Li, D., Zhu, L. and Hu, S. (2015) Long non-coding RNAs and their biological roles in plants. *Genomics, Proteomics & Bioinformatics* 13(3), 137–147.

López Sánchez, A., Stassen, J.H.M., Furci, L., Smith, L.M. and Ton, J. (2016) The role of DNA (de)methylation in immune responsiveness of *Arabidopsis*. *The Plant Journal* 88(3), 361–374.

Maier, T.R., Hewezi, T., Peng, J. and Baum, T.J. (2013) Isolation of whole esophageal gland cells from plant-parasitic nematodes for transcriptome analyses and effector identification. *Molecular Plant–Microbe Interactions* 26(1), 31–35.

Makarevitch, I., Waters, A.J., West, P.T., Stitzer, M., Hirsch, C.N. *et al.* (2015) Transposable elements contribute to activation of maize genes in response to abiotic stress. *PLoS Genetics* 11(1), e1004915.

Marin, E., Jouannet, V., Herz, A., Lokerse, A.S., Weijers, D. *et al.* (2010) MiR390, *Arabidopsis* TAS3 tasiRNAs, and their AUXIN RESPONSE FACTOR targets define an autoregulatory network quantitatively regulating lateral root growth. *The Plant Cell* 22(4), 1104–1117.

Medina, C., da Rocha, M., Magliano, M., Ratpopoulo, A., Revel, B. *et al.* (2017) Characterization of microRNAs from *Arabidopsis* galls highlights a role for miR159 in the plant response to the root-knot nematode *Meloidogyne incognita*. *The New Phytologist* 216(3), 882–896.

Medina, C., de Rocha, M., Magliano, M., Ratpopoulo, A., Marteu, N. *et al.* (2018) Characterization of siRNAs clusters in *Arabidopsis thaliana* galls induced by the root-knot nematode *Meloidogyne incognita*. *BMC Genomics* 19(1), 943.

Mirouze, M. and Paszkowski, J. (2011) Epigenetic contribution to stress adaptation in plants. *Current Opinion in Plant Biology* 14(3), 267–274.

Nejat, N. and Mantri, N. (2018) Emerging roles of long non-coding RNAs in plant response to biotic and abiotic stresses. *Critical Reviews in Biotechnology* 38(1), 93–105.

Nikovics, K., Blein, T., Peaucelle, A., Ishida, T., Morin, H. *et al.* (2006) The balance between the MIR164A and CUC2 genes controls leaf margin serration in *Arabidopsis*. *The Plant Cell* 18(11), 2929–2945.

Noon, J.B. and Baum, TJ. (2016) Horizontal gene transfer of acetyltransferases, invertases and chorismate mutases from different bacteria to diverse recipients. *BMC Evolutionary Biology* 16, 74.

Noon, J.B., Hewezi, T., Maier, T.R., Simmons, C., Wei, J.-Z. *et al.* (2015) Eighteen new candidate effectors of the phytonematode *Heterodera glycines* produced specifically in the secretory esophageal gland cells during parasitism. *Phytopathology* 105(10), 1362–1372.

Noon, J.B., Hewezi, T. and Baum, T.J. (2019) Homeostasis in the soybean miRNA396-GRF network is essential for productive soybean cyst nematode infections. *Journal of Experimental Botany* 70(5), 1653–1668.

Noureddine, Y., da Rocha, M., An, J., Médina, C., Mejias, J. *et al.* (2023) AUXIN RESPONSIVE FACTOR8 regulates development of the feeding site induced by root-knot nematodes in tomato. *Journal of Experimental Botany* 74(18), 5752–5766.

Palatnik, J.F., Wollmann, H., Schommer, C., Schwab, R., Boisbouvier, J. *et al.* (2007) Sequence and expression differences underlie functional specialization of *Arabidopsis* microRNAs miR159 and miR319. *Developmental Cell* 13(1), 115–125.

Pan, X., Nichols, R.L., Li, C. and Zhang, B. (2019) MicroRNA-target gene responses to root knot nematode (*Meloidogyne incognita*) infection in cotton (*Gossypium hirsutum* L.). *Genomics* 111(3), 383–390.

Phani, V., Gowda, M.T. and Dutta, T.K. (2023) Grafting vegetable crops to manage plant-parasitic nematodes: A review. *Journal of Pest Science* 97(2), 539–560. DOI: 10.1007/s10340-023-01658-w.

Piya, S., Kihm, C., Rice, J.H., Baum, T.J. and Hewezi, T. (2017) Cooperative regulatory functions of miR858 and MYB83 during cyst nematode parasitism. *Plant Physiology* 174(3), 1897–1912.

Piya, S., Liu, J., Burch-Smith, T., Baum, T.J. and Hewezi, T. (2020) A role for *Arabidopsis* growth-regulating factors 1 and 3 in growth-stress antagonism. *Journal of Experimental Botany* 71(4), 1402–1417.

Rai, M.I., Alam, M., Lightfoot, D.A., Gurha, P. and Afzal, AJ. (2019) Classification and experimental identification of plant long non-coding RNAs. *Genomics* 111(5), 997–1005.

Rambani, A., Rice, J.H., Liu, J., Lane, T., Ranjan, P. *et al.* (2015) The methylome of soybean roots during the compatible interaction with the soybean cyst nematode. *Plant Physiology* 168(4), 1364–1377.

Rambani, A., Hu, Y., Piya, S., Long, M., Rice, J.H. *et al.* (2020) Identification of differentially methylated miRNA genes during compatible and incompatible interactions between soybean and soybean cyst nematode. *Molecular Plant–Microbe Interactions* 33(11), 1340–1352.

Ripoll, J.-J., Bailey, L.J., Mai, Q.-A., Wu, S.L., Hon, C.T. *et al.* (2015) MicroRNA regulation of fruit growth. *Nature Plants* 1(4), 15036.

Ruiz-Ferrer, V., Cabrera, J., Martinez-Argudo, I., Artaza, H., Fenoll, C. *et al.* (2018) Silenced retrotransposons are major rasiRNAs targets in *Arabidopsis* galls induced by *Meloidogyne javanica*. *Molecular Plant Pathology* 19(11), 2431–2445.

Schommer, C., Palatnik, J.F., Aggarwal, P., Chételat, A., Cubas, P. *et al.* (2008) Control of jasmonate biosynthesis and senescence by miR319 targets. *PLoS Biology* 6(9), e230.

Secco, D., Wang, C., Shou, H., Schultz, M.D., Chiarenza, S. *et al.* (2015) Stress induced gene expression drives transient DNA methylation changes at adjacent repetitive elements. *eLife* 4, e09343.

Simon, S.A. and Meyers, B.C. (2011) Small RNA-mediated epigenetic modifications in plants. *Current Opinion in Plant Biology* 14(2), 148–155.

Tian, B., Wang, S., Todd, T.C., Johnson, C.D., Tang, G. *et al.* (2017) Genome-wide identification of soybean microRNA responsive to soybean cyst nematodes infection by deep sequencing. *BMC Genomics* 18(1), 572.

Tirnaz, S. and Batley, J. (2019) DNA methylation: Toward crop disease resistance improvement. *Trends in Plant Science* 24(12), 1137–1150.

Vijayapalani, P., Hewezi, T., Pontvianne, F. and Baum, T.J. (2018) An effector from the cyst nematode *Heterodera schachtii* derepresses host rRNA genes by altering histone acetylation. *The Plant Cell* 30(11), 2795–2812.

Wang, X., Weigel, D. and Smith, L.M. (2013) Transposon variants and their effects on gene expression in *Arabidopsis*. *PLoS Genetics* 9(2), e1003255.

Xie, W., Huang, A., Li, H., Feng, L., Zhang, F. *et al.* (2017) Identification and comparative analysis of microRNAs in *Pinus massoniana* infected by *Bursaphelenchus xylophilus*. *Plant Growth Regulation* 83(2), 223–232.

Xu, M., Li, Y., Zhang, Q., Xu, T., Qiu, L. *et al.* (2014) Novel miRNA and phasiRNA biogenesis networks in soybean roots from two sister lines that are resistant and susceptible to SCN race 4. *PLoS ONE* 9(10), e110051.

Ye, D., Jiang, Y., Wang, C. and Roberts, P.A. (2020) Expression analysis of microRNAs and their target genes in *Cucumis metuliferus* infected by the root-knot nematode *Meloidogyne incognita*. *Physiological and Molecular Plant Pathology* 111, 101491.

Ye, D., Qi, Y., Cao, S., Duan, Y. and Huynh, B.L. (2024) Identification and characterization of microRNA396 and its targets in *Cucumis metuliferus* infected with *Meloidogyne incognita*. *South African Journal of Botany* 165, 417–427.

Yu, A., Lepère, G., Jay, F., Wang, J., Bapaume, L. *et al.* (2013) Dynamics and biological relevance of DNA demethylation in *Arabidopsis* antibacterial defense. *Proceedings of the National Academy of Sciences of the United States of America* 110(6), 2389–2394.

Yu, Y., Jia, T. and Chen, X. (2017) The "how" and "where" of plant microRNAs. *The New Phytologist* 216, 1002–1017.

Zhang, H. and Li, L. (2013) SQUAMOSA promoter binding protein-like7 regulated microRNA408 is required for vegetative development in *Arabidopsis*. *The Plant Journal* 74(1), 98–109.

Zhang, B., Wang, L., Zeng, L., Zhang, C. and Ma, H. (2015) *Arabidopsis* TOE proteins convey a photoperiodic signal to antagonize CONSTANS and regulate flowering time. *Genes & Development* 29(9), 975–987.

Zhang, Y., Wang, Y., Xie, F., Li, C., Zhang, B. *et al.* (2016) Identification and characterization of microRNAs in the plant parasitic root-knot nematode *Meloidogyne incognita* using deep sequencing. *Functional & Integrative Genomics* 16(2), 127–142.

Zhao, W., Li, Z., Fan, J., Hu, C., Yang, R. *et al.* (2015) Identification of jasmonic acid-associated microRNAs and characterization of the regulatory roles of the mir319/TCP4 module under root-knot nematode stress in tomato. *Journal of Experimental Botany* 66(15), 4653–4667.

Zhu, Q.H., Shan, W.X., Ayliffe, M.A. and Wang, M.B. (2016) Epigenetic mechanisms: An emerging player in plant–microbe interactions. *Molecular Plant–Microbe Interactions* 29(3), 187–196.

Index

Printed and bound by CPI Group (UK) Ltd, Croydon, CR0 4YY

13/09/2026

14961113-0007